AF344560

CORONAVIRUSES AND ARTERIVIRUSES

ADVANCES IN EXPERIMENTAL MEDICINE AND BIOLOGY

Recent Volumes in this Series

Volume 433
RECENT ADVANCES IN PROSTAGLANDIN, THROMBOXANE, AND LEUKOTRIENE RESEARCH
Edited by Helmut Sinzinger, Bengt Samuelsson, John R. Vane, Rodolfo Paoletti, Peter Ramwell, and Patrick Y-K Wong

Volume 434
PROCESS-INDUCED CHEMICAL CHANGES IN FOOD
Edited by Fereidoon Shahidi, Chi-Tang Ho, and Nguyen van Chuyen

Volume 435
GLYCOIMMUNOLOGY 2
Edited by John S. Axford

Volume 436
ASPARTIC PROTEINASES: Retroviral and Cellular Enzymes
Edited by Michael N. G. James

Volume 437
DRUGS OF ABUSE, IMMUNOMODULATION, AND AIDS
Edited by Herman Friedman, John J. Madden, and Thomas W. Klein

Volume 438
LACRIMAL GLAND, TEAR FILM, AND DRY EYE SYNDROMES 2: Basic Science and Clinical Relevance
Edited by David A. Sullivan, Darlene A. Dartt, and Michele A. Meneray

Volume 439
FLAVONOIDS IN THE LIVING SYSTEM
Edited by John A. Manthey and Béla S. Buslig

Volume 440
CORONAVIRUSES AND ARTERIVIRUSES
Edited by Luis Enjuanes, Stuart G. Siddell, and Willy Spaan

Volume 441
SKELETAL MUSCLE METABOLISM IN EXERCISE AND DIABETES
Edited by Erik A. Richter, Bente Kiens, Henrik Galbo, and Bengt Saltin

Volume 442
TAURINE 3: Cellular and Regulatory Mechanisms
Edited by Stephen Schaffer, John B. Lombardini, and Ryan J. Huxtable

CORONAVIRUSES AND ARTERIVIRUSES

Edited by

Luis Enjuanes

National Center for Biotechnology, CSIC
Campus Universidad Autónoma
Madrid, Spain

Stuart G. Siddell

Institute of Virology
University of Würzburg
Würzburg, Germany

and

Willy Spaan

Institute of Medical Microbiology
State University of Leiden
Leiden, The Netherlands

PLENUM PRESS • NEW YORK AND LONDON

Library of Congress Cataloging in Publication Data

Coronaviruses and arteriviruses / edited by Luis Enjuanes, Stuart G. Siddell, and Willy Spaan.
       p.   cm.—(Advances in experimental medicine and biology; v. 440)
     "Proceedings of the VIIth International Symposium on Coronaviruses and Arteriviruses, held May 10–15, 1997, in Segovia, Spain"—T.p. verso.
     Includes bibliographical references and index.
     ISBN 0-306-45910-8
     1. Coronaviruses—Congresses. 2. RNA viruses—Congresses. I. Enjuanes, Luis. II. Siddell, S. III. Spaan, Willy. IV. International Symposium on Coronaviruses and Arteriviruses (7th: 1997: Segovia, Spain) V. Series.
     QR399.C64   1998
     579.2′5—dc21
                                                                              98-8316
                                                                                  CIP

Proceedings of the VIIth International Symposium on Coronaviruses and Arteriviruses, held May 10–15, 1997, in Segovia, Spain

ISBN 0-306-45910-8

© 1998 Plenum Press, New York
A Division of Plenum Publishing Corporation
233 Spring Street, New York, N.Y. 10013

http://www.plenum.com

10 9 8 7 6 5 4 3 2 1

Printed in the United States of America

PREFACE

Coronaviruses and arteriviruses have emerged as major virus families relevant in animal and human health. The order *Nidovirales*, which includes the *Coronaviridae* and *Arteriviridae*, was recognized by the International Committee for the Taxonomy of Viruses (ICTV) in 1996 based on the striking features common to the two virus families. In accordance with this decision, the organizers of the coronavirus meeting, held every three years since 1980, invited colleagues working with both virus families to attend the VIIth International Symposium on Coronaviruses and Arteriviruses, held in Segovia, Spain, in 1997. This symposium provided a stimulating forum for assessing the progress made since the last meeting, which was held in Quebec, Canada. This volume is a collection of the scientific papers presented at this symposium attended by 170 scientists from academia and industry.

"Virus Entry" has been reviewed in the first section of this volume by studying the interaction between mutants of virus receptors and coronavirus spike protein. The neutralization of virus by soluble receptors has been used to study this interaction. Interestingly, the evolution of virus-receptor interaction leads to interspecies transfer of coronaviruses. Construction of chimeric aminopeptidase, the receptor of group I coronaviruses, has been useful in the identification of the virus-binding sites for coronaviruses of the porcine, canine, feline, and human species. The progress in this area has revealed that the feline aminopeptidase is a receptor for all group I coronaviruses. Early events in virus entry, such as increase in intracellular calcium, are also described.

The nidovirus replicase gene is the largest known among positive-strand RNA viruses, ranging from approximately 9–11,000 nucleotides for arteriviruses to approximately 20–22,000 nucleotides for coronaviruses. For this reason alone, the analysis of replicase gene organization, expression, and function has been a major challenge. However, in the last few years, considerable progress has been made. The nucleotide sequence of several arterivirus and coronavirus replicase genes is now known, and a common organization of putative functional domains within the polyproteins encoded by these genes has been recognized. The nidoviral replicase gene is comprised of two large open-reading frames, ORFs 1a and 1b, and expression of the downstream ORF 1b is mediated by (-1) ribosomal frame-shifting. Thus, two polyproteins, pp1a and pp1ab, are synthesized and, by a series of intricate co- and posttranslational proteolytic processing events, these polyproteins are converted to a functional complex responsible for both genomic RNA replication and subgenomic mRNA transcription. The details of these proteolytic cascades are now being worked out for several nidoviruses and they form the focus of the papers presented in "Replication I: The Polymerase."

The mechanism of coronavirus RNA synthesis has been under debate and recent advances are described in "Replication II: RNA Synthesis." Most experimental data are compatible with two models: leader-primed transcription and discontinuous transcription during negative-strand RNA synthesis. In the last model, the intergenic sequences of the genomic RNA at the mRNA start sites serve as termination or pausing signals for negative-strand synthesis. New insights in *Nidovirales* RNA replication are being facilitated by the construction of infectious cDNA clones of arteriviruses and by the engineering of the replicase gene. The role of nonstructural viral and cellular proteins in the replication of positive-strand RNA viruses is visited. The engineering of coronavirus minigenomes has contributed to the identification of sequence elements involved in the replication of viral RNA and has aided the study of coronavirus tropism.

A full-length cDNA clone-encoding infectious coronavirus RNA still has not been engineered. To overcome this limitation, the coronavirus genome is being manipulated by the selection of recombinants, or by using synthetic RNA minigenomes to identify transcription enhancing elements and to express heterologous genes. These approaches have identified secondary structure requirements in the 3' untranslated region of mouse hepatitis virus genome. Translation regulatory elements, such as the nucleocapsid (N) protein have also been described. These advances are presented in "Protein Expression and Assembly I: Expression."

Significant progress in the structure and assembly of coronaviruses has been achieved ("Protein Expression and Assembly II: Assembly") by describing the assembly of coronavirus envelopes with the large (M) and small (E) membrane proteins, and by the identification of a core shell within the coronaviruses. Now, in the three genera composing the *Nidovirales*, a core shell has been identified that protects the helical nucleocapsid. The protein-RNA interactions for RNA packaging and virion assembly have been described in detail and the biological properties of assembled viruslike particles, such as the induction of interferon, have been characterized.

The development of disease following virus infection represents a complex interaction between viral and cellular components, among which the immune system has a central role as described in "Pathogenesis I: The Immune System." To dissect the virus-host interaction is a difficult and sometimes frustrating process. Fortunately, the technologies that are now available—for example, transgenic mice with a null-phenotype in pivotal genes, and more sensitive methods of analysis, such as semi-quantitative RT-PCR—are increasingly sophisticated. Thus, it can be predicted that these advances will almost certainly provide us with many new insights into the immunopathology of coronavirus and arterivirus infections and the definition of parameters critical for the control of the disease process.

Several advances in the understanding of the pathogenesis of infections caused by coronaviruses, including MHV, IBV, PRRS, TGEV, and FECV, are reported in "Pathogenesis II: Pathology." The development of techniques that allow genetic manipulation of the coronavirus genome has progressed to the point that these techniques will be useful to those studying pathogenesis. As examples, direct recombination between MHV-2 and MHV-A59 has been used to map neurovirulence determinants. One common theme is the major role that the immune system plays in the pathogenesis of coronavirus-induced diseases. The role of MHV-3-induced prothrombinase activation and fulminant hepatitis in susceptible strains of mice is described. The difference between resistant and susceptible strains also correlates with the induction of apoptosis. The immune system also has a major role in demyelination induced by MHV strains. Mice in which the perforin gene is disrupted clear virus less well than normal mice, but also survive slightly longer, suggesting that reduced immune function ameliorates the disease. Cytotoxic T lymphocyte (CTL) escape mutants are also reported in MHV mice and their presence correlated with greater morbidity and mortality. The more extensive disease caused by these variants also appeared to have an immune component. Viruses with large deletions in the hypervariable region of the S gene were detected at late times after infection and mice infected with viruses containing these deletions had a more pronounced symptomatology than did mice in which wild type virus was predominantly detected. These deletions included the major CTL epitope recognized in this strain of mice. Host range variants were detected in infected tissue culture cells and their development was shown to correlate with a decrease in virus receptor on host cells. Overexpression of the receptor was shown to suppress virus growth, probably by trapping S protein intracellularly.

TGEV appears to be protected by a "coat" that is neuraminidase sensitive. Removal of this coat restores hemagglutinating activity but makes the virus detergent sensitive. Another role for glycoproteins is shown in a study using lactate dehydrogenase-elevating virus (LDV) in which neuropathogenicity and neutralization have been correlated with a decreased number of N-linked polylactosaminoglycan chains.

Studies designed to elucidate the pathogenesis of feline enteric coronaviruses (FECV) and their virulent relative, feline infectious peritonitis virus (FIPV), are also reported. From these studies and others using porcine epidemic diarrhea virus (PEDV), the presence of the protein encoded by ORF 3b appears to correlate with virulence.

Another major theme of investigation are the "Strategies to Control Coronavirus-Induced Diseases." Neutralizing antibodies can prevent disease if present prior to infection. With this goal in mind, the construction of transgenic animals secreting TGEV neutralizing antibodies in breast milk and their possible utility in preventing TGEV in young animals are described. Heterologous gene expression has been facilitated by the use of defective RNA minigenomes and by DNA immunization. Progress toward the identification of antigenic sites within structural proteins of PRRS virus, a pathogen with major economic impact, is also shown.

RNA virus evolution and the relationship between genetic variation and viral pathogenesis are important aspects of both coronavirus and arterivirus biology. Point mutation and recombination are thought to be the driving forces of nidoviral evolution and the papers comprising Symposium IX, "Variability and Evolution," describe the consequence of these processes in natural infection, experimental infection, and infection in tissue culture. Clearly, the evolution of the virus-receptor ligand, i.e., the surface glycoprotein, and the evolution of relevant domains in other structural proteins are major factors in determining the course of the infection and the outcome of the disease process. Interestingly, however, the co-evolution of the host cell, manifested at the level of the cell surface receptor, appears to play an important role in the balance between cytolytic and persistent infection and, possibly, the susceptibility of specific tissues or, indeed, species to coronavirus.

The organizers of the meeting wish to thank those persons who contributed their efforts to the symposium, particularly Dr. Kathryn Holmes for her collaboration to raise NIH funds to support the travel reimbursement for young American scientists, and Stanley Perlman for his contribution to this Preface. Also the organizers would like to thank those institutions and industries that made this symposium possible: Centro Nacional de Biotecnología (CNB), Consejo Superior de Investigaciones Científicas (CSIC), Comisión Interministerial de Ciencia y Tecnologia (CICYT), European Commission (Biotechnology Program), Sociedad Española de Microbiología, Federation of European Microbiological Societies, National Institute of Health, Segovia City Hall, Comunidad Autónoma de Castilla y León, Beckmab, Boehrinher Ingelheim, Cultek, Fort-Dodge Veterinaria S.A., Ingenasa, Leica España, S.A.

Finally, the success of the symposium and the completion of this book would not have been possible without the dedicated efforts of Sara Alonso, Joaquín Castilla, Jose M. González, Ander Izeta, María Muntión, Jose M. Sánchez, Carlos Sánchez, and Isabel Sola, members of the Coronavirus Laboratory at the Centro Nacional de Biotecnología, CSIC, Madrid. We would like to thank Verónica Hernández, Zoltan Penzes, and Victor Buckwold for their assistance in editing this book.

Luis Enjuanes
Stuart G. Siddell
Willy Spaan

CONTENTS

Replication I: The Polymerase

Replication II: RNA Synthesis

Protein Expression and Assembly I: Expression

Protein Expression and Assembly II: Assembly

Pathogenesis I: The Immune System

Pathogenesis II: Pathology

Strategies to Control Coronavirus-Induced Diseases

Variability and Evolution

Virus Entry

1

NEUTRALIZATION OF MHV-A59 BY SOLUBLE RECOMBINANT RECEPTOR GLYCOPROTEINS

Bruce D. Zelus,[1] David R. Wessner,[2] Gabriela S. Dveksler,[2] and Kathryn V. Holmes[1,2]

[1]Department of Microbiology
University of Colorado
Health Sciences Center
Denver, Colorado 80262
[2]Department of Pathology
Uniformed Services University of the Health Sciences
Bethesda, Maryland 20814

1. ABSTRACT

The interaction of viruses with specific receptors is an important determinant of viral tissue tropism and species specificity. Our goals are to understand how mouse hepatitis virus (MHV) recognizes its cellular receptor, MHVR, and how post-binding interactions with this receptor influence viral fusion and entry. Murine cells express a variety of cell surface molecule in the biliary glycoprotein (Bgp) family that are closely related to the MHVR. When these proteins are expressed at high levels in cell culture, they function as MHV receptors. We used a baculovirus expression system to produce soluble recombinant murine Bgp receptors in which the transmembrane and cytoplasmic domains have been replaced with a six-histidine tag. The soluble glycoproteins were purified to apparent homogeneity and shown to react with antisera to the native receptor. We compared the virus neutralizing activities of various soluble receptor glycoproteins. Soluble MHVR [sMHVR(1–4)] had 10–20 fold more virus neutralizing activity the soluble protein derived from the Bgp1[b] glycoprotein [sBgp1[b](1–4)], from MHV-resistant SJL mice. The sMHVR(1–4) glycoprotein was 60–100 fold more active than a truncated receptor molecule containing only the first two immunoglobulin-like domains, sMHVR(1,2). The observation that sMHVR lacking domains 3 and 4 neutralizes MHV-A59 very poorly suggests that these domains may influence virus binding or subsequent steps associated with neutralization.

Coronaviruses and Arteriviruses, edited by Enjuanes *et al.*
Plenum Press, New York, 1998

2. INTRODUCTION

Mouse hepatitis viruses (MHV) provide an excellent model system in which to study the role of receptor variation in virus biology and pathogenicity. This group of coronaviruses causes inapparent infection or a variety of murine diseases in mice including diarrhea, hepatitis, splenolysis, immunological dysfunction, thymic atrophy, and acute and chronic neurological disorders (Barthold, 1986; Wege *et al.*, 1982). MHV is efficiently spread within colonies of laboratory mice, and most MHV strains readily infect murine cell lines (including 17 Cl 1 cells, L2 cells and DBT cells) causing cell fusion, lysis and death (Frana *et al.*, 1985). Infection by MHV is initiated by binding of the viral attachment protein, S, to a specific virus receptor glycoprotein, MHVR (Dveksler *et al.*, 1991; Williams *et al.*, 1991; Collins *et al.*, 1982). Analysis of deletion mutant has shown that the viral S protein binds to the N-terminal domain (domain 1) of MHVR (Dveksler *et al.*, 1993b). The anti-MHVR monoclonal antibody MAb-CC1 also binds to domain 1, and blocks virus binding and infection (Dveksler *et al.*, 1993b). Binding is followed by S-mediated fusion of the viral envelope and host cell membranes, allowing the viral nucleocapsid to enter the cytoplasm (Sturman *et al.*, 1990).

MHVR, also referred to as Bgp1[a], is a biliary glycoprotein of the carcinoembryonic antigen (CEA) family of the immunoglobulin (Ig) superfamily (Brümmendorf *et al.*, 1994; Nedellec *et al.*, 1994; Rudert *et al.*, 1992). MHVR consists of four Ig-like extracellular domains, a transmembrane domain and either a long or short cytoplasmic tail (McCuaig *et al.*, 1993; Dveksler *et al.*, 1991). It is a cell adhesion protein that is expressed in fibroblasts and epithelial cells of many murine tissues (Godfraind *et al.*, 1995; Coutelier *et al.*, 1994; Benchimol *et al.*, 1989). The identification of MHVR as the receptor for MHV has led to the identification of MHV receptor activities for several related murine glycoproteins in the Bgp family (Figure 1) (Dveksler *et al.*, 1993a; Yokomori and Lai, 1992). These include a two domain splice variant of MHVR; an allelic protein called Bgp1[b] or mmCgm2, with four and two domain splice variants; and Bgp2; and brain CEA (Chen *et al.*, 1995; Nedellec *et al.*, 1994; McCuaig *et al.*, 1993).

SJL mice are homozygous for Bgp1[b], do not express MHVR, and are highly resistant to infection by MHV-A59. A cell line derived from SJL mice (PSJLSV) is also resistant to infection by MHV-A59 (Yokomori and Lai, 1992). Intestinal brush border membranes (BBM) isolated from SJL mice contain high levels of Bg1b glycoprotein but do not bind MHV-A59 as well as BBM isolated from MHV-susceptible BALB/c mice, which express MHVR (Williams *et al.*, 1991; Boyle *et al.*, 1987). This suggests that Bgp1[b] is a less efficient receptor for MHV-A59 than MHVR (Dveksler *et al.*, 1993a; Yokomori and Lai, 1992; Boyle *et al.*, 1987). When recombinant Bgp1[b] is expressed in PSJLSV or MHV-resistant hamster cells (BHK), these cells become susceptible to infection by MHV-A59, demonstrating that Bgp1[b] can serve as a receptor when expressed at high levels (Dveksler *et al.*, 1993a; Yokomori and Lai, 1992). Sequence comparison of the N-terminal, virus-binding domains of MHVR and Bgp1[b] revealed that 29 of the 108 amino acids differ, and mutational analysis of this region has shown that it is responsible for the observed differences in receptor activity (Rao and Gallagher, 1997; Wessner *et al.*, 1997; Dveksler *et al.*, 1993a; Yokomori and Lai, 1992). Other Bgp glycoproteins (Figure 1) also function as receptors for MHV-A59 when expressed at high levels in BHK cells, although less efficiently than MHVR or Bgp1[b] (Chen *et al.*, 1995; Nedellec *et al.*, 1994; Dveksler *et al.*, 1993a). These alternative receptors may be more efficient receptors for different MHV isolates. The combination of several alternative virus receptors, with differing tissue distributions and virus binding capabilities, with many well characterized and pathogenically

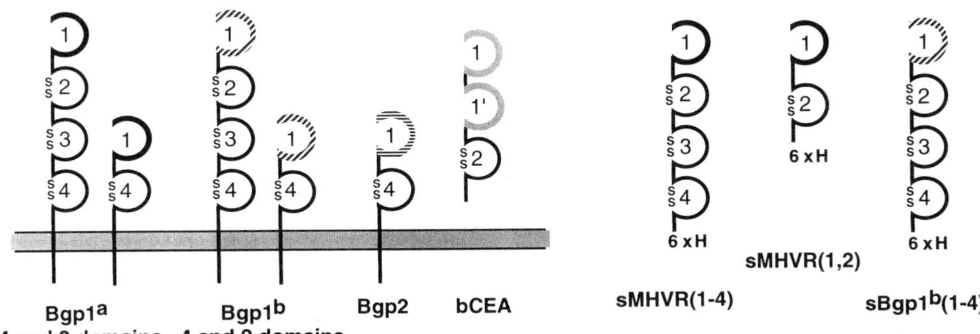

Figure 1. The various murine Bgp molecules are shown schematically, with the Ig-like domains numbered, beginning with the N-terminal domain. These glycoproteins differ markedly in the amino acid sequences of domain 1, the virus-binding domain, which is represented here by different shading patterns. Naturally occurring, membrane associated Bgps are shown on the left and the truncated, soluble Bgp molecules expressed using a baculovirus expression system are shown to the right. The names of the soluble proteins indicate which of the immunoglobulin-like domains are present, and the six-histidine tail present on each molecule is represented as 6xH.

distinct viruses makes the MHV system a fascinating arena for the investigation of virus-receptor interactions. The study of these virus-receptor interactions is important not only for understanding virus binding and the post-binding events of fusion, but also for elucidating the effects of receptor variation upon the pathogenesis of MHV infection.

3. RESULTS AND DISCUSSION

To pursue our investigations of MHV-receptor interactions, we developed a baculovirus expression system for production of large amounts of soluble, recombinant MHVR and Bgp1[b] glycoproteins (Figure 1, designated with the 's' prefix). Using PCR-mutagenesis techniques, we replaced the transmembrane and cytoplasmic domains of the glycoproteins with six histidine residues. The secreted glycoproteins were purified to apparent homogeneity by nickel affinity and ion exchange chromatography. The apparent molecular weights of the soluble proteins on SDS-PAGE gels were approximately double those predicted on the basis of their amino acid compositions, due to extensive glycosylation. A high degree of glycosylation is also seen in MHVR purified from Swiss Webster mouse liver and in two murine Bgps expressed in a baculovirus system (Ohtsuka *et al.*, 1996; Williams *et al.*, 1991). Immunoblot analysis demonstrated that sMHVR(1–4) and sMHVR(1,2) were recognized by the anti-MHVR MAb-CC1. Binding of MAb-CC1 is very dependent upon the proper tertiary configuration of domain 1 of MHVR as shown by mutagenesis studies (Wessner *et al.*, 1997; Dveksler *et al.*, 1996; Dveksler *et al.*, 1993b). These results indicate that the soluble, recombinant Bgps secreted from insect cells were properly folded and processed.

Soluble receptors for polioviruses, rhinoviruses and coronaviruses neutralize infectious virus in a concentration-dependant manner (Dveksler *et al.*, 1996; Ohtsuka *et al.*, 1996; Greve *et al.*, 1991; Kaplan *et al.*, 1990). Similar virus neutralization studies were conducted to compare the MHV receptor activities of the soluble four domain glycoproteins, sMHVR(1–4) and sBgp1[b](1–4) (Figure 2A). 5000 PFU of MHV-A59 was incubated with varying amounts of purified receptor for one hour at 37°C and the surviving virus was

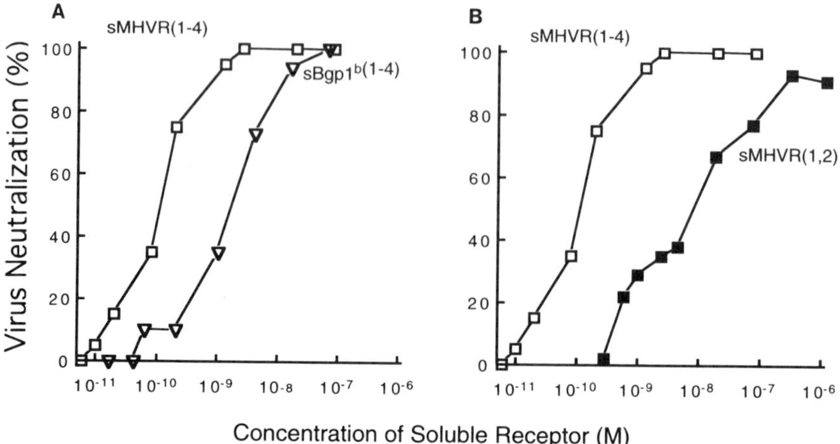

Figure 2. Neutralization of MHV-A59 by soluble receptors. A) The ability of the soluble four domain receptors, sMHVR(1–4) (□) and sBgp1[b](1–4) (∇), to neutralize MHV-A59 was determined by plaque assay on 17 Cl 1 cells. B) The neutralizing activity of the soluble two-domain sMHVR(1,2) (■) is compared to that of the four domain sMHVR(1–4) (□).

quantitated by plaque assay using murine 17 Cl 1 cells. Control reactions contained virus incubated with buffer alone. For the experiments presented in Figure 2, the 50% inhibitory dose (ID_{50}) is used to compare the neutralizing activities of the soluble receptors; the results of many experiments established the ranges given in the discussion. sMHVR(1–4) glycoprotein had the most neutralizing activity (ID_{50}=0.1 nM), while in repeated experiments, sBgp1[b](1–4) had 10–20 fold less virus neutralizing activity (ID_{50}=1.7 nM). The higher neutralization activity of sMHVR(1–4) compared to sBgp1[b](1–4) correlates with the higher receptor activity of anchored MHVR compared to Bgp1[b] when expressed in murine tissues and transfected BHK cells. Rao *et al.* (1997) created a chimeric receptor in which domain 1 of MVHR was replaced with domain 1 of Bgp1[b] and compared its receptor activity to that of MHVR when these two receptors were expressed in HeLa cells. MHVR had 10–100 fold more virus receptor activity than the Bgp1[b] chimera, a result comparable to the neutralizing activity of the soluble glycoproteins. Our results are quite different from those of Ohtsuka *et al.* (1996), who have found a greater difference between the receptor activities of soluble Bgps comprised of the first and fourth domains of MHVR and Bgp1[b], sMHVR(1,4) and sBgp1[b](1,4) in the nomenclature used here. Anchored two domain (1,4) proteins are naturally occurring isoforms generated by alternative RNA splicing (Figure 1) which have receptor activity when expressed at high levels in hamster cells (Dveksler *et al.*, 1993a). Ohtsuka *et al.* (1992) incubated soluble sMHVR(1,4) and sBgp1[b](1,4) glycoproteins in concentrated insect cell culture supernatants with MHV-JHM virus, and plaqued the survivors on DBT cells. In this system, the sMHVR(1,4) neutralized MHV-JHM 500 times more effectively than soluble sBgp1[b](1,4). One possible explanation for much larger difference in virus neutralizing activity may be their use of MHV-JHM virus, while we used MHV-A59. MHV-JHM has been reported to utilize the Bgp1[b] receptor less efficiently than MHV-A59 (Pasick *et al.*, 1992; Yokomori and Lai, 1992). MHV-JHM also has different receptor preferences than MHV-A59 in that MHV- JHM cannot infect cells transfected with brain CEA, while MHV-A59 can (Chen *et al.*, 1995). Taken together, these results show that these two strains of MHV have different interactions with Bgp1[b]. This supports the hypothesis that different strains of MHV may preferentially use different subsets of the MHVR "family" of receptors, which may affect the pathogenesis of individual viral strains.

The soluble two domain MHVR [sMHVR(1,2)] was constructed for biophysical and crystallographic studies. When purified, sMHVR(1,2) appeared to be properly processed and glycosylated, as demonstrated by its immunoreactivity and apparent molecular weight. However, it had 60 to 100 fold less virus neutralizing activity than the four domain sMHVR(1–4) (Figure 2B, ID_{50} = 8 nM) even though both molecules contain the same domain 1, the site of virus binding (Dveksler *et al.*, 1993b). The observation that sMHVR(1,2) has very low virus neutralizing activity is in agreement with the fact that membrane-anchored MHVR(1,2) transiently expressed in BHK cells is a very inefficient receptor for MHV-A59 (Wessner *et al.*, 1997). The low neutralizing activity of sMHVR(1,2) relative to sMHVR(1–4) is due to the absence of domains 3 and 4, which may alter the conformation of domain 1 and reduce its ability to bind the virus. Bgp glycoproteins appear to be signaling molecules that transmit information from the N-terminus outside the membrane to the C-terminus inside the cell via conformational changes. Thus, changes in distal regions of the protein might affect the conformation and the virus neutralizing activity of domain 1. Similarly, for the measles receptor, CD46, removal of the third and fourth of the seven short consensus repeats decreased the ability of the two N-terminal repeats to interact with measles virus (Devaux *et al.*, 1997). Virus neutralization assays reflect not only virus binding, but also post-binding events associated with membrane penetration and uncoating (Colston and Racaniello, 1995; Greve *et al.*, 1991; Kaplan *et al.*, 1990). For MHVR(1–4) and MHVR(1,4), we suggest that there may be an important structural and functional significance for the domain X-domain 4 linkage. This linkage may resemble an immunoglobulin hinge, which may play an important role in post-binding conformational changes associated with virus uncoating or penetration. This structural feature would be absent from the sMHVR(1,2) protein, and may explain its low virus binding and neutralizing activities.

4. CONCLUSIONS

We have produced and purified soluble Bgps that can neutralize MHV-A59, although with very different efficiencies: sMHVR(1–4) is 15–20 times more active than sBgp1[b](1–4) and 60–100 times more active than the two domain sMHVR(1,2). These differences in virus neutralization activity reflect the relative abilities of the anchored molecules to support MHV infection *in vivo* and when over expressed in BHK cells. These results, show that the soluble Bgps produced in the baculovirus system accurately reflect the behavior of Bgps produced by mammalian cells and are therefore a realistic model system for studying the biophysical parameters of virus-receptor interactions. Additional studies are underway to determine if the differences in virus neutralization are due to differences in virus binding affinities, or if post-binding events associated with membrane fusion and viral uncoating also affect neutralizing capability. The soluble receptors will also be used to investigate the receptor preferences of the various MHV strains to investigate the basis for their different biological properties and tissue tropisms.

ACKNOWLEDGMENTS

We thank David Wentworth and Dianna Blau for helpful discussions and review of this manuscript. We thank Fenna Tanner for excellent technical assistance and Kurt Christiansen of the UCHSC Cancer Center for assistance with the baculovirus expression system. This work was supported by NIH grants AI25231 and AI26075. The opinions in this paper are those of the authors and do not represent official views of the Uniformed Services University of the Health Sciences or the Department of Defense.

REFERENCES

Barthold, S.W., 1986, Mouse hepatitis virus: biology and epizootiology, in: *Viral and mycoplasma infections of laboratory rodents. Effects on biomedical research,* (P.N. Bhatt, R.O. Jacoby, H.C. Morse III and A.E. New, eds.), Academic Press, Orlando, FL, pp. 571–601.

Benchimol, S., Fuks, A., Jothy, A., Beauchemin, N., Shirota, K., and Stanners, C.P., 1989, Carcinoembryonic antigen, a human tumor marker, functions as an intercellular adhesion molecule, *Cell* **57**:327–334.

Boyle, J.F., Weismiller, D.G., and Holmes, K.V., 1987, Genetic resistance to Mouse Hepatitis Virus correlates with absence of virus-binding activity on target tissues, *J. Virol.* **61**:185–189.

Brümmendorf, T., and Rathjen, F.G., 1994, Introduction, in: *Cell Adhesion Molecules 1: immunoglobulin superfamily,* (P. Sheterline, ed.), Academic Press, London, pp. 951–962.

Chen, D.S., Asanaka, M., Yokomori, K., Wang, F., Hwang, S.B., Li, H.P., and Lai, M.M., 1995, A pregnancy-specific glycoprotein is expressed in the brain and serves as a receptor for mouse hepatitis virus, *Proc. Natl. Acad. Sci. USA* **92**:12095–12099.

Collins, A.R., Knobler, R.L., Powell, H., and Buchmeier, M.J., 1982, Monoclonal antibodies to murine hepatitis virus-4 (strain JHM) define the viral glycoprotein responsible for attachment and cell-cell fusion, *Virology* **119**:358–371.

Colston, E.M. and Racaniello, V.R., 1995, Poliovirus variants selected on mutant receptor-expressing cells identify capsid residues that expand receptor recognition, *J. Virol.* **69**:4823–4829.

Coutelier, J., Godfraind, C., Dveksler, G.S., Wysocka, M., Cardellichio, C.B., and Nöel, H., 1994, B lymphocyte and macrophage expression of carcinoembryonic antigen-related adhesion molecules that serve as receptors for murine coronoavirus, *Eur. J. Immunol.* **24**:1383–1390.

Devaux, P., Buchholz, C.J., Schneider, U., Escoffier, C., Cattaneo, R., and Gerlier, D., 1997, CD46 Short Consensus Repeats III and IV enhance Measles Virus binding but impair soluble hemagglutinin binding, *J. Virol.* **71**:4157–4160.

Dveksler, G.S., Pensiero, M.N., Cardellichio, C.B., Williams, R.K., Jiang, G.S., Holmes, K.V., and Dieffenbach, C.W., 1991, Cloning of the mouse hepatitis virus (MHV) receptor: expression in human and hamster cell lines confers susceptibility to MHV, *J. Virol.* **65**:6881–6891.

Dveksler, G.S., Dieffenbach, C.W., Cardellichio, C.B., McCuaig, K., Pensiero, M.N., Jiang, G.S., Beauchemin, N., and Holmes, K.V., 1993, Several members of the mouse carcinoembryonic antigen-related glycoprotein family are functional receptors for the coronavirus mouse hepatitis virus-A59, *J. Virol.* **67**:1–8.

Dveksler, G.S., Pensiero, M.N., Dieffenbach, C.W., Cardellichio, C.B., Basile, A.A., Elia, P.E., and Holmes, K.V., 1993, Mouse hepatitis virus strain A59 and blocking antireceptor monoclonal antibody bind to the N-terminal domain of cellular receptor, *Proc. Natl. Acad. Sci. USA* **90**:1716–1720.

Dveksler, G.S., Gagneten, S.E., Scanga, C.A., Cardellichio, C.B., and Holmes, K.V., 1996, Expression of the recombinant anchorless N-terminal domain of mouse hepatitis virus (MHV) receptor makes hamster of human cells susceptible to MHV infection, *J. Virol.* **70**:4142–4145.

Frana, M.F., Behnke, J.N., Sturman, L.S., and Holmes, K.V., 1985, Proteolytic cleavage of the E2 glycoprotein of murine coronavirus: host-dependent differences in proteolytic cleavage and cell fusion, *J. Virol.* **56**:912–920.

Godfraind, C., Langreth, S.G., Cardellichio, C.B., Knobler, R., Coutelier, J.P., Dubois-Dalcq, M., and Holmes, K.V., 1995, Tissue and cellular distribution of an adhesion molecule in the carcinoembryonic antigen family that serves as a receptor for mouse hepatitis virus, *Lab. Invest.* **73**:615–627.

Greve, J.M., Forte, C.P., Marlor, C.W., Meyer, A.M., Hoover-Litty, H., Wunderlich, D., and McClelland, A., 1991, Mechanisms of receptor-mediated rhinovirus neutralization defined by two soluble forms of ICAM-1, *J. Virol.* **65**:6015–6023.

Kaplan, G., Freistadt, M.S., and Racaniello, V.R., 1990, Neutralization of Poliovirus by Cell Receptors Expressed in Insect Cells, *J. Virol.* **64**:4697–4702.

McCuaig, K., Rosenberg, M., Nedéllec, P., Turbide, C., and Beauchemin, N., 1993, Expression of the Bgp gene and characterization of mouse colon biliary glycoprotein isoforms, *Gene* **127**:173–183.

Nedellec, P., Dveksler, G.S., Daniels, E., Turbide, C., Chow, B., Basile, A.A., Holmes, K.V., and Beauchemin, N., 1994, Bgp2, a new member of the carcinoembryonic antigen-related gene family, encodes an alternative receptor for mouse hepatitis viruses, *J. Virol.* **68**:4525–4537.

Ohtsuka, N., Yamada, Y.K., and Taguchi, F., 1996, Difference in virus-binding activity of two distinct receptor proteins for mouse hepatitis virus, *Journal of General Virology* **77**:1683–1692.

Pasick, J.M., Wilson, G.A., Morris, V.L., and Dales, S., 1992, SJL/J resistance to mouse hepatitis virus-JHM-induced neurologic disease can be partially overcome by viral variants of S and host immunosuppression, *Microbial Pathogenesis.* **13**:1–15.

Rao, P.V., and Gallagher, T.M., 1997, Identification of a contiguous 6-residue determinant in the MHV receptor that controls the level of virion binding to cells, *Virology* **229**:336–348.

Rudert, F., Saunders, A.M., Rebstock, S., Thompson, J.A., and Zimmerman, W., 1992, Characterization of murine carcinoembryonic antigen gene family members, *Mam. Genome* **3**:262–273.

Sturman, L.S., Ricard, C.S., and Holmes, K.V., 1990, Conformational change of the coronavirus peplomer glycoprotein at pH 8.0 and 37 degrees C correlates with virus aggregation and virus-induced cell fusion, *J. Virol.* **64**:3042–3050.

Wege, H., Siddell, S., and ter Meulan, V., 1982, The biology and pathogenisis of coronaviruses, *Curr. Top. in Microbiol. Immunol.* **99**:165–200.

Wessner, D.R., Shick, P.C., Lu, J.-H., Cardellichio, C.B., Gagneten, S.E., Beauchemin, N., Holmes, K.V., and Dveksler, G.S., 1997, Mutational Analysis of the Virus and Monoclonal Antibody Binding Sites in MHVR, the Cellular Receptor of the Murine Corona Virus MHV-A59, *J. Virol.* In Press.

Williams, R.K., Jiang, G.S., and Holmes, K.V., 1991, Receptor for mouse hepatitis virus is a member of the carcinoembryonic antigen family of glycoproteins, *Proc. Natl. Acad. Sci. USA* **88**:5533–5536.

Yokomori, K., and Lai, M.M., 1992, The receptor for mouse hepatitis virus in the resistant mouse strain SJL is functional: implications for the requirement of a second factor for viral infection, *J. Virol.* **66**:6931–6938.

2

ISOLATION AND CHARACTERIZATION OF MURINE CORONAVIRUS MUTANTS RESISTANT TO NEUTRALIZATION BY SOLUBLE RECEPTORS

Keiichi Saeki, Nobuhisa Ohtsuka, and Fumihiro Taguchi

Division of Animal Models for Human Diseases
National Institute of Neuroscience, NCNP
4-1-1 Ogawahigashi, Kodaira, Tokyo 187, Japan

1. ABSTRACT

Murine coronavirus mutants resistant to neutralization with soluble receptors were isolated to study the receptor-binding site on the S proteins since such mutants were expected to have mutations in an important site for receptor-binding. We have isolated five soluble receptor-resistant (srr) mutants which had mutations of a single amino acid at 3 different positions in S protein. Srr mutant 11 with an amino acid change at position 65 (Leu to His) in the S1 subunit showed an extremely reduced binding by virus overlay protein blot assay. However srr mutants with a mutation at 1114 (Leu to Phe) (srr mutants 3, 4 and 7) or 1163 (Cys to Phe) (srr mutant 18) in the S2 subunit had receptor-binding activity similar to that of wild type cl-2. These results suggest that an amino acid at position 62 located in a conserved region among MHV strains is in particular important for receptor binding. We also discuss why srr mutants with a mutation in S2 showed high resistance to neutralization by soluble receptor, irrespective of their binding to MHV receptors.

2. INTRODUCTION

The spike projecting from the virion surface of coronavirus is composed of two or three molecules of the spike (S) protein, each of which is a heterodimer consisting of two non-covalently bound subunits, S1 and S2 (Sturman and Holmes, 1984). These S1 and S2 subunits are derived from cleavage of the N-terminal and C-terminal halves of the precursor S protein by a host cell derived protease (Sturman and Holmes, 1984). The S1 subunit

Coronaviruses and Arteriviruses, edited by Enjuanes *et al.*
Plenum Press, New York, 1998

is believed to form the globular part of the spike and the S2 its stalk portion (De Groot et al., 1988). The S protein has many important biological functions. One of them is the fusion of cultured cells (Taguchi et al., 1992; Collins et al., 1982). Uncleaved S protein has a fusion activity (Stauber et al., 1993; Taguchi, 1993), suggesting that the mechanism of MHV fusion differs from that of other fusogenic RNA viruses (White, 1990). The S protein is the major target of the neutralizing antibodies induced in mice. It also elicits cytotoxic T cells (Flory et al., 1993). Furthermore, the S protein is suggested to be a major determinant of viral virulence in animals (Matsubara et al., 1991; Dalziel et al., 1986; Fleming et al., 1986).

The binding to the receptor proteins (Williams et al., 1991) is another important biological function of the MHV S protein. The receptor-binding domain on the S protein of MHV is located in the N-terminal region of the S1 composed of 330 amino acids [S1N(330)] (Suzuki and Taguchi, 1996; Kubo et al., 1994) but not on the S2 (Taguchi 1995). Recently, we have identified amino acid residues responsible for receptor-binding using various mutant S1N(330) generated by site-directed mutagenesis (Suzuki and Taguchi, 1996). To further identify the receptor-binding site in the S protein, we have utilized a different approach, the isolation and characterization of mutant viruses that escaped from neutralization by soluble receptor protein. Since such resistant viruses are expected to carry the mutation defective in the receptor binding (Colston and Racaniello, 1994), the mutations found in the S protein of such mutants appeared to be most useful tools to identify the amino acids responsible for receptor binding.

3. MATERIALS AND METHODS

3.1. Cells and Viruses

Cell lines (DBT and RK 13) were grown as previously reported (Kubo et al., 1994). The MHV strain JHMV cl-2 (Taguchi et al., 1985) and the recombinant vaccinia virus vTF7.3 were prepared and assayed as described elsewhere (Fuerst et al., 1987).

3.2. Preparation of Soluble Receptor Protein

The MHV receptor gene of MHVR1 (Kubo et al., 1994) or Bgp C was used for expression. The soluble form of the MHV receptor protein (soMHVR1) with the transmembrane and intracytoplasmic domains deleted was tagged with influenza HA epitope (Ohtsuka et al., 1996) and was expressed by recombinant vaccinia virus vTF7.3 (Fuerst et al., 1987). The soMHVR1 expressed and secreted into the culture was concentrated by ultrafiltration (Millipore) after removing vaccinia virus by centrifuging at 20,000 rpm for 2 h.

3.3. Isolation and Sequencing of the Srr Mutant S Genes

Total RNA (0.2 μ g) from DBT cells infected with virus was reverse transcribed and cDNAs were used for PCR using a pair of primers (5'-CGCAAGCTTCTAAA-CATGCTGTTCGTC-3' and 5'-ATCTTGGGACCGATGA-GGGCCAT-3' corresponding to the cl-2 S gene initiation codon and termination codon, respectively. The S cDNAs were directly sequenced on both strands using a series of S gene specific primers (Taguchi et al., 1992) labeled with FITC and Thermo Sequenase fluorescent labeled primer cycle sequencing kit (Amersham) in a DNA sequencer (Pharmacia, Uppsala, Sweden).

3.4. Western Blotting

Western blotting analysis was carried out using the S1-specific monoclonal antibody 30 B (Routledge et al., 1991) as previously reported (Taguchi, 1993).

3.5. Virus Overlay Protein Blot Assay (VOPBA)

The virus overlay protein blot assay (VOPBA) was used to detect the binding of the S proteins to the receptor protein, as described previously (Ohtsuka et al., 1996; Kubo et al., 1994).

4. RESULTS

4.1. Isolation of Srr Mutants Resistant to Neutralization by soMHVR1

To identify the amino acid residues of the S protein involved in the virus-receptor interaction, we isolated virus mutants resistant to neutralization by soluble receptor protein. In order to isolate the srr mutants, we used the concentrated soMHVR1 expressed by recombinant vaccinia virus vTF7.3. A mixture of 100 µl of cl-2 titered virus (2×10^5 PFU) and 500 µl of soMHVR1 was incubated at room temperature for 60 min and then inoculated onto DBT cells. The plaques produced were isolated 1 to 2 days later and subjected to 3 further plaque-purifications. As shown in Table 1, five virus clones (designated as srr 3, srr 4, srr 7, srr 11, and srr 18) were highly resistant to neutralization by soMHVR1. Wild type virus titer was reduced by 1/50 by incubation with soMHVR1 solution, while the five srr mutants were hardly affected. This result suggested that the five mutant viruses were selected though the resistance to neutralization with soMHVR1. Srr mutants were isolated with a frequency of about 1 per 1×10^5 PFU of the wild-type cl-2 population.

4.2. Sequence Analysis of S Gene in Srr Mutants

To identify the mutated amino acid residues in the S proteins of srr mutants, we determined the sequence of the entire S protein coding region. Sequence analysis of S cDNA containing 4,131-nucleotide revealed that each srr mutant has a single nucleotide mutation which creates an amino acid change in S1 or S2 subunit (Fig. 1). Srr 11 had an amino acid

Table 1. Neutralization of MHV-JHM cl-2 and srr mutants
by soMHVR1

Virus	Titer (pfu/0.1ml)		Relative neutralization efficiency
	+SoMHVR1	-SoMHVR1	
Srr 3	2.5×10^5	2.2×10^5	1.14
Srr4	2.3×10^5	2.4×10^5	0.96
Srr7	1.2×10^5	1.4×10^5	0.86
Srr 11	1.0×10^5	1.0×10^5	1.00
Srr 18	1.1×10^5	1.1×10^5	1.00
JHM cl-2	2.3×10^3	4.8×10^5	0.005

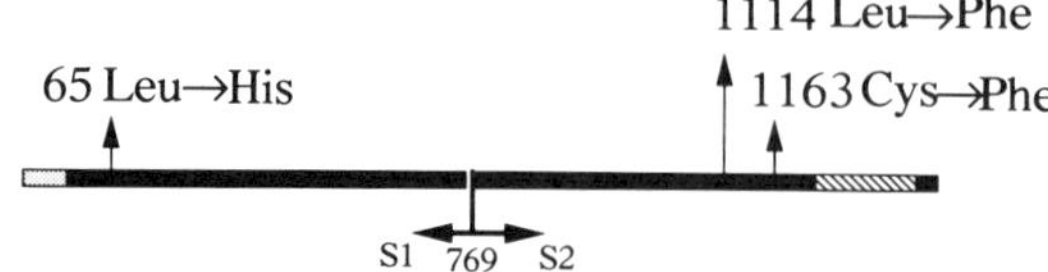

Figure 1. Distribution of amino acid changes on srr mutant S proteins. The S protein (1376 residues) is depicted by a horizontal line. Arrows indicate amino acid alterations in the mutant S proteins. Signal and transmembrane peptides are indicated by dotted and dashed bars respectively.

change in S1 subunit at position 65 (Leu to His) in a highly conserved region (at positions between 49 and 71) among MHV strains (Suzuki and Taguchi, 1996). Other alternations at positions 1114 (Leu to Phe) (srrs 3, 4 and 7) and 1163 (Cys to Phe) (srr 18) were present between two heptad repeat domains in the S2 subunit (De Groot et al., 1988). We chose three srr mutants (srr 7, srr 11, and srr 18) for further characterization because each of them had a single amino acid substitution in different positions of the S protein.

4.3. VOPBA of Srr Mutant S Proteins of soMHVR1

To compare the receptor-binding activities of wild type and srr S proteins, we assessed the ability of mutant S proteins to bind to the soMHVR1 by using VOPBA. Equivalent amounts of Srr mutant and wild-type S proteins were subjected to VOPBA. Srr 11 with a substitution at position 65 (Leu to His) in S1 subunit resulted in more than 98% loss of soMHVR1 binding activity compared to that of wild type (Fig. 2, lanes 1 and 3), while the S protein mutated at position 1114 (Leu to Phe, mutant 7) or 1163 (Cys to Phe, mutant 18) in the S2 subunit showed receptor-binding similar to that of wild type cl-2 (Fig. 2, lanes 1, 2 and 4). We have also assayed the binding activity of the mutant S proteins under several different S protein concentrations. However, the binding patterns for the srr 7 and srr 18 S proteins were similar to that of wild type S protein (data not shown). These results indicate that the substitution at position 65 in the S1 subunit remarkably reduced the capacity to bind to soMHVR1. However, the alteration at position 1114 or 1163 in the S2 subunit did not affect the binding to soMHVR1.

5. DISCUSSION

We have previously demonstrated by site-directed mutagenesis that the amino acids at positions 62 and a region consisting of amino acids 212 to 216 are important for receptor-binding (Suzuki and Taguchi, 1996). In the present study, we have used another procedure

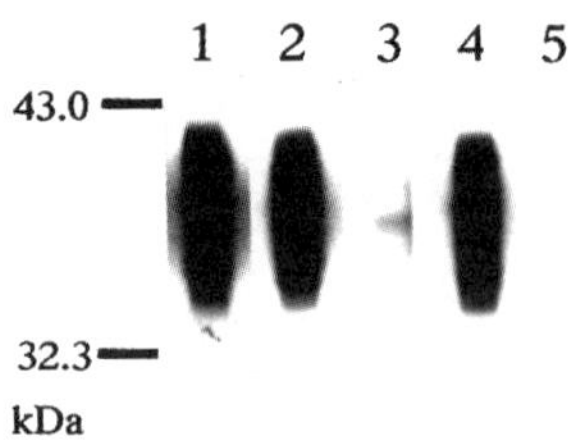

Figure 2. Analysis of receptor-binding activities by VOPBA. Wild type (Lane 1); srr 7 mutant position 1114 (Leu to Phe) (Lane 2); srr 11 mutant position 65 (Leu to His) (Lane 3); srr 18 mutant position 1163 (Cys to Phe) (Lane 4) and control DBT culture medium (Lane 5) were incubated at RT for 1 h with soMHVR1 transferred to the membrane paper by Western blotting. The binding of viruses were evaluated with the S1 specific monoclonal antibody No. 7 and horseradish peroxidase labeled anti-mouse IgG by enhanced chemiluminescence.

to define the region involved in receptor-binding. The srr mutants that are resistant to neutralization by the soluble receptor protein are supposed to contain amino acid changes in the receptor-binding site, as studied in detail for poliovirus (Colston and Racaniello, 1994). Three different srr mutants were isolated, each of which contained only one amino acid change in the S protein. Srr 11 contained the mutation at amino acid position 65 (Leu to His) in S1 subunit. This mutant showed high resistance to neutralization by soluble receptor and its binding activity to the soluble receptor was dramatically decreased. Thus, this particular amino acid at a position 65 appeared to be responsible for the receptor binding activity. This amino acid is located in the vicinity the amino acid 62 which was previously found to be important for receptor binding (Suzuki and Taguchi, 1996). These facts suggest that a region composed of amino acids from 62 to 65 is critical for receptor binding. Our previous study has shown that the mutations at different amino acids between 49 to 70 of the S protein, conserved in all MHV strains examined, did not influence the receptor-binding activity with the exception of the amino acid at position 62 (Suzuki and Taguchi, 1996). These data indicate that the amino acids at positions 62 and 65 are particularly important and that not necessarily all of the amino acids in this region are involved in the receptor-binding activity.

Srr 7 and srr 18 contained an amino acid mutation in the S2 subunit, at positions 1114 (Leu to Phe) and 1163 (Cys to Phe), respectively. We first speculated that the mutations at 1114 and 1163 reduce the receptor-binding activity of the S1, since the mutation in S2 subunit has been reported to influence the conformation of the S1 (Grosse and Siddell, 1994). However, that was not the case. Although these two mutants showed resistance to neutralization with soluble receptor protein, the receptor-binding capacity was not different from that of wild type virus. This means that the mutant S proteins bind to the soluble receptor protein as efficiently as did wild type S protein and that such soluble receptor-bound virus particles are still infectious. At present, no experimental data is available to account for this phenomenon. However, a difference was found between the mutant S proteins and that of wild type in the stability of the S1-S2 association. Wild type S1 and S2 proteins could easily be dissociated after binding to the receptor protein, but the S1-S2 association of srr 7 and srr 18 was more stable than that of wild type (data not shown). Such a stable association between S1 and S2 after binding to the receptor protein may account for the characteristics of srr of these two mutant viruses. Recently, it was reported that the S1 of wild type JHM with a large S protein can easily be dissociated from S2 after binding to receptor, however, the JHMX which contained a large amino acid deletion in the S1 subunit showed a stable association of S1 and S2 (Gallagher, 1997). We have found that JHMX is resistant to neutralization by soMHVR1, that is , JHMX has characteristics similar to our srr mutants (data not shown). These data suggest that the stable association of the S1 and S2 subunits after binding to the receptor protein may be related to the characteristics of the srr mutant.

ACKNOWLEDGMENTS

We thank K. Yamakawa for excellent technical assistance. This work was supported by grants from the Science and Technology Agency of Japan and Inoue Foundation for Science.

REFERENCES

Collins, A. R., Knobler, R. L., Powell, H., and Buchmeier, M. J., 1982, Monoclonal antibodies to murine hepatitis virus-4 (strain JHM) define the viral glycoprotein responsible for attachment and cell fusion, *Virology* **119**: 358–371.

Colston, E., and Racaniello, V. R., 1994, Soluble receptor-resistant poliovirus mutants identify surface and internal capsid residues that control interaction with the cell receptor, *EMBO J.* **13**: 5855–5862.

Dalziel, R. G., Lampert, P. W., Talbot, P. J., and Buchmeier, M. J., 1986, Site-specific alteration of murine hepatitis virus type 4 peplomer glycoprotein E2 results in reduced neurovirulence, *J. Virol.* **59**: 463- 471.

De Groot,R. J., Luytjes, W., Horzinek, M. C., Van der Zeijst, B. A. M., Spaan, W. J. M., and Lenstra, J. A.,1988, Evidence for a coiled-coil structure in the spike of coronaviruses, *J. Mol. Biol.* **196**: 963–966

Flory, E., Pfleiderer, M., Stuhler, A, and Wege, H., 1993, Induction of protective immunity against coronavirus-induced encephalomyelitis: evidence for an important role of CD8+ T cells in vivo, *Eur. J. Immunol.* **23**: 1757–1761.

Fuerst, T. R., Earl, P. L., and Moss, B., 1987, Use of hybrid vaccinia virus-T7 RNA polymerase system for the expression of target genes, Mol. Cell. Biol. 7: 2538–2544.

Fleming, J. O., Trousdale, M. D., El-Zaatari, F. A. K., Stohlman, S. A., and Weiner, L. P., 1986, Pathogenicity of antigenic variants of murine coronavirus JHM selected with monoclonal antibodies, *J. Virol.* **58**: 869–875.

Gallagher, T. M. 1997, A role for naturally occurring variation of the murine coronavirus spike protein in stabilizing association with the cellular receptor, *J. Virol.* **71**: 3129–3137.

Grosse, B., and Siddell, S. G., 1994, Single amino acid changes in the S2 subunit of the MHV surface glycoprotein confer resistance to neutralization by S1-specific monoclonal antibody, *Virology* **2202**: 814–824.

Kubo, H., Yamada, Y. K., and Taguchi, F., 1994, Localization of neutralizing epitopes and the receptor-binding site within the amino-terminal 330 amino acids of the murine coronavirus spike protein, *J. Virol.* **68**: 5403–5410.

Matsubara, Y., Watanabe, R., and Taguchi, F., 1991, Neurovirulence of six different murine coronavirus JHMV variants for rats, *Virus Res.* **20**: 45–58.

Ohtsuka, N., Yamada, Y. K., and Taguchi, F., 1996, Difference in virus-binding activity of two distinct receptor proteins for mouse hepatitis virus, *J. Gen. Virol.* **77**: 1683–1692.

Routledge, E., Stauber, R., Pfleiderer, M., and Siddell, S. G., 1991, Analysis of murine coronavirus surface glycoprotein functions by using monoclonal antibodies, *J. Virol.* **65**: 254–262.

Stauber, R., Pfleiderer, M., and Siddell, S., 1993, Proteolytic cleavage of the murine coronavirus surface glycoprotein is not required for fusion activity, *J. Gen. Virol.* **74**: 183–191.

Sturman, L. S., and Holmes, K. V., 1984, Proteolytic cleavage of peplomer glycoprotein E2 of MHV yields two 90 K subunits and activates cell fusion, *Adv. Exp. Med. Biol.* **173**: 25–35.

Suzuki, H., and Taguchi, F., 1996, Analysis of the receptor binding site of murine coronavirus spike glycoprotein, *J. Virol.* **70**: 2632–2636.

Taguchi, F., 1993, Fusion formation by uncleaved spike protein of murine coronavirus JHMV variant cl-2, *J. Virol.* **67**: 1195–1202.

Taguchi, F., 1995, The S2 subunit of the murine coronavirus spike protein is not involved in receptor binding, *J. Virol.* **69**: 7260–7263.

Taguchi, F., Ikeda, T., and Shida, H., 1992, Molecular cloning and expression of a spike protein of neurovirulent murine coronavirus JHMV variant cl-2, *J. Gen. Virol.* **73**: 1065–1072.

Taguchi, F., Siddell, S. G., Wege, H., and ter Meulen, V., 1985, Characterization of a variant virus selected in rat brain after infection by coronavirus mouse hepatitis virus JHM, *J. Virol.* **54**: 429–435.

White, J. M., 1990, Viral and cellular membrane fusion proteins, *Annu. Rev. Physiol.* **52**: 675–697.

Williams, R. K., Jiang, G. S., and Holmes, K. V., 1991, Receptor for mouse hepatitis virus is a member of the carcinoembryonic antigen family of glycoproteins, *Proc. Natl. Acad. Sci. USA* **88**: 5533–5536.

3

MUTATIONAL ANALYSIS OF FUSION PEPTIDE-LIKE REGIONS IN THE MOUSE HEPATITIS VIRUS STRAIN A59 SPIKE PROTEIN

Zongli Luo and Susan R. Weiss

Department of Microbiology
University of Pennsylvania School of Medicine
Philadelphia, Pennsylvania 19104

1. ABSTRACT

The coronavirus peplomer protein S is responsible for attachment and fusion during viral entry as well as for the induction of cell to cell fusion. While several regions within S have been shown to influence the ability to induce fusion, the region of the protein actually responsible for fusion, the fusion peptide, has not yet been identified. We identified two hydrophobic peptides (peptides #1 and #2) within MHV-A59 S2 as possible fusion domains. This was based on hydrophobicity, conservation among coronavirus S proteins and the prediction of a sided helix conformation. Using site directed mutagenesis and an *in vitro* cell to cell fusion assay we showed that substitution of hydrophobic amino acids with charged amino acids, within the predicted hydrophobic face of either of these two peptides eliminated fusion. Within peptide #1 substitution of the same hydrophobic amino acids with other hydrophobic amino acids or substitution of polar amino acids with charged or polar amino acids had little effect on fusion. Thus peptides #1 and #2 remain likely candidates for the MHV fusion peptide. A third previously identified peptide within S2 (Chambers *et al.*, 1990) is unlikely as a fusion peptide as it is not well conserved among coronaviruses and substitution within the hydrophobic face with charged amino acids does not effect fusion.

2. INTRODUCTION

Studies on viral fusion proteins indicate that their fusion peptides have certain common characteristics: (i) a relatively hydrophobic stretch of 16–26 amino acids which generally shows an amphipathic pattern if modeled as an alpha helix; (ii) they are rich in

Coronaviruses and Arteriviruses, edited by Enjuanes *et al.*
Plenum Press, New York, 1998

alanine and/or glycine; (3) they are found in a membrane anchored subunit and often near (but not within) the heptad repeat regions; and 4) they are highly conserved within but not between virus families (White, 1992; White, 1990). Fusion peptides can be divided into two groups based on their locations relative to the N-terminus of the membrane anchored subunit: amino terminal and internal fusion peptides. Fusion proteins such as paramyxovirus F proteins, influenza HA proteins, and the envelope protein of several retroviruses have amino terminal fusion peptides (White, 1990). Putative internal fusion peptides have been documented for togaviruses Sindbis and SFV and the retrovirus RSV (White, 1990). Besides having the common features of viral fusion peptides, these internal fusion peptides contain a proline residue in the center and a charged residue on both N- and C-ends. Sequence analysis of coronavirus fusion proteins suggests that they possess internal rather than aminoterminal fusion peptides. The coronavirus fusion peptide is presumed to be within S2 because S2 is highly conserved among coronaviruses and contains the membrane anchor and heptad repeats domains (White, 1992; White, 1990).

A prevalent hypothesis regarding the role of fusion peptides in the mechanism of viral fusion is that these peptides act as sided insertional helices (White, 1992; White, 1990). A typical feature of such helices is that distribution of the bulky hydrophobic residues is asymmetric. The more hydrophobic face of the helix is believed to insert into the membrane with an oblique orientation with respect to the lipid-water interface, thus inducing a local disorder of the bilayer structure.The other face of the helix is hydrophilic and contains polar and charged residues. We have analyzed the sequence of S2 of MHV-A59 and we have identified two conserved peptides that fit the description of fusion peptides as sided helices (White, 1992; White, 1990). These are amino acids 971–989 within heptad repeat #1 and 1098–1114 between the heptad repeat regions (see Figure 1). We are using site directed mutagenesis to analyze the effects of mutation of these peptides on the ability of S to induce fusion.

Mutagenesis studies presented here are consistent with either of these functioning as a fusion domain.We have tentatively ruled out a third peptide (amino acids 911–948), suggested previously to be a candidate fusion domain (Chambers *et al.*, 1990). This region, encoded just upstream of heptad repeat #1, predicted earlier to be a fusion peptide because this domain is less conserved and mutation of this domain does not seem to affect the ability of S to induce cell to cell fusion in an *in vitro* assay.

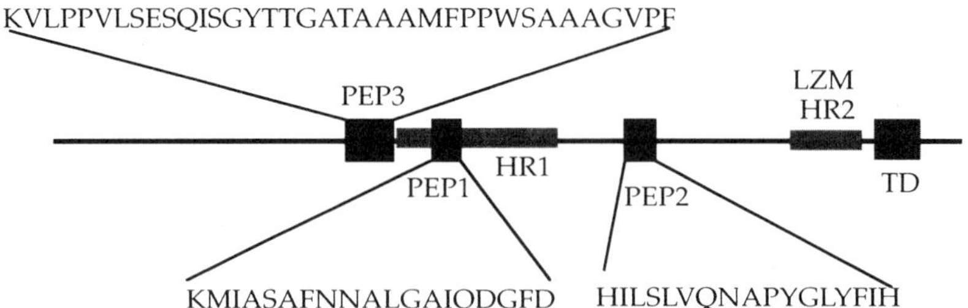

Figure 1. Schematic diagram of the S2 subunit of the MHV-A59 spike protein. Distinctive structural motifs in the S2 subunit are shown as boxes whereas the rest of S2 is shown as a thin line. PEP1, PEP2, PEP3: candidate fusion peptide # 1, 2, 3; HR1, HR2: heptad repeat region 1 and 2; LZM: leucine zipper motif; TD: transmembrane domain. The primary sequence of each candidate fusion peptide is shown.

3. METHODS

3.1. Plasmid Construction and Mutagenesis

For construction of a plasmid from which to express the S protein, the entire wild type MHV-A59 S gene was amplified by RT/ PCR from MHV-A59 intracellular RNA, using a 5' primer containing a SacI site and 3' primer containing a Bam HI site. This DNA was cloned into SacI/BamHI cut pBluescript II KS+ behind a T7 promoter and called pJG10. This plasmid was modified by Dr. Paul Masters to correct a mutation in the cleavage site and to introduce a silent mutation at nucleotide 517 which destroyed an internal HindIII site. The final clone, pINT 2, encodes an S protein with the identical amino acid sequence as our wild type MHV-A59 S (Hingley *et al.*, 1994). For mutagenesis, we followed published procedures (Landt *et al.*, 1990) in which the mutation (either point mutation or deletion) is introduced via a primer and PCR is used to amplify a DNA fragment containing the mutation. The product of the PCR amplification is gel purified, and subcloned into pBS II KS+. The presence of specific mutations in the resulting plasmid clones is verified by DNA sequencing.

3.2. Fusion Assay

We adapted the assay of Nussbaum et al., (1994) in which expression of the E. coli LacZ gene is used as the "readout" to measure S protein induced cell fusion. In this assay a group of DBT cells are infected with vaccinia vTF7–3 (expressing the T7 RNA polymerase), followed by transfection of the wild type (pINT2) or mutant S cDNA clone in the presence of Lipofectin. Another group of DBT cells is transfected also in the presence of Lipofectin with pG1NTβGal, a plasmid that expresses β galactosidase under control of the T7 promoter. After four hours the cells containing the S plasmid and vTF7–3 are removed from the plastic and plated onto a monolayer of the pG1NTβGal transfected cells and incubated overnight. Thus when the cells express fusion competent S, fusion occurs between the two cell types and the LacZ gene is expressed in the syncytia. Cells are fixed with 2% formaldehyde/0.2% glutaraldehyde and incubated in the presence of X-gal. A blue color is observed in the syncytia Fusion can be measured quantitatively by lysing the cells, adding the lysates to a solution of chlorophenol red-β-D-galactopyranoside (CPRG) and measuring the rate of hydrolysis compared to that of purified E. coli β galactosidase, in a microplate absorbance reader. The rate of hydrolysis is then converted to the amount of β galactosidase in ng/well background (after subtraction of the amount of activity for vTF7–3 infection alone) and then expressed as a percentage of wild type. When we quantitate fusion we carry out three independent transfections and assay each of these in duplicate. In Figure 2, the results of the fusion assay are designated: - (<10%); + (10–50%); and ++ (>50%) where wild type is 100%.

4. RESULTS

4.1. Identification of Putative Fusion Peptides

Using the computer program, Clustal, to analyze MHV-A59 S2, we have identified two peptides that fit the description of fusion peptides as they can be modeled as sided helices (White, 1992; White, 1990). The peptides identified are KMIASAFNNALGAIQDGFD (amino acids 971–989) and HILSLVQNAPYGLYFIH (amino acids 1098–1114). (See Figure

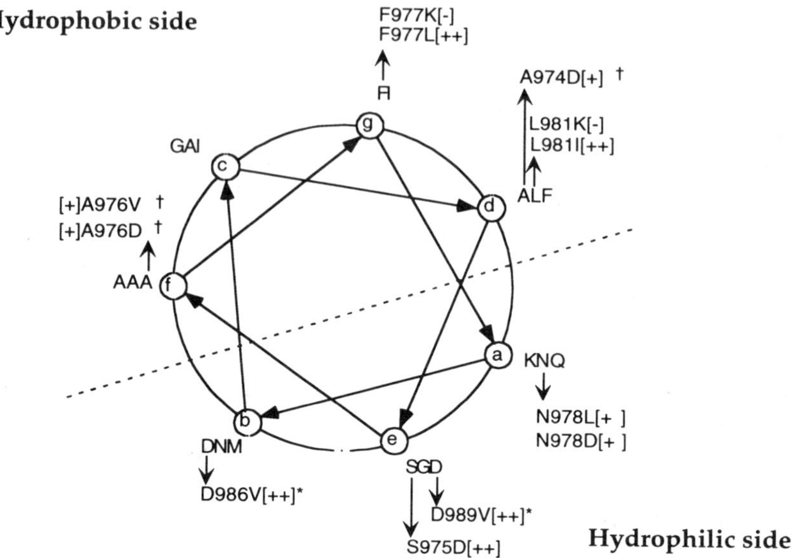

Figure 2. Candidate fusion peptide # 1 shown as a hypothetical α helix. The dotted line indicates the sidedness of the helix. Amino acid substitutions performed inside this region are indicated. The effect of each amino acid substitution on the fusogenic ability, as determined by measuring β-gal activity, is shown in the brackets as percentage of the wild type activity. "++": >50%; "+": 10–50%; "-": <10%; "*"; results shown are referred to the D986V-D989V double mutant; "†": Results shown are referred to the corresponding single mutants. Results for double mutants A974D-A976D and A974D-A976V are [-].

1 for location within S2.) The hydrophobicity indices of the predicted hydrophobic faces of peptides #1 and #2 are 0.91 and 0.96, respectively. These peptides are conserved among coronaviruses, an important criterion for fusion peptides (White, 1992; White, 1990) and nearly identical among several stains of MHV (data not shown). There is another region of S2 (peptide #3, amino acids 911–948, see Figure 1) that was suggested previously to be a candidate fusion domain based on its hydrophobicity and its proximity to the heptad repeats (Chambers *et al.*, 1990). We think this is a less likely candidate for a fusion peptide because it is less conserved than either peptide #1 and #2. It furthermore is not predicted to form a sided helix.

4.2. Effects of Mutagenesis of Putative Fusion Peptide Domains on Cell to Cell Fusion

4.2.1. Fusion Assay. In order to use site directed mutagenesis to map the fusion peptide we needed both a plasmid encoding S and a quantitative *in vitro* assay to measure cell to cell fusion. We constructed a cDNA clone (pINT2) with which to express the S protein under control of the T7 RNA polymerase promoter (see Materials and Methods). Using pINT2, transfected into cells infected with vaccinia virus expressing T7 polymerase (vTF7–3), we confirmed the results of others (Taguchi *et al.*, 1992; Gallagher *et al.*, 1991) that expression of S protein alone is sufficient to induce cell to cell fusion (data not shown). However for the analysis of S mutants, we wanted to establish a more quantitative assay. Thus we adapted the assay of Nussbaum et al., (1994) in which expression of the E. coli LacZ gene is used as the "readout" to measure S protein induced cell fusion. This is described in the Materials and Methods section. Using this assay, fusion mediated by S can be assayed qualitatively using an *in situ* assay by staining for LacZ activity or quantitatively by measuring the amount of LacZ activity compared to a standard curve for bacterial β-galactosidase all as described in the Materials and Methods.

4.2.2. Mutagenesis of Peptide #1. We analyzed the effects of mutations primarily in peptide #1. Deletion of the entire peptide eliminated cell fusion in this assay. We then introduced charged residues into the predicted hydrophobic face of peptide #1. These results are shown in Figure 2, in which peptide #1 is modeled as a sided helix. Replacement of hydrophobic residues (for example L981K or F977K) reduces fusion to background level. Replacement of the same residues with other hydrophobic residues (L981I or F977L), however, does not reduce the amount of fusion. These results suggest that maintenance of the hydrophobicity in these residues is important for fusogenicity, consistent with the sided helix model. We also introduced mutations into two alanine residues as alanine is a common feature of fusion peptides (White, 1992; White, 1990). While mutation of A976 or A974 alone reduced somewhat the levels of fusion, replacement of both of these residues eliminates fusion completely, suggesting that alanines may be important for fusion.

In general, substitutions within the predicted hydrophilic portion of the helix had less impact on fusion. For example, mutation of the polar residue N978 to either a charged (N879D) or hydrophobic (N879L) residue reduced but did not eliminate fusion and mutation of S975D had little, if any, effect on fusion. The double replacement of charged for hydrophobic residues (D986V/D989V) once again reduced but did not eliminate fusion.

In these initial experiments we only assayed the ability of mutant proteins to induce fusion. For fusion negative proteins, we need to determine whether these proteins are misfolded and never reach the surface of the cell or whether they are present on the cell surface and are defective in the fusion process itself. We have begun to examine the synthesis and processing of these mutant S proteins to determine whether they do, in fact, arrive at the surface of the cell. We have detected full length intracellular proteins for several of the fusion impaired mutants (D986V/D989V, A976D/A974D and A976V/A974D). Preliminary results of fluorescent activated cell sorting (FACS) analysis suggest that A974D, S975D, F977K and A976D/A974D do reach the cell surface at similar levels to wild type S.

4.2.3. Mutagenesis of Peptide #2. We have shown that deletion of peptide #2 eliminates fusion activity. We have introduced substitutions of hydrophobic residues with charged lysine residues in several positions within the predicted hydrophobic face of peptide #2. These substitutions all eliminate fusion. Furthermore substitution of the central proline residue with lysine also eliminates fusion. Thus this peptide remains a candidate fusion peptide.

4.2.4. Mutagenesis of Peptide #3. Mutation of a nonpolar methionine to either leucine of lysine had no effect on fusion. Furthermore mutation of an internal proline to leucine or lysine did not effect fusion. These results along with the fact that this region of S2 is not very conserved among coronaviruses makes it unlikely that this is indeed a fusion peptide.

5. DISCUSSION

Both peptides #1 and #2 are conserved, hydrophobic regions of S that can be modeled into sided helices. Both of these peptides have charged residues at the boundaries, another predicted common feature of internal fusion peptides (White, 1992; White, 1990). We tested the hypothesis for each of these peptides that the hydrophobic residues forming the predicted face are necessary for the induction of cell fusion.

We concentrated on peptide #1. We constructed a group of plasmids to express S proteins with substitution of the nonpolar amino acids predicted to form the hydrophobic face with charged amino acids such as aspartic acid or lysine; the prediction would be that these substitutions would greatly decrease or eliminate fusion. In the case of the HIV fusion peptide one valine to glutamic acid mutation completely eliminated fusion (Pereira *et al.*, 1995). The data in Figure 2 (for example F977K, L981K, A974D/A976D) conform to this prediction. We also mutated amino acids believed to be on the hydrophilic side of the helix; mutations in this region were less detrimental to the induction of fusion. However the substitution of two hydrophobic amino acids for two charged amino acids (D986V/D989V) did substantially reduce fusion. Thus the most drastic effect on fusion was the replacement of bulky hydrophobic residues within the predicted hydrophobic face with charged residues, consistent with the requirement for a hydrophobic face for the induction of fusion. Thus peptide #1 remains a likely candidate fusion peptide.

Peptide #2 has an additional feature thought to be common to internal fusion peptides, that is the central proline residue White, 1990; White 1992). Mutation of this residue or several of the hydrophobic residues on the predicted hydrophobic face of the helix with charged residues eliminates fusion. Thus from the limited analysis we have carried out, peptide #2 conforms to the sided helix model and remains a candidate fusion peptide.

It is not clear, however, that all fusion peptides form sided helices. Indeed, studies on the fusion peptides of PH-30, measles virus, and HIV have suggested that fusion peptides can assume conformations other than alpha-helix when inserted into lipid bilayers and it has been suggested that fusion peptides may be more generally hydrophobic regions (Gallaher *et al.*, 1992). By this criterion, as well, peptides #1 and #2 are likely candidate fusion peptides in that they are the most conserved, hydrophobic regions of S2 other than the transmembrane domain. We have included in our analysis, a third possible fusion peptide (amino acids 911–948), suggested previously to be a candidate fusion domain (Chambers *et al.*, 1990). Mutation of a methionine or central proline residue to either charged or polar residues had no effect on fusion. We will carry out other mutations within this peptide. However, based on our limited analysis and the fact that this region is not highly conserved among coronavirus S protein, this is not a likely fusion domain.

The analysis we have carried out so far examines fusion as an end point. Some of the fusion mutants may well be conformational mutants that never reach the cell surface. We are in the process of distinguishing processing mutants from those defective in fusion. Among those defective in fusion there will be some likely in domains other than the fusion peptide that effect fusion. For example, It has been shown previously that mutations in other domains of S, for example the heptad repeats (Gallagher *et al.*, 1991), hypervariable domain (Gallagher *et al.*, 1990) and the cleavage site (Gombold *et al.*, 1993; Bos *et al.*, 1996) affect cell to cell fusion. Thus once candidate fusion peptide are defined by mutagenesis, they must be assayed in an *in vitro* assay with liposomes (Pereira *et al.*, 1995) to conclusively show that they are involved in the fusion process.

ACKNOWLEDGMENTS

This work was supported by public health service grants NS-21954 and NS-30606. We thank Drs. Jim Gombold and Paul Masters for construction of pINT2.

REFERENCES

Bos, E. C. W., Heijnen, L., Luytjes, W., and Spaan, W. J. M., 1996, Mutational analysis of the murine coronavirus spike protein: effect on cell to cell fusion, *Virology* **214**: 453–463.

Chambers, P., Pringle, C. R., and Easton, A. J., 1990, Heptad repeat sequences are located adjacent to hydrophobic regions in several types of virus fusion glycoproteins, *J. Gen. Virol.* **71**: 3075–3080.

Gallagher, T. M., Parker, S. E., and Buchmeier, M. J., 1990, Neutralization resistant variants of a neurotropic coronavirus are generated by deletions within the amino terminal half of the spike glycoprotein, *J. Virol.* **64**: 731–741.

Gallagher, T. M., Escarmis, C., and Buchmeier, M. J., 1991, Alteration of pH dependence of coronavirus-induced cell fusion: Effect of mutations in the spike glycoprotein, *J. Virol.* **65**: 1916–1928.

Gallaher, W. R., Segrest, J. P., and Hunter, E., 1992, Are fusion peptides really "sided" insertional helices? *Cell* **70**: 531–532.

Gombold, J. L., Hingley, S. T., and Weiss, S. R., 1993, Fusion-defective mutants of mouse hepatitis virusA59 contain a mutation in the spike protein cleavage signal, *J. Virol.* **67**: 4504–4512.

Hingley, S. T., Gombold, J. L., Lavi, E., and Weiss, S. R., 1994, MHV-A59 fusion mutants are attenuated and display altered hepatotropism, *Virology* **200**: 1–10.

Landt, O., Grunert, H. P., and Hahn, U., 1990, A general method for rapid site directed mutagenesis, *Gene* **96**: 125–128.

Nussbaum, O., Broder, C. C., and Berger, E. A., 1994, Fusogenic mechanisms of enveloped-virus glycoproteins analyzeed by a novel recombinant vacciniavirus-based assay qunatitating cell fusion-dependent reporter gene activation, *J. Virol.* **68**: 5411–5422.

Pereira, F. B., Goni, F. M., and Nieva, J. L., 1995, Liposome destabilization induced by the HIV fusion peptide: Effect of single aminom acid changes, *FEBS Letters* **362**: 2243–246.

Taguchi, F., Ikeda, T., and Shida, H., 1992, Molecular cloning and expression of a spike protein of neurovirulent murine coronavirus JHMV variant cl-2, *J. Gen. Virol.* **73**: 1065–1072.

White, J., 1992, Membrane fusion, *Science* **258**: 917–924.

White, J. M., 1990, Viral and cellular membrane fusion proteins, *Ann. Rv. Physiol.* **52**: 675–697.

4

INTERACTIONS OF ENTEROTROPIC MOUSE HEPATITIS VIRUSES WITH Bgp2 RECEPTOR PROTEINS

Susan R. Compton

Section of Comparative Medicine
Yale University School of Medicine
New Haven, Connecticut 06520-8016

1. ABSTRACT

Enterotropic mouse hepatitis virus (MHV) infections are limited to the intestinal mucosa, rarely disseminate to other tissues and cause disease only in neonatal mice.The role of virus-host cell receptor interactions in the limited tissue tropism of enterotropic MHV infections is unclear. Previous studies have shown that enterotropic MHV-Y can infect BHK cells stably transfected with either the MHVR or the mmCGM2 receptor gene. In contrast, enterotropic MHV-RI infects BHK cells stably transfected with the MHVR but not the mmCGM2 receptor gene. Studies to determine whether MHV-Y and -RI can utilize the Bgp2 receptor isoform were performed. Both MHV-Y and -RI infected Vero cells transiently transfected with the Bgp2 receptor gene, though only MHV-Y infected CHO cells stably transfected with the Bgp2 receptor gene. Additionally, pretreatment with anti-MHVR monoclonal antibody (CC1) did not prevent MHV-Y and -RI infection of CMT93 cells. In contrast, pretreatment with CC1 prevented MHV-A59 infection of CMT93 cells. It is likely that MHV-Y and -RI use the Bgp2 receptor to infect CC1 pretreated CMT93 cells, since CMT93 cells are known to possess high levels of the Bgp2 receptor mRNAs, but it is also possible that they use an unidentified receptor.

2. INTRODUCTION

Mouse hepatitis virus (MHV), a singular name for several murine coronaviruses, causes a wide spectrum of diseases ranging from mild enteritis or rhinitis to fatal hepatitis

Coronaviruses and Arteriviruses, edited by Enjuanes *et al.*
Plenum Press, New York, 1998

or encephalitis. MHV strains can be divided into two biotypes, respiratory and enterotropic, on the basis of their initial site of replication. Following oronasal inoculation, respiratory MHV strains initiate replication in the upper respiratory tract and then disseminate to multiple organs if the mouse is sufficiently susceptible due to age, genotype or immune status (Barthold and Smith, 1987; Compton *et al.*, 1993). On the other hand, replication of enterotropic MHV strains, such as MHV-Y, is largely restricted to the intestinal mucosa, with minimal or no dissemination to other organs (Barthold, 1987; Barthold *et al.*, 1993). All ages and genotypes of mice are susceptible to infection with enterotropic MHV-Y but only young mice develop disease in the form of enteritis (Barthold, 1987; Barthold *et al.*, 1993). Unlike respiratory MHV strains in which viral titers mirror the severity of lesions, the titers of MHV-Y produced in the intestines do not reflect the level of lesion formation (Barthold and Smith, 1987; Barthold *et al.*, 1993).

Several members of the biliary glycoprotein (Bgp) subfamily of the murine carcinoembryonic antigen (mmCGM) family, have been previously shown to function as cell surface receptors for MHV (Boyle *et al.*, 1987; Compton, 1994; Dveksler *et al.*, 1991; Dveksler *et al.*, 1993a; Nedellec *et al.*, 1994; Williams *et al.*, 1990; Williams *et al.*, 1991; Yokomori *et al.*, 1992a; Yokomori *et al.*, 1992b). Both genes (MHVR and mmCGM2) at the Bgp1 genomic locus encode proteins that have a similar structure with a leader peptide, four immunoglobulin-like domains, a transmembrane domain and a cytoplasmic domain. Differential splicing of these two genes in different mouse tissues generates proteins containing either two or four of the immunoglobulin-like domains (Dveksler *et al.*, 1991; Dveksler *et al.*, 1993a;Yokomori *et al.*, 1992a; Yokomori *et al.*, 1992b). The MHVR(4d) and MHVR(2d) isoforms are present in tissues from BALB/c, C57BL/6 and C3H mice while the mmCGM$_2$(4d) and mmCGM$_2$(2d) isoforms are present in tissues from SJL mice (Dveksler *et al.*, 1991;Yokomori *et al.*, 1992a; Yokomori *et al.*, 1992b). Outbred CD-1 mice express both the MHVR and mmCGM$_2$ isoforms (Beauchemin *et al.*, 1989; McCuaig *et al.*, 1992; McCuaig *et al.*, 1993; Turbide *et al.*, 1991). MHVR and mmCGM$_2$ have extensive differences in domain 1, the domain to which binding site for the S protein has been mapped (Dveksler *et al.*, 1993b; Dveksler *et al.*, 1995). Only one isoform of the Bgp2 gene has been characterized. It contains a leader peptide, two immunoglobulin-like domains, a transmembrane domain and a cytoplasmic domain and is expressed in tissues of BALB/c, C57BL/6, C3H, SJL and outbred strains of mice (Nedellec *et al.*, 1994).

Using stably transfected cells expressing the MHVR(4d) isoform, it was shown that MHVR can act as a functional cell surface receptor for MHV-A59, -JHM, -Y and -RI (Compton, 1994; Dveksler *et al.*, 1991; Yokomori *et al.*, 1992a) . Using stably transfected cells expressing the mmCGM2(2d) isoform, mmCGM2 was been shown to act as a functional cell surface receptor for MHV-A59, -JHM and -Y but not MHV-RI (Compton, 1994; Dveksler *et al.*, 1993a; Yokomori *et al.*, 1992b). Using stably transfected cells expressing high levels of the Bgp2(2d) isoform, Bgp2 was been shown to act as a functional cell surface receptor for MHV-A59 and -JHM (Nedellec *et al.*, 1994). In contrast, Bgp2 did not serve as a functional receptor on CMT93 cells, a cell line expressing normal levels of Bgp2 (Nedellec *et al.*, 1994).

MHV-RI is capable of infecting SJL mice, producing very little clinical symptoms but high levels of virus. At the initiation of this study, a functional receptor for MHV-RI in SJL mice had not been identified. This study addressed the question of whether, the Bgp2 receptor isoform, expressed on many SJL tissues, could serve as a functional receptor for enterotropic MHV-RI and -Y.

3. MATERIALS AND METHODS

3.1. Viruses and Cells

MHV-Y was originally isolated in NCTC 1469 cells from the intestine of an infant mouse with acute typhlocolitis (Barthold *et al* , 1982). MHV-RI was originally isolated in CMT-93 cells from a nude mouse intestine (Barthold *et al.*, 1985). MHV-A59 was obtained from American Type Culture Collection (Rockville, MD). Virus stocks used in these studies were generated in J774A.1 cells. J774A.1 cells were obtained from American Type Culture Collection (Rockville, MD) and maintained in RPMI medium 1640 with 10% fetal bovine serum. CMT93 cells were obtained from American Type Culture Collection (Rockville, MD) and were maintained in Dulbecco's Modified Essential Medium/Leibowitz Medium with 10% fetal bovine sera. CHO-Bgp2 cells were obtained from G. Dveksler (Uniformed Services University of the Health Sciences, Bethesda, MD) and were maintained in alpha-Modified Essential Medium with 10 % fetal bovine sera. Vero cells were obtained from American Type Culture Collection (Rockville, MD) and were maintained in Dulbecco's Modified Essential Medium/Leibowitz Medium with 10% fetal bovine sera. Cells were transiently transfected with the pRCMV-Bgp2(2d) plasmid (obtained from G. Dveksler, Uniformed Services University of the Health Sciences, Bethesda, MD) using Lipofectamine as per the manufacturers instructions.

3.2. Antibodies and Immunofluorescence

Monolayer cultures of cells were grown on four chambered glass slides, mock- or MHV-infected and fixed in acetone. Viral antigens were detected with mouse anti-sialodacryoadenitis virus serum and FITC-conjugated goat anti-mouse IgG (Antibodies Inc, Davis CA) and counter-stained with Evan's blue. Cells were mounted in 80% glycerol and examined with an Olympus BH2 microscope. Anti-MHVR monoclonal antibody CC1 was obtained from K. Holmes (Uniformed Services University of the Health Sciences, Bethesda, MD).

4. RESULTS

CHO cells stably transfected with pRCMV plasmid expressing the Bgp2(2d) isoform of the MHV receptor were challenged with MHV-A59, -Y or MHV-RI using MOIs of 2–7. At 24 hours post-inoculation, cells were fixed and examined by immunofluorescence. Both MHV-A59 and -Y infected CHO-BGP2 cells and caused a low level of cell fusion (data not shown). MHV-RI did not infect CHO-Bgp2 cells and MHV-A59, -Y and -RI did not infect receptor-negative CHO-pRCMV cells based on the lack of visible cytopathic effect and viral antigen as detected by immunofluorescence at 24 hours post-inoculation. It was unclear whether the failure of MHV-RI to infect CHO-Bgp2 cells was due the inability of the Bgp2 isoform to be utilized as a functional receptor or whether CHO cells lack a factor necessary for MHV-RI replication. Since previous work (Compton, 1994) has shown that BHK cells possess all the factors necessary to uncoat and replicate MHV-RI, infection of BHK cells transiently transfected with the pRCMV-Bgp2(2d) plasmid was attempted. Infection levels with MHV-A59, -Y and -RI were very low, with only a few isolated cells expressing MHV antigens (data not shown). Therefore additional cell lines were tested. MHV-A59, -Y and -RI were capable of infecting Vero cells transiently transfected with the pRCMV-Bgp2(2d) plasmid. They produced numerous syncytia expressing MHV antigen as detected by immunofluorescence (data not shown).Therefore, the Bgp2(2d) isoform can serve as a functional receptor for the enterotropic MHV strains -Y and -RI.

Nedellec *et al.* (1994) has shown that even though the Bgp2(2d) molecule can serve as a functional cell surface receptor for MHV-A59, -JHM and -3 when expressed at high levels, it does not serve as a functional receptor for these strains on CMT93 cells which express normal levels of Bgp2 molecules. Given that CMT93 cells were used in the original isolation of MHV-RI from an infected nude mouse intestine and therefore probably express a receptor with high affinity for MHV-RI, it was of interest to determine if this receptor was Bgp2. CMT93 cells were pretreated with anti-MHVR monoclonal antibody for 1 hour and then inoculated with MHV-A59

(MOI = 0.1), MHV-Y (MOI = 3) or MHV-RI (MOI = 3). Similar to results published by Nedellec *et al.* (1994), CC1 pretreatment of CMT93 cells completely blocked MHV-A59 infection. In contrast CC1 pretreatment of CMT93 cells did not completely block infection by MHV- RI or -Y (75–85% inhibition). Additionally infection of CC1 pretreatment of CMT93 cells with an infant mouse intestinal stock of a wild type strain (wt2) isolated at Yale was only inhibited by 25–50%. Therefore at normal levels of expression, a receptor to which CC1 can not bind, Bgp2 or possibly an alternative unidentified receptor serves as functional receptor for MHV-RI and other enterotropic MHV strains in CMT93 cells.

5. DISCUSSION

Initiation of an MHV infection requires interactions between S proteins on the virus particle membrane or virus-infected cell membranes and cell surface receptors. The S proteins of MHV strains are highly variable and differences in the S protein may account for the different binding affinities observed by several labs to a single receptor isoform. The binding affinity of the S protein to different receptor isoforms vary greatly as demonstrated by Ohtsuka et al. (1996) with MHV-JHM and the MHVR(2d) and mmCGM2(2d) receptor isoforms. Many carcinoembryonic antigen-related family members have been identified which can serve as receptors for respiratory MHV strains when expressed at high levels in vitro in mon-murine cells. Initially, respiratory MHV strains were shown to use several members of the murine Bgp family (MHVR, mmCGM2, Bgp2) as cell surface receptors (Dveksler *et al.*, 1991; Dveksler *et al.*, 1993a; Nedellec *et al.*, 1994; Yokomori *et al.*, 1992a; Yokomori *et al.*, 1992b). More recently, several respiratory MHV strains (MHV-A59, -2, and -3) have also been shown to be able to utilize the murine pregnancy-specific glycoprotein, another member of the carcinoembryonic antigen-related family, as a receptor when it is expressed in COS7 cells (Chen *et al.*, 1995). MHV-A59 and -2 also have been shown to be able to utilize the human carcinoembryonic antigen and human biliary glycoprotein molecules as receptors when they are expressed in COS7 cells (Chen *et al.*, 1997). The affinity of MHV S proteins for these each of receptors seems to differ substantially though definitive studies have not been done to quantitatively measure these differences in affinity. At normal levels of receptor expression in murine cells, some S protein-Bgp interactions, such as the interaction of the MHV-A59 S protein with Bgp2 molecules on CC1 treated CMT93 cells, are not strong enough to initiate an infection (Nedellec *et al.*, 1994). Also, infected cells, which may express higher concentrations of S protein than virus particles, can be used to initiate infection in non-murine cells such as

BHK cells (Gallagher, 1992; Nash and Buchmeier, 1996). This form of cell-associated infection probably occurs following the interaction of S proteins with a unidentified low affinity hamster homolog of the murine Bgp molecule.

Given the expanding number of carcinoembryonic antigen-related molecules which have been identified as MHV receptors in vitro, it will be important to determine the role of each of these receptors in vivo for each MHV strain. Several murine cells lines which express functional receptors are not infectable by MHV, indicating that the presence of a functional receptor is only the first step at which infection can be restricted (Yokomori and Lai, 1992b; Yokomori *et al.*, 1993). Similar restrictions in uncoating/ or replication factors may also occur in vivo since several tissues which express substantial levels of MHV receptor molecules are not target tissues in vivo (Godfraind *et al.*, 1995; Nedellec *et al.*, 1994). Many cell types express multiple receptor isoforms and it is possible that receptors are present on cclls as both homodimers and heterodimers (Dveksler, *et al.*, 1996). The presence of heterodimers on cells may explain why studies in CMT93 cells pretreated with monoclonal antibody CC1, which would bind to both MHVR/MHVR homodimers and MHVR/Bgp2 heterodimers, gave partial inhibition of enterotropic MHV infection. The fact that infection of CMT93 cells with an intestinal stock of a wild type enterotropic MHV strain, containing a set of quasi species which are adapted to the receptor distribution present in the mouse intestine, was inhibited only 2–4 fold by pretreatment with CC1 indicates that experiments with cell culture-adapted viruses may result in a picture of receptor usage which is not reflective of receptor usage in the animal. It appears likely that MHVR and mmCGM2 are the primary receptors for respiratory MHV strains in BALB/c and SJL mice respectively. In contrast, Bgp2 may be the primary receptor for enterotropic MHV- RI in SJL mice, and possibly in other mouse strains. The receptor binding domain of several MHV strains has been mapped to the amino terminal 330 amino acids of the S protein (Kubo *et al.*, 1994; Suzuki and Taguchi, 1996). This and other studies with enterotropic MHV strains, have shown that a small number of changes (14 changes in 330 amino acids between MHV-RI and MHV-4) in the receptor binding domain of the S protein may be responsible for large differences in receptor usage patterns (Compton,1994; Kunita *et al.*, 1995). Studies using anti-Bgp2 antibodies and S protein chimeras are being performed to map these difference in receptor utilization to one or more of the 14 amino acid changes in the amino terminal region of the S protein and to determine the role of Bgp2 receptors in enterotropic MHV strains in vivo.

ACKNOWLEDGMENTS

I am grateful for the technical assistance of Gail Lizarraga. This work was supported by Public Health Service grant RR2039 from the National Institutes of Health.

REFERENCES

Barthold, S.W., 1987, Host age and genotypic effects on enterotropic mouse hepatitis virus infection, *Lab. Anim. Sci.* **37**:36–40.

Barthold, S.W. and A.L. Smith., 1987, Response of genetically susceptible and resistant mice to intranasal inoculation with mouse hepatitis virus JHM, *Virus Res.* **7**:225–239.

Barthold, S.W., A.L. Smith, P.F.S. Lord, P.N. Bhatt, R.O. Jacoby, and A.J. Main., 1982, Epizootic coronaviral typhlocolitis in suckling mice, *Lab. Anim. Sci.* **32**:376–383.

Barthold, S.W., A.L. Smith and M.L. Povar., 1985, Enterotropic mouse hepatitis virus infection in nude mice, *Lab. Anim. Sci.* **35**:613–618.

Barthold, S.W., D.S. Beck and A.L. Smith., 1993, Enterotropic coronavirus (mouse hepatitis virus) in mice: influence of host age and strain on infection and disease, *Lab. Anim. Sci.* **43**:276–284.

Beauchemin, N., C. Turbide, D. Afar, J. Bell, M. Raymond, C.P. Stanners and A. Fuks., 1989, A mouse analogue of the human carcinoembryonic antigen, *Cancer Res.* **49**:2017–2021.

Boyle, J.F., Weismiller, D.G. and Holmes, K.V., 1987, Genetic resistance to mouse hepatitis virus correlates with absence of virus-binding activity on target tissues, *J. Virol.* **61**:185–189.

Chen, D.S., Asanaka, M., Yokomori, K., Wang, F.-I., Hwang, S.B., Li, H.-P. and Lai, M.M.C., 1995, A pregnancy-specific glycoprotein is expressed in the brain and serves as a receptor for mouse hepatitis virus, *Proc. Natl. Acad. Sci. USA* **92**:12095–12099.

Chen, D.S., Asanaka, M., Chen, F.S., Shively, J.E. and Lai, M.M.C., 1997, Human carcinoembryonic antigen and biliary glycoprotein can serve as mouse hepatitis virus receptors, *J.Virol.* **71**:1688–1691.

Compton, S.R., S.W. Barthold, and AL. Smith., 1993, The cellular and molecular pathogenesis of coronaviruses, *Lab. Anim. Sci.* **43**:1–14.

Compton, S.R., 1994, Enterotropic strains of mouse coronavirus differ in their use of murine carcinoembryonic antigen-related glycoprotein receptors, *Virology* **203**:197–201.

Dveksler, G.S., M.N. Pensiero, C.B. Cardellichio, R.K. Williams, G-S. Jiang, K.V. Holmes, and C.W. Dieffenbach., 1991, Cloning of the mouse hepatitis virus (MHV) receptor: expression in human and hamster cell lines confers susceptibility to MHV, *J. Virol.* **65**:6881–6891.

Dveksler, G.S., C.W. Dieffenbach, C.B. Cardellichio, K. McCuaig, M.N. Pensiero, G-S. Jiang, N. Beauchemin, and K.V. Holmes., 1993a, Several members of the mouse carcinoembryonic antigen-related glycoprotein family are functional receptors for mouse hepatitis virus -A59, *J. Virol.* **67**:1–8.

Dveksler, G.S., M.N. Pensiero, C.W. Dieffenbach, C.B. Cardellichio, A.A. Basile, P.E. Elia and K.V. Holmes., 1993b, Mouse hepatitis virus strain A59 and blocking antireceptor monoclonal antibody bind to the N-terminal domain of cellular receptor, *Proc. Natl. Acad. Sci. USA*. **90**:1716–1720.

Dveksler, G.S., A.A. Basile, C.B. Cardellichio and K.V. Holmes., 1995, Mouse hepatitis virus receptor activities of an MHVR/mph chimera and MHVR mutants lacking N-linked glycosylation of the N-terminal domain. *J. Virol.* **69**:543–546.

Dveksler, G.S., S.E. Gagneten, C.A. Scanga, C.B. Cardellichio and K.V. Holmes., 1996, Expression of the recombinant anchorless N-terminal domain of mouse hepatitis virus (MHV) receptor makes hamster or human cells susceptible to MHV infection, *J. Virol.* **70**:4142–4145.

Gallagher, T.M., Buchmeier, M.J. and Perlman, S., 1992, Cell receptor-independent infection by a neurotropic murine coronavirus, *Virology* **191**:517–522.

Godfraind, C., Langreth, S.G., Cardellichio, C.B., Knobler, R.L., Coutelier, J.P., Dubois-Dalcq, M. and Holmes, K.V., 1995, Tissue and cellular distribution of an adhesion molecule in the carcinoembryonic antigen family that serves as a receptor for mouse hepatitis virus, *Lab. Invest.* **73**:615–627.

Kubo, H., Yamada, Y.K. and Taguchi, F., 1994, Localization of neutralizing epitopes and receptor-binding site within the amino-terminal 330 amino acids of the murine coronavirus spike protein, *J. Virol.* **68**:5403–5410.

Kunita, S., Zhang, L., Homberger, F.R. and Compton, S.R., 1995, Molecular characterization of the S proteins of two enterotropic murine coronavirus strains, *Virus Res.* **35**:277–289.

McCuaig, K., C. Turbide and N. Beauchemin., 1992, mmCGM1a: a mouse carcinoembryonic antigen gene family member, generated by alternative splicing, functions as an adhesion molecule, *Cell Growth Differ.* **3**:165–174.

McCuaig, K., M. Rosenberg, P. Nedellec, C. Turbide and N. Beauchemin., 1993, Expression of the Bgp gene and characterization of mouse colon biliary glycoprotein isoforms, *Gene* **127**:173–183.

Nash, T.C. and Buchmeier, M.J., 1996, Spike glycoprotein-mediated fusion in biliary glycoprotein-independent cell-associated spread of mouse hepatitis virus infection, *Virology* **223**:68–78.

Nedellec, P., G.S. Dveksler, E. Daniels, C. Turbide, B. Chow, A.A. Basile, K.V. Holmes and N. Beauchemin., 1994, Bgp2, a new member of the carcinoembryonic antigen-related gene family, encodes an alternative receptor for mouse hepatitis viruses, *J. Virol.* **68**: 4525–4537.

Ohtsuka, N., Yamada, Y.K. and Taguchi, F., 1996, Difference in virus-binding activity to two distinct receptor proteins for mouse hepatitis virus, *J. Gen. Virol.* **77**:1683–1692.

Suzuki, H. and Taguchi, F., 1996, Analysis of the receptor-binding site of murine coronavirus spike protein, *J. Virol.* **70**:2632–2636.

Turbide, C., M. Rojas, C.P. Stanners and N. Beauchemin., 1991, A mouse carcinoembryonic antigen gene family member is a calcium-dependent cell adhesion molecule, *J. Biol. Chem.* **266**:309–315.

Williams R.K., G-S. Jiang , S.W. Snyder, M.F. Frana, and K.V. Holmes., 1990, Purification of the 110-kilodalton glycoprotein receptor for mouse hepatitis virus (MHV)-A59 from mouse liver and identification of a non-functional homologous protein in MHV-resistant SJL/J mice, *J. Virol.* **64**:3817–3823.

Williams R.K., G-S. Jiang and K.V. Holmes., 1991, Receptor for mouse hepatitis virus is a member of the carcinoembryonic antigen family of glycoproteins, *Proc. Natl. Acad. Sci. USA* **88**:5533–5536.

Yokomori, K. and M.M.C. Lai., 1992a, Mouse hepatitis virus utilizes two carcinoembryonic antigens as alternative receptors, *J. Virol.* **66**:6194–6199.

Yokomori, K. and M.M.C. Lai., 1992b, The receptor for mouse hepatitis virus in the resistant mouse strain SJL is functional: implications for the requirement of a second factor for viral infection, *J. Virol.* **66**:6931–6938.

Yokomori, K., Asanaka, M., Stohlman, S.A. and Lai, M.M.C., 1993, A spike protein-dependent cellular factor other than the viral receptor is required for mouse hepatitis virus entry, *Virology* **196**:45–56.

5

VIRUS-RECEPTOR INTERACTIONS AND INTERSPECIES TRANSFER OF A MOUSE HEPATITIS VIRUS

Lisa E. Hensley,[1] Kathryn V. Holmes,[2] Nicole Beauchemin,[3,4] and Ralph S. Baric[1,5]

[1]Department of Epidemiology
[5]Department Of Microbiology And Immunology
University of North Carolina at Chapel Hill
Chapel Hill, North Carolina 27599
[2]Department of Microbiology
University of Colorado Health Sciences Center
Denver, Colorado
[3]Department of Biochemstry
[4]Department of Medicine
McGill Cancer Centre
McGill University
Montreal, Quebec, Canada

1. ABSTRACT

Molecular mechanisms regulating virus xenotropism and cross-species transmission are poorly understood. Host range mutants (MHV-H2) of mouse hepatitis virus (MHV) strains were isolated from mixed cultures containing progressively increasing concentrations of nonpermissive Syrian baby hamster kidney (BHK) cells and decreasing concentrations of permissive murine astrocytoma (DBT) cells. MHV-H2 was polytrophic, replicating efficiently in normally nonpermissive BHK cells, Syrian and Chinese hamster (DDT-1 and CHO) cells, human adenocarcinoma (HRT), primate kidney (VERO) and in murine 17Cl-1 cell lines. Little if any virus replication was detected in feline kidney (CRFK), and porcine testicular (ST) cell lines. To study the effects of xenotropic spread on virus receptor-interactions in the original host, murine DBT cells were pretreated with a monoclonal antibody (MAb) CC1, directed against the MHV receptor, MHVR, a biliary glycoprotein (Bgp1[a]). Under treatment conditions that completely ablated the replication of the parental MHV strains, CC1 antireceptor antibodies did not block MHV-H2 replica-

Coronaviruses and Arteriviruses, edited by Enjuanes *et al.*
Plenum Press, New York, 1998

tion. Following expression of MHVR in normally nonpermissive ST and CRFK cells, infection with the parental MHV strains, but not MHV-H2 was observed. To characterize the molecular basis preventing the interaction between MHV-H2 and MHVR, revertants of MHV-H2 (MHV-H2R6, MHV-H2R11) were isolated following a persistent MHV-H2 infection in DBT cells. These revertant viruses efficiently recognized MHVR, however infection of murine cells was resistant to MAb CC1 blockade. In addition, MHV-H2 and the revertant viruses efficiently recognized other Bgp receptors for docking and entry. These data suggest that interspecies transfer may remodel normal virus-receptor interactions that may result in altered virulence, tropism or pathogenesis in the original host.

2. INTRODUCTION

Emerging virus are frequently defined as either newly recognized viruses, or viruses that are rapidly increasing in incidence or expanding in geographic range (Morse, 1995). Although some new viral diseases may have resulted from mutations that altered tissue tropism, virulence or pathogenesis in the normal host, most new human viruses probably arose by cross species transmission from animal reservoirs (Morse, 1995; Kilbourne 1991). Recent examples include equine morbillivirus, human immunodeficiency viruses, hantavirus, hemorrhagic fever viruses, arboviruses, bovine spongiform encephalopathy, canine distemper, and influenza viruses. Factors that precipitate zoonotic virus transference to alternative hosts are extremely difficult to predict, hampering efforts to prevent the emergence and dissemination of new viral diseases in animals, humans, and plants (Baric et al., 1997).

MHV is highly species specific and tissue tropic, and is therfor an excellent model to investigate the fundamental molecular mechanisms that regulate virus xenotropism and cross-species transmission. Similar to many emerging RNA viruses, MHV host range is predominantly mediated at the level of entry as the MHV genomic RNA is infectious in nonpermissive baby hamster kidney cells and also human cells, and expression of MHVR permits viral docking and entry in nonpermissive hosts (Dveskler et. al., 1991; Schochetman et. al., 1977). In this paper, we investigate whether cross-species transmission of MHV has altered virus interactions in the original host.

3. MATERIALS AND METHODS

3.1. Viruses and Cells

Briefly, murine astrocytoma (DBT), murine 17Cl-1, baby hamster kidney (BHK), Chinese hamster ovary (CHO), hamster smooth muscle (DDT-1), feline kidney (CRFK), porcine testicular (ST), primate kidney (VERO), and human colorectal (HRT) cells were maintained as previously described (Baric et al., 1997). CHO cells stably expressing recombinant Bgp2 (CHO-Bgp2) and BHK cells stably expressing Bgp1^b (BHK-1^b) were maintained in standard tissue culture medium containing genetecin (800 µg/ml). These Bgp genes contain the Ig-like domains 1 and 4, the transmembrane domain, and the cytoplamic domain.

MHV-A59, MHV-JHM, MHV-H2, H2R6 and H2R11 were plaque purified and propagated in DBT cells or BHK cells. The TRSB strain of sindbis virus (kindly provided by Dr. Robert E. Johnson, University of North Carolina at Chapel Hill) was propagated in DBT cells.

3.2. Virus Growth Curves

Cultures of BHK, DDT-1, CHO, 17Cl-1, VERO, CRFK, MDCK, ST, HRT, BHK-1[b] and CHO-Bgp2 cells in $60mm^2$ dishes or in LabTek chamber slides (Nunc. Inc., Naperville, Ill.) were infected at an MOI of 10. After 1 hour at room temperature the innocula were removed and the monolayers were washed 3 times with phosphate buffered saline (PBS). The cultures were overlaid with complete media and maintained at 37°C. Samples of virus were harvested at various times post-infection and stored at -70°C. Virus titers were determined by plaque assay in DBT cells.

3.3. Detection of Viral Antigens by Immunofluorescence

Different cell lines grown on LabTek chamber slides (Nunc. Inc., Naperville, Ill.) were infected with MHV-A59 or MHV-H2 at room temperature at a MOI of 10 for 1 hour.. The innocula were removed and the monolayers were washed 3 times with PBS. Fresh media were added and the cells were maintained at 37°C. Infected cells were fixed and prepared for examination by Immunofluorescence as previously described (Chen, W. et. al., 1997).

3.4. Stable Expression of MHVR

A plasmid containing MHVR under the control of the cytomegalovirus (CMV) promoter in the pcDNA3 (Invitrogen) expression vector was transfected into CRFK and ST cells and stable expressing cells were selected with geneticin (Chen, W. et al., 1997). Stable cell lines were infected with H2R6, H2R11, MHV-JHM, MHV-H2 and MHV-A59.

3.5. Blockade of the Bgp1[a] Receptor

Blockade of MHV entry using the anti-MHVR MAb CC1 was performed as previously described (Chen et. al., 1997). As controls, additional blockade experiments were performed using two commercially antibodies, a monoclonal (kat4c), directed against human CEA gene family members: Bgp, CEA, CGM6, CGM1a and NCA, and a polyclonal (CEA) directed against all human CEA gene family members (DAKO).

3.6. Transient Expression of Bgp2

Bgp2 under the control of the CMV promoter in the prCMV expression system (5μg) (Invitrogen), was cotransfected with the pHook2 expression vector (2.5μg) (Invitrogen) by electroporation. Transfected cells were selected at 40 hours following electroporation. Briefly the cells were incubated with pHook magnetic beads for 1–2 hours, scraped and separated with a magnet. The procedure was repeated 24 hours later. Selected cells were seeded onto LabTek slides (Nunc Inc.). Cells were innoculated with MHV-H2 at an MOI of 10 for 1 hour and examined by immunofluorescence at 30 hours post-infection.

3.7. Scanning Electron Microscopy

Cultures of DBT cells were infected with MHV-A59, MHV-JHM, MHV-H2, H2R6 and H2R11 at an MOI of 10 for 1 hour at room temperature. At 14 hours post infection cultures were prepared for examination according to standard techniques (Bozzola, 1992).

Table 1. Virus replication in different mammalian cell lines

Cell lines	Virus Titer (Pfu / ml)		
	MHV - A59	MHV-JHM	MHV - H2
Murine DBT	5.0×10^7	1.0×10^7	9.5×10^6
Murine 17Cl-1	7.4×10^7	9.5×10^6	2.8×10^6
Hamster BHK	$< 1.0 \times 10^2$	$< 1.0 \times 10^2$	2.0×10^8
Hamster DDT-1	$< 1.0 \times 10^2$	$< 1.0 \times 10^2$	7.5×10^7
Hamster CHO	1.0×10^3	$< 1.0 \times 10^2$	7.5×10^6

4. RESULTS

4.1. Host Range Phenotype of MHV-A59 and MHV-H2

As previously reported MHV-H2 replicated efficiently in cultures of hamster (BHK, and CHO) and murine cells (DBT and 17Cl-1) (Baric et al., 1997). Under identical conditions MHV-A59 and MHV-JHM failed to replicate in the hamster cell lines. Although the replication of MHV-H2 was less efficient than the parental viruses in DBT cells (Table 1) significant synctium formation were noted within 16 hours of MHV-A59, MHV-JHM and MHV-H2 infection (Figure 1). These data demonstrated that the cross-species transmission of MHV-2 had not ablated the capacity of the virus to replicate in cells derived from the original murine host. Similar findings have been reported with primate-adapted strains of MHV (Murray, 1992).

4.2. MHV-H2 Resistance to MHVR Receptor Blockade

As coronavirus host range specificity is mediated at the level of entry, it seemed likely that expanded xenotropism may have altered specific interactions between the cellular receptor and the MHV-H2 viral attachment protein. To address this question murine DBT cells were pretreated with CC1 anti-MHVR MAb for 1 hour prior to infection. Anti-MHVR MAb CC1 has been demonstrated to block infection by binding to the N-terminus of MHVR (Dveksler et. al., 1993b). Although CC1 pretreatment significantly reduced MHV-A59 viral titers and the percentage of infected cells as observed by immunofluorescence, pretreatment failed to block MHV-H2 infection and did not reduce the percentage of infected cells (Table 2 and Table 3). As a control pretreatment of DBT cells with commercially available antibodies kat4c and a polyclonal CEA antibody directed against all human CEA glycoproteins did not block MHV-A59 or MHV-H2 infection (Table 2). These data suggested that MHV-H2 had either remodeled its normal interactions with MHVR so that it was no longer susceptible to blockade by the anti-MHVR MAb CC1 or suggested that MHV-H2 was no longer efficiently utilizing MHVR as a receptor for entry.

4.3. MHV-H2 Does Not Efficiently Utilize MHVR for Entry

To distinguish among these possibilities, MHVR in the pcDNA3 vector was transfected into nonpermissive ST cells. Although the parental strains MHV-A59 and MHV-JHM replicated efficiently in MHVR expressing ST cells, with titers approaching 1×10^7 PFU/ml in 24 hours, MHV-H2 failed to replicate above background levels (Table 4).

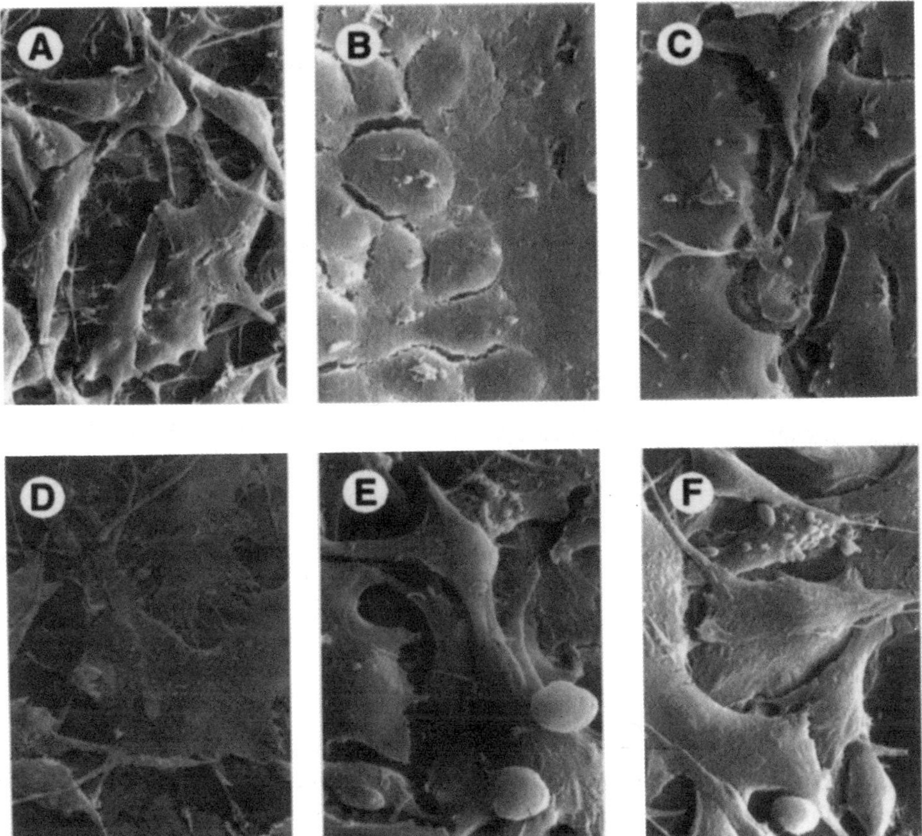

Figure 1. Synctium formation in DBT cells. DBT cells were infected with MHV-A59, MHV-JHM, MHV-H2, H2R6 and H2R11 at an MOI of 10 for 1 hour. At 16 hours post infection cells were examined with a scanning electron microscope. Panel A: Uninfected DBT cells; Panel B: DBT cells infected with MHV-A59; Panel C: DBT cells infected with MHV-JHM; Panel D: DBT cells infected with MHV-H2; Panel E: DBT cells infected with H2R6; Panel F: DBT cells infected with H2R11.

Table 2. Blockade of MHV infection in DBT cells

Virus	Virus Titers (Pfu / ml)			
	Untreated	Anit-MHVR CC1*	CEA$^\Delta$	Kat4c$^+$
MHV-A59	2.5×10^6	1.5×10^2	5.6×10^6	5.6×10^6
MHV-JHM	1.5×10^6	$< 1.0 \times 10^2$	N/D‡	N/D‡
MHV-H2	1.5×10^6	6.5×10^5	2.7×10^5	1.3×10^6

*A monoclonal antibody directed against MHVR.
$^\Delta$A commercially available (DAKO©) polyclonal antibody directed against human CEA and CEA related glycoproteins.
$^+$A commercially available (DAKO©) monoclonal antibody directed against human cd66abde glycoproteins.
‡ Not done.

Table 3. MHV-H2 resistance to CC1 blockade

Virus	Percentage of infected cells[2]	
	Untreated	Anit- MHVR CC1
MHV-A59	> 98.0 %	< 1.0 %
MHV-H2	> 98.0 %	> 98.0 %

[1]DBT cells were pretreated for 1hour with the Mab CC1. Cells were then challenged with either MHV-A59 or MHV-H2 at an MOI of 10. Cells were fixed at 22 hours post infection and examined by immunofluorescence.
[2]As measured by immunofluorescence at 22 hours post infection.

While significant numbers of infected cells were observed in cultures infected with MHV-A59 by immunofluorescence, there was no observable increase in the number of MHV-H2 infected cells (Table 5). Importantly, CC1 treatment of the ST-MHVR cell lines blocked MHV-A59 and MHV-JHM infection (Table 4). In addition, infection of cells expressing Bgp1^b allowed for low levels of MHV-A59 and MHV-H2 replication. These data indicated that MHV-H2 no longer efficiently recognized MHVR as a receptor for entry.

4.4. MHV-H2 Usage of Bgp2

DBT cells, which were derived from outbred CD1 mice, express MHVR, Bgp1^b and encode Bgp2 (Nedellec et al., 1994). In order to determine if MHV-H2 utilized different Bgp genes for entry into DBT cells, Bgp2 was transiently transfected into nonpermissive ST cells. Transfected cells, selected by the pHook system, were susceptible to infection by both MHV-A59 and MHV-H2 (Table 5). These data suggested that MHV-H2 may enter into murine cells by recognizing Bgp2, and to a lesser extent the N terminus of Bgp1b as a receptor.

4.5. Bgp Receptor Usage by the H2R6 and H2R11 Revertants

To further investigate the inability of MHV-H2 to utilize MHVR for entry into cells, revertants (H2R6 and H2R11) were isolated on day 25 and day 93 post-infection respectively, from a culture of DBT cells persistently-infected with MHV-H2. Importantly these revertants replicated efficiently in DBT cells but little if any replication was observed in

Table 4. MHV replication in ST-MHVR cells

	Virus Titer (Pfu / cell)[1]		
	MHV - A59	MHV-JHM	MHV - H2
ST cells (prior to transfection)	2.5×10^2	$< 1.0 \times 10^2$	2.5×10^2
ST-MHVR cells	7.1×10^7	1.7×10^5	$< 1.0 \times 10^2$
ST-MHVR cells + Mab CC1	4.1×10^4	2.5×10^2	$< 1.0 \times 10^2$

[1] At 22 hours post infection

Table 5. Virus infection in ST cells expressing MHVR or Bgp2[1]

		Percentage of infected cells	
Virus	ST cells	MHVR stable expressing ST cells	ST cells transiently expressing BGP2
MHV-A59	< 0.01%	40 − 50%	~30 %
MHV-H2	< 0.01 %	< 0.01%	~25 %

[1] At 30 hours post infection

BHK and CHO cells. Thus persistent MHV-H2 infection in DBT cells selected for variants which had lost their capacity to replicate efficiently in hamster cells. Interestingly, these virus also failed to produce syncytium, rather significant cell rounding was noted in infected cultures (Figure 1). While these revertants infected less than 5% of BHK cells under normal conditions, >95% of the BHK cell lines stably expressing Bgp1b or MHVR were susceptible to H2R6 and H2R11 infection. Similarly, while less than 5% of the control CHO cells were infected under normal conditions, > 95% of CHO cell lines expressing high levels of Bgp2 were susceptible to infection with H2R6 and H2R11 (Table 6). These data suggested that H2R6 and H2R11 had remodeled their interactions with the murine Bgp glycoprotein receptors and now utilized MHVR, Bgp1[b] and Bgp2 receptors for entry into murine cell lines.

5. DISCUSSION

Gene-environment interactions that govern virus cross-species transmission and xenotropism are poorly understood. For RNA viruses, host range specificity is often mediated at level of the virus-receptor interaction. For example, the host range expansion of canine parvovirus type 2 from a feline parvovirus in 1978 was likely attributed to a few changes in the region of the virus capsid where three protein monomers interact (Parish, 1994). We have suggested that mutations affecting virus xenotropism may result in new diseases by either promoting virus adaptation and dissemination into a new host species, or by altering normal virus-receptor interactions that result in new tissue tropisms, altered virulence or pathogenicity in the original host. In this paper we have used MHV as a model to study the consequence of virus interspecies traffic and xenotropism on virus-receptor interactions in the original host.

Table 6. MHV-H2 revertants H2R6 and H2R11
replication in murine and hamster cells and
Bgp receptor usage

	Virus Titer (Pfu / cell)	
Cell line	H2R6	H2R11
DBT	$1.0\ 10^8$	1.5×10^8
BHK	1.0×10^4	1.0×10^4
CHO	$< 1.0 \times 10^4$	$< 1.0 \times 10^4$
BHK-MHVR	1.0×10^8	1.1×10^8
BHK - Bgp1	1.5×10^8	1.0×10^8
CHO - Bgp2	2.5×10^6	1.0×10^7

The host range mutant, MHV-H2, was resistant to CC1 antibody blockade, suggesting that MHV-H2 may either be recognizing a portion of MHVR which was not accessible to CC1 blockade, had a decreased affinity for MHVR, or was recognizing alternative receptors for entry into murine cells. The failure of MHVR expression in nonpermissive cells to support MHV-H2 infection, while allowing for MHV-A59, MHV-JHM infection suggested that MHV-H2 no longer efficiently recognized MHVR as a receptor for docking and entry into murine cells. Although it remains possible that ST cells may not express some unidentified cofactors necessary for MHV-H2 entry and replication, the ability of MHV-A59 and MHV-JHM to replicate efficiently in ST cells expressing MHVR make this unlikely. To our knowledge this is the first MHV variant described which has lost its capacity to efficiently recognize MHVR as a receptor for docking and entry.

In addition to demonstrating that MHVR was no longer recognized efficiently for MHV-H2 entry we have shown that different Bgp genes or alleles expressed in DBT cells are likely acting as the principle receptors for virus docking and entry. These data suggest that interspecies traffic of viruses may subtly alter virus-receptor interactions which may result in altered tissue tropism, pathogenesis or virulence in the original host. Interestingly, virus adaptation to alternative host species in vitro has been used to isolate attenuated vaccine variants of measles and polioviruses. While attenuating mutations in these viruses have been identified, in most cases the functions of these mutations in virus replication in the original and the adopted hosts has not been elucidated. It will be of interest to determine whether the virulence, tissue tropism and pathogenesis of MHV-H2 has been altered in B6, Balb-c and SJL mice. If so, the MHV-H2 host range variant may provide an interesting model system to investigate the mechanisms by which virus-host interactions attenuate virulence and pathogenicity.

In conclusion, the isolation of MHV host range variants has selected for viruses which have remodeled their normal interactions with cellular receptors. The availability of these mutants may allow us to identify virus residues which promote or abrogate S glycoprotein-MHVR interactions and promote high affinity interactions with Bgp2.

REFERENCES

Baric, R. S., Yount, B., Hensley, L., Peel, S., and Chen W., 1997, Episodic Evolution and Interspecies Transfer of a Murine Coronavirus, *J. Virol.* **71**:1946–55.

Bozzola, Russel. Electron Microscopy, Jones and Bartlett Publishers, Boston. 1992.

Chen, C.S., Asanaka, M., Yokomori, K., Wang, F-I., Hwang, S.B., Li, H-P., and Lai, M.M.C., 1995, Pregnancy-specific glycoprotein is expressed in the brain and serves as a receptor for mouse hepatitis virus, *Proc. Natl. Acad. Sci. USA.* **71**:1688–1691.

Chen, D.S., Asanaka, M., Chen, F.S., Shively, J.E., and Lai, M.M.C., 1997, Human carcinoembryonic antigen and biliary glycoprotein can serve as mouse hepatitis virus receptors, *J. Virol.* **71**:1688–91.

Chen, W., and Baric, R.S., 1996, Molecular anatomy of mouse hepatitis virus persistence: coevolution of increased host cell resistance and virus virulence, *J. Virol.* **70**:3947–3960

Dveksler, G.S., Pensiero, Cardellichio, C.B. Williams, J.-K. Jiang, G.S. Holmes, K.V.,and Dieffenbach, C.W., 1991, Cloning of the mouse hepatitis virus (MHV) receptor: expression in human and hamster cell lines confers susceptibility to MHV, *J. Virol.* **65**:6881–6891.

Dveksler, G.S., Pensiero, C.W., Cardellichio, C.B., McCuaig, K., Pensiero, M.N., Jiang, G.-S., Beauchemin, N., and K.V. Holmes, 1993a, Several members of the mouse carcinoembryonic antigen-related glycoprotein family are functional receptors for the coronavirus mouse hepatitis virus-A59, *J. Virol.* **67**:1–8.

Dveksler, G.S., Pennsiero, M.N., Dieffenbach, C.W., Cardellichio, C.B., Basile, A.A., Elia, P.E., and Holmes, K.V., 1993b, Mouse hepatitis virus strain A59 and blocking antireceptor monoclonal antibody bind to the N-terminal domain of cellular receptor, *Proc. Natl. Acad. Sci. USA.* **90**:1716–1720.

Kilbourne, ED., 1991, New viruses and new disease: mutation, evolution and ecology, *Curr. Opin. Immunol.*, **3**:518–524.

Morse, SS., 1995, Factors in the emergence of infectious diseases, *Emerging Infect. Dis.* **1**:7–15.

Murray, Cai, G., Hoel, K. Zhang, J.-Y., Soike, K. F., Cabirac, G.F., 1992, Coronavirus Infects and Causes Demylination in Primate Central Nervous System, *Virology*, **188**: 274–278.

Nedellec., Dveksler, G. S., Daniels., Turbide., Chow., Basile., Holmes, K.V., Beauchemin. N., 1994, Bgp2, a New Member of the Carcinoembryonic Antigen-Related Gene Family, Encodes an Alternative Receptor for Mouse Hepatitis Viruses, *J. Virol*, **64**:4525–37.

Schochetman G., Stevens, RH., Simpson, W., 1977, Presence of infectious polyadenylated RNA in the coronavirus avian infectious bronchitis virus, *Virology* **77**: 772–778.

Parish, C.R. 1994, The emergence and evolution of canine parvovirus- an example of recent host range mutation, *Semin. Virol.* **5**:121–132.

Williams, R.K., Jiang, G-S. and Holmes, K.V., 1991, Receptor for mouse hepatitis virus is a member of the carcinoembryonic antigen family of glycoproteins, *Proc. Natl. Acad. Sci. USA.*, **88**:5533–5536.

Yokomori, K., and Lai, M.M.C., 1992a, Mouse hepatitis virus utilizes two carcinoembryonic antigens as alternative receptors, *J. Virol.*, **66**:6149–6199.

Yokomori, K., and Lai, M.M.C., 1992b, The receptor for mouse hepatitis virus in the resistant mouse strain SJL is functional· Implications for the requirement of a second factor for viral infection, *J. Virol.* **66**:6931–6938.

6

HUMAN BILIARY GLYCOPROTEINS FUNCTION AS RECEPTORS FOR INTERSPECIES TRANSFER OF MOUSE HEPATITIS VIRUS

Lisa E. Hensley[1] and Ralph S. Baric[1,2]

[1]Department of Epidemiology
[2]Department Of Microbiology And Immunology
University Of North Carolina At Chapel Hill
Chapel Hill, North Carolina 27599

1. ABSTRACT

A variant Mouse Hepatitis virus (MHV), designated MHV-H2, was isolated by serial passage in mixed cultures of permissive DBT cells and nonpermissive Syrian Hamster Kidney (BHK) cells. MHV-H2 replicated efficiently in hamster, mouse, primate kidney (Vero, Cos 1, Cos 7), and human adenocarcinoma (HRT) cell lines but failed to replicate in porcine testicular (ST), feline kidney (CRFK), and canine kidney (MDCK) cells. To understand the molecular basis for coronavirus cross-species transfer into human cell lines, the replication of MHV-H2 was studied in hepatocellular carcinoma (HepG2) cells which expressed high levels of the human homologue of the normal murine receptor, biliary glycoprotein (Bgp). MHV-H2 replicated efficiently in human HepG2 cells, at low levels in breast carcinoma (MCF7) cells, and poorly, if at all, in human colon adenocarcinoma (LS 174T) cell lines which expressed high levels of carcinoembryonic antigen (CEA). These data suggested that MHV-H2 may utilize the human Bgp homologue as a receptor for entry into HepG2 cells. To further study MHV-H2 receptor utilization in human cell lines, blockade experiments were performed with a panel of different monoclonal or polyclonal antiserum directed against the human CEA genes. Pretreatment of HepG2 cells with a polyclonal antiserum directed against all CEA family members, or with a monoclonal antibody, Kat4c (cd66abde), directed against Bgp1, CGM6, CGM1a, NCA and CEA, significantly reduced virus replication and the capacity of MHV-H2 to infect HepG2 cells. Using another panel of monoclonals with more restricted cross reactivities among the human CEA's, Col-4 and Col-14, but not B6.2, B1.13, Col-1, Col-6 and Col-12 blocked MHV-H2 infection in HepG2 cells. These antibodies did not block sindbis virus (SB) replication in HepG2 cells, or block SB, MHV-A59 or MHV-H2 replication in DBT cells. Monoclonal antibodies Col-4, Col-14, and Kat4c (cd66abde) all reacted

Coronaviruses and Arteriviruses, edited by Enjuanes *et al.*
Plenum Press, New York, 1998

strongly with human Bgp and CEA, but displayed variable binding patterns with other CEA genes. Following expression of human Bgp in normally nonpermissive porcine testicular (ST) and feline kidney (CRFK) cells, the cells became susceptible to MHV-H2 infection. These data suggested that phylogenetic homologues of virus receptors represent natural conduits for virus xenotropism and cross-species transfer.

2. INTRODUCTION

The exact mechanisms of virus cross-species transmission remain largely unexplored and few model systems exist to study the sites of virus host interactions which regulate the host range of animal viruses. Given the high degree of species specificity and tissue tropism, coronaviruses provide an excellent biochemical and genetic resource to study the mechanisms of virus xenotropism and cross-species transmission. MHV was passaged in progressively decreasing concentrations of permissive murine DBT cells and increasing concentrations of nonpermissive BHK cells, a setting which may be reflective of *in vivo* conditions present in heavily immunosuppressed humans following xenotransplantation (Baric et al., 1997). The variant viruses isolated from this procedure replicated in murine and hamster cells and surprisingly in primate and human cell lines (Baric et al., 1997). Since the MHV genomic RNA is infectious in nonpermissive hosts and expression of the MHV receptor for entry (MHVR), a biliary glycoprotein, converts nonpermissive cells into MHV susceptible hosts it is likely that MHV host species specificity is mediated at receptor-binding, docking and entry (Dveskler et al., 1991; Schochetman et al., 1977). In this paper we explore the molecular mechanisms mediating the interspecies transfer of a murine virus (MHV) into human cells lines in vitro.

3. MATERIALS AND METHODS

3.1. Viruses and Cells

Briefly, murine astrocytoma (DBT), and 17CL-1 cells, Syrian baby hamster kidney (BHK), Chinese hamster ovary (CHO), Syrian hamster smooth muscle (DDT-1), feline kidney (CRFK), porcine testicular (ST), primate kidney (VERO), and human colorectal (HRT) cells were maintained as previously described (Baric et al., 1997). Human hepatocellular carcinoma (HepG2) cells were maintained in Eagles minimum essential medium containing 8% fetal calf serum and 1% geneticin and kanamycin. Human colon adenocarcinoma (LS174T) cells were maintained in minimum essential medium with 8% fetal calf serum, 1% geneticin and kanamycin, and 1% non-essential amino acids. Breast carcinoma (MCF7) cells were maintained in Eagles minimum essential medium with 8% fetal calf serum, 1% geneticin and kanamycin, 1% non-essential amino acids, 1% sodium pyruvate and 10 µg/ml of insulin. Canine kidney (MDCK) cells were maintained in DMEM-H minimum essential medium containing 8% fetal calf serum and supplemented with 10% tryptose phosphate broth and 1% geneticin and kanamycin. Primate kidney cells, Cos 1 and Cos 7, were maintained in Eagle's minimum essential medium with 8% fetal calf serum and 1% geneticin and kanamycin.

MHV-A59 and MHV-JHM were plaque purified and propagated in DBT cells. The MHV-H2 variant was propagated in BHK cells as previously described. The TRSB strain of sindbis virus (kindly provided by Dr. Robert E. Johnson, University of North Carolina at Chapel Hill) was propagated in DBT cells.

3.2. Virus Growth Curves and Immunofluorescence

Different cell lines grown on LabTek chamber slides (Nunc. Inc., Naperville, Ill.) were infected with MHV-A59 or MHV-H2 at an MOI of 10 for 1 hour at room temperature. The inocula were removed and the monolayers were washed 3 times with PBS. Samples of virus were harvested at various times postinfection and stored at -70(C for plaque assay. Immediately following the last sample cells were fixed and prepared for examination by immunofluorescence as previously described (Chen et al., 1997).

3.3. Transmission Electron Microscopy

HepG2 cells were seeded on $60mm^2$ dishes and infected with MHV-H2 at an MOI of 10 for 1 hour. At 36 hours post infection these cells were then prepared for examination by transmission electron microscopy using standard techniques (Bozzola, 1992).

3.4. Blockade of MHV Infection

HepG2 or DBT cells were seeded at densities of 2×10^4 on LabTek 8 well chamber slides (Nunc. Inc., Naperville, Ill.). Cells were pretreated with 100 µl of a 1:4 dilution of either kat4c (cd66abde), a commercially available monoclonal antibody directed against the human BgpA, CGM6, CGM1a, NCA and CEA glycoproteins, or a commercially available polyclonal antibody which recognizes all human CEA gene family glycoproteins for 1 hour at room temperature. The antibodies were removed and the cells were infected with either MHV-A59, MHV-H2, or TRSB at an MOI of 10 for one hour at room temperature. Cells were washed 3 times with phosphate buffered saline (PBS) to remove residual virus, and complete media containing a 1:16 dilution of antibody were added to the individual chambers. Additionally, these experiments were repeated using a panel of monoclonal antibodies with more restrictive specificities to CEA gene family members at an original dilution of 1:3. As a control, cells were pretreated with an equivalent amount of R501, an irrelevant MAb (with the same IgG isotype as CC1) directed against the anti-E2 glycoprotein of Sindbis virus.

3.5. Transient Expression of Human Biliary Glycoprotein

The human biliary glycoprotein (hBgp1) was cloned from HepG2 cells using RT-PCR, and expressed in the $pcDNA_3$ vector system. Using electroporation, transient cotransfections were performed with 6 µg $pcDNA_3$ vector (Invitrogen) containing the human Bgp1 (hBgp1) gene and 3 µg of the pHook2 vector. Briefly the cells were incubated with pHook magnetic beads for 1–2 hours, scraped and separated with a magnet. The procedure was repeated 24 hours later. Selected cells were seeded onto LabTek slides (Nunc Inc.). Cells were infected with either MHV-H2, MHV-A59 or MHV-JHM at an MOI of 10 and examined for the presence of viral antigens at 32 or 36 hours post infection.

4. RESULTS

4.1. Virus Growth Curves in Various Mammalian Cell Lines

Previous work has demonstrated that MHV-H2 replicated efficiently in hamster (BHK, CHO, and DDT-1), murine (DBT and 17CL-1), and human (HRT) cell lines. Feline

Table 1. Host range phenotype of MHV-H2

	MHV-A59	MHV-JHM	MHV-H2
Mouse			
DBT	+	+	+
17Cl1	+	+	+
Hamster			
BHK	–	–	+
CHO	–	–	+
Feline			
CRFK	–	–	–
Canine			
MDCK	–	–	–
Pig			
ST	–	–	–
Primate			
Vero	–	–	+
Cos1	–	–	+
Cos7	–	–	+
Human			
HRT	–	–	+
HepG2	–	–	+
LS174T	–	–	–
MCF7	–	–	+/–

+ indicate evidence of viral growth; - indicates no evidence of viral growth by plaque assay and detection of viral proteins by radiolabelling or flourescence; + / – indicates low levels of immunofluorecence (< 4.0%).

kidney (CRFK) and porcine testicular cells (ST), however, were resistant to MHV-H2 infection (Baric et al., 1997). Infection of the canine kidney, MDCK, cell line with MHV-H2 also resulted in a nonproductive infection. In contrast, MHV-H2 infection of various primate cells, (VERO, COS 1 and COS 7), resulted in a productive infection with virus titers ranging between $1 \times 10^4 - 1 \times 10^5$ PFU/ml (Table 1).

MHV-H2 failed to replicate efficiently in human MCF7 cells and in human LS174T cell lines which express high levels of carcinoembryonic antigen (CEA) (Tom et al., 1976). In contrast MHV-H2 infection of HepG2 cells, which express high levels of human biliary glycoprotein (Hauck et al., 1994), resulted in virus titers approaching 1.0×10^8 (Table 1 and Figure 1). In addition, when viewed in a scanning electron microscope, significant CPE were visible in HepG2 cells infected with MHV-H2 (data not shown). Not surprisingly, the parental strains MHV-A59 and MHV-JHM only replicated in murine cells (Table 1). The ability of MHV-H2 to replicate in select cell lines suggested that one or more specific cellular factors were necessary for a productive MHV-H2 infection in human cell lines.

4.2. Pretreatment of HepG2 Cells with CEA Antibodies

Both human and mouse species have been shown to express members of the carcinoembryonic antigen gene family, including Bgp (Barnett et al., 1993; Rudert et al., 1992). Sequence comparisons have revealed that the human Bgp1 gene is highly homologous with the murine Bgp1 gene and to a lesser extent with the Bgp2 gene (data not shown). In addition human Bgp and CEA glycoproteins may function as receptors for

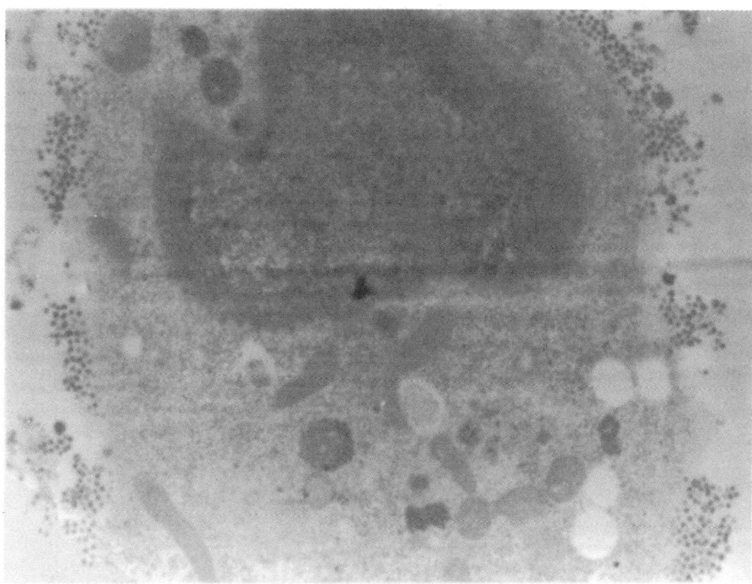

Figure 1. MHV-H2 infection in HepG2 cells. HepG2 cells were infected with MHV-H2 at an MOI of 20 for 1 hour. Cells were fixed and prepared for examination by electron microscopy 36 hours post infection. Magnification: 7,750X.

wildtype MHV entry into nonpermissive when expressed at high levels in non-permissive cells (Chen *et al.,* 1997). Consequently, it appeared likely that virus scanning for receptor homologues in different species may mediate cross-species transmission of MHV-H2 into human cells. A variety of polyclonal and monoclonal antibody were available that recognized the human CEA genes. Pretreatment of HepG2 cells with two commercially available antibodies, kat4c (cd66abde) and a polyclonal CEA (DAKO) significantly reduced MHV-H2 titers (Figure 2) and the number of infected HepG2 cells as observed by immunofluorescence (data not shown) Pretreatment with an irrelevant antibody, R501, failed to block MHV-H2 infection (Table 2). Consequently, pretreatment of HepG2 cells with these antibodies also failed to block Sindbis virus infection or inhibit MHV-H2 and MHV-A59 infections in VERO or DBT cells (data no shown). These data suggested that the blockade was specific and not due to antibody toxicity or other nonspecific effects.

Human CEA genes are highly homologous and the polyclonal CEA and kat4c (cd66abde) monoclonals' have broad crossreactivities and bind many CEA glycoproteins. Therefor an additional panel of monoclonal antibodies with more discrete ranges of CEA crossreactivities were obtained. Pretreatment of HepG2 cells with this panel of anti-CEA monoclonal antibodies (kindly provided by Dr. Jeffrey Schlom, NIH, Bethesda MD) produced mixed results (Table 2). Two antibodies, Col-4 and Col-14, which have been reported to react with hBgp1, successfully blocked MHV-H2 infection in HepG2 cells. Col-1, Col-6, and B6.2 which did not recognize hBgp1, all failed to block infection. A sixth antibody, Col-12 whose crossreactivity with hBgp1 has yet to be determined failed to block MHV-H2 infection in HepG2 cells. Pretreatment of HepG2 cells with these antibodies did block TRSB infection in these cells or block MHV-H2 (Table 2) or MHV-infection of murine cells (data not shown). These data indicated that human CEA genes likely functioned to initiate MHV-H2 docking and entry into human cells.

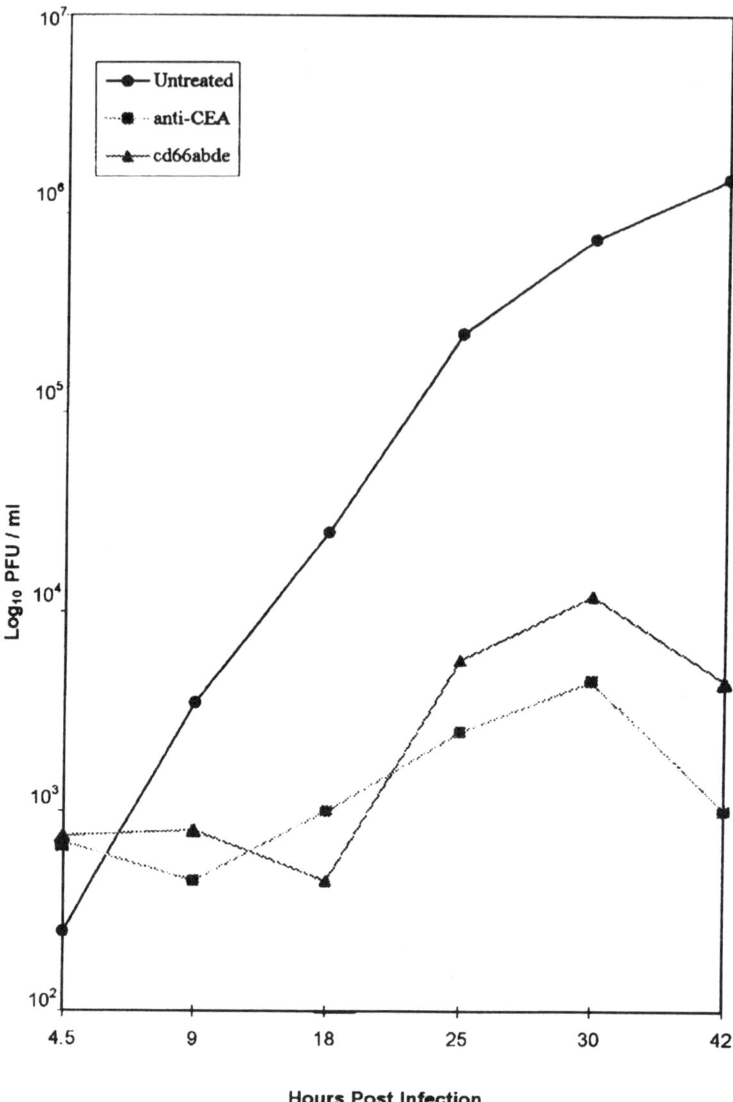

Figure 2. Anti-CEA and MAb cd66abde blockade of MHV-H2 in HepG2 cells. Cultures of HepG2 cells were pre-treated with the polyclonal anti-CEA antibody, which is directed against all human CEA members and with the MAb kat4c (cd66abde) directed against cd66abde for 1 hour. Cells were challenged with MHV-H2 at an MOI = 10 PFU/ml for one hours. Virus samples were taken at various times post infection and analyzed by plaque assay in DBT cells.

4.3. Expression of hBgp1 Promotes MHV-H2 Infection

The blockade data coupled with the observation that cell lines expressing high levels of human Bgp were susceptible to MHV-H2 infection, suggested that human Bgp1 may serve as a receptor for entry into human cells. To investigate this possibility, the human Bgp under the control of the CMV promoter was expressed in non-permissive ST cells. Following MHV-H2 infection of hBgp1-transfected cells, a significant increase in the

Table 2. Antibody blockade of MHV infections in HepG2 cells

Virus	Antibody	Virus Titer Pfu /ml
MHV-H2	Untreated	2.4×10^7
MHV-H2	TRSB	4.0×10^7
MHV-H2	Col-1	1.0×10^6
MHV-H2	Col - 4	5.5×10^3
MHV-H2	Col -14	7.0×10^3
MHV-H2	B1.13	3.0×10^6
MHV-H2	B6.2	5.0×10^5
MHV-H2	Col-6	1.5×10^6
MHV-H2	Col-12	5.0×10^5
TRSB	Untreated	1.5×10^7
TRSB	R501	6.0×10^7
TRSB	Col - 4	2.0×10^6
TRSB	Col - 14	2.5×10^6
TRSB	Polyclonal CEA	3.5×10^7
TRSB	kat 4c	5.0×10^6

number of fluorescent cells was observed by immunofluorescence. Under identical conditions 1–2% of the cells were susceptible to MHV-H2 and <0.05% of the cells to MHV-A59 infection. When transfected cells were sorted using the pHook system (Invitrogen) prior to infection, approximately 15–20% of the transfected cells were susceptible to MHV-H2 infection (see Figure 3) while only 1–2% of transfectants were permissive to MHV-A59 as observed by immunofluorescence (data not shown).

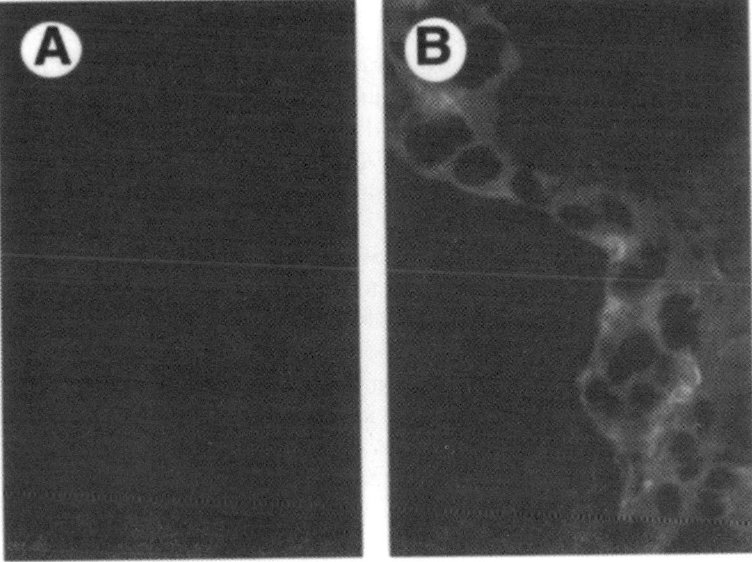

Figure 3. Virus infection in ST cells expressing hBgp1. ST cells transiently expressing hBgp1 were infected with MHV-H2 for 1 hour at an MOI of 10. Cells were fixed 36 hours later and examined by immunofluorescence. Panel A: ST cells infected with MHV-H2. Panel B: ST cells expressing hBgp1 infected with MHV-H2.

5. DISCUSSION

Host range specificity and the evolution of new viral diseases are complex phenomena involving interactions between the virus, the host and the environment (Morse, 1990). The cellular receptor is an essential component for virus entry and a major determinant of host range specificity, tissue tropism and pathogenesis (Wimmer, 1994). Cellular receptors for virus docking and entry have been demonstrated to represent a major site of virus-host interaction that regulate virus xenotropic spread. Other sites may include additional cofactors for entry, transcription factors, and host factors for virus assembly and release (Morse, 1994). Since little information is available regarding receptor utilization among phylogenetically related viruses that replicate in distinct species (Haywood, 1994; Wimmer 1994), our knowledge of virus cross-species transmission is incomplete. Focusing at the level of entry, the available literature suggests that new viral diseases may either emerge by a homologue scanning or a receptor augmentation mechanism (Baric *et al.*, 1997; Haywood, 1994; Wimmer 1994). The homologue scanning model predicts that phylogenetic homologues of the normal receptor function as natural conduits for the cross-species transmission and entry of viruses into alternative host species. For example, HIV likely evolved from SIV. Both viruses utilize CD4 and chemokine genes as receptors and co-receptors for entry into cells and target similar tissues in humans and primates (Signoret et al., 1993). Among the group I coronaviruses, transmissible gastroenteritis virus (TGEV), and human coronavirus 229E use the aminopeptidase N glycoprotein as a receptor, and produce similar infections in their respective hosts (Delmas *et al.*, 1992; Yeager et al., 1992). It appears likely that host range expansion among these virus families, in part, evolved along phylogenetically related receptor molecules. In contrast, the receptor augmentation model proposes that host range expansion is mediated by virus recognition of entirely new receptor molecules. For example, murine adapted strains of poliovirus do not recognize the murine homologue of the poliovirus receptor (PVR), rather, these variants recognize unique receptor moieties for docking and entry into murine cells (Morrison *et al.*, 1992; Morrison *et al.*, 1994).

The results presented in this paper strongly support the hypothesis that MHV-H2 utilizes human CEA glycoproteins, most likely hBgp1, for entry into permissive cell lines. MHV-H2 was able to replicate efficiently in human cell lines expressing high levels of hBgp, but was not able to replicate efficiently in LS174T cells which express high levels of CEA. In addition pretreatment of HepG2 cells with antibodies that crossreact with hBgp1 were successful in blocking infection. Finally expression of hBgp1 permitted MHV-H2 infection in non-permissive cells. These data indicate that MHV cross-species transmission evolved by virus recognition of phylogenetic homologues of the natural receptor. Expression of hBgp also appeared to permit low levels of MHV-A59 infection. This observation is consistent with previously published data, and is suggestive of an extremely low affinity or inefficient utilization of the hBgp1 homologues by MHV-A59 (Chen, D.S., et al., 1997). Since the human CEA genes are highly conserved it remains possible, if not likely, that MHV-H2 may also recognize other CEA gene family members for docking and entry. This is currently under investigation.

Transplantation of organs has become the preferred treatment for end-stage organ failure. Chronic shortages in human organs coupled with improved immunosuppressive therapies have established the feasibility of using animal organs in human transplantation as a life-saving alternative treatment (Michler, 1996). Using an in vitro model, which may be reflective of mixed species tissues that are present in human xenograph recipients, we have demonstrated that host range mutants emerge which replicate efficiently in the new host species. While further studies are warranted, xenograph recipients may represent excellent ecosystems for the emergence and cross-species transmission of animal viruses into the human host.

REFERENCES

Baric, R. S., Yount, B., Hensley, L., Peel, S., and Chen, W., 1997, Episodic Evolution and Interspecies Transfer of a Murine Coronavirus, *J. Virol.* **71**:1946–55.

Barnett, Drake, Pickle II., 1993, Human Biliary Glycoporein Gene: Characterization of a family of novel Alternatively Spliced RNAs and their Expressed proteins, *Molec. Cell Biol.*, **13**:1273–82.

Bozzola, Russel. Electron Microscopsy, Jones and Bartlett Publishers, Boston. 1992.

Chen, C.S., Asanaka, M., Yokomori, K., Wang, F-I, Hwang, S.B., Li, H-P., and M.M.C. Lai., 1995, Pregnancy-specific glycoprotein is expressed in the brain and serves as a receptor for mouse hepatitis virus, *Proc. Natl. Acad. Sci. USA*, **71**:1688–1691.

Chen, D.S., Asanaka, M., Chen, F.S., Shively, J.E., and Lai, M.M.C., 1997, Human carcinoembryonic antigen and biliary glycoprotein can serve as mouse hepatitis virus receptors, *J. Virol* **71**:1688–91.

Chen, W., and Baric, R.S., 1996, Molecular anatomy of mouse hepatitis virus persistence: coevolution of increased host cell resistance and virus virulence, *J. Virol.* **70**:3947–3960.

Delmas, B., Gelfi, J., L'Haridon, R., Vogel, L.K., Sjostrom, H., Noren, O., and Laude, H., 1992, Aminopeptidase N is a major receptor for the enteropathogenic coronavirus TGEV, *Nature*, **357**:417–420.

Hauck, Nédellec, Turbide, Stanners, Barnett., 1994, Transcriptional Control of the human biliary glycoprotein gene, a CEA gene family member down-regulated in colorectal carcinomas, *Euro. J. Biochem.*, **223**: 529–41.

Haywood, A.M., 1994, Virus receptors: Binding, adhesion strengthening, and changes in viral structure, *J. Virol.*, **68**: 1–5.

Hinoda, Y., Neumaier, M., Hefta, S.A., Drzeniek, Z., Wagner, C., Shively, L., Hefta, L.J., Shively, J.E., Paxton, R.J., 1988, Molecular cloning of a cDNA coding biliary glycoprotein I: Primary structure of a glycoprotein immunologically crossreactive with carcinoembryonic antigen, *Proc. Natl. Acad. Sci. USA*, **85**:6959–6963.

Hirano, N., Murakami, T., Fugiwara, K., and Matsumoto, M., 1978, Utility of the cell line DBT for propagation and assay of mouse hepatitis virus, *Jph. J. Exp. Med.*, **48**:71–75.

Holmes, K.V., 1994, in: *Cellular receptors for animal viruses.* E.Wimmer eds., Coldspring Harbor Laboratory Press, pgs 1–14.

Kilbourne, E.D., 1991, New viruses and new disease: mutation, evolution and ecology. *Curr. Opin. Immunol.*, **3**:518–524.

Michler., 1996, Xenotransplantation: risks, clinical potentail and future prospects, *Emerging Infect. Dis.*, **2**:64–70

Morrison, M.E., and Racaniello, V.R., 1992, Molecular cloning and expression of a murine homolog of a the human poliovirus receptor gene, *J. Virol.* **66**:2807–2813.

Morrison, M.E., He, Y-H., Wien, M.W. Hogle, J.M., and Racaniello, V.R., 1994, Homolog-scanning mutagenesis reveals poliovirus receptor residues important for virus binding and replication, *J. Virol*, **68**:2578–2588.

Morse, S.S., and Schluederberg, A., 1990, Emerging viruses: The evolution of viruses and viral diseases, *J. Infect. Dis.*, **162**:1–15.

Morse, S.S., 1994, The evolutionary biology of viruses, New York, Raven Press.

Murphy, F.A., 1994, New emerging and reemerging infectious diseases, *Adv. Vir. Res.* **43**:1–52.

Murray, Cai, G., Hoel, K., Zhang, J.-Y., Soike, K. F., Cabirac, G.F., 1992, Coronavirus Infects and Causes Demylination in Primate Central Nervous System, *Virology* **188**: 274–278.

Rudert, F., Saunders, A. Rebstock ,S., Thompson, J.A., Zimmermann, W., 1992, Characterization of murine carcinoembryonic antigen gene family members, *Mammalian Genome.* **3**: 262–273.

Schochetman, G., Stevens, RH., Simpson, W., 1977, Presence of infectious polyadenylated RNA in the coronavirus avian infectious bronchitis virus, *Virology* **77**: 772–778.

Signoret, N., Poignard, P., Blanc, D., Sattentau, Q.J., 1993, Human and simian immunodeficiency viruses: virus-receptor interactions, *Trends Microbiol.* **1**:328–33.

Tom, B.H., Rutzkey, L.P., Jakstys, M., 1976, Human colon adenocarcinoma cells I. Establishment and description of a new line, *In Vitro*, **12**: 180.

Williams, R.K., Jiang, G -S. and Holmes, K.V., 1991, Receptor for mouse hepatitis virus is a member of the carcinoembryonic antigen family of glycoproteins, *Proc. Natl. Acad. Sci. USA*, **88**:5533 5536.

Yeager, C.L., Ashmun, R.A., Williams, R.K., Cardellichio, C.B., Shapiro, L.H., Look, A.T. Holmes, K.V., 1992, Human aminopeptidase N is a receptor for human coronavirus 229E, *Nature.* **357**:420–2.

Yokomori, K., and Lai, M.M.C., 1992a, Mouse hepatitis virus utilizes two carcinoembryonic antigens as alternative receptors, *J. Virol.* **66**:6149–6199.

Wimmer, B., 1994, An Introduction in: *Cellular receptors for animal viruses.* E. Wimmer eds., Coldspring Harbor Laboratory Press, pgs 1–14.

7

OBTENTION OF PORCINE AMINOPEPTIDASE-N TRANSGENIC MICE AND ANALYSIS OF THEIR SUSCEPTIBILITY TO TRANSMISSIBLE GASTROENTERITIS VIRUS

L. Benbacer, M.-G. Stinackre, H. Laude, and B. Delmas

Unité de Virologie et Immunologie Moléculaires
Institut National de la Recherche Agronomique
F-78350 Jouy-en-Josas, France

1. ABSTRACT

To obtain a laboratory animal model for transmissible gastroenteritis virus (TGEV) infection, transgenic mice (Tg) were produced by introducing two porcine aminopeptidase-n (APN) cDNA-derived constructs into the mouse genome. In the first construct, the APN cDNA was fused in 5' with the 1kb upstream region of the APN gene and in 3' with the SV40 small intron and polyadenylation site. In the second construct, the 5' end of the APN cDNA was replaced by the corresponding domain of the APN gene comprising the three first introns, an additional intron (the rabbit β-like globine intron 2) was inserted at the 3' extremity of the construct and the resulting DNA stretch was placed under the control of the rat intestinal fatty acid-binding protein (I-FABP) gene promoter. Transgenes were obtained with these two constructs, and RNA expression was evidenced by RT-PCR with the second construct in a transgene lineage. Using two different immunoassays, expression of the porcine APN protein was not detected in the transgenic intestines of animals of the RT-PCR positive lineage. Northern blot analyses did not revealed TGEV replication in infected adult mice. Additional assays will be carried out on young animals to detect potential TGEV susceptibility.

2. INTRODUCTION

TGEV enters the target cells by binding to a specific surface receptor, porcine APN, (Delmas et al., 1992), followed by the fusion between viral and cellular membranes to re-

Coronaviruses and Arteriviruses, edited by Enjuanes *et al.*
Plenum Press, New York, 1998

lease the genomic RNA into the cytoplasm. In susceptible animals, TGEV infection starts with ingestion of the virus followed by a primary infection of enterocytes of the jejunum and, to a lesser extent, of the duodenum and the ileon (reviewed by Enjuanes and Van der Zeijst, 1995). TGEV also replicates with a low efficiency in the upper respiratory and in alveolar macrophages. In contrast, porcine respiratory coronavirus, a natural variant of TGEV which also uses APN as a receptor, replicates efficiently in the respiratory tract, but poorly in the gut. The mechanism for this host cell restriction is not known. In addition, TGEV and PRCV replication is mainly restricted to these cell types and systemic infection were never clearly evidenced (reviewed by Laude et al., 1993). APN is present in almost all tissues, is highly expressed in kidney proximal tubule cells and in enterocytes but also, in less amount, in other epithelial cells (reviewed by Norén et al., 1997).

The pig species is the only natural host for TGEV. Some other species such as canine and feline species naturally infectable by antigenically-related viruses have been experimentally infected by TGEV, but infections were found to be asymptomatic (reviewed by Holmes and Compton, 1995). Identification of APN as a cellular receptor for TGEV has opened new perspectives in developing transgenic laboratory animal models to study TGEV pathogenesis. To this purpose, we have constructed two porcine APN cDNA-derived minigenes to target expression in epithelial cells of the gut and of the respiratory tract. Transgenic mice obtained with the two minigenes were analyzed for their susceptibility to TGEV infection.

3. METHODS AND MATERIALS

3.1. Plasmids Construction

For minigene 1, the 2.8-kb EcoRI/EcoRI fragment of porcine APN cDNA was inserted downstream of the porcine APN gene DNA fragment HindIII/EcoRI made of 1.2-kb upstream of the transcription initiation site and part of the exon I. The polyadenylation signal and intron sequences of SV40 small T antigen were amplified by PCR using primers that create a BglII restriction enzyme site at the 5' end and a XbaI restriction enzyme site at the 3' end. This PCR product was then cloned at the 3' end of the construct. For minigene 2, the 3.3-kb BamHI/XhoI fragment of the cDNA encoding porcine APN was cloned downstream of the rat I-FABP promoter. APN intron I (provided by J. Olsen, Copenhagen) was inserted by using EcoRI and AccI enzyme restriction sites present in the cDNA. Introns 2 and 3 were amplified by PCR with genomic porcine DNA as template using primers with AccI and BspE1 restriction enzyme sites present in the cDNA. The PCR product was cloned in the relevant restriction sites of the construct. A 2.2-kb BamHI/BspEI DNA fragment containing the rabbit ß-globin 3' intron and polyadenylation signal was inserted downstream of the APN coding region. The structure of the final plasmids was confirmed by restriction enzyme digestion. For microinjection, minigene 1 (5.3-kb) was cleaved by XhoI and XbaI, minigene 2 (7.5-kb) by NotI and XhoI to remove vector sequences. The inserts were purified by agarose gel electrophoresis and diluted before microinjection.

3.2. Microinjection and Production of Transgenic Mice

DNA inserts were microinjected in fertilized eggs of the F1 generation of C57Bl6 x CBA mice and transgenic mouse foundators were identified by PCR assay of tail DNA, using adequate primers. Southern blot analysis was also performed to evaluate the number of copies per genome.

3.3. DNA and RNA Isolation

Tails of newborn or 3-week-old mice were cut and digested in 300mM sodium acetate (pH 8), 1mM EDTA, 1% sodium dodecyl sulfate, and 150 μg of proteinase K per 750 μl. Undigested debris were removed by centrifugation. After two phenolic extractions, DNA was precipitated with isopropanol and dissolved in distilled water. Total RNA was isolated from small intestine by homogenization in 4M guanidium thiocyanate with a homogenizer followed by acid phenol extraction. RNA was precipitated with isopropanol and dissolved in distilled water.

4. RESULTS

4.1. Minigene 1

Transgenic mice were generated by using a construct in which porcine APN cDNA was linked to the porcine APN promoter region containing 1.2 kb of upstream sequence and, in the 3' part of the minigene, the intron and polyadelylation signal of SV40 t antigen (Fig. 1). This construct, named minigene 1, was expected to drive APN expression in epithelial cells. To ensure that minigene 1 will produce a functional TGEV receptor, plasmid DNA was transiently transformed into non-permissive MDCK cells followed by TGEV infection. Neosynthesized viral antigens were revealed in cells transformed by minigene 1 plasmid, but not in cells transformed with the vector alone (not shown). The minigene 1 was excised and microinjected into fertilized mouse eggs. Transgenic mice were screened for the presence of the transgene by PCR and Southern blotting analysis of tail DNA (Fig. 2). Four transgenic founders were identified, and two of them (founders 2 and 11) were used to further study.

Northern blot hybridization analysis was used to determine whether porcine APN RNA is expressed in the small intestine and in the liver of minigene 1 transgenic offspring

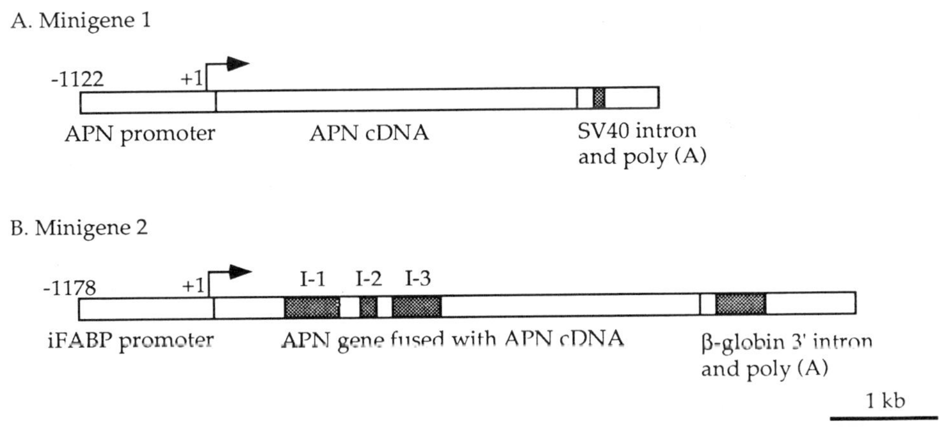

Figure 1. Structure of the minigenes used to generate transgenic mice. The introns are indicated in grey. I-1 to -3 are the first 5' introns of the porcine APN gene. Arrows indicate the site of initiation of transcription.

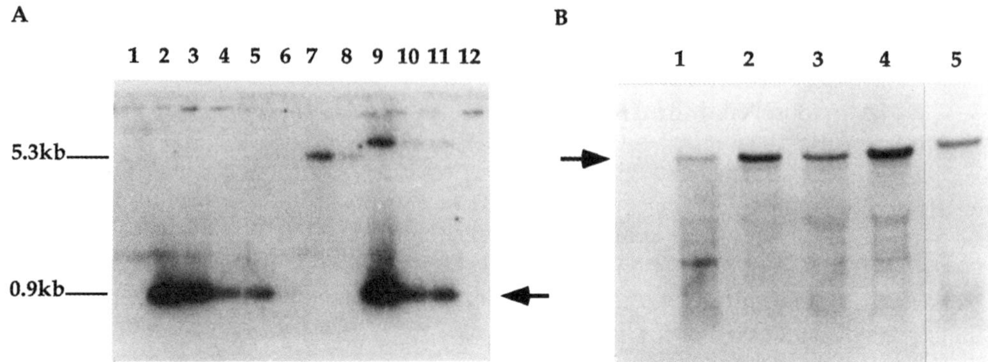

Figure 2. (A) Southern blot analysis of transgenic founders for minigene 1. Tail DNA of normal (lane 1), Tg2 (lane 2), Tg11 (lane 3), Tg13 (lane 4), Tg24 (lane 5) and non-Tg mice (lanes 6, 9, 10, 11 and 12) was digested by NcoI enzyme. All these lanes contained equal amounts of DNA, as judged by ethidium bromide staining. Lanes 7 and 8 contain the equivalent of 10 and 2 copies/genome of minigene 1 cut XhoI-XbaI (5.3-kb), respectively. In lanes 9, 10, 11 and 12, the equivalent of 20, 10, 5 and 0 copies/genome were added bebore DNA digestion. The DNA probe was the 0.9-kb NcoI-NcoI fragment of the cDNA. The arrow indicates the expected position of the porcine APN DNA of transgenic mice. (B) Northern blot analysis of total cell RNA from the swine ST cell line (lane 1) and from the small intestine of a normal mouse (lane 2), of theTg2 offspring 2.15 (lane 3), of pig (lane 4) and of the Tg11 offspring 11.09 (lane 5). The arrow indicates the expected position of the endogenous swine or mouse APN mRNA.

of founders 2 and 11. Analyses carried out with probes deriving from the porcine APN RNA revealed the 3.3-kb endogeneous mRNA encoding for the murine APN, but a 3.5-kb band consistent with that predicted from the structure of minigene 1 was never detected (Fig. 2). In the same way, a 3.5-kb band was not revealed by using as a probe the DNA stretch deriving from the SV40 DNA stretch present in the construct (not shown).

However, since a low concentration of receptor at the cell surface of target cell, at least in vitro, was found to be sufficient to confer susceptibility in cell culture (Delmas et al., 1995), infection assays were performed. Adult and new-born mice of the lineage 2 were orally infected with around 10^6 PFU of TGEV to analyse their permissivity. No enteritis was observed up to 3 days after infection. Viral replication was assayed by Northern blot analysis, by indirect immunofluorescence experiments and by quantification of interferon induction. No TGEV replication could be evidenced by these different assays (not shown).

4.2. Minigene 2

We designed another minigene by using a promoter which was previously validated in transgenic animals to direct expression in enterocytes and by adding introns in the cDNA for efficient mRNA maturation. To direct expression of porcine APN to the enterocytes of transgenic mice, the promoter for rat I-FABP was chosen. The I-FABP gene product is expressed in differentiated, villus-associated, small intestine epithelial cells (Sweetser et al, 1987). This promoter was validated to direct expression of diverse proteins in the enterocytes of transgenic animals (Sweetser et al., 1988, Zhang and Racaniello, 1997). Genomic sequences for porcine APN introns 1–3 were included in the full-length porcine APN cDNA which was cloned downstream of the I-FABP promoter. In addition, a fragment containing the rabbit β-globin intron 2 and poly(A) polyadenylation signal was placed immediately downstream of the APN cDNA to ensure posttranscriptional maturation and mRNA stability. To ensure that this construct, named minigene 2, produces a functional TGEV receptor, plasmid DNA was transiently transformed into receptor-negative BHK cells followed by TGEV infection. Neosynthesized viral antigens were revealed in cells transformed by minigene 2 plasmid, but not in cells transformed with the vector alone (not shown). Transgenic mice were produced with minigene 2 and screened for the presence of the transgene by PCR and Southern blotting analyses of tail DNA (not shown). Five transgenic founders, numbered 56, 58, 68, 116 and 126, were identified and used to further study.

Northern blot and RT-PCR analyses were carried out to determine whether porcine APN mRNA is expressed in transgenic animals of the lineages 56, 58, 68 and 126. Transgenic RNA was only detected in the small intestine of offspring of the founder mouse 126 by reverse transcription-polymerase chain reaction (RT-PCR) with primers specific of the APN cDNA and of the β-globin exon 3 (Fig. 3). Cloning and sequencing of the 480 bases pair RT-PCR product revealed a correct fusion between the two DNA domains and the expected splicing of β-globin intron 2.

To determine if offspring of founder 126 was susceptible to TGEV infection, 2×10^6 PFU were inoculated into the jejunum of five two month-old mice of this lineage. No enteritis was observed up to 3 days after infection. Northern blot analysis of intestine RNA extracted 8, 24 and 48h post infection were used to vizualize TGEV replication. No viral subgenomic mRNA were detected in the mice (Fig. 4).

5. DISCUSSION

A susceptible mouse model for TGEV infection would have obvious advantages over the use of the porcine species, its natural host. This laboratory model could be easily used to study the pathology and the intestinal immune response associated to the infection and would facilitate the development of antiviral approaches. For this purpose, we constructed two different minigenes to obtain transgenic mice expressing porcine APN in en-

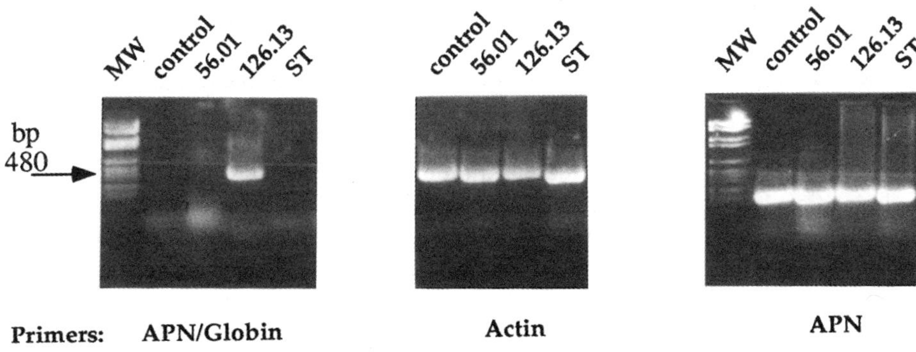

Figure 3. RT-PCR analysis of transgenic minigene 2 mRNA in the intestine of mice from transgenic lines 126 and 56. Reverse transcription was done on total tissue RNA with an oligo(dT) primer. Beta-actin and murine APN cDNAs were used as positive controls. PCR of the APN/globin fragment was done with primers 5'-TAATAGTGCCTGGTCCTTCCCGCCACCTGG and 5'-ATGAGACAGCACAACAACCAGCACGTTGCC. MW, molecular weight marker VI (Boehringer), control shows a negative control with a normal mouse and ST the PCR amplification obtained with the ST porcine cell line. Gels were stained with ethidium bromide.

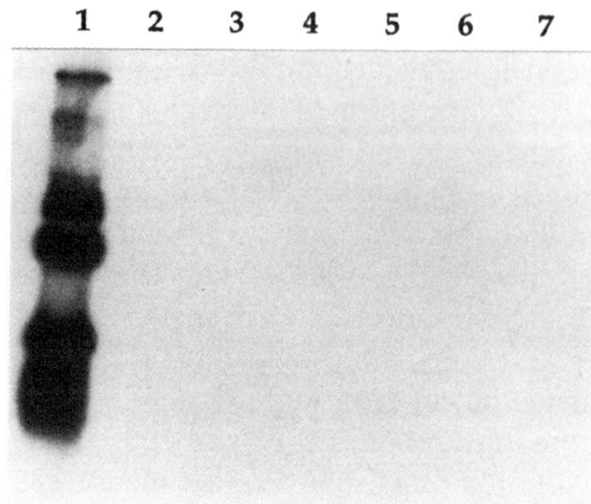

Figure 4. Northern blot analysis of total cell RNA from TGEV-infected mice of the line 126. Lane 1: Swine testis (ST) infected cell line, lanes 2 to 6: Two month-old infected transgenic animals, sacrified 8 h post infection (p.i.) (lanes 2 and 3), 20 h p.i. (lanes 4 and 5) and 48 h p.i. (lane 6). Lane 7, normal infected mouse sacrified 8 h p.i. The DNA probe was an HindIII-KpnI domain in the coding region of the N TGEV gene. Subgenomic TGEV RNAs were labeled in lane 1.

terocytes of the small intestine. In minigene 1, the APN cDNA was fused with the 1kb upstream region of the APN gene. We did not detect a transcriptional activity in the offsprings of two transgenic founders. This result could be due to absence of transcription, a defect of mRNA maturation or to the low number of lineages we analyzed. In minigene 2, the I-FABP promoter was found to be functional in one out the four lineages studied. A transcript was evidenced by RT-PCR in the lineage 126, but adult mice of this lineage were found not susceptible to TGEV infection as determined by Northern blot analysis.

There are several possible explanations for the failure of TGEV to replicate in the intestine of mice of the lineage 126. First, we were unable to detect the porcine APN in the transgenic animals by immunostaining of intestine sections (not shown) whereas in the mammal intestine, APN represent 5 to 8% of the total proteins present at the enterocyte brush border membranes. The concentration of receptor on the cell surface is possibly a critical factor for virus entry in the digestive tract. Secondly, mouse enterocytes possibly lack a functional host factor for TGEV entry or replication. Such a hypothesis has been recently proposed for the unability of poliovirus to replicate in enterocytes of transgenic mice expressing the poliovirus receptor (Zhang and Racaniello, 1997). In the same way, transgenic mice ubiquitously expressing human CD46, the receptor for measles virus, were not susceptible to virus infection (Horvat et al., 1996). Thirdly, we only infected adult mice and it would be of interest to infect newborns which are expected to be more susceptible than adults, as it is observed following experimental and natural infection of the swine.

The recent finding that feline APN can act as a functional receptor for TGEV in cell culture (Tresnan et al., 1996) raises the question of the capacity of TGEV to adapt to other mammals species. Indeed, TGEV was never reported to cause enteritis and to multiply efficiently in the cats. Thus, host factors other that the fixation of TGEV on APN may govern the spread of the virus and its ability to cause disease.

REFERENCES

Delmas, B., Gelfi, J., L'Haridon, R., Vogel, L.K., Sjöström, H., Norén, O., and Laude, H., 1992, Aminopeptidase N is a major receptor for the enteropathogenic coronavirus TGEV, *Nature.* **357**: 417–419.

Delmas, B., Kut, E., Gelfi, J., and Laude, H., 1995, Overexpression of TGEV cell receptor impairs the production of virus particles, in: *Corona- and related viruses*, Volume 380 (P. Talbot, and G. Levy, eds.), Plenum Press, New York, pp. 379–385.

Enjuanes, L., Van der Zeijst, B.A.M., 1995, Molecular basis of transmissible gastroenteritis virus epidemiology, in: *The Coronaviridae* (S.G. Siddell, ed.), Plenum Press, New York and London, pp.337–376.

Holmes, K.V., and Compton, S.R., 1995, Coronavirus receptors, in: *The Coronaviridae* (S.G. Siddell, ed.), Plenum Press, New York and London, pp. 55–71.

Horvat, B., Rivailler, P., Varior-Krishnan, G., Cardoso, A., Gerlier, D., and Rabourdin-Combe, C., 1996, Transgenic mice expressing human measles virus (MV) receptor CD46 provide cells exhibiting different permissivities to MV infection, *J. Virol.* **70**: 6673–6681.

Laude, H., Vanreeth, K., and Pensaert, M., 1993, Porcine respiratory coronavirus-Molecular features and virus host interactions, *Vet. Res.* **24**: 125–150.

Norén, O., Sjöström, H., and Olsen, J., 1997, Aminopeptidase N, in: *Cell-surface peptidases in health and disease* (A.J. Kenny, and C.M. Boustead, eds.), BIOS Scientific Publishers, pp. 175–191.

Sweetser, D.A., Birkenmeier, E.H., Klisak, I.J., Zollman, S., Sparkes, R.S., Mohandas, T., Lusis, A.J., and Gordon, J.I., 1987, The human and rodent intestinal fatty acid binding protein gene, *J. Biol. Chem.* **262**: 16060–16071.

Sweetser, D.A., Hauft, S.M., Hoppe, P.C., Birkenmeier, E.H., and Gordon, J.I., 1988, Transgenic mice containing intestinal fatty acid-binding protein-human growth hormone fusion genes exhibit correct regional and cell-specific expression of the reporter gene in their small intestine, *Proc. Natl. Acad. Sci. USA* **85**: 9611–9615.

Tresnan, D.B., Levis, R., and Holmes, K.V., 1996, Feline aminopeptidase N serves as a receptor for feline, canine, porcine, and human coronaviruses in serogroup I, *J. Virol.* **70**: 8669–8674.

Zhang, S., and Racaniello, V.R., 1997, Expression of the poliovirus receptor in intestinal epithelial cells is not sufficient to permit poliovirus replication in the mouse gut, *J. Virol.* **71**: 4915–4920.

8

MOLECULAR ANALYSIS OF THE CORONAVIRUS-RECEPTOR FUNCTION OF AMINOPEPTIDASE N

Andreas F. Kolb, Annette Hegyi, Julia Maile, Angelien Heister,
Margitta Hagemann, and Stuart G. Siddell

Institute of Virology and Immunology
University of Würzburg, Versbacherstr. 7
D-97078 Würzburg, Germany

1. ABSTRACT

Aminopeptidase N (APN) is a major cell surface for coronaviruses of the serogroup I. By using chimeric APN proteins assembled from human, porcine and feline APN we have identified determinants which are critically involved in the coronavirus-APN interaction. Our results indicate that human coronavirus 229E (HCV 229E) is distinct from the other serogroup I coronaviruses in that determinants located within the N-terminal parts of the human and feline APN proteins mediate the infection of HCV 229E, whereas determinants located within the C-terminal parts of porcine, feline and canine APN mediate the infection of transmissible gastro-enteritis virus (TGEV), feline infectious peritonitis virus (FIPV) and canine coronavirus (CCV), respectively. A further analysis of the mapped amino acid segments by site directed mutagenesis revealed that a short stretch of 8 amino acids in the hAPN protein plays a decisive role in mediating HCV 229E reception.

2. INTRODUCTION

On the basis of serological and genetical data, coronaviruses are classified into three different serogroups (Siddell, 1995). The members of serogroup I utilise the cell surface molecule aminopeptidase N (APN) as a receptor (Delmas et al., 1992; Yeager et al., 1992; Tresnan et al., 1996). Human APN and porcine APN act species specifically in that they only mediate the infection of the human coronavirus 229E and the porcine coronavirus TGEV, respectively (Delmas et al., 1994; Kolb et al., 1996). In contrast, the feline APN protein is able to mediate infection of FIPV, TGEV, HCV 229E and CCV (Tresnan et al.,

Coronaviruses and Arteriviruses, edited by Enjuanes *et al.*
Plenum Press, New York, 1998

1996). However, the feline coronavirus FIPV can only utilise fAPN as a receptor. By using chimeric molecules assembled from hAPN and pAPN the determinants which play an essential role in the infection of HCV 229E and TGEV have been identified (Delmas et al., 1994; Kolb et al., 1996). The region encompassing pAPN amino acids 717–813 has been delimited as the protein segment which is critically involved in TGEV infection via APN (Delmas et al., 1994). In contrast, a region located in the amino-terminal part of hAPN (hAPN amino acids 260–353) has been defined as being essential for the infection of HCV 229E via hAPN (Kolb et al., 1996). Crandell feline kidney (CRFK) cells are susceptible to FIPV, TGEV and HCV 229E. We anticipated that infection of HCV 229E into these cells would be mediated by a feline homologue of hAPN. We have isolated a fAPN cDNA from CRFK cells and by using chimeric molecules assembled from human, porcine and feline APN we have delimited the fAPN amino acid regions, which are involved in the HCV 229E, TGEV and FIPV receptor function. Within these regions we have analysed the amino acid residues which play an essential role in mediating coronavirus infections.

3. MATERIALS AND METHODS

3.1. Cells and Viruses

The HCV 229E isolate used in these studies was described previously. TGEV strain Purdue 46 was obtained from G. Herrler, University of Marburg, Germany. FIPV strain 79–1146 was obtained from R. de Groot, University of Utrecht, The Netherlands. CRFK cells (ECACC 86093002) and 293 HEK cells (ECACC 85120602) were cultivated according to standard techniques. Transfections were performed as described (Kolb et al., 1996).

3.2. Recombinant DNA

The construction of cDNA clones and chimeras used during these studies are described elsewhere (Delmas et al., 1994; Kolb et al., 1996; Kolb et al., submitted; Hegyi et al., in preparation).

3.3. Protein Detection

Detection of de novo synthesised HCV 229E or TGEV antigens was performed by Western blotting (Kolb et al., 1996). Detection of de novo synthesised FIPV antigens was done by immune-precipitation of viral proteins from infected and metabolically labelled cells as described (Grosse and Siddell, 1994). The enzymatic activity of APN proteins was analysed as described (Kolb et al., 1996).

4. RESULTS

A feline APN cDNA was isolated from CRFK cells by reverse transcription-PCR. The primers used for the amplification were derived from nucleotide sequences which are conserved between the known mammalian APN genes. Two PCR products encompassing the complete fAPN cDNA were cloned into an expression vector and transfected into 293 HEK cells. Cells transfected with the fAPN expression construct became susceptible to infection with FIPV, TGEV and HCV 229E (Table 1). In accordance with published data

Table 1. Receptor activity of mammalian APN proteins and chimeric proteins derived thereof[1]

Chimera	Backbone	Inserted residues	Susceptibility to HCV229E	Susceptibility to TGEV	Susceptibility to FIPV
pAPN (Delmas et al., 1992)			−	+	−
hAPN (Yeager et al., 1992)			+	−	−
fAPN [Fcwf] (Tresnan et al., 1996)			+	+	+
fAPN [CRFK] (Kolb et al., subm.)			+	+	+
AP24 (Delmas et al., 1994)	hAPN	pAPN 717-813	+	+	−
AP18-hbh (Benbacer et al., 1997)	hAPN	bAPN 643-841	nd	+	−
AP18-hch (Benbacer et al., 1997)	hAPN	cAPN 643-841	nd	+	+
AP34 (Kolb et al., subm.)	pAPN	fAPN 135-297	+	+	−
AP33 (Hegyi et al., in prep.)	hAPN	fAPN 670-840	+	+	+

nd: not determined

[1]The susceptibility of transfected cells to the infection with HCV 229E, TGEV and FIPV is indicated. Susceptibility was analysed by detection of viral antigens in infected cells.

(Delmas et al., 1993), transfection of pAPN or hAPN expression vectors failed to induce susceptibility to FIPV infection (Table 1). The sequence of the fAPN cDNA derived from CRFK cells is highly homologous to the fAPN cDNA recently isolated from Fcwf cells (Tresnan et al., 1996). However, the predicted CRFK fAPN protein differs from the Fcwf fAPN protein by 16 amino acids (Kolb et al., submitted).

In order to define the determinants in the fAPN protein which mediate the infection of HCV 229E a chimeric protein was generated (AP34) which carries fAPN amino acids 135–297 incorporated into a pAPN backbone (Table 1; Fig. 1). This chimera was able to act as a receptor for HCV 229E, indicating that the critical determinants required for the infection of HCV 229E via feline APN are present within fAPN amino acids 135–297 (Kolb et al., submitted). The sequences of human, feline and porcine APN were aligned by using the CLUSTAL method of the MegAlign program (DNAStar, Madison, WI, USA). This alignment reveals that the chimeras AP32, which carries hAPN amino acids 260–353 inserted into a pAPN backbone (Kolb et al., 1996), and AP34 overlap by as little as 39 amino acids (Fig. 1). These 39 amino acids contain a sequence which is highly divergent between the 3 APN proteins (Fig. 1). Within this short sequence the hAPN and fAPN are more related to each other that to the pAPN protein. A number of site directed mutants was therefore generated to more closely characterise this amino acid segment. In one of these mutants the pAPN amino acids 283–290 (QSVNETAQ) were replaced by the hAPN amino acids 288–295 (DYVEKQAS) (mutant AP35, Table 2). This replacement of 6 amino acids of pAPN by the corresponding hAPN residues was sufficient to convert pAPN into a functional receptor for HCV 229E (Kolb et al., submitted). At the same time this mutant was still able to mediate the infection of TGEV, suggesting that the exchanged amino acid residues do not play an essential role for the TGEV-pAPN interaction (Table 2).

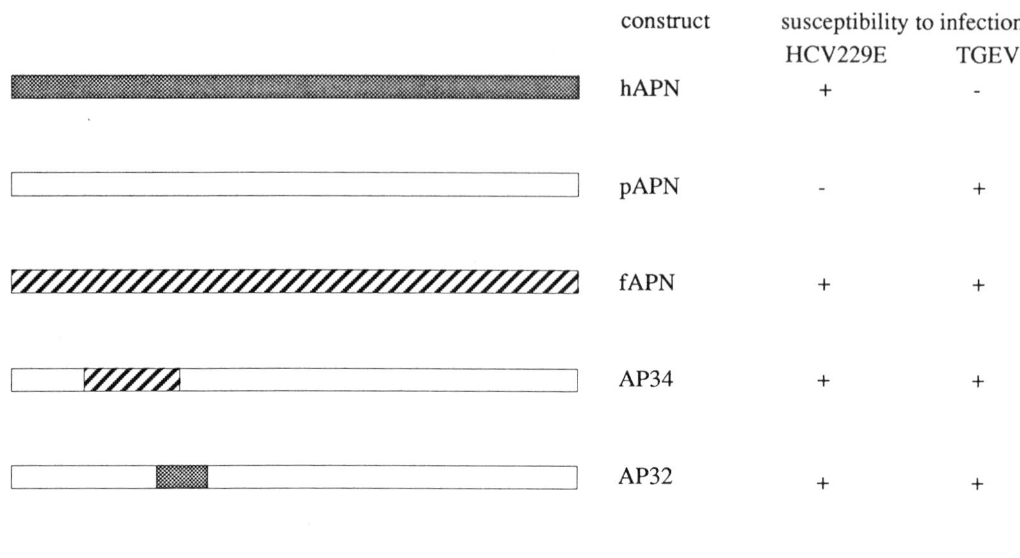

Figure 1. Schematic representation of the proteins hAPN, pAPN, fAPN and the chimeric proteins AP34 and AP32. The susceptibility of 293 HEK cells transfected with the corresponding expression constructs to infection with HCV229E and TGEV is indicated. Sequence alignment of hAPN amino acids 260 to 299 with the corresponding regions of the fAPN and pAPN proteins. The alignment was done using the CLUSTAL method of the MegAlign program (DNAStar, Madison, WI, USA). Residues diverging from the hAPN sequence are boxed.

The hAPN amino acid segment 288–295 was analysed for features that are similar in the fAPN sequence and different in the pAPN sequence to establish a molecular basis for the ability of hAPN and fAPN to act as receptors for HCV 229E. A first obvious difference between the pAPN sequence on the one side and the hAPN and fAPN sequences on the other is the presence of a N-linked glycosylation signal in the pAPN protein (residues: 283–290 QSV*NETA*Q) which is absent from hAPN and fAPN. The glycosylation signal was removed by site directed mutagenesis (converting the pAPN amino acids 283–290 to

Table 2. Receptor activity of mutated pAPN proteins. The susceptibility of transfected cells to the infection with HCV 229E and TGEV is indicated

Mutant	Mutated sequence: pAPN 283-290: QSVNETAQ	Susceptibility to HCV 229E	Susceptibility to TGEV
AP35	DYVEKQAS	+	+
AP36	QSVETRAQ	−	+
AP37	QYVNERAQ	−	+
AP38	QYVNKTAQ	−	+

QSVETRAQ). A mutant carrying this sequence alteration (AP36), however, failed to mediate HCV 229E infection (Table 2), suggesting that the absence of the glycosylation signal within this protein domain is not sufficient to convert pAPN into a receptor for HCV 229E. The mutant AP36 was still able to act as a receptor for TGEV, again indicating that the mutated sequence does not play an essential role in the pAPN-TGEV interaction.

Another prominent property which distinguishes hAPN and fAPN from pAPN is the presence of a tyrosine residue (hAPN: Y-289; fAPN Y-287) and a positively charged residue (hAPN: K-292; fAPN: R-291) within the critical amino acid segment defined by mutant AP35 (Fig. 1). A computer prediction using the predict-protein program (Rost, 1996) also indicates that this amino acid segment is located within an alpha-helical domain of the APN protein. The tyrosine residue and the positively charged residue are predicted to lie on the same surface of the helix. These two residues may therefore play a critical role in the HCV 229E receptor function of hAPN and fAPN. Two further mutants were thus generated in which the pAPN sequence QSVNETAQ (residues 283–290) was converted to QYVNERAQ (mutant AP37) or QYVNKTAQ (mutant AP38). Both of these mutants, however, failed to convert pAPN into a receptor for HCV 229E (Table 2). Again, both mutants could still act as receptors for TGEV (Table 2).

The amino acid residues which are critically involved in mediating the HCV 229E receptor function of both hAPN and fAPN are located within the amino-terminal part of the protein (Fig. 1). The mutant AP34, however, failed to mediate FIPV infection, suggesting that the determinants which are essential for the FIPV receptor function of fAPN are not present within this chimera (Kolb et al., submitted). The determinants mediating TGEV and CCV infection have been mapped to the carboxy-terminal part of porcine and canine APN, respectively (Delmas et al., 1994; Benbacer et al., 1997). To investigate whether the determinants mediating the FIPV receptor function are also located within the C-terminal part of fAPN, we generated a chimera (AP33) in which fAPN amino acids 670–840 were integrated into a hAPN backbone (Table 1, Fig. 2). Transfection/infection experiments revealed that AP33 can mediate infection of FIPV and TGEV (Hegyi et al., in prep.). This indicates that a) the determinants which are essential for the fAPN-FIPV interaction are indeed located within the C-terminal segment of fAPN, b) TGEV and FIPV interact with similar determinants within their respective receptor proteins and c) as in hAPN and pAPN, HCV 229E and TGEV interact with different amino acid segments of the fAPN protein.

In order to define critical amino acid residues within the domain involved in TGEV, FIPV and CCV reception we aligned the amino acid sequences of APN proteins which have been analysed for their coronavirus receptor function (Fig. 2). Chimeric proteins carrying fragments of the C-terminal part of porcine, canine, bovine and feline APN (encoding the amino acids indicated in Table 1) within an hAPN backbone can all act as functional receptors for TGEV (Delmas et al., 1994; Benbacer et al., 1997). Only the corresponding hAPN sequence is unable to mediate the TGEV-receptor function, suggesting that residues critically involved in the TGEV-APN interaction are not present within the hAPN sequence. As few as 9 amino acids in the 717–813 (pAPN) region are specific for the human APN sequence (Fig. 2).

By using site directed mutagenesis we converted two stretches of amino acids within the C terminal part of the hAPN protein (alone and in combination) to the corresponding porcine sequence (Table 3). None of the mutations, however, converted hAPN into a functional receptor for TGEV. The EEE->QDQ mutation in the mutant AP42 rather rendered the APN protein non-functional in that the expressed protein neither showed any HCV 229E receptor activity nor retained its aminopeptidase function (Table 3). The catalytic centre of the hAPN protein as well as the stretch of amino acids which plays an essential

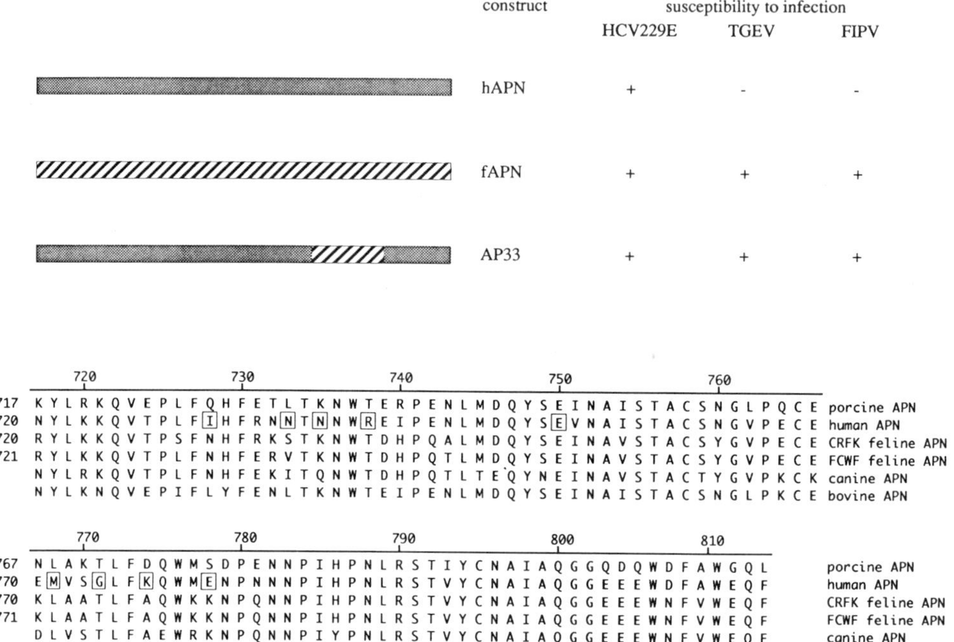

Figure 2. Schematic representation of the proteins hAPN, fAPN and the chimeric protein AP33. The susceptibility of cells transfected with the corresponding expression constructs to infection with HCV229E, TGEV and FIPV is indicated. Sequence alignment of pAPN amino acids 717 to 813 with the corresponding regions of the human, feline (from FCWF and CRFK cells) canine and bovine APN proteins. The alignment was done using the CLUSTAL method of the MegAlign program (DNAStar, Madison, WI, USA). Residues unique to the hAPN sequence are boxed.

role in the HCV 229E receptor function of hAPN is distant to the site of the EEE->QDQ mutation in the primary structure. This result may thus suggest that the determinants identified as being essential for the HCV 229E- and TGEV-receptor function of human, porcine and feline APN are not independent protein domains. Most of the sequenced mammalian APN proteins carry the EEE sequence at the position corresponding to pAPN amino acids 806–808, whereas the QDQ sequence is unique to the porcine APN protein (Table 3). As the introduction of the QDQ sequence into hAPN destroys the HCV 229E receptor function and the enzymatic activity, a second site in the pAPN protein must have a compensatory effect to ensure the proper biological function of pAPN.

Table 3. Receptor activity of mutated hAPN proteins. The susceptibility of transfected cells to the infection with HCV 229E and TGEV and the enzymatic activity of the mutated proteins are indicated

Mutant	Original hAPN sequence (residues)	Mutated sequence (residues)	Susceptibility to HCV 229E	Susceptibility to TGEV	Enzymatic activity
AP40	IHFRNN (731-736) + EEE (806-808)	QHFETL (731-736) + QDQ (806-808)	−	−	−
AP41	IHFRNN (731-736)	QHFETL (731-736)	+	−	+
AP42	EEE (806-808)	QDQ (806-808)	−	−	−

5. DISCUSSION

We have identified, by using chimeric APN proteins and site directed mutagenesis, a short amino acid segment in the hAPN protein which is critically involved in mediating the HCV 229E receptor function (Kolb et al., submitted). Additionally, we have demonstrated that the determinants in the fAPN protein which mediate reception of HCV 229E and TGEV are located in similar parts of the protein as the corresponding determinants in hAPN and pAPN (Kolb et al., 1996; Delmas et al., 1994). The utilisation of different determinants of the receptor protein may also reflect the serological and genetical differences between HCV 229E and the other group I coronaviruses like TGEV, FIPV and CCV (Siddell, 1995). In the absence of an APN crystal structure, however, it is difficult to predict whether these determinants are really independent protein domains or rather assemble into a single epitope which mediates coronavirus infections.

REFERENCES

Benbacer, L., Kut, E., Besnardeau, L., Laude, H., and Delmas, B., 1997, Interspecies aminopeptidase-N chimeras reveal species-specific receptor recognition by canine coronavirus, feline infectious peritonitis virus, and transmissible gastroenteritis virus, *J. Virol.* **71**: 734–737.

Delmas, B., Gelfi, J., L'Haridon, R., Vogel, L.K., Sjostrom, H., Noren, O., and Laude, H., 1992, Aminopeptidase N is a major receptor for the entero-pathogenic coronavirus TGEV, *Nature* **357**: 417–20.

Delmas, B., Gelfi, J., Sjostrom, H., Noren, O., and Laude, H., 1993, Further characterization of aminopeptidase-N as a receptor for coronaviruses, *Adv. Exp. Med. Biol.* **342**: 293–298.

Delmas, B., Gelfi, J., Kut, E., Sjostrom, H., Noren, O., and Laude, H., 1994, Determinants essential for the transmissible gastroenteritis virus-receptor interaction reside within a domain of aminopeptidase-N that is distinct from the enzymatic site, *J. Virol.* **68**: 5216–24.

Grosse, B. and Siddell, S.G., 1994, Single amino acid changes in the S2 subunit of the MHV surface glycoprotein confer resistance to neutralization by S1 subunit-specific monoclonal antibody, *Virology* **202**: 814–824.

Kolb, A.F., Maile, J., Heister, A., and Siddell, S.G., 1996, Characterization of functional domains in the human coronavirus HCV 229E receptor, *J. Gen. Virol.* 2515–2521.

Rost, B., 1996, PHD: predicting one dimensional protein structure by profile based neural networks., *Meth. Enzymol.* **266**: 525–539.

Siddell, S.G., 1995, The Coronaviridae: an introduction, in: *The Coronaviridae* (S. G. Siddell), Plenum Press, New York, pp. 1–10.

Tresnan, D.B., Levis, R., and Holmes, K.V., 1996, Feline aminopeptidase N serves as a receptor for feline, canine, porcine, and human coronaviruses in serogroup I, *J .Virol .* **70**: 8669–8674.

Yeager, C.L., Ashmun, R.A., Williams, R.K., Cardellichio, C.B., Shapiro, L.H., Look, A.T., and Holmes, K.V., 1992, Human aminopeptidase N is a receptor for human coronavirus 229E, *Nature* **357**: 420–2.

9

FELINE AMINOPEPTIDASE N IS A RECEPTOR FOR ALL GROUP I CORONAVIRUSES

Dina B. Tresnan and Kathryn V. Holmes

Department of Microbiology
University of Colorado Health Sciences Center
Denver, Colorado 80262

ABSTRACT

Human coronavirus HCV-229E and porcine transmissible gastroenteritis virus (TGEV), both members of coronavirus group I, use aminopeptidase N (APN) as their cellular receptors. These viruses show marked species specificity in receptor utilization as they can only use APN of their respective species to initiate virus infection. Feline and canine coronaviruses are also group I coronaviruses. To determine whether feline APN could serve as a receptor for feline coronaviruses (FCoVs), we cloned the cDNA encoding feline APN (fAPN) by PCR from feline cells and stably expressed it in FCoV-resistant mouse or hamster cells. These became susceptible to infection with either of several biotypes of FCoVs. The expression of recombinant fAPN also made hamster and mouse cells susceptible to infection with other group I coronaviruses, including several canine coronavirus strains, transmissible gastroenteritis virus (TGEV), and human coronavirus HCV-229E. Thus, fAPN served as a functional receptor for each of these coronaviruses in group I. As expected, fAPN could not serve as a receptor for mouse hepatitis virus (MHV), a group II coronavirus which uses murine biliary glycoproteins as receptors. Thus, fAPN acts as a common receptor for coronaviruses in group I, in marked contrast to human and porcine APN glycoproteins which serve as receptors only for human and porcine coronaviruses, respectively. These observations suggest that cats could serve as a "mixing vessel" in which simultaneous infection with several group I coronaviruses could lead to recombination of viral genomes.

INTRODUCTION

Coronaviruses in group I cause disease in many different species of animals and humans, generally in a host specific manner (Wege, 1982). The members of group I include

Coronaviruses and Arteriviruses, edited by Enjuanes *et al.*
Plenum Press, New York, 1998

two feline coronaviruses (FCoVs), feline infectious peritonitis virus (FIPV) and feline enteric coronavirus (FeCV); two porcine coronaviruses, transmissible gastroenteritis virus (TGEV) and porcine respiratory coronavirus (PRCV); canine coronavirus (CCV); and human coronavirus 229E (HCV-229E) (Holmes and Lai, 1996). Except for FIPV, which causes systemic disease, infections with these viruses are generally limited to the enteric or respiratory tracts.

The feline coronaviruses cause disease in domestic and exotic *Felidae*. Isolates of FIPV from diseased cats vary markedly in virulence *in vivo* and cytopathogenicity *in vitro*.

While some isolates called FeCV cause inapparent or mild infection in the enteric tract in kittens (Pedersen, 1987), other strains known as FIPV cause a chronic, systemic immune-mediated disease called feline infectious peritonitis (FIP) (Scott, 1986a). This disease is characterized by fibrinonecrotic and pyogranulomatous peritonitis and pleuritis and is usually fatal (Barlough and Stoddart, 1990; Pedersen and Boyle, 1980).

Based on serological studies, FCoVs have been subdivided into biotypes I and II (Pedersen *et al.*, 1984). Type II FCoVs are antigenically more closely related to CCV than to type I strains and also more closely resemble CCV in growth in tissue culture (Pedersen *et al.*, 1984). Comparative analysis of nucleotide sequences of feline and canine coronaviruses suggests that type II FCoVs may have arisen by RNA recombination between type I FCoV and CCV (Herrewegh, *et al.*, 1995).

Although, in general, the group I coronaviruses can only infect cultured cells of their normal host species, feline cells in culture can be infected with CCV, TGEV and HCV-229E as well as the FCoVs (Levis *et al.*, 1995). All of these viruses bind to brush border membranes isolated from cat intestines (Holmes and Compton, 1995). In addition, cats can be experimentally infected with non-feline coronaviruses in group I. The infected animals develop anti-viral antibody and may shed some virus, but they do not exhibit any clinical disease and infection does not protect them from subsequent infection with FCoVs (Scott, 1986b; Barlough *et al.*, 1985; Barlough *et al.*, 1984; Reynolds and Garwes, 1979).

Aminopeptidase N (APN) is a 150-kDa plasma membrane-associated type II glycoprotein with metalloprotease activity. It is expressed on the plasma membranes of epithelial cells of the intestinal brush border, renal proximal tubules, and respiratory tract, the synaptic membranes of the central nervous system and cells of the granulocyte and monocyte lineages (Look *et al.*, 1989; Kenny and Maroux, 1982). Cellular entry of both HCV-229E and the porcine coronaviruses TGEV and PRCV is mediated by the same APN receptor (Delmas *et al.*, 1994a; 1992; Yeager *et al.*, 1992). However, HCV-229E cannot utilize porcine APN (pAPN) as a receptor, and TGEV cannot utilize human APN (hAPN) as a receptor (Levis, *et al.*, 1995; Delmas *et al.*, 1994b; 1992). Hamster (BHK) cells transfected with pAPN were not susceptible to FIPV or CCV, suggesting that the pAPN expressed in these cells could not function as a receptor for these cat and dog coronaviruses (Delmas *et al.*, 1994a). Thus, in these instances, receptor specificity appears to be a major determinant of the observed species specificity of HCV-229E and TGEV infection.

Since APNs serve as receptors for several group I coronaviruses, we investigated whether feline APN (fAPN) could serve as a receptor for FCoVs. Using primers that recognize consensus sequences in APN genes from other species, a cDNA encoding feline APN was amplified by PCR from feline cell cDNA, cloned, and then expressed in mouse or hamster cells. Cells transfected with the feline APN gene were challenged with FCoVs and other group I coronaviruses to determine whether fAPN was a functional receptor for these viruses.

MATERIALS AND METHODS

Methods were as described in Tresnan *et al.*, 1996. Feline coronavirus strain DF-2 was obtained from American Type Culture Collection (ATCC) (Gaithersburg, MD). Canine coronavirus strain TN-449 was obtained from ATCC, and CCV/K378 was kindly provided by Leland E. Carmichael (James A. Baker Institute, Cornell University, Ithaca, NY). To increase the level of expression of recombinant fAPN, a new construct called fAPN-K was engineered using PCR. This construct contained an improved Kozak consensus sequence (Kozak, 1987) in comparison with the original fAPN construct (Tresnan, *et al.*, 1996). Hamster or mouse cells transfected with fAPN-K became susceptible to infection with feline coronaviruses at a higher rate than cells transfected with the original fAPN construct.

RESULTS

Our approach to studying the virus receptor activity and species specificity of fAPN was to clone and express a full-length cDNA clone of fAPN in FCoV-resistant cells, and then test them for susceptibility to coronavirus infection. The complete nucleotide sequence of fAPN cDNA was determined from a full length cDNA and three independently cloned overlapping segments of the APN cDNA (Tresnan *et al.*, 1996). Pairwise alignment of the amino acid sequence of fAPN with the sequences of hAPN or pAPN revealed 78 and 77% identity, respectively, using the human (M22324) and porcine (Z29522) APN sequences obtained from the GenBank database.

To determine whether the fAPN glycoprotein is a functional receptor for FCoVs, the full length cDNA encoding fAPN was stably expressed in hamster (BHK) or mouse (NIH3T3) cell lines, which are normally resistant to FCoV infection. The stably transfected cells were inoculated with either of several strains of FIPV including 79–1146, UCD-1, and DF-2, or mock infected with medium. Antigens of all three strains of FIPV were detected in the cytoplasm of the fAPN-transfected mouse or hamster cells by 6 hours post inoculation (p.i.), but no viral antigens were detected in hamster or mouse cells stably transfected with the vector alone and then inoculated with FIPV (Table 1 and 2; Tresnan *et al.*, 1996). In hamster cells transfected with the fAPN-K construct, which contained the improved Kozak consensus sequence, and inoculated with FIPV 79–1146, a higher per cent of the cells was infected and there was more extensive fusion than in cells transfected with the original

Table 1. Feline coronavirus strains that can utilize fAPN as a receptor

Strain	Biotype[a]	Growth in cell culture	Disease	Use fAPN as a receptor[b]
FIPV				
79-1146	II	+	FIP	+
DF-2	II	+	FIP	+
UCD 1	I	poor	FIP	+
FeCV				
79-1683	II	+	Enteritis	+

[a] Biotypes of different strains were reviewed by Herrewegh *et al.*, 1995.
[b] The ability to use fAPN as a receptor was demonstrated by the production of viral antigens in the cytoplasm of hamster and or mouse cells transfected with fAPN after inoculation with each strain of FCoV (Tresnan *et al.*, 1996).

fAPN construct (Fig. 1). Hamster and mouse cells transfected with fAPN or with vector alone were also inoculated with FeCV, an enterotropic feline coronavirus closely related to FIPV. FeCV viral antigens were detected in the cytoplasm of fAPN-transfected NIH3T3 cells and BHK, but not in cells transfected with vector alone. Thus, fAPN serves as a receptor for all feline coronaviruses tested, including several biologically different strains of FIPV and one of FeCV. Expression of recombinant fAPN in mouse or hamster cells is sufficient to allow FCoV binding and entry and the synthesis of viral antigens.

To determine whether fAPN-transfected hamster cells could produce infectious FIPV or FeCV, these cells were inoculated with FIPV 79–1146 or FeCV 79–1683, and the infectious virus released into the supernatant medium was titered by plaquing on Fcwf cells. With both FCoVs, infectious virions were produced and released into the medium from fAPN-transfected BHK cells, but not from BHK cells transfected with vector alone (data not shown).

Because the Fcwf line of feline cells is susceptible to infection by other coronaviruses in group I including CCV, TGEV and HCV-229E in addition to FCoVs (Table 2 and Levis *et al.*, 1995), we tested these viruses for infectivity on fAPN-transfected BHK and fAPN-transfected NIH3T3 cells. The fAPN-transfected BHK and NIH3T3 cells or control cells transfected with vector alone were inoculated with HCV-229E, TGEV, or with MHV-A59 and with several strains of CCV including 1–71, K378 and TN-449. Transfected cells were infected in two separate experiments and multiple cover slips were analyzed to determine the percentage of infected cells by immunolabeling with anti-viral antibody. The fAPN-transfected hamster and mouse cells were susceptible to infection

BHK-fAPN

FIPV 79-1146

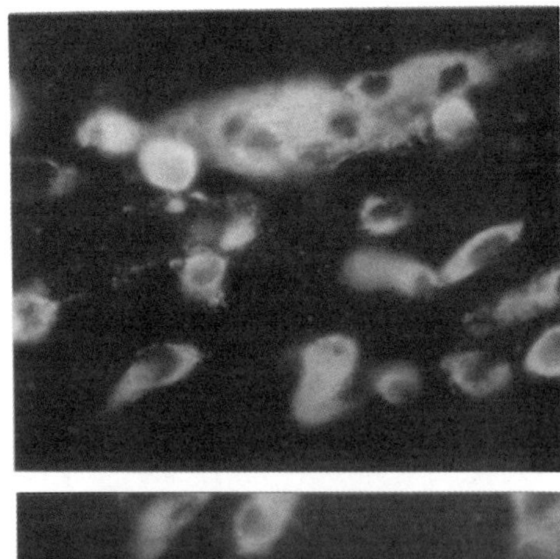

CCV TN-449

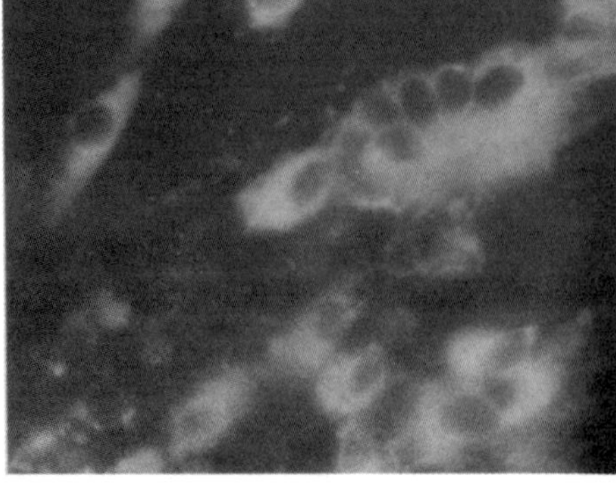

Figure 1. Recombinant feline aminopeptidase N can serve as a receptor for both feline and canine coronaviruses. Hamster cells transfected with feline aminopeptidase N were inoculated with feline coronavirus FIPV 79–1146 or canine coronavirus TN-449. Viral antigens were detected in infected cells by immunofluorescent labeling.

Table 2. Receptor activity of fAPN for various coronaviruses

	FIPV 79-1146	FeCV 79-1683	CCV 1-71	TGEV	HCV-229E	MHV-A59
Fcwf	+[a]	+	+	+	+	−
BHK-pCR3	−	−	−	−	−	−
BHK-fAPN	+	+	+	+	+	−

[a]Susceptibility to coronavirus infection was determined by the detection of viral antigens in infected cells using an indirect immunofluorescence assay (Tresnan *et al.*, 1996).

with each of the coronaviruses in group I as shown by viral antigen production in the cytoplasm of some of the transfected cells (Table 2; Fig.1, and Tresnan *et al.*, 1996). Neither mouse or hamster cells transfected with vector alone was infected by any of the group I coronaviruses tested. Thus, fAPN served as a receptor for CCV, HCV-229E and TGEV as well as for FCoVs. Because these experiments were done using pools of cells stably transfected with fAPN, we plan further studies on cloned cell lines expressing known levels of fAPN in order to determine whether fAPN is a more efficient receptor for feline coronaviruses than for other group I coronaviruses.

The fAPN-transfected BHK cells were not susceptible to infection with murine coronavirus MHV-A59. As expected, both NIH3T3 cells transfected with vector alone (data not shown) and fAPN-transfected NIH3T3 cells were susceptible to MHV infection because these mouse cells express the biliary glycoproteins that serve as receptors for MHV (Holmes and Dveksler, 1994).

DISCUSSION

We have shown that fAPN serves as a receptor for strains of FCoVs that differ markedly in their ability to grow in cultured feline cells and in their virulence in cats. In addition, we found that fAPN can serve as a receptor for all of the non-feline coronaviruses in group I that were tested (Tresnan, *et al.*, 1996). The common use of APN as a receptor for related coronaviruses that infect different species has important implications for the evolution of this group of coronaviruses. The phylogenetic relationships between group I coronaviruses were examined by comparing the nucleotide sequences of their spike protein genes. This work showed that canine coronaviruses are more similar to FCoVs than to the porcine coronaviruses, while the spike protein sequences of HCV-229E clearly place that virus in a separate cluster (Wesseling, *et al.*, 1994). The close relationships among these related coronaviruses and their use of a common molecule as a receptor suggests that they may have evolved from a common ancestor. Perhaps FCoV was the first ancestral group I coronavirus that exploited APN as a receptor. The virus could then have adapted to different species by a series of mutations in the S gene that optimized the usage of species-specific determinants on their APN receptors as shown for both porcine and human APN (Delmas *et al.* 1994a). Mutations in the spike protein that could optimize adaptation to species specific APN receptors may include altering the receptor binding site of S, its ability to undergo a conformational change, and/or its ability to mediate fusion of the viral envelope with the host membranes during virus entry.

Although APN molecules from various species have been shown to function as receptors for some group I coronaviruses, the specific functional domains that bind to the receptor glycoprotein have not yet been completely defined. Using chimeras between pAPN and hAPN, Delmas and co-workers identified a region on pAPN between amino acids 717 and 813, located 330 amino acids downstream from the active site of the enzyme, that determined its ability to serve as a receptor for TGEV (Delmas *et al.*, 1994a). Recently, a chimera containing amino acids 643 to 841 from canine APN in the framework of human APN was shown to act as a receptor for CCV, FIPV and TGEV (Benbacer *et al.*, 1997). The amino acid sequences of the APN glycoproteins of different species differ markedly in this region. Unexpectedly, using these same chimeras, the corresponding region of hAPN was not responsible for the species specificity of HCV-229E infection. Instead, a small region containing amino acids 260–353 of human APN was shown to determine the species specificity of HCV-229E infection (Kolb *et al.*, 1996). Although chimeric APN glycoproteins help to define domains that affect the species specificity of group I coronavirus receptors, they do not define the precise areas of interaction of the S glycoproteins with the receptors. Some contributions to the functions of the receptor glycoprotein may be provided by the framework of the APN glycoprotein into which a segment from another species is inserted, in addition to the functions provided by the inserted segment.

We have shown that the full length fAPN glycoprotein can serve as a common receptor for all of the coronaviruses in group I. This is in marked contrast to the porcine and human APN glycoproteins which function as receptors only for coronaviruses that cause disease in a single species (Kolb *et al.*,1996; Delmas, *et al.*1994a). The non-feline coronaviruses in group I can infect cats without causing disease (Barlough *et al.*, 1985; 1984). These viruses probably infect cats *in vivo* by using feline APN as a receptor, as they can do *in vitro* (Tresnan, *et al.* 1996). If all group I coronaviruses can infect cat cells *in vivo* using their common fAPN receptor, then it is possible that recombination between different coronaviruses in group I could occasionally occur in a cell of a cat that is simultaneously infected with several of these viruses. Thus, the cat could serve as a "mixing vessel," allowing recombination and shuffling of genes from different group I coronaviruses. Strong support for this hypothesis is provided by the observation that type I FCoVs may have arisen naturally by recombination between a type II FCoV and CCV (Herrewegh, *et al.*, 1995). While most recombinant coronaviruses might be poorly suited for transmission from animal to animal, rarely a recombinant virus might arise that can replicate to high titers *in vivo* and be transmitted efficiently. Such a transmissible recombinant coronavirus would be a candidate for causing an emerging virus disease either in cats or another host species.

ACKNOWLEDGMENTS

We are grateful to the excellent technical assistance of John Schneider and James Ahn and to Bruce Zelus, David Wentworth and Dianna Blau for thoughtful comments on the manuscript. We also are grateful to the many colleagues who provided us with virus strains and cell lines. This work was supported by U.S. Public Health Service grant AI26075 and by Physician Scientist Award K11 AI01151 to D.B.T.

REFERENCES

Barlough, J. E., and Stoddart, C. A., 1990, Feline coronaviral infections, in: *Infectious diseases of the dog and cat*, (C. E. Greene, ed.), W. B. Saunders Co., Philadelphia, pp. 300–312.
Barlough, J. E., Stoddart, C. A., Sorresso, G. P., Scott F. W., and Jacobson, R. H.,1984, Experimental inoculation of cats with canine coronavirus and subsequent challenge with feline infectious peritonitis virus, *Lab. Anim. Sci.* **34:** 592–597.

Barlough, J. E., Johnson-Lussenburg, C. M., Stoddart, C. A., Scott, F. W., and Jacobson, R. H., 1985, Experimental inoculation of cats with human coronavirus 229E and subsequent challenge with feline infectious peritonitis virus, *Can. J. Comp. Med.* **49:** 303–307.

Benbacer, L., Kut, E., Besnardeau, L., Laude, H., and Delmas, B., 1997, Interspecies aminopeptidase-N chimeras reveal species specific receptor recognition by CCV, FIPV, and TGEV, *J. Virol.* **71:** 734–737.

Delmas, B., Gelfi, J., L'Haridon, R., Vogel, L. K., Sjöström, H., Norén, O., and Laude, H., 1992, Aminopeptidase N is a major receptor for the enteropathogenic coronavirus TGEV, *Nature* (London) **357:** 417–419.

Delmas, B., Gelfi, J., Kut, E., Sjöström, H., Norén, O., and Laude, H., 1994a, Determinants essential for the transmissible gastroenteritis virus-receptor interaction reside within a domain of aminopeptidase N that is distinct from the enzymatic site, *J. Virol.* **68:** 5216–5224.

Delmas, B., Gelfi, J., Sjöström, H., Norén, O., and Laude, H., 1994b, Further characterization of aminopeptidase N as a receptor for coronaviruses, *Adv. Exp. Med. Biol.* **342:** 293–298.

Herrewegh, A. A. P. M., Vennema, H., Horzinek, M.C., Rottier, P.J.M., and de Groot, R.J., 1995, The molecular genetics of Feline coronaviruses: Comparative sequence analysis of the ORF7a/7b transcription unit of different biotypes, *Virology* **212:** 622–631.

Holmes, K.V., and Dveksler, G.S., 1994, Specificity of coronavirus/receptor interactions, in: *Cellular Receptors for Animal Viruses*, (E. Wimmer, ed.), Cold Spring Harbor Laboratory Press, pp. 403–443.

Holmes, K. V., and Compton, S. R., 1995, Coronavirus receptors, in: *The Coronaviruses*, (S. Siddell, ed.), Plenum Press, New York, pp. 55–72.

Holmes, K. V., and Lai, M. M. C., 1996, Coronaviridae: The viruses and their replication, in: *Virology* (Fields), 3rd ed., (B. N. Fields, D. M. Knipe and P. M. Howley, eds.), Lippincott-Raven Publishers, Philadelphia, pp. 1075 - 1093.

Kenny, A. J., and Maroux, S., 1982, Topology of microvillar membrane hydrolases of kidney and intestine, *Physiol. Rev.* **62:** 91–128.

Kolb, A., Maile, J., Heister, A., and Siddell, S., 1996, Characterization of functional domains in the human coronavirus 229E receptor, *J. Gen. Virol.* **77:** 2515–2521.

Kozak, M., 1987, An analysis of 5´-noncoding sequences from 699 vertebrate messenger RNAs, *Nucleic Acids Res.* **15:**8125–8148.

Levis, R., Cardellichio, C.B., Scanga, C.A., Compton, S.R., and Holmes, K.V., 1995, Multiple receptor-dependent steps determine the species specificity of HCV-229E infection, *Adv. Exp. Med. Biol.* **380:** 337–344.

Look, A. T., Ashmun, R. A., Shapiro, L. H., and Peiper, S. C., 1989, Human myeloid plasma membrane glycoprotein CD13 (gp150) is identical to aminopeptidase N, *J. Clin. Invest.* **83:** 1299–1307.

Pedersen, N. C., 1987, Virologic and immunologic aspects of feline infectious peritonitis virus infection, *Adv. Exp. Med. Biol.* **218:** 529–550.

Pedersen, N. C., and Boyle, J. F., 1980, Immunologic phenomena in the effusive form of feline infectious peritonitis, *Am. J. Vet. Res.* **41:** 868–876.

Pedersen, N.C., Black, J.W., Boyle, J.F., Evermann, J.F., McKeirnan, A.J., and Ott, R.L., 1984, Pathogenic differences between various feline coronavirus isolates, *Adv. Exp. Med. Biol.* **173** :337–344.

Reynolds, D. J., and Garwes D. J., 1979, Virus isolation and serum antibody responses after infection of cats with transmissible gastroenteritis virus, *Arch. Virol.* **60:**161–166.

Scott, F. W., 1986a, Feline infectious peritonitis and other feline coronaviruses, in: *Current veterinary therapy IX*, (R. W. Kirk, ed.), W. B. Saunders Co., Philadelphia, pp. 1059–1062.

Scott, F. W., 1986b, Immunization against feline coronaviruses, *Adv. Exp. Med. Biol.* **218:**569–576.

Tresnan, D.B., Levis, R., and Holmes, K.V., 1996, Feline aminopeptidase N (fAPN) serves as a receptor for feline, canine, porcine and human coronaviruses in serogroup I, *J. Virol.* **70:**8669–8674

Wege, H., Siddell, S., and ter Meulen, V., 1982, The Biology and Pathogenesis of Coronaviruses, *Curr. Topics Microbiol. Immunol.* **99:** 165–200.

Wesseling, J.G., Vennema, H., Godeke, G.J., Horzinek, M.C., Rottier, P.J., 1994, Nucleotide sequence and expression of the spike (S) gene of canine coronavirus and comparison with the S proteins of feline and porcine coronaviruses, *J. Gen. Virol.* **75:** 1789–1794.

Yeager, C. L., Ashmun, R. A., Williams, R. K., Cardellichio, C. B., Shapiro, L. H., Look, A.T., and Holmes, K. V., 1992, Human aminopeptidase N is a receptor for human coronavirus 229E, *Nature* (London) **357:** 420–422.

10

DIFFERENTIAL RECEPTOR-FUNCTIONALITY OF THE TWO DISTINCT RECEPTOR PROTEINS FOR MOUSE HEPATITIS VIRUS

N. Ohtsuka,[1] Y. K. Yamada,[2] K. Saeki,[1] and F. Taguchi[1]

[1]National Institute of Neuroscience, NCNP
4-1-1 Ogawahigashi, Kodaira
Tokyo 1871, Japan
[2]National Institute of Infectious Diseases
4-7-1 Gakuen, Musashimurayama
Tokyo 190-122, Japan

1. ABSTRACT

We compared the virus-binding activity and receptor-functionality of the receptor proteins isolated from mouse hepatitis virus (MHV)-susceptible BALB/c mice (MHVR1) and MHV-resistant SJL mice (MHVR2). By using a soluble receptor protein which lacked the transmembrane and intracytoplasmic domains, virus overlay protein blot assay and neutralization tests showed that MHVR1 bound to JHM cl-2 virus with 300–500 times higher efficiency than to MHVR2. MHVR1 was revealed to have 10–30 fold higher receptor-functionality than MHVR2 when examined by measuring virus-binding to the receptor expressed on the cell surface. These findings suggested that the differences in susceptibility between BALB/c and SJL mice may depend upon the genotype of the MHV receptor.

2. INTRODUCTION

The receptor proteins for mouse hepatitis virus (MHV) isolated from MHV-susceptible BALB/c mice was reported to interact with the virus particles in virus overlay protein blot assay (VOPBA), although such proteins were not detected in SJL mice resistant to MHV (Boyle et al., 1987). This receptor was thereafter identified as Bgp1[a] with four domains (the splice variant of Bgp1[a] containing two domains is designated as MHVR1 in this paper) which is classified as a carcinoembryonic antigen gene family member in the immunoglobulin super family and is a functional receptor protein for MHV-A59 (Dveksler et al., 1991). This finding could imply that the difference in susceptibility between BALB/c and

Coronaviruses and Arteriviruses, edited by Enjuanes *et al.*
Plenum Press, New York, 1998

SJL mouse strains to MHV infection may be due to the presence or absence of the functional receptor. In SJL mice, however, the homologous protein to MHVR1, named Bgp1[b] with two domains (called as MHVR2 in this paper), was found to serve as a functional receptor as well (Yokomori and Lai, 1992a, b), These results imply that the difference in susceptibility to MHV could not be simply explained by the presence or absence of functional receptors. To understand how the resistance of mice to MHV is controlled, we have compared the virus-binding activity and receptor-functionality of these proteins.

3. RESULTS AND DISCUSSION

To examine the virus-binding activity, we utilized soluble forms of the receptor proteins which lacked the transmembrane domain and intracytoplasmic tail and were tagged with influenza HA epitope. The genes for these soluble receptor proteins were prepared by PCR as described previously (Yamada et al., 1993) (Fig. 1). These genes were transfected into RK13 cells by electroporation and soluble receptors were expressed with the vaccinia virus T7 system (Fuerst et al., 1986). Soluble forms of MHVR1 and MHVR2 secreted in the culture fluid were concentrated by ultrafiltration (millopore) and used to compare the virus-binding activity by VOPBA (Kubo et al., 1994; Taguchi, 1995). As has been reported (Boyle et al., 1987), we have observed that JHM cl-2 virus (Taguchi et al., 1985) bound to MHVR1 but not to MHVR2 (data not shown). Furthermore we performed a similar experiment by dot-blot hybridization to analyze the virus-binding activity quantitatively. The result was that MHVR1 had 300 times higher virus-binding activity than did MHVR2 (data not shown). Neutralization tests of JHM with the soluble form of the receptor proteins revealed that MHVR1 has a 500-fold higher neutralizing activity than MHVR2 (Fig. 2). These findings by VOPBA and neutralization assay indicate that there is a marked difference between MHVR1 and MHVR2 in virus-binding activity.

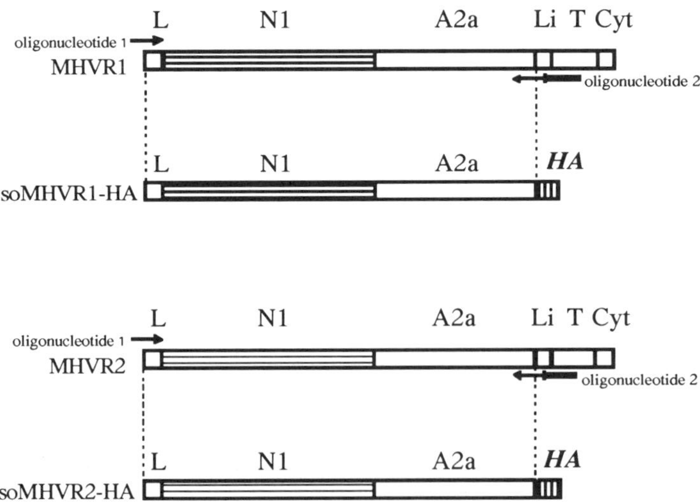

Figure 1. Structure of the genes used to express soluble MHVR1 and MHVR2. MHVR1 and MHVR2 are composed of two major domains, N and A2. L, Li, T, and Cyt indicate leader sequence, linker region, transmembrane domain and intracytoplasmic domain, respectively. The genes were designed to delete the transmembrane and intracytoplasmic domains and were tagged with HA epitope by PCR.

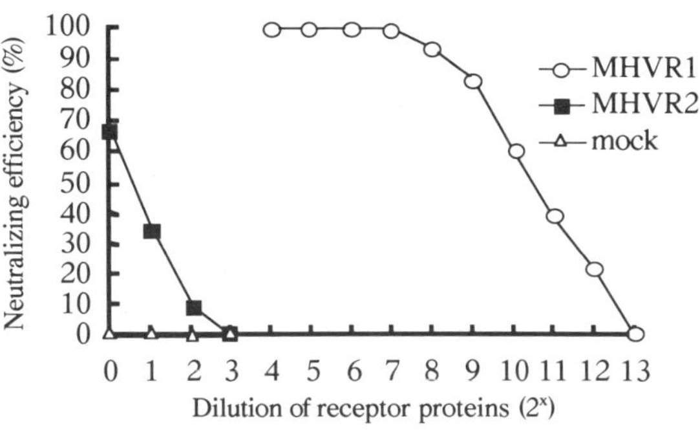

Figure 2. Neutralization of JHM by soluble MHVR1 and MHVR2. Soluble MHVR1 and MHVR2 as well as control culture fluids diluted two fold were mixed with JHM and incubated at RT for 1 h. The virus titres in the mixtures were plaque assayed on DBT cells. The rate of inhibition of virus infectivity is shown as neutralizing efficiency.

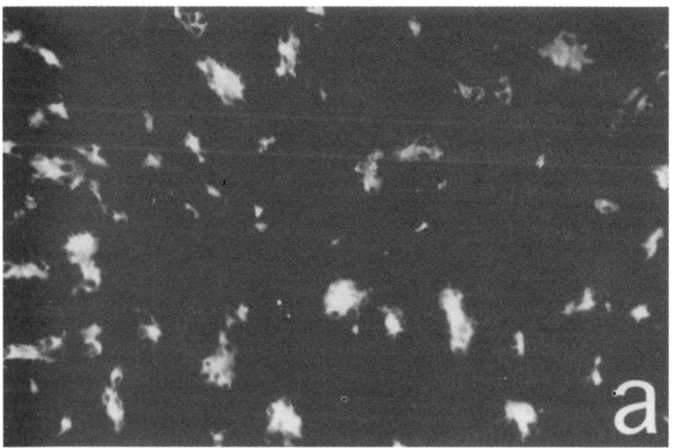

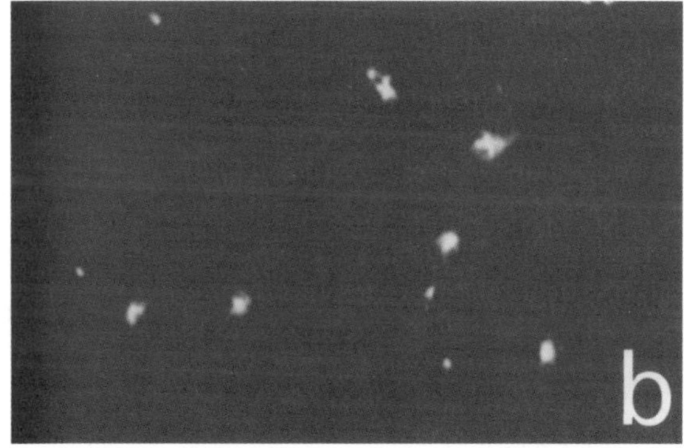

Figure 3. Immunofluorescence of BHK-MHVR1 and BHK-MHVR2 cells infected with JHM. BHK-MHVR1 (a), BHK-MHVR2 (b) and control BHK (c) cells infected with JHM at an m.o.i. of 0.5 were fixed with acetone at 10 h post-inoculation and examined for the presence of JHM antigen with a JHM-specific MAb.

We also compared the receptor-functionality of these proteins using MHVR expressed on cell membranes. MHVR1 and MHVR2 with transmembrane and intracytoplasmic domains were expressed in BHK-21 cells as described previously (Suzuki and Taguchi, 1996). There was a significant difference in the susceptibility to JHM between BHK cells expressing MHVR1 (BHK-MHVR1) and BHK cells expressing MHVR2 (BHK-MHVR2) (Fig.3). The numbers of fused BHK-MHVR1 cells were 10-fold more than that BHK-MHVR2 after infection at a multiplicity of infection (m.o.i.) of 0.5 (Fig. 3), and 30-fold higher at a m.o.i. of 0.005. We also investigated virus growth in these cells. JHM grew more efficiently in the BHK-MHVR1 cells than BHK-MHVR2. The difference was 6–60 times (data not shown). These results indicate that MHVR1 served as a highly efficient receptor for JHM, 10–30 fold higher as compared with MHVR2.

In the present study, a quantitative difference was shown in virus-binding ability between MHVR1 and MHVR2. Such a difference could account for the difference in susceptibility of cells to JHM infection. The finding that BHK-MHVR1 was 10–30 fold more susceptibility than BHK-MHVR2 cells could indicate the possibility that the resistance of SJL mice to JHM could be accounted for by the low virus-affinity of MHVR2 as compared with the MHVR1 of the susceptible BALB/c mouse strain. A 10 to 30-fold difference in virus affinity would be amplified into a huge difference in virus growth after repeated cycles of infection within a few days after the initial infection, which could result in the fatal disease of susceptible BALB/c mice and the survival of SJL mice. To investigate this possibility, susceptibilities of BALB/c mice whose MHVR1 gene has been replaced by the MHVR2 gene and SJL mice with the MHVR1 gene should be studied.

REFERENCES

Boyle J.F. , Weismiller D.G., and Holmes K.V., 1987, Genetic resistance to mouse hepatitis virus correlates with absence of virus-binding activity on target tissues, *J. Virol.* **61**: 185–189.

Dveksler G.S., Pensiero M.N., Cardellichio C.B., Williams R.K., Jiang G., Holmes K.V. and Diffenbach, C.W., 1991, Cloning of the mouse hepatitis virus (MHV) receptor: expression in human and hamster cell lines confers susceptibility to MHV, *J. Virol.* **65**: 6881–6891.

Fuerst, T.R., Niles, E.G., Studier, F.W., and Moss, B., 1986, Eukaryotic transient expression system based on recombinant vaccinia virus that synthesizes T7 RNA polymerase, *Proc. Natl. Acad. Sci. USA* **83**: 8122–8126.

Kubo, H., Yamada, Y.K. and Taguchi, F., 1994, Localization of neutralizing epitopes and the receptor-binding site within the amino-ternimal 330 amino acids of the murine coronavirus spike protein, *J. Virol.* **68**: 5403–5410.

Suzuki, H., and Taguchi, F., 1996, Analysis of receptor-binding site of murine coronavirus spike protein, *J. Virol.* 70: 2632–2636.

Taguchi, F., Siddell, S.G., Wege, H. and ter Meulen, V., 1985, Characterization of a variant virus selected in rat brains after infection by coronavirus mouse hepatitis virus JHM, *J. Virol.* **54**: 429–435.

Yamada, Y.K., Abe, M., Yamada, A. and Taguchi, F., 1993, Detection of mouse hepatitis virus by the polymerase chain reaction and its application to the rapid diagnosis of infection, *Lab. Anim. Sci.* **43**: 285–290.

Yokomori K. and Lai. M.M.C., 1992a, Mouse hepatitis virus utilizes two carcionoembryonic antigens as alternative receptors, *J. Virol.* **66**: 6194–6199.

Yokomori, K., and Lai. M.M.C., 1992b, The receptor for mouse hepatitis virus in the resistant mouse strain SJL is functional: Implications for the requirement of a second factor for viral infection, *J. Virol.* **66**: 6931–6938.

11

PORCINE REPRODUCTIVE AND RESPIRATORY SYNDROME VIRUS INFECTION OF ALVEOLAR MACROPHAGES CAN BE BLOCKED BY MONOCLONAL ANTIBODIES AGAINST CELL SURFACE ANTIGENS

X. Duan, H. J. Nauwynck, H. Favoreel, and M. B. Pensaert

Laboratory of Veterinary Virology
Faculty of Veterinary Medicine
University of Gent
Salisburylaan 133
B-9820 Merelbeke, Belgium

1. ABSTRACT

PRRSV has a restricted macrophage tropism. To explore if the difference in susceptibility of porcine alveolar macrophages (PAM) and peripheral blood mononuclear cells (PBMC) to PRRSV is correlated with certain cellular surface antigens which may serve as a virus receptor, polyclonal antibodies against PAM and PBMC were prepared. Anti-PAM but not anti-PBMC antibodies protected PAM from PRRSV infection suggesting that specific receptor(s) may exist on PAM. Furthermore, monoclonal antibodies (MAbs) against putative receptor(s) were produced. Balb/c mice were firstly immune-tolerized with freshly isolated PBMC after which they were immunized with PAM. Two MAbs (41D3 and 41D5) which blocked PRRSV infection of PAM were obtained. MAb 41D3 and 41D5 prevented the attachment of purified PRRSV to PAM. Both MAbs bound to the cellular membrane of PAM but not to that of porcine peritoneal macrophages, PBMC and three porcine cell lines (SK, ST and PK-15) as revealed by flow cytometry. This membrane reactivity correlates well with the susceptibility of these cells to a PRRSV infection. Taken together, these data suggest that MAb 41D3 and 41D5 recognize a potential cellular receptor for PRRSV on PAM.

Coronaviruses and Arteriviruses, edited by Enjuanes *et al.*
Plenum Press, New York, 1998

2. INTRODUCTION

The porcine reproductive and respiratory syndrome virus (PRRSV) is a new member of the arterivirus group (Meulenberg et al., 1992; Conzelmann et al., 1992). One common but peculiar characteristic of arteriviruses is that macrophages are the primary if not the only host cells supporting virus replication in their respective host (Plagemann and Moennig, 1992).

PRRSV demonstrates a high tropism for cells of the monocyte/macrophage lineage both "in vivo" and "in vitro". Of many porcine cell systems tested, porcine alveolar macrophages (PAM) and some cultivated porcine peripheral blood monocytes (PBM) support a productive replication of PRRSV (Duan et al., 1997; Voicu et al., 1994, Kim et al., 1993; Bautista et al., 1992; Benfield et al., 1992; Collins et al., 1992; Yoon et al., 1992; Wensvoort et al., 1991). However, the replication of PRRSV is subject to several types of restriction. Firstly, only some subsets of monocyte/macrophage lineage cells are susceptible to PRRSV. Monocytes/macrophages from different anatomic origin show different susceptibility to PRRSV in vitro (Duan et al., unpublished data). PRRSV replication is confined to monocyte/macrophage lineage cells in specific tissues such as lungs and lymphoid tissues in vivo (Duan et al., 1997; Halbur et al., 1995; Pol et al., 1991) and was not detected in PBMC and bone marrow cells, which contain the precursors of macrophages (Duan et al., 1997). Secondly, only relatively few cells of a certain susceptible population become infected, this number was for example less than two percent for lung lavaged PAM during any time point after natural and experimental infection (Duan et al., 1997; Mengeling et al., 1995). Thirdly, some specific states of differentiation and activation of macrophages may affect their susceptibility to PRRSV. This was particularly evident in PAM, in which virus replication significantly increased after one day in vitro cultivation and in which the infection can be completely blocked by activation of PAM with phorbol esters (Duan et al., unpublished data). Thus, the replication of the virus seems to be closely associated with certain specific biological properties of the monocyte/macrophage cell type. However, the mechanism which determines the PRRSV tropism and restriction of virus replication is unknown.

Since the receptor is often the key and in some cases the sole determinant of viral tropism (for recent reviews see Norkin, 1995 and Haywood, 1994), we have attempted to identify the PAM receptor which may determine the susceptibility of macrophage to PRRSV by producing anti-PAM monoclonal antibodies (MAb) which block the virus infection. The production of such MAb(s) will not only lead to the direct identification of the putative virus receptor but may provide a useful tool for understanding the pathogenesis of PRRSV by studying the specific cellular tropism of the virus in vivo.

3. MATERIALS AND METHODS

3.1. Cells

Porcine alveolar macrophages (PAM) and porcine peripheral blood mononuclear cells (PBMC) were obtained from conventional Belgian Landrace pigs from a PRRSV negative herd. They were cultivated according to the methods previously described by Wensvoort et al. (1991) and by Nauwynck et al. (1994).

The PRRSV permissive cell line MARC-145 was used as the source of PRRSV in all studies requiring purified virus. MARC-145 and the porcine PRRSV non-permissive cell lines PK15, ST and SK were cultivated by standard methods.

Two virus isolates were used: the Lelystad strain of PRRSV (kindly provided by Dr. Wensvoort) and a Belgian isolate of PRRSV designed 94V360. The latter was adapted to MARC-145 cells.

3.2. Virus Purification and Biotinylation

A fifth passage of 94V360 was purified through sucrose gradient centrifugation. The resulting preparation had a titre of 5 to 10×10^7 TCID$_{50}$/ml as determined in PAM and 1 to 2×10^{10} virus particles per ml as quantified by electron microscopy (containing 2.5 mg protein per ml).

Purified PRRSV was labelled with biotin by using a protein biotinylation kit (Amersham International, Buckinghamshire, UK). After biotinylation, the titre of PRRSV was reduced by 50%. Biotinylated virions were stored at -70°C.

3.3. Antibodies

Polyclonal antibodies against PAM and PBMC were produced in 10-week-old Balb/c mice immunised intraperitoneally with 1×10^7 cells per mouse of one day cultivated PAM or freshly isolated PBMC.

To produce monoclonal antibodies (MAb), immuno-tolerization was first performed in 10 weeks old female Balb/c mice by injecting freshly isolated PBMC and cyclophosphamide intraperitoneally (Sigma, St. Louis, MO, USA) as earlier described (Matthew and Sandrock, 1987). Hybridomas were produced by fusing the spleen cells from one immune-tolerized mouse with SP2/0 myeloma cells according to standard methods. The hybridomas were first screened with the cell-ELISA. The positive hybridomas were tested afterwards for their ability to protect PAM from PRRSV infection with a cell protection assay. MAbs were sub-cloned twice by limiting dilution from positive hybridomas and their isotypes were determined using a kit (Innogenetics, Zwijnaarde, Belgium).

MAbs 74–22–15, 517.2 reacting with porcine monocytes/macrophages and irrelevant MAbs 18E8, 13D12 and 4G3 of the isotype IgG1 were used as controls.

3.4. Cell-ELISA

For screening MAbs against porcine macrophage membrane antigens, a cell-ELISA assay was established. Briefly, plates with PAM were washed and fixed with cold 4% paraformaldehyde. After blocking with 1% (W/V) bovine serum albumin (BSA) in 0.05 M sodium carbonate buffer, pH 9.6, serum or ascites fluid or hybridoma supernatant was added and incubated for 60 min at 4°C. The plates were subsequently washed three times with washing buffer (0.01 M sodium phosphate buffer, pH7.2 containing 0.15 M NaCl and 0.05% (V/V) Tween 20). The bound antibodies were detected by an alkaline phosphatase detecting system (Sigma, St. Louis, MO, USA).

Cell protection assays: In order to exame if anti-PAM antibodies in serum, ascites fluid or hybridoma supernatant were able to protect PAM from PRRSV infection, a cell protection assay was established. PAM were cultivated in 96-well cell culture plate and subsequently incubated during 60 min with serum or ascites fluid or hybridoma supernatant on ice. The cells were then washed and inoculated with 100 TCID$_{50}$ PRRSV per well. After inoculation, the cells were washed and refed with medium containing the antibodies. The plate was fixed and stained after 48 hours post inoculation using an immunoperoxidase monolayer assay (IPMA) as previously described (Wensvoort et al, 1991).

3.5. Flow Cytometry Assay

The reactivity of MAbs with cell surface antigen(s) on PAM and PBMC, MARC-145, PK15, ST and SK was analysed by flow cytometry. Briefly, 1×10^6 cells were subsequently incubated with 50 μl of goat serum, MAbs and goat anti-mouse antibody FITC conjugate (Amersham International, Buckinghamshire, UK) for 60 min on ice. Three washings were done with cold PBS before and after each incubation. Finally, the cells were scanned with the flow cytometer FACSCalibur™ (Becton Dickinson Immunocytometry Systems, USA).

Virus attachment blocking assay: To further determine if the effect of the MAbs which block PRRSV infection of PAM is due to prevention of the virus binding to the cells, a virus attachment blocking assay was established. Briefly, 2×10^5 PAM were first incubated on ice with MAbs, later with biotinylated virus and finally with streptavidin-FITC in PBS plus 2% FCS for 60 min. The cells were washed three times with PBS plus 2% FCS after each incubation. The cells were resuspended in 0.5 ml of PBS plus 2% FCS plus 1 μg/ml of propidium iodide (PI). The relative fluorescence intensity of each sample was determined by a flow cytometry analysis, in which dead cells labelled with PI were excluded. The inhibition of the virus binding by antibodies was calculated using the formula: % inhibition = [median fluorescence (MFI) of cells incubated with control MAb - MFI of cells incubated with MAb 41D3 and 41D5] / (MFI of cells incubated control MAb) x 100.

4. RESULTS

Anti-PAM but not anti-PBMC polyclonal antibodies blocked PRRSV infection of PAM whereas both antibodies had a similar cell-ELISA titre to PAM (Table 1).

For the production of MAbs, 814 hybridomas were obtained after fusion. 148 hybridomas were PAM cell-ELISA positive. Two of these hybridomas (41D3 and 41D5) produced antibodies which blocked the replication of PRRSV in PAM. MAb 41D3 and 41D5 were classified as IgG1.

The ability of MAbs 41D3 and 41D5 to protect PAM from PRRSV infection was confirmed by using affinity purified antibodies. The protection was concentration dependent as shown in Fig. 1. Five μg/ml 41D5 and 0.6 μg/ml 41D3 completely blocked PRRSV infection of 1×10^5 one day cultivated PAM. None of the control MAbs affected PRRSV replication (Figure 1).

Table 1. Antibody titres in sera of mice immunised with either
porcine alveolar macrophages (PAM) or porcine blood
mononuclear cells (PBMC) as determined with the
PAM cell-ELISA and cell protection assay

		Antibody titre	
Mouse no.	Immunised with	PAM cell-ELISA	PAM cell protection assay
1	PAM	1600	160
2	PAM	1600	320
3	PBMC	1600	<10
4	PBMC	1600	<10

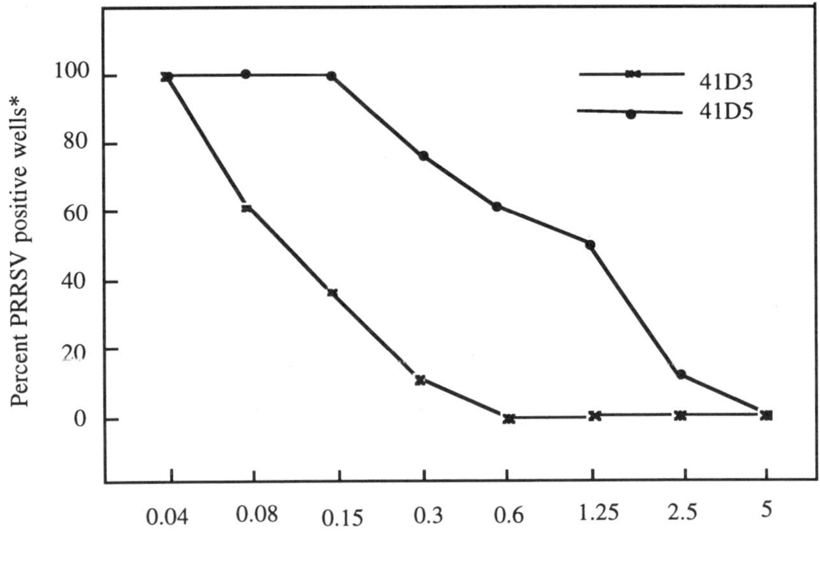

Figure 1. Ability of monoclonal antibodies 41D3 and 41D5 to inhibit PRRSV infection of porcine alveolar macrophages (PAM). *Mean value of three experiments using PAM originating from three pigs.

Preincubation of PAM with MAb 41D3 and 41D5 inhibited PRRSV attachment to PAM and the effect was concentration dependent (Figure 2).

About 40 to 65 % reduction of virus binding was observed when 2×10^5 PAM were pre-incubated with 50µl of 1mg/ml MAbs. The reduced binding of PRRSV was not observed when PAM were pre-incubated with one of the control MAbs.

MAb 41D3 and 41D5 reacted with the cellular membrane of PAM but not with that of porcine peritoneal macrophages, PBMC and three porcine cell lines (SK, ST and PK-15) or MARC-145 cell lines. (Table 2.)

5. DISCUSSION

This paper describes the production of specific monoclonal antibodies against PAM, which block PRRSV infection of PAM. They recognise epitope(s) on protein(s) expressed at the outer cell membrane, which seem to be involved in PRRSV attachment and infection.

Some cellular molecules on the macrophage membranes may serve as virus receptors and may be responsible for the restricted macrophage tropism of PRRSV. This was indicated by our finding that anti-PAM antibodies but not anti-PBMC antibodies protected PAM from PRRSV infection even though both antibodies had the same titre in a PAM cell-ELISA.

MAbs 41D3 and 41D5 block PRRSV infection of PAM. This blocking effect is unlikely to be non-specific since all isotype-matched control MAbs had no effect on the course of infection. The level at which virus replication is interrupted by both MAbs is virus attachment. This has been shown in the present study by the virus attachment blocking assay. This assay was not influenced by the phagocytosis of macrophages as it was per-

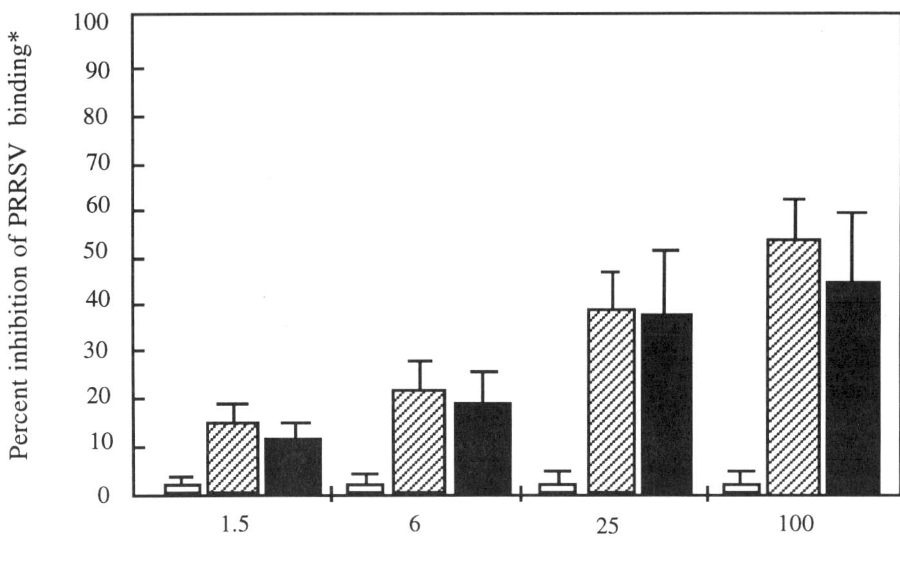

Figure 2. Ability of monoclonal antibodies 41D3 and 41D5 to prevent PRRSV attachment to the porcine alveolar macrophages (PAM). 50 µl of various dilutions of affinity purified anti-PAM monoclonal antibodies 41D3 (■), 41D5 (hatched), or anisotype matched irrelevant monoclonal antibody 18E8 (□) were pre-incubated with 2×10^5 PAM before incubation with biotinylated PRRSV. The fluorescence intensity was measured by flow cytometry and expressed as a percentage of the control (without antibodies). *Mean value of three experiments using PAM originating from three pigs.

formed on ice which completely inhibits phagocytosis. All the control antibodies did not prevent virus binding, which confirms the specificity of the blocking effect of both 41D3 and 41D5. Thus, MAb 41D3 and 41D5 appear to recognise the PRRSV receptor site. This finding is substantiated through the results that the protein(s) reacting with 41D3 and 41D5 is expressed on the membrane of porcine cells which are susceptible to PRRSV, such as PAM, but not on those which are refractory, such as peritoneal macrophages, peripheral blood mononuclear cells, and SK, ST, PK-15 cells.

Of interest is the observation that 41D3 and 41D5 did not react with the cell line MARC-145 derived from foetal rhesus monkey kidney (Kim et al., 1993). These results

Table 2. Expression of protein containing 41D3 and 41D5 reactive
epitopes on the cell membrane of different cell types

Cells	Reactivity with monoclonal antibody	
	41D3	41D5
Porcine alveolar macrophages	+	+
Peritoneal macrophages	–	–
Porcine peripheral blood mononuclear cells	–	–
SK	–	–
PK-15	–	–
ST	–	–
MARC-145	–	–

suggest that PRRSV may use a route of entry for MARC-145 cells which is different from that for macrophages. The difference in susceptibility of MARC-145 cells to different PRRSV isolates and the necessary adaptation of some isolates to these cells provide further evidence for different ways of virus entry (Drew, 1996; Bautista et al., 1993).

The PRRSV receptor presumably has a certain cellular function. Initial studies demonstrated that the 41D3 and 41D5 epitope(s) is only expressed on the surface of porcine alveolar macrophages. This suggests that 41D3 and 41D5 react with a macrophage differentiation antigen which seems to be expressed only in a sub-population of porcine monocytes/macrophages. The nature of this protein is still unclear. The isolation of the protein may become possible with the generation of suitable monoclonal antibodies. Identification and characterisation of the receptor for PRRSV will help to evaluate the importance of receptor-mediated events in the regulation of virus tropism.

ACKNOWLEDGMENTS

We thank Dr. G. Charlier (National Veterinary Research Institute, Brussels, Belgium) for the electron-microscopic quantification of purified PRRSV, Ms. L. Sys for her technical assistance in the fusion process for the production of monoclonal antibodies, and Dr. G. Wensvoort for the supply of the Lelystad isolate of PRRSV.

REFERENCES

Bautista, E.M., Goyal, S.M., Yoon, I.J., Joo, H.S., Collins, J.E., 1993, Comparison of porcine alveolar macrophages and CL 2621 for the detection of porcine reproductive and respiratory syndrome (PRRS) virus and anti-PRRS antibody, *J. Vet. Diagn. Invest,*.**5:** 163–5.

Benfield, D.A., Nelson, E., and Collins, J.E., 1992, Characterisation of swine infertility and respiratory syndrome (SIRS) virus (isolate ATCC VR-2332), *J. Vet. Diagn. Invest.* **4:** 127–133.

Conzelmann, K.K., Visser, N., van Woensel, P., and Thiel, H.J., 1993, Molecular characterisation of porcine reproductive and respiratory syndrome virus, a member of the arterivirus group, *Virology* **193:** 329–339.

Drew, T.W., 1996, Studies on the genome and proteins of porcine reproductive and respiratory syndrome virus. Ph.D. Thesis, The Open University, London

Duan, X., Nauwynck, H.J., and Pensaert, M.B., 1997, Virus quantification and identification of cellular targets in the lungs and lymphoid tissues of pigs at different time intervals after inoculation with porcine reproductive and respiratory syndrome virus (PRRSV), *Vet. Microbiol.* In press.

Duan, X., Nauwynck, H.J. and Pensaert, M.B., 1997, Effects of origin and state of differentiation and activation of monocytes/macrophages on their susceptibility to porcine reproductive and respiratory syndrome virus (PRRSV), Unpublished data.

Halbur, P.G., Miller, L.D., Paul, P.S., Meng, X.J., Huffman, E.L., and Andrews, J.J., 1995, Immunohistochemical identification of porcine reproductive and respiratory syndrome virus (PRRSV) antigen in the heart and lymphoid system of three-week-old colostrum-deprived pigs, *Vet. Pathol.* **32:** 200–204.

Haywood, A.M.,1994, Virus receptors: binding, adhesion strengthening, and changes in viral structure, *J. Virol.* **68:** 1–5.

Kim, H.S., Kwang, J., Yoon, I., Joo, H.S., and Frey, M.L., 1993, Enhanced replication of porcine reproductive and respiratory syndrome (PRRS) virus in a homogeneous subpopulation of MA-104 cell line, *Arch. Virol.* **133:** 477–483.

Matthew, W.D. and Sandrock, A.W., 1987, Cyclophosphamide treatment used to manipulate the immune response for the production of monoclonal antibodies. J. Immunol, *Methods* **100:** 73 82.

Mengeling, W.L., Lager, K.M., and Vorwald, A.C., 1995, Diagnosis of porcine reproductive and respiratory syndrome, *J. Vet. Diagn. Invest.* **7:** 3–16.

Meulenberg, J.J.M., Hulst, M.M., de Meijer, E.J., Moonen, P.L.J.M., de Besten, A., de Kluyver, E.P., Wensvoort, G., and Moormann, R.J.M., 1993, Lelystad virus, the causative agent of porcine epidemic abortion and respiratory syndrome (PEARS), is related to LDV and EAV, *Virology* **192:** 62–72.

Nauwynck, H.J. and Pensaert, M.B., 1994, Virus production and viral antigen expression in porcine blood monocytes inoculated with pseudorabies virus, *Arch. Virol.* **137:** 69–79.

Norkin, L.C.V., 1995, Virus receptors: implications for pathogenesis and the design of antiviral agents, *Clin. Microbiol. Rev.* **8:** 293–315.

Plagemann, P.G.W. and Moennig, V., 1992, Lactate dehydrogenase-elevating virus, equine arteritis virus, and simian hemorrhagic fever virus: a new group of positive-strand RNA viruses, *Adv. Virus. Res.* **41:** 99–192.

Pol, J.M.A., van Dijk, J.E., Wensvoort, G., and Terpstra, C., 1991, Pathological, ultrastructural, and immunohistochemical changes caused by Lelystad virus in experimentally induced infections of mystery swine disease, *Vet. Q.* **13:** 137–143.

Voicu, I.L., Silim, A., Morin, M., Elazhary, M.A.S.Y., 1994, Interaction of porcine reproductive and respiratory syndrome virus with swine monocytes, *Vet. Rec.* **134:** 422–423.

Wensvoort, G., Terpstra, C., Pol, J. M. A., ter Laak, E. A., Bloemraad, M., de Kluyver, E. P., Kragten, C., van Buiten, L., den Besten, A., Wagenaar, F., Broekhuijsen, J. M., Moonen, P. L. J. M., Zetstra, T., de Boer, E. A., Tibben, H. J., de Jong, M. F., van't Veld, P., Groenland, G. J. R., van Gennep, J. A., Voets, M. T., Verheijden, J. H. M., and Braamskamp, J., 1991, Mystery swine disease in The Netherlands: the isolation of the Lelystad virus, *Vet. Q.* **13:** 121–130.

Yoon, I.J., Joo, H.S., Christianson, W.T., Morrison, R.B., and Dial, G.D., 1992, Isolation of a cytopathic virus from weak pigs on farms with a history of swine infertility and respiratory syndrome, *J. Vet. Diagn. Invest*, **4:** 139–143.

12

REQUIREMENT OF PROTEOLYTIC CLEAVAGE OF THE MURINE CORONAVIRUS MHV-2 SPIKE PROTEIN FOR FUSION ACTIVITY

Yasuko K. Yamada,[1] Kazuhiro Takimoto,[1] Mikiko Yabe,[1] and
Fumihiro Taguchi[2]

[1]National Institute of Infectious Diseases
4-7-1 Gakuen, Musashimurayama
Tokyo 208, Japan
[2]National Institute of Neuroscience, NCNP
4-1-1 Ogawahigashi, Kodaira
Tokyo 187, Japan

1. ABSTRACT

The spike (S) protein of a non-fusogenic murine coronavirus, MHV-2, was compared to that of a variant, MHV-2f, with fusion activity. Two amino acids differed between the S proteins of these viruses; one was located in the signal sequence (amino acid 12) and the other in the putative cleavage site (amino acid 757). To determine which one of these amino acid changes is important for the alteration of fusogenicity, chimeric S proteins between MHV-2 and -2f were constructed and expressed in DBT cells by a vaccinia virus expression system. The results revealed that one amino acid change (Ser to Arg) at position 757 is responsible for the acquisition of fusogenicity of the MHV-2f S protein. This change also altered the susceptibility to proteolytic cleavage of the MHV-2 S protein which was originally uncleavable. We concluded that the non-fusogenic activity of MHV-2 results from the lack of cleavage of its S protein.

2. INTRODUCTION

The S protein of murine coronavirus (MHV) mediates syncytium formation. Other fusogenic viruses, such as orthomyxoviruses, paramyxoviruses and retroviruses, require proteolytic cleavage of the surface glycoprotein for fusion activity, since the cleavage exposes the fusion protein on the N-terminus of the membrane-anchored subunit of the surface

Coronaviruses and Arteriviruses, edited by Enjuanes *et al.*
Plenum Press, New York, 1998

glycoprotein (White, 1990). It is still controversial, however, as to whether S protein cleavage of MHV is required for syncytium formation. Examination of amino acid sequences of the S proteins of MHV-JHMV and -A59 mutants and an MHV-A59 variant suggested that cleavage of the S protein may not be absolutely necessary for syncytium formation, since the S proteins of these mutants were not cleaved, but still had fusion activity (Bos et al., 1995; Gombold et al., 1993; Stauber et al., 1993; Taguchi, 1993).

MHV-2 is the only strain with a non-fusogenic character among various MHV strains. From MHV-2 virus stock, we recently isolated a variant, MHV-2f, which induces syncytium formation like other fusogenic MHV strains. The comparison of amino acid sequences showed that there were two amino acid differences between the S proteins of these two viruses. The amino acid at position 12 in the signal sequence was Ser in MHV-2 and Cys in MHV-2f. The amino acid sequence of the cleavage site of MHV-2 was HRARS, while that of MHV-2f was HRARR showing one amino acid change at position 757. Cleavability of the S protein is also different between MHV-2 and -2f; the S protein of MHV-2 is not cleaved in infected DBT cells, while the S protein of MHV-2f is cleaved (Yamada et al., 1997). In order to determine the amino acid involved in the alteration of fusogenicity and cleavability, we constructed chimeric S proteins in which each amino acid change out of two positions was introduced in the MHV-2 S protein and expressed them in DBT cells by a vaccinia virus expression system.

3. MATERIALS AND METHODS

3.1. Construction of Chimeric S Proteins

The full length coding regions of the S proteins of MHV-2 and -2f amplified by RT-PCR from infected DBT cells using primers around the initiation and the stop codons (2S-PH and 2S-NB, respectively) (Yamada et al., 1997) were cloned into pGEM4Z (Promega) and designated as pS-2 and pS-2f, respectively. To introduce single mutation at nucleotide 35 in the MHV-2 S protein (GenBank Database Accession No. U72635), the S1 subunit of MHV-2f, containing the MHV-2 type cleavage site was amplified from pS-2f using 2S-PH and 2S-2283N (antisense nucleotide 2264–2283 of MHV-2) primers. The S2 subunit of MHV-2 was amplified from pS-2 using 2S-2264P (nucleotide 2264 to 2283 of MHV-2) and 2S-NB primers. The full length coding region was then amplified from the mixture of these subunits using 2S-PH and 2S-NB primers, and cloned into pGEM4Z (pS-2f/2). To make the MHV-2 S protein with a single mutation at nucleotide 2271, the S1 subunit of the MHV-2 containing the MHV-2f type cleavage site amplified from pS-2 using 2S-PH and 2fS-2283N (antisense nucleotide 2264–2283 of MHV-2f) primers, and the S2 subunit of MHV-2f amplified from pS-2f using 2fS-2264P (nucleotide 2264–2283 of MHV-2f) and 2S-NB primers were mixed, amplified using 2S-PH and 2S-NB primers, and cloned into pGEM4Z (pS-2/2f). The nucleotide sequences of pS-2, -2f, -2f/2, and -2/2f were analyzed by the dideoxy termination labeling method according to the manufacture's instructions (Applied Biosystems).

3.2. Expression of Chimeric S Proteins

DBT cells were infected with a recombinant vaccinia virus encoding bacteriophage T7 RNA polymerase (vTF7–3, kindly provided by Dr. B. H. Moss) and then transfected with plasmids, pS-2, -2f, -2f/2, and -2/2f with DOTAP transfection reagent (Boehringer

Mannneim). At 16 hr posttransfection, cytopathic effects were determined. After fixation with chilled acetone, expression of the S protein was determined by indirect immunofluorescent staining. In other series, transfected cells were washed with PBS and lysed with PBS containing 1% Nonidet-P40. The lysed samples were centrifuged at 15,000 rpm for 10 min. The supernatants were collected and stored at -20°C.

3.3. Western Blot Analysis

The lysates were boiled in a sample buffer and electrophoresed in a 7.5% SDS-polyacrylamide gel. Western blotting analysis was carried out as described previously (Yamada et al., 1997) using an S2 subunit-specific monoclonal antibody (kindly provided by Dr. S. Siddell) and horseradish peroxidase-conjugated anti-mouse IgGs (TAGO). The bands were visualized with PBS containing 0.05% diaminobenzidine and 0.015% hydrogen peroxide.

4. RESULTS

We previously demonstrated two amino acid differences between fusion-negative MHV-2 and its fusion-positive variant, MHV-2f. One change at position 12 is located in the signal sequence and the other at position 757 is in the putative cleavage site. In order to determine which amino acid change was responsible for the alteration of fusogenicity, chimeric S proteins were constructed (Figure 1) and cloned into a pGEM vector. Compared with the MHV-2 S protein, pS-2f/2 contained a single amino acid change at position 12 (Ser to Cys), while pS-2/2f had a change at position 757 (Ser to Arg). These chimeras along with the cloned MHV-2 and -2f S proteins were transiently expressed in DBT cells by a vaccinia virus expression system. At 16 hr posttransfection, syncytium formation was observed in cells transfected with pS-2f and with pS-2/2f. On the other hand, syncytium formation was detected in neither cells transfected with pS-2 nor in those transfected with pS-2f/2, even though the expression of the S protein was detectable by immunofluorescent staining with an S protein specific monoclonal antibody (Figure 2). These results suggested that one amino acid change at position 757 was responsible for the alteration of the fusogenicity. Although some other amino acid changes were introduced during the process of cloning and constructing chimeras (shown in parentheses in Figure 1), these changes did not seem to affect the fusogenicity.

Since the amino acid at position 757 is located in the putative cleavage site, this mutation could influence the cleavage of the S protein. We analyzed the cleavability of chimeric S proteins by Western blotting with S2-subunit specific monoclonal antibody (Figure 3). Two bands at 150 and 180 kDa were detected in pS-2- and -2f/2-transfected cells, while the 150- and 90-kDa bands were detected in cells transfected with pS-2f and -2/2f. The 150-kDa protein may represent a nonglycosylated form of the mature 180-kDa

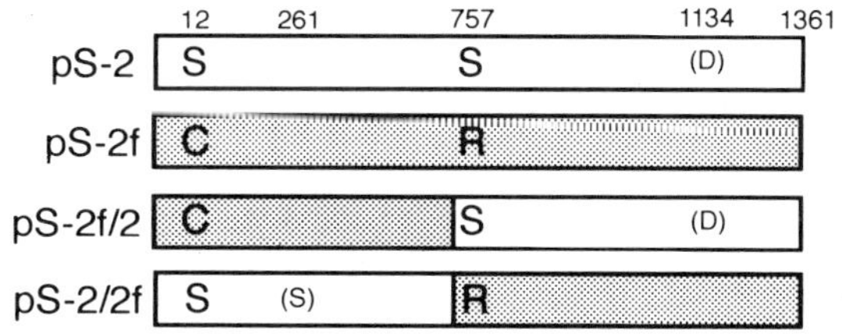

Figure 1. Schematic representation of chimeric S proteins. Open and shaded columns show parts of the S protein derived from MHV-2 and -2f, respectively.

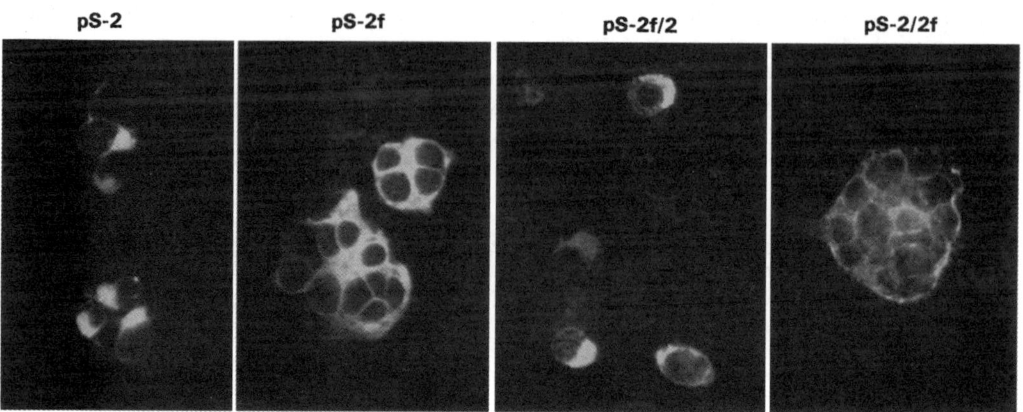

Figure 2. Immunofluorescent staining of DBT cells. Cloned and chimeric S proteins were expressed by a vaccinia virus expression system. The S protein was detected with anti-S protein specific monoclonal antibody and FITC-conjugated anti-mouse IgGs (TAGO) after fixation with chilled acetone.

protein. The mature 180-kDa protein may be cleaved into two subunits of which S2 subunit can be detected as the 90-kDa band by its specific antibody. In cells transfected with pS-2 and with pS-2f/2, faint bands of around 90 kDa were seen; these probably do not represent S protein cleavage products, because similar faint bands were also seen in control DBT cells infected with vTF7–3. These data showed that the S proteins expressed by pS-2 and -2f/2 were not cleaved, while those by pS-2f and -2/2f were cleaved.

These results demonstrated that MHV-2f acquired the fusion activity as a result of one amino acid change which altered the cleavability of the non-fusogenic MHV-2 S protein. We conclude that the non-fusogenic activity of MHV-2 results from the lack of cleavage of its S protein.

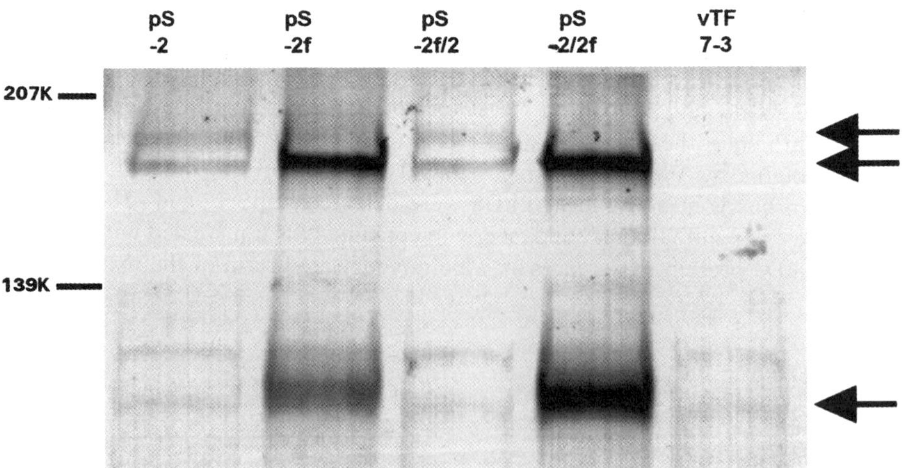

Figure 3. Western blot analysis of S proteins. Cloned and chimeric S proteins were expressed in DBT cells by a vaccinia virus expression system. Cell lysates at 16 hr posttransfection were analyzed as mentioned in Materials and Methods. Arrows indicate from top to bottom 180, 150 and 90 kDa.

5. DISCUSSION

Fusogenic MHV strains JHMV and A59, do not absolutely require cleavage of the S protein for their fusion activity, because the S proteins of these viruses with mutations in the putative cleavage site can induce syncytium, yet are not cleaved (Bos et al., 1995; Gombold et al., 1993; Stauber et al., 1993; Taguchi, 1993). On the other hand, MHV-2 requires cleavage of the S protein to acquire fusogenicity. Cleavage of the MHV-2 S protein may alter its conformation thereby activating the otherwise non-fusogenic character of this protein. Such a process would be very similar to that of other fusogenic orthomyxo-, paramyxo- and retroviruses (White, 1990). In these viruses, the newly appearing hydrophobic N-terminal region of the membrane-anchored subunit is postulated to work as a fusion peptide. In the N-terminus of the MHV-2 S2 subunit, however, a similar fusion peptide with a stretch of apolar amino acids was not found. We speculated that the region affecting fusion activity is possibly located inside S2, as is postulated for SFV, Sindbis and RSV viruses (White, 1990), because fusion activity is thought not to reside in the N-terminus of the MHV S protein (Keck et al., 1988). The finding that the uncleaved form of the MHV-2 S protein is inactive with respect to fusion is inconsistent with the previously reported result that the S proteins of A59 and JHMV are actively fusogenic without cleavage. Precise examination of the amino acid difference between MHV-2 and other MHV strains would be helpful in defining the molecular basis for the mechanisms of fusion induction which might be different among them.

REFERENCES

Bos, E. C. W., Heijnen, L., Luytjes, W., and Spaan, W. J. M., 1995, Mutation analysis of the murine coronavirus spike protein: Effect on cell-to-cell fusion, *Virology* **214**: 453–463.

Gombold, J. L., Hingley, S. T., and Weiss, S. R., 1993, Fusion-defective mutants of mouse hepatitis virus A59 contain a mutation in the spike protein cleavage signal, *J. Virol.* **67**: 4504–4512.

Keck, J. G., Soe, L. H., Makino, S., Stohlman, S. A., and Lai, M. M. C., 1988, RNA recombination of murine coronaviruses: Recombination between fusion-positive mouse hepatitis A59 and fusion-negative mouse hepatitis virus 2, J. Virol. 62: 1989–1998.

Stauber, R., Pfleiderera, M., and Siddell, S., 1993, Proteolytic cleavage of the murine coronavirus surface glycoprotein is not required for fusion activity, *J. Gen. Virol.* **74**: 183–191.

Taguchi, F., 1993, Fusion formation by the uncleaved spike protein of murine coronavirus JHMV variant cl-2, *J. Virol.* **67**: 1195–1202.

White, J. M., 1990, Viral and cellular membrane fusion proteins, *Annu. Rev. Physiol.* **52**: 675–697.

Yamada, Y. K., Takimoto, K., Yabe, M., and Taguchi, F., 1997, Acquired fusion activity of a Murine Coronavirus MHV-2 variant with mutations in the proteolytic cleavage site and the signal sequence of the S protein, *Virology* **227**: 215–219.

Replication I: The Polymerase

13

THE ARTERIVIRUS REPLICASE

The Road from RNA to Protein(s), and Back Again

Eric J. Snijder

Department of Virology
Leiden University Medical Center
AZL P4–26, Postbus 9600
2300 RC Leiden, The Netherlands
Telephone: ++ 31 715261657; fax: ++ 31 715266761; email:
  snijder@virology.azl.nl.

1. INTRODUCTION

About seven years ago, the identification of arteriviruses (den Boon *et al.*, 1991) and toroviruses (Snijder *et al.*, 1990) as distant relatives of "traditional" coronaviruses incited a discussion on the taxonomic position of these three virus groups (Cavanagh *et al.*, 1994). As a first result, the *genus torovirus* was included in the *Coronaviridae* family (Cavanagh *et al.*, 1993). The taxonomic debate ended at the 1996 International Congress of Virology in Jerusalem with the establishment of the *Arteriviridae* family and the order of the *Nidovirales*, containing the *Coronaviridae* and *Arteriviridae* families (Cavanagh, 1997). These re-classifications acknowledged both the many unique properties of arteriviruses and coronaviruses as well as their intriguing ancestral relationship at the level of replicase genes, genome organization, and replication strategy (Snijder and Horzinek, 1993; Snijder and Spaan, 1995; Cavanagh, 1997; de Vries *et al.*, 1997; Snijder and Meulenberg, 1998).

At the center of nidovirus molecular biology is a nonstructural gene which encodes a polyprotein of between 3175 (for the arterivirus equine arteritis virus (EAV); den Boon *et al.*, 1991) and approximately 7200 amino acids (for the coronavirus mouse hepatitis virus (MHV); Lee *et al.*, 1991; Bonilla *et al.*, 1994; Brown and Brierley, 1995). This size variation of the replicase again illustrates how both conservation and variation have played a part in nidovirus evolution. In addition to a number of highly conserved domains/functions which can be considered to form the "core" of the nidovirus replicase, each virus (or cluster of viruses) appears to have developed its own set of accessory nonstructural functions (see also below). However, the basic genome expression strategy is the same for all nidoviruses (Figure 1): from the incoming genomic RNA they generate two large replicase

Coronaviruses and Arteriviruses, edited by Enjuanes *et al.*
Plenum Press, New York, 1998

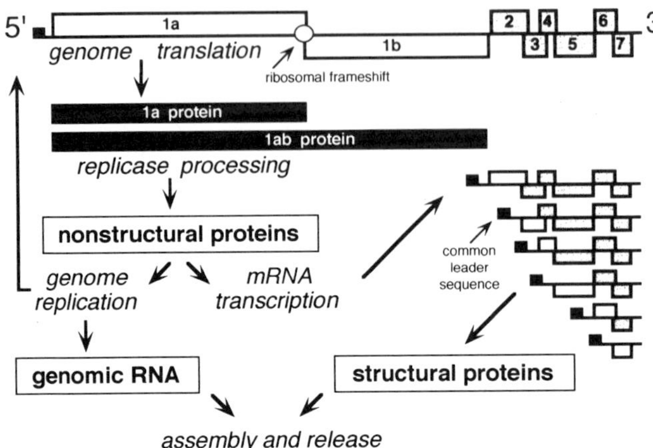

Figure 1. Overview of the nidovirus life cycle, illustrating the central role of the viral nonstructural proteins. EAV was used as an example in this figure.

polyproteins, which cleave themselves into smaller functional subunits. Subsequently, these nonstructural proteins can carry out the processes of genome replication and subgenomic mRNA transcription, the latter leading to the expression of a set of (mostly structural) genes from the 3' end of the genome. As expected from the common ancestry of the viral enzymes involved, the basic features of arterivirus mRNA transcription have recently been shown to be identical to those of coronaviruses (den Boon *et al.*, 1995b; den Boon *et al.*, 1996).

This chapter will focus on what is currently known about the structure and function of the arterivirus replicase polyprotein. Most of the data have been generated in Leiden using the arterivirus prototype EAV, but references to other arteriviruses and coronaviruses will be made when appropriate.

2. ORGANIZATION OF THE NIDOVIRUS REPLICASE

All nidovirus replicase genes consist of two open reading frames (ORFs), ORF1a and ORF1b. ORF1a is translated directly from the genomic RNA The expression of ORF1b requires a -1 ribosomal frameshift just upstream of the ORF1a termination codon, which leads to the synthesis of an ORF1ab polyprotein. In all nidovirus replicase genes two frameshift-promoting RNA signals have been identified (Jacks *et al.*, 1988; Brierley, 1995): a so-called "slippery" sequence, which is the actual frameshift site, and a downstream RNA pseudoknot structure. In EAV (den Boon *et al.*, 1991) the putative shift site 5' GUUAAAC 3' is followed immediately by the ORF1a termination codon and thus all 1727 residues of the ORF1a polyprotein are also present in the ORF1ab frameshift product. In many other nidoviruses, additional codons (up to 25) are present between frameshift site and ORF1a termination codon. This results in an ORF1a protein that contains unique C-terminal sequences which are lacking in the ORF1ab frameshift protein.

The arrangement of conserved domains (Figure 2) within the nidovirus replicase polyprotein is unique, and clearly separates the nidoviruses from other groups of plus-stranded RNA viruses. Sequence comparison between arterivirus and coronaviruses replicases revealed up to 30% amino acid sequence identity in the most conserved domains, a percentage that cannot be due to convergent evolution. The conservation of the putative RNA-dependent RNA polymerase and the putative RNA helicase, two domains which are

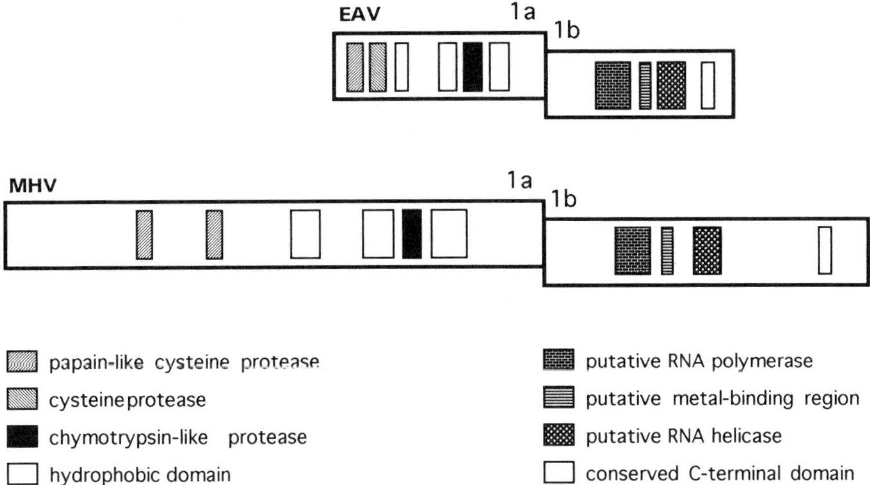

Figure 2. Comparison of the organization of the smallest (EAV) and largest (MHV) nidovirus replicase polyproteins.

common to positive-stranded RNA viruses, is not very surprising. It is remarkable, however, that only in nidovirus replicases the helicase domain is located downstream of the polymerase motif. The polymerase motif also carries another nidovirus trademark: the substitution of the canonical GDD in the core of the motif by SDD. Additional domains were identified that are conserved in nidovirus replicases (both in sequence and in position): e.g. a conserved domain in the C-terminal part of the ORF1b protein (den Boon *et al.*, 1991; Snijder *et al.*, 1990) and a Cys/His-rich domain upstream of the helicase motif. The latter was previously proposed to have metal-binding properties (Gorbalenya *et al.*, 1989b; Lee *et al.*, 1991) and was recently implicated in a remarkable mRNA transcription defect (van Dinten *et al.*, 1997).

The nidovirus replicase polyproteins are all processed by multiple ORF1a-encoded proteases. Despite the fact that our understanding of coronavirus replicase processing is far from complete, a common pattern for coronaviruses and arteriviruses is emerging (Snijder and Spaan, 1995; de Vries *et al.*, 1997; Snijder and Meulenberg, 1998). In comparable positions, the ORF1a proteins of both virus groups contain a "main protease", which is responsible for most replicase processing steps. These proteases belong to the superfamily that comprises the chymotrypsin-like and picornavirus 3C-like proteolytic enzymes (Fig. 2) (Bazan and Fletterick, 1988; Gorbalenya *et al.*, 1989a). Although the catalytic nucleophile of the coronavirus 3C-like protease (Cys; Ziebuhr *et al.*, 1995; Lu *et al.*, 1995; Liu and Brown, 1995) differs from that in the arterivirus nsp4 protease (Ser; Snijder *et al.*, 1996), these domains may still have a common ancestry (Snijder *et al.*, 1996). The exchange of Cys for Ser (or vice versa) at the active site of the enzyme is considered to be feasible (Bazan and Fletterick, 1988; Gorbalenya *et al.*, 1989a; Bazan and Fletterick, 1990; Dougherty and Semler, 1993).

The functions encoded from the central region of ORF1a to the 3'-end of ORF1b appear to form the "core" of the nidovirus replicase polyprotein: the well-conserved domains (main protease – polymerase – metal-binding domain – helicase – C-terminal ORF1b domain) are within this area, and only small insertions and deletions in this part of the replicase can be detected within the coronavirus or arterivirus groups. The presence of hydrophobic domains on either side of the main protease domain is another property shared by arteriviruses and coronaviruses. The N-terminal half of the ORF1a protein is very variable, both in size and in sequence. A comparison between coronaviruses and arteriviruses in this region does not yield any significant similarities, and even within the coronavirus and arterivirus groups there is little conservation. However, one striking observation can be made. Both corona- and arteriviruses contain protease domains in the amino-terminal half of the ORF1a protein which have been shown, or are predicted, to belong to the papain-like cysteine protease superfamily (Fig. 2) (Gorbalenya *et al.*, 1991; Gorbalenya and Snijder, 1996). A number of these nidovirus papain-like proteases have recently been characterized experimentally (Gao *et al.*, 1996; Bonilla *et al.*, 1995; Baker *et al.*, 1993; Lee *et al.*, 1991; Snijder *et al.*, 1995; den Boon *et al.*, 1995a; Snijder *et al.*, 1992).

3. ARTERIVIRUS PROTEASES

Proteolytic processing of nonstructural proteins fulfils a key role in the life cycle of most viruses. In the course of virus evolution, highly specific virus-encoded proteases have evolved and their importance for the regulation of virus replication is becoming more and more evident (Dougherty and Semler, 1993; Gorbalenya and Snijder, 1996). For EAV, protease domains have been identified in three replicase cleavage products: nonstructural protein 1 (nsp1), nsp2, and nsp4 (Figure 3). These proteases and their corresponding cleavage sites are well-conserved in other arteriviruses (Meulenberg *et al.*, 1993; Godeny *et al.*, 1993; Snijder and Spaan, 1995; de Vries *et al.*, 1997; Snijder and Meulenberg, 1998).

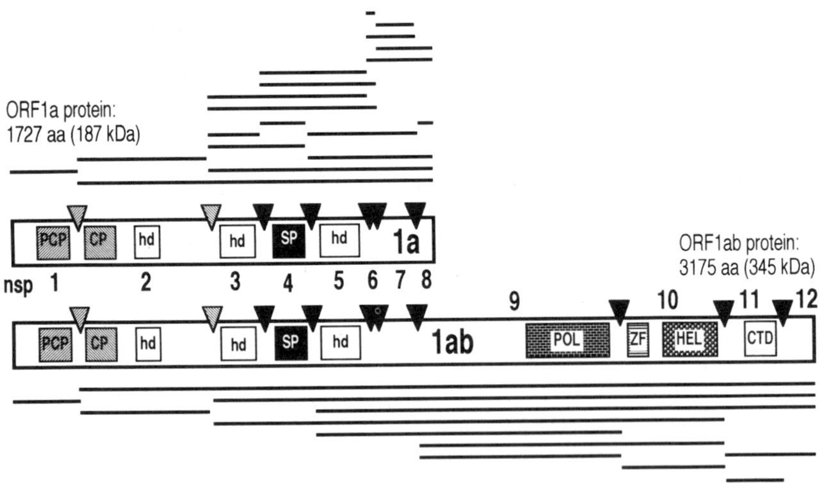

Figure 3. Proteolytic processing of the EAV replicase polyproteins. The processing schemes for the EAV ORF1a and ORF1ab proteins are depicted. The lines above and below the polyproteins represent the processing end products and intermediates which are currently known to be present in infected cells. Most of these protein species have been identified in immunoprecipitation analyses (Snijder *et al.*, 1994; van Dinten *et al.*, 1996; Wassenaar *et al.*, 1997). The three protease domains in the ORF1a polyprotein and their corresponding cleavage sites are indicated: PCP, nsp1 papainlike cysteine protease; CP, nsp2 cysteine protease; SP, nsp4 serine protease; hd, hydrophobic domain. In the ORF1b-encoded polypeptide the four major domains conserved in nidoviruses have been depicted: POL, putative RNA-dependent RNA polymerase; ZF, putative zinc finger domain; HEL, putative RNA helicase; CTD, conserved C-terminal domain specific for nidoviruses.

The internal papain-like cysteine protease (PCP) in EAV nsp1 is responsible for the production of this 29 kDa amino-terminal ORF1a cleavage product (Snijder *et al.*, 1992). Residues Cys-164 and His-230 have been proposed to form the catalytic dyad of the nsp1 PCP, which cleaves between Gly-260 and Gly-261 (Snijder *et al.*, 1992). Also the production of the next amino-terminal cleavage product, nsp2 (61 kDa), was shown to be an autoproteolytic event, which is mediated by a cysteine protease (CP) in the nsp2 N-terminal region (putative active site residues in EAV: Cys-270 and His-332; putative cleavage site: between Gly-831 and Gly-832). Although the nsp2 CP is most similar to the papain-like group of viral proteases, it possesses a number of unique properties and has been proposed to belong to a new subgroup of viral cysteine proteases (Snijder *et al.*, 1995). Two other arteriviruses, lactate dehydrogenase-elevating virus (LDV) and porcine reproductive and respiratory syndrome virus (PRRSV), have been found to use even a third N-terminal cysteine protease domain (Meulenberg *et al.*, 1993; Godeny *et al.*, 1993; den Boon *et al.*, 1995a). During arterivirus evolution, EAV appears to have lost this proteolytic activity, which is responsible for the production of an additional amino-terminal cleavage product from the nsp1 region of PRRSV and LDV (den Boon *et al.*, 1995a).

The arterivirus main protease, the nsp4 serine protease (SP), is a member of a relatively rare group of proteolytic enzymes, the 3C-like serine proteases (Snijder *et al.*, 1996). It combines the catalytic triad of classical chymotrypsinlike proteases (His-1103, Asp-1129, and Ser-1184 in EAV) with the substrate specificity of the 3C-like cysteine proteases, a subgroup of chymotrypsinlike enzymes named after the picornavirus 3C proteases. The latter property, which is assumed to be determined by a number of residues in the substrate-binding region of the nsp4 SP, explains the specificity for cleavage sites containing a Glu↓Gly/Ser dipeptide (Wassenaar *et al.*, 1997; Snijder *et al.*, 1996). Five of these cleavage sites have now been identified in the ORF1a protein: Glu-1064↓Gly-1065, Glu-1268↓Ser-1269, Glu-1430↓Gly-Gly1431, Glu-1452↓Ser-1453, and Glu-1677↓Gly-1678. The presence of three additional SP cleavage sites in the ORF1b-encoded polyprotein has been predicted (van Dinten *et al.*, 1996), but remains to be confirmed experimentally.

4. PROTEOLYTIC PROCESSING OF THE EAV REPLICASE

The post-translational fate of the EAV ORF1a (1727 aa) and ORF1ab (3175 aa) polyproteins has now been studied extensively and an apparently complete processing scheme has recently been obtained (Fig. 3) (Snijder *et al.*, 1994; van Dinten *et al.*, 1996; Wassenaar *et al.*, 1997). The ORF1ab polyprotein is cleaved ten times by the three ORF1a-encoded proteases described above. In combination with the ribosomal frameshift, this leads to the generation of 12 processing end products (named nsp 1 to 12). In the case of EAV, the C-terminal cleavage product of the ORF1a protein (nsp8) is identical to the N-terminal domain of the first ORF1b-encoded subunit (nsp9). In addition to the processing end products, a large number of intermediates has been detected, many of which have considerable half-lives and may therefore fulfil a specific role in the viral life cycle. The processing analysis of other viral replicases has revealed that such intermediates can be functional subunits themselves, e.g. in the replication of poliovirus (Ypma-Wong *et al.*, 1988; Jore *et al.*, 1988) and Sindbis virus (de Groot *et al.*, 1990; Lemm and Rice, 1993a; Lemm and Rice, 1993b; Shirako and Strauss, 1994; Lemm *et al.*, 1994).

An interesting recent observation was the fact that two alternative pathways can be followed for the processing of the C-terminal part of the ORF1a protein (Wassenaar *et al.*,

1997). After the rapid autoproteolytic release of nsp1 and nsp2 from the ORF1a polyprotein, a 96kDa nsp3–8 processing intermediate remains which contains the viral main protease, the nsp4 SP. It has now been found that nsp3–8 can cleave itself following either a "major pathway" (i.e. the pathway used most abundantly in infected cells) or following a "minor pathway". Co-expression experiments in the vaccinia virus/T7 system strongly suggested that the association with cleaved nsp2 is required for nsp3–8 to enter the major pathway. This leads to cleavage of the nsp4/5 and nsp7/8 junctions, but the nsp5/6 and nsp6/7 sites are not processed. When nsp2 is lacking the nsp4/5 site is no longer cleaved and instead the nsp5/6 and nsp6/7 sites are processed. This yields a set of "alternative" processing products which are not produced from the major pathway. An interesting consequence of the use of the two pathways is the generation of multiple SP-containing proteins. In a number of intermediates (e.g. nsp3–4 and nsp4–5) the SP is linked to a hydrophobic domain and may therefore be membrane-associated. The fully cleaved nsp4 is probably cytoplasmic and was shown to be efficient in processing other parts of the replicase polyprotein *in trans* (van Dinten *et al.*, unpublished observations).

Cofactors that strongly influence polyprotein processing have previously been identified in several other animal positive-stranded RNA virus systems. In most of these cases, the cofactor probably interacts (more or less) directly with the protease domain. In the case of EAV, however, the nsp2 cofactor role appears to be limited to just one of the eight cleavages carried out by the nsp4 SP. It may rather affect the conformation of the region containing the nsp4/5 junction than the activity of the nsp4 protease domain itself, since the SP was perfectly capable of processing other sites (nsp3/4, nsp5/6, nsp6/7, and nsp7/8) in the absence of nsp2. Nsp2 was previously reported to interact very strongly with nsp3 and nsp3-containing intermediates (Snijder *et al.*, 1994). Possibly, this interaction is the basis for the cofactor role of nsp2 in the "major" pathway for SP-directed nsp3–8 processing. After processing of the nsp4/5 junction in nsp3–8 (yielding nsp3–4 and nsp5–8), the conformation of nsp5–8 (51 kDa) appears to be such that cleavage of its internal nsp5/6 and nsp6/7 sites is prevented. However, the Glu-1677↓Gly-1678 site in the nsp5–8 C-terminal region is accessible to the SP, which results in the production of nsp5–7 and nsp8. Both nsp2 and nsp3–8 (and its derivatives) are complex molecules containing several clusters of conserved Cys residues (Snijder *et al.*, 1994) and multiple hydrophobic domains. The latter have recently been implicated in the membrane-association of the EAV replication complex (van Dinten *et al.*, 1996; van der Meer *et al.*, unpublished observations). Together, these results suggest that translation, processing, and membrane-association of the nsp2 to nsp8 region are interconnected processes.

5. MEMBRANE-ASSOCIATION OF THE EAV REPLICATION COMPLEX

The first indications for the membrane-association of the EAV replication complex were obtained when the rabbit antisera directed against various replicase regions (Snijder *et al.*, 1994; van Dinten *et al.*, 1996) were used for immunofluorescence assays on EAV-infected cells. With the exception of the anti-nsp1 serum, all antisera which could be used in this assay gave the same perinuclear staining (Figure 4). The co-localization of most of the replicase processing products, including those that contain the putative RNA polymerase (nsp9) and helicase (nsp10) functions (van Dinten *et al.*, 1996), indicated that these proteins assemble into a large replication/transcription complex. Their concentration in the

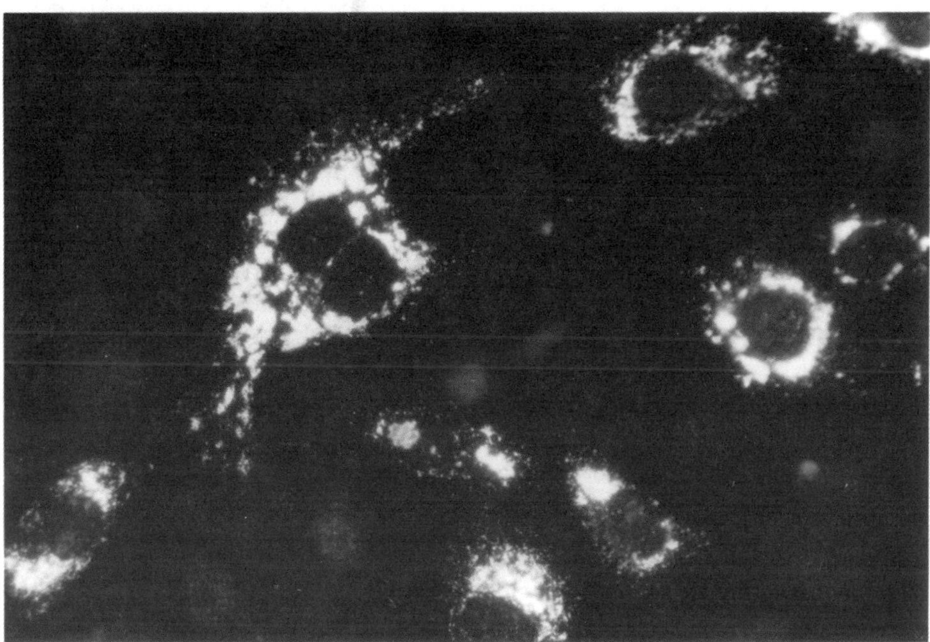

Figure 4. Immunofluorescence staining of the EAV replication complex in infected Vero cells using an antiserum directed against the C-terminal 220 residues of the ORF1a protein (Snijder *et al.*, 1994).

perinuclear region suggested association with intracellular membranes, which was subsequently confirmed by double-labeling experiments with marker antibodies directed against cellular membrane proteins (van der Meer *et al.*, unpublished observations). Using a metabolic RNA labeling, it was recently confirmed that the localization of the replicase proteins indeed coincides with the site of viral RNA transcription.

The ORF1a cleavage products nsp2, nsp3, and nsp5 contain hydrophobic domains which are likely to play a role in anchoring the EAV replication complex to intracellular membranes (Snijder *et al.*, 1994; van Dinten *et al.*, 1996). To confirm this biochemically, we recently isolated the membrane fraction from EAV-infected cells and compared it with the cytoplasmic fraction for the presence of replicase cleavage products (van der Meer *et al.*, unpublished observations). A number of major processing products (nsp2, nsp3–8, nsp3–4, nsp5–8, nsp5–7) was indeed predominantly recovered from the membrane fraction. This was also the case at pH 11, which suggests that these proteins probably contain transmembrane domains (Fujiki *et al.*, 1982). Upon Triton X-114 extraction (Bordier, 1981), the major part of the same proteins was retrieved from the detergent phase, which again confirmed their affinity for membranes. As expected, a number of cleavage products that lack hydrophobic domains displayed no affinity for the Triton X-114 detergent phase.

A typical feature of arterivirus replication is the formation of paired membranes and double membrane vesicles (DMV) at 3 to 6 hours post infection (Wood *et al.*, 1970; Breese and McCollum, 1970; Stueckemann *et al.*, 1982; Pol *et al.*, 1997). Although the origin and function of these membrane structures are unclear, they do not appear to be involved in virus assembly. Their possible relationship to the viral replication complex is currently under investigation using immuno electron microscopy and EAV replicase-specific antisera.

6. DEVELOPMENT OF AN INFECTIOUS cDNA CLONE FOR EAV

The molecular analysis and experimental manipulation of RNA virus genomes can be greatly facilitated by the availability of full-length cDNA clones from which functional RNA transcripts can be generated *in vitro*. This is especially true for positive-stranded RNA viruses since their genomes function as the mRNA for viral replicase translation. Thus, a virus infection can be initiated by the "simple" introduction of an infectious recombinant RNA molecule into a susceptible host cell. The availability of an infectious cDNA clone and the subsequent application of recombinant DNA technology ("reverse genetics") can enormously increase our understanding of many (molecular) aspects of the viral life cycle. This is e.g. illustrated by the advances which were made in the research on several animal RNA virus groups after infectious cDNA clones became available (5 to 15 years ago). Furthermore, full-length cDNA clones have proven to be valuable tools for pathogenesis studies and the development of vaccines and RNA virus expression vectors (Bredenbeek and Rice, 1992; Boyer and Haenni, 1994).

Since 1992, the development of a nidovirus infectious cDNA clone has been a major goal in our Department, since it was expected to be an essential tool to study (among other things) the common features of the replication and transcription of coronaviruses and arteriviruses. Also the development of nidovirus-based expression vectors would be very interesting because of the polycistronic genome organization and the "natural" generation of a set of subgenomic mRNAs. Compared to coronaviruses, the arteriviruses offered the obvious advantage of their considerably smaller genome size. In 1996, researchers in three Institutes in the Netherlands independently assembled arterivirus infectious cDNA clones. Full-length EAV cDNA clones were generated at Leiden (van Dinten *et al.*, 1997) and Utrecht (de Vries *et al.*, 1997), whereas an infectious PRRSV cDNA clone was developed at Lelystad (Meulenberg *et al.*, 1997).

The EAV infectious clone from Leiden (pEAV030; EMBL database accession number Y07862) was assembled from fully sequenced cDNA clones which had previously been used to determine the EAV genome sequence (den Boon *et al.*, 1991). The 17 most 5' nucleotides, which were lacking in the cDNA library, were determined (van Dinten *et al.*, 1997) and attached to the 5' end using a PCR. In the same PCR, a T7 RNA polymerase promoter was placed upstream of nucleotide G-1 of the EAV genome.

Unfortunately, RNA transcribed from the initial reconstruction (pEAV030F) was not infectious. When differences between two cDNA clones had been detected during the sequence analysis of the EAV genomic library, a third cDNA clone had always been analyzed to determine the consensus sequence. However, for two of these positions, nucleotide 5,899 and nucleotide 7,508, a third clone had not been available. One of these point mutations (C instead of T at position 7,508), which induces a Ser-2429 to Pro substitution in the ORF1b-encoded replicase part, was eventually identified as the reason why pEAV030F-transcripts are not infectious (van Dinten *et al.*, 1997). Upon correction of this error, BHK-21 cells could be infected by transfection (electroporation) of pEAV030 RNA transcripts and the first recombinant arterivirus, carrying a translationally silent mutation to discriminate it from the wild-type virus, was generated (van Dinten *et al.*, 1997). Subsequent experiments confirmed the potential for heterologous gene expression from arterivirus RNA vectors. It was shown to be possible to express a foreign gene (chloramphenicol acetyl transferase; CAT) from two different EAV subgenomic mRNAs. These CAT-expressing vectors were not infectious, due to the fact that (in both cases) the insertion prevented the expression of one of the viral structural proteins. Modified expression vectors which should retain their infectivity are currently under construction.

Surprisingly, the non-infectious RNA from the pEAV030F clone (which carried the T-7,508 to C mutation) turned out to be capable of efficient self-replication. In contrast, subgenomic mRNA synthesis (and therefore also structural protein expression) did not occur at a detectable level, explaining why virus particles were not produced (van Dinten *et al.*, 1997). This intriguing phenotype is most likely due to the Ser-2429 to Pro mutation in the EAV replicase. Residue Ser-2,429 is conserved in the three available arterivirus replicase sequences. It is located between two well-conserved domains in the nidovirus replicase: the putative zinc finger region (residues 2,374–2,426 in EAV) and the putative RNA helicase domain (residues 2,500–2,800 in EAV). These two domains are both part of the 50 kDa nsp10 subunit of the EAV replicase (van Dinten *et al.*, 1996). This mutant shows that the requirements for nidovirus genome replication and discontinuous mRNA synthesis are (at least in part) different. Since the mutation is located only 3 residues downstream of the most C-terminal conserved Cys residue of the putative zinc finger domain, it is currently being tested whether replacement of conserved Cys and His residues in this region yields the same phenotype (genome replication without subgenomic mRNA transcription).

7. REVERSE GENETICS USING THE EAV INFECTIOUS cDNA CLONE

In addition to the development of arterivirus-based expression vectors, the EAV full-length cDNA clone is currently being used for a number of fundamental studies into the arterivirus life cycle. This work includes an analysis of the functions of various replicase subunits, the role of replicase processing in the regulation of arterivirus replication, and the details of the discontinuous transcription mechanism used to generate the subgenomic mRNAs.

Obviously, the fortuitously detected phenotype of the Ser-2429 to Pro mutant is receiving considerable attention, both at the RNA (van Marle *et al.*, unpublished observations) and at the protein level (van Dinten *et al.*, unpublished observations). The effects of a number of internal deletions in the full-length cDNA clone have been examined and these studies have yielded preliminary information about the RNA sequences required for genome replication. At the 3' end not more than 200 nt upstream of the poly(A) tail are required for RNA replication. However, transcripts containing longer 3'-terminal sequences seem to replicate considerably more efficient. Deletion of ORFs 2 to 7 yielded an RNA that could replicate and produce a subgenomic mRNA from the single remaining mRNA promoter (the RNA2 promoter in the 3' part of ORF1b). This showed that the products of EAV ORFs 2 to 7 (three known viral envelope proteins, the nucleocapsid protein, and two as yet unidentified proteins) are not required for genome replication or mRNA transcription. At the 5' end, deletion of the nsp1-encoding region yielded an RNA that could replicate but did not produce subgenomic mRNAs. Whether the deletion of certain RNA sequences from the 5'-terminal region of the genome or the absence of nsp1 forms the basis for this phenotype is currently being investigated.

8. CONCLUDING REMARKS

With the development of infectious cDNA clones, arterivirus (and nidovirus) research has entered a new era. The possibility to genetically modify two economically relevant arteriviruses could lead to the rapid development of (marker) vaccines and arterivirus-based expression vectors. At the fundamental level, the functional dissection of the EAV replicase polyprotein and the RNA transcription processes are important challenges, especially in view of the similarities and differences between arteriviruses and coronaviruses. At present, the set of structural proteins employed by arteriviruses and the structure of the virion into which they assemble, appear to be unique. Still, common features of the assembly of coronaviruses and arteriviruses may emerge, once the structural proteins of the latter group are characterized in more detail. It is striking that both groups of viruses assemble at intracellular membranes and contain a triple-spanning membrane (M) protein, which is assumed to play a crucial role in this process. Since arteriviruses are now known to both replicate and assemble at intracellular membranes, one could speculate on the possibility of an integrated process of genome replication, genome encapsidation, and virus budding. In any case, the analysis of the interactions between the viral replication machinery and organelles and molecules of the host cell will be an important novel aspect of future nidovirus research.

ACKNOWLEDGMENTS

This chapter would not have been written without the continuous experimental, theoretical, and/or moral support of: Johan den Boon, Leonie van Dinten, Sasha Gorbalenya, Yvonne van der Meer, Sietske Rensen, Willy Spaan, Hans van Tol, and (most of all) Fred Wassenaar.

REFERENCES

Baker, S. C., Yokomori, K., Dong, S., Carlisle, R., Gorbalenya, A. E., Koonin, E. V., andLai, M. M., 1993, Identification of the catalytic sites of a papain-like cysteine proteinase of murine coronavirus, *J. Virol.* **67**:6056–6063.

Bazan, J. F., and Fletterick, R. J., 1988, Viral cysteine proteases are homologous to the trypsin-like family of serine proteases: structural and functional implications, *Proc. Natl. Acad. Sci. USA* **85**:7872–7876.

Bazan, J. F., and Fletterick, R. J., 1990, Structural and catalytic models of trypsin-like viral proteases, *Semin. Virol.* **1**:311–322.

Bonilla, P. J., Gorbalenya, A. E., and Weiss, S. R., 1994, Mouse hepatitis virus strain A59 RNA polymerase gene ORF 1a: heterogeneity among MHV strains, *Virology* **198**:736–740.

Bonilla, P. J., Hughes, S. A., Pinon, J. D., and Weiss, S. R., 1995, Characterization of the leader papain-like proteinase of MHV A-59: identification of a new in vitro cleavage site, *Virology* **209**:489–497.

Bordier, C., 1981, Phase separation of integral membrane proteins in Triton X-114 solution, *J. Biol. Chem.* **256**:1604–1607.

Boyer, J. C. and Haenni, A. L., 1994, Infectious transcripts and cDNA clones of RNA viruses, *Virology* **198**:415–426.

Bredenbeek, P. J., and Rice, C. M., 1992, Animal RNA virus expression systems, *Semin. Virol.* **3**:297–310.

Breese, S. S.,Jr. and McCollum, W. H., 1970, *Proceedings of the 2nd International Conference on Equine Infectious Diseases* (Bryans, J. T. and Gerber, H. Eds.) S. Karger, Basel. pp.133–139.

Brierley, I., 1995, Ribosomal frameshifting on viral RNAs, *J. Gen. Virol.* 76:1885–1892. Brown, T. D. K., and Brierley, I., 1995, *The Coronaviridae* (Siddell, S. G. Ed.) Plenum Press, New York. pp. 191–217.

Cavanagh, D., 1997, Nidovirales: a new order comprising Coronaviridae and Arteriviridae, *Arch. Virol.* **142**:629–633.

Cavanagh, D., Brian, D. A., Brinton, M. A., Enjuanes, L., Holmes, K. V., Horzinek, M. C., Lai, M. M. C., Laude, H., Plagemann, P. G. W., Siddell, S. G., Spaan, W. J. M., Taguchi, F., and Talbot, P. J., 1993, The coronaviridae now comprises two genera, coronavirus and torovirus: report of the coronaviridae study group, *Adv. Exp. Med. Biol.* **342**:255–257.

Cavanagh, D., Brian, D. A., Brinton, M. A., Enjuanes, L., Holmes, K. V., Horzinek, M. C., Lai, M. M. C., Laude, H., Plagemann, P. G. W., Siddell, S. G., Spaan, W. J. M., Taguchi, F., and Talbot, P. J., 1994, Revision of the taxonomy of the Coronavirus, Torovirus, and Arterivirus genera, *Arch. Virol.* **135**:227–237.

de Groot, R. J., Hardy, W. R., Shirako, Y., and Strauss, J. H., 1990, Cleavage-site preferences of Sindbis virus polyproteins containing the non-structural proteinase. Evidence for temporal regulation of polyprotein processing in vivo, *EMBO J.* **9**:2631–2638.

de Vries, A. A. F., Horzinek, M. C., Rottier, P. J. M., and de Groot, R. J., 1997, The genome organization of the Nidovirales: similarities and differences between arteri-, toro-, and coronaviruses, *Semin. Virol.*, in press.

den Boon, J. A., Snijder, E. J., Chirnside, E. D., de Vries, A. A. F., Horzinek, M. C., and Spaan, W. J. M., 1991, Equine arteritis virus is not a togavirus but belongs to the coronaviruslike superfamily, *J. Virol.* **65**:2910–2920.

den Boon, J. A., Faaberg, K. S., Meulenberg, J. J. M., Wassenaar, A. L. M., Plagemann, P. G. W., Gorbalenya, A. E., and Snijder, E. J., 1995a, Processing and evolution of the N-terminal region of the arterivirus replicase ORF1a protein: identification of two papainlike cysteine proteases, *J. Virol.* **69**:4500–4505.

den Boon, J. A., Spaan, W. J. M., and Snijder, E. J., 1995b, Equine arteritis virus subgenomic RNA transcription: UV inactivation and translation inhibition studies, *Virology* **213**:364–372.

den Boon, J. A., Kleijnen, M. F., Spaan, W. J. M., and Snijder, E. J, 1996, Equine arteritis virus subgenomic mRNA synthesis: analysis of leader-body junctions and replicative-form RNAs, *J. Virol.* **70**:4291–4298.

Dougherty, W. G., and Semler, B. L., 1993, Expression of virus-encoded proteinases: functional and structural similarities with cellular enzymes, *Microbiol. Rev.* **57**:781–822.

Fujiki, Y., Hubbard, A. L., Fowler, S., and Lazarow, P. B., 1982, Isolation of intracellular membranes by means of sodium carbonate treatment: application to endoplasmic reticulum, *J. Cell Biol.* **93**:97–102.

Gao, H. Q., Schiller, J. J., and Baker, S. C., 1996, Identification of the polymerase polyprotein products p72 and p65 of the murine coronavirus MHV JHM, *Virus Res.* **45**:101–109.

Godeny, E. K., Chen, L., Kumar, S. N., Methven, S. L., Koonin, E. V., and Brinton, M. A., 1993, Complete genomic sequence and phylogenetic analysis of the lactate dehydrogenase-elevating virus (LDV), *Virology* **194**:585–596.

Gorbalenya, A. E., and Snijder, E. J., 1996, Viral cysteine proteases, *Perspect. Drug Discov. Design* **6**:64–86.

Gorbalenya, A. E., Donchenko, A. P., Blinov, V. M., and Koonin, E. V., 1989a, Cysteine proteases of positive strand RNA viruses and chymotrypsin-like serine proteases. A distinct protein superfamily with a common structural fold, *FEBS Lett.* **243**:103–114.

Gorbalenya, A. E., Koonin, E. V., Donchenko, A. P., and Blinov, V. M., 1989b, Coronavirus genome: prediction of putative functional domains in the non-structural polyprotein by comparative amino acid sequence analysis, *Nucleic Acids Res.* **17**:4847–4861.

Gorbalenya, A. E., Koonin, E. V., and Lai, M. M., 1991, Putative papain-related thiol proteases of positive-strand RNA viruses. Identification of rubi- and aphthovirus proteases and delineation of a novel conserved domain associated with proteases of rubi-, alpha- and coronaviruses, *FEBS Lett.* **288**:201–205.

Jacks, T., Madhani, H. D., Masiarz, F. R., and Varmus, H. E., 1988, Signals for ribosomal frameshifting in the Rous sarcoma virus gagpol region, *Cell* **55**:447–458.

Jore, J., De Geus, B., Jackson, R. J., Pouwels, P. H., and Enger Valk, B. E., 1988, Poliovirus protein 3CD is the active protease for processing of the precursor protein P1 in vitro, *J. Gen. Virol.* **69**:1627–1636.

Lee, H. J., Shieh, C. K., Gorbalenya, A. E., Koonin, E. V., La Monica, N., Tuler, J., Bagdzhadzhyan, A., and Lai, M. M. C., 1991, The complete sequence (22 kilobases) of murine coronavirus gene 1 encoding the putative proteases and RNA polymerase, *Virology* **180**: 567–582.

Lemm, J. A., and Rice, C. M., 1993a, Roles of nonstructural polyproteins and cleavage products in regulating Sindbis virus RNA replication and transcription, *J. Virol.* **67**: 1916- 1926.

Lemm, J. A., and Rice, C. M., 1993b, Assembly of functional Sindbis virus RNA replication complexes: requirement for coexpression of p123 and p34, *J. Virol.* **67**:1905–1915.

Lemm, J. A., Rumenapf, T., Strauss, E. G., Strauss, J. H., and Rice, C. M., 1994, Polypeptide requirements for assembly of functional Sindbis virus replication complexes: a model for the temporal regulation of minus- and plus-strand RNA synthesis, *EMBO J.* **13**:2925–2934.

Liu, D. X. and Brown, T. D. K., 1995, Characterisation and mutational analysis of an ORF 1a-encoding proteinase domain responsible for proteolytic processing of the infectious bronchitis virus 1a/1b polyprotein, *Virology* **209**:420–427.

Lu, Y., Lu, X., and Denison, M. R., 1995, Identification and characterization of a serine-like proteinase of the murine coronavirus MHV-A59, *J. Virol.* **69**:3554–3559.

Meulenberg, J. J. M., Hulst, M. M., de Meijer, E. J., Moonen, P. L., den Besten, A., de Kluyver, E. P., Wensvoort, G., and Moormann, R. J. M., 1993, Lelystad virus, the causative agent of porcine epidemic abortion and respiratory syndrome (PEARS), is related to LDV and EAV, *Virology* **192**:62–72.

Meulenberg, J. J. M., Bos-de Ruijter, J. N. A., Wensvoort, G., and Moormann, R. J. M., 1997, Infectious transcripts from cloned genome-length cDNA of porcine reproductive respiratory syndrome virus, *J. Virol.*, submitted.

Pol, J. M., Wagenaar, F., and Reus, J. E. G., 1997, Comparative morphogenesis of three PRRS virus strains, *Vet. Microbiol.* **55**:203–208.

Shirako, Y., and Strauss, J. H., 1994, Regulation of Sindbis virus RNA replication: uncleaved P123 and nsP4 function in minus-strand RNA synthesis, whereas cleaved products from P123 are required for efficient plus-strand RNA synthesis, *J. Virol.* **68**:1874–1885.

Snijder, E. J., and Meulenberg, J. J. M., 1998, The molecular biology of arteriviruses, *J. Gen. Virol.*, submitted.

Snijder, E. J., and Spaan, W. J. M., 1995, *The Coronaviridae* (Siddell, S. G. Ed.) Plenum Press, New York. pp. 239–255.

Snijder, E. J., den Boon, J. A., Bredenbeek, P. J., Horzinek, M. C., Rijnbrand, R., and Spaan, W. J., 1990, The carboxyl-terminal part of the putative Berne virus polymerase is expressed by ribosomal frameshifting and contains sequence motifs which indicate that toro- and coronaviruses are evolutionarily related, *Nucleic Acids Res* **18**:4535–4542.

Snijder, E. J., Wassenaar, A. L. M., and Spaan, W. J. M., 1992, The 5' end of the equine arteritis virus replicase gene encodes a papainlike cysteine protease, *J. Virol.* **66**:7040–7048.

Snijder, E. J., Wassenaar, A. L. M., and Spaan, W. J. M., 1994, Proteolytic processing of the replicase ORF1a protein of equine arteritis virus, *J. Virol.* **68**:5755–5764.

Snijder, E. J., Wassenaar, A. L. M., Spaan, W. J. M., and Gorbalenya, A. E., 1995, The arterivirus nsp2 protease. an unusual cysteine protease with primary structure similarities to both papain-like and chymotrypsin-like proteases, *J. Biol. Chem.* **270**:16671–16676.

Snijder, E. J., Wassenaar, A. L. M., van Dinten, L. C., Spaan, W. J. M., and Gorbalenya, A. E. 1996. The arterivirus nsp4 protease is the prototype of a novel group of chymotrypsin-like enzymes, the 3C-like serine proteases, *J. Biol. Chem.* **271**:4864–4871.

Snijder, E. J. and Horzinek, M. C, 1993, Toroviruses: replication, evolution and comparison with other members of the coronavirus-like superfamily, *J. Gen. Virol.* **74**:2305–2316.

Stueckemann, J. A., Holth, M., Swart, W. J., Kowalchyk, K., Smith, M. S., Wolstenholme, A. J., Cafruny, W. A., and Plagemann, P. G. W., 1982, Replication of lactate dehydrogenase-elevating virus in macrophages. 2. mechanism of persistent infection in mice and cell culture, *J. Gen. Virol.* **59**:263–272.

van Dinten, L. C., Wassenaar, A. L. M., Gorbalenya, A. E., Spaan, W. J. M., and Snijder, E. J., 1996, Processing of the equine arteritis virus replicase ORF1b protein: identification of cleavage products containing the putative viral polymerase and helicase domains, *J. Virol.* **70**:6625–6633.

van Dinten, L. C., den Boon, J. A., Wassenaar, A. L. M., Spaan, W. J. M., and Snijder, E. J., 1997, An infectious arterivirus cDNA clone: identification of a replicase point mutation which abolishes discontinuous mRNA transcription, *Proc. Natl. Acad. Sci. USA* **94**:991–996.

Wassenaar, A. L. M., Spaan, W. J. M., Gorbalenya, A. E., and Snijder, E. J., 1997, Alternative proteolytic processing of the arterivirus ORF1a polyprotein: evidence that nsp2 acts as a cofactor for the nsp4 serine protease, *J. Virol.*, submitted.

Wood, O., Tauraso, N. M., and Liebhaber, H. 1970. Electron microscopic study of tissue cultures infected with simian haemorrhagic fever virus, *J. Gen. Virol.* **7**:129–136.

Ypma-Wong, M. F., Dewalt, P. G., Johnson, V. H., Lamb, J. G., and Semler, B. L,. 1988, Protein 3CD is the major poliovirus proteinase responsible for cleavage of the P1 capsid precursor, *Virology* **166**:265–270.

Ziebuhr, J., Herold, J., and Siddell, S. G., 1995, Characterization of a human coronavirus (strain 229E) 3C-like proteinase activity, *J. Virol.* **69**:4331–4338.

14

REPLICATION AND TRANSCRIPTION OF HCV 229E REPLICONS

Volker Thiel, Stuart G. Siddell, and Jens Herold

Institute of Virology and Immunology
University of Wuerzburg
Versbacherstr. 7
97078 Wuerzburg, Germany

1. ABSTRACT

Replicons based upon the human coronavirus 229E (HCV 229E) genome were transfected into HCV 229E infected cells. We demonstrate that a synthetic RNA comprised of 646 nucloetides from the 5' end and 1465 nucloetides from the 3' end of the HCV 229E genome is replication competent. We conclude that the cis-acting elements necessary for replication are located in these 5' and 3' genomic regions. Furthermore, we inserted the intergenic region of the HCV 229E nucleocapsid protein gene into this basic construct and were able to demonstrate the transcription of "subgenomic" RNAs

2. INTRODUCTION

The study of coronavirus replication has been greatly facilitated by the use of defective interfering (DI) RNAs that can replicate in coronavirus infected cells (Makino and Lai, 1989; Van der Most et al., 1991). The analysis of several different DI RNAs has revealed that cis-acting elements required for replication are located at the 5' and 3' end of the coronavirus genome. Furthermore, the insertion of an intergenic sequence into these DI RNAs leads to the synthesis of "subgenomic" RNA transcripts (Makino et al., 1991). In order to establish a similar system to study HCV 229E replication, we have constructed an HCV 229E replicon containing 5' and 3' sequences of the HCV 229E genome behind a T7 RNA polymerase promoter. Furthermore, we have inserted the intergenic sequence of the HCV 229E nucleocapsid protein gene at different positions within this construct and tested for replication and transcription. The potential use of this repli-

Coronaviruses and Arteriviruses, edited by Enjuanes *et al.*
Plenum Press, New York, 1998

cation and transcription assay in combination with a recombinant expressed HCV 229E RNA polymerase is discussed.

3. METHODS

3.1. DNA Construction

A PCR fragment was generated using DNA of the plasmid T35D5 (Herold et al., 1993) as template and the primers 35D5(up) (5'-ATCAGGGAATTCTAATACGACTCAC TATAGGGACTTAAGTACCTTATCTATCTACAGATAGAAAAGTTGCTTTTTA-3') and 35D5(down) (5'-ATTTGCGGGTACCCTCTGGTCAGTGTATGTCACATT-3'). The primer 35D5(up) contains an EcoRI restriction site followed by the T7 RNA polymerase promoter, three G bases and 44 nucleotides corresponding to the 5' end of the HCV 229E genome. The primer 35D5(down) contains a KpnI restriction site preceded by 21 nucleotides complementary to the nucloetides 626 - 646 of the HCV 229E genome. A second PCR fragment was generated using the plasmid pBH1 as template and the primers BH1(up) (5'-CCA-CAACGTGGTCGTCAGGGT-3') and BH1(down) (5'-ACTAGTGGATCCCCCCGA-CATCGATT-3'). The plasmid pBH1 contains the HCV 229E genomic sequence from nt. 23868 to the 3' end of the genome (Herold, unpublished). The primer BH1(up) corresponds to the nucloetides 25719–25739 of the HCV 229E RNA sequence and the primer BH1(down) contains a BamHI restriction site preceded by 12 nucleotides that correspond to the sequence downstream of the poly(A) stretch in the construct pBH1. Both PCR fragments were cleaved with KpnI and ligated. The ligation product was cleaved with EcoRI and BamHI and inserted into the EcoRI and BamHI cleaved vector pBR322. The resulting construct is designated as pRep1.1. In order to insert an intergenic sequence, two annealed oligonucleotides with KpnI overhangs were cloned into the KpnI site of pRep1.1. The inserted sequence corresponds to the nucleotides 25656–25684 of the HCV 229E sequence (including the intergenic sequence of the HCV 229E nucleocapsid protein gene) flanked by 6 nucleotides upstream and 30 nucleotides downstream that were introduced for further cloning procedures. The resulting construct is designated as pRep6.1. To obtain the constructs pRep6.2, pRep6.3 and pRep6.4, we inserted the same sequence into pRep1.1 using the Bsu36I, BstXI and PacI sites, respectively. The intergenic sequence of the constructs pRep6.1, pRep6.2, pRep6.3 and pRep6.4 is located 1495, 1030, 802 and 277 nucleotides, respectively, upstream of the poly(A) stretch representing the 3' end of the HCV 229E genomic sequence.

3.2. RNA Replication and Transcription Assay

In vitro transcription with T7 RNA polymerase was done with BamHI-linearised plasmid DNA templates using the MEGAscript-Kit (Ambion, Germany). 1×10^5 HeLa cells were infected with HCV 229E (m.o.i. 10) and incubated at 33°C. At 1h post infection, 5 μg in vitro synthesised RNA was transfected using 12.5 μl Lipofectin (Life Technologies, Germany) according to the manufacturer's instructions. At 30 min post transfection the cells were washed twice with medium and incubated for further 16 h at 33°C. Poly(A) RNA was prepared using oligo $(dT)_{25}$ dynabeads (Dynal) according to the manufacturer's instructions and electrophoresed on 2.2 M formaldehyde-1% agarose gels. The gels were dried and hybridised with a ^{32}P-(5'-end)-labelled oligonucleotide, as described previously (Meinkoth and Wahl, 1984).

4. RESULTS

4.1. The Structure of HCV 229E Replicons

Previous studies on coronavirus replication have shown that RNAs containing 5' and 3' genomic sequences can replicate in coronavirus infected cells. However, there has been no analysis of the RNA sequences that are necessary for replication in HCV 229E infected cells. To identify the cis-acting sequences that are sufficient for the replication of RNAs by the HCV 229E RNA polymerase, we cloned 646 nucleotides from the 5' end and 1465 nucleotides from the 3' end of the HCV 229E genome behind a T7 RNA polymerase promoter. To allow efficient in vitro transcription with T7 RNA polymerase, we inserted three additional G bases between the T7 promoter and the 5' end of the HCV 229E sequence. A BamHI restriction site, 11 nucleotides downstream of the poly(A) sequence of the HCV 229E 3' end, was used to linearise plasmid DNA prior to in vitro transcription. The structure of this construct, designated as pRep1.1, is shown in Figure 1. In order to analyse transcription, we inserted the intergenic sequence of the HCV 229E nucleocapsid protein gene at different positions in the pRep1.1 sequence as illustrated in Figure 1. The resulting constructs are designated as pRep 6.1, 6.2, 6.3 and 6.4.

The structure of the construct pRep 1.1 relative to the HCV 229E genome is shown. The position of the insertion of the intergenic sequence in the constructs pRep 6.1, 6.2, 6.3 and 6.4 are indicated.

4.2. Replication and Transcription

To study replication and transcription by the HCV 229E RNA polymerase, we synthesised in vitro transcripts of the HCV 229E replicons using T7 RNA polymerase. Then, 5 μg of these synthetic RNAs were transfected into 1×10^5 HeLa cells that had been previously infected with HCV 229E. After 16 h incubation at 33 °C, the cells were lysed and the poly(A) RNA was isolated. The RNA samples were separated by gel electrophoresis

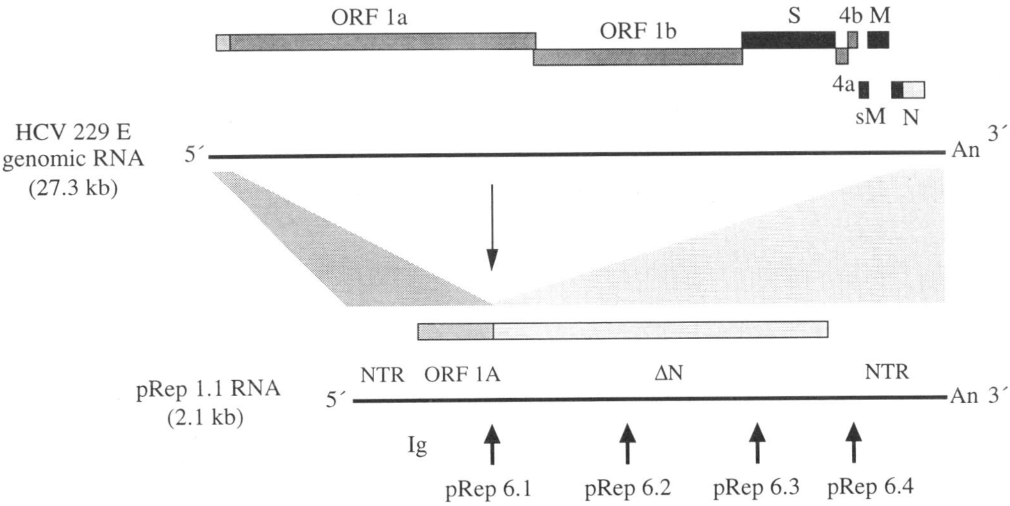

Figure 1. Schematic representation of HCV 229E replicons.

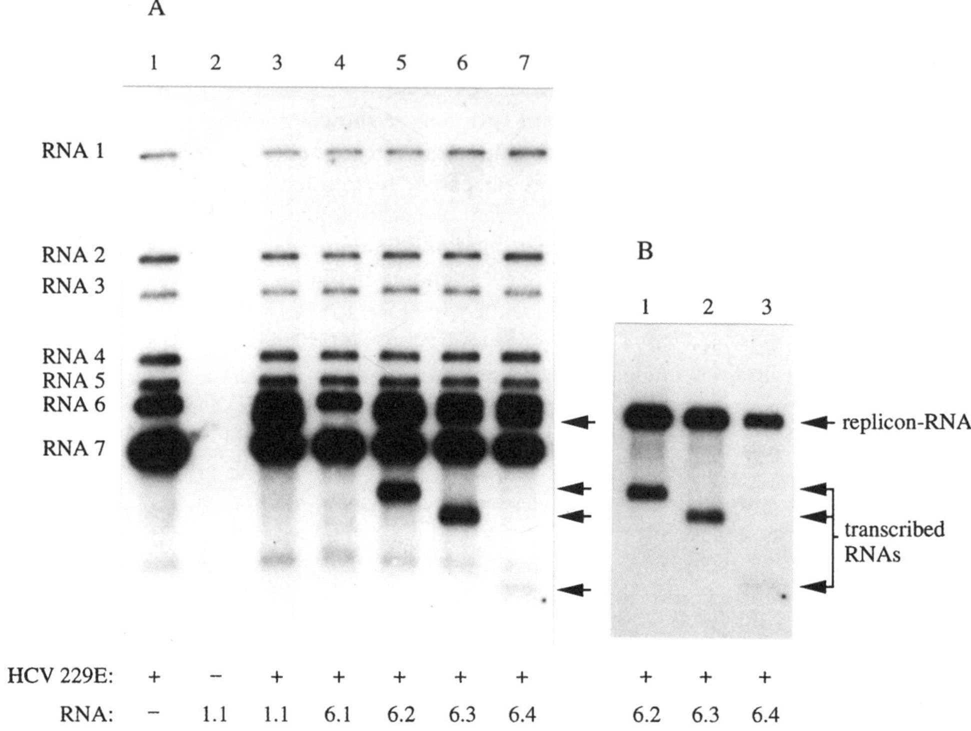

Figure 2. Replication and transcription assay.

and hybridised with a HCV 229E nucleocapsid protein gene specific ^{32}P labelled oligonucleotide. The result of this analysis is shown in Figure 2a.

Lane 1 shows poly(A) RNA from HCV 229E infected HeLa cells. The HCV 229E genomic RNA and six subgenomic mRNA can be detected. Lane 2 shows that transfected RNA derived from pRep1.1 cannot be detected in unifected cells. In contrast, as is shown in lane 3, RNA derived from pRep1.1 is replicated in HCV 229E-infected cells. Lanes 5, 6 and 7 show that this is also true for RNA derived from pRep6.2, pRep6.3 and pRep6.4. RNA derived from pREp6.1 did not replicate in infected cells (lane 4) and the reason for this difference remains to be determined. Importantly, we were also able to detect "subgenomic" RNAs transcribed from the RNA derived from pRep6.2, pRep6.3 and pRep6.4. Depending on the position of the intergenic sequence in the replicon construct, "subgenomic" RNAs of 1.0 kb, 0.8 kb and 0.3 kb, respectively, were detected (lanes 5, 6 and 7).

Poly(A) RNAs were separated by electrophoresis. Viral RNAs were hybridised with a) an HCV 229E nucleocapsid gene-specific or b) a replicon-specific, 5'-end-labelled oligonucleotide. Viral and replicon specific RNAs are indicated.

To further confirm the identity of the replicated and transcribed RNAs, we did a second hybridisation analysis. In this case, the nucleocapsid protein gene specific ^{32}P labelled oligonucleotide was replaced by a ^{32}P labelled oligonucleotide specific for the insertion of the HCV 229E intergenic sequence. The result is shown in Figure 2b. As expected, the viral RNAs can no longer be detected, whereas the replicons (with a size of approximately 2.1 kb) are clearly visible. Furthermore, the detection of the "subgenomic RNAs (1.0, 0.8

and 0.3 kb) demonstrate transcription from these replicating RNAs. Again the sizes of the transcribed RNAs correlate to the position of the intergenic sequence in the replicon construct.

5. DISCUSSION

In this study, we have established an assay to detect HCV 229E RNA polymerase activity using HCV 229E derived replicating RNA. We have tested several synthetic RNAs comprised of 5' and 3' genomic sequences for their ability to replicate in HCV 229E infected cells. This analysis revealed that 646 nts from the 5' end and 1465 nts from the 3' end of the HCV 229E genome are sufficient for replication. We therefore conclude that all cis-acting sequences necessary for replication are located in these 5' and 3' genomic regions. Furthermore, we have demonstrated transcription of "subgenomic" RNAs from replicating RNA, under the control of an intergenic region.

The use of assays similar to those reported here have greatly enhanced our understanding of, in particular, cis-acting elements and their role in coronaviral RNA replication and transcription. However, the structure-function relationships of the coronavirus RNA polymerase still remain, by and large, an enigma. The RNA polymerase gene is comprised of two overlapping ORFs with a length of about 20 kb. The primary polymerase translation products are extensively processed into smaller polypeptides and it is not yet known which polymerase gene products participate in the many and complex functions of the replication/transcription process. To approach such questions, it will be necessary to clone a functional RNA polymerase gene and establish a system that allows for the use of reverse genetics. One possibility is to express the RNA polymerase in eucaryotic cells using recombinant vaccinia viruses. The replication/transcription assay presented in this study represents an important step towards achieving this goal.

REFERENCES

Herold, J., Raabe, T., Schelle-Prinz, B., and Siddell. S.G., 1993, Nucleotide sequence of the human coronavirus 229E RNA polymerase locus, *Virology* **195**: 680–691.

Makino, S., and Lai. M.M.C., 1989, High-frequency leader sequence switching during coronavirus defective interfering RNA replication, *J. Virol.* **63**: 5285–5292.

Makino, S., Joo, M., and Makino. J.K., 1991, A system for study of coronavirus mRNA synthesis: a regulated, expressed subgenomic defective interfering RNA results from intergenic site insertion, *J. Virol.* **65**: 6031–6041.

Meinkoth, J., and Wahl. G., 1984, Hybridization of nucleic acids immobilized on solid supports, *Anal.Biochem.* **138**: 267–284.

Van der Most, R.G., Bredenbeck, P.J., and Spaan. W.J.M., 1991, A domain at the 3' end of the polymerase gene is essential for encapsidation of coronavirus defective interfering RNAs, *J. Virol.* **65**: 3219–3226.

15

SUBSTRATE SPECIFICITY OF THE HUMAN CORONAVIRUS 229E 3C-LIKE PROTEINASE

John Ziebuhr, Gerhard Heusipp, Anja Seybert, and Stuart G. Siddell

Institute of Virology
University of Würzburg
Versbacher Strasse 7
97078 Würzburg, Germany

1. ABSTRACT

Coronavirus gene expression involves proteolytic processing of the gene 1-encoded polyproteins and a key enzyme in this process is the virus-encoded 3C-like proteinase. In this study, we describe the biosynthesis of the human coronavirus 229E 3C-like proteinase in *Escherichia coli* and the substrate specificity of the purified protein. Using immunofluorescence microscopy, we have also investigated the subcellular localization of the 3C-like proteinase and have found a punctate, perinuclear distribution of the proteinase in virus-infected cells.

2. INTRODUCTION

The human coronavirus (HCV) 229E genome is a positive-strand RNA of approximately 27,000 nucleotides. The replicase gene or gene 1, which is located at the 5' end of the genome, is comprised of two large, overlapping open reading frames (ORFs), ORF 1a and ORF 1b. The upstream ORF, ORF 1a, encodes a polyprotein, pp1a, with a calculated molecular mass of 454 kDa. In vitro studies suggest that the downstream ORF, ORF 1b, is expressed as a fusion protein with the ORF 1a-derived polyprotein pp1a by a process of (-1) ribosomal frameshifting. The fusion protein, pp1ab, has a calculated molecular mass of 754 kDa (Herold et al., 1993; Herold and Siddell, 1993).

There is clear evidence that the functional polypeptides involved in coronaviral RNA replication are released from the replicase gene-encoded polyproteins by extensive proteolytic processing. A key enzyme in this process appears to be the virus-encoded 3C-like proteinase (Liu et al., 1997; Heusipp et al., 1997a, 1997b; Grötzinger et al., 1996; Lu

Coronaviruses and Arteriviruses, edited by Enjuanes *et al.*
Plenum Press, New York, 1998

et al., 1996; Tibbles et al., 1996; Ziebuhr et al., 1995; Liu et al., 1994). Recently, considerable progress has been made in characterizing the enzymatic properties of the coronavirus 3C-like proteinase and it appears that the enzyme has a catalytic system which is similar to the picornavirus 3C proteinases. However, the coronavirus enzyme has also several additional, unique features (Ziebuhr et al., 1997; Lu and Denison, 1997; Seybert et al., 1996; Lu et al., 1995; Liu and Brown, 1995; Ziebuhr et al., 1995).

In infected cells, the 3C-like proteinases of mouse hepatitis virus (MHV) A59 and HCV 229E have been identified as polypeptides with apparent molecular masses of 27 kDa and 34 kDa, respectively (Lu et al., 1996; Ziebuhr et al., 1995). However, to date nothing is known about the subcellular localization of the coronavirus 3C-like proteinase.

3. MATERIALS AND METHODS

3.1. Bacterial Expression and Purification of the HCV 229E 3C-Like Proteinase

The construction of plasmid pMalc2–3CL and the bacterial expression of the HCV 229E 3C-like proteinase have been described previously (Ziebuhr et al., 1995). Briefly, a fusion protein, MBP-3CLpro, containing the maltose-binding protein (MBP) of *E. coli* and the 3C-like proteinase domain of HCV 229E (amino acids 2966–3267 of pp1a and pp1ab) was synthesized in bacteria and then purified by amylose-affinity chromatography. The authentic 3C-like proteinase was released from the fusion protein by factor Xa cleavage and the protein mixture was loaded onto a Phenylsepharose HP column (Pharmacia Biotech, Freiburg, Germany) that had been pre-equilibrated with 12.5 mM Bis-Tris-HCl, pH 7.0, 300 mM NaCl, 1 mM dithiothreitol, 0.1 mM EDTA. After extensive washing of the column, the recombinant 3C-like proteinase was eluted with 12.5 mM Bis-Tris-HCl, pH 7.0, 1mM dithiothreitol, 0.1 mM EDTA. The fractions containing the 3C-like proteinase were pooled and concentrated to 3 mg/ml using Centricon-3 concentrators (Amicon, Beverly, Mass.). The concentrated material was purified further by chromatography on a Superdex 75 pg column (Pharmacia Biotech) run under isocratic conditions with 10 mM Tris-HCl, pH 7.3, 200 mM NaCl, 0.1 mM EDTA, 1 mM dithiothreitol. Finally, the purified protein was concentrated to 10 mg/ml using Centricon-3 concentrators and stored at -80°C.

3.2. Trans-Cleavage Assays

3.2.1. Peptide Substrates. Synthetic peptides were prepared by solid-phase chemistry and purified by high-performance liquid chromatography (HPLC) on a reversed-phase C_{18} silica column (Jerini Bio-Tools, Berlin, Germany). The identity and homogeneity of the peptides were confirmed by mass spectrometry and analytical reversed-phase chromatography. Cleavage reactions were incubated for 2 h at 25°C and contained 20 mM Bis-Tris-HCl, pH 7.0, 1 μM recombinant 3C-like proteinase and 0.5 mM substrate peptide in a total volume of 20 μl. The reactions were terminated by the addition of 80 μl 0.1% trifluoroacetic acid and the mixture was centrifuged for 5 min at 14 000 x *g* prior to analysis by reversed-phase HPLC on a 3.9 x 150 mm Delta Pak C_{18} column (Waters, Milford, Mass.). Cleavage products were resolved using a 22-min, 5–90% linear gradient of acetonitrile in 0.1% trifluoroacetic acid. The absorbance was determined at 215 nm. Under these conditions, the conversion of functional substrate peptides was complete. Nonfunctional substrate peptides did not show detectable hydrolysis under the same conditions.

3.2.2. Fusion Protein Substrates. Fusion proteins consisting of MBP and various replicase gene-derived peptide sequences were synthesized in *E. coli* TB1 cells and purified by amylose-affinity chromatography as described previously (Grötzinger et. al., 1996; Heusipp et al., 1997a; Heusipp et al., 1997b). In this way, five fusion proteins were purified that contain the following gene 1-encoded amino acids (aa): (i) aa 3934–4237, (ii) aa 4774–5259, (iii) aa 5500–5771, (iv) aa 5981–6602, and (v) aa 5981–6290. These substrate proteins were incubated with recombinant 3C-like proteinase as described by Grötzinger et al. (1996). The cleavage products were separated by SDS-polyacrylamide gel electrophoresis, transferred electrophoretically to polyvinylidene difluoride membranes, and the amino-terminal sequences of the carboxyl-terminal cleavage products were determined by automated Edman degradation as previously described (Ziebuhr et al., 1995).

3.3. Immunofluorescence Assay

MRC-5 cells were grown on coverslips, infected with HCV 229E (multiplicity of infection of 10 pfu per cell), and incubated at 33°C. After 15 h, the cells were fixed for 10 min with 4% paraformaldehyde in PBS. After washing with PBS, the cells were permeabilized with PBS-0.2% Triton X-100 and washed with PBS-1% NP-40. The indirect immunofluorescence assays were carried out with the HCV 229E 3C-like proteinase-specific antiserum K17 (Ziebuhr et al., 1995) at a 1:100 dilution in 10 mM Tris-HCl, 150 mM NaCl, 1% NP-40 containing 5% normal goat serum to reduce the background fluorescence. A FITC-conjugated goat anti-rabbit immunoglobulin G (1:100 dilution, Dianova, Hamburg, Germany) was used as the secondary antibody.

4. RESULTS AND DISCUSSION

4.1. Substrate Specificity of the HCV 229E 3C-Like Proteinase

Recently, considerable progress has been made in characterizing the primary structure of substrates that are recognized and cleaved by the HCV 229E 3C-like proteinase (Ziebuhr et al., 1997; Heusipp et al., 1997a; Heusipp et al., 1997b; Grötzinger et al., 1996, Ziebuhr et al., 1995). In order to substantiate the conclusions that have been drawn from these experiments, we have extended, in the present study, the analysis of putative functional substrates of the HCV 229E 3C-like proteinase. We have used both synthetic peptides and bacterially synthesized fusion protein substrates in the trans-cleavage assays with recombinant 3C-like proteinase. These data, combined with published results on the substrate specificity of the HCV 229E 3C-like proteinase, are summerized in Table 1 and lead us to the following conclusions. The Gln residue in the P1 position is absolutely conserved among all 3C-like proteinase substrates identified to date. The P2 and P1´ positions are occupied by Leu/Ile and Ser/Ala/Gly, respectively. However, as Table 1 also shows, the presence of the conserved Leu/Ile-Gln-Ser/Ala/Gly sequence is not sufficient for cleavage by the 3C-like proteinase: Two peptides containing either Ile-Gln-Ser or Ile-Gln-Gly tripeptides are not cleaved. Clearly, proteolysis is not exclusively determined by the P2 through P1' positions. It is also interesting to note, that the HCV 229E 3C-like proteinase cleaves efficiently an MHV-A59 derived peptide substrate indicating that the substrate specificity of coronavirus 3C-like enzymes is, at least to some extent, conserved.

Table 1. Substrate specificity of recombinant HCV 229E 3C-like proteinase

Substrate[a]	Sequence																Polyprotein
P	K	T	L	I	F	T	L	Q	↓	A	A	F	G	N	A	G	HCV 229E pp1a/1ab
B/P	V	S	Y	G	S	T	L	Q	↓	A	G	L	R	K	M	A	HCV 229E pp1a/1ab
P	Q	M	F	G	V	N	L	Q	↓	S	G	K	T	T	S	M	HCV 229E pp1a/1ab
B	T	C	D	R	T	A	I	Q	↓	S	F	D	N	S	Y	L	HCV 229E pp1a/1ab
B	Y	E	K	S	T	V	L	Q	↓	A	A	G	L	C	V	V	HCV 229E pp1ab
B/P	E	I	T	M	T	D	L	Q	↓	S	E	S	S	C	G	L	HCV 229E pp1ab
B	T	F	T	E	V	N	L	Q	↓	G	L	E	N	I	A	F	HCV 229E pp1ab
B	A	T	F	Y	P	Q	L	Q	↓	S	A	E	W	K	C	G	HCV 229E pp1ab
P	L	C	T	T	S	F	L	Q	↓	S	G	I	V	K	M	V	MHV-JHM pp1a/1ab
P	C	F	L	V	T	K	F	R		R	M	F	G	D	L	S	HCV 229E pp1a/1ab
P	D	S	F	C	K	T	I	Q		S	A	L	S	V	V	S	HCV 229E pp1a/1ab
P	K	Q	R	I	T	T	I	Q		G	P	P	G	S	G	K	HCV 229E pp1ab

[a]P, synthetic peptide; B, bacterially synthesized fusion protein.
The cleavage of a specific peptide bond is indicated by an arrow.

4.2. Intracellular Localization of the HCV 229E 3C-Like Proteinase

In the case of coronaviruses, there is a paucity of information on the intracellular localization of replicase gene-encoded polypeptides. Therefore, we decided to study the intracellular localization of the 3C-like proteinase by immunofluorescence microscopy. MRC-5 cells were infected with HCV 229E at a multiplicity of infection of 10 pfu per cell, and 15 h postinfection we performed an indirect immunofluorescence analysis using the proteinase-specific rabbit antiserum K17 (Ziebuhr et al., 1995). The results of this experiment are shown in Figure 1. We observed a punctate, perinuclear immunofluorescence

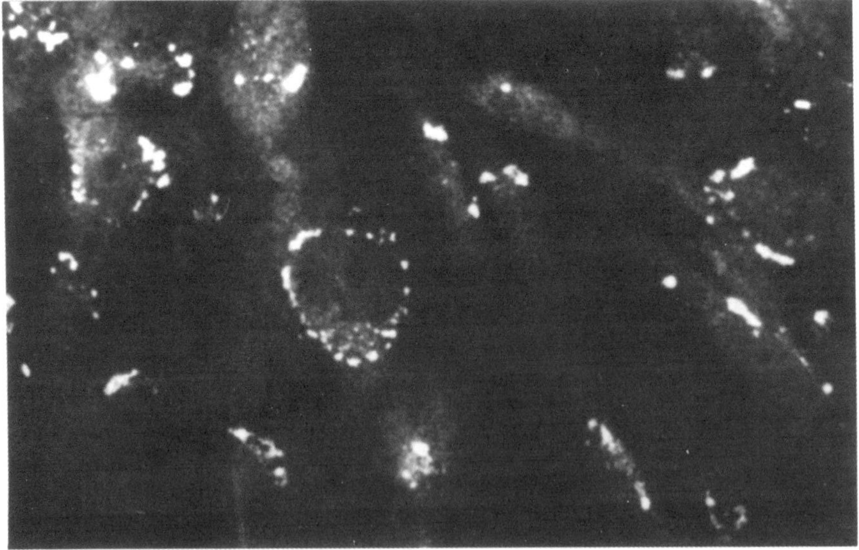

Figure 1. Immunofluorescence analysis of the subcellular localization of the 3C-like proteinase in HCV 229E-infected cells. MRC-5 cells were infected with HCV 229E (panel A) or mock infected (panel B) and analyzed 15 h postinfection.

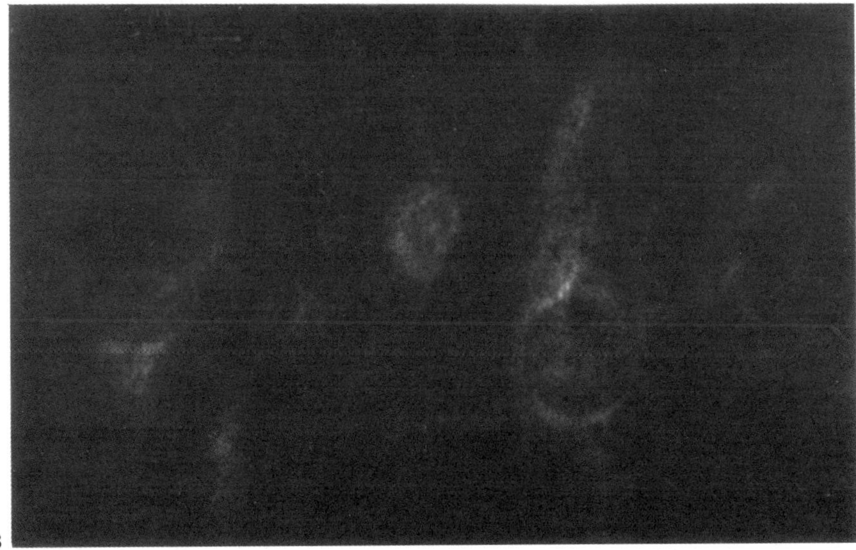

Figure 1B.

in virus-infected cells. A very similar staining pattern was also observed in HCV 229E-infected HeLa-CD13 cells (data not shown). These data provide the first experimental evidence that the coronavirus 3C-like proteinase is located in a perinuclear, presumably membraneous compartment. Interestingly, a similar immunofluorescence pattern has also been reported for the replicase gene-encoded putative helicase and RNA polymerase proteins of the equine arteritis virus (EAV), a virus that is closely related to coronaviruses (van Dinten et al., 1996).

REFERENCES

Grötzinger, C., Heusipp, G., Ziebuhr, J., Harms, U., Süss, J., and Siddell, S. G., 1996, Characterization of a 105-kDa polypeptide encoded in gene 1 of the human coronavirus HCV 229E, *Virology* **222:** 227–235.

Herold, J. and Siddell, S. G., 1993, An elaborated pseudoknot is required for high frequency frameshifting during translation of HCV 229E polymerase mRNA, *Nucleic Acids Res.* **21:** 5838–5842.

Herold, J., Raabe, T., Schelle-Prinz, B. and Siddell, S. G., 1993, Nucleotide sequence of the human coronavirus 229E RNA polymerase locus, *Virology* **195:** 680–691.

Heusipp, G., Grötzinger, C., Herold, J., Siddell, S. G., and Ziebuhr, J., 1997a, Identification and subcellular localization of a 41-kDa, polyprotein 1ab processing product in human coronavirus 229E-infected cells, *J. Gen. Virol.*, in press

Heusipp, G., Harms, U., Siddell, S. G., and Ziebuhr, J., 1997b, Identification of an ATPase activity associated with a 71-kDa polypeptide encoded in gene 1 of the human coronavirus 229E, *J. Virol.*, in press

Liu, D. X. and Brown, T. D. K., 1995, Characterization and mutational analysis of an ORF 1a-encoding proteinase domain responsible for proteolytic processing of the infectious bronchitis virus 1a/1b polyprotein, *Virology* **209:** 420–427.

Liu, D. X., Brierley, I., Tibbles., K. W., and Brown, T. D. K., 1994, A 100-kilodalton polypeptide encoded by open reading frame (ORF) 1b of the coronavirus infectious bronchitis virus is processed by ORF 1a products, *J. Virol.* **68:** 5773–5780.

Liu, D. X., Xu, H. Y., and Brown, T. D. K., 1997, Proteolytic processing of the coronavirus infectious bronchitis virus 1a polyprotein: identification of a 10-kilodalton polypeptide and determination of its cleavage sites, *J. Virol.* **71:** 1814–1820.

Lu, Y. and Denison M. R., 1997, Determinants of mouse hepatitis virus 3C-like proteinase activity, *Virology* **230**: 335–342.

Lu, X., Lu, Y., and Denison, M. R., 1996, Intracellular and in vitro-translated 27-kDa proteins contain the 3C-like proteinase activity of the coronavirus MHV-A59, *Virology* **222**: 375–382.

Lu, Y., Lu, X., and Denison, M. R., 1995, Identification and characterization of a serine-like proteinase of the murine coronavirus MHV-A59, *J. Virol.* **69**: 3554–3559.

Seybert, A., Ziebuhr, J., and Siddell, S. G., 1997, Expression and characterization of a recombinant murine coronavirus 3C-like proteinase, *J. Gen. Virol.* **78**: 71–75.

Tibbles, K. W., Brierley, I., Cavanagh, D., and Brown, T. D. K., 1996, Characterization in vitro of an autocatalytic processing activity associated with the predicted 3C-like proteinase domain of the coronavirus avian infectious bronchitis virus, *J. Virol.* **70**: 1923–1930.

Van Dinten, L. C., Wassenaar, A. L. M., Gorbalenya, A. E., Spaan, W. J. M., and Snijder, E. J., 1996, Processing of the equine arteritis virus replicase ORF1b protein: identification of cleavage products containing the putative viral polymerase and helicase domains, *J. Virol.* **70**: 6625–6633.

Ziebuhr, J., Herold, J., and Siddell, S. G., 1995, Characterization of a human coronavirus (strain 229E) 3C-like proteinase activity, *J. Virol.* **69**: 4331–4338.

Ziebuhr, J., Heusipp, G., and Siddell, S. G. (1997). Biosynthesis, purification, and characterization of the human coronavirus 229E 3C-like proteinase, *J. Virol.* **71**: 3992–3997.

16

PROCESSING OF THE MHV-A59 GENE 1 POLYPROTEIN BY THE 3C-LIKE PROTEINASE

M. R. Denison,[1,2,3] A. C. Sims,[2,3] C. A. Gibson,[2,3] and X. T. Lu[1,3]

[1]Department of Pediatrics
[2]Department of Microbiology and Immunology
[3]Elizabeth B. Lamb Center for Pediatric Research
Vanderbilt University Medical Center
Nashville, Tennessee

1. ABSTRACT

The 3C-like proteinase of mouse hepatitis virus (MHV-3CLpro) is predicted to cleave at least 10 sites in the gene 1 polyprotein, resulting in processing of proteinase, polymerase and helicase proteins from the polyprotein. We have used *E. coli* expressed recombinant 3CLpro (r3CLpro) to define cleavage sites in carboxy-terminal region of the ORF 1a polyprotein. Polypeptides containing one or more putative 3CLpro cleavage site were translated *in vitro* from subcloned regions of gene 1, and the polypeptides were incubated with r3CLpro. Analysis of the cleavage products confirmed several putative cleavage sites, as well as identifying cleavage sites not previously predicted by analysis of the MHV sequence. Antibodies directed against a portion of the ORF 1a polyprotein were used to probe virus infected cells, and detected proteins that correspond to the cleavage sites used by 3CLpro *in vitro*. These results suggest that MHV 3CLpro cleaves at least 7 sites in the ORF 1a polyprotein, and that the specificity of 3CLpro for cleavage site dipeptides may be broader than previously predicted.

2. INTRODUCTION

Gene 1 of MHV-A59 encodes a fusion polyprotein with a predicted mass of 803 kDa (Gorbalenya *et al.*, 1989; Lee *et al.*, 1991; Bonilla *et al.*, 1994). Expression of the entire gene 1 polyprotein requires translation of two open reading frames, ORFs 1a and 1b. Since these ORFs are in different reading frames, ORF 1b can only be expressed if a ribosomal frameshift occurs at the end of ORF 1a (Brierley, 1989). The ORF 1a portion of the gene 1 polyprotein encodes two experimentally confirmed proteinases, PLP-1 and 3CLpro, as well as an additional proteinase, PLP-2, for which no activity has yet been identified. The MHV 3CLpro has been shown to autoproteolytically liberate itself from the nascent polyprotein, both *in vitro* and in virus-infected cells (Lu *et al.*, 1995; Lu *et al.*, 1996). Multiple other cleavage sites have been predicted for 3CLpro, all of which are carboxy terminal to 3CLpro in the polyprotein (Lee *et al.*, 1991). At least five sites appear to be highly conserved among four sequenced coronaviruses.

We were interested in the region between the end of the 3CLpro encoding region and the end of ORF 1a (nt 11115–13589). This region is predicted to encode the predominantly hydrophobic MP-2 domain, as well as the remainder of the ORF 1a polyprotein with 4 predicted 3CLpro cleavage sites. We therefore undertook a study to define the cleavage sites in this portion of the ORF 1a polyprotein. In addition we sought to determine if any of the putative protein products could be identified in virus-infected cells.

3. METHODS

3.1. Cloning and Expression of Recombinant 3CLpro (r3CLpro)

Active MHV 3CLpro was cloned and expressed in *E. coli* as described in the accompanying paper. Briefly, the 3CLpro domain was expressed as a fusion protein with maltose binding protein (MBP-r3CLpro), partially purified on an amylose affinity column, and separated from MBP using factor Xa. The partially purified r3CLpro preparation was used for all experiments.

3.2. Cloning and Expression of Polypeptides Containing Putative 3CLpro Cleavage Sites

Subclones of the region of ORF 1a between nt 11991 and 13118 were constructed using PCR from a cDNA of the 3' 2.5 kb of the ORF 1a region of gene 1 (Figure 2). Left primers included an Nco 1 site and right primers an Xho 1 site, allowing for ligation into pET-23d (Novagen). This vector contains an optimized AUG at the insertion site, allowing for full-length expression of all constructs. The recombinant plasmids were used to program expression of polypeptides in an combined *in vitro* transcription/translation reticulocyte lysate system (TnT-Promega) in the presence of [^{35}S]methionine and [^{35}S]cysteine. Reactions were terminated by addition of SDS-PAGE sample buffer, or in the case of *trans* cleavage assays, by the addition of cycloheximide.

3.3. *Trans* Cleavage Assays

Radiolabeled polypeptides were incubated with elutate from the amylose affinity column containing r3CLpro, maltose binding protein and factor Xa at 30°C for 4 hours. Typically, 1–2 ul or *in vitro* translation reaction was incubated with 5 ul of r3CLpro containing solution. Reactions were terminated by addition of SDS-PAGE sample buffer and boiling for 5 minutes, prior to electophoresis on SDS- 5–18% gradient polyacrylamide gels.

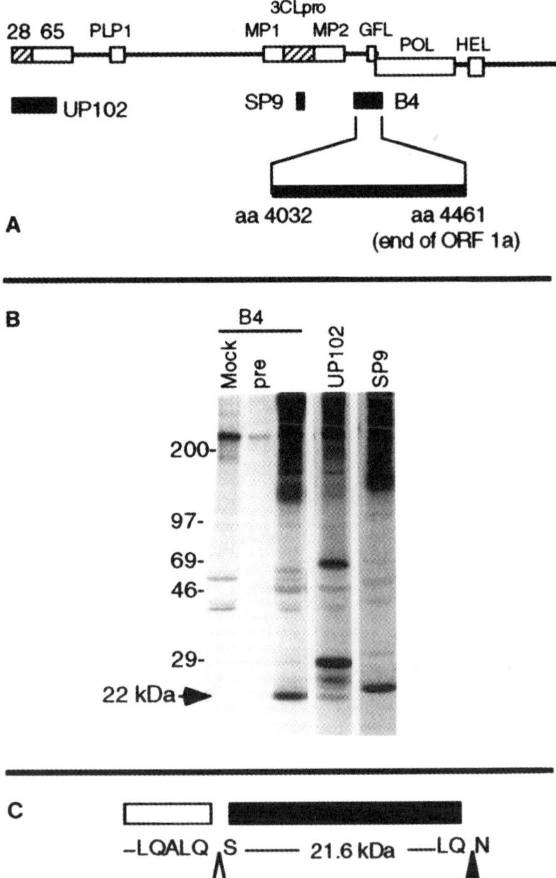

Figure 1. Identification of gene 1 proteins in MHV infected DBT cells. A. Scheme of gene 1 showing location of fusion proteins or peptides used to induce antisera UP102, SP9 and B4. The expression of the B4 fusion protein and antibody induction is described in the methods section. B. Immunoprecipitation of radiolabeled gene 1 proteins from virus infected cells. Mock infected (Mock) or infected cells were radiolabeled with [^{35}S]methionine. Lysates were immunoprecipitated with preimmune serum (pre) or immune sera (B4, UP102, or SP9). Molecular mass markers are to the left of the gel and the location of the B4-precipitated 22 kDa protein is noted. C. The location of the confirmed amino terminal cleavage site of the 22 kDa protein is shown (open arrowhead) and the putative carboxy terminal site is indicated (black arrowhead).

3.4. Antibody Production and Immunoprecipitation

The B4 protein was expressed in *E. coli* in the pQE-30 vector (Quiagen). the B4 subclone extended from the Xba 1 site at nt 12302 to the Kpn 1 site at 13901 (aa 4032 - 4461) (Figure 1). This clone contains the ORF 1a/ ORF 1b frameshift and potentially encodes 114 aa of ORF 1b; however the size of the *E. coli* expressed protein indicated that only the ORF 1a portion of this construct was translated. The 46 kDa protein was excised from SDS-polyacrylamide gels and used to induce polyclonal antibodies in rabbits. This antiserum was called B4. Antisera UP102 and SP9 have been previously described (Denison *et al.*, 1995; Lu *et al.*, 1996). Virus infections and immunoprecipitation experiments were performed as previously described (Denison *et al.*, 1992).

3.5. Amino Terminal Radiosequencing

The 22 kDa protein immunoprecipitated from MHV-infected DBT cells by the B4 antiserum was excised from the polyacrylamide gel following localization by autoradiography, and was transferred to a polyvinylidene difluoride (PVDF) membrane. Transfer of the protein and radiosequencing was performed on a ABI 470 sequencer as previously described (Lu *et al.*, 1995).

4. RESULTS

4.1. Detection of Gene 1 Proteins in MHV-Infected DBT Cells

The B4 antiserum was used to detect proteins at the carboxy terminus of ORF 1b in MHV-A59 infected DBT cells, along with the previously characterized UP102 and SP9 antisera (Figure 1). UP102 detected the amino terminal p28 and p65 proteins; SP9 precipitated 3CLpro (p27) from virus-infected cells. B4 detected a product of 22 kDa that was not detected by preimmune serum or in mock infected cells. Because of the extent of the B4 protein used to induce the B4 antiserum, it was not possible to define the amino and carboxy termini based on the antibody specificity alone.

Thus the [^{35}S]met labeled 22 kDa protein was immunoprecipitated, blotted to PVDF membrane, and used for radioterminal sequencing. The result indicated that the amino terminus of the B4 precipitated 22 kDa protein was cleaved at its amino terminus between Glutamine$_{4013}$ and Serine$_{4014}$ residues in the ORF 1a polyprotein. This Q_S dipeptide is within the sequence NTVLQALQ_SEFVN, which also contains the only directly adjacent possible 3CLpro cleavage sites in the gene 1 polyprotein, LQ_ALQ_S. The LQ_S is conserved in the gene 1 sequence of infectious bronchitis virus, transmissible gastroenteritis virus, and the human coronavirus 229E, whereas the LQA is not present in any of these sequences (Boursnell *et al.*, 1987; Herold *et al.*, 1993; Eleouet *et al.*, 1995).

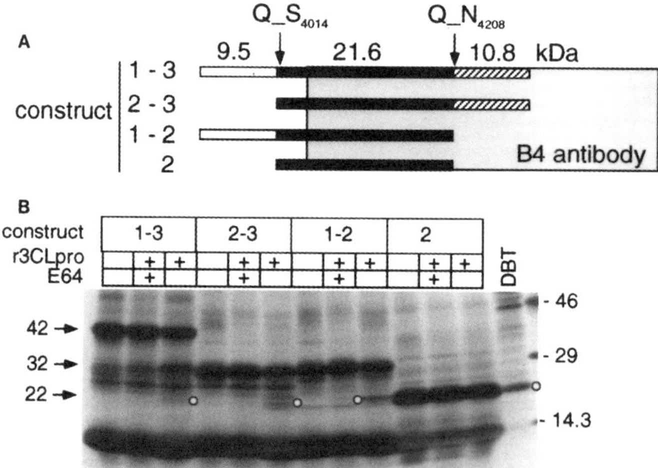

Figure 2. ORF 1a subclones, *in vitro* translation and 3CLpro *trans* cleavage. A. The ORF 1a subclones were constructed as described in Methods. All constructs had an optimal AUG at the 5' end in the construct. The constructs are labeled by the portions of ORF 1a potentially expressed: 1- 9.5 kDa fragment amino terminal to the LQ_S$_{4014}$; 2- 22 kDa protein precipitated by B4 antibody (location shown by grey box); 3 - 10.8 kDa fragment carboxy terminal to putative LQ_N$_{4208}$. B. *In vitro* translation and *trans* cleavage assay. The translation of the ORF 1a subclones is shown. Incubation of IVT products with recombinant 3CLpro (r3CLpro) and the cysteine proteinase inhibitor E64d is indicated by (+). Molecular mass markers are to the right of the gel and the location of specific translation products and cleavage products is indicated by numbers with arrows to the left of the gel. Location of 22 kDa proteins translated (construct 2) or cleaved are noted by clear circles. DBT indicates immunoprecipitation of proteins from MHV infected DBT cells by B4 antiserum.

4.2. Identification of a New 3CLpro Cleavage Site and Determination of the Carboxy Terminus of the 22 kDa Protein

Based on the apparent mass of the B4 precipitated protein and the determined amino terminus, we predicted that the 22 kDa protein must have its carboxy terminal cleavage at the sequence $VVLQ_{4207}\_N_{4208}NEL$. The LQNNE sequence is completely conserved in MHV, IBV, HCV 229E and TGEV, and the LQ_N has recently been shown to be a cleavage site for the IBV 3CLpro (Liu *et al.*, 1997). We therefore constructed a series of subclones of ORF 1b containing 1 or 2 cleavage sites in this region, in order to determine if cleavage in fact could occur *in trans* at the LQ_N peptide by recombinant 3Clpro (Figure 2).

The construct 2 contained the precise region between the amino-terminus of the B4-precipitated 22 kDa protein (Q_S_{4014}) and the predicted LQ_N_{4208} cleavage site. *In vitro* translation of this construct resulted in a 22 kDa protein that had the exact migration of the protein isolated from virus-infected cells. Incubation of this polypeptide with r3CLpro did not result in any cleavage of the protein. The construct 1–2 contained the 22 kDa protein, the amino-terminal LQ_S site and an amino terminal flanking fragment with a predicted mass of 9.5 kDa. Expression of this protein resulted in a dominant protein of 32 kDa, consistent with the total mass predicted for this polypeptide. Incubation with r3CLpro resulted in cleavage of a 22 kDa protein. Similarly, the 2–3 construct contained the 22 kDa protein, the putative carboxy terminal LQ_N cleavage site and a 10.8 kDa carboxy terminal fragment. *In vitro* translation of this construct also resulted in an approximately 32 kDa protein, and a 22 kDa cleavage product was seen after incubation with r3CLpro. Finally, the 1–3 construct containing both cleavage sites and amino and carboxy terminal fragments was translated to a 42 kDa precursor, and could be cleaved, albeit less efficiently, into 32 and 22 kDa products. All of the cleavages observed in this experiment could be blocked by the addition of the proteinase inhibitor E64. Thus the cleavages were specific, were mediated by 3CLpro, and occurred at the predicted sites.

5. DISCUSSION

We have previously shown that 3CLpro mediates autoproteolytic cleavage of the proteinase from the polyprotein (Lu *et al.*, 1995; Lu *et al.*, 1996; Lu and Denison, 1997), and similar results have been obtained for 3CLpro of HCV-229E and IBV (Ziebuhr *et al.*, 1995; Liu and Brown, 1995). In this study we have confirmed two additional cleavage sites, including one, LQ_N4208, that had not previously been predicted for MHV. We have also shown the product of cleavage seen *in vitro* is also a stable mature product of polyprotein processing in MHV infected cells. This suggests that we will be able to determine with accuracy the boundaries of mature proteins processed from the remainder of gene 1 polyprotein. It also suggests that additional sites may be used by the proteinase that do not conform to the predicted Q_ (S,A,G) consensus cleavage sites.

Based on the sites we have confirmed, along with the remaining predicted cleavage sites for r3CLpro, it is likely that up to 12 proteins are processed by 3CLpro during gene 1 translation. (Figure 3).

It is particularly intriguing that the region of the ORF 1a polyprotein between the MP-2 domain and the end of ORF 1a is likely to incorporate four mature proteins of 10, 22, 12, and 15 kDa. Only the putative 15 kDa protein has been predicted to have any similarity to known proteins, in this case a similarity to murine epidermal growth factor recep-

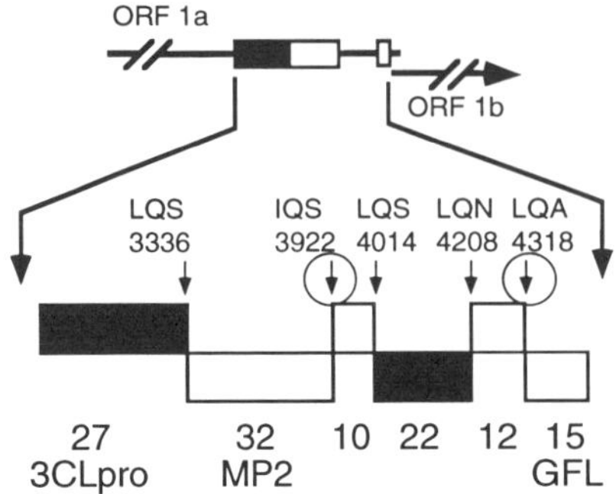

Figure 3. Model of 3CLpro ORF 1a polyprotein cleavage sites and protein products. A schematic of gene 1 with predicted or confirmed functional domains is shown at the top of the figure. The enlarged section shows the location of putative (black arrows) or confirmed (circled arrowheads) 3CLpro cleavage sites. Proteins that have been detected both *in vitro* and in virus-infected cells are shown as black boxes.

tor binding ligands (GFL). Thus it is likely that these proteins, including the confirmed 22 kDa protein, will mediate unique functions during MHV replication.

ACKNOWLEDGMENTS

This work was supported by Public Health Service grant AI-26603 from the National Institute of Allergy and Infectious Diseases. Protein sequencing was performed in the shared resource of the Vanderbilt University Cancer Center (IP30CA68485).

REFERENCES

Bonilla, P. J., Gorbalenya, A. E., and Weiss, S. R., 1994, Mouse hepatitis virus strain A59 RNA polymerase gene ORF 1a: heterogeneity among MHV strains, *Virology* **198**: 736–40.

Boursnell, M. F. G., Brown, T. D. K., Foulds, I. J., Green, P. F., Tomley, F. M., and Binns, M. M.,1987, Completion of the sequence of the genome of the coronavirus avian infectious bronchitis virus, *J. Gen. Virol.* **68**: 57–77.

Brierley, I., Digard, P., and Inglis, S. C., 1989, Characterization of an efficient coronavirus ribosomal framshift signal: requirement for an RNA pseudoknot, *Cell* **57**: 537–547.

Denison, M. R., Hughes, S. A., and Weiss, S. R., 1995, Identification and characterization of a 65-kDa protein processed from the gene 1 polyprotein of the murine coronavirus MHV-A59, *Virology* **207**: 316–20.

Denison, M. R., Zoltick, P. W., Hughes, S. A., Giangreco, B., Olson, A. L., Perlman, S., Leibowitz, J. L., and Weiss, S. R., 1992, Intracellular processing of the N-terminal ORF 1a proteins of the coronavirus MHV-A59 requires multiple proteolytic events, *Virology* **189**: 274–84.

Eleouet, J. F., Rasschaert, D., Lambert, P., Levy, L., Vende, P., and Laude, H., 1995, Complete sequence (20 kilobases) of the polyprotein-encoding gene 1 of transmissible gastroenteritis virus, *Virology* **206**: 817–22.

Gorbalenya, A. E., Koonin, E. V., Donchenko, A. P., and Blinov, V. M., 1989, Coronavirus genome: prediction of putative functional domains in the nonstructural polyprotein by comparative amino acid sequence analysis, *Nucleic Acids Res.* **17**: 4847–4861.

Herold, J., Raabe, T., Schelle, P. B., and Siddell, S. G., 1993, Nucleotide sequence of the human coronavirus 229E RNA polymerase locus, *Virology* **195**: 680–91.

Lee, H.-J., Shieh, C.-K., Gorbalenya, A. E., Koonin, E. V., LaMonica, N., Tuler, J., Bagdzhadhzyan, A., and Lai, M. M. C., 1991, The complete sequence (22 kilobases) of murine coronavirus gene 1 encoding the putative proteases and RNA polymerase, *Virology* **180**: 567–582.

Liu, D., Xu, H. and Brown, T., 1997, Proteolytic processing of the coronavirus infectious bronchitis virus 1a polyprotein: identification of a 10-kilodalton polypeptide and determination of its cleavage sites, *J Virol* **71**: 1814–1820.

Liu, D. X. and Brown, T. D. K., 1995, Characterisation and mutational analysis of an ORF 1a-encoding proteinase domain responsible for proteolytic processing of the infectious bronchitis virus 1a/1b polyprotein, *Virology* **209**: 420–427.

Lu, X., Lu, Y. and Denison, M. R., 1996, Intracellular and in vitro translated 27-kDa proteins contain the 3C-like proteinase activity of the coronavirus MHV-A59, *Virology* **222**: 375–382.

Lu, Y., Lu, X. and Denison, M. R., 1995, Identification and characterization of a serine-like proteinase of the murine coronavirus MHV-A59, *J Virol* **69**: 3554–9.

Lu, Y. Q. and Denison, M. R., 1997, Determinants of mouse hepatitis virus 3C-like proteinase activity, *Virology* **230**: 335–342.

Ziebuhr, J., Herold, J., and Siddell, S. G., 1995, Characterization of a human coronavirus (strain 229E) 3C-like proteinase activity, *J Virol* **69**: 4331–8.

17

EXPRESSION, PURIFICATION, AND ACTIVITY OF RECOMBINANT MHV-A59 3CLpro

A. C. Sims,[1,3] X. T. Lu,[2,3] and M. R. Denison[1,2,3]

[1]Department of Microbiology and Immunology
[2]Department of Pediatrics
[3]Elizabeth B. Lamb Center for Pediatric Research
Vanderbilt University Medical Center
Nashville, Tennessee

1. ABSTRACT

The 3C-like proteinase (3CLpro) of MHV-A59 is predicted to mediate the majority of proteolytic processing events within the gene 1 polyprotein. We have overexpressed 3CLpro in *E. coli* as a fusion protein with maltose binding protein (MBP). The MBP-3CLpro fusion protein was purified from contaminating *E. coli* proteins by amylose column chromatography, and r3CLpro was cleaved from the fusion protein by factor Xa. Recombinant 3CLpro (r3CLpro) was able to cleave a polypeptide substrate containing mutated inactive 3CLpro and portions of the flanking domains. R3CLpro cleaved substrate completely within 5 minutes and the activity of r3CLpro was sensitive to inhibition by serine and cysteine proteinase inhibitors; however, it was not inhibited by EDTA, suggesting that metal ions were not critical for 3CLpro activity.

2. INTRODUCTION

Expression of viral proteins mediating RNA transcription and replication is presumed to occur primarily if not exclusively from gene 1 of MHV. Translation of the 22 kb gene 1 has been demonstrated *in vitro* and *in cyto*, resulting in expression of a number of proteins (Denison and Perlman, 1986; Denison *et al.*, 1992). Gene 1 is composed of two overlapping open reading frames, ORF 1a and ORF 1b. Translation of ORF 1a is calcuated to result in expression of a polyprotein of 495 kDa (pp1a) and translation of ORFs 1a and 1b together via a ribosomal frameshift would result in expression of an 803 kDa polyprotein (pp1ab). The ORF 1a portion of the polyprotein contains two experimentally

Coronaviruses and Arteriviruses, edited by Enjuanes *et al.*
Plenum Press, New York, 1998

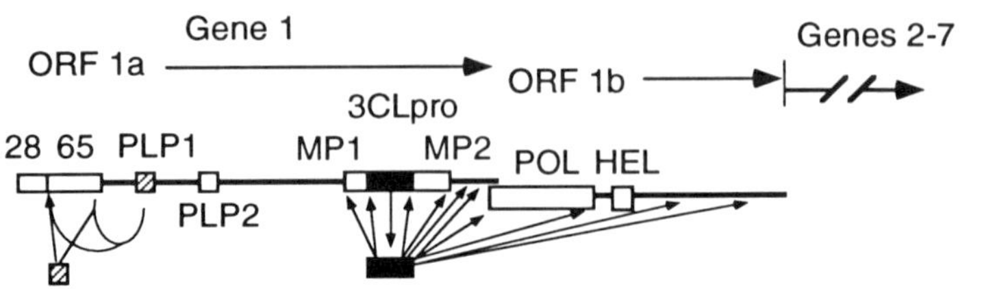

Figure 1. Location of proteinase domains and cleavage sites in the gene 1 polyprotein. This schematic shows the location of known (PLP-1, 3CLpro) and predicted (PLP-2) proteinases in the gene 1 polyprotein. Both PLP-1 and 3CLpro have been shown to be capable of acting *in trans*. Predicted cleavage sites for 3CLpro (Lee *et al.*, 1991) are indicated by arrows.

confirmed proteinase activities; a papain-like proteinase (PLP-1) and a 3C-like proteinase (3CLpro). PLP-1 is known to cleave the amino-terminal gene 1 proteins p28 and p65 (Dong and Baker, 1994; Hughes *et al.*, 1995). 3CLpro has been shown to cleave itself from the gene 1 polyprotein (Lu *et al.*, 1995). In addition 3CLpro is predicted to mediate all of the gene 1 polyprotein cleavages in pp1ab carboxy terminal to 3CLpro; these include processing of the putative polymerase and helicase proteins encoded in ORF 1b (Lee *et al.*, 1991) (Figure 1). Thus the activity of MHV 3CLpro is presumed to be critical to virus replication.

In this study we have cloned the precise domain of 3CLpro and have overexpressed the proteinase in *E. coli* as a fusion protein with maltose binding protein. Following purification and cleavage of r3CLpro, we have demonstrated its activity against cleavage sites in the ORF 1a polyprotein and have characterized the determinants of 3CLpro activity against *in vitro* translated substrate. Our results demonstrate that MHV 3CLpro can be expressed in *E. coli*, and that the proteinase has unique features of activity compared with other 3C or 3C-like proteinases.

3. METHODS

3.1. Cloning and Expression of Recombinant 3CLpro (r3CLpro) in *E. coli*

The region of MHV gene 1 encoding the functional 3CLpro was amplified by PCR from the existing clone pGpro (Lu *et al.*, 1995). The left primer began at nt 10209 and the right at nt 11112; neither primer had additional nucleotides. The amplified fragment was ligated into Xmn 1 digested pMAL-c2 (New England Biolabs) and the ligated plasmid was transformed into nm522 *E. coli* cells. The recombinant plasmid (pMAL MPB-r3Clpro) was transformed into DH5a cells for protein expression. Following induction, *E. coli* were lysed per manufacturer's instructions and the clarified lysate passed over an amylose column. The bound MBP-r3CLpro fusion protein was eluted from the column and cleaved with factor Xa (obtained from Paul Bock) at 0.1% for 12 hours at 4°C. Cleavage of MBP and r3CLpro was confirmed by visualization of proteins on SDS-PAGE and Coomassie Blue staining (Figure 2). The elutate containing separated MBP and r3CLpro was used for all experiments without additional purification.

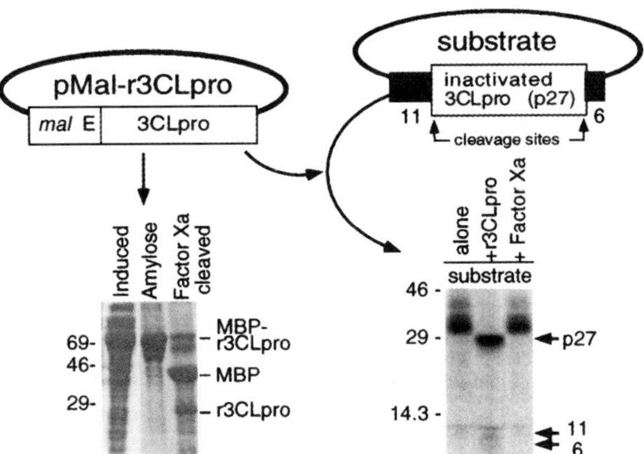

Figure 2. Expression of r3CLpro in E. coli and *trans* cleavage of substrate by r3CLpro. The pMAL MBP-r3CLpro construct is shown in the upper left. The gel below the construct shows results of expression in *E. Coli*. Induced lane shows total *E. coli* proteins following induction. The MBP-r3CLpro was purified on amylose resin (amylose) and then cleaved with factor Xa (factor Xa). The location of MBP-r3CLpro, MBP alone, and r3CLpro alone are shown. The substrate construct shows the organization and apparent molecular masses (see Methods). Gel shows electrophoresis of substrate (alone), substrate incubated with r3CLpro (+r3CLpro), and substrate incubated with factor Xa (+Factor Xa). Locations of cleavage products are shown to the right of the gel.

3.2. Cloning and Expression of Substrate for 3CLpro

The plasmid pGproH41Q was used to program *in vitro* translation in a combined transcription/translation reticulocyte lysate system, as previously described (Lu *et al.*, 1995). pGproH41Q expresses a polypeptide containing mutated inactive 3CLpro and 11 kDa and 6 kDa fragments of the flanking amino and carboxy terminal domains respectively (Figure 2). Although the protein expressed from pGproH41Q (hereafter referred to as "substrate") is unable to autoproteolytically cleave itself, the cleavage sites flanking the mutated proteinase are intact and can be cleaved in trans. Translation was performed in the presence of [^{35}S]methionine for 90 minutes.

3.3. *Trans* Cleavage Assays

Elutate containing r3CLpro, MBP and factor Xa was added directly to reticulocyte lysate reactions containing [^{35}S]met labeled substrate. Usually 2 ul of substrate was incubated with r3CLpro in a total volume of 10 ul. Incubations were performed at 30°C for the times indicated in each experiment. *Trans* cleavage experiments were terminated by addition of 2X Laemlli buffer or by snap freezing in dry ice/ethanol. Reaction products were separated on SDS- gradient polyacrylamide gels and analyzed by fluorography.

4. RESULTS

4.1. Cloning, Expression, and Activity of r3CLpro

The pMAL-r3CLpro plasmid directed expression of large amounts of soluble MBP-r3CLpro fusion protein in DH5a cells following induction with IPTG for 4 hours at 32°C

(Figure 2). Typically 20–50 mg/l of fusion protein was obtained under these conditions. Cleavage of MBP-r3CLpro by factor Xa was never complete, despite prolonged incubation at a variety of temperatures, even with the addition of fresh aliquots of factor Xa.

The substrate for 3CLpro was translated in a combined *in vitro* transcription/translation reticulocyte lysate system (TnT) and radiolabeled with [^{35}S]met. Radiolabeled substrate was incubated in the absence of proteinase (alone), in the presence of r3CLpro (+ r3CLpro) or in the presence of factor Xa alone for 1 hour at 30°C (Figure 2). Under these conditions r3CLpro completely processed the substrate to p27 and fragments of 11 and 6 kDa. These data demonstrated that 3CLpro was able to cleave *in trans* at its amino and carboxy termini. and that this process was rapid and complete.

4.2. Determinants of r3CLpro Activity *in Vitro*

We next defined the extent of r3CLpro activity against the *in vitro* translated substrate (Figure 3). Radiolabled substrate was incubated with excess r3CLpro at 30°C and samples were taken a various times from 0.5 to 240 minutes.

Since no intermediate cleavage products were identified, even at very early times, it appeared that r3CLpro was able to cleave at both cleavage sites concurrently. Substrate was 50% cleaved at the earliest time point examined, 0.5 min. indicating that proteinase/substrate interaction was quite rapid. Complete cleavage of substrate into the "mature" components of 11, 27 and 6 kDa occurred by 4.5 min. The rapidity and extent of cleavage was much greater than we have previously observed with 3CLpro expressed *in vitro* or purified from virus-infected cells.

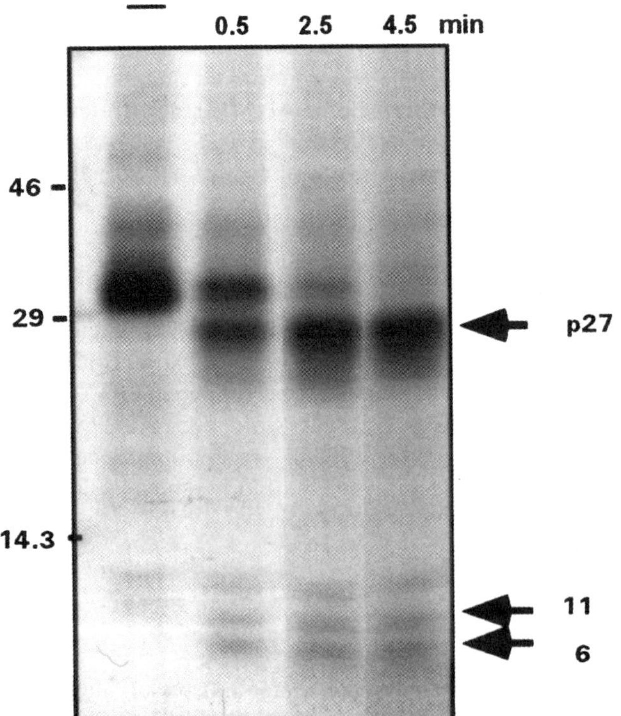

Figure 3. r3CLpro substrate cleavage. Substrate was incubated alone (-) or in the presence of r3CLpro (0.5–4.5 min.). Molecular mass markers are to the left of the gel and location of cleavage products are noted to the right of the gel.

Table 1. Inhibition r3CLpro activity

Inhibitor	Inhibitor class	Minimal inhibitory concentration (mM)	Inhibition
Leupeptin	Serine/ cysteine	0.2	+
PMSF	Serine	0.1	+
Zinc Chloride	Cysteine	3.0	+
E64	Cysteine	2.0	+
EDTA	Metallo/ Thiol		−
1,10 phenanthroline	Metallo		−

4.3. Inhibition of r3CLpro Activity

We examined the sensitivity of r3CLpro to inhibition by proteinase inhibitors (Table 1). Substrate was incubated with r3CLpro for 1 hr. at 30°C in the presence of several different classes of proteinase inhibitors. The serine proteinase inhibitor PMSF was the most active in blocking cleavage of substrate, but r3CLpro activity also was inhibited by leupeptin, albeit requiring high concentrations. $ZnCl_2$ has been shown to be a specific inhibitor of cysteine proteinases, and was able to inhibit r3CLpro in our assays.

The irreversible cysteine proteinase inhibitor E64 inhibited cleavage activity of r3CLpro, similar to results obtained in virus infected cells and during *in vitro* translation of 3CLpro containing precursors. EDTA and metalloproteinase inhibitor 1,10 phenanthroline did not affect activity, even at high concentrations (not shown) indicating that metal ions are not critical for activity. Together these results suggest some differences in the activity and specificity of MHV 3CLpro in comparison with other proteinases in this family.

5. DISCUSSION

In this report we have demonstrated that the active 3CLpro can be overexpressed in *E. coli* and that that partially purified proteinase is able to cleave substrate. A similar result has been reported for MHV-JHM (Seybert *et al.*, 1997). However, the 229E proteinase was tested against a small synthetic peptide, whereas we have demonstrated activity at both flanking cleavage sites in the polyprotein context. Our results are also consistent with those we previously reported for 3CLpro expressed *in vitro* or in virus-infected cells; specifically that the "mature" 3CLpro has no requirement for membranes. We were surprised at the efficiency of cleavage of substrate, since autoproteolytic liberation of mature proteinases in other virus systems is a regulated process in which precursors play a role in polyprotein processing as well as in other replication functions. However, our results may represent the final steps in activity of 3CLpro, with amplification of processing by mature 3CLpro. In order to address this more carefully, it will be necessary to express r3CLpro along with both flanking domains to assess the degree of cleavage. This may be a challenge in *E. coli*, because of the profoundly hydrophobic nature of these domains (Pinon *et al.*, 1997).

We have previously demonstrated that addition of the irreversible cysteine proteinase inhibitor E64 to virus infected cells results in rapid shutoff of MHV RNA synthesis and virus replication. Our demonstration that r3CLpro is also inhibited by E64 corroborates our previous results *in vitro* and *in cyto*. Other cysteine substituted chymotrypsin-like proteinases such as the 3Cpro of the picornaviruses are not inhibited by E64. Thus the sensitivity to E64 may represent a significant difference in MHV 3CLpro mechanism. The requirement for high concentrations of E64 and leupeptin may reflect the amount of proteinase used in our assays. Since we have only partially purified the enzyme, we did not quantitate precise amounts added to the assays. Kinetic analyses using highly purified r3CLpro and defined peptide substrates will be necessary to determine the inhibitor profile with more precision.

The partially purified r3CLpro is active at two defined cleavage sites. The availability of the recombinant enzyme will allow us to define other cleavage sites in the gene 1 polyprotein, and will also allow us to pursue structure function analyses and define the kinetics of this novel proteinase.

ACKNOWLEDGMENTS

This work was supported in part by NIH grant RO1-AI 26603 from the National Institutes of Allergy and Infectious Diseases.

REFERENCES

Denison, M. R. and Perlman, S., 1986, Translation and processing of mouse hepatitis virus virion RNA in a cell-free system, *J. Virol* **60**: 12–8.

Denison, M. R., Zoltick, P. W., Hughes, S. A., Giangreco, B., Olson, A. L., Perlman, S., Leibowitz, J. L., and Weiss, S. R., 1992, Intracellular processing of the N-terminal ORF 1a proteins of the coronavirus MHV-A59 requires multiple proteolytic events, *Virology* **189**: 274–84.

Dong, S. and Baker, S. C., 1994, Determinants of the p28 cleavage site recognized by the first papain-like cysteine proteinase of murine coronavirus, *Virology* **204**: 541–9.

Hughes, S. A., Bonilla, P. J., and Weiss, S. R., 1995, Identification of the murine coronavirus p28 cleavage site, *J. Virol* **69**: 809–13.

Lee, H.-J., Shieh, C.-K., Gorbalenya, A. E., Koonin, E. V., LaMonica, N., Tuler, J., Bagdzhadhzyan, A., and Lai, M. M. C., 1991, The complete sequence (22 kilobases) of murine coronavirus gene 1 encoding the putative proteases and RNA polymerase, *Virology* **180**: 567–582.

Lu, Y., Lu, X., and Denison, M. R., 1995, Identification and characterization of a serine-like proteinase of the murine coronavirus MHV-A59, *J. Virol* **69**: 3554–9.

Pinon, J., Mayreddy, R., Turner, J., Khan, F., Bonilla, P., and Weiss, S., 1997, Efficient autoproteolytic processing of the MHV-A59 3C-like proteinase from the flanking hydrophobic domains requires membranes, *Virology* **230**: 309–322.

Seybert, A., Ziebuhr, J. and Siddell, S. G., 1997, Expression and characterization of a recombinant murine coronavirus 3C-like proteinase, *J. Gen. Virol.*

18

MATURATION OF THE POLYMERASE POLYPROTEIN OF THE CORONAVIRUS MHV STRAIN JHM INVOLVES A CASCADE OF PROTEOLYTIC PROCESSING EVENTS

Jennifer J. Schiller and Susan C. Baker

Department of Microbiology and Immunology
Loyola University of Chicago
Stritch School of Medicine
2160 South First Ave., Bldg. 105
Maywood, Illinois 60153

1. ABSTRACT

The RNA polymerase gene of the murine coronavirus mouse hepatitis virus (MHV) encodes a polyprotein of greater than 750 kDa. The amino-terminal cleavage product of the MHV polymerase polyprotein, p28, has been shown to be cleaved from the polyprotein by the virus-encoded protease PCP-1. We aim to identify the MHV-JHM proteolytic products downstream of p28 and to determine which viral proteinase domains are responsible for generating each of them. To this end, we have generated antisera directed at specific MHV-JHM ORF1a regions and have used these antisera to identify six viral proteins, representing a large portion of ORF1a, from MHV-JHM-infected cells. These proteins include p28, p72, p65, p250, p210, and p27.

2. INTRODUCTION

Gene 1 at the 5' end of the genome of mouse hepatitis virus (MHV) encodes a large polyprotein precursor to the viral RNA polymerase complex. Gene 1 is 22 kilobases in length, contains two overlapping open reading frames (ORF1a and ORF1b), and is believed to encode a polyprotein of greater than 750 kDa (Lee *et al.*, 1991; Pachuk *et al.*, 1989). This polyprotein precursor is thought to be proteolytically processed by two viral papain-like cysteine proteinase domains, PCP-1 and PCP-2, and a poliovirus 3C-like pro-

Coronaviruses and Arteriviruses, edited by Enjuanes *et al.*
Plenum Press, New York, 1998

teinase domain, 3C-pro, to generate the mature protein products associated with virus replication (Gorbalenya *et al.*, 1989).The release of the amino terminal cleavage product, p28, is mediated by PCP-1 (Hughes *et al.*, 1995; Dong and Baker, 1994; Baker *et al.*, 1993, 1989). This cleavage event is detectable both in vitro and in vivo. In addition, a 65 kDa protein encoded immediately downstream in MHV strain A59 is also believed to be released by PCP-1 activity (Bonilla *et al.*, 1997; Denison *et al.*, 1995).

Our goal is to identify the cleavage products generated downstream of p28 from cells infected with the neurotropic strain of MHV, strain JHM. We have recently published the in vivo identification of p72 and p65, two proteins immediately downstream of p28, and documented their precursor and product relationship (Gao *et al.*, 1996). Here we provide a preliminary report of the in vivo detection of three additional gene 1 products representing a large portion of the polyprotein encoded by ORF1a. These three proteins were detected by immunoprecipitation of viral proteins from infected cell lysates using polyclonal rabbit antisera generated against fusion proteins of GST with various regions of ORF1a.

3. MATERIALS AND METHODS

3.1. Infection of Cells and Preparation of Lysates

Monolayers of 80–100% confluent DBT cells were infected with MHV-JHM-x at a multiplicity of infection of 5–10 plaque-forming units/cell, and lysates prepared as described by Gao *et al.* (1996). Briefly, proteins were metabolically labeled between 6.0–6.5 hr post-infection with [^{35}S]methionine at a final concentration of 100 uCi/ml. Cells were then washed and incubated in MEM media containing a 10-fold excess of methionine for the indicated periods of time. Cells were lysed in a buffer containing 4% SDS, 3% DTT, 40% glycerol and 0.0625 M Tris, pH 6.8, and proteins solubilized by multiple passages through a tuberculin syringe.

3.2. Generation of Rabbit Antisera for Immunoprecipitation

A panel of antisera directed against GST-fusion proteins representing multiple domains within the MHV-JHM ORF1a were generated. Each fusion protein domain was RT-PCR amplified from total RNA of MHV-JHM-x infected DBT cells and directionally cloned into the bacterial expression vector pGEX-KG, kindly provided by Dr. Steven Broyles (Purdue University, West Lafayette, IN). Fusion proteins were induced and isolated following the method of Guan and Dixon (1991), and injected into rabbits intramuscularly and subcutaneously. Polyclonal immune sera were collected 7–10d following the final (3rd or 4th) injection.

3.3. Immunoprecipitation and SDS-PAGE

Virus-encoded proteins generated during infection of DBT cells were identified by immunoprecipitation from whole cell lysates prepared as described above. Lysates from approximately 1×10^6 cells were immunoprecipitated with 5–10 ul of immune serum in a total volume of 1 ml RIPA buffer, as previously described (Gao *et al.*, 1996). Proteins were electrophoresed on 5–15% gradient SDS-PAGE gels and analyzed by fluorography.

4. RESULTS AND DISCUSSION

To identify MHV-JHM ORF1a gene products, a series of antisera directed against ORF1a-GST fusion proteins (FPs) were generated. The fusion proteins were generated by cloning ORF1a RT-PCR products into the vector pGEX-KG (Figure 1A). Constructs were transformed into DH5 alpha cells, and FP production was induced by addition of IPTG to growth media. Bacterial cells were lysed by sonication, GST-FPs were purified by affinity chromatography and used for rabbit immunizations (Figure 1B).

Three polyclonal antisera were generated: anti-D3, anti-D5 and anti-D12. The D3-FP encompasses amino acids (aa's) 784–1037 (nt 2563–3326) of the polymerase polyprotein within the p72/p65 domains as well as the region downstream. The D5-FP encompasses aa's 1212–1365 (nt 3847–4393), including the catalytic His residue of the PCP-1 domain, and the D12-FP encompasses aa's 3340–3470 (nt 10231–10246), covering the catalytic His residue of the 3C-pro domain (Figure 2A).

Temporal processing of the MHV-JHM ORF1a polyprotein was analyzed in a pulse-chase experiment. Specific viral proteins p72, p65, p250, p210 and p27 were detected using two of the antisera described above (Figure 2B). P250 and p210 were also detected in experiments using the anti-D5 serum (data not shown). Based on the electrophoretic mobilities of the proteins and locations of the FPs used to generate the antisera (as well as previously published results identifying p72 and p65 using the anti-647 serum [Gao *et al.*, 1996]), the predicted locations of the protein products are depicted schematically in Figure 2A. Proteins p28, p72, p250 and p27 are detected immediately following pulse labeling of cells, whereas proteins p65 and p210 represent proteolytic products of the larger precursors. Two additional predicted protein products, p7 and p40, have not yet been detected from MHV-JHM-infected cells. This experiment also confirms our previously pub-

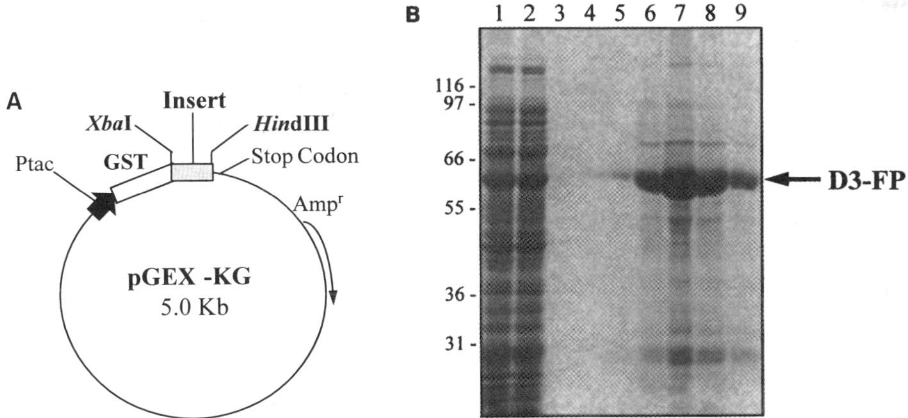

Figure 1. Production and purification of GST-fusion proteins. A) Schematic representation of the vector pGEX-KG provided by Dr. Steven Broyles (Purdue Univ., IN). RT-PCR products representing various domains within gene 1 were directionally cloned in frame with GST using *Xba*I and *Hind*III. B) Fusion proteins were isolated from sonicated bacterial lysates by affinity chromatography using glutathione-sepharose columns. Protein purity was assessed on Fast-Stained (Zoion Research, Newton, MA) SDS-10% polyacrylamide gels, as shown here for the D3-FP. Lane 1, bacterial lysates; lane 2, column flow-through; lanes 3–5, column washes; lanes 6–9, D3-FP eluted by competition with glutathione. Molecular weight markers are labeled in kilodaltons on the left of the gel. Quantitation of proteins was achieved using the Bradford method (Bio-Rad, Richmond, CA) prior to injection into rabbits (200–600 ug per injection).

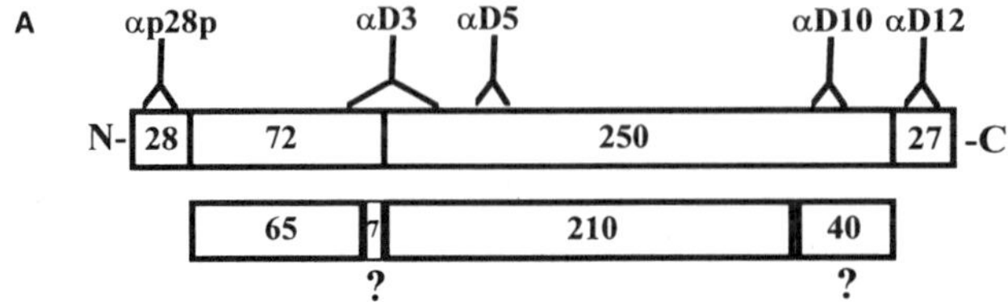

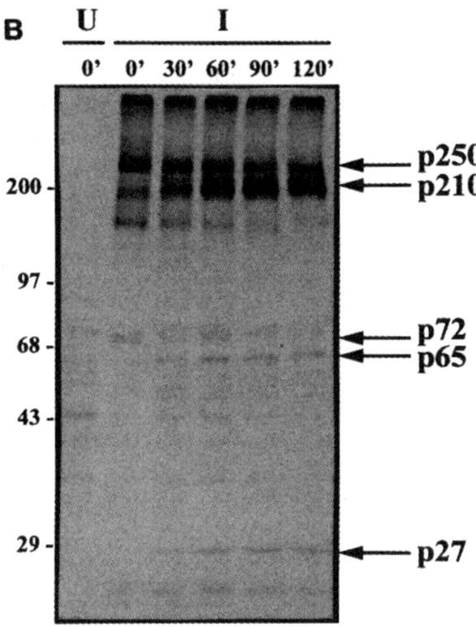

Figure 2. Synthesis and processing of MHV-JHM ORF1a polyproteins in vivo. A) The MHV-JHM ORF1a polyprotein precursor is processed into at least 8 protein products depicted here along with the antisera used to detect them. Putative protein products predicted based on the electrophoretic mobilities of other proteins are indicated by the (?). B) Temporal processing of the MHV-JHM ORF1a polyprotein. Pulse-chase immunoprecipitation analysis using a combination of the anti-D3 and anti-D12 antisera detects precursor-product relationships between p250-p210 and p72-p65, and an accumulation of p27 (3C-pro) during the chase. Molecular weight markers are labeled in kilodaltons on the left of the gel.

lished identification of p72 and p65 and their precursor-product relationship (Gao *et al*, 1996). Furthermore, p27 (representing 3C-pro) is processed even at the earliest time point in the chase, and accumulates for the duration of the experiment. This implies that p27 is cleaved rapidly from the precursor as it becomes available, and unlike the HCV-229E 3C-like pro domain (Ziebuhr *et al.*, 1995), it is quite stable in vivo. This result is also consistent with the generation of a 27 kDa 3C-pro by MHV-A59 (Lu *et al.*, 1995), which has activity in vivo and in vitro (Seybert *et al.*, 1997; Lu *et al.*, 1996).

Further immunoprecipitation analyses of gene 1 proteolytic processing with additional antisera, including the indicated anti-D10 serum (Figure 2A), are needed to identify all of the viral products associated with viral replication, as are functional assays to determine the role of each subunit in the unique coronavirus discontinuous transcription mechanism. Such experiments are currently in progress.

ACKNOWLEDGMENTS

We would like to thank Amornrat Kanjanahaluethai for her assistance in the production of the fusion protein D3, and John Zaryczny for his assistance with all rabbit injections and collection of sera.

REFERENCES

Baker, S.C., Shieh, C.K., Soe, L.H., Chang, M.F., Vannier, D.M., and Lai, M.M.C., 1989, Identification of a domain required for autoproteolytic cleavage of murine coronavirus gene A polyprotein, *J. Virol.* **63**:3693–3699.

Baker, S.C., Yokomori, K., Dong, S., Carlisle, R., Gorbalenya, A.E., Koonin, E.V., and Lai, M.M.C., 1993, Identification of the catalytic sites of a papain-like cysteine proteinase of murine coronavirus, *J. Virol.* **67**:6056–6063.

Bonilla, P.J., Hughes, S.A., and Weiss, S.R., 1997, Characterization of a second cleavage site and demonstration of activity in *trans* by the papain-like proteinase of the murine coronavirus mouse hepatitis virus strain A59, *J. Virol.* **71**:900–909.

Denison, M.R., Hughes, S.A., and Weiss, S.R., 1995, Identification and characterization of a 65-kDa protein processed from the gene 1 polyprotein of the murine coronavirus MHV-A59, *Virology* **207**:316–320.

Dong, S., and Baker, S.C., 1994, Determinants of the p28 cleavage site recognized by the first papain-like cysteine proteinase of murine coronavirus, *Virology* **204**:541–549.

Gao, H.-Q., Schiller, J.J., and Baker, S.C., 1996, Identification of the polymerase polyprotein products p72 and p65 of the murine coronavirus MHV-JHM, *Virus Research* **45**:101–109.

Gorbalenya, A.E., Koonin, E.V., Donchenko, A.P., and Blinov, V.M., 1989, Coronavirus genome: Prediction of putative functional domains in the non-structural polyprotein by comparative amino acid sequence analysis, *Nucleic Acids Res.* **17**:4847–4861.

Guan, K., and Dixon, J.E., 1991, Eukaryotic proteins expressed in Escherichia coli: an improved thrombin cleavage and purification procedure for fusion proteins with glutathione S-transferase, *Anal. Biochem.* **192**:262–267.

Hughes, S.A., Bonilla, P., and Weiss, S.R., 1995, Identification of the murine coronavirus p28 cleavage site, *J. Virol.* **69**:809–813.

Lee, H.-J., Shieh, C.-K., Gorbalenya, A.E., Koonin, E.V., La Monica, N., Tuler, J., Bagdzhadzhyan, A., and Lai, M.M.C., 1991, The complete sequence (22 kilobases) of murine coronavirus gene 1 encoding the putative proteases and RNA polymerase, *Virology* **180**:567–582.

Lu, Y., Lu, X., and Denison, M.R., 1995, Identification and characterization of a serine-like proteinase of the murine coronavirus MHV-A59, *J. Virol.* **69**:3554–3559.

Lu, X., Lu, Y., and Denison, M.R., 1996, Intracellular and *in vitro*-translated 27-kDa proteins contain the 3C-like proteinase activity of the coronavirus MHV-A59, *Virology* **222**:375–382.

Pachuk, C.J., Bredenbeek, P.J., Zoltick, P.W., Spaan, W.J.M., and Weiss, S.R., 1989, Molecular cloning of the gene encoding the putative polymerase of mouse hepatitis coronavirus, strain A59, *Virology* **171**:141–148.

Seybert, A., Ziebuhr, J., and Siddell, S.G., 1997, Expression and characterization of a recombinant murine coronavirus 3C-like proteinase, *J. Gen. Virol.* **78**:71–75.

Ziebuhr, J., Herold, J., and Siddell, S.G., 1995, Characterization of a human coronavirus (strain 229E) 3C-like proteinase activity, *J. Virol.* **69**:4331–4338.

19

CHARACTERIZATION OF A PAPAIN-LIKE CYSTEINE-PROTEINASE ENCODED BY GENE 1 OF THE HUMAN CORONAVIRUS HCV 229E

Jens Herold, Volker Thiel, and Stuart G. Siddell

Institute of Virology and Immunology
University of Würzburg
Versbacher Strasse 7
97078 Würzburg, Germany

1. ABSTRACT

Expression of the coronaviral gene 1 polyproteins, pp1a and pp1ab, involves a series of proteolytic events that are mediated by virus-encoded proteinases similar to cellular papain-like cysteine-proteinases and the 3C-like proteinases of picornaviruses. In this study, we have characterized, *in vitro,* the human coronavirus HCV 229E papain-like cysteine-proteinase PCP 1. We show that PCP 1 is able to mediate cleavage of an aminoterminal polypeptide, p9, from *in vitro* translation products representing the aminoproximal region of pp1a/pp1ab. Mutagenesis studies support the prediction of Cys1054 and His1278 as the catalytic amino acids of the HCV 229E PCP 1, since mutation of these residues abolishes the proteolytic activity of the enzyme.

2. INTRODUCTION

The HCV 229E genome is a single-stranded, positive sense RNA comprised of approximately 27000 nucleotides. Gene 1 encompasses the 5'-two thirds of the genome and encodes for proteins thought to be involved in viral replication and transcription (Herold et al., 1993). Gene 1 contains two large overlapping open reading frames, ORF 1a and ORF 1b, with the potential to encode for polypeptides of approximately 450 kDa and 300 kDa, respectively (Herold et al., 1993). In vitro studies suggest that the ORF 1b gene product is expressed as a fusion protein by a (-1) ribosomal frameshifting event mediated by an element located at the junction of ORF 1a and ORF 1b (Herold and Siddell, 1993).

Coronaviruses and Arteriviruses, edited by Enjuanes *et al.*
Plenum Press, New York, 1998

Computer-assisted analyses of the HCV 229E gene 1 sequence has revealed a number of putative functional domains within pp1a and pp1ab. These include an RNA-dependent RNA-polymerase domain and a metal-binding/helicase domain, both encoded in ORF 1b, as well as domains indicative of papain-like cysteine-proteinases and a chymotrypsin-like cysteine-proteinases (3C-like proteinases) encoded by ORF 1a (Herold et al., 1993).

Recently, we have shown that an autoproteolytic activity is encoded within the aminoterminal 1315 amino acids of the HCV 229E pp1a/pp1ab, a region that includes the predicted PCP 1 domain (Herold, 1995). The proteolytic activity could be blocked *in vitro* by $ZnCl_2$ but not by leupeptin.

In this study, we have produced a polyclonal rabbit antiserum specific for the amino acids 41 to 250 of HCV 229E pp1a/pp1ab. Using this antiserum, we have been able to show that the HCV 229E PCP 1 proteinase is responsible for the release of the aminoterminal polypeptide, p9, from pp1a/pp1ab and we have determined, by mutagenesis analysis, the putative catalytic amino acids of this proteinase.

3. MATERIALS AND METHODS

3.1. Preparation of Antigen and Antiserum

A 632 bp SphI/KpnI cDNA fragment, representing the nucleotides 412–1043 of the genomic RNA of HCV 229E, was excised from the plasmid J12E6 (Herold et al., 1993) and ligated with Sph/KpnI digested DNA of the bacterial expression vector plasmid QE30 (Diagen, Germany). The resulting plasmid, I1a.1, was transformed into competent JM109 bacteria and characterized by restriction enzyme analysis and sequencing of the cloning sites.

For bacterial expression of the recombinant polypeptide, plasmid I1a.1 was transformed into M15/pRep4 bacteria. The recombinant protein expressed from this plasmid comprises 230 amino acids: 12 amino acids at the aminoterminus that are encoded by the expression vector (including 6 consecutive histidines), 210 amino acids encoded by the HCV 229E RNA polymerase gene, corresponding to amino acids 41–250 of pp1a/pp1ab and two vector-derived amino acids at the carboxy terminus. Expression and purification of this fusion protein, pHis(41–250), and immunization of rabbits was done as described previously (Ziebuhr et al., 1995). The resulting pHis(41–250) protein-specific antiserum is referred to as IS1720.

3.2. Plasmids

The construction of plasmid Pap has been described (Herold et al., 1996). Briefly, a cDNA representing the nucleotides 224–4793 of the genomic RNA of HCV 229E was amplified by RT-PCR from poly(A) selected RNA from HCV 229E infected C16 cells and cloned into a BluescriptIISK+ vector. The nucleotide sequence of Pap was determined and several PCR-derived nucleotide misincorporations were identified. In the context of this study, the change of a lysine codon AAA (Lys-1316) to a termination codon TAA is relevant.

3.3. PCR

Pap DNA (1ng) was used as a template for four different PCR reactions using the Elongase amplification system (Life Technologies, USA) as recommended by the manufacturer. Oligonucleotide T7 (5'-AATACGACTCACTATAG-3') was the upstream primer, and the oligonucleotides I (5'-CACAAGTCACAGTGGTTGG-3'), II (5'-GTGCTGATTGAATAGTCTTAC-3'), III (5'-GTTAGTCTGGTAATGACCAC-3') and IV (5'-GCAAGGTTCTCATTAGCA-3') were used as downstream primers. The cycle conditions were: initial denaturation, 94 °C for 30 s; 30 cycles at 94 °C for 30 s, 50 °C for 30 s, 68 °C for 75 s/1 kb to be amplified; final elongation step 72 °C for 10 min. The PCR DNAs are referred to as PCR I to PCR IV and the encoded polypeptides are called, according to their length, pI-613, pII-956, pIII-1209 and pIV-1309.

3.4. *In Vitro* Transcription and Translation

RNAs were synthesized in vitro using the MEGAscript T7 Kit (Ambion, USA) and translated (100 ng per μl translation reaction) in a reticulocyte lysate in the presence of [^{35}S]-methionine as described earlier (Herold et al., 1996). Aliquots (2 μl) of the translation reaction were used for immunoprecipitation with IS1720 as described previously (Ziebuhr et al., 1995).

3.5. Site-Directed Mutagenesis

Site specific mutations were introduced into pPap by *in vivo*-recombination PCR as described (Herold and Siddell., 1993).

4. RESULTS

We have expressed an antigen, pHis(41–250), representing amino acids 41–250 of the HCV 229E pp1a/pp1ab, which allows for a fast, single-step purification of the recombinant protein by affinity chromatography on Ni-NTA-agarose columns (Figure 1). The polypeptide has been used to immunize rabbits, resulting in the antiserum IS1720.

A recombinant plasmid, Pap, containing a T7 RNA polymerase promotor followed by the coding sequence of the aminoterminal 1315 amino acids of pp1a/pp1ab was constructed. A series of PCR-DNAs (I-IV) were derived from pPap and subsequently transcribed and translated *in vitro*.. The translation products were then immunoprecipitated with IS1720. As expected, and as is shown in Figure 2, RNA derived from PCR I, PCR II and PCR III encoded for proteins with apparent molecular masses of 66 kDa (pI-613), 104 kDa (pII-956) and 133 kDa (pIII-1209), respectively. PCR IV encodes for the aminoterminal 1.309 amino acids of the HCV 229E pp1a/pp1ab and contains the predicted catalytic amino acids of the first papain-like cysteine-proteinase domain PCP 1. Upon translation of PCR IV-RNA a 9 kDa polypeptide was detected, together with a number of higher molecular mass polypeptides. These polypeptides include products with apparent molecular masses of 143 kDa, the expected size of the PCR IV-RNA primary translation product, and 134 kDa, the expected size of the processed product.

Our interpretation of the data shown in Fig. 2 is that PCP 1 recognizes the full length translation product of PCR IV-RNA, releasing p9 and a processed polypeptide of 134 KDa. Due to the position of pHis(41–250), relative to pp1a/pp1ab, we concluded that p9 represents the aminoterminal polypeptide of pp1a/pp1ab. Using this *in vitro* approach, we are also able to position the carboxyterminal border of the active PCP 1 domain between the amino acids 1.209 and 1.309 of pp1a/pp1ab.

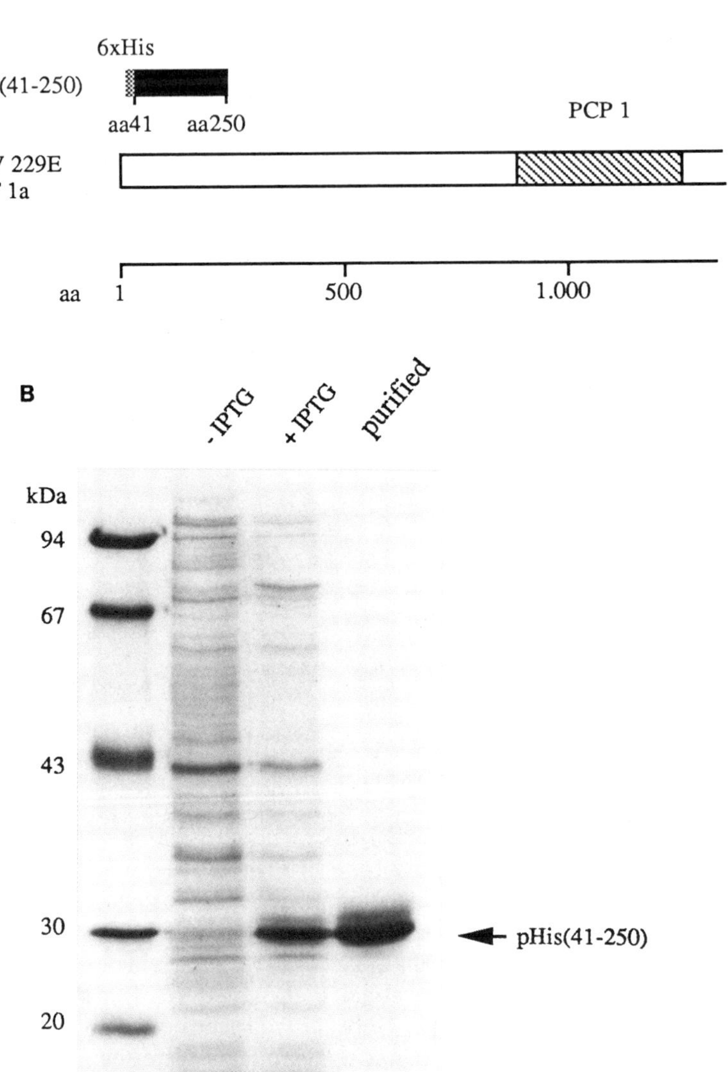

Figure 1. Expression and purification of pHis(41–250). (A) Schematic representation of pHis(41–250) in relation to the aminoproximal region of the HCV 229E pp1a/pp1ab. (B) Cell lysates from non-induced and IPTG-induced bacteria and purified pHis(41–250) were separated by SDS-PAGE in a 15% polyacrylamide gel.

In a further series of experiments, the codons for the cysteine residues Cys962 and Cys1054 and the histidine residues His1205 and His1278 have been mutated and the effect of these changes on proteolytic activity have been monitored (Figure 3). Upon transcription and translation of pPap-DNA, p9 and a processed form of the full length translation product (primary translation product 144 kDa, processed form 135 kDa) are detected . When Cys962 is changed to Gly (Cys962Gly) or His1278 is changed to Gly (His1278Gly) or Val (His1278Val), the proteolytic processing remained unaffected. In contrast, when Cys1054 was changed to Arg (Cys1054Arg), Gly (Cys1054Gly), or Ser (Cys1054Ser) and His1205 was changed to Ala (His1205Ala) or Gly (His1205Gly), the generation of p9 was

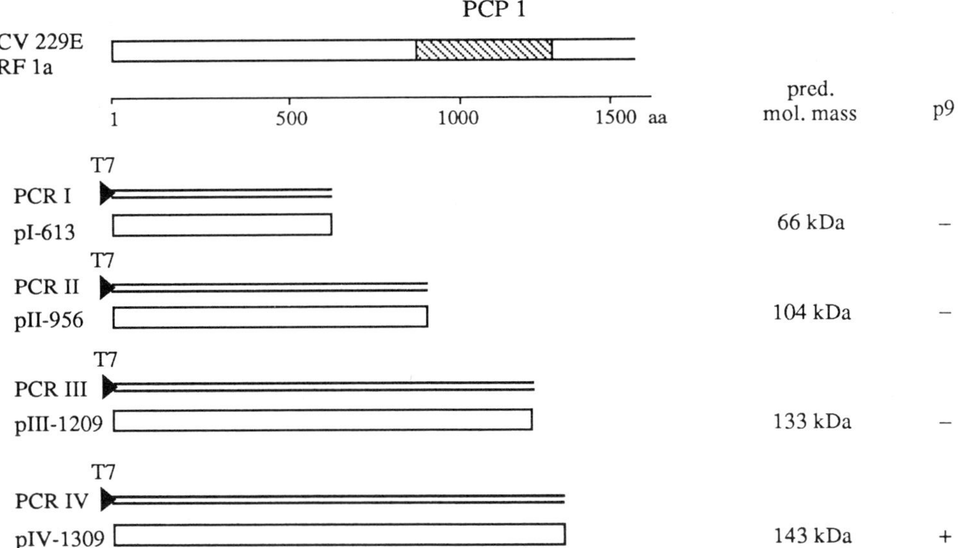

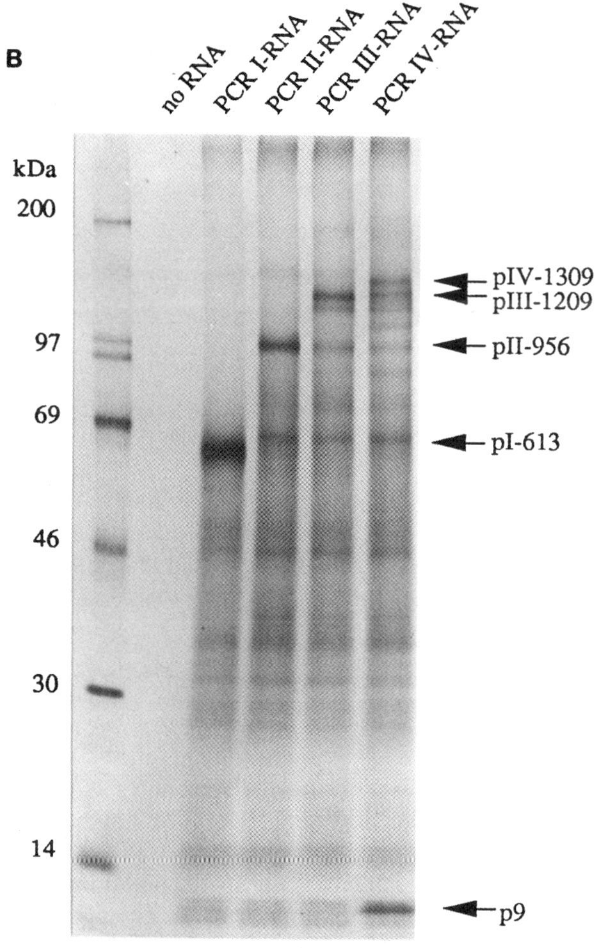

Figure 2. Mapping of the domain necessary for p9 processing. (A) Schematic representation of the HCV 229E ORF 1a, the PCR-DNAs and their predicted primary translation products. (B) In vitro transcription/translation of PCR-DNAs and immunoprecipitation of the translation products. The immunoprecipitated poylpeptides were separated by SDS-PAGE in 10–17.5%-polyacrylamide gels. Protein molecular weight markers are indicated.

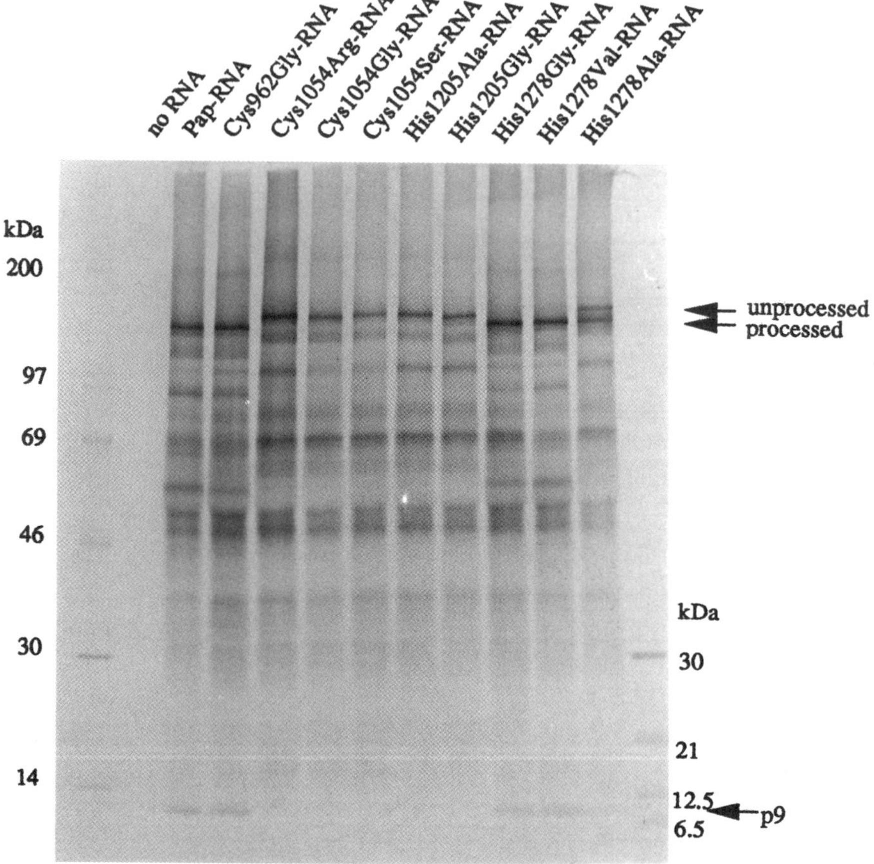

Figure 3. Mutagenic analysis of the HCV 229E-PCP 1. Pap RNA and RNAs derived from mutated DNAs were translated in reticulocyte lysates and the translation products were immunoprecipitated with IS1720. The proteins were analyzed by electrophoresis in a 10–17.5% SDS-polyacrylamide gradient gel. Protein molecular weight markers are indicated.

completely abolished. The change His1278Ala decreased the extent of proteolytic processing. These data support the previous prediction of Cys1054 and His1205 as the catalytic residues of HCV 229E PCP 1.

5. DISCUSSION

Proteolytic processing of non-structural gene products is thought to play a major role in the genesis of funtional replication complexes for many positive stranded RNA viruses (Dougherty and Semmler, 1993; Gorbalenya and Snijder, 1996). Although an adequate processing map of the coronaviral replicase polyproteins has not yet been established, considerable progress has been made in identifying the proteinases involved and their cleavage sites within the polyproteins. In this study, we have investigated proteolytic processing events within the aminoterminal part of the human coronavirus 229E pp1a/pp1ab.

Broadly, our results confirm a number of earlier studies that investigated the MHV PCP1 activity (Baker et al., 1989). It appears that the coronavirus PCP1 activity is responsible at least for the cleavage of a small polypeptide from the aminoterminus of pp1a/pp1ab. In the case of MHV this polypeptide is p28. In the case of HCV 229E it is p9. At the present time, however, there is no indication of the possible function of these proteins. Immunofluorescence assays of HCV 229E-infected cells using the IS1720 antiserum has revealed a punctated pattern of staining in the perinuclear region. This may suggest that at least one of these proteins recognized by this serum remains associated with the virus replication complex (Heusipp et al., 1997).

REFERENCES

Baker, S. C., Shieh, C. K., Soe, L. H., Chang, M. F., Vannier, D. N., and Lai, M. M. C., 1989, Identification of a domain required for autoproteolytic cleavage of murine coronavirus gene A polyprotein, *J. Virol.* **63**:3693–3699.

Dougherty, W. G., and Semler, B. L., 1993, Expression of virus-encoded proteinases: functional and strutural similarities with cellular enzymes, *Micribiol. Rev.* **57**:781–822.

Gorbalenya, A. E., and Snijder, E. J., 1996, Viral cysteine proteinases, *Perspectives in Drug Discovery and Design* **6**: 64–86.

Herold, J., and Siddell, S.G., 1993, An elaborated pseudoknot is required for high frequency frameshifting during translation of HCV 229E polymerase mRNA, *Nucleic Acids Res.* **21**:5838–5842.

Herold, J, 1995, Organization and Expression of RNA Polymerase Locus of the Human Coronavirus 229E, Ph.D. Thesis., University of Würzburg, Germany.

Herold, J., Siddell, S.G., and Ziebuhr, J., 1996, Characterization of coronavirus RNA polymerase gene products, *Methods Enzymol.* **275**:68–89.

Herold, J., Raabe, T., Schelle-Prinz, B., and Siddell, S.G., 1993a, Nucleotide sequence of the human coronavirus 229E RNA polymerase locus, *Virology* **195**:680–691.

Heusipp, G., Grötzinger, C., Herold, J., Siddell, S. G., and Ziebuhr, J., 1997, Identification and subcellular localization of a 41-kDa, polyprotein 1ab processing product in human coronavirus 229E-infected cells, *J. Gen. Virol.* in press.

Ziebuhr, J., Herold, J., and Siddell, S. G., 1995, Characterization of a human coronavirus (strain 229E) 3C-like proteinase activity, *J. Virol.* **69**:4331–4338.

20

PROTEOLYTIC PROCESSING OF THE POLYPROTEIN ENCODED BY ORF1b OF THE CORONAVIRUS INFECTIOUS BRONCHITIS VIRUS (IBV)

D. X. Liu,[1,2] S. Shen,[1] H. Y. Xu,[1] and T. D. K. Brown[2]

[1]Institute of Molecular Agrobiology
59A The Fleming
1 Science Park Drive
Singapore 118240
[2]Division of Virology
Department of Pathology
University of Cambridge
Tennis Court Road
Cambridge, CB2 1QP, United Kingdom

1. ABSTRACT

We present here evidence demonstrating that four previously predicted Q-S(G) cleavage sites, encoded by the IBV sequences from nucleotide 15129 to 15134, 16929 to 16934, 18492 to 18497, and 19506 to 19511, respectively, can be recognised and *trans*-cleaved by the 3C-like proteinase. Five mature products with sizes of approximately 100 kDa, 65 kDa, 63 kDa, 42 kDa and 35 kDa are released from the ORF1b polyprotein by the 3C-like proteinase-mediated cleavage at these positions. Meanwhile, expression of plasmids containing only the ORF1b region showed no autocleavage of the polyproptein encoded, suggesting that the 3C-like proteinase may be the sole proteinase involved in processing of the 1b polyprotein. These data may therefore represent a complete processing map of the polyprotein encoded by ORF1b of mRNA1.

2. INTRODUCTION

Three proteinase domains, including two overlapping papain-like proteinase domains encoded by the IBV sequences from nucleotide 4242 to 5553 and one trypsin-like proteinase domain of the picornavirus 3C proteinase group (3C-like proteinase) encoded between nucleotides 8937 and 9357, have been predicted to be encoded by ORF 1a (Gorbalenya *et al.*, 1989; Lee *et al.*, 1991). The papain-like proteinase domains were demonstrated to be involved in proteolytic processing of the 1a polyprotein to an 87 kDa mature viral product (Liu *et al.*, 1995). However, it is likely that only a limited number of cleavages may be mediated by this proteinase. More important role in proteolytic processing of the mRNA1-encoding polyprotein may be played by the 3C-like proteinase domain, as more than ten Q-S(G) dipeptide bonds located in the 1a/1b polyproteins are predicted to be the cleavage sites for this proteinase (Gorbalenya *et al.*, 1989). Indeed, we have reported the identification of a 100 kDa polymerase domain-containing protein, released from the 1a/1b polyprotein by the 3C-like proteinase-mediated cleavage at two predicted Q-S dipeptide bonds (Liu *et al.*, 1994; Liu and Brown, 1995). More recently, a 10 kDa mature viral product encoded by ORF 1a from nucleotide 11545 to 11878 was shown to be cleaved from the 1a polyprotein by the same proteinase (Liu *et al.*, 1997).

In this communication, we report experiments designed to map the 3C-like protienase-dependent cleavage sites in the polyprotein encoded by the ORF 1b region of mRNA1. Three ORF 1b-specific antisera, V58, V60 and V17 were used in the study. Cotransfection, deletion and site-directed mutagenesis studies demonstrated that four previously predicted Q-S(G) cleavage sites, encoded by the IBV sequences from nucleotide 15129 to 15134, 16929 to 16934, 18492 to 18497 and 19506 to 19511, respectively, were recognised and cleaved by the 3C-like proteinase to release five mature products with sizes of approximately 100 kDa, 65 kDa, 63 kDa, 42 kDa and 35 kDa. Information gained from this study could be used to guide further identification and characterisation of the ORF 1b-specific viral products.

3. MATERIALS AND METHODS

3.1. Transient Expression of IBV Sequences in Vero Cells Using a Vaccinia-T7 Expression System

Open reading frames placed under control of the T7 promoter were expressed transiently in eukaryotic cells using the vaccinia virus-T7 expression system (Fuerst *et al.*, 1986).

3.2. Polymerase Chain Reaction (PCR)

Appropriate primers and template DNAs were used in amplification reactions with Pfu DNA polymerase (Stratagene) under standard buffer conditions with 2 mM $MgCl_2$. PCR reaction conditions were 30 cycles of 95°C for 45 seconds, X$^\circ$C for 45 seconds and 72°C for X minutes. The annealing temperature and the extension time were adjusted according to the melting temperature of the primers used and the length of the PCR fragments synthesised.

3.3. Site-Directed Mutagenesis

Site-directed mutagenesis was carried out, as previously described (Liu *et al.*, 1994), using single stranded DNA templates prepared from pIBV1b5 and appropriate oligonucleotide primers.

Substitution mutation of the $Q^{1491(1b)}$-$G^{1492(1b)}$ dipeptide bond was introduced, as previously described (Liu *et al.*, 1997), by two rounds of PCR with two pairs of primers.

3.4. Radioimmunoprecipitation

Plasmid DNA-transfected Vero cells were lysed with RIPA buffer (50 mM Tris HCl, pH 7.5, 150 mM NaCl, 1% sodium deoxycholate, 0.1% SDS) and pre-cleared by centrifugation at 12,000 rpm for 5 minutes at 4°C in a microfuge. Immunoprecipitation was carried out as described previously (Liu *et al.*, 1994).

3.5. SDS-Polyacrylamide Gel Electrophoresis

SDS-polyacrylamide gel electrophoresis (SDS-PAGE) of virus polypeptides was carried out with 10% polyacrylamide gels (Laemmli, 1970). Labelled polypeptides were detected by autoradiography or fluorography of dried gels.

3.6. Construction of Plasmids

Plasmid pIBV1b3, which covers the IBV sequence from nucleotide 15135 to 16931, was constructed by cloning an NcoI/BamHI digested-PCR fragment into NcoI/BamHI digested pKT0 (Liu *et al.*, 1994). An artificial AUG initiation codon in an optimal context (ACC**AUG**G) located immediately upstream of the viral sequence and a UAG termination codon located immediately downstream of the viral sequence were meanwhile introduced by PCR. The sequence of the upstream PCR primer is 5'-CGACTTCC**ATG** GCTTGTGGCGTT-3', and the sequence of the downstream primer is 5'-CCAAAG-GATC**CTA**TTGCAGACTTG-3'. Plasmid pIBV1b4 was constructed by cloning an SnaBI/DraI fragment containing the IBV sequence from nucleotide 18930 to 20874 into EcoRV/SmaI digested pKT0, and selected by nucleotide sequencing. Plasmid pIBV1b5 was made by cloning an NcoI/BamHI digested PCR fragment, which covers the IBV sequence from nucleotide 16932 to 20490, into NcoI-BamHI digested pKT0. An artificial AUG initiation codon in an optimal context (ACC**AUG**G) located immediately upstream of the viral sequence was introduced by PCR. The sequence of the upstream PCR primer is 5'-ACAAGTCC**ATG**GGTACAGGTT-3', and the sequence of the downstream primer is 5'-GCACCCCCGGGATCCTGCCAAC-3'.

Plasmids pIBV3C and pIBV5, which cover the IBV sequence from nucleotide 8871 to 9786 and from 10752 to 16980, respectively, were described previously (Liu *et al.*, 1994; Liu *et al.*, 1997).

Plasmid pIBV20 was made by cloning a 3,665 bp PCR fragment, which contains the IBV sequence from nucleotide 16841 to 20506, into EspI- and BamHI-digested pIBV1b3 (EspI cuts the IBV sequence at nucleotide position 16840, the BamHI site was located immediately downstream of the IBV sequence and was introduced by PCR with primer 5'-GCACCCCCGGGATCCTGCCAAC-3'.

Plasmid pIBV14Æ1Q$^{891(1b)}$-E, which covers the IBV sequences from nucleotide 8693 to 16980 with deletions of the regions from nucleotide 9911 to 12227 and from 15537 to 16788, respectively, and contains a Q$^{891(1b)}$ to E mutation, was described previously (Liu and Brown, 1995). Plasmid pIBV21, which covers the IBV sequences from nucleotide 8693 to 17630 with the two deletions in pIBV14Æ1Q$^{891(1b)}$-E and contains substitution mutations at both the $Q^{891(1b)}$-$S^{892(1b)}$ and $Q^{1491(1b)}$-$S^{1492(1b)}$ sites, was constructed as follows. A 1,242 bp PCR fragment, which covers the IBV sequence from nucleotide 16788 to 17630 and contains a $Q^{1491(1b)}$ to $M^{1491(1b)}$ mutation, was generated by two rounds of PCR. The sequence of the oligonucleotide primer used to introduce the mutation is 5'-ACAAGT**CCCATG**GGTACAGGTT-3', and the sequence of the complementary primer is 5'-AACAAACCTGTACCCATGGGAC-3'. This PCR fragment was then digested with PstI (which cuts the IBV sequence at nucleotide 15537), gel-purified and ligated into PstI-SmaI digested pIBV14Æ1Q$^{891(1b)}$-E, giving plasmid pIBV21. The construct was selected by restriction digestion with NcoI and confirmed by nucleotide sequencing.

Two mutant constructs, pIBV1b5Q$^{2350(1b)}$-E and pIBV1b5Q$^{2012(1b)}$-E, with alterations at the putative $Q^{2350(1b)}$-$S^{2351(1b)}$ and $Q^{2012(1b)}$-$S^{2013(1b)}$ cleavage sites, respectively, were made by site-directed mutagenesis with the single stranded DNA templates prepared from plasmid pIBV1b5. The sequences of the oligonucleotide primers used to introduce the mutations are: 5'-TCCACAGCTT**GAA**TCAGCATG-3' and 5'-TTCAGCTCT**CGAG**TCTATCGAC-3', respectively.

4. RESULTS

4.1. Characterisation of Two Antisera Raised against Bacterial-Viral Fusion Proteins Containing the C-Terminal Regions of the ORF 1b-Encoding Polyprotein

We have recently reported the identification of a 100 kDa protein encoded by the IBV sequence from nucleotide 12313 to 15131 in IBV-infected Vero cells with an ORF 1b-specific antiserum V58 (Liu *et al.*, 1994; Liu and Brown, 1995). No viral products, however, were detected by using two other ORF 1b-specific antisera, V60 and V17 (Liu *et al.*, 1994; Liu, unpublished observations). Antisera V60 and V17 were raised in rabbits against β-galactosidase-IBV fusion proteins containing the IBV sequences encoded from nucleotide 16066 to 16783 and from 19154 to 20649, respectively (Figure 1a). The specificity and affinity of these two antisera for their target proteins were tested by immunoprecipitation against *in vitro* synthesised ORF 1b products.

As the results shown in Figure 1b, expression of plasmid pIBV1b3, which contains the IBV sequence from nucleotide 15132 to 16932, led to the synthesis of a polypeptide of approximately 65 kDa, consistent with the calculated molecular weight of 69 kDa of the full-length product encoded by this construct. Immunoprecipitation study showed that the 65 kDa protein can be efficiently and specifically precipitated by antiserum V60 (Figure 1b).

To characterise antiserum V17, plasmid pIBV1b5, which contains IBV sequence from nucleotide 16930 to 20874, was linearised at different positions by restriction enzymes (Figure 1a), and transcribed and translated in the TnT system. As shown in Figure 1c, transcription and translation of SacI-digested pIBV1b5 led to the synthesis of a polyprotein of approximately 140 kDa, representing the full-length product encoded by this construct. Transcription and translation of BstBI-, MboII- and StuI-digested pIBV1b5 resulted in the synthesis of proteins of approximately 110 kDa, 98 kDa and 85 kDa, respectively (Figure 1c). Immunoprecipitation studies showed that the 140 kDa full-length product as well as the 110 kDa and 98 kDa proteins could be efficiently precipitated by antiserum V17 (Figure 1c). However, the 85 kDa protein could not be precipitated by the same antiserum. These results confirm the specific recognition of the target products by the antiserum.

4.2. Transient Expression of ORF 1b in Eukaryotic Cells

Plasmids covering the ORF1b region were then expressed in Vero cells by using the vaccinia virus-T7 expression system, and analysed by immunoprecipitation with three ORF-1b-specific antisera. As can be seen, expression of plasmid pIBV5, which contains the IBV sequence from nucleotide 10752 to 16980, resulted in the detection of two major protein species with apparent molecular weights of approximately 60 kDa and 200 kDa, respectively, representing the ORF 1a termination product and the full-length product encoded by the construct (also see Liu *et al.*, 1994). The 200 kDa full-length protein can be immunoprecipitated by both antisera V58 and V60 (Figure 2). No 100 kDa and any other cleavage products were detected (Figure 2). Similarly, expression of pIBV20, which covers the IBV sequence from nucleotide 15135 to 20506, led to the detection of a product with an apparent molecular weight of approximately 190 kDa, representing the full-length polyprotein encoded by this construct. This 190 kDa protein can be immunoprecipited by both antisera V60 and V17 (Figure 2). No other products were detected by either antiserum (Figure 2), confirming that there is no autoprocessing of the polyprotein encoded by this region of ORF 1b.

4.3. Determination of the C-Terminal Cleavage Site of the Putative Metal-Binding and RNA-Helicase Protein

Data presented above suggest that there is no autocleavage of the polyprotein encoded by ORF1b. In a previous report, we confirmed that a previously predicted QS ($Q^{891(1b)}S^{892(1b)}$) dipeptide bond is the C-terminal cleavage site of the 100 kDa protein (Liu and Brown, 1995). Three other QS(G) dipeptide bonds in the 1b polyprotein were also predicted to be the cleavage sites of the 3C proteinase (Gorbalenya *et al.*, 1989). The first of these sites is the $Q^{1491(1b)}G^{1492(1b)}$ dipeptide bond. Substitution of the $Q^{1491(1b)}$ residue with an M was therefore carried out to test if this is a scissile bond of the 3C-like proteinase. As shown in Figure 3, expression of pIBV14 1$Q^{891(1b)}$-E led to the detection of a protein of 125 kDa (see also Liu and Brown, 1995). Expression of pIBV21, however, resulted in the detection of a protein migrating in SDS-PAGE much more slowly than the 125 kDa protein species (Figure 3). The apparent molecular weight of approximately 155 kDa of this product suggested that it is a fusion protein comprising the 125 kDa protein and the product encoded by the IBV sequence from nucleotide 16932 to 17630. This result indicates that mutation of the $Q^{1491(1b)}$ residue to an M blocked the cleavage occurred at this position.

4.4. Analyses of the Putative Q^{2350}-S^{2351} and Q^{2012}-S^{2013} Cleavage Sites

During the course of studying the ORF1b expression, we have consistently observed a very low level of expression of constructs covering the 3C-like proteinase domain and the complete ORF1b sequence. In order to analyse the two putative cleavage sites (Q^{2350}-S^{2351} and Q^{2012}-S^{2013}) located in the C-terminal one third region of the 1b polyprotein, co-transfection approaches were used. As shown in Figure 4a, expression of plasmid pIBV1b5, which covers the IBV sequences from nucleotide 16932 to 20490, led to the synthesis of a 140 kDa polyprotein, representing the full-length product encoded by this construct. Once again, no processing of this polyprotein to smaller products were observed (Figure 4a). However, co-transfection of pIBV1b5 with pIBV3C, a plasmid containing the whole 3C-like proteinase domain, resulted in the detection of five protein species. In addition to the 140 kDa full-length product, four other protein species with apparent molecular weights of approximately 105 kDa, 77 kDa, 42 kDa and 35 kDa, respectively, were also detected (Figure 4a). To reveal the identities of these products, plasmid pIBV1b4, which contains the IBV sequence from nucleotide 18930 to 20874, was expressed in Vero cells. As shown in Figure 4a, transfection of pIBV1b4 alone led to the synthesis of a polypeptide of approximately 60 kDa, consistent with the calculated full-length product of 60 kDa

A

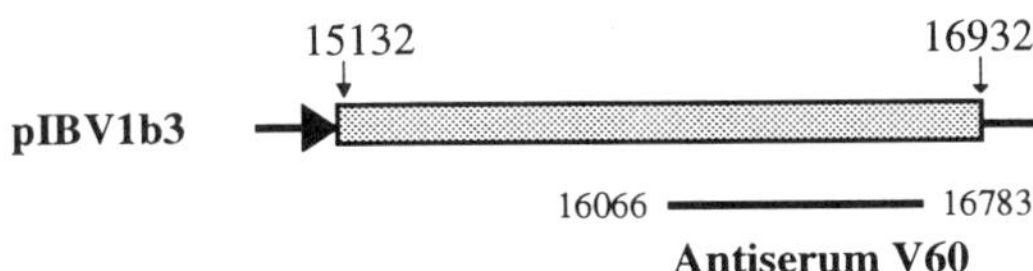

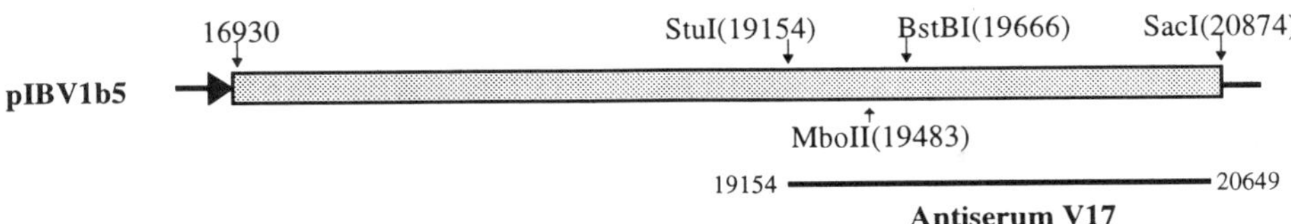

B

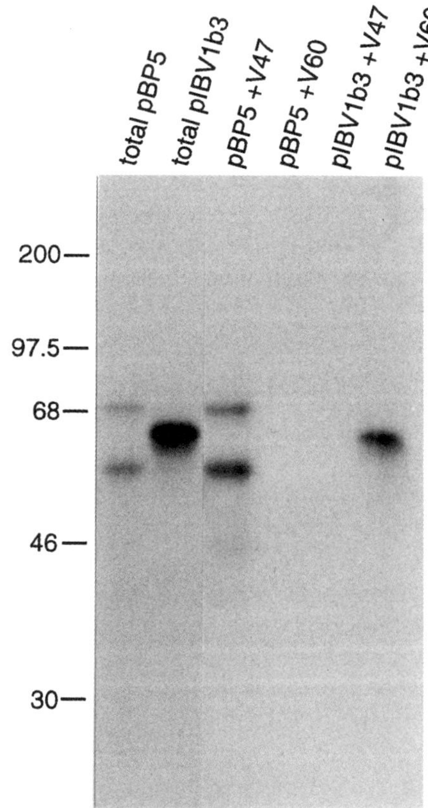

Figure 1. (A). Diagram showing IBV sequences present in plasmids pIBV1b3 and pIBV1b5, and the sequences used to raise antisera V17 and V60. Also shown are the restriction sites used to linearise pIBV1b5. (B). Testing of the reactivities of antiserum V60 to the *in vitro* synthesised target polypeptides. The polypeptides were transcribed and translated from pBP5 and pIBV1b3 in reticulocyte lysate with the TnT coupled translation system (Promega). Plasmid DNA was added to reticulocyte lysate at approximately 200 µg/ml. [^{35}S] methionine-labelled translation products were separated on an SDS-12.5% polyacrylamide gel directly or after immunoprecipitation with either antiserum V47, which recognises the IBV sequence encoded between nucleotides 11488 and 12600 (Liu *et al.*, 1994), or antiserum V60, and detected by fluorography. Numbers indicate molecular mass in kilodaltons. Plasmid pBP5 covers the IBV sequence from nucleotide 10752 to 12600 (Liu *et al.*, 1994). (C). Testing of the reactivities of antiserum V17 to the *in vitro* synthesised target polypeptides. The polypeptides were transcribed and translated from SacI-, BstBI-, MboII- and StuI-digested pIBV1b5 in reticulocyte lysate with the TnT translation system, and were separated on an SDS-12.5% polyacrylamide gel directly or after immunoprecipitation with antiserum V17.

C

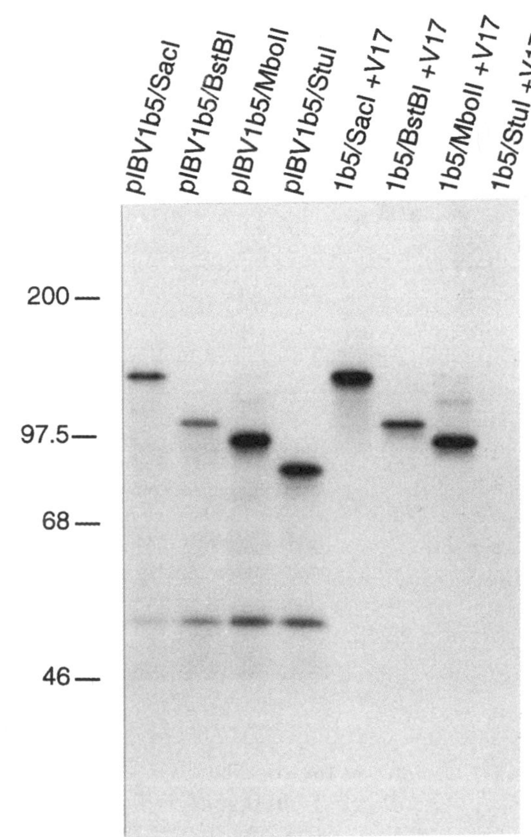

Figure 1C.

encoded by this construct. As expected, two additional products with apparent molecular weights of 25 kDa and 35 kDa, respectively, were produced from co-expression of pIBV1b4 with pIBV3C (Figure 4a). The 35 kDa protein co-migrates with the 35 kDa protein produced from the co-expression of pIBV1b5 and pIBV3C (Figure 4a), indicating that they are the same cleavage products from the C-terminal region of the 1b polyprotein.

To confirm further if cleavage did occur at the Q^{2350}-S^{2351} cleavage site, substitution of the Q^{2350} residue with an E was produced by site-directed mutagenesis using the DNA template prepared from pIBV1b5, giving a mutant construct pIBV1b5$Q^{2350(1b)}$-E. Expression of pIBV1b5$Q^{2350(1b)}$-E in Vero cells resulted in the synthesis, once again, of the full-length 140 kDa polyprotein (Figure 4a). Co-expression of pIBV1b5$Q^{2350(1b)}$-E and pIBV3C led to the formation of a polypeptide co-migrating with the 77 kDa protein detected from co-transfection of pIBV1b5 and pIBV3C (Figure 4a). No other cleavage products were detected (Figure 4a), suggesting that mutation of the Q^{2350} residue to E did block the cleavage occurred at this position. These results confirm that the Q^{2350}-S^{2351} dipeptide bond is responsible for releasing the N-terminus of the 35 kDa and the C-terminus of the 42 kDa proteins. The 77 kDa protein detected from co-expression of pIBV1b5 and pIBV3C is therefore representing an intermediate cleavage product encoded by the IBV sequence from nucleotide 18496 to 20414.

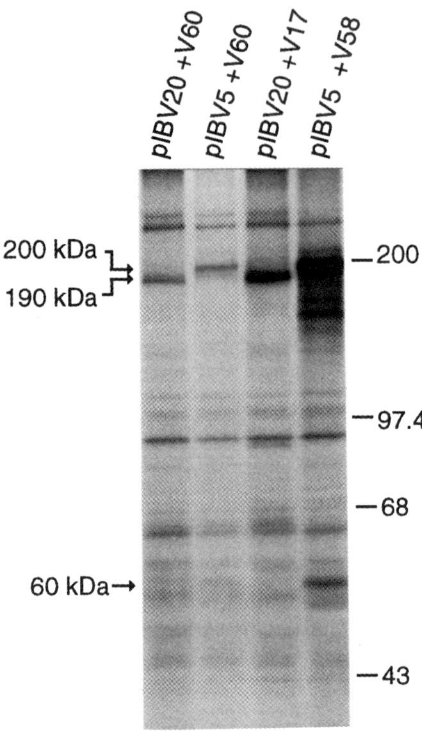

Figure 2. Analysis of transiently expressed mRNA1 products from transfection with plasmids pIBV5 and pIBV20, using the vaccinia virus-T7 expression system. Cells were labelled with [^{35}S] methionine, lysates were prepared, and polypeptides were either analysed directly, or immunoprecipitated with either antiserum V17, V58 or V60. Polypeptides were separated on an SDS-12.5% polyacrylamide gel, and detected by fluorography. HMW-high molecular weight markers (numbers indicate kilodaltons).

Substitution of the Q^{2012} residue of the predicted Q^{2012}-S^{2013} cleavage site with an E was subsequently made by site-directed mutagenesis, giving the mutant construct pIBV1b5Q^{2012}-E. Co-expression of this plasmid with pIBV3C led to the synthesis of the full-length 140 kDa protein (Figure 4b). In addition, two polypeptides co-migrating, respectively, with the 105 kDa and the 35 kDa proteins detected from the co-expression of pIBV1b5 and pIBV3C, were also detected (Figure 4b). No 42 kDa and 77 kDa proteins, however, were observed (Figure 4b). These results indicate that mutation of the Q^{2012} residue to E abolished the cleavage occurred at this position.

5. DISCUSSION

We have recently reported that an ORF 1a-encoding proteinase of the picornavirus 3C proteinase group is responsible for cleavage of the polyproteins encoded by mRNA1 to a 10 kDa and a 100 kDa mature viral products (Liu *et al.,* 1994; Liu *et al.,* 1997). Deletion and site-directed mutagenesis studies demonstrated that the 100 kDa protein was encoded by IBV sequence from nucleotide 12313 to 15131 (Liu *et al.,* 1994; Liu and Brown, 1995). Two Q-S dipeptide bonds, encoded by the ORF 1a sequence from nucleotide 12310 to 12315, and by ORF 1b from nucleotide 15129 to 15134, respectively, were identified to be the cleavage sites responsible for releasing the 100 kDa protein from the 1a/1b fusion polyprotein (Liu and Brown, 1995). In this report, we show that the 3C-like proteinase is also able to mediate cleavage at three other sites located in the ORF 1b polyprotein, resulting in the releasing of four more cleavage products approximately 65 kDa, 63 kDa, 42 kDa and 35 kDa, respectively.

Among the five protein species identified in this study, only the 100 kDa protein has actually been detected from IBV-infected cells (Liu *et al.*, 1994). Attempts were made to detect the rest four protein species in virus-infected cells, but were unsuccessful so far. It is currently uncertain if these proteins represent genuine viral products. However, as important functional domains, such as the metal-binding and RNA helicase domains, were predicted to be located in these proteins, some of them may be essential for the replication of viral RNA. Failure to detect these products may simply reflect the rapid degradation of these products in virus-infected cells during the virus infection cycle. Alternatively, complex mechanisms may be involved in regulation of the expression and accumulation of these products. Further investigations are underway to explore these possibilities.

Our previous mutagenesis data demonstrated that the Q residue of a Q-S cleavage site is very sensitive to substitution mutations. Substitution of the Q residue with an E totally block the cleavage occurred at the corresponding site (Liu and Brown, 1995). Substitution of the Q residue with an M also blocks the cleavage occurred at a Q-G dipeptide bond. The S residue, however, is more tolerant to similar substitution mutations. Substitution of the S residue with either an A or S does not inhibit the cleavage (Liu and Brown, 1995). It seems that a wide range of Q-X (X represents any amino acid residue) dipeptide bonds might be used by the 3C-like proteinase domain to release mature viral products. To support this hypothesis, we have recently identified a Q-N dipeptide bond responsible for the release of a 10 kDa mature viral product from the 1a polyprotein (Liu *et al.*, 1997).

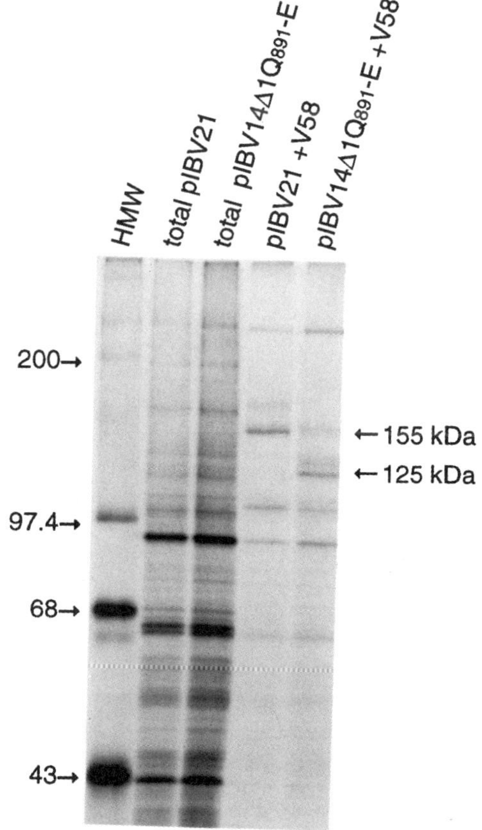

Figure 3. Mutational analysis of the C-terminal cleavage site of the putative metal-binding and RNA helicase protein. Plasmids were transiently expressed in Vero cells, using the vaccinia virus-T7 expression system. The transfected cells were labelled with [^{35}S] methionine, lysates were prepared, and polypeptides were either analysed directly, or immunoprecipitated with antiserum V58. Polypeptides were separated on an SDS-12.5% polyacrylamide gel, and detected by fluorography. HMW-high molecular weight markers (numbers indicate kilodaltons).

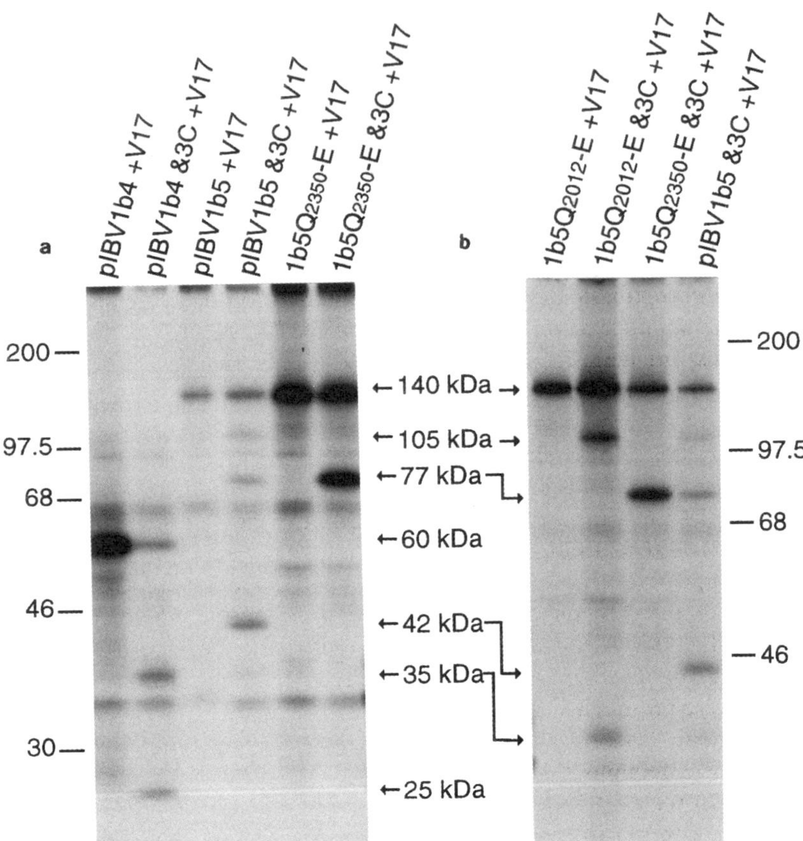

Figure 4. Analysis of transiently expressed products from pIBV1b4- and pIBV1b5-transfected Vero cells (a), and mutational analyses of the predicted $Q^{2350(1b)}S^{2351(1b)}$ (a) and $Q^{2012(1b)}S^{2013(1b)}$ cleavage sites (b). The transfected cells were labelled with [^{35}S] methionine, lysates were prepared, and polypeptides were either analysed directly, or immunoprecipitated with antiserum V17. Polypeptides were separated on an SDS-12.5% polyacrylamide gel, and detected by fluorography. HMW-high molecular weight markers (numbers indicate kilodaltons).

Systematic mutagenesis of the cleavage dipeptides and their context is required to address this issue further.

REFERENCES

Fuerst, T. R., Niles, E. G., Studier, F. W., and Moss, B., 1986, Eukaryotic transient-expression system based on recombinant vaccinia virus that synthesizes bacteriophage T7 RNA polymerase, *Proc. Natl. Acad. Sci. USA* **83**: 8122–8127.

Gorbalenya, A. E., Koonin, E. Y., Donchenko, A. P., and Blinov, V. M., 1989, Coronavirus genome: prediction of putative functional domains in the non-structural polyprotein by comparative amino acid sequence analysis, *Nuc. Ac. Res.* **17**: 4847–4860.

Laemmli, U. K., 1970, Cleavage of structural proteins during the assembly of the bacteriophage T4, *Nature (London)* **227**: 680–685.

Lee, H-J., Shieh, C-K., Gorbalenya, A. E., Koonin, E. V., Monica, N. L., Tuler, J., Bagdzhadzhyan, A., and Lai, M. M. C., 1991, The complete sequence (22 kilobases) of murine coronavirus gene 1 encoding the putative proteases and RNA polymerase, *Virology* **180**: 567–582.

Liu, D. X., and Brown, T. D. K., 1995, Characterization and mutational analysis of an ORF-1a-encoding proteinase domain responsible for proteolytic processing of the infectious bronchitis virus 1a/1b polyprotein, *Virology* **209**: 420–427.

Liu, D. X., Brierley, I., Tibbles, K. W., and Brown, T. D. K., 1994, A 100-kilodalton polypeptide encoded by open reading frame (ORF) 1b of the coronavirus infectious bronchitis virus is processed by ORF 1a products, *J. Virol.* **68**: 5772–5780.

Liu, D. X., Tibbles, K. W., Cavanagh. D., Brown, T. D. K., and Brierley, I., 1995, Identification, expression, and processing of an 87-kDa polypeptide encoded by ORF 1a of the coronavirus infectious bronchitis virus, *Virology* **208**: 48–57.

Liu, D. X., Xu, H. Y., and Brown, T. D. K., 1997, Proteolytic processing of the coronavirus infectious bronchitis virus 1a polyprotein: identificaiton of a 10 kDa polypeptide and determination of its cleavage sites, *J. Virol.* **71**: 1814–1820.

21

FURTHER CHARACTERISATION OF THE CORONAVIRUS IBV ORF 1a PRODUCTS ENCODED BY THE 3C-LIKE PROTEINASE DOMAIN AND THE FLANKING REGIONS

Lisa F. P. Ng and D. X. Liu

Institute of Molecular Agrobiology
59A The Fleming
1 Science Park Drive
Singapore 118240

1. ABSTRACT

Coronavirus IBV encodes a piconarvirus 3C-like proteinase. In a previous report, this proteinase was shown to undergo rapid degradation *in vitro* in reticulocyte lysate due to a posttranslational event involving ubiquitination of the protein. Several lines of evidence presented here indicate that the proteinase itself is stable. Translation of the IBV sequence from nucleotide 8864 to 9787 resulted in the synthesis of a 33 kDa protein, representing the full-length 3C-like proteinase. Pulse-chase and time-course experiments showed that this protein was stable in reticulocyte lysate for up to 2 hours. However, a 45 kDa protein encoded by the IBV sequence from nucleotide 8693 to 9911 underwent rapid degradation in reticulocyte lysate, but was stable in wheat germ extract, suggesting that an ATP-dependent protein degradation pathway may be involved in the turnover of the 45 kDa protein. To identify the IBV sequence responsible for the instability of this 45 kDa protein species, the region from nucleotide 8693 to 9787 was translated both *in vitro* and *in vivo* , leading to the synthesis of a stable 43 kDa protein. These results suggest that a destabilising signal may be located in the IBV sequences between the nucleotides 9787 and 9911. Meanwhile, protein aggregation was observed when the product encoded by the IBV sequence from nucleotide 9911 to 10510 was boiled for 5 minutes before being analysed in SDS-PAGE; when the same product was treated at 37°C for 15 minutes, however, protein aggregation was not detected. Deletion studies indicate that the presence of a hydrophobic domain downstream of the 3C-like proteinase-encoding region may be the cause for the aggregation of the product encoded by this region of ORF 1a.

Coronaviruses and Arteriviruses, edited by Enjuanes *et al.*
Plenum Press, New York, 1998

2. INTRODUCTION

The 5'-unique region of IBV mRNA1 comprises two large ORFs (1a and 1b), overlapping by 42 nucleotides. ORF 1a potentially encodes a polypeptide of 441 kDa and ORF 1b potentially encodes a polypeptide of 300 kDa (Boursnell *et al.*, 1987). The downstream ORF 1b is produced as a fusion protein of 741 kDa with 1a by a mechanism involving ribosomal frame shifting (Brierley *et al.*, 1987). The 1a/1b fusion polyprotein is expected to be cleaved by viral or cellular proteinases to produce functional products associated with viral RNA replication. Three proteinase domains have been predicted to be encoded by ORF 1a, including two overlapping papain-like proteinase domains, and a trypsin-like proteinase domain of the picornavirus 3C proteinase group (Gorbalenya *et al.*, 1989). The papain-like proteinase domains are involved in proteolytic processing of the 1a polyprotein to an 87 kDa viral product (Liu *et al.*, 1995). More important role in the processing and maturation of the ORF 1 proteins is likely to be played by the 3C-like proteinase domain. In fact, processing of the 1a / 1b polyprotein to a 10 kDa and a 100 kDa protein species has been reported (Liu *et al.*, 1994, 1997). Meanwhile, internal deletion and substitution mutation studies have shown that the predicted nucleophilic cysteine residue (Cys^{2922}) and the histidine 2820 residue are essential for the activity of this proteinase, confirming that this proteinase belongs to the picornavirus 3C proteinase supergroup (Liu and Brown, 1995).

Recently, Tibbles *et al.* have reported that the 3C-like proteinase is subjected to rapid turnover when expressed in rabbit reticulocyte lysate due to a posttranslational event involving ubiquitination of the protein and degradation by an ATP-dependent system (Tibbles *et al.*, 1995). In this paper, we present experiments designed to verify the stability of the 3C-like proteinase. Data presented demonstrated that the 3C-like proteinase itself is stable both in reticulocyte lysate and in intact cells. A destabilising signal may be located downstream of the region encoding the 3C-like proteinase. In addition, deletion studies have confirmed that the extremely hydrophobic nature of the products encoded by the IBV sequence from nucleotide 10375 to 10450 causes protein aggregation.

3. MATERIALS AND METHODS

3.1. Transient Expression of the IBV Sequences in Vero Cells by Using a Vaccinia Virus-T7 Expression System

ORFs placed under control of the T7 promoter were expressed transiently in eukaryotic cells as described in Fuerst *et al.*, 1986. Semiconfluent monolayers of Vero cells were infected with 10 pfu per cell of a recombinant vaccinia virus (vTF7–3) which expresses the T7 phage RNA polymerase and then transfected with appropriate plasmid DNA with Dosper (Boehringer Mannheim) according to manufacturer's instructions. After incubation of the cells at 37°C overnight (12 hours), the cells were incubated in methionine-free medium for 30 min prior to being labelled. After 4 hours of labelling with [^{35}S]methionine, the cells were lysed with radioimmunoprecipitation assay buffer (50 mM Tris-HCl [pH7.5], 150 mM NaCl, 1% NP-40, 0.1% sodium dodecyl sulphate [SDS]) and removed from the dishes. The cells were then precleared by centrifugation at 15,000 rpm for 5 min at 4°C and stored at -70°C.

3.2. Radioimmunoprecipitation

Radioimmunoprecipitation with polyclonal rabbit antisera was carried out as described previously (Liu *et al.*, 1994).

3.3. Polymerase Chain Reaction (PCR)

Appropriate primers and template DNAs were used in amplification reactions with *Pfu* DNA polymerase (Stratagene) under standard buffer conditions with 2mM $MgCl_2$.

3.4. Cell-Free Transcription and Translation

Plasmid DNA (5µg) was linearised at an appropriate restriction site downstream of the IBV sequence (unless stated otherwise) before they were transcribed and translated using the TNT transcription coupled translation system (Promega). Alternatively, linearised plasmid DNA was transcribed *in vitro* with T7 phage RNA polymerase as described before (Liu *et* al., 1994), with the dinucleotide ^{7m}GpppG incorporated to provide a 5' cap structure (Contreras, 1982). Product mRNA was recovered from the reaction mixtures by extraction with phenol-chloroform (1:1) and precipitation with ethanol before the RNA was translated in either the rabbit reticulocyte lysate system (Promega) or the wheat germ extract system (Promega) in the presence of [^{35}S]methionine as described previously (Liu *et al.*, 1994). Translation reactions were carried out at 30°C, typically for 90 min (unless stated otherwise), and were stopped by incubation with equal volume of translation stop buffer (1 mg/ml Rnase A, and 0.5M EDTA) for 20 min at room temperature. During pulse-chase experiments, excess cold methionine (5mM) was added at the times indicated for the individual experiments. Reaction products were separated by SDS-PAGE and detected by autoradioagraphy.

3.5. SDS-Polyacrylamide Gel Electrophoresis

SDS-polyacrylamide gel electrophoresis (SDS-PAGE) of virus polypeptides was carried out using 10% and 12.5% polyacylamide gels (Laemmli, 1970). Labelled polypeptides were detected by autoradiography or fluorography of dried gels.

3.6. Construction of Plasmids

Plasmids pIBV4, pIBV6, pIBV9 (formerly called pKT205), pIBV10, pIBV11 and pIBV3C, which cover the IBV sequences from nucleotides 10752 to 13896, 10510 to 12700, 8693 to 10925, 11952 to 13896, 13896 to 16980, and 8864 to 9787 respectively, were described before (Liu *et al.*, 1994, 1997).

Plasmid pBV9ÆE2, which contains the IBV sequence from nucleotide 8693 to 9787 was made by cloning a *Bgl* II/*Spe* I restriction fragment covering the IBV sequence from nucleotide 8693 to 9608, into *Bgl* II/*Spe* I digested pIBV3C. *Spe* I cuts the IBV sequence at nucleotide 9608 and the *Bgl* II site is located between the T7 promoter and the IBV sequence.

Plasmid pIBV9Q^{2779}-E covering the IBV sequence from nucleotide 8693 to 10637 and containing a mutation of the Gln^{2779} residue to a Glu was created by PCR with pIBV9 as the template. The sequence of the mutation primer is 5'-GTTAGTAGATTA-GAGTCTGGTTTTAAG-3', and the sequence of the complementary primer is 5'-GTCACCACCAATTCCCTCTATAAGTAT-3'.

Plasmid pCAT-205, which covers the IBV sequence from nucleotides 9911 to 10925 tagged downstream to a gene coding for the bacterial chloramphenicol acetyltransferase (CAT), was made by cloning a 516 bp *Bam* HI/*Sca* I digested CAT fragment into *Bgl* II/Pvu II digested pIBV9.

4. RESULTS

4.1. *In Vitro* Expression of IBV Sequence Containing the 3C-Like Proteinase Domain and the Flanking Regions

We have previously reported that translation of the IBV sequence containing the 3C-like proteinase domain and the surrounding regions in an *in vitro* expression system led to the poor expression of the viral proteins (Liu *et al.*, 1994). To extend this study, four plasmids, pIBV9, pIBV6, pIBV10 and pIBV11, containing the IBV sequences from nucleotide 8693 to 10925, 8693 to 12700, 8693 to 13896 and 8693 to 16980 respectively, together with pIBV4 (10752 to 13896) were translated *in vitro* by using the TNT transcription cou-

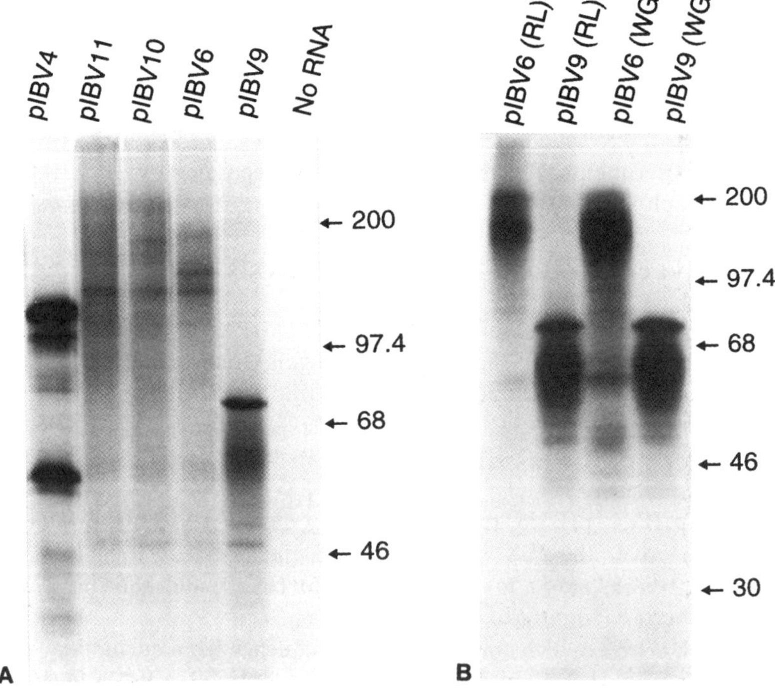

Figure 1. (A). Analysis of *in vitro* translation products produced from plasmids pIBV9, pIBV6, pIBV10, pIBV11, and pIBV4 in reticulocyte lysate by using the TNT coupled translation system (Promega). Plasmid DNA was added to reticulocyte lysate at 200 µg/ml, and the ^{35}S-labelled products were separated on an SDS-12.5%-polyacrylamide gel and detected by fluorography. (B). Comparison of *in vitro* translation products produced from plasmids pIBV9 and pIBV6 in both reticulocyte lysate and wheat germ extract systems (Promega). Capped transcripts were generated from the mentioned plasmids and then translated in either reticulocyte lysate or wheat germ extract. Translation reactions were stopped after 60 min incubation by the addition of translation stop buffer. The samples were analysed on an SDS-12.5%-polyacrylamide gel.

pled translation system (Promega). As shown in Figure 1a, a full-length product of 82 kDa from translation of pIBV9 was observed; transcription and translation of pIBV6, pIBV10, and pIBV11 led to the synthesis of a 138 kDa protein, representing the ORF1a stop product. Two proteins of approximately 150 and 195 kDa, representing 1a-1b fusion products, were also detected from the expression of pIBV6 and pIBV10, respectively (Figure 1a). No full-length product, however, was seen from the translation of pIBV11(Figure 1a). A heterogenous smear of polypeptides covering a large size range was also seen from expression of plasmids pIBV6, pIBV10 and pIBV11(Figure 1a), indicating that this region may contain elements that affect the expression of downstream sequence or accumulation of the gene products.

To determine if the poor expression of pIBV6 and pIBV9 was due to the instability of the translated products in reticulocyte lysate system, plasmids pIBV6 and pIBV9 were expressed in wheat germ extract. As can be seen in Figure 1b, transcription and translation of pIBV6 and pIBV9 led to the synthesis of full-length proteins of approximately 138 kDa and 82 kDa, respectively. No significant differences in the translational pattern of the two constructs were observed from wheat germ extract, although more efficient translation of both plasmids was seen in this system (Figure 1b). Once again, translation of both plasmids showed the formation of the heterogenous high molecular mass species. These results indicate that products expressed from reticulocyte lysate may be unstable, and tend to aggregate.

4.2. Products Encoded by the IBV Sequences Flanking the 3C-Like Proteinase Domain Are Unstable

The data presented suggest that proteins expressed from plasmid pIBV6 and derivatives may be unstable in reticulocyte lysate. To investigate this further, pIBV6 was linearised separately with two restriction enzymes, *Hinc* II and *Sma* I, which digest the IBV sequences at nucleotide positions 10438 and 12677, respectively, and analysed by a pulse-chase experiment. Aliquots were taken from the translation reaction mixture after incubation for 0, 5, 10, 15, 20, 30, 40, 50, and 60 minutes, and the polypeptides were analysed by 12.5% SDS-PAGE. As shown in Figure 2a, translation of *Sma* I digested pIBV6 showed that no product of the predicted full-length of 160 kDa was detected. Instead, premature termination products of approximately 48 kDa and 65 kDa were observed. The products were seen to degrade rapidly and convert into high molecular mass forms, becoming trapped at the top of the gel. Translation of *Hinc* II digested pIBV6 led to the detection of the full-length product of 67 kDa; however, once again accumulation of high molecular mass protein species and decreasing of the 67 kDa protein species were seen over the 60 minute time course (Figure 2b). These results indicate that some regions flanking the 3C-like proteinase may contain destabilising sequences, which promote rapid protein degradation.

To define further the destabilising signals-containing region, plasmid pIBV9 was linearised with restriction enzyme *Pvu* II which cuts the IBV sequence at nucleotide position 9911, and analysed by pulse-chase experiment in both reticulocyte lysate and wheat germ extract systems. As presented in Figure 3, the full-length protein of 45 kDa was detected after 10 min chase; it then underwent rapid degradation in reticulocyte lysate. Interestingly, the protein was stable in wheat germ extract over the time course, suggesting that an ATP-dependent proteolytic system present in reticulocyte lysate may be responsible for the rapid degradation of this protein. These results indicate that a protein destabilising signal may be located in the IBV sequence between nucleotides 8693 and 9911.

 L. F. P. Ng and D. X. Liu

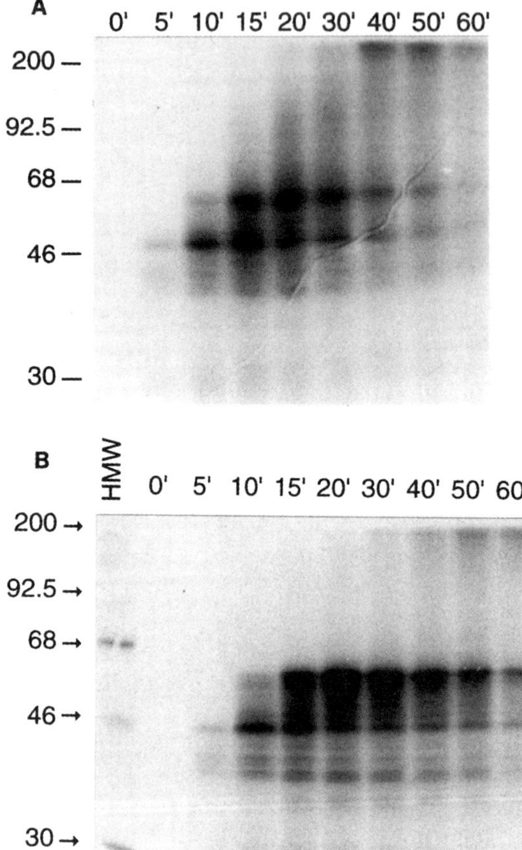

Figure 2. (A). Pulse-chase analysis of the accumulation of [^{35}S]methionine-labelled products from *Sma* I digested pIBV6. 50µl reaction was performed and 10-fold excess of unlabelled methionine was added to the reaction after 10 min incubation at 30°C. Aliquots were taken at intervals (min) as indicated. Polypeptide products were separated on an SDS-12.5%-polyacrylamide gel and detected by fluorography. (B). Pulse-chase experiment of *Hinc* II digested pIBV6. Reaction was performed as described above and polypeptide products were separated on an SDS-12.5%-polyacrylamide gel and detected by fluorography.

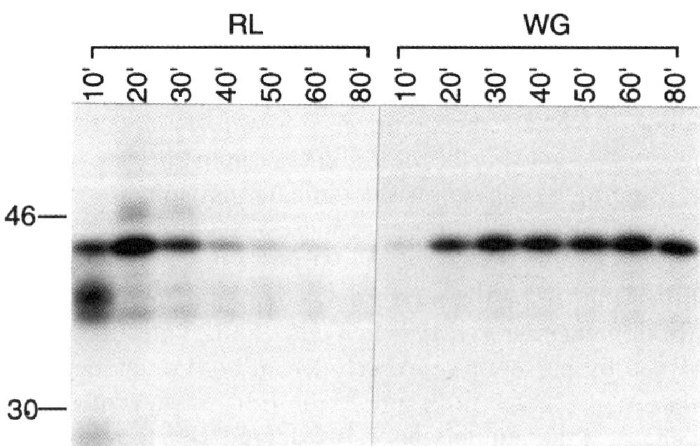

Figure 3. Pulse-chase analysis of cell-free translation products of mRNA obtained by *in vitro* transcription of *Pvu* II digested pIBV9. RNA was added to reticulocyte lysate and wheat germ extract as indicated. Samples were taken from the translation reaction mixtures after incubation for 10, 20, 30, 40, 50, 60, and 80 min, and separated on an SDS-12.5%-polyacrylamide gel and detected by fluorography.

4.3. Stable Expression of the 3C-Like Proteinase *in Vitro* and *in Vivo*

To define further the destabilising signal, pIBV9Δ2, which covers the IBV sequence from 8693 to 9787, was cloned and analysed by *in vitro* pulse-chase assay. As shown in Figure 4a, a full-length protein of 43 kDa was detected after 10 min incubation, and was seen to remain stable till the end of the pulse-chase experiment. The stability of the 43 kDa protein was further studied by *in vivo* expression of plasmid pIBV9Q^{2779}-E, which has

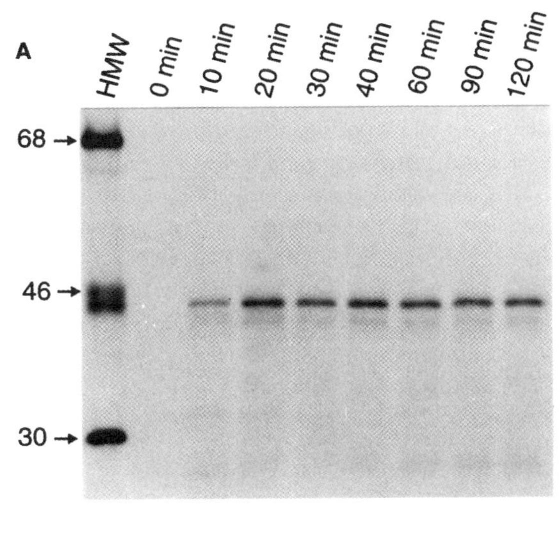

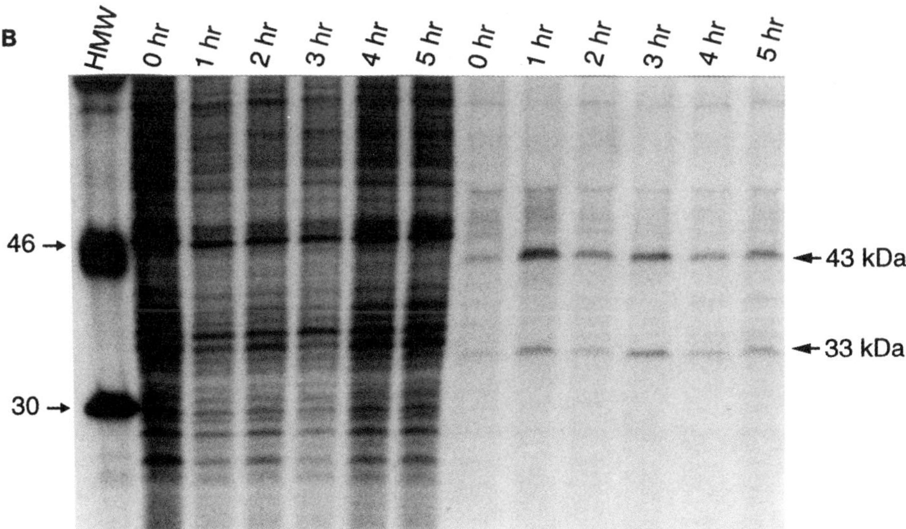

Figure 4. (A). *In vitro* pulse-chase analysis of pIBV9Æ2. Aliquots were removed at intervals as indicated, and polypeptide products were separated on an SDS-12.5%-polyacrylamide gel and detected by fluorography. (B). *In vivo* expression and stability analysis of pIBV9 ^{Q2779}-E in Vero cells. At 13 hours postinfection, infected cells were incubated in methionine-free medium for 30 min prior to being labelled with [^{35}S]methionine for 4 hours. 10-fold excess unlabelled methionine was added and cells were harvested at intervals (hour) as indicated. Total cell lysates were immunoprecipitated with anti-3C antiserum, and separated on an SDS-12.5%-polyacrylamide gel.

the $Q^{2779}S$ site mutated to ES. As expected, the 43 kDa protein was detected and seen to remained stable 4 hours after addition of 10-fold excess of cold methionine. These results demonstrate that product encoded by the IBV sequence from nucleotide 8693 to 9787 is stable. It is therefore suggested that the 3C-like proteinase, predicted to be encoded by the IBV sequence from nucleotide 8864 to 9787 should be stably expressed (Figure 4b).

To support this conclusion further, plasmid pIBV3C, which contains the sequence encoding the 3C-like proteinase, was analysed by pulse-chase experiment. As shown in Figure 5a, a 33 kDa protein representing the full-length product encoded by this construct, was observed after incubation for 10 min and remained stable in the reticulocyte lysate up to 120 min after adding 10-fold excess of cold-methionine in the reaction mixture.

The stability of the 3C-like proteinase was finally assessed by expression of pIBV9 *in vivo*. Expression of pIBV9 *in vitro* showed the synthesis of a product of 82 kDa (Figure 1b and 1c). *In vivo* expression of this plasmid, however, showed processing of full-length protein to smaller products. As shown in Figure 5b, instead of the 82 kDa protein, a product with an apparent molecular weight of approximately 33 kDa was immunoprecipitated by anti-3C antiserum. This 33 kDa protein co-migrated with the 33 kDa expressed from

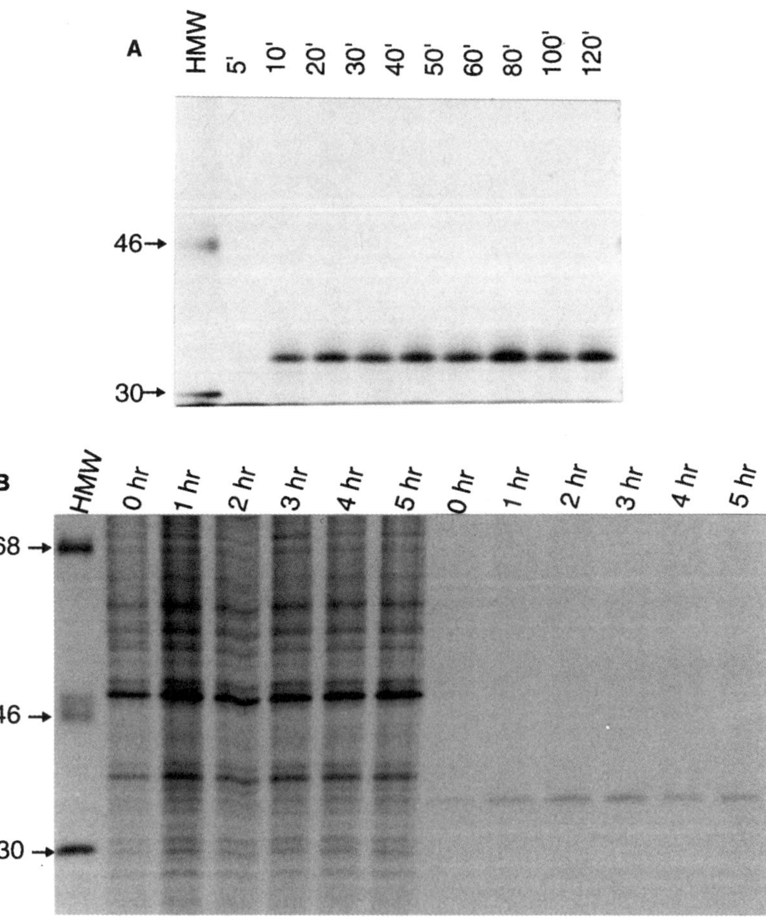

Figure 5. (A). *In vitro* pulse-chase analysis of pIBV3C. (B). *In vivo* expression, processing and stability analysis of pIBV9 in Vero cells.

pIBV3C and was shown to be released from the 1a polyprotein by autocleavage at two QS dipeptide bonds ($Q^{2779}S$ and $Q^{3086}S$) (Ng and Liu, unpublised observations), suggesting that it is the protein associated with the 3C-like proteinase activity. Time-course and pulse-chase experiments showed that this protein remained stable after chasing with 10-fold excess of cold-methionine up to 5 hours (Figure 5b).

4.4. Products Encoded by the IBV Sequence from 10438 to 10900 Tend to Aggregate

During the expression of plasmid pIBV6 and its derivatives, heterogenous high molecular mass protein species were consistently observed when the IBV sequence from 10438 to 10900 were included (Figure 2 and unpublished data). Examination of the deduced amino acid sequence showed that this region encoded a stretch of 39 hydrophobic amino acid residues. To investigate the possibility that the formation of the heterogenous smear of high molecular mass protein species is due to protein aggregation caused by the hydrophobic amino acid sequences, the IBV sequence from nucleotide 9911 to 10925 was tagged to the C-terminal end of the CAT gene, giving plasmid pCAT-205. Plasmid pCAT-205 was then linearised with *Hinc* II, *Sau* I, *Hin*d III, and *Bam* HI, which cut the IBV sequences at 10438, 10510, 10721 and 10925 respectively, and transcribed and translated. The polypeptides were treated separately with two different temperatures, at 100°C for 4 minutes, and at 37°C for 15 minutes before subjected to SDS-PAGE analysis. Data presented in Figure 6 showed that products translated from *Hin*d III and *Bam* HI digested

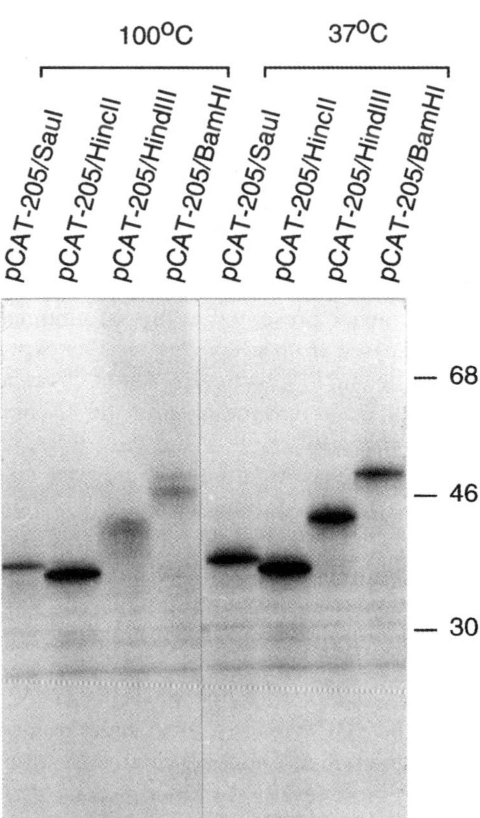

Figure 6. C-terminal deletion analysis of translation products of pCAT-205. Plasmid pCAT-205 was linearised at the restriction sites described in the text and transcribed *in vitro* with T7 RNA polymerase. Translation was performed in the presence of [^{35}S]methionine in reticulocyte lysate, and labelled products were analysed on an SDS-10%-polyacrylamide gel after treatment either at 37°C for 15 min or at 100°C for 4 min.

pCAT-205 tended to aggregate after treatment at 100°C for 4 min; no aggregation was observed after the samples were treated at 37°C. However, protein aggregation was not observed when products translated from *Sau* I and *Hinc* II digested pCAT-205 were analysed either after treatment at 37°C for 15 min or at 100°C for 4 min. These results confirm that the hydrophobic amino acid stretch encoded by the IBV sequence from 10375 to 10450 caused the protein aggregation at 100°C.

5. DISCUSSION

Coronavirus gene expression involves proteolytic processing of the mRNA 1-encoding polyproteins by viral and cellular proteinases. Recent reports have demonstrated that an ORF 1a-encoding proteinase domain of the picornavirus 3C proteinase group is responsible for proteolytic processing of the IBV 1a/1b fusion polyprotein to a 100 kDa and a 10 kDa protein species (Liu *et al.*, 1994 ,1997). It is likely that this proteinase may play a major role in processing of the 1a/ 1b polyprotein during the viral replication cycle. In an attempt to characterise this proteinase, Tibbles *et al.* observed that products encoded by the 3C-like proteinase domain and the flanking regions were unstable. They suggested that the 3C-like proteinase domain may be a target for the ubiquitin-mediated ATP-dependent degradation present in the reticulocyte lysate (Tibbles *et al.*, 1995). In this report, systematic analysis of the stability of the 3C-like proteinase was carried out using constructs covering the 3C-like proteinase domain and the surrounding regions. It was observed that *in vitro* transcription and translation of plasmids covering both upstream and downstream regions of the 3C-like proteinase were poorly expressed in reticulocyte lysate system. Products encoded by this region of ORF 1a tend to be aggregated to form heterogenous smear of high molecualr mass protein species. Construction and expression of a series of deletion constructs confirmed that a 33 kDa protein, representing the 3C-like proteinase is stably expressed in reticulocyte lysate and in Vero cells. However, the IBV sequence between 9787 to 9911 may code for a destabilising signal promoting rapid protein degradation. The protein aggregation may be caused by the presence of a stretch of 39 hydrophobic amino acid residues encoded by the IBV sequence from 10375 to 10450.

Evidence presented in this communication shows that the 33 kDa protein expressed and processed from pIBV9 as well as expressed from pIBV3C is stable both in reticulocyte lysate and in intact cells. As the 33 kDa protein is actually the 3C-like proteinase, this study suggested that the level of the 3C-like proteinase may be stably maintained during, at least, the early stage of the viral RNA replication cycle. This is understandable as the 3C-like proteinase is a major proteinase responsible for cleaving the polyprotein encoded by ORF 1b. Considering that the physical distance between the 3C-like proteinase-encoding region and the last two scissile bonds encoded by ORF 1b is approximately 10 kb, it is reasonable to assume that stable maintenance of the proteinase is a prerequisite for efficient cleavage of the target product at these positions. Kinetic study of the synthesis and accumulation of the 3C-like proteinase in virus-infected cells is underway to address this issue further.

The observation that products encoded by the IBV sequence between nucleotides 10375 and 10450 tend to form aggregation highlights a classical difficulty in analysis of the extremely hydrophobic proteins. As attempts to raise antisera against proteins encoded by this region of ORF 1a failed, we are currently lacking tools to extend our study on this region to intact cells.

REFERENCES

Boursnell, M. E. G., T. D. K. Brown, I. J. Foulds, P. F. Green, F. M. Tomley, and M. M. Binns., 1987, Completion of the sequence of the genome of the coronavirus avian infectious bronchitis virus, *J. Gen. Virol.* **68:** 57–77.

Brierley, I., M. E. G. Boursnell, M. M. Binns, B. Billimoria, V. C. Blok, T. D. K. Brown, and S. C. Inglis., 1987, An efficient ribosomal frame-shifting signal in the polymerase-encoding region of the coronavirus IBV, *EMBO. J.* **6:**3779–3785.

Contreras, R., H. Cheroutre, W. Degrave, and W. Fiers., 1982, Simple efficient *in vitro* synthesis of capped RNA useful for direct expression of cloned DNA, *Nucleic Acids Res.* **10:**6353–6362.

Fuerst, T. R., E. G. Niles, F. W. Studier, and B. Moss., 1986, Eukaryotic transient-expression system based on recombinant vaccinia virus that synthesizes bacteriophage T7 RNA polymerase, *Proc. Natl. Acad. Sci. USA* **83:**8122–8127.

Gorbalenya, A. E., E. Y. Koonin, A. P. Donchenko, and V. M. Blinov., 1989, Coronavirus genome: prediction of putative functional domains in the non-structural polyprotein by comparative amino acid sequence analysis, *Nucleic Acids Res.* **17:**4847–4860.

Laemmli, U. K., 1970, Cleavage of structural proteins during the assembly of the head of bacteriophage T4, *Nature* (London) **227:**680–685.

Liu D. X., I, Brierley, K. W. Tibbles, and T. D. K. Brown, 1994, A 100 kilodalton polypeptide encoded by open reading frame (ORF) 1b of the coronavirus infectious bronchitis virus is processed by ORF 1a products, *J. Virol.* **68:**5772–5780.

Liu D. X., K. W. Tibbles, D. Cavanagh, T. D. K. Brown, and I. Brierley., 1995, Identification, expression, and processing of an 87 kDa polypeptide encoded by ORF 1a of the coronavirus infectious bronchitis virus, *Virology* **208:**48–57.

Liu D. X., and T. D. K. Brown., 1995, Characterisation and mutational analysis of an ORF-1a-encoding proteinase domain responsible for proteolytic processing of the infectious bronchitis virus 1a/1b polyprotein, *Virology* **209:**420–427.

Liu D. X., H. Y. Xu, and T. D. K. Brown., 1997, Proteolytic processing of the coronavirus infectious bronchitis virus 1a polyprotein: identification of a 10 kilodalton polypeptide and determination of its cleavage sites, *J. Virol.* **71:**48–57.

Tibbles, K. W., I. Brierley, D. Cavanagh, and T. D. K. Brown., 1995, A region of the infectious bronchitis virus 1a polyprotein encoding the 3C-like protease domain is subject to rapid turnover when expressed in rabbit reticulocyte lysate, *J. Gen. Virol.* **76:**3059–3070.

22

CHARACTERISATION OF A PAPAIN-LIKE PROTEINASE DOMAIN ENCODED BY ORF1a OF THE CORONAVIRUS IBV AND DETERMINATION OF THE C-TERMINAL CLEAVAGE SITE OF AN 87 kDa PROTEIN

K. P. Lim and D. X. Liu

Institute of Molecular Agrobiology
59A The Fleming
1 Science Park Drive
Singapore 118240

1. ABSTRACT

Our previous studies have shown that two overlapping papain-like proteinase domains (PLPDs) encoded by the IBV sequence from nucleotides 4155 to 5550 is responsible for cleavage of the ORF 1a polyprotein to an 87 kDa protein. In this study, we demonstrate that only the more 5' one of the two domains, PLPD-1 encoded between nucleotides 4155 and 5031, is required for processing to the 87 kDa protein. Site-directed mutagenesis studies have shown that the Cys^{1274} and His^{1435} residues are essential for the PLPD-1 activity, suggesting that they may be the components of the catalytic centre of this proteinase. Coexpression and immunoprecipitation studies have further revealed that PLPD can interact with the 87 kDa protein. Meanwhile, data obtained from the construction and expression of a series of deletion mutants have indicated that the 87 kDa protein is encoded by the 5'- most 2600 bp part of ORF1a. Further deletion and mutagenesis studies are underway to determine precisely the C-terminal cleavage site of the 87 kDa protein.

2. INTRODUCTION

Proteolytic processing of the 741 kDa polyprotein encoded by mRNA 1 is one of the strategies employed by the Coronavirus IBV in replication of its genome. Two overlapping viral papain-like proteinase domains (PLPDs) belonging to the class of cysteine pro-

Coronaviruses and Arteriviruses, edited by Enjuanes *et al.*
Plenum Press, New York, 1998

teinases and a trypsin-like proteinase domain belonging to the picornavirus 3C proteinase supergroup are predicted to be encoded by the IBV sequence from nucleotide 4155 to 5550 and from 8864 to 9787, respectively (Gorbalenya *et al.*, 1989). The two overlapping PLPDs have been demonstrated to be responsible for cleavage of the 1a polyprotein to a novel 87 kDa protein, as the release of this 87 kDa product was observed only when the IBV sequence from nucleotide 4859 to 5753 was included in the expression plasmids (Liu *et al.*, 1995).

In this communication we report that only the first PLPD encoded between nucleotides 4242 and 5031 is required for *cis*-cleavage of the 87 kDa protein at its C-terminus. Substitution of either of the two putative core residues, Cys^{1274} and His^{1435}, totally abolished the proteinase activity, thereby allowing us to conclude that they may be components of the catalytic centre of PLPD-1. The C-terminal cleavage site of the 87 kDa protein was meanwhile determined by construction and expression of a series of deletion constructs. It was shown that the C-terminal scissile bond of the 87 kDa protein was most likely located in the vicinity of Gly^{673} residue. In addition, evidence presented indicated that the 87 kDa protein may interact with products encoded by the IBV sequence from nucleotide 2548 to 5753.

3. METHODS AND MATERIALS

3.1. In Vivo Expression of IBV Sequences

IBV sequences placed under the control of a T7 promoter were transiently expressed in Vero cells using the vaccinia recombinant virus-T7 expression system as described before (Liu *et al.*, 1994).

3.2. Polymerase Chain Reaction

Amplification reactions of the respective template DNAs with appropriate primers were performed with Native Pfu DNA polymerase (Stratagene) under standard buffer conditions with 2 mM $MgCl_2$.

3.3. Site-Directed Mutagenesis Studies

Site-directed mutagenesis was performed as previously described (Liu *et al.*, 1997).

3.4. Radioimmunoprecipitation

Plasmid DNA-transfected Vero cells were lysed with RIPA buffer (50 mM Tris HCl, pH 7.5, 150 mM NaCl, 1% sodium deoxycholate, 0.1% SDS) and precleared by centrifugation at 12,000 rpm for 5 minutes at 4°C. Immunoprecipitation was performed as described previously (Liu *et al.*, 1994).

3.5. Cell-Free Transcription and Translation

Plasmid DNA was transcribed and translated *in vitro* with the TnT T7-coupled reticulocyte lysate system (TnT system) according to the instructions of the manufacturer (Promega). Reaction products were separated by SDS-PAGE and detected by autoradiography.

3.6. SDS-Polyacrylamide Gel Electrophoresis

SDS-polyacrylamide gel electrophoresis (SDS-PAGE) of viral polypeptides was performed using 10% polyacrylamide gels (Laemmli, 1970). The [^{35}S]-labelled polypeptides were detected by autoradiography of the dried gels.

3.7. Construction of Plasmids

Plasmid pIBV1a2 (formerly referred to as pKT1a2) which covers the IBV sequence from nucleotide 362 to 5753 (Liu *et al.*, 1995), was used to construct pIBV1a2Δ1, pIBV1a2Δ4, pIBV1a2Δ5 and pIBV1a2Δ8. Plasmid pIBV1a2Δ1 which covers the IBV sequence from nuclcotide 1311 to 5753, was constructed by cloning a BamHI/MluI digested-PCR fragment into BglII/MluI digested pIBV1a2. BglII digests pIBV1a2 at a position immediately upstream of the viral sequence and MluI digests the IBV sequence at nucleotide position 3997. An artificial AUG initiation codon in an optimal context (AC-C*AUG*G) was introduced immediately upstream of the viral sequence. The sequences of the upstream primer (LDX-3) and downstream primer (LDX-4) used for the construction of this plasmid are indicated in Table 1. Plasmid pIBV1a2Δ4 was constructed by cloning a 2.7 kb BamHI/MluI digested PCR fragment containing the IBV sequence from nucleotides 362 to 3047 into BglII/MluI digested pIBV1a2, resulting in the deletion of the region between nucleotides 3047 and 3997. The upstream and downstream primers used are LKP-1 and LKP-2 respectively (Table 1). Plasmid pIBV1a2Δ5 contains the IBV sequence from nucleotide 362 to 5027. It was constructed by ligation of a 1.03 kb DNA fragment, obtained by digestion of pIBV1a2 with MluI and SpeI, into MluI/XbaI digested pIBV1a1 (Liu *et al.*, 1995). SpeI cuts the IBV sequence at nucleotide position 5027.A second series of deletion mutants was made by introducing a UAA termination codon into different positions of the viral sequences. The first plasmid, pIBV1a2Δ8 which contains a stop codon at the nucleotide position 2856, was constructed by cloning a BamHI/MluI digested PCR fragment into BglII/MluI digested pIBV1a2. The upstream primer was LKP-1 and the downstream primer was LKP1a2Δ8. Plasmids pIBV1a2Δ10, pIBV1a2Δ11, and pIBV1a2Δ12 contain an UAA codon at nucleotide positions 2601, 2547 and 2517 respectively. The primers used to introduce these mutations are LKP1a2Δ10, LKP1a2Δ11 and

Table 1. Primers used for PCR amplification and site-directed mutagenesis

Primer name	Nucleotide sequences	ORF1a position
Deletion mutagensis primers		
LDX-3	5' -ACG (CGGATCCAC) **CATG**GGTTCTAAG- 3'	1304-1323
LDX-4	5' -TCTGTTTGCAAGTTACATCG- 3'	4060-4041
LKP-1	5' - (TCTCAGGGA) TCCCCCCACATACC - 3'	370-383
LKP-2	5' -ACGCTA (ACGCGT) TCATCAAGAGGCAG- 3'	3047-3022
LKP1a2Δ8	5' -CTCTCATAGACGCG*TTA*GATCAAATGGC- 3'	2870-2843
LKP1a2Δ10	5' -ATAGG (CCCGGG*TTA*) AGGTGGTATCT - 3'	2612-2584
LKP1a2Δ11	5' -CACT (CCCGGG*TTA*) TGCTTTGCAAACCAC - 3'	2557-2530
LKP1a2Δ12	5' -CATTATT (CCCGGG*TTA*) TTGAGACATTGGTG - 3'	2530-2501
Catalytic residue mutagenesis primers		
LKP1a2Δ4[C1274-S]	5' -CGATGGAAAT (AGC) TGGATTAGTTCAGC- 3'	4338-4364
LKP1a2Δ4[H1435-K]	5' -CTAATAGTGGC (AAG) TGTTATACACAAGC- 3'	4826-4853

LKP1a2Δ12. The PCR fragments obtained for these constructs were digested with BamHI and SmaI, and then cloned into BglII/SmaI-digested pKT0 (Liu *et al.*, 1994). Two mutants with alterations at the putative catalytic residues Cys^{1274} (pIBV1a2Δ4$^{C1274-S}$) and His^{1435} (pIBV1a2Δ4$^{H1435-K}$) were created by PCR amplification using pIBV1a2Δ4 as templates. The respective mutation primers used were LKP1a2Δ4$^{C1274-S}$ and LKP1a2Δ$^{H1435-K}$ (Table 1). All constructs made by site-directed mutagenesis were selected and confirmed by automated sequencing.

Regions of the IBV sequences presented in these constructs are illustrated in Fig. 1.

4. RESULTS

4.1. Further Analysis of the Involvement of the Two Papain-Like Proteinase Domains in Processing of the 1a Polyprotein to the 87 kDa Protein

It was previously shown that when a construct (pIBV1a1) containing the IBV sequence from nucleotide 362 to 4858 was expressed, only the full-length product of 220 kDa was observed (Liu *et al.*, 1995). However, expression of the IBV sequence up to nucleotide 5753 (pIBV1a2), resulted in the detection of three protein species: a 250 kDa full-length product and two cleavage products of 160 kDa and 87 kDa (Liu *et al.*, 1995), implying that processing of the full length polypeptide had occurred. These results suggest strongly that the region

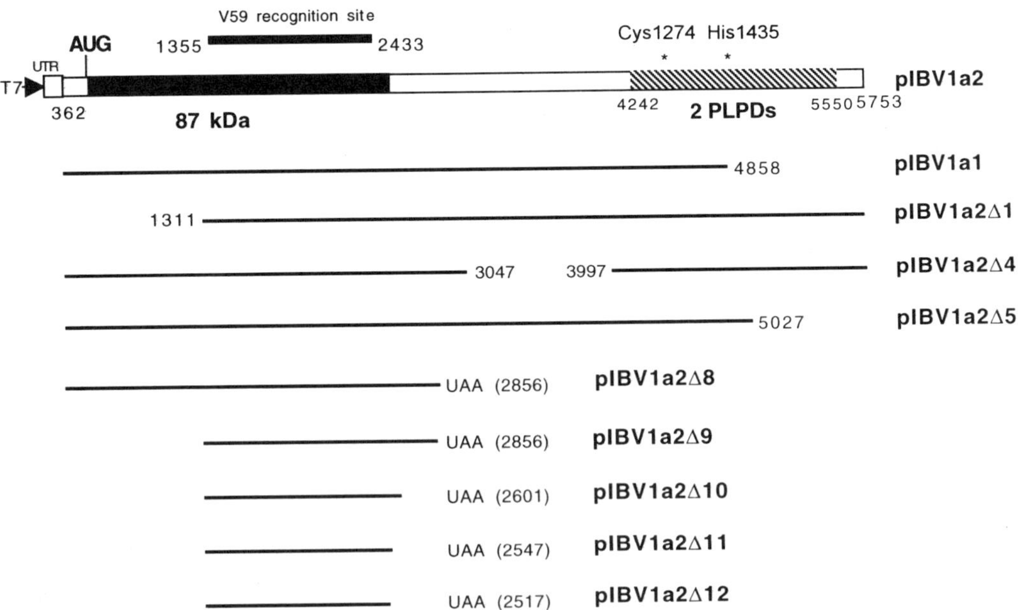

Figure 1. Diagram of the ORF1a region presented in plasmid pIBV1a2, showing the locations of the 87 kDa protein, the two PLPDs and the region recognised by antiserum V59. The positions of the putative C-terminal cleavage site (Gly^{673}↓Gly^{674} dipeptide) of the 87 kDa protein and the catalytic residues of the PLPD-1 (Cys^{1274} and His^{1435}) are also indicated. Also shown are the regions of ORF1a present in plasmids pIBV1a1 and the various deletion mutants. The nucleotide positions of the UAA termination codon introduced are indicated in the brackets.

from nucleotide 4859 to 5753 is essential for the processing of the 1a polyprotein to the 87 kDa protein. As computer prediction indicates that two putative overlapping PLPDs may be encoded from nucleotide 4155 to 5550 (Lee *et al.*, 1991), it is highly likely that this cleavage is mediated by the two proteinase domains. However, it was not clear if both domains were required for the cleavage. To clarify this issue, plasmid pIBV1a2Δ5 (Fig. 1) which contains the IBV sequence from nucleotide 362 to 5027 was constructed and expressed in Vero cells using the vaccinia virus-T7 expression system (Fuerst, 1986). The deletion showed little, if any, effect on the processing to the 87 kDa protein. As can be seen from Figure 2, expression of pIBV1a2Δ5 resulted in the detection of the 87 kDa protein as well as the full length 225 kDa protein and 136 kDa cleavage product. This data indicates that only the first PLPD is responsible for the release of the 87 kDa protein from its precursor.

4.2. Mutational Analysis of the Putative Catalytic Dyad of PLPD-1

Data presented above demonstrated that only the first of the two overlapping papain-like proteinase domains is required for processing of the 1a polyprotein to the 87 kDa

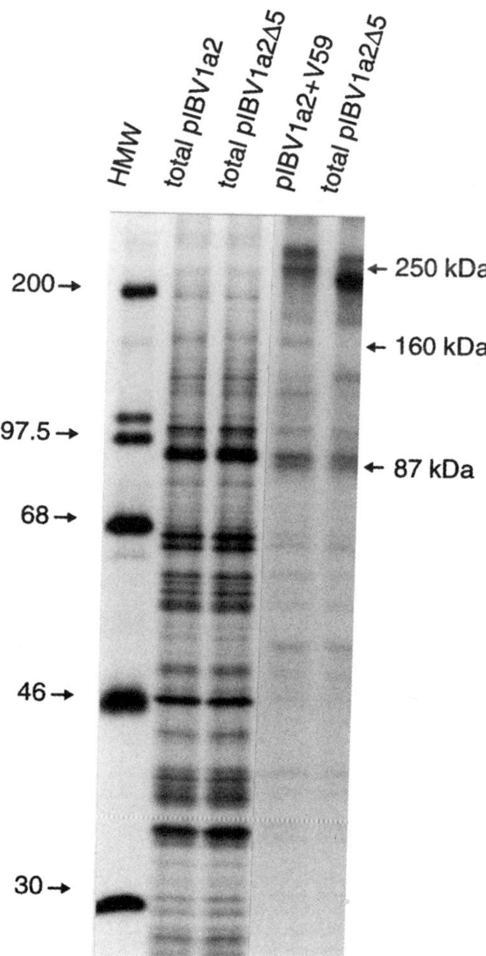

Figure 2. In vivo expression of pIBV1a2 and pIBV1a2Δ5 in Vero cells using the vaccinia recombinant virus-T7 expression system. The transfected cells were labelled with [^{35}S] methionine and the cell lysates were subjected to immunoprecipitation with antiserum V59 as indicated above each lane. SDS-PAGE was performed on an SDS-10% polyacrylamide gel. HMW-high molecular weight markers (numbers indicate kilodaltons).

product. We then sought to test if the two predicted residues, Cys^{1274} and His^{1435}, is associated with the catalytic activity of the proteinase (Lee *et al.*, 1991). To facilitate this study, plasmid pIBV1a2Δ4 (Fig. 1), containing an internal deletion of 0.96 kb from nucleotide 3047 to 3997, was constructed and used to make the mutant constructs. By monitoring the production of the 87 kDa protein *in vivo*, we were able to test the effect of the mutation on the proteinase activity. Using pIBV1a2Δ4 as the template, the putative nucleophilic residue Cys^{1274} was mutated to a Ser residue ($pIBV1a2Δ4^{C1274-S}$) and the predicted His^{1435} residue was mutated to a Lys ($pIBV1a2Δ4^{H1435-K}$). Fig. 3 shows that expression of pIBV1a2Δ4 yielded both the 190 kDa full length polypeptide and the 87 kDa protein. However, expression of $pIBV1a2Δ4^{C1274-S}$ and $pIBV1a2Δ4^{H1435-K}$ resulted in the detection of only the 190 kDa full-length product by antiserum V59 (Fig. 3). No 87 kDa protein was detected from the two mutant constructs (Fig. 3). This result demonstrated that substitutions of Cys^{1274} with a Ser and His^{1435} with a Lys disrupt the proteinase activity required for the re-

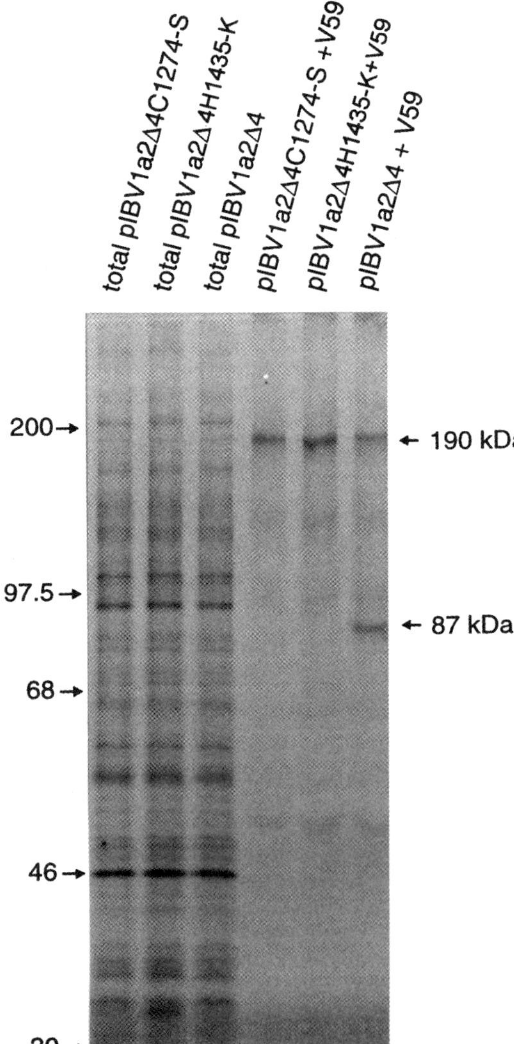

Figure 3. Mutational analysis of the putative catalytic residues Cys^{1274} and His^{1435} of PLPD-1. Plasmids pIBV1a2Δ4, $pIBV1a2Δ4^{C1274-S}$ and $pIBV1a2Δ4^{H1435-K}$ were transiently expressed in Vero cells using the vaccinia recombinant virus-T7 expression system. The plasmid transfected cells were labelled with [^{35}S] methionine and the lysates were subjected to immunoprecipitation with V59. The polypeptides were separated on an SDS-10% polyacrylamide gel. HMW-high molecular weight markers (numbers indicate kilodaltons).

lease of the 87 kDa protein. It is therefore suggested that the Cys^{1274} and His^{1435} may be core residues of the catalytic centre of PLPD-1.

4.3. Deletion Analysis of the C-Terminal Cleavage Site of the 87 kDa Protein

The target cleavage site of PLPD-1 was next analysed by deletion analysis. The first deletion mutant, pIBV1a2Δ8, was constructed by introducing a UAA termination codon into pIBV1a2 at the nucleotide position 2856. Expression of this plasmid led to the synthesis of a product with an apparent molecular mass of 93 kDa, which appeared much larger than the expected size of 87 kDa in SDS-PAGE (Fig. 4a). This implies that the actual cleavage site responsible for releasing the C-terminus of the 87 kDa protein is located at least 250 bp upstream of this position.

To rule out the possibility that the size difference observed was caused by N-terminal cleavage of the pIBV1a2-encoded polyprotein, pIBV1a2Δ1 was constructed by deletion of the IBV sequence from nucleotide 362 to 1310. Expression of this construct resulted in the detection of three protein species of 220 kDa, 160 kDa and 58 kDa, representing the full-length and the cleavage products, respectively (Fig. 4b). The migration of the 58 kDa protein in SDS-PAGE suggested that it may be the 5'-truncated version of the 87 kDa protein. Based on pIBV1a2Δ1, plasmids pIBV1a2Δ9, pIBV1a2Δ10, pIBV1a2Δ11 and pIBV1a2Δ12 (Fig. 1) were then constructed by inserting the UAA stop codon into pIBV1a2Δ1 at nucleotide positions 2856, 2601, 2547 and 2517 respectively, and were expressed in Vero cells. As shown in Fig. 4b, transfection of pIBV1a2Δ9 and pIBV1a2Δ10 resulted in the synthesis of the full-length stop products migrating in SDS-PAGE more slowly than the 58 kDa protein. Their approximate sizes were 65 kDa and 61kDa, respectively. Plasmids pIBV1a2Δ11 and pIBV1a2Δ12 encoded proteins with apparent molecular masses of 58 kDa and 52 kDa. The product encoded by pIBV1a2Δ11 had an electrophoretic mobility in SDS-PAGE almost identical to that of the 58 kDa protein encoded by pIBV1a2Δ1 (Fig. 4b). As the UAA termination codon in pIBV1a2Δ11 was inserted immediately downstream of the deduced amino acid position 673, it is likely that the C-terminal cleavage site of the 87 kDa protein is in the vicinity of this residue.

4.4. Interaction of the 87 kDa Protein with Products Encoded by the Papain-Like Proteinase Domains

When pIBV1a2 and its derivatives were expressed *in vivo* , detection of a 160 kDa protein was consistently observed (Fig. 2 and see Liu *et al.*, 1995). It was previously postulated that the 160 kDa protein was an intermediate cleavage product containing the 87 kDa protein, as it could be immunoprecipitated by antiserum V59. However, this possibility was ruled out by the expression of the N- and C- terminal deletion constructs pIBV1a2Δ1 and pIBV1a2Δ5. Expression of pIBV1a2Δ1 showed the detection of the 160 kDa protein (Fig. 4b). Expression of pIBV1a2Δ5, however, resulted in the synthesis of a 136 kDa product; no 160 kDa protein was detected (see Fig. 2). It is therefore likely that the 160 kDa product was a C-terminal cleavage product released from the 250 kDa polyprotein encoded by pIBV1a2. Detection of the 160 kDa protein by the N-terminal specific antiserum V59 may be due to the interaction of this protein with the 87 kDa protein species.

To investigate this possibility, C-terminal specific antiserum V16 which recognised the IBV sequence between nucleotides 4864 and 5576, was used in immunoprecipitation

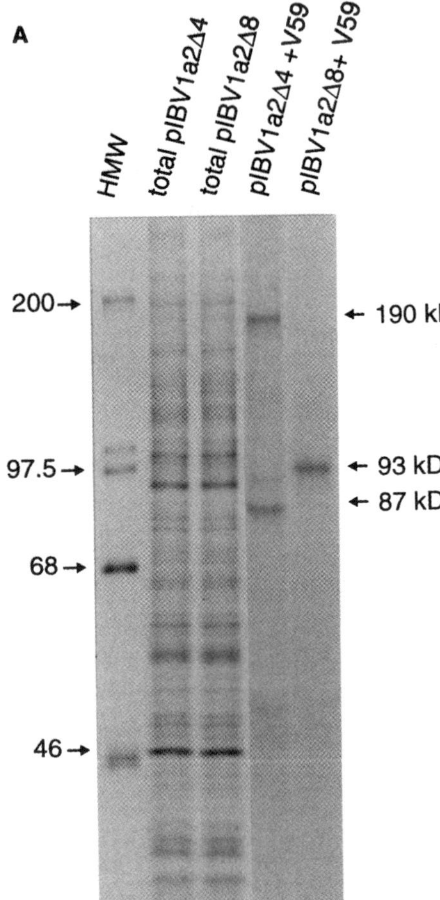

Figure 4. Determination of the C-terminal cleavage site of the 87 kDa protein by deletional studies. A) *In vivo* expression of pIBV1a2Δ8 and pIBV1a2Δ4 in Vero cells using the vaccinia recombinant virus-T7 expression system. The cells were labelled with [^{35}S] methionine and the plasmid-transfected cell lysates were subjected to immunoprecipitation with antiserum V59 as indicated above each lane. The proteins synthesised were separated on an SDS-10% polyacrylamide gel. HMW-high molecular weight markers (numbers indicate kilodaltons). B) Immunoprecipitation of the proteins encoded by pIBV1a2Δ1, pIBV1a2Δ9, pIBV1a2Δ10, pIBV1a2Δ11, and pIBV1a2Δ12 were done with V59.

studies. As shown in Figure 5a, immunoprecipitation of pIBV1a2Δ1-transfected cell lysate with V16 led to the detection of the 220 kDa, 160 kDa and 58 kDa protein species. The same three protein species could be immunoprecipitated by V59 (Fig. 5a).

To test the specificities of antisera V59 and V16, plasmids pIBV1a2Δ8 (covers the IBV sequences between nucleotides 362 and 2856) and pIBV2P (contains the IBV sequences between nucleotides 3832 and 5757) was expressed in vitro. As can be seen from Figure 5b, pIBV1a2Δ8 encoded a polypeptide with an apparent molecular weight of 93 kDa. This protein could only be immunoprecipitated by antiserum V59 (Fig 5b). Similarly, pIBV2P encoded a 74 kDa protein, which could only be recognised by antiserum V16 (Fig. 5b).

5. DISCUSSION

A number of positive-stranded RNA viruses employs proteolytic processing as one of the strategies for their gene expression. Post-translational processing of the 741 kDa polyprotein encoded by the 27.6 kb genome-length mRNA 1 of the Coronavirus IBV into

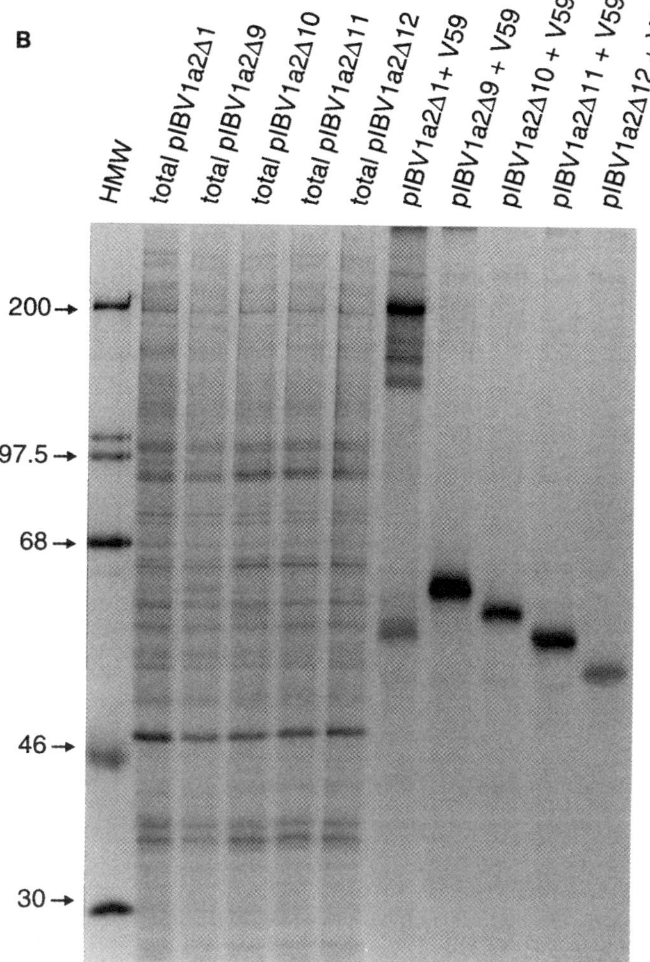

Figure 4B.

smaller functional proteins is probably mediated by both viral and cellular proteinases. One of such cleavage product is an 87 kDa protein identified previously from IBV-infected Vero cells by using a region-specific antiserum V59 (Liu *et al.*, 1995). Data obtained from expression of IBV sequence up to nucleotide 5753 have shown that inclusion of the predicted sequence encoding the two putative overlapping papain-like proteinase domains was required for releasing the 87 kDa protein from the full length precursor. Through the expression of a construct lacking the second of the overlapping papain-like proteinase domains, we demonstrated that this domain was not essential for the *in vivo* processing of the 1a polyprotein to the 87 kDa product. Therefore, only PLPD-1 encoded between nucleotides 4242 and 5031 was required for the *cis* -cleavage of the 87 kDa protein from the 1a polyprotein precursor. Site-directed mutagenesis studies confirmed the previously predicted Cys^{1274} and His^{1435} residues as members of the catalytic centre of PLPD-1. The lack of proteolytic activity when either of the catalytic residues was substi-

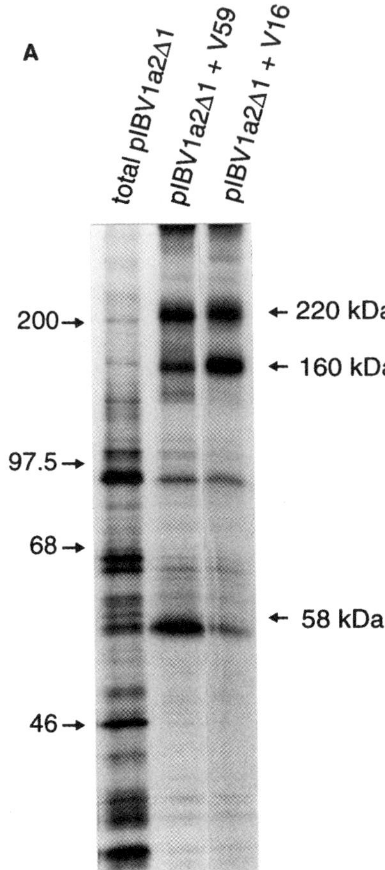

Figure 5. A) Analysis of polypeptides synthesised from *in vivo* expression of pIBV1a2Δ1. The cells were labelled with [^{35}S] methionine and the lysate was subjected to immunoprecipitation with antisera V59 and V16 as indicated above each lane. Electrophoresis of the polypeptides were performed on an SDS-10% polyacrylamide gel. HMW-high molecular weight markers (numbers indicate kilodaltons). B) Immunoprecipitation of *in vitro* translation products of RNAs cotranscribed from pIBV1a2Δ8 and pIBV2P in reticulocyte lysate using the transcription-coupled-translation system (Promega). Approximately 200μg/ml of plasmid DNA was added to reticulocyte lysate and the proteins were labelled with [^{35}S] methionine. The polypeptides were separated on an SDS-10% PAGE either directly or after immunoprecipitation with antisera V59 and V16 and detected by autoradiography. Numbers indicate molecular mass in kilodaltons.

tuted even though the second proteinase domain was still intact further reinforces the conclusion that only the first PLPD is required for the processing to the 87 kDa protein.

Unlike MHV, processing of the ORF1a polyprotein into smaller products by the PLPDs was not observed during the *in vitro* translation of constructs containing this region of ORF1a (Liu *et al.*, 1995). Microsequencing of the C-terminal cleavage site of the 87 kDa protein could not be performed as insufficient templates could be generated by using the *in vivo* expression system. In order to define the C-terminal boundary of the 87 kDa protein, we constructed and expressed a series of deletion constructs by introducing a UAA triplet at different positions of the IBV sequences. By comparing the migration of the termination products in SDS-PAGE with the cleavage product, this approach allows us to define Gly^{673}-Gly^{674} dipeptide as a potential cleavage site for releasing the C-terminus of the 87 kDa protein. Mutagenesis studies are underway to confirm this speculation.

We have previously reported that the 160 kDa species is probably the intermediate cleavage product covering the 87 kDa protein, as it could also be immunoprecipitated by region-specific antiserum V59. By construction and expression of N- and C-terminal deletion constructs, we showed that the 160 kDa protein species was actually not encoded from the 5'-end of ORF1a, as the full-length 160 kDa protein was observed even when the 5'-end of ORF1a up to nucleotide 1311 was deleted. Therefore, the 160 kDa protein species probably represents the C-terminal cleavage product resulted from cleavage of the

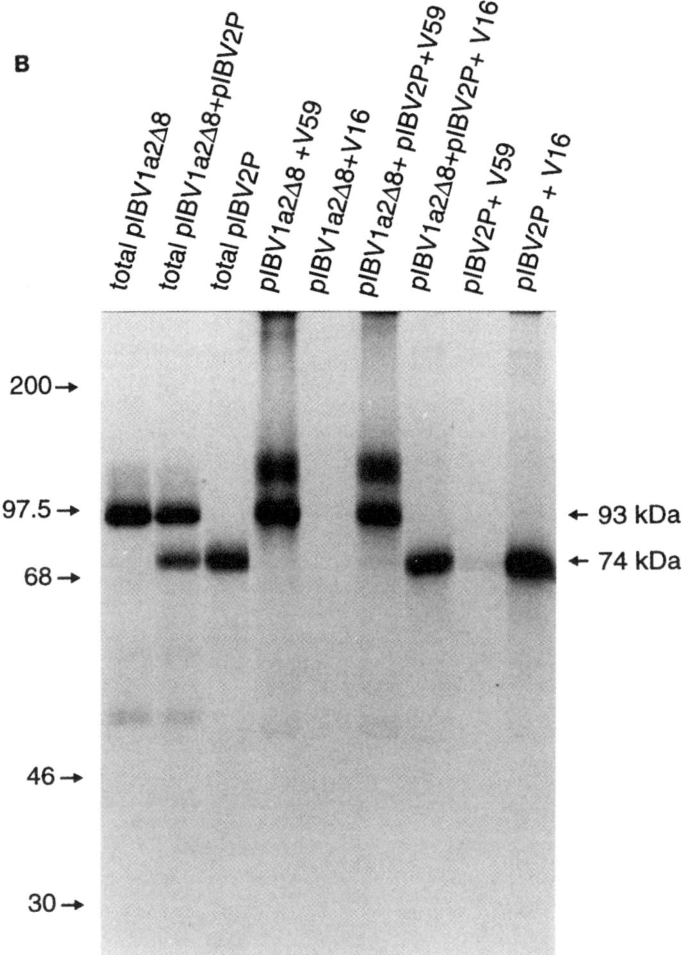

Figure 5B.

250 kDa polypeptide encoded by pIBV1a2, at the Gly^{673}-Gly^{674} dipeptide bond. This result meanwhile rules out the possibility that a second PLPD cleavage site may be located in this portion of the 1a polyprotein.

Evidence presented confirmed that the detection of the 160 kDa protein by N-terminal specific antiserum is due to the interaction of the 160 kDa protein with the 87 kDa protein. We are currently uncertain if this interaction is essential for the functions of the products encoded by this region of the viral RNA.

REFERENCES

Boursnell, M. E. G., T. D. K. Brown, I. J. Foulds, P. F. Green, F. M. Tomley and M. M. Binns., 1987, Completion of the sequence of the genome of the coronavirus avian infectious bronchitis virus, *J. Gen. Virol.* **68**:57–77.
Fuerst, T. R., E. G. Niles, F. W. Studier and B. Moss., 1986, Eukaryotic transient-expression system based on recombinant vaccinia virus that synthesises bacteriophage T7 RNA polymerase, *Proc. Natl. Acad. Sci. USA* **83**:8122–8127.

Gorbalenya, A. E., E. Y. Koonin, A. P. Donchenko and V. M. Blinov., 1989, Coronavirus genome: prediction of putative functional domains in the non-structural polyprotein by comparative amino acid sequence analysis, *Nucleic Acids Research* **17**:4847–4860.

Hughes, S. A., P. J. Bonilla and S. R. Weiss. 1995. Identification of the murine coronavirus p28 cleavage site, *J. Virol.* **69**:809–813.

Laemmli, U. K, 1970, Cleavage of structural proteins during the assembly of the bacteriophage T4. *Nature* (London) **227**:680–685.

Lee, H-J., C-K. Shieh, A. E. Gorbalenya, E. V. Koonin, N. L. Monica, J. Tuler, A. Bagdzhadzhyan and M. M. C. Lai, 1991, The complete sequence (22 kilobases) of murine coronavirus gene 1 encoding the putative proteases and RNA polymerase, *Virology* **180**:567–582.

Liu, D. X., I. Brierley, K. W. Tibbles and T. D. K. Brown., 1994, A 100-kilodalton polypeptide encoded by open reading frame (ORF) 1b of the coronavirus IBVis processed by ORF 1a products, *J. Virol.* **68**:5772–5780.

Liu, D. X., K. W. Tibbles, D. Cavanagh, T. D. K. Brown and I. Brierley., 1995, Identification, expression and processing of an 87 kDa polypeptide encoded by ORF1a of the coronavirus infectious bronchitis virus, *Virology* **208**:48–57.

Liu, D. X., H. Y. Xu, T. D. K. Brown., 1997, Proteolytic processing of the coronavirus IBV1a polyprotein: identification of a 10-kilodalton polypeptide and determination of its cleavage sites, *J. Virol* **208**:48–57.

Replication II: RNA Synthesis

23

ROLE OF THE NONSTRUCTURAL POLYPROTEINS IN ALPHAVIRUS RNA SYNTHESIS

Dorothea L. Sawicki and Stanley G. Sawicki

Department of Microbiology and Immunology
Medical College of Ohio
Toledo, Ohio 43699

1. INTRODUCTION

Alphaviruses are of general interest because they replicate in such a broad range of animals, including invertebrates as well as mammals and birds. This broad replicative potential is an essential part of their transmission strategy. The successful synthesis of many infectious progeny starts with the translation of the genome into the viral nonstructural proteins (nsPs) that form an unstable viral RNA-dependent RNA polymerase activity, which copies the infecting genome into a complementary, or minus-strand, RNA template of genome-length. Complexes composed of the minus-strand templates and long-lived viral replicases or transcriptases then produce throughout the rest of the infectious cycle large amounts of genomes or via internal initiation of a 3'-co-terminal, subgenomic mRNA. The 26S mRNA serves as a template for translation of the viral capsid and envelope proteins. Thus, the crucial step after an alphavirus enters the cell is the formation of the initial replication complex that produces minus-strand RNA since the number of minus strand templates that form during the early part of the infectious cycle determines the amount and rate of viral plus-strand RNA synthesis. This review will highlight research efforts directed toward understanding the synthesis of alphavirus RNA and its regulation.

2. ALPHAVIRUS NONSTRUCTURAL PROTEINS

During the past decade, there have been important advances made using alphaviruses as a model system. One of the most significant was the successful construction of an infectious clone of the heat resistant strain of Sindbis virus (SIN HR) by Charlie Rice and his colleagues (Rice et al., 1987) and subsequently of Semliki Forest virus (SFV)

Coronaviruses and Arteriviruses, edited by Enjuanes *et al.*
Plenum Press, New York, 1998

by Liljestrom et al. (1991). This has allowed mutations in many of the well characterized temperature-sensitive (ts) mutants of SIN HR to be mapped and the functions of the four nsPs and of their conserved motifs or domains to be determined. The original collection of ts mutants of SIN HR was isolated by Burge and Pfefferkorn (1966a, b) and was added to by Strauss et al. (1976). The ts mutants of SIN HR were assigned originally into seven complementation groups, of which four (A, B, F and G) were RNA-negative and failed to synthesize viral RNA when infection was initiated and maintained at the nonpermissive temperature (40°C). These mutants and the infectious clone of SIN HR have provided powerful tools for studying alphavirus RNA synthesis.

Analysis of conditionally-lethal mutants has shown that all four nsPs are essential for viral replication and transcription. The nsPs are numbered from the 5' end of the genome according to their order in the genome and are synthesized initially as two polyproteins, P1234 and P123 (Figure 1). SIN polyprotein P1234 results from readthrough of an opal termination codon located between the coding sequence of nsP3 and nsP4 (Strauss et al., 1983) and its efficiency is affected by a single C residue immediately downstream of the termination codon (Li and Rice, 1993). Cleavage of P1234 and P123 is regulated and is catalyzed by a viral thiol protease located in nsP2 (Ding and Schlesinger, 1989; Hardy and Strauss, 1989). In SFV, nsP4 may encode a second protease (Takkinen et al., 1990). The nsPs are present in membrane-associated complexes expressing replicase or transcriptase activities and these complexes retain their activity in vitro (Barton et al., 1991). The viral replicases synthesize genome-length plus and minus-strand RNA, and the transcriptase synthesizes the subgenomic 26S mRNA.

2.1. Functions of the Nonstructural Proteins

The following functions are associated with the nsPs. nsP1 has methyltransferase and guanylyltransferase activities (Ahola and Kääriäinen, 1995; Mi and Stollar, 1991) and is thought to function in capping of the genome and 26S mRNA as well as in the initiation of minus-strand RNA synthesis. This latter property may involve the binding of a N-terminal Y residue of nsP4 into a hydrophobic pocket in the C-domain of nsP1 since substitutions in nsP1 at amino acid 348 led to ts minus strand initiation (Wang et al., 1991; Hahn et al., 1989b; Sawicki et al., 1981) and at amino acid 349 led to rescue of a replication defective nsP4 mutant that possessed A instead of Y as its N-terminus (cited in Strauss and Strauss, 1994). nsP2 also possesses several functional domains. The N-terminal region of nsP2 contains sequence motifs found in NTP-binding proteins/helicases, binds ATP and GTP, and its ATPase activity is stimulated by RNA (reviewed in Rikkonen et al., 1994);

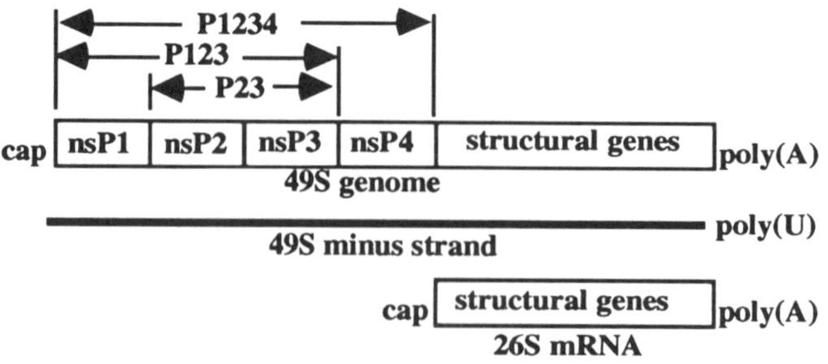

Figure 1. Alphavirus gene arrangement and RNA species.

whether it functions as a nucleic acid chaperone is not known. The C-terminal region of nsP2 contains the thiol protease activity and processes nonstructural polyprotein precursors (Gorbalenya and Koonin, 1993; Ding and Schlesinger, 1989; Hardy and Strauss, 1989) and encodes a nuclear localization signal (Rikkonen et al., 1992). Although the phosphoprotein nsP3 is essential for replicase and transcriptase activities (LaStarza and Rice, 1994; Wang et al., 1994; Peränen, 1991; Li et al., 1990), its functional role is not clear. nsP4 contains the conserved GDD motif found in RNA-dependent RNA polymerases (Gorbalenya and Koonin, 1993) and is the polymerase subunit. Recently, Gerhke and coworkers (Ansel-McKinney et al., 1996) identified a second conserved sequence (PT/NXRS) in the N-domain of nsP4 based on its homology with a PTXRS sequence in a RNA-binding motif in the capsid proteins of ilarviruses and alfalfa mosaic virus (AMV). Binding of capsid proteins or peptides containing this motif to the 3' end of AMV templates requires the R residue and is essential for RNA transcription. Since capsid proteins coisolate with the viral replicase, Ansel-McKinney et al. (1996) propose capsid binding serves to stabilize the viral replicase-RNA complex.

Sequence analysis indicated that only nsP1, the N-domain of nsP2, and nsP4 of alphaviruses and rubiviruses share homology with nsPs of plus-stranded RNA plant members of the SIN super-family of viruses (Koonin and Dolja, 1993; Ahlquist et al., 1985). Among these, only the carla, capillo and tymoviruses are known presently to encode proteases similar to nsP2 (Koonin and Dolja, 1993). Of interest, group 1 helicases and papain-like proteases found in alpha- and related viruses are present in coronaviruses and arteriviruses (Gorbalenya et al., 1991) and the coronavirus protease is next to an "X" domain as in alphaviruses (Koonin and Dolja, 1993). The alpha-like viruses possess a unique group 1 methyltransferase that may reflect its unusual ability to methylate guanine before it is linked to nascent plus strand transcripts (Ahola and Määriäinen, 1995).

2.2. nsP Interaction with Host Components

Alphavirus replication is dependent on host factors, one or more of which interacts with nsP4 to determine host range (Lemm et al., 1990). Calreticulin was shown by gel-shift assays (Atreya et al., 1995) to bind the 3' sequence of rubella and SIN genome RNA and may serve as a host component of the minus strand replicase or act to mark the genome as a template for minus strand synthesis. La or a La-related protein binds to the 5' end of plus strands (Pardigon et al., 1996; Pogue et al., 1996). Solubilization and immunoprecipitation of replication complexes from alphavirus infected cells late in infection (RC_{stable}) was accomplished by Barton et al. (1991), who found detergent treatment released complexes from infected cell membranes that were composed of cleaved nsPs and retained transcriptive activity after immunoprecipitation. A host protein of approximately 120–130 kDa molecular weight co-immunoprecipitated with the complex in proportion to the amount of nsP1 and may be an additional required host factor.

2.3. Role of Nonstructural Polyproteins in Minus Strand Synthesis

Genetic (Sawicki et al., 1990; 1981; Sawicki and Sawicki, 1987; 1985; Wang et al., 1994; 1991) and biochemical (Lemm et al., 1994; 1993; Shirako and Strauss, 1994) evidence supports proposals (de Groot et al, 1990; Hardy et al., 1990; Sawicki and Sawicki, 1980) of a specific processing pathway whereby the nascent P1234 and P123 polyproteins are formed into the various alphavirus polymerase activities (Figure 2). Complexes of polyproteins and fully cleaved nsPs are generated first. Specifically, cleavage at the 3/4

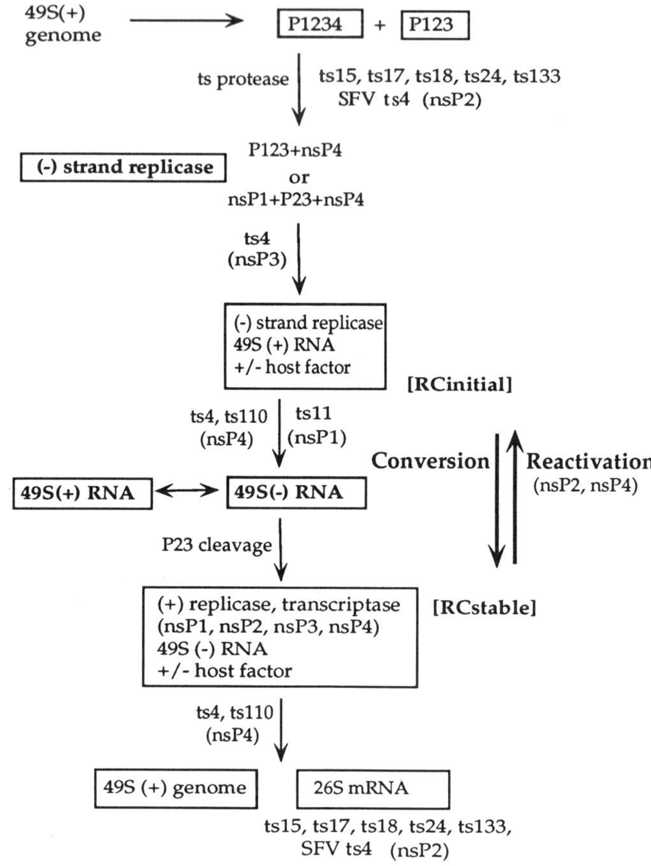

Figure 2. A model for alphavirus replication and the role of polyproteins in minus strand synthesis.

site to release nsP4 occurs rapidly, probably co-translationally, and is essential to the formation of functional polymerase activity. Polymerase complexes containing P123+nsP4 bind to the 3' end of genome RNA and elongate minus strands. When overproduced from two separate mRNAs using a vaccinia vector expression system, uncleaved P123+nsP4 complexes could function as plus strand replicases and formed complexes in vivo in the absence of template RNAs that associated with membranes pelleting at 15,000 x g (Lemm et al., 1997; 1994; 1993), similar to authentic alphavirus complexes. Subsequent cleavage at the 1/2 site forms replicases composed of nsP1+P23+nsP4 that are active in minus and plus strand synthesis (Lemm et al., 1997; 1994). Of interest, functional polymerases formed in cells when the nsPs were translated from one mRNA as P1234 or when separate mRNAs encoded P123 and nsP4 or nsP1, P23 and nsP4 but not when mRNAs encoded each nsP separately. Shirako and Strauss (1994) obtained similar results when combinations of nsPs and stable forms of the polyproteins were expressed from infectious RNA transcripts following transfection into cells. One difference found in this assay system compared to overexpression of nsPs and template RNAs in the vaccinia system was the inefficient synthesis of plus strands by P123+nsP4 $RC_{initial}$ compared to nsP1+P23+nsP4 $RC_{initial}$. This suggests nsP1+P23+nsP4 may be the more active of these two initial complexes in infected cells, even though its minus strand replicase activity also is short-lived.

Table 1. Characterization of SIN HR conditionally lethal RNA-negative mutants and identification of their functional lesions

Virus	Mutant protein	Predicted change	Specific ts phenotypes[d]
ts11 (B)[a]	nsP1	Ala-348>Thr[b]	minus strand synthesis
ts14 (AII)	nsP2	Glu-163>Lys[c]	? conversion
ts16 (AII)	nsP2	Val-275>Ile[c]	? conversion
ts21 (AII)	nsP2	Cys-304>Tyr[b]	? conversion
ts19 (AII)	nsP2	Leu-416>Ser[c]	? conversion
ts118 (F)	nsP2	Val-425>Ala[b]	? conversion
ts15 (AI)	(nsP2)		protease and 26S
ts18 (G)	nsP2	Phe-509>Leu[b]	protease and 26S
ts17 (AI)	nsP2	Ala-517>Thr[c]	protease, 26S, template specificity
ts7 (G)	nsP2	Asp-522>Asn[b]	pfu[b] and template specificity[c]
ts133 (AI)	nsP2	Asn-700>Lys[c]	protease, 26S, template specificity
ts24 (AI)	nsP2	Gly-736>Ser[b]	protease and 26S
ts138 (AII)	nsP3	Ala-68>Gly[c]	? conversion
ts4 (AII)	nsP3	Ala-248>Val[c]	minus strand synthesis
ts7 (G)	nsP3	Phe-312>Ser[b]	RNA synthesis[b], ? conversion
ts118 (F)	nsP4	Gln-93>Arg[b]	ts requires both lesions
ts6 (F)	nsP4	Gly-153>Glu[b]	elongation
ts24 (AI)	nsP4	Gln-191>Lys[c]	template specificity
ts110 (F)	nsP4	Gly-324>Glu[b]	elongation

[a]Complementation group

[b]Mapping was reported by Hahn et al (1989a, b)

[c]Mapping was reported by Dé et al. (1997; 1996); Wang et al. (1994); Sawicki and Sawicki (1993) or Sawicki et al. (1990).

[d]Abbreviations of ts phenotypes exhibited after shift of infected cultures to 40°C refer to reactivation of minus strand synthesis by RC_{stable} (template specificity) that does not confer a ts RNA-negative phenotype; ts nsP2 protease and failure to cleave P1234 or P123 (protease); ts selective loss of 26S mRNA synthesis (26S); and failure to synthesize RNA at nonpermissive temperature (RNA synthesis).

Finally, cleavage of P23 polyproteins at the 2/3 site results in long-lived complexes composed of nsP1, nsP2, nsP3, and nsP4 (RC_{stable}) that make single-stranded 49S genomes and 26S mRNAs at high rates (Shirako and Strauss, 1994; Lemm et al., 1994, 1993; Barton et al., 1991). Thus, P23 cleavage inactivates the minus strand replicase activity and forms a new transcriptase activity but retains genome replicase activity.

That polyproteins played a role in minus strand synthesis was indicated also by genetic complementation analyses and by the mapping of ts mutants of SIN HR (Table 1). While complementation occurs among SIN HR mutants, it is inefficient and alphavirus mutants other than those from SIN HR fail to complement. This may be because their formation as polyproteins allows the nsPs to more efficiently form a complex with one another and, once formed, these complexes do not readily dissociate. In this regard, plant viruses belonging to the SIN superfamily produce homologous nsPs that actually function as polyproteins. As expected from classical genetic complementation, mutants in complementations groups B or F rescued mutants assigned to the other three complementation groups and were defective in individual and different nsPs. The sole member of the B group, ts11 was selectively defective in minus strand synthesis (Sawicki et al., 1981) and its defect mapped to nsP1 (Table 1). This indicated nsP1 functions in the initiation of minus strands in a manner that is likely to be independent of its activity in capping of plus strands since overall plus strand synthesis is unaffected. The mutations in ts6, ts110, ts118 that had been assigned to the F complementation group mapped to nsP4, as did several host range mutants (Lemm et al., 1990). Of interest was the finding that, although ts118 had a complementation pattern of an nsP4 mutant, it was in fact a double mutant and possessed changes in both nsP4 and nsP2 genes; each mutation by itself did not lead to ts virus production. The implied interaction between the two nsPs is reminiscent of the need for specific sequences in BMV homologues, 1a and 2a, which are produced from separate mRNAs, for complex formation and transcriptional activation (Dinant et al., 1993). Wang et al. (1994) concluded that the ability to rescue mutants assigned to complementation groups other than B and F argued that nsP4 and nsP1 were diffusable and active in trans and thus that their cleavage from nascent P1234 and P123 occurred early in the process of forming the replication complex.

Mutants assigned originally to either the A or G complementation groups had defects in nsP2 or nsP3 (Table 1) and none of these mutants complemented each other (Wang et al., 1994). This was an unexpected and striking result as it was a general property of 12 different mutations that covered most of the nsP2 sequence and the conserved N-domain of nsP3. nsP2 and nsP3 apparently act initially as a single cistron. This may be because they function as polyprotein P23 or they form a stable association while in the precursor stage that is retained after cleavage at the 2/3 site. Because Barton et al. (1991) found P23 was not a major component of stable, plus strand replicases and transcriptases, we proposed P23 functioned early in the formation or activity of the initial replication complex.

The notion that cleaved nsP1 was a functional replicase component and that polyprotein forms of nsP2 and nsP3 were essential for SIN minus strand replicase activity was confirmed recently by Lemm et al. (1997), who reconstituted active replication complexes using mRNAs and template RNAs produced from vaccinia virus vectors. Stable P23 forms were created either by mutating the 2/3 cleavage site or by replacing the protease active site C with a S residue. nsP1+P23+nsP4 complexes transcribed wildtype levels or more of minus strand RNA and about twice as much genome RNA but were blocked for 26S mRNA transcription. This overproduction of 49S plus strands is similar to what is seen at 40°C in ts 26S RNA mutant-infected cells and is thought to result from the recruitment for 49S genome synthesis of minus strand templates previously engaged in 26S mRNA synthesis (Sawicki et al., 1978).

In summary, genetic and biochemical approaches lead to the conclusion that nsP4 must be cleaved to be active as a polymerase, that P34 is an inactive form, and that nsP2 and nsP3 function initially as polyprotein P123 or P23 for minus strand synthesis. Cleavage of P123 forms nsP1+P23+nsP4 that has both efficient plus and minus strand replicase activities and final, full cleavage to nsP1+nsP2+nsP3+nsP4 activates 26S mRNA transcription and inhibits the minus strand replicase function. Interestingly, this latter cleavage apparently does not diminish plus strand replicase activity of the less mature complex, just as 1/2 cleavage does not diminish minus strand replicase activity of the initial P123+nsP4 complex. It is clear that nsP polyproteins are essential, active forms. P23 polyproteins play a unique role in template specificity for minus strand synthesis, while nsP2 with free C-termini (and/or nsP3 with free N-termini) have altered promoter recognition, leading to internal initiation of 26S mRNA synthesis.

3. ALPHAVIRUS MINUS STRAND SYNTHESIS

In alphavirus replication, minus strand synthesis occurs only early in infection and ceases at a time when the rate of plus strand synthesis reaches a maximum and constant level (Sawicki and Sawicki, 1987). Its cessation fixes the number of replication and transcription complexes making genome and subgenomic 26S mRNA (Wang et al., 1991). The synthesis of minus strand RNA only early in infection may be a common feature of SIN superfamily viruses since this is also a strategy in replication of the plant viruses TMV (Ishikawa et al., 1991) and BMV (Kroner et al, 1990). For coronaviruses, both plus and minus strand polymerases are unstable (Sawicki and Sawicki, 1986c); for alphaviruses, only the minus strand replicases are unstable and require continuous protein synthesis (Sawicki and Sawicki, 1980). Their products are found in partially double-stranded RIs where they serve as templates for genome and 26S mRNA synthesis and, in certain ts mutants, the same minus strand template can be switched between these two activities in a temperature-dependent manner (Sawicki et al., 1978). The plus strand replicase and transcriptase are stable activities and function throughout infection. Thus, it was apparent that alphavirus replication required formation of two distinct forms of polymerase complexes: Initial and short-lived $RC_{initial}$ that produce minus strands and long-lived RC_{stable} that yield 49S or 26S mRNAs. The two complexes were not independently formed from nascent polyproteins. Rather, $RC_{initial}$ expressing minus strand replicase activity was formed first from nascent polyproteins and was then converted by a process that included P23 cleavage to RC_{stable} expressing plus strand replicase or transcriptase. We failed to identify mutants or conditions forming RC_{stable} without $RC_{initial}$ formation; we found the number of minus strands determined the number of RC_{stable} and the rate of overall plus strand synthesis; and, although mature nsPs compose active RC_{stable} complexes, their translation initially as polyproteins was found essential for their activity (see section 2.3).

3.1. Mutants Defective in Synthesis of Minus Strands

As expected for the RNA-dependent RNA polymerase subunit, nsP4 functions in the elongation of nascent viral transcripts and is an activity necessary but not selective for minus strand synthesis (Barton et al, 1988). Two mutants were identified whose phenotypes upon shift to nonpermissive temperature (40°C) reproduced the effect of translation inhibition: ts11 and ts4 rapidly and selectively shut off the synthesis of minus strand RNA at 40°C, in contrast to the normal cessation that occurred later at 30°C (Table 1). The mutation in ts11 was mapped to nsP1 (Table 1) and downstream of the methyltransferase domain located within the first 100 amino acids of nsP1. The defect in nsP1 did not affect production of nsP1, nsP2, nsP3 and nsP4 at 40°C and it functioned in trans to inhibit minus strand synthesis on other wildtype, viral templates, consistent with it functioning via the nsP1 protein. Like many of these mutants, mutant nsP1 made at 40°C in ts11 infected cells failed to regain activity upon return to 30°C. They also were not dominant negative and did not interfere with nsP's newly synthesized at 30°C that were responsible for the resumption of minus strand synthesis that occurred at 30°C after the 2.5 h incubation at 40°C.

The mutation in ts4 mapped to a change of A-268 to V in the conserved N-half of nsP3 (Table 1). nsP3 is a 549-amino acid, phosphorylated protein whose N-terminal 325 residues are highly conserved among the alphaviruses, while the C-terminal 225 residues are variable or can be deleted. The N-half of nsP3 contains essential sequences since deletion of amino acids 218–273 is lethal (LaStarza et al, 1994) as well as the "X" conserved sequences found also in rubiviruses, hepatitis E virus and coronaviruses, flanking their papain protease domains (Gorbalenya and Koonin, 1993; Gorbalenya et al, 1991). It is apparent that nsP3 (as P123 or P23) has a role in the formation of $RC_{initial}$ as opposed to nsP1 that functions both early or late in the initiation of minus strands. When present in previously formed replication complexes together with a mutant nsP4, V268 nsP3 did not block reactivation of minus strand synthesis but T-348 nsP1 did (Wang et al., 1994). The V268 nsP3 phenotype also was consistent with a lesion that was expressed in a precursor of nsP3 (e.g. P23) that would be part of the initial complex but not a stable complex that contained cleaved nsP3. The defect in V268 nsP3 may be related to a defect in phosphorylation at 40°C as revertants restored both normal replicative and phosphorylation phenotypes (Dé et al., 1997). Finally, absence of nsP3 or "X" domain homologues in many alpha-like plant viruses that synthesize minus strands and its presence in coronaviruses make its role of mutual interest.

3.2. Mutants Defective in Cessation of Minus Strand Synthesis

Normally, minus strands are synthesized only early in infection and their synthesis stops when the rate of plus strand synthesis achieves a maximum rate or if protein synthesis is inhibited during the early phase of the replication cycle (Sawicki and Sawicki, 1980; 1986b). We identified a class of mutants with ts defects in the shut-off of minus strand synthesis; these also allowed minus strand synthesis to continue at 40°C or to restart at 40°C once it had ceased at 30°C. One mutation was in the nsP4 gene, predicting a change from Q191 to K in ts24 and its revertants (not Q195; see Strauss and Strauss, 1994) that allowed minus strand synthesis to resume in the absence of protein synthesis and at late time after infection if the infected cells were shifted to 40°C (Sawicki et al., 1990). Exposure to 40°C early in infection prevented the normal cessation of minus strand synthesis by these cultures. Such minus strands accumulated in replicative structures and could be shown to serve as templates for plus strand synthesis; however, they did not increase the rate of plus strand synthesis. This suggested that reactivation differed from early minus strand synthesis and did not lead to significant formation of new replication complexes. We argued reactivation was due to template switching by preformed complexes that had been engaged in plus strand synthesis until the moment of shift to 40°C (i.e. RC_{stable} replaced their minus strand templates with plus strand templates at 40°C). This phenomenon also indicated nsP4 functioned normally in promoter recognition to commit the complex to plus strand synthesis by fixing the minus strand as the preferred template. The substitution at residue 191 introduces a positive charge near the region affecting host range (Lemm et al., 1990). Whether at 40°C Q191 nsP4 shows reduced binding of host factors essential for plus strand synthesis or enhanced binding of host factors for minus strand synthesis is not known but clearly of interest.

Recently, certain SIN and SFV nsP2 mutants also were found to reactivate minus strand synthesis at 40°C in a temperature-reversible manner and without the need for new viral proteins (Suopanki et al., 1997; Sawicki and Sawicki, 1993; Wang et al, 1994)). Intriguingly, the four nsP2 mutants map to the C-domain, overlapping both protease sequences and those that function in some unknown manner to affect initiation of 26S subgenomic mRNA synthesis. The two nsPs may perform different steps in the reactivation process as their effects were additive when both were part of the same RC_{stable} (Sawicki and Sawicki, 1993). Reversibility suggested the ts mutations induced reversible changes in protein interactions or conformation and modified the polymerase so that it resembled a minus strand replicase at 40°C or affected association of a strand-specific host factor. As shown in Figure 2, we proposed that these alterations caused nsP2 and nsP4 to switch from recognizing minus strands as templates to recognize plus strands and thereby mimic the conformation of $RC_{initial}$. If so, the normal actions of nsP2 and nsP4 result in the conversion of the complex from minus strand to plus strand synthesis and lead to the overproduction of plus strands and the shut off of minus strand synthesis if no new $RC_{initial}$ forms. In addition, reactivation of minus strand synthesis by stable nsP2 or nsP4 proteins made earlier in infection and the recovery of the normal pattern of RNA synthesis in cultures shifted-down to 30°C in the absence of new protein synthesis was one factor arguing against the hypothesis by de Groot et al. (1990) that P34 was the only active form of the polymerase late in infection (Sawicki and Sawicki, 1993).

3.3. Mutants Defective in Conversion to RC_{stable}

Minus strand reactivation led us to propose that $RC_{initial}$ were converted to RC_{stable} after or during the synthesis of the minus strand (Wang et al., 1994). Analysis of ts RNA-negative mutants also support such a regulatory pathway (Figure 2). While lesions in each of the four SIN nsPs yielded RNA-negative mutants unable to produce functional RC_{stable} at 40°C, only ts6 and ts110 were defective in elongation and only the nsP3 mutant ts4 and the nsP1 mutant ts11 were defective solely in the formation or activity of $RC_{initial}$, respectively. Only some of the others had ts nsP2 protease and failed to process P1234 into functional replicases (Hardy et al., 1990; Keränen and Kääriäinen, 1979; Sawicki and Sawicki, 1985) and many of these were ts for 26S mRNA sythesis. Other alphavirus RNA-negative ts mutants might be blocked after the formation of $RC_{initial}$ and before the switch to plus strand synthesis, e.g., conversion of $RC_{initial}$ to RC_{stable}. The phenotype of conversion defective mutants would include the ability to continue synthesis of minus strand RNA after shift to 40°C without an increase in the overall rate of plus strand transcription because new RC_{stable} were not formed. Mutations whose phenotypes appear to fullfill these criteria are in nsP2 and nsP3 (Dé et al., 1996) and also in SIN HR nsP4 in a conserved PNIRS motif identified originally in AMV by Ansel-McKinney et al. (1996). Substitution of its essential R residue led to loss of the RNA-binding function in AMV (Ansel-McKinney et al., 1996) and resulted in a SIN HR variant ts for RNA synthesis that exhibited properties expected of a conversion defective mutant (Fata et al., 1997).

In summary, translation of the alphavirus genome produces P1234 and P123 polyproteins. Cleavage at the 3/4 site is essential, as cleaved nsP4 is the RNA-dependent RNA polymerase early and late, not P34. The minus strand replicase requires P23 polyproteins and both P123+nsP4 and nsP1+P23+nsP4 complexes are minus strand replicases. The latter complex is more efficient in genome plus strand synthesis. P23 cleavage is responsible for the short-lived nature of the minus strand replicase and converts this activity to a 26S transcriptase while retaining the initial plus strand replicase activity. Thus, the number of $RC_{initial}$ (P123+nsP4 or nsP1+P23+nsP4) that form early in infection directly determine the number of RC_{stable} and the maximum rate of plus strand synthesis late in infection. Why then does all minus strand synthesis cease late in infection?

4. TEMPORAL CESSATION OF ALPHAVIRUS MINUS STRAND SYNTHESIS

Several models to explain the temporal synthesis of minus strands have been proposed. Two of them evoke the action of a viral regulatory protein such as the nsP2 protease, whose accumulation during infection leads to the premature and rapid cleavage of newly synthesized P1234 and P123 and thereby inhibits minus strand synthesis. de Groot et al (1990) proposed that accumulation of the nsP2 protease altered polyprotein cleavage late in infection to form P34 but not nsP4 and assumed for this model that P34 would be active only in plus strand synthesis. This model was disproved by analysis of ts protease mutants (Sawicki and Sawicki, 1993) and by the direct demonstration that P34 was not active as a polymerase subunit (Shirako and Strauss, 1994; Lemm et al., 1993). An alternative is that $RC_{initial}$ would not be made if P23 is cleaved prematurely and in trans by accumulated nsP2 protease. This second model is supported by finding that a hypercleaving nsP2 variant failed to make detectable P23 in vitro and was lethal as virus (Strauss et al, 1992). Earlier studies of SFV ts1, however, suggest this model may be unlikely. Even though SFV ts1 overproduced the nsPs as a result of reduced 26S mRNA synthesis and increased genome synthesis, infected cells did not more rapidly shut-down minus strand synthesis as would be predicted from such a model. Instead, they showed the normal timing of synthesis and cessation of minus strand synthesis and accumulated similar numbers of RIs and minus strands as wildtype SFV infected cells (Sawicki and Sawicki, 1986a).

Therefore, we continue to favor a third model. Minus strand synthesis occurs as a result of the formation of new $RC_{initial}$ complexes but their conversion to RC_{stable} limits their activity in minus strand synthesis to a short time. Early in infection, new $RC_{initial}$ can form; late in infection they cannot and thus minus strand synthesis stops. Because temporal cessation of minus strand synthesis was not the result of a failure to continue to synthesize nsPs late in infection and was not due to encapsidation of plus strand templates (Sawicki and Sawicki, 1980), we suggested minus strand synthesis would cease when no further $RC_{initial}$ complexes could be formed because essential host factors or membrane sites had become depleted (Dé et al., 1996; Sawicki et al, 1990; Sawicki and Sawicki, 1987, 1986a). Coronavirus minus strand synthesis also occurs predominantly during the early phase of virus replication, requires continued protein synthesis, and results in the formation of increased numbers of replication complexes making plus strand RNA and thus in greater rates of plus strand synthesis; plus strand synthesis also requires continued protein synthesis (Sawicki and Sawicki, 1986c and cited therein). The translation of coronavirus and arterivirus nsPs initially as polyproteins makes it likely that they or their intermediates may function in these short-lived activities. Which additional features of alphavirus replication may be shared with coronaviruses or arteriviruses await future investigation and discoveries.

REFERENCES

Ahlquist, P., Strauss, E.G., Rice, C.M., Strauss, J.H., Haseloff, J., and Zimmern, D., 1985, Sindbis virus proteins nsP1 and nsP2 contain homology to nonstructural proteins from several RNA plant viruses, *J. Virol.* **53**:536–542.

Ahola, T. and Kääriäinen, L. 1995. Reaction in alphavirus mRNA capping: Formation of a covalent complex of nonstructural protein nsP1 with 7-methyl-GMP, *PNAS* **92**:507–511.

Ansel-McKinney, P., Scott, S.W., Swanson, M., Ge, X., and Gehrke, L., 1996, A plant viral coat protein RNA-binding consensus sequence contains a crucial arginine, *EMBO J.* **15**:5077–5084.

Atreya, C.D., Singh, N.K., and Nakhasi, H.L., 1995, The rubella virus RNA bindign activity of calreticulin is localized to the N-terminal domain, *J. Virol.* **69**:3848–3851.

Barton, D.J., Sawicki, S.G., and Sawicki, D.L., 1988, Demonstration in vitro of temperature-sensitive elongation of RNA in Sindbis virus mutant ts6, *J. Virol.* **62**:3597–3602.

Barton, D.J., Sawicki, S.G., and Sawicki, D.L., 1991, Solubilization and immunoprecipitation of alphavirus replication complexes, *J. Virol.*, **65**:1496–1506

Burge, B.W., and Pfefferkorn, E.R., 1966a, Isolation and characterization of conditional-lethal mutants of Sindbis virus, *Virology* **30**:204–213.

Burge, B.W., and Pfefferkorn, E.R., 1966b, Complementation between temperature-sensitive mutants of Sindbis virus, *Virology.* **30**:214–223.

de Groot, R.J., Hardy, W.R., Shirako, Y., and Strauss, J.H., 1990, Cleavage-site preferences of Sindbis virus polyproteins containing the nonstructural proteinase: evidence for temporal regulation of polyprotein processing in vivo, *EMBO J.* **9**:2631–2638.

Dé, I., Sawicki, S.G., and Sawicki, D.L., 1996, Sindbis virus RNA-negative mutants that fail to convert from minus-strand to plus-strand synthesis: role of the nsP2 protein, *J. Virol.* **70**:2706–2719.

Dé, I., Sawicki, S.G., Fata, C., and Sawicki, D.L., 1997, unpublished results.

Dinant, S., Janda, M., Kroner, P.A., and Ahlquist, P., 1993, Bromovirus RNA replication and transcription require compatibility between the polymerase and helicase-like viral RNA synthesis proteins, *J. Virol.* **67**:7181–7189.

Ding, M., and Schlesinger, M.J. 1989. Evidence that Sindbis virus nsP2 is an autoprotease which processes the virus nonstructural polyprotein, *Virology.* **171**:280–284.

Fata, C., Sawicki, S.G., Gerhke, L., Ansel-McKinney, P., and Sawicki, D.L., 1997, In preparation.

Gorbalenya, A.E. and Koonin, E.V. 1993. Comparative analysis of amino-acid sequences of key enzymes of replication and expression of positive-strand RNA viruses: validity of approach and functional and evolutionary implications, *Soc. Sci. Rev. D. Physicochem. Biol.* **11**:1–84.

Gorbalenya, A.E., Koonin, E.V., and Lai, M.M.C. 1991. Putative papain-related thiol proteases of positive-strand RNA viruses, *FEBS* **288**:201–205.

Hahn, Y.S., Grakoui, A., Rice, C.M., Stauss, E.G., and Strauss, J.H., 1989a, Mapping of RNA⁻ temperature-sensitive mutants of Sindbis virus: complementation group F mutants have lesions in nsP4, *J. Virol.* **63**:1194–1202.

Hahn, Y.S., Strauss, E.G., and Strauss, J.H. 1989b. Mapping of RNA temperature-sensitive mutants of Sindbis virus: assignment of complementation groups A,B, and G to nonstructural proteins, *J. Virol.* **63**:3142–3150.

Hardy, W.R., and Strauss, J.H. 1989. Processing the nonstructural polyproteins of Sindbis virus: the nonstructural proteinase is in the C-terminal half of nsP2 and functions both in cis and in trans, *J. Virol.* **63**:4653–4664.

Hardy, W.R., Hahn, Y.S., de Groot, R.J., Strauss, E.G. and Strauss, J.H., 1990, Synthesis and processing of the nonstructural polyproteins of several temperature-sensitive mutants of Sindbis virus, *Virology* **177**:199–208.

Ishikawa, M., Meshi, T., Ohno, T., and Okada, Y., 1991, Specific cessation of minus strand RNA accumulation at an early stage of tobacco mosaic virus infection, *J. Virol.* **65**:861–868.

Keränen, S., and Kääriäinen, L., 1979, Functional defects of RNA-negative temperature-sensitive mutants of Sindbis and Semliki Forest virus, *J. Virol.* **32**:19–29.

Koonin, E.V., and Dolja, V.V., 1993, Evolution and taxonomy of positive-strand RNA viruses: implications of comparative analysis of amino acid sequences, *Crit. Rev. Biochem. Mol. Biol.* **28**:375–430.

Kroner, P.A., Young, B.M., and Ahlquist, P., 1990, Analysis of the role of brome mosaic virus 1a protein domains in RNA replication, using linker insertion mutagenesis, *J. Virol.* **64**:6110–6120.

LaStarza, M., and Rice. C.M., 1994, Deletion and duplication mutations in the C-terminal nonconserved region of Sindbis virus nsP3: effects on phosphorylation and on virus replication in vertebrate and invertebrate cells, *Virology* **202**:224–232.

Lemm, J.A., and Rice, C.M., 1993, Assembly of functional Sindbis virus replication complexes: requirement for coexpression of P123 and P34, *J. Virol.* **67**:1905–1915.

Lemm, J.A., and Rice, C.M., 1997, In preparation.

Lemm, J.A., Durbin, R.K., Stollar, V., and Rice, C.M., 1990, Mutations which alter the level or structure of nsP4 can affect the efficiency of Sindbis virus replication in a host-dependent manner, *J. Virol.* **64**:3001–3011.

Lemm, J.A., Rumenapf, T., Strauss, E.G., Strauss, J.H., and Rice, C.M., 1994, Polypeptide requirements for assembly of functional Sindbis virus replication complexes: a model for the temporal regulation of minus-strand and plus-strand RNA-synthesis, *EMBO J.* **13**:2925–2934.

Li, G., and Rice, C.M., 1993, The signal for translational readthrough of a UGA codon in Sindbis virus RNA involves a single cytidine residue immediately downstream of the termination codon, *J. Virol.* **67**:5062–5067.

Li, G., LaStarza, M.W., Hardy, W.R., Strauss, J.H., and Rice, C.M., 1990, Phosphorylation of Sindbis virus nsP3 in vivo and in vitro, *Virology* **179**:416–427.

Liljestrom, P., Lusa, S., Huylebroeck, D. , and Garoff, H., 1991, In vitro mutagenesis of a full-length cDNA clone of Semliki Forest virus: the small 6000-molecular weight membrane protein modulates virus release, *J. Virol.* **65**:4107–4113.

Mi, S. and Stollar, V., 1991, Expression of Sindbis virus nsP1 and methyltransferase activity in E. coli, *Virology* **184**:423–427.

Pardignon, N., and Strauss, J.H., 1996, Mosquito homologue of the La autoantigen binds to Sindbis virus RNA, *J. Virol.* **70**:1173–1181.

Peranen, J., 1991, Localization and phosphorylation of Semliki Forest virus non-structural protein nsP3 expressed in COS cells from a cloned cDNA, *J. Gen. Virol.* **72**:195–199.

Pogue, G.P., Hofmann, J., Duncan, R., Best, J.M., Etherington, J., Sontheimer, R.D., and Nakhasi, H.L., 1996, RNA-protein complexes containing rubella virus RNA cis-acting elements are recognized by sera from patients with Sjogren's Syndrome and Systemic Lupus Erythematosus, *J. Virol.* **70**:6269–6277.

Rice, C.M., Levis, R. , Strauss, J.H., and Huang, H.V., 1987, Production of infectious RNA transcripts from Sindbis virus cDNA clones: mapping of lethal mutations, rescue of a temperature-sensitive marker, and in vitro mutagenesis to generate defined mutants, *J. Virol.* **61**:3809–3819.

Rikkonen, M., Peränen, J., and Kääriäinen, L., 1992, Nuclear and nucleolar targeting signals of Semliki Forest virus nonstructural protein nsP2, *Virology* **189**:462–473.

Rikkonen, M., Peränen, J., and Kääriäinen, L., 1994, ATPase and GTPase activities associated with Semliki Forest virus nonstructural protein nsP2, *J. Virol.* **68**:5804–5810.

Sawicki, D.L., and Sawicki, S.G., 1980, Short-lived minus strand polymerase for Semliki Forest virus, *J. Virol.* **34**:108–118.

Sawicki, D.L., and Sawicki, S.G., 1985, Functional analysis of the A complementation group mutants of Sindbis HR virus, *Virology* **144**:20–34.

Sawicki, S.G., and Sawicki, D.L., 1986a, The effect of overproduction of nonstructural proteins on Semliki Forest virus plus and minus strand RNA synthesis, *Virology* **152**:507–512.

Sawicki, S.G., and Sawicki, D.L., 1986b, The effect of loss of regulation of minus strand RNA synthesis on Sindbis virus replication, *Virology* **151**:339–349.

Sawicki, S.G., and Sawicki, D.L., 1986c, Coronavirus minus strand RNA synthesis and the effect of cycloheximide on coronavirus RNA synthesis, *J. Virol.* **57**:328–334.

Sawicki, D.L., and Sawicki, S.G., 1987, Alphavirus plus strand and minus strand RNA synthesis, in: *Positive Strand RNA Viruses*, (M. Brinton and R. Rueckert, eds), Alan R. Liss, New York, pp251–259.

Sawicki, D.L., and Sawicki, S.G., 1993, A second nonstructural protein functions in the regulation of alphavirus negative-strand RNA, *J. Virol.* **67**:3605–3610.

Sawicki, D.L., Barkhimer, D.B., Sawicki, S.G., Rice, C.M., and Schlesinger, S., 1990, Temperature sensitive shutoff of alphavirus minus strand RNA synthesis maps to a nonstructural protein, nsP4, *Virology* **174**:43–52.

Sawicki, D.L., Kääriäinen, L., Lambek, C., and Gomatos, P.J., 1978, Cleavage defect in the nonstructural polyprotein of Semliki Forest virus has two separate effects on viral RNA synthesis, *J. Virol.* **25**:19–27.

Sawicki, D.L., Sawicki, S.G., Keränen, S., and Kääriäinen, L., 1981. Specific Sindbis virus coded function for minus-strand RNA synthesis, *J. Virol.* **39**:348–358.

Shirako, Y., and Strauss, J.H., 1994, Regulation of Sindbis virus RNA replication: uncleaved P123 and nsP4 function in minus strand RNA synthesis whereas cleaved products from P123 are required for efficient plus strand RNA synthesis, *J. Virol.* **68**:1874–1885.

Suopanki, J., Sawicki, D.L., Sawicki, S.G., and Kääriäinen, L., 1997, In preparation.

Strauss, E.G., de Groot, R.J., Levinson, R., and Strauss, J.H., 1992, Identification of the active site residues in the nsP2 proteinase of Sindbis virus, *Virology* **191**:932–940.

Strauss, E.G., Lenches, E.M., and Strauss, J.H., 1976, Mutants of Sindbis virus. I. Isolation and partial characterization of 89 new temperature-sensitive mutants, *Virology* **74**:154–168.

Strauss, E.G., Rice, C.M., and Strauss, J.H., 1983, Sequence coding for the alphavirus nonstr-uctural proteins is interrupted by an opal termination codon, *PNAS* **80**:5271–5275.

Strauss, J.H., and Strauss, E.G., 1994, The Alphaviruses: Gene expression, replication and evolution, *Microbiological Reviews* **58**:491–562.

Takkinen, K., Peränen, J., Keranen, S., Soderlung, H., and Kääriäinen, L., 1990, The Semliki Forest virus-specific nonstructural protein nsP4 is an autoproteinase, *Eur. J. Biochem.* **189**:33–38.

Wang, Y.F., Sawicki, S.G., and Sawicki, D.L., 1991, Characterization of a single amino acid substitution in nsP1 that confers temperature-sensitive negative strand synthesis, *J. Virol.* **65**: 985–988.

Wang, Y.F., Sawicki, S.G., and Sawicki, D.L., 1994, Alphavirus nsP3 functions to form replication complexes transcribing negative-strand RNA, *J. Virol.* **68**:6466–6475.

24

AN INFECTIOUS cDNA CLONE OF PORCINE REPRODUCTIVE AND RESPIRATORY SYNDROME VIRUS

J. J. M. Meulenberg, J. N. A. Bos-de Ruijter, G. Wensvoort, and R. J. M. Moormann

Institute for Animal Science and Health
P.O. Box 365
NL-8200, AJ Lelystad, The Netherlands

1. SUMMARY

A plasmid containing a full-length cDNA copy of the Lelystad virus isolate (LV) of porcine reproductive and respiratory syndrome virus was constructed. When RNA that was transcribed in vitro from this full-length cDNA clone was transfected to BHK-21 cells, infectious LV was produced and secreted. The virus was rescued by passage to porcine alveolar lung macrophages or CL2621 cells. When infectious transcripts were transfected to porcine alveolar lung macrophages or CL2621 cells no infectious virus was produced due to the poor transfection efficiency of these cells. The growth properties of the viruses produced by BHK-21 cells transfected with infectious transcripts of LV cDNA resembled the growth properties of the parental virus from which the cDNA was derived. The infectious clone of LV enables us to mutagenize the viral genome at specific sites and thus will be useful for detailed molecular characterization of the virus, as well as for the development of a safe and effective live vaccine for use in pigs.

2. INTRODUCTION

The causative agent of a new disease, now known as porcine reproductive and respiratory syndrome (PRRS), was first identified in 1991 by Wensvoort et al. (1991) and was named Lelystad virus (LV). The main symptoms of the disease are respiratory problems in pigs and abortions in sows. Although major outbreaks, such as those observed at first in the United States in 1987 and in Europe in 1991 have diminished, this virus still causes significant economic losses in herds in the United States, Europe, and Asia. The

Coronaviruses and Arteriviruses, edited by Enjuanes *et al.*
Plenum Press, New York, 1998

virus, now generally called porcine reproductive and respiratory syndrome virus (PRRSV) is a member of the *Arteriviridae* family, which also comprises equine arteritis virus (EAV), lactate dehydrogenase-elevating virus (LDV), and simian hemorrhagic fever virus (SHFV; Meulenberg *et al.*, 1993). Recently, the International Committee on the Taxonomy of Viruses has decided to incorporate this family in a new order of viruses, the *Nidovirales*, together with the *Coronaviridae*, and *Toroviridae* (Cavanagh, 1997). PRRSV preferentially grows in porcine alveolar lung macrophages (PAMs; Wensvoort *et al.*, 1991). A few cell lines, such as CL2621 and other cell lines cloned from the monkey kidney cell line MA-104 are also susceptible to the virus (Collins *et al.*, 1992; Kim *et al.*, 1993). The genomic cDNA sequence of LV and other isolates of PRRSV was determined (Meulenberg *et al.*, 1993; Murtaugh *et al.*, 1995). In addition to the RNA-dependent RNA polymerase (ORFs 1a and 1b), the genomic sequence encodes four envelope glycoproteins named GP_2 (ORF2), GP_3 (ORF3), GP_4 (ORF4) and GP_5 (ORF5) as well as a nonglycosylated membrane protein M (ORF6) and the nucleocapsid protein N (ORF7; Meulenberg *et al.*, 1995; Meulenberg *et al.*, 1996; van Nieuwstadt *et al.*, 1996; Mardassi *et al.*, 1996).

The production of cDNA clones from which infectious RNA can be transcribed in vitro has become an essential step in the molecular genetic analysis of positive-strand RNA viruses (for a review see Boyer and Haenni, 1994). Here we describe for the first time the construction of an infectious clone of the LV isolate of PRRSV.

3. MATERIALS AND METHODS

3.1. Cells and Viruses

The *Ter Huurne* strain of LV was isolated in 1991 (Wensvoort *et al.*, 1991) and grown in porcine alveolar macrophages (PAMs). Passage 6 of the *Ter Huurne* strain (TH) was used in this study as well as a derivative of this strain, LV4.2.1, which was adapted for growth on CL2621 cells by serial passage. Virus titers ($TCID_{50}$/ml) were determined on PAMs or CL2621 cells by endpoint dilution, as described previously (Wensvoort *et al.*, 1986). BHK-21 cells were maintained and transfected as described by Liljeström and Garoff (1991).

3.2. Isolation of Viral RNAs

Intracellular LV RNA was isolated as described previously (Meulenberg *et al.*, 1993). In order to isolate virion genomic RNA, virions were purified on sucrose gradients as described by van Nieuwstadt *et al.* (1996). The virions were treated with Proteinase K, the RNA was extracted with phenol/chloroform and precipitated with ethanol.

3.3. Cloning of the 5' and 3' Termini of the LV Genome

The 5' end of the viral genome of LV was cloned using a modified single strand ligation to single-stranded cDNA procedure according to the protocol of the 5'-Amplifinder RACE kit (CLONTECH). Details of this procedure will be publised elsewhere. A similar strategy was used to clone the 5' terminus of the LV genome from intracellular LV RNA. A 3' end cDNA clone containing a poly(A) tail of 109 A's was constructed by reverse transcription and PCR of LV RNA.

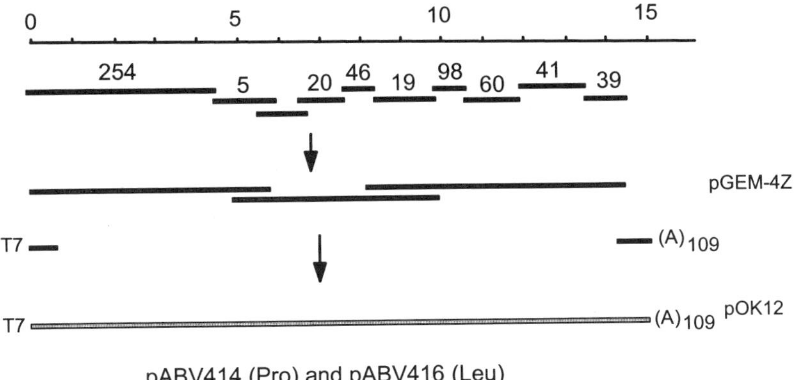

Figure 1. Construction of genome-length cDNA clones of LV. The upper part shows the ligation of cDNA clones, which were previously sequenced (Meulenberg *et al.*, 1993), in pGEM-4Z. The pABV numbers of the clones that were used are indicated. Black boxes are cDNA clones newly generated by reverse transcription and PCR, or PCR only. The lower part shows the assembly of the larger cDNA clones with the 5' end clone, containing a T7 RNA polymerase promoter, and the 3' end clone, containing a poly(A) tail, in low copy number vector pOK12.

3.4. Construction of Full-Length Genomic cDNA Clones of LV

cDNA clones generated earlier to determine the nucleotide sequence of the genome of LV (Meulenberg *et al.*, 1993) were ligated together at convenient restriction sites in high copy number plasmid pGEM-4Z as shown in Figure 1. Since further ligation of cDNA fragments in pGEM-4Z resulted in instable clones, the inserts were ligated to each other and to the 5' and 3' cDNA fragments in low copy number vector pOK12 (Viera and Messing, 1991; Figure 1). Plasmids were transformed to *Escherichia coli* strain DH5α, and were grown at 32⁰C in the presence of 5 to 15 µg/ml Kanamycin to keep their copy number as low as possible. A T7 RNA polymerase promoter was fused to the exact 5' end of the viral genome by PCR, whereas a poly(A) tail of 109 residues was incorporated at the 3' end. In this way, we obtained two genome-length cDNA clones, which were designated pABV414 and pABV416. These genome-length cDNA clones encode identical viral protein sequences except for one amino acid at position 1084 in ORF1a, which is a Pro in pABV414 and a Leu in pABV416.

3.5. In Vitro Transcription and Transfection of RNA

Full-length genomic cDNA clones were linearized with *Pvu*I, which is located directly downstream of the poly(A) stretch. Plasmid pABV296, Semliki Forest virus vector pSFV1 expressing the GP$_4$ protein encoded by ORF4 of LV (Meulenberg *et al.*, 1997), served as control for in vitro transcription and transfection experiments and was linearized with *Spe*I. The linearized plasmids were precipitated with ethanol and 1.5 µg of these plasmids was used for in vitro transcription with T7 RNA polymerase (full-length cDNA clones) or Sp6 RNA polymerase (pABV296), by the methods described for SFV by Liljeström and Garoff (1991). The in vitro transcribed RNA was precipitated with isopropanol, washed with 70% ethanol and stored at -20⁰C until use.

BHK-21 cells were seeded in 35-mm wells (approximately 10^6 cells/well) and were transfected with 2.5 µg in vitro transcribed RNA or 2.5 µg intracellular LV RNA mixed with

10 µl lipofectin in optimem as described by Liljeström and Garoff (1991). Alternatively, RNA was introduced in BHK-21 cells by electroporation. In this case, 10 µg in vitro transcribed RNA or 10 µg intracellular LV RNA was transfected to approximately 10^7 BHK-21 cells using the electroporation conditions of Liljeström and Garoff (1991). The electroporated cells were seeded in 4 wells (35-mm). The medium was harvested 24 h after transfection, and transferred to CL2621 cells or PAMs to rescue infectious virus. Transfected and infected cells were tested for expression of LV proteins by an immunoperoxidase monolayer assay (IPMA), essentially as described by Wensvoort *et al.* (1986). Monoclonal antibodies (MAbs) 122.13, 122.59, 122.9, and 122.17, directed against respectively the GP_3, GP_4, M and N protein (van Nieuwstadt *et al.*, 1996) were used for staining in this assay.

4. RESULTS

4.1. Infectivity of LV RNA

Since PAMs and cell line CL2621 or other clones derived from the monkey kidney cell line MA-104 are the only cells, which were shown to propagate LV and other isolates of PRRSV (Collins *et al.*, 1992; Kim *et al.*, 1993), we first tested these in transfection experiments to demonstrate the infectivity of LV RNA. When intracellular RNA isolated from CL2621 cells infected with LV was transfected to PAMs or CL2621 cells at different doses using different methods, such as lipofectin, lipofectamin, DEAE-dextran and electroporation, no cythopathic effect or plaques were observed nor could the production of structural proteins be detected in IPMA using LV-specific MAbs (Table 1). RNA transcribed in vitro from pABV296 (Semliki Forest virus vector pSFV1 expressing GP_4; Meulenberg *et al.*, 1997) was used as control in these experiments. This RNA was transfected most efficiently by electroporation. However, still only 0.01% of the CL2621 cells stained with GP_4-specific MAbs in IPMA. In contrast, when BHK-21 were electroporated under similar conditions, 90–100% of the cells stained. Since these results indicated that BHK-21 cells were much more efficiently transfected than CL2621 cells, we used them to test the infectivity of LV RNA. Thus, intracellular LV RNA (2.5 µg which was estimated to contain approximately 1–2 ng of LV genomic RNA) was transfected to 10^6 BHK-21 cells with lipofectin. At 24 h after transfection approximately 5–15 individual cells were stained with LV-specific MAbs, but no infectious centers or plaques were observed, indicating that LV did not spread to neighbouring cells (Fig. 2A). The number of positive cells increased 2- to 4-fold when the RNA was transfected to BHK-21 cells by electroporation

Table 1. Transfection of RNA to different cells by electroporation

Cells	LV RNA[1]	SFV-ORF4[2]
Macrophages	–	–
CL2621	–	+ (0.01%)
BHK-21	+[3]	+ (90-100%)

[1] Transfected cells were stained in IPMA with MAb 122.17 directed against the N protein.
[2] Transfected cells were stainded in IPMA with MAb122.59 directed against GP_4.
[3] A few positive cells.

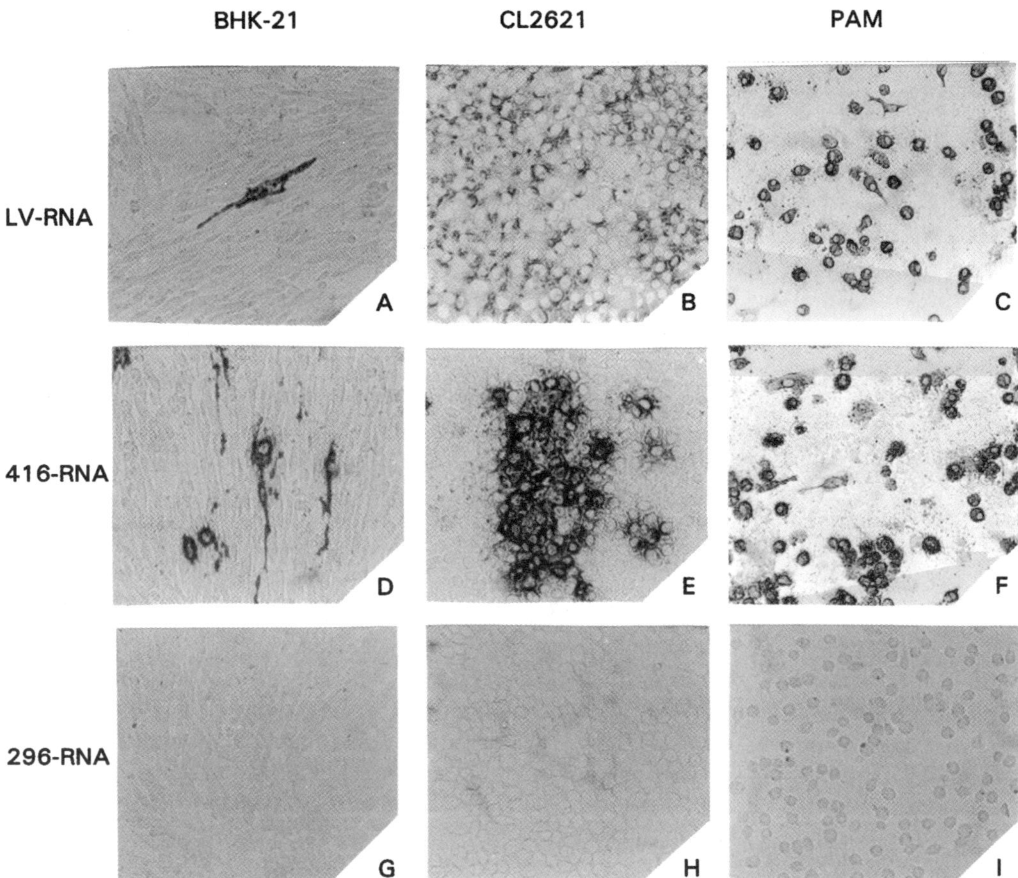

Figure 2. Infectivity of RNA. BHK-21 cells were transfected with LV intracellular RNA (A), transcripts from pABV416 (D) or pABV296 (G) using lipofectin and were stained at 24 h post transfection with N-specific MAb122.17 in IPMA. The supernatant of these transfected BHK-21 cells was used to infect CL2621 cells (B, E, and H, respectively) and PAMs (C, F, and I respectively). These PAMs and CL2621 cells were stained after two and three days, respectively, with N-specific MAb 122.17.

(Table 1). At 24 h after transfection, the supernatant of the BHK-21 cells transfected with intracellular LV RNA or pABV296 RNA was transferred to PAMs and CL2621 cells. Cythopathic effect was observed in PAM cultures at 2 days and in CL2621 cultures at 3 to 4 days after inoculation with the supernatant from BHK-21 cells transfected with intracellular LV RNA. The infected PAMs and CL2621 cells were positively stained with LV-specific MAbs in IPMA (Figure 2B and C). Similar results were obtained when RNA isolated from purified virions of LV was transfected to BHK-21 cells (data not shown). No cythopathic effect or staining with LV-specific MAbs directed against the N protein (Fig.2 H and I) or GP_4 (data not shown) was observed in PAMs or CL2621 cells incubated with the supernatant from BHK-21 cells transfected with pABV296 RNA. Thus, these results show that BHK-21 cells can be used to demonstrate the infectivity of LV RNA. Although LV cannot infect BHK-21 cells, probably because they lack the receptor for LV, once the genomic RNA has been introduced in BHK-21 cells, new infectious virus particles are produced and excreted into the medium.

4.2. Reconstruction of the 5' Terminal Sequence of the Genomic RNA of LV

To clone the 5' end of the LV genome, a modified single strand ligation to single-stranded cDNA (SLIC; Edwards *et al.*, 1991) procedure was used. Twelve clones, obtained from two independent PCRs on ligated products derived from LV intracellular RNA and 14 clones derived from two independent PCRs on ligated products derived from virion RNA were sequenced. From these 26 cDNA clones, 22 clones contained an extension of 10 nucleotides (5' ATGATGTGTA 3') compared to the cDNA sequence, that was reported in our previous study (Meulenberg *et al.*, 1993). The other 4 clones lacked 1 to 3 nucleotides at the 5' end of this additional sequence. This led us to conclude that these ten nucleotides represent the utmost 5' end of the LV genome and they were therefore incorporated in the genome-length cDNA clone.

4.3. Construction of Genome-Length cDNA Clones of LV

A genome-length cDNA clone of LV was constructed according to the strategy depicted in Figure 1. A T7 RNA polymerase promoter for in vitro transcription was directly linked to the newly determined 5' terminus of the genome of LV by PCR and inserted in the genome-length cDNA clone. Resequencing of nucleotides 3420 to 3725 of six newly generated and independent cDNA clones indicated that at nucleotide 3472 a C and T were present at a ratio of 1:1, resulting in a Pro or Leu at amino acid residue 1084 in ORF1a. Since we could not predict the influence of the amino acid substitution at this position on the infectivity of the RNA transcribed from the final genome-length cDNA clone, we constructed two genome-length cDNA clones encoding either a Leu or Pro at this position. At the 3' end a poly(A) stretch of 109 A residues was incorporated in the genome-length cDNA clone.

We tried to ligate the larger cDNA fragments in pGEM-4Z, but this resulted in the accumulation of deletions. Therefore we finally fused these clones to each other in low copy number vector pOK12 and obtained the genome-length cDNA clones pABV414 (Pro) and pABV416 (Leu). These could be stably propagated in *E. coli* under the growth conditions used.

4.4. In Vitro Synthesis of Infectious RNA

Transcripts of genome-length cDNA clones pABV414 and pABV416 were transfected to BHK-21 to test their infectivity. These transcripts, synthesized in vitro using T7 RNA polymerase, were expected to contain 2 non viral nucleotides (CG) at the 3' end. In addition, transcripts were expected to contain a nonviral G at the 5' end, which is the transcription start site of T7 RNA polymerase. Approximately 2.5 µg of RNA transcribed in vitro from pABV414 or pABV416 was transfected to BHK-21 cells using lipofectin and at 24 h after transfection 800 to 2700 cells stained positive with N-specific MAb122.17 in IPMA (Fig. 2D). PAMs that were inoculated with the supernatant derived from BHK-21 cells that were transfected with transcripts of pABV414 or pABV416 displayed cythopathic effect after 2 days. Individual plaques were produced after 3 to 4 days in CL2621 cultures that were inoculated with either of the supernatants. The infected PAMs and CL2621 cells stained in the IPMA with MAb122.17 directed against the N protein (Fig. 2E and F) and with MAbs directed against the M, GP$_4$, and GP$_3$ protein (data not shown), confirming that these proteins were all properly expressed. Therefore, these results clearly show that when RNA transcribed from genome-length cDNA clones pABV414 or pABV416 is transfected to BHK-21 cells, infectious LV is produced and excreted.

The initial transfection and infection experiments suggested that the rescued recombinant viruses, designated vABV414 and vABV416, infect and grow equally well in PAMs, but grow slower on CL2621 cells than the virus rescued from BHK-21 cells transfected with intracellular LV RNA (compare Fig. 2C with Fig. 2F, and Fig. 2B with Fig. 2E). This intracellular LV RNA was isolated from CL2621 cells infected with LV4.2.1, which has been adapted for growth on CL2621. This observation was confirmed by growth curves of vABV414 and vABV416, wild-type LV that has only been passaged on PAMs (TH) and LV4.2.1 performed in PAMs and CL2621 cells. The TH virus had similar growth characteristics on PAMs and CL2621 cells as the recombinants (data not shown). In contrast, the CL2621-adapted virus LV4.2.1 grew faster on CL2621 cells than the viruses vABV414, vABV416 and TH. This confirmed the larger plaque size of this virus compared to vABV414, vABV416 and TH observed in CL2621 cells. In summary, these results demonstrate that the growth properties of the recombinant viruses are similar to those of the TH virus. This was expected, since the cDNA sequence used to construct the infectious clones was derived from the parental "non-adapted" TH virus.

5. DISCUSSION

In this study, we generated for the first time an infectious clone of the LV isolate of PRRSV. The exact 5' end of the LV genome was determined. This sequence ligated to cDNA fragments covering the entire LV genome resulted in a genome-length cDNA clone of 15.2 kb, from which infectious transcripts could be produced. The infectious clone, described here, is to our knowledge the longest infectious clone of a positive-strand RNA virus thusfar developed. Recently, an infectious clone of another arterivirus, EAV, has been reported which is 12.7 kb in length (van Dinten *et al.*, 1997). Transcripts of full-length cDNA of LV lacking a cap structure were not infectious (data not shown). This indicated that the cap structure is most likely essential for translation of the genomic RNA. The infectivity of genomic RNA or transcripts of infectious cDNA clones of other positive-strand RNA viruses have always been tested in cell lines that are susceptible to the virus. This was not possible for LV, due to the poor transfection efficiency of CL2621 cells and PAMs. However, transfection of transcripts from full-length cDNA clones, intracellular LV RNA, and virion RNA to BHK-21, a cell line which is not susceptible to infection with LV, resulted in the production and release of infectious virus, which could be rescued in CL2621 cells and PAMs. Since infectious virus is produced when LV RNA is transfected into BHK-21 cells, but not when these cells are inoculated with LV particles, we hypothesize that it is impossible for LV to enter BHK-21 cells because these cells lack the receptor of LV. A similar difference in the infectivity of LDV and its virion RNA in various cell types has been observed (Inada *et al.*, 1993). The growth properties of the two recombinant viruses vABV414 and vABV416, which differ only at one amino acid at position 1084 in ORF1a (Pro versus Leu), were similar to those of the parental TH strain, from which the cDNA was originally derived. The wild-type nature of these viruses still has to be confirmed by experimental infection of pigs.

The infectious clone of LV is an excellent tool for site-directed mutagenesis and is an important finding for future projects whose aim is to construct new live vaccines against PRRSV. In addition, the infectious clone of the LV isolate of PRRSV might provide a model system to study and unravel the intriguing mechanism of transcription and replication of arteriviruses and coronaviruses.

ACKNOWLEDGMENTS

This work was supported by Boehringer Ingelheim, Germany. We thank J. Castrop for critical reading of the manuscript and F. van Poelwijk for fruitful discussions.

REFERENCES

Boyer, J., and Haenni, A., 1994, Infectious transcripts and cDNA clones of RNA viruses, *Virology* **198**:415–426.

Cavanagh, D., 1997, Nidovirales: a new order comprising Coronaviridae and Arteriviridae, *Arch. Virology* **142**:629–633.

Collins, J.E., Benfield, D.A., Christianson, W.T., Harris, L., Hennings, J.C., Shaw, D.P., Goyal, S.M., McCullough, S., Morrison, R.B., Joo, H.S.,. Gorcyca, D.E., and Chladek, D.W., 1992, Isolation of swine infertility and respiratory syndrome virus (Isolate ATCC-VR-2332) in North America and experimental reproduction of the disease in gnotobiotic pigs, *J. Vet. Diagn. Invest.* **4**:117–126.

Edwards, J.B.D.M., Delort, J., and Mallet, J., 1991, Oligodeoxyribonucleotide ligation to single-stranded cDNAs; A new tool for cloning 5' ends of mRNAs and for contructing cDNA libraries by in vitro amplification, *Nucleic Acids Res.* **19**:5227–5232.

Inada, T., Kikuchi, H., and Yamazaki, S., 1993, Comparison of the ability of lactate dehydrogenase-elevating virus and its virion RNA to infect murine leukemia virus-infected or -uninfected cell lines, *J. Virol.* **67**:5698–5703.

Kim, H.S., Kwang, J.,and Yoon, I.Y., 1993, Enhanced replication of porcine reproductive and respiratory syndrome virus in a homogeneous subpopulation of MA-104 cell line, *Arch. Virol.* **133**:477–483.

Liljeström, P., and Garoff, H., 1991, A new generation of animal cell expression vectors based on the Semliki Forest virus replicon, *Biotechnol.* **9**: 1356–1361.

Mardassi, H., Massie, B., and Dea, S., 1996, Intracellular synthesis, processing and transport of proteins encoded by ORFs 5 to 7 of porcine reproductive and respiratory syndrome virus, *Virology* **221**:98–112.

Meulenberg, J.J.M., Hulst, M.M., de Meijer, E.J., Moonen, P.L.J.M., den Besten, A., de Kluyver, E.P., Wensvoort, G., and Moormann, R.J.M., 1993, Lelystad virus, the causative agent of porcine epidemic abortion and respiratory syndrome (PEARS) is related to LDV and EAV, *Virology* **192**:62–74.

Meulenberg, J.J.M., and Petersen-den Besten, A., 1996, Identification and characterization of a sixth structural protein of Lelystad virus: the glycoprotein GP_2 encoded by ORF2 is incorporated in virus particles, *Virology* **225**:44–51.

Meulenberg, J.J.M., Petersen-den Besten, A., de Kluyver, E.P., Moormann, R.J.M., and Wensvoort, G., 1995, Characterization of proteins encoded by ORFs 2 to 7 of Lelystad virus, *Virology* **206**:155–163.

Meulenberg, J.J.M., van Nieuwstadt, A.P., van Essen-Zandbergen, A., Langeveld, J.P.M., 1997, Post-translational processing and identification of a neutralization site of the GP_4 protein encoded by ORF4 of Lelystad virus, *J. Virol.* **71**: 6061–6067.

Murtaugh, M.P., Elam, M.R., and Kakach, 1995, Comparison of the structural protein coding sequences of the VR-2332 and Lelystad virus strains of the PRRS virus, *Arch. Virol.* **140**:1451–1460.

van Dinten, L.C., den Boon, J.A., Wassenaar, A.L..M., Spaan, W.J.M., and E.J. Snijder, E.J., 1997, An infectious arterivirus cDNA clone: Identification of a replicase point mutation that abolishes discontinuos mRNA transcription, *Proc. Natl. Acad. Sci. USA* **94**:991–2952.

van Nieuwstadt, A.P., Meulenberg, J.J.M., van Essen-Zandbergen, A., Petersen-den Besten, A., Bende, R.J., Moormann, R.J.M., and Wensvoort, G., 1996, Proteins encoded by ORFs 3 and 4 of the genome of Lelystad virus (Arteriviridae) are structural proteins of the virion, *J. Virol.* **70**:4767–4772.

Viera, J., and Messing, J., 1991, New pUC-derived cloning vectors with different selectable markers and DNA replication origins, *Gene* **100**, 189–194.

Wensvoort, G., Terpstra, C., Boonstra, J., Bloemraad, M., and van Zaane, D.,1986, Production of monoclonal antibodies against swine fever virus and their use in laboratory diagnosis, *Vet. Microbiol.* **12**:101–108.

Wensvoort, G., Terpstra, C., Pol, J.M.A., ter Laak, E.A., Bloemraad, M., de Kluyver, E.P., Kragten, C., van Buiten, L., den Besten, A., Wagenaar, F., Broekhuijsen, J.M., Moonen, P.L.J.M., Zetstra, T., de Boer, E.A., Tibben, H.J., de Jong, M.F., van 't Veld, P., Groenland, G.J.R., van Gennep, J.A., Voets, M.T., Verheijden, J.H.M., and Braamskamp, J., 1991, Mystery swine disease in the Netherlands: the isolation of Lelystad virus, *Vet. Q.* **13**:121–130.

25

THE SPIKE PROTEIN OF TRANSMISSIBLE GASTROENTERITIS CORONAVIRUS CONTROLS THE TROPISM OF PSEUDORECOMBINANT VIRIONS ENGINEERED USING SYNTHETIC MINIGENOMES

A. Izeta,[1] C. M. Sanchez,[1] C. Smerdou,[1] A. Mendez,[1] S. Alonso,[1]
M. Balasch,[2] J. Plana-Durán,[2] and L. Enjuanes[1]

[1]Department of Molecular and Cell Biology
Centro Nacional de Biotecnología, CSIC
Campus Universidad Autónoma, Canto Blanco
28049 Madrid, Spain
[2]Fort Dodge Veterinaria
Vall de Bianya
17813 Girona, Spain

1. ABSTRACT

The minimum sequence required for the replication and packaging of transmissible gastroenteritis virus (TGEV)-derived minigenomes has been determined. To this end, cDNAs encoding defective RNAs have been cloned and used to express heterologous spike proteins, to determine the influence of the peplomer protein in the control of TGEV tropism.

A TGEV defective interfering RNA of 9.7 kb (DI-C) was isolated, and a cDNA complementary to DI-C RNA was cloned under the control of T7 promoter. *In vitro* transcribed DI-C RNA was replicated *in trans* upon transfection of helper virus-infected cells. A collection of DI-C deletion mutants (TGEV minigenomes) was generated and tested for their ability to be replicated and packaged. The size of the smallest minigenome replicated *in trans* was 3.3 kb. The rescue system was used to express the spike protein of an enteric TGEV isolate (C11) using as helper virus a TGEV strain (C8) that replicates very little in the gut. A mixture of two pseudorecombinant viruses containing either the helper virus genome or the minigenome was obtained. These pseudorecombinants display in the sur-

Coronaviruses and Arteriviruses, edited by Enjuanes *et al.*
Plenum Press, New York, 1998

face the S proteins from the enteric and the attenuated virus, and showed 10^4-fold increase in their gut replication levels as compared to the helper isolate (C8). In addition, the pseudorecombinant virus increased its enteric pathogenicity as compared to the C8 isolate.

2. INTRODUCTION

TGEV is a member of the *Coronaviridae* family (Lai and Cavanagh, 1997; Enjuanes and Van der Zeijst, 1995) with a plus-stranded, polyadenylated RNA genome of 28.5 kb (Eleouet *et al.*, 1995). Defective RNAs, i.e. deletion mutants maintaining the *cis*-signals required for replication and packaging but dependent on other viral functions that can be supplied *in trans* by a helper virus, may arise in coronavirus replication cycle. To study the molecular basis of coronavirus replication, evolution and tropism, it would be convenient to have a full-length infectious cDNA clone. Since this is not available, an alternative approach is the cloning of subgenomic-sized defective RNAs to manipulate the coronavirus genome. Recently, cDNAs encoding small coronavirus defective RNAs have been successfully used to express heterologous genes (Zhang *et al.*, 1997; Liao *et al.*, 1995; Liao and Lai, 1994; Lin and Lai, 1993).

In our laboratory, three TGEV defective interfering RNAs of 22, 10.6 and 9.7 kb (DI-A, DI-B and DI-C, respectively) were isolated (Mendez *et al.*, 1996). This chapter reports the construction of a full-length cDNA clone from the smallest TGEV DI RNA, DI-C, which is replicated and packaged upon transfection of helper virus-infected cells. The construction and rescue of a collection of TGEV minigenomes has provided a TGEV-derived expression system, in which several heterologous genes have been expressed.

In order to identify the molecular bases of TGEV enteropathogenicity, the S gene of the virulent strain C11 of PUR46 virus, which grew to high levels in the gut, was cloned into a minigenome and it was rescued (i.e. replicated and packaged *in trans*) using an attenuated strain, clone C8 of PUR46, which grew poorly in the enteric tract. Pseudorecombinant viruses derived from the attenuated C8 clone, carrying the S protein from the enteric strain C11, increased 10^4-fold their gut replication levels, confirming data from our laboratory (Ballesteros *et al.*, 1997; Sanchez *et al.*, 1992) that demonstrated that the tropism of TGEV depends on the origin of the S protein. To our knowledge, this constitutes the first report of a tropism change being engineered in a coronavirus.

3. MATERIALS AND METHODS

3.1. Cells and Viruses

Viruses were grown in swine testis (ST) cells (McClurkin and Norman, 1966). TGEV PUR46MAD (Sanchez *et al.*, 1990) and PUR46 C8 and C11 strains (Sanchez *et al.*, 1997) have been described.

3.2. Construction of cDNAs Encoding RNA Minigenomes

(i) pDI-C. Four overlapping cDNA fragments of DI-C RNA obtained by RT-PCR amplification (Mendez *et al.*, 1996) were corrected for point mutations introduced by RT-PCR procedure and they were assembled into plasmid pSL1190 (Pharmacia), under the control of T7 promoter. A cDNA encoding HDV ribozyme and T7 terminator was cloned by PCR from transcription vector 2,0, kindly provided by A. Ball, University of Alabama (Pattnaik *et al.*, 1992), and placed immediately downstream of the cDNA complementary to DI-C RNA plus a synthetic poly(A) tract. (ii) Plasmid pDI-C-derived deletion mutants. Sequence analysis of pDI-C was performed and several DI-C internal deletion mutants, i.e. M54, were made by deleting the sequences between two restriction endonuclease sites using standard procedures (Sambrook *et al.*, 1989). (iii) Plasmid M54-S$_{C11}$. Five RT-PCR overlapping fragments of strain C11 S gene were assembled into pGEM-T (Promega), and sequenced with an Applied Biosystems 373A automated DNA sequencer. The complete S gene, including its intergenic sequences, was then inserted into pDI-C deletion mutant M54 in ORF1b gene.

3.3. *In Vitro* Transcription

In vitro transcription of linearized DNA templates was performed with T7 RNA Polymerase (Promega), according to the manufacturer instructions.

3.4. Electroporation of Helper Virus-Infected ST Cells

ST cells were grown to confluence and infected with helper virus with a m.o.i. of 10 PFU/cell. At 4–6 h p.i., cells were trypsinized, collected and resuspended in ice-cold PBS. The cells were then electroporated with *in vitro* transcribed RNA in a Gene Pulser apparatus (BioRad). The electroporated cells were incubated at 37ºC for 12h. Supernatants of these cultures were used to infect fresh ST cell monolayers, and at least six passages were performed to amplify the RNA.

3.5. RNA Analysis by Northern Hybridization and Northern Blot

Cytoplasmic RNA was extracted from helper virus infected and RNA transfected ST cells at different passages, as described previously (Mendez *et al.*, 1996). Northern hybridization was performed as described elsewhere (Penzes *et al.*, 1996), using a leader-specific oligonucleotide complementary to nt 66 to 91 of TGEV genome. Northern Blot analysis was performed using a 3'-UTR specific DNA probe, complementary to nt 28299 to 28543, and following standard procedures (Sambrook *et al.*, 1989).

3.6. Virus Tropism Analysis

The *in vivo* growth of the pseudorecombinant virus was determined after oro-nasal and intragastric inoculation of two to three days-old NIH miniswine (Lunney *et al.*, 1986; Sachs *et al.*, 1976) with doses of 5×10^9 PFU. Piglets were sacrificed at days 1, 2, 3, and 4 post-inoculation, and the virus present in jejunum, ileum, and intestinal content was determined. Tissues were processed as previously described (Ballesteros *et al.*, 1997).

4. RESULTS

4.1. Replication and Packaging of Synthetic DI-C RNA in TGEV-Infected ST Cells

A cDNA complementary to DI-C RNA was cloned under the control of T7 promoter. Four overlapping fragments comprising the full length DI-C RNA where corrected to have the consensus virus sequence, and assembled. DI-C cDNA was cloned immediately downstream of T7 promoter with only two extra bases of non viral origin (GG) between the promoter and DI-C cDNA (Figure 1A and 1B). The 3' end of the poly(A) tail (25A) was self-processed to give perfect 3' ends, by inserting after the poly(A) a cDNA encoding the hepatitis delta virus (HDV) ribozyme and a T7 transcription terminator (results not shown).

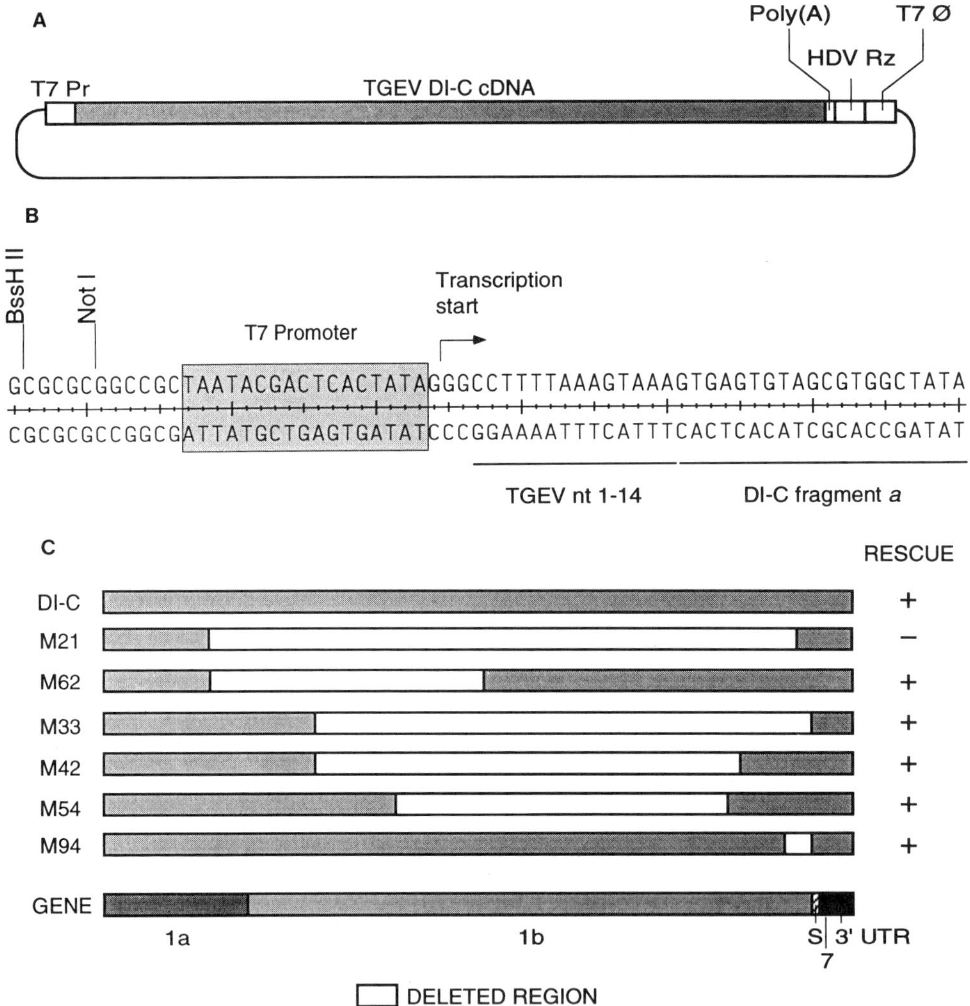

Figure 1. Rescue of TGEV minigenomes. (A) General structure of the plasmid-encoding DI-C RNA. A cDNA complementary to DI-C RNA was assembled downstream of T7 promoter. At DI-C 3' end, a synthetic poly(A) tract was added. HDV Rz, Hepatitis delta virus ribozyme; T7Ø, T7 transcription terminator. (B) Sequence of DI-C cDNA 5' end. Two restriction endonuclease sites, T7 Promoter and TGEV nucleotides 1–14 were placed upstream of DI-C fragment *a*, which started at TGEV nt 15 (Mendez *et al.*, 1996). TGEV nt 1–14 and DI-C fragment *a* 20-first nucleotides are underlined. (C) The structure of DI-C deletion mutants and their ability to be rescued by PUR46MAD helper virus is shown. Cytoplasmic RNAs were analyzed by Northern hybridization using a oligonucleotide complementary to nt 66 to 91 of TGEV genome. The presence of bands with the expected size for each of the minigenomes is indicated (+), while the absence of the band was interpreted as the absence of rescue by the helper virus (-). Grey boxes, sequence of the minigenomes. Empty boxes, deleted fragments.

TGEV PUR46MAD-infected ST cells were electroporated with *in vitro* transcribed RNA from linearized pDI-C. Supernatants of these cultures were passed six times onto fresh ST cell monolayers and total intracellular RNA was extracted. In Northern hybridization analysis, synthetic DI-C RNA was clearly detected and remained stable in RNA transfected cultures at least from passage 3 to 6, but not in the non-transfected ones (results not shown).

4.2. Effects of Internal Deletions on the Replication and Packaging of DI-C

A collection of TGEV minigenomes was generated and their ability to be replicated and packaged using PUR46MAD as helper virus was tested (Figure 1C). Minigenomes are named by a number that indicates their size in hundreds of nucleotides. All minigenomes but one were rescued, some of them to higher levels than the naturally selected DI C RNA. The non-rescued RNA (M21) of 2.1 kb, was the smallest RNA generated. The smallest that was rescued (M33) had 3.3 kb. By comparing the structure of the non-rescued RNA with those of the rescued minigenomes, it was concluded that at least one of the two ORF1 fragments of 4.1 and 1.3 kb present in M62 and M33 but not in M21 was required for rescue, but none of these fragments was essential. Whether these fragments are required for replication or packaging is being investigated.

4.3. Expression of S_{C11} Heterologous Protein in M54 TGEV Minigenome

The rescue system was used to generate pseudorecombinant viruses expressing the spike protein from two viruses, with low and high replication level in the enteric tract (clones C8 and C11 of TGEV, respectively). Using clone C8 as helper virus, a recombinant M54 minigenome expressing the spike gene of C11 strain was rescued (Figure 2A). Northern blot analysis of cytoplasmic RNA of helper virus infected ST cells showed the standard set of viral mRNAs (Figure 2B, left panel). Analysis of the RNAs from ST cells infected with the helper virus and transfected with M54 RNA (with no heterologous insert) showed the standard RNA pattern plus an RNA with the size expected for the minigenome, which was rescued to high levels. Finally, when RNAs from cells infected with the helper virus and transfected with the M54-S_{C11} minigenome were analyzed, the mRNA with the expected size for the RNA encoding the C11 spike protein was observed. Interestingly, the C11 spike gene was transcribed to levels similar to those of the mRNA encoding the S protein of the helper virus.

4.4. Pseudorecombinant Virus Tropism

Three groups of 4 piglets were infected either with TGEV strains C8, C11, or with the pseudorecombinant virus (Figure 2B, right panel). One piglet of each group was sacrificed at different days post-inoculation, and the titer of the virus recovered from intestinal homogenates was determined. The pseudorecombinant virus grew up to titers of 10^6 PFU/g of tissue, in contrast to the C8 isolate, that grew to a limited extent (10^2 PFU/g of tissue). In addition, all the animals infected with the pseudorecombinant virus presented enteritis 2 days post-infection, a pathology not observed in the animals infected with strain C8. These results indicate that the pseudorecombinant virus, carrying the spike proteins from clones C8 and C11, increased 10.000-fold its replication level in the gut.

5. DISCUSSION

The rescue of synthetic TGEV derived minigenomes has been shown. The minimum size of the minigenomes that have been rescued is 3.3 kb. This minigenome requires for replication and packaging a 1.3 kb fragment, not present in a 2.1 kb minigenome that is not rescued with the same helper virus (M21). Alternatively it may require a 4.1 kb fragment present in M62 but not in M21. Both fragments are required for rescue, although none of these fragments is essential, since defective minigenomes with either one or the other fragment could be replicated and passaged at least six times in ST cells. Whether these fragments are required for replication or packaging is not known at present.

The synthetic minigenomes with the helper virus were useful to express a heterologous antigen in large amounts. In fact, pseudorecombinant viruses expressing the S protein from the helper virus plus the S protein of a virus clone that replicates to high levels in the enteric tract increased ten thousand-fold their replication in the enteric tissues. This result strongly suggests two important conclusions: (i) The S protein of the enteric virus provided *in trans* was sufficient to increase the enteric tropism of the helper virus, and (ii) the expression levels of the enteric virus spike protein were probably as high as those coming from the helper virus S protein expression, in order to form pseudorecombinants with a significantly increased enteric tropism. This conclusion is also supported by the high levels of the mRNA corresponding to the S protein of the enteric virus, encoded by the minigenome.

The possibility of modifying the proteins assembled into coronaviruses by engineering coronavirus derived minigenomes opens important avenues to study the role of the different structural proteins of this virus family.

ACKNOWLEDGMENTS

This work has been supported by grants from the Consejo Superior de Investigaciones Científicas, the Comisión Interministerial de Ciencia y Tecnología (CICYT), La Consejería de Educación y Cultura de la Comunidad de Madrid, and Fort Dodge Veterinaria from Spain, and the European Communities (Projects Biotechnology and FAIR). AI and SA received fellowships from the Department of Education, University and Research of the Gobierno Vasco.

Figure 2. Generation of pseudorecombinant viruses with increased enteric tropism. (A). Scheme of pseudorecombinant virus generation. The role of S protein in tropism was studied by constructing pseudorecombinant viruses based on the helper virus C8, with a reduced enteric tropism, carrying the S protein from the highly enteric isolate C11. The S_{C11} gene was cloned into M54 minigenome. *In vitro* transcription with T7 RNA polymerase led to the synthesis of a minigenome RNA that was transfected into ST cells previously infected with helper virus C8. A mixture of pseudorecombinant viruses containing either the helper virus genome or the minigenome was obtained. Dark and empty peplomers represent C8 and C11 spike proteins, respectively. (B). Analysis of the pseudorecombinant virus RNA replication and infectivity in the enteric tract. Left panel: supernatants from ST cells infected with isolate C8 and transfected with different minigenome RNAs were amplified by virus passage on ST cell monolayers. Cytoplasmic RNAs were analyzed by Northern Blot using an anti-3'UTR ^{32}P-labeled probe. Numbers on the left indicate the viral mRNAs. Arrows on the right show the position of the transfected M54-S_{C11} and M54 RNAs, as well as the mRNA transcribed from M54-S_{C11}. UK, unknown RNAs. Right panel: three groups of 4 piglets each were infected with PUR46 strains C8, C11 and with the indicated pseudorecombinant virus. One piglet of each group was sacrificed at days 1, 2, 3, and 4 post-inoculation, and the titer of the virus recovered from intestinal homogenates (jejunum, ileum, and intestinal content) was determined.

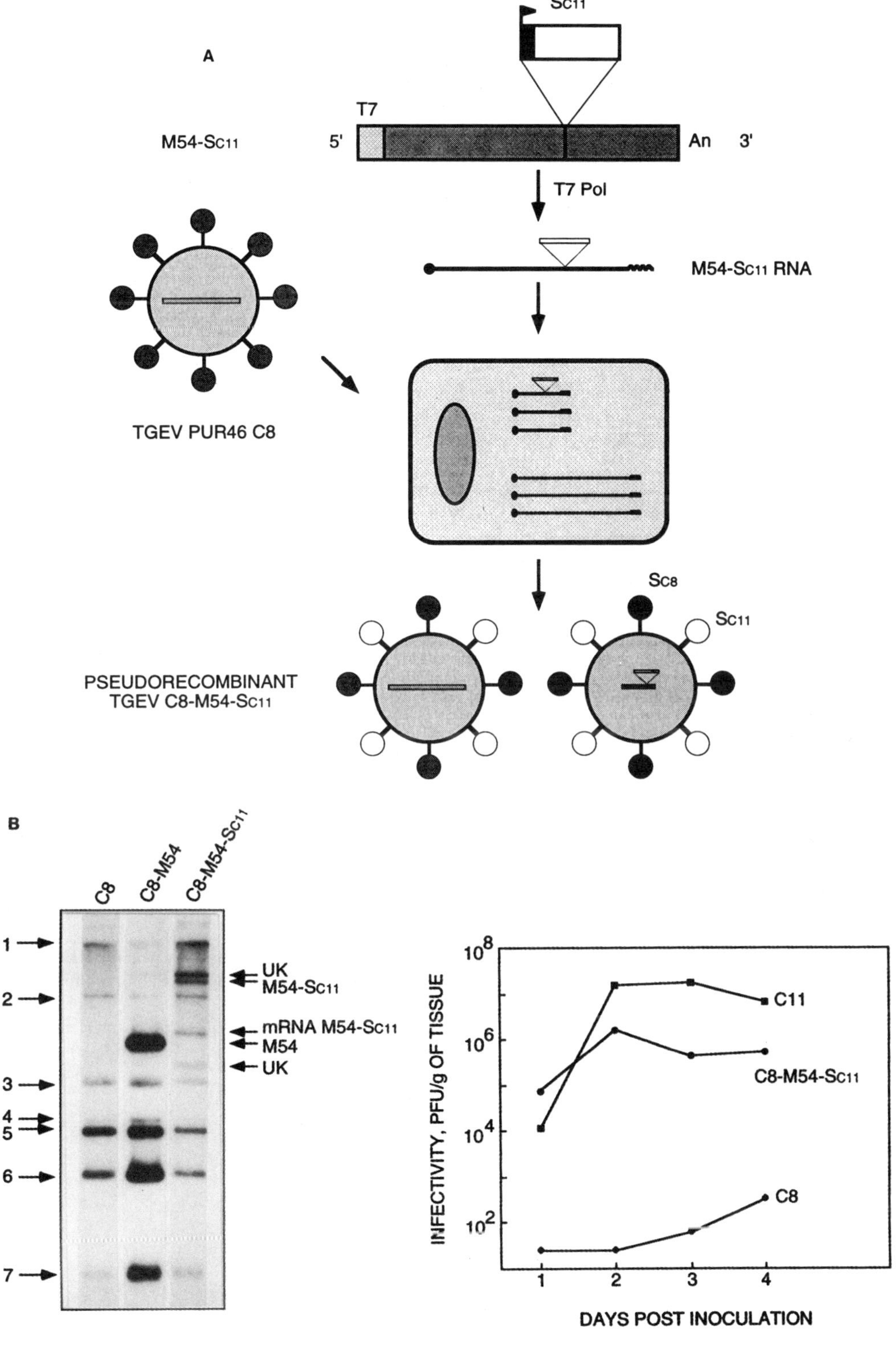

A
Sc11
T7
5'
An   3'
T7 Pol
M54-Sc11
M54-Sc11 RNA
TGEV PUR46 C8
Sc8
Sc11
PSEUDORECOMBINANT
TGEV C8-M54-Sc11
B
C8
C8-M54
C8-M54-Sc11
1
2
3
4
5
6
7
UK
M54-Sc11
mRNA M54-Sc11
M54
UK
INFECTIVITY, PFU/g OF TISSUE
C11
C8-M54-Sc11
C8
DAYS POST INOCULATION

REFERENCES

Ballesteros, M. L., Sanchez, C. M., and Enjuanes, L., 1997, Two amino acid changes at the N-terminus of transmissible gastroenteritis coronavirus spike protein result in the loss of enteric tropism, *Virology* **227:** 378–388.

Eleouet, J. F., Rasschaert, D., Lambert, P., Levy, L., Vende, P., and Laude, H., 1995, Complete sequence (20 kilobases) of the polyprotein-encoding gene 1 of transmissible gastroenteritis virus, *Virology* **206:** 817–822.

Enjuanes, L., and Van der Zeijst, B. A. M., 1995, Molecular basis of transmissible gastroenteritis coronavirus epidemiology, in: *The Coronaviridae* (S. G. Siddell, Ed.), Plenum Press, New York, pp. 337–376.

Lai, M. M. C., and Cavanagh, D., 1997, The molecular biology of coronaviruses, *Adv. Vir. Res.* **48:** 1–100.

Liao, C.-L., and Lai, M. M. C., 1994, Requirement of 5'-end genomic sequence as an upstream *cis*-acting element for coronavirus subgenomic mRNA transcription, *J. Virol.* **68**(8): 4727–4737.

Liao, C.-L., Zhang, X., and Lai, M. M. C., 1995, Coronavirus defective-interfering RNA as an expression vector: the generation of a pseudorecombinant mouse hepatitis virus expressing hemagglutinin-esterase, *Virology* **208:** 319–327.

Lin, Y. J., and Lai, M. M. C., 1993, Deletion mapping of a mouse hepatitis virus defective interfering RNA reveals the requirement of an internal and discontiguous sequence for replication, *J. Virol.* **67:** 6110–6118.

Lunney, J. K., Pescovitz, M. D., and Sachs, D. H., 1986, The swine major histocompatibility complex: its structure and function, in: *Swine in biomedical research* (M. E. Tumbleson, Ed.), Plenum Press, New York, pp. 1821–1836.

McClurkin, A. W., and Norman, J. O., 1966, Studies on transmissible gastroenteritis of swine. II. Selected characteristics of a cytopathogenic virus common to five isolates from transmissible gastroenteritis, *Can. J. Comp. Vet. Sci.* **30:** 190–198.

Mendez, A., Smerdou, C., Izeta, A., Gebauer, F., and Enjuanes, L., 1996, Molecular characterization of transmissible gastroenteritis coronavirus defective interfering genomes: Packaging and heterogeneity, *Virology* **217:** 495–507.

Pattnaik, A. K., Ball, L. A., LeGrone, A. W., and Wertz, G.W., 1992, Infectious defective interfering particles of VSV from transcripts of a cDNA clone, *Cell* **69:** 1011–1020.

Penzes, Z., Wroe, C., Brown, T. D. K., Britton, P., and Cavanagh, D., 1996, Replication and packaging of coronavirus infectious bronchitis virus defective RNAs lacking a long open reading frame, *J. Virol.* **70:** 8660–8668.

Sachs, D., Leight, G., Cone, J., Schwarz, S., Stuart, L., and Rosemberg, S., 1976, Transplantation in miniature swine. I. Fixation of the major histocompatibility complex, *Transplantation* **22:** 559–567.

Sambrook, J., Fritsch, E. F., and Maniatis, T., 1989, *Molecular cloning: a laboratory manual*, 2nd ed., Cold Spring Harbor Laboratory, Cold Spring Harbor, New York.

Sanchez, C. M., Jiménez, G., Laviada, M. D., Correa, I., Suñé, C., Bullido, M. J., Gebauer, F., Smerdou, C., Callebaut, P., Escribano, J. M., and Enjuanes, L., 1990, Antigenic homology among coronaviruses related to transmissible gastroenteritis virus, *Virology* **174:** 410–417.

Sanchez, C. M., Gebauer, F., Suñé, C., Mendez, A., Dopazo, J., and Enjuanes, L., 1992, Genetic evolution and tropism of transmissible gastroenteritis coronaviruses, *Virology* **190:** 92–105.

Sanchez, C. M., Ballesteros, M. L., and Enjuanes, L., 1997, Tropism, virulence and primary genome structure in a cluster of transmissible gastroenteritis coronavirus derived from the Purdue isolate, Submitted for publication.

Zhang, X., Hinton, D. R., Cua, D. J., Stohlman, S. A., and Lai, M. M. C., 1997, Expression of interferon-γ by a coronavirus defective-interfering RNA vector and its effect on viral replication, spread, and pathogenicity, *Virology* **233:** 327–338.

26

A NEW MODEL FOR CORONAVIRUS TRANSCRIPTION

S. G. Sawicki and D. L. Sawicki

Department of Microbiology and Immunology
Medical College of Ohio
Toledo, Ohio 43699

1. ABSTRACT

Coronaviruses contain an unusually long (27–32,000 ribonucleotide) positive sense RNA genome that is polyadenylated at the 3' end and capped at the 5' end. In addition to the genome, infected cells contain subgenomic mRNAs that form a 3' co-terminal nested set with the genome. In addition to their common 3' ends, the genome and the subgenomic mRNAs contain an identical 5' leader sequence. The transcription mechanism that coronaviruses use to produce subgenomic mRNA is not known and has been the subject of speculation since sequencing of the subgenomic mRNAs showed they must arise by discontinuous transcription. The current model called leader-primed transcription has subgenomic mRNAs transcribed directly from genome-length negative strands. It was based on the failure to find in coronavirus infected cells subgenome-length negative strands or replication intermediates containing subgenome-length negative strands. Clearly, these structures exist in infected cells and are transcriptionally active. We proposed a new model for coronavirus transcription which we called 3' discontinuous extension of negative strands. This model predicts that subgenome-length negative strands would be derived directly by transcription using the genome RNA as a template. The subgenome-length templates would contain the common 5' leader sequence and serve as templates for the production of subgenomic mRNAs. Our findings include showing that: 1. Replication intermediates (RIs) containing subgenome-length RNA exist in infected cells and are separable from RIs with genome-length templates. The RFs with subgenome-length templates are not derived by RNase treatment of RIs with genome-length templates. 2. The subgenome-length negative strands are formed early in infection when RIs are accumulating and the rate of viral RNA synthesis is increasing exponentially. 3. Subgenome-length negative strands contain at their 3' ends a complementary copy of the 72 nucleotide leader RNA that is found in the genome only at their 5' end. 4. RIs with subgenomic templates

Coronaviruses and Arteriviruses, edited by Enjuanes *et al.*
Plenum Press, New York, 1998

serve immediately as templates for transcription of subgenomic mRNAs. Because subgenomic mRNAs are not replicated, i.e., copied into negative strands that in turn are used as templates for subgenomic mRNA synthesis, we propose that the subgenome-length negative strands must arise directly by transcription of the genome and acquire their common 3' anti-leader sequence after polymerase jumping from the intergenic regions to the leader sequence at the 5' end of the genome. This would make negative strand synthesis discontinuous and subgenomic mRNA synthesis continuous, which is the opposite of what was proposed in the leader primed model.

2. RESULTS AND DISCUSSION

At the VIth International Symposium on Corona and Related Viruses in Quebec, Canada, we first presented a new model for coronavirus transcription, which we called discontinuous extension of negative strands, to explain how subgenome-length negative strands would be derived directly from the genome (Sawicki and Sawicki, 1995). A new model was needed because the current model called leader-primed transcription (see review by Lai, 1990) was based on data which claimed to demonstrate that only genome-length negative strands existed in coronavirus infected cells (Lai, et al. 1982; Baric et al., 1983). Therefore, subgenomic mRNAs had to be derived from genome-length negative strands. The seminal observation made in David Brian's laboratory (Sethna et al., 1989) that transmissible gastroenteritis virus (TGEV) infected cells contained subgenome-length negative strands invalidated the premise on which the leader-primed model was based, mainly that only genome length-negative strands existed in coronavirus infected cells. Shortly afterward we demonstrated (Sawicki and Sawicki, 1990) that RIs containing subgenome-length negative strands existed and were transcriptionally active in mouse hepatitis virus (MHV) infected cells.

Because subgenomic mRNAs are not themselves able to be copied directly into negative strands (Brian et al., 1994; Makino et al., 1991; Masters et al., 1994), we suggested (Sawicki and Sawicki, 1995) that subgenome-length negative strands would be generated directly from the genome. We are able to separate RIs with subgenome-length templates from RIs with genome length templates, either by velocity gradient centrifugation on sucrose gradients or by agarose gel electrophoresis. The RIs with subgenomic templates incorporate ^{3}H-uridine into positive and negative strands with the same kinetics as RIs with genome-length templates. The RIs with subgenome-length templates were formed quickly after infection and were detectable by 1.5 -2 hours post infection; but we could not determine whether they were formed directly from transcription of the genome or if they were derived by copying subgenomic mRNAs. The latter possibility seems unlikely based on the experiments cited above that subgenomic mRNAs are not replicated when transfected into cells. Sequences downstream of the leader are required for replication. Therefore, we concluded that subgenome-length negative strand templates arose directly from copying the genome RNA and derived their 3' ends, which are complementary to the leader sequence at the 5' end of the genome, by discontinuous transcription.

As we proposed three years ago the transcription apparatus of coronaviruses shows similarities to the E.coli RNA polymerase which transcribes DNA in a discontinuous fashion (Komissarova and Kashlev, 1997; Nudler et al., 1997). Nudler et al. (1997) stated that the basic features of the ternary complex of RNA polymerase are conserved in all living organisms but its detailed structure is unknown. We suggest that the coronavirus RNA polymerase also has an inherent capacity to transcribe its RNA template monotonously but the intergenic sequences (IS elements) in the genome are sites of discontinuous elongation. Pausing at these sites would allow for detachment of the polymerase with its nascent negative strand. The polymerase with its nascent negative strand would reattach in cis or in trans at the same site or translocate specifically to the 5' end of the genome. The movement of the following polymerase may cover the site of detachment and force the detached polymerase to reattach at the 5' end of the genome. We predict that the genome would have multiple polymerase molecules copying it into negative strands. One possibility is that the initial or leading polymerase would be committed to produce a genome-length negative strand template and, thereby, ensure replication of the genome and the survival of the virus. Following polymerases would be subject to pausing. Loading of the genome with polymerases would force pausing of the following polymerase by the leading polymerase and account for the relationship of the relative abundance of the smaller compared to the larger subgenomic mRNAs with their position relative to the 3' of the genome.

The generation of RIs with subgenome-length templates would result from polymerase reattachment at the 5' end of the genome, either the same genome (cis) or another genome (trans). The high rate of recombination observed in coronaviruses would result from reattachment in trans to another genome but at the same site on the genome from which the polymerase had detached. Reattachment at the 5' end of the genome would be facilitated, as we suggested, by protein-protein interaction between the polymerase and a protein bound between the leader sequence at the 5' end of the genome and first open reading frame, which would be a region necessary for the genome to serve as a template for negative strand synthesis. The report by Chang et al. (1996) demonstrated that leader conversion of defective interfering (DI) RNA of bovine coronavirus was not guided by the IS element at the 5' end of the DI RNA but by a recombination event during negative strand synthesis. In their case the polymerase detached during negative strand synthesis before copying the leader sequence at the 5' end of the DI RNA and reattached in trans at the 5' end of the helper virus genome but downstream of the leader sequence. The recombinant DI RNA possessed the leader sequence including the IS element of the helper virus genome. Since the DI RNA did not contain an IS element at its 5' end the polymerase paused and detached up to 25 nucleotides before the leader sequence in the DI RNA. In this case the sequence downstream of the leader sequence in the DI RNA and in the helper virus were nearly identical. This allowed the detached negative strand to reinitiate RNA synthesis downstream of the IS element at the 5' end of the helper virus genome. In the case of generating subgenome-length templates, only the IS elements in the intergenic regions of the genome would be homologous with the IS element at the 5' end of the genome. Therefore, if the polymerase detaches at a site other than the IS elements when copying the genome it cannot easily elongate and complete negative strand synthesis at the 5' end of the genome because of base pair mismatch. Only when there are matching base pairs at the 3' end of the the detached nascent negative strand and the genome template will elongation of the nascent negative strands occur. This would also explain the heterogeneity of leader-body fusion sites on subgenomic RNA (see van der Most et al., 1994).

The IS elements in the genome of coronaviruses can be viewed as analogous to the intergenic dinucleotide and transcription start sequence in the genome of vesicular stomatitis virus, which are required for the efficient termination and initiation of transcription (Stillman and Whitt, 1997). It appears that like DNA-dependent RNA polymerases, RNA-dependent RNA polymerases transcribe a template in monotonous or in discontinuous fashion. The structural features of a template that determine discontinuous transcription remain to be elucidated.

At the last Symposium on Corona and Related viruses we stated that to our knowledge there is no experimental data that disproves our model. This remains true today. Although the details of the mechanism of discontinuous synthesis of negative strands remain a mystery, we can summarize the state of our understanding of coronavirus transcription as follows:

1. During the early phase of coronavirus infection, negative strand synthesis occurs coincidentally with positive strand synthesis.
2. Negative strand synthesis requires continued protein synthesis and is immediately halted by the addition of inhibitors of translation.
3. The rate of positive strand synthesis is determined by the number of negative strand templates.
4. Positive strand synthesis, in contrast to negative strand synthesis, is more stable and continues unabated for about 30 min. after inhibition of translation and then declines slowly. This suggests that a protein factor is required for the activity of the coronavirus positive strand RNA polymerase. There appears to be a pool of this factor sufficient to sustain the positive strand polymerase activity for 30 min.; and, if not replenished by protein synthesis, positive strand synthesis declines.
5. The decline in negative strand synthesis at about 5 hours post infection at 37°C in 17Cl-1 cells infected with the A59 strain of mouse hepatitis virus is correlated with leveling of positive strand synthesis. Late in the infection cycle, the rate of positive strand synthesis gradually declines.
6. The template for subgenome-length negative strands is the genome. The subgenome-length negative strands are produced by discontinuous transcription and are found in partially double stranded RIs, not as single stranded RNAs.
7. At all times post infection RIs with genome- and subgenome-length templates are actively producing genomes and subgenomic mRNAs.
8. The ratio of genomes to subgenomic mRNAs produced throughout infection does not vary and is determined by the ratio of RIs with genome- and subgenome-length templates.

REFERENCES

Baric, R. S., Stohlman, S. A., Lai, M. M. C., 1983, Characterization of replicative intermediate RNA of mouse hepatitis virus: presence of leader RNA sequences on nascent chains, *J. Virol.* **48**:633.

Brian, D.A., Chang, R.-Y., Hofmann, M.A., Sethna, P.B., 1994, Role of subgenomic minus-strand RNA in coronavirus replication, *Arch. Virol.* (Suppl.) **9**:173–180.

Chang, R.Y., Krishnan, R., and Brian, D.A., 1996, The UCUAAAC promoter motif is not required for high frequency leader recombination in bovine coronavirus defective interfering RNA, *J. Virol.* **70**:2720–2729.

Komissarova, N., and Kashlev, M., 1997, Transcription arrest: Escherichia coli RNA polymerase translocates backward, leaving the 3' end of the RNA intact and extruded, *Proc. Natl. Acad. Sci. USA* **94**:1755–1760.

Lai, M. M. C., Patton, C. D., and Stohlman, S. A., 1982, Replication of mouse hepatitis virus: negative-stranded RNA and replicative form RNA are of genome length, *J. Virol.* **44**:487–492.

Lai, M.M.C., 1990, Coronavirus-organization, replication and expression of genome, *Annu. Rev. Microbiol.* **44**:303–333.

Makino, S., Joo, M. , and Makino, J.K., 1991, A system for study of coronavirus mRNA synthesis: a regulated, expressed subgenomic defective interfering RNA results from intergenic site insertion, *J. Virol.* **65**:6031–6041.

Masters, P.S., Koetzner, C.A., Kerr, C.A., and Heo, Y., 1994, Optimization of targeted RNA recombination and mapping of a novel nucleocapsid gene mutation in the coronavirus mouse hepatitis, *J. Virol.* **68**:328–337.

Nudler, E., Mustaev, A., Lukhtanov, E., and Goldfarb, A., 1997, The RNA-DNA hybrid maintains the register of transcription by preventing backtracking on RNA polymerase, Cell. **89**:33–41.

Sawicki, S.G, and Sawicki, D.L., 1990. Coronavirus transcription: subgenomic mouse hepatitis virus replicative intermediates function in RNA synthesis, *J. Virol.* **64**:1050–1056.

Sawicki, S.G.and Sawicki, D.L., 1995, Coronaviruses use discontinuous extension for synthesis of subgenome-length negative strands, in: *Corona and Related Viruses* (P.J. Talbot and G.A. Levy, eds) Plenum Press, New York, pp. 499–505.

Sethna, P.B., Hung, S-L., and Brian, D.A., 1989, Coronavirus subgenomic minus-strand RNAs and the potential for mRNA replicons, *Proc. Natl. Acad. Sci. USA* **86**:5626–5630.

Stillman, E.A., and Whitt, M.A., 1997, Mutational analysis of the intergenic dinucleotide and the transcriptional start sequence of vesicular stomatitis virus (VSV) define sequences required for efficient termination and initiation of VSV transcripts, *J. Virol.* **71**:2127–2137.

van der Most, R.G., de Groot, R.J., and Spaan, W.J.W., 1994, Subgenomic RNA synthesis directed by a synthetic defective interfering RNA of mouse hepatitis virus: a study of coronavirus transcription initiation, *J. Virol.* **68**:3656–3666.

27

NEGATIVE STRAND RNA SYNTHESIS BY TEMPERATURE-SENSITIVE MUTANTS OF MOUSE HEPATITIS VIRUS

D. R. Younker and S. G. Sawicki

Department of Microbiology and Immunology
Medical College of Ohio
Toledo, Ohio 43699

1. ABSTRACT

Mutants of the A59 strain of mouse hepatitis virus (MHV) were studied to determine the effects of temperature shift on negative and positive strand RNA synthesis. Mutant LA 9 was originally reported to have a selective temperature sensitive defect in negative strand synthesis. We found that this mutant continued to synthesize negative strands for at least one hour after temperature shift. LA 6 was found to possess a temperature sensitive defect in negative strand synthesis. Following temperature shift, negative strand synthesis rapidly declined. The effect of temperature shift on negative strand synthesis by LA 6 was similar to the effect of cycloheximide treatment of the parental A59 virus. Temperature shift of Alb 16 infected cells did not stop negative strand synthesis but did prevent a corresponding rise in the rate of positive strand synthesis. Therefore, we suggest that Alb 16 is a conversion mutant because it cannot convert newly synthesized negative strands into templates for positive strand synthesis at the nonpermissive temperature.

2. INTRODUCTION

Coronaviruses are positive strand RNA viruses and their genome must be translated after entry of the virus into the cytoplasm. The translation products of the genome are non-structural proteins, which have been characterized only partially, and must include the viral RNA-dependent RNA polymerase. The coronavirus RNA polymerase would be responsible for producing both negative and positive strand templates. Previously we had shown that the rate of positive strand RNA synthesis in MHV infected cells was coupled to negative strand synthesis and the number of negative stands controlled the rate of posi-

Coronaviruses and Arteriviruses, edited by Enjuanes *et al.*
Plenum Press, New York, 1998

tive strand synthesis (Sawicki and Sawicki, 1986). During the early phase of infection, negative strand synthesis occurred coincidentally with positive strand synthesis. After the rate of both positive and negative strand synthesis reached a maximum at 5–6 hours p.i. at 37°C the rate of negative strand synthesis declined and the rate of positive strand synthesis became relatively constant. If translation was inhibited by cycloheximide before the maximum rate of viral RNA synthesis was attained, negative strand synthesis ceased abruptly and the rate of positive strand synthesis leveled and then declined beginning about 30 min. after translation had halted. We interpreted these data to mean that new polymerase was required for the production of negative strands. Once negative strands were produced they functioned as templates for positive strand synthesis. However, positive strand synthesis continued unabated for only a limited time after inhibition of translation. Therefore, positive strand synthesis required the production of either new polymerase or a protein factor that is necessary for polymerase activity. We have been studying RNA negative, temperature sensitive mutants of MHV/A59. These mutants fail to produce viral RNA when cells are infected and maintained at the nonpermissive temperature. Schaad et al. (1990) suggested that the rate of positive strand synthesis could be increased in the absence of negative strand synthesis in the temperature sensitive mutant LA 9. After shifting to the nonpermissive temperature after viral RNA synthesis had started at the permissive temperature, it appeared that positive strands, but not negative strands accumulated. In fact, they did not measure the rate of negative strand synthesis, e.g., the incorporation of ^{3}H-uridine into positive or negative strands, but looked for the accumulation of negative strands after temperature shift. We obtained LA 9 and LA 6 from Dr. Ralph Baric and Alb 16 from Dr. Larry Sturman (Sturman et al. 1987) and determined whether or not they had a specific defects in negative strand RNA synthesis.

3. RESULTS

All of the mutants possessed a ts RNA negative phenotype, i.e., they did not make viral RNA if the infection was initiated and maintained at 39.5°C the nonpermissive temperature. To study the effect of temperature on negative strand synthesis by these mutants, we allowed viral RNA synthesis to begin and reach about 20% of the maximum rate at 30°C, the permissive temperature. Then we shifted the infected cells to 40°C and labeled the viral RNA with ^{3}H-uridine in the presence of dactinomycin. To determine negative strand synthesis we followed our published procedure (Sawicki and Sawicki, 1986) of first isolating RF RNA by cellulose CF-11 chromatography and subjecting denatured RF RNA to hybridization with an excess of unlabeled virion RNA.

When the parental A59 virus was shifted from 30°C to 40°C at 9 hours post infection and pulse labeled with ^{3}H-uridine for two half hour periods, the rate of positive strand synthesis increased about two fold during the second half hour pulse period compared to the first half hour pulse (Fig. 1A). Similarly, the rate of negative strand synthesis increased about two-fold in the second half-hour relative to the first half hour after shift (Fig. 1B). The radioactivity in RF RNA also increased about two fold but the percent labeled RF RNA as negative strands remained constant. We interpret this to mean that the negative strands produced after shift were utilized as templates for positive strand synthesis to increase the rate of positive strand synthesis. With the addition of cycloheximide at the time of shift, positive strand synthesis remained the same during the second half hour pulse period compared to the first half hour pulse period (Fig. 1A). However, cycloheximide caused a dramatic reduction in negative strand synthesis (Fig. 1B). We reported previously

(Sawicki and Sawicki, 1986) that cycloheximide immediately inhibited negative strand synthesis whereas positive strand synthesis began to decline only after 30–60 min. after its addition. Therefore, the inhibition of negative strand synthesis after the addition of cycloheximide prevented the increase in the rate of positive strand synthesis which was observed after shift in the absence of cycloheximide.

Schaad et al. (1990) reported that the ts mutant LA 9 of the A59 strain MHV possessed a specific temperature sensitive defect in negative strand synthesis: Positive strands and infectious virus, but not negative strands, accumulated after temperature shift. Since they had not measured negative strand synthesis directly, we determined if negative strand synthesis by LA 9 was inhibited immediately by temperature shift. When LA 9 infected cells were shifted from 30°C to 40°C at 9 hours post infection and pulse labeled with ^{3}H-uridine for two half hour periods, the rate of positive strand synthesis increased about two fold during the second half hour pulse period compared to the first half hour pulse (Fig. 1C). The rate of negative strand synthesis also increased about two-fold in the second half-hour relative to the first half hour after shift (Fig. 1D). When cycloheximide was added at the time of shift, LA 9 positive strands synthesis declined slightly during the second half hour pulse period compared to the first half hour pulse period (Fig. 1C); and, cycloheximide caused a dramatic reduction in negative strand synthesis (Fig. 1D). If LA 9 had a specific defect in the synthesis of negative strands, we would have observed a similar reduction in the incorporation of ^{3}H-uridine into negative strands in the presence as in the absence of cycloheximide. Clearly, this was not the case. Rather, we observed that the increase in positive strand synthesis that occurred immediately after temperature shift was coupled to an increase in negative strand synthesis. Therefore, we could not confirm that LA 9 had a specific temperature sensitive defect in negative strand synthesis.

When shifted between 2 and 10 hours post infection to the nonpermissive temperature, LA 9 infected cells increased their rate of positive strand synthesis approximately five-fold (data not shown). With LA 9 we found the time of temperature shift to be critical. Shifting early when RNA synthesis is below detectable levels had little or no effect on viral RNA synthesis or virus yield. Shifting too late, when viral synthesis had reached its maximum rate, produced no immediate reduction in viral RNA synthesis or viral yield. We confirmed the original observation (Schaad et al., 1990) that LA 9 increased the rate of viral RNA synthesis and virus yield if the time of temperature shift was after six hours post infection.

We also screened two other ts mutants (LA 6 and Alb 16) for defects in negative or positive strand synthesis. LA 6 infected cells were shifted from 30°C to 40°C at 9 hours post infection and pulse labeled with ^{3}H-uridine for three twenty minutes periods. The rate of positive strand synthesis remained the same for the three pulse periods (Fig. 1E). Notice that the rate of ^{3}H-uridine incorporation into viral RNA was less with LA 6 than with the parental A59 or LA 9. LA 6 at the permissive temperature showed three to six fold less viral RNA synthesis compared to the parental A59 virus or revertants of LA 6. With LA 6 negative strand synthesis decreased dramatically after shift (Fig. 1F). With the addition of cycloheximide at the time of shift, positive strands synthesis remained the same during the three twenty minutes pulse (Fig. 1E) and cycloheximide caused a dramatic reduction in negative strand synthesis (Fig. 1F). We concluded that with LA 6 synthesis of negative strands, but not positive strands, was temperature sensitive. The failure to produce more negative strands after shift resulted in the failure to increase positive strand synthesis. Therefore, temperature shift in the presence and absence of cycloheximide produced nearly the same results.

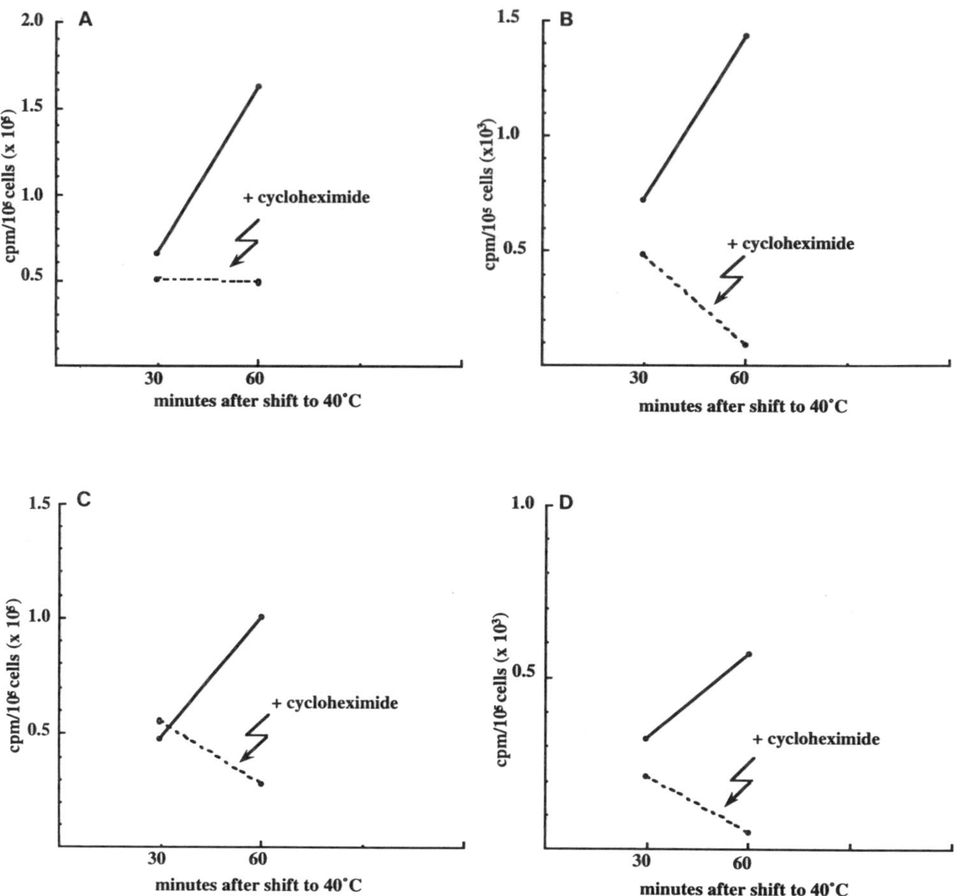

Figure 1. Viral and negative strand synthesis by parental A59 and temperature sensitive mutants after temperature shift. 17Cl-1 cells were infected and maintained at 30°C and shifted to 40°C at 8 hours for Alb 16 or 9 hours for A59, LA 9, and LA 6 viruses and labeled for 30 or 20 minute pulses with 200 μCi/ml of [³H]uridine and 20 μg/ml dactinomycin in the presence or absence of 100 μg/ml cycloheximide. Viral RNA was determined by acid precipitation from cell lysates. RF RNA was purified by column chromatography on CF-11 after RNase T1 digestion. Labeled negative strands were obtained from the RF RNA after they were denatured, annealed to a 100-fold excess unlabeled virion RNA, and treated with RNase A and T1. (A) A59, viral RNA synthesis. (B) A59 negative strand synthesis. (C) LA 9, viral RNA synthesis. (D) LA 9, negative strand synthesis. (E) LA 6, viral RNA synthesis. (F) LA 6, negative strand synthesis. (G) Alb 16, viral RNA synthesis. (H) Alb 16, negative strand synthesis.

When Alb 16 (Sturman et al., 1987) was shifted at 8 hours post infection and pulse labeled with ³H-uridine for three twenty minute periods, the rate of positive strand synthesis increased slightly or remained constant (Fig. 1G). The rate of negative strand synthesis remained relative constant for one hour after temperature shift (Fig. 1H). With the addition of cycloheximide at the time of shift, positive strand synthesis remained the about the same (Fig. 1G), whereas negative strand synthesis declined (Fig. 1H) as it did for the parental A59 virus. Therefore, Alb 16 has an interesting phenotype with regard to negative strand synthesis. Negative strand synthesis is not as sensitive to temperature shift as it is to cycloheximide. Nevertheless, positive strand synthesis did not increase significantly during the first hour after temperature shift, which mimics the effect of cycloheximide.

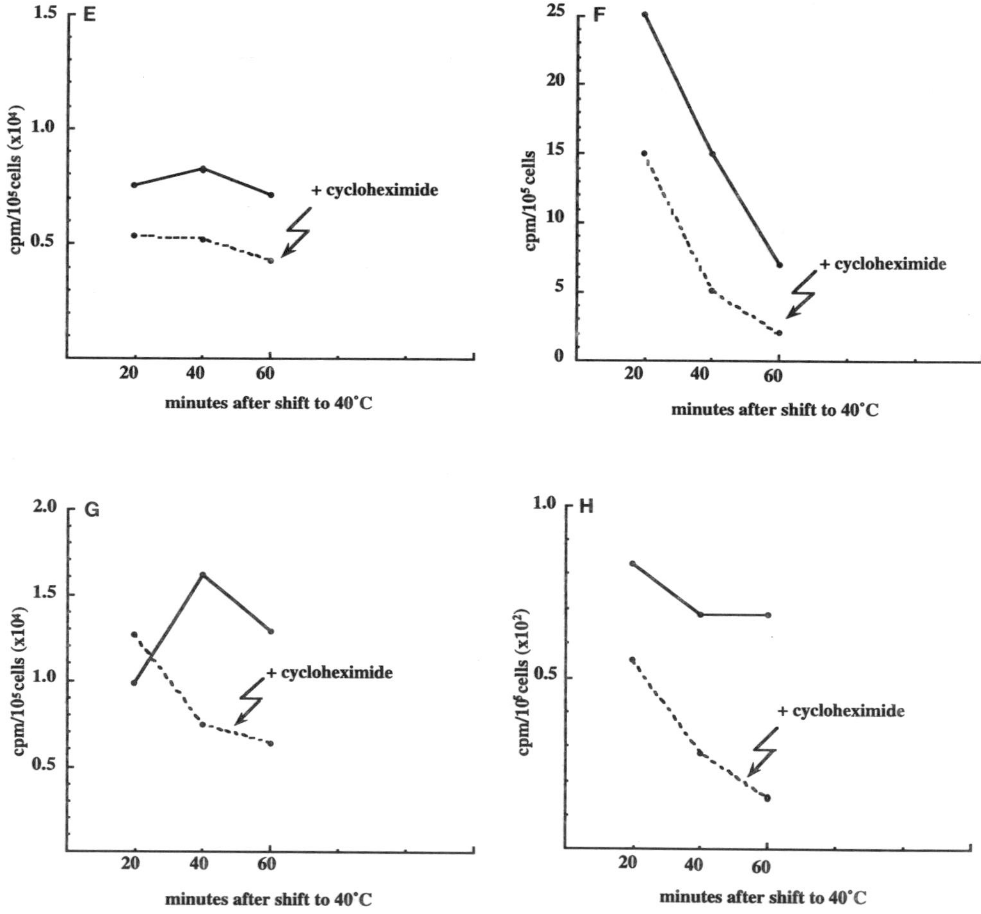

Figure 1E-H.

4. DISCUSSION

The use of RNA-negative temperature sensitive mutants should give some insight into mechanisms coronaviruses use to synthesize RNA. Previously, Schaad et al. (1990) had concluded that LA 9 had a temperature sensitive defect in negative strand synthesis. We could not confirm their results for negative strand synthesis but could confirm their results for positive strand synthesis and virus yield. Temperature shift with this mutant allowed negative strands to be synthesized for about one hour before the synthesis of negative strands decreased. This was in turn followed by a five fold increase in the rate of positive strand synthesis. The timing of the shift was critical. If the rate of viral RNA synthesis was 20% of the maximum or higher at the time of the shift, then a full burst of viral RNA synthesis and virus yield occurred. If the shift was earlier, then less viral RNA and virus were made.

The mutant LA 6 had a temperature sensitive defect in producing polymerase that would synthesize negative strands, but the polymerase producing positive strands was unaffected by shift to the nonpermissive temperature. Since negative strand synthesis was not more sensitive to temperature shift than was the parental A59 virus to cycloheximide, we concluded the mutation in LA 6 affected the initiation of negative strand synthesis or the formation of the negative strand replicase. Without the continued production of negative strands the rate of positive strand synthesis remained constant after the shift to the nonpermissive temperature.

The mutant Alb 16 had an interesting phenotype. Negative strand synthesis was not inhibited immediately by temperature shift; however, the rate of positive strand synthesis did not increase. We (see Sawicki, D.L., and Sawicki, S.G., this volume) called alphavirus mutants with this phenotype conversion mutants because the newly formed polymerase which has negative strand activity cannot be converted to a polymerase with positive strand activity.

REFERENCES

Sawicki, S.G., and Sawicki, D.L., 1986, Coronavirus minus-strand RNA synthesis and effect of cycloheximide on coronavirus RNA synthesis, *J. Virol.* **57**:328–334.

Schaad, M.C., and Baric, R.S., 1994, Genetics of mouse hepatitis virus transcription: evidence that subgenomic negative strands are functional templates, *J. Virol.* **68**:8169–8179.

Schaad, M.C., Stohlman, S.A., Egbert, J., Lum, K., Fu, K., Wei, Jr., T., and Baric, R.S., 1990, Genetics of mouse hepatitis virus transcription: identification of cistrons which may function in positive and negative strand RNA synthesis, *Virology* **177**:634.

Sturman, L.S., Eastwood, C., Frana, M.F., Duchala, C., Baker, F., Ricard, C.S., Sawicki, S.G. and Holmes, K.V., 1987, Temperature-sensitive mutants of MHV-A59, in: *Coronaviruses* (M.M.C. Lai and S.A. Stohlman, eds) Plenum Press, New York, pp. 159–168.

28

CELLULAR PROTEIN hnRNP-A1 INTERACTS WITH THE 3'-END AND THE INTERGENIC SEQUENCE OF MOUSE HEPATITIS VIRUS NEGATIVE-STRAND RNA TO FORM A RIBONUCLEOPROTEIN COMPLEX

Xuming Zhang,[1,4] Hsin-Pai Li,[2] Wenmei Xue,[2] and Michael M. C. Lai[1,2,3]

[1]Department of Neurology
[2]Department of Molecular Microbiology and Immunology
[3]Howard Hughes Medical Institute
University of Southern California School of Medicine
Los Angeles, California 90033
[4]Department of Microbiology and Immunology
University of Arkansas for Medical Sciences
Little Rock, Arkansas 72205

1. ABSTRACT

We previously showed that several cellular proteins specifically bind to the 3'-end and the intergenic sequences of the negative-strand RNA of mouse hepatitis virus (MHV), and proposed that these distant RNA sequences can be brought together by cellular and viral proteins (Furuya and Lai, 1993; Zhang et al., 1994; Zhang and Lai, 1995). The cellular protein p35 has been identified as a heterogeneous nuclear ribonucleoprotein (hnRNP) A1. We have now expressed hnRNP-A1 as a glutathione-*S*-transferase (GST) fusion protein and demonstrated that the amino terminal two-thirds of hnRNP-A1 interacted with the two MHV regulatory RNA sequences (3'-end and intergenic sequences) through protein-RNA interaction while its carboxy-terminal glycine-rich domain mediated homomeric (protein-protein) interactions. In a partially reconstituted reaction, in which the two MHV RNA fragments and the purified GST-hnRNP-A1 fusion protein were mixed, an RNP complex was formed. Depletion of either hnRNP-A1 or one of the RNA components abolished the complex formation. These results indicate that hnRNP-A1 can mediate the formation of an MHV RNP complex, which includes both the negative-strand leader and intergenic sequences. Site-directed mutagenesis revealed that mutations in the MHV intergenic sequences, which inhibited MHV RNA transcription, also inhibited the RNP complex formation. Deletion analysis showed that the amino terminal RNA-binding domains of hnRNP-A1 is essential for the RNP complex formation while the carboxy-terminal protein-binding domain enhanced the complex formation by 90-fold. These findings provide direct evidence demonstrating that the negative-strand leader RNA and intergenic sequences can form an RNP complex mediated by cellular protein hnRNP-A1.

2. INTRODUCTION

Heterogeneous nuclear ribonucleoprotein (hnRNP) is a collective term representing a number of proteins (termed A to U) that are associated with pre-mRNAs or hnRNAs to form RNP complexes in mammalian cells (Dreyfuss et al. 1993). hnRNP A1 is the first and best characterized hnRNP protein. It contains 320 amino acids deduced from the cDNA sequence (Buvoli et al., 1988) with an estimated molecular weight of 34 kDa. Its N-terminal two-thirds (residues 1 to 196) forms two tandemly arranged RNA-binding domains (RBD) which bind to the 5'- and 3'-splicing consensus sequences of pre-mRNAs (Dreyfuss et al., 1993). The C-terminal domain (residues 197 to 320) is very rich in glycine (40%), and is designated as a glycine-rich domain. This domain contains a 36 residue sequence analogous to an "RGG box", an RNA-binding motif first identified in hnRNP U (Kiledjian and Dreyfuss, 1992). Recently, it has been shown that the glycine-rich domain mediates the protein-protein interaction of A1 with itself and with other hnRNP basic "core" proteins and some SR-proteins in vitro (Cartegni et al., 1996). Downstream of the RGG box is the M9 domain, which was shown to mediate both nuclear import and export of the A1 protein (Siomi and Dreyfuss, 1995; Weighardt et al., 1995, Pollard et al., 1996). A1 has also been shown to be involved in RNA-reannealing and alternative RNA-splicing. Its diverse and specific protein-RNA and protein-protein interactions have been implicated to function in mRNA maturation in mammalian cells (Dreyfuss et al., 1993).

We previously showed that two cellular proteins with molecular weights of 35 and 38 kDa (p35/38) specifically bound to the 3'-end and the intergenic (IG) sequences of the negative-strand RNA of mouse hepatitis virus (MHV) (Furuya and Lai, 1993; Zhang and Lai, 1995). By performing site-directed mutagenesis of the IG sequence, we further demonstrated that p35/38 is possibly involved in the regulation of MHV RNA transcription, since the efficiency of p35/38 binding to the IG sequence correlated with the amount of mRNAs transcribed (Zhang and Lai, 1995). We hypothesized that these regulatory RNA sequences of MHV can be brought together by cellular and viral proteins to form a transcription initiation complex (Zhang et al, 1994; Zhang and Lai, 1995). Recently, we have isolated the cellular protein p35 and identified it as hnRNP-A1 (Li et al., 1997). Protein expression and UV cross-linking experiments further confirmed that hnRNP-A1 had the same RNA-binding properties to MHV RNAs as p35/38.

In the present study, we investigated the role of the cellular protein hnRNP-A1 (previously designated p35) in the formation of an MHV RNP complex by analyzing the interactions between hnRNP-A1 and MHV RNAs. We present direct evidence demonstrating that both protein-RNA and protein-protein interactions are involved in the formation of an MHV RNP complex. Our results demonstrated that the RNA-binding domain of hnRNP-A1 mediate the binding of hnRNP-A1 to both the 3'-end and the IG sequence of the negative-strand RNA of MHV, whereas the carboxy-terminal glycine-rich domain exclusively involved in protein-protein interaction. Furthermore, mutations of the intergenic sequence of MHV RNA, which inhibited the protein-RNA interaction, also inhibited the RNP complex formation. Deletion analyses of hnRNP-A1 protein indicated that protein-RNA interaction is required for an RNP complex formation while the protein-protein interaction significantly enhanced the complex formation. This study establishes the important components in the proposed model involving protein-RNA interaction, and thus contributes to the understanding of the complex mechanism of coronavirus mRNA transcription.

3. MATERIALS AND METHODS

3.1. Construction of cDNA Clones

The cDNA encoding the murine hnRNP A1 gene was synthesized from RNAs isolated from a murine astrocytoma cell line (DBT) by reverse transcription (RT) and polymerase chain reaction (PCR) according to the procedures described previously (Zhang et al., 1994). It was then cloned into the *Bam*HI and *Eco*RI sites of pBluescript and pGEX 4–1, resulting in pBS-mA1 and pGST-mA1, respectively. To generate deletion constructs of A1, pGST-mA1 DNA was used as a template for PCR amplification using various sets of primers. These PCR fragments were then digested with *Bam*HI and *Eco*RI and directionally cloned into the pGEX4–1, resulting in GST-fusion plasmids containing various domains of A1.

3.2. Expression of GST Fusion Proteins

The procedure for expressing GST fusion proteins was as described previously (Smith and Johnson, 1988; Ausubel et al., 1989).

3.3. In Vitro Transcription and in Vitro Translation

In vitro transcription and translation were carried out according to the methods recommended by the manufacturer (Promega).

3.4. UV Cross-Linking of RNA-Protein Complex

The UV crosslinking study was carried out to determine the interaction of hnRNP-A1 protein with MHV RNA sequences. The purified GST-fusion proteins were used as a protein source, and the ^{32}P-labeled, *in vitro*-transcribed MHV RNA was used as a probe. The procedure for UV crosslinking experiments was essentially as described previously (Zhang and Lai, 1995).

3.5. Protein Binding Assay

A protein binding assay was performed to determine the protein-protein interaction as described previously (Ausubel et al., 1989).

3.6. In Vitro Formation and Detection of the RNP Complex

The *in vitro* formation and detection of the RNP complex was based on the procedures described by Wills (1996) and Lamond and Sproat (1996) with slight modifications. Briefly, three basic components [the *in vitro*-transcribed cold RNA fragment of the 3'-end

182 nt sequence, XN182(-); the *in vitro*-transcribed ^{32}P-labeled intergenic sequence of 118 nt, ^{32}P-IG7-SM118(-); and the purified GST-A1 fusion protein] were mixed in a reaction buffer of 100 µl containing 300 mM potassium chloride, 1.5 mM ATP, 5 mM creatine phosphate, 10 mM creatine phosphokinase, 2 mM magnesium chloride, and 40 U of RNasin. The mixture was incubated at 30 oC for 1 h. To identify RNP complex formation, the biotin-labeled riboprobe [biotin-56(+)], which is complementary to the 3'-end of XN182(-) RNA, was added to the reaction, and the incubation continued for an additional hour. Preblocked streptavidin-agarose beads (Sigma) were added to the reaction and incubated at 4 oC on a rocking platform for 1 h, followed by 4 washes with wash buffer WB300 (20 mM HEPES-potassium hydroxide, pH7.9, 300 mM potassium chloride, and 0.1% nonidet P-40). After adding 200 µl of Bonner's RNA extraction buffer (7 M urea, 350 mM sodium chloride, 10 mM Tris, pH8.0, 10 mM ethylenediamine tetraacetic acid, 1% sodium dodecyl sulfate) and 20 µl of 0.2 mM glycogen, RNA was extracted with phenol/chloroform, and analyzed by denaturing PAGE containing 6% polyacrylamide as described (Lamond and Sproat, 1996). The dried gel was autoradiographed and/or scanned in a PhosphorImager and the amount of radioactivity of specific bands was determined.

4. RESULTS AND DISCUSSION

4.1. Mapping the RNA- and Protein-Binding Domains of hnRNP-A1

Although several functional domains of hnRNP-A1 have been identified, it is not known which domain(s) specifically binds MHV RNA. To understand how an MHV RNP complex is formed, it will be necessary to determine which domains of hnRNP-A1 bind MHV regulatory RNA sequences. For this purpose, we employed PCR to generate 9 deletion fragments by deleting one or more domains of hnRNP-A1, and expressed them in *E. coli* as GST-fusion proteins. We then used purified GST fusion proteins and one of the two ^{32}P-labeled riboprobes [NX182(-) RNA or IG7-SM118(-) RNA] for UV crosslinking studies. As shown in Fig. 1, the full-length GST-A1 and five carboxy-terminal deletion GST-fusion proteins (ΔIX, ΔCT, 2xRBD, RBD I, RBD II) bound to both XN182(-) and IG7-SM118 (-) RNAs. In contrast, all GST fusion proteins with amino-terminal deletions (Gly, RGG, CT, and VIII) did not bind to the two RNAs. These results are consistent with the previous findings on the general RNA-binding properties of hnRNP-A1 (Dreyfuss et al., 1993). Surprisingly, the RGG box, which has been shown to contribute to RNA-binding and to be an RBD in hnRNP-U (Kiledjian and Dreyfuss, 1992), did not bind to MHV RNA. We conclude from these results that the amino-terminal two RBDs are the only domains capable of binding the complementary strand of MHV 5'-untranslated region and the IG sequence.

We further examined the protein-binding sites of hnRNP-A1 using the full-length and truncated GST-fusion proteins and the *in vitro*-translated, ^{35}S-labeled full-length hnRNP-A1 in a protein-binding assay. As shown in Fig. 1, GST-A1 fusion proteins containing the complete or portions of the glycine-rich domain retained the protein-binding activity, whereas GST-A1 fusion proteins lacking this domain did not. These results demonstrate that the carboxy terminal glycine-rich domain of hnRNP-A1 possesses the protein-binding activity with itself, consistent with the published data on human hnRNP-A1 (Cartegni et al., 1996). The binding specificity was further confirmed by using GST protein as a negative control. No background binding activity was observed between GST and hnRNP-A1 under this experimental condition (data not shown).

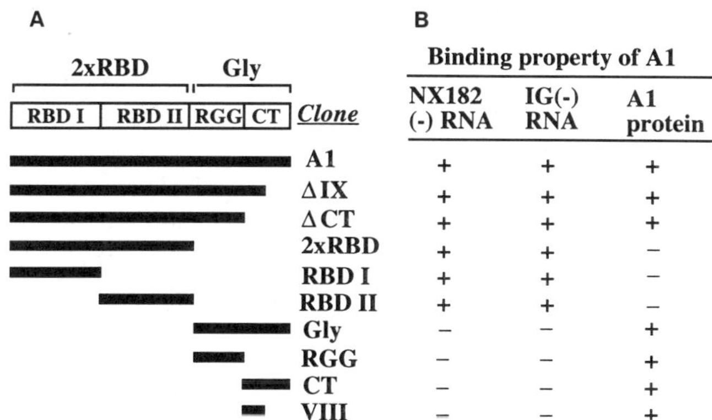

Figure 1. Summary of deletion analyses on protein-RNA and protein-protein interactions. (A) Schematic diagram of the structure of hnRNP-A1 and its deletion derivatives, which are expressed as glutathione-*S*-transferase (GST)-fusion proteins. RBD, RNA-binding domain; Gly, glycine-rich domain; RGG, arginine-glycine-glycine box; CT, C-terminus. The name of each construct is shown under *Clone* which reflects its functional domain(s). (B) Interactions between GST-A1 fusion proteins and NX182(-) RNA or IG(-) RNA, and between GST-A1 proteins and in vitro-translated A1 protein were determined by UV cross-linking and protein-binding assays, respectively. Bound proteins were analyzed by SDS-PAGE containing 10% polyacrylamide. (+), positive reaction; (-), negative reaction.

4.2. In Vitro Formation of an RNP Complex

We previously hypothesized that cellular and viral proteins and coronavirus regulatory RNA sequences form a transcription initiation complex by protein-RNA and protein-protein interactions, thus regulating viral RNA transcription (Zhang et al., 1994). To test this hypothesis, we developed an in vitro assay to determine whether the two distant *cis*-acting regulatory sequences of MHV RNA can form a complex with cellular protein hnRNP-A1. When the two RNA fragments [IG7-SM118(-) RNA was labeled with ^{32}P and XN182(-) RNA was unlabeled] were mixed with purified GST-A1 protein, an RNP complex containing these three basic components was formed (Fig. 2, lane 1). This complex was then isolated by affinity selection with a biotin-labeled antisense riboprobe, biotin-56(+), which is complementary to the 3'-end 56 nt of the unlabeled XN182(-) RNA and subsequently with streptavidin agarose beads. When GST-A1 was replaced with GST or omitted, no specific RNP complexes were formed (Fig. 2, lane GST; data not shown). Similarly, when either XN182(-) RNA or biotin-56(+) was omitted in the reaction, no specific RNP complexes were formed (data not shown). We conclude that the two MHV RNA sequences can form an RNP complex by cellular protein hnRNP-A1. This is the first demonstration of such an RNP complex in vitro.

4.3. Mutations of the IG Sequence Affect RNP Complex Formation

To determine whether protein-RNA interaction is critical in RNP complex formation, we chose two mutant MHV IG sequences, MG(-) and MGG(-), which have been shown previously to abrogate the protein-RNA interaction and transcription efficiency (Zhang and Lai, 1995). MG(-) has one G mutation in the consensus sequence (UCUAAA*C* to UCUAAA*G*) while MGG(-) has two G mutations (UC*U*AAA*C* to UC*G*AAA*G*). The amounts of the RNP formed in MG(-) and MGG(-) were 16 and 9% of the wild-type, re-

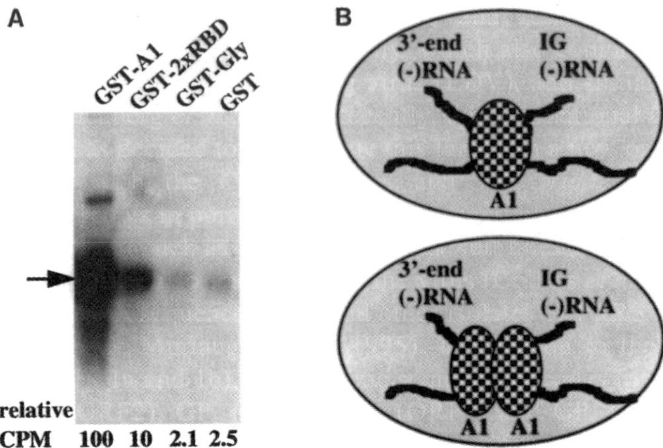

Figure 2. Protein-RNA and protein-protein interactions are involved in the formation of an MHV RNP complex in vitro. (A) In vitro RNP complex formation. The full-length and two deletion constructs of hnRNP-A1 were expressed in *E. coli* as GST-fusion proteins, which were then purified with glutathione-Sepharose beads and used for studies of RNP complex formation. After PAGE separation, gels were autoradiographed. The ^{32}P-labeled IG7(-) RNA recovered from the complex is indicated by an arrow. The radioactivity of this band in each lane was determined by a PhosphorImager (Life Technology), and the absolute CPMs were converted to relative CPMs to the value of the full-length hnRNP-A1, which was set to 100 (shown at the bottom). The proteins used in each reaction are indicated at the top. (B) Two possible models of the RNP complex formation in vitro.

spectively (data not shown), indicating that protein-RNA interaction is important for the RNP complex formation.

4.4. Effects of Deletion of hnRNP-A1 on RNP Complex Formation

We further determined whether both protein- and RNA-binding activities of hnRNP-A1 are required for the formation of an RNP complex. To do this, we performed the RNP reconstitution reaction by mixing the truncated GST-fusion proteins with the two RNA fragments [NX182(-) RNA and ^{32}P-IG7(-) RNA] in a reconstitution reaction buffer. As shown in Fig. 2A, the RNP complex was significantly reduced (10 % of the full-length) when the complete carboxy-terminal glycine-rich domain was deleted (lane GST-2xRBD), while almost no RNP complex was formed (2.1% of the full-length), when the amino-terminal RBD was deleted (lane GST-Gly). The radioactivity of the complex formed by GST-Gly was comparable to that by GST (lane GST, 2.5% of the full-length). These results indicate that the RBDs are essential for the formation of the RNP complex, again suggesting the importance of the protein-RNA interaction in RNP complex formation. They also indicate that the glycine-rich domain is not absolutely required for, but enhances the RNP complex formation, suggesting that protein-protein interaction is an auxiliary function in RNP complex formation in vitro.

These results suggest two possible senarios in the formation of an RNP complex (Fig. 2B). One is that a single hnRNP-A1 protein is able to bind to the two separate RNA sequences (3'-end and intergenic sequences of the negative-strand RNA) to form a complex without the involvement of protein-protein interaction. This is most likely to be mediated by the two RNA-binding domains of hnRNP-A1 through direct protein-RNA interaction, as it was shown that each RNA-binding domain independently binds the 3'-

end and intergenic sequences (Fig. 1). On the other hand, the formation of such a complex may require two hnRNP-A1 proteins through protein-protein interaction, as evidenced by the deletion analysis (Fig. 2A). Both possibilities may operate in vitro and in vivo. It should be noted that in our *in vitro*-RNP reaction, the amounts of RNAs and hnRNP-A1 protein may be significantly large as compared to the intracellular milieu, in which viral RNAs are restricted to certain compartments and are surrounded by diverse macromolecules in excess; the cis-acting sequences of MHV RNA might not be easily accessible to a single hnRNP-A1 protein molecule as shown here *in vitro*. Thus, we can envisage that if such a complex were formed *in vivo*, protein-protein interactions of hnRNP-A1 would have a greater role. This assumption is consistent with the findings that protein interactions of hnRNP-A1 are essential in the assembly of a splicing complex in mammalian cells by associating with other hnRNP and SR proteins (Dreyfuss et al., 1993). Whether other MHV RNA sequences and viral and cellular proteins are also involved in the formation of an MHV RNP complex remains to be determined.

ACKNOWLEDGMENTS

This work was supported in part by Public Health Service Research Grants NS14168 and AI16144 from the National Institutes of Health. We thank Dr. Tom Kienzle for critical reading the manuscript. M.M.C.L. is an Investigator of the Howard Hughes Medical Institute.

REFERENCES

Ausubel, F.M., Brent, R., Kingston, R.E., Moore, D.D., Seidman, J.G., Smith, J.A., and Struhl, K., 1994, *Current protocols in molecular biology*. John Wiley & Sons Inc., New York.

Buvoli, M., Biamonti, G., Tsoulfas, P., Bassi, M. T., Ghetti, A., Riva, S., and Morandi, C., 1988, cDNA cloning of human hnRNP protein A1 reveals the existence of multiple mRNA isoforms, *Nucl. Acids Res.* **16**:3751–3770.

Cartegni, L., Maconi, M., Morandi, E., Cobianchi, F., Riva, S., and Biamonti, G., 1996, hnRNP A1 selectively interacts through its Gly-rich domain with different RNA-binding proteins, *J. Mol. Biol.* **259**:337–348.

Dreyfuss, G., Matunis, M.J., Pinol-Roma, S., and Burd, C.G, 1993, hnRNP proteins and the biogenesis of mRNA, *Annu. Rev. Biochem.* **62**:289–321.

Furuya, T. and Lai, M.M.C., 1993, Three different cellular proteins bind to the complementary sites on the 5'-end positive- and 3'-end negative-strands of mouse hepatitis virus RNA, *J. Virol.* **67**:7215–7222.

Kiledjian, M. and Dreyfuss, G., 1992, Primary structure and binding activity of the hnRNP U protein: binding RNA through RGG box, *EMBO J.* **11**: 2655–2664.

Lamond, A., and Sproat, B.S., 1996, Isolation and characterization of ribonucleoprotein complexes. p103–140. In: *RNA processing: A practical approach* (Higgins, S.J., and Hames, B.D., eds.). New York:Oxford University Press.

Li, H.-P., Zhang, X.M., Comai, L., Duncan, R., and Lai, M.M.C., 1997, Identification of the hetergeneous nuclear ribonucleoprotein (hnRNP) A1 as the cellular protein binding to the regulatory RNA regions of mouse hepatitis virus, in: *Coronaviruses and Arteriviruses* (eds. L. Enjuanes, S. Siddell, and W. Spaan), Plenum, New York.

Pollard, V.W., Michael, W.M., Nakielny, S., Siomi, M.C., Wang, F., and Dreyfuss, G., 1996, A novel receptor-mediated nuclear protein import pathway, *Cell* **86**, 985–994.

Simoi, H., and Dreyfuss, G., 1995, A nuclear localization domain in the hnRNP A1 protein, *J. Cell Biol.* **129**:551–560.

Smith, D.B. and Johnson, K.S., 1988, Single-step purification of polypeptides expressed in Escherichia coli as fusions with glutathione S-transferase, *Gene* **67**:31–40.

Weighardt, F., Biamonti, G., and Riva, S, 1995, Nucleo-cytoplasmic distribution of human hnRNP proteins: a search for the targeting domains in hnRNP A1, *J. Cell Sci.* **108**: 545–555.

Will, C.L., Kastner, B., Luehrmann, R., 1996, Analysis of ribonucleoprotein interactions. p141–177. In: *RNA processing: A practical approach* (Higgins, S.J., and Hames, B.D., eds.). New York:Oxford University Press.

Zhang, X.M, Liao, C.-L. and Lai, M.M.C., 1994, Coronavirus leader RNA regulates and initiates subgenomic mRNA transcription, both in trans'and in cis, *J. Virol.* **68**:4738–4746.

Zhang, X.M., and Lai, M. M. C., 1995, Interactions between the cytoplasmic proteins and the intergenic (promoter) sequence of murine hepatitis virus RNAs: Correlation with the amounts of subgenomic mRNA transcribed, *J. Virol.* **69:** 1637–1644.

29

CELL PROTEINS BIND TO A 67 NUCLEOTIDE SEQUENCE WITHIN THE 3' NONCODING REGION (NCR) OF SIMIAN HEMORRHAGIC FEVER VIRUS (SHFV) NEGATIVE-STRAND RNA

You-Kyung Hwang and Margo A. Brinton

Department of Biology
Georgia State University
Atlanta, Georgia 30303

1. ABSTRACT

The 3'NCR of the SHFV negative-strand RNA [SHFV 3'(-)NCR RNA] is thought to be the initiation site of full-length and possibly also subgenomic positive-strand RNA and so is likely to contain *cis*-acting signals for viral RNA replication. Cellular and viral proteins may specifically interact with this region to form replication complexes. When *in vitro* transcribed SHFV 3'(-)NCR RNA was used as a probe in gel mobility shift assays, two RNA-protein complexes were detected with MA104 S100 cytoplasmic extracts. The specificity of these RNA-protein interactions was demonstrated by competition gel mobility shift assays. Four MA104 proteins (103, 86, 55, and 36 kDa) were detected by UV-induced cross-linking assays and three proteins (103, 55, and 36 kDa) were detected by northwestern blotting assays. The binding sites for these proteins were mapped to the region between nucleotides 117 to 184 on the SHFV 3'(-)NCR RNA. Four cellular proteins with identical molecular masses to those of the proteins that bind to the SHFV 3'(-)NCR RNA were detected by the 3'(-) NCR of another arterivirus, LDV-C, suggesting that divergent arteriviruses utilize the same set of conserved cell protein domains.

2. INTRODUCTION

Simian hemorrhagic fever virus (SHFV) causes a fetal hemorrhagic disease in macaque monkeys. SHFV-infected macaques develop early fever, mild facial edema, anorexia, dehydration, proteinuria, cyanosis, skin petechia, bloody diarrhea, nose bleeds,

Coronaviruses and Arteriviruses, edited by Enjuanes *et al.*
Plenum Press, New York, 1998

hemorrhages in the skin, and occasional bleeding in the orbits of the eyes. Death occurs within 1 to 2 weeks after SHFV infection (London, 1977). In contrast, SHFV causes a persistent infection but no disease in chimpanzees, African green monkeys, and baboons (Gravell et al., 1980; London, 1977). These species are thought to be natural hosts for SHFV.

SHFV is a positive-strand RNA virus and a member of the family *Arteriviridae*. Although the organization and replication strategy of the arterivirus genome are similar to those of coronaviruses, arteriviruses differ from coronaviruses in their virion morphology and genome size (Plagemann, 1996). The genome of SHFV is approximately 14.5 kilobases in length (Sagripanti, 1984; Brinton et al., unpublished data; Smith, 1997) and contains a type I cap at the 5' end (Sagripanti et al., 1986) and a poly (A) tract at the 3' end (Sagripanti, 1984). The SHFV genome encodes a large, nonstructural polyprotein ORF at the 5' end and 9 structural ORFs at the 3' end (Brinton et al., unpublished data; Zeng et al., 1995; Godeny et al., 1995; Smith, 1997). Both coronaviruses and arteriviruses produce a 3' co-terminal nested set of mRNAs during their replication cycle (Plagemann, 1996; Lai, 1990). Each SHFV subgenomic mRNA contains a leader sequence at its 5' end that is identical to the one located near the 5' end of the genome (Zeng et al., 1995). Short conserved sequences, designated junction sequences, precede each ORF (Zeng et al., 1995; Smith, 1997). SHFV junction sequences are shorter and less redundant than coronavirus intergenic sequences.

It was previously reported that a 35/38 kDa protein in DBT cytoplasmic extracts binds to the 3' end of the negative-strand RNA of the coronavirus, MHV (Furuya and Lai, 1993) and that three cellular proteins, 70, 48, and 35/38 kDa, bind to the MHV negative-strand intergenic region RNA located between genes 6 and 7 (Zhang and Lai, 1995). In this study the 3' NCR of the SHFV negative-strand RNA was used as a probe in gel mobility shift, UV-induced cross-linking, and northwestern blotting assays with cytoplasmic extracts from uninfected and SHFV-infected MA104 cells to determine whether any cellular or viral proteins bind to this RNA.

3. MATERIALS AND METHODS

3.1. Cells and Virus

Cytoplasmic extracts were prepared as described previously (Blackwell and Brinton, 1995) from confluent monolayers of MA104 cells that had been cultured in Dulbecco's minimal essential medium supplemented with 10% fetal bovine serum. For preparation of infected cell extracts, MA104 cells were infected with SHFV, strain LVR 42–0/M6941, at a multiplicity of infection of 5 for 6 hr prior to cell lysis . Virus stock pools were prepared in MA104 cells and contained titers of about 10^8 PFU/ml.

3.2. Construction of cDNA Templates and *in Vitro* Synthesis of RNA Transcripts

SHFV RNA was purified from virions using Catrimox-14 Surfactant (Iowa Biotechnology Corp.). SHFV 3'(-)NCR RNA cDNA was synthesized from viral RNA by reverse transcription (RT), amplified by PCR and cloned into the pCR 2.1 plasmid (Invitrogen). Different primer pairs were used to generate the PCR templates for the various truncated SHFV 3'(-) RNAs used as probes. All antigenomic sense primers contained the T7 promoter. Negative-strand RNA transcripts were synthesized *in vitro* from the various PCR templates with T7 RNA polymerase. *In vitro* RNA transcription reactions were performed as described by the manufacturer (Ambion).

3.3. Gel Mobility Shift Assay

[32]P-labeled SHFV 3'(-)NCR RNA and uninfected MA104 S100 cytoplasmic extracts were incubated for 30 min at room temperature in a volume of 10 μl and then analyzed on nondenaturing polyacrylamide gels as described previously (Blackwell and Brinton, 1995).

3.4. UV-Induced Cross-Linking Assay

Binding reactions were done as described above and then the reaction mixtures were irradiated for 30 min with UV light on ice as described previously (Blackwell and Brinton, 1995). The cross-linked products were digested with RNase A, precipitated with 50% acetone/methanol, pelleted by microcentrifugation, and visualized by autoradiography.

4. RESULTS

4.1. Analysis of Interactions between SHFV 3'(-)NCR RNA and MA104 Cytoplasmic Proteins

To determine whether any MA104 cytoplasmic proteins bind specifically to the 3' end of the SHFV negative-strand RNA, gel mobility shift assays were performed. The RNA probe used in the initial experiments was 209 nucleotides (nt) in length and contained the entire 3' noncoding region of SHFV 3'(-)NCR (Zeng et al., 1995). Two RNA-protein complexes were detected in both mock- and SHFV-infected MA104 cytoplasmic extracts in gel mobility shift assays. The number and eletrophoretic mobility of the complexes formed between SHFV 3'(-)NCR RNA and MA104 proteins in mock- and SHFV-infected cytoplasmic extracts were identical, suggesting that these RNA-protein complexes contained only cellular proteins. To determine the specificity of these RNA-protein interactions, competition gel mobility shift assays were performed. Unlabeled SHFV 3'(-)NCR RNA was used as the specific competitor and poly (I)- poly (C), yeast tRNA, and an RNA synthesized *in vitro* from plasmid pCR 2.1 vector DNA were used as non-specific competitors. While 350 ng of the specific competitor completely inhibited RNA-protein complex formation, no inhibition was detected with much higher amounts of poly (I)-poly (C), yeast tRNA, or plasmid RNA.

4.2. Determination of the Molecular Masses of the Cellular Proteins That Bind to the SHFV 3'(-) NCR RNA

UV-induced cross linking and Northwestern blotting assays were utilized to determine the molecular masses of the cellular proteins that bind to the SHFV 3'(-)NCR RNA. Four MA104 cytoplasmic proteins (103, 86, 55, and 36 kDa) were detected with the ^{32}P-labeled SHFV 3'(-)NCR by UV-induced cross-linking assays (Figure 1). Four cellular proteins with identical molecular masses were detected with the SHFV 3'(-)NCR probe in uninfected and SHFV-infected cells extracts.

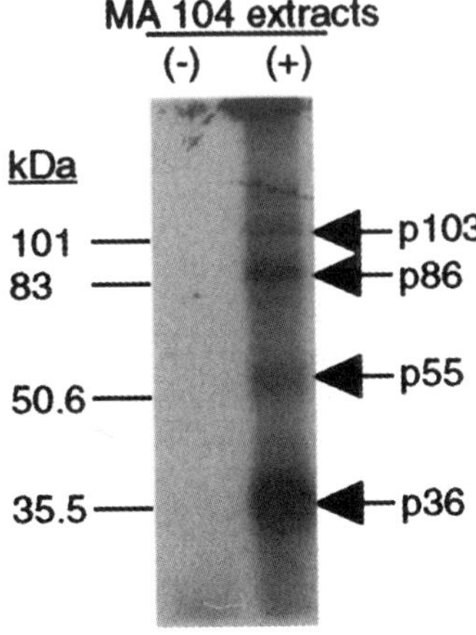

Figure 1. UV-induced cross-linking analysis of cellular proteins that bind to the SHFV 3'(-)NCR RNA. ^{32}P-labeled SHFV 3'(-)NCR RNA probe was incubated with MA104 S100 cytoplasmic extract (5 µg) for 30 min at room temperature and the RNA-protein complexes were subjected to UV-induced cross-linking. The arrows indicate the cellular protein bands (103, 86, 55, and 36 kDa). (-), no cellular extracts added; (+), MA104 S100 cytoplasmic extracts added to the reaction. Bio-Rad protein standards were used as markers.

Only three of the cellular proteins, 103, 55, and 36 kDa, were detected in northwestern blotting assays. The 86 kDa protein may not renature after separation by SDS-PAGE, may not transfer efficiently, may require a co-factor to bind to the SHFV 3'(-)NCR RNA or may require another protein for cooperative binding.

4.3. Mapping the Binding Sites of the Four Cellular Proteins on the SHFV 3'(-)NCR RNA

SHFV 3'(-)5–184 RNA and SHFV 3'(-)45–184 RNA (Figure 2) were generated and used as competitors in competition gel mobility shift assays with the ^{32}P-labeled SHFV 3'(-)209 RNA probe. The data obtained indicated that the 5' portion of the SHFV 3'(-) NCR RNA contained binding sites for each of the four cellular proteins. Additional truncated SHFV 3'(-)NCR RNAs were then generated and used as probes in UV-induced cross-linking assays to more precisely locate the binding sites for the individual cellular proteins. ^{32}P-labeled SHFV 3'(-)5–184 RNA and SHFV 3'(-)137–202 RNA probes were used in initial UV-induced cross-linking assays. While the binding sites for the four proteins (103, 86, 55, and 36 kDa) were detected with the SHFV 3' (-)5–184 RNA probe, only the binding sites for the 103 and 86 kDa proteins and a partial binding site of the 55

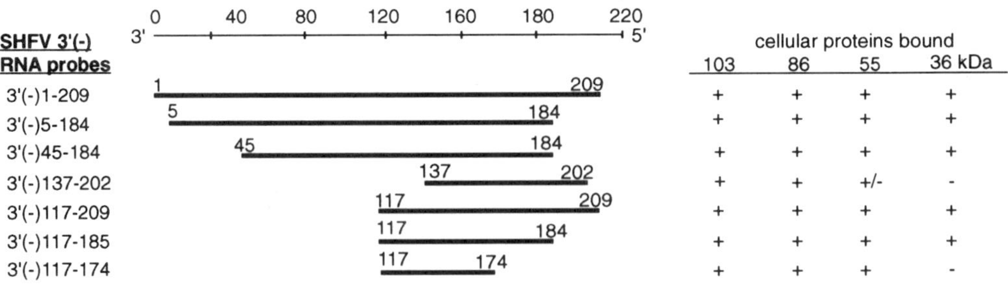

SHFV 3'(-) RNA probes	103	86	55	36 kDa
3'(-)1-209	+	+	+	+
3'(-)5-184	+	+	+	+
3'(-)45-184	+	+	+	+
3'(-)137-202	+	+	+/-	-
3'(-)117-209	+	+	+	+
3'(-)117-185	+	+	+	+
3'(-)117-174	+	+	+	-

Figure 2. Schematic diagram of the truncated SHFV 3'(-)NCR RNAs used for mapping the binding sites of the cellular proteins on SHFV 3'(-) RNA. cDNAs representing the various truncated SHFV 3'(-) RNAs were generated by PCR and used as templates for in vitro RNA transcription. The nucleotides were numbered starting from the 3' end of the minus-strand RNA. The proteins detected by each of the deleted RNAs were assayed by UV-induced cross-linking. A summary of the data obtained is presented on the right.

kDa protein were detected with the SHFV 3'(-)137–202 RNA probe. These data suggested that the region between nt 184 and 202 contains contact sites for the 36 and 55 kDa proteins. An RNA which contained 20 additional 3' nt [SHFV 3'(-)117–209] bound efficiently to all four of the cellular proteins. To define the 5' boundary of the binding region for the cellular proteins on the SHFV 3'(-)NCR RNA, SHFV 3'(-)117–184 RNA and SHFV 3'(-)117–174 RNAs were generated and tested. While the SHFV 3'(-)117–184 RNA contained the binding sites for the all four of the proteins, SHFV 3'(-)117–174 RNA contained binding sites for only the 103, 86, and 55 kDa proteins. Therefore, the binding sites for the 103, 86, 55, and 36 kDa proteins were mapped to the region between nt 117 and 184 on the SHFV 3'(-)NCR RNA. Data obtained from the UV-induced cross-linking assays are summarized in Figure 2.

4.4. Conservation of the MA104 Cellular Protein Binding Sites in Another Arterivirus 3'(-)NCR RNA

The family *Arteriviridae* is currently composed of four members, equine arteritis virus (EAV), porcine reproductive and respiratory syndrome virus (PRRSV), lactate dehydrogenase-elevating virus (LDV) and SHFV. Natural infections with each of the arteriviruses are host restricted. To determine whether a divergent arterivirus can bind to the same MA104 cellular proteins as the SHFV 3'(-)NCR RNA, the 3' NCR of LDV-C negative-strand RNA [LDV-C 3'(-)NCR RNA] was used as an unlabeled competitor in competition gel mobility shift assays and as a labeled probe in UV-induced cross-linking assays with MA104 S100 cytoplasmic extracts (data not shown). The formation of SHFV RNA-protein complexes was inhibited by the unlabeled LDV-C 3'(-)NCR RNA competitor. In UV-induced cross-linking assays, the LDV-C 3'(-)NCR RNA probe detected four cellular proteins with identical molecular masses to those detected by the SHFV 3'(-)NCR RNA.

5. DISCUSSION

We have demonstrated that cellular proteins in MA104 cytoplasmic extracts bind specifically to the SHFV 3'(-)NCR RNA. The molecular masses of the cellular proteins that bind to the SHFV 3'(-)NCR RNA in mock-infected and SHFV-infected MA104 cytoplasmic extracts are 103, 86, 55, and 36 kDa, indicating that the proteins involved in the formation of these RNA-protein complexes are cellular proteins and not viral proteins. Using a series of 3' and 5' deleted, *in vitro* transcribed SHFV 3'(-)NCR RNAs, the binding sites for the four cellular proteins were mapped to the region between nucleotides 117 and 184. Competition gel mobility shift and UV-induced cross-linking analyses with LDV-C 3'(-)NCR RNA showed that this RNA contains binding sites for the same four cellular proteins as the SHFV 3'(-)NCR RNA. These results suggest that the 3'(-)NCR RNAs of divergent arteriviruses bind to the same conserved domains on cellular proteins in their respective host cells. The functions provided by the cellular proteins that bind to the SHFV 3'(-)NCR RNA are not yet known.

Recently, we have developed an *in vivo* SHFV replication system based on that of Hiscox et al. (1995) to determine whether the region containing the cellular protein binding sites (nt 117 to 184) on the SHFV 3'(-)NCR RNA functions as a *cis*-acting sequence for positive-strand RNA synthesis. Previous data obtained with other viral systems have been shown that the RNA region to which cellular proteins bind is required for viral multiplication (Sriskanda et al., 1996; Yu and Leibowitz, 1995). Preliminary data obtained from our system indicates that the SHFV 3'(-)NCR RNA is required in *cis* for *in vivo* replication of an *in vitro* transcribed SHFV-CAT negative-strand RNA. Studies are underway to map the *cis*-acting sequence elements on the SHFV 3'(-)NCR.

ACKNOWLEDGMENTS

This research was supported by the Georgia State University Research Foundation.

REFERENCES

Blackwell J. L., and Brinton M. A., 1995, BHK cell proteins that bind to the 3' stem-loop structure of the West Nile virus genome RNA, *J. Virol.* **69**: 5650–5658.

Furuya T., and Lai M. M. C., 1993, Three different cellular proteins bind to complementary sites on the 5'-end-positive and 3'-end-negative strands of mouse hepatitis virus RNA, *J. Virol.* **67**: 7215–7222.

Godeny E. K., Zeng L., Smith S. L., and Brinton M. A., 1995, Molecular characterization of the 3' terminus of the simian hemorrhagic fever virus genome, *J. Virol.* **69**: 2679–2683.

Gravell M., London W.T., Rodriguez M., Palmer A.E., and Hamilton R.S., 1980, Simian hemorrhagic fever (SHF): new virus isolate from a chronically infected patas monkey, *J. Virol.* **51**: 99–106.

Hiscox J.A., Mawditt,K.L., Cavanagh D., and Britton P., 1995, Investigation of the control of coronavirus sub-genomic mRNA transcription by using T7-generated negative-sense RNA, *J.Virol* **69**: 6219–6227.

Lai, M.M.C., 1990, Coronavirus: organization, replication, and expression of genome. *Annu. Rev. Microbiol.* **44**: 303–333.

London W.T., 1977, Epizootology, transmission and approach to prevention of fatal simian hemorrhagic fever virus in rhesus monkeys, *Nature* **268**: 344–345.

Plagemann P.G.W., 1996, Lactate dehydrogenase-elevating virus and related viruses, in: *Virology*, Third Edition, (Fields, B.N., Knipe, D.M., Howley P.M., et al., eds.), Lippincott-Raven Publishers, Philadelphia, PA, pp 1075–1093.

Sagripanti J.L., 1984, The genome of simian hemorrhagic fever virus, *Arch. Virol.* **82**: 61–72.

Sagripanti J.L, Zandomeni R.O., and Weinmann R., 1986, The cap structure of simian hemorrhagic fever viron RNA, *Virology* **151**: 146–150.

Smith, S. L., Wang, X. C., and Godeny, E. K., 1997, Organization of the simian hemorrhagic fever virus (SHFV)genome and identification of the sgRNA junction sequences, *Gene*. In press.

Sriskanda V.S., Pruss G., Ge X., and Vance V.B., 1996, An eight-nucleotide sequence in the potato virus X 3'untranslated region is required for both host protein binding and viral multiplication, *J. Virol.* **70**: 5266–5271.

Yu W., and Leibowitz J.L., 1995, A conserved motif at the 3' end of mouse hepatitis virus genomic RNA is required for host protein binding and viral RNA replication, *Virology* **214**: 128–138.

Zeng L., Godeny E.K., Methven S.L, and Brinton M.A., 1995, Analysis of simian hemorrhagic fever virus (SHFV) subgenomic RNAs, junction sequences, and 5' leader, *Virology* **207**: 543–548.

Zhang, X., and Lai M.M.C., 1995, Interaction between cytoplasmic proteins and the intergenic (promoter) sequence of mouse hepatitis virus RNA: correlation with the amounts of subgenomic mRNAs transcribed, *J. Virol.* **69**: 1637–1644.

30

STUDIES OF MURINE CORONAVIRUS DI RNA REPLICATION FROM NEGATIVE-STRAND TRANSCRIPTS

S. Banerjee and S. Makino

Department of Microbiology, and Institute of Cellular and Molecular Biology
The University of Texas at Austin
Austin, Texas 78712

1. ABSTRACT

The positive-strand transcripts as well as negative-strand transcripts of mouse hepatitis virus (MHV) defective interfering (DI) RNA, when introduced into MHV-infected cells, resulted in DI RNA replication and accumulation. The leader sequence of the majority of DI RNAs that accumulated from the expression of negative-strand DI RNA transcripts with no extra non-MHV nucleotides at the 3' end switched to that of helper virus, whereas this leader sequence switching did not occur in most of the positive-strand DI RNAs that accumulated from the expressed negative-strand DI RNA transcripts with extra non-MHV nucleotides at the 3' end. These data demonstrated that the extra 4 nucleotides at the 3'-end of negative-strand DI RNA transcripts affected leader sequence switching on DI RNA, and indicated that the leader switching probably occurred during positive-strand DI RNA synthesis.

2. INTRODUCTION

Mouse hepatitis virus (MHV), the prototypic coronavirus, contains a single-strand, positive-sense RNA that is approximately 31 kb in length (Lee *et al.*, 1991; Lai and Stohlman, 1978). MHV-infected cells generate seven to eight species of virus-specific mRNAs whose sequences comprise a 3'-co-terminal nested set structure (Lai *et al.*, 1981; Leibowitz *et al.*, 1981). At their 5'-ends all of the mRNAs are fused to a 72- to 77- nucleotide-long leader sequence (Lai *et al.*, 1984; Spaan *et al.*, 1983). Genomic sized and subgenomic sized negative-strand RNAs are present in coronavirus-infected cells in amounts that are significantly lower than the amounts of corresponding positive-strand RNAs (Sethna *et al.*, 1989). The subgenomic sized negative-strand RNAs may be template RNAs

Coronaviruses and Arteriviruses, edited by Enjuanes *et al.*
Plenum Press, New York, 1998

for subgenomic mRNAs (Schaad and Baric, 1994; Sethna *et al.*, 1991; Sawicki and Sawicki, 1990) or they may be dead-end transcription products (Jeong and Makino, 1992).

Complete cDNA clones of coronavirus defective interfering (DI) RNAs under control of the T7 promoter have been used to study various aspects of coronavirus replication. Transfection of in vitro-synthesized positive-strand MHV DI RNA transcripts into MHV-infected cells as well as expression of positive-strand DI RNA transcripts by a recombinant T7 vaccinia virus expression system in MHV-infected cells results in DI RNA replication. The leader sequence of DI RNAs that contain a nine-nucleotide stretch of sequence (UUUAUAAAC) at the junction between the leader and the remaining DI sequence efficiently switches to that of helper virus during replication (Makino and Lai, 1989). This high-frequency leader sequence switching is not observed when the nine-nucleotide are deleted (Makino and Lai, 1989). Similar leader sequence switching is found in other coronavirus DI RNAs (Chang *et al.*, 1996) yet the mechanism of leader switching remains unclear.

MHV DI RNA accumulation occurs in MHV-infected cells when the DI RNA negative-strand transcripts are expressed using a vaccinia virus T7 expression system; DI RNA replication depends on expression of T7 polymerase and on the presence of the T7 promoter (Joo *et al.*, 1996). In this study, we examined the leader sequence of DI RNAs that accumulated from expression of two different negative-stranded DI RNA transcripts, and found that the sequence difference at the very 3' end of the negative-strand transcripts affected the leader switching of the accumulated DI RNAs. The data shown in this study also indicated that the leader switching in DI RNA probably occurred during positive-strand DI RNA synthesis.

3. MATERIAL AND METHODS

3.1. Viruses and Cells

The plaque-cloned A59 strain of MHV (Lai *et al.*, 1981) was used as a helper virus. Mouse DBT cells were used for MHV growth and DNA transfection. Recombinant vaccinia virus (VV) vTF7–3 that expresses T7 RNA polymerase (Fuerst *et al.*, 1986) were grown and titered in RK13 cells.

3.2. DNA Transfection

DNA transfection into vTF7–3-infected DBT cells was described previously (Joo *et al.*, 1996). A plasmid, pS5A, containing the Chloramphenicol Acetyl Transferase (CAT) gene (Woo *et al.*, 1997) was co-transfected with the DNA.

3.3. Preparation of Virus-Specific Intracellular RNA

Viral RNAs from virus-infected cells were extracted as previously described (Makino *et al.*, 1984). In some experiments poly(A) containing RNA was collected by oligo(dT) column chromatography. RNA samples were treated with RNase-free DNase I (Promega) at 37^{0}C.

3.4. Northern (RNA) Blotting

Northern blot analysis using a random-primed probe corresponding to the 0.55 kb Stu I - Sph I fragment of DE 25 was performed as previously described (Jeong and Makino, 1992). For detection of positive-stranded MHV DI-RNA, oligonucleotide 10080 (5' GGCAAGCCGTCCTCTTCTTGGGTATCGGC 3'), which binds at nucleotides 931–960 of DE25 from the 5'-end was 5'-end labeled and used as a probe.

3.5. Reverse Transcriptase (RT)-PCR

Positive-strand DI RNA specific-cDNAs were synthesized by incubating intracellular RNA species with oligonucleotide 10120 (5' CTTTAGACAACGCCAGTT 3'), which binds to positive-strand DI RNA at nucleotides 1594 to 1611 nucleotide from 5' end, as described previously (Makino *et al.*, 1988); this oligonucleotide binds at the junction site of domains II and III of the positive strand of DIssE (Makino *et al.*, 1988). RT-PCR products were synthesized by incubating the cDNAs with oligonucleotide 10066 (5'TATAA-GAGTGATTGGCGTCCG 3'), which binds to the 5' end of leader sequence (nucleotides 1–21). PCR was carried out at 94°C for 5 min (to inactivate RT), 94°C for 30 sec, 50°C for 1 min. 30 sec, and 72°C for 1 min. 30 sec., for 30 cycles.

3.6. Sequence Analysis

The gel-purified, DI-specific, RT-PCR products were sequenced by cloning the RT-PCR products into a TA cloning vector (Invitrogen). To sequence the leader region of these clones oligonucleotide 10258 (5' CTGGCGCCGAATGGACACGTC 3') was used. The oligonucleotide binds to nucleotides 168–188 of DIssE, from the 5'-end.

4. RESULTS

Previously we constructed and characterized two plasmids, pDER and pDER4, to study DI RNA replication from MHV DI RNA negative-strand transcripts (Joo *et al.*, 1996). In pDER a complete cDNA of MHV DI RNA, DIssE, is inserted between the T7 promoter and the hepatitis delta virus ribozyme domain such that negative-strand DI RNA transcripts are synthesized from the T7 promoter. The leader sequence of pDER has 4 repeats of UCUAA at the junction of the leader and body sequences and lacks the nine-nucleotide sequence (UUUAUAAAC) downstream of the UCUAA repeats (on the positive-sense), while MHV-A59 has 2 UCUAA repeats and contains the nine-nucleotide sequence (Makino and Lai, 1989). The leader sequence of pDER also has five substituted nucleotides relative to the MHV leader sequence. The structure of pDER4 is nearly identical to that of pDER, except that pDER4 has four extra nucleotides, AGCT, between the 3'-end of the leader sequence and the ribozyme domain; negative-strand transcripts of pDER4 should have the corresponding (UCGA) additional non-MHV sequence at the 3'-end (Fig. 1). When the negative-strand transcripts of the DI RNA are expressed in MHV-infected cells by a vaccinia virus T7 expression system, positive-strand DI RNAs accumulate in the plasmid-transfected cells (Joo *et al.*, 1996). DI RNA replication depends on expression of T7 polymerase and on the presence of the T7 promoter (Joo *et al.*, 1996).

To test whether leader switching occurred in the accumulated DI RNAs from the expressed negative-strand DI RNA transcripts, we examined the leader sequence of accumulated DI RNAs from the cells expressing pDER transcripts and pDER4 transcripts. Intracellular viral RNA was extracted from plasmid-transfected cells at 10 h p.i. with MHV and incubated with DNase I. We amplified the 5'-region of positive-strand DI RNAs

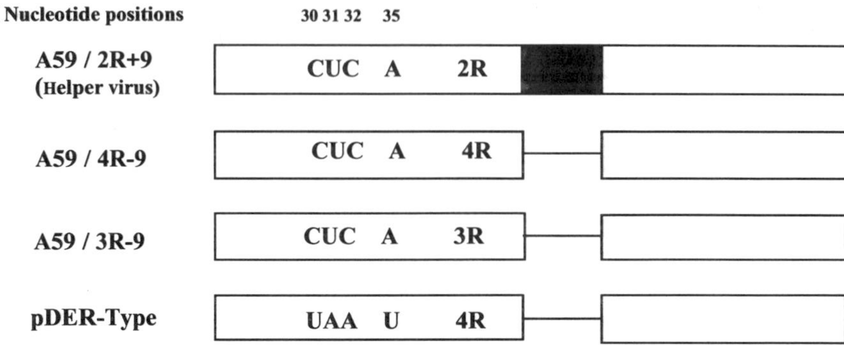

Figure 1. Diagram of the 5'-end sequences of the DI RNA transcripts in the positive sense. A59/2R+9 represents MHV genomic RNA. The shaded box represents the nine-nucleotide sequence (5' UUUAUAAAC 3'). 2R, 3R and 4R represent two, three and four repeats of UCUAA sequence, respectively. The nine-nucleotide sequence that is deleted is shown as a thin line. The nucleotide sequences which are not denoted share the same sequence as described previously (Makino and Lai, 1989). pDER-type represents both pDER and pDER4. The only difference between pDER and pDER4 is the presence of four extra nucleotides (5' UCGA 3') at the 3'-end of pDER4.

using RT-PCR as described in the Materials and Methods. We detected the expected RT-PCR product from DNA-transfected, MHV-infected cells, whereas we did not detect this product in the DNA-transfected, mock-infected cells, or in mock-transfected, MHV-infected cells (data not shown), indicating that the RT-PCR product was indeed derived from positive-strand DI RNAs. The PCR product was purified on an agarose gel and cloned into a plasmid vector. Sequencing analysis of five out of a total of six clones derived from accumulated positive-strand DI RNAs in the pDER-transfected cells had helper virus-specific leader sequence with two repeats of the UCUAA sequence and the nine-nucleotides, while one clone had helper virus-specific leader sequence with four repeats of UCUAA and deletion of the nine-nucleotide sequence. Sequencing analysis of DI RNAs from pDER4-transfected cells showed that five out of nine clones had the pDER4-specific leader sequence. We do not know whether those accumulated DI RNAs maintained the extra 4 nucleotides present at the extreme 3' end of the pDER4 transcripts, because this RT-PCR analysis did not identify the very 5' end of the leader sequence. The remaining four clones showed the helper virus-specific leader sequence, yet all of the clones lacked the nine-nucleotide sequence. Half of the clones contained three UCUAA repeats, while the remaining two clones had four UCUAA repeats (Table 1).

The leader switching does not occur after transfection of positive-strand DI RNAs lacking the nine-nucleotides (Makino and Lai, 1989), whereas we showed here that most of the accumulated DI RNAs underwent the leader switching in cells expressing pDER transcripts, which lacked the nine-nucleotides. Furthermore, our data indicated that the presence of 4 extra nucleotides at the very 3'-end of the negative-strand DI RNA transcripts affected leader switching; it seemed that the extra nucleotides prevented the leader switching.

5. DISCUSSION

The present data provided further information about the leader switching mechanism. The leader switching does not occur when positive-strand DI RNA transcripts lacking the nine-nucleotide sequence are transfected into MHV-infected cells (Makino and

Table 1. Leader sequences of cloned PCR products of DI RNAs from MHV-infected pDER-expressing cells, and from MHV-infected pDER4-expressing cells

Transfected clones[b]	Leader type[a]			
	A59/2R+9	A59/3R-9	A59/4R-9	pDER-type
pDER + pS5A[c]	5	–	1	–
pDER4 + pS5A[d]	–	2	2	5

[a]Leader sequences as described in Figure 1.

[b]Positive-strand DI RNA-specific RT-PCR products were cloned into TA cloning vector. Sequence analysis of leader regions of the cloned PCR products was performed.

[c]Five clones out of six had helper virus leader sequences. One clone had helper virus type leader with four UCUAA repeats and lacking the nine-nucleotide sequence.

[d]Out of nine clones sequenced, two clones had A59/3R-9 leader, two had A59/4R-9 leader and five retained the pDER-type leader.

Lai, 1989). Accordingly, accumulated negative-stranded DI RNAs also most likely lack the nine-nucleotide sequence. Most MHV negative-strand RNAs appear to exist in double-strand (ds) forms (Sawicki and Sawicki, 1986), therefore, the majority of negative-strand DI RNAs and some positive-strand DI RNA probably formed ds RNAs. We assume that efficient positive-strand RNA synthesis occurs from those ds RNAs. If the positive-strand DI RNA molecules that were synthesized from the expressed pDER transcripts are faithful copies of the pDER transcripts, then they should have lacked the nine-nucleotides. If this is the case, the ds RNAs made of newly synthesized positive-strand DI RNA and expressed negative-strand pDER DI RNA transcripts are structurally very similar to the ds RNAs that are produced after transfection of positive-strand DI RNA transcripts lacking the nine-nucleotides. Accordingly, one might expect that the structure of nascent positive-strand DI RNAs from both of those ds DI RNAs is the same. However, positive-strand DI RNAs accumulated in the cells expressing pDER transcripts contain helper virus-derived leader sequence, whereas those accumulated after transfection of the positive-strand DI RNA transcripts do not undergo leader switching. Therefore, we speculate that initial positive-strand DI RNA molecules synthesized from the expressed pDER transcripts were not faithful copies of the pDER. It is more likely that the helper virus-derived RNA(s) containing the leader sequence was used as a primer for the initiation of positive-strand DI RNA (Makino and Lai, 1989); this would result in leader switching on DI RNAs after expression of pDER transcripts. We further speculate that if the template RNAs exist as ds RNAs, this primer-dependent initiation of positive-strand DI RNA does not occur. This speculation may explain why those DI RNAs that accumulate after transfection of the positive-strand DI RNAs do not undergo leader switching. The ds DI RNAs produced after transfection of the positive-strand DI RNA transcripts consist of in-put DI RNA transcripts and faithful copies of the negative-strand DI RNAs. Subsequent positive-strand DI RNA synthesis does not require the helper virus-derived RNA containing the leader sequence as a primer, and the accumulated DI RNAs maintain the original leader sequence.

The leader switching did not occur in the majority of DI RNAs that accumulated in the cells expressing pDER4 transcripts, indicating that the presence of the extra 4 nucleotides at the very 3'-end of pDER4 transcripts was crucial for the leader switching. Why did most of the positive-strand DI RNAs that accumulated from the expressed pDER4 transcripts not undergo leader switching? The extra 4 nucleotides at the very 3' end of pDER4 transcripts and sequence immediately upstream may form a specific secondary structure, which may facilitate the initiation of positive-strand DI RNA synthesis. If the very 3' end negative-strand transcript lacks extra nucleotides, i.e., pDER transcripts, the very 3'-end of the expressed transcripts may not form the secondary structure that is optimal for the initiation of positive-strand DI RNA synthesis. If nascent positive-strand DI RNAs are not initiated from the very 3'-end of the negative-strand DI RNA transcripts, then MHV replication mechanism may use helper virus-derived leader sequence as a primer for the initiation of positive-strand DI RNA synthesis.

ACKNOWLEDGMENTS

This work was supported by Public Health Service grants AI29984 and AI32591 from the National Institutes of Health.

REFERENCES

Chang, R.-Y., Krishnan, R. and Brian, David, 1996, The UCUAAAC promoter motif is not required for high frequency leader recombination in bovine coronavirus defective interfering RNA, *J. Virol.* **70**:2720–2729.

Fuerst, T.R., Niles, E.G., Studier, F.W., and Moss, B., 1986, Eukaryotic transient-expression system based on recombination vaccinia virus that synthesizes bacteriophage T7 polymerase, *Proc. Natl. Acad. Sci. USA* **83**:8122–8126.

Jeong, Y.S., and Makino, S., 1992, Mechanism of coronavirus transcription:duration of primary transcription initiation activity and effect of subgenomic RNA transcription on RNA replication, *J. Virol.* **66**:3339–3346.

Lai, M.M.C., and Stohlman, S.A., 1978, RNA of mouse hepatitis virus, *J. Virol.*. **26**:236–242.

Lai, M.M.C., Baric, R.S., Brayton, P.R. and Stohlman, S.A., 1984, Characterization of leader RNA sequences on the virion and mRNA of mouse hepatitis virus, a cytoplasmic virus, *Proc. Natl. Acad. Sci. USA* **81**:3626–3630.

Lai, M.M.C., Brayton, P.R., Armen, R.C., Patton, C.D., Pugh, C., and Stohlman, S.A., 1981, Mouse hepatitis virus A59: mRNA structure and genetic localization of the sequence divergence from hepatotropic strain MHV-3, *J. Virol.* **39**:823–834.

Lee, H.-J., Shieh, C.-K., Gorbalenya, A.E., Eugene, E.V., La Monica, N., Tuler, J., Bagdzhadzhyan, A., and Lai, M.M.C., 1991, The complete sequence (22 kilobases) of murine coronavirus gene1 encoding the putative proteases and RNA polymerase, *Virology* **180**:567–582.

Leibowitz, J.L., Wilhelmsen, K.C., and Bond, C.W., 1981, The virus-specific intracellular RNA species of two murine coronaviruses: MHV-A59 and MHV-JHM, *Virology* **114**:39–51.

Makino, S., and Lai, M.M.C., 1989, High frequency leader sequence switching during coronavirus defective interfering RNA replication, *J. Virol.* **63**:5285–5292.

Makino, S., Taguchi, F., Hirano, N., and Fujiwara, K., 1984, Analysis of genomic and intracellular viral RNAs of small plaque mutants of mouse hepatitis virus, JHM strain, *Virology* **139**:138–151.

Makino, S., Shieh, C.-K., Soe, L.H., Baker, S., and Lai, M.M.C., 1988, Primary structure and translation of a defective interfering RNA of murine coronavirus, *Virology* **166**:550–560.

Sawicki, S.G., and Sawicki, D.L., 1990, Coronavirus transcription: subgenomic mouse hepatitis virus replicative intermediates function in RNA synthesis, *J. Virol.* **64**:1050–1056.

Sawicki, S.G., and Sawicki, D.L., 1986, Coronavirus minus-strand mRNA synthesis and effect of cyclohexamide on coronavirus RNA synthesis, *J. Virol.* **57**:328–334.

Schaad, M.C., and Baric, R.S., 1994, Genetics of mouse hepatitis virus transcription: evidence that subgenomic negative strands are functional templates, *J. Virol.* **68**:8169–8179.

Sethna, P.B., Hofmann, M.A., and Brian, D.A., 1991, Minus-strand copies of replicating coronavirus mRNA contain antileaders, *J. Virol.* **65**:320–325.

Sethna, P.B., Hung, S.-L., and Brian, D.A., 1989, Coronavirus subgenomic minus-strand RNAs and the potential for mRNA replicons, *Proc. Natl. Acad. Sci. USA* **86**:5626–5630.

Spaan, W., Delius, H., Skinner, M., Armstrong, J., Rottier, P., Smeekens, S., van der Zeijst, B.A.M., and Siddell, S.G., 1983, Coronavirus mRNA synthesis involves fusion of non- contiguous sequences, *EMBO J.* **2**:1939–1944.

Woo, K., Joo, M., Narayanan, K., Kim, K.H., and Makino, S., 1997, Murine coronavirus packaging signal confers packaging to nonviral RNA, *J. Virol.* **71**:824–827.

31

THE EFFECT OF DELETION OF A CONSERVED 11 NUCLEOTIDE SEQUENCE ON MOUSE HEPATITIS VIRUS RNA REPLICATION

J. F. Repass and S. Makino

Department of Microbiology, and Institute of Cellular and Molecular Biology
The University of Texas at Austin
Austin, Texas 78712

1. ABSTRACT

A conserved 11 nt sequence present at near the 3' end of mouse hepatitis virus (MHV) genomic RNA binds to host proteins and is important for MHV RNA replication (Yu and Leibowitz, 1995). To better understand the role of this 11 nt sequence in positive-strand MHV RNA replication, we examined whether positive-strand MHV DI RNAs from negative-strand DI RNA transcripts lacking the 11 nt sequence were synthesized in MHV-infected cells. Positive-strand DI RNAs did not accumulate efficiently, indicating that the conserved 11 nt sequence was necessary for positive-strand MHV RNA synthesis.

2. INTRODUCTION

An infectious RNA virus genome contains cis-acting replication elements that are recognized by virus-specific proteins and probably by host factors. Defective interfering (DI) RNAs of various RNA viruses, including coronavirus mouse hepatitis virus (MHV), have been used for the identification of these cis-acting replication elements; the coronavirus cis-acting element is called the cis-acting RNA replication signal. MHV, the prototypic coronavirus, contains a single-strand of positive-sense RNA that is approximately 31 kb in length (Lai and Stohlman, 1978). Deletion analyses of MHV DI RNAs showed that the cis-acting replication signals of all the MHV DI RNAs so far examined include both of the terminal sequences of MHV genomic RNA; the 5' end element spans approximately 0.5 kb and the 3' end about 0.45 kb. DI RNAs derived from the JHM strain of MHV (MHV-JHM) require an additional 57 nt-long internal sequence for DI RNA synthesis (Kim and Makino, 1995). Lin et al. (1994) demonstrated that only the 3' end 55 nt plus a poly A sequence are necessary for negative-strand DI RNA synthesis. Accordingly, other identified regions of the cis-acting sig-

Coronaviruses and Arteriviruses, edited by Enjuanes *et al.*
Plenum Press, New York, 1998

nals are probably important for positive-strand DI RNA synthesis. Indeed, the 57 nt-long internal sequence of MHV-JHM DI RNA functions in positive-strand DI RNA synthesis (Kim and Makino 1995). Although the very 3'-end 55 nt of MHV RNA and poly (A) are necessary for negative-strand MHV RNA synthesis (Lin et al., 1994), whether this region plays a role in positive-strand RNA synthesis is not known.

Yu and Leibowitz (1995) demonstrated that host proteins bind to a conserved sequence of 11 nt (5'-UGAAUGAAGUU-3') (the 11 nt-region) spanning nucleotide 36 to 26 from the 3' end of genomic RNA; this region is within the very 3'-end 55 nt sequence that is necessary for negative-strand DI RNA synthesis (Lin et al. 1994). DI RNAs containing nucleotide substitutions within the 11 nt-region replicate less efficiently than wild type (wt) DI RNA, indicating that the 11 nt-region is important for MHV RNA synthesis (Yu and Leibowitz, 1995). In some cases, mutations introduced into input DI RNA transcripts are repaired by RNA recombination between DI RNA and helper virus. Unfortunately, RNA recombination-mediated sequence repair makes straightforward characterization of the 11 nt-region difficult.

We wanted know how the 11 nt-region functioned in DI RNA replication. Particularly we are interested in whether the 11 nt-region is involved in positive-strand DI RNA synthesis. Previously we demonstrated that expression of negative-strand DI RNA transcripts using a vaccinia virus T7 expression system in MHV-infected cells results in replication of DI RNA (Joo et al., 1996). In this study we expressed negative-strand DI RNA transcripts lacking the 11 nt-region in MHV-infected cells and examined accumulation of the positive-strand DI RNAs. We did not observe the accumulation of positive-strand DI RNA from the expressed negative-strand DI RNA transcripts, indicating involvement of the 11 nt-region in positive-strand RNA synthesis.

3. MATERIALS AND METHODS

3.1. Viruses and Cells

The plaque-cloned A59 strain of MHV (Lai et al., 1981) was used as a helper virus in all experiments. Recombinant vaccinia virus vTF7–3 (Fuerst et al., 1986) was the source of T7 polymerase for DNA transfection experiments. Mouse cell line DBT (Hirano et al., 1974) was used as a host cell for DNA transfection experiments as well as for growth and titer determination of A59.

3.2. Plasmid Construction

A recombinant PCR procedure was used to make deletion mutants (Higuchi, 1990).

3.3. DNA Transfection and Northern Blot Analysis

DNA transfections were performed as previously described (Joo et al., 1996). Northern blots were performed as previously described (Joo et al., 1996).

4. RESULTS

To examine the role of the 11 nt-region sequence in positive-strand DI RNA synthesis, we constructed a plasmid pDERΔ11, in which a 2.2 kb-long MHV DI RNA DIssE cDNA sequence lacking the 11 nt-region was inserted between the T7 promoter and a hepatitis delta virus ribozyme domain; T7 polymerase-mediated transcripts represented negative-strand DI RNAs.

Yu and Leibowitz (1995) found that DI RNAs accumulate after transfection of in vitro-synthesized positive-strand DI RNA transcripts lacking the 11 nt-region into MHV-infected cells, yet accumulated DI RNAs contained the 11 nt-region (Yu and Leibowitz, 1995); deletion of the 11 nt-region was most likely repaired by RNA recombination between DI RNAs and helper virus RNAs. We assumed that some of the expressed pDERΔ11 transcripts would also acquire an 11 nt-region by a RNA recombination-mediated sequence repair mechanism, and that DI RNAs containing the 11 nt-region would accumulate. We also assumed that repair by RNA recombination of the 11 nt deletions present in all of the expressed RNA transcripts would be highly unlikely, and that the majority of expressed transcripts would maintain the 11 nt-region deletion. If the 11 nt-region is not involved in positive-strand DI RNA synthesis, then positive-strand DI RNAs lacking the 11 nt-region should be synthesized from the expressed negative-stranded DI RNA transcripts. The 11 nt-region is located within the cis-acting replication signal that is necessary for negative-strand RNA synthesis (Lin and Lai, 1994). Accordingly, if positive-strand DI RNAs lacking the 11 nt-region are synthesized from the expressed pDERΔ11 RNA transcripts, they would not undergo subsequent negative-strand DI RNA synthesis.

We used Northern blot analysis to detect the synthesis of the positive-strand DI RNAs lacking the 11 nt-region from the expressed pDERΔ11 transcripts in MHV-infected cells. To detect positive-strand DI RNAs lacking the 11 nt-region, we used 5'-end labeled oligonucleotide 10293 (5'-CCATGATCTTTACTA-3'); this oligonucleotide specifically hybridized with the positive-strand RNAs lacking the 11 nt region, but not with those containing the 11 nt-region. Oligonucleotide 10283 (5'-TCAACTTCATTCATT-3') was used to detect positive-strand DI RNA containing the 11 nt-region. As shown in Fig. 1A, oligonucleotide 10293 specifically hybridized with the in vitro-synthesized positive-strand transcripts lacking the 11 nt-region (DE25Δ11), while oligonucleotide 10283 specifically hybridized with in vitro-synthesized DE25 transcripts, which contain the 11 nt region. Both probes demonstrated complete specificity, even upon overexposure (data not shown).

To examine the positive-strand DI RNA synthesis from expressed pDERΔ11, pDERΔ11 was transfected into vTF7–3-infected cells, and then superinfected with MHV. For the control plasmids, we used DE25, which expresses positive-strand DIssE transcripts (Makino and Lai, 1989), and pDER, which encodes a negative-strand DIssE transcript (Joo et al., 1996). Transfection of DE25 transcripts and expression of pDER transcripts by a recombinant T7 vaccinia virus expression system in MHV-infected cells results in the accumulation of DI RNA (Joo et al., 1996). We also prepared another control plasmid, pDERΔ135, which expresses negative-strand DI RNA lacking the 135 nt-long internal cis-acting replication signal. Because the internal cis-acting replication signal is important for positive-strand DI RNA synthesis (Kim and Makino, 1995), positive-strand DI RNA synthesis should not occur from pDERΔ135 transcripts. We extracted intracellular RNAs at 12 h postinfection of MHV and examined them for the presence of positive-strand DI RNA by Northern blot analysis. In repeated experiments, we could not detect an appreciable amount of positive-strand DI RNA synthesis from pDERΔ11 transcripts.

In some experiments, after long exposure of the autoradiogram, a faint band corresponding to pDERΔ11 was seen, indicating a very low level of positive-strand DI RNA synthesis occurred in some experiments. We detected the accumulation of DE25 using oligonucleotide probe 10283. Positive-strand DI RNA synthesis from pDER was not obvious using this probe due to the strong signals of helper virus mRNAs 6 and 7 (Fig. 1A). We exam-

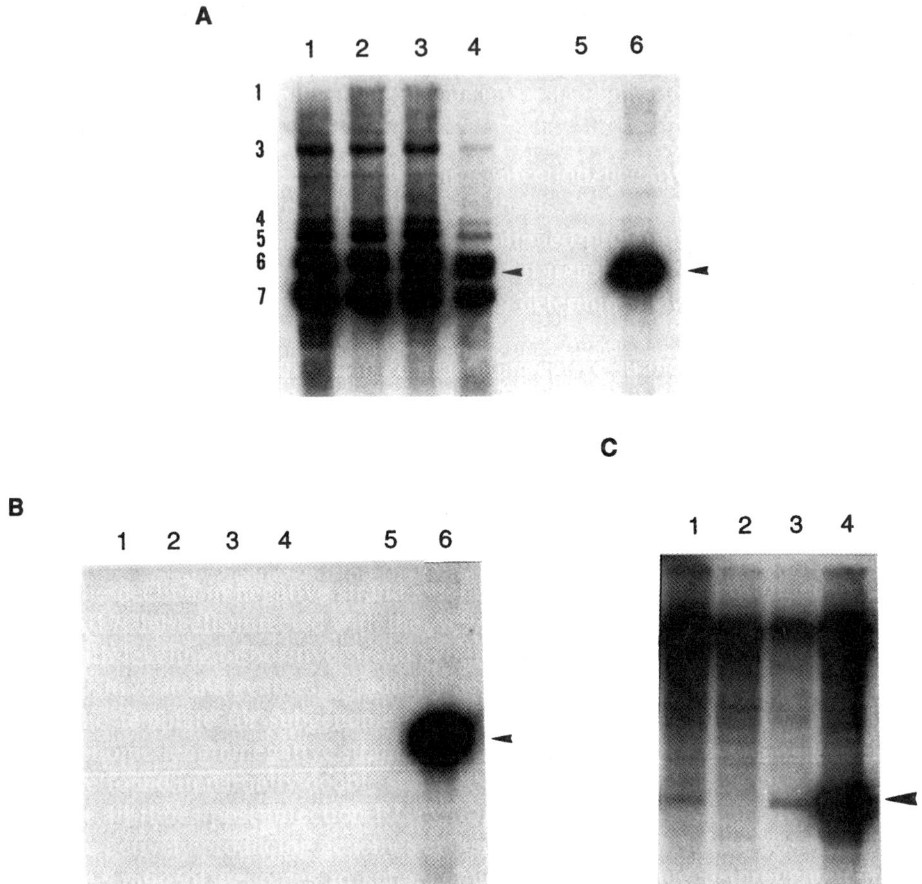

Figure 1. Positive-strand DI RNA synthesis in DNA-transfected, MHV-infected cells. Plasmid DNA (8–10μg) was transfected into vTF7–3-infected cells at 2 h postinfection. After 6 h postinfection, cells were superinfected with MHV. Intracellular RNA was extracted at 12 h after MHV infection and analyzed by Northern blot analysis. (A) The 5'-end-labeled oligonucleotide 10283, which detects positive-sense transcripts containing the 11 nt sequence, was used as a probe. RNAs from pDERΔ135-transfected cells (lane 1), from pDERΔ11-transfected cells (lane 2), from pDER-transfected cells (lane 3), and from DE25-transfected cells (lane 4). Lanes 5 and 6 contain in vitro-transcribed DE25Δ11 and DE25 RNA, respectively, which were run as controls. Numbers to the left of figure represent the MHV mRNAs. Arrowheads represent in vitro-synthesized DI RNAs (lanes 5 and 6) and replicating DI RNA (lane 4). (B) The 5'-end-labeled oligonucleotide 10293, which detects positve-sense transcripts without the 11 nt sequence, was used as a probe. Lanes 1–4 were loaded with the same RNA as in Fig. 1A. Lanes 5 and 6 contain in vitro-transcribed DE25 and DE25Δ11 RNA, respectively. Arrowhead represents in vitro-synthesized DI RNAs (lanes 5 and 6). (C) A probe which specifically hybridizes near the 5'-end of MHV genomic RNA was used. RNAs from pDERΔ135-transfected cells (lane 1), from pDERΔ11-transfected cells (lane 2), from pDER-transfected cells (lane 3), and from DE25-transfected cells (lane 4). Arrowhead represents DI RNA.

ined accumulation of DI RNAs using an oligonucleotide probe, 10250, that specifically hybridizes at the 5'-region (487–517 nt from the 5'-end of MHV genomic RNA), and found that positive-strand DI RNAs from pDER and DE25 transcripts accumulated (Fig. 1C). Nor could we detect the accumulation of positive-strand DI RNA of the expected size in the cells expressing pDERΔ135, yet we detected accumulation of DI RNAs that had the same size as DIssE; this DI RNAs most probably represented the DI RNAs, in which deletion of the 135 nt-long internal cis-acting replication was repaired by RNA recombination (Fig. 1C).

These data demonstrated that positive-strand DI RNA synthesis did not occur from the negative-strand DI RNA transcripts lacking the 11 nt-region, indicating that the 11 nt region was important for positive-strand DI RNA synthesis.

5. DISCUSSION

Transfection and expression of MHV DI RNA deletion mutants and expression of MHV DI RNA positive-strand fragments in MHV-infected cells revealed a cis-acting replication signal (Kim et al., 1993; Kim and Makino, 1995; Lin and Lai, 1993) and a signal that is required for negative-strand RNA synthesis (Lin et al., 1994). In this study we examined whether a short 11 nt sequence, which is present within the signal required for negative-strand RNA synthesis is involved in positive-strand DI RNA synthesis. Because we did not detect efficient positive-strand DI RNA synthesis from pDER∆11, we concluded that the 11 nt-region was also important for positive-strand DI RNA synthesis; the signal required for positive-strand DI RNA synthesis overlapped the signal that is required for negative-strand RNA synthesis.

We previously directly demonstrated that the internal cis-acting region is important for positive-strand DI RNA synthesis (Kim and Makino, 1995). Our data indicated that RNA secondary structure is important for biological function, and further characterization of the internal cis-acting replication signal indeed supported this speculation (Repass, J. F., and Makino, S. unpublished data). Here we showed the sequence present at the very 3'-region of DI RNA is also involved in positive-strand DI RNA synthesis. These data revealed that RNA elements required for positive-strand DI RNA synthesis consist of various regions of DI RNA, which is in striking contrast to the requirement for negative-strand RNA synthesis of only the very 3'-end 55 nt and poly A sequence.

How these various regions of MHV DI RNA sequence function in positive-strand DI RNA synthesis remains to be studied. Most of the MHV negative-strand RNAs are present as double-strand (ds) RNAs in MHV-infected cells (Lin et at., 1994; Sawicki and Sawicki, 1986). We speculated that negative-strand DI RNAs forming the ds RNAs are better templates for positive-strand DI RNA synthesis than those present as signal-strand negative-strand DI RNAs (Joo et al., 1996). Certain regions of ds RNAs may exist as a single-strand and these single-strand regions may interact with each other to form specific tertiary structure(s). These structures may be recognized by a complex of viral-specific proteins and host proteins that initiate RNA synthesis and elongation of nascent RNA molecules.

ACKNOWLEDGMENTS

This work was supported by Public Health Service grants AI29984 and AI32591 from the National Institutes of Health.

REFERENCES

Fuerst T.R., Niles E.G., Studier F.W. and Moss B., 1986, Eukaryotic transient-expression system based on recombinant vaccinia virus that synthesizes bacteriophage T7 RNA polymerase, *Proc. Natl. Acad. Sci. USA* **83**:8122–8126.

Higuchi R., 1990, Recombinant PCR, in *"PCR Protocols"* (Michael A. Innis, David H. Gelfland, John J. Sninsky and Thomas J. White, Eds.), pp. 177–183. Academic Press, San Diego.

Hirano N., Fujiwara K., Hino S. and Matsumoto M., 1974, Replication and plaque formation of mouse hepatitis virus (MHV-2) in mouse cell line DBT culture, *Arch. Gesamte Virusforsch* **44**:298–302.

Joo M., Banerjee S. and Makino S., 1996, Replication of murine coronavirus defective interfering RNA from negative-strand transcripts, *J. Virol.* **70**:5769–5776.

Kim Y.-N., and Makino S., 1995, Characterization of a murine coronavirus defective interfering RNA internal cis-acting replication signal, *J. Virol.* **69**:4963–4971.

Kim Y.-N., Jeong Y.S. and Makino S., 1993, Analysis of cis-acting sequences essential for coronavirus defective interfering RNA replication, *Virology* **197**:53–63.

Lai M.M.C. and Stohlman S.A., 1978, RNA of mouse hepatitis virus, *J. Virol.* **26**:236–242.

Lai M.M.C., Brayton P.R., Armen R.C., Patton C.D., Pugh C. and Stohlman S.A., 1981, Mouse hepatitis virus A59: mRNA structure and genetic localization of the sequence divergence from hepatotropic strain MHV-3l, *J. Virol.* **39**:823–834.

Lin Y.J. and Lai M.M.C., 1993, Deletion mapping of a mouse hepatitis virus defective interfering RNA reveals the requirement of an internal and discontiguous sequence for replication, *J. Virol.* **67**:6110–6118.

Lin Y.J., Liao C.-L. and Lai M.M.C., 1994, Identification of the cis-acting signal for minus- strand RNA synthesis of a murine coronavirus: implications for the role of minus- strand RNA in RNA replication and transcription, *J. Virol.* **68**:8131–8140.

Makino S., Fujioka N. and Fujiwara K., 1985, Structure of the intracellular defective viral RNAs of defective interfering particles of mouse hepatitis virus, *J. Virol.* **54**:329–336.

Makino S. and Lai M.M.C., 1989, High-frequency leader sequence switching during coronavirus defective interfering RNA replication, *J. Virol.* **63**:5285–5292.

Sawicki S.G. and Sawicki D.L., 1986, Coronavirus minus-strand RNA synthesis and effect of cycloheximide on coronavirus RNA synthesis, *J. Virol.* **57**:328–334.

Yu W. and Leibowitz J., 1995, A conserved motif at the 3' end of mouse hepatitis virus genomic RNA required for host protein binding and viral RNA replication, *Virology* **214**:128–138.

32

SEQUENCE ELEMENTS INVOLVED IN THE RESCUE OF IBV DEFECTIVE RNA CD-91

Kevin Dalton,[1] Z. Penzes,[1] C. Wroe,[2] K. Stirrups,[1] S. Evans,[1] K. Shaw,[1] T. D. K. Brown,[2] P. Britton,[1] and D. Cavanagh[1]

[1]IAH Compton RG20 7NN England
[2]Division of Virology
Department of Pathology
University of Cambridge, England

1. ABSTRACT

Deletion mutagenesis has been used to identify essential regions for rescue of coronavirus defective RNAs (D-RNAs). Using this technique on a cloned IBV D-RNA CD-91, we have identified a region potentially important in its rescue. Comparing the sequence of D-RNAs rescued with those not rescued we have deduced that a 72 base region corresponding to base number 13824 to 13896 in the viral genome is required for rescue. This may be an IBV D-RNA packaging signal or a cis-acting element involved in replication. Further experiments and modification of our techniques will be required to differentiate between the two processes.

2. INTRODUCTION

The cloning of a naturally occurring IBV D-RNA, CD-91 and the development of a technique to detect rescue of RNA transcripts of this D-RNA has allowed us to use deletion mutagenesis to distinguish regions essential for rescue from non-essential regions (Penzes *et al.*, 1994; Penzes *et al.*, 1996). Identification of regions required for replication, transcription and packaging of D-RNA can be used to construct D-RNAs for use as direct expression vectors of foreign genes and/or vehicles to drive recombination between genes of choice. D-RNAs may not be replicated in the same manner as genome; they may act as a mini-genome or as messages which are efficiently packaged. However they can still be used to help identify sequences necessary for recognition by the polymerase and in the process of packaging.

Coronaviruses and Arteriviruses, edited by Enjuanes *et al.*
Plenum Press, New York, 1998

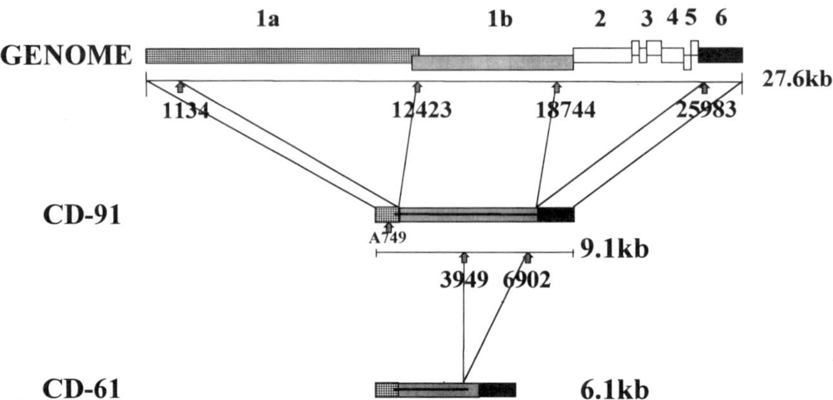

Figure 1. Diagram showing structure of the cloned naturally occurring IBV D-RNA CD-91 and the deletion mutant CD-61 derived from it. The open reading frame is indicated by the black bar.

CD-91 comprises three discontinuous regions of the IBV genome (Figure 1) (Penzes *et al.*, 1994; Penzes Z. *et al.*, 1996). Both 5' and 3' ends (1133 and 1622 nucleotides (nts) respectively) include terminal untranslated regions (UTRs) and a portion of the neighbouring message and a 6kb region of the ORF 1b bases 12423–18744.

Deletion mutagenesis has been used to identify essential regions of coronavirus D-RNAs, including those of MHV strains JHM and A59, and BCV (Chang and Brian, 1996; Chang *et al.*, 1994; Kim *et al.*, 1993; Kim and Makino, 1995; Lin and Lai, 1993; Lin *et al.*, 1994). Recently isolated TGEV D-RNAs are also under investigation (Mendez *et al.*, 1996). Results from each group have differed, suggesting the uniqueness of each of the D-RNAs studied and or differences in the techniques used to study them.

A 61 base stem loop has been identified in MHV JHM strain which is essential for packaging of D-RNA and can cause non-viral RNA to be packaged (Fosmire *et al.*, 1992; Woo K., *et al.*, 1997). This structure is within ORF 1b and is base number 20356–20416. This structure has since been shown to exist in MHV A59 D-RNAs, that are efficiently packaged. No corresponding structure has yet been identified in BCV D-RNA, which requires no internal 1b region for rescue (Chang and Brian, 1996; Chang *et al.*, 1994). The D-RNAs of TGEV and IBV both contain internal regions from pol 1b but packaging signals have yet to be defined (Mendez *et al.*, 1996; Penzes *et al.*, 1994; Penzes *et al.*, 1996). If the IBV D-RNA CD-91 contains such a packaging signal it is in a different position to that of MHV JHM, as the corresponding bases are missing from CD-91.

Sequences at the termini are required for the rescue of D-RNAs . Approximately 450 has been shown to be required at both ends of MHV D-RNAs (Kim *et al.*, 1993; Kim and Makino, 1995; Lin *et al.*, 1994). However, BCV D-RNA requires the whole of the 3'UTR and the N gene virtually intact for rescue to occur (Chang and Brian, 1996; Chang *et al.*, 1994). The requirement for 55nts at the 3' end for the synthesis of negative strand MHV JHM D-RNA suggests the interaction of both ends in the replication of this D-RNA but this has yet to be shown for other D-RNAs.

MHV JHM D-RNAs DIssE and DIssF have been shown to require an internal 135nt which is said to act as a replication enhancer (Kim and Makino, 1995). A 57nt stem loop structure is present within this region and may play a role in the function of this sequence (Kim and Makino, 1995). Other D-RNAs of BCV and MHV A59 have been shown not to require any such internal region (Chang and Brian, 1996; Chang *et al.*, 1994; Kim and Makino, 1995).

The requirement of these D-RNAs to contain a long ORF also differs. It has been shown for MHV and BCV D-RNAs that a long ORF is required for replication (De Groot, *et al.*, 1992; Kim *et al.*, 1993). Using a deletion mutant of the IBV D-RNA CD-91, CD-61 it has been shown that its long ORF can be interrupted by inserting stop codons, shortening it to just 11 amino acids with no noticeable effect on rescue (Penzes *et al.*, 1996).

From these results it is clear that in all coronavirus D-RNAs studied the terminal regions are essential for rescue and these may or may not require an internal region; the latter appears to be dependent on the particular D-RNA or strain of helper virus. Results found with one D-RNA do not necessarily imply the same for another D-RNA.

3. METHODS

Cloned D-RNA CD-91 under the control of a T7 promoter were used to construct all deletion mutants in this paper using convenient restriction sites.

RNA transcripts were prepared using T7 polymerase and a transcription kit supplied by Promega. IBV Beaudette infected primary chick kidney cells were electroporated with D-RNA and up to 6 passages performed at 20–24 hour intervals. Cellular RNA was harvested and analysed by Northern blotting. Rescued D-RNAs were then reverse transcribed and the cDNA amplified by PCR, so that deleted junctions could be sequenced.

4. RESULTS

Deletion mutants of CD-91 were created using existing restriction sites. (Figure 2).

Northern blot analysis was used to detect rescue of the deletion mutants. Probes used corresponded to the 5' end of genome (excluding leader, specific for genome and D-RNAs.) and /or the 3' end of genome which would hybridize to all viral messages including D-RNA. Results of the rescue experiments are indicated in Figure 2.

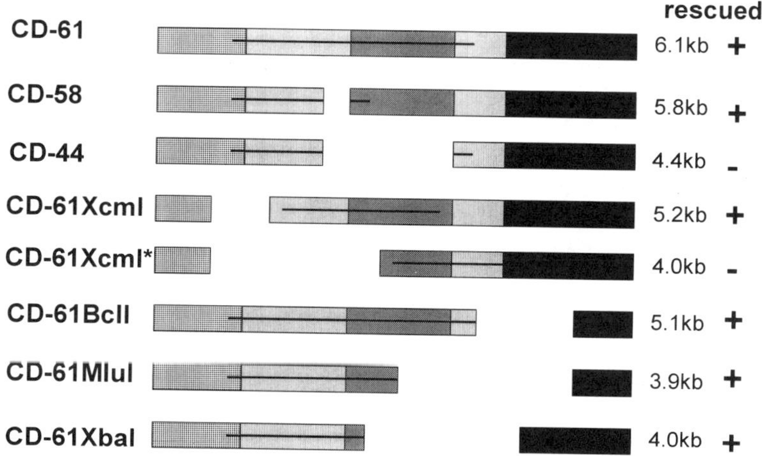

Figure 2. Diagram showing CD-61 and deletion mutants constructed to identify regions essential for rescue. Rescue was determined by Northern blotting.

Initial deletion mutants of CD-91 pointed to the importance for rescue of the central part of the 1b region. CD-61 and CD-58 were both rescued in IBV-infected CKCs. CD-44, which differs from CD-58 by the removal of a central continuous 1.4kb region was not rescued, thus indicating some importance for rescue of this 1.4kb region. This domain corresponds to nucleotide 13824 to 15239 in the viral genome (Penzes *et al.*, 1996).

Further deletion mutagenesis of CD-61 was carried out to investigate which part of the 1.4kb region was essential for rescue (Figure 2).

CD-61XcmI* has a 2.1 kb region (801 -2955) deleted and lacks the 5' 421 bases of the 1.4kb region. CD-61XcmI has an 880 base deletion from 803–1683, with the 1. 4kb region intact.

CD-61XcmI was rescued but CD-61XcmI* was not. This suggested that the 5' 421 bases of the 1.4kb region might contain an essential region or that deletion of this region had disrupted essential sequence of other regions.

Deletion mutants were constructed to remove 3' regions of CD-61 (Figure 2). CD-61BclI left the 1.4kb region untouched and therefore acted as a 3' deletion control, to confirm that sufficient sequence remained at the 3' end to allow rescue. CD-61MluI lacked the 3' 769 bases of the 1.4kb region and CD-61XbaI lacked virtually all of the 1.4kb region, leaving only the 5' 72 bases. Experiments showed all three constructs were rescued (Figure 2). D-RNA CD-61MluI is 3.9 kb, the smallest IBV D-RNAs to be rescued to date.

5. DISCUSSION

A 61 base stem loop structure has been identified in MHV JHM strain which enables D-RNAs to be packaged efficiently (Fosmire, Hwang, andMakino, 1992). This stem loop is located from base 20356 to base 20416 within the ORF 1b of the viral genome. A 1.4kb region of ORF1b is required for the rescue of the IBV D-RNA CD-91. Deletion mutants created to lack parts of the 1.4kb region indicated that the 5' most 72 bases of this 1.4kb region were sufficient for rescue. This may contain an IBV packaging signal or a cis-acting sequence essential for replication. Further experiments and modification of our techniques are required to answer this question fully. Rescue is a consequence of both replication and packaging. Our system of detection is unable to differentiate between these two processes. It may be possible to use a reporter gene inserted into a non-essential region as a tool for discriminating between packaging and replication.

ACKNOWLEDGMENTS

I would like to thank Kathy Shaw, Clive Naylor and Karen Mawditt for helpful advice in the laboratory, and also the BBSRC for funding this work.

REFERENCES

Chang, R. Y., and Brian, D. A., 1996, *cis* Requirement for N-specific protein sequence in bovine coronavirus defective interfering RNA replication, *J. Virol.* **70**: 2201–7.

Chang, R. Y., Hofmann, M. A., Sethna, P. B. and Brian, D. A., 1994, A *cis*-acting function for the coronavirus leader in defective interfering RNA replication, *J. Virol.* **68**: 8223–31.

De Groot, R. J., van der Most, R. G., and W. J. M., S., 1992, The fitness of defective interfering murine coronavirus DI-a and its derivatives is decreased by nonsense and frameshift mutations, *J. Virol.* **66**: 5898–5905.

Fosmire, J. A., Hwang, K., and Makino, S., 1992, Identification and characterization of a coronavirus packaging signal, *J. Virol.* **66**: 3522–30.

Kim, Y., Lai, M. M. C., and Makino, S., 1993, Generation and selection of coronavirus defective interfering RNA with large open reading frame by RNA recombination and possible editing, *Virology* **194**: 244–253.

Kim, Y. N., Jeong, Y. S., and Makino, S., 1993, Analysis of cis-acting sequences essential for coronavirus defective interfering RNA replication, *Virology* **197**: 53–63.

Kim, Y. N., and Makino, S., 1995, Characterization of a murine coronavirus defective interfering RNA internal cis-acting replication signal, *J. Virol.* **69**:4963–71.

Lin, Y. J., and Lai, M. M., 1993, Deletion mapping of a mouse hepatitis virus defective interfering RNA reveals the requirement of an internal and discontiguous sequence for replication, *J. Virol.* **67**: 6110–6118.

Lin, Y. J., Liao, C. L., and Lai, M. M., 1994, Identification of the *cis*-acting signal for minus-strand RNA synthesis of a murine coronavirus: implications for the role of minus-strand RNA in RNA replication and transcription, *J. Virol.* **68**: 8131–40.

Mendez, A., Smerdou, C., Izeta, A., Gebauer, F., and Enjuanes, L., 1996, Molecular characterization of transmissible gastroenteritis coronavirus defective interfering genomes: Packaging and heterogeneity, *Virology* **217**: 495–507.

Penzes, Z., Tibbles, K., Shaw, K., Britton, P., Brown, T.D.K., and Cavanagh, D., 1994, Characterization of a replicating and packaged defective RNA of avian coronavirus infectious bronchitis virus, *Virology* **203**: 286–93.

Penzes Z., Wroe C., Brown T.D.K., Britton P., Cavanagh D., 1996, Replication and packaging of coronavirus infectious bronchitis virus defective RNAs lacking a long open reading frame. *J. Virol.* **70**: 8660–8668.

Woo K., J. M., Narayanan K., Kim K. H., Makino S., 1997, Murine coronavirus packaging signal confers packaging to nonviral RNA, *J. Virol.* **71**: 824–827.

33

RESCUE OF IBV D-RNA BY HETEROLOGOUS HELPER VIRUS STRAINS

Kathleen Stirrups, Kathleen Shaw, Sharon Evans, Kevin Dalton, David Cavanagh, and Paul Britton

Division of Molecular Biology
Institute for Animal Health
Compton Laboratory
Compton, Newbury, RG20 7NN
United Kingdom

1. ABSTRACT

Coronavirus defective RNA (D-RNA) vectors could be developed to deliver selected genes for the production of recombinant coronavirus vaccines. An IBV D-RNA, CD-61, derived from a naturally occurring IBV Beaudette D-RNA, CD-91, is being developed as a D-RNA vector for IBV. In order to use CD-61 as a vector it will require rescue by heterologous strains in addition to Beaudette. Rescue will be determined by recognition of replication and packaging signals within the D-RNA by the helper virus. The 5' and 3' UTRs are believed to contain sequences involved in replication and transcription. The 5' and 3' UTRs of six strains of IBV have been sequenced and experiments performed using six strains of helper virus for rescue of CD-61 to determine whether rescue correlates with sequence conservation within the 5' and 3' UTRs. Results indicate that all strains of helper virus rescued the D-RNA to varying degrees. Sequence comparisons show a high degree of sequence identity in the UTRs, but enough strain differences exist to be used as markers. The 5' and 3' UTRs of the D-RNAs rescued by the heterologous strains were also sequenced and leader switching between the helper virus and the Beaudette leader on the D-RNAs was observed.

2. INTRODUCTION

Coronaviruses have large non-segmented RNA genomes making the production of a cDNA clone difficult, hence targeted recombination is a potential method for the introduction

Coronaviruses and Arteriviruses, edited by Enjuanes *et al.*
Plenum Press, New York, 1998

of site-specific mutations into the coronavirus genome. Recombination has been shown to occur in MHV in vitro and in vivo (Lai, 1990) and in IBV following a mixed infection (Kottier *et al.,* 1995). It should therefore be possible to utilise recombination between genomic and replicating defective-RNA (D-RNA) to create IBV targeted recombinants, as demonstrated recently for MHV (Zhang *et al.,* 1997). The D-RNA vector available for IBV is CD-61 derived from the naturally occurring D-RNA CD-91 from the Beaudette strain (Penzes *et al.,* 1996) (Fig. 1). One possible use of recombination would be to replace the spike gene of a given strain of IBV with one of a strain with different properties such as nephropathogenicity and determine the effect on pathogenicity, antigenicity and tissue tropism. Alternatively, the spike gene of the current vaccine strain could be exchanged with that predominant in the field. The Beaudette strain grows atypically in chickens and therefore it is desirable to use another strain of IBV in the recombination experiments. The D-RNA vector, CD-61 can be rescued (replicated and packaged) by the parental strain, Beaudette, but for use as a potential vector for recombination it will require rescue by heterologous strains of IBV. Rescue of the D-RNA should occur if replication and packaging signals within the D-RNA are recognised by the helper strain of IBV. The 5' and 3' UTRs are believed to contain sequences involved in the replication of coronavirus RNAs and the leader sequence is required for the discontinuous transcription of coronavirus subgenomic mRNAs.

The rescue of CD-61 has been investigated using six different heterologous helper strains to determine which may be suitable for rescue of CD-61. The six strains were chosen because they were either used as vaccines, capable of being differentiated by the use of neutralising monoclonal antibodies, consisted of different spike serotypes which may indicate sequence differences in the 5' and 3' UTRs or because of known deletions in the 3' UTR. Both the 5' and 3' UTRs of all seven strains were sequenced and analysed to determine the level of sequence conservation, resulting in potential differences between the replication and transcription signals and to look for nucleotide differences to act as markers for identification of any recombination between the D-RNA and helper virus.

In this paper, we describe that heterologous virus can rescue the D-RNAs derived from another strain and when this occurred leader switching was also observed.

3. MATERIALS AND METHODS

Virus stocks were grown in chick kidney (CK) cells and embryonated fowl eggs.

Rescue of T7-transcribed CD-61 RNA in IBV-infected CK cells and isolation and analysis of IBV-derived RNA as described previously by Penzes *et al.* (1996).

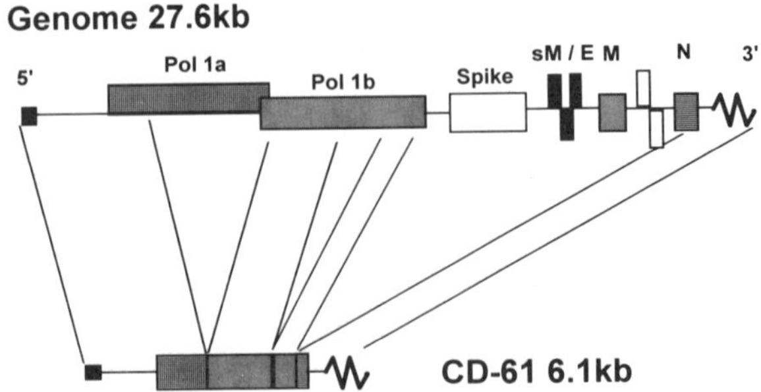

Figure 1. Comparison of the IBV genome with the IBV D-RNA CD-61.

PCR products from either the 5' or 3' UTR were generated by RT-PCR using Gibco Reverse transcriptase and Pharmacia Taq DNA polymerase. Oligonucleotides were designed using published IBV Beaudette sequence (Boursnell *et al.*, 1987).

The PCR-derived DNAs were blunt-ended and ligated into the EcoRV site of pBluescript II SK(+) (Stratagene). Dideoxy sequencing was carried out using a USB Sequenase version 2.0 DNA sequencing kit. Sequence data was compiled and analysed using the Staden package.

DNA fragments specific for CD-61 D-RNA were produced by RT-PCR using an oligonucleotide complementary to a region within the IBV polymerase 1b sequence present in CD-61 and an oligonucleotide corresponding to the first 23 bases of the Beaudette genome sequence. Sequencing of the PCR products was carried out using the USB 70170 sequenase PCR product sequencing kit.

4. RESULTS

4.1. Rescue of CD-61 D-RNA by Heterologous Strains

Rescue of the D-RNA CD-61 occurred with all six heterologous strains of IBV in chicken kidney cells (CKs). This has been demonstrated from Northern blot analysis of RNA extracted from the cells and PCR analysis looking for CD-61 specific junction regions and a missing A residue at base 749, a feature of CD-61. As can be seen from Fig. 2 although rescue occurs with all six heterologous strains it did not always occur with the same efficiency as observed by the passage number or by the amount detected when compared to the mRNAs.

4.2. Sequence Comparisons of 5' and 3' UTRs of the Seven IBV Strains

Sequence analysis of the 5' and 3' UTRs of seven strains of IBV shows a high level of conservation when compared to average level of conservation of other regions of the

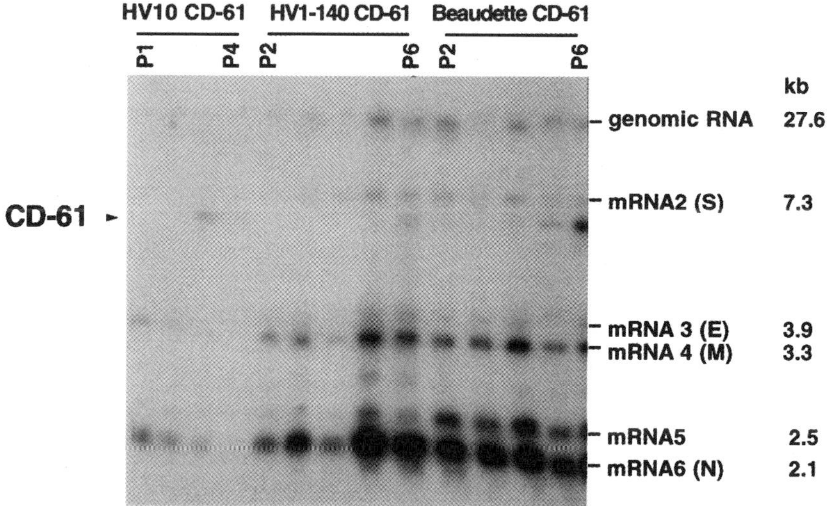

Figure 2. Analysis of IBV RNAs. RNA was extracted from CK cells infected with three different strains (HV10, HVI-140 and Beaudette) of IBV and electroporated with D-RNA CD-61. The passage number in the CKs from which the RNA was extracted is mentioned. The IBV derived RNAs were detected by northern blot analysis using ^{32}P-labelled PCR products from the N gene and 3' UTR.

Table 1. Average percentage nucleotide differences between the six
heterologous IBV strains and Beaudette genomic sequence
for the 5' and 3' UTRs

5' UTR			
Whole UTR 529 bases	Leader only 68 bases	UTR without leader 461 bases	First 350 bases of polymerase
2.64%	10.4%	2.0%	5.7%

3' UTR			
Whole UTR	Bases 1-206[*]	Bases 207 – 517	Nucleocapsid gene[1]
6.1%	13.6%	0.7%	5.9%

[*]Bases 1-188 of the 3'UTR are missing from M41 strain
[1]Data from Zwaagstra *et al.,* (1991) - % nucleotide differences between 20 strains of IBV compared to strain D1466 for a 360 base region of the nucleocapsid gene

genome. However, there are a greater number of changes in the leader and first 206 bases of the 3' UTR than the rest of the UTRs.

4.3. Leader Switching in D-RNAs Rescued by Heterologous Strains

Any evidence of sequence changes in the UTRs of CD-61 rescued by heterologous helper strains of IBV was investigated. This was done by a junction specific PCR of CD-61 and subsequent sequence analysis of the product. Results of the analysis revealed that the leader sequence of the rescued D-RNA has been preferentially derived from the helper virus genome rather than from the input D-RNA (Fig. 3).

5' Leader sequence

```
CD-61   ACTTAAGATAGATATTAATATATATC TATT A C ACTAGCCTTG C GCTAGAT
HCD-61  ACTTAAGATAGATATTAATATATATC ATAC A T ACTAGCCTTG T GCTAGAT
HV10    ACTTAAGATAGATATTAATATATATC ATAC A T ACTAGCCTTG T GCTAGAT
```

Leader junction site

```
CD-61   TT T TAA CTTAACAA AACGGACTTAAATACCTACAGCTGGTCC T CATAGGT
HCD-61  TT C TAA CTTAACAA AACGGACTTAAATACCTACAGCTGGTCC T CATAGGT
HV10    TT C TAA CTTAACAA AACGGACTTAAATACCTACAGCTGGTCC C CATAGGT
```

```
CD-61   GTTCCATTGCAGTGCACTTTAGTGCCCTGGATGGCACCTGGCCACCTGTC
HCD-61  GTTCCATTGCAGTGCACTTTAGTGCCCTGGATGGCACCTGGCCACCTGTC
HV10    GTTCCATTGCAGTGCACTTTAGTGCCCTGGATGGCACCTGGCCACCTGTC
```

```
CD-61   AGGTTTTTGTTATTAAAAT C T T ATTGTTGCTGGTATCACTGCTTGTTTTG
HCD-61  AGGTTTTTGTTATTAAAAT C T T ATTGTTGCTGGTATCACTGCTTGTTTTG
HV10    AGGTTTTTGTTATTAAAAT A T C ATTGTTGCTGGTATCACTGCTTGTTTTG
```

```
CD-61   CCGTGTCTCACTTTATACATC T GTTGCTTGGGCTACCTAGT G TCC
HCD-61  CCGTGTCTCACTTTATACATC T GTTGCTTGGGCTACCTAGT G TCC   245
HV10    CCGTGTCTCACTTTATACATC C GTTGCTTGGGCTACCTAGT A TCC
```

Figure 3. Sequence comparison of the 5' UTRs. The sequence of the 5' UTRs (nucleotides 1–245) derived from CD-61, HV10 genome and CD-61 rescued by HV10 (HCD-61) were determined and aligned. The sequence differences are boxed and the leader junction site is underlined.

5. DISCUSSION

5.1. Rescue of CD-61 D-RNA by Heterologous Strains

The IBV D-RNA CD-61 was rescued by all six heterologous strains of virus but to varying degrees. There are a number of potential reasons for the observed differences in rescue. The heterologous strains can only be propagated in the primary cell line of CKs a heterogeneous and variable mix of cells which can affect the passage number of rescue of the CD-61. Alternatively rescue of CD-61 could be affected by differences in the ability of each helper virus to replicate in CKs, some strains may not replicate as well as others in the 24 hour span allowed for each passage. Although the observed levels of mRNAs detected for each virus does not seem to vary significantly by northern blot analysis. Also the level of input virus may differ and future work will include a determination of virus titre of each strain to determine if this is the case. A final reason could be due to slight incompatibility of replication, or packaging signals between strains affecting rescue of CD-61.

5.2. Sequence Comparisons of 5' and 3' UTRs of the Seven IBV Strains

No striking sequence differences were observed between the UTRs of the strains that could be assigned to potential differences in replication signals. Our data agrees with that of Williams *et al.* (1993) on the comparison of the 3' UTRs showing that nucleotides 1–206 of the 3' UTR are hypervariable. One of the strains, M41, is lacking this region of the 3'UTR. M41 is able to rescue CD-61 indicating that any signals involved in rescue are unlikely to be located in this region. In contrast, nucleotides 207–517 of the 3'UTR are highly conserved, with only the occasional base difference occurring. The 5' UTR is also well conserved, though a greater percentage of the differences occur in the 68 base leader sequence. The data therefore suggests that if there are constraints on any potential changes throughout the 5' and conserved part of the 3' UTRs that are involved in recognition by the polymerase they occur at very specific nucleotides. Enough strain differences were found to identify the six strains from each other.

5.3. Leader Switching in D-RNAs Rescued by Heterologous Strains

The phenomenon of leader switching for D-RNAs rescued by heterologous helper viruses of MHV and BCV has previously been demonstrated (Makino and Lai, 1989; Chang *et al.,* 1996). Therefore, any evidence of sequence changes in the UTRs of CD-61 rescued by heterologous helper strains of IBV were investigated. This was done by a junction specific PCR of CD-61 and subsequent sequence analysis of the product. Results of the analysis revealed that the leader sequence of the rescued D-RNA has been preferentially derived from the helper virus genome rather than from the input D-RNA. This indicated a recombination event, leader switching had occurred analogous to the discontinuous mechanism involved in the transcription of subgenomic mRNAs. No sequence differences derived from the genome of the helper virus were found in the 3' UTR of the D-RNA. Our results indicated that the alternative mechanism for replication of D RNAs, involving the direct copying and subsequent amplification of the D-RNA in an analogous manner to the replication of the genomic RNA does not take place. Our results also indicate an intimate interaction between the replicating D-RNA and genomic RNA, an essential requirement for recombination, and that CD-61 can be used with heterologous strains of IBV for the production of recombinant viruses.

REFERENCES

Boursnell, M.E.G., Brown, T.D.K., Foulds, I.J., Green, P.F., Tomley, F.M. and Binns, M.M., 1987, Completion of the sequence of the genome of the coronavirus infectious bronchitis virus, *J. Gen. Virol.* **68**: 57–77.

Chang, R.-Y., Krishnan, R. and Brian, D.A., 1996, The UCUAAAC Promoter motif is not required for high efficiency leader recombination in bovine coronavirus defective interfering RNA, *J. Virol.* **70**: 2720–2729.

Kottier, S.A., Cavanagh, D. and Britton, P., 1995, Experimental evidence of recombination in coronavirus infectious bronchitis virus, *Virology* **213**: 569–580.

Lai, M.M.C., 1990, Coronavirus: Organization, replication and expression of genome, *Ann. Rev. Microbiol.* **44**: 303–333.

Makino, S. and Lai M.M.C., 1989, High frequency leader switching during coronavirus defective interfering RNA replication, *J. Virol.* **63**: 5285–5292.

Penzes, Z., Tibbles, K., Shaw, K., Britton, P., Brown, T.D.K. and Cavanagh, D., 1994, Characterization of a replicating and packaged defective RNA of Avian coronavirus Infectious Bronchitis Virus, *Virology.* **203**: 286–293.

Penzes, Z., Wroe, C., Brown, T.D.K., Britton, P. and Cavanagh, D., 1996, Replication and packaging of coronavirus infectious bronchitis virus defective RNAs lacking a long open reading frame, *J. Virol.* **70**: 8660–8668.

Williams, A.K., Wang, L., Sneed, L.W. and Collisson, E.W., 1993, Analysis of a hypervariable region in the 3' non-coding region of the infectious bronchitis virus genome, *Virus Res.* **28**: 19–27.

Zhang, L., Homberger, F., Spaan, W. and Luytjes, W., 1997, Recombinant genomic RNA of Coronavirus MHV-A59 after coreplication with a DI RNA containing the MHV-RI spike gene, *Virology* **230**: 93–102.

Zwaagstra, K. A., Van Der Zeijst, B. A. M. and Kusters J.G., 1991, Rapid detection and identification of avian infectious bronchitis virus, *J. Clin. Microbiol.* **30**: 70–84.

34

A STRATEGY FOR THE GENERATION OF INFECTIOUS RNAs AND AUTONOMOUSLY REPLICATING RNAs BASED ON THE HCV 229E GENOME

J. Herold, V. Thiel, and S. G. Siddell

Institute for Virology and Immunology
University of Würzburg
Versbacher Strasse 7
97078 Würzburg, Germany

1. ABSTRACT

A strategy to generate in vitro transcripts representing infectious RNAs and autonomously replicating RNAs based on the HCV 229E genome is presented. PCR-DNAs were ligated *in vitro,* resulting in 27 kbp and 22 kbp ligation products. These DNAs can now be transcribed *in vitro* and the RNAs tested for infectivity and their ability to replicate.

2. INTRODUCTION

The generation of infectious transcripts from cloned DNAs equivalent to coronaviral genomes has been hampered by a variety of problems. For example, vectors carrying coronaviral polymerase sequences are often unstable upon propagation in bacteria. Also, the size of the coronaviral genome prevents conventional cloning techniques in the common plasmid or bacteriophage vectors.

We have developed a system that allows for the generation of exceptionally long HCV 229E specific RT-PCR derived DNAs, starting with polyA-RNA isolated from HCV 229E infected cells. These DNAs are ligated in vitro to generate DNAs representing replicationally active RNA molecules.

Coronaviruses and Arteriviruses, edited by Enjuanes *et al.*
Plenum Press, New York, 1998

3. METHODS AND RESULTS

3.1. The RT-PCR Derived DNAs

We are able to generate RT-PCR derived DNA molecules that together represent either a full length DNA copy of the HCV 229E genome or a DNA encoding the 5- and 3′-cis acting signals necessary for replication and transcription combined with the coronaviral RNA polymerase gene i.e. a DNA copy of an autonomously replicating HCV 229E RNA.

The DNA fragments (AI, AII and AIII or BI, BII and BIII, Fig. 1) were obtained using primers containing coronavirus specific sequences and either a T7 RNA polymerase

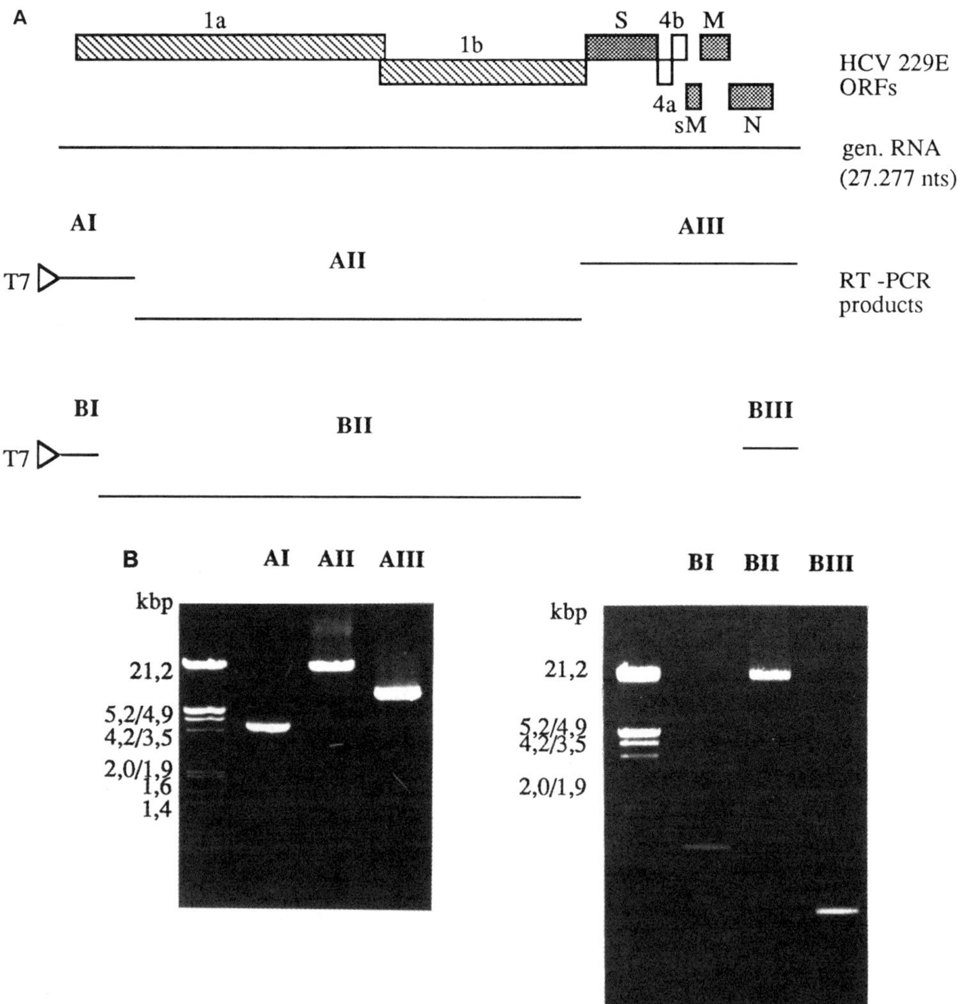

Figure 1. RT-PCR DNAs representing either (A) the genome of HCV 229E or (B) an autonomously replicating RNA. In the upper scheme, the position of the DNAs, relative to the HCV 229E genome, is shown. The DNAs were separated on 0.7% agarose gels to verify the correct length and homogeneity of the RT-PCR products.

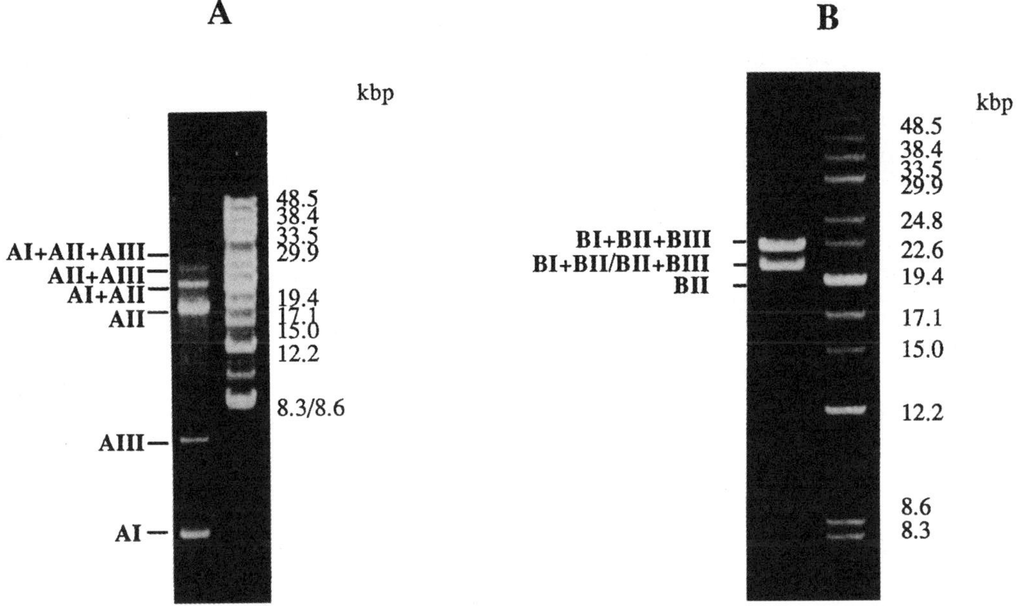

Figure 2. Ligation reaction products representing (A) the genome of HCV 229E (AI+AII+AIII) or (B) an autonomously replicating RNA (BI+BII+BIII). The ligated DNAs have been separated on a 0.8% agarose gel using the pulsed-field electrophoresis. The RT-PCR DNAs, intermediate reaction products and the full-length reaction products are shown.

promotor (fragments AI and BI, upstream primer), specific restriction enzyme recognition sites (all fragments) and marker mutations that allow for the identification of recombinant genomic RNA(all fragments).

3.2. In Vitro Ligation

The RT -PCR DNAs were treated with restriction enzymes to generate compatible ends. Subsequently the DNAs were purified and ligated in vitro using T4 DNA ligase. The reaction products were separated by PFGE (Fig. 2). Although the amounts of the full length reaction products are low, we could clearly detect the ligation products AI+AII+AIII and BI+BII+BIII. It is obvious that complete digestion of the RT-PCR DNA ends is the most critical parameter for complete ligation. In the future we will use this in vitro ligation technology to ligate quantitative amounts of PCR DNAs that can be subsequently used for in vitro transcription.

3.3. In Vitro Transcription

We have also tested the ability of T7 RNA polymerase to generate long HCV 229E-specific transcripts in vitro. A series of RT-PCR DNAs, representing the first 2 to 10 kb of the HCV 229E genome, with a T7 RNA polymerase promotor at the 5′-end has been synthesized and transcribed in vitro using T7 RNA polymerase.

Fig. 3 shows that T7 RNA polymerase is able to transcribe efficiently RNA using RT-PCR DNAs as substrates (approx. 3–10 μgRNA/μg DNA). Allthough the efficiency de-

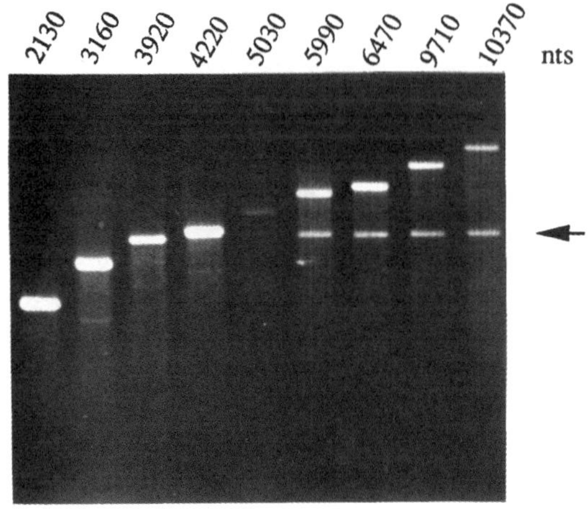

Figure 3. RT-PCR DNAs containing a T7 RNA polymerase promotor at the 5′end were transcribed in vitro using T7 RNA polymerase. After transcription, the template DNA was digested with RQI-DNase and the RNA was isolated by LiCl$_2$ precipitation. The RNAs (0.5 µg) were then separated by electrophoresis in a 1% agarose gel containing 0.1% SDS. The arrow indicates the RNA species derived by premature termination of the T7 RNA polymerase.

creases with the length of the DNA template sufficient amounts of 10 kb RNA molecules are synthesized. In this experiment, we have also identified a T7 RNA polymerase termination signal encoded in the RT-PCR DNA, located between nt 3.900 and 4.200 of the HCV 229E genome. A possible candidate is the heptanucleotide sequence ATCTGTT starting at position 4039. This heptanucleotide has been described recently to be a T7 RNA polymerase termination signal in vitro (Hartvig and Christiansen, 1996).

4. DISCUSSION

We intend to use the above strategy to produce quantitative amounts of RNAs equivalent to both the HCV 229E genome [containing marker mutations] and an autonomously replicating RNA. For this purpose, we will have to optimize at least two conditions: (i) in order to increase the amount of the full length in vitro ligation products, we will have to control the extent of endonucleolytic digestion of the RT-PCR DNAs and (ii) we will have to compare bacteriophage T7, T3, and SP6 RNA polymerases for their ability to synthesize long transcripts in vitro. It will also be necessary to avoid the generation of additional RNA species due to cryptic RNA polymerase termination signals. These RNAs will then be transfected into eukaryotic cells and screened for their ability to replicate and, eventually, to initiate an infectious cycle. We hope that the ability to produce full length infectious transcripts will greatly facilitate both molecular and pathogenesis studies on coronaviruses. In the long term, autonomously replicating coronavirus RNAs could also be used to establish a system for the expression of foreign genes in eukaryotic cells.

REFERENCES

Hartvig, L., and Christiansen, J., 1996, Intrinsic termination of T7 RNA polymerase mediated either by RNA or DNA, *EMBO J.* **15**:4767–4774.

35

LONG DISTANCE RT-PCRs OF HUMAN CORONAVIRUS 229E RNA

Volker Thiel, Jens Herold, and Stuart G. Siddell

Institute of Virology and Immunology
University of Wuerzburg
Versbacherstr. 7
97078 Wuerzburg, Germany

1. ABSTRACT

The generation and cloning of cDNA fragments longer than 10 kb is often a difficult and time consuming task. In this study, we have analysed the conditions necessary to produce reverse transcripts longer than 10 kb that can be amplified by polymerase chain reaction. Thus, we isolated poly(A)-RNA from human coronavirus 229E infected MRC-5 cells and did reverse transcription using a sequence-specific primer. Subsequently, we amplified PCR products of varying length upstream of the primer position. Optimisation of the poly(A)-RNA preparation, the reverse transcription protocol and the polymerase chain reaction cycle conditions enabled us to successfully amplify regions of the human coronavirus 229E genome between 11.5 and 20.3 kb in length.

2. INTRODUCTION

Polymerase chain reaction (PCR) and related techniques have become an almost indispensible tool in biological research and medicine. Consequently, there is a constant effort to improve the sensitivity and specificity of the procedure and to extend the application of PCR and related techniques to an increasing range of problems. For example, PCR technology has been adapted to the amplification of long DNA templates and to DNA templates of a more complex nature, i.e. human genomic DNA and mitochondrial DNA (Barnes, 1994; Cheng et al., 1994a; Cheng et al., 1994b). PCR, combined with reverse transcription (RT), is also a suitable method for the production of DNA from RNA templates. However only a few reports describe RT-PCR protocols that enable the amplifcation of DNAs longer than about 7 kilobases (Kb) (Fakhfakh et al., 1996; Martinez et al., 1996; Tellier et al., 1996). Here we report our studies on the conditions necessary to produce RT-PCR products longer than 10 kb using HCV 229E RNA as a template for reverse

Coronaviruses and Arteriviruses, edited by Enjuanes *et al*.
Plenum Press, New York, 1998

transcription. We focused, especially, on the requirements of the RNA template, the reverse transcription and the amplification of the cDNA. To carry out these studies, we used the HCV 229E genomic RNA as template and HCV 229E specific oligonucleotides to prime the reverse transcription. Then we tried to amplify products of varying length upstream of the RT-primer position.

3. METHODS

3.1. Preparation of Polyadenylated RNA

MRC-5 cells were infected with HCV 229E at a m.o.i. of 10, incubated at 33°C and polyadenylated RNA was prepared 18 h p.i. using two different methods. First, cytoplasmic RNA was prepared by phenol extraction and ethanol precipitation and poly(A)-RNA was selected by chromatography on poly(U)-Sepharose. The second method is based on preparation of poly(A)-RNA using oligo-(dT)$_{25}$-dynabeads (Dynal, Germany).

3.2. RT-PCR

Polyadenylated RNA (200 ng) from HCV 229E infected MRC-5 cells and 30 ng of oligonucleotide primer were used for reverse transcription in a volume of 20 µl with 200 U Superscript II reverse transcriptase (Life Technologies). The reactions were carried out for 90 min at 42°C, then heated for 2 min at 94°C and chilled on ice.

Aliquots (0.5 µl) of the reverse transcription reaction were used for the PCR. Reactions were performed in a total volume of 50 µl using the Elongase Enzyme Mix (Life Technologies) according to the manufacturer's instructions. Unless otherwise indicated, the PCR cycles were: 1 min 94°C, followed by 30 cycles of 20 s denaturation at 94°C, 30 s annealing at 50°C and elongation for 1 min per kb of expected product length at 68°C. During the last 18 cycles the elongation time was increased by 30 s per cycle. The reaction was terminated by a 10 min elongation at 72°C.

4. RESULTS

MRC-5 cells were infected with HCV 229E and the poly(A)-containing RNA was prepared using poly(U)-Sepharose or oligo-(dT)$_{25}$ coupled to magnetic beads. The poly(A)-containing RNAs were separated by gel elecrophoresis and the viral mRNAs were visualised by hybridisation analysis (Figure 1). In both preparations, it is possible to identify the genomic RNA and the six subgenomic mRNAs that are characteristic of HCV 229E infection. The viral RNAs are more abundant in the material isolated by poly(U)-Sepharose, however, the hybridisation analysis indicates that the material isolated with oligo(dT)$_{25}$ magnetic beads is less degraded.

In order to generate HCV 229E specific cDNAs we performed reverse transcriptions with HCV 229E specific oligonucleotides and the RNA templates shown in Figure 1. To amplify DNA products from the HCV 229E cDNAs, we inserted aliquots of the RT reaction into the PCR. Up to a distance of 4.8 kb from the RT priming site, we were able to obtain the expected PCR products regardless of the poly(A)-containing RNA preparation that we used as template for the RT reaction. However, when we tried to synthesise longer PCR products, we only succeeded with the poly(A)-RNA template prepared by the Dyna-

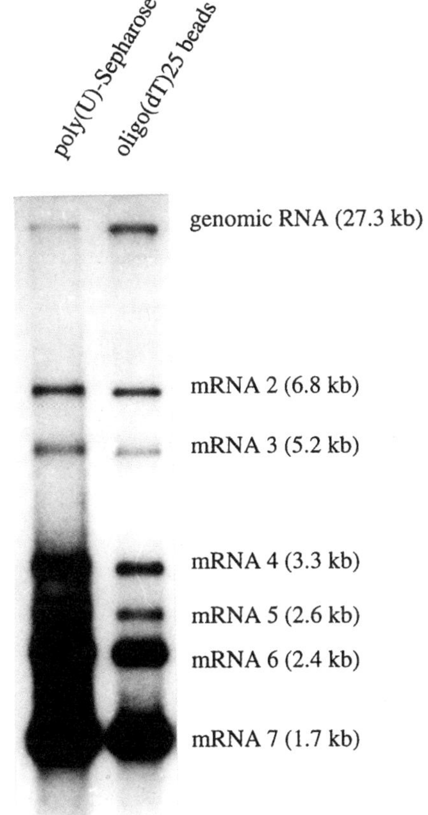

Figure 1. Hybridisation analysis of HCV 229E specific RNAs. Poly(A)-RNA (0.5 µg) prepared from HCV 229E-infected cells using poly(U)-Sepharose or oligo-(dT)$_{25}$ magnetic beads was separated by electrophoresis. Viral RNAs were hybridised with a HCV 229E nucleocapsid gene-specific, 5'-end-labelled oligonucleotide.

beads method. This result indicates that the quality of the RNA preparation is important when RT-PCR products longer than 4.8 kb are desired.

We then established an RT-PCR protocol that enabled us to generate a DNA product with a size of 12.6 kb. As RNA template we used the poly(A)-containing RNA that was prepared from HCV 229E infected cells using oligo-(dT)$_{25}$ magnetic beads. We primed the reverse transcription with the oligonucleotide 85 (Figure 2a). Two microliter of this reverse transcription reaction then served as template for a subsequent PCR reaction with the primers 159 and 89. As is shown in Figure 2b (lane 2), we obtained a PCR product with the expected size of 12.6 kb.

Our next goal was to amplify cDNAs longer than 12.6 kb. Therefore, we primed the reverse transcription with the oligonucleotide 32 (Figure 2a). First a PCR was done with 0.5 µl of the RT reaction as template and the primers 27 and 11 . As shown in Figure 2b, it was possible to amplify a DNA fragment with the expected size of 11.5 kb (lane 1). Using the same protocol, we successfully amplified DNA fragments of 16.7 kb and 17.5 kb in length (Figure 2b, lanes 3 and 4). It was also possible to synthesise DNA products of 19.0 and 20.3 kb, however the yields of these products were relatively poor. We therefore optimised the PCR conditions by varying the cycle profile. As is shown in Figure 2b, lanes 5 and 6, a profile of 5 cycles with 5s at 94°C, 20s at 50°C, 23 min at 68°C and 25 cycles with 5s at 94°C, 23 min at 68°C plus 30s per cycle increased the specificity of these PCRs and greatly enhanced the synthesis of the 19.0 and 20.3 kb RT-PCR products.

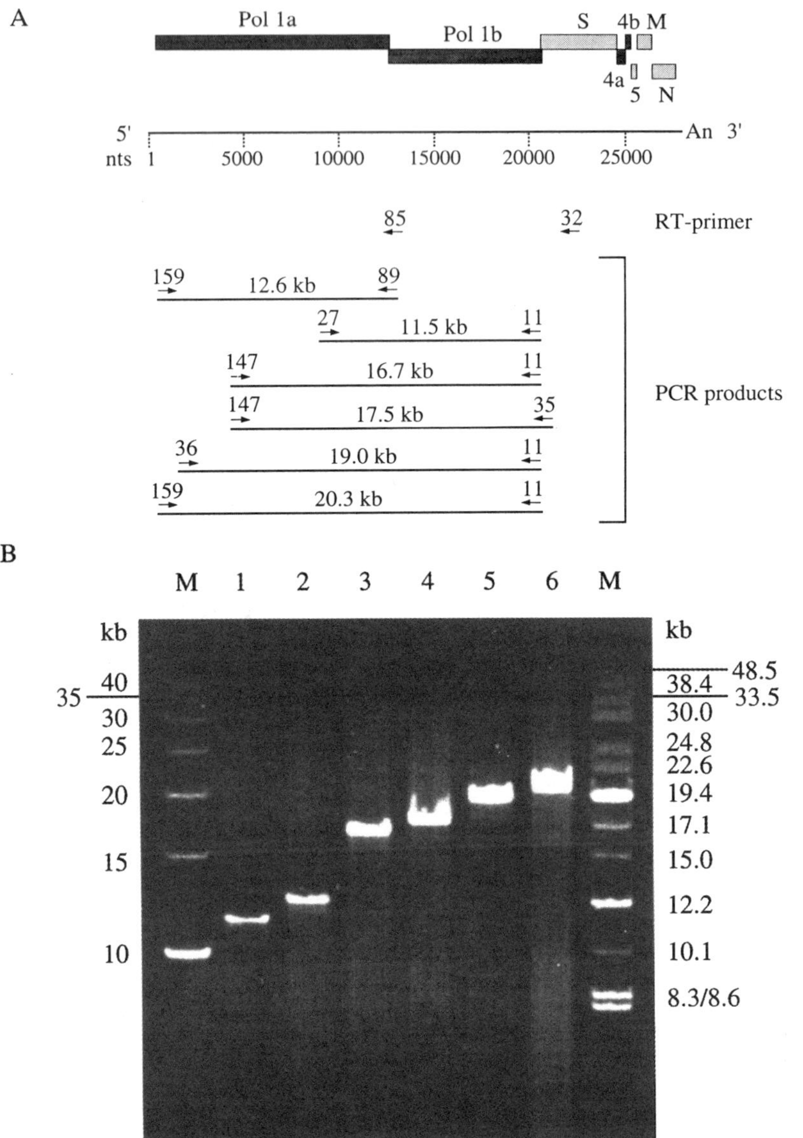

Figure 2. RT-PCRs of HCV 229E RNA. A) Schematic representation of HCV 229E specific oligonucleotides and RT-PCR products. The oligonucleotides are indicated as arrows according to their orientation and position relative to the HCV 229E genomic RNA. The expected sizes of the RT-PCR products are indicated. B) RT-PCR products of HCV 229E RNA. Five microliters of the PCR reactions were separated by PFGE along with a 5kb DNA ladder and a high molecular weight marker. The sizes of the RT-PCR products and markers are indicated.

5. DISCUSSION

In this study, we have combined the concept of long PCR technology to reverse transcription PCR. We have demonstrated that there is no limitation concerning the ability of reverse transcriptase to synthesise cDNAs of up to 20 kb. However, to achieve this goal, a number of critical parameters have to be kept in mind. First the integrity of the RNA template is an important prerequisite to perform long RT-PCRs. Second the conditions of the reverse transcription reaction strongly influence the outcome of the subsequent PCR. Thus, "non-stringent" priming during the RT reaction can be responsible for background amplification but this problem can be overcome by minimising the RT-primer concentration in the RT reaction. Finally, as is the case for all PCRs, the cycle conditions have to be optimised according to the amount of template, the PCR primers and the cycle profile.

The system that we have described has been applied to the amplification of viral RNA sequences. There are a number of RNA viruses with genome lengths above 10 kb, and for many of them it would be very desirable to develop a simple and reproducible system for the generation of genomic cDNA clones or RT-PCR DNA that represents a quasi-species population. An approach using long RT-PCRs to generate full length cDNAs in vitro could solve many of the problems encountered with conventional cloning techniques. The ability to obtain RT-PCR products up to 20 kb, in microgram amounts, greatly extends the number of RNA viruses that become amenable to this approach.

REFERENCES

Barnes, W.M., 1994, PCR amplification of up to 35-kb DNA with high fidelity and high yield from l bacterio-phage templates, *Proc. Natl. Acad. Sci. USA* **91:** 2216–2220.

Cheng, S., C. Fockler, W.M. Barnes, and R. Higuchi., 1994a, Effective amplification of long targets from cloned inserts and human genomic DNA, *Proc. Natl. Acad. Sci. USA* **91:** 5695–5699.

Cheng, S., R. Higuchi, and M. Stoneking, 1994b, Complete mitochondrial genome amplification. *Nature Genet.* **7:** 350–351.

Chumakov, K.M., 1996, PCR Engeneering of Viral Quasispecies: a New Method To Preserve and Manipulate Genetic Diversity of RNA Virus Populations, *J. Virol.* **70:** 7331–7334.

Fakhfakh, H., F. Vilaine, M. Makni, and C. Robaglia., 1996, Cell-free cloning and biolistic inoculation of an infectious cDNA of potato virus Y, *J. Gen. Virol.* **77:** 519–523.

Martinez, J.M., H.H. Breidenbach, and R. Cawthon, 1996, Long RT-PCR of the Entire 8.5-kb *NF1* Open Reading Frame and Mutation Detection on Agarose Gels, *Genome Res.* **6:** 58–66.

Tellier, R., J. Bukh, S.U. Emerson, and R.H. Purcell., 1996, Amplification of the full-length hepatitis A virus genome by long reverse transcription-PCR and transcription of infectious RNA directly from the amplicon, *Proc. Natl. Acad. Sci. USA* **93:** 4370–4373.

36

SUBGENOMIC RNA7 IS TRANSCRIBED WITH DIFFERENT LEADER-BODY JUNCTION SITES IN PRRSV (STRAIN VR2332) INFECTION OF CL2621 CELLS

Kay S. Faaberg, Margaret R. Elam, Chris J. Nelsen, and Michael P. Murtaugh

Department of Veterinary PathoBiology
University of Minnesota
1971 Commonwealth Avenue
St. Paul, Minnesota 55108

1. ABSTRACT

Porcine reproductive and respiratory syndrome virus (PRRSV), like all members of the order Nidoviridae, is expressed in the infected cell as a nested set of subgenomic (sg) RNAs with a common 5'-leader sequence. We have determined that the 5'-leader sequence for the US prototype strain (VR2332, Collins, *et al.*, 1992) is distinct from the European prototype strain [Lelystad (LV); Wensvoort, *et al.*, 1991, Meulenberg *et al.*, 1993a], yet these two strains use almost the same sequence for downstream sites of 5'-leader-body junction formation. Analysis of VR2332 genomic sequence identified several potential 5'-leader-body junction sequences upstream of open reading frame (ORF) 7, coding for the nucleocapsid protein, that could be used for generation of VR2332 sgRNA7 transcripts. Sequence determinations of RT-PCR-generated cDNA clones of sgRNA7 identified two species of RNA7 transcripts in infected cells, one utilizing a leader-body junction sequence (AUAACC) 123 nucleotides upstream of the AUG start site and one utilizing a sequence (UAAACC) 9 nucleotides upstream of the AUG start site for ORF7 translation.

2. INTRODUCTION

PRRSV causes severe respiratory problems in young pigs and reproductive failures in infected sows. This viral disease first appeared in North America in 1987 (Keffaber,

Coronaviruses and Arteriviruses, edited by Enjuanes *et al.*
Plenum Press, New York, 1998

1989) and in Europe in 1990 (Paton *et al.*, 1991), and has since spread worldwide (Halbur *et al.*, 1996). The positive sense, single stranded RNA genomes of both a North American strain (VR2332, approximately 15 kilobases (kb) in length, Collins, *et al.*, 1992) and a European strain (LV, 15.1 kb in length, Meulenberg *et al.*, 1993b) of PRRSV encode at least 8 ORFs. Both PRRSV genomes are transcribed in the infected cell into a nested set of 7 RNAs with a common 5′-leader sequence. Each RNA transcript, except sgRNA7, encodes more than one ORF, but is believed to only express the 5′ terminal ORF sequence. Protein sequence comparison of ORFs 2–7 of these two viruses reveal that VR2332 and LV are markedly different, however. Amino acid homologies for the ORFs range from 59% (ORF 5) to 70% (ORF 6) identity (Murtaugh, *et al.*, 1995).

We report that these two strains of PRRSV code for 5′-leader sequences of only moderate nucleotide similarity and that two leader-junction sites are utilized to produce sgRNA7 transcripts for strain VR2332. Previously, only one leader-junction site for sgRNA 7 had been identified for the LV strain (Meulenberg, *et al.*, 1993a) and for a strain isolated in Japan (EDRD-1), which is more closely related to a North American strain (VR2385) than LV in the region encoding ORFs 5–7 (Saito, *et al.*, 1996).

3. MATERIALS AND METHODS

A genomic library of the VR2332 strain of PRRSV (Collins, *et al.*, 1992) produced cDNA clones representing partial VR2332 sgRNAs with VR2332 5′-leader sequence (data not shown). Forward primer 658P1/ (CAGGAGCTGTGACCATTGGCA) was constructed from a leader sequence clone (c658). 658P1/ and 3′ RACE primers (Qt, CCAGTGAGCA-GAGTGACGAGGACTCGAGCTCAAGCTTTTTTTTTTTTTTTTTT, and Qi, GAGGAC-TCGAGCTCAAGC) were used to amplify species of sgRNA7 from VR2332 infected CL2621 cell total RNA using standard RT-PCR techniques. The total RNA from these infected cells (RNAeasy, Qaigen Inc., Santa Clarita, CA) was reversed transcribed to obtain almost full-length cDNAs representing sgRNA7. The cDNA was amplified by PCR and the RT-PCR (GeneAmp RNA PCR, Perkin Elmer, Foster City, CA) DNA products were analyzed by gel electrophoresis (data not shown). The DNA products of the approximate size were TA-cloned and sequenced. Taq DyeDeoxy™ Terminator Cycle Sequencing Kit (Applied Biosystems, Inc., Foster City, CA) was utilized to generate the sequence of 10 clones of VR2332 sgRNA7. Resulting sequences were analyzed and aligned using the AS-SEMBLE and LINEUP programs in the software package by the Genetics Computer Group (GCG), University of Wisconsin, USA.

4. RESULTS

4.1. VR2332 5′-Leader Sequence

The 5′-leader sequence determined to date for strain VR2332 is 170 nucleotides in length and will be published elsewhere (manuscript in preparation). Primer extension analysis has demonstrated that 20 nucleotides at the 5′-end of the sequence remain uncloned (data not shown). Accounting for the missing sequence, we predict that the VR2332 5′-leader contains approximately 190 nucleotides. The 5′-leader sequence of the LV strain is 221 nucleotides in length (Meulenberg, *et al.*, 1993a; personal communication), which is about 31 nucleotides longer than the VR2332 leader sequence. Sequence comparison indicates that PRRSV 5′-leader sequence for strain VR2332 (incomplete) and strain LV leader sequence exhibit 81% sequence similarity [Needleman-Wunsch comparison (Gap creation penalty:3, Gap extension penalty:0.1)].

4.2. VR2332 sgRNA7 Leader-Body Junction Sequences

Computer analysis of strain VR2332 identified 4 possible leader-body junction sites located downstream of the initiation AUG codon for ORF6 and upstream of the ORF7 AUG codon. These sites were located at nucleotides 2549–2554 (UUCACC), 2772–2777 (AUAACC), 2871–2876 (UUAAAC), and 2886–2891 (UAAACC) of the submitted VR2332 sequence (GENBANK submission U00153). Sequence analysis of 10 DNA clones of sgRNA7 indicated that at least two leader-body junction sites were utilized in the production of VR2332 ORF 7 transcripts in CL2621 cells. The most abundant 5′-leader-body junction site, representing 7 clones, identified the AUAACC sequence 123 nucleotides proceeding the starting AUG of ORF7 (nucleotides 2901–2903). Another sequence (UAAACC), located 9 nucleotides proceeding the ORF7 AUG (3 clones), was also used for sgRNA7 transcription. Each set of similar clones displayed identical leader-body junction sequences.

4.3. PRRSV Strain Comparison of Utilized 5′-Leader-Body Junction Sequences

A comparison was made between two other PRRSV strains for which leader-body junction sequences have been determined. Two clones of sgRNA7 for a Japanese strain, EDRD-1, have been analyzed. ERDR-1 was determined to be 94.3% similar in ORF7 to strain VR2385 (Saito, *et al.*, 1996), a North American strain. The investigators found that only the sequence (AUAACC), 123 nucleotides between the end of the leader-body junction sequence and the iniation codon for ORF7, was used for sgRNA7 transcription. These results directly correspond with the major sgRNA7 leader-body junction sequence (AUAACC) for strain VR2332. Six clones derived from sgRNA7 for LV, a European strain, have been reported. All of the LV clones possessed a leader-body junction site (UUAACC), spanning nucleotides 14573–14578 (GENBANK accession number M96262), which is located 9 nucleotides upstream of the ORF7 initiating AUG. This sequence varies by one nucleotide from the second sequence (UAAACC) used in VR2332 sgRNA7 transcription., and corresponds to the same number of nucleotides between the junction sequence and the start site for translation of ORF7.

5. DISCUSSION

All arteriviruses have been shown to utilize similar leader-body junction sequences [Saito, *et al.*, 1996 (EDRD-1); Meulenberg, *et al.*, 1993a (LV); Chen *et al.*, 1993 (lactate dehydrogenase-elevating virus(LDV)); Zeng, *et al.*, 1995 (simian hemorrhagic fever virus (SHFV)); den Boon, *et al.*, 1996 (equine arteritis virus (EAV))]. The consensus sequence (U/A)(U/A)AACC is used in transcription of sgRNA7 for all strains of PRRSV, and minor variations on this basic hexanucleotide sequence is used in production of all PRRSV sgRNAs. However, although the 5′-leader-junction sequences are similar, other regions of the PRRSV genome are dissimilar in sequence. The 3′-end of the genomes of VR2332 and LV are quite distinct (Murtaugh, *et al.*, 1995) and as related in this abstract, the North

American prototype (VR2332) and the European prototype (LV) display dissimalarity in both the length and the sequence of their 5′-leader regions. In addition, sequences upstream of the reported structural regions show considerable genotypic divergence (manuscript in preparation).

Two leader-junction motifs were shown to be utilized in VR2332 sgRNA7 (encoding the nucleocapsid (N) protein) in infected CL2621 cells. No other arterivirus, including PRRSV strains EDRD-1 and LV, has been shown to exhibit a variation in sgRNA expression for a critical structural protein, although EAV can express multiple forms of sgRNA3 (den Boon, *et al.*, 1996). Our evidence suggests that the predominant VR2332 sgRNA7 utilizes the site 123 bp upstream of ORF7, and the second site is used less frequently. We are presently determining whether infection of porcine alveolar macrophages, the natural host cell for PRRSV infection, will produce similar results and if other PRRSV sgRNA7 5′-leader-body junctions sites may be utilized at a low frequency. LV sgRNA7 appears to utilize the sequence 9 bp upstream of LV ORF 7 exclusively (Meulenberg, *et al.*, 1993a). It will be interesting to explore the functional sugnificance of the differences in 5′-leader-body junction sequences utilized by the North American and European strains of PRRSV.

ACKNOWLEDGMENTS

The authors wish to thank Judy Laber and Dan Strom of the University of Minnesota Advance Genetic Analysis Center for sequencing expertise and Thy M. Truong and Sara Proman for technical assistance. Boehringer Ingelheim Animal Health, Inc. provided financial support for this project.

REFERENCES

Chen, Z., Kuo, L., Rowland, R.R.R., Even, C., Faaberg, K.S., and Plagemann, P.G.W., 1993, Sequences of 3′ end of genome and of 5′ end of open reading frame 1a of lactate dehydrogenase-elevating virus and common junction motifs between 5′ leader and bodies of seven subgenomic mRNAs, *J. Gen. Virol.* **74**: 643–660.

Collins, J.E., Benfield, D.A., Christianson, W.T., Harris, L., Hennings, J.C., Shaw, D.P., Goyal, S.M., McCullough, S., Morrison, R.B., Joo, H.S., Gorcyca, D., and Chladek, D., 1992, Isolation of swine infertility and respiratory syndrome virus (isolate ATCC VR-2332) in North America and experimental reproduction of the disease in gnotobiotic pigs, *J. Vet. Diagn. Invest.* **4**: 117–126.

den Boon, J.A., Kleijnen, M.F., Spaan, W.J.M., and Snijder, E.J., 1996, Equine arteritis virus subgenomic mRNA synthesis: Analysis of leader-body junctions and replicative-form RNAs, *J. Virol.* **70**: 4291–4298.

Halbur, P.G., Paul, P.S., Meng, X.J., Lum, M.A., Andrews, J.J., and Rathje, J.A., 1996, Comparative pathogenicity of nine US porcine reproductive and respiratory syndrome virus (PRRSV) isolates in a five-week-old cesarean-derived, colostrum-deprived pig model, *J. Vet. Diagn. Invest.* **8**: 1120.

Keffaber, K.K., 1989, Reproductive failure of unknown etiology, *Am. Assoc. Swine Pract. Newsl.* **1**: 1–9.

Meulenberg, J.J., de Meijer, E.J., and Moormann, R.J., 1993a, Subgenomic RNAs of Lelystad virus contain a conserved leader-body junction sequence, *J. Gen. Virol.* **74**: 1697–1701.

Meulenberg, J.J., Hulst, M.M., de Meijer, E.J., Moonen, P.L., den Besten, A., de Kluyver, E.P., Wensvoort, G., Moormann, R.J., 1993b, Lelystad virus, the causative agent of porcine epidemic abortion and respiratory syndrome (PEARS), is related to LDV and EAV, *Virology* **192**: 62–72.

Murtaugh, M.P., Elam, M.E., and Kakach, L.T., 1995, Comparison of the structural protein coding sequences of the VR-2332 and Lelystad virus strains of the PRRS virus, *Arch. Virol.* **140**: 1451–1460.

Paton, D. J., Brown, I.H., Edwards, S., and Wensvoort, G., 1991, Blue ear disease of pigs, *Vet. Record* **128**: 617.

Saito, A., Kanno, T., Murakami, Y., Muramatsu, M., Yamaguchi, S., 1996, Characteristics of major structural protein coding gene and leader-body sequence in subgenomic mRNA of porcine reproductive and respiratory syndrome virus isolated in Japan, *J. Vet. Med. Sci.* **58**: 377–80.

Wensvoort, G., Terpstra, C., Pol, J.M.A., ter Laak, E.A., Bloemraad, M., de Kluyver, E.P., Kragten, C., Van Buiten, L., den Besten, A., Wagenaar, F., Broekhuijsen, J.M., Moonen, P.L.J.M., Zetstra,T., de Boer, E.A., Tibben, H.J., de Jong, M.F., van't Veld, P., Groenland, G.J.R., van Gennep, J.A., Voets, M.T., Verheijden, J.H.M., and Braamskamp, J., 1991, Mystery swine disease in the Netherlands: the isolation of Lelystad virus, *Vet. Quarterly* **13**: 121–130.

Zeng, L., Godeny, E.K., Methven, S.L., and Brinton, M.A., 1995, Analysis of simian hemorrhagic fever virus (SHFV) subgenomic RNAs, junction sequences, and 5′ leader, *Virology* **207**: 543–548.

37

ORGANIZATION OF THE SIMIAN HEMORRHAGIC FEVER VIRUS GENOME AND IDENTIFICATION OF THE sgRNA JUNCTION SEQUENCES

X. C. Wang, S. L. Smith, and E. K. Godeny

Department of Veterinary Microbiology and Parasitology
School of Veterinary Medicine
Louisiana State University
Baton Rouge, Louisiana 70803

1. ABSTRACT

SHFV is a member of the *Arteriviridae* family. Viruses within this family encode eight open reading frames (ORFs), two of which are translated from the full-length genome RNA. The remaining six ORFs are translated from a nested set of six or seven 3′ co-terminal, subgenomic RNAs (sgRNAs). We have cloned and sequenced approximately 6000 nucleotides (nt) from the 3′ end of the SHFV genome. Eleven ORFs, numbered ORFs 1a, 1b, 2a, 2b, 3, 4, 5, 6, 7, 8, and 9, were identified, three more than the other arteriviruses. The characteristics of the peptides encoded by ORFs 2a through 9 were determined from their computer-generated amino acid sequences. We also amplified the junction sequences from each of the SHFV subgenomic RNAs (sgRNAs) using RT-PCR analysis . Eight separate junction sequences were found which suggests that SHFV produces eight sgRNAs during replication. ORFs 2a and 2b appear to be encoded on the same sgRNA implying that RNA 2 is polycistronic. Sequence analysis identified the conserved SHFV junction sequence as 5′-(U/C)(C/U)N(U/C)(U/C)(A/C/G)AC(C/U)-3′. Since SHFV encodes additional ORFs and produces additional sgRNAs during replication, these data suggest that SHFV may be more complex than the other arteriviruses.

2. INTRODUCTION

Simian hemorrhagic fever virus (SHFV), along with equine arteritis virus (EAV), lactate dehydrogenase-elevating virus (LDV) and porcine reproductive and respiratory

Coronaviruses and Arteriviruses, edited by Enjuanes *et al.*
Plenum Press, New York, 1998

syndrome virus (PRRSV), was recently reclassified into a new virus family, the *Arteriviridae*. These viruses are morphologically similar to the togaviruses; however, their genome organization and replication strategy is similar to the coronaviruses. During replication, the arteriviruses produce six or seven subgenomic mRNAs (sgRNAs) which are nested at the 3′ end of the viral genome (Snijder and Spaan, 1995). The arterivirus genomes consists of eight overlapping open reading frames (ORFs; den Boon *et al.*, 1991; Godeny *et al.*, 1993; Meulenberg *et al.*, 1993b). Due to a frame-shifting mechanism, the first two ORFs, ORFs 1a and 1b, are translated from the full length genome RNA as one large polypeptide (Snijder and Spaan, 1995). The remaining ORFs are translated from the sgRNAs which are 3′ co-terminal. Only the 5′-most ORF is translated from each sgRNA.

Each of the sgRNAs contains a leader sequence at the 5′ end which is encoded at the 5′ terminus of the viral genome. The 5′ leader sequence is joined to the sgRNAs at junction or intergenic sequences. These junction sequences are highly conserved among the sgRNAs of a specific virus and are similar among all of the arteriviruses thus far sequenced. The conserved junction sequences of the other arteriviruses are: 5′-UCAAC-3′ (EAV; den Boon *et al.*, 1991); 5′-U(A/G)(U/A)A(A/-)CC-3′ (LDV; Chen *et al.*, 1993) and 5′-GNUNAAC-3′ (PRRSV; Meulenberg *et al.*, 1993a).

It had previously been reported that SHFV produces six sgRNAs during replication and that these RNAs are nested and 3′ co-terminal (Godeny *et al.*, 1995; Zeng *et al.*, 1995). The SHFV capsid (C) and nonglycosylated membrane (M) proteins map to the ultimate and penultimate 3′ ORFs, respectively, suggesting that these proteins are translated from the two smallest sgRNAs (Godeny *et al.*, 1995). The SHFV 5′ leader was sequenced and the conserved junction sequences of the two smallest SHFV sgRNAs had been determined to be 5′-U(U/C)AACC-3′ (Zeng *et al.*, 1995). We have recently obtained the 3′ sequence of the SHFV genome (Smith *et al.*, 1997). The genome sequence shows that SHFV encodes three additional ORFs at the 3′ end of the genome as compared to the other arteriviruses. The purpose of this study was to identify the characteristics of the peptides encoded by the SHFV 3′ ORFs and to sequence and identify the conserved junction sequence for the remaining SHFV sgRNAs.

3. MATERIALS AND METHODS

3.1. Computer Analysis

The 3′ end of the genome sequence from the prototype strain, LVR 42–0/M6941, of SHFV was reported previously (Smith *et al.*, 1997) and can be accessed through GenBank (accession number U63121). Translation of the SHFV ORFs and the characteristics of the deduced peptides were determined by the Translation and Protein Analysis programs, respectively, supplied in the University of Wisconsin Genetics Computer Group (GCG) software (Program Manual for the Wisconsin Package, Version 8, August 1994, Genetics Computer Group, 575 Science Drive, Madison, WI, 53711).

3.2. Isolation, Cloning, and Sequencing of the SHFV sgRNA Junction Sequences

Using the method of Sawicki *et al.* (1981), intracellular RNA was isolated from MA104 cells twenty hours post-infection with the prototype strain of SHFV. The SHFV sgRNA junction sequences were reverse transcribed from the intracellular RNA using cDNA primers complementary to specific SHFV genome ORF sequences. The resulting SHFV cDNA was amplified in a thermocycler; the forward primer used was identical to nucleotides 60 through 77 of the SHFV 5′ leader sequence (Zeng *et al.*, 1995) and the reverse primers were those used to reverse transcribe the viral RNA. The amplified products were cloned into the pCRII plasmid vector supplied in the TA Cloning Kit (Invitrogen Corp., San Diego, CA). The resulting clones were sequenced by the dideoxy chain-termination method using the Sequenase™ DNA Sequencing Kit (U.S. Biochemical Corp., Cleveland. OH).

4. RESULTS

4.1. Organization of the SHFV Genome

The 3′ end sequence of the SHFV genome, beginning at the 3′ poly(A) tract and continuing into the helicase domain in ORF 1b, has been reported (Smith *et al.*, 1997). As shown in Figure 1, this sequence contains 9 complete ORFs. Although similar in organization to the other arteriviruses, SHFV has three additional ORFs at the 3′ end (Fig. 1). With the exception of the beginning of ORFs 4 and 7, all of the SHFV ORFs overlap adjacent ORFs. Interestingly, the 5′ end of ORF 2a overlaps the 3′ end of ORF 1b. This is a unique property among the arteriviruses as the 3′ ends of ORFs 1b of EAV, LDV, and PRRSV do not overlap their adjacent ORFs.

ORF 9, which encodes a peptide 111 amino acids in length, is the smallest SHFV ORF and ORF 2a is the largest ORF encoding a 281 amino acid peptide. The deduced peptides encoded by the SHFV ORFs have slightly acidic to very basic pI values ranging from 6.2 for peptides 3 and 5 through 11.7 for peptide 9. All of the SHFV ORFs encode peptides with at least one potential N-linked glycosylation site. However, the utilization of these potential glycosylation sites during viral protein processing has yet to be determined.

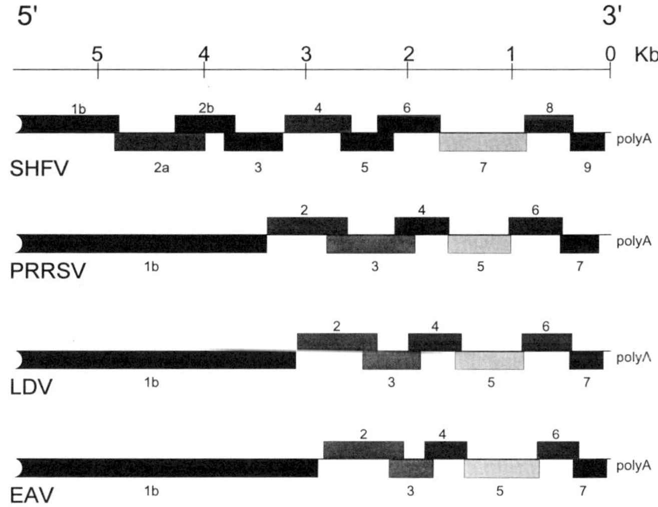

Figure 1. Comparison of the genome organization of the 3′ SHFV ORFs with those of the other arteriviruses. The size and location of each of the SHFV ORFs was determined from the SHFV genome sequence using the GCG Translation program. ORFs are drawn approximately to scale.

4.2. SHFV sgRNA Junction Sequences

The junction sequences on the SHFV sgRNAs between the leader sequence and the 5′-most ORF on that sgRNA were determined and are shown in Figure 2. Although SHFV encodes nine ORFs at the 3′ end of its genome, only eight SHFV sgRNA junction sequences were found. ORFs 2a and 2b appear to share the same junction sequence. There is a junction sequence located upstream of each of the remaining seven SHFV ORFs. The consensus junction sequence among the sgRNAs is 5′-(U/C)(C/U)N(U/C)(U/C)(A/C/G)AC(C/U)-3′ (Fig.

```
                            5'                                        3'
    5' Leader      ...GCAGACCCUCCUUAACCAUGUUCUGUGAGU...

    ORF 2a sgRNA   ...GCAGACCCUCCUUAACUUCUG-81nt-AUG...
    Genome         ...AGCCAAGGUCUUUAACUUCUG-81nt-AUG...

    ORF 2b sgRNA   ...GCAGACCCUCCUUAACUUCU-574nt-AUG...
    Genome         ...AGCCAAGGUCUUUAACUUCU-574nt-AUG...

    ORF 3 sgRNA    ...GCAGACCCUCCUUCACCCUGA-51nt-AUG...
    Genome         ...UGCCUUAACCUUUCACCCUGA-51nt-AUG...

    ORF 4 sgRNA    ...GCAGACCCUCCUUAACCAAA-124nt-AUG...
    Genome         ...UAGGAUUUCUAUUAACCAAA-124nt-AUG...

    ORF 5 sgRNA    ...GCAGACCCUCACUAACCCAUGGAUGUCCGU...
    Genome         ...CAGUUAUCUCACUAACCCAUGGAUGUCCGU...

    ORF 6 sgRNA    ...GCAGACCCUCCUUGACCAAA-175nt-AUG...
    Genome         ...CAACGUUGUCAUUGACCAAA-175nt-AUG...

    ORF 7 sgRNA    ...GCAGACCCUCCUUAACUACCUAAUUAUGUA...
    Genome         ...GACUCCGCUCCUUAACUACCUAAUUAUGUA...

    ORF 8 sgRNA    ...GCAGACCCUCCUCAACCACG-123nt-AUG...
    Genome         ...UAGAUUAUUUGUCAACCACG-123nt-AUG...

    ORF 9 sgRNA    ...GCAGACCCUCCUUAACCUGAGGAAGUAUGG...
    Genome         ...AAGGGGUCUUGUUAACCUGAGGAAGUAUGG...


                            CU  CCC   U
    CONSENSUS             -UCNUUAACC-
                                 G
```

Figure 2. The 5′ nucleotide sequences of the SHFV sgRNAs encoding ORFs 2a, 2b, 3, 4, 5, 6, 7, 8 and 9. Nucleotides in bold-type represent the 5′ leader sequence which is encoded upstream of ORF 1a. ":" indicates nucleotide identity between the sgRNA and the genome sequences. Initiation codons for the respective ORFs are underlined. The consensus junction sequence is shown at the bottom.

2). The distance between the junction sequence and the initiation codon for the respective ORF varies from 1 nt (ORF 5) to 177 nts (ORF 6).

5. DISCUSSION

SHFV is one of four members of the new virus family, *Arteriviridae*. The genomes of the other arteriviruses, EAV, LDV and PRRSV, encode six ORFs in the region between ORF 1b and the 3′ terminus (Godeny *et al.*, 1993; Meulenberg *et al.*, 1993b; den Boon *et al.*, 1991). The ultimate and penultimate ORFs at the 3′ end of the arterivirus genomes encode the capsid and membrane proteins, respectively (Meulenberg *et al.*, 1995; de Vries *et al.*, 1992; Godeny *et al.*, 1990). For EAV and LDV, ORFs 2 and 5 have been shown to encode the small and large envelope glycoproteins (de Vries *et al.*, 1992; Faaberg and Plagemann, 1995). The PRRSV ORF 5 product was also shown to encode an envelope glycoprotein but the ORF 2 product could not be detected in purified virions (Meulenberg *et al.*, 1995). The PRRSV ORFs 3 and 4 gene products were also reported as viral envelope glycoproteins (van Nieuwstadt *et al.*, 1996); however, these gene products in EAV and LDV have not been identified.

The 3′ end of the SHFV genome RNA, beginning in the helicase domain within ORF 1b and ending at the 3′ poly(A) tract, has been cloned and sequenced (Smith *et al.*, 1997). Nine complete ORFs were found in this genome region, which is three additional ORFs as compared to the other arteriviruses. Although the gene products of ORFs 8 and 9 have been identified as the membrane and capsid proteins, respectively (Godeny *et al.*, 1995), the remaining SHFV ORF products remain to be identified.

Previously, SHFV has been shown to produce six sgRNAs during replication (Godeny *et al.*, 1995; Zeng *et al.*, 1995). However, this study suggests that SHFV produces at least eight sgRNAs, since separate junction sequences have been obtained for all of the 3′ ORFs except ORF 2b. These sgRNAs are diagramed in Figure 3. Interestingly, sgRNA 2 contains two complete ORFs at the 5′ end and therefore has the potential to en-

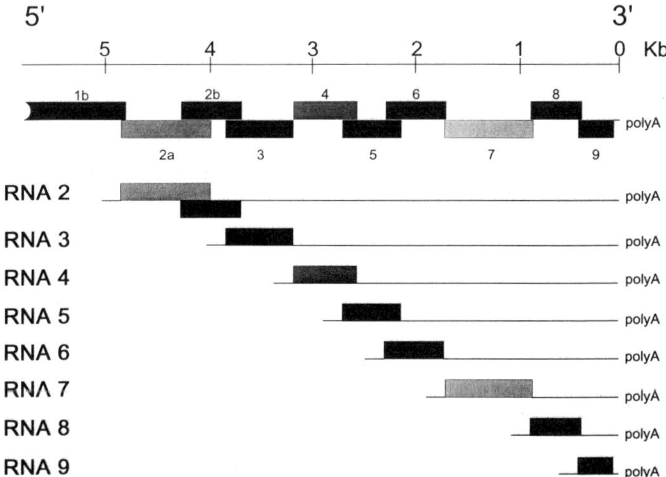

Figure 3. Replication strategy of SHFV. The deduced SHFV subgenomic mRNAs and the ORFs translated from each RNA species are shown.

code two peptides. Although this characteristic has been observed for a few of the coronavirus sgRNAs (Liu and Inglis, 1992; Senanayake *et al.*, 1992; Liu *et al.*, 1991), it is a unique feature among the arteriviruses.

From the eight junction sequences, we determined the conserved junction sequence on the SHFV sgRNAs as 5'-(U/C)(C/U)N(U/C)(U/C)(A/C/G)AC(C/U)-3'. This sequence is similar to the conserved junction sequences of EAV (den Boon *et al.*, 1991), LDV (Chen *et al.*, 1993) and PRRSV (Meulenberg *et al.*, 1993b). Each of the arterivirus junction sequences contains a form of the nucleotide sequence -AAC- towards the 3' end of this sequence. The SHFV junction sequence is longer and more diverse than the junction sequences of the other arteriviruses. The SHFV sequence is also rich in uridine and cytidine residues; whereas, the EAV, LDV and PRRSV junction sequences are rich in adenine and uridine (Snijder and Spaan, 1995). Interestingly, the junction sequence found on the sgRNAs after the 5' leader sequence was attached is not the same for all of the sgRNAs. The first two nucleotides at the 5' end of the junction sequence were derived from the leader sequence whereas the last six nucleotides were derived from the genome RNA sequence. The third nucleotide is the most variable in the sequence and may be derived from either the leader or genome sequences. These data suggests that a switching event occurs between the 5' leader sequence and position 3 of the junction sequence during the replication of the SHFV sgRNAs.

ACKNOWLEDGMENTS

This work was supported by a Public Health Service grant from the NCRR, RR06841.

REFERENCES

Chen, Z., Kuo, L., Rowland, R. R. R., Even, C., Faaberg, K. S. and Plagemann, P. G. W., 1993, Sequences of 3' end of genome and of 5' end of open reading frame 1a of lactate dehydrogenase-elevating virus and common junction motifs between 5' leader and bodies of seven subgenomic mRNAs, *J. Gen. Virol.* **74:** 643–660.

de Vries, A. A. F., Chirnside, E. D., Horzinek, M. C. and Rottier, P. J. M., 1992, Structural proteins of equine arteritis virus, *J. Virol.* **66:** 6294–6303.

den Boon, J., Snijder, E. J., Chirnside, E. D., de Vries, A. A. F., Horzinek, M. C. and Spaan, W. J. M., 1991, Equine arteritis virus is not a togavirus but belongs to the coronavirus-like family, *J. Virol.* **65:** 2910–2920.

Faaberg, K. S. and Plagemann, P. G. W., 1995, The envelope proteins of lactate dehydrogenase-elevating virus and their membrane topography, *Virology* **212:** 512–525.

Godeny, E. K., Chen, L., Kumar, S. N., Methven, S. L., Koonin, E. V. and Brinton, M. A., 1993, Complete genomic sequence and phylogenetic analysis of the lactate dehydrogenase-elevating virus (LDV), *Virology* **194:** 585–596.

Godeny, E. K., Speicher, D. W. and Brinton, M. A., 1990, Map location of lactate dehydrogenase-elevating virus (LDV) capsid protein (Vp1) gene, *Virology* **177:** 768–771.

Godeny, E. K., Zeng, L., Smith, S. L. and Brinton, M. A., 1995, Molecular characterization of the 3' terminus of the simian hemorrhagic fever virus genome, *J. Virol.* **69:** 2679–2683.

Liu, D. X., Cavanagh, D., Green, P. and Inglis, S. C., 1991, A polycistonic mRNA specified by the coronavirus infectious bronchitis virus, *Virology* **184:** 531–544.

Liu, D. X. and Inglis, S. C., 1992, Identification of two new polypeptides encoded by mRNA 5 of the coronavirus infectious bronchitis virus, *Virology* **186:** 342–347.

Meulenberg, J. J. M., de Meijer, E. J. and Moormann, R. J. M., 1993a, Subgenomic RNAs of Lelystad virus contain a conserved leader-body junction sequence, *J. Gen Virol.* **74:** 1693–1701.

Meulenberg, J. J. M., Hulst, M. M., de Meijer, E. J., Moonen, P. L. J. M., den Besten, A., de Kluyver, E. P., Wensvoort, G. and Moormann, R. J. M., 1993b, Lelystad virus, the causative agent of porcine epidemic abortion and respiratory syndrome (PEARS) is related to LDV and EAV, *Virology* **192**: 62–72.

Meulenberg, J. J. M., Petersen-Den Besten, A., De Kluyver, E. P., Moormann, R. J. M., Schaaper, W. M. M. and Wensvoort, G., 1995, Characterization of proteins encoded by ORFs 2 to 7 of Lelystad virus, *Virology* **206**: 155–163.

Sawicki, S. G., Sawicki, D. L., Kääriänen, L. and Keranen, S., 1981, A sindbis virus mutant temperature sensitive in the regulation of minus strand RNA synthesis, *Virology* **115**: 161–172.

Senanayake, S. D., Hofmann, M. A., Maki, J. L. and Brian, D. A., 1992, The nucleocapsid gene of bovine coronavirus is bicistronic, *J. Virol.* **66**: 5277–5283.

Smith, S., Wang, X. and Godeny, E., 1997, Sequence of the 3' end of the simian hemorrhagic fever virus genome, *Gene* (In Press).

Snijder, E. J. and Spaan, W. J. M., 1995, The coronaviruslike superfamily, in: *The Coronaviridae* (S. G. Siddell, ed.), Plenum Press, New York, pp. 239–255.

van Nieuwstadt, A. P., Meulenberg, J. J. M., van Essen-Zandbergen, A., Petersen-den Besten, A., Bende, R. J., Moormann, R. J. M. and Wensvoort, G., 1996, Proteins encoded by open reading frames 3 and 4 of the genome of Lelystad virus (Arteriviridae) are structural proteins of the virion, *J. Virol.* **70**: 4767–4772.

Zeng, L., Godeny, E. K., Methven, S. L. and Brinton, M. A., 1995, Analysis of simian hemorrhagic fever virus (SHFV) subgenomic RNAs, junction sequences, and 5' leader, *Virology* **207**: 543–548.

Protein Expression and Assembly I: Expression

38

CONSTRUCTION OF A MOUSE HEPATITIS VIRUS RECOMBINANT EXPRESSING A FOREIGN GENE

Françoise Fischer, Carola F. Stegen, Cheri A. Koetzner, and Paul S. Masters

David Axelrod Institute
Wadsworth Center for Laboratories and Research
New York State Department of Health
Albany, New York 12201-2002

ABSTRACT

The genome of the coronavirus mouse hepatitis virus (MHV) contains genes which have been shown to be nonessential for viral replication and which could, in principle, be used as sites for the introduction of foreign sequences. We have inserted heterologous genetic material into gene 4 of MHV in order (i) to test the applicability of targeted RNA recombination for site-directed mutagenesis of the MHV genome upstream of the N gene; (ii) to develop further genetic tools for mutagenesis of structural genes other than N; and (iii) to examine the feasibility of using MHV as an expression vector. A DI-like donor RNA vector containing the MHV S gene and all genes distal to S was constructed. Initially, a derivative of this was used to insert a 19-nucleotide tag into the start of ORF 4a of MHV-A59 using the N gene deletion mutant Alb4 as the recipient virus. Subsequently, the entire gene for the green fluorescent protein (GFP) was inserted in place of gene 4. This heterologous gene was shown to be expressed by recombinant viruses but not at levels sufficient to allow detection of fluorescence of viral plaques. Northern blot analysis of transcripts of GFP recombinants showed the expected displacement of the mobility, relative to those of wild-type, of all subgenomic mRNAs larger than mRNA5. An unexpected result of the Northern analysis was the observation that GFP recombinants also produced an RNA species the same size as that of wild-type mRNA4. RT-PCR analysis of the 5' end of this species revealed that it was actually a collection of mRNAs originating from a cluster of 10 different sites, none of which possessed a canonical intergenic sequence. The finding of these aberrant mRNAs, all of nearly the same size as wild-type mRNA4, suggests that long range structure of the MHV genome can sometimes be the sole determinant of the site of initiation of transcription.

Coronaviruses and Arteriviruses, edited by Enjuanes *et al.*
Plenum Press, New York, 1998

INTRODUCTION

The exceptionally large genomic sizes of coronaviruses currently places them beyond the reach of the preferred method for the engineering of positive-strand RNA viruses - the transcription of infectious RNA from a cDNA clone of the entire genome. Our laboratory has used the method of targeted RNA recombination to generate a number of MHV mutants containing site-specific mutations in the N gene and the 3' UTR of the viral genome (Koetzner *et al.*, 1992; Masters *et al.*, 1994; Peng *et al.*, 1995a; Peng *et al.*, 1995b; Fischer *et al.*, 1997a). This technique takes advantage of recombination between a synthetic DI RNA and a thermolabile N gene deletion mutant (Alb4) that can be selected against when we seek to identify recombinants that have repaired the deletion. We are interested in extending the applicability of targeted recombination to all the structural protein genes of MHV. In the present study we have tested this by inserting heterologous genetic material into gene 4 of MHV, which is nonessential for the virus (Weiss *et al.*, 1993; Yokomori and Lai, 1991). An unexpected consequence of this work was our finding of a novel aspect of coronavirus transcription: under certain conditions leader-body fusion can occur in the absence of any proximal intergenic sequence (IGS), presumably mediated by long range genomic interactions.

MATERIALS AND METHODS

The methods used in these studies have been described in detail elsewhere (Fischer *et al.*, 1997b). In brief, transcription vectors encoding DI RNAs were constructed by standard techniques, and mutations or substitutions were made in these by PCR-based methods or by direct insertion of a restriction fragment containing heterologous material of interest. Recombination experiments were performed by infection followed by transfection of synthetic RNA. Transfections were carried out by electroporation of mouse L2 cells, which were then plated onto monolayers of mouse 17 clone 1 cells. Progeny virus were titered on mouse L2 cells at 39°C, and candidate recombinants able to form large plaques were purified and analyzed further. Initial characterization of recombinants was by RT-PCR analysis or by sequencing of RNA from infected cells; final verification of recombinants was accomplished through sequencing of genomic RNA from purified virions.

RESULTS AND DISCUSSION

To carry out targeted recombination upstream of the MHV N gene, we constructed an extended version of our original N gene-containing DI RNA vector. This enlarged donor RNA vector, designated pFV1 contained a 5' genomic segment fused to the genes for S, ns4, ns5a, E, M, N, and the 3' UTR. Thus, from the S gene 3'-ward, its encoded RNA was identical to the wild-type MHV-A59 genome, except for a coding-silent restriction site polymorphism (*Hin*dIII to *Ase*I) introduced some 500 nt from the 5' end of the S gene.

Initially, we used a derivative of pFV1 to insert a 19-nt tag into the start of gene 4 of MHV (Fig. 1). This established that targeted recombination could be used to create mutations upstream of the N gene, and created the possibility that in future work we may be able to identify recombinants by screening for this tag rather than by selection.

We next inserted an entire heterologous gene, that for GFP (Chalfie *et al.*, 1994), into the pFV1 vector in place of most of gene 4, and carried out targeted recombination

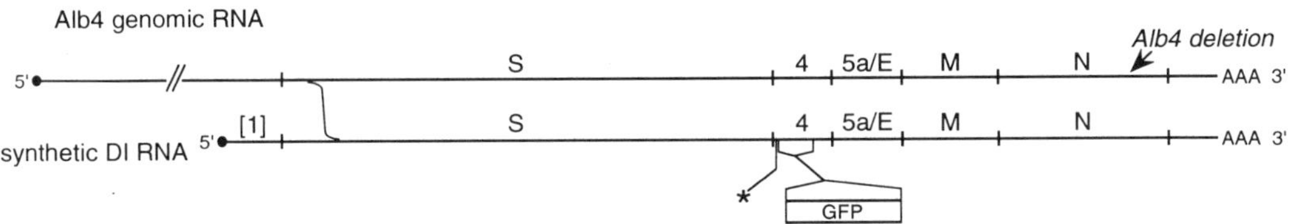

Figure 1. Targeted RNA recombination between the genome of the temperature-sensitive, thermolabile mutant Alb4 and synthetic DI RNA containing either a 19-nt tag (asterisk) at the start of gene 4 or the entire gene for the green fluorescent protein (GFP) substituted for most of gene 4. The locus of the N gene deletion responsible for the Alb4 phenotype is indicated with an arrow.

with the N gene deletion Alb4 mutant as the recipient virus (Fig. 1). Eight candidate recombinant viruses were analyzed by RT-PCR. Use of a primer pair crossing the N gene deletion revealed that all 8 were recombinants that had repaired the lesion in Alb4. A primer pair crossing the region of the insertion showed that 5 of these 8 viruses also contained the GFP gene, and all of these 5 had an expected internal *Dra*I site within the GFP gene. Finally, PCR analysis of the 5' end of S gene showed that 1 of the 5 GFP recombinants had the *Ase*I/*Hind*III polymorphism that marked the donor RNA vector. Therefore, the crossover that generated this particular recombinant had to have occurred more than 6.8 kb from the 3' end of the MHV genome. Final confirmation of the presence of the GFP gene was accomplished through direct RNA sequencing of purified genomes of two of the recombinants.

Western blot analysis of lysates of GFP recombinant-infected cells and wild-type-infected and mock-infected controls demonstrated that GFP was, indeed, expressed by the recombinants. However, the level of expression of GFP was not sufficiently high to allow detection of fluorescing plaques.

Northern blot analysis of RNA from GFP recombinant-, wild-type-, and mock-infected cells showed the expected pattern of displacement of the mobilities of mRNA4 and larger mRNAs for the GFP recombinants when cellular RNAs were hybridized with a probe specific for the N gene. Also, as expected, a probe specific for the 5' end of the GFP gene hybridized only to mRNA4 and larger mRNAs from GFP-infected cells. To our surprise, however, the N gene probe also detected an RNA species the same apparent size as that of wild-type mRNA4 in cells infected with the GFP recombinants, and this was roughly equimolar with the larger, GFP-containing mRNA4. This species was not seen with the probe for the 5' end of the GFP gene, implying that it originated in the middle of that gene.

To understand the source of this unexpected transcript, we performed RT-PCR analysis to determine the leader-body junctions of gene 4-related RNAs in both GFP recombinant-infected cells and wild-type-infected controls. Products of the expected sizes were obtained from primers specific for the MHV leader paired with primers for either the middle of the GFP gene or the 3' end of gene 4. These products were cloned in bulk and then sequenced. We thus found that the leader-body junctions of both wild-type mRNA4 and the upstream GFP recombinant mRNA4 are canonical and homogeneous (Fig. 2), having the MHV consensus common region motif 5'AAUCUAAAC3'. By contrast, the aberrant mRNA originating from the middle of the GFP gene turned out to be a collection of at least 11 species, the leader-body junctions of which are heterogeneous and noncanonical. In the most extreme case, the leader-body common region was found

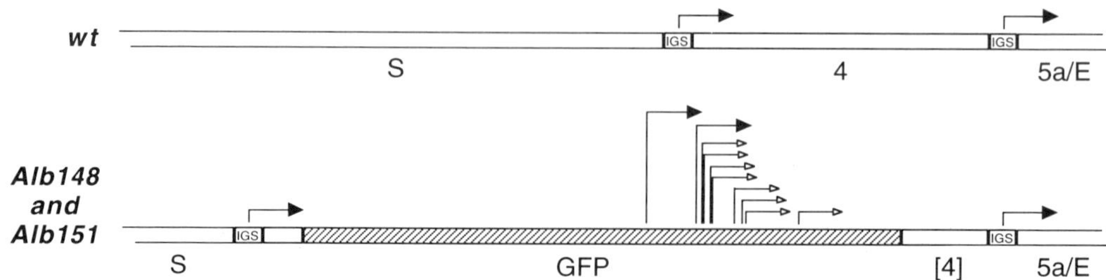

Figure 2. Loci of normal and aberrant mRNA fusion sites produced in the gene 4 / gene 5 regions of wild-type (wt) and GFP recombinant viruses Alb148 and Alb151. The two major and eight minor leader-body fusion sites originating within the GFP gene are denoted by solid arrows and open arrows, respectively. The shown portions of the mutant and wild-type viral genomes are aligned at their 3' ends, and consensus intergenic sequences (IGS) are indicated.

to be as small as one nucleotide. Remarkably, all the points of leader-body fusion occur within a small window corresponding to the distance, measured from the 3' end, where wild-type-size mRNA4 would originate (Fig. 2). There are no identifiable cryptic IGSs at or near the aberrant sites of fusion in the GFP gene, and we contend that the transcription from this region is IGS-independent and must be caused by genomic elements located outside of the GFP gene. This observation has some resemblance to the broad spectrum of fusion site heterogeneity seen by Zhang and Lai (1994), who examined mRNAs produced by a DI containing either an orthodox IGS or an atypical IGS upstream of a reporter gene. However, the anomalous fusion sites they observed were correlated with variations in leader RNA composition (number of UCUAA repeats and presence or absence of a 9-nt motif downstream of the leader). Also, in their system, all fusions occurred within, or adjacent to, authentic IGS motifs. Neither of those two conditions pertain in our results.

In conclusion, we have established that targeted RNA recombination can be used to insert mutations as far as 6.8 kb upstream from the 3' end of the MHV genome (and more than 6.3 kb upstream of the selected marker). Furthermore, a heterologous gene can be inserted into the genome of MHV, replacing most of the nonessential gene 4, and the inserted gene is expressed. This, incidentally, has produced what is currently the largest known RNA virus genome (31.8 kb). Finally, we found that, as expected, in the GFP-MHV recombinant, a separate subgenomic RNA4(GFP) was initiated at the (now displaced) IGS upstream of the inserted GFP gene. However, the virus also somehow retained a "memory" of where the IGS used to be, and it initiated synthesis of wild-type-sized transcripts from a cluster of sites in the middle of the GFP gene, where there is no canonical IGS. Thus, in some cases, long-range RNA (or RNP) interactions can be the sole determinants of the sites of leader-body mRNA fusion in MHV.

ACKNOWLEDGMENTS

We are grateful to James Gombold for providing a clone of the MHV-A59 S gene and to Monica Parker for one of the precursor clones incorporated into pFV1. This work was supported in part by Public Health Service grants AI 31622 and AI 39544 from the National Institutes of Health.

REFERENCES

Chalfie, M., Tu, Y., Euskirchen, G., Ward, W.W., and Prasher, D.C., 1994, Green fluorescent protein as a marker for gene expression, *Science* **263**:802–805.

Fischer, F., Peng, D., Hingley, S.T., Weiss, S.R., and Masters, P.S., 1997a, The internal open reading frame within the nucleocapsid gene of mouse hepatitis virus encodes a structural protein that is not essential for viral replication, *J. Virol.* **71**:996–1003.

Fischer, F., Stegen, C.F., Koetzner, C.A., and Masters, P.S., 1997b, Analysis of a recombinant mouse hepatitis virus expressing a foreign gene reveals a novel aspect of coronavirus transcription, *J. Virol.* **71**:5148–5160.

Koetzner, C.A., Parker, M.M., Ricard, C.S., Sturman, L.S., and Masters, P.S., 1992, Repair and mutagenesis of the genome of a deletion mutant of the coronavirus mouse hepatitis virus by targeted RNA recombination, *J. Virol.* **66**:1841–1848.

Masters, P.S., Koetzner, C.A., Kerr, C.A., and Heo, Y., 1994, Optimization of targeted RNA recombination and mapping of a novel nucleocapsid gene mutation in the coronavirus mouse hepatitis virus, *J. Virol.* **68**:328–337.

Peng, D., Koetzner, C.A., and Masters, P.S., 1995a, Analysis of second-site revertants of a murine coronavirus nucleocapsid protein deletion mutant and construction of nucleocapsid protein mutants by targeted RNA recombination, *J. Virol.* **69**:3449–3457.

Peng, D., Koetzner, C.A., McMahon, T., Zhu, Y., and Masters, P.S., 1995b, Construction of murine coronavirus mutants containing interspecies chimeric nucleocapsid proteins, *J. Virol.* **69**:5475–5484.

Weiss, S.R., Zoltick, P.W., and Leibowitz, J.L., 1993, The ns 4 gene of mouse hepatitis virus (MHV), strain A59 contains two ORFs and thus differs from ns 4 of the JHM and S strains, *Arch. Virol.* **129**:301–309.

Yokomori, K., and Lai, M.M.C., 1991, Mouse hepatitis virus S RNA sequence reveals that nonstructural proteins ns4 and ns5a are not essential for murine coronavirus replication, *J. Virol.* **65**:5605–5608.

Zhang, X., and Lai, M.M.C., 1994, Unusual heterogeneity of leader-mRNA fusion in a murine coronavirus: implications for the mechanism of RNA transcription and recombination, *J. Virol.* **68**:6626–6633.

39

AN ESSENTIAL SECONDARY STRUCTURE IN THE 3' UNTRANSLATED REGION OF THE MOUSE HEPATITIS VIRUS GENOME

Bilan Hsue and Paul S. Masters

David Axelrod Institute
Wadsworth Center for Laboratories and Research
New York State Department of Health
Albany, New York 12201-2002

ABSTRACT

The 3' untranslated regions (3' UTRs) of coronaviruses contain the signals necessary for negative strand RNA synthesis and may also harbor elements essential for positive strand replication and subgenomic RNA transcription. The 3' UTRs of mouse hepatitis virus (MHV) and bovine coronavirus (BCV) are more than 30% divergent. In an effort to learn what parts of these regions might be functionally interchangeable, we attempted to replace the 3' UTR of MHV with its BCV counterpart by targeted RNA recombination. Initially, we tried to substitute the 3' 267 nucleotides (nt) of the 301 nt MHV 3' UTR with the corresponding region of the BCV 3' UTR. This exchange did not yield viable recombinant viruses, and the donor DI RNA was shown to be unable to replicate with MHV as a helper virus. Subsequent analysis revealed that the entire BCV 3' UTR could be inserted into the MHV genome in place of the entire MHV 3' UTR. It resulted that the failure of the initial attempted substitution was due to the inadvertent disruption of an essential conserved bulged stem-loop secondary structure in the MHV and BCV 3' UTRs immediately downstream of the N gene stop codon.

INTRODUCTION

At the outset of this study, we sought to examine the sequence or structural requirements of the MHV 3' UTR by attempting to replace it in the viral genome with the BCV 3' UTR. We took this phylogenetic approach in order to complement the DI RNA deletion analyses that have been performed by other groups. These prior studies have established

Coronaviruses and Arteriviruses, edited by Enjuanes *et al.*
Plenum Press, New York, 1998

that some 378–463 nt at the 3' end of the MHV genome is required for DI RNA replication (Kim *et al.*, 1993; Lin and Lai, 1993; Van der Most *et al.*, 1995) but that the minimal *cis*-acting signal essential for minus-strand RNA synthesis is contained within only the last 55 nt at the 3' terminus of the genome (Lin *et al.*, 1994).

We have previously shown that it is possible to incorporate mutations into the nucleocapsid gene and the 3' UTR of MHV by targeted recombination (Koetzner *et al.*, 1992; Masters *et al.*, 1994; Peng *et al.*, 1995a; Peng *et al.*, 1995b; Fischer *et al.*, 1997). This technique takes advantage of recombination between a synthetic DI RNA and a thermolabile N gene deletion mutant (Alb4) that can be selected against when we seek to identify recombinants that have repaired the deletion. We sought to use the same method to exchange portions of the MHV and BCV 3' UTRs. Although the 3' UTRs of different MHV strains are more than 99% conserved (Parker and Masters, 1990), we set out to examine the possibility of replacing the MHV 3' UTR with the corresponding region of the BCV genome, which has only 69% sequence homology to its MHV counterpart. In the course of accomplishing this aim, we discovered that the 5' extreme end of the MHV 3' UTR contains a functionally essential RNA secondary structure.

MATERIALS AND METHODS

The methods used in these studies have been described in detail elsewhere (Hsue and Masters, 1997). In brief, transcription vectors encoding DI RNAs were constructed by standard techniques, and mutations or substitutions were made in these by a number of current PCR-based methods or by direct exchange of restriction fragments containing portions of the BCV 3' UTR. The parent vector for all constructs was the replicating DI vector pB36 (Masters *et al.*, 1994), which contains the 5' 467 nt of the MHV genome connected, via a short linker, to the entire nucleocapsid (N) gene and 3' UTR of MHV. Recombination experiments were performed by infection followed by transfection of synthetic RNA. Transfections were carried out by electroporation of mouse L2 cells, which were then plated onto monolayers of mouse 17 clone 1 cells. Progeny virus were titered on mouse L2 cells at 39°C, and candidate recombinants able to form large plaques were purified and analyzed further. Initial characterization of recombinants was by RT-PCR analysis or by sequencing of RNA from infected cells; final verification of recombinants was accomplished through sequencing of genomic RNA from purified virions. Replicating DI RNA was metabolically labeled with either [5,6–^{3}H]uridine or ^{32}P$_i$ essentially as described previously (Masters *et al.*, 1994).

RESULTS AND DISCUSSION

For our initial experiments, we constructed a transcription vector for DI donor RNA containing almost the entire BCV 3' UTR in place of the MHV 3' UTR. In this vector, designated pBL34, the 3' 267 nt of the 301 nt MHV 3' UTR were replaced with the corresponding region of the BCV 3' UTR. Because this construct was made on the basis of available restriction sites, the MHV/BCV junction formed was not precisely homologous with respect to the alignment of the two sequences, resulting in an 8-nt insertion. Vector pBL34 also contained a phenotypically silent tag in the spacer B region of the N gene. When targeted recombination was carried out with pBL34 donor RNA, many candidate recombinants were recovered. However, although all of these were found to have the tag sequence repairing the N gene deletion in the recipient virus, none contained BCV 3' UTR sequences. Therefore, they must have been generated by double crossover events. We tentatively concluded from these experiments that the substitution of the BCV 3' UTR is lethal to MHV.

This result raised the question of whether the BCV-MHV chimeric pBL34 RNA was a functional DI RNA. Metabolic labeling of RNA in MHV-infected DI RNA-transfected cells revealed that, unlike its purely MHV parent, pBL34-derived RNA was not able to replicate with MHV as a helper virus. We next tested whether it was possible to evolve pBL34 RNA into a replicating DI RNA species. To accomplish this, we used the DNA shuffling technique developed by Stemmer (1994a, 1994b). Transfection of MHV-infected cells with a pool of pBL34-derived RNAs mutagenized by this method resulted in the detection of replicating RNA species. To analyze these, we amplified total cytoplasmic RNA from these cells by RT-PCR with primers specific for the DI RNA, and we cloned the obtained products. Despite a high frequency of clones resulting from mispriming, we were able to identify two independent clones that contained regions of the BCV 3' UTR. These both differed from their nonreplicating progenitor, pBL34, in three respects. First, they crossed over from BCV 3' UTR sequence to MHV 3' UTR sequence upstream of nt 212 (numbering from the 3' end of the genome). Second, they had MHV nucleotides rather than BCV nucleotides at positions 22 and 25 at the extreme 3' end of the 3' UTR (these were generated by one of the primers used in the DNA shuffling strategy). Third, they contained multiple, apparently random point mutations.

This finding led us to analyze which of the three classes of changes were responsible for allowing these DI RNAs with BCV-MHV chimeric 3' UTRs to replicate. In summary, numerous experiments allowed us to conclude that the latter two features were not crucial determinants allowing or preventing replication of DI RNAs or viruses containing chimeric 3' UTRs. Consequently, we next focussed on the 5' end of the 3' UTR, since the composition of the evolved replicating DI RNAs implied that the 5' 60–88 nt in a chimeric 3' UTR must come from MHV in order for it to be functional.

Examination of this region revealed that it could be folded into a potential bulged stem-loop structure extending from nt 234 to 301 of the MHV 3' UTR (Fig. 1). The same secondary structure is predicted for the homologous segment of the BCV 3' UTR, and this is the only predicted folding in the entire 3' UTR that is conserved between BCV and MHV. Eight of the ten nucleotides that differ between MHV and BCV in this region form four covariant base pairs in the stem portions of the structure. Moreover, two other coronaviruses in the same antigenic group, HCV-OC43 and BECV, have the same eight covariant stem nucleotides as BCV, as well as additional covariant base-pairs (Fig. 1). This provided convincing phylogenetic evidence that the computer-generated structure was of actual functional significance.

The original chimeric BCV-MHV 3' UTR that we had constructed in pBL34 would have disrupted the stem structure by attempting to pair the MHV left stem arm with the BCV right stem arm. It also would have created an 8-nt insertion in the loop. One or both of these changes might have been lethal to the function of the 3' UTR. To test this hypothesis, we constructed a series of DI RNA templates, derived from the original nonreplicating pBL34 template, in which all combinations of MHV or BCV left and right arms of the stem were paired, including the two possible cases incorporating the 8-nt insert within a homogeneous MHV or BCV stem-loop (Fig. 2). The replicative ability of these various DI RNA constructs was then tested. We observed that all RNAs containing left and right stem arms from the same source, whether BCV or MHV, were able to replicate in the presence of MHV helper virus. By contrast, RNAs with mixed pairs of stem arms failed to rep-

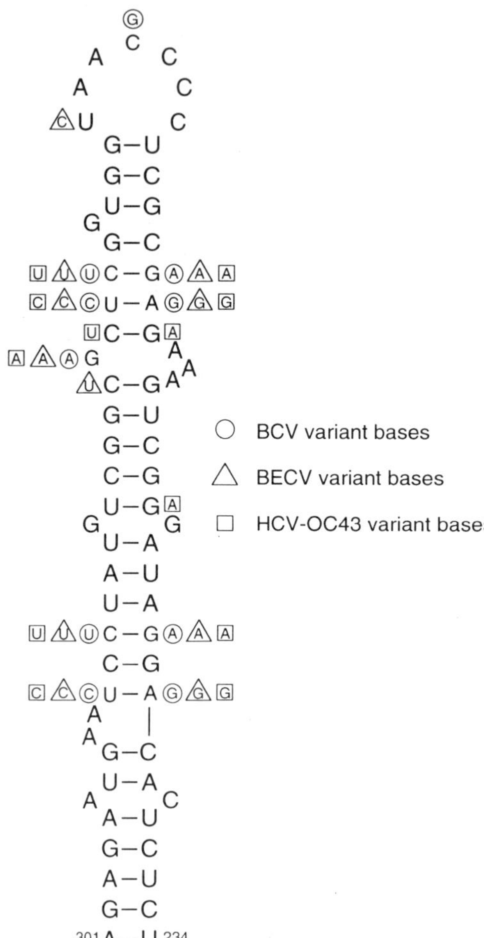

Figure 1. A phylogenetically conserved RNA secondary structure in the MHV 3' UTR, between nucleotides 234 and 301 (numbered from the 3' end of the genome). The immediately upstream N gene stop codon is boxed. Nonconserved nucleotides in the BCV, bovine enteric coronavirus (BECV), and human coronavirus OC43 (HCV-OC43) genomes are indicated by circles, triangles, and squares, respectively.

licate (Fig. 2). This suggested that the bulged stem-loop structure is functionally essential for DI RNA replication. In addition, the data showed that the presence of the 8-nt insertion in the loops of those DI RNAs harboring a homogeneous stem-loop did not abolish their replicative ability.

We next attempted to recombine each of the synthetic DI RNAs into the MHV genome. It was found that only those DI RNAs that contained homogeneous stem arms of either MHV or BCV origin could recombine their 3' UTRs into the MHV genome, irrespective of the presence of the 8-nt loop insert. DI RNAs containing mixed stem-loops gave double-crossover recombinants that did not contain any portion of the BCV 3' UTR, similar to those obtained with the original pBL34 RNA. Thus, there was complete congruence between the ability of a given 3' UTR to support DI replication and its ability to function within the intact viral genome.

In conclusion, we have found that the entire 3' UTR of MHV can be replaced by the 3' UTR of BCV, both in DI RNAs and in the actual virus. Additionally, the 3' UTR of MHV contains a 68-nt bulged stem-loop immediately downstream of the N gene stop codon, and this secondary structure is essential for replication, both in DI RNAs and in the virus.

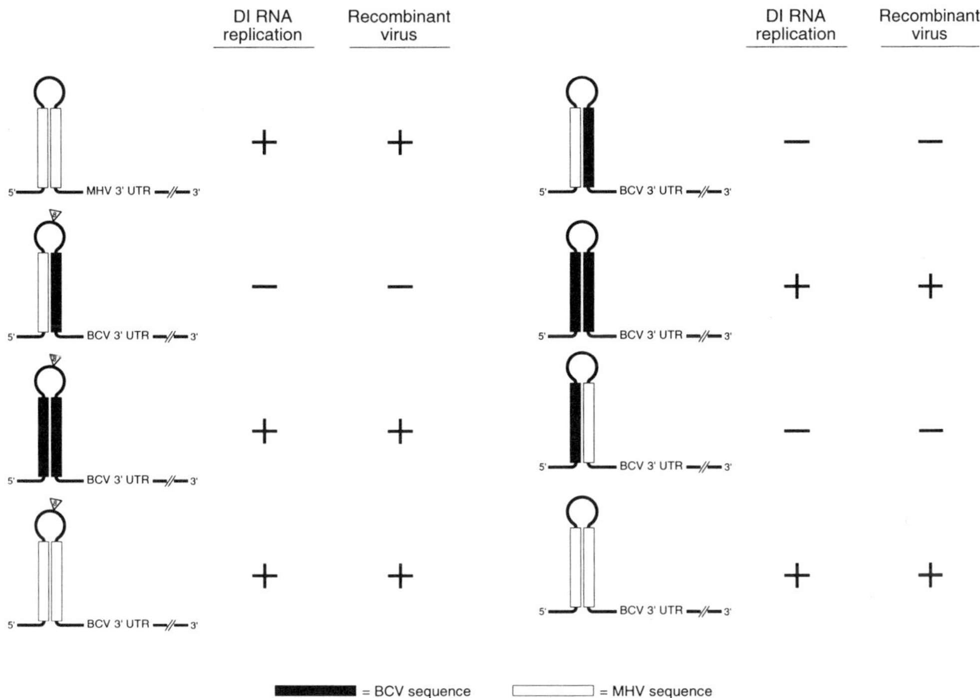

Figure 2. DI RNAs containing different combinations of left and right arms of the putative bulged stem-loop structure, with or without the 8-nt insert (triangle) in the loop. Shaded and open rectangles represent BCV sequence and MHV sequence, respectively. Except for the first construct, the remainder of each 3' UTR corresponds to that of BCV. The ability of each DI RNA to replicate or to donate its 3' UTR to a viable recombinant virus is indicated.

ACKNOWLEDGMENTS

We are grateful to Savithra Senanayake and David Brian (University of Tennessee) for generously providing the BCV clone pLN. This work was supported in part by Public Health Service grants AI 31622 and AI 39544 from the National Institutes of Health.

REFERENCES

Fischer, F., Peng, D., Hingley, S.T., Weiss, S.R., and Masters, P.S., 1997, The internal open reading frame within the nucleocapsid gene of mouse hepatitis virus encodes a structural protein that is not essential for viral replication, *J. Virol.* **71**:996–1003.

Hsue, B., and Masters, P.S., 1997, A bulged stem-loop structure in the 3' untranslated region of the genome of the coronavirus mouse hepatitis virus is essential for replication, Submitted to *J. Virol.*

Kim, Y.-N., Jeong, Y.S., and Makino, S., 1993, Analysis of *cis*-acting sequences essential for coronavirus defective interfering RNA replication, *Virology* **197**:53–63.

Koetzner, C.A., Parker, M.M., Ricard, C.S., Sturman, L.S., and Masters, P.S., 1992, Repair and mutagenesis of the genome of a deletion mutant of the coronavirus mouse hepatitis virus by targeted RNA recombination, *J. Virol.* **66**:1841–1848.

Lin, Y.-J., and Lai, M.M.C., 1993, Deletion mapping of a mouse hepatitis virus defective interfering RNA reveals the requirement of an internal and discontiguous sequence for replication, *J. Virol.* **67**:6110–6118.

Lin, Y.-J., Liao, C.-L., and Lai, M.M.C., 1994, Identification of the *cis*-acting signal for minus-strand RNA synthesis of a murine coronavirus: implications for the role of minus-strand RNA in RNA replication and transcription, *J. Virol.* **68**:8131–8140.

Masters, P.S., Koetzner, C.A., Kerr, C.A., and Heo, Y., 1994, Optimization of targeted RNA recombination and mapping of a novel nucleocapsid gene mutation in the coronavirus mouse hepatitis virus, *J. Virol.* **68**:328–337.

Parker, M.M., and Masters, P.S., 1990, Sequence comparison of the N genes of five strains of the coronavirus mouse hepatitis virus suggests a three domain structure for the nucleocapsid protein, *Virology* **179**:463–468.

Peng, D., Koetzner, C.A., and Masters, P.S., 1995a, Analysis of second-site revertants of a murine coronavirus nucleocapsid protein deletion mutant and construction of nucleocapsid protein mutants by targeted RNA recombination, *J. Virol.* **69**:3449–3457.

Peng, D., Koetzner, C.A., McMahon, T., Zhu, Y., and Masters, P.S., 1995b, Construction of murine coronavirus mutants containing interspecies chimeric nucleocapsid proteins, *J. Virol.* **69**:5475–5484.

Stemmer, W.P.C. 1994a, Rapid evolution of a protein *in vitro* by DNA shuffling, *Nature* **370**:389–391.

Stemmer, W.P.C. 1994b, DNA shuffling by random fragmentation and reassembly: *In vitro* recombination for molecular evolution, *Proc. Natl. Acad. Sci. USA* **91**:10747–10751.

Van der Most, R.G., Luytjes, W., Rutjes, S., and Spaan, W.J.M., 1995, Translation but not the encoded sequence is essential for the efficient propogation of defective interfering RNAs of the coronavirus mouse hepatitis virus, *J. Virol.* **69**:3744–3751.

40

REGULATION OF mRNA 1 EXPRESSION BY THE 5'-UNTRANSLATED REGION (5'-UTR) OF THE CORONAVIRUS INFECTIOUS BRONCHITIS VIRUS (IBV)

D. X. Liu, H. Y. Xu, and K. P. Lim

Institute of Molecular Agrobiology
National University of Singapore
59A The Fleming
1 Science Park Drive
Singapore 118240

1. ABSTRACT

In this report, we show that expression of the coronavirus IBV mRNA1 is regulated by its 5'-UTR. Evidence presented domonstrates that the IBV sequence from nucleotide 1 to 1904 directs very inefficient synthesis of a product of approximately 43 kDa. Deletion of either the first 362 bp or the whole part of the 5'-UTR, however, dramatically increased the expression of the 43 kDa protein species. The mechanisms involved were investigated by two different approaches. Firstly, translation of the same construct in the presence of $[^3H]$-leucine ruled out the possibility that initiation of small reading frames from non-AUG codons located in the 5'-UTR may compete with the authentic AUG initiation codon, and therefore inhibit the expression of ORF 1a. Secondly, expression and deletion analyses of a dicistronic construct showed that translation of the 43 kDa protein was initiated by ribosome internal entry mechanism. These studies suggest that a 'weak' ribosome internal entry signal is located in the 5'-UTR and is involved in the regulation of mRNA1 expression.

2. INTRODUCTION

Six mRNA species, including the genome-length mRNA (mRNA1) of 27.6 kb and five subgenomic mRNA species (mRNAs 2–6) with sizes ranging from 2 to 7 kb, are produced in cells infected with the prototype virus of the Coronaviridae, avian infectious

Coronaviruses and Arteriviruses, edited by Enjuanes *et al.*
Plenum Press, New York, 1998

bronchitis virus (IBV). Nucleotide sequencing of the genomic RNA has shown that the 5'-terminal unique region of mRNA 1 contains two large ORFs (1a and 1b), with potential to encode a fusion polyprotein of 741 kDa by a ribosomal frameshift (Boursnell *et al.*, 1987; Brierley *et al.*, 1987, 1989). This 1a/1b fusion polyprotein is expected to be cleaved by viral or cellular proteinases to produce functional products associated with viral RNA replication (Gorbalenya *et al.*, 1989).

Examination of the nucleotide sequence of mRNA 1 shows that a 527bp untranslated region is located at its 5' end. During the studies of the expression of mRNA1, it was consistently observed that this messenger RNA was very poorly expressed both in *in vitro* and *in vivo* expression systems unless the majority of the 5'-UTR was deleted from the constructs. In this report, we present evidence demonstrating that the 5'-UTR may contain a 'weak' ribosome internal entry signal, which is likely involved in the regulation of mRNA 1 expression.

3. MATERIALS AND METHODS

3.1. SDS-Polyacrylamide Gel Electrophoresis

SDS-polyacrylamide gel electrophoresis (SDS-PAGE) of virus polypeptides was carried out with a range of polyacrylamide concentrations from 12.5 to 17.5 % (Laemmli, 1970). Labelled polypeptides were detected by autoradiography or fluorography of dried gels.

3.2. Cell-Free Transcription and Translation

Plasmid DNA was transcribed and translated *in vitro* by using the TnT T7-coupled reticulocyte lysate system (TnT system) according to the instructions of the manufacturer (Promega). Reaction products were separated by SDS-PAGE and detected by autoradiography.

3.3. Polymerase Chain Reaction (PCR)

Appropriate primers and template DNAs were used in amplification reactions with Pfu DNA polymerase (Stratagene) under standard buffer conditions with 2 mM $MgCl_2$. The PCR conditions were 30 cycles of 92°C for 30 sec, 56°C for 30 sec and 72°C for 6 min.

3.4. Construction of Plasmids

Plasmid pIBV1a5, which covers the IBV sequence from nucleotide 1 to 1904, was constructed by cloning an ApaI- and EcoRI-digested PCR fragment into ApaI/EcoRI digested pPCRII vector (Invitrogen). The ApaI site is introduced by the upstream PCR primer and EcoRI cuts the IBV sequence at nucleotide 1904. The sequence of the upstream primer is 5'-ACTAGGGCCCACTTAAGATAGATATTAA-3', and the sequence of the downstream primer is 5'-TTCCATATGCAAGCTTCCAGA-3', which is complementary to the IBV sequence from nucleotide 1909 to 1929.

Plasmid pBP5Δ5, which covers the IBV sequence from nucleotide 11306 to 11877, was described previously (Liu *et al.*, 1997).

For construction of pBP5–1a5, a 1904 bp fragment covering the IBV sequence from nucleotide 1 to 1904 was generated by digestion of pIBV1a5 with ApaI, end-repair with Klenow and re-digestion with EcoRI. This fragment was then cloned into pBP5Δ5, which was firstly digested with BamHI (the BamHI site is located immediately downstream of the viral sequence), end-repaired with Klenow and re-digested with EcoRI, giving pBP5–1a5.

Plasmid pBP5–1a5Δ1 was constructed by cloning a BamHI- and EcoRI-digested PCR fragment into BamHI/EcoRI-digested pBP5Δ5. This PCR fragment covers the IBV sequence from nucleotide 140 to 1904 and was generated by using pIBV1a5 as the template and oligonucleotide 5'-CTGGATGGATCCTGGCCACCTG-3' as the upstream primer. The same downstream primer used to clone pIBV1a5 was used to generate this PCR fragment as well as PCR fragments for constructing pBP5–1a5Δ2 and pBP5–1a5Δ3. Plasmid pBP5–1a5Δ2 was constructed by cloning a BamHI- and EcoRI-digested PCR fragment, covering the IBV sequence from nucleotide 528 to 1904, into BamHI/EcoRI-digested pBP5Δ5. The sequence of the upstream primer is 5'-ACGCGGATCCAC-CATGGCTTCAAGC-3'. Plasmid pBP5–1a5Δ3 was constructed by cloning a BamHI- and EcoRI-digested PCR fragment, covering the IBV sequence from nucleotide 288 to 1904, into BamHI/EcoRI-digested pBP5Δ5. The sequence of the upstream primer is 5'-GTGCGAGGATCCTCTGGTTCA-3'. Plasmid pBP5–1a5Δ4 was constructed by digestion of pBP5–1a5Δ1 with BamHI and BstEII (BstEII cuts the IBV sequence at nucleotide 364), end-repair with Klenow and re-ligation with T4 DNA ligase.

Plasmid pIBV1a2, which was formerly called pKT1a2 and covers the IBV sequence from nucleotide 362 to 5753, was described previously (Liu *et al.*, 1995). Plasmid pIBV1a2Δ2, which covers the IBV sequence from nucleotide 528 to 5753, was constructed as follows. A BamHI/MluI digested PCR fragment covering the IBV sequence from nucleotide 528 to 3997 was cloned into BglII/MluI digested pIBV1a2, giving the deletion construct pIBV1a2Δ2. MluI cuts the IBV sequence at nucleotide 3997. The upstream PCR primer is the same one used to construct pBP5–1a5Δ2; the sequence of the downstream primer is 5'-TCCAAATTGACCCAATGAGTGTC-3', which is complementary to the IBV sequence from nucleotide 4117 to 4140.

4. RESULTS

4.1. Very Inefficient Expression of Constructs Containing the 5'-Most 2 kb Part of mRNA1

During the course of studying the expression of IBV mRNA1, it was consistently observed that constructs containing the 527 bp 5'-UTR of mRNA1 exhibited very low expression efficiency. Figure 1 shows an example. As can be seen, expression of EcoRI-digested plasmid pIBV1a5, which covers the IBV sequence from nucleotide 1 to 1904, resulted in very inefficient expression of a polypeptide of approximately 43 kDa. Only a trace amount of the 43 kDa protein was observed after prolonged exposure of the gel (Figure 1). Deletion of the 5'-UTR, however, dramatically increased the expression of the 43 kDa protein species. As shown in Figure 1, expression of plasmids pIBV1a2 and pIBV1a2Δ2, which contain deletions of the IBV sequences from nucleotide 1 to 346 and from 1 to 526, respectively, led to much more efficient expression of the 43 kDa protein. It was of interest to investigate the mechanisms.

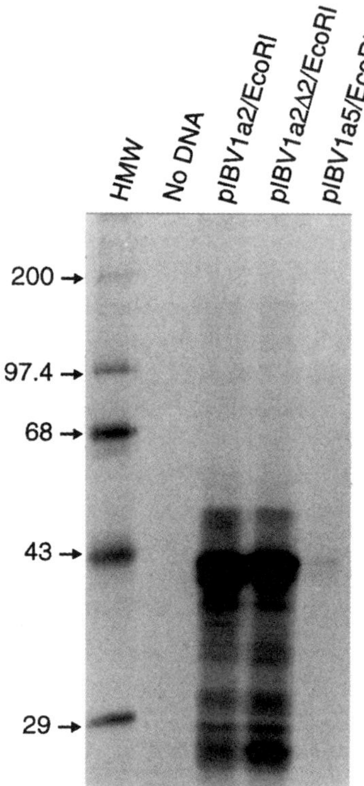

Figure 1. Analysis of *in vitro* translation products of RNAs obtained by *in vitro* transcription of EcoRI-digested pIBV1a2, pIBV1a2Δ2 and pIBV1a5 with T7 RNA polymerase. RNA was added to reticulocyte lysate as indicated above each lane. [^{35}S] methionine-labelled translation products were separated on an SDS-12.5% polyacrylamide gel and detected by fluorography. HMW-high molecular weight markers (numbers indicate molecular mass in kilodaltons).

4.2. No Initiation of Small Reading Frames from Non-AUG Initiation Codons Located in the 5'-UTR

Examination of the 5'-UTR sequence shows that several small reading frames (over 50 amino acids) are located in the IBV sequence from nucleotide 1 to 450. A prominent example is a 261 bp reading frame from nucleotide 138 to 398. Translation of this reading frame would result in the synthesis of a polypeptide with 86 amino acid residues, if a CUG triplet (from nucleotide 138 to 140) could be used as a translation initiation codon. This possibility was investigated by *in vitro* translation of pIBV1a5 in the presence of [^{3}H]-Leucine. For this purpose, pIBV1a5 was linearised by digestion with AlwNI and EcoRI, which cut the IBV sequences at nucleotide positions 486 and 1904, respectively, and was translated in reticulocyte lysate in the presence of either [^{35}S]-methionine or [^{3}H]-leucine. As shown in Figure 2, expression of EcoRI-linearised pIBV1a5 led to the detection of the 43 kDa protein species. No specific product was detected from expression of AlwNI-digested pIBV1a5. This result rules out the possibility that the inefficient expression of mRNA1 is caused by expression of small reading frames located in the 5'-UTR.

4.3. *In Vitro* Translation of Capped and Uncapped Transcripts Containing the 5'-UTR

In vitro synthesised capped and uncapped RNAs were then translated *in vitro* in reticulocyte lysate to compare the translatability of the two transcripts. For this purpose, pIBV1a5 was linearised with EcoRI and transcribed *in vitro* with T7 bacteriophage RNA

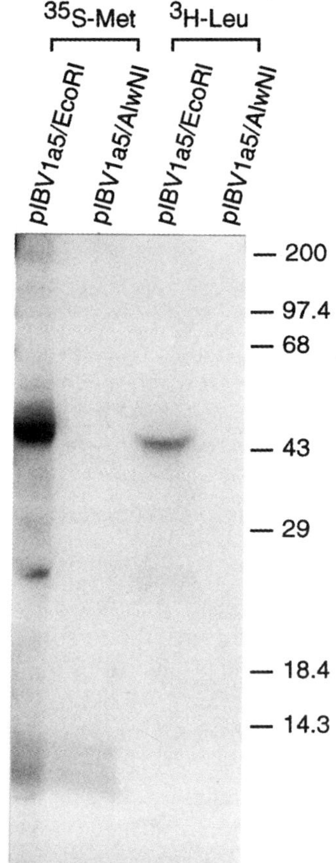

Figure 2. Analysis of *in vitro* translation products of RNAs obtained by *in vitro* transcription of EcoRI- and AlwNI-digested pIBV1a5 with T7 RNA polymerase. RNA was added to reticulocyte lysate and translated in the presence of [^{35}S]-methionine and [^{3}H]-leucine as indicated above each lane. The radio-labelled translation products were separated on an SDS-17.5% polyacrylamide gel and detected by fluorography. Numbers indicate molecular mass in kilodaltons.

polymerase. Equal amount of *in vitro* synthesised RNAs were used to translate *in vitro* in reticulocyte lysate. As shown in Figure 3, both transcripts direct inefficient expression of the 43 kDa protein. Furthermore, the amounts of the 43 kDa protein species produced from both transcripts were very similar. This result indicates that the cap structure of mRNA1 may not be essential for its expression and suggests that translation of mRNA1 may be initiated by cap-independent mechanism.

4.4. Cap-Independent Translation of mRNA1

To study further if expression of mRNA1 is cap-independent, the IBV sequence from nucleotide 1 to 1,904 was cloned downstream of a reporter gene covering the IBV sequence from nucleotide 11306 to 11877, giving rise to a dicistronic construct pBP5–1a5 (Figure 4a). Translation of plasmids pIBV1a5 and pBP5Δ5 in reticulocyte lysate led to the detection of two proteins with apparent molecular weights of 43 kDa and 25 kDa, respectively, representing the full-length products encoded by the two constructs (Figure 4b). Expression of EcoRI-linearised pBP5–1a5 also resulted in the detection of two protein species of approximately 25 kDa and 43 kDa. The 25 kDa protein co-migrates with the 25 kDa protein translated from pBP5Δ5 and the 43 kDa product co-migrates with the 43 kDa protein expressed from pIBV1a5 (Figure 4b), suggesting that they may represent the prod-

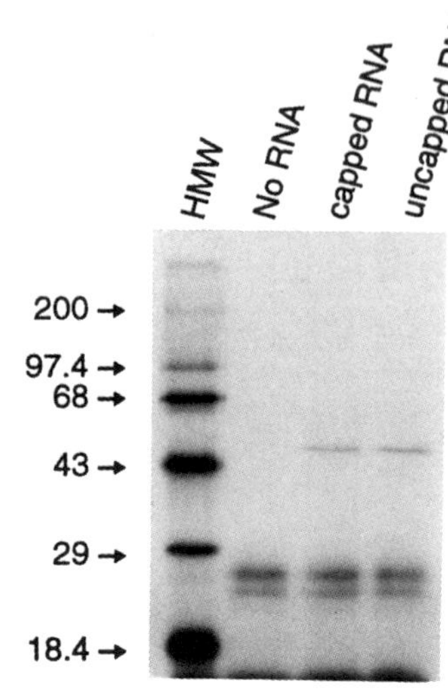

Figure 3. Analysis of *in vitro* translation products of capped and uncapped RNAs obtained by *in vitro* transcription of EcoRI-digested pIBV1a5 with T7 RNA polymerase. The capped RNA was prepared by incorporating the dinucleotide ^{7m}GpppG to provide a 5' cap structure (Contreras *et al.*, 1982). RNA was added to reticulocyte lysate and translated in the presence of [^{35}S]-methionine. The radio-labelled translation products were separated on an SDS-12.5% polyacrylamide gel and detected by fluorography. HMW-high molecular weight markers (numbers indicate molecular mass in kilodaltons).

ucts translated from the upstream and downstream ORFs, respectively. These results indicate that both the upstream and downstream ORFs were expressed from the dicistronic construct. Deletion of the 5'-UTR from pBP5–1a5, however, abolished the expression of downstream ORF expression. As can be seen from Figure 4b, only the 43 kDa protein was observed from transcription and translation of pBP5–1a5Δ2. Deletion of the 5'-most 140 bp of the 5'-UTR, however, did not affect the expression of the downstream ORF. Only slight reduction of the 43 kDa protein expression was observed from transcription and translation of pBP5–1a5Δ1 (Figure 4b). These results confirm that expression the 43 kDa protein from the dicistronic construct is dependent on the presence of the 5'-UTR.

Two more deletion constructs were made to define the sequence required for the internal translation of the 43 kDa protein. Plasmids pBP5–1a5Δ3 contains a deletion of the IBV sequence from nucleotide 1 to 288, and pBP5–1a5Δ4 contains a deletion from nucleotide 1 to 364 (Figure 4a). Expression of both constructs showed that very similar amounts of the 43 kDa protein was expressed (Figure 4b).

4.5. Expression of the Full-Length Product Encoded by the 5'-Most 5753 bp Part of ORF1a Is Increased by the Presence of the Second Half of the 5'-UTR

Data presented above showed that a weak internal initiation signal is located in the 5'-UTR. In an attempt to increase the expression efficiency of the ORF 1a product, plasmid pIBV1a2Δ2 was constructed. This construct covers the IBV sequence from nucleotide 528 to 5753 and therefore excludes the whole 5'-UTR. Expression of the construct in reticulocyte lysate, however, resulted in the synthesis of much less full-length product than that from pIBV1a2. As can be seen from Figure 5, when equal amount of MluI- and BamHI-digested pIBV1a2 and pIBV1a2Δ2 were expressed *in vitro*, both plasmids can direct synthesis of the full-length products of approximately 180kDa and 250 kDa, respectively. However, significantly more efficient expression of the 180 kDa and 250 kDa products was observed from translation of pIBV1a2 than that from pIBV1a2Δ2 (Figure 5). Meanwhile, when EcoRI-digested pIBV1a2 and pIBV1a2Δ2 were expressed, no obvious difference in the synthesis of the 43 kDa products was detected.

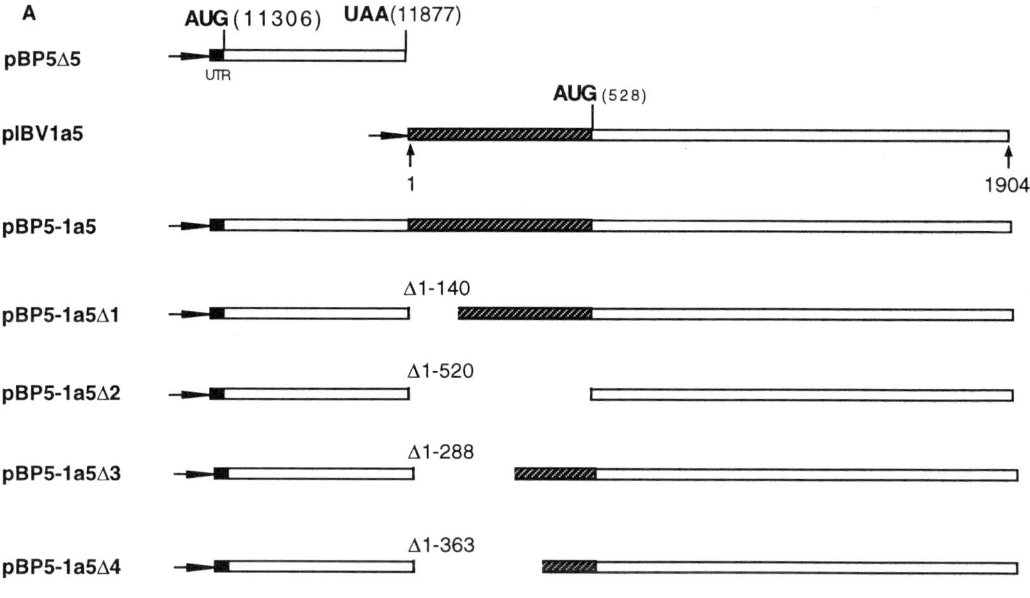

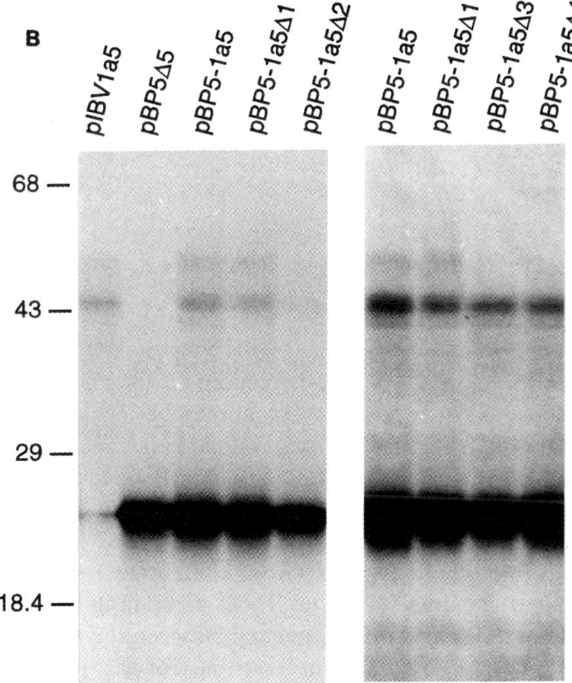

Figure 4. (A). Diagram showing the structures of plasmids pBP5Δ5, pIBV1a5, pBP5–1a5, pBP5–1a5Δ1, pBP5–1a5Δ2, pBP5–1a5Δ3 and pBP5–1a5Δ4. (B). Analysis of *in vitro* translation products of RNAs obtained by *in vitro* transcription of EcoRI-digested pIBV1a5, pBP5Δ5, pBP5–1a5, pBP5–1a5Δ1, pBP5–1a5Δ2, pBP5–1a5Δ3 and pBP5–1a5Δ4 with T7 RNA polymerase. RNA was added to reticulocyte lysate and translated in the presence of [^{35}S]-methionine. The radio-labelled translation products were separated on an SDS-12.5% polyacrylamide gel and detected by fluorography. Numbers indicate molecular mass in kilodaltons.

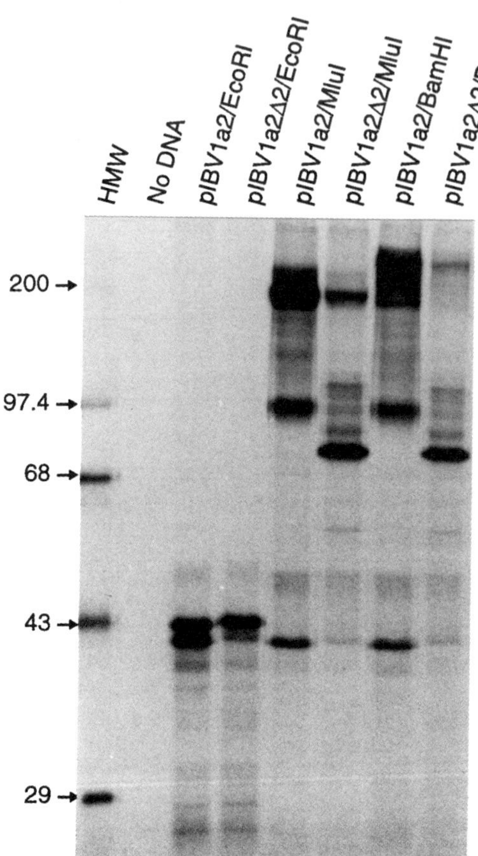

Figure 5. Analysis of *in vitro* translation products of RNAs obtained by *in vitro* transcription of EcoRI-, MluI- and BamHI-digested pIBV1a2 and pIBV1a2Δ2 with T7 RNA polymerase. RNA was added to reticulocyte lysate and translated in the presence of [^{35}S]-methionine. The radio-labelled translation products were separated on an SDS-10% polyacrylamide gel and detected by fluorography. HMW-high molecular weight markers (numbers indicate molecular mass in kilodaltons).

5. DISCUSSION

The genome-length mRNA, mRNA 1, contains a long 5'-UTR. In this report, we show that a 'weak' ribosome internal entry signal may be located in this region and is involved in the regulation of IBV mRNA 1 expression. This conclusion is based on the following observations. Firstly, *in vitro* translation of the IBV sequence from nucleotide 1 to 1904 resulted in very inefficient synthesis of a 43 kDa protein. Expression of this protein, however, was dramatically increased by deletion of either 362 bp from the 5'-UTR (from nucleotide 1 to 362) or the whole 5'-UTR (from nucleotide 1 to 527). Secondly, *in vitro* translation of capped and uncapped artificial RNAs containing the whole 5'-UTR showed no obvious difference in the translation efficiency between the two RNAs, indicating that the cap structure of mRNA 1 may be not essential for the initiation of translation of ORF1a and further suggesting that ribosomes may bypass the cap structure and bind to some internal sites of the 5'-UTR. Furthermore, construction and expression of a dicistronic plasmid showed that the 5'-UTR can direct the expression of the downstream ORFs from the dicistronic constructs. Deletion studies confirmed that the expression of the downstream ORF was dependent on the presence of the 5'-UTR.

Like most other eukaryotic mRNAs, IBV mRNA1 is naturally capped. In this report, we show that this cap structure may be not essential for translation initiation of the mRNA1 expression. Instead of using the conventional cap-dependent translation initiation mechanism, expression of this mRNA may be initiated by ribosome internally binding to the 5'-UTR. Cap-independent translation initiation of other naturally capped mRNAs of coronaviruses was also reported. These include translation of the third ORF of IBV mRNA3 and the second ORF of mouse hepatitis virus mRNA5 (Liu and Inglis, 1992; Thiel and Siddell, 1994). It is currently unclear why coronaviruses utilise both cap-dependent and cap-independent mechanisms to initiate the translation of their individual messenger RNAs, but this may reflect a subtle regulatory mean employed by virus to control the individual gene expression during the virus life cycle.

It is intriguing to observe that when the IBV sequence up to nucleotide 5753 was expressed *in vitro*, much more full-length product was synthesised from the construct containing the 3' end 160 bp part of the 5'-UTR (pIBV1a2) than that from the construct without the 5'-UTR (pIBV1a2Δ2). However, very similar amount of the full-length product was synthesised from both constructs when transcripts covering the IBV sequence only up to nucleotide 1904 was expressed. These results indicate that there is no obvious differences in the stage of translation initiation between the two constructs. It is therefore likely that the presence of the 160 bp leader sequence in the run-off transcripts transcribed from pIBV1a2 may stabilize the *in vitro* synthesised RNA, and consequently increase the synthesis of the full-length product. Further investigations are underway to address this possibility.

REFERENCES

Boursnell, M. E. G., T. D. K. Brown, I. J. Foulds, P. F. Green, F. M. Tomley, and M. M. Binns., 1987, Completion of the sequence of the genome of the coronavirus avian infectious bronchitis virus, *J. Gen. Virol.* **68**:57–77.

Brierley, I., M. E. G. Boursnell, M. M. Binns, B. Bilimoria, V. C. Blok, T. D. K. Brown, and S. C. Inglis., 1987, An efficient ribosomal frame-shifting signal in the polymerase-encoding region of the coronavirus IBV, *EMBO. J.* **6**:3779–3785.

Brierley, I., P. Digard, and S. C. Inglis., 1989, Characterization of an efficient coronavirus ribosomal frameshifting signal: requirement for an RNA pseudoknot, *Cell* **57**:537–547.

Contreras, R., H. Cheroutre, W. Degrave, and W. Fiers., 1982, Simple efficient *in vitro* synthesis of capped RNA useful for direct expression of cloned DNA, *Nucleic Acids Res.* **10**: 6353–6362.

Gorbalenya, A. E., E. Y. Koonin, A. P. Donchenko, and V. M. Blinov., 1989, Coronavirus genome: prediction of putative functional domains in the non-structural polyprotein by comparative amino acid sequence analysis, *Nucleic Acids Research* **17**:4847–4860.

Laemmli, U. K., 1970, Cleavage of structural proteins during the assembly of the head of bacteriophage T4, *Nature (London)* **227**:680–685.

Liu, D. X., and S. C. Inglis., 1992, Internal entry of ribosomes on a tricistronic mRNA encoded by infectious bronchitis virus, *J. Virol.* **66**: 6142–6154.

Liu, D. X., K. W. Tibbles, D. Cavanagh, T. D. K. Brown, and I. Brierley., 1995, Identification, expression and processing of an 87 kDa polypeptide encoded by ORF1a of the coronavirus infectious bronchitis virus, *Virology* **208**: 48–57.

Liu, D. X., and T. D. K. Brown. 1997. Proteolytic processing of the coronavirus infectious bronchitis virus 1a polyprotein: identification of a 10-kilodalton polypeptide and determination of its cleavage sites, *J. Virol.* **71**: 1814–1820.

Thiel, V., and S. G. Siddell. 1994. Internal ribosome entry in the coding region of murine hepatitis virus mRNA 5, *J. Gen. Virol.* **75**: 3041–3046.

41

MOUSE HEPATITIS VIRUS NUCLEOCAPSID PROTEIN AS A TRANSLATIONAL EFFECTOR OF VIRAL mRNAs

Stanley M. Tahara,[2] Therese A. Dietlin,[1] Gary W. Nelson,[1] Stephen A. Stohlman,[2] and David J. Manno[1]

[1]Department of Molecular Microbiology and Immunology
[2]Department of Neurology
USC School of Medicine
Los Angeles, California 90033

1. ABSTRACT

The mouse hepatitis virus (MHV) nucleocapsid protein stimulated translation of a chimeric reporter mRNA containing an intact MHV 5'-untranslated region and the chloramphenicol acetyltransferase (CAT) coding region. The nucleocapsid protein binds specifically the tandemly repeated -UCYAA- of the MHV leader. This RNA sequence is the same as the intergenic motif found in the genome RNA.

Preferential translation of viral mRNA in MHV infected cells is stimulated in part by this interaction and represents a specific, positive translational control mechanism employed by coronaviruses.

2. INTRODUCTION

An acute effect of mouse hepatitis virus (MHV) infection is the shut-off of host mRNA translation (Siddell et al., 1981; Hilton et al., 1986; Tahara et al., 1994). This phenomenon is observed for many cytolytic virus infections and is believed to be a means of ensuring optimum viral protein expression (Schneider and Shenk, 1987). Steady-state mRNA levels are not globally downregulated by MHV infection (Hilton et al., 1986; Kyuwa et al., 1994) so increases in viral protein expression are the sum of increased mRNA translation efficiency and increased synthesis of viral mRNAs. We reported earlier that increased translation efficiency is a property of the 5'-untranslated region UTRs of MHV mRNAs (Tahara et al., 1994). Since all MHV mRNAs of a given strain have virtually identical 5'-UTRs in an infected cell, this mechanism ensures that increased translation of viral mRNAs occurs in concert.

Coronaviruses and Arteriviruses, edited by Enjuanes *et al.*
Plenum Press, New York, 1998

We identified a 13 base sequence of the viral 5'-UTR, which includes the intergenic sequence, as the cis element primarily responsible for the increase in viral mRNA utilization (Tahara et al., 1994). The 3'-proximal region of the UTR, where this 13 nt sequence is located, was previously shown to be important for binding of MHV nucleocapsid protein to leader RNA (Stohlman et al., 1988). Thus we hypothesized that the viral nucleocapsid protein is the *trans*-acting factor responsible for increased translation efficiency of MHV mRNAs. In this report we show experimentally that it has high affinity for the intergenic sequence motif of MHV genomic and subgenomic mRNAs and acts as a *trans*-acting positive regulator of translation.

3. METHODS AND MATERIALS

3.1. Cells and Virus

Mouse DBT cells were maintained in MEM medium containing 7% newborn calf serum (heat inactivated, Gemini), 10% tryptose phosphate broth (Difco), 2 mM glutamine, 100 µg/ml streptomycin and 100 U/ml penicillin. Media and supplements were from Life Technologies.

The A59 strain of MHV was used.

3.2. Plasmid Constructions

Construction of pMCAT and pCAT transfection vectors was performed by insertion of CAT genes between the unique HindIII and XhoI sites of pBC/CMV/SEAP (Cullen and Malim, 1992). pCAT coding region was prepared by PCR amplification with primers containing unique HindIII and XhoI restriction sites. pMCAT was constructed by PCR amplification of the gene 6 leader sequence from phαGL-1 (Tahara et al., 1994), followed by co-amplification with CAT coding region. The resulting chimeric gene contained the MHV 5'-UTR fused directly to the start codon of the CAT reporter.

pRcCMV-N encodes the MHV-A59 nucleocapsid protein and was a gift from Dr. John Polo.

Construction of mutVI plasmid was performed by PCR methods such that an XbaI site was introduced to disrupt the intergenic sequence.

All constructs were sequenced by the dideoxynucleotide method (Sanger et al., 1977) with Sequenase II (Amersham) for verification.

3.3. DNA Transfections and CAT Assay

DBT cells were seeded at 0.5–1 X 10^6 cells per 35 mm cell culture well at least 12 h prior to transfection. Preparation of DNA-Ca-PO$_4$ coprecipitates, transfection conditions and glycerol shock were as described previously (Lin et al., 1986). Reporter and effector plasmids were transfected at levels indicated in the text. pCMV-SPORT-βGAL (0.5 µg/well; Life Technologies) and carrier DNA (pRcCMV; Invitrogen) were added to maintain a total DNA input of 1.7 µg/well. Cells were maintained at 37°C for 24 h prior to harvest and assay for CAT activity.

Stable transformants were prepared by cotransfecting DBT cells with 1–2 µg of pMCAT or pCAT and pSV2neo (Southern and Berg, 1982) at a 10:1 mass ratio. After transfection, cells were selected in 400 µg/ml G418 (Geneticin, Life Technologies) for 3 weeks.

Foci of G418 resistant cells were trypsinized and cloned separately in multi-well culture dishes. Continued selection in G418 was performed for an additional 3 weeks. All drug resistant clones were assayed for presence of CAT gene expression prior to any further use.

3.4. Enzyme Assays

Cell pellets were lysed by four, freeze-thaw cycles; cell-free extracts were assayed for CAT activity as described earlier (Gorman et al., 1982). Acetylated products were resolved by thin-layer chromatography and dried chromatograms were quantitated in a radiographic plate scanner (AMBIS). All conditions were assayed in duplicate.

β-Galactosidase activity was measured by a chemiluminescence assay (Galacto-Light, Tropix, Inc.) using conditions recommended by the vendor. Luminescence was measured in a Berthold luminometer. β-galactosidase activities for each transfectant were used to normalize transfection efficiency and observed CAT activity.

Protein measurements were determined by the method of Bradford using bovine gamma globulin as the standard (Bradford, 1976).

3.5. RNA Binding Assay

Construction of recombinant glutathione-S-transferase (GST) fusion proteins of JHMV nucleocapsid protein and fragments was as described earlier (Nelson, 1996). Data in Fig. 1A were obtained with N protein without GST fusion protein. There was no difference in kinetic constants for the GST fusion proteins compared to their counterparts without GST (Nelson, 1996).

RNA binding assays were performed with ^{32}P-labeled, *in vitro* transcribed pBSL (Nelson and Stohlman, 1993) as described previously (Nelson, 1996). Transcription of pBSL after linearization with HindIII using T7 RNA polymerase yields a 153 nt RNA molecule which has the first 114 bases of mRNA 6 of JHMV.

Binding data were analyzed using the EZ-Fit program, (written by Dr. Frank Perrella, E. I. DuPont de Nemours & Co.). Data were fitted to a single substrate binding curve.

4. RESULTS

Using cell-free translation extracts, we previously established that reporter mRNAs with intact MHV 5'-UTRs were specifically stimulated in cell-free extracts prepared from MHV infected cells (Tahara et al., 1994). In order to determine whether this type of stimu-

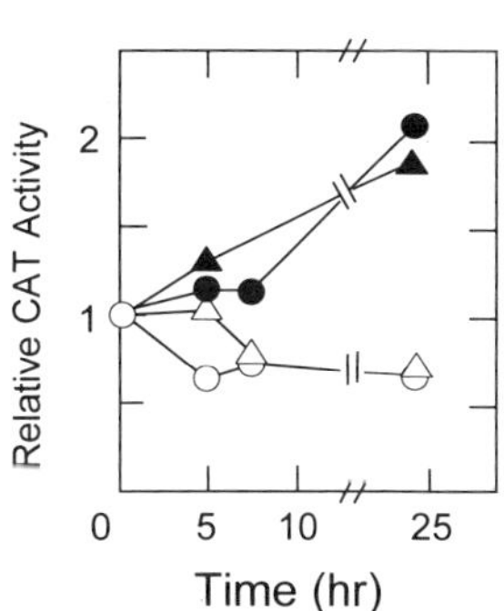

Figure 1. MHV infection of stably transformed DBT cells expressing MHV-CAT and CAT. Cell lines expressing pMCAT or pCAT were infected with MHV-A59 at an m.o.i.~1 using standard conditions. Cells were harvested at the indicated times and assayed for CAT activity. Relative CAT expression is shown in the figure. pCAT: O, clone A1; Δ, clone B6. pMCAT: ●, clone A5; ▲, clone B6.

lation was physiologically relevant in the intact cell we continued these studies using a transfection approach in DBT cells. We established stable transformants expressing either a control CAT gene or an intact MHV 5'-UTR fused to the CAT reporter. These stable transformants were in turn infected with MHV-A59 and assayed for CAT expression vs. time. We expected that only CAT mRNAs with an MHV 5'-UTR (MHV-CAT) would show increased activity, based on our previous *in vitro* studies (Tahara et al., 1994). The results of such an experiment are shown in Fig. 1. Infection of clonal lines expressing MHV-CAT genes showed a general two-fold stimulation of CAT expression over a 24 hour period of infection. By contrast, MHV infection of cell lines expressing the control CAT gene showed no such increase in CAT activity. Indeed, a decline in activity was observed which was consistent with the general shut-off of translation due to MHV infection. This result confirmed the previous result obtained *in vitro* and demonstrated that the effect of MHV leader was independent of the reporter gene.

Previous experiments implicated the MHV nucleocapsid gene product as the viral gene product which binds leader RNA. We examined the biochemical properties of JHMV N protein interactions with viral leader RNA. In these studies, recombinant N protein was synthesized as a GST fusion protein in *E. coli* and assayed for binding activity with the synthetic MHV leader RNA as ligand. We observed that the N protein exhibited saturable binding kinetics when assayed against the full length leader RNA molecule. The apparent K_d for this reaction was 14 nM. Subfragments of N protein were also prepared as GST fusions consisting of N^{1-168} (A), $N^{176-230}$ (B1) and $N^{309-454}$ (C). These were tested for leader RNA binding. As shown in Fig. 2B only the GST-B1 fusion protein showed RNA binding activity.

In order to test the hypothesis that the N protein recognizes 3'-proximal sequences of the MHV leader, specifically the intergenic motif, we assayed ligands which had one to three copies of the pentamer repeat sequence. As shown in Table 1, the binding of these three ligands to GST-N was compared. We found that there was no difference in the affin-

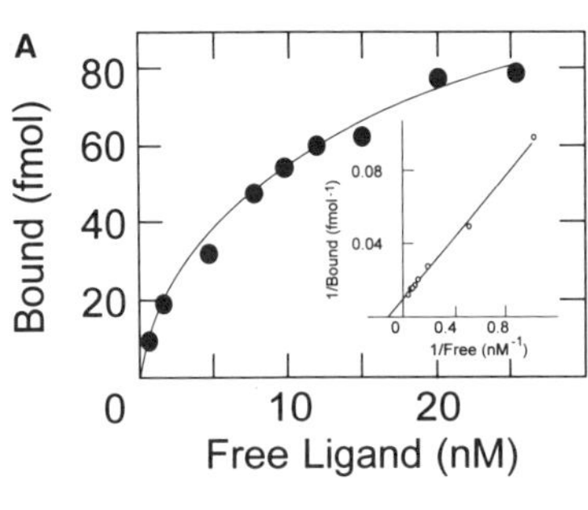

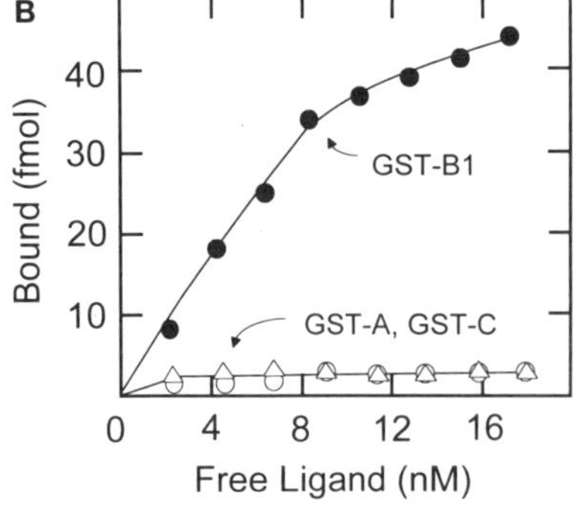

Figure 2. Leader RNA binding activity of GST-N protein. RNA binding was assayed for GST-N protein as indicated in Materials and Methods. A. Primary plot and double-reciprocal plot (inset) of leader RNA binding activity of N protein. B. Binding activity of N protein subfragments to leader RNA.

Table 1. Effect of UCYAA copy number on N protein binding

Ligand[a]	Apparent K_d
- UCUAA -	14.7 ± 3.1 nM
- UCUAAUCCAA -	18 ± 5.4 nM
- UCUAAUCUAAUCUAA -	14 ± 4 nM

[a] RNA ligands containing one, two or three copies of the UCYAA motif were tested as for binding to N protein. For the two copyligand, RNA from pBSL was used. K_d values were determined as in Fig. 2.

ity of N protein for the multimers of the UCUAA pentamer. All of the UCUAA repeat ligands showed the same experimental K_d for binding to N protein. This was a somewhat surprising result because a change in the number of tandem repeats is observed to alter the rates of viral mRNA transcription (Zhang et al., 1994; van Marle et al., 1995). Our results indicate then that the interaction between N protein and the UCUAA repeat is not capable of distinguishing between tandem copies of this sequence, thus N protein likely is not involved in this type of discrimination.

In order to demonstrate a trans effect of the JHMV nucleocapsid protein on translation of 5'-leader containing mRNA, we performed a cotransfection experiment using a nucleocapsid expression plasmid as the effector with reporter plasmids containing the CAT gene and either an intact JMHV leader sequence (...*UCUAAUCCAA*C...; MHV-CAT) or a mutated intergenic sequence (...*UCUUCUAGAA*C...; mutVI). As shown in Table 2, inclusion of the effector plasmid in the transfection experiment resulted in an increase in CAT expression from MHV-CAT in a dose-dependent fashion. Expression of the MHV-CAT reporter was consistently two-fold higher than the activity of mutVI which showed at best modest increases in activity. This was the first clear indication that the nucleocapsid protein stimulates translation of mRNAs which contain the viral 5'-UTR. It is important to note that the effector to reporter ratios varied from 1:20 to 1:10 in this experiment. The results are indicative that N protein was acting in a rate limiting process, *e.g.* initiation of translation, rather than in a process requiring stoichiometric ratios of mRNA and protein.

5. DISCUSSION

Identification of the MHV nucleocapsid protein as a factor important for translation of MHV mRNAs underscores its multifunctionality. This protein is thought to have roles in tran-

Table 2. Co-transfection of nucleocapsid protein gene with chimeric MHV-CAT reporter

Nucleocapsid DNA	Relative reporter activity[a]	
	mutVI	MHV-CAT
none	1	1
50 ng	0.86	1.81
100 ng	1.79	4.45

[a] Activity is normalized to control CAT reporter plasmid which completely lacks viral 5'-UTR sequences.

scription and replication in addition to encapsidation; to this list we propose to add its function in translation of MHV mRNAs. All of these processes are tied to N protein via its RNA binding activity. Our identification of the RNA sequence bound by N protein as well as the corresponding domain of the N protein itself clearly establishes this RNA-protein interaction. This information can now be used as a starting point for an investigation into N protein domains which are responsible for the effector activities in the processes listed above.

The RNA binding domain of nucleocapsid protein was identified in earlier studies as domain B ($N^{169-308}$) (Nelson and Stohlman, 1993) or domain II ($N^{163-380}$) . In this report we show that the B1 region ($N^{176-230}$) of the nucleocapsid protein is responsible for binding to the intergenic region of the viral leader RNA. Our results do not address a mechanism of N protein as a translational activator, *i.e.* whether it has intrinsic activity or whether it facilitates the activity of the canonical initiation factors. Regardless of its mechanism this phenomenon represents one of few examples of positive regulators of translation activity and as such deserves further study.

ACKNOWLEDGMENTS

This work supported by grants from the Public Health Service: NS30880 (SMT), NS18146 (SAS). DJM was supported by training grant NS07149.

REFERENCES

Bradford, M., 1976, A rapid and sensitive method for the quantitation of microgram quantities of proteins utilizing the principle of dye binding, *Anal. Biochem.* **72**:248–254.

Cullen, B., and Malim, M. H., 1992, Secreted placental alkaline phosphatase as a eukaryotic reporter gene, *Meth. Enzymol.* **216**:362–368.

Gorman, C. M., Moffat, L. F., and Howard, B. H., 1982, Recombinant genomes which express chloramphenicol acetyltransferase in mammalian cells, *Molec. Cell. Biol.* **2**:1044–1051.

Hilton, A., Mizzen, L., Macintyre, G., Cheley, S., and Anderson, R., 1986, Translational control in murine hepatitis virus infection, *J. Gen. Virol.* **67**:923–932.

Kyuwa, S., Cohen, M., Nelson, G., Tahara, S. M., and Stohlman, S. A., 1994, Modulation of cellular macromolecular synthesis by coronavirus: implication for pathogenesis, *J. Virol.* **68**:6815–6819.

Lin, A. Y., Chang, S. C., and Lee, A. S., 1986, A calcium ionophore-inducible cellular promoter is highly active and has enhancerlike properties, *Molec. Cell Biol.* **6**:1235–1243.

Nelson, G. W., 1996, [Ph.D. Dissertation]. (University of Southern California, Los Angeles, CA).

Nelson, G. W., and Stohlman, S. A., 1993, Localization of the RNA-binding domain of mouse hepatitis virus nucleocapsid protein, *J. Gen. Virol.* **74**:1975–1979.

Sanger, F., Nicklen, S., and Coulson, A. R., 1977, DNA sequencing with chain-terminating inhibitors, *Proc. Natl. Acad. Sci. USA* **74**:5463–5467.

Schneider, R., and Shenk, T., 1987, Impact of virus infection on host cell protein synthesis, *Annu. Rev. Biochem.* **56**:317–332.

Siddell, S., Wege, H., Barthel, A., and ter Meulen, V., 1981, Intracellular protein synthesis and the in vitro translation of coronavirus JHM mRNA, *Adv. Exptl. Biol. Med.* **142**:193–207.

Southern, P. J., and Berg, P., 1982, Transformation of mammalian cells to antibiotic resistance with a bacterial gene under control of the SV40 early region promoter, *J. Mol. Appl. Gen.* **1**:327–341.

Stohlman, S. A., Baric, R. S., Nelson, G. N., Soe, L. H., Welter, L. M., and Deans, R. J., 1988, Specific interaction between coronavirus leader RNA and nucleocapsid protein, *J. Virol.* **62**:4288–4295.

Tahara, S. M., Dietlin, T. A., Bergmann, C. C., Nelson, G. W., Kyuwa, S., Anthony, R. P., and Stohlman, S. A., 1994, Coronavirus translational regulation: leader affects mRNA efficiency, *Virology* **202**:621–630.

van Marle, G., Luytjes, W., van der Most, R. G., van der Straaten, T., and Spaan, W. J. M., 1995, Regulation of coronavirus mRNA transcription, *J. Virol.* **69**:7851–7856.

Zhang, X., Liao, C.-L., and Lai, M. M.-C., 1994, Coronavirus leader RNA regulates and initiates subgenomic mRNA transcription both in trans and in cis, *J. Virol.* **68**:4738–4746.

42

PROGRESS TOWARDS THE CONSTRUCTION OF A TRANSMISSIBLE GASTROENTERITIS CORONAVIRUS SELF-REPLICATING RNA USING A TWO-LAYER EXPRESSION SYSTEM

Zoltán Pénzes,* José Manuel González, Ander Izeta, María Muntión, and Luis Enjuanes

Department of Molecular and Cell Biology
Centro Nacional de Biotecnología, CSIC
Campus Universidad Autónoma
Cantoblanco, 28049 Madrid, Spain

1. ABSTRACT

Three transmissible gastroenteritis coronavirus (TGEV) defective interfering RNAs of 21, 10.6 and 9.7 kb (DI-A, DI-B and DI-C, respectively) were isolated. Dilution experiments showed that the largest DI RNA, DI-A, is a self-replicating RNA (replicon), and thus codes for a functional RNA polymerase and all the necessary replication signals. In order to engineer a cDNA encoding the RNA replicon a strategy based on the cloning of DI-C cDNA, followed by the insertion of the sequences required to complete the DI-A sequence has been developed. A cDNA complementary to DI-C RNA was cloned under the control of the CMV promoter (pDI-C-CMV) and rescued with a helper virus. In the ORF 1a of polymerase gene pDI-C-CMV contained a 10 kb deletion and in ORF 1b a 1.1 kb deletion. The consensus sequence corresponding to the deleted regions was cloned, and the deletions in pDI-C-CMV were replaced to yield a complete cDNA clone of DI-A, pDI-A-21-CMV, containing a full-length TGEV polymerase, driven by a CMV promoter. Expression of a functional TGEV polymerase is being investigated.

---

* Z. Pénzes is on sabbatical from the Veterinary Medical Research Institute of the Hungarian Academy of Sciences, Budapest, Hungary.

Coronaviruses and Arteriviruses, edited by Enjuanes *et al.*
Plenum Press, New York, 1998

2. INTRODUCTION

Defective RNAs of TGEV PUR46-MAD, DI-A, DI-B, and DI-C of 21, 10.4, and 9.7 kb, respectively were generated by undiluted passages and their structure was determined previously (Méndez et al., 1996), indicating that DI-A RNA contains the information required to encode the viral polymerase, thus it is a potential replicon, similarly to DIssA RNA of MHV which is synthesized in the absence of helper functions (Makino et al., 1988).

In this paper we show that, in fact, DI-A RNA is self-replicating. Our aims were to assemble a full-length cDNA clone of DI-A RNA, and later to study the expression of synthetic DI-A. Since DI-A RNA length is higher than 21 kb, its in vitro expression probably would be difficult, suggesting that the expression of this RNA in the cell nucleus under the CMV promoter could provide a procedure leading to a full length DI-A RNA, unless splicing would occur.

A full-length clone of DI-C (9.7 kb) was assembled after the CMV promoter (pDI-C-CMV) and a synthetic DI-C RNA was rescued and stably passaged, indicating that the intracellular synthesis of DI-A RNA under CMV promoter may be an efficient strategy to rescue a TGEV derived replicon.

3. MATERIALS AND METHODS

3.1. Cells and Virus

Plaque-cloned TGEV PUR46-MAD (Sánchez et al., 1990) was used. The passage and growth of the virus on swine testis (ST) cells has been described (Méndez et al, 1996).

3.2. RNA Analysis

Cytoplasmic RNA was extracted from virus-infected ST cells as described previously (Méndez et al., 1996). Northern hybridizations were performed as described (Sambrook et al., 1989) using a ^{32}P-labeled cDNA probe complementary to the 3' end of the TGEV sequence. RNA was quantified on a BioRad phosphorimager.

3.3. Plasmid Construction

Standard recombinant DNA techniques were used for the construction of all plasmids. The cDNA clone of 9.7 kb DI-C RNA behind a T7 promoter, pDI-C (Izeta et al., 1997) was used as a starting point to construct the pDI-A-21. The required fragments of genes 1a and 1b were synthesized by RT-PCR as described (Méndez et al., 1996), cloned and all but one mutation introduced by the PCR were corrected (see Results). pDI-A-21 with a total length of 23.5 kb, was assembled in plasmid pACNR1180, kindly provided by J.D. Tratschin (Ruggli et al., 1996). For the construction of plasmids carrying a cDNA encoding DI-C and DI-A under the control of the CMV promoter (pDI-C-CMV and pDI-A-21-CMV, respectively), the CMV promoter was amplified from pcDNA3.1 (Invitrogen) and was directly joined to the 5' end of the TGEV sequence by PCR-directed mutagenesis (Dubensky et al., 1996). A cDNA coding for the bovine growth hormone (BGH) terminator sequences was extracted by PCR from plasmid pcDNA3.1 and placed immediately downstream of the cDNAs complementary to DI-C and DI-A RNAs plus a synthetic polyA tract and a hepatitis delta ribozyme (HDVR) to generate perfect 3' ends.

3.4. DNA Transfection

Plasmid DNA was transfected into ST cells using DOTAP liposomal transfection reagent (Boehringer Mannheim) according to the manufacturer's instructions.

4. RESULTS

4.1. Rescue of TGEV Derived Minigenomes Using the CMV Promoter and Cellular Polymerase II

The use of the CMV promoter for the expression of TGEV minigenomes was studied. pDI-C, a full-length cDNA clone of DI-C RNA containing a T7 promoter, was modified by PCR-mediated site-directed mutagenesis to include the CMV promoter joined precisely to the TGEV 5' end. Accordingly, the 3' end of the DI-C cDNA was engineered to contain the polyA, followed by the HDVR and the BGH terminator sequences. Plasmid DNA was transfected into ST cells and cells were infected with TGEV helper virus. The virus was passaged undiluted and total cellular RNA was analyzed by Northern hybridization in passage 4. DI-C RNA was clearly detected (Fig. 1). The expression of the virus containing the DI-C minigenome was stable for more than ten passages (results not shown).

To study the expression of a foreign gene by the TGEV DI vector system, a DI-C-derived minigenome, M39, containing the beta-glucuronidase (GUS) reporter gene under a TGEV subgenomic promoter (Izeta et al., 1997) was constructed and the expression of the minigenome was engineered under the control of CMV promoter. Upon lipofection of the plasmid DNA into ST cells and infection with helper virus, high amounts GUS activity were detected in cell lysates prepared from passage zero, indicating the expression of GUS (data not shown).

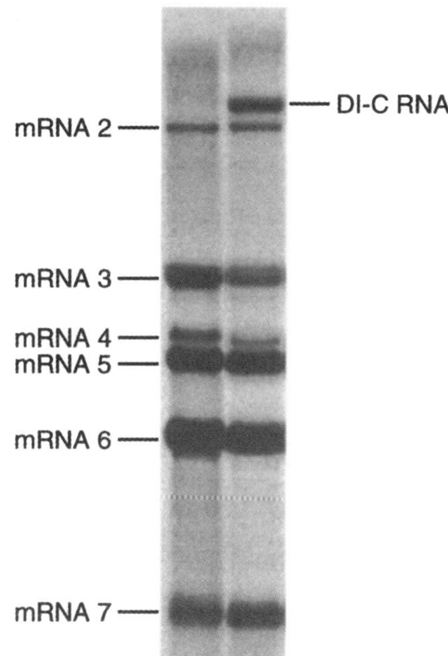

Figure 1. Rescue of DI-C RNA expressed from transfected plasmid in the nucleus under the control of the CMV promoter. Northern blot analysis of the RNAs expressed after plasmid pDI-C-CMV was transfected into ST cells and infected with helper virus. Following serial undiluted passages, total cellular RNA was analyzed from passage 4. Rescue of DI-C RNA was detected as a strong band migrating above the S mRNA. The left lane shows the standard pattern of mRNAs after infection with the helper virus. The lane on the right side shows the mRNA pattern plus a band with the size expected for the DI-C RNA.

4.2. TGEV DI-A Minigenome Is a Self-Replicating RNA

To study TGEV DI-A RNA replication, ST cells were infected with twofold serially diluted DI-containing TGEV. At 7 hr p.i. total cellular RNA was extracted, electrophoresed in a formaldehyde agarose gel, transferred to a nylon membrane by vacuum blotting and hybridized to a ^{32}P-labeled cDNA probe complementary to the 3' untranslated region. The RNA bands corresponding to the genomic RNA (gRNA), DI-A, DI-B, and the spike (S) mRNA were quantified on a phosphorimager and the graphs were fitted by linear regression analysis. The effects of the virus dilution on the synthesis of the gRNA, DI-A, DI-B, and mRNA S were evaluated (Fig. 2). The levels of gRNA, DI-A and mRNA S decreased linearly with the virus dilution, following one-hit kinetics, while DI-B synthesis decreased more rapidly, following two-hit kinetics. These data indicated that DI-A RNA was synthesized independently of a helper function, while for the replication of DI-B RNA, in addition to the DI-B RNA, a second component was required, most probably the RNA corresponding to the helper virus which encoded the viral replicase.

4.3. Progress towards the Construction of a Synthetic TGEV Derived Replicon

The complete cDNA clone of DI-C RNA, pDI-C, (Izeta et al., 1997) was used as a starting point to construct a potential TGEV replicon of 21 kb, that codes for a full TGEV

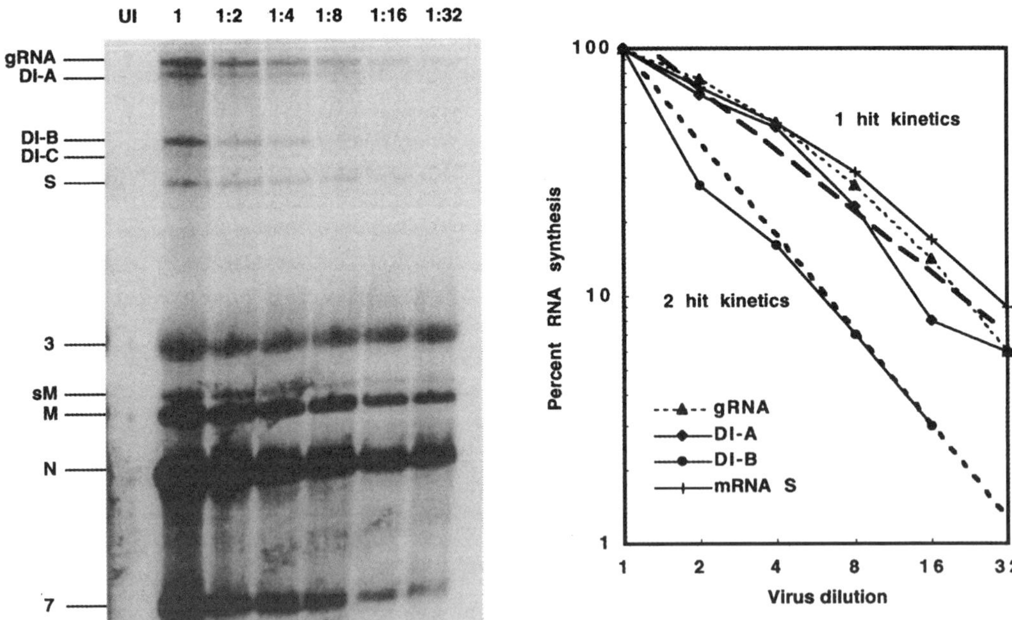

Figure 2. Effects of virus dilution on TGEV RNA synthesis. ST cells were infected with twofold serially diluted DI containing TGEV virus. At 7 hr p.i. total cellular RNA was extracted, electrophoresed on a formaldehyde agarose gel, transferred to a nylon membrane and hybridized to a ^{32}P-labeled cDNA probe complementary to the 3' untranslated region. The bands corresponding to the gRNA, DI-A, DI-B, and mRNA S were quantified on a phosphorimager and plotted. The theoretical one-hit kinetics and two-hit kinetics are shown as discontinuous lines. The gRNA, DI-A, and mRNA S fit the one-hit kinetics, while DI-B fits the two-hit kinetics curve.

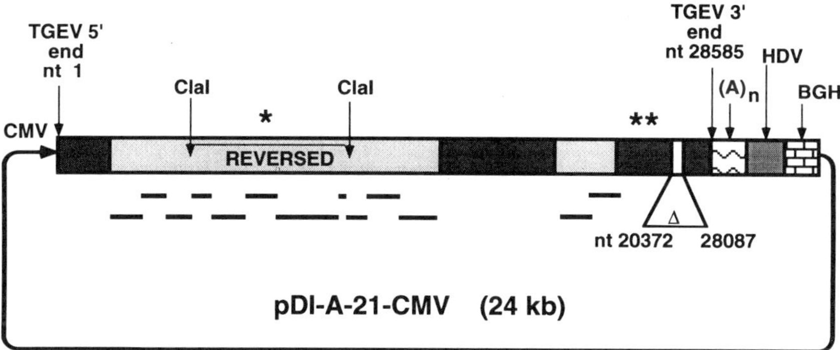

Figure 3. Construction of a TGEV self-replicating RNA cDNA clone. The complete polymerase gene (ORF 1a-b) of TGEV PUR46-MAD and a 0.5 kb region of the 3' end of the genome has been cloned, sequenced and assembled in a single plasmid (pDI-A-21-CMV of 24 kb). All errors introduced by the DNA polymerases during cloning have been corrected with the exception of two marker mutations (*, nucleotide 6752: A to G; **, nucleotide 18997: T to C) which do not change the aminoacids. The dark boxes represent sequences retained from the DI-C plasmid. The light gray boxes represent the filled deletion. The short black lines represent the individual clones used in the construction. (A)$_n$, polyA; HDV, hepatitis delta ribozyme; BGH, bovine growth hormone terminator; Δ, deleted sequences.

RNA polymerase and also contains sequences from the 5' and 3' ends of the TGEV genome as replication signals. The 10 kb region of gene 1a, and a 1.1 kb region in gene 1b, which were missing from pDI-C, were amplified by RT-PCR in various fragments using a viral RNA preparation of TGEV PUR46-MAD as a template. The PCR fragments were cloned, sequenced and all nucleotide differences encountered between the PUR46-MAD sequences and the cDNA clones were corrected with the exception of a silent marker mutation at nucleotide 6752 (A to G, leading to a Ser) (Fig. 3). The complete TGEV replicon cDNA clone was assembled in a low copy number plasmid pACNR1180 behind CMV promoter. The replicon cDNA clone, named pDI-A-21-CMV (24 kb), and the 13 clones that were used for the filling up of pDI-C are shown in Fig. 3. To delete an unwanted ClaI site at position 18997 a silent T to C change has been introduced by PCR-mediated site-directed mutagenesis. This mutation facilitated the assembly of the full-length clone and will allow us to easily distinguish a future infectious clone from the wild type virus.

Plasmid pDI-A-21-CMV contains a 5.2 kb ClaI gene 1a fragment in the reverse orientation. Due to toxicity problems in bacteria, stable cloning of this fragment in the correct orientation was not possible. To overcome this problem, an *in vitro* ligation strategy has been developed, which includes the release and *in vitro* religation of the previously reversed 5.2 kb fragment in virus orientation and its transformation into mammalian cells. The presence of the correct assembly of the full length clone encoding the replicase has been shown by PCR analysis (results not shown).

To detect the expression of the replicase activity by the engineered cDNA construct, co-transformation with a plasmid encoding a minigenome carrying a reporter gene (GUS) under the control of a TGEV promoter is being performed. GUS activity would be only detected if a functional polymerase is being expressed, that in turn will led to the mRNA encoding the GUS activity.

5. DISCUSSION

In this chapter we have studied the expression of TGEV minigenomes DI-C and M39 containing a foreign gene, GUS, under the nuclear CMV promoter. Rescue of DI-C RNA expressed from CMV promoter was detected when plasmid DNA transfected cells were infected with TGEV helper virus. The rescue of DI-C RNA expressed from CMV promoter was more efficient than the rescue performed with the *in vitro* transcribed (T7 promoter) and electroporated DI-C RNA (results not shown). This is probably due to the two amplification steps, in which first high amounts of DI-C RNA are transcribed in the nucleus under CMV promoter, followed by a specific amplification by the TGEV polymerase in the cytoplasm. A TGEV minigenome cloned behind CMV promoter was used to express the GUS reporter gene. In the transfected cells (passage 0), the GUS activity was significantly higher within the cells transfected with the DNA construct under CMV promoter control, than when the cells were electroporated with minigenome RNA transcribed *in vitro* (data not shown). The detection of gene expression in the cells directly transfected with synthetic minigenomes, without the need for amplification throughout the passage in tissue culture, may be a useful tool to study the replication of coronavirus RNA.

For the expression of the generally large synthetic coronavirus RNAs the use of the CMV promoter could be more advantageous than the T7 promoter, since it would eliminate the need for the *in vitro* transcription and the manipulation of large synthetic RNAs, and provides a method to synthesize large RNA transcripts in the cells.

The mechanism of synthesis of TGEV DI-A in TGEV infected cells was investigated. Dilution experiments indicated that the synthesis of DI-A RNA followed the same kinetics as gRNA, i.e., the synthesis of DI-A RNA decreased linearly with virus dilution, following one-hit kinetics. Méndez et al. (1996) showed that TGEV DIs are packaged in separate particles. All these data suggest that DI-A was synthesized independently from helper functions, thus codes for a functional polymerase as previously described for DIssA, a large MHV DI RNA that previously has also been shown to replicate independently from helper functions (Makino et al., 1988). In contrast, the synthesis of DI-B RNA decreased exponentially, following two-hit kinetics.

Encouraged by the autoreplication of DI-A RNA, we have constructed a full-length, 21 kb, cDNA clone of DI-A RNA by conventional bacterial cloning. Since stable insertion of a 5.2 kb ClaI fragment of gene 1a was not possible in the DI-A context, this fragment was inserted in reverse orientation into pDI-A-21-CMV. Then, *in vitro* ligation technology was used to release and religate this 5.2 kb fragment that subsequently was transfected directly into mammalian cells, to prevent the toxicity in bacteria. A cDNA with the correct sequence of the viral replicase has been assembled, as shown by PCR analysis. Current research efforts are being directed to show that the engineered clone encodes a functional replicase.

ACKNOWLEDGMENTS

This work has been supported by grants from the Consejo Superior de Investigaciones Científicas, the Comisión Interministerial de Ciencia y Tecnología (CICYT), La Consejería de Educación y Cultura de la Comunidad de Madrid, and Fort-Dodge Veterinaria from Spain, and the European Communities (Projects Biotechnology and FAIR).

REFERENCES

Dubensky, J., T. W., Driver, D. A., Polo, J. M., Belli, B. A., Latham, E. M., Ibanez, C. E., Chada, S., Brumm, D., Banks, T. A., Mento, S. J., Jolly, D. J., and Chang, S. M. W., 1996, Sindbis virus DNA-based expression vectors: utility for in vitro and in vivo gene transfer, *J. Virol.* **70**: 508–519.

Izeta, A., Smerdou, C., Mendez, A., Alonso, S., Pénzes, Z., and Enjuanes, L., 1997, Replication and packaging of synthetic minigenomes derived from defective transmissible gastroenteritis coronavirus, J. Virol. Manuscript submitted.

Makino, S., Shieh, C. K., Keck, J. G., and Lai, M. M. C., 1988, Defective-interfering particles of murine coronaviruses: mechanism of synthesis of defective viral RNAs, *Virology* **163**: 104–111.

Méndez, A., Smerdou, C., Izeta, A., Gebauer, F., and Enjuanes, L., 1996, Molecular characterization of transmissible gastroenteritis coronavirus defective interfering genomes: Packaging and heterogeneity, *Virology* **217**: 495–507.

Ruggli, N., Tratschin, J. D., Mittelholzer, C., Hofmann, M. A., 1996, Nucleotide sequence of classical swine fever virus strain Alfort/187 and transcription of infectious RNA from stably cloned full-length cDNA, *J. Virol.* **70**: 3478–3487.

Sambrook, J., Fritsch, E. F., and Maniatis, T., 1989, *Molecular cloning: a laboratory manual*, Cold Spring Harbor Laboratory, Cold Spring Harbor, New York.

Sánchez, C. M., Jiménez, G., Laviada, M. D., Correa, I., Suñé, C., Bullido, M. J., Gebauer, F., Smerdou, C., Callebaut, P., Escribano, J. M., and Enjuanes, L., 1990, Antigenic homology among coronaviruses related to transmissible gastroenteritis virus, *Virology* **174**: 410–417.

Protein Expression and Assembly II: Assembly

43

MOLECULAR EVENTS IN THE ASSEMBLY OF RETROVIRUS PARTICLES

Michael Sakalian and Eric Hunter

University of Alabama at Birmingham
Department of Microbiology and The UAB AIDS Center
UAB Station
Birmingham, Alabama 35294

1. ABSTRACT

Retrovirus assembly results from the ability of a single gene product, the gag polyprotein precursor, to coalesce into a spherical particle capable of release from the cell. In conjunction with this primary process of capsid formation additional viral gene products such as the replicative enzymes and envelope glycoproteins as well as the genomic RNA are incorporated to form an infectious virus.

2. INTRODUCTION

Retrovirus assembly involves a process in which a large number of chemically distinct macromolecules migrate through different transport pathways to the plasma membrane of the cell (Figure 1) where they are assembled into a nascent viral particle. The capsid of the virus is assembled from a large number of polyprotein precursors that must be transported through the cytoplasm to the underside of the plasma membrane. The membrane-spanning viral glycoproteins, on the other hand, must be transported through the secretory pathway of the cell to the cell surface where they co-localize with the nascent, membrane-extruding capsid. At a point still undetermined in the capsid assembly process, genome-length viral RNA molecules, along with necessary cell-derived tRNAs, must become associated with both capsid and polymerase components. By understanding the molecular basis of the interaction involved in assembly of the pathogenic human retroviruses, it may be possible to develop methods to therapeutically intervene in their replication.

Coronaviruses and Arteriviruses, edited by Enjuanes *et al.*
Plenum Press, New York, 1998

3. CAPSID ASSEMBLY

3.1. The *gag* Gene Product Directs Virus Assembly

All replication competent retroviruses contain four genes that encode the structural and enzymatic components of the virion (Figure 2). These are *gag* (capsid protein), *pro* (aspartyl proteinase), *pol* (reverse transcriptase and integrase enzymes) and *env* (envelope glycoprotein). Unlike most other enveloped RNA viruses in which the viral glycoproteins appear to catalyze virus particle formation, assembly and release of retrovirus particles occurs when capsid proteins are produced in the absence of the other gene products. Thus the product of this gene has the necessary structural information to mediate intracellular transport, to direct self-assembly into the capsid shell and to catalyze the process of membrane extrusion known as budding (Craven and Parent, 1996; Kräusslich and Welker, 1996; Hunter, 1994).

The *gag* gene product, a polyprotein precursor, is translated on free polyribosomes from an unspliced, genome length mRNA (Eisenman et al., 1974). In most retroviruses, the nascent Gag polyproteins are transported directly to the plasma membrane where as-

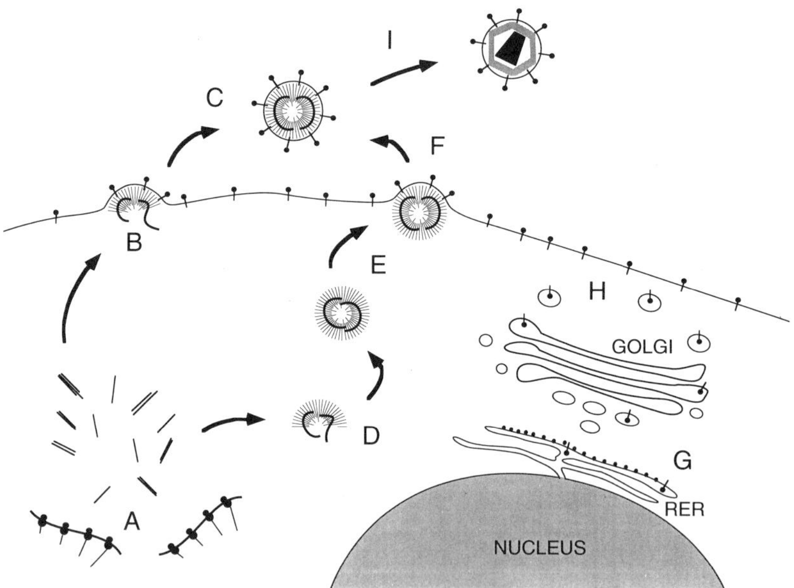

Figure 1. Schematic representation of retrovirus assembly. The morphogenesis of both C-type (A to B) and B/D-type (A to D-F) retroviruses is depicted. The Gag and Gag-Pol precursors are synthesized on free polyribosomes (A). In the case of C-type morphogenesis (i.e. HIV), the capsid precursors are transported, either individually or in small aggregates, directly to the plasma membrane where capsid assembly and budding (B) occur simultaneously. In contrast, for the B/D-type viruses, the capsid precursors are first transported to an intracytoplasmic assembly site (D). Immature capsids are then transported (E) to the plasma membrane where they induce budding to occur (F). The envelope glycoproteins of these viruses are translated on membrane bound polyribosomes and then traverse a common route to the plasma membrane (G-H), the secretory pathway of the cell. During the budding process, the viral glycoproteins presumably migrate to the same site in the membrane and are incorporated into the virus. For both morphogenetic types of retroviruses, an immature virus particle is released from the cell (C) and rapidly undergoes maturation (I). The mature virus particles have dense cores with a variable morphology characteristic of individual virus species.

sembly of the capsid shell and membrane extrusion occur simultaneously (Figure 1). Viruses that undergo this form of morphogenesis are known as type-C viruses and include the avian and mammalian leukemia/sarcoma viruses (e.g. Rous sarcoma, avian leukosis, and murine leukemia virus). The pathogenic human viruses, human T-cell leukemia virus and human immunodeficiency virus (HTLV-I and HIV), assemble their capsids in a similar fashion. In the second morphogenic class of viruses, the Gag precursors appear to be targeted to an intracytoplasmic site where capsid assembly occurs. These preassembled immature (A-type) particles are then transported to the plasma membrane where they undergo budding and envelopment (Figure 1). Viruses that undergo this process of assembly and release include the type-B, mouse mammary tumor virus (MMTV), the type-D, Mason-Pfizer monkey virus (M-PMV) and related simian retroviruses (SRV.1–5) as well as members of the spumavirus family. Despite the different morphogenic pathways, the process by which Gag precursors self-assemble into capsids is probably similar for the type-C and type-B/D viruses since a single amino acid change within the *gag* gene product of M-PMV was shown to divert it to the type-C morphogenic pathway (Hunter, 1994). Irrespective of the pathway to virus release, the newly budded-off virions have a common immature morphology. In thin-section electron microscopy the capsid shell appears as an electron opaque band in tight apposition to the membrane with an electron lucent center. During the process of virus maturation, the capsid polyprotein precursors are cleaved by the virus-encoded aspartyl proteinase, and this structure collapses into an electron dense core with a morphology characteristic of the virus family. (Reviewed in Nermut and Hockley, 1996; Gelderblom, 1991.)

3.2. Organization and Functions of the *gag* Gene Products in Virion Particles

The Gag polyprotein precursor functions as the primary building block in virus capsid assembly and during maturation it is cleaved by the viral proteinase to yield a number of individual proteins that make up the mature virion (Figure 2B). The translation of these products as a precursor protein thus ensures that equimolar amounts of each of the struc-

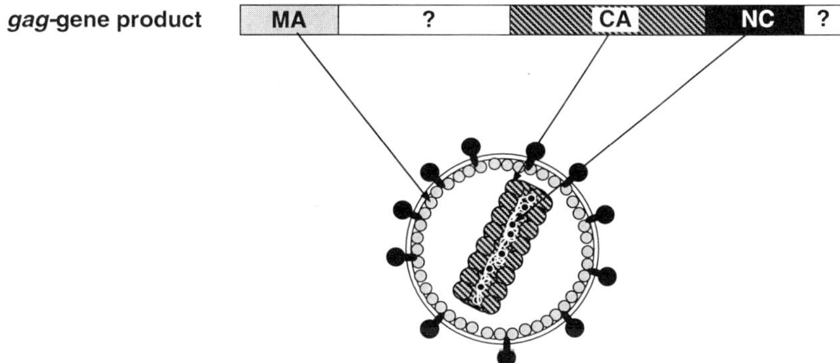

Figure 2. Organization of a generic retrovirus *gag* gene product. The Gag polyprotein precursor consists of the conserved matrix (MA), capsid (CA) and nucleocapsid (NC) protein domains. Unshaded boxes represent regions for which no common defined function has been established. Cleavage by the viral protease results in production of the matured Gag proteins the locations of which are shown in the schematic representation of a retrovirus particle.

tural proteins are incorporated into the virus. While the size and protein content of the precursor varies between different retroviral families, at least three *gag*-encoded proteins are found in all retroviruses; these are the matrix protein (MA), the capsid protein (CA) and the nucleocapsid protein (NC) (Figure 2) (Nermut and Hockley, 1996; Leis et al., 1988). The matrix protein is closely associated with the viral membrane. In most retroviruses, MA is modified co-translationally by the N-terminal addition of a myristic acid residue which is critical for its function. The CA protein forms the major protein component of the electron dense core in mature virions, where it appears to form a protein shell into which the virion RNA genome and replicative enzymes are packed (Gelderblom, 1991). The NC protein is located within the CA-derived shell (Figure 2C) where it is found associated with the viral RNA genome (Linial and Miller, 1990). This domain of the Gag precursor, in all retroviruses but the spumaviridae, contains a conserved cysteine-histidine rich, zinc-finger-like, region (Cys-X2-Cys-X4-His-X4-Cys) that is thought to play an important role in the specific packaging of viral RNA into the assembling virus (Katz and Jentoft, 1989).

The arrangement of the proteins on the precursor (NH$_2$-MA--CA--NC-COOH) reflects their position in both the immature capsid and the matured virion, where they appear to form concentric shells of protein after cleavage from the precursor (reviewed in Hunter, 1994). Despite a common organization, Gag precursors from different retroviruses share little amino acid sequence homology. Functional homologies must thus be reflected at the level of three-dimensional structure, as has been observed between retroviral proteinases (Weber, 1990) and among the MA protein structures of HIV, SIV, BLV (Matthews et al., 1996) and M-PMV (Conte et al., 1997) which maintain very similar three-dimensional conformations.

In addition to the functionally conserved MA, CA, and NC proteins, the Gag precursor can, depending on the virus encoding it, contain additional peptide sequences (Figure 2) whose function in virus assembly and future cycles of infection is for the most part obscure. Several of the undesignated peptide domains are short and may represent spacer peptides that allow the correct folding of the precursor protein for assembly and subsequent processing into the mature virion proteins. Alternately, since the precursor protein is the assembly unit for capsid construction, such regions could represent novel domains of the precursor that function transiently during the process of virus assembly and which are then dispensed with during the process of virus maturation.

3.3. Biosynthesis and Assembly of *pro* and *pol* Gene Products

The viral proteinase and reverse-transcriptase/integrase enzymes pose a translational problem for most retroviruses since their coding sequences are located in the middle of the genome and they also must be targeted intracellularly for assembly into virions. These problems are solved in most retroviruses by translating the enzymes from an unspliced, genome-length mRNA as long polyprotein precursors that represent C-terminal extensions of the *gag* gene product itself. The Gag sequence present at the amino-terminus of the Pro and Pol precursors is thought to target the molecules to the site of virus assembly and to facilitate their inclusion into the nascent virion. In many retroviruses synthesis of these Gag-Pro-Pol fusion proteins is achieved by way of a ribosomal frameshift mechanism. This process which occurs with an efficiency of approximately 5% also provides a mechanism for the virus to regulate the level of enzymatic proteins incorporated into virions. For some viruses, such as M-PMV and HTLV, *pro* lies within its own reading frame between *gag* and *pol*. Two extended Gag precursors, Gag-Pro, and Gag-Pro-Pol, are therefore produced as the result of one or two frameshift events respectively (Jacks, 1990).

MuLV, utilizes a subtly different mechanism and expresses *pro* and *pol* through the use of a suppressor tRNA which allows ribosomes to read past the stop codon. Foamy viruses, exemplified by the human spumaretrovirus (HSRV), are unique in that they express their *pro* and *pol* genes by translation of a separate spliced message (Bodem et al., 1996). Since the human spumavirus does not express *pro* and *pol* as a fusion with *gag* it is unclear how these enzymes are incorporated into virus particles.

3.4. Transport of Capsid Proteins to the Site of Assembly

The mechanism(s) by which the capsid precursor proteins are directed to the site of assembly remains for the most part unknown but is primarily mediated by the MA domain of the Gag precursor (Kräusslich and Welker, 1996). In the C-type retroviruses, the Gag and Gag-Pol precursors are transported directly to the plasma membrane where self-assembly into capsids occurs. In B- and D-type retroviruses, where capsid assembly and virus budding are discrete events, transport involves two stages: in the first capsid precursors are directed to an intracytoplasmic assembly site; in the second preassembled capsids are transported to the plasma membrane (Hunter, 1994). A recent genetic dissection of this process suggests that Gag-containing precursors of D-type viruses express a dominant sorting signal that targets the proteins to the initial assembly site (Figure 1). A point mutation within the MA coding domain of M-PMV abrogated this targeting and resulted in the direct, type-C-like, transport of precursors to the plasma membrane where efficient capsid assembly occurred. This result demonstrates that while the morphogenic pathway of a retrovirus can be altered the assembly of particles utilizes a common mechanism.

Transport of the preassembled capsids of D-type viruses to the plasma membrane appears to be dependent on an active process rather than passive diffusion since a variety of mutations within the MA domain of M-PMV result in the accumulation of discrete clusters of immature capsids deep within the cytoplasm—apparently unable to exit the site of assembly. This phenotype was initially observed for mutations that blocked the co-translational addition of myristic acid to the amino-terminus of the Gag precursors. However, two additional transport-defective M-PMV mutants with single amino acid changes in MA have similar phenotypes even though their Gag precursors are myristylated normally.

In the C-type retroviruses, the MA domain of the Gag precursor also plays a major role in targeting capsid components to the plasma membrane site of assembly. Mutations which block myristylation interfere with this process (reviewed in Kräusslich and Welker, 1996; Hunter, 1994). Mutations in MA can redirect Gag precursors to a new site for assembly, as with the morphogenesis mutant of M-PMV described above. Similarly, a deletion within the MA protein of HIV-1 results in the targeting of capsid precursors to the endoplasmic reticulum where they assemble and bud into the lumen of this compartment (Kräusslich and Welker, 1996).

By what pathway(s) might preassembled capsids of the B- and D-type viruses or the Gag-containing precursors of the C-type viruses be transported to the plasma membrane? Several studies have suggested that cytoskeletal elements might be involved in this process. MuLV, HIV, and HSRV Gag have been shown to be associated with cytoskeletal elements. Moreover, electron micrographs of MMTV- and RSV-infected cells have shown microfilaments in close association with budding viruses (reviewed in Kräusslich and Welker, 1996). Recent experiments indicate that the process is an active rather than a passive one since it appears to require energy. In cells treated with agents that block ATP synthesis M-PMV capsids are unable to migrate to the plasma membrane from sites deep within the cytoplasm (Weldon and Hunter, 1997). A requirement for energy suggests a possible role for chaperonins in the transport process.

3.5. Assembly of the Capsid Shell

Whether the Gag precursor proteins are first transported to an intracytoplasmic site or to the plasma membrane they must associate with other precursors in a regular and symmetrical fashion to self-assemble into the nascent capsid. Mutational analyses of *gag* genes have attempted to define what portions of the Gag precursor are required for particle formation itself and which domains might have other functions in the virus replication cycle. It is important to recognize that assembly domains within the Gag precursor may not necessarily reside within the boundaries of the mature cleavage products of Gag but may span cleavage sites and thus would be destroyed during the process of virus maturation. In this way the process of maturation would define the transition from an assembly function for Gag to an entry/infection function where there is a requirement for efficient disassembly and release of a transcriptionally active core upon infection of a new cell.

Deletion analysis has identified three essential assembly domains designated M, L, and I to reflect their specific functions for Membrane association, Late budding release, and Gag-Gag Interaction (Craven and Parent, 1996). Assembly domain M lies at the N-terminus of Gag and appears to be necessary for membrane association . Domain L has been identified for both RSV and M-PMV to contain a PPPY motif that is required for final release of budding particles. In RSV this sequence is located in the p2b spacer peptide between MA and p10. In M-PMV the same sequence is found within the pp16 region. Mutations in this sequence of either virus result in budding particles that remain attached to the plasma membrane (Yasuda and Hunter, 1997; Craven and Parent, 1996). In HIV the equivalent role is played by the carboxy-terminal p6 domain. Chimeric Gag proteins containing L domains from HIV or RSV in either the p2b or carboxy-terminal position have been shown to assemble and bud normally supporting the idea that the mechanisms by which retroviruses assemble are fundamentally equivalent.

The Gag-Gag interaction or I domain is essential for the production of particles with the correct density and size. For RSV, the I domain has been narrowed down to the carboxy-terminal end of the CA domain and half of the NC domain (Craven and Parent, 1996). The Gag proteins of HIV and MuLV also require equivalent regions for the production of particles with the correct density. This domain could establish the correct protein-protein interactions which allow the tight packing of Gag molecules during assembly or since this region is implicated in RNA packaging, it may influence particle density by directly mediating RNA encapsidation. Thus RNA could serve as a necessary scaffold upon which Gag proteins tightly pack during particle assembly—an idea supported by recent *in vitro* experiments (see below).

In some retroviruses, additional domains of Gag or other gene products may influence the efficiency of particle assembly without being essential to the process itself. For example, in M-PMV the p12 coding region can be deleted completely without affecting capsid assembly if Gag precursors are expressed at high levels. In contrast, such mutant Gag precursors are defective in assembly when expressed at levels typical of a normal virus infection (Sommerfelt et al., 1992).

How might Gag precursors become organized into a virus particle? Some researchers have hypothesized that Gag precursors pack together in a regular fashion to form a fullerine-like structure similar to those of icosahedral viruses (Nermut and Hockley, 1996). However, recent cryoelectron microscopy studies have demonstrated a lack of icosahedral symmetry as well as a large heterogeneity in particle size raising questions as to the organization of Gag molecules in capsids.

3.6. Assembly of Gag Proteins *in Vitro*

Recently, the possibility of establishing *in vitro* systems for the assembly of retroviral capsids has been explored. Gag proteins produced in bacteria have been successfully assembled into capsid-like structures. Fragments of either the RSV or HIV Gag precursor equivalent to the CA plus NC domains can be assembled into rod-shaped coils when combined with RNA (Campbell and Vogt, 1995). A larger fragment of RSV Gag comprising the entire precursor minus the proteinase domain can be similarly assembled in the presence of RNA, but in this case, forms spherical particles like those produced by proteinase-defective mutants of RSV (Campbell and Vogt, 1997). For either rods or spheres the assembly process appears to be independent of the RNA sequence since random *E. coli* RNA can serve as a suitable co-ssembly substrate. Studies of M-PMV *gag* expression in bacteria have demonstrated that capsids can assemble in *E. coli*. Capsid-like structures can be seen by thin-section electron microscopy within cellular inclusion bodies. Most importantly, capsid structures can be reassembled from these inclusion bodies after solubilization in urea (Klikova et al., 1995)

A second *in vitro* approach has been to produce Gag precursors in *in vitro* translation systems. Production of the Gag and Gag-Pro precursors of M-PMV in a reticulocyte system leads to the formation of immature capsid-like structures (Sakalian et al., 1996). In support of the authenticity of this system M-PMV mutant *gag* gene products which are defective for capsid assembly *in vivo* are also assembly incompetent *in vitro*. While the ability of HIV Gag to assemble into particles in a reticulocyte system remains controversial (Sakalian et al., 1996; Spearman and Ratner, 1996) efficient assembly of such particles was recently demonstrated in a wheat germ translation system (Lingappa et al., 1997). In addition, non-hydrolysable ATP analogs can block the formation of both HIV (Lingappa et al., 1997) and M-PMV (Sakalian and Hunter, 1997) particles in these *in vitro* systems, thus establishing a requirement for energy for the process of assembly itself. Furthermore, recent results with M-PMV have demonstrated that this in vitro assay system can distinguish true assembly mutants from intracellular transport mutants. The D- to C-type MA mutant, R55W, can assemble *in vitro* demonstrating that it does not contain an assembly defect while p12 deletion mutants which fail to assemble in cells likewise fail to assemble *in vitro* demonstrating the defect is not merely one in transport (Sakalian and Hunter, 1997). The further development of these cell-free assembly systems will not only advance the molecular dissection of capsid assembly but, as has been demonstrated by the use of anti-Gag monoclonal antibodies, could facilitate the analysis of potential inhibitors of retrovirus capsid assembly (Sakalian et al., 1996).

3.7. Incorporation of RNA

During assembly, in a process that is not understood but for which no viral gene product other then Gag is necessary, two copies of the viral RNA genome are packaged (Berkowitz et al., 1996). The retrovirus RNA genome contains in the 5' untranslated region a sequence element that mediates packaging of viral RNA. In most retroviruses, this element is near to, or perhaps coincident with, the dimer linkage site, the sequence through which the two identical RNA subunits in each virion are held together. In many retroviruses, stem-loop structures have been identified as important components the packaging element.

What corresponding determinants in Gag recognize and bind to Ψ RNA? Extensive studies of the NC protein, as a part of the Gag precursor, have established the importance of this domain in the packaging of genomic RNA (Berkowitz et al., 1996). Analysis of specific RNA packaging by a series of RSV Gag deletion mutants pointed to a possible role for sequences corresponding to the I assembly domain (Sakalian et al., 1994) leading again, as already mentioned, to the idea that RNA, whether viral or cellular, is required for assembly.

3.8. Membrane Association and Extrusion

Irrespective of the pathway followed by the Gag precursors, at a point either during assembly or following assembly in the cytoplasm, the nascent capsid must associate with the plasma membrane and induce the lipid bilayer to extrude; a process termed budding. It is clear that the viral glycoproteins are not necessary for this process to occur, since viruses lacking the env gene efficiently assemble and release membrane-enveloped virions.

It seems likely that during the process of membrane association and virus budding, conformational changes in the capsid precursors will occur. For the preassembled capsids of the type-D retroviruses, it is probable that the hydrophobic myristic acid modification of MA is buried inside the capsid structure during its transport through the hydrophilic environment of the cytoplasm. However, once the capsid has associated with the inner surface of the lipid bilayer, a conformational change which results in the exposure of this hydrophobic fatty acid on its surface might be expected to provide a more stable Gag-lipid interaction and could even provide the force for driving the budding process itself. A similar conformation-change driven process might also operate during C-type retroviral budding.

The final step in virus particle release requires the 'pinching-off' from the cell membrane of the immature particle—a process that involves a membrane fusion event. Although we now know that the L domain is necessary for this process the mechanism by which this occurs is a mystery. It is difficult to reconcile the positionally independent function of this domain with a hypothesis for a single cellular factor. However, it has been suggested recently that the L domain could constitute a binding site for "WW" proteins which contain a 38 amino acid sequence motif with two widely spaced tryptophan residues (Garnier et al., 1996). How such WW motif proteins, which include cell signaling molecules and cytoskeletal proteins, might function in releasing enveloped virions is unknown.

4. ASSEMBLY OF ENVELOPE GLYCOPROTEINS

While the capsid precursor proteins are transported through the cytoplasm by ill-defined pathways, the surface components of the virion, the envelope glycoproteins, are directed to the site of assembly through the relatively well characterized 'secretory pathway' of the cell (reviewed in Einfeld, 1996). The envelope glycoprotein complex of replication competent retroviruses is comprised of two polypeptides, an external, glycosylated, hydrophilic protein (SU) and a membrane-spanning protein (TM), that form a knob or knobbed spike on the surface of the virion. Both polypeptides are encoded by the env gene in the order - NH_2--SU--TM--COOH - and are synthesized in the form of a glycosylated, polyprotein precursor that oligomerizes in the rough endoplasmic reticulum. This assembled complex is then transported to the Golgi where the attached oligosaccharides are processed and the precursor is proteolytically cleaved to its mature subunits. The glycoprotein oligomers must then be transported from the Golgi to the surface of the cell in order to participate in the assembly process. While the glycoproteins are not required for the assembly of enveloped virus particles, they do play a critical role in the virus replication cycle by recognizing and binding to specific receptors and by mediating the fusion of viral and cell membranes; virus particles lacking envelope glycoproteins are thus non-infectious.

What directs these nascent glycoproteins into budding virions? It has been generally accepted that they are preferentially incorporated into virus while the bulk of the host cell membrane proteins are excluded from the assembling particle. There is biochemical and genetic evidence for an interaction between the MA domain of capsid and the cytoplasmic domain of the TM protein (reviewed in Hunter, 1994). Mutations within both the MA protein and the TM protein coding domains of HIV-1 and M-PMV have been shown to reduce the efficiency of glycoprotein incorporation into virions indicating that interactions between these components of the virus might be important during the assembly process. Moreover, in polarized epithelial cells, the HIV glycoprotein can specifically target HIV capsid assembly and release to the basolateral plasma membrane . This data suggests that retroviral glycoproteins are positively selected from the mixture of molecules in the plasma membrane.

While such a positive selection model is attractive, it does not explain the facility with which foreign glycoproteins can be incorporated into retrovirus particles or the lack of effect deletion of the cytoplasmic domain of the TM protein can have on glycoprotein incorporation in some retroviral systems. In the case of RSV and SIV, truncation of the cytoplasmic domain does not block glycoprotein incorporation . Furthermore, a glycolipid anchored form of the HIV-1 glycoprotein complex was also efficiently incorporated into virus particles, suggesting that the protein anchor and cytoplasmic domain are not critical signals for incorporation (reviewed in Einfeld, 1996; Hunter, 1994).

The fact that a variety of non-retroviral glycoproteins can be assembled efficiently into retrovirus particles and in some viruses removal of the cytoplasmic domain does not prevent glycoprotein incorporation, is inconsistent with a specific requirement for strong capsid-glycoprotein interactions. How then might viral proteins be preferentially incorporated into virions? Given an intimate MA/membrane association, it is possible that as virus assembly occurs, the associated areas of the plasma membrane are cleared of cellular glycoproteins that have connections with cytoplasmic counterparts (cytoskeletal elements or associated kinases for example). This active exclusion of cellular proteins might then allow mobile viral (and cellular) membrane proteins to freely diffuse into this region and be passively incorporated into virions. Such an active exclusion/passive inclusion mechanism would not preclude interactions between the capsid and glycoprotein, but neither would it require this interaction to take place for the glycoprotein complex to be incorporated into virus. Support for this hypothesis comes from more recent studies of HIV glycoprotein incorporation. In the context of mutations in the MA domain of Gag the envelope glycoprotein is excluded from virions; however, truncated Env species can be incorporated (Freed and Martin, 1995).

Indeed it may be possible to reconcile the apparent contradiction of a requirement for specific capsid/glycoprotein interactions during assembly of HIV and the 'passive incorporation' of foreign glycoproteins. Internalization signals have been found in the cytoplasmic domain of the TM proteins of HIV and SIV which can decrease the level of a glycoprotein on the cell surface (LaBranche et al., 1995). This same internalization signal, consisting of a membrane proximal tyrosine residue in the context of a "tight turn" motif has also been identified as the signal for basolateral membrane targeting of both the envelope glycoprotein and virion budding (Lodge et al., 1997). The efficiency of glycoprotein incorporation could therefore reflect an equilibrium between competing interactions of viral glycoprotein endocytosis and viral glycoprotein/capsid association. Thus, mutations which interfere with the capsid/viral glycoprotein interaction reduce the efficiency with which glycoprotein is incorporated into virus due to internalization while mutations which remove endocytosis signals from the TM cytoplasmic domain result in higher levels of virion-associated glycoprotein. Recent results with HIV are consistent with this in that the Gag precursor can indeed suppress the endocytosis of the envelope glycoprotein while a mutant Gag that is unable to mediate glycoprotein incorporation cannot (Egan et al., 1997).

5. VIRUS MATURATION

The final stage in the assembly of an infectious retrovirus particle occurs following release of the immature virion and involves its conversion to a mature virus. The process of maturation requires proteolysis of the Gag and Gag-related precursors by the viral protease and a reorganization of the mature protein products and viral nucleic acids (reviewed in Vogt, 1996). During this process an electron dense core forms (Figure 1), the diploid viral genome forms a stable hydrogen-bonded dimer RNA structure and the virus becomes an infectious entity. What mechanism regulates the activation of the viral protease remains unknown. However, because the type-D viruses preassemble an immature capsid containing an inactive protease precursor, activation by concentrating the precursors in an assembling capsid is unlikely. It is possible that conformational changes in the capsid precursor during association with the membrane trigger protease activity. Mutations or inhibitors that block proteolysis prevent the formation of a mature virus and block infectivity (reviewed in Vogt, 1996). This process of maturation, which results in a relatively fragile, more easily disrupted structure, irreversibly commits the virus to the infection pathway and can be considered both the finale and the beginning of the virus life cycle.

REFERENCES

Berkowitz, R., Fisher, J., and Goff, S. P., 1996, RNA packaging, *Curr. Top. Microbiol. Immunol.* **214**: 177–218.
Bodem, J., Löchelt, M., Winkler, I., Flower, R. P., Delius, H., and Flügel, R. M., 1996, Characterization of the spliced *pol* transcript of feline foamy virus: the splice acceptor site of the *pol* transcript is located in *gag* of foamy viruses, *J. Virol.* **70**: 9024–9027.
Campbell, S., and Vogt, V. M., 1995, Self assembly in vitro of purified CA-NC proteins from Rous sarcoma virus and human immunodeficiency virus type1, *J. Virol.* **69**: 6487–6497.
Campbell, S., and Vogt, V. M., 1997, In vitro assembly of virus-like particles with Rous sarcoma virus gag deletion mutants: Identification of the p10 domain as a morphological determinant in the formation of spherical particles, *J. Virol.* **71**: 4425–4435.
Conte, M. R., Klikova, M., Hunter, E., Ruml, T., and Matthews, S., 1997, The three-dimensional solution structure of the matrix protein from the type D retrovirus, the Mason-Pfizer Monkey virus, *submitted.*
Craven, R. C., and Parent, L. J., 1996, Dynamic interactions of the Gag polyprotein, *Curr. Top. Microbiol. Immunol.* **214**: 65–94.
Egan, M. A., Carruth, L. M., Rowell, J. F., Yu, X., and Siliciano, R. F., 1997, Human immunodeficiency virus type 1 envelope protein endocytosis mediated by a highly conserved intrinsic internalization signal in the cytoplasmic domain of gp41 is suppressed in the presence of the Pr55gag precursor protein, *J. Virol.* **70**: 6547–6556.
Einfeld, D., 1996, Matruation and assembly of retroviral glycoproteins, *Curr. Top. Microbiol. Immunol.* **214**: 133–176.

Eisenman, R. N., Vogt, V. M., and Diggelmann, H., 1974, Synthesis of avian RNA tumor virus structural proteins, *Cold Spring Harbor Symp. Quant. Biol.* **39**: 1067–1075.

Freed, E. O., and Martin, M. A., 1995, Virion incorporation of envelope glycoproteis with long but not short cytoplasmic tails is blocked by specific, single amino acid substitutions in the human immunodeficiency virus type 1 matrix., *J. Virol.* **69**: 1984–1989.

Garnier, L., Wills, J. W., Verderame, M. F., and Sudol, M., 1996, WW domains and retrovirus budding, *Nature (London)* **381**: 744–745.

Gelderblom, H. R., 1991, Assembly and morphology of HIV: potential effect of structure on viral function, *AIDS* **5**: 617–638.

Hunter, E., 1994, Macromolecular interactions in the assembly of HIV and other retroviruses, *Sem. in Virology* **5**: 71–83.

Jacks, T., 1990, Translational suppression in gene expression in retroviruses and retrotransposons, *Curr. Top. Microbiol. Immunol.* **157**: 93–124.

Katz, R. A., and Jentoft, J. E., 1989, What is the role of the *cys-his* motif in retroviral nucleocapsid (NC) proteins?, *BioEssays* **11**: 176–181.

Klikova, M., Rhee, S. S., Hunter, E., and Ruml, T., 1995, Efficient *in vivo* and *in vitro* assembly of retroviral capsids from Gag precursor proteins expressed in bacteria, *J. Virol.* **69**: 1093–1098.

Kräusslich, H.-G., and Welker, R., 1996, Intracellular transport of retroviral capsid components, *Curr. Top. Microbiol. Immunol.* **214**: 25–64.

LaBranche, C. C., Sauter, M. M., Haggarty, B. S., Vance, P. J., Romano, J., Hart, T. K., Bugelski, P. J., Marsh, M., and Hoxie, J. A., 1995, A single amino acid change in the cytoplasmic domain of SIVmac transmembrane molecule increases envelope glycoprotein expression on infected cells, *J. Virol.* **69**: 5217–5227.

Leis, J., Baltimore, D., Bishop, J. M., Coffin, J., Fleissner, E., Goff, S. P., Oroszlan, S., Robinson, H., Skalka, A. M., Temin, H. M., and Vogt, V., 1988, Standardized and simplified nomenclature for proteins common to all retroviruses, *J. Virol.* **62**: 1808–1809.

Lingappa, J. R., Hill, R. L., Wong, M. L., and Hegde, R. S., 1997, A multistep, ATP-dependent pathway for assembly of human immunodeficiency virus capsids in a cell-free system, *J. Cell Biol.* **136**: 567–581.

Linial, M. L., and Miller, A. D., 1990, Retroviral RNA packaging: Sequence requirements and implications, *Curr. Top. Microbiol. Immunol.* **157**: 125–152.

Lodge, R., Lalonde, J.-P., Lemay, G., and Cohen, E. A., 1997, The membrane-proximal intracytoplasmic tyrosine residue of HIV-1 envelope glycoprotein is critical for basolateral targeting of viral budding in MDCK cells, *EMBO J.* **16**: 695–705.

Matthews, S., Mikhailov, M., Burny, A., and Roy, P., 1996, The solution structure of the bovine leukaemia virus matrix protein and similarity with lentiviral matrix proteins, *EMBO J.* **15**: 3267–3274.

Nermut, M. V., and Hockley, D. J., 1996, Comparative morphology and structural classification of retroviruses, *Curr. Top. Microbiol. Immunol.* **214**: 1–24.

Sakalian, M., and Hunter, E., 1997, Unpublished results.

Sakalian, M., Parker, S. D., Weldon, R. A., Jr., and Hunter, E., 1996, Synthesis and assembly of retrovirus gag precursors into immature capsids in vitro, *J. Virol.* **70**: 3706–3715.

Sakalian, M., Wills, J. W., and Vogt, V. M., 1994, Efficiency and selectivity of RNA packaging by Rous sarcoma virus Gag deletion mutants, *J. Virol.* **68**: 5969–5981.

Sommerfelt, M. A., Rhee, S. S., and Hunter, E., 1992, Importance of the p12 protein in Mason-Pfizer monkey virus assembly and infectivity, *J. Virol.* **66**: 7005–7011.

Spearman, P., and Ratner, L., 1996, Human immunodeficiency virus type 1 capsid formation in reticulocyte lyates, *J. Virol.* **70**: 8187–8194.

Vogt, V. M., 1996, Proteolytic processing and particle maturation, *Curr. Top. Microbiol. Immunol.* **214**: 95–131.

Weber, I., 1990, Comparison of the crystal structures and intersubunit interactions of human immunodeficiency and Rous sarcoma virus proteases, *J. Biol. Chem.* **265**: 10492–10496.

Weldon, R. A. J., and Hunter, E., 1997, Unpublished results.

Yasuda, J., and Hunter, E., 1997, Unpublished results.

44

STRUCTURE AND INTRACELLULAR ASSEMBLY OF THE TRANSMISSIBLE GASTROENTERITIS CORONAVIRUS

C. Risco, I. M. Antón, M. Muntión, J. M. González, J. L. Carrascosa, and L. Enjuanes

Centro Nacional de Biotecnología (CSIC)
Campus Universidad Autónoma
28049 Madrid, Spain

1. ABSTRACT

Coronaviruses have been described as pleomorphic, round particles with a helical nucleocapsid as the unique internal structure under the virion envelope. Our studies on the organization of the transmissible gastroenteritis coronavirus (TGEV) have shown that the structure of these viruses is more complex. Different electron microscopy techniques, including cryomicroscopy of vitrified viruses, revealed the existence of an internal core, most probably icosahedral, in TGEV virions. Disruption of these cores induced the release of elongated ribonucleoprotein complexes. Ultrastructural analysis of freeze-substituted TGEV-infected swine testis (ST) cells showed characteristic intracellular budding profiles as well as two types of virions. While large virions with an electron-dense internal periphery are seen at perinuclear regions, smaller viral particles exhibiting compact internal cores of poligonal contours are more abundant in areas closer to the plasma membrane of the cell. These data strongly suggest that maturation events following the budding process are responsible for the formation of the internal core shell, the new structural element that we have recently described in extracellular infectious TGEV virions.

2. INTRODUCTION

The morphogenesis of coronaviruses has been defined as a complex process, that starts with the interaction of a viral nucleocapsid with membranes of the endoplasmic reticulum-Golgi intermediate compartment (ERGIC). This compartment is an specialized domain of the endoplasmic reticulum where the coronavirus M glycoprotein accumulates,

Coronaviruses and Arteriviruses, edited by Enjuanes *et al.*
Plenum Press, New York, 1998

defining the cellular location of viral assembly (Krijnse-Locker *et al.*, 1994). The assembled virions pass through the Golgi apparatus and finally use the constitutive secretory pathway to exit the cell. Our recent description of a new structural element in coronavirus particles, the internal core (Risco *et al.*, 1996), introduced a new component in the sequence of assembly. After visualizing the different structural elements that form the infectious extracellular virion, we have looked for them inside the infected cells. The use of cryomethods at the EM level is being decisive for the detailed characterization of the coronavirus structure and assembly.

3. MATERIALS AND METHODS

3.1. Electron Microscopy Characterization of Purified Extracellular Virions and Viral Cores

TGEV virions (PUR46-MAD strain) were purified from infected cultures of swine testis (ST) cells (Jiménez *et al.*, 1986). Negative staining of both intact and detergent-treated virions and viral cores was done with 2 % uranyl acetate or 2 % sodium phosphotungstate according to standard procedures (Risco *et al.*, 1996). Cryoelectron microscopy of vitrified, unstained hydrated virions was performed as described (Booy, 1993). Immunogold labeling with specific monoclonal antibodies (MAbs) (Gebauer *et al.*, 1991) and a 5 - nm colloidal gold conjugate was performed following previously stablished procedures (Risco *et al.*, 1995a).

3.2. Structural Analysis of TGEV-Infected Cells

TGEV-infected ST cells were fixed at different post-infection times before cryoprotection and rapid freezing in liquid propane. Vitrified samples were then submitted to freeze-substitution, embedding, ultrathin sectioning and immunogold labeling as described (Grief *et al.*, 1994; Risco *et al.*, 1995b). Samples were studied in a JEOL 1200 EX II electron microscope.

4. RESULTS

Our structural studies, based of different electron microscopy methods (Risco *et al.*, 1996) have revealed the existence of an internal core in TGEV virions. In particular, cryomicroscopy of vitrified viruses provided a direct view of the different viral components (Figure 1A-C). Since these specimens are hydrated and no chemical fixation or staining was used, the information obtained is very close to the native state of the structure. Purification of the viral cores was done after finding an adequate treatment with detergents, that releases the internal core after removing the envelopes (Figure 1D-F). The cores released by this treatment contained both M and N proteins, as well as RNA (Risco *et al.*, 1996). Disruption of the viral cores induced the release of an internal component that heavily reacted with anti-N MAbs and that, once separated from the core, exhibited an elongated structure covered by N protein molecules, according to immunogold detection (Figure 2).

The search for the generation of all these viral structural elements has been done in ultra-thin sections of infected cells processed by freeze-substitution. Using this method a considerable improvement of fine structural details (when compared with conventional

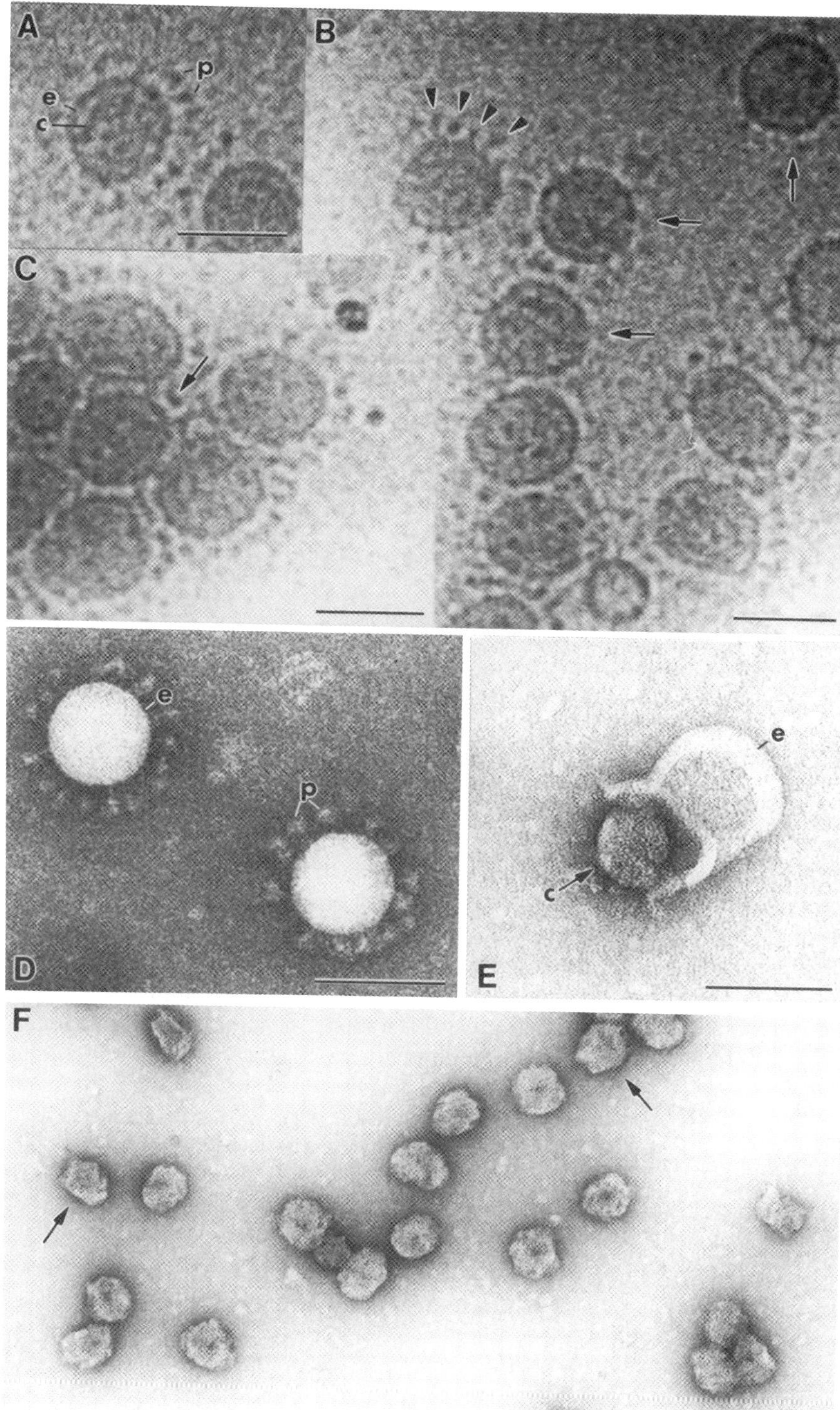

Figure 1. Electron microscopy visualization of intact and detergent-treated TGEV virions, as well as viral cores. A) to C) vitrified virions show well-extended peplomers (p) and a compact internal core (c), clearly separated from the viral envelope (e). D) and E) show intact and NP40-treated negative stained TGEV virions, respectively. The treatment with the detergent induces the release of the internal core (c) from the disrupted envelope (e). In F) a group of released viral cores is shown. The arrows point to particles that exhibit a clear polygonal contour. Bars: 100 nm.

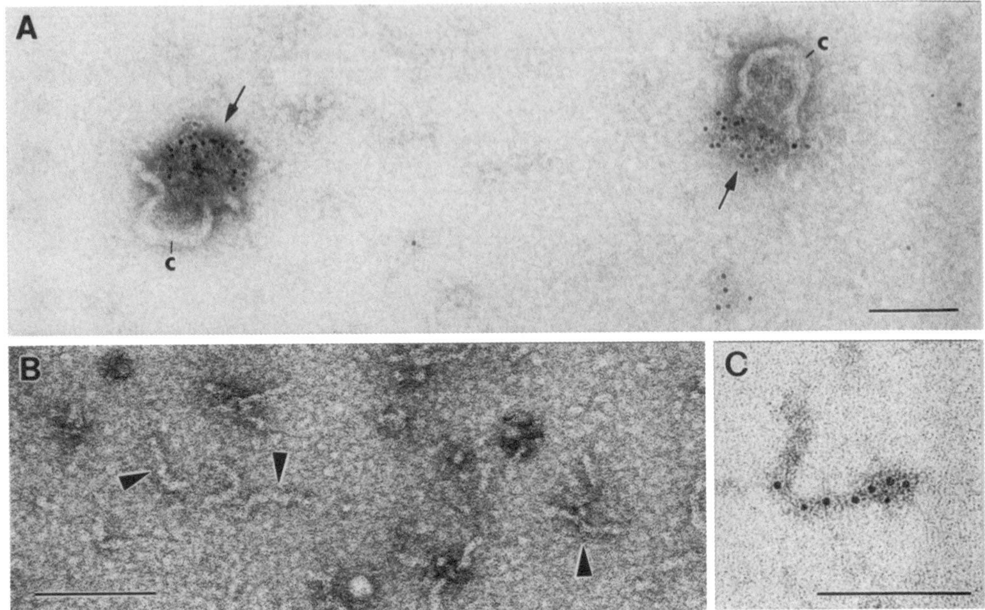

Figure 2. Release of N protein–containing aggregates from disrupted viral cores. A) Triton X-100-treated cores (c) release a material that strongly reacts with monoclonal antibodies against the nucleocapsid protein (arrows). B) The released material corresponds to elongated structures (arrowheads) of rather homogeneous length. These structures are covered by N protein molecules, according to immunogold detection with anti-N MAbs (C). Bars: 100 nm.

embedding procedures) has been obtained in ST cells. In freeze-substituted infected cells we have distinguished two types of viral particles. Large virions (apparently generated from budding profiles) as well as smaller viral particles with dense cores are clearly distinguished (Figure 3A). These smaller virions are similar in dimensions and internal organization to the extracellular purified TGEV viruses characterized in the first part of this study. Furthermore, and in spite of their great difference in volume and organization, both types of virions react with anti-TGEV antibodies (Figure 3B).

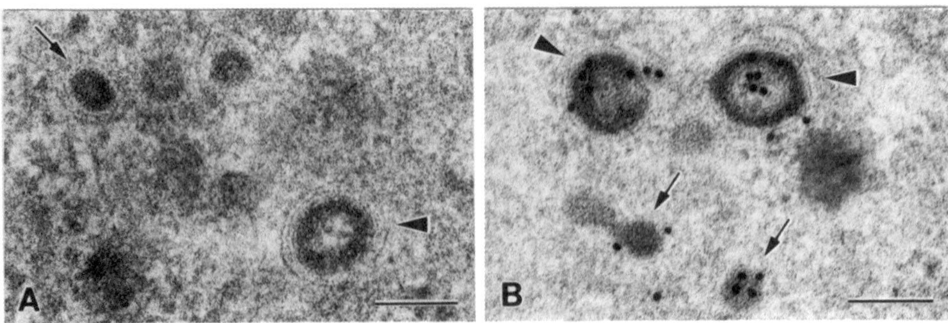

Figure 3. Ultra-thin sections of freeze-substituted TGEV-infected ST-cells. A) Small (arrow) and large (arrowhead) virions are distinguished. They both react with anti-TGEV antibodies, as shown by immunogold labeling (B). Bars: 100 nm.

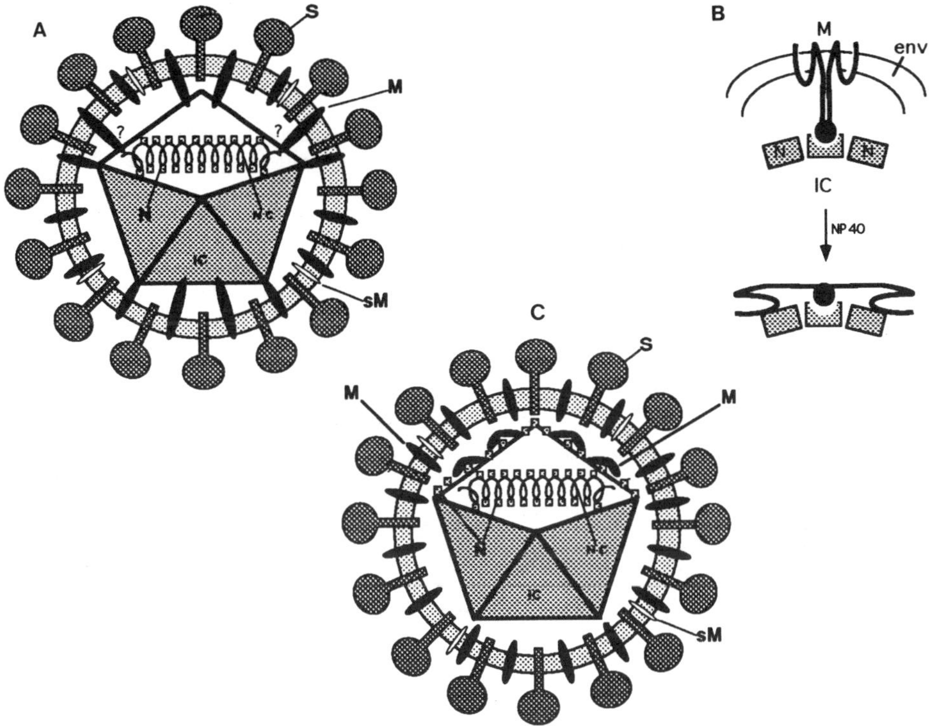

Figure 4. Structural models of TGEV showing two different possibilities to explain the association of M protein molecules to the surface of the purified viral core. A) Internal domains of envelope M protein molecules could interact with the viral core, remaining associated to it once the lipids are removed by the detergent (B). C) Hypothetically, two different populations of M protein molecules (that could exhibit or not different modifications) could separate during viral assembly. IC: internal core; NC: nucleocapsid; env: viral envelope.

5. DISCUSSION

The new model for the organization of coronaviruses (Figure 4), supported by our structural evidences has raised a number of questions concerning the organization of the different structural proteins within the viral particle. The envelope M protein, for example, was consistently found on the surface of purified viral cores. It is well documented that subviral particles prepared by NP-40 disruption of purified MHV, HEV, IBV, or BCV, still contained M protein associated with the nucleocapsids (Rottier, 1995). The significance of these interactions remains to be assessed. Envelope M protein molecules could interact with the internal core through an intravirion domain (Figure 4A), but a separation of two populations of M protein molecules (envelope and core-associated molecules) during viral assembly cannot be discarded (Figure 4B). High-resolution studies are necessary to define these aspects. Our structural analysis of TGEV-infected cells strongly suggest that assembly starts with the formation of large viral particles that undergo a dramatic reorganization to render smaller particles with assembled icosahedral cores. Studies on the cellular and viral factors involved in the sequence of assembly are now in progress. For this purpose, EM cryomethods, that provide an adequate preservation of fine structural details, are of key importance for the definition of the coronavirus morphogenesis.

REFERENCES

Booy, F.P., 1993, Cryoelectron microscopy, in: *Viral fusion mechanisms* (J. Bentz, ed.), CRC Press, Inc., Boca Raton, Fla., pp. 21–54.

Gebauer, F., Posthumus, W.A.P., Correa, I., Suñé, C., Sánchez, C.M., Smerdou, C., Lenstra, J.A., Meloen, R., and Enjuanes, L.,1991, Residues involved in the formation of the antigenic sites of the transmissible gastroenteritis coronavirus, *Virology* **183:** 225–238.

Grief, C., Nermut, M.V., and Hockley, D.J., 1994, A morphological study of freeze-substituted human and simian immunodeficiency viruses, *Micron.* **25:** 119–128.

Jiménez, G., Correa I., Melgosa, M.P., Bullido, M.P., and Enjuanes, L., 1986, Critical epitopes in transmissible gastroenteritis virus neutralization, *J. Virol.* **60:** 131–139.

Krijnse-Locker, J., Ericsson, M., Rottier, P.J., and Griffiths, G., 1994, Characterization of the budding compartment of mouse hepatitis virus: evidence that transport from the RER to the Golgi complex requires only one vesicular transport step, *J. Cell Biol.* **124:** 55–70.

Risco, C., Antón, I.M., Suñé, C., Pedregosa, A.M., Martín-Alonso, J.M., Parra, F., Carrascosa, J.L., and Enjuanes, L., 1995a, Membrane protein molecules of the transmissible gastroenteritis coronavirus also expose the carboxy-terminal region on the external surface of the virion, *J. Virol.* **69:** 5269–5277.

Risco, C., Menéndez-Arias, L., Copeland, T.D., Pinto da Silva, P., and Oroszlan, S., 1995b, Intracellular transport of the murine leukemia virus during acute infection of NIH 3T3 cells: nuclear import of nucleocapsid protein and integrase, *J. Cell Sci.* **108:** 3039–3050.

Risco, C., Antón, I.M., Enjuanes, L., and Carrascosa, J.L., 1996, The transmissible gastroenteritis coronavirus contains a spherical core shell consisting of M and N proteins, *J. Virol.* **70:** 4773–4777.

Rottier, J.M., 1995, The coronavirus membrane glycoprotein, in: *The Coronaviridae* (S.G. Sidell, ed.), Plenum Press, New York and London, pp.115–139.

45

CHARACTERIZATION OF CORONAVIRUS DI RNA PACKAGING

K. H. Kim, K. Narayanan, and S. Makino

Department of Microbiology, and Institute of Cellular and Molecular Biology
The University of Texas at Austin
Austin, Texas 78712

ABSTRACT

Studies of defective interfering (DI) RNAs of mouse hepatitis virus (MHV), suggest that a 69 nt-long packaging signal, which is located about 20 kb from the 5'-end of the 31 kb-long MHV genomic RNA, is necessary and sufficient for MHV genomic RNA packaging into MHV particles. We demonstrated that use of a low pH culture medium combined with subsequent ultrafiltration increased MHV infectivity about 60 times over MHV preparations grown in neutral medium. Using this virus concentration procedure, we successfully prepared DI particle-rich MHV preparations. Characterization of virus samples released from the cells infected with DI particle-rich MHV revealed that infectious MHV genomic RNA was not required for packaging of DI RNAs. These data suggested that interaction of the DI packaging signal with an unidentified region(s) of helper virus genomic RNA is unlikely, and therefore unlikely to facilitate the packaging of MHV DI RNA into the MHV virion. Rather, both DI RNA and MHV genomic RNA probably use the packaging signal for RNA packaging.

INTRODUCTION

Mouse hepatitis virus (MHV), contains three envelope proteins, S, M, and E proteins and an internal N protein. N protein and a 31 kb-long MHV genomic RNA (Lee *et al.*, 1991; Pachuk *et al.*, 1989) form a helical nucleocapsid (Macnaughton *et al.*, 1978). M protein and E protein are both necessary for packaging of viral nucleocapsid (Kim *et al.*, 1997) and MHV assembly (Kim *et al.*, 1997; Vennema *et al.*, 1996; Bos *et al.*, 1996), while S protein is dispensable for both (Kim *et al.*, 1997; Vennema *et al.*, 1996; Bos *et al.*, 1996). Of the 7–8 species of MHV mRNAs, which consist of a 3' coterminal nested-set

Coronaviruses and Arteriviruses, edited by Enjuanes *et al.*
Plenum Press, New York, 1998

structure, only mRNA 1 is packaged into MHV virions (Lai and Stohlman, 1978), indicating that there is a mechanism which allows specific packaging of mRNA 1 into MHV.

Studies of cloned defective interfering (DI) RNAs of MHV identified an MHV RNA signal (packaging signal) that is necessary for MHV DI RNA packaging into MHV particles (Fosmire *et al*., 1992). The 69 nt-long packaging signal is sufficient for RNA packaging into MHV particles (Woo *et al*., 1997). This packaging signal resides about 20 kb from the 5'-end of the MHV genome (Fosmire *et al*., 1992), and presumably is necessary for packaging of MHV mRNA 1 into MHV particles.

MHV DI particles are not separable from infectious MHV particles by sucrose gradient centrifugation (Fosmire *et al*., 1992; Makino *et al*., 1990), therefore, whether or not MHV genomic RNA is required for packaging of MHV DI RNAs into MHV particles is not known. If MHV genomic RNA is always required for packaging of DI RNAs, there is a possibility that the packaging signal in DI RNAs is not a signal for MHV genomic RNA, but is a signal that interacts with the helper virus genomic RNA; DI RNAs may "piggyback" MHV genomic RNA and be packaged into MHV particles. If this is the case, then the mechanism of MHV genomic RNA packaging should differ from that of DI RNA packaging; the identified 69-nt-long packaging signal should be necessary only for packaging of MHV DI RNAs, but not for packaging of MHV genomic RNA. If copackaging of MHV genomic RNA is not always necessary for DI RNA packaging, then the identified packaging signal is most likely one of the elements that is required for MHV RNA packaging.

To further understand the mechanism of coronavirus RNA packaging, we asked whether MHV genomic RNA is required for packaging of DI RNAs into virions. To address this question, we first needed to establish a procedure which significantly increases MHV infectivity. Our study demonstrated that copackaging of genomic RNA with MHV DI RNA was not required for MHV DI RNA packaging.

MATERIALS AND METHODS

Viruses and Cells

Plaque-cloned MHV-A59 was used as a helper virus. Mouse DBT cells (Hirano *et al*., 1974) were used for MHV growth.

Concentration of MHV

For ammonium sulfate precipitation, MHV-infected DBT cell culture fluid was brought to 50% saturation by ammonium sulfate followed by centrifugation at 8,000 rpm for 30 min using a Beckman JA-14 rotor. Virus was suspended in NTE buffer (0.1 M NaCl, 0.01 M Tris-HCl, pH 7.2, and 0.001 M EDTA) at 1/10–1/20 of the original volume and transferred into two different dialysis membranes, one with a molecular weight cut off of 10 kDa (MWCO 10 k membrane) and the other with a molecular weight cut off of 500 kDa (MWCO 500 k membrane). The virus suspension was then dialyzed against 1.5 liters of phosphate buffered saline (PBS) at 4°C for overnight. To the dialysate, we added kanamycin to a final concentration of 1 mg/ml and fetal calf serum to 2% and stored the samples at -80°C. For volume concentration by powdered polyethylene glycol 8000 (PEG), the culture fluid was transferred into either a 10 k molecular weight cutoff (MWCO) membrane or a 500 k MWCO membrane, and both membranes were buried in

PEG powder and placed at 0°C for 3–5 h. The volume of culture fluid was reduced through adsorption of water by the PEG powder. Once the total volume decreased to about 1/10–1/20 of the original volume, the virus samples were dialyzed against PBS and stored as described above. We used centricon 100 (Amicon) for ultrafiltration. Virus samples were placed into the top chamber of a centricon 100 unit, and centrifuged at 3,000 rpm using a Beckman JA-14 rotor. After 1 h of centrifugation, the volumes of the virus samples were reduced to about 1/20 th of the original volume. Concentrated virus samples were then stored at -80°C.

RNA Transcription and Transfection

MHV DI cDNA FA1 (Fosmire *et al.*, 1992) was linearized by Xba I digestion and RNA was synthesized in vitro as previously described (Makino and Lai, 1989). The lipofection method was used for RNA transfection as previously described (Felgner *et al.*, 1987).

Purification of MHV

MHV particles were purified by sucrose gradient centrifugation as described previously (Kim *et al.*, 1997).

Preparation and Characterization of Virion RNA and Intracellular RNAs

Synthesis of virus-specific RNAs was examined by labeling of virus RNA with ^{32}P-orthophosphate in the presence of 2.5 µg/ml actinomycin D for 2 h from 5 to 7 h postinfection (p.i.). Virion RNA and the intracellular RNA was extracted (Makino *et al.*, 1984) and analyzed by agarose gel electrophoresis (McMaster and Carmichael, 1977). Northern blot analysis was performed as described previously (Makino *et al.*, 1991). A 1 kb-long Stu I-Spe I fragment that corresponds to the 0.5–1.5 kb region from the 5' end of MHV DIssF clone, DF1–2 (Makino *et al.*, 1990), was used as a probe.

RESULTS

If MHV genomic RNA is always required for DI RNA packaging, a reasonable amount of MHV genomic RNAs should be detected in DI virion RNA preparations. If co-packaging of MHV genomic RNA is not required for DI RNA packaging, then the amount of MHV genomic RNA may be very minute in the DI particle-rich virus preparations. To examine whether the presence of MHV genomic RNA in virus particles was required for DI RNA packaging, we needed to prepare DI particle-rich MHV samples, in which most of the viruses were DI particles. Generally, virus samples containing large amounts of MHV DI particles were prepared from cell cultures in which most of cells were coinfected with DI particles and infectious helper virus. We attempted to enhance the concentration of DI particles by using various procedures that may increase infectivities of MHV.

We first tested the effect of culture medium pH on MHV replication, because some coronaviruses grow to higher titers in culture media with low pH than in neutral pH media (Alexander and Collins, 1975; Pocock and Garwes, 1975). In agreement with this, we found that MHV grew to a higher titer using the low pH (pH 6.0) medium than the neutral

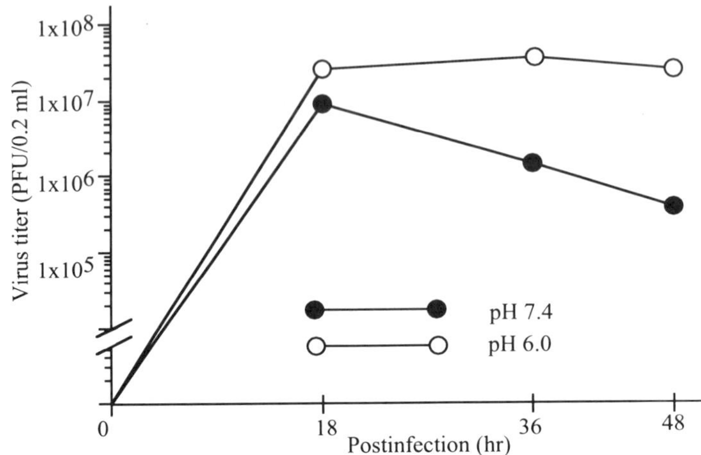

Figure 1. Growth of MHV in the neutral pH medium and low pH medium. DBT cells were infected with MHV at a multiplicity of infection (m.o.i.) of 5, and incubated with the neutral pH medium (black circles) or the low pH medium (open circles). Culture fluids were harvested at p.i., 18 h, 36 h and 48 h and virus infectivities were examined by plaque assay.

pH (pH 7.4) medium (Fig. 1). We then tested three different procedures for concentrating MHV. Ammonium sulfate precipitation, concentration by PEG powder, and ultrafiltration all increased MHV infectivity, however, ultrafiltration consistently showed excellent enhancement of infectivity and specific infectivity (Table 1). Enhancement of infectivity when using ammonium sulfate precipitation and PEG concentration was about half to two thirds that of the ultrafiltration procedure (data not shown). Inhibition of MHV RNA synthesis did not occur in the cells that were infected with MHV concentrated by ultrafiltration (Fig. 2A). MHV RNAs accumulated efficiently in the cells infected with MHV grown in the low pH medium (Fig. 2B). Although viral RNA was efficiently produced in the neutral pH cells at 18 h p.i., it was significantly lower at 36 h p.i. (Fig. 2B, lanes 3, 4); this lower level of MHV RNA synthesis at 36 h p.i. was not so obvious in the MHV samples

Table 1. Enhancement of MHV infectivities after concentration of viruses by using ultrafiltration

Concentration methods	MHV titer PFU/0.2 ml	Volume reduction (VR)[a]	Enhancement of infectivity (EI)[b]	Specific infectivity (VR×EI)
Exp. 1				
No concentration	1.5×10^6			
After concentration	2.3×10^7	1/20	15.3	0.77
Exp. 2				
No concentration	4.3×10^6			
After concentration	3.2×10^7	1/20	7.4	0.37
Exp. 3				
No concentration	5.0×10^5			
After concentration	4.4×10^6	1/20	8.8	0.44

[a]Final volume of virus samples after dialysis or ultrafiltration.

[b]Fold increase of MHV infectivities after concentration/dialysis or ultrafiltration as compared with MHV infectivities in culture fluid.

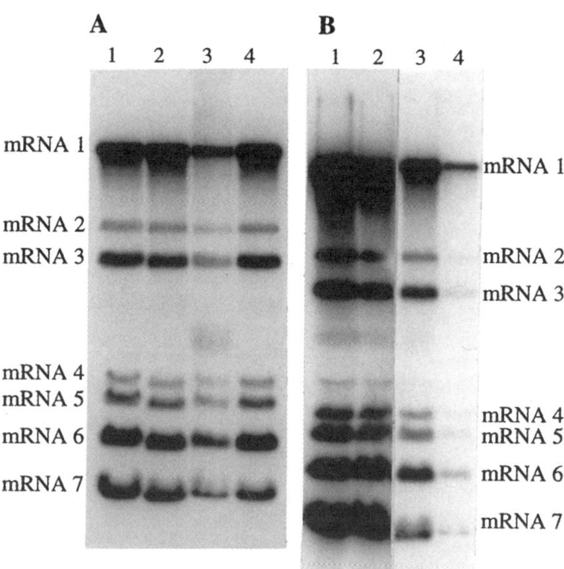

Figure 2. Synthesis of MHV-specific intracellular RNAs. (A) MHV samples that were concentrated by ultrafiltration were used as inoculum. Lanes 1 and 3 show MHV-specific intracellular RNAs after infection of untreated virus samples. Lanes 2 and 4 show MHV-specific intracellular RNAs after infection of concentrated virus (Centricon 100) samples. The m.o.i. of inoculum were 1, 15, 0.3 and 3 for lanes 1, 2, 3 and 4 respectively. (B) MHV-specific intracellular RNA synthesis after infection of MHV samples that were grown in the low pH (pH 6.0) medium (lanes 1, 2) or the neutral pH (pH 7.4) medium (lanes 3, 4). The m.o.i. of inoculum were 17, 23, 6 and 1 for lanes 1, 2, 3 and 4 respectively. RNA was extracted at 18 h p.i. for lanes 1 and 3, and 36 h p.i. for lanes 2 and 4. All MHV samples were inoculated without dilution. Species and location of seven MHV mRNAs are shown by mRNA 1–mRNA 7.

grown in the low pH medium (Fig. 2B, lanes 1, 2). To obtain MHV with the highest infectivity, we combined these two procedures. MHV-infected DBT cells were incubated in the low pH medium or neutral pH medium and harvested at 18 h p.i. The culture fluid from the low pH medium was concentrated by ultrafiltration. Plaque assay analysis showed that MHV released into the neutral pH medium had a titer of 3.5×10^6 PFU/0.2 ml, while the virus sample isolated after growth in cells cultured in the low pH medium followed by 20-fold ultrafiltration concentration yielded 2.2×10^8 PFU/0.2 ml. Combination of the two procedures increased MHV infectivity about 60 times.

FA1 DI RNA (Fosmire *et al.*, 1992), which contains the packaging signal, was synthesized in vitro and transfected into MHV-infected cells. Cells were grown in the low pH medium and the culture fluid was collected after overnight incubation. DBT cells were infected with this virus sample or virus samples that were concentrated by using ultrafiltration. Cells were cultured for 12 h in the presence of ^{32}Pi. Agarose gel electrophoresis of ^{32}P-labeled virion RNAs extracted from purified MHV particles showed that the majority of virion RNAs were DI RNAs (Fig. 3, lane 2). Only a trace amount of infectious virus genomic RNA was found in the virus sample containing FA1 DI particles, whereas a large amount of infectious virus genomic RNA was present in the helper virus preparation (Fig. 3, lane 1). If unconcentrated inoculum was used, we detected a higher amount of genomic RNA in some experiments (data not shown), while virus samples obtained from the cells infected with concentrated virus sample consistently showed the presence of a m-

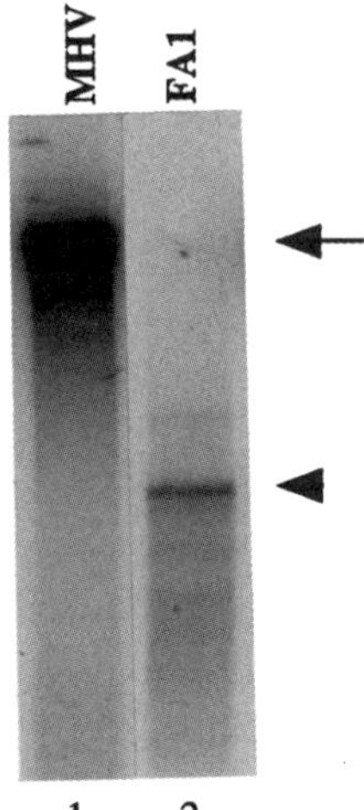

Figure 3. Agarose gel electrophoresis of ^{32}P-labeled virion RNA extracted from purified virus particles. ^{32}P-labeled MHVs were purified from cells infected with MHV-A59 (lane 1) or from cells infected with MHV-A59 containing FA1 DI particles (lane 2). Extracted virion RNAs were separated by 1% agarose-formaldehyde gel electrophoresis. Arrow indicates MHV-A59 genomic RNA. Arrowhead indicates FA1 DI RNA.

inute amount of helper virus genomic RNA (data not shown). Densitometric analysis of the autoradiogram showed that the molar ratio of helper virus genomic RNA to FA1 DI RNA was approximately 1 to 1,000. If packaging of MHV DI RNAs requires the presence of at least one infectious virus genomic RNA in a virion, then this data would have indicated that each MHV DI particle must contain, on average, one infectious genomic RNA plus 1,000 DI RNA molecules. However, the buoyant density of DI particles and that of helper virus is the same (Fosmire *et al.*, 1992; Makino *et al.*, 1990). It is highly unlikely that MHV particles that contained one infectious genomic RNA molecule plus 1,000 DI RNA molecules had the same density as infectious MHV. These data indicated not only that most DI RNA molecules were packaged into MHV virion in the absence of the infectious virus genome, but also that most DI particles contained only DI RNAs. Therefore, the possibility that the packaging signal in DI RNAs is a signal that interacts with the helper virus genomic RNA is highly unlikely. Packaging of MHV DI RNA and MHV genomic RNA probably underwent the same mechanism.

DISCUSSION

In the present study we found that the combination of ultrafiltration and low pH culture medium significantly increased MHV infectivity. This concentration procedure will be useful for the preparation of MHV samples with high infectivity.

The data shown in Fig. 3 indicated that MHV DI particles contained mostly DI RNAs; if DI particles that contain MHV genomic RNA exist at all, they exist only as a very minor population. That the packaging signal of DI RNA interacts with an unidentified region(s) of helper virus genomic RNA, and that such an interaction facilitates the packaging of MHV DI RNA into the virion is less likely than the possibility that both DI RNA and MHV genomic RNA probably use the packaging signal for RNA packaging. The data shown in this study was consistent with the observation that the kinetics of intracellular DI RNA synthesis after infection of the serially diluted DI particle-containing MHV preparation follows two hit-kinetics, while kinetics of helper virus mRNA synthesis follows one-hit kinetics based on virus dilution experiments (Makino *et al.*, 1988). We obtained a similar result using the efficiently packaged DI RNA, DIssF (Makino, unpublished data). If DI RNAs were always copackaged with helper virus genomic RNA, then DI RNA and helper virus genomic RNA should be cointroduced into host cells, and the virus dilution experiment would have resulted in one-hit kinetics of DI RNA.

ACKNOWLEDGMENTS

This work was supported by Public Health Service grants AI29984 and AI32591 from the National Institutes of Health.

REFERENCES

Alexander, D.J., and Collins, M.S., 1975, Effect of pH on the growth and cytopathogenicity of avian infectious bronchitis virus in chicken kidney cells, *Arch. Virol.* **49**:339–348.

Bos, E.C.W., Luytjes, W., van der Muelen, H., Koerten, H.K., and Spaan, W.J.M., 1996, The production of recombinant infectious DI-particles of a murine coronavirus in the absence of helper virus, *Virology* **218**:52–60.

Felgner, P.L., Gadek, T.R., Holm, M., Roman, R., Chan, H.W., Wenz, M., Northrop, J.P., Ringgold, G.M., and Danielson, M., 1987, Lipofection: a high efficient, lipid- mediated DNA-transfection procedure, *Proc. Natl. Acad. Sci. USA* **84**:7413–7417.

Fosmire, J.A., Hwang, K., and Makino, S., 1992, Identification and characterization of a coronavirus packaging signal, *J. Virol.* **66**:3522–3530.

Hirano, N., Fujiwara, K., Hino, S., and Matsumoto, M., 1974, Replication and plaque formation of mouse hepatitis virus (MHV-2) in mouse cell line DBT culture, *Arch. Gesamte. Virusforch.* **44**:298–302.

Kim, K.H., Narayanan, K., and Makino, S., 1997, Assembled coronavirus from complementation of two defective interfering RNAs, *J. Virol.* **71**:3922–3931.

Lai, M.M.C., and Stohlman, S.A., 1978, RNA of mouse hepatitis virus, *J. Virol.* **26**:236- 242.

Lee, H.-J., Shieh, C.-K., Gorbalenya, A.E., Eugene, E.V., La Monica, N., Tuler, J., Bagdzhadzhyan, A., and Lai, M.M.C., 1991, The complete sequence (22 kilobases) of murine coronavirus gene 1 encoding the putative proteases and RNA polymerase, *Virology* **180**:567–582.

Macnaughton, M.R., Davies, H.A., and Nermut, M.V., 1978, Ribonucleoprotein-like structures from coronavirus particles, *J. Gen. Virol.* **39**:545–549.

Makino, S., unpublished data.

Makino, S., and Lai, M.M.C., 1989, High-frequency leader sequence switching during coronavirus defective interfering RNA replication, *J. Virol.* **63**:5285–5292.

Makino, S., Taguchi, F., Hirano, N., and Fujiwara, K., 1984, Analysis of genomic and intracellular viral RNAs of small plaque mutants of mouse hepatitis virus, JHM strain, *Virology* **139**:138–151.

Makino, S., Shieh, C.-K., Keck, J.G., and Lai, M.M.C., 1988, Defective-interfering particles of murine coronavirus: mechanism of synthesis of defective viral RNAs, *Virology* **163**:104–111.

Makino, S., Yokomori, K., and Lai, M.M.C., 1990, Analysis of efficiently packaged defective interfering RNAs of murine coronavirus: localization of a possible RNA- packaging signal, *J. Virol.* **64**:6045–6053.

Makino, S., Joo, M., and Makino, J.K., 1991, A system for study of coronavirus mRNA synthesis: a regulated, expressed subgenomic defective interfering RNA results from intergenic site insertion, *J. Virol.* **65**:6031–6041.

McMaster, G.K., and Carmichael, G.G., 1977, Analysis of single- and double-stranded nucleic acids on polyacrylamide and agarose gels by using glyoxal and acridine orange, *Proc. Natl. Acad. Sci. USA* **74**:4835–4838.

Pachuk, C.J., Bredenbeek, P.J., Zoltick, P.W., Spaan, W.J.M., and Weiss, S.R., 1989, Molecular cloning of the gene encoding the putative polymerase of mouse hepatitis virus, strain A59, *Virology* **171**:141–148.

Pocock, D.H., and Garwes, D.J., 1975, The influence of pH on the growth and stability of transmissible gastroenteritis virus in vitro, *Arch. Virol.* **49**:239–247.

Vennema, H., Godeke, G.-J., Rossen, J.W.A., Voorhout, W.F., Horzinek, M.C., Opstelten, D.-J.E., and Rottier, P.J.M., 1996, Nucleocapsid-independent assembly of coronavirus-like particles by co-expression of viral envelope protein genes, *EMBO J.* **15**:2020–2028.

Woo, K., Joo, M., Narayanan, K., Kim, K.H., and Makino, S., 1997, Murine coronavirus packaging signal confers packaging to nonviral RNA, *J. Virol.* **71**:824–827.

46

CORONAVIRUS NUCLEOCAPSID PROTEIN

RNA Interactions

Raymond Cologna[1] and Brenda G. Hogue[1,2]

[1]Division of Molecular Virology
[2]Baylor College of Medicine
Houston, Texas

1. ABSTRACT

The coronavirus nucleocapsid protein (N) is involved in encapsidation and packaging of viral RNA. In this study we investigated the ability of the bovine coronavirus (BCV) N protein to interact with RNA. Histidine-tagged BCV N (his-N) protein was expressed in bacteria. A filter binding assay was established to quantitatively measure the binding efficiency of purified his-N to different RNAs. The results indicate that bacterially expressed N bound both BCV and mouse hepatitis coronavirus (MHV) RNAs. Binding to in vitro generated BCV and MHV RNA transcripts was significantly higher than binding to a non-coronavirus RNA. Similar binding efficiencies were measured for a BCV defective genome, pDrep, and a transcript that contained the MHV packaging signal. Interestingly, the entire MHV DI, pMIDI-C, was bound at a higher efficiency than the packaging signal alone. This is one of the first reports to show that N interacts with the MHV packaging signal.

2. INTRODUCTION

The coronavirus N protein interacts with the genomic RNA to form a helical nucleocapsid (for review see Cavanagh, et al., 1994; Siddell, 1995). The helical nature of the nucleocapsid is unique for a positive-strand RNA animal virus, since all others have icosahedral symmetry. A new structural model for the coronavirus core was recently proposed which suggests that the transmissible gastroenteritis coronavirus has a spherical internal core composed of the membrane protein (M), N and genomic RNA (Risco et al., 1996). Further study will be necessary to determine if the description of the coronavirus

Coronaviruses and Arteriviruses, edited by Enjuanes *et al.*
Plenum Press, New York, 1998

nucleocapsid should be modified. Despite this, it is clear that N encapsidates the coronavirus genomic RNA, but the mechanism of encapsidation is not well defined.

A 69-nucleotide (nt) packaging signal that maps within open reading frame 1b (ORF 1b) was previously identified within MHV defective interfering (DI) RNAs (Fosmire, et al., 1992; Van der Most et al., 1991; Makino et al., 1990). Non-MHV transcripts containing the packaging signal are packaged into MHV virions (Woo et al., 1997). Furthermore, a subgenomic MHV RNA containing the packaging signal is sufficient for packaging, even though the subgenomic is packaged less efficiently than the DI genomic RNA (Bos et al., 1997).

A BCV defective RNA, pDrep, is replicated and packaged in BCV infected cells (Chang and Brian, 1996). The defective RNA contains 288 nts from the 5' end of the genome (ORF 1a of gene 1), the entire 1,344 nt coding region of the N gene, all of the 3'-noncoding region (291 nts) and the 68 nt poly(A) tail (Chang and Brian, 1996). Since pDrep lacks the ORF 1b region of the genome where the MHV packaging signal maps, what specifies packaging of the replicon? It was previously reported that, in addition to genomic RNA, BCV subgenomic mRNAs are packaged during virus infection (Hofmann et al., 1990). Collectively, this suggests that BCV may contain an alternative packaging signal.

Whatever sequence(s) functions as a packaging signal, we expect that there must be a specific interaction between that signal and the N protein. We currently do not understand how N recognizes and interacts with the RNA to encapsidate the viral genome. A direct interaction between N and the MHV packaging signal has not been demonstrated. As part of a study to understand the mechanism by which N encapsidates viral RNA we have examined the ability of BCV N to bind RNA in vitro.

3. MATERIALS AND METHODS

3.1. Bacterial Expression and Purification

The coding sequence for the BCV N protein was subcloned into the pQE.31 vector (Qiagen). Induction and denaturing purification using 8 M urea were carried out basically as described by Qiagen. Purified histidine tagged N (his-N) was renatured by dialysis, concentrated and the protein concentration was determined. Protein purity was monitored by SDS-PAGE and Coomassie blue staining.

3.2. Filter Binding Assay

Purified protein was incubated in reaction buffer with non-specific competitor RNAs, heparin, and ^{32}P labeled in vitro generated transcripts (Fig. 1) for 20 min at room temperature. Reactions were applied to pre-wetted nitrocellulose filters, washed, air dried and Cerenkov counted. Binding was calculated using the following formula: % RNA bound = (counts per minute (cpm) retained on the filter/total cpm added to the reaction) (100).

4. RESULTS AND DISCUSSION

Purified bacterially expressed, his-N was used to measure the ability of N to bind and retain ^{32}P labeled in vitro transcripts on nitrocellulose membranes in a filter binding assay. Six different RNAs have been analyzed at this time. A schematic representation of the transcripts is shown in Figure 1.

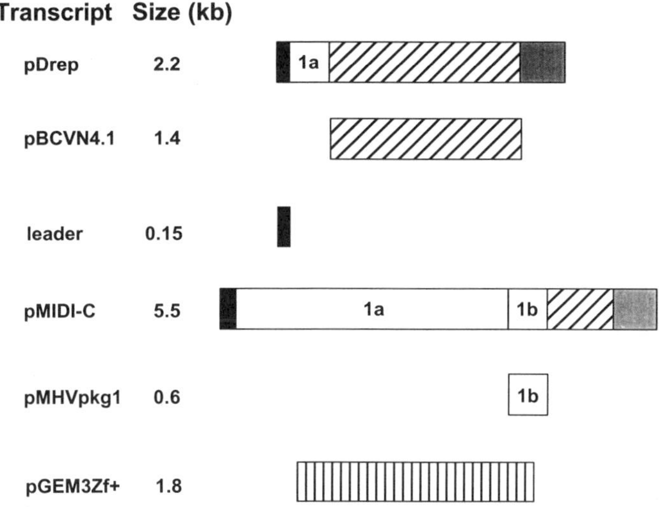

Figure 1. Schematic representation of RNA transcripts used in filter binding experiments. Leader sequences (black boxes), the nucleocapsid coding sequences (hatched boxes) and the 3' non-coding sequences (shaded boxes) are indicated for the BCV (pDrep, pBCVN4.1 and leader) and MHV (pMIDI-C and pMHVpkg1) transcripts. The non-coronavirus RNA, pGEM3Zf+, is represented as a vertically striped box. The size in kilobases of each transcript is indicated.

Initially we analyzed the interactions between his-N and the BCV transcripts. ^{32}P-labeled in vitro generated RNAs were incubated with increasing amounts of purified his-tagged N (Fig. 2). Both pDrep, the BCV defective RNA, and pBCVN4.1, the coding region of the BCV N gene, were bound at similar efficiencies. The BCV transcripts were bound more efficiently than a non-coronavirus RNA, pGEM3Zf+. The higher binding efficiency indicates that there is binding specificity for the viral RNAs.

We also tested the binding efficiency between a leader-containing transcript and his-N. The transcript contained 150 nts from the 5' end of the BCV genome, including the leader and intergenic region. The leader containing RNA was retained on the filter at a

Figure 2. His-tagged BCV N protein binding efficiencies for BCV RNAs and a non-coronavirus RNA. Increasing amounts of his-N were incubated for 20 min with 0.1 nM ^{32}P labeled in vitro transcripts under optimal binding conditions. Percent binding values represent the mean of three separate filter binding experiments for each transcript. Error bars reflect the standard deviation.

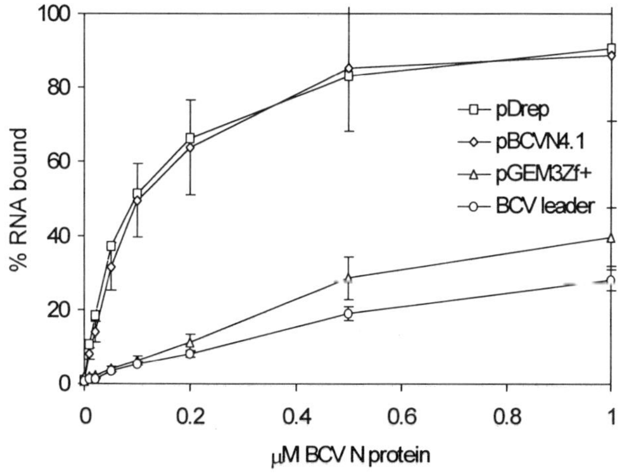

level lower than the background binding of the non-coronavirus RNA, suggesting that there is little binding affinity for the leader-containing transcript. We were surprised at this result since MHV N binds the leader, both in vitro and in vivo (Baric, et al., 1988; Stohlman, et al., 1988). We have also determined that genomic and all subgenomics are coimmunoprecipitated with N from BCV infected cells (Cologna and Hogue, unpublished data). We assumed until we obtained the filter binding results that N was interacting with the leader. It is possible that the 150 nt transcript is not sufficient to allow for formation of a structure that may be necessary for N binding. Further mapping will be required to re-solved this. Based on our limited mapping at this time, we hypothesize that the packaging signal for pDrep will map within the nucleocapsid coding region.

Interactions between N and a coronavirus packaging signal have not been described. We have not yet identified a packaging signal for pDrep. Therefore, we did not know what kind of binding efficiency to expect from N interactions with such a signal in the filter binding assay. Since the packaging signal has been identified for MHV, we used the filter binding assay to determine how efficiently his-N bound the MHV DI, pMIDI-C (De Groot et al., 1992), and have compared this with the binding efficiency of a transcript, pMHVpkg1, that contained the MHV packaging signal (Fig. 3). We previously deter-mined, using other assays, that BCV N and MHV N reciprocally interact with pDrep and pMIDI-C (Cologna and Hogue, unpublished data). For comparison pDrep was analyzed in parallel with the MHV transcripts. His-N bound pDrep and MHVpkg1 at similar efficien-cies. Surprisingly, the binding efficiency to the entire transcript from MIDI-C was signifi-cantly higher (Fig. 3). It is possible that pMIDI-C contains multiple sequences which interact with N at different efficiencies. The additional N interactions may account for the more efficient packaging of the MHV DI genome in the recent study by Bos and col-leagues (Bos et al., 1997). It was suggested that cis-acting enhancement elements or struc-ture may be important for efficient packaging (Bos et al, 1997).

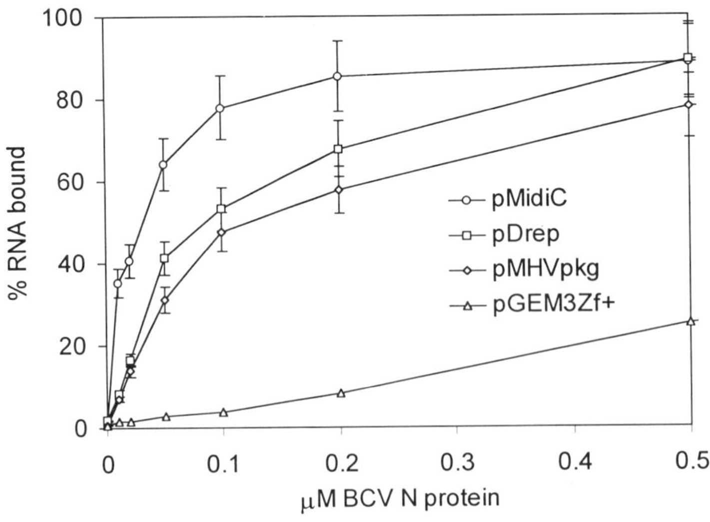

Figure 3. Comparison of his-N binding efficiencies for pMIDI-C, MHV packaging signal transcripts and pDrep. Binding efficiencies were measured as described in the legend for Fig. 2. Results from the mean of three separate experiments are shown for all transcripts except pGEM3Zf+.

The measurement of the binding efficiency between N and the MHV packaging signal is one of the first reports demonstrating that N can interact with the packaging signal. It is the first report that provides information on N:BCV RNA interactions. This provides a first step toward the goal of understanding the mechanism of coronavirus encapsidation. Further analysis will be required to more precisely define the N:RNA interactions that play an important role in encapsidation of MHV and BCV genomic RNAs.

ACKNOWLEDGMENTS

We thank Willy Spaan for generously providing the pMIDI-C clone. This research was supported by Public Health Service grant AI33500 from the National Institutes of Health. R. C. was supported in part by National Institutes of Health training grant AI07471.

REFERENCES

Baric, W. R., Nelson, G. W., Fleming, J. O., Deans, R. J., Keck, J. G., Casteel, N., and Stohlman, S. A., 1988, Interactions between coronavirus nucleocapsid protein and viral RNAs: implications for viral transcription, *J. Virol.* **62**:4280–4287.

Bos, E. C. W., Dobbe, J. C., Luytjes, W., and Spaan, W. J. M., 1997, A subgenomic mRNA transcript of the coronavirus mouse hepatitis virus strain A59 defective interfering (DI) RNA is packaged when it contains the DI packaging signal, *J. Virol.* **71**:5684–5687.

Cavanagh, D., and The Coronaviridae Study Group of the International Committee on Taxonomy of Viruses, 1994, Revision of the taxonomy of the Coronavirus, Torovirus, and Arterivirus general, *Arch. Virol.* **135**:226–237.

Chang, R.-Y., and Brian, D. A., 1996, Cis requirement for N-specific protein sequence in bovine coronavirus defective interfering RNA replication, *J. Virol.* **70**:2201–2207.

De Groot, R. J., Van der Most, R. G., and Spaan, W. J. M., 1992, The fitness of defective interfering murine coronavirus Di-1 and its derivatives is decreased by nonsense and frameshift mutations, *J. Virol.* **66**:5898–5905.

Fosmire, J., Hwang, K., Makino, S., 1992, Identification and characterization of a coronavirus packaging signal, *J. Virol.* **66**:3522–3530.

Makino, S., Yokomori, K. and Lai, M. M. C., 1990, Analysis of efficiently packaged defective interfering RNAs of murine coronavirus: localization of a possible RNA-packaging signal, **64**:6045–6053.

Siddell, S. G. 1995, The Coronaviridae. An introduction, p 1–10. In S. G. Siddell (ed), *The Coronaviridae*, Plenum Press, New York.

Stohlman, S. A., Baric, R. S., Nelson, G. N., Soe, L. H., Welter, L. M. and Deans, R. J., 1988, Specific interactions between coronavirus leader RNA and nucleocapsid protein, *J. Virol.* **62**: 4288–4295.

Van der Most, R. G., Bredenbeek, P. J., and Spaan, W. J. M., 1991, A domain at the 3' end of the polymerase gene is essential for encapsidation of coronavirus defective interfering RNAs, *J. Virol.*, **65**:3219–3226.

Woo, K., Joo, M, Narayanan, K., Kim, K. H., and Makino, S., 1997, Murine coronavirus packaging signal confers packaging to nonviral RNA, *J. Virol.* **71**:824–827.

47

CORONAVIRUS ENVELOPE GLYCOPROTEIN ASSEMBLY COMPLEXES

Vinh-Phuc Nguyen[1] and Brenda G. Hogue[1,2]

[1]Division of Molecular Virology
[2]Department of Microbiology and Immunology
Baylor College of Medicine
Houston, Texas 77030

1. ABSTRACT

Protein:protein interactions, and their subcellular localization, play important roles in coronavirus assembly. In this study, we have identified similar envelope glycoprotein complexes that are present in mouse hepatitis coronavirus A59 (MHV-A59) and bovine coronavirus (BCV) infected cells. Complexes consisting of the spike (S) and membrane (M) proteins were identified in cells infected with MHV-A59 or BCV. Kinetic analyses demonstrated that S and M quickly associated after translation, and suggested that both initially interacted in a pre-Golgi site. In addition, the hemagglutinin esterase (HE) was identified as part of a complex with M and S in BCV infected cells. Taken together, our data indicate that similar glycoprotein complexes are present in cells infected with two different coronaviruses, and thus likely represent important prerequisite complexes involved in virus assembly.

2. INTRODUCTION

All coronaviruses contain two major structural envelope glycoproteins, the membrane protein (M) and the spike protein (S). A few molecules of the envelope protein (E) are also found in purified virions. BCV is the prototype of coronavirus strains which, in addition to S and M, also express a third major envelope glycoprotein, the hemagglutinin esterase (HE) (King, 1982). An interesting feature of the coronavirus life cycle is their assembly at the internal membranes (Tooze, 1984; Krijinse-Locker, 1994). This feature distinguishes them from other enveloped, RNA-containing viruses, like influenza virus and vesicular stomatitis virus which assemble and bud at the plasma membrane. In addition,

Coronaviruses and Arteriviruses, edited by Enjuanes *et al.*
Plenum Press, New York, 1998

coronaviruses do not possess a typical membrane-associated matrix protein. Thus, their assembly is thought to be mediated by direct interactions between the viral glycoproteins and the nucleocapsid at the membranes where virions bud. However, the precise molecular mechanism driving coronavirus assembly is not fully defined.

To investigate how the coronavirus glycoproteins are assembled into virions, we sought to identify protein complexes that are present in virus infected cells. Part of the data from our analyses of both MHV-A59 and BCV infected cells are presented here. Specific viral protein complexes that likely represent prerequisite components for the assembly of virions were detected by coimmunoprecipitations. Pulse-chase experiments were used to follow the kinetics of protein association in MHV-A59 infected cells. We then extended our study to include BCV since this virus expresses and incorporates HE into virions, in addition to S and M. HE was identified in a complex with M and S in virus infected cells.

3. MATERIALS AND METHODS

3.1. Viruses and Cell Lines

MHV-A59 and BCV Mebus strains were grown in DBT and HCT-8 cells, respectively.

Both cell types were grown in Dulbecco's modified Eagle medium (Gibco BRL) supplemented with 10% fetal calf serum.

3.2. Virus Infection, Radiolabeling, and Immunoprecipitation

Cells were infected at a multiplicity of infection (moi) of 5–10. At 15 h post-infection with BCV or 5 h post-infection with MHV-A59, cells were starved for 30 min in methionine-deficient medium, labeled for 10 min with 200 (Ci/ml of EXPRE35S35S (NEN), washed and harvested immediately or chased in the presence of excess methionine and cysteine for the indicated times.

Cells were lysed on ice in buffer containing 50 mM Tris (pH 8.0), 62.5 mM EDTA, 0.5% Nonidet P40 (NP40), 0.5% sodium deoxycholate (DOC) and 1 mM phenylmethylsulfonyl fluoride (PMSF). Nuclei and cellular debris were removed by centrifugation at 4°C for 10 min at 13K x g. Clarified lysates were precleared with protein A Sepharose CL-4B, and immunoprecipitated with specific antibodies. Antigen-antibody complexes were collected with protein A Sepharose and washed four times with RIPA buffer [50 mM Tris-HCl (pH 7.5), 150 mM NaCl, 1% NP40, 0.5% DOC, 0.1% SDS] and once in RIPA without detergents. Proteins were eluted in sample buffer, resolved on 5–20% gradient gels by SDS-PAGE, and fluorographed.

3.3. Antibodies

All of the antibodies against the BCV proteins that were used in this study, with the exception of a polyclonal anti-BCV N, were previously described (Deregt and Babiuk, 1987; Deregt et al., 1987). The anti-BCV N polyclonal was made against histidine-tagged BCV N protein (Cologna and Hogue, unpublished data). The monoclonals against the MHV proteins were previously described (Fleming et al., 1983).

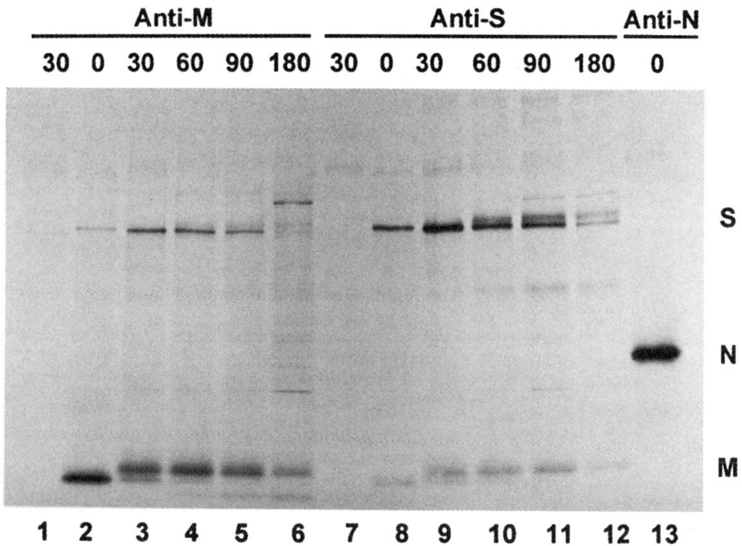

Figure 1. Kinetics of protein complex formation between MHV-A59 glycoproteins. At 5 hours post-infection, MHV-A59 infected DBT cells were pulsed with a mix of ^{35}S methionine/cysteine for 10 min, and chased in medium containing an excess of methionine and cysteine. At the indicated time points, cells were lysed, and immunoprecipitated with a mix of anti-M monoclonal antibodies (MAbs) (J1.3 and J2.7) (lanes 1–6), an anti-S MAb (J7.5) (lanes 7 to 12), or a mix of anti-N MAbs (J3.1 and J3.3) (lane 13). Eluted proteins were resolved by 5–20% gradient SDS-PAGE, and fluorographed. Mock-infected cell lysates were used as negative controls (lanes 1 and 7). Letters denote the MHV-A59 protein designations: S (spike), N (nucleocapsid), and M (membrane).

4. RESULTS

To begin our studies on virus assembly we initially identified protein complexes that were present in MHV-A59 infected cells. Mock- and virus infected cells were pulse labeled for 10 minutes with ^{35}S methionine/cysteine and chased in the presence of excess label. Complexes consisting of S and M were detected immediately after the pulse when lysates were immunoprecipitated with either anti-M or anti-S antibodies (Fig. 1, lanes 2 and 8). The MHV-A59 M has been shown to be unglycosylated in the endoplasmic reticulum (ER), and becomes O-glycosylated in the Golgi (Krijinse-Locker, 1992). After a 10 min of pulse, the S-associated M appeared to be unglycosylated, and subsequently acquired sugar chains through the chase (Fig. 1, compare lane 12 to 8). Similar post-translation modifications, namely N-linked glycosylation, also appeared to occur on the M-associated S (Fig. 1, compare lane 6 to 2). This suggests that S and M initially associated in a pre-Golgi site, most likely in the ER. The S-M complexes were subsequently transported along the exocytic pathway, probably as part of assembled virions.

Next we extended our studies on coronavirus assembly to include BCV, with an interest in the hemagglutinin esterase (HE) glycoprotein. To date, the molecular mechanism of incorporation of HE into virions has not been addressed. HCT cells were infected with BCV Mebus strain at a moi of 10. At 15 h post-infection, virus infected cells were metabolically labeled for 20 min, chased for 1 hour, lysed and immunoprecipitated with specific anti-BCV antibodies. Similar to MHV-A59 infected cells, complexes consisting of S and M were also identified (Fig. 2, lanes 3–4). Furthermore, HE was also found in a complex with S-M (Fig. 2,

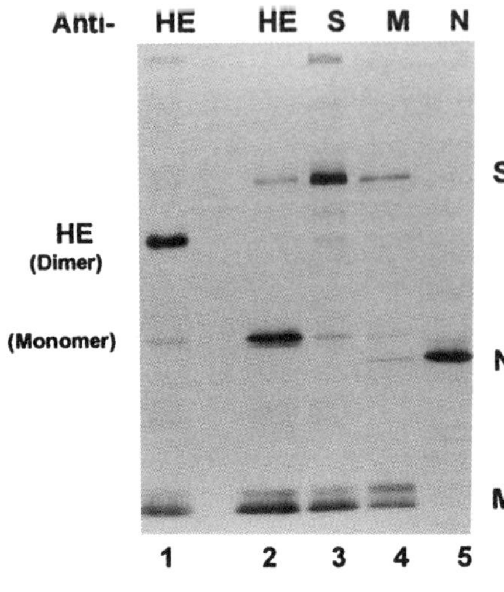

Figure 2. Detection of protein complexes in BCV infected cells. At 15 hours post-infection, BCV infected HCT cells were metabolically pulsed for 20 min, chased for 1 h, lysed and immunoprecipitated wit`h specific anti-BCV protein antibodies. To visualize HE dimers, immunoprecipitated proteins were eluted in sample buffer without β-mercaptoethanol (nonreducing conditions) (lane 1).

lane 2–4). When proteins were eluted in the absence of reducing agent, the majority of immunoprecipitated HE was the dimeric form (Fig. 2, lane 1). Parallel immunoprecipitation with a rabbit anti-N antiserum showed that HE and S are not in a complex with N, even though a small amount of N-M complexes were coimmunoprecipitated (Fig. 2, compare lane 4–5).

5. DISCUSSION

Both protein:protein and protein:RNA interactions, as well as the subcellular localization of protein complexes, are thought to be important for coronavirus assembly. To date, the molecular mechanism of coronavirus assembly remains largely undefined. This study aimed at examining the interactions between the viral envelope proteins that ultimately lead to the formation of mature particles in virus infected cells. In this report, we showed that similar protein complexes were detected by coimmunoprecipitation from cells infected with either MHV-A59 or BCV. The complexes are likely important prerequisite components for assemby of virions.

Our data showed that in MHV-A59 infected cells, complexes of S and M were readily detected after a 10 min pulse. This suggests that these glycoproteins quickly associate after synthesis. Shorter pulse-labeling yielded identical results (data not shown). The kinetics of M-S complex formation differ from those previously reported for MHV (Opstelten et al., 1995). Different cells and antibodies may account for this. Otherwise, our data are consistent with the previously reported results.

The M glycoprotein has been shown to be transported to the Golgi when it is expressed in the absence of other coronavirus proteins (Klumperman, 1994). Nevertheless, it remains the only viral protein that is retained in the same general area where coronaviruses bud. Therefore it has been proposed that M, through self-oligomerization and interactions with other viral proteins, retains and organizes the glycoproteins in the membrane so the ribonucleocapsid complex can bind and initiate virion budding (Krijinse-Locker, 1995). Based on their glycosylation patterns, S and M appear to initially associate at a pre-Golgi site, presumably in ER, and the formed complexes are subsequently transported to the Golgi. Our results correlate well with the proposed scheme of events that lead to the formation of coronavirus particles.

The HE glycoprotein is a major BCV envelope glycoprotein that is absent from MHV A59 virions (Hogue, 1984). To date, the molecular mechanism of HE incorporation in BCV particles has yet to be addressed. This is the first report describing the presence of HE in a complex with S and M. HE does not directly interact with S, but initially interacts with M and later becomes part of complex consisting of S-M-HE (Nguyen and Hogue, submitted). HE has been shown to rapidly dimerize after synthesis (Hogue, 1989). It is unknown at present which form of HE, monomeric or dimeric, initially associates with M; however, only dimers are found as part of S-M-HE complexes (Nguyen and Hogue, submitted).

Taken altogether, our data showed that the viral envelope glycoproteins associate to form stable complexes. These complexes are present in both MHV and BCV infected cells. Since common complexes are detected in cells infected with two different coronaviruses, this supports the idea that these are important preassembly complexes. An additional glycoprotein, HE, is readily accommodated as part of the glycoprotein complexes. These complexes presumably create a favorable micro-environment which excludes cellular proteins, and allows the nucleocapsid to bind and form budding progeny virions.

ACKNOWLEDGMENTS

We thank John Fleming and Lorne Babiuk for generously providing antibodies. This work was supported by Public Health Service grant AI33500 from the National Institutes of Health.

REFERENCES

Deregt, D. and L. A. Babiuk., 1987, Monoclonal antibodies to bovine coronavirus: characteristics and topographical mapping of neutralizing epitopes on E2 and E3 glycoproteins, *Virology* **161**:410–420.

Deregt, D., M. Sabara, and L. A. Babiuk, 1987, Structural proteins of bovine coronavirus and their intracellular processing, *J. Gen. Virol.* **68**:2863–2877.

Fleming, J. O., S. A. Stohlman, R. C. Harmon, M. M. Lai, J. A. Frelinger, L. P. Weiner., 1983, Antigenic relationships of murine coronaviruses: analysis using monoclonal antibodies to JHM (MHV-4) virus, *Virology* **131**:296–307

Hogue, B. G., B. King, and D. A. Brian., 1984, Antigenic relationship among proteins of bovine coronavirus, human respiratory coronavirus OC43, and mouse hepatitis coronavirus A59, *J. Virol.* **51**:384–388.

Hogue, B. G., T. E. Kienzle, and D. A. Brian., 1989, Synthesis and processing of the bovine enteric hemagglutinin protein, *J. Gen. Virol.* **70**:345–352.

King, B., and D. A. Brian., 1982, Bovine coronavirus structural proteins, *J. Virol.* **42**:700–707.

Klumperman, J., J. Krijinse-Locker, A. Meijer, M. C. Horzinek, H. J. Geuze, and P. J. M. Rottier., 1994, Coronavirus M proteins accumulate in the Golgi complex beyond the site of virus budding, *J. Virol.* **68**:6523–6534.

Krijinse-Locker, J., G. Griffiths, M. C. Horzinek, and P. J. M. Rottier., 1992, O-glycosylation of the coronavirus M protein, *J. Biol. Chem.* **267**:14094–14101.

Krijinse-Locker, J., M. Ericsson, P. J. M. Rottier, and G. Griffiths., 1994, Characterization of the budding compartment of mouse hepatitis virus: Evidence that transport from the RER to the Golgi complex requires only one vesicular transport step, *J. Cell. Biol.* **124**:55–70.

Krijinse-Locker, J., D.-J. E. Opstelten, M. Ericsson, M. C. Horzinek, and P. J. M. Rottier., 1995, Oligomerization of a trans-Golgi/trans-Golgi network retained protein occurs in the Golgi complex and may be part of its retention, *J. Biol. Chem.* **270**:8815–8821.

Opstelten, D.-J. E., M. J. B. Raasmsman, K. Wolfs, M. C. Horzinek, and P. J. M. Rottier., 1995, Envelope glycoprotein interactions in coronavirus assembly, *J. Cell. Biol.* **131**:339–349.

Tooze, J., S. A. Tooze, and G. Warren., 1984, Replication of coronavirus MHV-A59 in Sac- cells: determination of the first site of budding of progeny virions, *Eur. J. Cell. Biol.* **33**:281–293.

48

CORONAVIRUS ENVELOPE ASSEMBLY IS SENSITIVE TO CHANGES IN THE TERMINAL REGIONS OF THE VIRAL M PROTEIN

C. A. M. de Haan, H. Vennema, and P. J. M. Rottier

Institute of Virology
Department of Infectious Diseases and Immunology
Faculty of Veterinary Medicine, and Institute of Biomembranes
Utrecht University, Yalelaan 1
3584 CL Utrecht, The Netherlands

1. ABSTRACT

Recently we demonstrated that the co-expressed coronavirus membrane proteins have the capacity to assemble viral envelopes which are similar to normal virus particles in dimensions and appearance, and which can form independent of a nucleocapsid (Vennema *et al.*, 1996). For the formation of these particles only the M and the E protein are required; the S protein is dispensable but is incorporated when present. As we illustrate here, this virus-like particle assembly system is an ideal tool to study the interactions between the essential assembly partners M and E in molecular detail. Taking a mutagenetic approach we demonstrate that envelope assembly is critically sensitive to changes in the primary structure of both terminal domains of the M protein. The effects were most dramatically observed after mutation of the carboxy-terminal domain where the deletion of just one single amino acid at the extreme terminus abolished particle formation almost completely. But also some subtle mutations in the amino-terminal domain were severely inhibitory to the assembly process. Interestingly, mutant M proteins that were themselves incompetent to support particle formation appeared to inhibit, in a concentration dependent manner, the assembly of particles by wild-type M and E protein.

2. INTRODUCTION

Enveloped viruses assemble by budding through cellular membranes. The viral nucleocapsid or core, present in the cytoplasm, meets the viral envelope proteins at a specified cellular membrane and their interactions lead to the formation of a virion. Often, as is

Coronaviruses and Arteriviruses, edited by Enjuanes *et al.*
Plenum Press, New York, 1998

the case for togaviruses (Suomalainen *et al.*, 1992; Hobman *et al.*, 1994), these interactions are an absolute requirement for budding to occur. For several viruses, however, the envelope proteins are not required: retroviruses (Delchambre *et al.*, 1989; Gheysen *et al.*, 1989) and rhabdoviruses (Mebatsion *et al.*, 1996) allow budding of their nucleocapsids without the involvement of these proteins, giving rise to noninfectious virions. Conversely, a number of other viruses have been found to assemble "empty" particles just from their envelope proteins, i.e. without the need for a nucleocapsid. Usually these capsidless viral particles are smaller than the normal virions, as observed with flaviviruses (Schalich *et al.*, 1996), and morphologically different, as seen with hepatitis B virus (Simon *et al.*, 1988). This is, however, not the case for coronaviruses. The membrane proteins of these viruses can drive the formation of envelopes which, when viewed by electron microscopy, look perfectly similar to authentic virions (Vennema *et al.*, 1996).

Of the four coronaviral envelope proteins only two are actually essential for budding: the membrane (M) protein and the envelope (E) protein. Coronavirions without the spike (S) protein, and thus without spikes, have been demonstrated already a long time ago (e.g. Rottier *et al.*, 1981). The haemagglutinin-esterase (HE) protein occurs only in a subset of coronaviruses. The triple-spanning M protein is the major component of the viral envelope. In mouse hepatitis virus (MHV) it has its carboxy-terminus inside the particle while the amino-terminus, which carries an oligosaccharide side chain linked to threonine, protrudes from the outside of the envelope (see Rottier, 1995). Of the E protein the carboxy-terminus is also facing the virion inside but little if anything of this protein seems to be exposed on the outside (M. Raamsman, H. Vennema and P. Rottier, unpublished data).

Coexpression of M and E in cells gives rise to the synthesis of the virus-like particles (VLPs) mentioned above. This provides us with a very simple assay system to investigate the primary structure requirements of these two proteins for envelope formation. In this study we have focussed on the terminal parts of the M protein.

3. MATERIALS AND METHODS

3.1. Expression Vectors and Site-Directed Mutagenesis

Expression construct pTM5ab contains the MHV mRNA 5 open reading frames (ORFs) 5a and 5b, the latter encoding the E protein, in pTUG3 (Vennema *et al.*, 1996). Expression construct pTUMM contains the MHV M gene cloned as a XhoI fragment into pTUG3. All mutant M proteins were generated by standard polymerase chain reaction (PCR) cloning methods and cloned into pTUG3, except mutants ΔN, $G_{11}N_{13}$ and ΔC. These were made as described previously (Rottier *et al.*, 1990), and expressed from the transcription vector pTZ19R (Mead *et al.*, 1986).

3.2. Expression, Metabolic Labeling, and Immunoprecipitation

Subconfluent monolayers of OST-7 cells were infected with recombinant vaccinia virus encoding the T7 RNA polymerase (vTF7–3) and subsequently transfected with expression constructs as described before (Vennema *et al.*, 1996). Metabolic labeling of the transfected cells with ^{35}S methionine was performed essentially as described by de Vries et al. (1995) for three hours starting at t=5. Rabbit anti-MHV serum #134 (Rottier *et al.*, 1981) was used at a 500-fold dilution for immunoprecipitation of MHV proteins from cell lysates and cell culture media as described before (de Vries *et al.*, 1995).

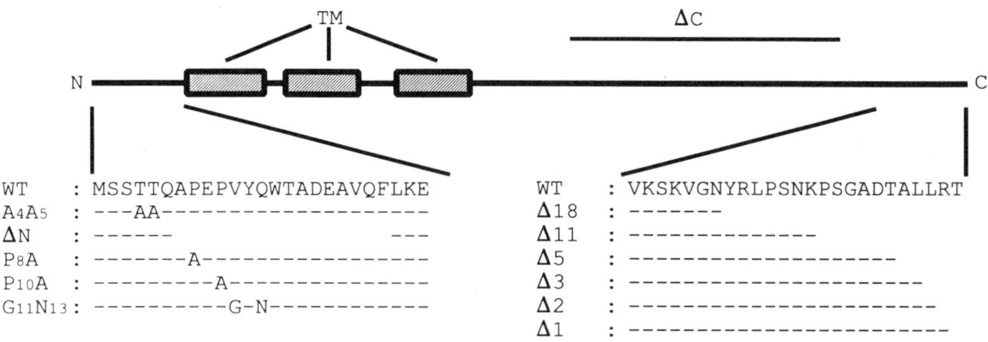

Figure 1. Overview of mutant M proteins. On top a schematic representation of the structure of the M protein. The three transmembrane (TM) domains are indicated by boxes. The left and right panel show amino acid sequences of the amino-terminal and carboxy-terminal domain, respectively, and the mutations in these domains. Dashes represent unchanged residues, a gap refers to a deletion. The domain deleted in mutant ΔC (residues E_{121} through D_{195}) is indicated by a horizontal line.

4. RESULTS

4.1. The MHV M Cytoplasmic Domain in VLP Assembly

In this study we analyzed the primary structure requirements of the M protein for MHV envelope assembly. To this aim we constructed a number of M mutants.

Our first set of mutants (Fig. 1) consisted of a panel of molecules with progressively smaller deletions at the carboxy-terminus, ranging from 18 till 1 amino acids (aa), as well as one internal deletion mutant (ΔC), from which E_{121} through D_{195}, which comprise most of the amphiphilic domain, were removed. The ability of these mutant M molecules to function in assembly was tested by coexpression of each of the mutant genes together with the E protein gene. Genes were expressed using the recombinant vaccinia virus/bacteriophage T7 RNA polymerase system in OST-7 cells and labeled with ^{35}S methionine/cysteine from 5 to 8 h postinfection. Cells and media were collected separately and viral proteins were immunoprecipitated with MHV-specific antibodies followed by SDS-PAGE in a 15% gel (Fig. 2). The analysis of the cell lysates of the single expressions demonstrates that all mutant constructs were expressed, yielding products of expected sizes. In all cases the M proteins appear as a set of proteins differing in apparent molecular weight, due to different extents of O-glycosylation. The patterns of the glycosylated species of M mutants are not much different from that of the wild-type (WT), indicating that the mutations in the C-terminus did not affect the ability of the N-terminus to become glycosylated, nor did they affect the ability of the proteins to be transported to the Golgi complex. In the double expressions, the presence of the E protein did not seem to affect the synthesis of M quantitatively or qualitatively under the experimental conditions used. The E protein is not resolved in a 15% gel, but was synthesized as was shown using an E-specific serum (not shown). Particle assembly and secretion was assayed by measuring the release of the M protein into the culture medium. The E protein is extremely difficult to detect in VLPs, due to its small size and very low abundance. When looking at WT-M it is clear that M release into the medium is critically dependent on the presence of the E protein, consistent with our earlier findings (Vennema *et al.*, 1996). However, all mutant M pro-

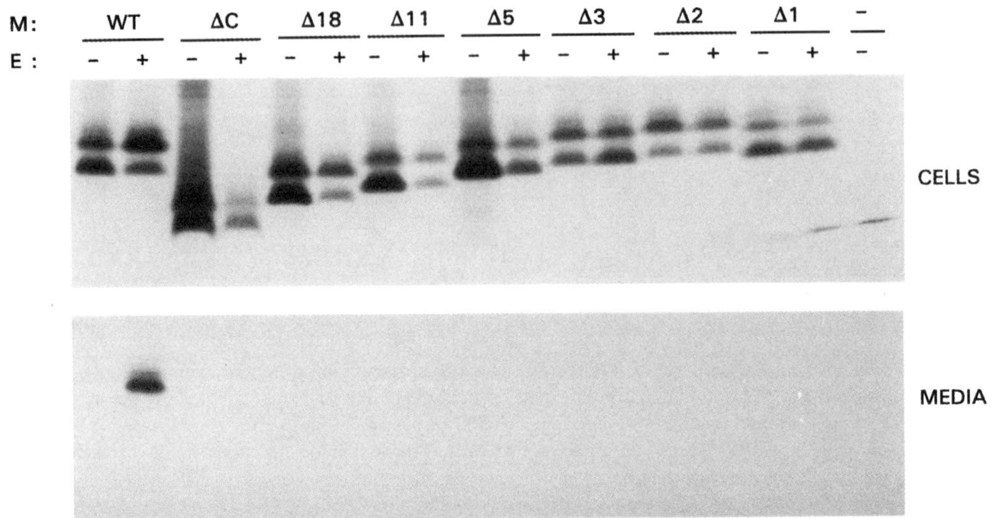

Figure 2. Effect of deletions in the M cytoplasmic domain on VLP assembly. Recombinant vaccinia virus vTF7–3 infected OST7–1 cells were transfected with a plasmid containing the wild-type (WT) or mutant M gene either alone or in combination with a plasmid containing the E protein gene, each gene behind a T7 promoter. Cells were labeled for 3 h with ^{35}S-labeled amino acids. Both cells (upper panel) and the culture medium (lower panel) were prepared and used for immunoprecipitation and the precipitates were analyzed by SDS-PAGE. The different M genes expressed are indicated above each set.

teins fail to be secreted into the culture medium when expressed in combination with the E protein. Not only an internal deletion in the cytoplasmic domain but also deletions at the C-terminus are fatal for VLP assembly. Even the deletion of only one single amino acid at the extreme C-terminus abolished VLP formation almost completely (Fig. 2).

4.2. The MHV M Amino-Terminal Domain in VLP Assembly

In order to investigate the role of the amino-terminal domain of M in VLP formation, a second set of mutants was constructed (Fig. 1), consisting of a panel of molecules with several mutations in this domain. In mutant A_4A_5 threonines at position 4 and 5 are substituted by alanines. Mutant ΔN lacks A_7 through F_{22}, an internal deletion of 16 aa. Three mutants have substitutions within this deleted sequence. In mutant P_8A the proline residue at position 8 is substituted by alanine; $P_{10}A$ has a similar substitution of the proline at position 10. Mutant $G_{11}N_{13}$ has substitutions of valine and glutamine at positions 11 and 13 into glycine and asparagine, respectively. The mutant constructs were expressed alone and in combination with the E protein gene. Cells and media were processed and analyzed as in the previous experiment. The analysis of the cell lysates (Fig. 3) shows, that all mutants were expressed to yield products of expected sizes. Mutant $G_{11}N_{13}$ shows the same glycosylation pattern as WT-M indicating that this mutant is transported at least to the Golgi complex. Mutants P_8A and $P_{10}A$ also become glycosylated but to a lesser extent. This is, however, not a reflection of the inability of the mutants to be transported to the Golgi apparatus, as was verified by immunofluorescence (not shown), but substitution of the prolines results in less efficient glycosylation at the N-terminal end. Mutants A_4A_5 and ΔN do not become glycosylated at all and, again, this is not a reflection of their inability

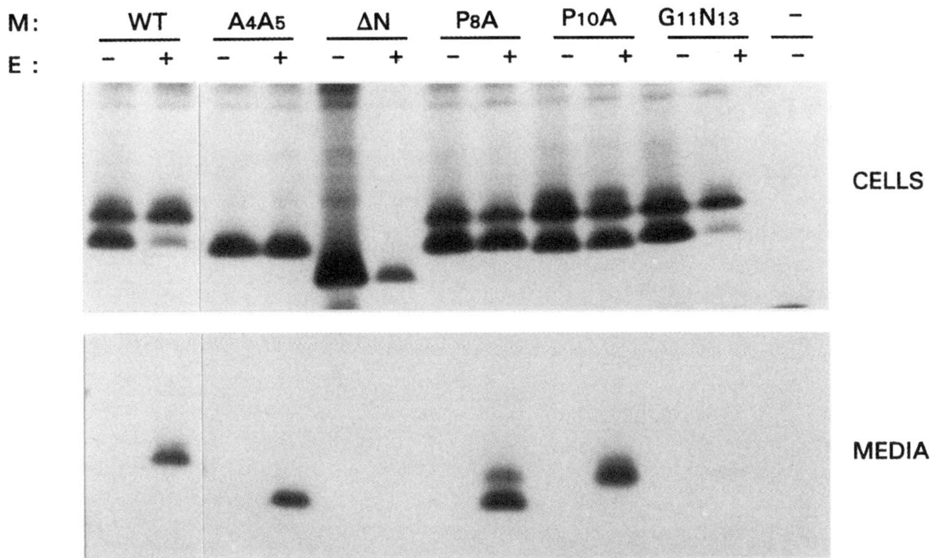

Figure 3. Effect of mutations in the amino-terminal domain of M on VLP assembly. M and E genes were expressed as described in the legend to Fig. 2.

to be transported to the Golgi complex, but results from their mutations being fatal for glycosylation at the N-terminus. The presence of the E protein does not seem to affect the synthesis of mutant M quantitatively or qualitatively under the experimental conditions used. All mutant M proteins fail to be secreted into the medium in the single expressions. When coexpressed with the E protein, mutant A_4A_5 is secreted into the medium with the same efficiency as WT-M. This result indicates that the threonines are not a primary structure requirement for VLP formation. Furthermore, O-glycosylation of the M protein is also not a prerequisite for VLP assembly and release. Mutant ΔN is not secreted into the medium at all; the 16 aa deletion for some reason abolishes the M protein's ability to be assembled into VLPs. Since P_8A and $P_{10}A$ are secreted efficiently the prolines do not seem important for VLP assembly. In contrast, secretion of $G_{11}N_{13}$ is decreased significantly, indicating that envelope-assembly is sensitive to changes in the region downstream of the prolines (Fig. 3).

4.3. Inhibition of VLP Formation by Assembly-Incompetent Mutant M Proteins

Our next aim was to investigate whether mutants of M which are themselves incompetent in VLP formation have an effect on assembly of VLPs by WT-M and E. Therefor, a triple expression experiment was performed, expressing WT-M, E and mutant M. Fixed amounts of plasmid DNA coding WT-M and E were used in the transfections, while of the plasmid DNA coding for mutant M an amount equal to or smaller than that of WT-M was used. Cells and media were processed and analyzed similarly as in the previous experiments. The analysis of the cell lysates shows the differently glycosylated M species (Fig. 4). Mutant Δ1 can not be distinguished from WT-M, due to their small size difference. Of mutant Δ18 the unglycosylated form can clearly be discriminated from WT-M;

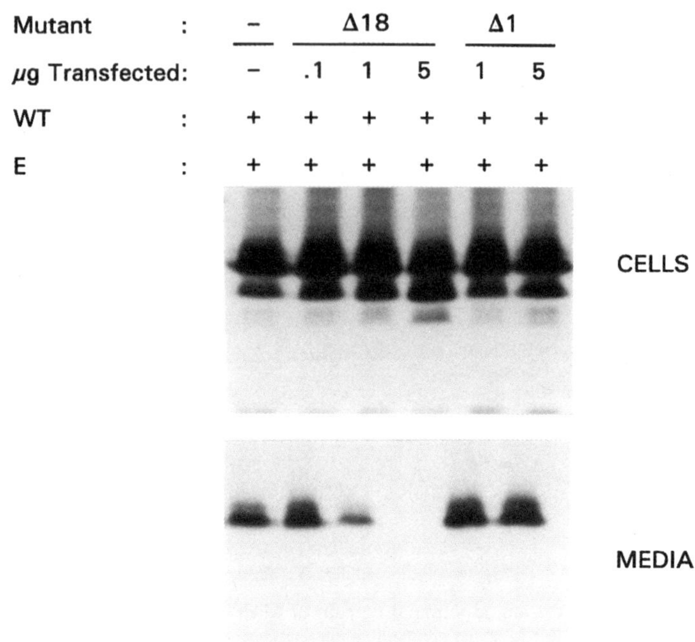

Figure 4. Inhibition of VLP formation by carboxy-terminal deletion mutant M proteins. After infection of OST7–1 cells with recombinant vaccinia virus vTF7–3, cells were transfected with plasmid DNA encoding WT M and E (5 μg each) together with 0.1, 1 or 5 μg plasmid DNA encoding mutant M. The different M mutants expressed are indicated above each set. Cells were labeled for 3 h with ^{35}S-labeled amino acids. Both cells (upper panel) and the culture medium (lower panel) were prepared for immunoprecipitation and the precipitates were analyzed by SDS-PAGE.

the glycosylated Δ18 species, however, comigrates with the unglycosylated form of WT-M. The glycosylated species of WT-M run slower and can thus be discriminated from the Δ18 proteins. When equal amounts of plasmid DNA of WT-M and Δ18 were co-transfected, products from both were clearly detected in the cell lysate, although the expression level of the mutant seems much lower than that of WT. Using lower amounts of mutant Δ18 plasmid DNA led to lower amounts of the unglycosylated Δ18 species that were immunoprecipitated, while the amounts of glycosylated WT-M forms was not noticeably changed. Analysis of the culture media showed that that the amount of WT-M secreted was reduced when Δ18 was coexpressed. The effect was dependent on the expression level of mutant Δ18, VLP release being completely inhibited at the higher level. Surprizingly, no effect was observed in the case of mutant Δ1, not even after transfection with the higher amount (Fig. 4).

5. DISCUSSION

Coronaviruses have the capacity to assemble uniform envelopes from their membrane proteins (Vennema et al. 1996). In this study we have exploited this feature to start an analysis of the primary structure requirements of the M protein for envelope formation. Together with the N protein, M is the most abundant virion component. The protein is considered to be the key element in virion assembly. It is essential not only for the formation of the viral envelope but also for drawing the nucleocapsid into the nascent particle by its interactions with the N protein. Its presumed role in localizing the budding of coronaviruses to membranes of the early exocytic pathway became questionable when the protein expressed by itself appeared to be transported to the Golgi complex (Klumperman *et al.*, 1994). This role might be played by the E protein, which seems to be retained in pre-Golgi membranes when expressed independently (M. Raamsman, H. Vennema and P. Rottier, unpublished observations).

Our preliminary observations presented here demonstrate that assembly of virus like particles is critically sensitive to changes in either of the terminal domains of the M protein. Mutations not only in the carboxy-terminal tail but also in its amino-terminal domain can interfere with particle formation. The carboxy-terminal part of M protrudes into the cytoplasm of cells. In infected cells it supposedly interacts with the nucleocapsid and ends up in the interior of the budded virion. Any truncation of this domain led to severe inhibition of particle formation in the VLP assay. The mere removal of just one terminal residue abolished assembly almost completely while no particles were detectable when two or more residues of the tail were lacking. Similarly fatal was the deletion of an internal domain, located in the amphiphilic region of the M protein (mutant ΔC). These effects are not likely to be the result of incorrect membrane integration or folding of the mutant M proteins or of an inability to be transported as judged from their normal O-glycosylation patterns and their transport to the Golgi complex and beyond (mutant $\Delta 18$; data not shown).

How then do these mutations exert their effects? When considering the different interactions which the carboxy-terminal tail might entertain, a number of possibilites come up. The terminal residues might be involved in interactions between M molecules, as well as with E, with membrane lipids or with host proteins somehow involved in the assembly process. Mutations, even the deletion of just one residue, could frustrate such interactions either directly or, when the mutation affects intramolecular interactions and, consequently, the proper formation of a necessary secondary structure in the M protein, indirectly. None of the various possibilities can as yet be excluded except probably the involvement of the carboxy-terminus in homotypic M-M interactions. Such interactions have been demonstrated earlier by Krijnse Locker et al. (1996) by showing the appearance in cells of large homomeric complexes of the expressed M protein, using sucrose density gradient analysis. In that study we found that such complexes, which presumably result from lateral interactions between M molecules, were also formed by a mutant M protein lacking its 22 carboxy-terminal residues.

The amino-terminus of M faces the lumen after the polypeptide has assembled in the ER membrane; in virions this domain is exposed on the outside (Rottier *et al.*, 1984). Deleting the middle part of this domain (mutant ΔN) rendered the M protein unproductive in VLP formation. More subtle changes within this middle part gave divergent results. Replacement of the two conspicuous proline residues by alanines was without effect. In contrast, mutant $G_{11}N_{13}$, meant to introduce a.o. an N-glycosylation consensus sequence (which, by the way, appeared not to be functional), was strongly crippled in VLP assembly. Neither the deletion in mutant ΔN nor the substitutions in mutant $G_{11}N_{13}$ had affected the protein's fitness as judged from its normal exit from the ER and transport to the Golgi complex (immunofluorescence data mutant ΔN: Rottier *et al.*, 1990; normal O-glycosylation mutant $G_{11}N_{13}$, Fig.3). This suggests that the effects of these mutations are not caused by drastic changes in the secondary or tertiary structure of the M molecule. An alternative possibility is that the amino-terminal domain is involved in intermolecular interactions, most likely between M molecules because little if anything of the membrane-integrated E protein is available on the luminal side (M.Raamsman, H.Vennema and P.Rottier, unpublished observations). For the human coronavirus (HCV-229E) M protein such interactions have indeed been demonstrated by Arpin and Talbot (1990) who showed the occurrence of covalently linked homodimers apparently formed by the single cysteine present in that molecule's ectodomain.

Two longstanding questions in the field are what the function is of the glycosylation of the M protein and why the protein in some coronaviruses carries N-linked sugars as opposed to O-linked oligosaccharides in other viruses. Here we show that glycosylation *per se* is not required for the formation of the coronaviral envelope. Substitutions of the threonines at positions 4 and 5 completely abolished the O-glycosylation of the MHV M protein but did not affect VLP formation. This result eliminates one possible answer; the old questions, however, still remain.

An interesting observation made in our studies was that a mutant M protein that was itselve negative in VLP formation could interfere with the assembly of viral envelopes by wild-type M protein. The mutant $\Delta 18$ appeared to poison the assembly process in a concentration-dependent way. Similar results were obtained with other cytoplasmic tail mutants but the effects became gradually weaker as the truncations were shorter (not shown), mutant $\Delta 1$ showing no detectable interference (Fig.3). These findings confirm our earlier conclusion that lateral interactions between M molecules are an important driving force in coronavirus particle formation (Krijnse Locker *et al.*, 1995). They are also consistent with the finding that the carboxy-terminal tail is not required to form and stabilize these oligomeric interactions (*ibid.*). Oligomerization of M proteins is, however, obviously not sufficient for VLP formation; additional interactions involving the cytoplasmic domain are apparently required. The possible nature of these interactions was discussed above. Our observations are reminiscent of findings reported for Semliki Forest virus where p62E1 heterodimers, nucleocapsid binding-deficient due to mutations in the cytoplasmic domain of the p62 protein, were found to inhibit normal virus budding also in a concentration dependent manner (Ekström *et al.*, 1994).

It is clear from this study that the VLP system has great potential for the analysis of the structural requirements of the individual coronaviral membrane proteins as well as for the investigation of their interactions during coronavirus envelope assembly. Whether, and to what extent, observations made in this system can be extrapolated to the assembly of coronavirions is still unclear. This will require the introduction of selected mutations into the coronaviral genome, a major challenge for the future.

ACKNOWLEDGMENTS

We are gratefull to Peggy Roestenberg and Marèl de Wit for their assistence in part of the experimental work. The investigations were supported by the Netherlands Foundation for Chemical Research (SON) with the financial aid from the Netherlands Organization for Scientific Research (NWO).

REFERENCES

Arpin, N., and Talbot, P.J., 1990, Molecular characterization of the 229E strain of human coronavirus, *Adv. Exp. Med. Biol.* **276**:73–80.

de Vries, A.A.F., Raamsman, M.J.B., van Dijk, H.A., Horzinek, M.C., and Rottier, P.J.M., 1995, The small envelope glycoprotein (Gs) of equine arteritis virus folds into three distinct monomers and a disulfide-linked dimer, *J. Virol.* **69**:344–3448.

Delchambre, M., Gheysen, D., Thines, D., Thiriart, C., Jacobs, E., Verdin, E., Horth, M., Burny, A., and Bex, F., 1989, The Gag precursor of simian immunodeficiency virus assembles into virus-like particles, *EMBO J.* **8**:2653–2660.

Ekström, M., Liljeström, P., and Garoff, H., 1994, Membrane protein lateral interactions control Semliki Forest virus budding, *EMBO J.* **13**:1058–1064.

Gheysen, D., Jacobs, E., de Foresta, A., Thiriart, C., Francotte, M., Thines, D., and De Wilde, M., 1989, Assembly and release of HIV-1 precursor Pr55gag virus-like particles from recombinant baculovirus-infected cells, *Cell* **59**:103–112.

Hobman, T.C., Lundstrom, M.L., Mauracher, C.A., Woodward, L., Gilliam, S., and Farquhar, M.G., 1994, Assembly of rubella virus structural proteins into virus-like particles in transfected cells, *Virology* **115**:574–585.

Klumperman, J., Krijnse Locker, J., Meijer, A., Horzinek, M.C., Geuze, H.J., and Rottier, P.J.M., 1994, Coronavirus M proteins accumulate in the Golgi complex beyond the site of virion buddin, *J. Virol.* **68**:6523–6-534.

Krijnse Locker, J., Opstelten, D.-J.E., Ericsson, M., Horzinek, M.C., and Rottier, P.J.M., 1995, Oligomerization of a trans-Golgi/trans-Golgi network retained protein occurs in the Golgi complex and may be part of its retention, *J. Biol. Chem.* **270**:88158821.

Mead, D.A., Szczesna-Skorupa, E., and Kemper, B., 1986, Single-stranded DNA 'blue' T7 promotor plasmids: A versatile tandem promotor system for cloning and protein engineering, *Prot. Engineering* **1**:67–74.

Mebatsion, T., König, M., and Conzelman, K.-K., 1996, Budding of rabies virus particles in the absence of the spike glycoprotein, *Cell* **84**:941–951.

Rottier, P.J.M., Horzinek, M.C., and van der Zeijst, B.A.M., 1981, Viral protein synthesis in mouse hepatitis virus strain A59-infected cells: effect of tunicamycin, *J. Virol.* **40**:350-357.

Rottier, P.J.M., Brandenburg, D., Armstrong, J., van der Zeijst B.A.M., and Warren, G., 1984, Assembly in vitro of a spanning membrane protein of the endoplasmatic reticulum: the E1 glycoprotein of coronavirus mouse hepatitis virus A59, *Proc. Natl. Acad. Sci. USA* **81**:1421–1425.

Rottier, P.J.M., Krijnse Locker, J., Horzinek, M.C., and Spaan, W.J.M., 1990. Expression of MHV-A59 M glycoprotein: Effects of deletions on membrane integration and intracellular transport, *Adv. Exp. Med. Biol.* **276**:127–135.

Rottier, P.J.M., 1995, The coronavirus membrane protein, p. 115–139. *In* S. G. Siddell (ed.), *The Coronaviridae,* Plenum Press, New York.

Schalich, J., Allison, S.L., Stiasny, K., Mandl, C.J., Kunz, C., and Heinz, F.X., 1996, Recombinant subviral particles from tick-borne encephalitis virus are fusogenic and provide a model system for studying flavivirus envelope glycoprotein functions, *J. Virol.* **70**:4549–4557.

Simon, K., Lingappa, V.R., and Ganem, D., 1988, Secreted hepatitis B surface antigen polypeptides are derived from a transmembrane precursor, *J. Cell Biol.* **107**:2163–2168.

Suomalainen, M., Liljeström, P., and Garoff, H., 1992, Spike protein-nucleocapsid interations drive the budding of alphaviruses, *J. Virol.* **66**:4737-4747.

Vennema, H., Godeke, G.-J., Rossen, J.W.A., Voorhout, W.F., Horzinek, M.C., Opstelten, D.-J.E., and Rottier, P.J.M., 1996, Nucleocapsid-independent assembly of corona virus-like particles by co-expression of viral envelope protein genes, *EMBO J.* **15**:2020-2028.

49

INTERFERON ALPHA INDUCING PROPERTY OF CORONAVIRUS PARTICLES AND PSEUDOPARTICLES

P. Baudoux,[1] L. Besnardeau,[1] C. Carrat,[1] P. Rottier,[2] B. Charley,[1] and H. Laude[1]

[1]Unité de Virologie Immunologie Moléculaires
INRA, Jouy-en-Josas, France
[2]Institute of Virology
Veterinary Faculty
Utrecht University, The Netherlands

1. ABSTRACT

Previous work in our laboratory have provided evidence that the membrane glycoprotein M of TGEV is centrally involved in efficient induction of alpha interferon (IFN-α) synthesis by non-immune peripheral blood mononuclear cells incubated with fixed, TGEV-infected cells or inactivated virions. Here we report recent completion of studies initiated to get a better understanding of the nature of the interferogenic determinant(s). Transfected cells expressing TGEV M together with the minor structural component E (formerly called sM) were found to trigger IFN-α synthesis. Co-expression of these two proteins was shown to be necessary and sufficient for assembly and release of pseudoparticles resembling TGEV virions. Purified pseudoparticles exhibited an interferogenic activity close to that of authentic virions. Chimeric recombinant particles expressing BCV M ectodomain also induced IFN. Examination of cell cultures infected by viruses representative of the three *Nidovirales* genera revealed that the capacity to act as an efficient IFN-α inducer is a common feature of viral particles of the coronavirus genus. Altogether these data bring new insights regarding the putative nature of the viral structure involved in IFN-α induction.

2. INTRODUCTION

The replication of transmissible gastroenteritis virus (TGEV), a highly enteropathogenic virus of swine, primarily takes place in the differenciated enterocytes of the small

Coronaviruses and Arteriviruses, edited by Enjuanes *et al.*
Plenum Press, New York, 1998

intestine. In infected animals, however, high titers of interferon (IFN) are detected at both the local and systemic levels, and the IFN activity essentially belongs to the alpha type (La Bonnardière and Laude, 1981). Accordingly, the infected enterocytes are unlikely to represent the major IFN-α-producing cells. Interfaced investigations involving two groups of our laboratory were being pursued in an attempt to elucidate the nature of the IFN-α-producing cells and the mechanism by which TGEV triggers such an efficient synthesis.

An ex vivo induction system was set up in which swine peripheral blood mononuclear cells (PBMC) were used as a source of IFN-producing cells. Incubation of these cells with fixed, TGEV-infected cultures or with UV-inactivated, purified virions results in a strong and early IFN-α synthesis (Charley and Laude, 1988). This led to the conclusion that the IFN-α synthesis is triggered by non-replicating, inert viral structures, a mechanism also shown to be operative for a number of other viruses (for a review, see Fitzgerald-Bocarsly, 1993). The characterization of the IFN-α-producing cells is still in progress, but there is now clear evidence that they represent a very minor subpopulation of leukocytes (Riffault et al., 1997 and ref. thererin). The available information concerning the nature of the interferogenic determinants is overall scarce. In the case of TGEV, several lines of evidence support the view that the M protein plays a key role in IFN induction. Firstly, within a collection of antibodies directed to the structural proteins, only two blocked the IFN induction. Both of them recognized the ectodomain of the M protein (Charley and Laude, 1988), which is about 35 amino acid-long after cleavage of the signal peptide (Laude et al., 1987). Secondly, two of the epitope mutants selected towards the above antibodies exhibited a markedly decreased interferogenic activity. Sequencing of the mutant M genes revealed that the substitutions all mapped within the N-terminal, ectodomain of the protein and impaired its glycosylation without altering the virus infectivity (Laude et al., 1992).

The aim of the present paper is to briefly report our research progress regarding this phenomenon. Two major issues which have been addressed during the past few years are: i) would the single or combined expression of recombinant TGEV M protein induce IFN synthesis ?; ii) is the strong IFN inducer property a unique feature of TGEV or is it shared by other coronaviruses ?

3. MATERIALS AND METHODS

The procedures for preparation of swine PBMC, induction of IFN by TGEV-infected cultures and IFN-α titration by ELISA have been described earlier (Charley and Laude, 1988; Riffault et al., 1997). Purification of TGEV pseudoparticles and immunoprecipitation of labeled viral proteins were performed as already reported (Laude et al., 1986). The pcDNA1 or pTM1-based plasmid constructs used to express viral proteins under control of T7 promotor are described in a separate publication (Baudoux et al., submitted). Tetracycline-controlled expression of TGEV genes was done by using the pUHD15.1 and pUHD10.3 plasmid vectors designed by M. Gossen & H. Bujard (1992).

The origin of viruses and cell lines introduced in the laboratory for IFN induction assays is as follows: FIPV and FEA cells: VIRBAC R & D Biologie (Le Carros, France); MHV A59 and L cells, EAV and BHK21 cells, BEV and Equine. Dermis cells: Institute of Virology (Utrecht, The Netherlands); IBV and Vero cells, CCV: Rhône-Mérieux (Lyon, France). The origin of the other viruses and cells listed in Table 1 are mentioned in our earlier publications.

Table 1. Induction of alpha interferon in swine leucocytes incubated with cell monolayers infected by coronaviruses and other members of the *Nidovirales*

Genus	Virus	Strain	Cell line	Fixation[1]	IFN-α titre[2]
Coronavirus (group 1)	TGEV	Purdue 115	ST	–	17500
				+	16400
	FIPV	1146 (type 2)	FEA	–	4800
				+	3300
			FCWF	–	1800
				+	400
	CCV	IRC5	FEA	+	1330
			FCWF	–	550
				+	510
	HCV	229E	FCWF	–	5400
				+	330
	PEDV	CV777	VERO	–	680
				+	680
Coronavirus (group 2)	MHV	A59	L	–	5400
				+	1100
	BCV	F15	HRT18	–	620
				+	50
Coronavirus (group 3)	IBV	Mass.41	VERO	–	11800
				+	2400
		Beaud.		–	7300
				+	1500
Torovirus	BEV		Eq. Dermis	–	≤10
				+	≤10
Arterivirus	EAV	Bucyrus	BHK21	–	≤10
				+	≤10

[1] The cell sheets were fixed (+) or not (–) with glutaraldehyde.
[2] IFN-α titre was ≤10 IU/ml in mock-infected cells.

4. RESULTS

4.1. IFN Induction by Cells Co-Expressing TGEV M and Other Viral Proteins

The vaccinia virus/T7-driven system was used to express M protein in different combinations with the other structural proteins: S (spike), N (nucleocapsid) and the minor small membrane protein (Godet et al., 1992). For the latter, the designation sM used in previous publications was changed to E according to a recommendation of the ICTV study group made during the Segovia symposium. Fig. 1 shows the IFN-α activity induced by these overexpressing cultures following glutaraldehyde fixation and incubation with PBMC. As a main result, IFN induction occurred only with monolayers co-transfected with the M and E genes at least. This finding is consistent with results from our earlier experiments in which stable or transient expression of M protein in the absence of other viral components failed to induce IFN (Baudoux et al., 1995). The additional expression of S and/or N did not result in appreciably higher IFN levels. Altogether these data led to the conclusion that the co-expression of M and E was necessary and sufficient to generate in-

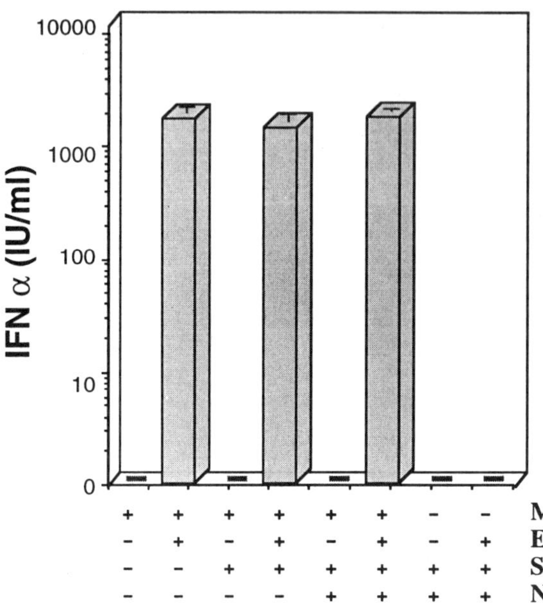

Figure 1. Induction of IFN-α by cell cultures expressing TGEV structural proteins. Vaccinia virus vT7–3-infected RK13 cells were lipofected with one or a combination of plasmids each encoding one of the TGEV genes, as indicated. The monolayers were fixed by 0.25 % glutaraldehyde at 10 hrs p.i. then incubated overnight with PBMC (10^7/ml). Interferon titer in the maintenance medium was measured by an ELISA assay specific for IFN-α.

terferogenic material. The measured IFN titres reached values ^{3} 10^3 UI/ml, which are close to the titer induced by TGEV-infected, fixed ST cells used for comparison in this assay; in addition, IFN induction appeared to be in correlation with an increased expression of M antigen as detected by an ELISA test using Mabs directed to the M ectodomain (data not shown).

4.2. TGEV Pseudoparticles Are Released from Cells Co-Expressing the M and E Proteins

The above data strongly suggested that a direct or indirect interaction with E, a protein shown to be present at a low copy number in TGEV virions (Godet et al., 1992), was absolutely required for the M protein to acquire an interferogenic structure and/or to gain access to IFN-producing cells. This possibility was reinforced by the recent demonstration by Vennema and coworkers (1996) that co-expression of the MHV M and E proteins led to the assembly and release of particles resembling authentic coronavirions. Experiments were therefore designed to formally assess this point in the case of TGEV. As shown in Fig. 2A, M material was immunoprecipitated from the supernatant of metabolically labelled cell cultures in which M and E were expressed through the VV/T7 vector (right panel), thus demonstrating that soluble M material was released from cells co-expressing these two proteins. As it can be noted, only a low proportion of the M material synthesized intracellularly was secreted in this experiment, and the level of E protein expression appeared to be a limiting factor as the amount of extracellular M increased with the concentration of E-encoding plasmid used for transfection. In subsequent experiments the assay was improved but the proportion of released material never exceeded 10 % of the synthesized M. This may indicate that the assembly and/or release in this model system is less efficient than in infected cells, where other viral components such as non-structural polypeptides could play a role so as to enhance this process. Investigations are currently being performed in order to address this point.

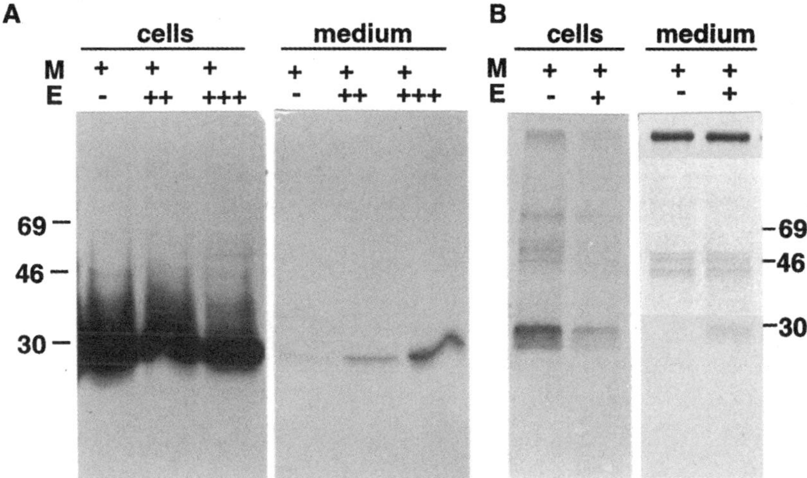

Figure 2. Secretion of TGEV M protein through co-expression of M and E gene constructs. (A) vT7–3-infected monolayers were transfected with pcsM (E) and pTM2 (M) plasmids as indicated, and labelled with [^{35}S]methionine. Viral proteins from the cell lysates and culture media were immunoprecipitated by anti-M monoclonal antibodies and analyzed by SDS-PAGE electrophoresis. (B) Cells were transfected with constructs allowing a tetracycline controlled expression of M and E genes. The resulting material was processed as above. The medium lanes are shown after a 10-fold prolonged exposure.

To rule out the possibility that release of M protein occurred merely as a consequence of the infection by vaccinia virus, an additional experiment was performed in which TGEV M and E proteins were expressed through an alternative system based on the use of tetracycline-dependent plasmid expression vectors. As shown in Fig. 2B, release of M protein occurred as well, thus supporting the view that it was a relevant phenomenon rather that an artifact resulting from the use of the VV/T7 vector. Finally, ultracentrifugation in sucrose gradient revealed that isotopically labelled M material released from transfected, VV/T7-infected cells sedimented as a discrete peak, at a slightly slower rate than TGEV virions (not shown). Overall these data led us to conclude that TGEV E and M proteins interactions allow the formation of pseudoparticles, similar to that reported for MHV.

4.3. Purified TGEV Pseudoparticles Are Potent IFN-α Inducers

A key issue was whether the specific interferogenic activity of the so obtained recombinant particles compared that of TGEV virions. To address it, a stock of pseudoparticles produced in transfected, VV/T7 infected RK13 cells was prepared and purified following the same procedure as used for TGEV virions. When examined in electron microscopy, the resulting preparation was shown to contain particles which looked like authentic virions except that they were devoid of prominent, typical coronavirus spikes (Fig. 3) It was also noted that, consistent with the absence of a nucleocapsid core, the pseudoparticles had a tendency to collapse, which implies a real flexibility of the coronavirus envelope.

Suspensions of purified pseudoparticles and virions having the same M antigen content were allowed to incubate with PBMC. The IFN titers induced by the two types of material were found to be of the same order of magnitude, thus establishing that pseudoparticles

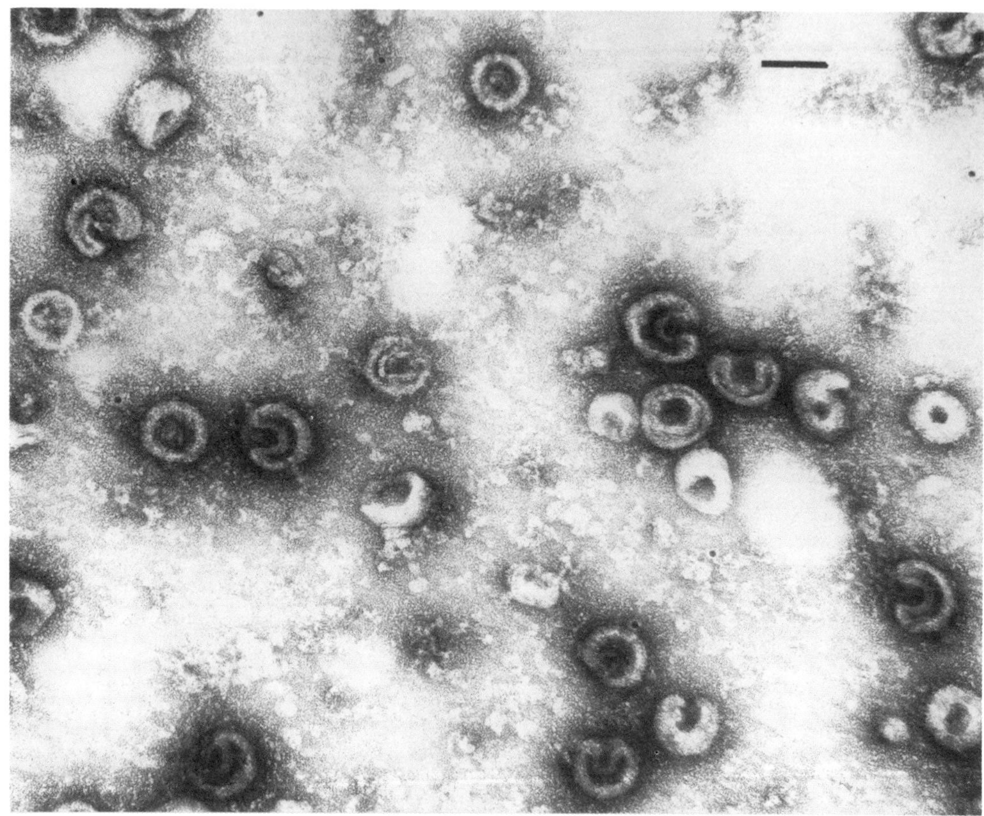

Figure 3. Electron microscopy of TGEV pseudoparticles. Particles released from RK13 cells expressing M and E genes through the VV/T7 vector were purified in a sucrose gradient. Particles were viewed after negative staining (bar: 100nm).

were nearly as potent IFN inducer as TGEV virions (Fig. 4). Moreover, the addition of appropriate anti-M ectodomain monoclonal antibodies during the incubation strongly inhibited the IFN induction by pseudoparticles, similar to that already observed with TGEV. This confirmed that identical determinants were involved in both cases.

4.4. IFN Induction by a TGEV/BCV Chimeric M Protein

The possibility of generating fully interferogenic, TGEV-like particles from engineered genes opened the way to closer investigations regarding the nature of the molecular determinants responsible for such an activity. As a first approach, it was decided to generate TGEV/BCV chimeric particles by taking advantage of BCV M genes previously cloned in the laboratory (Savoysky et al., 1990; Woloszyn et al., 1990). As will be described in a separate paper (Baudoux et al., submitted), a general outcome of these experiments was that the co-expression experiments using constructs encoding chimeric either M or E, or both, did not allow particle formation at a workable level. A noticeable exception was the M chimera called BTT, in which the BCV ectodomain was swapped in the TGEV M sequence. Not only the BTT chimera lead to relatively efficient particle forma-

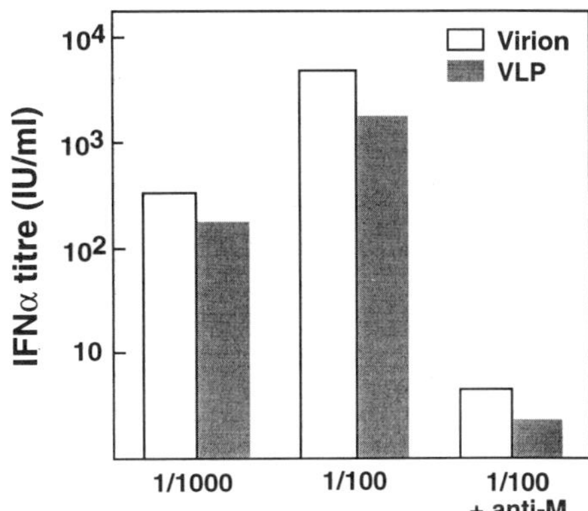

Figure 4. IFN-α inducing activity of pseudoparticles and virions of TGEV. Purified preparations of particles were adjusted to have the same content in M antigen as judged by ELISA assay, and incubated with PBMC at the indicated dilutions. The blocking effect of anti-M monoclonal antibodies on IFN induction is shown on the right.

tion, but also it exhibited a limited specificity of interaction with E (data not shown). When analyzed for IFN induction, BTT was found to promote IFN synthesis as efficiently as homologous TGEV M when coexpressed with TGEV E (Fig. 5). Co-expression of BTT with BCV E also induced IFN, although at an appreciably lower level, which was correlated to a less efficient particle formation. Last but not least, cultures co-expressing the homologous BCV M + E pair exhibited also substantial interferogenic activity.

4.5. Replication-Independent Induction of IFN-α Is a Common Feature of Coronaviruses

The finding that, like TGEV, BCV pseudoparticles were interferogenic raised the question of whether this property could be shared by other coronaviruses or by other members of the *Nidovirales*. To answer it, cultures infected by one each of the ten different viruses listed in Table 1 were produced using appropriate permissive cell systems. After onset of the cytopathic effect, fixed or unfixed monolayers were allowed to incubate with swine leukocytes as a source of IFN-producing cell, and the IFN-a titre associated with the resulting supernatants was determined by ELISA assay. As a striking result, all the seven coronaviruses tested, which are representative from the three genetic subsets of the genus, were able to trigger a substantial IFN synthesis, like TGEV. Importantly, IFN

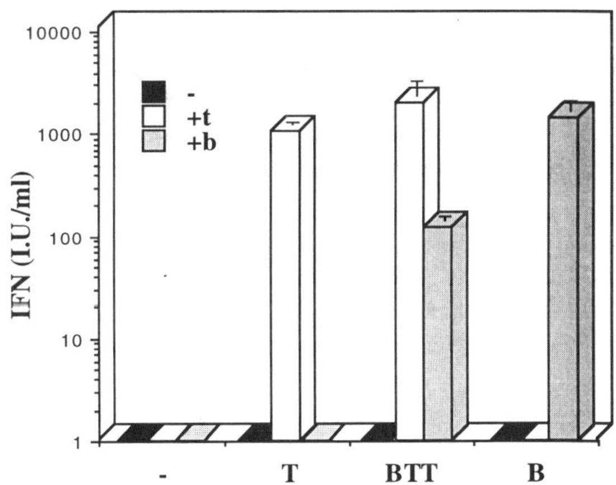

Figure 5. Induction of IFN-α by cell cultures expressing homologous or chimeric pseudoparticles of TGEV and BCV. The experiment was performed as in Fig. 1. The expressed viral genes are designated as uppercase letters on abscissa axis for M constructs, and lowercase letters in insert for E constructs. T and t for TGEV, B and b for BCV. BTT designates a chimeric M gene (see text).

production persisted after glutaraldehyde fixation of the monolayers prior to incubation with the leukocytes. For several infected cultures, the IFN titer decreased after fixation, possibly due to the fact that synthesis of interferogenic material was blocked by fixation while it proceeded during incubation with unfixed cultures. An inconstant effect of fixation on the IFN titer has been recognized in the course of our experiments with TGEV. Alternatively, the integrity of the interferogenic determinants may be affected by glutaraldehyde to variable extents differing with the virus tested. In any case, these data demonstrate that the property of triggering IFN synthesis in leukocytes by recognition of inert viral structures is shared by many coronaviruses. In contrast, Berne virus (BEV) and equine arteritis virus (EAV), representative of the two other nidovirus genera, failed to induce IFN in two separate assays in the same above conditions. In an independent study, the porcine reproductive respiratory syndrome virus (PRRSV, arterivirus) failed to induce an in vitro IFN-α response in swine PBMC (E. Albina; personnal communication). It remains possible, however, that latter viruses would act as inducers for leukocytes from other species than the swine.

The experiments using infected cultures as inducers do not allow to draw valid conclusions regarding the respective interferogenic activity of the different coronaviruses since the amount of viral material produced may vary significantly among the infected cultures and was not quantified. Moreover, PBMC from different batches were used for induction. The only way to address this point is to use suspensions of purified virions at identical concentrations as inducers in the same assay, which was done for two viruses in addition to TGEV. β-propiolactone-inactivated FIPV virions incubated with swine PBMC (1 μg / 2 $\times$ 10^{6}cells) induced IFN titers of 4–5 $\times$ 10^3 IU/ml, close to what UV-inactivated TGEV virions induced in the same assay (7–8 $\times$ 10^3 IU/ml). BCV, however, appeared to be a less potent inducer than the two above viruses, with IFN titres of 3–4 $\times$ 10^2 UI/ml, a result which can be correlated to the relatively low interferogenic activity of BCV-infected, fixed cultures. This last result was not expected since cultures producing BCV and TGEV recombinant pseudoparticles induced comparable IFN titers. Although this finding would require further examination, one can speculate that the presence on BCV virions of a second layer of short spikes (HE protein) may result in a partial masking of interferogenic determinants.

5. DISCUSSION

The major achievement of these studies was the demonstration that recombinant TGEV-like particles, lacking spikes, nucleocapsid and genomic RNA, may substitute to authentic virions for triggering an efficient IFN-α synthesis by swine leukocytes. The pseudoparticles were generated through the expression of the major virion component, the polytopic transmembrane glycoprotein M, which was previously proposed by us to play a key role as IFN inducer (see introduction), together with the minor, small transmembrane protein E (formerly called sM). These results nicely confirmed the observations recently reported on MHV by Vennema and coworkers and further support the view that, through an interaction whose precise nature remains to elucidate, the M and E proteins represent both necessary and sufficient partners for coronavirus assembly. The characteristics of the induction of IFN-α by TGEV pseudoparticles closely resembled those previously established for virions: i) their intrinsic interferogenic activity nearly approached that of TGEV virions, as estimated on the basis of their respective M antigen content; it should be noted in this respect that, for practical reasons, the two types of particles were produced in distinct cell systems, and thus may differ to some extent in terms of carbohydrate and lipid composition; ii) antibodies directed to the M ectodomain were able to prevent the IFN induction; iii) co-expression of S spike protein had no discernible effect on the interferogenic activity, in agreement with our earlier finding that anti-S or anti-APN receptor antibodies did not affect IFN stimulation.

Another relevant result of our work was that the ability to act as a potent IFN-α inducer for PBMC is not restricted to TGEV but appears to be a common, so far unrecognized trait of coronaviruses. Indeed, cultures infected by any of the members of the genus tested so far, as well as FIPV and BCV purified virus suspensions and BCV pseudoparticles-producing cells, were found to induce an efficient IFN synthesis by swine leukocytes. There is circumstantial evidence that the strong IFN response observed in TGEV-infected animals is raised through this virus RNA-independent induction pathway. Hence, it might be worth investigating the interferogenic activity of other viruses of the genus towards leukocytes from their respective, natural host species, in order to evaluate any possible biological relevancy of this phenomenon in coronavirus infections.

The mechanism by which coronavirus particles, either free or associated with the cell surface, commit the IFN-producing cells to express the related genes is far from being understood. However, our findings altogether brought truly new insights in this respect. The early step of the induction presumably involves the recognition of a surface-exposed, viral structure. Both theoretical and experimental reasons led us to propose the ectodomain of the TGEV M protein, as a candidate for this function. It appeared, however, that engineered TGEV chimeric pseudoparticles carrying the BCV M ectodomain, which differs from its TGEV counterpart in terms of size, amino acid sequence, and glycosylation type, were as interferogenic as non-chimeric particles. Furthermore, as outlined above, this property is potentially a general feature of the coronavirus particle. Inspection of the amino acid sequences failed to reveal any candidate motif at least in the assumedly exposed regions of the protein. Thus an attractive view emerging from these studies is that the induction could be mediated by a tridimensionnal, possibly repetitive, structure rather than by a sequence motif. Such a hypothetical IFN-inducing determinant should be unique to the coronavirus particle since other Nidoviruses do not appear to share this intriguing biological property.

ACKNOWLEDGMENTS

We express our deep thanks to Joëlle Cronier and Véronique Duquesne (Virbac) for providing virus, cells and FIPV-inactivated suspension for purification; Gert-Jan Godeke, Twan de Vries and Lisette Cornelissen from Peter Rottier's lab for kindly making available to us MHV, EAV and BEV viruses, and L, BHK21 and Equine Dermis cell lines; Jean-François Bouquet and Gilles Chappuis (Rhône-Mérieux) for IBV and CCV virus strains; Pascal Boireau for providing the cloned M and E genes.

REFERENCES

Baudoux P., Charley B., Laude H., 1995, Recombinant expression of the TGEV membrane protein, in "Corona- and related viruses" (P. Talbot & G. Levy Eds.), *Adv. Exp. Med. Biol.* **380**: 305–310.
Charley B., Laude H., 1988, Induction of interferon alpha by transmissible gastroenteritis coronavirus : role of transmembrane glycoprotein E1, *J. Virol.* **62**: 8–11.
Fitzgerald-Bocarsly P., 1993, Human Natural Interferon-a producing cells, *Pharmacol. Ther.* **60**: 39–62.

Godet M., L'Haridon R., Vautherot J.F., Laude H., 1992, TGEV coronavirus ORF4 encodes a membrane protein that is incorporated into virions, *Virology* **188**: 666–675.

La Bonnardière C., Laude H., 1981, High interferon titer in newborn pig intestine during experimentally induced viral enteritis, *Infec. Immun.* **32**: 28–31.

Laude H., Chapsal J.M., Gelfi J., Labiau S., Grosclaude J., 1986, Antigenic structure of transmissible gastroenteritis coronavirus I. Properties of monoclonal antibodies directed against virion proteins, *J. Gen. Virol.* **67**: 119–130.

Laude H., Gelfi J., Lavenant L., Charley B., 1992, Single amino acid changes in the viral glycoprotein M affect induction of alpha interferon by coronavirus TGEV, *J. Virol.* **66**: 743–749.

Laude H., Rasschaert D., Huet J.C., 1987, Sequence and N-terminal processing of the transmembrane protein El of the coronavirus transmissible gastroenteritis Virus, *J. Gen. Virol.* **68**: 1687–1693.

Riffault S. Carrat C. Besnardeau L., La Bonnardiere C., Charley B., 1997, In vivo induction of interferon-a in pig by non-infectious coronavirus: tissue localization and in situ phenotypic characterization of interferon-a producing cells, *J. Gen. Virol.* **78**: 2483–2487.

Savoysky E., Boireau P., Finance C., Laporte, J.,1990, Sequence and analysis of BECV F15 matrix protein, *Res. Virol.* **141**: 411–425.

Vennema H., Godeke G-J., Rossen J.W.A., Vorhout W.F., Horzinek M.C., Opstelten D.-J., Rottier P.J.M., 1996, Nucleocapsid-independent assembly of coronavirus -like particles by co-expression of viral envelope protein genes, *EMBO J.* **15**: 2020–2028.

Woloszyn N., Boireau P., Laporte J., 1990, Nucleotide sequence of the bovine enteric coronavirus F15 mRNA 5 and mRNA 6 unique regions, *Nucl. Ac. Res.* **18**: 1303.

50

EXPRESSION AND PROCESSING OF NONSTRUCTURAL PROTEINS OF THE HUMAN ASTROVIRUSES

C. A. Gibson,[1,3] J. Chen,[4] S. A. Monroe,[4] and M. R. Denison[1,2,3]

[1]Department of Microbiology and Immunology
[2]Department of Pediatrics
 Vanderbilt University
[3]Elizabeth B. Lamb Center for Pediatric Research
 Vanderbilt University Medical Center
 Nashville, Tennessee
[4]Centers for Disease Control
 Atlanta, Georgia

1. ABSTRACT

The human astroviruses (HAst) are increasingly recognized as an important cause of gastroenteritis. These viruses contain a 6.8-kb positive-sense, single-stranded RNA molecule that is infectious when transfected into permissive cells. The HAst gene 1 is composed of two open reading frames (ORFs 1a and 1b) connected by a ribosomal frameshift. Gene 1 is predicted to encode two nonstructural polyproteins (pp1a and pp1ab), and analysis of the HAst gene 1 sequence has resulted in predictions of a serine proteinase within the ORF1a polyprotein. However, none of the gene 1 proteins have been identified. To examine the expression and processing of the HAst2 gene 1 polyprotein, we have translated pp1a and pp1ab *in vitro*. These ongoing studies will provide the foundation for correlating gene 1 expression *in vitro* with proteins expressed in virus-infected cells.

2. INTRODUCTION

Astroviruses were first described in 1975 as star-shaped particles in electron micrographs from diarrheal stools of infants (Madeley *et al*). Members of the Astroviridae are now known to be involved in endemic, epidemic and common-source outbreaks of diarrhea, and may especially impact immunocompromised individuals. Of the eight as-

Coronaviruses and Arteriviruses, edited by Enjuanes *et al.*
Plenum Press, New York, 1998

trovirus serotypes that cause human disease, types 1 and 2 (HAst1 and HAst2) are associated with the majority of infections and clinical gastroenteritis. Complete sequences for both of these serotypes are known.

The astrovirus virion particle contains a 6.8kb single-stranded, positive-sense RNA molecule (RNA 1) that is infectious when directly transfected into permissive cells (Geigenmuller *et al*, 1997). During infection a single subgenomic RNA (RNA 2) is generated; this RNA is 3' coterminal with RNA1 (Monroe *et al*, 1991; Monroe *et al*, 1993). The HAst2 gene 1 consists of two out of frame open reading frames (ORFs 1a and 1b) that are connected by a ribosomal frameshift (Figure 1) (Jiang *et al*, 1993; Marczinke *et al*, 1995; Lewis *et al*, 1996). The gene 1 polyprotein is predicted to be cleaved by the putative serine proteinase (Spro). Sequence analyses have also predicted a nuclear localization signal (NLS) and an immune response element (IRE) as well as membrane spanning domains (MB) and an RNA-dependent-RNA polymerase (Pol) (Matsui *et al*, 1993; Carter, 1994; Lewis *et al*, 1994). RNA 2 is translated as a 90 kDa capsid precursor that is processed to mature capsid proteins, but the mechanism of subgenomic RNA 2 transcription is unknown.

Nothing is known about the pattern of gene 1 polyprotein expression and processing, nor of the role of viral nonstructural proteins in initiating virus replication. We have combined *in vitro* and *in cyto* approaches to begin to define the steps in replication occurring immediately after virus entry. These studies will allow us to compare astrovirus strategies of replication with those of other positive–sense RNA viruses sharing similar genome organization.

3. MATERIALS AND METHODS

3.1. Construction and Expression of the HAst2 Gene 1 Clones

Fragments incorporating ORFs 1a and 1ab were obtained by single step RT-PCR of HAst2-infected cell lysates and were cloned into pBluescript II SK- to generate the Mono

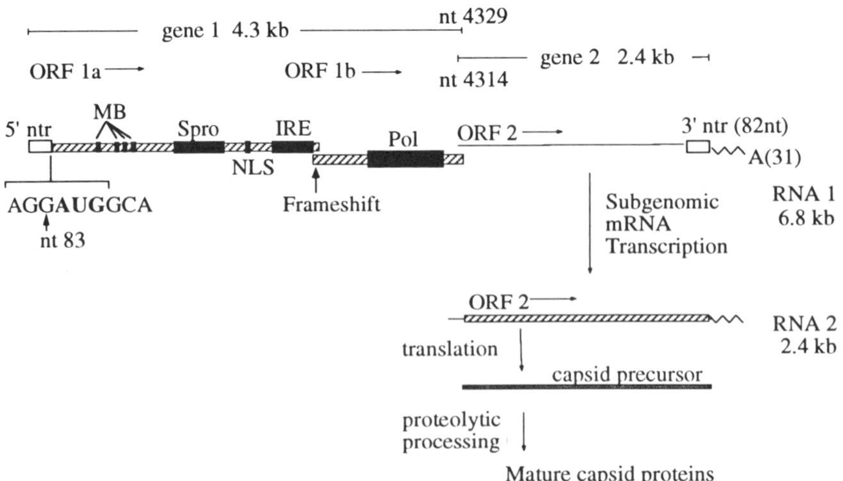

Figure 1. HAst2 genome organization and predicted replication strategy.

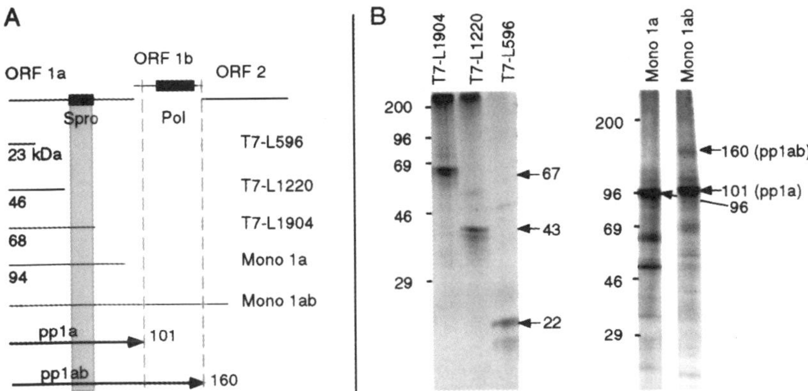

Figure 2. *In vitro* expression of HAst2 gene 1 clones.

1a and Mono 1ab constructs (Figure 2A). The "T7" clones were subcloned from Mono 1ab cDNA using a common 5' primer immediately upstream of the 5' nontranslated region. The masses of the expected proteins (kDa) are shown beneath each clone. The vertical gray box demarcates the putative Spro region. The predicted polyproteins (pp1a and pp1ab) resulting from translation of ORF 1a or ORFs1a/b are shown below the Mono 1ab clone.

The partial and full-length gene 1 cDNA constructs were used to express proteins, using a combined transcription/translation rabbit reticulocyte lysate system (TnT, Promega) with [^{35}S]-methionine (Figure 2B). Numbers to the left of the gels indicate the molecular mass of the marker proteins. Apparent mass (kDa) of individual proteins is shown to the right of the gels.

3.2. Induction of Antibodies against Gene 1 Proteins

Antigenic domains in the gene 1 polyprotein sequence were identified, and synthetic peptides were synthesized (Figure 3). Peptides were coupled to keyhole limpet hemocyanin

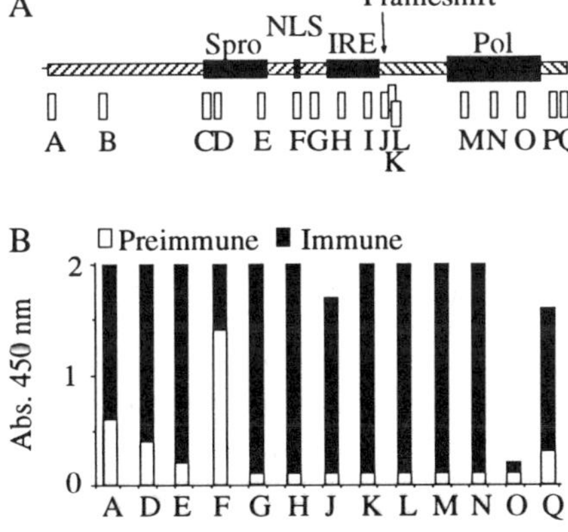

Figure 3. Antigenic domains in gene 1, synthesis of peptides, and immune response in rabbits.

and were used to induce antibodies in rabbits. The HAst2 peptides were used as capture antigens ELISAs with the corresponding preimmune and immune sera (Figure 3B). Sera were diluted 10^2 to 10^7, and the 450nm absorbance reading at the 10^2 dilution was plotted.

3.3. Construction of a Full Length HAst2 Clone

A full length 7kb HAst2 cDNA was generated by single step RT-PCR of HAst2-infected cell lysates. The insert was cloned into pBluescript II SK- by restriction digest at the 5' Cla 1 and 3' BamH1 sites to generate the pAst2 construct.

4. RESULTS AND DISCUSSION

4.1. Expression of the HAst2 Gene 1 Clones

The clone extending from the 5' end of the genome to 214 nucleotides upstream from the ribosomal frameshift (Mono 1a) directed expression of a 96 kDa protein, the expected size for full-length translation of this clone (Fig 2). The Mono 1ab clone expressed a 101 kDa protein consistent with translation termination at the frameshift and also expressed a less prominent protein of 160 kDa, identical to the predicted size of the frameshifted gene 1 fusion polyprotein. These data indicate that ORF 1b is expressed only as a 1ab polyprotein (pp1ab) and that frameshifting occurs in the context of all of gene 1 with an efficiency similar to that observed in subclones containing the frameshift region (Lewis and Matsui, 1996).

Since several of the clones expressed the putative Spro domain, it was surprising that full-length translation products were detected. The reason for the lack of processing is not known but may be due to a need for higher concentrations of viral proteinase than can be synthesized *in vitro*. Alternatively, this result could indicate that *in vitro* processing of the polyprotein is delayed, or that cellular factors not present in the reticulocyte lysate, such as membranes or cellular proteinases, may be needed to initiate or facilitate polyprotein processing.

4.2. Specificity of the Polyclonal Antisera Directed against HAst2 Gene 1 Peptides

Antisera raised to peptides A, D, G, and H when diluted up to 10^5 retained optical density readings at least 3-fold higher than readings for the corresponding preimmune sera. Antisera raised to peptides K, L, M, and N were detectable over background when diluted up to 10^3. Antisera raised against the D peptide are able to detect E. coli-expressed recombinant Spro (rSpro) by Western blot analysis (data not shown).

4.3. Generation of a Full Length HAst2 Clone

In parallel with the *in vitro* expression studies, we are developing an infectious clone for the human serotype 2 astroviruses in order to further characterize the role of the nonstructural polyproteins in HAst2 replication. A full-length HAst2 cDNA clone has been obtained. Preliminary nucleotide sequencing data have revealed several independent point mutations among the clones, which may yield new insights into regions of the genome that are essential for infectivity. The panel of polyclonal antisera will be a tremendous asset to these studies.

REFERENCES

Carter, M.J., 1994, Genomic organization and expression of astroviruses and caliciviruses, *Arch. Virol. Supp.* **9**: 429–439.

Geigenmuller, U., Ginzton, N., and Matsui, S., 1996, Construction of a genome-length cDNA clone for human astrovirus serotype 1 and synthesis of infectious RNA transcripts, *J. Virol.* **71**: 1713–1717.

Jiang, B., Monroe, S.S., Koonin, E.V., Stine, S.E., and Glass, R.I., 1993, RNA sequence of astrovirus: distinctive genomic organization and a putative retrovirus-like ribosomal frameshifting signal that directs the viral replicase synthesis, *Proc. Nat. Acad. Sci.* **90**: 10539–10543.

Lewis, T., and Matsui, S., 1996, Astrovirus ribosomal frameshifting in an infection-transfection transient expression system, *J. Virol.* **70**: 2869–2875.

Madeley, C.R., Cosgrove, B.P., and Bell, E.J., Stool viruses in babies in Glasgow2: Investigation of normal newborns in hospital, *J. Hygiene* **81**: 285–294.

Matsui, S.M., Kim, J.P., Greenberg, H.B., Young, L.M., Smith, L.S., Lewis, T.L., Herrmann, J.E., Blacklow, N.R., Dupuis, K., and Reyes, G.R., 1993, Cloning and characterization of human astrovirus immunoreactive epitopes, *J. Virol.* **67**: 1712–1715.

Marczinke, B., Bloys, A.J., Brown, T.D., Willcocks, M.M., Carter, M.J., and Brierley, I., 1994, The human astrovirus RNA-dependent RNA polymerase coding region is expressed by ribosomal frameshifting, *J. Virol.* **68**: 5588–5595.

Monroe, S.S., Stine, S.E., Gorelkin, L., Herrmann, J.E., Blacklow, N.R., and Glass, R.I., 1991, Temporal synthesis of proteins and RNAs during human astrovirus infection of cultured cells, *J. Virol.* **65**: 641–648.

Monroe, S.S., Jiang, B., Stine, S.E., Koopmans, M., and Glass, R.I., 1993, Subgenomic RNA sequence of human astrovirus supports classification of Astroviridae as a new family of RNA viruses, *J. Virol.* **67**: 3611–3614.

Pathogenesis I: The Immune System

51

EXTRACELLULAR ENVELOPED VACCINIA VIRUS

Entry, Egress, and Evasion

Geoffrey L. Smith and Alain Vanderplasschen

Sir William Dunn School of Pathology
University of Oxford
South Parks Road
Oxford OX1 3RE United Kingdom

1. ABSTRACT

Vaccinia virus is a large and complex virus that produces two types of infectious virus particles, termed intracellular mature virus (IMV) and extracellular enveloped virus (EEV). EEV contains an extra lipid envelope and ten associated proteins that are absent from IMV. Although EEV represents less than 1% of infectious progeny it is very important biologically. First, it mediates virus dissemination and second, it is the virus against which protective immune responses are directed. This article reviews the genes known to encode EEV proteins and their functions, describes recent data showing that the cellular receptors for IMV and EEV are different, and demonstrates that EEV, in contrast to IMV, is resistant to neutralisation by antibody.

2. INTRODUCTION

Vaccinia virus is the live vaccine that was used to eradicate smallpox (Fenner *et al.*, 1988). It is a member of the Poxviridae, a group of large and complex DNA viruses that replicate within the cytoplasm and encode many of their own enzymes for transcription and DNA replication (Moss, 1996). Strains of vaccinia virus and variola virus (the cause of smallpox) have been completely sequenced and contain approximately 200 genes (Goebel *et al.*, 1990; Massung *et al.*, 1993; Shchelkunov *et al.*, 1995). Genes near the centre of the virus genome are generally important for virus replication, encoding for instance, enzymes needed for transcription and DNA replication and for virus structural proteins. In

Coronaviruses and Arteriviruses, edited by Enjuanes *et al.*
Plenum Press, New York, 1998

contrast, the genes near the genomic termini are mostly non-essential for virus replication and encode proteins which affect virus host range or virulence. These include many proteins which interfere with the host response to infection such as complement, interferon, inflammation, cytokines and chemokines (Alcamí and Smith, 1995; Smith, 1996; Spriggs, 1996).

Although vaccinia virus is the only vaccine to have been used to achieve the global eradication of a human disease, the origin of the virus and its natural host are unknown (Baxby, 1981; Alcamí and Smith, 1996). Vaccinia virus is not cowpox, the virus believed to have been used by Jenner for vaccination in 1796, but is a related orthopoxvirus which, like cowpox, is an effective vaccine against variola virus.

3. VIRUS REPLICATION

The replication cycle of vaccinia virus has been recently reviewed (Moss, 1996) and with the exception of entry and release is considered only briefly here.

3.1. Binding and Entry

Before considering how vaccinia virus binds and enters cells, it is important to realise that there are two structurally different infectious vaccinia virus particles (Appleyard *et al.*, 1971; Ichihashi and Dales, 1971). The majority of infectious progeny is called intracellular mature virus (IMV) and, as the name implies, this virus remains inside the infected cells and is only released upon cell lysis. With most strains of vaccinia virus IMV may represent more than 99% of infectivity (Payne, 1980). The other form of virus is called extracellular enveloped virus (EEV) and is released from the cell. The details of how these particles are formed and released are described below under Morphogenesis (section 4.5). Although EEV is much less abundant than IMV, it is important because it mediates virus spread. Strains of virus producing more EEV form a comet shaped plaque where the primary plaque forms the head of the comet and the secondary plaques form the tail (see Fig. 9) (Appleyard *et al.*, 1971; Boulter and Appleyard, 1973; Payne, 1980; Payne and Kristensson, 1985). EEV is also the form of virus against which the protective immune response is directed (Boulter, 1969; Appleyard *et al.*, 1971; Boulter *et al.*, 1971; Boulter and Appleyard, 1973; Appleyard and Andrews, 1974; Payne, 1980). IMV and EEV differ in that EEV contains an additional lipid envelope and approximately 10 associated proteins which are absent from IMV (Appleyard *et al.*, 1971; Payne and Norrby, 1976; Payne, 1978). These proteins give EEV different immunological and biological properties and also make it likely that IMV and EEV bind to different cell receptors, although until recently (Vanderplasschen and Smith, 1997) there was no direct evidence for this. To permit any discussion of how vaccinia virus binds to or enters cells and of what the cellular receptors for vaccinia virus are, the form of virus being considered must first be defined.

3.1.1. Cellular Receptors for Vaccinia Virus Attachment. The cellular receptors for both the IMV and EEV are unknown. Shortly after the discovery that vaccinia virus expresses a growth factor (VGF) (Stroobant *et al.*, 1985) with amino acid similarity to epidermal growth factor, it was claimed that the cell receptor for vaccinia virus was the epidermal growth factor receptor (EGFR) (Eppstein *et al.*, 1985). However, VGF is a soluble protein that is released from cells and is not known to form part of either IMV or EEV, and viruses lacking the VGF gene are equally as infectious as wild type virus (Buller *et al.*, 1988). Moreover, the binding of vaccinia virus to cells which do or do not express EGFR and the infectivity on these cells were identical (Buller *et al.*, 1988).

A step towards the identification of receptors for vaccinia virus was made with the isolation of a monoclonal antibody (mab), called B2, which recognises a trypsin-sensitive cell surface protein and inhibits the binding of IMV to cells (Chang *et al.*, 1995). This antibody inhibits IMV binding sterically but the inhibition is incomplete, suggesting that there is more than one IMV receptor (Vanderplasschen and Smith, 1997). Importantly, mab B2 provided a tool for the molecular cloning of an IMV receptor. Preliminary results with a cDNA clone encoding a protein recognised by mab B2, indicated that the protein is unrelated to EGFR and other GenBank entries and is likely to be a large membrane-associated glycoprotein (Wen Chang, personal communication). Little information about EEV receptors is available other than they are different from those for IMV, see below, and that they bind EEV better after treatment of cells with trypsin, pronase or neuraminidase (Fig. 7) (Vanderplasschen and Smith, 1997).

3.1.2. Penetration. Whatever the cellular receptors are for IMV and EEV, once either virus has bound to the cell it has to penetrate the cell membrane. Both IMV and EEV are surrounded by lipid membranes and, like other enveloped viruses, each form of the virus enters by membrane fusion (Janeczko *et al.*, 1987; Doms *et al.*, 1990). However, for both IMV and EEV, fusion of the outer membrane with a cellular membrane would not automatically release the virus core into the cytosol because each virus contains more than one membrane (Fig. 1). For EEV, which has three membranes, the particle released into the cytosol after fusion at the plasma membrane would represent an IMV which is still surrounded by two tightly juxtaposed membranes (Sodeik *et al.*, 1993). For IMV, fusion

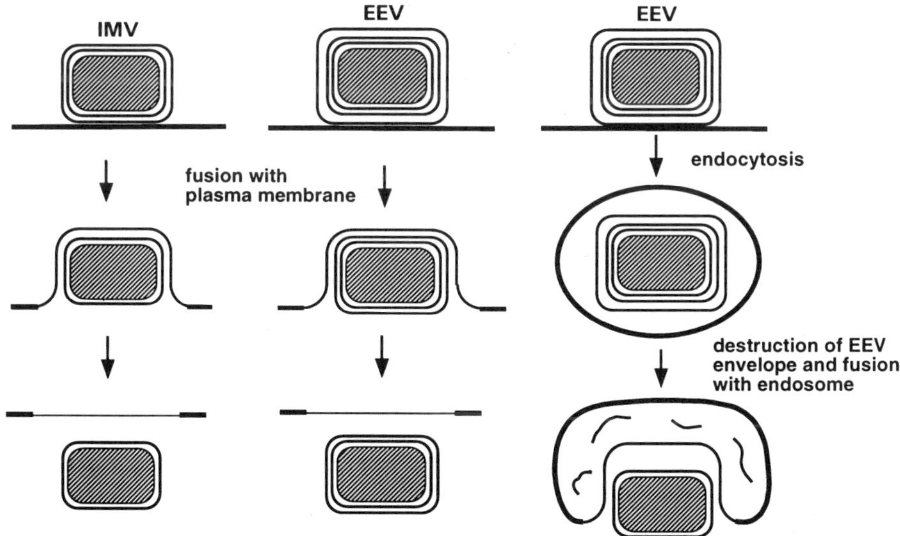

Figure 1. Models for vaccinia virus entry. IMV entry: after binding of IMV to cells, the outer IMV membrane fuses with the plasma membrane, releasing a virus particle with one membrane into the cytoplasm; how the second membrane is lost is unknown. EEV entry: EEV either fuses at the cell surface releasing an IMV into the cytosol (centre) or is endocytosed into an intracellular vesicle e.g. endosome (right). The EEV membrane may then be destroyed (Ichihashi, 1996), releasing an IMV which then fuses with the endosome releasing a core with one membrane into the cytosol.

with the plasma membrane will release the virus core still surrounded by the inner membrane into the cytosol. Without disruption or removal of these additional membranes transcription and virus replication could not commence. Clearly these problems are overcome by the virus in ways we do not yet understand. It seems very probable that the entry mechanisms of IMV and EEV may be different because they have different receptors (Vanderplasschen and Smith, 1997) and because they have to shed different numbers of membranes before entry is complete.

Until recently, studies on the entry of VV into cells have focused on IMV. Whereas early studies, which relied largely on electron microscopy, suggested that endocytosis is the more common route of IMV entry (Dales and Kajioka, 1964; Payne and Norrby, 1978), others have shown that fusion can occur directly at the cell surface (Armstrong *et al.*, 1973; Chang and Metz, 1976). Later biochemical evidence for the fusion of IMV (Janeczko *et al.*, 1987) at the plasma membrane was provided by the demonstration that fusion was pH-independent. An independent study which monitored fusion of fluorescent-labelled IMV reached the same conclusion (Doms *et al.*, 1990).

Considerably less data are available about the entry of EEV into cells. It has been suggested that EEV enters cells by fusion at the plasma membrane, like IMV, but at twice the rate and efficiency of IMV (Payne, 1979; Doms *et al.*, 1990). However, more recently it was proposed that EEV enters cells by endocytosis (Ichihashi, 1996) (Fig. 1). In this model, EEV entry consists of binding to cells, endocytosis and disruption of the EEV envelope within endosomes, releasing an infectious IMV particle into the endosome. This IMV particle then fuses with the endosomal membrane releasing the core surrounded by the inner membrane into the cytosol. Based on this model the function of the EEV membrane is to protect the internal IMV from antibody and complement neutralisation and to mediate the binding onto the cell surface. Further studies are needed to confirm the mechanisms of IMV and EEV entry and to solve the associated topological problems.

3.2. Virus Gene Expression

Vaccinia virus genes are of three types (early, intermediate and late) which are expressed in a co-ordinated cascade within the cell cytoplasm with each class being dependent upon prior expression of the previous class. Early genes are expressed immediately after the virus has entered the cell and the virus core is exposed to the cytosol, while the intermediate and late gene classes are expressed after DNA replication has commenced.

Early mRNAs are synthesised by the virus DNA-dependent RNA polymerase and associated transcription factors and capping and polyadenylating enzymes which are all packaged within virus particles. Once formed, the mRNAs are extruded from virus cores into the cytoplasm and are translated. Early genes encode proteins that further uncoat the virus core, enzymes for nucleoside metabolism and DNA replication, factors for intermediate and late transcription, and many virulence factors some of which are secreted from the cell to interfere with the host response to infection. Approximately half the virus genes (100) are expressed early in infection.

Intermediate genes are less numerous and encode mostly factors required for late transcription. During virus infection intermediate genes are only expressed once DNA replication has commenced, but this is not an absolute requirement because intermediate gene promoters on plasmids can be transcribed in infected cells in the presence of inhibitors of DNA replication. Interestingly, there is a host cell protein that is required for intermediate transcription.

Late genes are expressed once the late transcription factors are available. They encode the structural proteins of the virus, enzymes that are to be packaged into virions for early transcription, and additional virulence factors. There are two unusual features of late mRNAs: first, they are long and heterogeneous in length due to a failure to terminate at the early transcription termination motif (TTTTTNT), and second, they contain a poly A tract of approximately 35 bases immediately downstream from the 5' cap structure and before the open reading frame. The function of the 5' poly A is not understood. In general, late proteins are expressed at higher levels than early proteins due to their requirement in stoichiometric rather than catalytic amounts.

3.3. IMV Surface Proteins

Many IMV surface proteins have been characterised and the encoding gene identified (Jensen *et al.*, 1996). Although a role for some proteins in the binding and entry of IMV has been suggested, viruses lacking some of these proteins individually are still infectious. For instance, the 32 kDa IMV surface protein (gene D8L) can bind to the surface of cells (Maa *et al.*, 1990), but is non-essential for IMV infectivity because interruption of the D8L open reading frame, does not reduce virus infectivity (Niles and Seto, 1988). Similarly, the 14 kDa IMV fusion protein (gene A27L) (Rodriguez and Esteban, 1987; Rodriguez *et al.*, 1987) is not required for entry because a virus in which the expression of this gene is repressed by the *E. coli* lac repressor still forms infectious IMV (Rodriguez and Smith, 1990). The infectivity of virus with mutant D8L protein could be explained if the virus has several binding proteins, which is perhaps not surprising given its structural complexity. Likewise, in the absence of the 14 kDa fusion protein, the IMV particle must have another fusion protein or use a different mechanism to enter cells. Despite this, antibodies to 14 kDa neutralise virus infectivity (Rodriguez *et al.*, 1985; Czerny and Mahnel, 1990).

3.4. EEV Specific Proteins

Ten proteins that are unique to EEV envelope have been described (Payne, 1978; Payne, 1979) and six genes encoding these have been mapped (Table 1). The genes are: A56R, encoding the virus haemagglutinin (HA), a heavily glycosylated protein of 86 kDa (Payne and Norrby, 1976; Shida, 1986); F13L, encoding a 37 kDa protein, p37 (Hirt *et al.*, 1986); A34R, encoding a triplet of glycoproteins of 22–24 kDa, gp22–24 (Duncan and Smith, 1992); B5R, encoding a 42 kDa glycoprotein, gp42 (Engelstad *et al.*, 1992; Isaacs *et al.*, 1992); A36R, encoding a 45–50 kDa protein, p45–50 (Parkinson and Smith, 1994); and A33R, encoding a 23–28 kDa glycoprotein, gp23–28 (Roper *et al.*, 1996). The function of these proteins is being investigated by the construction of mutant viruses in which the target gene is either deleted or repressed by the *E. coli* lac repressor. Such mutants are

Table 1. Vaccinia virus EEV proteins

Gene	Protein	Membrane topology	Protein superfamily
A33R	gp24-28	type II	
A34R	gp22-24	type II	C-type lectin
A36R	p45-50	type II	
A56R	gp86 (HA)	type I	immunoglobulin
B5R	gp42	type I	complement control
F13L	p37	integral, inner face of EEV outer membrane	phospholipase D

Table 2. Properties of vaccinia virus mutants lacking specific EEV proteins

Mutant virus	Plaque size	IMV synthesis	CEV synthesis	Actin tails	EEV synthesis	Attenuation
ΔA34R	tiny	normal	yes	no	⇑ x 25, infectivity ⇑ x 5	yes
ΔA36R	small	normal	yes	no	⇑ x 3	yes
ΔA56R	syncytial	normal	yes	yes	normal	yes, intracranial no, intranasal
ΔB5R	tiny	normal	no	no	⇑ x 10	yes
ΔF13L	tiny	normal	no	no	⇑ x 20	yes

described for each gene except A33R. In each case, the loss of the EEV protein does not affect the formation or infectivity of IMV, however, the proteins have differing roles in EEV formation and infectivity, plaque formation and virulence (Table 2).

3.4.1. HA (A56R). The HA is a type I membrane protein with the majority of the protein within the lumen of the intracellular membrane compartments and on the surface of the infected cell and EEV particle. The gene is expressed early and late during infection (Brown *et al.*, 1991) and is heavily glycosylated with N- and O-linked oligosaccharides that represent more than half the mass of the mature HA (Shida and Dales, 1981). Loss of the HA protein does not affect virus plaque size, but induces cell-to-cell fusion so that large syncytia are formed (Ichihashi and Dales, 1971; Shida and Dales, 1982). Therefore, the HA inhibits cell fusion. Viruses with amino acid substitutions within HA have been isolated and characterised and these studies have defined amino acid residues that are required for either the inhibition of cell fusion or for agglutination of erythrocytes (Shida and Matsumoto, 1983; Seki *et al.*, 1990). The HA is a member of the immunoglobulin superfamily (IgSF) with a single Ig domain near the amino terminus (Jin *et al.*, 1989). This domain is more conserved than the remainder of the protein when comparing either strains of vaccinia virus or vaccinia and variola viruses (Aguado *et al.*, 1992). Deletion of the HA gene caused virus attenuation when the virus was administered by intracranial injection (Flexner *et al.*, 1987) but had no effect on the outcome of intranasal infection (GL Smith, unpublished data). The deletion of HA does not greatly affect EEV production but the infectivity of HA negative EEV needs careful investigation.

3.4.2. p37 (F13L). The p37 EEV protein is expressed late during infection and is a hydrophobic membrane protein that is acylated and accumulates in the trans-Golgi network (Hiller and Weber, 1985; Hirt *et al.*, 1986). The protein has no conventional signal sequence and is present on the inner surface of EEV outer membrane (Schmutz *et al.*, 1995). Mutations within the p37 protein confer resistance to the drug N_1-isonicotinoyl-N_2-3-methyl-4-chloro-benzoylhydrazine (IMCBH) (Schmutz *et al.*, 1991) which blocks IMV envelopment by membranes derived from the trans-Golgi network or early tubular endosomes (Tooze *et al.*, 1993; Schmelz *et al.*, 1994). Similarly, a virus mutant lacking the F13L gene makes normal IMV particles but these are not enveloped and very little EEV is released. This mutant forms only a very small plaque and is severely attenuated (Blasco and Moss, 1991). Palmitylation of p37 is required for the correct location of p37 in cells and for normal plaque formation (Grosenbach *et al.*, 1997). Interestingly, p37 shows amino acid similarity to phospholipase D (Koonin, 1996; Ponting and Kerr, 1996).

3.4.3. gp22-24 (A34R). The A34R protein is expressed late during infection and is modified by N-linked carbohydrate to forms a triplet of proteins of 22–24 kDa (Duncan and Smith, 1992). The protein has amino acid similarity to members of the C-type lectin super-family and is a type II membrane glycoprotein in which the N-terminus remains within the cytosol and inside the EEV membrane, while the lectin domain is within the lumen of intra-cellular cisternae and on the surface of EEV (Duncan and Smith, 1992). Proteins of 23–28 kDa that are recognised by a mab (Mab 4) were originally thought to be the A34R gene products (Payne, 1992) but in fact recognise the A33R gene product (Roper *et al.*, 1996), see below. Virus mutants lacking the A34R protein form very small plaques (Duncan and Smith, 1992), are severely attenuated (McIntosh and Smith, 1996) and are unable to induce actin tail formation on intracellular enveloped virus (IEV) (Wolffe *et al.*, 1997). Deletion of the A34R gene also causes a 25 fold increase in EEV release, but the A34R negative EEV has a five to six fold reduced specific infectivity (McIntosh and Smith, 1996). This implies that the A34R protein is required for some aspect of the re-entry process. In this regard, cells infected with ΔA34R virus, unlike wild type virus, are unable to fuse after brief expo-sure to low pH (Duncan, 1992; Wolffe *et al.*, 1997). A single amino acid substitution (K151D) within the lectin domain of A34R causes the majority of EEV, which is otherwise retained on the cell surface as cell-associated enveloped virus (CEV) (Blasco and Moss, 1992), to be released from the cell (Blasco *et al.*, 1993). The A34R protein is highly con-served between different vaccinia virus strains (McIntosh and Smith, 1996) and in variola virus (Aguado *et al.*, 1992). Surprisingly, the predominant amino acid at position 151 is K rather than D, and consequently most orthopoxviruses are predicted to retain the majority of enveloped virus on the surface as CEV (McIntosh and Smith, 1996).

3.4.4. gp42 (B5R). The B5R gene is transcribed early and late during infection and encodes a 42 kDa type I membrane protein which has N-linked carbohydrate and can form higher molecular weight complexes (Engelstad *et al.*, 1992; Isaacs *et al.*, 1992; Payne, 1992). Some of the B5R protein is proteolytically cleaved to release a 35 kDa fragment (Martinez-Pomares *et al.*, 1993). Like the HA and gp22–24 EEV proteins, B5R is a mem-ber of a protein superfamily and contains four copies of a 50–70 amino acid motif called the short consensus repeat (SCR) that is found in members of the regulators of comple-ment activation (RCA) proteins (Takahashi-Nishimaki *et al.*, 1991; Engelstad *et al.*, 1992; Isaacs *et al.*, 1992). Claims that B5R interacts with HA and p37 have been made (Oie *et al.*, 1990); Payne, 1992 #70] and refuted (Schmutz *et al.*, 1995). Recently, it was shown that the distribution of B5R in infected cells is not altered by mutations which prevent p37 palmitylation and alter p37 localisation (Grosenbach *et al.*, 1997). Virus mutants which lack the B5R protein form a small plaque, fail to envelope IMV particles, release less than 10% of the normal amount of EEV and are highly attenuated in animal models (Taka-hashi-Nishimaki *et al.*, 1991; Engelstad and Smith, 1993; Wolffe *et al.*, 1993). Consistent with this attenuation, a highly attenuated vaccine strain, LC16m8, that was used for small-pox vaccination in Japan was found not to synthesise B5R due to a frameshift mutation (Takahashi-Nishimaki *et al.*, 1991). When the transmembrane domain and short cytoplas-mic tail of B5R were joined to the C-terminus of the extracellular domain of HIV gp120, the latter protein was expressed on the surface of EEV particles (Katz *et al.*, 1997). This suggested that the amino acids targeting B5R to the EEV envelope are present in the C-terminal region (Katz *et al.*, 1997).

3.4.5. p45-50 (A36R). Gene A36R is predicted to encode a type II membrane protein with a molecular weight of 25 kDa, but the polypeptide migrates as a 45–50 kDa protein during sodium dodecyl sulphate polyacrylamide gel electrophoresis (Parkinson and Smith, 1994). The protein contains 7 potential N-linked glycosylation sites but apparently these are not used because the protein has the same electrophoretic mobility after synthesis in the presence of tunicamycin (Parkinson and Smith, 1994). Deletion of A36R results in a modest reduction in plaque size, a 3 fold reduction in EEV formation and considerable virus attenuation (Parkinson and Smith, 1994).

3.4.6. gp23-28 (A33R). The most recently discovered EEV protein is encoded by gene A33R. Like A34R and A36R, the A33R protein has a type II membrane topology and is glycosylated to form multiple species that have sizes of 23–28 kDa (Roper *et al.*, 1996). A mab (Mab 4) that was originally thought to recognise the A34R gene products (Payne, 1992), in fact recognises the A33R protein (Roper *et al.*, 1996). This is both acylated and phosphorylated and forms disulphide-linked dimers (Payne, 1992). The function of the protein in the virus life cycle is unknown and a deletion mutant has not yet been reported.

There is as yet no data indicating which of the EEV proteins are required for virus attachment, but it is noteworthy that the three protein superfamilies to which the HA (A56R), gp22–24 (A34R) and gp42 (B5R) EEV proteins belong (IgSF, C-type lectin and complement control proteins, respectively) are all characterised by molecules which mediate interactions between cells or between soluble molecules and cells. Possibly, vaccinia virus and other orthopoxviruses have acquired such proteins during evolution that allow the virus to bind to a broad range of surface molecules and so give the virus a wide tropism.

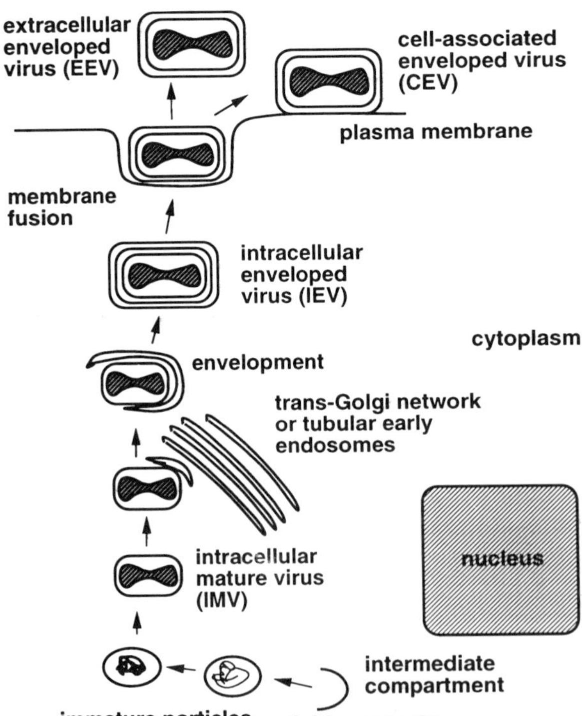

Figure 2. Vaccinia virus morphogenesis. See text for details. Note that the membrane drawn around the immature virus particles and IMV is thought to be a double lipid bilayer, although for simplicity it is only drawn as a single line here. IEV particles may move to the cell surface either by polymerising actin (Fig. 3) or by another pathway.

3.5. Morphogenesis

Morphogenesis has been extensively studied by electron microscopy and occurs within eosinophilic cytoplasm inclusion bodies (B type inclusions) often called virus factories. The first sign of new virus particles is the appearance of crescent shaped structures (Fig. 2). These are derived from membranes of the intermediate compartment between the endoplasmic reticulum and the Golgi stack (Sodeik *et al.*, 1993) which have been modified by addition of several virus membrane-associated or membrane-spanning proteins. Crescents are thought to contain two lipid bilayers that are very closely juxtaposed and their formation is reversibly inhibited by rifampicin (Moss *et al.*, 1969). The crescents then extend to form a complete oval into which the DNA-nucleoprotein complex is packaged forming immature virus (IV). The DNA-nucleoprotein condenses into an electron dense core structure and thereafter the virus is infectious and is called IMV. During IMV morphogenesis several of the major core proteins are proteolytically cleaved, for review see (VanSlyke and Hruby, 1990).

Some IMV particles become enveloped with additional cisternae derived from the early tubular endosomes or trans-Golgi network (Tooze *et al.*, 1993; Schmelz *et al.*, 1994). Most of the early IMV particles made become enveloped, but later during infection, as more IMV particles mature, a smaller proportion associate with these additional membranes (Ulaeto *et al.*, 1996). Envelopment of IMV must involve the specific interaction between one of more IMV surface proteins and the cytosolic domain of one of more virus EEV protein which become enriched in the these wrapping membranes. The details of this interaction are not understood, but this process is inhibited by brefeldin A (Ulaeto *et al.*, 1995) and the IMV 14 kDa fusion protein is required, since in its absence, envelopment is restricted and less EEV is formed (Rodriguez and Smith, 1990). As described above and summarised in Table 2, deletion of the B5R gene (gp42) (Engelstad and Smith, 1993; Wolffe *et al.*, 1993) or F13L gene (p37) (Blasco and Moss, 1991) drastically inhibits envelopment. Once envelopment is complete, the virus particles are surrounded by four membranes and are called intracellular enveloped virus (IEV).

Movement of IEV to the cell surface can occur in at least two ways. Firstly, in a process that is reminiscent of the actin polymerisation and motility induced by some intracellular bacteria such as Listeria and Shigella (Tilney and Portnoy, 1989), IEV particles can induce the polymerisation of actin on one side of the IEV particle so that an actin tail is formed which propels the IEV particle to the cell surface (Cudmore *et al.*, 1995; Cudmore *et al.*, 1996). Once the IEV particle has reached the plasma membrane, the actin tail continues to grow, thereby pushing the IEV particle out of the cell on the tip of specialised microvilli that are easily visible by scanning electron microscopy (Fig. 3). Recent studies into the proteins required for actin polymerisation on the IEV particle have shown a number of EEV proteins are involved. Loss of either the gp22–24 (A34R) (Wolffe *et al.*, 1997) or p45–50 (A36R) proteins (C.A. Sanderson, M. Way and G.L. Smith unpublished data) prevents actin tail formation on IEV particles. The requirement for these proteins in actin tail formation is somewhat surprising because each is predicted to have very few amino acids on the cytosolic side of the membrane where actin nucleation occurs. Possibly these proteins are essential components of a larger complex which mediates actin polymerisation. Importantly, the A36R deletion mutant still makes approximately 30% of wild type levels of EEV (Parkinson and Smith, 1994), and the A34R mutant makes 25 fold more EEV (McIntosh and Smith, 1996). These observations clearly demonstrate that IEV can reach the cell surface independent of actin tails.

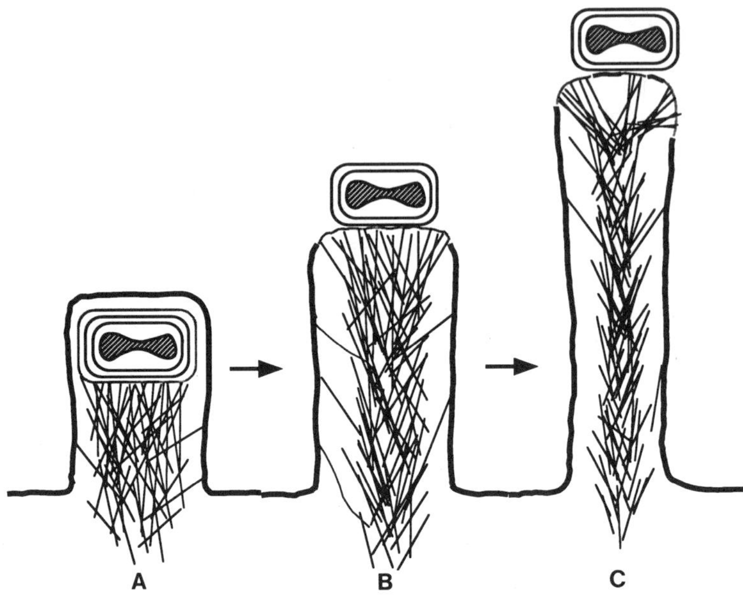

Figure 3. Formation of CEV (and EEV) via actin tails and microvilli. (A). An IEV particle with its associated actin tail reaches the plasma membrane and continues to grow outwards forming a specialised microvillus. (B). The EEV outer membrane (thin grey line) fuses with the plasma membrane (thick black line) at the tip of the microvillus thereby exposing a CEV on the cell surface. (C). The actin tail continues to grow from the actin-nucleating proteins which have been transferred to the plasma membrane. The CEV particle may detach from the cell forming EEV.

3.6. Release of EEV

Upon reaching the cell surface the outer membrane of IEV fuses with the plasma membrane forming a virus with three membranes (Fig. 2). With most vaccinia virus strains the majority of this enveloped virus remains on the cell surface and has been called cell associated enveloped virus (CEV). This virus can be released by mild trypsin treatment (Blasco and Moss, 1992), by a K151D substitution within the gp22–24 (A34R) (Blasco *et al.*, 1993) or by deletion of the A34R gene (McIntosh and Smith, 1996). Why it is advantageous to vaccinia virus to retain most of the enveloped virus on the cell, rather than have it released, is unclear. Some other viruses such as influenza virus express a receptor destroying enzyme (neuraminidase) to enable virus release from cells and prevent virus aggregation.

4. RESULTS

4.1. Vaccinia Virus IMV and EEV Bind to Different Receptors

The different proteins present on the surface of IMV and EEV make it likely that these viruses bind to different receptors, but direct evidence for this has been lacking until recently. To study virus binding and entry, the IMV and EEV forms need to be purified and analysed separately. IMV can be purified in a relatively straightforward manner from cytoplasmic extracts (Joklik, 1962), but the situation with EEV is more complex because it is 100 fold less abundant than IMV and its outer envelope is fragile. Consequently, if only 1% of the infected cells lyse, the EEV in the supernatant becomes overwhelmingly contaminated with released IMV. EEV can be separated from IMV by caesium chloride density gradient centrifugation because of its lower buoyant density (1.23 g/ml, compared with 1.27 g/ml for IMV) or by sucrose gradients (Doms *et al.*, 1990). However, the EEV envelope is easily ruptured by either freeze-thawing, sonication or ultracentrifugation, so that although the particles purify as EEV, biologically they behave as IMV due to exposure of IMV surface proteins (Payne and Norrby, 1976; McIntosh and Smith, 1996, Ichihashi, 1996 #58; Vanderplasschen and Smith, 1997).

To assess the degree of IMV contamination in an EEV preparation, we used a mab, 5B4/2F2, which recognises the 14 kDa fusion protein of IMV and efficiently neutralises IMV particles (Czerny and Mahnel, 1990). Interestingly, this antibody does not prevent IMV binding to cells but inhibits a subsequent step of the infection process. Figure 4 shows that the proportion of total infectivity neutralised by mab 5B4/2F2 increases during EEV purification. IMV represents up to 25% of infectivity in the supernatant of infected cells but this increases to up to 65% after gradient centrifugation. Therefore for all subsequent measurements of EEV binding and infectivity we used fresh EEV from the clarified supernatant of infected cells and neutralised any contaminating IMV by mab 5B4/2F2 (Vanderplasschen and Smith, 1997).

To distinguish IMV and EEV within a mixed population, we developed a method using mabs which stain either only EEV (rat mab 19C2 against B5R (Schmelz *et al.*, 1994)) or both IMV and EEV (mouse mab AB1.1 against D8L (Parkinson and Smith, 1994)) and visualised individual virus particles bound to cells by confocal microscopy (Vanderplasschen and Smith, 1997). Figure 5 shows a cell to which IMV and EEV particles had bound at 4°C, and which was then stained with mabs against the B5R and D8L gene products of the EEV and IMV surface, respectively. After addition of appropriate secondary antibod-

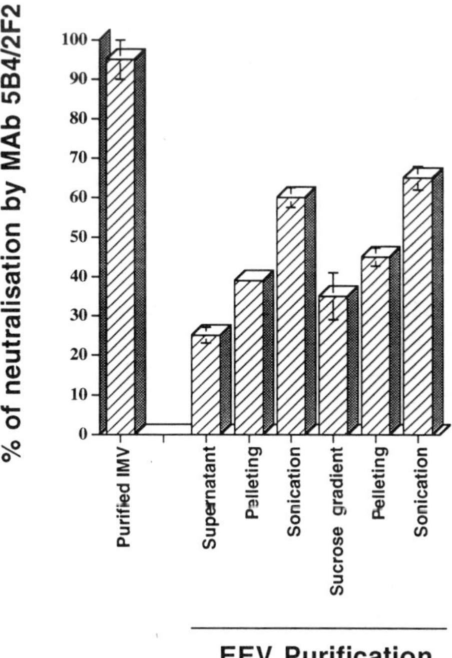

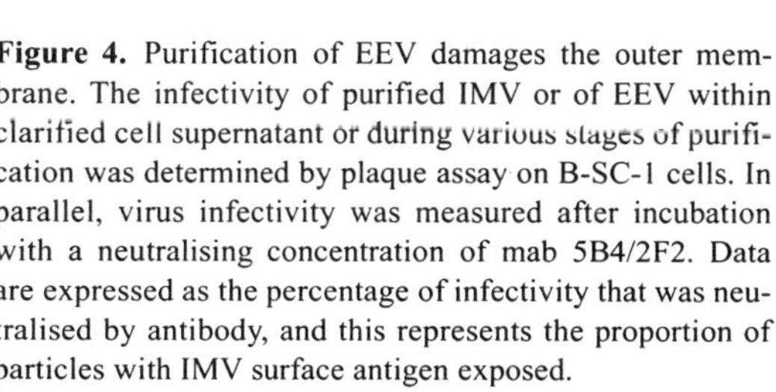

Figure 4. Purification of EEV damages the outer membrane. The infectivity of purified IMV or of EEV within clarified cell supernatant or during various stages of purification was determined by plaque assay on B-SC-1 cells. In parallel, virus infectivity was measured after incubation with a neutralising concentration of mab 5B4/2F2. Data are expressed as the percentage of infectivity that was neutralised by antibody, and this represents the proportion of particles with IMV surface antigen exposed.

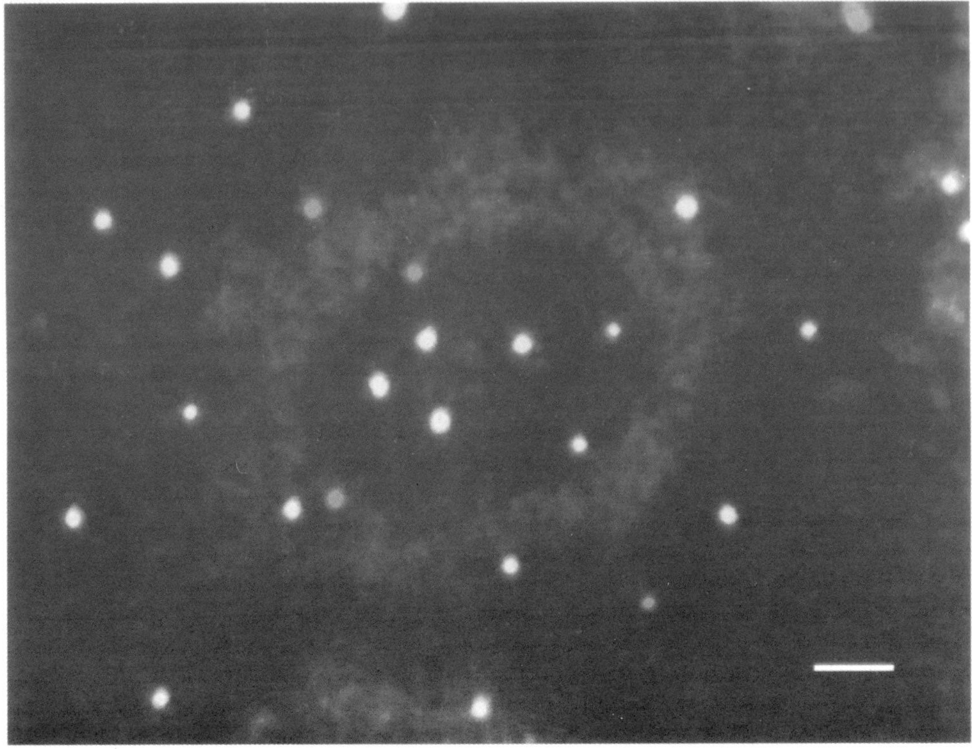

Figure 5. Distinguishing IMV and EEV particles bound to cells with mabs and confocal microscopy. A mixture of IMV and EEV was adsorbed to RK-13 cells at 4°C and then fixed and permeabilised. Cells were then stained with biotinylated mouse mab AB1.1 against the IMV surface protein D8L (Parkinson and Smith, 1994) and rat mab 19C2 against EEV surface protein B5R (Schmelz *et al.*, 1994). Bound antibodies were detected with streptavidin-rhodamine and FITC-conjugated goat anti-rat respectively. The photograph shows a merged image; bar equals 2 μm.

ies, IMV particles appeared red while EEV particles were green, and in the merged image shown, EEV particles appear yellow.

This method allowed the number of IMV and EEV particles within a virus preparation to be determined. Parallel measurements of infectivity (plaque assay) in the presence or absence of IMV neutralising antibody 5B4/2F2, revealed the infectivity associated with IMV or EEV particles and hence the particle/plaque forming unit (pfu) ratios were determined. These data showed that fresh IMV had a particle/pfu ratio of 45 ± 11.1 (mean ± SD, n=3) and this increased to 64.6 ± 16.5 (n=3) during purification, while EEV had a lower particle/pfu ratio of 12.7 ± 6.1 (n=3).

Having established a method that distinguishes IMV and EEV particles bound to cells, it was possible to assess if the binding or infectivity of the two forms of virus varied in the same way under different conditions. Binding to different cells was examined first. If IMV and EEV used a common receptor, the particles would be expected to bind to different cells in the same way. Figure 6A shows that IMV bound equally to BS-C-1 and RK-13 cells, whereas EEV bound much better to BS-C-1 cells than RK-13. The small amount of IMV contaminating the fresh EEV preparation provided a perfect internal control and showed that the different binding of EEV to these cells was not due to any artefact of the

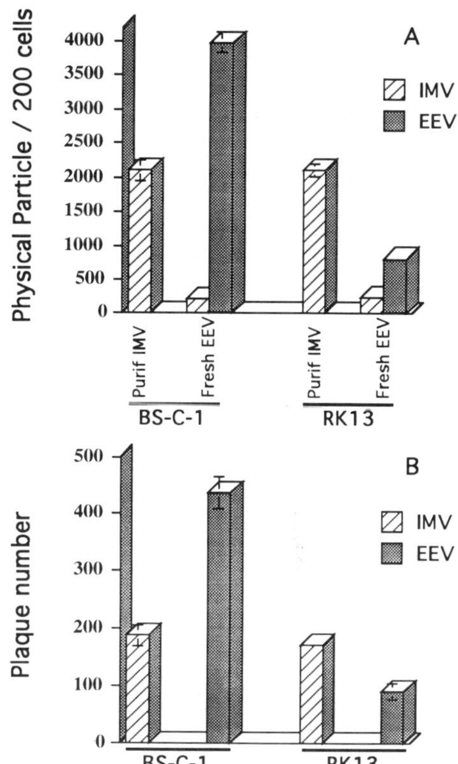

Figure 6. IMV and EEV particles bind differently to different cell types. (A). Purified IMV or fresh EEV were bound to B-SC-1 or RK-13 cells at 4°C and then detected with mabs and confocal microscopy as described in Fig. 5. The data show the number of IMV or EEV particles bound to 200 cells. (B). The infectivity of the same preparations of virus as used in (A) were determined by plaque assay on B-SC-1 or RK-13 cells. For infectivity determination of EEV, any contaminating IMV was neutralised with mab 5B4/2F2.

virus preparation. The infectivity of these virus preparations was determined in parallel, including mab 5B4/2F2 to neutralise any contaminating IMV within the EEV preparation, and showed exactly the same ratios (Fig. 6B).

Next we examined whether treatment of RK-13 cells with different enzymes would affect the binding of IMV and EEV in different ways. (Fig. 7). These data showed that IMV binding was increased slightly by trypsin and neuraminidase but almost abolished by pronase. In contrast, EEV binding was increased twofold by trypsin, threefold by pronase and 1.6 fold by neuraminidase. Examination of the infectivity of IMV and EEV on these cells showed exactly the same result (Fig. 7B). These data indicated the receptors for IMV and EEV were different.

It was reported that IMV binding to cells could be inhibited by a mouse IgM mab (B2) (Chang *et al.*, 1995), however the effect on EEV binding was not examined. This was investigated and the data are shown in Fig. 8. While IMV binding and infectivity were reduced by 50 to 80%, depending on the cell type (data not shown), EEV binding and infectivity were unaffected. This might have been explained if IMV and EEV bound to different regions of the same molecule and mab B2 was able to block only IMV binding. To address this point, capping of the B2 reactive molecule was induced. However, this did not increase the inhibition of IMV binding and had no effect on EEV binding (Vanderplasschen and Smith, 1997). Moreover, the EEV particles that were bound to cells on which capping of the B2 reactive epitope had been induced, were clearly localised away from the B2 reactive molecule (Vanderplasschen and Smith, 1997). These data showed that the IMV and EEV receptor(s) are different and, because mab B2 did not inhibit all IMV binding, there is more than one IMV receptor.

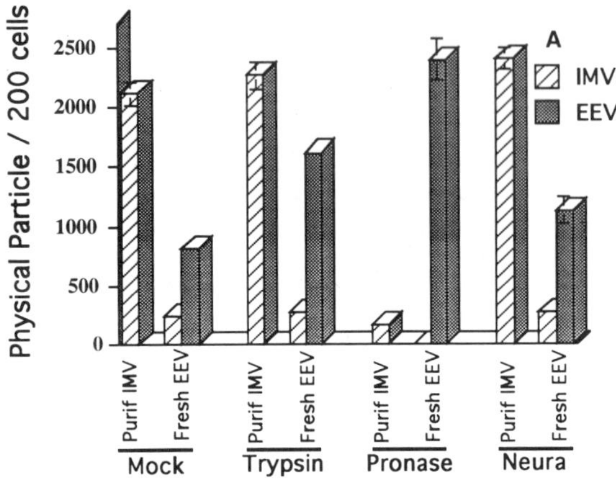

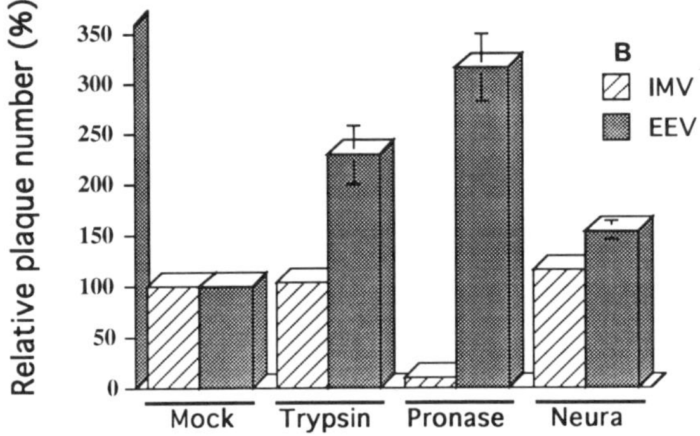

Figure 7. Effect of enzyme treatment on IMV and EEV binding and infectivity on RK-13 cells. Cells were either mock treated or treated with trypsin, pronase or neuraminidase as described elsewhere (Vanderplasschen and Smith, 1997). (A). Purified IMV or fresh EEV were bound to cells as in Fig. 6 and the number of IMV and EEV particles bound to 200 cells was determined. (B). Parallel infectivity measurement of IMV and EEV on enzyme-treated cells. Any contaminating IMV in the EEV preparation was neutralised by mab 5B4/2F2.

4.2. EEV Is Resistant to Neutralisation by Antibody

Many studies have demonstrated the importance of antibody raised against EEV proteins in protection against poxviruses and that these antibodies prevent comet formation (see introduction). However, Ichihashi reported recently that EEV was resistant to antibody neutralisation (Ichihashi, 1996). This was examined in more detail using the methodology developed above (section 5.1) for the simultaneous analysis of IMV and EEV. After repeated immunisation with live vaccinia virus, sera from man, mouse, rat and rabbit were tested for their ability to neutralise IMV and EEV infectivity. The data showed that while IMV was inhibited well by all sera, none of the sera were able to inhibit EEV infectivity and nor were mabs against EEV proteins either alone or in combination (Table 3) (Vanderplasschen et al., 1997).

Despite its failure to neutralise EEV, the polyclonal rabbit serum (which had the highest anti-vaccinia virus titre, Table 3) was able to efficiently inhibit comet formation (Fig. 9). Since EEV is required for comet formation, this result presented a paradox: on the one hand, antibodies cannot neutralise EEV infectivity, while on the other hand, they block comet formation. This paradox was explained by electron microscopic analysis of infected cells in the presence or absence of anti-vaccinia virus serum. In the presence, but not the absence, of antibody, large numbers of enveloped virus particles were found on the cell surface (Fig. 10). Consistent with this result, the number of physical particles and infectious particles in the cell supernatant were reduced by 3 to 4 $\log_{10}$ units in the presence of antibody (Vanderplasschen et al., 1997). Thus antibodies to EEV provide immunological protection by restricting virus release rather than preventing infection.

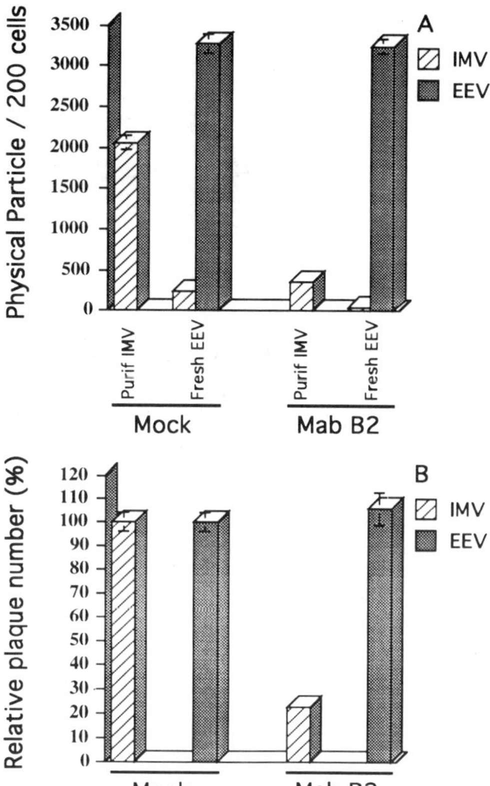

Figure 8. Mab B2 inhibits IMV but not EEV binding and infectivity. B-SC-1 cells were either mock treated or treated with mab B2. (A). Purified IMV or fresh EEV were bound to cells as in Fig. 6 and the number of IMV and EEV particles bound to 200 cells was determined. (B). Parallel infectivity measurement of IMV and EEV on mab B2-treated or mock-treated cells. Any contaminating IMV in the EEV preparation was neutralised by mab 5B4/2F2.

Table 3. Antibody neutralisation of vaccinia virus

Sera and Mabs	Titre[†]	Neutralisation assay[*]	
		IMV	EEV
Human	1/200	20.9%	98.3%
Rabbit	1/800	1.8%	101.5%
Rat	1/500	3.9%	97.3%
Mouse	1/250	28.2%	96.0%
Mab 19C2 (α B5R)	1/16	106.7%	102.2%
Mab 1H831 (α HA)	1/8	103.1%	106.8%
Mab 19C2 plus Mab 1H831 (1:1)	—	99.5%	103.5%

[*]The infectivity of IMV and EEV preparations were determined by plaque assay after incubation with hyperimmune serum from different species or with the mabs against EEV proteins. Sera and mabs were tested at dilutions of 1/100 and 1/10, respectively. The results are expressed as the percentage of the controls (pre-immune or non-immune serum from the same species, giving between 150-200 pfu/well). Each value represents the average for triplicate cultures. SD of all values were below 0.05.

[†]The arbitrary titre of antibody against VV of the different sera and mabs was determined by indirect immunofluorescent staining of vaccinia virus infected cells. The titre represents the highest dilution which gave an optimal fluorescent signal.

Reproduced with permission from Vanderplasschen, A., Hollinshead, M. and Smith, G. L. (1997). Antibodies against vaccinia virus do not neutralise extracellular enveloped virus but prevent virus release from infected cells and comet formation. *J. Gen. Virol.* **78**, in press.

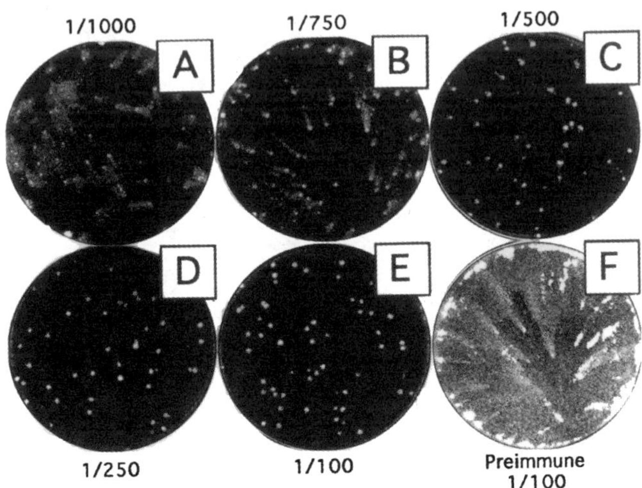

Figure 9. Inhibition of comet formation by rabbit anti-serum raised against a live vaccinia virus infection. RK-13 cells were infected with approximately 50 plaque forming units of vaccinia virus strain IHD-J and then incubated under liquid overlay containing either a 1/100 dilution of rabbit pre-immune serum (F) or various dilutions of immune serum (A-E). After 48 hours the medium was removed and the cells were stained with 0.1% crystal violet in 15% ethanol. Note the appearance of comet shaped plaques in panel F. Reproduced with permission from Vanderplasschen, A., Hollinshead, M. and Smith, G. L. (1997). Antibodies against vaccinia virus do not neutralise extracellular enveloped virus but prevent virus release from infected cells and comet formation. *J. Gen. Virol.* **78**, in press.

This result raises the question of how EEV that has been opsonised by antibody remains infectious. One possibility is that the virus enters cells via Fc receptors which bind to the Fc region of immunoglobulins attached to the virus particle. Whatever the mechanism, it is likely that free virus and virus complexed with antibody may use different cell receptors which may lead to a change in cell tropism.

The EEV evasion of neutralisation by antibody is another example of a growing list of strategies used by vaccinia virus to evade the host response to infection. In addition to this method of evading neutralisation by antibodies, vaccinia virus is involved in processes that oppose apoptosis, counteract complement, capture chemokines, interfere with interferon, intercept interleukins and synthesise steroids. The study of vaccinia virus therefore provides an excellent means of studying virus pathogenesis and of learning fundamental aspects of the host response to infection.

ACKNOWLEDGMENTS

The work in this laboratory is supported by The Wellcome Trust and the Medical Research Council. A.V. is a senior research assistant of the Fonds National Belge de la Recherche Scientifique (F.N.R.S.) and is the laureate of NATO and Huynen research fellowships. We thank Christopher Sanderson, Henriette van Eijl, Liz Mathew and Nicola Price for critical reading of the manuscript. G.L.S. thanks Luis Enjuanes, Willy Spaan and Stuart Siddel for the kind invitation to attend the VIIth International Symposium on Coronaviruses and Arteriviruses in Segovia.

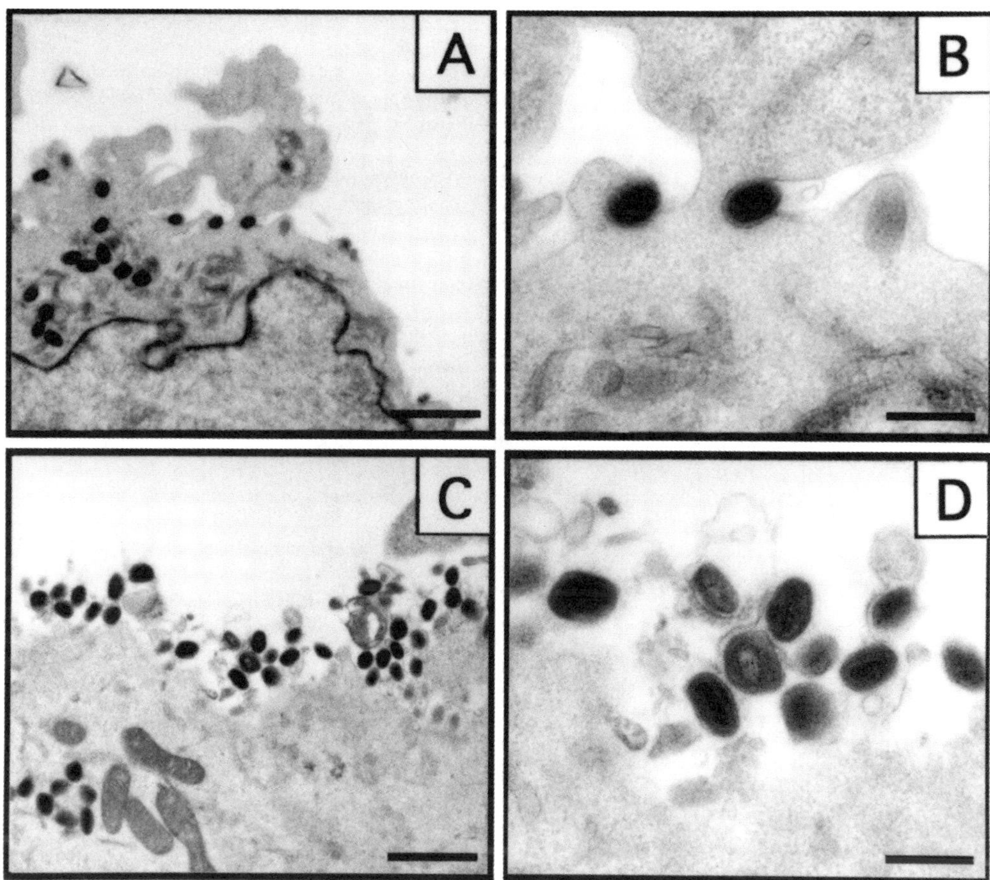

Figure 10. Electron micrographs showing aggregated virus particles on the surface of infected cells. RK-13 cells were infected with vaccinia virus strain WR at 0.1 pfu/cell and incubated for 24 hours in the presence of either pre-immune serum (A and B) or with rabbit antiserum raised against a live vaccinia virus infection (C and D). Cells were then processed for electron microscopy. Bar in panels A and C = 1.25 μm and in panels B and D = 0.4 μm. Reproduced with permission from Vanderplasschen, A., Hollinshead, M. and Smith, G. L. (1997). Antibodies against vaccinia virus do not neutralise extracellular enveloped virus but prevent virus release from infected cells and comet formation. *J. Gen. Virol.* **78**, in press.

REFERENCES

Aguado, B., Selmes, I. P., and Smith, G. L., 1992, Nucleotide sequence of 21.8 kbp of variola major virus strain Harvey and comparison with vaccinia virus, *J. Gen. Virol.* **73**: 2887–2902.

Alcamí, A. and Smith, G. L., 1995, Cytokine receptors encoded by poxviruses: a lesson in cytokine biology, *Immunol. Today* **16**: 474–478.

Alcamí, A. and Smith, G. L., 1996, Receptors for gamma-interferon encoded by poxviruses: implications for the unknown origin of vaccinia virus, *Trends Microbiol.* **4**: 321–326.

Appleyard, G. and Andrews, C., 1974, Neutralizing activities of antisera to poxvirus soluble antigens, *J. Gen. Virol.* **23**: 197–200.

Appleyard, G., Hapel, A. J., and Boulter, E. A., 1971, An antigenic difference between intracellular and extracellular rabbitpox virus, *J. Gen. Virol.* **13**: 9–17.

Armstrong, J. A., Metz, D. H., and Young, M. R., 1973, The mode of entry of vaccinia virus into L cells, *J. Gen. Virol.* **21**: 533–537.

Baxby, D., 1981, *"Jenner's smallpox vaccine: the riddle of the origin of vaccinia virus."* Heineman Educational Books, London.

Blasco, R. and Moss, B., 1991, Extracellular vaccinia virus formation and cell-to-cell virus transmission are prevented by deletion of the gene encoding the 37,000-Dalton outer envelope protein, *J. Virol.* **65**, 5910–5920.

Blasco, R. and Moss, B., 1992, Role of cell-associated enveloped vaccinia virus in cell-to-cell spread, *J. Virol.* **66**: 4170–4179.

Blasco, R., Sisler, J. R., and Moss, B., 1993, Dissociation of progeny vaccinia virus from the cell membrane is regulated by a viral envelope glycoprotein: effect of a point mutation in the lectin homology domain of the A34R gene, *J. Virol.* **67**: 3319–3325.

Boulter, E. A.,1969, Protection against poxviruses, *Proc. R. Soc. Med.* **62**: 295–297.

Boulter, E. A. and Appleyard, G., 1973, Differences between extracellular and intracellular forms of poxviruses and their implications, *Prog. Med. Virol.* **16**: 86–108.

Boulter, E. A., Zwartouw, H. T., Titmuss, D. H. I., and Maber, H. B., 1971, The nature of the immune state produced by inactivated vaccinia virus in rabbits, *Am. J. Epidemiol.* **94**: 612–620.

Brown, C. K., Turner, P. C. and Moyer, R. W., 1991, Molecular characterization of the vaccinia virus hemagglutinin gene, *J. Virol.* **65**: 3598–606.

Buller, R. M. L., Chakrabarti, S., Cooper, J. A., Twardzik, D. R., and Moss, B., 1988, Deletion of the vaccinia virus growth factor gene reduces virus virulence, *J. Virol.* **62**: 866–874.

Chang, A. and Metz, D. H., 1976, Further investigations on the mode of entry of vaccinia virus into cells, *J. Gen. Virol.* **32**: 275–282.

Chang, W., Hsiao, J.-C., Chung, C.-S., and Bair, C.-H., 1995, Isolation of a monoclonal antibody which blocks vaccinia virus infection, *J. Virol.* **69**: 517–522.

Cudmore, S., Cossart, P., Griffiths, G., and Way, M., 1995, Actin-based motility of vaccinia virus, *Nature* **378**: 636–638.

Cudmore, S., Reckmann, I., Griffiths, G., and Way, M., 1996, Vaccinia virus: a model system for actin-membrane interactions, *J. Cell Sci.* **109**: 1739–1747.

Czerny, C. P. and Mahnel, H., 1990, Structural and functional analysis of orthopoxvirus epitopes with neutralizing monoclonal antibodies, *J. Gen. Virol.* **71**: 2341–2352.

Dales, S. and Kajioka, R., 1964, The cycle of multiplication of vaccinia virus in Earles strain L cells. I. Uptake and penetration, *Virology* **24**: 278–294.

Doms, R. W., Blumenthal, R., and Moss, B., 1990, Fusion of intra- and extracellular forms of vaccinia virus with the cell membrane, *J. Virol.* **64**: 4884–4892.

Duncan, S. A., 1992, D. Phil. *Analysis of three vaccinia virus genes, one of which is essential for plaque formation.* University of Oxford, Oxford.

Duncan, S. A. and Smith, G. L., 1992, Identification and characterization of an extracellular envelope glycoprotein affecting vaccinia virus egress, *J. Virol.* **66**: 1610–1621.

Engelstad, M., Howard, S. T. and Smith, G. L., 1992, A constitutively expressed vaccinia virus gene encodes a 42 kDa glycoprotein related to complement control factors that forms part of the extracellular envelope, *Virology* **188**: 801–810.

Engelstad, M. and Smith, G. L., 1993, The vaccinia virus 42 kDa envelope protein is required for envelopment and egress of extracellular virus and for virulence, *Virology* **194**: 627–637.

Eppstein, D. A., Marsh, Y. V., Schreiber, A. B., Newman, S. R., Todaro, G. J. and Nestor, J. J., 1985, Epidermal growth factor receptor occupancy inhibits vaccinia virus infection, *Nature* **318**: 663–665.

Fenner, F., Anderson, D. A., Arita, I., Jezek, Z. and Ladnyi, I. D., 1988, *"Smallpox and Its Eradication."* World Health Organisation, Geneva.

Flexner, C., Hugin, A. and Moss, B., 1987, Prevention of vaccinia virus infection in immunodeficient mice by vector-directed IL-2 expression, *Nature* **330**: 259–262.

Goebel, S. J., Johnson, G. P., Perkus, M. E., Davis, S. W., Winslow, J. P. and Paoletti, E., 1990, The complete DNA sequence of vaccinia virus, *Virology* **179**: 247–266.

Grosenbach, D. W., Ulaeto, D. O. and Hruby, D. E., 1997, Palmitylation of the vaccinia virus 37-kDa major envelope antigen. Identification of a conserved acceptor motif and biological relevance, *J. Biol. Chem.* **272**: 1956–1964.

Hiller, G. and Weber, K., 1985, Golgi-derived membranes that contain an acylated viral polypeptide are used for vaccinia virus envelopment, *J. Virol.* **55**: 651–659.

Hirt, P., Hiller, G. and Wittek, R., 1986, Localization and fine structure of a vaccinia virus gene encoding an envelope antigen, *J. Virol.* **58**: 757–764.

Ichihashi, Y., 1996, Extracellular enveloped vaccinia virus escapes neutralization, *Virology* **217**: 478–485.

Ichihashi, Y. and Dales, S., 1971, Biogenesis of poxviruses: interrelationship between hemagglutinin production and polykaryocytosis, *Virology* **46**: 533–543.

Isaacs, S. N., Wolffe, E. J., Payne, L. G. and Moss, B., 1992, Characterization of a vaccinia virus-encoded 42-kilodalton class I membrane glycoprotein component of the extracellular virus envelope, *J. Virol.* **66**: 7217–7224.

Janeczko, R. A., Rodriguez, J. F. and Esteban, M., 1987, Studies on the mechanism of entry of vaccinia virus in animal cells, *Arch. Virol.* **92**: 135–150.

Jensen, O. N., Houthaeve, T., Shevchenko, A., Cudmore, S., Ashford, T., Mann, M., Griffiths, G. and Krijnse Locker, J., 1996, Identification of the major membrane and core proteins of vaccinia virus by two-dimensional electrophoresis, *J. Virol.* **70**: 7485–7497.

Jin, D., Li, Z., Jin, Q., Yuwen, H. and Hou, Y., 1989, Vaccinia virus hemagglutinin. A novel member of the immunoglobulin superfamily, *J. Exp. Med.* **170**: 571–576.

Joklik, W. K., 1962, The purification of four strains of poxvirus, *Virology* **18**: 9–18.

Katz, E., Wolffe, E. J. and Moss, B., 1997, The cytoplasmic domains of the vaccinia viru B5R protein target a chimeric human immunodeficiency virus type 1 glycoprotein to the outer envelope of nascent vaccinia virions, *J. Virol.* **71**: 3178–3187.

Koonin, E. V., 1996, A duplicated catalytic motif in a new superfamily of phosphohydrolases and phospholipid synthases that includes poxvirus envelope proteins, *Trends Biochem. Sci.* **21**: 242–243.

Maa, J. S., Rodriguez, J. F. and Esteban, M., 1990, Structural and functional characterization of a cell surface binding protein of vaccinia virus, *J. Biol. Chem.* **265**: 1569–77.

Martinez-Pomares, L., Stern, R. J. and Moyer, R. W., 1993, The ps/hr gene (B5R open reading frame homolog) of rabbitpox virus controls pock colour, is a component of extracellular enveloped virus, and is secreted into the medium, *J. Virol.* **67**: 5450–5462.

Massung, R. F., Esposito, J. J., Liu, L.-i., Qi, J., Utterback, T. R., knight, J. C., Aubin, L., Yuran, T. E., Parsons, J. M., Loparev, V. N., Selivanov, N. A., Cavallaro, K. F., Kerlavage, A. R., Mahy, B. W. J. and Venter, A. J., 1993, Potential virulence determinants in terminal regions of variola smallpox virus genome; *Nature* **366**: 748–751.

McIntosh, A. A. G. and Smith, G. L., 1996, Vaccinia virus glycoprotein A34R is required for infectivity of extracellular enveloped virus, *J. Virol.* **70**: 272–281.

Moss, B., 1996, Poxviridae: the viruses and their replication, in *"Fields Virology"* (B. N. Fields, D. M. Knipe, and P. M. Howley, Eds.), Vol. 2, pp. 2637–2671. 2 vols. Lippincott Raven Press, New York.

Moss, B., Rosenblum, E. N., Katz, E. and Grimley, P. M., 1969, Rifampicin: a specific inhibitor of vaccinia virus assembly, *Nature* **224**: 1280–1284.

Niles, E. G. and Seto, J., 1988, Vaccinia virus gene D8 encodes a virion transmembrane protein, *J. Virol.* **62**: 3772–3778.

Oie, M., Shida, H. and Ichihashi, Y., 1990, The function of the vaccinia hemagglutinin in the proteolytic activation of infectivity, *Virology* **176**: 494–504.

Parkinson, J. E. and Smith, G. L., 1994, Vaccinia virus gene A36R encodes a M_r 43–50 K protein on the surface of extracellular enveloped virus, *Virology* **204**: 376–390.

Payne, L., 1978, Polypeptide composition of extracellular enveloped vaccinia virus, *J. Virol.* **27**: 28–37.

Payne, L. G., 1979, Identification of the vaccinia hemagglutinin polypeptide from a cell system yielding large amounts of extracellular enveloped virus, *J. Virol.* **31**: 147–155.

Payne, L. G., 1980, Significance of extracellular enveloped virus in the *in vitro* and *in vivo* dissemination of vaccinia virus, *J. Gen.Virol.* **50**: 89–100.

Payne, L. G., 1992, Characterization of vaccinia virus glycoproteins by monoclonal antibody preparations, *Virology* **187**: 251–260.

Payne, L. G. and Kristensson, K., 1985, Extracellular release of enveloped vaccinia virus from mouse nasal epithelial cells in vivo, *J. Gen. Virol.* **66**: 643–646.

Payne, L. G. and Norrby, E., 1976, Presence of haemagglutinin in the envelope of extracellular vaccinia virus particles, *J. Gen. Virol.* **32**: 63–72.

Payne, L. G. and Norrby, E., 1978, Adsorption and penetration of enveloped and naked vaccinia virus particles, *J. Virol.* **27**: 19–27.

Ponting, C. P. and Kerr, I. D., 1996, A novel family of phospholipase D homologues that includes phospholipid synthases and putative endonucleases: identification of duplicated repeats and potential active site residues, *Protein Sci.* **5**: 914–922.

Rodriguez, J. F. and Esteban, M., 1987, Mapping and nucleotide sequence of the vaccinia virus gene that encodes a 14-kilodalton fusion protein, *J. Virol.* **61**: 3550–3554.

Rodriguez, J. F., Janezcko, R. and Esteban, M., 1985, Isolation and characterization of neutralizing monoclonal antibodies to vaccinia virus, *J. Virol.* **56**: 482–488.

Rodriguez, J. F., Paez, E. and Esteban, M., 1987, A 14,000-M_r envelope protein of vaccinia virus is involved in cell fusion and forms covalently linked trimers, *J. Virol.* **61**: 395–404.

Rodriguez, J. F. and Smith, G. L., 1990, IPTG-dependent vaccinia virus : identification of a virus protein enabling virion envelopment by Golgi membrane and egress, *Nucl. Acids Res.* **18**: 5347–5351.

Roper, R. L., Payne, L. G. and Moss, B., 1996, Extracellular vaccinia virus envelope glycoprotein encoded by the A33R gene, *J. Virol.* **70**: 3753–3762.

Schmelz, M., Sodeik, B., Ericsson, M., Wolffe, E. J., Shida, H., G., H. and Griffiths, G., 1994, Assembly of vaccinia virus: the second wrapping cisterna is derived from the trans golgi network, *J. Virol.* **68**: 130–147.

Schmutz, C., Payne, L. G., Gubser, J. and Wittek, R., 1991, A mutation in the gene encoding the vaccinia virus 37,000-M_r protein confers resistance to an inhibitor of virus envelopment and release, *J. Virol.* **65**: 3435–3442.

Schmutz, C., Rindisbacher, L., Galmiche, M. C. and Wittek, R., 1995, Biochemical analysis of the major vaccinia virus envelope antigen, *Virology* **213**: 19–27.

Seki, M., Oie, M., Ichihashi, Y. and Shida, H., 1990, Hemadsorption and fusion inhibition activities of hemagglutinin analyzed by vaccinia virus mutants, *Virology* **175**: 372–384.

Shchelkunov, S. N., Massung, R. F. and Esposito, J. J., 1995, Comparison of the genome DNA sequences of Bangladesh-1975 and India-1967 variola viruses, *Virus Res.* **36**: 107–118.

Shida, H., 1986, Nucleotide sequence of the vaccinia virus hemagglutinin gene, *Virology* **150**: 451–462.

Shida, H. and Dales, S., 1981, Biogenesis of vaccinia: carbohydrate of the hemagglutinin molecule, *Virology* **111**: 56–72.

Shida, H. and Dales, S., 1982, Biogenesis of vaccinia: molecular basis for the hemagglutinin-negative phenotype of the IHD-W strain, *Virology* **117**: 219–237.

Shida, H. and Matsumoto, S., 1983, Analysis of the hemagglutinin glycoprotein from mutants of vaccinia virus that accumulates on the nuclear envelope, *Cell* **33**: 423–434.

Smith, G. L., 1996, Virus proteins that bind cytokines, chemokines and interferons, *Curr. Opin. Immunol.* **8**: 467–471.

Sodeik, B., Doms, R. W., Ericsson, M., Hiller, G., Machamer, C. E., van 't Hof, W., van Meer, G., Moss, B. and Griffiths, G., 1993, Assembly of vaccinia virus: role of the intermediate compartment between the endoplasmic reticulum and the Golgi stacks, *J. Cell Biol.* **121**: 521–541.

Spriggs, M. K., 1996, One step ahead of the game: viral immunomodulatory molecules, *Ann. Rev. Immunol,* **14**: 110–130.

Stroobant, P., Rice, A. P., Gullick, W. J., Cheng, D. J., Kerr, I. M. and Waterfield, M. D., 1985, Purification and characterization of vaccinia virus growth factor., *Cell* **42**: 383–393.

Takahashi-Nishimaki, F., Funahashi, S., Miki, K., Hashizume, S. and Sugimoto, M., 1991, Regulation of plaque size and host range by a vaccinia virus gene related to complement system proteins, *Virology* **181**: 158–164.

Tilney, L. G. and Portnoy, D. A., 1989, Actin filaments and the growth, movement and spread of the intracellular bacterial parasite, *Listeria monocytogenes*, *J. Cell Biol.* **109**: 1597–1608.

Tooze, J., Hollinshead, M., Reis, B., Radsak, K. and Kern, H., 1993, Progeny vaccinia viruses and human cytomegalovirus particles utilize early endosomal cisternae for their envelopes; *Eur. J. Cell Biol.* **60**: 163–178.

Ulaeto, D., Grosenbach, D. and Hruby, D. E., 1995, Brefeldin A inhibits vaccinia virus envelopment but does not prevent normal processing and localization of the putative envelopment receptor P37, *J. Gen. Virol.* **76**: 103–111.

Ulaeto, D., Grosenbach, D. and Hruby, D. E., 1996, The vaccinia virus 4c and A-type inclusion proteins are specific markers for the intracellular mature virus particle, *J. Virol.* **70**: 3372–3377.

Vanderplasschen, A., Hollinshead, M. and Smith, G. L., 1997, Antibodies against vaccinia virus do not neutralise extracellular enveloped virus but prevent virus release from infected cells and comet formation, *J. Gen. Virol.* **78**: In press.

Vanderplasschen, A. and Smith, G. L., 1997, A novel virus binding assay using confocal microscopy: demonstration that the intracellular and extracellular vaccinia virions bind to different cellular receptors, *J. Virol.* **71**: 4032–4041.

VanSlyke, J. K. and Hruby, D. E., 1990, Posttranslational modification of vaccinia virus proteins, *Curr. Top. Microbiol. Immunol.* **163**: 185–206.

Wolffe, E., Katz, E., Weisberg, A. and Moss, B., 1997, The A34R glycoprotein gene is required for induction of specialized actin-containing microvilli and efficient cell-to-cell transmission of vaccinia virus, *J. Virol.* **71**: 3905–3915.

Wolffe, E. J., Isaacs, S. N. and Moss, B., 1993, Deletion of the vaccinia virus B5R gene encoding a 42-kilodalton membrane glycoprotein inhibits extracellular virus envelope formation and dissemination, *J. Virol.* **67**: 4732–4741.

52

RESISTANCE OF NAIVE MICE TO MURINE HEPATITIS VIRUS STRAIN 3 REQUIRES DEVELOPMENT OF A Th1, BUT NOT A Th2, RESPONSE, WHEREAS PRE-EXISTING ANTIBODY PARTIALLY PROTECTS AGAINST PRIMARY INFECTION

M. F. Liu,[1] Q. Ning,[1] M. Pope,[1] T. Mosmann,[2] J. Leibowitz,[3] J. W. Ding,[1] L. S. Fung,[1] O. Rotstein,[1] R. Gorczynski,[1] and G. A. Levy[1]

[1]The Toronto Hospital
621 University Ave.
10th Floor, Room 151
Toronto, Canada M5G 2C4
[2]Department of Medical Microbiology and Immunology
University of Alberta
Edmonton, Alberta, Canada
[3]Department of Pathology
Texas A & M University
College Station, Texas 77843

1. ABSTRACT

Murine hepatitis virus strain 3 (MHV-3) produces a host-strain-dependent spectrum of disease. The development of liver necrosis has been shown to be related to production of a unique macrophage procoagulant activity (PCA), encoded by the gene fgl-2, in susceptible mice. These studies were designed to examine the influence of Th1/Th2 cells on resistance/susceptibility and production of macrophage procagulant activity (PCA) in resistant (A/J) and susceptible (Balb/cJ) strains of mice following infection with MHV-3. Immuniza tion of A/J mice with MHV-3induced a Th1 cellular immune response and one Th1 cell line (3E9.1) protected susceptible mice and inhibited production of PCA by macrophages both in vitro and in vivo. In contrast, immunization of Balb/cJ mice with an attenuated variant of MHV-3 derived from passaging MHV-3 in YAC-1 cells resulted in a Th2 response. Transfer

Coronaviruses and Arteriviruses, edited by Enjuanes *et al.*
Plenum Press, New York, 1998

415

of spleen cells and T cell lines from immunized Balb/cJ mice failed to protect naive susceptible syngeneic mice from infection with MHV-3 and augmented production of IL-1β, TNF-α and PCA by macrophages to MHV-3 in vitro. Serum from immunized Balb/cJ mice contained high titered neutralizing antibody which protected naive Balb/cJ animals from lethal primary MHV-3 infection. These results demonstrate that susceptible Balb/cJ mice generate a Th2 response following MHV-3 infection and that these Th2 cells neither inhibit MHV-3-induced macrophage PCA production nor protect naive mice from MHV-3 infection.. The results suggest that antibody protects against primary infection, but could not eradicate ongoing infection. Ribavirin, a synthetic guanosine analogue prolonged survival to MHV-3 infection, inhibited production and transcription of the macrophage pro-inflammatory cytokines IL-1β and TNF-α and Th2 cytokines while preserving Th1 cytokine production. Thus, this data defines the differential role of Th1/Th2 lymphocytes in primary and secondary MHV-3 infection and further defines the importance of macrophage inflammatory mediators in the pathogenesis of MHV-3 infection.

2. INTRODUCTION

Murine hepatitis virus strain 3 (MHV-3) causes fulminant viral hepatitis with a mortality rate close to 80% and has been used as a murine model to study the pathogenesis of viral related hepatic failure. Although viral replication is similar in both susceptible and resistant strains, susceptible strains, such as BALB/cJ, develop acute liver hepatitis with obvious liver necrosis and die within 3 to 5 days following i.p. of MHV-3 while resistant A/J mice do not develop clinical or histologic evidence of hepatitis.

Different differentiation pathways of antigen specific Th cells are involved in the resistance and susceptibility of mice to pathogen. Th1 cells, through the production of IL-2, IFN-γ, and lymphotoxin, mediate cellular immunity, which is essential for clearance of viral and other intracellular pathogens such as Leishmania. Th2 cells, conversely, produce IL-4, IL-10, and IL-13 and are most effective in providing help for B cell differentiation to plasma cells (Mosmann et al., 1989). IL-12 production by APCs has been shown to preferentially stimulate a Th1 response (Sypek et al., 1993). While IFN-γ inhibits the proliferation of Th2 cells, treatment with neutralizing Abs to IFN-γ promotes the development of a Th2 response (Belosevic et al., 1989). IL-4 and IL-10 are regulatory cytokines, favoring the development of a Th2 response (Chatelain et al., 1992). IL-10 has been shown to act on the APC to inhibit IFN-γ production by Th1 clones (Fiorentino et al., 1991). Immunomodulatory treatments that have converted a Th1 to a Th2 response have resulted in the loss of resistance (Chatelain et al., 1992), and conversely, induction of a Th1 response has led to resolution of the infection (Romani et al., 1992).

Our laborotory has had a long interest in pathogenesis of MHV-3 induced liver failure. We have shown that the induction of a unique prothrombinase which encoded by Fgl-2 gene plays a critical role in virus-related acute murine hepatitis in susceptible and semi-susceptible mice. Once induced by MHV-3 infection, fgl-2 prothrombinase cleaves prothrombin to thrombin, which leads to fibrin deposition, intravascular thrombosis, and ischemic liver necrosis (Parr et al., 1995).

In this study, we examined the role of Th1 and Th2 cells in susceptibility and resistance to MHV-3 and regulation of macrophage PCA production in susceptible mice. Mice were immunized with an attenuated variant of MHV-3 (YAC-1-MHV-3) and CD4+ cell lines were isolated and studied for their effects on macrophage procoagulant production and on survival of MHV-3-infected susceptible BALB/cJ mice. Ribavirin (1-beta-D-ribofuranosyl-1,2,4-triazole-3-carbozamide), a synthetic guanosine analogue which has been shown to attenuate the course of acute liver failure induced by murine hepatitis virus (Idwell et al., 1977), was also studied for its effect on MHV-3 replication, cytokine production of Th cells and PCA induction by MHV-3.

3. MATERIAL AND METHODS

Animals: Female A/J (H-2a), C57BL/6J (H-2b), and BALB/cJ (H-2d) mice, 8 wk of age, were obtained from the Jackson Laboratory (Bar Harbor, ME). The recombinant inbred strain AXB8 (H-2a) was obtained from the Montreal General Hospital (Montreal, Canada). Lewis rats were obtained from Charles River Laboratories (Montreal, Canada).

Virus and cells: MHV-3 was plaque-purified on monolayers of DBT cells and grown to a titer of 1×10^7 PFU/ml in 17 CL1 cells. The attenuated variant of MHV-3, YAC-1-MHV-3, was grown by repeated passages of the virulent MHV-3 in YAC-1 cells as described (Lamontagne et al., 1991). Mononuclear cells from spleen or lymph nodes of mice or rats, peritoneal macrophages were obtained as described (Levy et al., 1981).

Serum: Susceptible BALB/cJ mice were immunized i.p. with 1000 PFU of attenuated MHV-3 (YAC-1-MHV-3) and boosted i.p, 14 days later with 1000 PFU of virulent MHV-3. Blood was collected 3, 7, 11, 14, and 18 days following primary immunization.

T cell lines were prepared from susceptible mice according to previously described method (Chung et al., 1994). The T cells obtained were subtyped into CD3+, CD4+, and CD8+ populations based upon indirect immunofluorescence analysis of surface membrane makers.

Lymphokine production of the T cell lines: Cells were cocultured in growth factor-free medium with MHV-3-infected, irradiated SMNC as APCs for 48 hours. The supernatants were assayed for IL-1,IL-2, IL-4, TNF-α and IFN-γ. Their activities were assessed as described (Chung et al., 1994). The cytokines and fgl-2 PCA production of macrophages also performed as above and the northern-blot for those factors were performed as described (Pope et al., 1995).

A Th1/Th2 cell line was derived from C3H/HeJ mice, which had been immunized with 1×10^8 irradiated (2000 rad) B10BR spleen cells and poly I:C i.p.. T cells were recovered from spleen, diluted and incubated with 2×10^5 irradiated B10BR cells. Cells were fed weekly by adding fresh medium and rechallenged every 10 days with freshly harvested and irradiated B10BR spleen cells. Cells showing proliferation were transferred to T-25 tissue culture flasks. Stable T cells were then passaged and redistributed at limiting dilution under the same condition .

4. RESULTS

An attenuated variant of MHV-3, derived by passaging the virus repeatedly in YAC-1 cells was used to immunize Balb/cJ mice. Infection of Balb/cJ mice with 1000 PFU of attenuated MHV-3 induced mild acute hepatitis, characterized by the presence of small discrete foci of hepatic necrosis with a sparse polymorphonuclear leukocyte infiltrate 72 to 96 h postinfection. However, all Balb/cJ mice survived a subsequent lethal i.p. dose of 1000 PFU of virulent MHV-3 administered 2 wk following the primary immunization (Fig.1). No infectious virus was detected by plaque assay in liver, serum, or spleen homogenates 4 days after infection with virulent MHV-3.

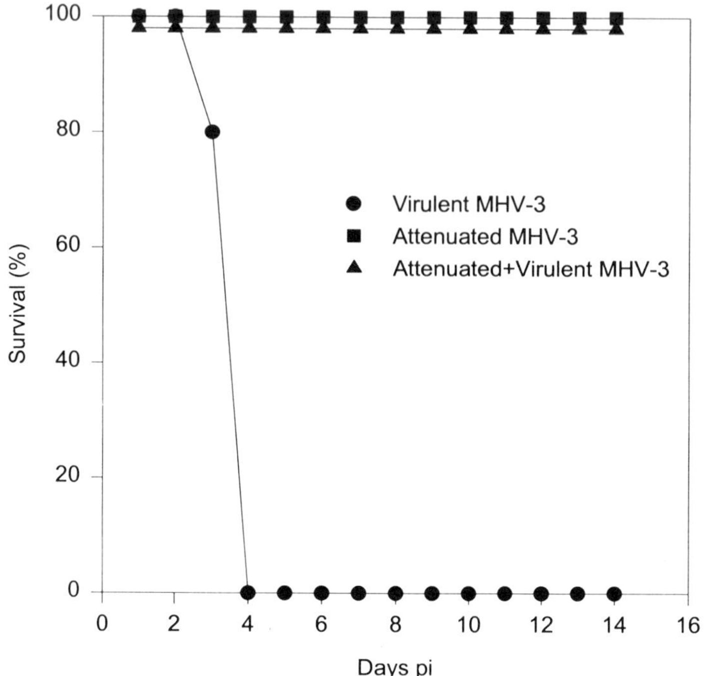

Figure 1. Attenuated YAC-1 MHV induces immunity to virulent MHV-3 in Balb/cJ mice. Groups of 10 naïve Balb/cJ mice were infected with either 10 PFU of virulent MHV-3 i.p. or 1000 PFU of attenuated MHV-3 i.p. The mice that received the attenuated MHV-3 were reinfected 2 wk later with 1000 PFU of virulent MHV-3. Pi indicates postinfection.

To test whether humoral or cell-mediated immunity conveyed resistance to BALB/cJ mice immunized with YAC-1-MHV, naive BALB/cJ mice were injected 200 ul of immune serum i.v. or 5×10^7 immune spleen cells i.v. 1 h before infection with 10 PFU of virulent MHV-3. While spleen cells from A/J mice which had been immunized with MHV-3 prevented mortality of naive A/J mice rendered susceptible to MHV-3 infection by irradiation (650 rad) (Fig.2B), the transfer of spleen cells from Balb/cJ mice which had been immunized with YAC-1-MHV-3 failed to protect naive susceptible Balb/cJ mice from MHV-3 infection (Fig. 2A). The transfer of spleen cells from naive mice to susceptible mice which were subsequently infected with MHV-3 similarly did not demonstrate any protective effect (not shown). In contrast, when susceptible mice received a single dose of serum from YAC-1-MHV immunized mice, survival was significantly prolonged compared with the mice receiving serum from non-immunized mice (Fig. 2A: p<0.05).

To examine the T cell response in Balb/cJ mice which has been infected with attenuated MHV-3, we stimulated spleen-derived mononuclear cell cultures with the attenuated virus. In these studies, analysis of cytokine production demonstrated significant levels of IL-4 (not shown). Twelve Th cell lines from Balb/cJ mice immunized with attenuated MHV-3 were established, all of which had cytokine profiles consistent with a Th2 phenotype. Four of these were studied in detail.

In contrast to the T cell lines from immunized A/J mice (Chung et al., 1994), which produced a Th1 cytokines profile, results showed that all four CD4+ Th cell lines isolated from BALB/cJ mice immunized with attenuated MHV-3 produced IL-4 and low amounts of IFN-γ or IL-2, consistent with a Th2 cytokine secretion profile.

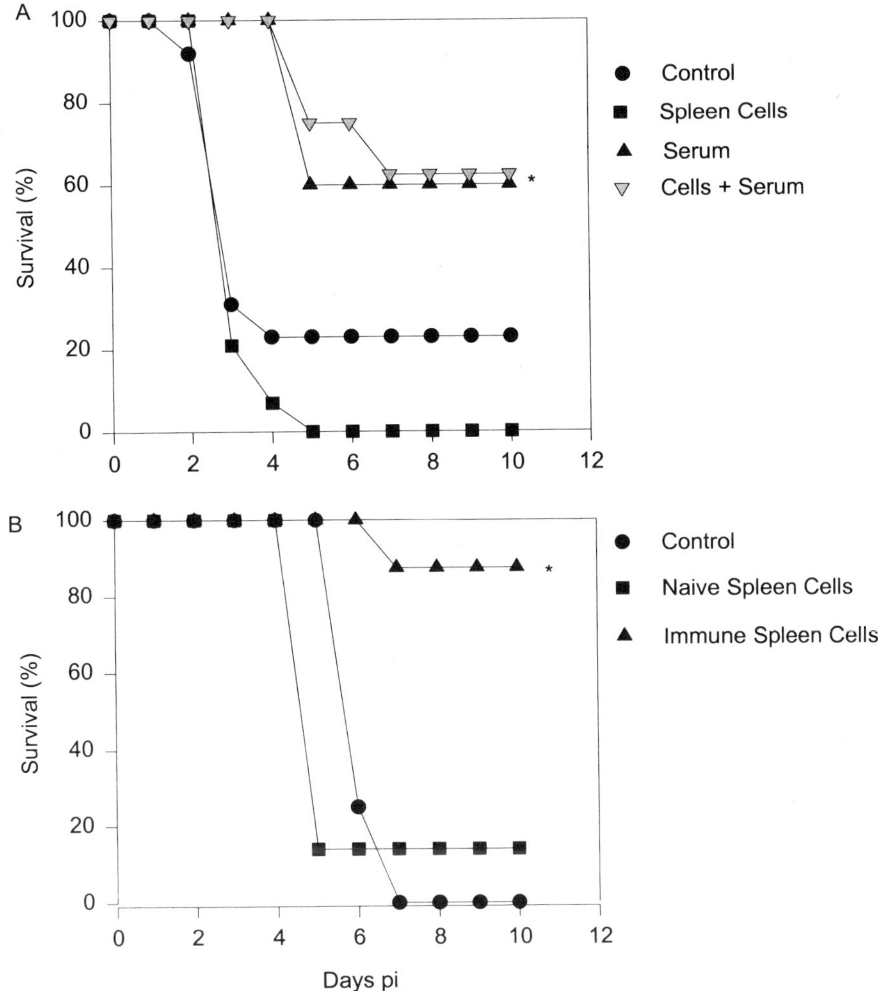

Figure 2. A) Groups of 15 naive Balb/cJ mice that had been immunized with attenuated MHV-3, following which they were infected with 10 PFU of virulent MHV-3 i.p. Fifteen control mice received 10 PFU of virulent MHV-3 i.p. alone. The mean survival times and 95% CIs for each group were compared by survival analysis using the Kaplan-Meier method (*indicates p<0.05) and the analysis methods are the same for the following studies. B) Groups of 10 naïve A/J mice were irradiated with 650 rad from a gamma source, following which they received 5×10^7 spleen cells i.v. from naïve or immunized A/J mice before infection with virulent MHV-3 (10 PFU i.p.). Ten control mice received 10 PFU of virulent MHV-3 i.p. only. Pi indicates postinfection.

To determine the effect of the T cell lines on PCA induction, 8×10^6 T cells from A/J or Balb/cJ mice were added to 5×10^5 peritoneal macrophages from susceptible AZB8 or Balb/cJ mice in the presence of 1×10^6 PFU of MHV-3. After incubation for 12 h, PCA was measured as described (Chung et al. 1994). Results of PCA assay showed that spleen cells from BALB/cJ mice immunized with attenuated MHV-3 augmented macrophage PCA production, while spleen cells from naive BALB/cJ mice had no effect on PCA (not shown). All four Th2 cell lines from immunized susceptible BALB/cJ mice also augmented MHV-3-induced macrophage procoagulant production, consistent with the effects

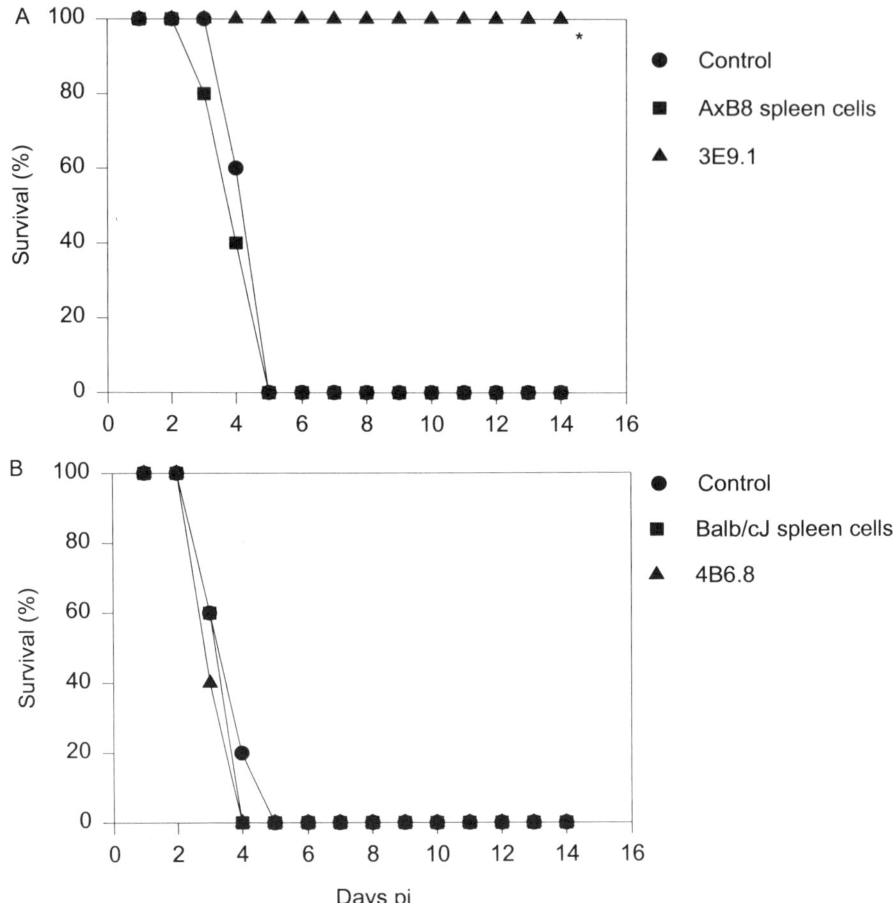

Figure 3. A) Groups of 10 AXB8 mice received either 5×10^6 T cells i.v. or 5×10^6 naïve AXB8 spleen cells i.v., following which they were infected with 10 PFU of virulent MHV-3 i.p. Ten control mice received 10 PFU of virulent MHV-3 alone. B) Groups of 15 naïve Balb/cJ mice received ether 5×10^6 T cells i.v. or 5×10^6 naïve Balb/cJ spleen cells i.v., following which they were infected with 10 PFU of virulent MHV-3 i.p. Fifteen control mice received 10 PFU of virulent MHV-3 alone.

seen using bulk spleen cell cultures. No PCA was detected when Th2 cells were incubated with macrophages in the absence of MHV-3 (not shown).

To assess the effects of Th cell lines for susceptible Balb/cJ mice on the course of MHV-3 infection in vivo, T cells were collected 8 days following antigen restimulation, and 5×10^6 cells were injected i.v. into naive BALB/cJ mice. The Th2 cell line 4B6.8 from immunized susceptible Balb/cJ mice, which increased PCA, did not protect naive BALB/cJ mice following MHV-3 infection (Fig. 3B). In contrast, as previously reported (Chung et al., 1994), the Th1, PCA-inhibitory 3E9.1 cell line from A/J mice prevented mortality in naive MHC-compatible and susceptible AXB8 mice (Fig. 3A).

Ribavirin, which has been shown to attenuate the course of acute hepatitis in mice induced by MHV-3, was also studied for its effects on MHV-3 replication, cytokine production of Th cells and the proinflammatory monokines IL-1, TNF and PCA. When cocultured with Balb/cJ macrophages in the presence of MHV-3, this drug specifically inhibited

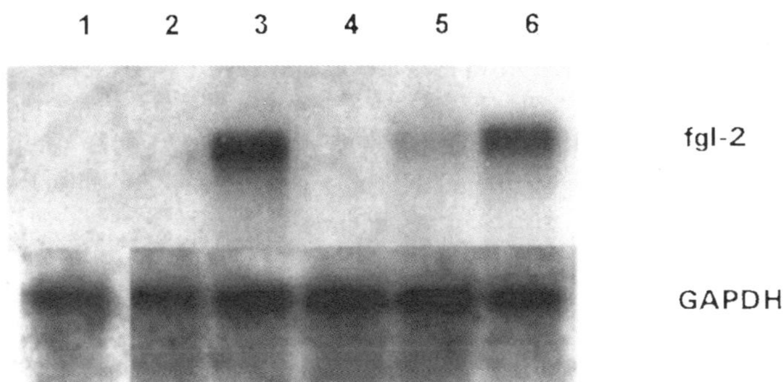

Figure 4. Northern-blot analysis of total RNA from Balb/cJ mice macrophage stimulated with MHV-3 in the presence of ribavirin. 1, macrophage alone; 2, macrophage + ribavirin (500 µg/ml); 3, macrophage + MHV-3; 4, macrophage + MHV+ ribavirin (500 µg/ml); 5, macrophage + MHV-3 + ribavirin (100 µg/ml); 6, macrophage + MHV-3 + ribavirin (1 µg/ml).

the transcription of fgl-2, IL-1 and TNF in a dose dependent manner. In addition, ribavirin at a concentration level higher than 100 ug/ml inhibited the Th2 cytokine IL-4, but not the TH1 cytokine IFN-γ (Fig. 4 and not shown).

5. DISCUSSION

The outcome of viral infection represents a balance between replication of the virus and the host immune response, which consists of an early nonspecific inflammatory response and a later acquired antigen-specific immune response. Studies on the pathogenesis of MHV-3 infection in our laboratory as well as by others have shown that the resistant A/J mice generate a protective cellular immune response following MHV-3 infection (Chung et al., 1994). Furthermore, spleen cells from resistant A/J mice can adoptively transfer resistance to susceptible neonatal mice (Levy-Leblond et al., 1977), and the resistance can be overcome following implementation of various immunosuppressive regiments that inhibit cell-mediated immunity (Dupuy et al., 1975). Also, the principal effector cells responsible for resistance to MHV-3 infection have been shown to be CD4+ Th cells (Korner et al., 1991).

Infection of naive susceptible BALB/cJ mice with virulent MHV-3 is rapidly fatal and associated with the induction of macrophage PCA, recently identified as musfiblp, encoded by the fgl-2 gene, which results in microvascular thrombosis and hepatocellular necrosis, with all mice dying within 5 days (Parr et al., 1995). It has been shown that induction of musfiblp is regulated by T cells in both susceptible and resistant mice (Chung et al., 1991 and 1994).

Lamontagne et al. demonstrated that infection of susceptible mice with an attenuated variant of MHV-3 in YAC-1 cells (YAC-1-MHV-3) rendered resistance to subsequent challenge with virulent MHV-3 (1991). The attenuated variant of MHV-3 is antigenically similar to the virulent MHV-3 and, in the present study, it induced neutralizing antibodies that protected susceptible mice against infection with a lethal dose of virulent MHV-3. While neutralizing Abs were produced in the susceptible Balb/cJ mice following infection with the attenuated variant of MHV-3, no neutralizing Abs were detected following infection with virulent MHV-3 as previously described (Dindzans et al., 1987).

Adoptive transfer of neutralizing Abs from YAC-1-MHV immunized mice confer immunity by preventing initial viral entry and replication. However, viral replication is not completely inhibited by antibody treatment, as suggested by the small foci of liver necrosis 3 days following challenge of YAC-MHV-3 immunized Balb/cJ mice with virulent MHV-3. Nonimmunized mice, which do not have neutralizing antibody and are unable to prevent initial viral entry and replication, produce PCA and develop lethal MHV-3 infection (not shown).

Adoptive transfer of spleen cells from YAC-1-MHV-3-immunized Balb/cJ mice failed to protect naive mice from MHV-3 infection. This was in contrast to the protective role of Tcells from immunized resistant A/J mice, suggesting differences in cell-mediated immunity in resistant and susceptible mice. MHV-3-specific Th cell lines isolated from resistant A/J mice had a cytokine profile consistent with a Th1 phenotype as evidenced by IL-2 and IFN-γ production (Chung et al., 1994). In contrast, all MHV-3 specific Th cell lines isolated from susceptible Balb/cJ mice which had been infected with YAC-1MHV were of the Th2 phenotype, as demonstrated by the production of IL-4. The studies with ribavirin support the hypothesis of Th1 cells being protective and Th2 cells being non protective of MHV-3 infection. (Fig. 4 and not shown).

Our results showed that treatment of naive susceptible BALB/cJ mice with the neutralizing anti-IL-4 mAb failed to improve survival following MHV-3 infection, and all mice died within 8 days of infection with a similar time course. However, this observation can not exclude a possible causative role of Th2 cells in the pathogenesis of MHV-3 infection. These results demonstrate significant differences in the cellular immune responses of resistant and susceptible mice following MHV-3 infection. Resistant A/J mice generate a protective Th1 cellular immune response. In contrast, while susceptible BALB/cJ mice are able to produce neutralizing anti-viral Abs following immunization with an attenuated variant of MHV-3, all anti-MHV-3 Th cell lines are of the Th2 phenotype and do not protect naive animals from the lethality of MHV-3 infection. Their failure to protect naive mice is consistent with the fact they augment PCA production. These studies provide new insights into our understanding of the pathogenesis of fulminant viral hepatitis by MHV-3, demonstrating a protective role for Th1 cells and a role for Th2 cells in the pathogensis of MHV-3-induced liver failure.

ACKNOWLEDGMENT

We are grateful to Charmaine Beal and Dawn Paluch for their excellent secretarial assistance. This work was suported by a group grant PG 11810 from the Medical Research Coucil of Canada.

REFERENCES

Belosevic, M., D. S. Finbloom, Pl H. Van der Meide, M. V. Slayter, and C. Nacy., 1989, Administration of monoclonal anti-IFN-gamma antibodies in vivo abrogates natural resistance of C3H/HeN mice to infection with Leishmania major, *J. Immunol.* **143**: 266–274.

Chatelain, R., K. Varkila, and R. L. Coffman., 1992, IL-4 induces a Th2 response in Leishmania major-infected mice, *J. Immunol.* **148**: 1182–1187.

Chung, S., R. Gorczynski, B. Cruz, R. Fingerote, E. Skamene, S. Perlman, J. L. Leibowitz, L. S. Fung, M. Flowers, and G. A. Levy., 1994, A Th1 helper cell line (3E9.1) from resistant A/J mice inhibits induction of macrophage procoagulatn activity (PCA) in vitro and protects against MHV-3 mortality in vivo, *Immunol.* **83**: 353–361.

Chung, S., S. Sinclair, J. Leibowitz, E. Skamene, L. S. Fung, and G. Levy., 1991, Cellular and metabolic requirements for induction of macrophage procoagulant activity by murine hepatitis virus strain 3 in vitro, *J. Immunol.* **146**: 271–278.

Dindzans, V. J., B. Zimmerman, A. Sherker, and G. A. Levy., 1987, Susceptibility to mouse hepatitis virus strain 3 in Balb/cJ mice: failure of immune cell proliferation and interleukin 2 production, *Adv. Exp. Med. Biol.* **218**: 411–420.

Dupuy, J. M., E. Levy-Leblond, and C. Le Prevost., 1975, Immunopathology of mouse hepatitis virus type 3 infection. II. Effect of immunosuppression in resistant mice, *J. Immunol.* **114**:226–230.

Fiorentino, D. F., A. Zlotnik, P. Vieira, T. R. Mosmann, M. Howard, K. W. Moore, and A. O'Garra., 1991, IL-10 acts on the antigen-presenting cell to inhibit cytokine production by Th1 cells, *J. Immunol.* **146**:3444–3451.

Idwell, R. W., J. H. Huffman, N. Campbell, L. B. Allen., 1977, Effect of ribavirin on viral hepatitis in laboratory animal,. *Annals of the New York Academy of Science.* **284**: 239–246.

Korner, H., A. Schliephake, J. Winter, F. Zimprich, H. Lassmann, J. Sedgwick, S. Siddell, and H. Wege., 1991, Nucleocapsid or spike protein-specific CD4+ T lymphocytes protect against coronavirus-induced encephalomyelitis in the absence of CD8+ cells., *J. Immunol.* **147**: 2317–2323.

Lamontagne, L., and P. Jolicoeur., 1991, Mouse hepatitis virus 3-thymic cell intreactions correlating with viral pathogenecity, *J. Immunol.* **146**: 3152–3159.

Levy, G. A., J. L. Leibowitz, and T.S. Edgington., 1981, Induction of monocyte rocagulant activity by murine hepatitis virus type 3 parallels disease susceptibility im mice, *J. Ecp. Med.* **154**: 1150–1163.

Levy-Leblond, E., and J. M. Dupuy., 1977, Neonatal susceptibility to MHV-3 infection in mice. I. Transfer of resistance, *J. Immunol.* **118**: 1219–1223.

Li, C., L. S. Fung, A. Crow, N. Myers-Mason, J. L. Leibowitz, E. Cole, and G. A. Levy., 1992,

Monoclonal anti-prothrombinase (3D4.3) prevents mortality from murine hepatitis virus infection (MHV-3), *J. Exp. Med.* **1763**: 689–697.

Mosmann, T. R., and R. L. Coffman., 1989, Th1 and Th2 cells: different patterns of lymphokine secretion lead to different functional properties, *Annu. Rev. Immunol.* **7**: 145–214.

Parr, R., J. L. Leibowitz, L. S. Fung, S. J. Reneker, N. Myers-Mason, and G. A. Levy., 1995, The association of mouse fibrinogen-like protein (Musfiblp) with murine hepatitis virus induced prothrombinase activity, *J. Virol.* **69**: 5033–5038.

Pope, M., O. Roststein, E. Cole, S. Sinclair, R. Parr, B. Cruz, R. Fingerote, S. Chung, R. Gorczynski, L. Fung, J. Leibowitz, Y. S. Rao, and G. Levy. 1995. Pattern of disease after murine hepatitis virus strain 3 infection correlates with macrophage activation and not viral replication, *J. Virol.* **69**: 5252–5260.

Romani, L., A. Mencacci, U. Grohmann, S. Mocci, P. Mosci, P. Puccetti, and F. Bistoni., 1992, Neutralizing antibody to interleukin 4 inducess systemic protection and T helper type 1- associated immunity in murine candidiasis, *J. Exp. Med.* **176**: 19–25.

Sypek, J. P., C. L. Chung, S. H. Mayeor, J. M. Subramanyam, S. J. Goldman, D. S. Sieburth, S. F. Wolf, and R. G. Schaub., 1993, Resolution of cutaneous leishmaniasis:interleukin-12 initiates a protective T helper type 1 immune response, *J. Exp. Med.* **177**: 1797–1802

53

APOPTOSIS OF JHMV-SPECIFIC CTL IN THE CNS IN THE ABSENCE OF CD4[+] T CELLS

S. A. Stohlman,[1,2] C. C. Bergmann,[1,2] D. J. Cua,[1] M. T. Lin,[3] S. Ho,[2] W. Wei,[2] and D. R. Hinton[3]

[1]Department of Neurology
[2]Department of Microbiology and Immunology
[3]Department of Pathology
University of Southern California
Los Angeles, California 90033

1. ABSTRACT

The role of CD4[+] T cells in altering the activity of cytotoxic T lymphocytes (CTL) during infection of the central nervous system (CNS) by the neuroptropic JHMV strain of mouse hepatitis virus was examined. Adoptive transfer of in vitro activated CTL into CD4-depleted and control recipients showed that CTL were not effective in reducing JHMV replication within the CNS. The distribution of CD4[+] and CD8[+] T cells within the CNS during JHMV infection showed that the CD4[+] T cells remained in perivascular and subarachnoid spaces and few entered the parenchyma. By contrast approximately half of the CD8[+] T cells entered the parenchyma. In CD4-depleted mice the trafficking of CD8[+] T cells was not inhibited; however, the majority of the cells were found to be apoptotic. These data suggested that CD4[+] T cells were not required for CTL induction but were required for the maintenance of CTL viability. The limited role of CD4[+] T cells in CTL induction was confirmed by comparison of CTL activity from CD4-depleted and control mice.

2. INTRODUCTION

JHMV induces both acute and chronic infections in the CNS of rodents (Kyuwa and Stohlman, 1990; Lane and Buchmeier, 1997). Although JHMV is rapidly cleared from the CNS following infection, incomplete clearance correlates with the establishment of chronic infection. Chronic infection is associated with chronic ongoing primary demyeli-

Coronaviruses and Arteriviruses, edited by Enjuanes *et al.*
Plenum Press, New York, 1998

nation, the persistence of viral mRNA and viral antigen (Kyuwa and Stohlman, 1990; Lane and Buchmeier, 1997). During persistent infection viral antigen is found in both astrocytes and oligodendroglia (Kyuwa and Stohlman, 1990). In some models of JHMV persistence and chronic demyelination infectious virus can not be demonstrated; however, in other models infectious virus is present in the CNS. The CD8$^+$ subset of T cells with virus-specific cytotoxic activity is not only recruited to the CNS during infection (Williamson, Sykes and Stohlman, 1991) but adoptive transfer experiments show that CTL are a major effector of virus clearance (Stohlman, *et al.*, 1995). CTL preferentially clear JHMV from astrocytes, microglia, and ependymal cells diminishing the overall virus load and preventing chronic demyelination (Stohlman, *et al.*, 1995). This ability to prevent chronic demyelination contrasts with the effects of both non neutralizing antibody and CD4$^+$ T cells. Previous data suggested that both CD4$^+$ T cells and CD8$^+$ T cells were effectors of virus clearance (Williamson and Stohlman, 1990). Furthermore, it appeared that CD4$^+$ T cells were required for induction of JHMV-specific CTL activity. This notion was based on the identification of a unique population of CD4$^+$ T cells which protected from acute disease and reduced virus replication in a MHC class I restricted manner (Sussman, *et al.*, 1989). In addition, the elimination of either CD4$^+$ or CD8$^+$ T cell using both antibody deletions and genetically deficient mice showed that both populations contribute to virus clearance (Williamson, and Stohlman, 1990; Houghton and Fleming, 1996; Gombold, *et al.*, 1995). Having demonstrated that the adoptive transfer of CTL could inhibit virus replication, we examined the role(s) of CD4$^+$ T cells in conjunction with CTL in the pathogenesis of JHMV infection.

3. MATERIALS AND METHODS

3.1. Mice

BALB/c By (H-2^d) mice were obtained from the Jackson Laboratory (Bar Harbor ME). Donors were immunized by intraperitoneal injection of 1 x 10^6 plaque forming units of JHMV. Recipient and control mice were infected intracranially with 100 plaque forming units of JHMV in 30 ul PBS. This inoculum is uniformly fatal within 7 to 9 days post infection. Recipients were depleted of CD4$^+$ T cells by interperitoneal injection of 200 ug of anti-CD4 mAb GK1.5 at -2, 0 and +2 days relative to infection as previously described (Williamson and Stohlman, 1990). This regimen resulted in ≥ 98% depletion of CD4$^+$ T cells as analyzed by FACS analysis at 4 days after the final injection.

3.2. Viruses

The DM strain of JHMV was propagated and plaque assayed using the murine DBT astrocytoma cell line. Virus titers were determined by homogenization of 1/2 the brain in 4.0 ml of Dulbecco's PBS, pH 7.4 using Tenbrock tissue homogenizers. Data presented are the average titer of groups of three or more mice.

3.3. Induction and Transfer of Bulk Effector CTL

CTL were prepared from mice immunized 3–8 wk earlier. Spleen cells (1 × 10^8) were cultured for 6 days at 37°C in 40 ml of RPMI 1640 complete medium supplemented with 10 % rat Con A supernatant (RCS) and 1 μM pN318–335 peptide as previously described (Stohlman, *et al.*, 1995). Viable cells were purified by centrifugation onto Lympholyte M (Cedar Lane) cushions prior to transfer. CD8[+] T cells, purified by positive selection using magnetic beds, were transferred intravenously (1–2 x 10[7]cells/recipient) either 4 - 6 h prior to infection or at 2 days post infection for BrdU-labeled cells. To label CD8[+] T cells, 50 nM BrdU was added *in vitro* for the final 48 hours incubation. BrdU-labeled CD8[+] T cells were purified by positive selection prior to transfer.

3.4. Cytotoxicity Assay

Cytolytic activity was measured in a 4-hour ^{51}Cr-release assay as previously described (Stohlman, *et al.*, 1995). Data are expressed as percent specific release defined as [(experimental release)-(spontaneous release)]/ [Total (detergent release)-(spontaneous release)]. Maximum spontaneous release values were ≤20% of total release values.

3.5. Histology and Cell Trafficking

Brains and spinal cords were fixed for 3 hours in Clark's solution (75% ethanol and 25% glacial acetic acid) and embedded in paraffin. Sections were stained with either hematoxylin and eosin or luxol fast blue for routine examination. The distribution of JHMV antigen was examined using immunoperoxidase staining (Vectastain-ABC kit, Vector Laboratories, Burlingame, CA) and anti-nucleocapsid mAb J.3.3. Distribution of BrdU[+] cells was determined on paraffin embedded sections by immunoperoxidase (Vectastains ABC Kit) using mouse mAb BU-33 (Sigma). To identify CD4[+] and CD8[+] T cells, immunostaining was performed on acetone fixed frozen sections using rat anti-CD4[+] (L3T4, PharMingen) and rat anti-CD8[+] (Ly-2, PharMingen) antibodies at 1:200 dilution. Anti-CD4[+] and CD8[+] antibodies were detected with biotinylated rabbit-anti-rat Ab preabsorbed with mouse serum (Vector Laboratory) and the Vetastain-ABC kit. Terminal deoxynucleotidyl transferase-mediated dUTP-biotin nick-end labeling (Tunnel) was used to examine the distribution of apoptotic cells according to the suppliers instructions (Oncor, Gaitherburg, MD).

4. RESULTS

Previous data suggested that during JHMV infection CD4[+] T cells were required for induction of CTL (Kyuwa and Stohlman, 1990). Therefore we reasoned that the adoptive transfer of JHMV-specific CTL into mice depleted of CD4[+] T cells should reduce virus replication. CTL were expanded in vitro and transferred to naive untreated recipients and recipients depleted of CD4[+] T cells by administration of GK1.5 as previously described (Williamson and Stohlman, 1990). CTL recipient and control mice were infected with JHMV and the amount of virus replication in the brains quantitated by plaque assay at 5 days post infection. Surprisingly, adoptively transferred CTL which reduce virus titers in the immunocompetent recipients were unable to mediate virus clearance from the CNS of CD4-depleted mice (Table 1).

To attempt and understand the inability of CTL to alter virus replication, we examined the distribution of CD4[+] and CD8[+] T cells within the CNS of normal and CD4-depleted mice. Figure 1 shows that during infection of immunocompetent mice, both CD4[+] and CD8[+] T cells enter the CNS. CD4[+] T cells localize to perivascular areas and subarachnoid spaces consistent with their localization following adoptive transfer of CD4[+] T cell clones (Stohlman *et al.*, 1986; Yamaguchi *et al.*, 1991). The absence of CD4[+] T cells

Table 1. CD8[+] T cells mediate JHMV clearance from the CNS

Treatment	#Cells transferred	Virus titer[a]
Experiment 1		
None	None	6.37 ± 0.25
C' only	2 × 10⁷	3.75 ± 0.52
Anti-CD8[+] C	2 × 10⁷	6.13 ± 0.31
Experiment 2		
None	None	7.28 ± 0.16
None	2 × 10⁷	4.61 ± 0.34
CD8[+b]	1 × 10⁷	4.35 ± 0.02

[a]Log_{10} ± SEM pfu/gm brain determined 5 d post infection. Average 3-4 mice per group.
[b]CD8[+] T cells isolated using magnetic beads.

from the parenchyma suggests that they may not constitute a primary effector population in controlling JHMV infection. In contrast to CD4[+] T cells, CD8[+] T cells were equally distributed between the perivascular and subarachnoid areas and the parenchyma, consistent with their ability to inhibit virus replication. Examination of CD8[+] T cells in the mice depleted of CD4[+] T cells showed a distribution similar to the distribution in normal mice although there was a substantial reduction in cell numbers. However, we found a large number of mononuclear cells which could not be stained with either anti-C4 or anti-CD8 in these sections. These cells all showed evidence of apoptosis by Tunnel assay, suggesting the reason CTL are ineffective within the CNS.

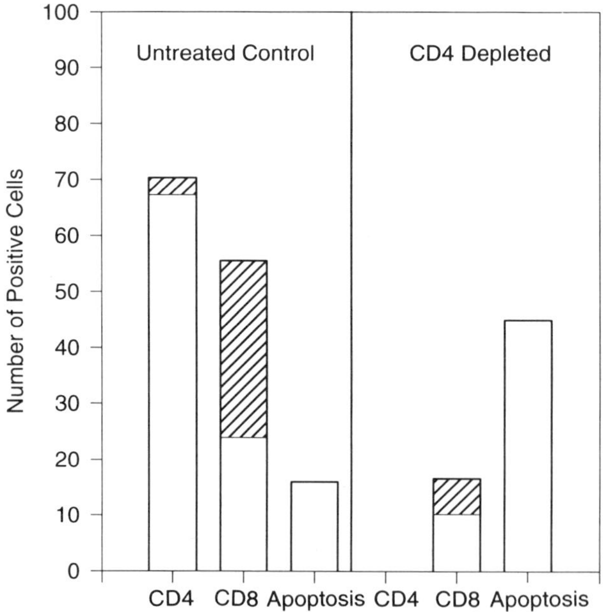

Figure 1. Localization of CD4[+] and CD8[+] T cells in the brains of immunocompetent and CD4-depleted mice.

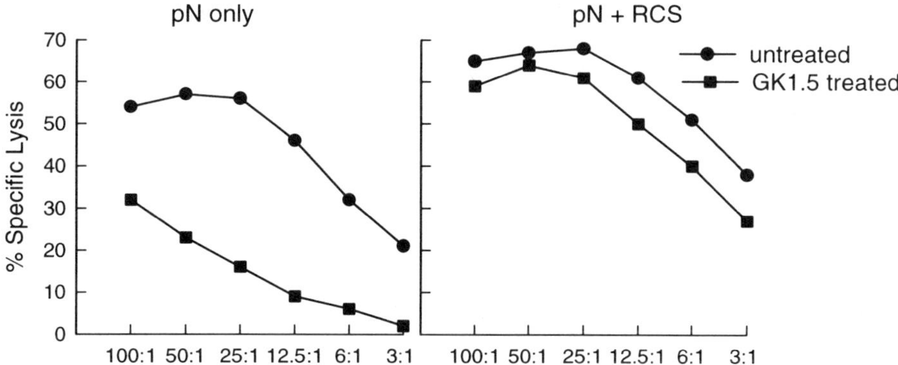

Figure 2. CTL activity of spleen cells from immunocompetent and CD4-depleted mice cultured 6 days with/without addition of RCS as a source of IL-2.

CD4-deficient mice is that CD4$^+$ T cells are required to maintain CTL viability once they enter the parenchyma. To confirm this impression we compared the ability of BrdU-labeled purified pN-specific CTL to traffic into the CNS of immunocompetent and CD4-depleted mice. No difference in the ability of these cells to traffic into the CNS was detected, supporting the notion that CTL can effectively enter the CNS parenchyma but are ineffective at reducing virus due to their untimely death.

These data suggested that CD4$^+$ T cells did not influence CTL trafficking into the CNS, therefore, we reexamined the question of whether or not CD4$^+$ T cells were required for the induction of CTL during JHMV infection. Immunocompetent and CD4-depleted mice were infected with JHMV and the spleens removed at 6 days post infection. CTL were expanded in vitro under identical conditions and tested for CTL activity. Figure 2 shows that in the absence of CD4$^+$ T cells CTL are indeed induced during JHMV infection albeit at a reduced level compared to immunocompetent mice. To provide suggestive evidence for apoptosis of the CTL found within the CNS of JHMV infected mice, parallel cultures were incubated in the presence of IL-2. These data show that there is no difference in the CTL activity derived from immunocompetent and CD4-depleted mice when cultured in the presence of IL-2.

5. DISCUSSION

These data demonstrate a number of interesting findings. First, CD4$^+$ T cells appear to localize within the CNS during JHMV infection in areas distinct from the majority of the virus infected cells. This localization is consistent with the distribution of inflammatory cells following adoptive transfer of CD4$^+$ T cells clones into JHMV infected (Stohlman, *et al.*, 1986). Analysis of the ability of CD4$^+$ T cells to directly alter virus replication in the CNS have yielded inconsistent results (Stohlman, *et al.*, 1986; Yamaguchi, *et al.* 1991) and include the suggestion that CD4-mediated anti viral activity is independent of CTL (Korner, *et al.*, 1991). CD4$^+$ T cells not only secrete IL-2 but also IFN-γ. Therefore, these data are consistent with a partial role of IFN-γ in suppressing virus replication (Lane, Paoletti, and Buchmeier, 1997). Second, CD8$^+$ CTL appear to migrate into the CNS parenchyma, consistent with their ability to reduce virus load within the CNS. These data suggest that the basis for the absence of CTL-mediated JHMV clearance from the CNS of mice deficient in CD4$^+$ T cells is not related to the absence of CTL, but rather the inability of the CTL to remain viable and exhibit effector function once they enter the parenchyma. Third, these data clearly show that CTL are induced during JHMV infection of the CNS in a CD4-independent manner. CTL induction is less efficient than in immunocompetent mice by approximately 16 fold. Whether this reduced induction plays a role in the inability of CTL to clear virus is not clear. However, infection of mice with lymphocytic choriomeningitis virus, reported to be CD4-independent (Zinkernagel, *et al.*, 1986), also results in a similar decrease in CTL induction. In contrast to JHMV infection of the CNS, lymphocytic choriomeningitis virus is cleared from the liver in CD4-depleted mice (Zinkernagel, *et al.*, 1986). This may reflect both access to liver versus the CNS (Ando, *et al.*, 1994) and the prominent role CTL play in eliminating this virus versus the only partial protective effect of CTL during JHMV infection (Stohlman, *et al.*, 1995). Finally, the increase in cytolytic activity of T cells derived from CD4-depleted mice cultured in the presence of IL-2 suggests that secretion of IL-2 by CD4$^+$ T cells may be important in preventing apoptosis of CTL once they enter the parenchyma.

REFERENCES

Ando, K., Gidotti, L. G., Cerny, A., Ishikawa, T., and Chisari, F. V., 1994, CTL access to tissue antigen is restricted in vivo, *J. Immunol.* **153:** 482–488.

Gombold, J., Sutherland, R., Lavi, E., Paterson, Y., and Weiss, S. R., 1995, Mouse hepatitis virus A59-induced demyelination can occur in the absence of CD8+ T cells, *Microb. Pathog.* **18:** 211–221.

Houtman, J. J., and Fleming, J. O., 1996, Dissociation of demyelination and viral clearance iln congenitally immunodeficient mice infected with murine coronavirus JHM, *J. Neurovirol.* **2:** 101–111.

Korner, H., Schliephake, A., Winter, J., Zimprich, F., Lassmann, H., Sedgwick, J., Siddell, S., and Wege, H., 1991, Nucleocapsid or spike protein-specific CD4+ T lymphocytes protect against coronavirus-induced encephalomyelitis in the absence of CD8+ T cells, *J. Immunol.* **147:** 2317–2323.

Kyuwa, S., and Stohlman, S. A., 1990, Pathogenesis of a neurotropic murine coronavirus strain JHM in the central nervous system, *Seminar Virol.* **1:** 273.

Lane, T. E., and Buchmeier, M. J., 1997, Murine coronavirus infection: a paradigm for virus-induced demyelinating disease, *Trends Microbiol.* 9–14.

Lane, T. E., Paoletti, A. D., and Buchmeier, M. J., 1997, Disassociation between the in vitro and in vivo effects of nitric oxide on a neurotropic murine coronavirus, *J. Virol.* **71:** 2202–2210.

Stohlman, S. A., Bergmann, C., van der Veen, R., and Hinton, D., 1995, Mouse hepatitis virus-specific cytotoxic T lymphocytes protect from lethal infection without eliminating virus from oligodendroglia, *J. Virol.* **69:** 684–694.

Stohlman, S. A., Matsushima, G. K., Casteel, N., and Weiner, L. P., 1986, In vivo effects of coronavirus-specific T cell clones: DTH inducer cells prevent a lethal infection but do not inhibit virus replication, *J. Immunol* **136:** 3052–3056.

Sussman, M., Shubin, R., Kyuwa, S., and Stohlman, S. A., 1989, T cell-mediated clearance of mouse hepatitis virus strain JHM from the central nervous system, *J. Virol.* **63:** 3051–3059.

Williamson, J., Sykes, S. P. K., and Stohlman, S., 1991, Charcterization of brain infiltrating mononuclear cells during infection with mouse hepatitis virus strain JHM, *J. Neuroimmunol.* **32:** 199–207.

Williamson, J. S. P., and Stohlman, S. A. S., 1990, Effective clearance of mouse hepatitis virus from the CNS requires both CD4+ and CD8+ T cells, *J. Virol.* **64:** 4590–4592.

Yamachuchi, K., Goto, N., Kyuwa, S., Hayami, M., and Toyoda, Y., 1991, Protection of mice from a lethal coronavirus infection in the central nervous system by adoptive transfer of virus-specific T cell clones, *J. Neuroimmunol.* **32:** 1–9.

Zinkernagel, R. M., Haenseler, M. E., Leist, T., Cerny, A., Hengartner, H., and Zlthage, A., 1986, T cell-mediated hepatitis in mice infected with lymphocytic choriomeningitis virus: liver cell destruction by H-2 class I-restricted virus-specific cytotoxic T cells as a physiological correlate of the 51 Cr-release assay?, *J. Exp. Med.* **164:** 1075–1092.

54

MECHANISMS OF VIRAL CLEARANCE IN PERFORIN-DEFICIENT MICE

M. T. Lin,[3] D. R. Hinton,[2,3] and S. A. Stohlman[1,2]

[1]Department of Molecular Microbiology and Immunology
[2]Department of Neurology
[3]Department of Pathology
University of Southern California
School of Medicine
Los Angeles, California 90033

1. ABSTRACT

The roles of CD4+ T cells, IFN-γ and TNF-α in viral clearance from the central nervous system (CNS) were examined in perforin gene deficient (PKO) mice. Depletion of CD4+ T cells from the PKO mice resulted in a significant 1 $\log_{10}$ PFU/gm increase in viral titer over control-treated PKO mice. PKO mice treated with anti-IFN-γ mAb also had a significant 1 $\log_{10}$ increase in infectious virus whereas inhibition of TNF-α did not alter viral clearance or clinical disease in the PKO mice. These data suggest, in addition to perforin-mediated cytolysis, CD4+ T cells and IFN-γ, but not TNF-α could contribute to JHMV clearance from the CNS.

2. INTRODUCTION

Infection of the CNS by JHMV (MHV4), a neurotropic strain of mouse hepatitis virus (MHV) induces primary demyelination and persistent infection in susceptible strains of mice (Lampert, Sims, and Kniazeff, 1973; Weiner, 1973). Although the mechanisms of viral persistence remains unclear, the complete elimination of virus from the CNS by the immune response during the acute infection appears to be the critical parameter in preventing viral persistence and chronic demyelination (Houtman and Fleming, 1996b). Viral clearance of JHMV from the CNS is influenced by both cellular and humoral responses (Compton, Barthold, and Smith, 1993; Fazakerley and Buchmeier, 1993; Kyuwa and Stohlman, 1990); however, since T cell recruitment and viral clearance occur prior to de-

Coronaviruses and Arteriviruses, edited by Enjuanes *et al.*
Plenum Press, New York, 1998

tectable serum neutralizing antibodies, cellular immunity appears to be most critical. Consistent with this concept, adoptive transfer of JHMV-specific CD8+ cytotoxic T lymphocytes (CTL) provides protection from both acute and chronic infection by limiting viral replication in astrocytes and microglia, but not major histocompatibility complex class I-negative oligodendrocytes (Stohlman et al., 1995a). The exact mechanisms involved in immune mediated clearance are not yet clear. Anti-viral effects of CD8+ CTL appear to be predominantly due to direct lysis of infected cells via perforin-mediated cytolysis since clearance of JHMV from the CNS in mice genetically deficient in perforin (PKO) is delayed, but not abolished (Lin, Stohlman, and Hinton, 1997). Moreover, these data suggest the coexistence of other anti-viral mechanisms along with perforin-mediated cytolysis such as CD4+ T cells or cytokines. CD4+ T cells participate in viral clearance by providing help for CD8+ CTLs (Williamson and Stohlman, 1990). However, adoptive transfer of some, but not all JHMV-specific CD4+ T cells also results in clearance of virus from the CNS (Korner et al., 1991; Stohlman et al., 1986; Wijburg et al., 1996; Yamaguchi et al., 1991). Furthermore, cytokine secretion by both CD8+ and CD4+ T cells may exert anti-viral activity either directly or indirectly. During JHMV infection, tumor necrosis factor (TNF)-α and interferon (IFN)-γ mRNA expression increased at the time of maximal decrease in virus replication (Parra et al., in press; Pearce et al., 1994). Consistent with IFN-γ as an anti-viral agent, mice treated with anti-IFN-γ are more susceptible to MHV infection, while mice treated with IFN-γ (Smith et al., 1991)or adoptively transferred with a CD4+ Th1 line secreting IFN-γ are protected (Pope et al., 1996). Unlike IFN-γ, anti-TNF-α treatment of mice infected with JHMV did not show any alteration in encephalitis or demyelination (Stohlman et al., 1995b). In this report, the presence of anti-viral mechanisms other then perforin-mediated cytolysis were examined in PKO mice depleted of either CD4+ T cells, IFN-γ or TNF-α.

3. MATERIALS AND METHODS

Breeding pairs of perforin +/+ and -/- mice of 129 x C57BL/6 background were kindly provided by Dr. W. Clarke (UCLA). Mice were breed and maintained SPF at University of Southern California vivaria. Groups of seven to eight week old mice were infected intracranially with 100 PFU of the DM strain of JHMV which inducs a uniformly fatal encephalomyelitis. JHMV replication in the CNS was determined by plaque assay as previously described (Stohlman et al., 1986). Data presented are the average of triplicate determinations for groups of three to seven mice. For anti-cytokine mAb treatments, mice were treated with 1 mg of anti-IFN-γ (XMG1.2) or anti-TNF-α (XT22.11) at -2, 0, +2 days post-infection (p.i.). To deplete CD4+ T cells, mice were treated with 200 ug of anti-CD4+ (GK1.5) mAb using the same time course. Control mice experiments were treated with isotype-matched rat-anti-galactosidase mAb (GL113). CNS penetration of these mAb's were confirmed using goat-anti-rat antibodies conjugated with peroxidase and visualized with AEC chromagen (Stohlman et al., 1995b). Flow cytometry analysis of splenocytes from anti-CD4 treated mice showed > 98 % depletion (data not shown).

4. RESULTS

It has been suggested that CD4+ T cells are involved in the clearance of JHMV from the CNS by providing help to CTL since depletion of CD4+ T cells prevented JHMV

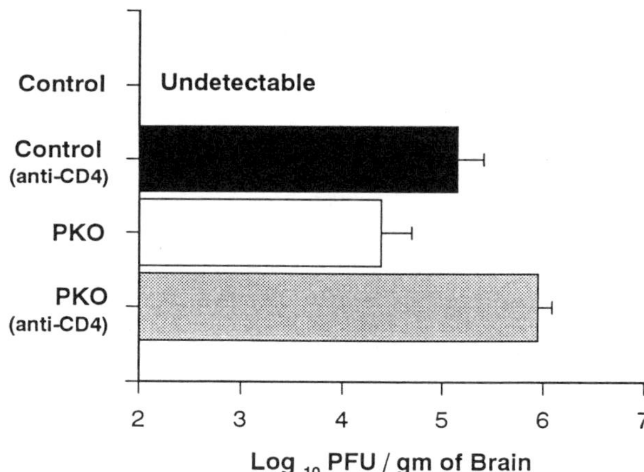

Figure 1. Depletion of CD4$^+$ T cells results in decreased viral clearance. Perforin wild-type (control) and perforin-deficient mice (PKO) were either treated with isotype-matched anti-β-galactosidase mAb or with anti-CD4$^+$ mAb and CNS virus titer assayed at 10 days post-infection.

clearance from the CNS (Williamson and Stohlman, 1990). However, viral clearance data from PKO mice indicated that, in addition to perforin-mediated anti-viral mechanism, CD4+ may participate directly in viral clearance (Lin, Stohlman, and Hinton, 1997). To determine whether CD4+ T cells contribute to viral clearance in the absence of perforin, viral replication in CD4 depleted PKO and control mice was compared. Figure 1 shows that, in agreement with previous experiments (Williamson and Stohlman, 1990), anti-CD4 treatment significantly (P < 0.05) increased virus titer in the control mice from undetectable to over 4 $\log_{10}$ PFU/gm at day 10 p.i.. Interestingly, depletion of CD4+ T cells from the PKO mice also resulted in a further increase of 1 $\log_{10}$ of infectious virus as compared to the isotype-treated control PKO mice (P < 0.05). These data suggest that in addition to its role in providing help for CTL, CD4+ T cell may contribute to viral clearance either directly or via secretion of anti-viral cytokines.

TNF-α has antiviral activity and can synergize with interferons in the induction of resistance to both RNA and DNA virus in diverse cell types (Wong and Goeddel, 1986). Since both CTL and CD4+ T helper cells secret TNF-α, the contribution of TNF-α to viral clearance in the absence of perforin-mediated cytolysis was addressed by examining JHMV clearance in control and PKO mice treated with anti-TNF-α mAb. In agreement with previous experiments, virus titers were undetectable in moribund control mice at day 10 p.i. Anti-TNF-α mAb did not inhibit viral clearance nor improve clinical disease (data not shown). Similarly, inhibition of TNF-α in JHMV infected PKO mice resulted in only a slight increase in virus titer over the untreated PKO mice. The ability of transferred mAb to cross the blood brain barrier was confirmed by the diffuse presence of rat immunoglobulin within the CNS (Figure 2). These data support previous data by showing that TNF-α does not contribute to JHMV clearance from the CNS even in the absence of perforin.

IFN-γ is secreted by CD8+ and CD4+ T cells and has been suggested to contribute to viral clearance. Although NK cells are thought to mediate anti-viral response via secretion of IFN-γ, the kinetics of IFN-γ mRNA expression in the CNS correlates with the infiltration of T lymphocytes into the CNS (Parra et al., in press). In addition, nude mice with intact NK response were unable to clear JHMV from the CNS (Houtman and Fleming, 1996a). To determine the potential anti-viral effect of IFN-γ, virus replication was compared in PKO and control mice were treated with anti-IFN-γ mAb (Figure 3). Similar to mice depleted of CD4+ T cells, treatment with anti-IFN-γ resulted in a significant (P < 0.05) delay in viral clearance in control mice. Anti-IFN-γ treatment of PKO mice also re-

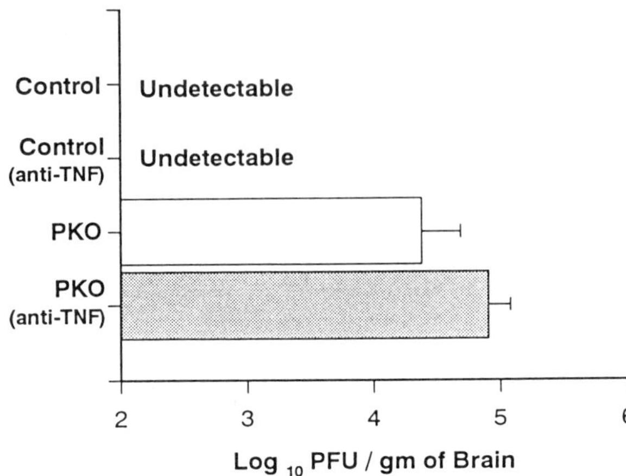

Figure 2. Inhibition of TNF-α do not inhibit viral clearance. Perforin wild-type (control) and perforin-deficient mice (PKO) were either treated with isotype-matched anti-β-galactosidase mAb or with anti-TNF-α mAb and CNS virus titer assayed at 10 days post-infection.

sulted in a significant ($p < 0.05$) 1 log increase in virus titer over the isotype-treated PKO mice. These data suggest that IFN-γ can provide anti-JHMV function in addition to perforin-mediated cytolysis.

5. DISCUSSION

Unlike many viral infections where a single anti-viral mechanism is sufficient for protection and clearance of virus, JHMV infection of the CNS had been shown to involve participation of many types of immune effector mechanisms (Kyuwa and Stohlman, 1990). This report attempts to further define the mechanisms involved in JHMV clearance from the CNS. The data presented are in agreement with previously published reports on the effect of anti-CD4, anti-IFN-γ and anti-TNF-α treatment in JHMV pathogenesis (Smith et al., 1991; Stohlman et al., 1995b; Williamson and Stohlman, 1990). Depletion of CD4+ T cells from mice resulted in significantly decreased viral clearance in the treated control mice over the untreated control mice while a smaller, but significant decrease in clearance was observed in the treated PKO mice over the untreated PKO mice. This data

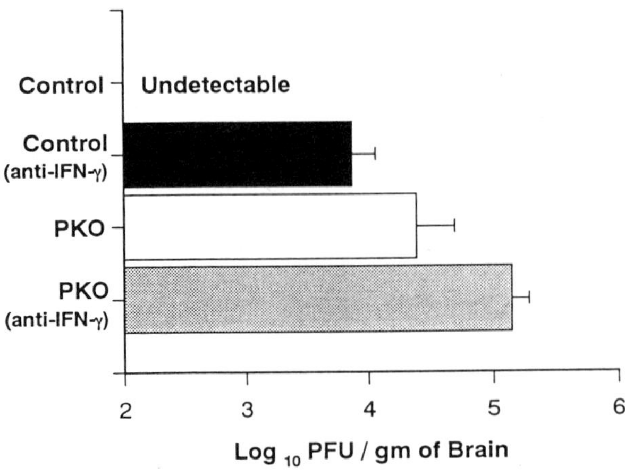

Figure 3. Anti-IFN-γ treatment increased JHMV replication in CNS. Perforin wild-type (control) and perforin-deficient mice (PKO) were either treated with isotype-matched anti-β-galactosidase mAb or with anti-IFN-γ mAb and CNS virus titer assayed at 10 days post-infection.

confirms the importance of CD4+ T cells in JHMV clearance from the CNS. Moreover, reduced virus clearance in the CD4-depleted PKO group support the hypothesis that CD4+ T cells have direct anti-viral roles in addition to providing help to CD8+ T cells. However, it remains unclear whether depletion of CD4+ T cells alters CTL development, trafficking into the CNS, cytokine secretion, or survival in the CNS.

CD4+ T cells could help eliminate virus from the CNS through direct lysis of infected MHC class II positive CNS cells via perforin, Fas/FasL, or TNF-mediated cytolysis. These data analyzing CD4-depleted PKO mice suggest the possibility of Fas/FasL or TNF-mediated involvement in viral clearance. Fas/FasL-mediated cytolysis has been implicated in the down-regulation of immune responses but has not been shown to mediate viral clearance (Hahn et al., 1995; Kagi et al., 1995; Scott, Grdina, and Shi, 1996; Van Parijs, Ibraghimov, and Abbas, 1996). TNF-α is a pleuripotent pro-inflammtory cytokine that has anti-viral activity and is expressed by many cell types, including CD4+ T cells, CD8+ T cells, microglia, astrocytes, and macrophages. In the absence of perforin, anti-TNF-α mAb treatment did not influence viral clearance or clinical disease, consistent with previous reports showing no effect of TNF-α inhibition on JHMV-induced demyelination or encephalitis (Stohlman et al., 1995b). TNF-α mRNA expression are associated with acute encephalitis and it's level increased in the CNS of JHMV-DM infected mice until death when no virus could be isolated (Parra et al., in press). Further, TNF-α mRNA is present in astrocytes of persistently infected mice (Sun et al., 1995). Although, TNF-α mRNA is not translated in JHMV-infected cells, it may be secreted by adjacent non-infected reactive CNS cells (Stohlman et al., 1995b). Together, these data suggest TNF-α has no direct role in the pathogenesis of JHMV-induced CNS disease and that its detection in the CNS may be a result of an on-going reactive gliotic process during viral infection.

Like TNF-α, IFN-γ is a pleuripotent cytokine that possesses direct anti-viral activity; however, the anti-viral activity of this cytokine is confounded by it's participation in the regulation of cell-mediated immunity. This may explain the discrepency in the amount of virus titer increases observed between the anti-IFN-γ treated control and PKO mice. Nonetheless, a direct but minor anti-viral role for IFN-γ is suggested in the PKO mice. These results are consistent with the in vitro anti-JHMV activity of IFN-α (Zhang et al., in press). As well, they are in agreement with the delayed clearance observed in IFN-γ deficient mice infected with a OBLV-60 strain of JHMV (Lane, Paoletti, and Buchmeier, 1997) and the kinetics of IFN-γ mRNA expression during infection (Parra et al., in press; Pearce et al., 1994). Further experiments are needed for clarification of IFN-γ's ability to influence cellular immunity.

In summary, the present study showed that CD4+ T cell are important for viral clearance either by a direct anti-viral effector mechanism or by secretion of IFN-γ. Although not directly demonstrated, these data indicate that CD4+ T cells may in fact utilize each of these mechanisms to help eliminate JHMV from the CNS.

REFERENCES

Compton, S. R., Barthold, S. W., and Smith, A. L., 1993, The cellular and molecular pathogenesis of coronaviruses [published erratum appears in Lab Anim Sci 1993 Apr;43(2):203], *Lab. Anim. Sci.* **43**: 15–28.

Fazakerley, J. K., and Buchmeier, M. J., 1993, Pathogenesis of virus-induced demyelination, *Adv. Virus. Res.* **42**: 249–324.

Hahn, S., Stalder, T., Wernli, M., Burgin, D., Tschopp, J., Nagata, S., and Erb, P., 1995, Down-modulation of CD4+ T helper type 2 and type 0 cells by T helper type 1 cells via Fas/Fas-ligand interaction, *Eur. J. Immunol.* **25**: 2679–85.

Houtman, J. J., and Fleming, J. O., 1996a, Dissociation of demyelination and viral clearance in congenitally immunodeficient mice infected with murine coronavirus JHM, *J. Neurovirol.* **2**: 101–111.

Houtman, J. J., and Fleming, J. O., 1996b, Pathogenesis of mouse hepatitis virus-induced demyelination, *J. Neurovirol.* **2**: 361–376.

Kagi, D., Seiler, P., Pavlovic, J., Ledermann, B., Burki, K., Zinkernagel, R. M., and Hengartner, H., 1995, The roles of perforin- and Fas-dependent cytotoxicity in protection against cytopathic and noncytopathic viruses, *Eur. J. Immunol.* **25**: 3256–62.

Korner, H., Schliephake, A., Winter, J., Zimprich, F., Lassmann, H., Sedgwick, J., Siddell, S., and Wege, H., 1991, Nucleocapsid or spike protein-specific CD4+ T lymphocytes protect against coronavirus-induced encephalomyelitis in the absence of CD8+ T cells, *J. Immunol.* **147**: 2317–2323.

Kyuwa, S., and Stohlman, S. A., 1990, Pathogenesis of neurotropic murine coronavirus strain JHM in the central nervous system, *Semin. Virol.* **1**: 273–280.

Lampert, P. W., Sims, J. K., and Kniazeff, A. J., 1973, Mechanism of demyelination in JHM virus encephalomyelitis, *Acta Neuropathol.* **24**, 76–85.

Lane, T. E., Paoletti, A. D., and Buchmeier, M. J., 1997, Disassociation between the in vitro and in vivo effects of nitric oxide on a neurotropic murine coronavirus, *J. Virol.* **71**: 2202–10.

Lin, M. T., Stohlman, S. A., and Hinton, D. R., 1997, Mouse hepatitis virus is cleared from the central nervous systems of mice lacking perforin-mediated cytolysis, *J. Virol.* **71**: 383–91.

Parra, B., Hinton, D. R., Lin, M. T., Cua, D. J., and Stohlman, S. A., Kinetics of cytokine mRNA expression in the central nervous system following lthal and nonlethal coronavirus-induced acute encephalomyelitis, *Virology*. In press.

Pearce, B. D., Hobbs, M. V., McGraw, T. S., and Buchmeier, M. J., 1994, Cytokine induction during T-cell-mediated clearance of mouse hepatitis virus from neurons in vivo, *J. Virol.* **68**: 5483–95.

Pope, M., Chung, S. W., Mosmann, T., Leibowitz, J. L., Gorczynski, R. M., and Levy, G. A., 1996, Resistance of naive mice to murine hepatitis virus strain 3 requires development of a Th1, but not a Th2, response, whereas pre-existing antibody partially protects against primary infection, *J. Immunol.* **156**: 3342–9.

Scott, D. W., Grdina, T., and Shi, Y., 1996, T cells commit suicide, but B cells are murdered!, *J. Immunol.* **156**: 2352–6.

Smith, A. L., Barthold, S. W., de Souza, M. S., and Bottomly, K., 1991, The role of gamma interferon in infection of susceptible mice with murine coronavirus, MHV-JHM, *Arch. Virol.* **121**: 89–100.

Stohlman, S. A., Bergmann, C., van der Veen, R., and Hinton, D., 1995a, Mouse hepatitis virus-specific cytotoxic T lymphocytes protect from lethal infection without eliminating virus from oligodendroglia, *J. Virol.* **69**: 684–694.

Stohlman, S. A., Hinton, D. R., Cua, D., Dimacali, E., Sensintaffar, J., Tahara, S., Hofman, F., and Yao, Q., 1995b, Tumor necrosis factor expression during mouse hepatitis virus induced demyelination, *J. Virol.* **69**: 5898–5903.

Stohlman, S. A., Matsushima, G. K., Casteel, N., and Weiner, L. P., 1986, In vivo effects of coronavirus-specific T cell clones: DTH inducer cells prevent a lethal infection but do not inhibit virus replication, *J. Immunol.* **136**: 3052–6.

Sun, N., Grzybicki, D., Castro, R. F., Murphy, S., and Perlman, S., 1995, Activation of astrocytes in the spinal cord of mice chronically infected with a neurotropic coronavirus, *Virology* **213**: 482–93.

Van Parijs, L., Ibraghimov, A., and Abbas, A. K., 1996, The roles of costimulation and Fas in T cell apoptosis and peripheral tolerance, *Immunity* **4**: 321–8.

Weiner, L. P., 1973, Pathogenesis of demyelination induced by a mouse hepatitis virus (JHM virus), *Acta. Neurol.* **18**: 298–303.

Wijburg, O. L., Heemskerk, M. H., Sanders, A., Boog, C. J., and Van Rooijen, N., 1996, Role of virus-specific CD4+ cytotoxic T cells in recovery from mouse hepatitis virus infection. *Immunology* **87**: 34–41.

Williamson, J. S., and Stohlman, S. A., 1990, Effective clearance of mouse hepatitis virus from the central nervous system requires both CD4+ and CD8+ T cells. *J Virol* **64**: 4589–92.

Wong, G. H., and Goeddel, D. V., 1986, Tumour necrosis factors alpha and beta inhibit virus replication and synergize with interferons. *Nature* **323**: 819–822.

Yamaguchi, K., Goto, N., Kyuwa, S., Hayami, M., and Toyoda, Y., 1991, Protection of mice from a lethal coronavirus infection in the central nervous system by adoptive trnsfer of virus-specific T cell clones. *J. Neuroimmunol.* **32**, 1–9.

Zhang, X., Hinton, D. R., Cua, D. J., Stohlman, S. A., and Lai, M. M., C. Expression of gamma interferon by a coronavirus defective-interfering RNA vector and its effect on viral replication, spread and pathogenicity. *Virology*. In press.

55

CORONAVIRUS INFECTION AND DEMYELINATION

Development of Inflammatory Lesions in Lewis Rats

Helmut Wege,[1] Hermann Schluesener,[2] Richard Meyermann,[2]
Vesna Barac-Latas,[3] Gerda Suchanek,[3] and Hans Lassmann[3]

[1]Institute of Diagnostic Virology
Federal Research Centre for Virus Diseases of Animals
Friedrich-Loeffler-Institutes, D-17498 Isle of Riems
[2]Institute of Brain Research
D-72076 Tübingen
[3]Institute of Neurology
A-1090 Vienna, Austria

1. ABSTRACT

Coronavirus infections of rodents can cause diseases of the central nervous system characterised by inflammatory demyelination. The lesions mimick in many aspects the pathology of multiple sclerosis in humans and of other neurological diseases. As an animal model for demyelination, we studied the MHV-JHM induced encephalomyelitis of Lewis rats. The pathomorphological analysis revealed patterns of lesions which developed in stages. Infected oligodendrocytes were first destroyed by necrosis. Later stages were characterized by demyelinated plaques. In the center of plaques, no virus antigen was found and oligodendrocytes were mainly destroyed by apoptosis. At the edge of plaques, virus antigen was expressed in parallel to infiltrations consisting of lymphocytes and macrophages. The prevailing mechanisms leading to demyelination may change individually and during defined stages of the disease. The transcriptional expression of chemoattractants and other mediators of inflammation was studied by semiquantitative RT-PCR. Virus induced inflammatory demyelination was accompanied by high expression of a relatively novel cytokine, the endothelial monocyte activating polypeptide II (EMAP II). By immunocytochemistry, EMAP II was detected in parenchymal microglia located both within the lesions and in unaffected areas. Furthermore, the level of transcriptional expression of the regulatory calcium binding S100 proteins MRP8, MRP14 and CP10 was associated with inflammatory demyelination and expression of IFN γ, IL-2, TNF α, and iNOS.

Coronaviruses and Arteriviruses, edited by Enjuanes *et al.*
Plenum Press, New York, 1998

2. INTRODUCTION

Inflammatory demyelination is an important feature characteristic for a number of chronic central nervous system (CNS) diseases in both animal and human for which the etiology and/or pathogenesis is not yet clear. The most enigmatic example is multiple sclerosis (MS) which is driven by an immune pathological process leading to selective destruction of myelin. Similar processes are induced by virus infections such as Visna in sheep, canine distemper or Theilers virus in mice (Wege et al., 1995; Fazakerley et al., 1993; Kyuwa et al., 1990). To analyse mechanisms of virus induced demyelination, the encephalomyelitis caused by murine coronaviruses in rodents has received much attention. Here we describe results of studies obtained by intracerebral infection of Lewis rats with MHV-JHM. The outcome is largely dependent on host factors and the virus variant employed for infection. Of major interest is the subacute demyelinating encephalomyelitis (SDE), which develops after intracerebral infection of Lewis-rats with MHV-JHM (Barac-Latas et al., 1997; Flory et al., 1995; Zimprich et al., 1991; Wege et al., 1984; Watanabe et al., 1983). Clinical signs such as ataxic gait and paresis develop several weeks to months p. i., animals may recover completely, or progress to a fatal outcome. The neuropathological hallmark of disease are lesions of primary demyelination in the white matter of the CNS. The pathogenic mechanisms leading to the histological changes are controversially discussed. In this context, in both autoimmune and virus induced experimental models the pathogenetic mechanisms driving inflammatory demyelination may be very similar. Increasing interest is focused to the role calcium binding S-100 proteins, chemoattractants an cytokines, which initiate and control inflammatory responses. As a first stage to analyse these parameters in more detail, we studied the transcriptional expression of some immune modulatory molecules which play an important role during inflammatory processes in the CNS.

3. MATERIALS AND METHODS

Specific pathogen free Lewis–rats between 3 to 8 weeks of age were intracerebrally inoculated with MHV-JHM. For histology, animals were perfused with 4% paraformaldehyde in 0.1 M phosphate buffer. Parts of brain and spinal cord were embedded in paraffin. To obtain specimens for RT-PCR, animals were dissected and left half of the brain frozen in liquid nitrogen before storage at -80 °C. The right half of the brain and spinal cord was further processed for histology. Combinations of histological stainings, immunocytochemistry and in situ hybridisation were employed as described previously (Barac-Latas et al., 1997; Gold et al., 1994; Breitschopf et al., 1992; Zimprich et al., 1991). Processing of tissue for RT-PCR was performed according to Stühler et al. (1997). Semiquantitative RT-PCR was evaluated by parallel processing of probes sampled from consecutive cycle numbers. The specific primer pairs were derived from gene banks and literature. The PCR was optimised for each primer pair and the identity of the amplification product was proven by Southern blot hybridisation.

4. RESULTS

The following analysis was obtained with CNS–tissue of rats, which displayed 3 to 5 weeks p. i. clinical signs and neuropathological changes of SDE. The specimens were selected out of a large number of rats characterised in detail during our previous studies (Barac-Latas et al., 1997; Stühler et al., 1997; Flory et al., 1995; Zimprich et al., 1991).

4.1. Pathomorphological Changes and Development of Lesions

By quantitative determination of virus antigen expressing cells, macrophages and T-lymphocytes three patterns of the demyelinating disease were differentiated (Barac-Latas et al., 1997). The first pattern was characterised by intensive virus antigen expression, although only mild inflammations were detected. Cells and myelin appeared normal, but the mRNA-expression of proteolipid-protein (PLP), a major constitutent of myelin, was reduced. By immunocytochmistry, both PLP, MBP and other myelin specific marker proteins remained at the same level as in uninfected white matter.

The second pattern represents the typical, sharply demarcated demyelinating plaques. These areas were surrounded by virus infected tissue which appeared otherwise unchanged. Within the plaques, only a few cells displayed viral antigen, PLP mRNA expression was reduced. Severe inflammatory infiltrations consisting of lymphocytes and high numbers of macrophages indicated the active demyelination process. The relation between necrotic versus apoptotic cells was analysed by morphological criteria and displaying the degree of DNA fragmentation by in situ tailing methods. Many T-cells appeard to be eliminated by apoptosis, whereas oligodendrocytes were affected by both cell destruction pathways. Necrosis of oligodendrocytes were in general associated with virus infection. Within the plaques, remyelination was associated with an increased number of oligodendrocytes displaying enhanced expression of PLP mRNA, virus antigen was not longer detectable in these lesions.

In the third type of lesions, virus antigen expression was very rarely, despite of ongoing demyelination and inflammation mainly characterised by macrophages containing myelin degradation products and cells dying from apoptosis. These lesions were surrounded by astroglial scar tissue.

4.2. Transcriptional Expression of Immunomodulatory Molecules in CNS Tissue

The study is based on rats for which neuropathological changes had been studied in detail. In total, we analysed CNS tissue from 30 rats with SDE. Furthermore, six uninfected controls, rats with EAE, rats at early times before clinical signs developed and animals, which recovered from SDE were included. A representive collection of specimens from rats with SDE was than further analysed by semiquantitative RT-PCR. These results are summarized in Table 1.

As a background to evaluate the degree of inflammation in CNS-tissue we monitored for the transcriptional expression of the cytokines IFN γ, TNF-α, IL-2 and for the nitric oxid syntheases iNOS and cNOS. Previous studies have shown, that these immunomodulatory substances are associated with diseased CNS infiltrated by immune cells. All rats which were diseased with SDE displayed relatively to the uninfected controls significant to strong upregulation of IFN γ, TNF-α, IL-2 and of the inducible nitric oxid synthease iNOS. The constituive expressed cNOS was detected in all CNS-samples at similar levels. Our major interest was to identify major calcium binding S 100 molecules, in particular the myeloid related proteins MRP8 and MRP14, the chemotactic S 100 protein CP-10 and the cytokine endothelial-monocyte activating protein EMAP II. The expression of MRP8, MRP14 and CP-10 varied to some extent among CNS tissue among individual rats. Considering all specimens investigated and taking into account the histological results a strong correlation between their expression and the inflammatory response was observed.

Table 1. Transcriptional expression of cytokines and immunomodulatory substances in CNS tissue of Lewis rats demonstrated by semiquantitative RT-PCR

Case No.	Organ	CP-10	MRP8	MRP 14	EMAP II	IFN-γ	TNF-α	IL-2	iNOS	cNOS
1 contr.	spleen	+++	++++	++++	++	+	±	-	-	-
2 contr.	brain	+	+	±	++	-	-	+	-	+++
3 EAE[1]	brain	++	++++	++	++++	++++	++	++	+	+++
4 SDE	brain	+	+	+	++++	+++	+	++++	+	+++
5 SDE	brain	++	+	+	++++	++++	+++	+++	++++	+++
6 SDE	brain	+++	++++	-	++++	+++	++++	+++	+++	++++
7 SDE	brain	+++	++++	+	++++	+++	+++	+	++	++++
8 SDE	brain	+	±	±	++++	++	+	++	±	+++
9 SDE	brain	+	++	±	+++	+	+	++	±	+++
10 SDE	brain	++++	+++	++	++++	+++	++	+++	+	++++

[1]EAE 11 days post sensitation with MBP

Furthermore, the transcriptional expression of EMAP II was strong in most cases. Recently, a monoclonal antibody suitable for immunocytochemical stainings had been established and was employed to locate EMAPII protein in tissue sections (Schluesener et al., 1997). The results are illustrated by Figure 1. In uninfected, healthy rats the expression of EMAP II is restricted to monocytes, macrophages and perivascular microglia. Parenchymal microglia of control rats is free of EMAP II by immunostaining. By contrast, parenchymal microglial cells of SDE lesions displayed EMAP II. The localisation of EMAP II expressing microglial cells is not limited to perivascular cuffs. EMAP II expressing microglial cells are only weakly positive for MHC class II antigens and not positive for ED1, which represents a marker of phagocytotic macrophages.

5. DISCUSSION

Destruction of myelin can involve a number of mechanisms (Fazakerley et al., 1993). Demyelination can be due to a cytolytic virus infection, promoted by immune reactions against virus infected cells or involve virus induced autoimmunity. In the model of the experimental allergic encephalomyelitis (EAE) the disease is mainly triggered by CD4+ - T-cells specific for CNS-antigens. The extent of demyelination and the intensity of inflammation depends from the balance between specific antibodies and cellular immune mechanisms (Linington et al., 1988). The pathomorphological changes observed in MS, EAE-models and virus induced demyelination are remarkably similar suggesting that identical mechnisms may operate in the evolution of demyelinating lesions (Barac-Latas et al., 1997; Ozawa et al., 1994). In vivo the mechanisms of oligodendrocyte damage appear to be related to the stage of the disease process. The virus infection during the early stage of SDE was cleared in many areas. It is conceivable, that the antiviral effect of certain cytokines and T-cell responses controls the spread of virus without destruction of infected glial cells (Stohlman et al., 1995; Pearce at al., 1994; Körner et al., 1991). Before any tissue damage was visible, the synthesis of PLP was downregulated. Therefore, the virus infection of oligodendrocytes may impair their function of producing and maintaining myelin proteins. However, this mechanism alone does not explain the inflammation process. In general, oligodendrocyte destruction by T-cell mediated cytotoxicity mediators released from macrophages is thought to result in apoptosis. Necrosis is the dominant

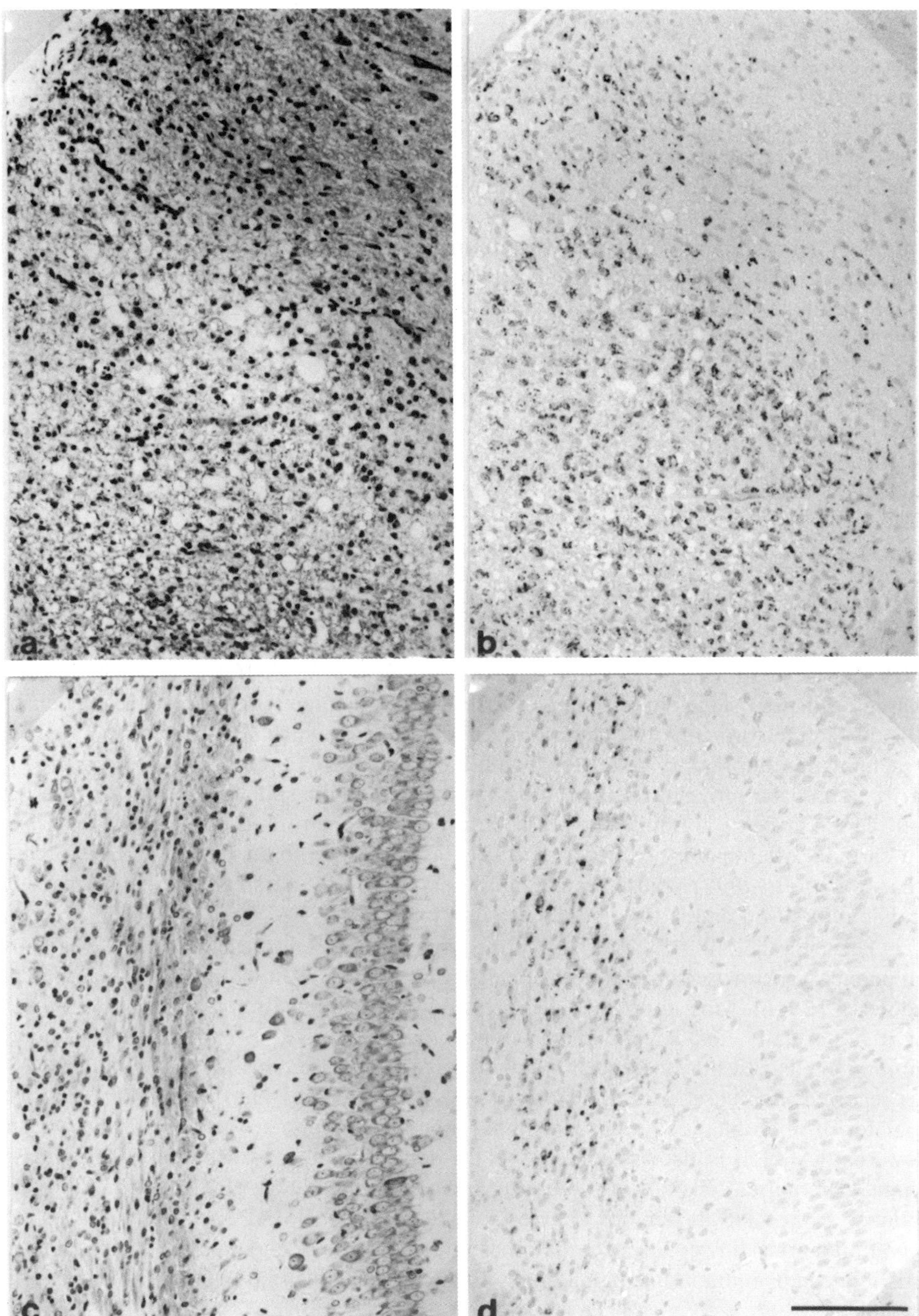

Figure 1. Demyelinating lesion in a Lewis rat with SDE and expression of EMAP II. The expression of EMAP II is not limited to macrophages within the demyelinated lesions. The cytokine is also expressed on microglia cells within the myelinated areas. a) Demyelinated area in the white matter of the spinal cord. b) Consecutive section, EMAP II expression represented by dark dots adjacent to the weakly stained nuclei. EMAP II is also found in ramified cells within still myelinated areas (upper part of the figure). c) Demyelinated area in the hippocampal region. Small bundle of myelinated fibers adjacent to the demyelinated area on the left side of the figure. d) Consecutive section, in both the demyelinated and myelinated areas phagocytes and microglia express EMAP II. A small bundle of myelinated fibers. Bar = 100 μm; a, c staining for myelin with luxol fast blue; b, d immunostaining with an anti-EMAP II mAb.

pattern of cell death in antibody—and complement mediated cytotoxicity. Many SDE-lesions displayed IgG and binding of complement C9. In these areas such complement fixing antibodies may promote demyelination (Piddlesen et al., 1993; Zimprich et al., 1991). In late stages virus antigen was detectable only in a few glial cells, in particular in astrocytes. However, destruction of myelin sheaths and oligodendrocytes continued. These lesions may involve a "bystander demyelination" maintained by macrophages. Individual patterns of demyelination may prevail at defined stages of the disease process. In brain tissue from MS patients, a similar variability in the pattern of demyelination and oligodendrocyte death has been observed (Ozawa et al., 1994).

Chemokines and cytokines involved in chemoattraction play a pivotal role for the state of the blood-brain barrier, attraction of macrophages and the regulation of inflammation. As a first step to study this network, we employed RT-PCR to monitor the transcriptional expression of the calcium binding S 100 molecules such as the myeloid related proteins MRP8, MRP14, CP-10 and the cytokine EMAP II. Further studies by immunocytochemistry and in situ hybridisation are indicated in order to localise and identify specific cell types within the morphological context. At present, for the rat system suitable antibodies for immunocytochemistry of CP-10, MRP8 and MRP14 are not yet available.

CP-10 is a potent chemotactic factor for murine and human myeloid cells (Lackman et al., 1993). In vivo CP-10 was described to locate at endothelial cells of the small blood vessels thus promoting the recruitment of monocytes and adhesion to the vessel walls. MRP8 and MRP14 are two calcium–binding proteins of the S-100 family which are expressed at distinct stages of monocytic differentiation (Imamichi et al., 1993). Their biologically active form involve heterodimeric complexes which had been associated with some human inflammatory disorders. In adult mice, MRP8/14 proteins were identified primarily in immature myeloid cells of the bone marrow, in the splenic red pulp and marginal zone. These proteins are highly expressed in recruited neutrophils and monocytes. Important chronic inflammatory reactions had been associated with an altered MRP expression (Bhardwaj et al., 1992). These proteins may serve as useful parameters to characterise the local activity of inflammations in the CNS.

The endothelial–monocyte activating polypeptide II represents a relatively novel and potent cytokine, which had been localised previously in monocytes and macrophages (Kao et al., 1994). Most pronounced expression was described for perivascular microglia, but it is normally not detectable in parenchymal microglia. In experimentally induced autoimmune diseases of rats, recently a widespread activation of EMAP II within the CNS was reported (Schluesener et al., 1997). These data indicate, that this cytokine may be a regulator of microglial cell reactivity in autoimmune inflammation of the CNS. Our data show, that EMAP II is also associated with the development of virus induced demyelinating lesions. Microglial cells are of increasing interest as regulatory elements of immunosurveillance in the CNS. It remains to be established whether EMAP II expression is an early activation marker for rat parenchymal microglia. The role of chemokines and cytokines in virus induced or immune mediated demyelination remains an interesting topic for further research.

ACKNOWLEDGMENTS

The work was supported by the Deutsche Forschungsgemeinschaft and Hertie Stiftung. We thank Regine Schröder and Hanna Wege for excellent technical assistance.

REFERENCES

Barac-Latas, V., Suchanek, G., Breitschopf, H., Stühler, A., Wege, H. and Lassmann, H., 1997, Patterns of oligodendrocyte pathology in coronavirus induced subacute demyelinating encephalomyelitis in the Lewis rat, *Glia* **19**: 1–12.

Bhardwaj, R.S., Zotz, C., Zwadlo-Klarwasser, G., Roth, J., Goebeler, M., Mahnke, K., Falk, M., Meinardus-Hager, G. and Sorg, C., 1992, The calcium-binding proteins MRP8 and MRP14 form a membrane-associated heterodimer in a subset of monocytes/macrophages present in acute but absent in chronic inflammatory lesions, *Eur. J. Immunol.* **22**: 1891–1897.

Breitschopf, H., Suchanek, G., Gould, R.M., Colman, D.R. and Lassmann, H., 1992, In situ hybridization with digoxigenin-labeled probes: Sensitive and reliable dedection method applied to myelinating rat brain, *Acta Neuropathol.* **84**: 581–587.

Fazakerley, J.K. and Buchmeier, M., 1993, Pathogenesis of virus-induced demyelination, *Adv. Virus Res.* **42**: 249–324.

Flory, E., Stühler, A., Barac-Latas, V., Lassmann, H. and Wege, H., 1995, Coronavirus induced encephalomyelitis: balance between protection and immune pathology depends on the immunization schedule with spike protein S, *J. Gen. Virol.* **76**: 873–879.

Gold, R., Schmied, M., Giegerich, G., Breitschopf, H., Hartung, H.P., Toyka, K.V. and Lassmann, H., 1994, Differentiation between cellular apoptosis and necrosis by combined use of in situ tailing and nick translation techniques, *Lab Invest.* **71**: 219–225.

Imamichi, T., Uchida, I., Wahl, S.M. and McCartney-Francis, N., 1993, Expression and cloning of migration inhibitory factor-related protein MRP 8 and MRP14 in arthritis-susceptible rats, *Biochem. Biophys. Res. Comm.* **194**: 819–825.

Kao, J., Houck, K., Fan, Y. Haehnel, I. Libutti, S.K., Kayton, M.L., Grikscheit, T., Chabot, J., Nowygrod, R., Greenberg, S., Kuang, W.-J., Leung, D.W., Hayward, J.R., Kisiel, W., Heath, M., Brett, J. and Stern, D.M., 1994, Characterization of a novel tumor-derived cytokine - endothelial-monocyte activating polypeptide II, *J. Biol. Chem.* **269**: 25106–25119.

Körner, H., Schliephake, A., Winter, J., Zimprich, F., Lassmann, H., Sedgwick, J., Siddell, S.G. and Wege, H., 1991, Nucleocapsid or spike protein-specific CD4+ T-lymphocytes protect against coronavirus-induced encephalomyelitis in the absence of CD8+ T-cells, *J. Immunol.* **147**: 2317–2323.

Kyuwa, S. and Stohlman, S.A., 1990, Pathogenesis of a neurotropic murine coronavirus, strain JHM in the central nervous system of mice, *Semin. Virol.* **1**: 273–280.

Lackmann, M., Rajasekariah, P., Lismaa, SE., Jones, G., Cornish, CJ., Hu, SP., Simpson, RJ., Moritz, RL. and Geczy, CL., 1993, Identification of a chemotactic domain of the proinflammatory S 100 protein CP-10, *J. Immunol.* **150**: 2981–2991.

Linington, C., Bradl, M., Lassmann, H., Brunner, C. and Vass, K., 1988, Augmentation of demyelination in rat acute allergic encephalomyelitis by circulating mouse monoclonal antibodies against myelin/oligodendroglia glycoprotein, *Am. J. Pathol.* **130**: 443–454.

Ozawa, K., Suchanek, G., Breitschopf, H., Brueck, W., Budka, H., Jellinger, K. and Lassmann, H., 1994, Patterns of oligodendroglia pathology in multiple sclerosis, *Brain.* **117**: 1311–1322.

Pearce, B.D., Hobbs, V., McGraw, T.S. and Buchmeier, M.J., 1994, Cytokine induction during T-cell-mediated clearance of mouse hepatitis virus from neurons in vivo, *J. Virol.* **68**: 5483–5495.

Piddlesden, S., Lassmann, H., Zimprich, F., Morgan, B.P. and Linington, C., 1993, The demyelinating potential of antibodies to myelin oligodendrocyte glycoprotein is related to their ability to fix complement, *Am. J. Pathol.* **143**: 555–564.

Stühler, A., Flory, E., Wege, H., Lassmann, H. and Wege, H., 1997, No evidence for quasispecies populations during persistence of the coronavirus mouse hepatitiis virus JHM: sequence conservationwithin the surface glycoprotein gene S in Lewis rats, *J. Gen. Virol.* **78**: 747–756.

Schluesener, H.J., Seid, K., Zhao, Y. and Meyermann, R., 1997, Localization of endothelial-monocyte-activating polypeptide II (EMAP II), a novel proinflammatory cytokine, to lesions of experimental autoimmune encephalomyelitis, neuritis and uveitis. *Glia.* In press

Stohlmann, S.A., Bergmann, C.C., van der Veen, R.C. and Hinton, D.R., 1995, Mouse hepatitis virus-specific cytotoxic T lymphocytes protect from lethal infection without eliminating virus from the central nervous system, *J. Virol.* **69**: 684–694.

Watanabe, R., Wege, H. and ter Meulen, V., 1983, Adoptive transfer of EAE-like lesions by BMP stimulated lymphocytes from rats with coronavirus-induced demyelinating encephalomyelitis, *Nature* **305**: 150–153.

Wege, H., 1995, Immunopathological aspects of coronavirus infections, *Springer Semin. in Immunopathol.* **17**: 133–148.

Wege, H., Watanabe, R. and ter Meulen, V., 1984, Relapsing subacute demyelinating encephalomyelitis in rats in the course of coronavirus JHM infection, *J. Neuroimmunol.* **6**: 325–336.

Zimprich, F., Winter, J., Wege, H. and Lassmann, H., 1991, Coronavirus induced primary demyelination: indications for the involvement of a humoral immune response, *Neuropathol. Appl. Neurobiol.* **17**: 469–484.

56

MHV-INDUCED FATAL PERITONITIS IN MICE LACKING IFN-γ

S. Kyuwa,[1] Y. Tagawa,[2] K. Machii,[3] S. Shibata,[4] K. Doi,[4] K. Fujiwara,[5] and Y. Iwakura[2]

[1]Department of Animal Pathology
[2]Laboratory of Animal Research Center
Institute of Medical Science
Univercity of Tokyo
Tokyo 108, Japan
[3]Department of Veterinary Public Health
Institute of Public Health
Tokyo 108, Japan
[4]Laboratory of Veterinary Pathology
Faculty of Agriculture
Univercity of Tokyo
Tokyo 113, Japan
[5]Department of Veterinary Pathobiology
Nihon Univercity
Fujisawa 252, Japan

1. ABSTRACT

IFN-γ gene was disrupted by homologous recombination in A3–1 embryonic stem cells. Germinally transmitted chimeric mice were successfully obtained and backcrossed with C57BL/6 (B6) mice 5 or 6 times. Deficiency of IFN-γ in homozygous mice was confirmed by northern blot analysis of spleen cells stimulated with phorbor esther and calcium ionophore and also by IFN-γ production in the culture supernatant of spleen cells stimulated with the same reagents. B6 mice lacking IFN-γ were infected intraperitoneally (ip) with 10^6 PFU of JHMV and monitored for their survival. Approximately 90% of the mice died at 50 days post-infection (pi) and the mean survival time was 28 days. Mice sacrificed at 3 weeks pi showed severe peritonitis accompanying the accumulation of a viscous fluid in the abdominal and thoracic cavities. Microscopically, the disease was characterized by disseminated granulomatous inflammation and exudative fibrinous serositis in the abdominal cavity. Infectious virus was isolated in most tissues including the

Coronaviruses and Arteriviruses, edited by Enjuanes *et al.*
Plenum Press, New York, 1998

liver, spleen, kidney, pancreas and lung during the experimental periods. The disease was not observed in wild-type or heterozygous littermates infected ip with JHMV. These results suggest that IFN-γ plays a critical role in MHV infection in mice. This experimental model may provide a unique opportunity to address the pathogenesis of virus-induced peritonitis such as feline infectious peritonitis in cats.

2. INTRODUCTION

MHV induce a variety of diseases including hepatitis, enteritis, encephalitis in mice, dependent on virus strain, infectious route and strain, age and immune status of the hosts. In particular, T cell-mediated immunity has been suggested to play an important role in the protection against ip infection with JHM strain (JHMV) in B6 mice (Kyuwa et al., 1989, Kyuwa et al., 1996). IFN-γ is an antiviral cytokine, which is produced by type 1 helper T cells, cytotoxic T lymphocytes and NK cells. IFN-γ activates macrophage activity, induces MHC class I, class II and adhesion molecules, and enhances IgG2a and IgG3 production. However, there is a discrepancy in the role of IFN-γ among virus infections in vivo (Baumgarth and Kelso, 1996; Sarawar et al., 1997). In the present study, we attempted to evaluate the role of IFN-γ in ip infection with JHMV in mice and found that JHMV induced subacute fatal peritonitis in mice lacking IFN-γ.

3. MATERIALS AND METHODS

3.1. Mice

Production of mice lacking IFN-γ was described previously (Tagawa et al., in press). A 129/SvJ mouse with heterozygously disrupted IFN-γ gene (IFN-γ$^{+/-}$) was backcrossed with B6 mice 5 or 6 times. Genotype of mice was determined by PCR as described previously (Tagawa et al., in press), and 8- to 12-week-old female mice were used. Breeding mice were maintained in a laminar flow rack and routinely checked forserologically free of MHV, Sendai virus, Mycoplasma pulmonis, Clostridium piriformis (Tyzzer's organism).

3.2. Virus

The DL variant of JHMV was propagated and plaque assayed on DBT cells as described previously (Kyuwa et al., 1996). Mice were infected ip with 110^6 plaque-forming units (PFU) of JHMV in a volume of 0.2 ml.

3.3. Serological Tests

Serum neutralizing antibodies were assayed as described elsewhere (Lin et al. 1997). Total anti-JHMV antibodies were determined by enzyme-linked immunosorbent assay by a commercial kit (Denka Seiken, Tokyo).

3.4. Depletion of CD4$^+$ or CD8$^+$ T Cells in Vivo

Purified anti-CD4 (GK1.5) or anti-CD8 (2.43) monoclonal antibodies were injected ip to deplete CD4$^+$ or CD8$^+$ T cells in vivo as previously described (Kyuwa et al., 1996).

3.5. Histopathology

Samples were fixed in 10% phosphate-buffered formalin and embedded in paraffin. Sections were prepared and stained with hematoxylin and eosin.

4. RESULTS

4.1. Production and Characterization of Mice Lacking IFN-γ

IFN-γ gene was disrupted by homologous recombination in A3–1 embryonic stem cells. The first exon was disrupted by inserting neomycine-resistant gene and cells homologously recombinated were selected in the presence of G418 (Tagawa et al., in press). Then, chimeric mice were produced and progenies were backcrossed with B6 mice 5 or 6 times.

The deficiency of IFN-γ in homozygous mice was confirmed by northern blot analysis of spleen cells stimulated with phorbor esther and calcium ionophore in vitro. Although spleen cells from wild type or heterozygous mice produced IFN-γ mRNA, those from homozygous mouse failed (Tagawa et al., in press). The lack of IFN-γ production in homozygous mice was also confirmed by protein level. IFN-γ protein was not detected in the culture supernatant of spleen cells from homozygous mouse stimulated with concanavalin A or phorbor esther and calcium ionophore (Tagawa et al., in press).

4.2. The Role of IFN-γ in the Acute Phase of JHMV Infection

IFN-$\gamma^{-/-}$, IFN-$\gamma^{+/-}$ and IFN-$\gamma^{+/+}$ mice were infected ip with 10^6 PFU of JHMV. After infection, survival of mice, viral growth and histopathological changes were examined. Although in heterozygous and wild type mice JHMV was cleared in the liver by 7 days after infection, the viral titers gradually decreased but persisted till 21 days after infection in homozygous mice. The result suggests that IFN-γ play a critical role in viral clearance during acute JHMV infection.

After ip infection with JHMV, a number of small lesions were observed in the liver in wild type mice 5 days after infection (Figure 1). In contrast, the liver lesions in IFN-$\gamma^{-/-}$ mice were larger than those in wild type mice, and infiltrated with abundant neutrophils (Figure 2).

4.3. Subacute Fatal Peritonitis in IFN-$\gamma^{-/-}$ mice

None of wild type and heterozygous mice died during the experimental periods. In contrast, some homozygous mice began to die from 16 days after infection, although they looked healthy till 10 days after infection. A 90% of the mice died at 50 days after infection. JHMV-infected homozygous mice were sacrificed at 21 days after infection. Most of mice showed severe peritonitis accompanying the accumulation of a viscous fluid in the abdominal and thoracic cavities. Microscopically, the disease was characterized by disseminated granulomatous inflammation and exudative fibrinous serositis in the abdominal and thoracic cavities. The deformation as well as adhesion of the abdominal organs was observed in affected animals (Figure 3). Although some neutrophils were infiltrated in the liver, hepatocyte injury was not observed and serum alanine aminotransferase (GPT) activity of the mice was normal. Lesions were observed not only in the abdominal cavity but also in the thoracic cavity. Infectious virus was recovered from most organs including the liver, spleen, mesenterium, kidney, pancreas and lung at 2 and 3 weeks after infection in IFN-$\gamma^{-/-}$ mice.

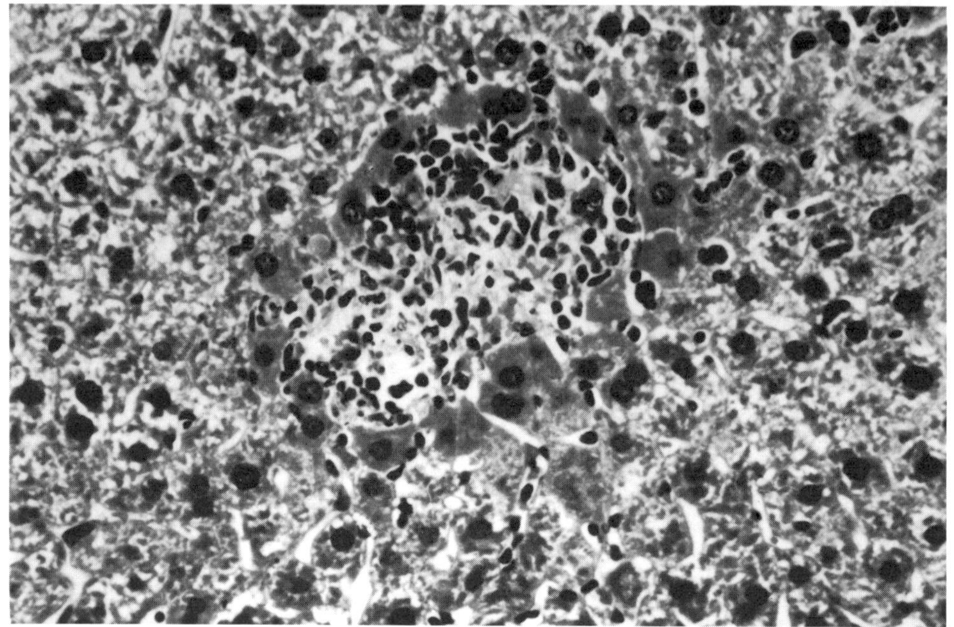

Figure 1. A representative lesion in the liver in JHMV-infected IFN-$\gamma^{+/+}$ mice. 5 days pi.

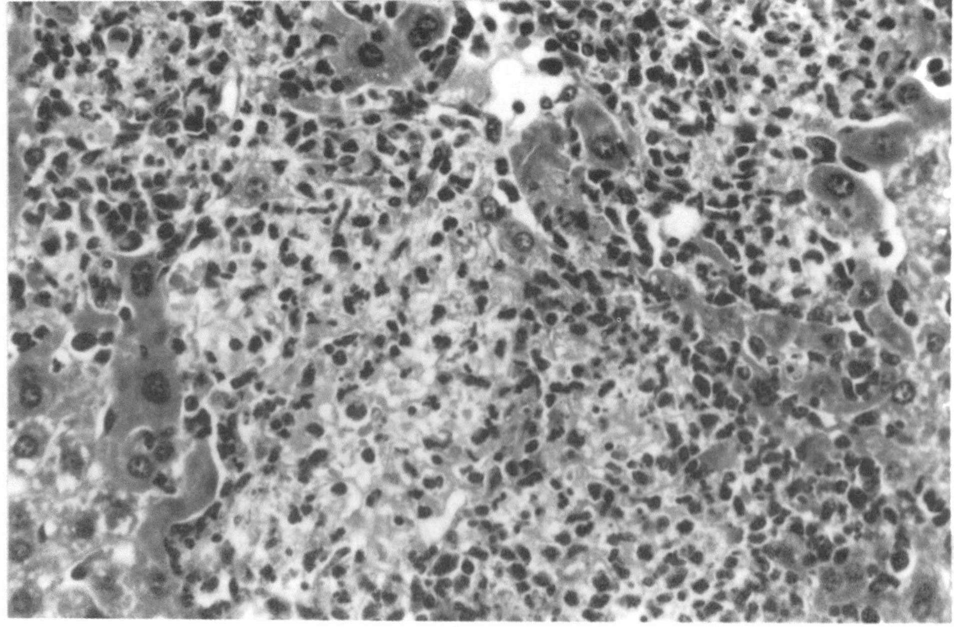

Figure 2. A representative lesion in the liver in JHMV-infected IFN-$\gamma^{-/-}$ mice. 5 days pi.

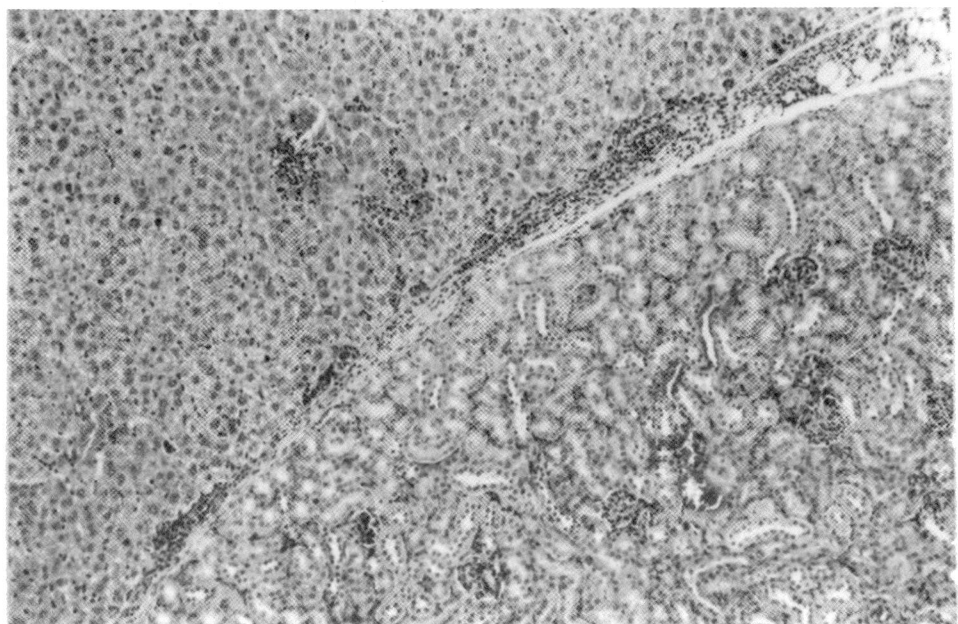

Figure 3. The adhesion of the liver and kidney in JHMV-infected IFN-γ$^{-/-}$ mice. 21 days pi.

Antiviral antibody responses in homozygous, heterozygous and wild type mice were examined. Sera were obtained from mice at 3 weeks after infection and total antiviral antibody titer was determined by a commercial ELISA kit. The antiviral antibody titers of homozygous mice were significantly higher (1:734) than those of heterozygous (1:165) or wild type (1:253) mice. Virus neutralization titer was also examined and the titers of homozygous mice were also higher than those of heterozygous or wild type mice.

4.4. The Role of CD4$^+$ and CD8$^+$ T Cells and Administration of Recombinant IFN-γ

Finally, effects of recombinant IFN-γ administration, depletion of either CD4$^+$T cells or CD8$^+$T cells on MHV-induced fatal peritonitis were examined. 3,000 units of recombinant IFN-γ or 200 μg of purified monoclonal antibodies were injected ip twice a week during the experimental periods. All of homozygous mice depleted of CD8$^+$T cells died within 10 days. Similarly, homozygous mice depleted of CD4$^+$T cells died within 2 weeks. These results suggest that protective effects of T cells are not solely mediated by IFN-γ. Recombinant IFN-γ administration partially inhibited MHV-induced fatal disease. At present, it remains obscure whether administration of higher dose of IFN-γ may completely prevent the disease or not.

5. DISCUSSION

In this study, we found some role of IFN-γ in MHV infection in mice using IFN-γ deficient mice. Complete viral clearance was not observed in IFN-γ deficient mice, suggesting that IFN-γ plays a critical role in viral clearance in ip JHMV infection in mice. However, the survival time of homozygous mice depleted of each T cell subset after JHMV infection was shorter than that of PBS-treated homozygous mice, suggesting that T cell-mediated antiviral effects except IFN-γ production also play a key role in the protection.

Intraperitoneal JHMV infection induces subacute fatal peritonitis in IFN-γ deficient mice. Previously, Yanagisawa et al. reported that MHV-NuU induced a similar disease in ICR-nude mice at a lower rate (Yanagisawa et al., 1985; Yanagisawa et al., 1986). Although the pathgenesis of this disease remains obscure at present, this experimental model may provide a unique opportunity to address virus-induced peritonitis such as feline infectious peritonitis in cats.

ACKNOWLEDGMENTS

This work was supported in part by a grant-in-aid from Ministry of Education, Science, Sport and Culture of Japan.

REFERENCES

Baumgarth, N. and Kelso, A., 1996, In vivo blockade of gamma interferon affects the influenza virus-induced humoral and the local cellular immune response in lung tissue, *J. Virol.* **70**:4411–4418.

Kyuwa, S., Yamaguchi, K., Toyoda, Y. and Fujiwara, K., 1989, Effect of sensitized T cell transfer on mouse hepatitis virus type 4 infection in athymic nude mice, *Jpn. J. Vet. Sci.* **51**:219–221.

Kyuwa, S., Machii, K. and Shibata, S., 1996, Role of CD4$^+$ and CD8$^+$ T cells in mouse hepatitis virus infection in mice, *Exp. Anim.* **45**:81–83.

Lin, M. T., Stohlman, S. A. and Hinton, D. R., 1997, Mouse hepatitis virus is cleared from the central nervous systems of mice lacking perforin-mediated cytolysis, *J. Virol.* **71**: 383–391.

Sarawar, S. R., Cardin, R. D., Brooks, J. W., Mehrpooya, M., Hamilton-Easton, A.-M., Mo, X. Y. and Doherty, P. C., 1997, Gamma interferon is not esssential for recovery from acute infection with murine gammaherpesvirus 68, *J. Virol.* **71**:3916–3921.

Yanagisawa, T., Nakanaga, K., Kyuwa, S. and Fujiwara, K., 1985, Ascitic disease in nude mice infected with mouse hepatitis virus, *Jpn. J. Vet. Sci.* **47**:171–174.

Yanagisawa, T., Nakanaga, K., Kyuwa, S. and Fujiwara, K., 1986, Ascitic disease in ICR-nude mice due to mouse hepatitis virus, *Jpn. J. Vet. Sci.* **48**:7–14.

57

CORONAVIRUS MHV-A59 CAUSES UPREGULATION OF INTERFERON-β RNA IN PRIMARY GLIAL CELL CULTURES

Q. Wang, J. A. Haluskey, and E. Lavi

Division of Neuropathology
Department of Pathology and Laboratory Medicine
University of Pennsylvania, School of Medicine
Philadelphiam, Pennsylvania 19104

1. ABSTRACT

Infection of mice with coronavirus mouse hepatitis virus strain MHV-A59 causes focal acute encephalitis, hepatitis and chronic demyelinating disease. To investigate host interferon (IFN) response to viral infection within the brain, RNA was extracted from A59-or MHV-2- infected and mock-infected primary astrocyte cultures from newborn mice, RT-PCR amplified RNA with primers specific for the various IFNs, transferred to nylon membranes and hybridized with IFN specific digoxigenin-labeled probes. Infection of primary astrocyte cultures from newborn mice with either A59 or MHV-2 caused upregulation of IFN-β RNA, but not IFN-γ or IFN-α. Thus, brain astrocytes are capable of producing a local IFN-β response upon infection with MHV. The response of the other IFNs following MHV infection may be derived from inflammatory cells.

2. INTRODUCTION

The pathologic hallmarks the chronic experimental disease induced by MHV infection in mice are similar to the human multiple sclerosis MS (Lavi et al., 1984a; Lavi et al., 1984b; Lavi et al., 1987a; Lavi and Weiss, 1989). As in MS, MHV-induced demyelination is believed to be immune-mediated (Wang et al., 1990). We initiated our investigation of primary glial cultures because astrocytes are considered as immune modulating cells in the CNS and one of the primary cell targets in MHV infection (Lavi et al., 1987a). We have previously shown that MHV infection causes upregulation of MHC class I expression in astrocytes and that astrocytes are the source of the MHC class I inducing factor

Coronaviruses and Arteriviruses, edited by Enjuanes *et al.*
Plenum Press, New York, 1998

(Lavi et al., 1987b; Lavi et al., 1989; Suzumura et al., 1988; Suzumura et al., 1986). Interferon response is an important antiviral, non-specific, host defense mechanisms. IFN-β is an antiviral and immune modulating factor, an MHC inducer, and is known to be secreted by astrocytes among other cells. IFN-β is used as one of the few treatment modalities in chronic demyelinating diseases such as MS although the mechanism of action of IFN-β in MS is not clear. In the present investigation we asked whether MHV infection causes upregulation of IFN-β in astrocytes. If so, what role may IFN-β play in MHV infection and CNS disease. We used two strains of MHV for this investigation. While MHV-A59 causes focal acute encephalitis, hepatitis and chronic demyelinating disease, (MHV-2), a closely related strain, causes acute meningitis and hepatitis without significant involvement of brain parenchyma. The study of the differences between these two strains may give us some insights into the mechanism of MHV-induced CNS disease.

3. METHODS AND MATERIALS

3.1. Preparation of Astrocyte Cultures

Late-term pregnant, virus-free, C57BL/6 mice were purchased from Jackson Laboratories, Bar Harbor Maine. Astrocyte-enriched glial cell cultures from 2–3-day-old neonatal mice were estabalished by the modification of the method of MaCarthy as described previously (McCarthy and de Vellis, 1980). Briefly, brains were removed and passed through 19G, 21G and 26G needles, suspended in DMEM supplemented with 20% FBS, and cultured in T75 flasks (Falcon). After 3 days, medium was changed and cells were maintained in DMEM with 10% FBS. The identification of the cells was confirmed by immunofluorescence analysis with an astrocytic marker anti glial fibrillary acidic protein (GFAP) antibodies. Cultures usually contained over 95% GFAP-positive cells. Infection was performed by 1 hour absorption of 1 pfu/cell of MHV at 10–14 days post plating.

3.2. Detection of IFNs by RT-PCR and Southern Blot Analysis

Total cellular RNA was isolated by the TRIzol Reagent (BRL) by following the manufacturer's protocol. Astrocytes were lysed directly in a culture dish by adding TRIzol Reagent (1ml per 10cm²), and passing the cell lysate through a pipette. Tissue samples were homogenized by mechanical disruption between two glass slides, then put in 1ml of TRIzol Reagent per 50–100 mg of tissue. Amplification grade DNAse I (BRL) was used for digestion of DNA before the RT-PCR reaction.

For RT-PCR reaction, 1μg of RNA was reverse transcribed using a random hexamer primer and SuperScript II reverse transcriptase (BRL), according to the manufacturer's protocol. Reagents of SuperScript Preamplification System for First Strand cDNA Synthesis Kit(BRL) were used in the PCR. The β-actin and IFNs primers were purchased from ClonTech. The expected size of β-actin , IFN-α, IFN-β and IFN-γ are 540bp, 294bp, 509bp and 365bp, respectively. The PCR products were further transblotted to the positive nylon membrane (BMB) in 20xSSC overnight at room temperature. Specific anti-sense probes (obtained from ClonTech or designed by MacVECTOR computer program) for β-actin and IFNs were labeled with Digoxigenin-11-ddUTP by using Genius 5 Oligonucleotide 3'-End Labeling kit (BMB). After 2 hours prehybridization in 65°C, the blot was sealed in a hybridization bag with 10 pmol/ml probe in hybridization solution. The bags were kept in a 65°C overnight. After post-hybridization wash, the membrane was blocked in blocking solution (BMB) for 30 min, then incubated with sheep anti-digoxigenin antibody (BMB), at a 1:10,000 dilution for 30 min. Detection of signal was done by applying chemiluminescent substrate Lumi-Phos 530 (BMB). The membrane was then exposed to a Kodak x-ray film.

Primers and probes for the RT-PCR reactions and Southern blot analysis

β-actin	5'primer	5'GTG GGC CGC TCT AGG CAC CAA 3'	
	3'primer	5'CTC TTT GAT GTC ACG CAC GAT TTC 3'	
	probe	5'GGG TGT TGA AGG TCT CAA ACA TGA TCT GGG 3'	(ClonTech)
IFN-α	5'primer	5'GAC TCA TCT GCT GCT TGG AAT GCA ACC CTC C 3'	
	3'primer	5'GAC TCA CTC CTT CTC CTC ACT CAG TCT TGC 3'	
	probe	5'GTT TCT TCT CTC TCA GGT ACA CAG TGA TCC 3'	(self-made)
IFN-β	5'primer	5'CAG CTC CAG CTC CAA GAA AGG ACG AAC ATT CG 3'	
	3'primer	5'CCA CCA CTC ATT CTG AGG CAT CAA CTG ACA GG 3'	
	probe	5'GTC TCA TTC CAC CCA GTG CTG GAG AAA TTG 3'	(self-made)
IFN-γ	5'primer	5'TGC ATC TTG GCT TTG CAG CTC TTC CTC ATG GC 3'	
	3'primer	5'TGG ACC TGT GGG TTG TTG ACC TCA AAC TTG GC 3'	
	probe	5'TTG TCT TTC AAG ACT TCA AAG AGT CTG AGG 3'	(ClonTech)

4. RESULTS

Uninfected and mock infected cultures did not express any of the three IFNs. Cultures always expressed detectable amounts of β-actin RNA. Following infection with MHV-A59 or MHV-2 cultures expressed detectable amounts of IFN-β RNA but not IFN-α or IFN-γ in addition to β-actin RNA. The PCR amplified reverse transcribed cDNA product was of the expected size. Southern blot analysis confirmed the specificity of the amplified PCR products showing positive hybridization only with the IFN-β and β-actin probes. PCR analysis at both 1 and 4 days post infection gave identical results.

5. DISCUSSION

The experiments shown here provide additional evidence for the role of astrocytes as a local immune cell in the CNS. Astrocytes are capable of producing IFN-β following viral infection when exposed to Newcastle Disease virus (NDV) (Lieberman et al., 1989). We now provide evidence that IFN-β can also be produced in cultured astrocytes after infection with the demyelinating coronavirus MHV. Although all three interferons RNAs are induced in response to MHV infection in the mouse brain (Lavi and Wang, 1995), upregulation of IFN-β RNA in the brain may be due to a local effect of the virus on CNS astrocytes. The other two interferons may be contributed by the inflammatory cells invading the brain during MHV infection. The exact role of IFN-beta in demyelination is not clear. However, the fact that both demyelinating and non-demyelinating strains of MHV produced the same effect of interferon induction in astrocytes may suggest that demyelination is not a function of interferon induction. Additional factors may be responsible for the inability of MHV-2 to produce demyelination.

ACKNOWLEDGMENTS

This work was supported by a grant from the National Multiple Sclerosis Society RG-26151/2.

REFERENCES

Lavi, E., Gilden, D. H., Highkin, M. K., and Weiss, S. R., 1984a, Persistence of MHV-A59 RNA in a slow virus demyelinating infection in mice as detected by in situ hybridization, *J. Virol.* **51:** 563–566.

Lavi, E., Gilden, D. H., Wroblewska, Z., Rorke, L. B., and Weiss, S. R., 1984b, Experimental demyelination produced by the A59 strain of mouse hepatitis virus, *Neurology* **34:** 597–603.

Lavi, E., Suzumura, A., Hirayama, M., Highkin, M. K., Dambach, D. M., Silberberg, D. H., and Weiss, S. R., 1987a, Coronavirus MHV-A59 causes a persistent, productive infection in glial cells, *Microbial Pathogenesis* **3:** 79–86.

Lavi, E., Suzumura, A., Lampson, L. A., Siegel, R. M., Murasko, D. M., Silberberg, D. H., and Weiss, S. R., 1987b, Expression of MHC class I genes in mouse hepatitis virus (MHV-A59) infection and in multiple sclerosis, *Adv. Exp. Med. Biol.* **218:** 219–222.

Lavi, E., Suzumura, A., Murray, E. M., Silberberg, D. H., and Weiss, S. R., 1989, Induction of MHC class I antigens on glial cells is dependent on persistent mouse hepatitis virus infection, *J. Neuroimmunol.* **22:** 107–111.

Lavi, E., and Wang, Q., 1995, The protective role of cytotoxic T cells and interferon against coronavirus invasion of the brain, *Adv. Exp. Med. Biol.* **380:** 145–149.

Lavi, E., and Weiss, S. R., 1989, Coronaviruses, in *"Clinical and molecular aspects of neurotropic viral infections"* (D. H. Gilden, and H. L. Lipton, Eds.), pp. 101–139. Kluwer, Academic Publishers, Boston.

Lieberman, A. P., Pitha, P. M., Shin, H. S., and Shin, M. L., 1989, Production of tumor necrosis factor and other cytokines by astrocytes stimulated with lipopolysaccharide or a neurotropic virus, *Proc. Natl. Acad. Sci. USA* **86:** 6348–6352.

McCarthy, K. D., and de Vellis, J., 1980, Preparation of separate astroglial and oligodendroglial cell cultures from rat cerebral tissue, *J. Cell Biol.* **85:** 890–902.

Suzumura, A., Lavi, E., Bhat, S., Murasko, D. M., Weiss, S. R., and Silberberg, D. H., 1988, Induction of glial cell MHC antigen expression in neurotropic coronavirus infection: characterization of the H-2 inducing soluble factor elaborated by infected brain cells, *J. Immunol.*, 2068–2072.

Suzumura, A., Lavi, E., Weiss, S. R., and Silberberg, D. H., 1986, Coronavirus infection induces H-2 antigen expression on oligodendrocytes and astrocytes, *Science* **232:** 991–993.

Wang, F. I., Stohlman, S.A., and Fleming, J. O., 1990, Demyelination induced by murine hepatitis virus JHM strain (MHV-4) is immunologically mediated, *J. Neuroimmunol.* **30:** 31–41.

58

CYTOTOXIC T LYMPHOCYTE RESPONSES TO INFECTIOUS BRONCHITIS VIRUS INFECTION

Sang Heui Seo and Ellen W. Collisson

Department of Veterinary Pathobiology
Texas A & M University
College Station, Texas 77843-4467

1. ABSTRACT

Cytotoxic T lymphocyte (CTL) activity to infectious bronchitis virus (IBV) was examined at regular intervals between 3 and 30 days post infection (p.i.). The maximal CTL lysis of target cells infected with IBV with 82% was detected at 10 days p.i. The specific CTL activity began to decrease only after viral loads, which peaked at day 8 p.i. in both kidneys and lungs, started to decline. Therefore, the CTL response correlated with elimination of acute infection. IgM antibody did not appear until day 10 and levels peaked at day 12 p.i. whereas IgG antibody titers were detectable only by day 15 p.i., but continued to increase exponentially until day 30 p.i., the last day examined. IBV specifc CTL epitope(s) were mapped within the carboxyl terminal 120 amino acids of nucleocapsid protein. In vivo inoculation of this fragment, as cDNA, induced protection against acute infection. The absence of viral neutralizing epitopes on the nucleocapsid protein would suggest that protection with known CTL eptiope(s) can be induced in the absence of neutralizing antibody.

2. INTRODUCTION

Infectious bronchitis virus (IBV), a prototype of the Coronaviridae family, causes a highly contagious respiratory disease in chickens. IBV infection in chicks is severe for young chicks in which signs of illness include tracheal rales, coughing, sneezing, and nasal discharge. Certain strains are also nephropathogenic, affect the reproductive system or cause enteric disease (Butcher et al, 1989; Collisson et al, 1990). Outbreaks of IBV infection in chicks can occur in vaccinated flocks. These vaccine failures are probably caused by antigenic variants of IBV arising from wild-type or vaccine viruses by point mutations

Coronaviruses and Arteriviruses, edited by Enjuanes *et al.*
Plenum Press, New York, 1998

455

or genomic RNA-RNA recombination (Wang et al, 1994). Whereas current IBV vaccines have targeted induction of viral neutralizing antibody, we have examined the cytotoxic T lymphocytes responses to IBV infection and correlated such immunity control of acute infection.

3. MATERIALS AND METHODS

3.1. Viral Source and Animals

The nephropathogenic Gray strain of IBV was propagated in the allantoic sac of 11-day old chicken embryos (Seo and Collisson, 1997). Embryonated eggs of inbred (B^{12}/B^{12}) and outbred (R68C) chickens were obtained from Hy-VAC (Adel, IA), and SPAFAS Incorporated (Preston, CT), respectively.

3.2. Generation of Recombinant Semliki Forest Virus (SFV) Vectors

IBV genomic RNA was prepared as described (Sneed et al, 1989). First strand cDNA synthesis was performed as described in the SUPERSCRIPT™ manual (GIBCO BRL) using 3' end primer (5'GTTGGATCCGAGTTCATTCTCACCTAAAGC3') of IBV genomic RNA before amplification. PCR was performed using the appropriate primers for synthesizing overlapping fragments of nucleocapsid gene with the following reaction conditions; primary denaturation at 94°C for 5 min, annealing at 55°C for 30s, extention at 72°C for 45s, and subsequent denaturation at 94°C for 1 min. A total of 35 cycles were used with a final extention step of 7 min at 72°C. Resulting PCR products were cloned into SFV1 plasmids (pSFV1). Packaging of SFV recombinant was carried out according to manufacturer's manual (GIBCOBRL) before infecting CK cells.

3.3. CTL Assay, ELISA, and EID50

Splenic T cells were evaluated for cytotoxic activity against syngeneic and allogenic target cells (Seo and Collisson, 1997). ELISA plates were prepared by adding Gray strain viral antigen to 96-well plates. Positive reactions were detected with ABTS peroxidase substrate (KPL, Gaithersburg, MD). Viral loads were determinded for lungs and kidneys from six chicks infected with Gray strain of IBV at the varying days p.i. (Cook et al, 1990; Seo and Collisson, 1997).

4. RESULTS

4.1. The Peak CTL Activity from IBV Infected Chicks Was at 10 Days p.i.

The CTL responses of splenic T cells from IBV infected syngeneic chicks were examined between 3 to 30 days p.i. using an E:T ratio of 100:1 (Fig 1a). The IBV specific CTL activity was detectable at day 3 p.i. with 22.9% release of radioactivity. The maximal lysis of target cells was observed at day 10 p.i. with 82.8% lysis. The CTL response decreased by days 15 and 20 with 46.30% and 38%, respectively. The CTL activity continued to decline to 19.23% on day 30 p.i., the last day examined. The CTL activity of splenic T cells from uninfected chicks was negligible.

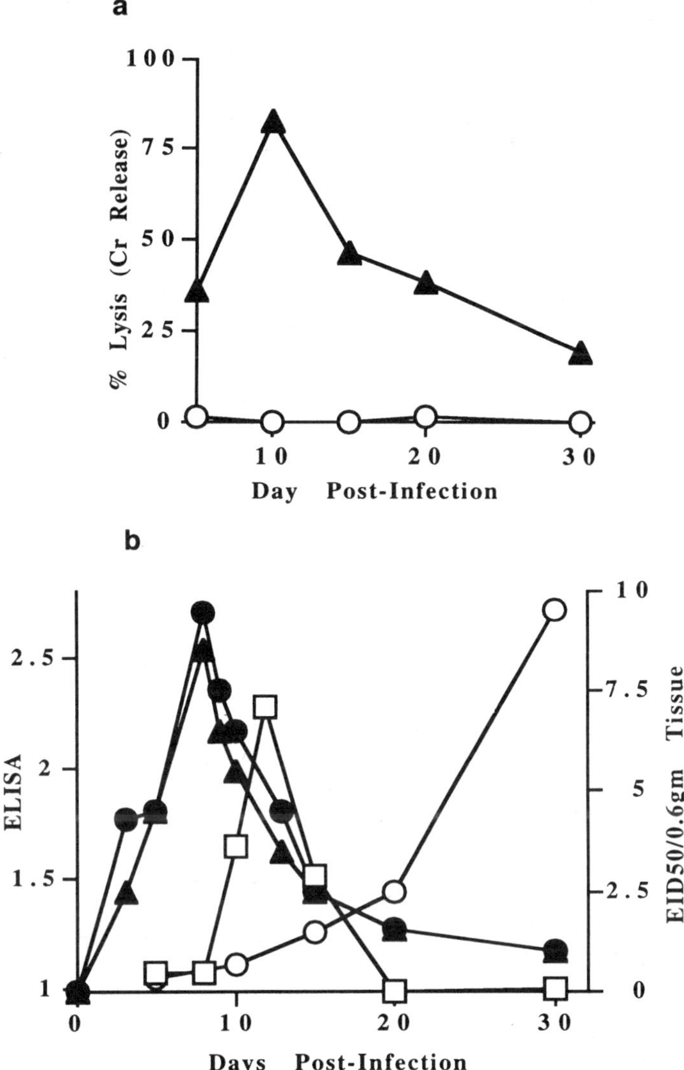

Figure 1. a) Time course of the responses to IBV infection. Triangles, target cells infected with IBV; and circles, control uninfected target cells. b) Viral and antibody titers. The EID_{50} in lungs is represented by closed circles, the EID_{50} in kidneys by closed triangles, open squares by IgM ELISA levels, and open circles by IgG levels.

4.2. Titers of Virus Isolated from Lungs and Kidneys Was Greatest at 8 Days p.i.

To examine the viral titers in chicks infected with IBV, lungs and kidneys were col lected from chicks at varying days p.i. and the viral titer was determined in EID_{50}. Virus was detected as early as day 3 p.i. in lungs with $10^{4.3}$ EID_{50} and kidneys with $10^{2.5}$ EID_{50}. Viral titers increased continuously until day 8 p.i. with a maximum of $10^{9.5}$ EID_{50} in lungs and $10^{8.5}$ EID_{50} in kidneys respectively. However, the viral titers started to decrease from day 9 until day 30 p.i. in either lungs or kidneys (Fig 1b).

4.3. IgG Antibody Titer Increased Exponentially in Contrast to CTL Activity

Sera were collected at varying days p.i. and the antibody titer was examined by ELISA. IgG antibody titer increased exponentially from day 10 p.i. when CTL activity was maximal, whereas, the highest titer of IgG antibody was observed at day 30 p.i., the last day examined when CTL activity dropped to the lowest level. IgM antibodis were detectable at 10 day p.i. and peaked at 12 day p.i (Fig 1b).

4.4. IBV Specific CTL Epitopes Were Mapped to the Carboxyl Terminal 120 Amino Acids of Nucleocapsid

To locate IBV specific CTL epitopes in the nucleocapsid, each overlapping fragments were cloned into SFV (Fig. 2a). The amino terminal 172 amino acids and carboxyl terminal 238 amino acids were expressed in target cells infected with Na-SFV and Nb-SFV. Splenic T cells collected from IBV infected chicks were stimulated in vitro with irradiated IBV infected CK cells for 8 days before using as CTL effector cells (Fig. 2b).

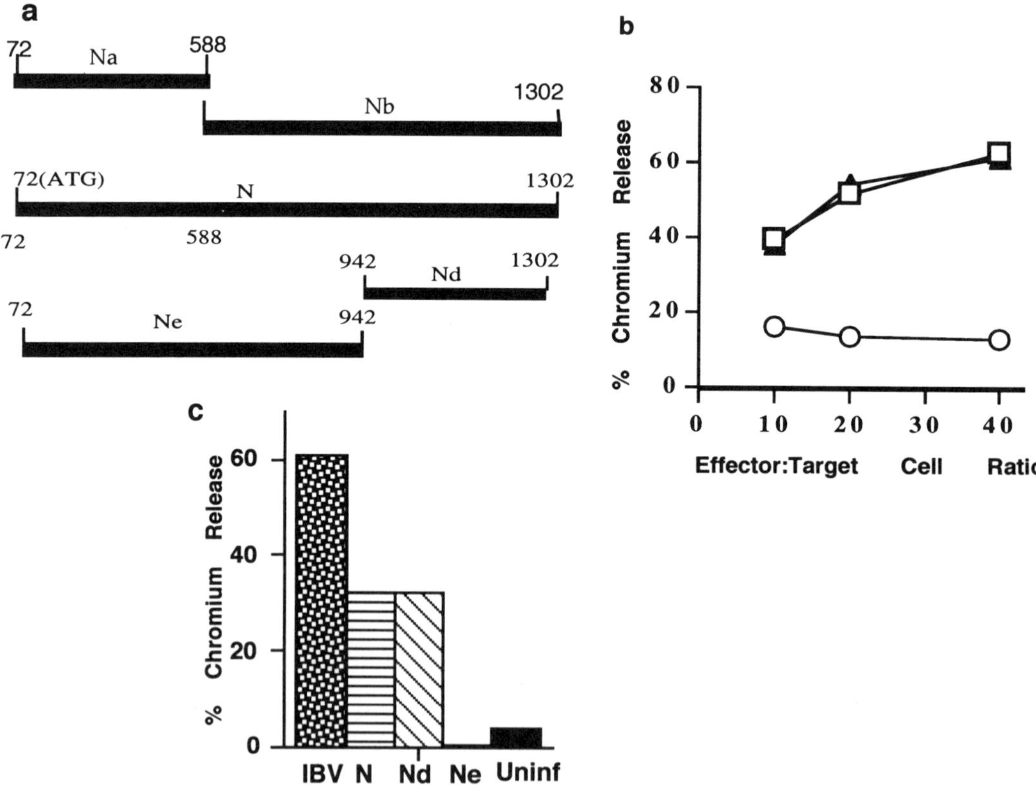

Figure 2. Mapping of CTL epitope to nucleocapsid protein. a) Map of the polypeptide of N. b) CTL responses to N. CK cells were infected with Na-SFV or Nb-SFV before using target cells. Lysis of target cells infected with Nb-SFV is represented by closed triangles, target cells infected with N-SFV by open squares, and target cells infected with Na-SFV by open circles. c) Further mapping of CTL epitopes. CK cells were infected with IBV, N-SFV, Nd-SFV, or Ne-SFV before using as target cells (effector:target ratio, 40:1).

Effector cells lysed both target cells infected with N-SFV and Nb-SFV with a maximum of 60% at the effector:target ratio of 40:1. Whereas, lysis of target cells infected with Na-SFV was minimal. Therefore, IBV specific CTL epitope(s) in nucleocapsid protein reside within carboxyl 238 amino acids.

To further map CTL determinants in the nucleocapsid, SFV vectors encoding the overlapping fragments Nc, Nd, or Ne were used to generate CK target cells. Splenic T cells from IBV infected chicks lysed syngeneic CK target cells infected with IBV, N-SFV, and Nd-SFV in a dose dependent manner but not the target cells expressing Nc-SFV, Ne-SFV, or the uninfected target cells (Fig. 2c). These results indicate that at least one IBV specific CTL epitope exists in Nd, the carboxyl terminal 120 amino acids of nucleocapsid.

4.5. IBV Nucleocapsid Specific CTL Were Generated in Vivo

In vivo induction of CTL with endogenously expressed nucleocapsid in the absence of other viral polypeptides was examined. pTargeT DNA plasmids encoding Nd or Ne of Gray strain of IBV were constructed for inoculation of chicks. Transfection and expression potential of pTargeT plasmids encoding Nd or Ne were first examined in vitro in CK cells. Splenic T cells collected from chicks inoculated with pTargeT plasmids were stimulated in vitro with the irradiated syngeneic CK cells infected with Nd-SFV or Ne-SFV for 8 days before using as CTL effector cells. Nd specific CTL were detected using target cells infected with Gray strain IBV or Nd-SFV in a dose responsive manner resulting in maximum lysis of 45% at 70:1 effector:target ratio (Fig. 3). In contrast, splenic T cells from uninoculated chicks or chicks inoculated with Ne-pTargeT did not lyse target cells infected with Gray strain of IBV.

Chicks inoculated with Nd-pTargeT or Ne-pTargeT were challenged with the Gray strain of IBV. Chicks immunized with Nd-pTargeT were protected from severe illness because only two out of six chicks experienced only mild respiratory signs of sneezing or nasal discharge. In contrast, all of six chicks immunized with Ne-pTargeT suffered a severe clinical illness, such as dyspnea, rales, lack of appetite or ataxia similar to unimmunized chicks. In addition, the amount of virus detected following IBV challenge in terms of EID50/0.6 g was reduced in the lungs by 7 logs from 9.5 in unimmunized chicks and reduced in kidneys to undetectable levels from 9.3 in unimmunized chicks (unpublished data).

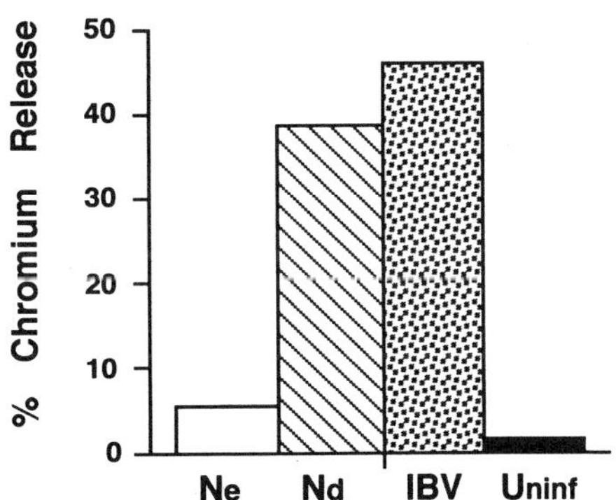

Figure 3. In vivo induction of nucleocapsid specific CTL by DNA inoculation. Splenic T cells collected from chicks inoculated as indicated were stimulated in vitro with irradiated syngeneic CK cells infected with N-SFV before using as effector cells. Effector:target cell ratio was 40:1.

5. DISCUSSION

Although chickens are a practical model for studying cellular immunity, few studies have been reported, describing CTL responses to viral infection (Omar and Schat, 1996). The CTL response of chickens to avian leukosis virus and the response of mice to MHV have been described but not correlated with acute viral infection (Thacker et al, 1995; Yamaguchi et al, 1988). We have characterized the CTL response to IBV antigens and correlated the response to viral infection and clinical illness. In this manuscript, we have shown that during acute IBV infection induces a vigorous specific CTL activity. This cellular response can be correlated more closely with the initial decline in viral load than the subsequent specific humoral responses.

The CTL response from IBV infected chicks was, at least in part, directed toward the carboxyl terminal 120 residue (Nd). Similarly, a CTL epitope for murine hepatitis virus was mapped to the 178 residue carboxyl terminus (Stohlman et al, 1992). SFV vectors were used to generate target cells expressing individual polypeptides. This system was prefered because IBV polypeptides were efficiently expressed in the cytoplasm in the absence of structural proteins from the vector.

In vivo inoculation of cDNA expressing the Nd fragment resulted in control of clinical illness and acute infection. Therefore, because the nucleocapsid gene does not have encode viral neutralizing epitopes, protection can be induced in the absence of neutralizing antibody. Although long-term protection may required a humoral response, this work suggests that development of effective IBV vaccines should include antigens which also induce CTL.

REFERENCES

Butcher, G.D., Winterfield, R.W., and Shapiro, D.P., 1989, An outbreak of nephropathogic H13 infectious bronchitis in commercial broilers, *Avian Dis.* **33**:823–826.

Collisson E.W., Li, J.Z., Sneed, L.W., Peters, M.L., and Wang, L., 1990, Detection of avian infectious bronchitis viral infection using in situ hybridization and recombinant DNA, *Vet. Microbiol.* **24**:261–271.

Cook, J.K.A., 1983, The classification of new serotypes of infectious bronchitis virus isolated from poultry flocks in Britain between 1981 and 1983, *Avian Pathol.* **13**:733–741.

Omar, A.R., and Schat, K.A., 1996, Syngeneic Marek's disease virus (MDV)-specific cell-mediated immune responses against immediate early, late, and unique MDV proteins, *Virology* **222**: 87–99.

Sneed, L., Butcher, G., Wang, L., Parr, R., and Collisson, E.W., 1989, Comparisons of the structural proteins of avian bronchitis virus as determined by western blot analyses, *Viral Immunol.* **2**: 221–227.

Seo, H.S., and Collisson, E.W., 1997, Specific cytotoxic T lymphocytes are involved in in vivo clearance of infectious bronchitis virus, *J Virol,* **71**:5173–5177

Thacker, E.L., Fulton, J.E., and Hunt, H.D., 1995, In vitro analysis of a primary, major histocompatibility complex (MHC)-restricted, cytotoxic T-lymphocyte response to avian leukosis virus virus (ALV), using target cells expressing MHC class I cDNA inserted into a recombinant ALV vector, *J. Virol.* **69**: 6439–6444.

Wang L., Junker, D., Collisson, E.W., 1993, Evidence of natural recombination within the S1 gene of infectious bronchitis virus, *Virology* **192**:710–716.

Yamaguchi, k., Kyuwa, S., Nakanaga, K., and Hayami, M., 1988, Establishment of cytotoxic T-cell specific for cells infected with mouse hepatitis virus, *J. Virol.* **62**: 2505–2507.

59

IN VIVO AND *IN VITRO* INTERFERON (IFN) STUDIES WITH THE PORCINE REPRODUCTIVE AND RESPIRATORY SYNDROME VIRUS (PRRSV)

W. Buddaert, K. Van Reeth, and M. Pensaert

Laboratory of Veterinary Virology
Faculty of Veterinary Medicine
University of Gent
Salisburylaan 133
B-9820 Merelbeke, Belgium

1. ABSTRACT

Some of the interactions between the porcine reproductive and respiratory syndrome virus (PRRSV) and the porcine interferon-α (IFN-α) system were studied. In a first experiment, it was shown that pretreatment of primary porcine alveolar macrophages (AMs) with recombinant porcine (rPo) IFN-α 1 resulted in significant reductions of PRRSV yield and numbers of antigen expressing cells. In a second experiment, sensitivity of PRRSV to IFN-α was confirmed *in vivo*. In pigs inoculated with porcine respiratory coronavirus (PRCV)—a potent inducer of endogenous IFN-α in the lungs of pigs—followed 2 days later by PRRSV—lung PRRSV titers were 1.7 to 2.9 $\log_{10}$ TCID$_{50}$ reduced compared to those in singly PRRSV inoculated pigs. It was concluded therefore that PRRSV has a fairly good sensitivity to the antiviral effects of IFN-α. A third experiment documented that *in vivo* PRRSV infection generally does not affect PRCV-induced IFN-α production in the lungs of pigs. In addition, it was shown that the IFN-inducing capacity of PRRSV is at least 159 times lower than that of PRCV. This finding suggests that cells other than AMs may be responsible for IFN production in the lungs of pigs.

2. INTRODUCTION

Infection with the porcine reproductive and respiratory syndrome virus (PRRSV) most frequently occurs via the respiratory route and the lungs are an important virus target

Coronaviruses and Arteriviruses, edited by Enjuanes *et al.*
Plenum Press, New York, 1998

organ. PRRSV replicates in the lungs during at least 3–4 weeks and titers up to $10^{5.5}$–$10^{5.9}$ TCID$_{50}$/g lung tissue are produced during the first 2 weeks. Macrophages of the alveolar spaces and alveolar septa play a prominent role in the pathogenesis of PRRSV infection. These cells make up 80–94% of the infected lung cells (Duan et al, 1997), are a major reservoir for virus in the lungs, and are a potential source of persistent viraemia. Cells of the macrophage type are quite intimately related tho the interferon (IFN) system. Macrophages are highly sensitive to the antiviral actions of interferons and, in turn, they are major IFN-producing cells in some animal species (Saksela et al., 1984).

Several porcine respiratory viruses are capable of inducing substantial amounts of biologically active interferon-α (IFN-α) in the lungs of pigs. One very potent IFN-inducer is the porcine respiratory coronavirus (PRCV), and IFN titers up to 4000–5000 biological units/ml are found in bronchoalveolar lavage fluids of experimentally infected pigs (Van Reeth and Pensaert, 1995). In dual infections with the PRCV followed by either H1N1-influenza virus or Aujeszky's disease virus (ADV), replication of the second virus in the lungs was significantly suppressed (Van Reeth and Pensaert, 1994; Van Reeth and Pensaert, 1996). This viral interference is likely mediated by IFN-α.

Until recently, little attention has been paid to the interactions between PRRSV and the porcine interferon-α system (Albina et al., 1997). In a first part of this study, we have investigated whether PRRSV is sensitive to the antiviral actions of IFN-α *in vitro* and *in vivo*. In the *in vitro* studies, primary porcine alveolar macrophages and recombinant porcine IFN-α 1 (rPo IFN-α) 1 (Lefèvre et al., 1990) were used. In the *in vivo* studies, PRCV was used as an endogenous inducer of natural IFN-α in the lungs of pigs. In a second part of this study, we examined if a prior PRRSV infection affects lung IFN production by PRCV.

3. METHODS AND MATERIALS

3.1. Antiviral Effect of IFN-α against PRRSV Replication *in Vitro*

Primary porcine alveolar macrophages (AMs) were obtained by post mortem lung lavage of 4 to 6-week-old conventional pigs that were negative for PRRSV antibodies. Freshly seeded AMs were pretreated for 18 h with recombinant porcine IFN-α 1 (generous gift from C. La Bonnardière, INRA, Jouy-en-Josas, France) at 1000 units/10^6 AMs or left untreated (virus controls). The IFN-treated and control wells were then rinsed with RPMI and inoculated with the Lelystad virus strain of PRRSV or with vesicular stomatitis virus

Table 1a. Kinetics of PRRSV production in alveolar macrophages pretreated with rPo IFN-α

		Virus titer[a] ($\log_{10}$ TCID$_{50}$/ml) at … h PI				
	Treatment	1	8	18	30	42
PRRSV	Control	3.2	3.2	4.5	5.0	5.6
	IFN-α	3.5	2.9	2.3	2.4	3.0
VSV	Control	4.5	6.3	7.3	7.8	8.5
	IFN-α	3.8	3.9	3.6	4.2	4.8

[a] Means of 2 experiments.

Table 1b. Kinetics of PRRSV antigen expression in alveolar macrophages pretreated with rPo IFN-α

		IPMA positive cells[a] at … h PI					
	Treatment	1	12	18	30	48	72
PRRSV	Control	0	15-26	56-173	264-301	CPE[b]	CPE
	IFN-α	0	0	0-4	1-5	146-212	CPE

[a]Number of positive cells in one representative microscopic field, a range of 6 wells is given.
[b]Complete cytopathic effect.

(VSV) (MOI 0.04). VSV was included because of its known high sensitivity to porcine IFN-α. After 2 h adsorption, the virus inoculum was removed and maintenance medium added. At different times post inoculation (PI), IFN-treated and control AM cultures were frozen (-70°C) and thawed (2 cycles) and infectious virus titers determined by titration in AM (PRRSV) and swine testicle (ST) (VSV) cell cultures.

In a parallel series of experiments, PRRS viral antigen expression in IFN-pretreated and control AM cultures was quantitated by an immunoperoxydase monolayer assay (IPMA) (Wensvoort et al., 1991).

3.2. Antiviral Effect of Interferon against PRRSV Replication in the Lungs of Pigs

Conventional 10-week-old pigs from herds free of PRRSV and PRCV were used. Two pigs were inoculated with PRCV (91V44 strain; 10^7 TCID$_{50}$/pig) followed 2 days later by PRRSV (Lelystad virus strain; $10^{5.0}$ TCID$_{50}$/pig) and euthanatized 2 and 3 days PI (DPI) with PRRSV respectively (Table 2). Two singly PRCV inoculated pigs and 2 singly PRRSV inoculated pigs were euthanatized at corresponding times PI. Two uninoculated control pigs were included. PRRSV and PRCV titers in the left apical and diaphragmatic lung lobes were determined by titrations in AMs and ST cells respectively. The right lung

Table 2. Virus titers and IFN-α titers in the lungs of dually PRCV-PRRSV and singly virus-inoculated pigs

	Euthanasia (DPI)[a]		Virus titer[b]			IFN-α titer (U/ml BAL fluid)
			PRCV	PRRSV		
Inoculation	PRCV	PRRSV	Apic[c]	Apic	Diaphr[d]	
none	–[e]	–	neg	neg	neg	neg
	–	–	neg	neg	neg	111
PRCV	4	–	6.7	neg	neg	406
	5	–	7.2	neg	neg	455
PRRSV	–	2	neg	5.5	5.2	219
	–	3	neg	5.9	4.9	98
PRCV-2d-	4	2	6.8	3.7	3.7	3281
PRRSV	5	3	7.5	2.7	2.7	7014

[a]Time of euthanasia in days post inoculation.
[b]Virus titer in $\log_{10}$ TCID$_{50}$/g lung.
[c,d]Apical, diaphragmatic lung lobes.
[e]Not applicable.

was lavaged and bronchoalveolar lavage (BAL) fluid IFN titers were determined in a virus CPE inhibition assay with Madin-Darby bovine kidney (MDBK) cells and VSV.

3.3. Effect of PRRSV Infection on the Ability of PRCV to Induce IFN-α in the Lungs

Caesarian-derived colostrum-deprived (CDCD) pigs were used to avoid eventual IFN production by concurrent natural infections. Pigs were inoculated at the age of 3 weeks by the intratracheal route. Four pigs were dually inoculated with PRRSV first followed 3 (2 pigs) or 7 (2 pigs) days later by PRCV (Table 3). One pig of each group was euthanatized 3 and 4 DPI with PRCV. Singly PRCV (2 pigs) and PRRSV (2 pigs) inoculated control pigs were killed at corresponding times PI. Two inoculated control pigs were included. IFN titers in BAL fluids and virus titers were determined as described higher.

4. RESULTS

4.1. *In Vitro* Sensitivity of PRRSV to IFN-α

Pretreatment of AMs with rPo IFN-α resulted in reduced PRRSV yields (Table 1a) and number of infected cells (Table 1b). Maximal reduction in virus yield was 2.6 $\log_{10}$ $TCID_{50}$ for PRRSV, compared to 3.7 $\log_{10}$ $TCID_{50}$ for VSV. In IFN-treated AMs, titers of both viruses gradually increased between 18 and 42 h PI. Numbers of PRRSV infected cells in IFN-treated AM cultures were at least 53 times reduced at 30 h PI and increased thereafter. On the whole, IFN-treated AMs showed a 24 h delay in PRRSV antigen expression and CPE onset.

4.2. *In Vivo* Sensitivity of PRRSV to IFN-α

Uninoculated control pigs were negative for PRCV and PRRSV. One of 2 pigs had an IFN-α titer of 111U/ml BAL fluid.

While singly PRRSV inoculated pigs were negative for PRCV, PRCV titers in dually PRCV-PRRSV and singly PRCV inoculated pigs were similar (Table 2). Prior inoculation

Table 3. Interferon-α titers in BAL fluids of dually PRRSV-PRCV and singly virus-inoculated pigs

	Euthanasia (DPI)[a]		IFN-α titer
Inoculation	PRRSV	PRCV	(U/ml BAL fluid)
PRCV	–[b]	3	27171
PRCV	–	4	9229
PRRSV	7	–	58
PRRSV	10	–	32
PRRSV-3d-PRCV	6	3	53248
PRRSV-3d-PRCV	7	4	649
PRRSV-7d-PRCV	10	3	28535
PRRSV-7d-PRCV	11	4	5114

[a]Time of euthanasia in days post inoculation.
[b]Not applicable.

of pigs with PRCV resulted in mean reductions in PRRSV titer of 1.7 and 2.9 $\log_{10}$ TCID$_{50}$ at 2 and 3 DPI with PRRSV respectively. IFN-α titers in the latter pigs were 15 and 72 times higher than those after inoculation with PRRSV only.

4.3. Effect of PRRSV Infection on the Ability of PRCV to Induce IFN-α in the Lungs

None of the uninoculated control pigs had any detectable IFN-α production.

All 3 groups of virus inoculated pigs were positive for IFN-α (Table 3). IFN-α titers reached up to 27171 U/ml BAL fluid in the singly PRCV inoculated pigs, but did not exceed 58 U/ml in the singly PRRSV inoculated pigs. As in singly PRCV inoculated pigs, IFN-α titers of dually PRRSV-3d-PRCV and PRRSV-7d-PRCV inoculated pigs were highest 3 DPI with PRCV. At that time, IFN-α titers amounted to 53248 and 28535 U/ml and were thus comparable to those detected after single PRCV inoculation.

5. DISCUSSION

The results presented here clearly demonstrate that PRRSV replication in AMs is sensitive to the antiviral effect of IFN-α *in vitro* and *in vivo*. The *in vitro* experimental conditions were selected on the basis of scientific and practical arguments. The natural virus host cell and species specific recombinant IFN-α were used. Interferon was used at a dose that can be achieved in the lungs of pigs by IFN induction or administration. While it could have been more appropriate to use an MOI of 1 to test the effect of IFN on one single cycle of virus replication, a relatively low MOI of 0.04 was used because of limitations in the PRRS virus stock titer. Furthermore, preliminary experiments had shown that IFN treatment before and after viral challenge was far more effective than pretreatment only. Such continuous treatment of AMs, however, hampered exact quantitations of infectious virus yield and viral antigen-expressing cells. Indeed, carry over of biologically active IFN from AMs in the antiviral assay into AM indicator cultures was demonstrated. Also, prolonged IFN treatment (> 48 h) caused a slight degree of toxicity in some AM cultures and thus interfered with interpretation of the IPMA staining. Consequently, IFN-pretreatment was selected for further experiments. All of these experimental conditions should be considered when making comparisons with previously determined sensitivities of other viruses. After all, compared to other antiviral studies with primary AMs and porcine (Esparza et al., 1988) or bovine viruses (Babiuk et al., 1985; Holland et al., 1991), PRRSV shows a fairly good sensitivity to IFN-α. A "normal high sensitivity" of PRRSV to IFN-α was recently also reported by Albina and coworkers (1997). *In vivo*, using PRCV as an IFN inducer in the lungs of pigs, a maximal PRRSV yield reduction of 2.9 $\log_{10}$ TCID$_{50}$ could be obtained. Previously, similar experiments with H1N1-influenza virus (Van Reeth and Pensaert, 1994) and ADV (Van Reeth and Pensaert, 1996) had shown maximal yield reductions of 2.7 and 1.5 $\log_{10}$ TCID$_{50}$ respectively. These *in vivo* differences between virus infections may depend largely upon their different sites of replication within the lungs.

Significant differences in IFN-α titers were found between the 2 *in vivo* experiments presented here. Maximal IFN titers detected were about 7000 U/ml in the first experiment and 53000 U/ml in the second experiment. In unpublished experiments with other viruses, we found that IFN titers following direct inoculation of virus into the trachea may be $\geq$ 10 times higher than those after aerosol or intranasal virus inoculation. Most likely, the difference in inoculation methods in the present experiments plays a major role in the observed differences in IFN production. Another difference between these 2 experiments is the sanitary status of the experimental pigs. In the PRCV-PRRSV dual infection experiment, we intended to mimic the field situation as closely as possible. Therefore, pigs originating from conventional breeding farms were used and these were transported to our isolation facilities 1 week before the start of the experiment. One of the control pigs, however, was positive for IFN-α. Artursson and coworkers (1989) have demonstrated that 25% of 10–12 week old conventional feeder pigs are positive for IFN-α in their serum within 5–10 days after regrouping in the fattening farm units. IFN production under these natural circumstances, as well as in our experiments, may result from virus infections or from stress caused by transport and adaptation to a new environment. Therefore, completely germ-free caesarian derived and colostrum deprived pigs, reared in Horsefall-type isolation units, were used in our second experiment.

One of the most interesting aspects of this study was that even though PRRSV was a very poor IFN inducer in the lungs of pigs, it did not appear to have an inhibitory effect on IFN production by a subsequent PRCV infection. To our knowledge, the effect of viruses with AM tropism on lung IFN production by a second inducer has never been studied *in vivo*. *In vitro*, on the other hand, ADV (Iglesias et al., 1992) as well as African swine fever virus (Powell et al., 1996) infection of primary porcine AMs have been shown to interfere with IFN-α production by synthetic inducers. Even more important, transmissible gastrenteritis virus (TGEV)-induced IFN-α secretion by AMs was totally inhibited when cells were infected by PRRSV before (Albina et al., 1997). Our present *in vivo* findings can be explained in 2 ways. If IFN-α in the pig lung is produced by AMs, PRRSV clearly does not interfere with the IFN-producing capacity of its target cell *in vivo*. Another and more likely possibility is that lung cells other than AMs are responsible for IFN production. In pigs, so-called "natural IFN-producing cells" or null lymphocytes have been found to constitute a major subpopulation of IFN-producing leukocytes upon *in vitro* exposure to several viruses (Charley and Lavenant, 1990). Studies of the IFN-α producing cells *in situ* in the lungs of pigs are needed to clarify these issues.

ACKNOWLEDGMENTS

We thank Lieve Sys for skillful technical assistance. This study was financially supported by the Belgian Ministry of Agriculture.

REFERENCES

Albina, E., Carrat, C., and Charley, B., 1997, Interferon alpha response to porcine reproductive and respiratory syndrome virus (PRRSV), *J. Gen. Virol.*, submitted.

Artursson, K., Wallgren, P., and Alm, G., 1989, Appearance of interferon-α in serum and signs of reduced immune function in pigs after transport and installation in a fattening farm, *Vet. Immunol. Immunopathol.* **23**: 345–353.

Babiuk, L. A., Bielefeldt Ohmann, H., Gifford, G., Czarniecki, C. W., Scialli, V. T., and Hamilton, E. B., 1985, *J. Gen. Virol.* **66**: 2383–2394.

Charley, B., and Lavenant, L.,1990, Characterization of blood mononuclear cells producing IFN-α following induction by coronavirus-infected cells (transmissible porcine gastroenteritis virus). *Res. Immunol.* **141**: 141.

Duan, X., Nauwynck, H.J., and Pensaert, M.B.,1997, Virus quantification and identification of cellular targets in the lungs and lymphoid tissues of pigs at different time intervals after inoculation with porcine reproductive and respiratory syndrome virus (PRRSV), *Vet. Microbiol.* **56**: 9–19.

Esparza, I., Gonzalez, J. C., and Vinuela, E., 1988, Effect of interferon-α, interferon-γ and tumour necrosis factor-α on African swine fever virus replication in porcine monocytes and macrophages, *J. Gen. Virol.* **69**: 2973–2980.

Holland, S. P., Fulton, R. W., Short, E. C., Wyckoff, J. H., and Fox, J. C., 1991, *In vitro* and *in vivo* 2', 5'-oligoadenylate synthetase activity induced by recombinant DNA-derived bovine interferon αI1 in bovine alveolar macrophages and blood mononuclear cells, *Am. J. Vet. Res.* **52**: 1779–1783.

Iglesias, G., Pijoan, C., and Molitor, T., 1992, Effects of pseudorabies virus infection upon cytotoxicity and antiviral activities of porcine alveolar macrophages, *Comp. Immunol. Microbiol. Infect. Dis.* **15**: 249–259.

Lefèvre, F., L'Haridon, R., Borras-Cuesta, F., and La Bonnardière, C., 1990, Production, purification and biological properties of an *Escherichia coli*-derived recombinant porcine alpha interferon, *J. Gen. Virol.* **71**: 1057–1063.

Powell, P.P., Dixon, L.K. and Parkhouse, R.M., 1996, An IkappaB homolog encoded by African swine fever virus provides a novel mechnanism for downregulation of proinflammatory cytokine responses in host macrophages, *J. Virol.* **70**: 8527–8533.

Saksela, E., Virtanen, I., Hovi, T., Secher, D.S., and Cantell, K., 1984, Monocyte is the main producer of human alpha interferons following Sendai virus induction, *Prog. Med. Virol.* **30**: 78.

Van Reeth, K., and Pensaert, M.B., 1994, Porcine respiratory coronavirus-mediated interference against influenza virus replication in the respiratory tract of feeder pigs, *Am. J. Vet. Res.* **55**: 1275–1281.

Van Reeth, K., and Pensaert, M., 1995, Production of interferon-α, tumor necrosis factor-α and interleukin-1 in the lungs of pigs infected with the porcine respiratory coronavirus. *Proc. 3rd Congress Europ. Soc. Vet. Virol.*, Interlaken, Switzerland, 197–201.

Van Reeth, K., and Pensaert, M., 1996, A clinical and virological study in pigs infected with Aujeszky's disease virus shortly after infection with porcine respiratory coronavirus. Proc. *14th Congress Int. Pig Vet. Soc.*, Bologna, Italy, 137.

Wensvoort, G., Terpstra, C., Pol, J.M.A., ter Laak, E.A., Bloemraad, M., de Kluyver, E.P., Kragten, C., van Buiten, L., den Besten, A., Wagenaar, F., Broekhuijsen, J.M., Moonen, P.L.J.M., Zetstra, T., de Boer, E.A., Tibben, H.J., de Jong, M.F., van Veld, P., Groenland, G.J.R., van Gennep, J.A., Voets, M.T., Verheijden, J.H.M., and Braanskamp, J., 1991, Mystery swine disease in The Netherlands: the isolation of the Lelystad virus, *Vet. Q.* **13**: 121–130.

60

IDENTIFICATION OF A COMMON ANTIGENIC SITE IN THE NUCLEOCAPSID PROTEIN OF EUROPEAN AND NORTH AMERICAN ISOLATES OF PORCINE REPRODUCTIVE AND RESPIRATORY SYNDROME VIRUS

J. I. Casal,[1] M. J. Rodriguez,[1] J. Sarraseca,[1] J. Garcia,[1] J. Plana-Duran,[2] and A. Sanz[1]

[1]INGENASA
 Hnos. Garcia Noblejas 41, 2º
 28037 Madrid, Spain
[2]Lab. Sobrino
 17037 Vall de Bianya, Spain

1. ABSTRACT

Porcine reproductive and respiratory syndrome virus (PRRSV) nucleocapsid (N) protein has been identified as the most immunodominant viral protein. The N protein genes from two PRRSV isolates Olot/91 (European) and Quebec 807/94 (North American) were cloned and expressed in *Escherichia coli* using the pET3x system. The antigenic structure of the PRRSV N protein was dissected using seven monoclonal antibodies (MAbs) and overlapping fragments of the protein expressed in *E.coli*. Three antigenic sites were found. Four MAbs recognized two discontinuous epitopes that were present in the partially folded protein or at least a large fragment comprising the first 78 residues, respectively. The other three MAbs revealed the presence of a common antigenic site localized in the central region of the protein (amino acids 50 to 66). This hydrophillic region is well conserved among different isolates of European and North American origin. However, since this epitope is not recognized by many pig sera, it is not adequate for diagnostic purposes. Moreover, none of the N protein fragments were able to mimic the antigenicity of the entire N protein.

Coronaviruses and Arteriviruses, edited by Enjuanes *et al.*
Plenum Press, New York, 1998

2. INTRODUCTION

Porcine reproductive and respiratory syndrome virus (PRRSV) causes an important pathology in pigs characterized by reproductive failure in sows and gilts, pneumonia and an increase in perinatal mortality (Terpstra et al., 1991). This arterivirus is now endemic in domestic pigs of the U.S. and Europe. The North American and European isolates of PRRSV represent two distinct genotypes (Suarez et al., 1996; Murtaugh et al., 1995; Meng et al., 1995; Mardassi et al., 1994), with important antigenic differences (Dea et al., 1996; Nelson et al., 1993; Wensvoort et al., 1992). However, even among different isolates of the same genotype there are considerable antigenic, genetic and pathogenic differences (Meng et al., 1996).

PRRSV is an enveloped RNA virus approximately 62 nm in diameter. The viral genome is a single stranded RNA of positive polarity, approximately 15 kb in size. The virion contains six structural proteins encoded by ORFs 2 to 7 (van Nieuwstadt et al., 1996; Meulenberg et al., 1996; Meulenberg et al., 1995). ORF7 encodes for the nucleocapsid (N) protein. Recently, we have shown that the majority of antibodies produced during the infection are specific for the N protein of PRRSV (Plana-Duran et al., 1997), which is, therefore, a suitable candidate for the detection of virus-specific antibodies and diagnosis of the disease. Since the important antigenic differences between both genotypes, the identification of antigenic sites in the N protein would help to prepare diagnostic assays in order to discriminate between PRRSV genotypes.

In this study we have cloned and expressed in *E.coli* the ORF 7 from two different PRRSV isolates: Olot/91 and Quebec 807/94. Then, we have characterized the antigenic map of the N protein of the European PRRSV Olot/91 isolate using a panel of monoclonal antibodies (MAbs) and pig antisera in combination with a collection of overlapping fragments. This study has allowed the identification of three major epitopes in the N protein, being one of them well conserved among the different isolates.

3. MATERIAL AND METHODS

3.1. Antisera and Monoclonal Antibodies

A collection of 48 European and 15 North American field pig sera positive for PRRSV, together with 7 MAbs specific for PRRSV Olot/91 N protein, were used for the antigenic characterization analysis of the N protein and its fragments. N protein specific MAbs were prepared from Balb/c mice immunized with semi-purified recombinant baculovirus-derived proteins (Plana-Durán et al., 1997). Protocols for immunization and the preparation of MAbs and ascitic fluids have been described previously (Sanz et al., 1985). Hybridoma supernatants were screened by ELISA and immunofluorescence assays. The isotype and subclass of each MAb were determined by ELISA using specific anti-mouse subtype antibodies (Sigma).

3.2. PRRSV Quebec 807/94 RNA Purification and cDNA Cloning

PRRSV Quebec 807/94 was grown in pig alveolar macrophages and semipurified as previously described (Plana-Durán et al., 1992). The semipurified virus was treated with 1% SDS and 500 µg/ml proteinase K for 1h at 37°C for extraction of total RNA. The ORF7 cDNA was obtained by RT-PCR as previously described (Rodriguez et al., 1997).

PCR products were cloned in phosphatase-treated *Sma*I-digested pMTL25. Plasmid DNA was sequenced by the dideoxynucleotide chain-termination method (Sanger et al., 1977). Protein sequence analyses were performed with PC/GENE software (Intelligenetics) and the PredictProtein mail server of EMBL (Rost, 1996).

3.3. Cloning of PRRSV Olot/91 ORF7 Fragments in the pET3x Expression Plasmid

The complete ORF7 coding sequences were obtained from the recombinant plasmid pPRRSV-8 (Plana-Durán et al., 1997) for Olot/91 and from pPRRSC-ORF7 for Quebec 807/94 and cloned in pET3x. The Olot/91 ORF7 fragments generated were the following: ORF 7 gene (*Hpa*I-*Afl*III), A (nt -16 to 237, *Hpa*I-*Taq*I), B (nt 54 to 291, PCR), C (nt 237 to 387, *Taq*I-*Afl*III), D (nt -16 to 126, *Hpa*I-*Eco*NI), E (nt 126 to 237, *Eco*NI-*Taq*I) and F (nt 151 to 201, PCR) (Fig.1). For all the PCR amplifications, 50 ng of template DNA was mixed with 800 ng of primers (Rodriguez et al., 1997). The samples were subjected to 25 rounds of amplification under the following reaction conditions: denaturation at 94°C for 1 min, annealing at 42–60°C for 1 min, and extension at 72°C for 1 min plus a final extension step of 7 min at 72°C. All of the fragments were recovered by *Bam*HI-compatible digestion and subcloned into I-digested phosphatase-treated pET3x (a,b,c) (Studier et al., 1990). The ligation mixtures were used to transform XL1 blue or DH5 competent cells. The resulting colonies were screened by digestion with appropriate restriction enzymes.

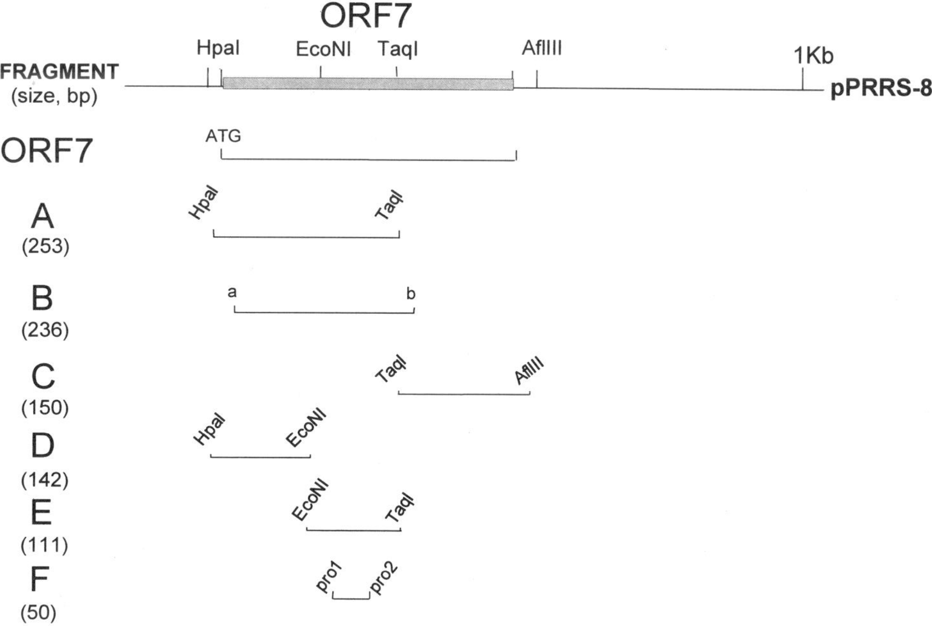

Figure 1. Map of PRRSV Olot/91 ORF 7 fragments. ORF 7 fragments were constructed as described in Materials and Methods, using conventional restriction sites or PCR. The ORF 7 is shown as a box with the restriction sites used in the fragmentation process. The lines from A to F show the restriction sites or the primers used to generate the fragments subsequently cloned in pET3x expression plasmids. The size of every fragment is indicated below the letter.

After verification of the sequence, *E.coli* BL21(DE3) pLysS competent cells were transformed with all the recombinant pET3x plasmids.

3.4. Growth, Induction, and Analysis of *E. coli*-Transformed Cells

Single clones of *E.coli* BL21 (DE3) pLysS cells containing the recombinant pET3x-derived plasmids were grown and induced with 0.4 mM isopropyl thiogalactopyranoside (Boehringer-Mannheim). After 3 h induction at 37°C, cells were centrifuged, washed twice with PBS and resuspended in a final volume of 500 µl of PBS for 50 ml of culture. Bacteria were freeze-thawed three times and then lysed by sonication. Detection of proteins was carried out by staining with Coomassie blue or by immunoblotting according to standard procedures (Harlow and Lane, 1988). Purification of the different fragments of the PRRSV N protein was carried out as described previously (Martinez-Torrecuadrada and Casal, 1995).

3.5. ELISAs

Microtiter 96-well plates (LabSystem) were coated overnight at 4°C with 100 µl of a dilution 1:100 of the soluble fraction obtained by guanidinium chloride-treatment of the recombinant N protein, or derived fragments, in 0.05M carbonate buffer pH: 9.6. Then, standard procedures were followed for the assay (Harlow and Lane, 1988). The reactivity caused by non-specific binding of sera to *E.coli* extracts was substracted from that of the recombinant proteins to obtain the PRRSV-specific value. The cut-off value was taken as (mean± 2 S.D.) of the 6 negative control sera.

For competitive ELISA, 50 µl of each MAb at 10-fold dilutions starting at 20 µg/well in ZPBS-0.05% Tween20 were mixed and incubated with 50 µl of biotin-labelled MAb for 1h at 37°C. After washing, peroxidase-labelled streptavidin (dilution 1:4,000) was added and incubated for 30 min at room temperature. In some cases, the competitor MAb was labelled with peroxidase instead of biotin.

For competitive ELISA among pig sera and MAbs, pig sera (dilution 1:5) were added for 1h at 37 °C and Mabs were added for 30 min at 37°C. Then, plates were washed and peroxidase-labelled anti-mouse IgG added for 30 min at 37°C. Finally, plates were washed and color was developed using ABTS as substrate.

4. RESULTS

4.1. Monoclonal Antibodies to PRRSV N Protein

Seven MAbs were produced that were positive for PRRSV by ELISA. The ascites fluid titers of each MAb are shown in Table 1, together with the isotype and other characteristics. All of them were positive for the native viral protein by indirect immunofluorescence and immunoperoxidase assays.

The ability of the MAbs to compete for binding to the N protein was tested by ELISA competition. The observed competition values were used to construct a preliminary antigenic map of the protein (Table 2). MAbs 1CH5 and 1DH10 showed a complete reciprocal competition, suggesting that the corresponding epitopes are very close or overlapping. The same complete competition was observed among 1EB9 and 1BD11 and among 1AC7, 1DA4 and 1AG11, respectively. In summary, the results were indicating the

Table 1. Characteristics of the N protein-specific MAbs

Hybridome	Isotype	ELISA (titer)		Immunoblotting	
		Virus	ORF7	Olot/91	Quebec 807/94
1DH10	IgG2a	10µg/ml	100ng/ml	–	–
1CH5	IgG2b	1µg/ml	10ng/ml	+/–	–
1EB9	IgG1	10ng/ml	1ng/ml	+	–
1BD11	IgG2a	10ng/ml	1ng/ml	+	–
1AG11	IgG1	1µg/ml	10ng/ml	+	+
1AC7	IgG1	1µg/ml	10ng/ml	+	+
1DA4	IgG1	1µg/ml	10ng/ml	+	+

existence of three different antigenic sites. These epitopes are quite distinct as indicated by the extremely low or null degree of competition among the MAbs of the different groups. Next step was to find the amino acid sequence corresponding to these epitopes.

4.2. Expression of Different PRRSV ORF7 Fragments in *E.coli*

A product of 370 bp containing the complete N protein coding sequence of PRRSV Quebec 807/94 isolate was obtained after RT-PCR amplification of the viral RNA. After cDNA sequencing and translation to protein, it showed a high amino acid identity (97.9%) with the Quebec isolate IAF-exp91. The only change consisted of Thr^5 by Asn^5. With the North American prototype isolate VR2332 (Collins et al., 1992) there was a homology of 95.2% (11 mismatches out of 123 residues). The homology with the European isolate Olot/91 was much lower (i.e. 61.2%).

Seven pET-derived recombinant plasmids were constructed containing the Olot/91 ORF7 and overlapping fragments of the N protein with various lengths from 51 to 237 nt, as shown in Fig.1. *E. coli* BL21 cells were transformed with the respective plasmids and induced for protein expression for 3h. The fusion proteins were all abundantly produced, from 0.02 mg of fragment C to 0.1 mg per ml of culture for most of the fragments. To con-

Table 2. Mapping of antigenic sites on the PRRSV N protein

PURIFIED MAbs	BIOTINILATED MAbs							ANTIGENIC SITE	N-FRAGMENT [2] REACTIVITY
	1DH10	1CH5	1EB9	1BD11	1AG11	1AC7	1DA4		
1DH10	+[1]	+						I	N PROTEIN
1CH5	+	+							
1EB9			+	+				II	N PROTEIN
1BD11			+	+					A, B
1AG11					+	+	+		
1AC7					+	+	+	III	N PROTEIN
1DA4					+	+	+		A, B, E, F

1. MAbs showing more than 80% competition
2. Reactivity was determined by western blot and ELISA.

firm the identity of the fusion proteins, fragments were probed with PRRSV-positive pig antisera by immunoblotting. The binding pattern of the pig sera was variable, most of the sera reacted equally with all the fragments, but not as well as with the complete N protein. However, in some cases the pig sera reacted strongly with fragments A, C and D, weakly with fragment B and failed to react with fragments E and F.

Previously, we have shown that only 5 out of 7 specific MAbs reacted strongly with the N protein of the European isolate by immunoblotting, MAb 1CH5 reacted weakly and 1DH10 did not recognize the recombinant N protein (Rodriguez et al., 1997). These two MAbs may recognize conformational epitopes not properly presented in the recombinant protein. Only MAbs 1AC7, 1AG11 and 1DA4 recognized the N protein from the North American isolate Quebec 807/94 expressed in *E.coli*. By indirect ELISA, the recombinant N protein Olot/91 was better recognized by all the MAbs, including 1DH10. Again, the Quebec 807/94 N protein was recognized by the same three MAbs. Since they recognize the N protein from both isolates, European and North American, they must be reacting with a common epitope.

4.3. Mapping of Antigenic Sites in the PRRSV N Protein

A fine epitope mapping of the PRRSV N protein was accomplished by probing the reactivity of the different fragments of the Olot/91 ORF7 gene expressed in *E.coli* with the MAbs. Using immunoblotting analyses, MAbs 1AC7, 1AG11, 1DA4 reacted with fragments A, B, E and F, defining antigenic site III. Fragment F, the smallest fragment recognized, comprises the amino acid sequence 50-PEKPHFPLAAEDDIRHH-66. MAbs 1BD11 and 1EB9 reacted with the complete N protein and with the larger fragments A and B, defining site II. They are probably recognizing a discontinuous epitope located in the first 78 residues of the N protein. Finally, MAbs 1CH5 and 1DH10 defined antigenic site I. The inability of these MAbs to bind the N protein by immunoblotting is suggesting that the recognized epitope must be discontinuous. By ELISA all of the MAbs reacted with the entire N protein. These data confirm that 1CH5 and 1DH10 recognize a discontinuous epitope and require the N protein to remain partially folded. Fragments C and D, containing the N and C terminus of the protein, respectively, were not recognized by any MAb, suggesting the absence of continuous antigenic sites in these places.

4.4. Application of the Recombinant N Protein, Fragments, and MAbs in Diagnosis of PRRSV

Finally, the utility of the recombinant polypeptides as diagnostic reagents was investigated. Both N proteins obtained from *E.coli* were well recognized by a collection of 48 European and 15 North American PRRSV-positive pig sera, respectively, by indirect ELISA. In all the cases the correlation with the routinely used immunoperoxidase monolayer assay (IPMA) was near 100%. In contrast, none of the N protein truncated fragments were recognized by the complete collection of positive sera.

To improve the specificity of the assay, a competitive binding assay was set up. The results of the blocking ELISA among the MAbs and several positive pig sera are shown in Table 3. Site I MAbs 1DH10 and 1CH5 binding was completely blocked by the pig sera. Site III MAbs, recognizing the conserved linear epitopes, were partially blocked (30–86%) and site II MAbs did not compete with the pig sera, suggesting that this site is very poorly recognized by the natural host of PRRSV. Since 1DH10 and 1CH5 are spe-

Table 3. Blocking ELISA among
MABs and PRRSV positive
pig sera

1DH10	100[1]
1CH5	99
1EB9	0
1BD11	6
1AG11	86
1AC7	30
1DA4	31

[1]Percentage of blocking

cific for the European strain, this assay would allow the discrimination among PRRSV
positive sera from different genotypes.

5. DISCUSSION

Two different antigenic groups have been described for PRRSV, which correspond
to the European and North American isolates. Newly developed vaccines based on live-at-
tenuated North American PRRS viruses are now being widely introduced in Europe. To
make a precise distinction of the origin of the virus, new diagnostic procedures have to be
developed in order to discriminate between both genotypes. These procedures will have to
rely on the use of marker proteins of PRRSV. The nucleocapsid protein has earlier been
demonstrated to be a suitable marker for PRRSV infection, since it is the most immunodominant antigen in the pig immune response to PRRSV (Plana-Durán et al., 1997). Unfortunately, the production and purification of PRRSV antigens from tissue culture is
troublesome and costly. To overcome this problem, we have described in this report the
expression of the N protein gene in *E.coli*, and their use in diagnosis, extending the analysis to map antigenic regions and to identify the precise sequence of the conserved epitopes
in both PRRSV genotypes.

The expression levels of the N protein were very high in *E.coli*, when expressed as a
large fusion protein of 46 kDa. After solubilization with 4M guanidinium hydrochloride
and without any further manipulation, the *E.coli*-derived protein could be used for diagnostic assays by simple dilution in PBS. Several truncated fragments of the N-protein
were also overexpressed in *E.coli* and used for the epitope mapping in combination with
seven specific MAbs.

From the results obtained, the MAbs could be divided into three groups, recognizing
at least three different epitopes on the PRRSV N protein. MAbs 1CH5 and 1DH10 recognized a discontinuous epitope (site I) on the N protein. 1CH5 and 1DH10 compete for the
same site and they were the only ones that compete efficiently with all the positive pig
sera in a competition ELISA, suggesting that this epitope is probably the most immunodominant PRRSV epitope in pigs. Group II MAbs, 1BD11 and 1EB9, may also recognize
a discontinuous epitope comprising several sequences inside the larger fragments A and B,
a region that covers the first 78 residues of the protein. Further attempts to delimitate this
epitope using shorter fragments (D, E) were unsuccessful. Site III is a continuous epitope,
as the MAbs react with the N protein after it has been denatured and with several fragments, including the short fragment F, which comprises residues 50–66. Neither site III

MAbs nor the group II MAbs were able to compete efficiently with all the PRRSV-positive pig sera in a competition ELISA assay.

Nelson et al. (1993), Wieczorek-Krohmer et al. (1996) and Dea et al., (1996) had reported common antigenic sites among different PRRSV isolates, but no attempts to define their amino acid composition were done. Here we define the sequence of the common antigenic site in the European and American PRRSV isolates, 3 out of 7 N protein-specific MAbs were directed to the epitope 50-PEKPHFPLAAEDDIRHH-66 in the middle of the N protein. Amino acid sequence comparisons of several PRRSV isolates show that the central region (aa 50 to 90) of the protein is well conserved. The prediction of secondary structure and solvent accessibility indicates that the epitope 50–66 shows a high probability of being part of a random structure with a good solvent accessibility, suggesting that this epitope is exposed on the surface of the N protein.

The reactivity of a collection of PRRSV-specific sera with the truncated fragments by indirect ELISA was weak in general, and about one third of the pig sera did not react with any of the fragments. Therefore fragment F, containing the common epitope, does not seem a suitable reagent for diagnostic purposes. The fact that the conserved, linear epitopes are not immunodominant would explain the lack of cross-reactivity for the N-protein among positive pig sera of different origin. This explanation is supported by the fact that MAbs 1CH5 and 1DH10, which compete efficiently with all the European-positive sera, recognize only the Olot/91 N protein, but not Quebec 807/94. The most immunodominant epitope, which is recognized by all the European pig sera and by the MAbs 1CH5 and 1DH10, is a variable and conformation-dependent epitope. As a consequence, only the entire N protein, expressed in *E.coli*, seems to be a suitable reagent for diagnostic purposes. It is expressed at very high levels and is easy to purify. In summary, the combination of the recombinant N protein and the MAbs 1DH10 and 1CH5 provides a powerful tool for a rapid and specific diagnosis of PRRSV and paves the way to a new test to differentiate infections caused by distinct PRRSV genotypes.

ACKNOWLEDGMENTS

This work was partially funded by a grant (351/96) of the Spanish Ministry of Industry and Energy.

REFERENCES

Collins, J.E., Benfield D.A., Christianson W.T., Harris J.C., Hennings, D.P., Shaw, S.M. Goyal, S.M., McCullough, S., Morrisson, R.B., Joo H.S, Gorcyca D. and Chladek D. W., 1992, Isolation of swine infertility and respiratory syndrome virus (isolate ATCC VR-2332) in North America and experimental reproduction in gnotobiotic pigs, *J. Vet. Diag. Inves.* **4**: 117–126.

Dea, S., Gagnon, C.A., Mardassi, H. and Milane G., 1996, Antigenic variability among North American and European strains of Porcine Reproductive and Respiratory Syndrome Virus as defined by monoclonal antibodies to the matrix protein, *J. Clin. Microbiol.* **34**: 1488–1493.

Harlow, E. and Lane, D., 1988, Antibodies, *A laboratory manual*, Cold Spring Harbor Laboratory, Cold Spring Harbor, New York.

Mardassi, H., Mounir, S. and Dea S., 1994, Identification of major differences in the nucleocapsid protein genes of a Québec strain and European strains of porcine and respiratory syndrome virus, *J. Gen. Virol.* **75**: 681–685.

Mardassi, H., Mounir S. and Dea S., 1994, Identification of major differences in the nucleocapsid protein genes of a Quebec strain and European strains of porcine reproductive and respiratory syndrome virus, *J. Gen. Virol.* **75**: 681–685.

Martinez-Torrecuadrada, J.L. and Casal J.I., 1995, Identification of a linear neutralization domain in the protein VP2 of African horsesickness Virus, *Virology* **210**: 391–399.

Meng, X.J., Paul, P.S, Halbur, P.G. and Lum, M.A., 1995, Phylogenetic analysis of the putative M (ORF6 and N (ORF7) genes of porcine reproductive and respiratory syndrome virus (PRRSV): implication for the existence of two genotypes of PRRSV in the US and Europe, *Arch. Virol.* **140**: 745–755.

Meng, X.J., Paul, P.S, Halbur, P.G. and Morozov I., 1996, Sequence comparison of ORFs 2 to 5 of low and high virulence United States isolates of porcine reproductive and respiratory syndrome virus, *J. Gen. Virol.* **76**:3181–3188.

Meulenberg, J.J., Hulst, M.M., De Meijer, E.J., Moonen, P.L.J.M., Den Besten, A.P., De Kluyver, E.P., Wensvoort, G. and Moormann, R.J.M., 1993, Lelystad Virus, the causative agent of porcine epidemic abortion and respiratory syndrome (PEARS), is related to LDV and EAV, *Virology* **192**: 62–72.

Meulenberg, J., Den Besten, A.P., De Kluyver, E.P., Moorman, R.J.M., Schaaper, W.M.M. and Wensvoort, G., 1995b, Characterization of proteins encoded by ORFs 2 to 7 of Lelystad Virus, *Virology* **206**: 155–163.

Meulenberg, J.J.M. and Petersen Den Besten, A., 1996, Identification and characterization of a sixth structural protein of Lelystad Virus: The glycoprotein GP$_2$ encoded by ORF2 is incorporated in virus particles, *Virology* **225**: 44–51.

Murtaugh, M.P., Elam, M.R. and Kakach L.T., 1995, Comparison of the structural protein coding sequences of the VR-2332 and Lelystad virus strains of the PRRS virus, *Arch. Virol.* **140**: 1451–1460.

Nelson, E. A., Christopher-Hennings, J., Drew, T., Wensvoort, G., Collins, J.E. and Benfield, D.A., 1993, Differentiation of United States and European isolates of porcine reproductive and respiratory syndrome virus by monoclonal antibodies, *J. Clin. Microbiol.* **31**: 3184–3189.

Plana-Duran, J., Vayreda, M., Vilarasa, J.L., Bastons, M., Rosell, R., Martinez, M., San Gabriel, A., Pujols, J., Badiola, J.L., Ramos, J.A and Domingo M., 1992, Porcine epidemic abortion and respiratory syndrome (mystery swine disease, Isolation in Spain of the causative agent and experimental reproduction of the disease, *Vet. Microbiol.* **33**, 203–211.

Plana-Duran, J., Climent, I., Sarraseca, J., Urniza, A., Cortés E., Vela C. and Casal, J.I., 1997, Baculovirus expression of proteins of Porcine Reproductive and Respiratory Syndrome Virus strain Olot/91. Involvement of ORF3 and ORF 5 proteins in protection, *Virus Genes*. **14**: 19–29.

Rodriguez, M.J., Sarraseca, J., Garcia, J., Sanz, A., Plana-Duran, J and Casal, J.I., 1997, Epitope mapping of the nucleocapsid protein of European and North American isolates of Porcine Reproductive and Respiratory Syndrome Virus, *J. Gen. Virol.* In press.

Rost, B., 1996, PHD: predicting one-dimensional protein structure by profile based neural networks, *Meth. Enzymol.* **266**: 525–539.

Sanger, F., Nicklen, S. and Coulson, A.R., 1977, DNA sequencing with chain terminating inhibitors, *Proc. Nat. Acad. Sci.* USA **74**: 5463–5467.

Sanz, A., Garcia-Barreno, B., Nogal, M.L., Viñuela, E. and Enjuanes, L., 1985, Monoclonal antibodies specific for African swine fever virus proteins, *J. Virol.* **54**: 199–206.

Studier, F.W., Rosenberg, A.H., Dunn, J.J. and Dubendorff, J.W., 1990, Use of T7 RNA polymerase to direct expression of cloned genes, *Met. Enzymol.* **185**: 60–89.

Suarez, P., Zardoya, R., Martín, M.J., Prieto, C., Dopazo, J., Solana, A. and Castro, J.M., 1996, Phylogenetic relationships of European strains of porcine reproductive and respiratory syndrome virus (PRRSV) inferred from DNA sequences of putative ORF-5 and ORF-7 genes, *Virus Res.* **42**: 159–165.

Terpstra, C., Wensvoort, G. and Pol, J.M.A., 1991, Experimental reproduction of porcine epidemic abortion and respiratory syndrome (mystery swine disease) by infection with Lelystad virus: KochĘs postulates fulfilled, *Vet. Q.* **13**: 131–136.

Van Nieuwstadt, A.P., Meulenberg, J.J.M., van Essen-Zanderbergen A., Petersen Den Besten, A., Bende, R.J., Moorman, R.J.M. and Wensvoort, G., 1996, Proteins encoded by open reading frames 3 and 4 of the genome of Lelystad virus (*Arteriviridae*) are structural proteins of the virion, *J. Virol.* **70**: 4767–4772.

Wensvoort, G., De Kluyver, E.P., Luijtze, E.A., Den Besten, A.P., Harris,L., Collins, J.E., Christianson, W.T. and Chladek, D., 1992, Antigenic comparison of Lelystad Virus and swine respiratory syndrome (SIRS) virus, *J. Vet. Diagn. Invest.* **4**: 134–138.

Wieczorek-Krohmer M., Weiland F., Conzelman, K., Kohl D., Visser N., Van Woensel, P., Thiel, H.-J. and Weiland, E., 1996, Porcine reproductive and respiratory syndrome virus (PRRSV): Monoclonal antibodies detect common epitopes on two viral proteins of European and U.S. isolates, *Vet. Microbiol.* **51**: 257–266.

61

INTRAHEPATIC $\alpha\beta$-TcR^{intermediate} LFA-1^{high} T CELLS ARE STIMULATED DURING MOUSE HEPATITIS VIRAL INFECTION

L. Lamontagne, E. Massicotte, and C. Page

Département des Sciences Biologiques
Université du Québec à Montréal
Montréal, Québec, Canada H3C 3P8

ABSTRACT

Mouse hepatitis virus type 3 (MHV3), a coronavirus, is an excellent animal model for the study of thymic and extrathymic T cell subpopulation disorders induced during the viral hepatitis. To understand local hepatic immune responses, the phenotypes of resident hepatic lymphocytes were determined and compared that of splenic and thymic T cell subpopulations during the acute viral hepatitis induced by MHV3 in susceptible C57BL/6 mice. Single positive (SP) CD4+ or CD8+ cells strongly increased in the liver. A specific cell population, the double positive (CD4+C8+) cells, normally present in liver and thymic cell preparations, decreased in C57BL/6 mice following the viral infection. $\alpha\beta$–TcR^{intermediate} T cells shifted toward $\alpha\beta$–TcR^{high} T cells in the liver and thymus of infected mice, but not in their spleen. The specific $\alpha\beta$–TcR$^{\text{int or high}}$ lymphocytes occurring in the liver of MHV3-infected mice expressed higher levels of leukocyte function antigen-1 (LFA-1) and Pgp-1 (CD44) activation markers, suggesting that they were either activated or antigen-experienced during the viral infection. No significant changes in T cell subpopulations were detected in the spleen. These observations suggest that MHV3 infection could induce an early *in situ* stimulation of resident hepatic T cells, despite a peripheral immunodeficiency in the thymus and spleen.

INTRODUCTION

Mouse hepatitis virus type 3 (MHV3), a coronavirus, is an excellent animal model for the study of hepatic T cell subpopulations during the virus-induced hepatitis. C57BL/6 mice, susceptible to MHV3 infection, develop an acute hepatitis and die within 3 days (Le Prévost *et al.*, 1975). We have previously reported the occurrence of lesions and cellular

Coronaviruses and Arteriviruses, edited by Enjuanes *et al.*
Plenum Press, New York, 1998

disorders in lymphoid organs, such as the thymus, spleen, and bone marrow of susceptible C57BL/6 mice acutely infected with MHV3 (Lamontagne *et al.*, 1989). Disorders in hepatic microcirculation induced by macrophage/monocyte procoagulant activity (PCA) under T cell control (Levy and Abecassis, 1989) have been proposed to explain the occurrence and extension of the hepatic lesions. No information, however, is available concerning the resident T cell subpopulations in the liver of MHV3-infected mice, or on their role in the hepatic pathogenic process.

Recent investigations have shown that the liver of normal adult mice contains both $\alpha\beta$–TcR and $\gamma\delta$–TcR cells, residing in hepatic sinusoids, characterized by an $\alpha\beta$–TcR of intermediate intensity ($\alpha\beta$–TcR^{int})(Ohteki *et al.*, 1992; Watanabe *et al.*, 1992). It was suggested that liver-resident $\alpha\beta$–TcR^{int} cells are important under pathological situations, such as in autoimmune diseases, malignancy or hepatic bacterial infections (Watanabe *et al.*, 1992; Abo *et al.*, 1991). However, no information is available concerning the role of $\alpha\beta$–TcR^{int} cell subpopulations during a virus-induced hepatitis.

In the present study, we analyzed the phenotypes of resident hepatic, splenic and thymic T cell subpopulations during the acute viral hepatitis induced by MHV3 in susceptible C57BL/6 mice. In contrast to that seen in the thymus or in the spleen, the number of hepatic $\alpha\beta$–TcR$^{low\ or\ int}$ shifted to $\alpha\beta$–TcR$^{int\ or\ high}$ T cells in infected mice. In addition, the specific hepatic $\alpha\beta$–TcR^{int}, LFA-1^{high} cell population increased in the liver of MHV3-infected mice.

MATERIALS AND METHODS

C57BL/6 mice were purchased from The Jackson Laboratory (Bar Harbor, ME). During the experiments, the female mice 8 to 12 weeks-old were housed in a sterile atmosphere (Forma Scientific, Marietta, OH). Groups of three mice were infected i.p. with 1000 TCID$_{50}$ of pathogenic MHV3, and sacrificed after 72 h p.i. Mock-infected mice received i.p. a similar volume of PBS. Mice were anesthetized with ketamine-sulphate (200mg/kg) and xylazine (10 mg/kg) by i.p. injection. Mice were bled by section of portal vein and aortic artery, as described by Watanabe *et al.* (1992). Liver, spleen and thymus were removed from dead mice after total bleeding.

The thymus and spleen were collected and pressed through a 70μm cell strainer (Becton-Dickinson, Lincoln Park, NJ) in RPMI with 20% FCS at room temperature. Cell preparations were electronically counted (Coulter Counter, Coulter Electronics, Hialeah, FL), and cell viability was assayed by the trypan blue exclusion test, ranging from 90 to 100%. To obtain MNC from the liver, groups of three livers were pressed through a 70μm cell strainer, which was then washed with 20mL of RPMI 1640 (GIBCO Laboratories, Grand Island, NY) contained 20% SVF and antibiotics. The suspension was then centrifuged on top of a discontinuous Percoll gradient (45%, 67% Percoll in PBS) for 1h at 2600 rpm. MNC were collected at the interface of the 45% and 67% Percoll layers. The cells collected at the interface of Percoll layers were then washed in RPMI 1640 containing 20% FCS and electronically counted.

Double immunolabeling was performed as follows: 10^{5}cells were resuspended in 500μL of RPMI 1640 containing 20% FCS, and incubated on ice for 1h with the optimal dilution of mAb anti-mouse CD4.FITC (clone RM-4) (Pharmingen, San Diego, CA), and mAb anti-mouse CD8.PE (clone 53–6.7)(Pharmingen). Procedure was similar for LFA-1 (mAb anti-mouse CD11a.FITC: clone M17/4, Pharmingen)/$\alpha\beta$–TcR (mAb anti-mouse $\alpha\beta$–TcR.PE: clone H57–597, Pharmingen) or CD44 (mAb anti-mouse CD44. PE: clone

IM7, Pharmingen)/αβ-TcR double labelings. Cytofluorometric analysis was done on a FacScan cytofluorometer with Lysis II Software (Becton-Dickinson). Gating was performed according to forward scatter *versus* 90° angle scatter (side scatter) so as to select the distinct mononuclear cell subpopulation. Percentage and absolute numbers were evaluated by the Student t test.

RESULTS

It was previously observed, in both human and mice, that percentages of hepatic T cell subpopulations differed from those of lymphoid organs (Watanabe *et al.*, 1992; Pham *et al.*, 1995). To verify if hepatitis outcome was associated with an intrahepatic immuno-disorder in the presence of inflammatory foci in the liver of infected mice, susceptible C57BL/6 mice were i.p. injected with 1000 $TCID_{50}$ of MHV3 virus (ten $TCID_{50}$ is equivalent to one LD_{50}). Mock-infected mice received i.p. a similar volume of PBS. Mice were anesthetized, bled, and the liver, spleen, and thymus were collected at 72 h p.i. The CD4-CD8 phenotypes of liver-resident lymphocytes from viral-infected or mock-infected C57BL/6 mice were analyzed and compared to those of splenic and thymic cell preparations. As shown in Table 1, absolute numbers of both SP CD4+ (p 0.001) and CD8+ (p 0.01) cells increased in the liver from MHV3-infected C57BL/6 mice whereas DP T cells decreased. Splenic SP CD4+ or CD8+ T cell subpopulations were not altered whereas thymic cells were strongly depleted (p 0.001) in spite of increase of SP CD4+ cells in infected animals (p 0.001).The number of cells isolated from the liver, however, is about 10 times less than in lymphoid organs. These observations suggest that hepatic MNC may be stimulated *in situ* rather than recruited from peripheral immune organs during viral hepatitis. In MHV3-infected animals, the hepatic CD4/CD8 ratio, normally higher than in peripheral lymphoid organs, was maintained through the increase of both CD4+ and CD8+ cells. Morever, viral infection slightly alters the CD4/CD8 ratio of splenic cells in infected C57BL/6 mice.

Matsumoto *et al.* (1994) have recently demonstrated the protective role of the hepatic αβ–TcR [int] T cells expressing a high level of LFA-1 marker during salmonellosis. To determine if αβ–TcR [int or high]LFA-1[high] T cells are induced in liver during MHV3 infection, C57BL/6 mice were infected with MHV3 for 72 hrs, and the hepatic, splenic and thymic mononuclear cells were double-labelled for the LFA-1 and αβ–TcR markers. As shown in Table 2, in mock-infected mice, the percentage of the αβ–TcR [int] LFA-1[high] cell population

Table 1. Absolute numbers (x 10^5) of CD4-CD8 phenotypes of hepatic, splenic and thymic T lymphocytes isolated from MHV3-infected C57BL/6 mice

Organ	Virus	CD4+CD8- *	CD4-CD8+	CD4+CD8+
Liver	Mock-infected	3.97 ± 0.32	0.99 ± 0.33	0.29 ± 0.01
	MHV3	13.16 ± 2.12***	2.67 ± 0.45***	0.17 ± 0.02***
Spleen	Mock-infected	92.06 ± 2.56	35.32 ± 2.47	N.A.
	MHV3	92.90 ± 5.31	25.29 ± 2.87	N.A.
Thymus	Mock-infected	81.19 ± 7.92	5.79 ± 0.54	743.3 ± 21.5
	MHV3	60.81 ± 9.87**	6.29 ± 2.01	161.9 ± 35.8***

*The cell preparations were double-labelled with FITC anti-CD4 mAb, and PE anti-CD8 mAb, and analyzed by flow cytometry. Data are means ± SD of three mice per group. These results are representative of three different experiments. **p 0.01; *** p 0.001; N.A. not applicable.

Table 2. Percentages of TcR-$\alpha\beta^{\text{int or high}}$ T cells, expressing high level of LFA-1 marker, isolated from liver, spleen and thymus from MHV3-infected C57BL/6 mice

Organ	Virus	TcR ^{int}	TcR $^{\text{high}*}$
Liver	Mock-infected	6.3 ± 2.9	15.2 ± 2.9
	MHV3	12.5 ± 1.2**	27.1 ± 6.5**
Spleen	Mock-infected	2.3 ± 0.9	10.4 ± 3.5
	MHV3	4.4 ± 2.7	17.0 ± 8.7
Thymus	Mock-infected	23.4 ± 9.6	2.4 ± 0.3
	MHV3	43.5 ± 6.8**	16.4 ± 3.9**

* The cell preparations were double-labelled with FITC anti-LFA-1 mAb, and PE anti-$\alpha\beta$TcR mAb, and analyzed by flow cytometry. Data are means ± SD of three mice per group. These results are representative of three different experiments. **p 0.001.

was higher among hepatic lymphocytes than in splenic cells. The $\alpha\beta$–TcR $^{\text{int or high}}$ LFA-1^{high} cell populations strongly increased after MHV3 infection in C57BL/6 mice (p 0.001). In addition, $\alpha\beta$–TcR^{high} LFA-1^{high} T cells was also seen in thymic cells from MHV3-infected mice (p 0.001). These results suggest that MHV3 infection induces an increase of $\alpha\beta$–TcR $^{\text{int or high}}$ LFA-1^{high} T cells in the liver, and to a lesser extent, in the thymus.

It was previously reported that many hepatic $\alpha\beta$–TcR ^{int} express IL-2Rβ and Pgp-1(CD44) cell surface markers, as seen on antigen-experienced T cells (Ohteki *et al.*, 1992), suggesting that these cells may be activated *in situ* . We have also analyzed CD44 expression in hepatic, splenic and thymic T cells from mock-infected and MHV3-infected C57BL/6 mice. Double staining for $\alpha\beta$-TcR and CD44 markers indicated that the hepatic TcR^{int}CD44^{high} cell population increased strongly in MHV3-infected mice (12.89 ± 3.52 to 20.30 ± 2.25%) (p 0.01), but not in the spleen or thymus (results not shown). These results suggest that activated T cells are generated *in situ* in the liver during viral infection.

DISCUSSION

We have demonstrated that MHV3 infection stimulates the intrahepatic CD4+ or CD8+ SP T cells in spite of immunodeficiencies in T lymphocytes in the thymus and spleen from susceptible C57Bl/6 mice. In addition, DP cell populations, present in liver and thymic cell preparations from mock-infected mice, decreased following viral infection in C57BL/6 mice. Such a stimulation results from a shift of TcR^{int} toward TcR^{high} T cells in the liver but not in the spleen from MHV3-infected mice. The specific hepatic $\alpha\beta$–TcR$^{\text{int or high}}$, LFA-1 ^{high} cell populations increased in the liver from MHV3-infected mice while LFA-1 molecules remained highly expressed in hepatic or thymic $\alpha\beta$–TcR$^{\text{int or high}}$ T cells.

MHV3 serotype possesses a tropism for several hepatic cells, such as hepatocytes, Kupffer, Ito and endothelial cells in susceptible mice (Pereira *et al.*, 1984). The pathogenic processes involved in the hepatitis are not clearly established but are not related directly to viral replication (Lamontagne *et al.*, 1989). On other hand, induction of procoagulant activity (PCA) correlated, rather, with resistance/susceptibility to infection in different mouse strains (Abecassis and Levy, 1989). We have also observed that MHV3 infection favors the expression of the LFA-1 and CD44 activation markers by hepatic lymphocytes indicating their ability to bind to ICAM-1 expressing cells. Some pathological observations suggest that the hepatocytes at inflammation sites express ICAM-1 antigen and that thrombin can also induce ICAM-1 expression in hepatic endothelial cells (Volpes *et al.*, 1990). Thus, the high level of CD44 expression in hepatic TcR[int or high] T cells from MHV3-infected mice indicates the ability of such hepatic lymphoid cells, possibly activated, to interact with endothelial cells (Shumizu *et al.*, 1992). Stromal thymic or hepatic endothelial cells or macrophages may therefore act as target cells for viral infection and, at the same time, act as antigen-presenting cells able to activate liver-resident T cells. Hepatic CD44 [high] T cells may thus be involved in the PCA activation.

REFERENCES

Abo, T., Ohteki, T., Seki, S., Koyamada, N., Yoshikai, Y., Masuda, T., Rikiish, H., and Kumagai, K., 1991, The appearance of T cells bearing self-reactive T cell receptor in the liver of mice injected with bacteria, *J. Exp. Med.* **174**:417–423.

Lamontagne, L., Descoteaux, J.P., and Jolicoeur, P., 1989, T and B lymphotropisms of mouse hepatitis virus 3 correlate with viral pathogenicity, *J. Immunol.* **142**:4458–4467.

Le Prévost, C., Levy-Leblond, B., Virelizier, J.L., and Dupuy, J.M., 1975, Immunopathology of mouse hepatitis virus type 3 infection. I. Role of humoral and cell-mediated immunity in resistance mechanism, *J. Immunol.* **114**:221–225.

Levy, G.A., and Abecassis, M., 1989, Activation of the immune coagulation system by murine hepatitis virus strain 3, *Rev. Infect. Dis.* **11**:712–727.

Matsumoto, Y., Emoto, M., Usami, J., Maeda, K., and Yoshikai, Y., 1994, A protective role of extrathymic ab TcR cells in the liver in primary murine salmonellosis, *Immunology* **81**:8–10.

Ohteki, T., Okuyama, R., Seki, S., Abo, T., Sugiura, K., Kusumi, A., Ohmori, T., Watanabe, H. and Kumagai, K., 1992, Age-dependent increase in extrathymic T cells in the liver and their appearance in the periphery of older mice, *J. Immunol.* **149**:1562–1567.

Pereira, C.A., Steffan, A.M., and Kirn, A., 1984, Interaction between mouse hepatitis virus and primary cultures of Kupffer and endothelial liver cells from resistant and susceptible inbred mouse strains, *J. Gen. Virol.* **65**:35–41.

Pham, B.N., Martinot-Peignoux, M., Mosnier, J.F., Njapoum, C., Marcellin, P., Bougy, F., Degott, C., Erlinger, S., Cohen, J.H., and Degos, F., 1995, CD4+/CD8+ ratio of liver-derived lymphocytes is related to viraemia and not to hepatitis C virus genotypes in chronic hepatitis C, *Clin. Exp. Immunol.* **102**:320–356.

Shimizu, Y., Newman, W., Tanaka, Y., and Shaw, S., 1992, Lymphocyte interactions with endothelial cells, *Immunol. Today* **13**: 106–108.

Volpes, R., Van Den Oord, J.J., and Desmet, V.J., 1990, Immunohistochemical study of adhesion molecules in liver inflammation, *Hepatology* **12**:59–68.

Watanabe, H., Ohtsuka, K., Kimura, M., Ikarashi, Y., Ohmori, K., Kusumi, A., Ohteki, T., Ski, S., and Abo, T., 1992, Details of an isolation method for hepatic lymphocytes in mice, *J. Immunol. Meth.* **146**:145–154.

62

CLONAL DELETION OF SOME Vβ+ T CELLS IN PERIPHERAL LYMPHOCYTES FROM C57BL/6 MICE INFECTED WITH MHV3

S. Gagne,[1,2] L. Thibodeau,[1] and L. Lamontagne[2]

[1]Centre de Recherche en Virologie
Institut Armand-Frappier
Laval, Québec, Canada H7N 4Z3
[2]Département des Sciences Biologiques
Université du Québec à Montréal
Montréal, Québec, Canada H3C 3P8

ABSTRACT

Mouse hepatitis virus type 3 infection is generally accompanied by a severe immune dysfunction involving thymic or splenic T cell subpopulations. We postulate that the peripheral lymphoid cell depletions were caused by a selective deletion of some Vβ subsets of mature T cells, as observed with superantigens. We have examined the expression of Vβ6, Vβ8 and Vβ14 in T cell subpopulations from the spleen and lymph nodes of pathogenic L2-MHV3-infected C57BL/6 mice. Cytofluorometric study showed decreases in splenic Vβ8+, Vβ6+, and Vβ14+ T cell subpopulations at 72 hrs post-infection. Single positive CD4+ T cells were diminushed but not the CD8+ cells. In contrast, the various Vβ splenic cell populations were not modified in mice infected with a non- pathogenic YAC-MHV3 variant. However, the Vβ8/CD4 ratio increased in splenic cells but decreased in lymphocytes from lymph nodes. The Vβ14/CD4 ratio decreased only in splenic cells while Vβ6/CD4 ratios were not modified. These results suggest that alterations in Vβ cell populations may play a role in the L2-MHV3-induced immunodeficiency.

INTRODUCTION

Mouse hepatitis virus type 3 (MHV3), a coronavirus, is an excellent animal model for the study of T cell subpopulation disorders during the virus-induced hepatitis. Susceptible C57BL/6 mice infected with pathogenic L2-MHV3 virus develop an acute hepatitis

Coronaviruses and Arteriviruses, edited by Enjuanes *et al.*
Plenum Press, New York, 1998

and die within 3 days whereas A\J mice support a subclinical infection with viral clearance occurring in the first week. The non-pathogenic YAC-MHV3 variant, generated during L2-MHV3 infection in YAC lymphoid T cell line, induces only a subclinical infection, with a rapid viral clearance (Lamontagne *et al.*, 1989). We have previously reported the occurrence of thymic atrophy and depletion of T cell subpopulation in the thymus and spleen from C57BL/6 mice acutely infected with the pathogenic L2-MHV3 strain but not with non-pathogenic YAC-MHV3 variant (Lamontagne and Jolicoeur, 1991).

Perturbations in the structural diversity of the TcR Vβ complex have been reported as altering the genetic sensitivity to viral diseases (Rodriguez *et al.*, 1992). Decreases by apoptosis or anergy of some specific Vβ families has been suggested as a mechanism to explain HIV-induced immunodeficiencies (Imberti *et al.*, 1991). This hypothesis is supported by the fact that murine retroviruses encode proteins acting as superantigens that induce activation and subsequent deletion of CD4+ T cell subsets expressing specific Vβ elements (Acha-Orbea *et al.*, 1991).

In this work, we have observed a correlation between the deletion of Vβ6, 8 or 14 T cell subpopulations in spleen and lymph nodes and pathogenicity of L2-MHV3 and YAC-MHV3 variants, suggesting a specific role of viral proteins in such deletions.

MATERIALS AND METHODS

C57BL/6 mice were purchased from The Jackson Laboratory (Bar Harbor, ME). During the experiments, the female mice of 8 to 12 weeks-old were housed in a sterile atmosphere (Forma Scientific, Marietta, OH). Groups of three mice were infected i.p. with 1000 $TCID_{50}$ of pathogenic L2-MHV3 or non-pathogenic YAC-MHV3, and sacrificed after 72 h p.i. Mock-infected mice received i.p. a similar volume of MEM. Mice were killed by CO_2 inhalation and spleen and lymph nodes were collected.

The spleen and lymph nodes were pressed through a 70μm cell strainer (Becton-Dickinson, Lincoln Park, NJ) in RPMI with 20% FCS at room temperature, and cell preparations were electronically counted (Coulter Counter, Coulter Electronics, Hialeah, FL). Double immunolabeling was performed as follows: 10^6 cells were resuspended in 500μL of RPMI 1640 containing 20% FCS, and incubated on ice for 1h with the optimal dilution of mAb anti-mouse Vβ6. FITC (clone RR4–7), Vβ8.FITC (clone MR5–2), or Vβ14.FITC (clone 14–2), and anti-mouse CD4.FITC (clone RM-4) (Pharmingen, San Diego, CA), or CD8.PE (clone 53–6.7)(Pharmingen). Cytofluorometric analysis was done on a FacScan cytofluorometer with Lysis II Software (Becton-Dickinson). Gating was performed according to forward scatter *versus* 90° angle scatter (side scatter) so as to select the distinct mononuclear cell subpopulation. Percentage and absolute numbers were evaluated by the Student t test.

RESULTS

To verify if the T cell immunodeficiency induced by the pathogenic L2-MHV3 virus in susceptible C57BL/6 mice results from a depletion of specific Vβ subpopulations, groups of three mice were infected with L2-MHV3 or non-pathogenic YAC-MHV3. At 72 hrs post-infection (p.i.), spleen and lymph nodes were collected and percentages of CD4+CD8- or CD4-CD8+ cells expressing Vβ6, Vβ8, or Vβ14 were analyzed. As shown in Table 1, splenic CD4+Vβ6, CD4+Vβ8, or CD4+Vβ14 T cells decreased in L2-MHV3-infected mice whereas no perturbation among these Vβ families was observed in YAC-

Table 1. Percentages of splenic CD4+ or CD8+ expressing Vβ6, Vβ8, or Vβ14 from L2-MHV3 or YAC-MHV3-infected C57BL/6 at 72 hrs p.i.

| Organ | T cell | Vβ | Pathogenic L2-MHV3 | | Non-pathogenic YAC-MHV3 | |
			Mock-infected mice	Infected mice	Mock-infected mice	Infected mice
Spleen	CD4+CD8-	Vβ6	1.29 ± 0.20^a	$0.42 \pm 0.02**$	1.29 ± 0.20	1.02 ± 0.45
		Vβ8	2.44 ± 0.31	$1.20 \pm 0.08**$	3.05 ± 0.19	2.35 ± 0.69
		Vβ14	1.63 ± 0.14	$0.34 \pm 0.02**$	2.06 ± 0.67	1.84 ± 0.23
	CD4-CD8+	Vβ6	0.40 ± 0.05	0.64 ± 0.28	0.65 ± 0.05	0.59 ± 0.15
		Vβ8	1.37 ± 0.54	1.40 ± 0.27	1.97 ± 0.19	1.89 ± 0.10
		Vβ14	0.29 ± 0.21	0.42 ± 0.21	0.51 ± 0.04	0.44 ± 0.07
Lymph nodes	CD4+CD8-	Vβ6	3.27 ± 0.50	$1.73 \pm 0.01**$	3.13 ± 0.16	2.82 ± 0.60
		Vβ8	6.83 ± 0.58	$2.93 \pm 0.30**$	6.69 ± 0.67	6.04 ± 0.79
		Vβ14	3.42 ± 1.03	$1.95 \pm 0.71*$	3.67 ± 0.29	3.54 ± 0.26
	CD4-CD8+	Vβ6	1.67 ± 0.67	1.27 ± 0.35	2.03 ± 0.30	1.56 ± 0.22
		Vβ8	4.06 ± 0.84	2.17 ± 1.66	3.97 ± 0.18	3.59 ± 0.37
		Vβ14	1.75 ± 1.39	0.67 ± 0.23	1.75 ± 0.51	$1.06 \pm 0.24*$

[a]The cell preparations were double-labelled with FITC anti-Vβ mAb, and PE anti-CD4 or CD8 mAbs, and analyzed by flow cytometry. Data are means $\pm$ SD of three mice per group. These results are representative of three different experiments. * p 0.05, ** p 0.001.

MHV3-infected mice. No decrease was found, however, in the CD8+Vβ6, Vβ8, or Vβ14 T cells from both L2-MHV3 or YAC-MHV3-infected mice.

Similar analysis of Vβ families in lymph nodes shown that CD4+Vβ6 or CD4+Vβ8 T cells decreased more than CD4+Vβ14 lymphocytes in L2-MHV3 infected mice (Table 2). CD8+ Vβ families were not altered in L2-MHV3-infected mice. In YAC-MHV3-infected mice, however, no decrease in various Vβ populations both in CD4+ or CD8+ T cells were observed.

Decreases in the studied Vβ families may reflect specific disturbances in T cell subpopulations in L2-MHV3-infected mice but may also be a consequence of the global decrease of CD4+ T cells.This hypothesis have been verified in analyzing the ratio of each Vβ tested on the total CD4+ or CD8+ cells. As shown in Table 2, the relative percentage of Vβ8 cells among the splenic CD4+ increased whereas it decreased among CD4+ cells in lymph nodes from L2-MHV3-infected mice. The percentages of Vβ14+ cells, however, decreased only in the spleen from L2-MHV3-infected mice. These results indicate that Vβ

Table 2. Percentages of Vβ6, Vβ8, and Vβ14 cell subpopulations in CD4+ cells in spleen or lymph nodes from L2-MHV3 infected C57BL/6 at 72 hrs p.i.

Organ	Vβ	Mock-infected mice	L2-MHV3-infected mice
Spleen	Vβ6	6.85 ± 0.69^a	7.38 ± 0.36
	Vβ8	12.77 ± 0.87	$15.47 \pm 0.86*$
	Vβ14	8.21 ± 0.61	$4.47 \pm 0.21**$
Lymph nodes	Vβ6	8.27 ± 0.57	8.37 ± 0.80
	Vβ8	17.43 ± 0.29	$14.26 \pm 0.13**$
	Vβ14	8.65 ± 2.03	8.61 ± 2.26

[a]The cell preparations were double-labelled with FITC anti-Vβ mAb, and PE anti-CD4 mAb, and analyzed by flow cytometry. Data are means $\pm$ SD of three mice per group. These results are representative of three different experiments. *p 0.05, ** p 0.001.

families are differently altered in spleen and lymph nodes in spite of decrease of total CD4+ cells.

DISCUSSION

The L2-MHV3 virus induces an immunodeficiency in susceptible C57BL/6 mice leading to an acute hepatitis. We have shown that CD4+ Vβ6, Vβ8 or Vβ14 specifically increased or decreased during the viral infection with the pathogenic virus but not with the non-pathogenic YAC-MHV3 variant suggesting a role for some Vβ gene families in the pathogenic process. Specific stimulation of some Vβ cell subsets followed by an anergy or a clonal deletion is a characteristic feature of microbial superantigen (Huber *et al.* 1996). Specific decrease of Vβ6, 8 or 14 cell subsets in lymph nodes and increase in spleen from mice infected with L2-MHV3 present some similarities with effect of staphyloccal enterotoxin B (SEB) (Marrack and Kappler, 1990). The SEB reacts with murine T cells expressing members of the some Vβ+ gene family and induces an anergic state preceded by expansion of Vβ+ T cells within three days following SEB administration (Kawabe and Ochi, 1990; Surman *et al.*, 1994). The presentation of a superantigen requires MHC class II proteins without processing but the T-cell response to superantigens is not class II restricted (Huber *et al.*, 1996). Similarities between MHV3 infection and SEB administration suggest also that some viral proteins would be able to bind specifically to TcR Vβ8 or/and Vβ14 chains and elicit a lymphocyte stimulation. The closely-related non-pathogenic YAC-MHV3 variant does not act as the L2-MHV3 virus with the Vβ gene family studied, suggesting an important role of the specific Vβ deletions in the pathogenic process of hepatitis. We did not find any differences in the molecular weight, by western blotting, or in the neutralization sites on S protein, in using monoclonal antibodies, between L2-MHV3 or YAC-MHV3 viruses (results not shown). However, YAC-MHV3 cannot infect B lymphocytes or thymic stromal cells, as seen with the L2-MHV3 virus, although it can replicate in peritoneal macrophages (Jolicoeur and Lamontagne, 1989; Lamontagne *et al.*, 1991), suggesting modifications in surface cell proteins involved in the attachment to the receptor or in the internalization. Absence of superantigen activity associated with this variant is another indication that the YAC-MHV3 could not bind to external part of the TcR Vβ6, 8 or 14 chains.

The superantigen activity may modulate some viral diseases, i.e. in faciliting the viral transmission, in inducing a self-tolerance in the thymus or in activating T or B cells to produce growth factors for amplification of viral replication (Huber *et al.*, 1996; Rodriguez *et al.*,1992). In other way, the anergy or cell deletion resulting from superantigen activity may contribute to the viral-induced immunodeficiency. We have previously demonstrated that the pathogenic L2-MHV3 induces a profound T cell immunodeficiency but the mechanism involved is not yet elucidated (Lamontagne *et al.*, 1989). Superantigen activity of L2-MHV3 virus would be a mechanism contributing to the T cell deletions or anergy without viral replication or without processing of viral proteins by antigen-presenting cells. Further work is in progress to evaluate the identify the viral protein involved in the superantigen activity and the importance of this mechanism in the outcome of acute or chronic hepatitis.

REFERENCES

Acha-Orbea, H., Shakhov, A.N., Scarpellino, L., and Kolb, E., 1991, Clonal deletion of V beta 14-bearing T cells in mice transgenic for mammary tumor virus, *Nature* **350**:207–210.

Huber, B.T., Hsu, P.N., and Sutkowski, N., 1996, Virus-encoded superantigens, *Microbiol. Rev.* **60:**473–482.

Imberti, L., Sottini, A., Bettinardi, A., Puoti, M., and Primi, D., 1990, Selective depletion in HIV infection of T cells that bear specific T cell receptor, *Science* **254:**860–863.

Kawabe, Y., and A. Ochi, 1990, Selective anergy of Vb8+ CD4+ T cells in *Staphylococcus* enterotoxin-B-primed mice, *J. Exp. Med.* **172:** 1065–1070.

Lamontagne, L., Descoteaux, J.P., and Jolicoeur, P., 1989, T and B lymphotropisms of mouse hepatitis virus 3 correlate with viral pathogenicity, *J. Immunol.* **142:**4458–4467.

Lamontagne, L., and Jolicoeur, P., 1991, Mouse hepatitis virus 3 thymic cell interactions correlating with viral pathogenicity, *J. Immunol.* **146:**3152–3162.

Marrack, P.E., and Kappler, L., 1990, The staphylococcal enterotoxins and their relatives, *Science* **248:**705–711.

Rodriguez, M., Patick, A.K., Pease, L.R., and David, C.S., 1992, Role of T cell receptor Vb genes in Theiler's virus-induced demyelination of mice, *J. Immunol.* **148:**921–927.

Surman, S., Deckhut, A.M., Blackman, M.A., and Woodland, D.L., 1994, MHC-specific recognition of a bacterial superantigen by weakly reactive T cells, *J. Immunol.* **152:**4893–4902.

63

A SEROLOGICAL SURVEY OF HUMAN CORONAVIRUS IN PIGS OF THE TOHOKU DISTRICT OF JAPAN

N. Hirano and K. Ono

Department of Veterinary Microbiology
Iwate University
Morioka, Japan

1. ABSTRACT

A total of 2496 swine sera from 60 farms in the Tohoku District of Japan was examined for hemagglutination inhibiting (HI) antibodies to human coronavirus (HCV), swine hemagglutinating encephalomyelitis virus (HEV) and bovine coronavirus (BCV). HI antibodies to HCV OC43 strain and HEV 67N strain were highly prevalent with positivity rates of 91.4 and 82.1%, respectively, while the BCV Kakegawa strain was 44.2% positive. Farm D in Miyagi Prefecture showed the highest antibody titers to HCV OC43 strain with geometric mean titers (GMT) of 1:200. These results suggest that pigs might be infected with HCV or an antigenetically related virus as well as HEV.

2. INTRODUCTION

Coronaviruses infect a wide variety of animal species, including avian and also human beings, and causes respiratory disease, enteritis, hepatitis or encephalitis. Among the swine coronaviruses, transmissible gastroenteritis virus, porcine epidemic diarrhea virus and HEV are well known. Notably, HEV infection in pigs has been reported as globally common by serological surveys. In Japan (Hirai et al., 1974) it was first reported that about 50% of the swine sera collected were positive for HI antibodies to HEV 67N strain. In 1987 in Japan HEV was first isolated from piglets in farms showing about 50% seropositivity to the virus (Hirahara., 1987). To see the prevalence of antibodies to HEV, we undertook serological surveys for HEV in 60 farms of the Tohoku District by HI tests. In addition to HEV, we also surveyed for HCV and BCV in swine sera. The results showed that HI antibodies to HCV was higher in seropositivities and titers than those of HEV or BCV.

Coronaviruses and Arteriviruses, edited by Enjuanes *et al.*
Plenum Press, New York, 1998

3. MATERIALS AND METHODS

3.1. Serum Samples

From 1985 to 1988, 2,469 sera were collected from pigs of 60 farms in Aomori, Iwate, Akita and Miyagi Prefectures. These sera were heated at 56^{0}C for 30 minutes, and mixed with an equal volume of 25% kaolin solution in phosphate buffered saline (PBS, pH 7.2). After centrifugation, the supernatant was mixed with an equal volume of 25% chicken red blood cell (CRBC) suspension in PBS and incubated at 20^{0}C for 30 min. After centrifugation, the serum samples were used as a 1:4 dilution of the original.

3.2. Virus

The virus strains used in this study were HCV OC436(HCV-OC43) , HEV 67N7(HEV-67N) and BCV Kakegawa (Hirano et al., 1985) (BCV-K) strains. As virus antigen, supernatant of infected mouse brain homogenates was used for HCV OC43, and infected culture fluid of SK-K for HEV 67N and of BEK-1 cells for BCV-K.

3.3. HI Test

The test was carried out in 96-well microplates using 0.5% CRBC in PBS. An HI titer of 1:8 or higher was recorded as positive.

4. RESULTS

HI Antibodies in Swine Sera to HCV-OC43, HEV-67N, and BCV-K

As shown in Table 1, 2,257 (91.4%), 2,028 (82.1%) and 1,092 out of 2,469 sera were positive for HCV-OC43, HEV-67N and BCV-K, respectively, in 4 Prefectures. The GMT for HCV-OC43, HEV-67N and BCV-K was 1:45, 1:43 and 1: 18. Among three viruses, HCV-OC43 showed the highest positive rate of 91.4% and a GMT of 1:45. The positive rates and the GMT for HCV-OC43 ranged from 86.2% (425/493) in Aomori to 94% (498/530) in Akita Prefecture, and from 1:37 in Akita to 1:59 in the Miyagi Prefecture, respectively.

Table 2 indicates swine farms showing the highest antibody titers for HCV-OC43 in each Prefecture. Farm A (Aomori), B (Iwate), C (Akita) and D (Miyagi Prefecture) had higher positive rates and antibody titers than those of HEV-67N or BCV-K. Farm D showed the highest GMT of 1:200 in 42 of 42 sera, whose antibody titers varied from 1:8 to 1:1028.

5. DISCUSSION

In 1974, Hirai et al conducted a serological survey on HEV infection in pigs in Japan, revealing that about 50% of all pigs in Japan and 27% those in the Tohoku District were positive for HEV-67N. In the present study, 82% of animals in the Toholu District were shown to be positive for HEV-67N. However, no clinical disease was observed in

Table 1. HI antibodies in swine sera to HCV-OC43, HEV-67N,
and BCV-K

Virus	Prefecture	Positive/tested	(%)	GMT
HCV-OC43	Aomori	425/493	(86.2)	52[a]
	Iwate	681/738	(92.3)	38
	Akita	498/530	(94.0)	37
	Miyagi	653/708	(92.2)	59
	Total	2257/2469	(91.4)	45
HEV-67N	Aomori	424/493	(86.0)	74
	Iwate	640/738	(86.2)	26
	Akita	337/530	(63.6)	29
	Miyagi	627/708	(82.1)	43
	Total	2028/2469	(82.1)	43
BCV-K	Aomori	320/493	(64.9)	25
	Iwate	258/783	(35.0)	13
	Akita	102/530	(19.2)	15
	Miyagi	412/708	(58.2)	17
	Total	1092/2469	(44.2)	18

[a]Reciprocal titer

any of the 60 farms examined. These results indicate that HEV might cause as inapparent form of infection in pigs.

Interestingly, HI antibodies to HCV-OC43 were detected in pigs at a higher incidence than antibodies to HEV-67N and BCV-K. Farm D in the Miyagi Prefecture showed the highest GMT of 1:200, suggesting that HCV or an antigenitically related virus might be responsible for HI antibodies to HCV-OC43 in 42 pigs. Kaye et al reported antibody responses to HCV-OC43 and HEV-67N in human and animal sera, and demonstrated that sera from veterinary students and meat producers showed higher titers of HI antibodies to HEV-67N than those of college students, suggesting the possibility of HEV exposure in humans.

The present study demonstrated that swine sera showed high antibody titers and seropositiviticates for HCV-OC43 as well as HEV-67N, suggesting the possibility that HCV or antigenitically related virus might infect pigs.

Table 2. Swine farms showing high antibody titers to HCV-OC43

Farm	(Prefecture)	Virus	Positive/tested (%)	GMT
A	(Aomori)	HCV-OC43	30/30 (100)	75
		HEV-67N	25/30 (83.3)	23
		BCV-K	13/30 (43.3)	9
B	(Iwate)	HCV-OC43	120/120 (100)	95
		HEV-67N	95/120 (79.2)	20
		BCV-K	39/120 (32.5)	10
C	(Akita)	HCV-OC43	19/19 (100)	59
		HEV-67N	15/19 (78.9)	22
		BCV-K	0/19	
D	(Miyagi)	HCV-OC43	42/42(100)	200
		HEV-67N	42/42(100)	36
		BCV-K	42/42(100)	19

[a]Reciprocal titers

REFERENCES

Hirai, K., Chang, C. N. and Shimakura, S., 1974, A serological survey on hemagglutinating encephalomyelitis virus infection in pigs in Japan, *Jpn. J. Vet. Sci.* **36**: 375–382.

Hirahara, T., Yasuhara, H., Kodama, K., Nakai, M. and Sasaki, N., 1987, Isolation of hemagglutinating encephalomyelitis virus from respiratory tract of pigs in Japan, *Jpn. J. Vet. Sci.* **49**: 85–93.

Hirano, N., Ono, K., Takasawa, H., Murakami, T. and Haga, S. J., 1990, Replication and plaque formation of swine hemagglutinating encephalomyelitis virus (67N) in swine cell line, SK-K culture, *J. Virol. Meth.* **27**: 91–100.

Hirano, N., Sada, F., Tsuchiya, K., Ono, K. and Murakami, T., 1985, Plaque assay of bovine coronavirus in BEK-1 cells, *Jpn. J. Vet. Sci.* **47**: 679–681.

Kaye, H. S., Yarbrough, W. R., Reed, C. J. and Harrison, A. K. J., 1977, Antigenic relationship between human coronavirus strain OC 43 and hemagglutinating encephalomyelitis virus strain 67N of swine: antibody responses in human and animal sera, *Inf. Dis.* **135**: 201–209.

McIntosh, K., Dees, J. H., Becker, W. B., Kapikian, A. Z. and Chanock, R. M., 1967, Recovery in tracheal organ cultures of novel viruses from patients with respiratory disease, *Proc. Natl. Acad. Sci. USA* **57**: 933–940.

Mengeling, W. L., Boothe, A. D. and Richte, A. E., 1972, Characteristics of a coronavirus (Strain 67N) of pigs, *Am. J. Vet. Res.* **33**: 297–308.

64

A MONOCLONAL ANTIBODY BLOCKING ELISA FOR THE DETECTION OF IBV ANTIBODIES IN FOWL

V. Moving,[1] G. Czifra,[2] and L. Renström[2]

[1]SVANOVA Biotech
The National Veterinary Institute
Uppsala, Sweden
[2]Department of Poultry
The National Veterinary Institute
Uppsala, Sweden

1. ABSTRACT

A murine monoclonal antibody (mAb) reacting with the spike protein of seven field and laboratory strains of Infectious Bronchitis Virus (IBV) was characterised. Neutralisation tests performed in chicken tracheal organ culture showed that the mAb is directed against a conserved neutralising epitope. A monoclonal antibody blocking ELISA (B-ELISA) was developed based on the mAb. The sensitivity and specificity of the test was evaluated by examining sera and egg yolk from IBV-free, vaccinated or naturally infected chickens from different European countries. The comparisons showed that the IBV blocking ELISA was very sensitive and more specific than the commercially available indirect ELISA and the haemagglutination-inhibition (HI) test. As part of the Swedish national health control program, more than 60.000 sera have been examined with the B-ELISA at the National Veterinary Institute (Sweden) since 1993. The test proved to be very specific in detecting the spread of disease during the Swedish IBV outbreaks in 1994.

2. INTRODUCTION

Our aim was to develop and evaluate a new serological test for the detection of IBV specific antibodies in blood, serum and egg yolk.

Coronaviruses and Arteriviruses, edited by Enjuanes *et al.*
Plenum Press, New York, 1998

VIRUS	**mAb recognise the virus**
IBV-M41	yes
IBV D274 (dutch strain)	yes
IBV D1466 (dutch strain)	yes
IBV CR (French strain)	yes
IBV PL (French strain)	yes
IBV Archansas	yes
IBV Conneticut	yes
IBV Swedish isolate	yes
Egg proteins (control)	no
Other coronaviruses:	
BCV	no
MHV	weakly
TGEV	no
PRCV	no
Other avian pathogens:	
NDV	no
APV	no
Mycoplasma gallisepticum	no
Mycoplasma synoviae	no

Figure 1. Specificity of the mAb.

3. MATERIAL AND METHODS

3.1. Monoclonal Antibody

The virus strain IBV M41 was used to produce murine monoclonal antibodies (mAb). The selection of IBV specific antibody-producing hybridomas was based on an indirect ELISA with IBV M41 coated plates. Species specificity of the IBV positive supernatants was confirmed by indirect ELISA using other avian virus antigens (Figure 1). Anti-IBV mAbs directed to relevant epitopes, that induce specific antibody response in the host, were identified using the blocking ELISA. The specificity of the mAbs was further studied by SDS-PAGE immunoblotting, neutralisation in chicken tracheal organ culture, HI and a commercially available indirect ELISA. The mAb used in the IBV blocking ELISA was horseradish peroxidase (HRP) conjugated (Figure 2).

3.2. Neutralisation in Chicken Tracheal Organ Culture

The tracheal organ culture (TOC) was prepared from 19 day old specific pathogen free (SPF) chicken embryos following the method of Cook et al. (1976). The sera and the monoclonal antibody were kept at a constant dilution while the IBV M41 virus strain was titrated. Results were recorded on day 3 as ciliostasis. When the culture is IBV infected the cilia does no longer move while in uninfected cultures the cilia maintains the capability of movement. Calculation of endpoint and neutralisation index (NI) were performed according to Reed and Muench (1938).

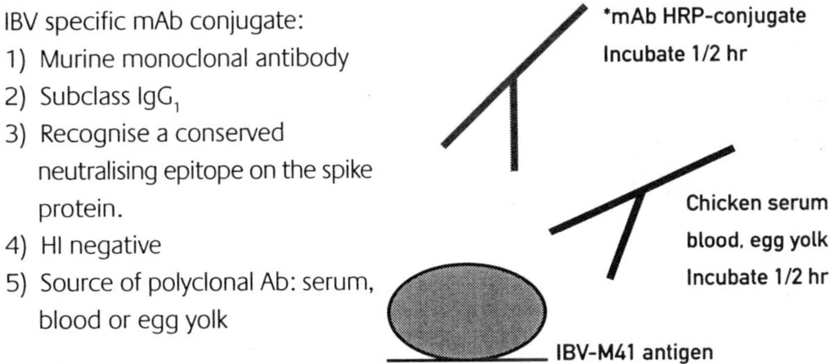

Figure 2. Diagram of the IBV blocking ELISA.

4. RESULTS

The IBV specific mAb in the IBV blocking ELISA kit recognise a conserved neutralising epitope on the spike protein. The neutralisation index of the monoclonal antibody was NI = 3.4, while the neutralisation index of a Swedish IBV positive serum pool was NI = 3.5. The mAb was classified as IgG_1 isotype using the Inno-Lia mouse mAb isotyping kit (Innogenetics, Belgium). The mAb was HI negative.

In indirect ELISA the mAb-conjugate recognised the ten IBV strains tested (Figure 1) and it did not recognise other avian pathogens as Newcastle (NDV), avian pneumovirus (APV), *M. gallisepticum, M. synoviae.* Of the coronaviruses tested it reacted weakly with MHV but did not recognise bovine coronaviurs (BCV), transmisible gastroenteritis virus (TGEV) and porcine respiratory coronaviurs (PRCV).

Blood, serum and egg yolk from Swedish infected poultry blocked the mAb conjugate very efficiently. The titration curves for a pool of Swedish IBV negative and positive sera are shown in Figure 3.

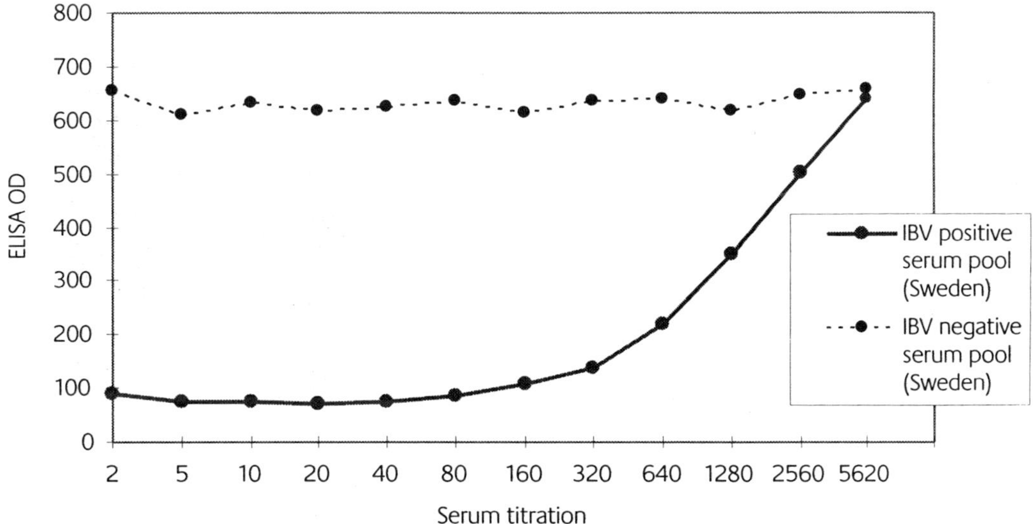

Figure 3. Titration of Swedish sera on IBV-M41 coated ELISA plates.

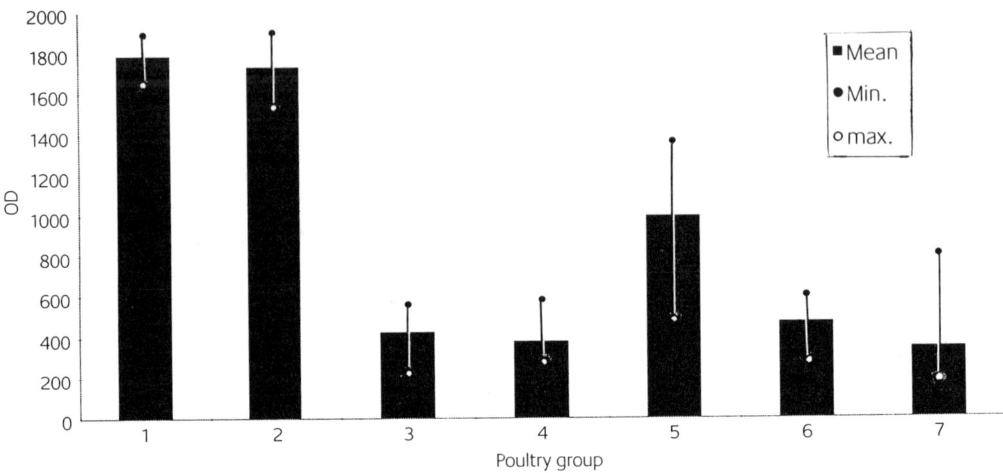

Figure 4. IBV blocking ELISA using sera from vaccination experiments in France. Sera source: 1, Swedish sera from 8 IBV negative chickens; 2, Hungarian sera from 6 NDV negative chickens; 3, Hungarian sera from 7 NDV positive and IBV vaccinated chickens; 4, French sera from 13 Adult layers vaccinated with IBV Live vaccine; 5, French sera from 20 SPF chickens vaccinated with IBV inactivated vaccine; 6, French sera from 20 Adult layers vaccinated with IBV inactivated vaccine; 7, French sera from 20 Field Adult Birds vaccinated with IBV Live and inactivated vaccine.

The results from a vaccination experiment in France are shown in Figure 4. Sera from the group of adult layers that received inactivated vaccine did not respond with an inhibition expresed in percentage (PI) as high as the other vaccinated groups.

When German chicken sera from different vaccination groups were tested only birds vaccinated with IBV had serum antibodies that inhibited the binding of the mAb conjugate in the IBV blocking ELISA (Figure 5). Sera from SPF chickens, NDV and *M. gallisep-*

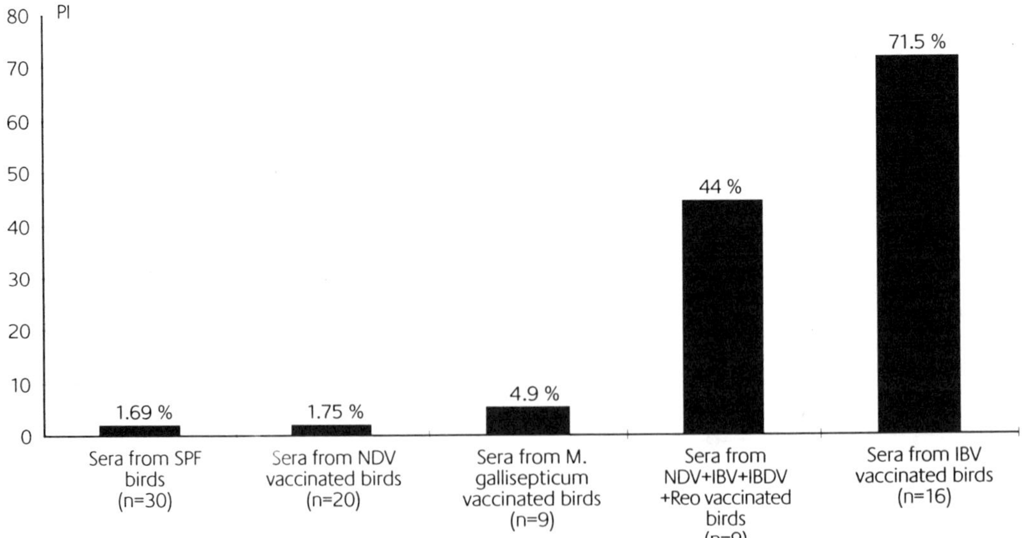

Figure 5. Analysis of German chicken sera from vaccinated and nonvaccinated groups tested in the Blocking ELISA.

	Blocking ELISA (SVANOVA)		
Frequency %	0	1	TOTAL
Indirect ELISA (IDEXX) 0	2 0.77	38 14.56	40 15.33
1	0 0.00	221 84.67	221 84.67
TOTAL	2 0.77	259 99.23	261 100.00

Figure 6. Comparison of the IDEXX-Group diagnosis test with the Blocking ELISA (SVANOVA) test.

ticum vaccinated birds only blocked the test by a percentage (PI) of 1 to 5 %. In birds that had received a mixture of NDV, IBV, IBDV, and Reo vaccine PI was 44%. In birds vaccinated with only IBV the PI was 71.5%.

During the 1994 IBV outbreaks in Sweden the specificity of the indirect ELISA from IDEXX was compared to the IBV blocking ELISA (Figure 6). Two sera out of the 261 sera tested were IBV negative and 221 sera IBV positive in both tests. No serum was IBV positive in IDEXX and negative in the blocking ELISA. On the other hand 38 sera were IBV positive in the blocking ELISA but negative in the IDEXX indirect ELISA. No correlation was observed between the IDEXX indirect ELISA and the IBV blocking ELISA, while a correlation between the HI test and the blocking ELISA was observed.

5. DISCUSSION

A mAb reacting with the spike protein of seven field and laboratory strains of IBV was characterised. The neutralisation test performed using chicken tracheal organ cultures showed that the mAb is directed against a conserved neutralising epitope.

The sensitivity and specificity of the blocking ELISA based on this mAb was evaluated with sera, blood and egg yolk from IBV-free, vaccinated or naturally infected chickens from different European countries. The comparison showed that the IBV blocking ELISA was more sensitive and specific than the commercially available indirect ELISA and the HI test.

As part of the Swedish National Health control program, more than 60.000 sera have been examined with the blocking ELISA at the National Veterinary Institute (Sweden) since 1993. The test proved to be very specific in detecting the spread of disease during the Swedish IBV outbreaks in 1994.

REFERENCES

Cook J. K. L., Darvyshire G. H. and Peters R. W., 1976, The use of chicken tracheal organ cultures for the isolation and assay of avian infectious bronchitis, *Arch. Virol.* **50**: 109–118.

Reed L. G. and Muench. H., 1938, A simple method for estimating fifty per cent endpoints, *Am. J. Hyg.* **27**: 493–496.

Pathogenesis II: Pathology

65

PATHOGENESIS OF CORONAVIRUS-INDUCED INFECTIONS

Review of Pathological and Immunological Aspects

S. Perlman

Department of Pediatrics and Microbiology and Interdisciplinary Program in
    Immunology
University of Iowa
Iowa City, Iowa 52242

1. ABSTRACT: MODELS OF CORONAVIRUS PATHOGENESIS

Coronaviruses and arteriviruses infect multiple species of mammals, including humans, causing diseases that range from encephalitis to enteritis. Several of these viruses infect domestic animals and cause significant morbidity and mortality, leading to major economic losses. In this category are included such pathogens as transmissible gastroenteritis virus, porcine respiratory and reproductive virus and infectious bronchitis virus. The feline coronaviruses (FECV) generally do not cause infections with high morbidity but in a small percentage of cases, the virus mutates to become more virulent. This virus, feline infectious peritonitis virus (FIPV), causes severe disease in young cats. This disease is in large part immunopathological and understanding it is a major goal of coronavirus research.

2. INTRODUCTION TO MHV-INDUCED NEUROLOGICAL DISEASE

The most commonly studied coronavirus is mouse hepatitis virus (MHV), in part because the natural host for this infection, the mouse, is more easily managed in the laboratory. Although, as illustrated throughout this symposium, much progress has been made in understanding the pathogenesis of the infections described above, this review will concentrate on MHV and in particular, the neurotropic strains of this virus. The first strain of mouse hepatitis virus was isolated in 1947 from a paralyzed mouse in a colony of Swiss

Coronaviruses and Arteriviruses, edited by Enjuanes *et al.*
Plenum Press, New York, 1998

white mice at Harvard Medical School (Cheever, et al., 1949). This virus initially caused hindlimb paralysis, but on repeated passage through mice, more virulent variants, which predominantly caused encephalitis were selected. This virus was named the JHM virus (JHMV) and subsequently shown to be a coronavirus, related to other MHV strains. The infection caused by this virus was not extensively investigated until Weiner described a study in which JHMV-induced demyelination was analyzed (Weiner, 1973).

JMHV is now used in many laboratories to study virus-induced neurological diseases, particularly demyelination, and in all cases, these viruses were derived from the virus originally isolated in 1947. Both mice, the natural host for this virus and rats, a species not naturally susceptible to MHV but which can be infected after intracerebral inoculation, are used. In both species, a hallmark of the infection is persistence in nearly all rodents that survive the acute infection. The JHMV used in different laboratories are not identical, however, as changes have occurred in the genome during the course of propagation at different geographical locales. In addition, different strains of mice and rats display different susceptibilities to infection with JMHV. This difference was most apparent in adult SJL mice, a strain that is resistant to JHMV (Stohlman and Frelinger, 1978). This resistance was later used to isolate the host cell receptor used by JHMV to infect susceptible cells (Williams, et al., 1991).

Other factors, such as route of inoculation and age of the host at the time of inoculation also affect disease manifestation and outcome. For example, after intracerebral inoculation with relatively large volumes of virus, JHMV spreads via neurons, the cerebrospinal fluid and even the blood. For the closely related and mildly neurotropic A59 strain of MHV (MHV-A59) intracerebral inoculation results in a hepatitis that may be severe enough to cause death (Lavi, et al., 1986). Delivery of the virus by this route also compromises the blood-brain barrier and this might affect the ultimate disease outcome. On the other hand, intranasal inoculation results in a CNS infection in which virus spreads only transneuronally within the CNS (Lavi, et al., 1988; Barnett and Perlman, 1993) but in which the initial process of inoculation may lead to an infection of the respiratory tract, resulting in pneumonia. These studies on route of inoculation also show that unless virus is injected directly into the brain, JHMV (and probably MHV-A59) enters the central nervous system (CNS) only via the olfactory nerve. Even after intraperitoneal inoculation in suckling mice, the pattern of virus spread within the brain is consistent with spread from the olfactory nerve (unpublished observations).

Further complicating studies using JHMV is the high rate of virus variation that arises during the course of an infection. Variability occurs as a consequence of the high error rate of the coronavirus polymerase, common to the polymerases of all RNA viruses since they lack proofreading ability and also as a consequence of the high recombination rate exhibited by all coronaviruses. The RNA molecules within a single cell are not identical but consist of a group of quasispecies. This distribution of similar, but not identical viruses, facilitates virus selection and allows the virus to adapt more readily to changes in the environment. These adaptive qualities of JHMV has been illustrated in several recent sets of studies (e.g., Chen and Baric, 1996).

3. MURINE MODELS OF MHV-INDUCED DEMYELINATION

As described above, the strain of JHMV that arose after many passages through the brains of susceptible mice was virulent and caused a high mortality. Demyelination was observed in a few survivors and since the pathogenesis of this process was deemed most interesting, efforts were made to increase the number of surviving mice with demyelination. The first approach was to isolate JHMV mutants that caused less encephalitis and more demyelination. The great variability exhibited by JHMV made this process relatively straightforward and several mutants were isolated that were attenuated in their ability to infect neurons (and thus caused less encephalitis) but continued to infect glial cells, resulting in the same amount or more demyelination. These mutant viruses include temperature-sensitive mutants, variants isolated from the original suckling mouse brain pool, monoclonal antibody-resistant variants and variants with deletions in the hypervariable region of the S protein (Haspel, et al., 1978; Stohlman, et al., 1982; Dalziel, et al., 1986). Some of these variants preferentially appeared to infect astrocytes or oligodendrocytes, resulting in different levels of persistence and demyelination. A second approach was to use virulent virus, but protect neurons from an acute infection by passive infusion of neutralizing antibodies or T cells (Buchmeier, et al., 1984; Stohlman, et al., 1986; Yamaguchi, et al., 1991). Under these conditions, the virus persisted in the white matter and was able to cause demyelination. In all of these models, mice are sickest at early times post inoculation (p.i.) and if they survive the acute infection, they recover clinically. Consistent with these clinical observations, infectious virus usually can not be cultured after 15–21 days p.i., but viral antigen or RNA can be detected, sometimes only using very sensitive methods such as PCR.

In one model, hindlimb paralysis develops several weeks after inoculation and infectious virus can be consistently isolated from these mice (Perlman, et al., 1987). In this case, suckling C57Bl/6 mice are inoculated intranasally with a virulent strain of JHMV. They are protected from acute encephalitis by nursing by dams previously immunized with live MHV-JHM. A variable fraction (40–90%) develop hindlimb paralysis with histological evidence of demyelination 3–8 weeks after inoculation. When suckling BALB/c mice are inoculated under similar conditions, they are protected from acute encephalitis but do not develop hindlimb paralysis at later times. This murine model differs from the ones described above in that mice remain well until they develop symptoms a few weeks after inoculation and in that infectious virus can be isolated. A recent set of experiments provided an explanation for some of these results (Pewe, et al., 1996). The CD8 T cell response is critical for clearance in most noncytopathic viral infections and although JHMV behaves as a lytic virus in tissue culture cells, in vivo it does not appear to be cytopathic in some cell types. As described below, the target epitopes for anti-MHV CD8 T cells has been identified in BALB/c and C57Bl/6 mice. In each strain, one epitope is dominant and in the case of C57Bl/6 mice, this epitope is located in a region of the S protein previously shown to be hypervariable (Parker, et al., 1989). When the virus isolated from the CNS of C57Bl/6 mice with hindlimb paralysis is sequenced, mutations are detected in every case in this CD8 T cell epitope. These changes abrogate recognition by CNS-derived lymphocytes in direct ex vivo cytotoxicity assays and thus behave like CTL escape mutants. Since these changes arise early during the infectious process, they are likely to contribute to the initiation of the process of persistence in this strain of mouse.

4. INFECTION OF RATS WITH JHMV

Although rats are not a natural host for JHMV, young rats can be infected if virus is delivered by intracerebral inoculation (Sorensen, et al., 1980; Watanabe, et al., 1987). In some strains, a fraction of these mice later develop subacute demyelinating encephalomyelitis (SDE) with clinical signs of paralysis. Several features of this model are unique. First, ongoing demyelination is detected in the presence of minimal amounts of viral antigen in Lewis rats. This is consistent with a previous report showing that T cells harvested from rats with SDE are able to cause neuropathological abnormalities after adoptive transfer into naive mice (Watanabe, et al., 1983). Second, other strains of rats inoculated with JHMV also develop SDE and in some cases, infectious virus can be isolated from symptomatic, but not asymptomatic animals. This ability to isolate infectious virus is the same as described above for the maternal antibody protection model.

5. SITES OF PERSISTENCE

Much of the experimental data obtained thus far suggest that the host's ability to clear MHV is a key step in the development of demyelination. Almost invariably, if an immunologically intact host is unable to clear the virus, demyelination results. Virus persistence is known to occur in the white matter of the spinal cord and brain although the molecular basis of this persistence is not understood. Virus can be identified in astrocytes in asymptomatic mice suggesting that this cell type serves as a reservoir in MHV-infected mice (Perlman and Ries, 1987). It is not known if microglia or oligodendrocytes can also serve as reservoirs for virus in asymptomatic mice. It will be particularly fruitful to investigate the role of microglia in this process. Identification of microglia is straightforward since good cell markers are available and there is precedent for microglia serving as a reservoir for virus in other models of virus-induced demyelination (Lipton, et al., 1995). Oligodendrocytes are readily identified as infected in MHV-infected rodents with either acute encephalitis or symptomatic, chronic demyelination because there is a large viral load under these conditions (Lampert, et al., 1973; Weiner, 1973; Sun, et al., 1995). However, in asymptomatic mice, in which virus load is low, the lack of reliable markers for oligodendrocytes makes determining whether this cell type is infected more difficult.

A summary of some features of commonly used MHV models of demyelination is shown in Table 1.

6. PATHOLOGICAL CHANGES IN MHV-INFECTED RODENTS

Although variations in virus and rodent strain used by different investigators studying JHMV-induced demyelination make a direct comparison of results difficult, several common themes emerge. First, virus persistence is a key element in the development of demyelination and, in general, the amount of demyelination and clinical disease appears to be proportional to the level of virus or its products. Demyelination could occur by one of several mechanisms. It could result from direct viral lysis of infected oligodendrocytes. Several early studies, based in part upon a lack of effect of immunosuppression on the disease process and in part on studies using electron microscopy, suggested that this mechanism was most important (Lampert, et al., 1973; Weiner, 1973). It is still believed that direct virus lysis of oligodendrocytes is important in mice with acute demyelination (Kyuwa and Stohlman, 1990). Another possibility is that demyelination results as a consequence of the immune response to the virus, either via direct cytolytic activity or as a consequence of cytokine activity. In support of this explanation, several more recent studies show that demyelination does not occur or occurs to a much lesser extent if mice are immunosuppressed prior to the initiation of the demyelinating process (Wang, et al., 1990; Houtman and Fleming, 1996). Alternatively, demyelination could result from an autoimmune response triggered by the initial viral infection. As mentioned above, there is evidence for the latter in rats infected with JHMV (Watanabe, et al., 1983) but this mechanism has not been identified thus far in infected mice.

Table 1. Common models of MHV-induced neurological disease[1]

Virus/strain	Persistence	Disease	Tropism	Comments
[3]MHV-3	yes	Vasculitis	Ependyma, Meninges	
[4]MHV-A59	yes	Encephalitis Hepatitis Demyelination	Neurons Glia	Common cause of death is hepatitis
[5]MHV-4 (JHM)		Encephalitis	Glia	100% mortality unless protected by anti-viral antibody or T cells[2]
		Demyelination	Neurons	
[6]JHM-W	yes	Encephalitis	Glia	Adopted by suckling mouse brain passage
		Demyelination	Neurons	
[7]JHM-DL	?	Encephalitis	Glia	Virulent large plaque variant from JHM-W
		Demyelination	Neurons	
[8]JHM 2.2v-1	yes	Demyelination	Glia	mAb derived mutant
[8]JHM 2.2/7.2-v-2	?	Minimal/No Demyelination	Glia	Double mAb escape from 2.2v-1
			Astrocytes?	
[7]JHM-DM	yes	Encephalitis Demyelination	Glia Neurons	Plaque variant of JHM-W
[7]JHM-DS	yes	Encephalitis Demyelination	Glia	Prominent demyelination From JHM-W
[9]JHM-Wurzburg	yes	Encephalitis Demyelination	Neurons Glia	SDE in rats
[10]JHM-cl2	yes	Encephalitis Demyelination	Glia Neurons	Virulent variant similar to MHV-4
[11]JHM-X	yes	Encephalitis Demyelination	Glia From ATCC	Deletion in S protein
[12]JHM V5A13.1	yes	Encephalitis Demyelination	Glia Neurons	mAb escape mutant
[13]JHM OBLV	no	Mild encephalitis	Olfactory neurons	
[14]JHM Ts8	yes	Demyelination	Glia	Prominent demyelination. Not readily available

[1]Many ts mutants and recombinants available are not included in Table 1.
[2]Persistent infection and Demyelination in maternal antibody-protection model (Perlman, 1987).
[3]Tardieu, et al., 1986; [4]Lavi, et al., 1984; [5]Cheever, et al., 1949; [6]Weiner, 1973; [7]Stohlman, et al., 1982; [8]Fleming, et al., 1987; [9]Nagashima, et al., 1978; [10]Taguchi, et al., 1985; [11]Nakanaga, et al., 1986; [12]Dalziel, et al., 1986; [13]Pearce, et al., 1994; [14]Haspel, et al., 1978.

A second major theme is that the same or similar pathological findings are present in all mice and rats with demyelination. In the earliest stages, viral antigen can be detected in the white matter with few histological changes observed. Soon thereafter, a large infiltration of lymphocytes and macrophages can be detected in areas of demyelination. The number of oligodendrocytes is decreased and the synthesis of mRNA specific for oligodendrocytes, such as myelin basic protein and proteolipid protein, is decreased at this time (Jordan, et al., 1989). Demyelination, with relative sparing of axons, occurs at this stage. Chemokines, particularly crg-2 and RANTES are expressed in areas of demyelination and attract lymphocytes (T Lane and M. Buchmeier, personal communication). These lymphocytes, in turn, secrete cytokines that attract macrophages and activate microglia These areas of demyelination become larger and are dominated histologically by astrogliosis and the presence of a large number of lipid-laden macrophages. The activated astrocytes stain for several cytokines and other immunomodulatory molecules, including TNF-α, IL-6, IL-1ß and the inducible form of nitric oxide synthase (NOS2) but not for MHC class I or class II antigen (Sun, et al., 1995). The macrophages present in these lesions are also highly activated, but unlike monocytes in other pathological settings do not express high levels of cytokines. They do, however, express both MHC class I and class II antigens on their surface (Figure 1). At present, it is not known if macrophages attack only virus-infected cells or if uninfected cells (and associated myelin) are also damaged. In later stages, the demyelinating plaques become well demarcated. Cellular infiltrates resolve, although a few macrophages and astrocytes may still be present in these areas. Remyelination is sometimes observed. Virus is cleared and cellular debris removed. If virus has not continued to replicate, the animals recover clinical function. If virus is not controlled however, new areas of demyelination develop in other areas of the white matter. In areas that do not show remyelination, inactive plaques can be detected (Barac-Latas, et al., 1997).

7. HOST IMMUNE RESPONSE TO JHMV AND MHV-A59

7.1. Humoral Response

Since a key part of the above model is that incomplete clearance of MHV is necessary for demyelination to develop and since demyelination is in large part an immu-

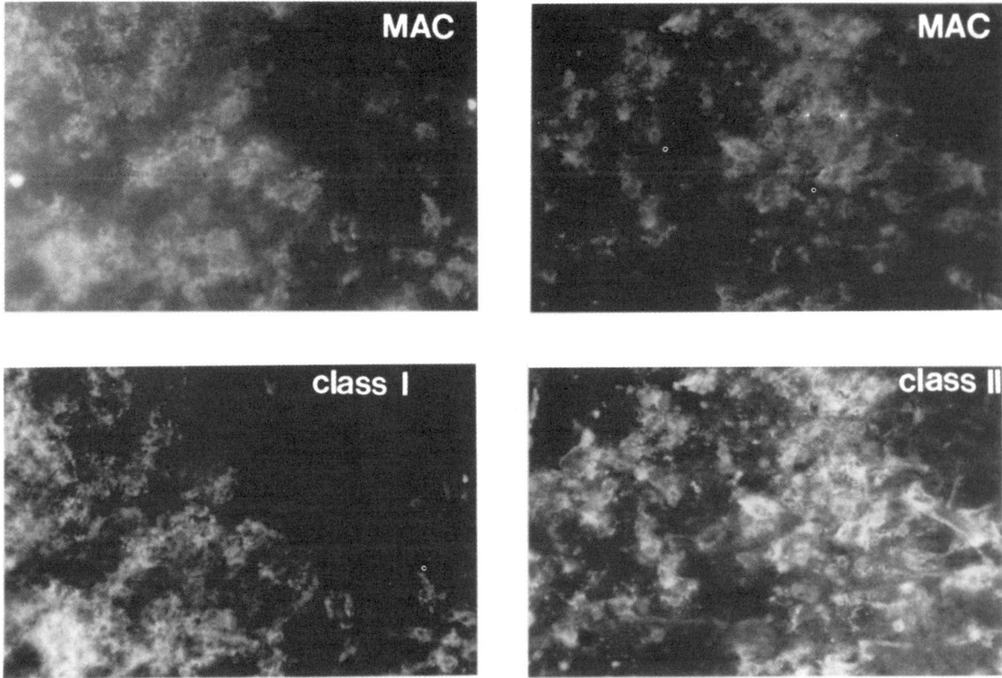

Figure 1. Macrophages/microglia express MHC class I and class II in mice persistently infected with JMHV. Suckling mice were infected with JHMV and nursed by dams previously immunized with live JHMV. The spinal cord was harvested from a mouse that developed hindlimb paralysis at 31 days p.i. Samples were simultaneously analyzed for expression of a macrophage marker (Mac-1) and MHC class I or class II antigen. FITC-conjugated goat anti-mouse antibody was used to detected MHC class I or class II antigen and Texas Red-conjugated goat-antirat antibody to detected Mac-1.

nopathological phenomenon, an understanding of the immune response to MHV is important in delineating the details of these processes. Initial studies concentrated on the humoral response to MHV. The administration of neutralizing antibody prior to intracerebral infection with MHV prevents acute encephalitis by protecting neurons from infection but does not prevent demyelination (Buchmeier, et al., 1984). Other studies showed that neutralizing antibody was present in the blood and cerebrospinal fluid of mice and rats with demyelination (Stohlman and Weiner, 1981; Sorensen, et al., 1984; Dorries, et al., 1986; Jacobsen and Perlman, 1990). Thus, once demyelination is underway, neutralizing antibody is unable to prevent the process from continuing although it is able to protect neurons from infection. Consistent with this, Brown Norway rats develop an asymptomatic chronic demyelinating disease induced by JHMV and in these animals, high levels of antibody can be detected in the CNS. These antibodies may prevent most virus spread within the CNS and contribute to the mild nature of the disease (Watanabe, et al., 1987).

7.2. Cell-Mediated Response

Other studies showed the importance of CD4 and CD8 T cells in virus clearance. In experiments in which CD4 or CD8 T cells are depleted prior to infection with JHMV, the kinetics of virus clearance is delayed (Williamson and Stohlman, 1990; Pearce, et al., 1994). Similar results were obtained with mice in which CD4 or CD8 T cell function was genetically disrupted. In mice deficient in CD8 function (ß2-microglobulin (-/-)), the LD_{50} for MHV-A59 is 0.001 that of normal mice (Gombold, et al., 1995).

Adoptive transfer experiments have also provided insight into the mechanisms of immune protection. Adoptive transfer of CD4 T cells prevented neuronal infection. In some cases, these cells also reduced virus replication and demyelination whereas in another report, virus replication and demyelination was not affected (Stohlman, et al., 1986; Körner, et al., 1991; Yamaguchi, et al., 1991). The difference in effect on virus replication may reflect differences in cytokine release or CD4 T cell cytotoxicity. The importance of CD8 T cells has also been shown in similar experiments. Adoptive transfer of CD8 T cells resulted in protection and enhanced virus clearance (Yamaguchi, et al., 1991; Stohlman, et al., 1995). In another set of experiments, rats immunized with recombinant vaccinia virus expressing the S protein were protected if exposed to virus 7 days after immunization; this is the time of maximal CD8 T cell response (Flory, et al., 1995). Interestingly, if rats were infected 21 days after immunization, they were not protected but rather became chronically infected with MHV. Since antibody can be detected at 21 days and not 7 days after immunization, these results are consistent with the idea, described above, that antibodies contribute to persistence and demyelination in these animals.

MHV-specific cytotoxic CD4 T cells may also have a role in virus clearance. No MHC class I-restricted CD8 T cells could be identified in mice infected with MHV-A59. However, MHC class II-restricted cytotoxic CD4 T cells could be identified in these animals. At present, it is not known if these cytotoxic T cells restricted by MHC class II antigen are uniquely important in mice with hepatitis or if they are generally important in MHV-infected mice (Heemskerk, et al., 1995).

In other studies, the CD8 and CD4 T cell epitopes recognized by JHMV-specific lymphocytes were identified. A summary of the epitopes identified thus far is shown in Table 2. These epitopes have been identified using either splenocytes harvested from immunized mice and stimulated in vitro, T cell clones developed from the spleens and CNS or lymphocytes harvested from the CNS of mice acutely or persistently infected with MHV and analyzed in direct ex vivo cytotoxicity assays. In BALB/c mice, a CD8 T cell epitope

Table 2. CD4 and CD8 T cell epitopes recognized in
MHV-infected rodents

Type of epitope	Source	MHV protein	Amino acids
[1]CD8 T cell	BALB/c mice	N	318-326
[2,3]CD8 T cell	C57Bl/6 mice	S	510-518
[2]CD8 T cell	C57Bl/6 mice	S	598-605
[4]CD4 T cell	Lewis rat	N	361-458
[5]CD4 T cell	C57Bl/6 mice	M	128-147
[6,7]CD4 T cell	C57Bl/6 mice	S	329-343
[6]CD4 T cell	BALB/c mice	S	329-343
[8]CD4 T cell	BALB/c mice	N	266-279
[7]CD4 T cell	C57Bl/6 mice	S	358-372
		S	408-422

[1]Bergmann, et al., 1993; [2]Castro and Perlman, 1995; [3]Bergmann, et al., 1996; [4]Wege, et al., 1993; [5]Xue, et al., 1995; [6]Heemskerk, et al., 1995; [7]Xue, S. and S.P., submitted for publication; [8]Van der Veen, 1996.

encompassing amino acids 318–326 of the N protein is immunodominant, although there is evidence that other JHMV-specific CD8 T cell epitopes are recognized (Bergmann, et al., 1993; Stohlman, et al., 1993). These epitopes appear to be present within one or more nonstructural protein and have not been further characterized. In C57Bl/6 mice, two CD8 T cell epitopes (S-510–518 and S-598–605) are recognized (Castro and Perlman, 1995; Bergmann, et al., 1996). Both are located within a region of the S protein which is commonly deleted in many JHMV variants and during the course of persistence in C57Bl/6 mice (Parker, et al., 1989; Rowe, et al., 1997). Deletions and missense mutations in this region are not lethal for the virus but result in CTL escape in C57Bl/6 mice as described above. CD4 T cell epitopes have also been described in both strains of mice. In BALB/c mice, CD4 T cell epitopes are present within the N protein (Van der Veen, 1996) whereas in C57Bl/6 mice such epitopes are located within the S and M but not the N proteins (Mobley, et al., 1992; Heemskerk, et al., 1995; Xue, et al., 1995). In no case are these epitopes located within the hypervariable region of the S protein and no mutations which result in escape from CD4 T cell surveillance have been identified.

8. ROLE OF CYTOKINES

Cytokines are likely to play a major role in MHV-induced demyelination but our understanding of this process is rudimentary. As in most acute encephalitides such cytokines as IL-1α, IL-1ß, TNF-α, IL-6 and IFN-γ are detected in RNA samples harvested from the CNS of infected mice and with the exception of IFN-γ are in part synthesized by resident CNS cells (Pearce, et al., 1994). IFN-γ is produced by infiltrating immune cells and depletion of IFN-γ, either with neutralizing antibody or using mice in which the gene for IFN-γ is disrupted leads to decreased virus clearance and greater mortality (Lane, et al., 1997). In contrast, neutralization of TNF-α did not appear to effect either the recruitment of T cells in the CNS, virus clearance or the development of demyelination (Stohlman, et al., 1995) and inhibition of NOS2 did not affect virus clearance (Lane, et al., 1997). Less is known about the role of cytokines in the chronic demyelinating process. The cytokines TNF-α, IL-1ß and IL-6 as well as NOS2 are expressed by astrocytes localized near to sites of demyelination in chronically infected spinal cords. These cytokines are synthesized for the most part by uninfected cells although infected astrocytes expressing these cytokines can occasionally be detected (Sun, et al., 1995). Some of these immunomodulatory molecules are directly toxic for oligodendrocytes or myelin and it will be important to determine the role of these cytokines in chronic demyelination.

9. CONCLUSIONS

Although progress has been made in understanding MHV persistence and the development of demyelination, much remains to be determined. The development of methods to genetically manipulate the MHV genome as well as the availability of mice in which the genes encoding one or more key immune functions are disrupted should facilitate progress in these areas. Several outstanding questions remain, including: 1) what enables virus to avoid clearance during the early stages of the infection? Virus is able to persist even if CTL escape mutants are not selected, albeit usually without the presence of infectious virus. 2) In what cells does virus persist, other than astrocytes, and what is the molecular form of the persistence? 3) What are the relative roles of virus, T cells, antibodies and cytokines in the demyelinating process? The answers to these questions will undoubtedly be complicated but the tools are available to start answering them.

ACKNOWLEDGMENTS

This research was supported in part by grants from the N.I.H. (NS24401, DC01711) and the National Multiple Sclerosis Society.

REFERENCES

Barac-Latas, V., Suchanek, G., Breitschopf, H., Stuehler, A., Wege, H., and Lassmann, H., 1997, Patterns of oligodendrocyte pathology in coronavirus-induced subacute demyelinating encephalomyelitis in the Lewis rat, *Glia* **19**: 1–12.

Barnett, E. and Perlman, S., 1993, The olfactory nerve and not the trigeminal nerve is the major site of CNS entry for mouse hepatitis virus, strain JHM, *Virology* **194**: 185–191.

Bergmann, C., McMillan, M., and Stohlman, S. A., 1993, Characterization of the Ld-restricted cytotoxic T-lymphocyte epitope in the mouse hepatitis virus nucleocapsid protein, *J. Virol.* **67**: 7041–49.

Bergmann, C. C., Yao, Q., Lin, M., and Stohlman, S. A., 1996, The JHM strain of mouse hepatitis virus induces a spike protein-specific Db-restricted CTL response, *J. Gen. Virol.* **77**: 315–325.

Buchmeier, M. J., Lewicki, H. A., Talbot, P. J., and Knobler, R. L., 1984, Murine hepatitis virus-4 (strain JHM)-induced neurologic disease is modulated in vivo by monoclonal antibody, *Virology* **132**: 261–270.

Castro, R. F. and Perlman, S., 1995, CD8+ T cell epitopes within the surface glycoprotein of a neurotropic coronavirus and correlation with pathogenicity, *J. Virol.* **69**: 8127–8131.

Cheever, F. S., Daniels, J. B., Pappenheimer, A. M., and Bailey, O. T., 1949, A murine virus (JHM) causing disseminated encephalomyelitis with extensive destruction of myelin, *J. Exp. Med.* **90**: 181–194.

Chen, W. and Baric, R., 1996, Molecular anatomy of MHV persistence: coevolution of increased host resistance and virus virulence, *J. Virol.* **70**: 3947–3960.

Dalziel, R. G., Lampert, P. W., Talbot, P. J., and Buchmeier, M. J., 1986, Site-specific alteration of murine hepatitis virus type 4 peplomer glycoprotein E2 results in reduced neurovirulence, *J. Virol.* **59**: 463–471.

Dorries, R., Watanabe, R., Wege, H., and ter Meulen, V., 1986, Murine coronavirus-induced encephalomyelitides in rats: Analysis of immunoglobulins and virus-specific antibodies in serum and cerebrospinal fluid, *J. Neuroimmunol.* **12**: 131–142.

Fleming, J. O., Trousdale, M. D., Bradbury, J., Stohlman, S. A., and Weiner, L. P., 1987, Experimental demyelination induced by coronavirus JHM (MHV-4): molecular identification of a viral determinant of paralytic disease, *Microb. Pathog.* **3**: 9–20.

Flory, E., Stuhler, A., Barac-Latas, V., Lassmann, H., and Wege, H., 1995, Coronavirus-induced encephalomyelitis: balance between protection and immunopathology depends on the immunization schedule with spike protein S, *J. Gen. Virol.* **76**: 873–879.

Gombold, J., Sutherland, R., Lavi, E., Paterson, Y., and Weiss, S. R., 1995, Mouse hepatitis virus A59-induced demyelination can occur in the absence of CD8+ T cells, *Microb. Pathog.* **18**: 211–221.

Haspel, M. V., Lampert, P. W., and Oldstone, M. B. A., 1978, Temperature-sensitive mutants of mouse hepatitis virus produce a high incidence of demyelination, *Proc. Natl. Acad. Sci. USA* **75**: 4033–4036.

Heemskerk, M., Schoemaker, H., De Jong, I., Schijns, V., Spaan, W., and Boog, C. J. P., 1995, Differential activation of mouse hepatitis virus-specific CD4+ cytotoxic T cells is defined by peptide length, *Immunology* **85**: 517–522.

Heemskerk, M., Schoemaker, H., Spaan, W., and Boog, C., 1995, Predominance of MHC class II-restricted CD4+ cytotoxic T cells against mouse hepatitis virus A59, *Immunology* **84**: 521–527.

Houtman, J. J. and Fleming, J. O., 1996, Dissociation of demyelination and viral clearance in congenitally immunodeficient mice infected with murine coronavirus JHM, *J. Neurovirology* **2**: 101–110.

Jacobsen, G. and Perlman, S., 1990, Localization of virus and antibody response in mice infected persistently with MHV-JHM, *Adv. Exp. Med. Biol.* **276**: 573–578.

Jordan, C., Friedrich, V., Godfraind, C., Caardellechio, C., Holmes, K., and Dubois-Dalcq, M., 1989, Expression of viral and myelin gene transcripts in a murine CNS demyelinating disease caused by a coronavirus, *Glia* **2**: 318–329.

Körner, H., Schliephake, A., Winter, J., Zimprich, F., Lassmann, H., Sedgwick, J., Siddell, S., and Wege, H., 1991, Nucleocapsid or spike protein-specific CD4$^+$ T lymphocytes protect against coronavirus-induced encephalomyelitis in the absence of CD8$^+$ T cells, *J. Immunol.* **147**: 2317–2323.

Kyuwa, S. and Stohlman, S. A., 1990, Pathogenesis of a neurotropic murine coronvirus, strain JHM in the central nervous system of mice, *Sem. Virol.* **1**, 273–280.

Lampert, P. W., Sims, J. K., and Kniazeff, A. J., 1973, Mechanism of demyelination in JHM virus encephalomyelitis, *Acta Neuropathol.* **24**: 76–85.

Lane, T. E., Paoletti, A. D., and Buchmeier, M. J., 1997, Disassociation between the in vitro and in vivo effects of nitric oxide on a neurotropic murine coronavirus, *J. Virol.* **71**: 2202–2210.

Lavi, E., Fishman, P. S., Highkin, M. K., and Weiss, S. R., 1988, Limbic encephalitis after inhalation of a murine coronavirus, *Lab. Invest.* **58**: 31–36.

Lavi, E., Gilden, D., Highkin, M., and Weiss, S. R., 1986, The organ tropism of mouse hepatitis virus A59 in mice is dependent on dose and route of inoculation, *Lab. Invest.* **36**: 130–135.

Lavi, E., Gilden, D., Wroblewska, Z., Rorke, L., and Weiss, S., 1984, Experimental demyelination produced by the A59 strain of MHV, *Neurology* **34**: 597–603.

Lipton, H. L., Twaddle, G., and Jelachich, M. L., 1995, The predominant virus antigen burden is present in macrophages in Theiler's murine encephalomyelitis virus-induced demyelinating disease, *J. Virol.* **69**: 2525–2533.

Mobley, J., Evans, G., Dailey, M. O., and Perlman, S., 1992, Immune response to a murine coronavirus: Identification of a homing receptor-negative CD4$^+$ T cell subset that responds to viral glycoproteins, *Virology* **187**: 443–452.

Nagashima, K., Wege, H., Meyermann, R., and ter Meulen, V., 1978, Coronavirus induced subacute demyelinating encephalomyelitis in rats: A morphological analysis, *Acta Neuropathol.* **44**: 63–70.

Nakanaga, K., Yamanouchi, K., and Fujiwara, K., 1986, Protective effect of monoclonal antibodies on lethal mouse hepatitis virus infection in mice, *J. Virol.* **59**: 168–171.

Parker, S. E., Gallagher, T. M., and Buchmeier, M. J., 1989, Sequence analysis reveals extensive polymorphism and evidence of deletions within the E2 glycoprotein gene of several strains of murine hepatitis virus, *Virology* **173**: 664–673.

Pearce, B. D., Hobbs, M. V., McGraw, T. S., and Buchmeier, M. J., 1994, Cytokine induction during T-cell-mediated clearance of mouse hepatitis virus from neurons in vivo, *J. Virol.* **68**: 5483–5495.

Perlman, S. and Ries, D., 1987, The astrocyte is a target cell in mice persistently infected with mouse hepatitis virus, strain JHM, *Microb. Pathog.* **3**: 309–314.

Perlman, S., Schelper, R., Bolger, E., and Ries, D., 1987, Late onset, symptomatic, demyelinating encephalomyelitis in mice infected with MHV-JHM in the presence of maternal antibody, *Microb. Pathog.* **2**, 185–194.

Pewe, L., Wu, G., Barnett, E. M., Castro, R., and Perlman, S., 1996, Cytotoxic T cell-resistant variants are selected in a virus-induced demyelinating disease, *Immunity* **5**: 253–262.

Rowe, C. L., Baker, S. C., Nathan, M. J., and Fleming, J. O., 1997, Evolution of mouse hepatitis virus: Detection and characterization of S1 deletion variants during persistent infection, *J. Virol.* **71**: 2959–2967.

Sorensen, O., Coulter-Mackie, M., Puchalski, S., and Dales, S., 1984, In vivo and in vitro model of demyelinating disease. IX. Progression of JHM virus infection in the central nervous system of the rat during overt and asymptomatic phases, *Virology* **137**: 347–357.

Sorensen, O., Perry, D., and Dales, S., 1980, In vivo and in vitro models of demyelinating diseases. III. JHM virus infection of rats, *Arch. Neurol.* **37**: 478–484.

Stohlman, S. A., Bergmann, C. C., van der Veen, R. C., and Hinton, D. R., 1995, Mouse hepatitis virus-specific cytotoxic T lymphocytes protect from lethal infection without eliminating virus from the central nervous system, *J. Virol.* **69**: 684–694.

Stohlman, S. A., Fleming, J. O., Brayton, P. R., Weiner, L. P., and Lai, M. M. C., 1982, Murine coronaviruses: isolation and characterization of two plaque morphology variants of the JHM neurotropic strain, *J. Gen. Virol.* **63**: 265–275.

Stohlman, S. A. and Frelinger, J. A., 1978, Resistance to fatal central nervous system disease by mouse hepatitis virus, strain JHM, *Immunogenetics* **6**: 277–281.

Stohlman, S. A., Hinton, D. R., Cua, D., Dimacali, E., Sensintaffar, J., Hofman, F. M., Tahara, S. M., and Yao, Q., 1995, Tumor necrosis factor expression during mouse hepatitis virus-induced demyelinating encephalomyelitis, *J. Virol.* **69**: 5898–5903.

Stohlman, S. A., Kyuwa, S., Polo, J. M., Brady, D., Lai, M. M. C., and Bergmann, C. C., 1993, Characterization of mouse hepatitis virus-specific cytotoxic T cells derived from the central nervous system of mice infected with the JHM strain, *J. Virol.* **67**: 7050–7059.

Stohlman, S. A., Matsushima, G. K., Casteel, N., and Weiner, L. P., 1986, In vivo effects of coronavirus-specific T cell clones: DTH inducer cells prevent a lethal infection but do not inhibit virus replication, *J. Immunol.* **136**: 3052–3056.

Stohlman, S. A. and Weiner, L. P., 1981, Chronic central nervous system demyelination in mice after JHM virus infection, *Neurology* **31**: 38–44.

Sun, N., Grzybicki, D., Castro, R., Murphy, S., and Perlman, S., 1995, Activation of astrocytes in the spinal cord of mice chronically infected with a neurotropic coronavirus, *Virology* **213**: 482–493.

Taguchi, F., Siddell, S., Wege, H., and ter Meulen, V., 1985, Characterization of a variant virus selected inrat brains after infection by coronavirus MHV JHM, *J. Virol.* **54**: 429–435.

Tardieu, M., Boespflug, O., and Barbe, T., 1986, Selective tropism of a neurotropic coronavirus for epndymal cells, neurons and meningeal cells, *J. Virol.* **60**: 574–582.

Van der Veen, R. C., 1996, Immunogenicity of JHM virus proteins: Characterization of a CD4+ T cell epitope on nucleocapsid protein which induces different T-helper cell subsets, *Virology* **225**: 339–346.

Wang, F., Stohlman, S. A., and Fleming, J. O., 1990, Demyelination induced by murine hepatitis virus JHM strain (MHV-4) is immunologically mediated, *J. Neuroimmunol.* **30**: 31–41.

Watanabe, R., Wege, H., and ter Meulen, V., 1983, Adoptive transfer of EAE-like lesions from rats with coronavirus-induced demyelinating encephalomyelitis, *Nature* **305**: 150–153.

Watanabe, R., Wege, H., and ter Meulen, V., 1987, Comparative analysis of coronavirus JHM-induced demyelinating encephalomyelitis in Lewis and Brown Norway rats, *Lab. Invest.* **57**: 375–384.

Wege, H., Schliephake, A., Korner, H., Flory, E., and Wege, H., 1993, An immunodominant CD4+ T cell site on the nucleocapsid protein of murine coronavirus contributes to protection against encephalomyelitis, *J. Gen. Virol.* **74**: 1287–1294.

Weiner, L. P., 1973, Pathogenesis of demyelination induced by a mouse hepatitis virus (JHM virus), *Arch. Neurol.* **28**: 298–303.

Williams, R. K., Jiang, G., and Holmes, K. V., 1991, Receptor for mouse hepatitis virus is a member of the carcinoembryonic antigen family of glycoproteins, *Proc. Natl. Acad. Sci. USA* **88**: 5533–5536.

Williamson, J. S. and Stohlman, S. A., 1990, Effective clearance of mouse hepatitis virus from the central nervous system requires both CD4$^+$ and CD8$^+$ T cells, *J. Virol.* **64**: 4589–4592.

Xue, S., Jaszewski, A., and Perlman, S., 1995, Identification of a CD4+ T cell epitope within the M protein of a neurotropic coronavirus, *Virology* **208**: 173–179.

Yamaguchi, K., Goto, N., Kyuwa, S., Hayami, M., and Toyoda, Y., 1991, Protection of mice from a lethal coronavirus infection in the central nervous system by adoptive transfer of virus-specific T cell clones, *J. Neuroimmunol.* **32**: 1–9.

66

ROLE OF CTL MUTANTS IN DEMYELINATION INDUCED BY MOUSE HEPATITIS VIRUS, STRAIN JHM

S. Perlman[1,2,3] and L. Pewe

[1]Department of Pediatrics
[2]Department of Microbiology and
[3]Interdisciplinary Program in Immunology
University of Iowa
Iowa City, Iowa 52242

1. ABSTRACT

Mouse hepatitis virus, strain JHM (MHV-JHM) is a well described cause of demyelination. C57Bl/6 (B6) mice infected at the suckling stage in the presence of protective antibodies remain asymptomatic initially but later develop clinical disease (hindlimb paralysis). Infectious virus can be isolated from these mice. Recently, two MHV-specific target epitopes for cytotoxic CD8 T cells have been identified in B6 mice. Our results show that in all mice with hindlimb paralysis, mutations can be detected in the RNA encoding the immunodominant of the two epitopes. These mutations result in a loss of recognition by MHV-specific cytotoxic T cells. These changes are not detected, for the most part, in mice that remain asymptomatic nor in mice with acute encephalitis. These results suggest that the development of CTL escape mutants is necessary for hindlimb paralysis to develop in this model.

2. INTRODUCTION

MHV-JHM causes acute encephalitis and acute and chronic demyelination in susceptible strains of mice and rats. Most strains of mice are susceptible to the virus and mice of all ages develop an acute fatal encephalitis after intranasal or intracerebral inoculation. This disease can be prevented if mice are infected with attenuated strains of virus or if they are protected by infusions of antiviral antibodies or T cells. Under most circumstances, animals are most ill for a few days after inoculation of virus and the highest mor-

Coronaviruses and Arteriviruses, edited by Enjuanes *et al.*
Plenum Press, New York, 1998

tality is recorded in this period. Evidence of demyelination can be detected in mice that survive the acute infection (Kyuwa and Stohlman, 1990; Houtman and Fleming, 1996; Lane and Buchmeier, 1997).

Over the past few years, a different version of this basic model has been studied in our laboratory. In this model, suckling B6 mice are inoculated intranasally with MHV-JHM. In the absence of any therapeutic interventions, all mice succumb to an acute encephalitis. This fatal disease can be prevented however, if mice are nursed by dams that have been previously immunized to MHV-JHM. The mice remain asymptomatic for several weeks, but 40–90% then develop clinical disease manifested by hindlimb paralysis. Infectious virus can be isolated from mice with clinical disease but not from mice that remain asymptomatic whereas viral antigen can be detected in all mice. In marked contrast to these results, suckling BALB/c mice similarly treated remain asymptomatic once protected from acute encephalitis (Perlman, et al., 1987; Castro, et al., 1994).

Recently, MHV-specific CD8 T cell targets have been identified in B6 and BALB/c mice (Bergmann, et al., 1993; Castro and Perlman, 1995; Bergmann, et al., 1996). The majority of MHV-specific CD8 T cells recognize the nucleocapsid (N) protein in BALB/c mice whereas the surface (S) glycoprotein is recognized by CD8 T cells harvested from B6 mice. Further analyses in B6 mice revealed the presence of two epitopes encompassing residues 510 to 518 (S-510–518) and 598–605 (S-598–605). S-510–518 is the more immunodominant of the two epitopes. The S protein contains a hypervariable region that appears to be readily deleted without loss of viability (Parker, et al., 1989) and both epitopes are located in this region of the S protein. In contrast, the N protein is highly conserved among the various species of MHV and the epitope is located in a region not prone to variability.

Mutations in CD8 T cell epitopes that lead to a loss of recognition by virus-specific T cells have been identified in patients infected with hepatitis B virus, human immunodeficiency virus, Epstein-Barr virus and hepatitis C and in mice infected with lymphocytic choriomeningitis virus (LCMV) (Koup, 1994; Franco, et al., 1995; Zinkernagel, 1996). In all cases, CTL escape mutants appear to arise most commonly and to be most important when the CTL response is nearly monospecific and strong. Since B6 mice infected with MHV-JHM mount such a response and the response is directed at an epitope located in a region prone to sequence variability, we postulated that CTL mutations in epitope S-510–518 might develop during the course of the infection and contribute to the development of hindlimb paralysis in mice that were initially well and later developed disease.

3. MATERIALS AND METHODS

The materials and experimental methods used in these experiments were previously described (Perlman, et al., 1987; Pewe, et al., 1996).

4. RESULTS

To determine if CTL escape mutants could be detected in mice that developed hindlimb paralysis several weeks after infection, infectious virus was isolated from these mice and propagated a minimal number of times in tissue culture cells so that sufficient material was available for analysis. The results of these initial analyses showed that in every case, mutations were detected in epitope S-510–518. The sequence of this epitope is

CSLWNGPHL and mutations were detected in residues 2 to 7 of the epitope. In each mouse, a single mutation was identified in virus harvested from infected CNS tissue (Pewe, et al., 1996) and different muttions were detected in different mice.

In previous studies, passage through tissue culture cells was shown to select for a subpopulation of viruses initially present in the infected host (Meyerhans, et al., 1989). To determine if this type of selection occurred in our experiments, RNA was isolated from the spinal cords and brains of chronically infected mice and the sequence of epitope S-510–518 determined. As before, mutations in epitope S-510–518 were detected in all samples (Pewe, et al., 1996). Additionally, cDNA clonal analysis was performed to determine if there was any heterogeneity in the sequence of epitope S-510–518 not easily detected by bulk analysis. In some samples, more than one mutation was present. Occasionally, a subpopulation of wild type sequence could also be detected. As controls for these experiments, the following additional samples were analyzed:

1. In mice with acute encephalitis, no changes in epitope S-510-518 were detected (Pewe, et al., 1996).

2. Mice destined to develop hindlimb paralysis do so by 80 days p.i. and the vast majority of mice develop disease by 60 days. Analysis of RNA harvested from the spinal cords of these mice revealed changes in epitope S-510–518 in one case and large deletions which included the epitope in 3/14 samples. Only wild type sequence was detected in 10/14 samples which were analyzed (Pewe, et al., 1996; Pewe, et al., in press).

3. CD8 T cells are not present in SCID mice, but mice readily become persistently infected with MHV-JHM. No changes in epitope S-510–518 were detected in these mice (Pewe, et al., in press).

4. S-510–518 is the immunodominant CTL epitope (Castro and Perlman, 1995). No changes were detected in the subdominant T cell epitope S-598–605. In addition, a major CD4 T cell epitope was identified in the M protein (M-135–143) and at least three CD4 T cell epitopes are present in the S glycoprotein (Heemskerk, et al., 1995; Xue, et al., 1995, unpublished observations). Mutations were not detected in the M-specific epitope nor in two of the three S-specific epitopes. In the third S-specific epitope, located in the region encompassing amino acids 328–347, a change was noted in a minority of cDNA sequenced from mice with hindlimb paralysis. This mutation, resulting in a change from alanine to threonine at position 337 did not appear to affect recognition by MHV-specific CD4 T cells in proliferation assays.

5. If variants containing mutations in epitope S-510–518 contribute to the pathogenesis of the chronic demyelinating encephalomyelitis, they should arise at early times in the infectious cycle before virus would normally be cleared. MHV is usually cleared by 12–21 days p.i. (Kyuwa and Stohlman, 1990; Houtman and Fleming, 1996; Lane and Buchmeier, 1997). Changes in epitope S-510–518 can be detected as early as 10–12 days p.i. and comprise the majority of the viral RNA detected in some mice at 15 days p.i. These data are consistent with previous analyses performed in mice infected with LCMV (Pircher, et al., 1990; Weidt, et al., 1995).

6. To show that the changes that we detected affect recognition by MHV-specific CD8 T cells, variant peptides were analyzed in direct ex vivo cytotoxicity assays using lymphocytes harvested from the CNS of mice with acute encephalitis. Mutations in positions 3, 4, 5, 6 and 7 were analyzed in these experiments. In

each case, the mutations resulted in a complete or major loss in recognition by the CNS-derived lymphocytes consistent with their role in the process of virus persistence and resultant demyelination (Pewe, et al., 1996).

Of note, mice inoculated with virus containing mutated epitope S-510-518 are unable to clear the virus and usually die by 20–25 d p.i. Extensive demyelination and large amounts of viral antigen are detected in the CNS of these mice (unpublished observations).

A summary of the changes that we have detected thus far in epitope S-510–518 is shown in the Table 1.

5. DISCUSSION

Cytotoxic T cells play a major role in controlling viral infections. CTL escape mutants have been identified in some persistent human viral infections as well as in mice infected with LCMV. The role of CTL escape mutants has been questioned, however, since in general, the CTL response to a pathogen is polyclonal and polyspecific (Franco, et al., 1995). Changes in a single epitope would not be expected, under these circumstances, to result in a loss of CTL recognition. However, CTL escape mutants would be expected to be important if the CTL response is functionally monospecific and oligoclonal so that a change in the epitope would abrogate recognition by all or a majority of CTLs.

This scenario occurs in B6 mice infected with MHV-JHM. The CTL response in these mice is important for virus clearance and is directed primarily to a single epitope present in a region of the S glycoprotein previously determined to be hypervariable. Mutations in this epitope arise readily and early in the infection and do not appear to affect virulence. They affect recognition by lymphocytes harvested from the CNS, the site of inflammation and would be expected to contribute to virus persistence. An increase in virus load directly contributes to an increase in clinical disease and histological evidence of demyelination. Clinical disease may result from direct viral lysis of glial cells or more likely, from the host response to the virus. Our results suggest that CTL escape mutants are necessary, but not sufficient, for these processes to occur.

Table 1. Summary of changes identified in epitopes S-510-518 (number)

Wild type:	CSLWNGPHL	Position 5:	CSLW*S*GPHL (10)
			CSLW*T*GPHL (1)
Position 2:	C*Y*LWNGPHL (1)		CSLW*H*GPHL (1)
	C*F*LWNGPHL (2)		CSLW*Y*GPHL (1)
	C*P*LWNGPHL (1)		CSLW*D*GPHL (1)
Position 3:	CS*R*WNGPHL (2)	Position 6:	CSLWN*V*PHL (2)
	CS*P*WNGPHL (2)		CSLWN*R*PHL (1)
	CS*F*WNGPHL (1)		
		Position 7:	CSLWNGL*L*HL (5)
Position 4:	CSL*R*NGPHL (3)		CSLWNG*H*HL (2)
Deletions:	C*F*WNGPHL (5)	Entire epitope deleted (7)	
	CSWNGPHL (1)		
	CSLNGPHL (2)		

ACKNOWLEDGMENTS

This research was supported in part by grants from the National Institutes of Health and the National Multiple Sclerosis Society (U.S.).

REFERENCES

Bergmann, C., McMillan, M., and Stohlman, S. A., 1993, Characterization of the Ld-restricted cytotoxic T-lymphocyte epitope in the mouse hepatitis virus nucleocapsid protein, *J. Virol.* **67**: 7041–49.

Bergmann, C. C., Yao, Q., Lin, M., and Stohlman, S. A., 1996, The JHM strain of mouse hepatitis virus induces a spike protein-specific Db-restricted CTL response, *J. Gen. Virol.* **77**: 315–325.

Castro, R. F., Evans, G. D., Jaszewski, A., and Perlman, S., 1994, Coronavirus-induced demyelination occurs in the presence of virus-specific cytotoxic T cells, *Virology* **200**: 733–743.

Castro, R. F. and Perlman, S., 1995, CD8+ T cell epitopes within the surface glycoprotein of a neurotropic coronavirus and correlation with pathogenicity, *J. Virol.* **69**: 8127–8131.

Franco, A., Ferrari, C., Sette, A., and Chisari, F. V., 1995, Viral mutations, TCR antagonism and escape from the immune response, *Curr. Opin. Immunol.* **7**: 524–531.

Heemskerk, M., Schoemaker, H., De Jong, I., Schijns, V., Spaan, W., and Boog, C. J. P., 1995, Differential activation of mouse hepatitis virus-specific CD4+ cytotoxic T cells is defined by peptide length, *Immunology* **85**: 517–522.

Houtman, J. J. and Fleming, J. O., 1996, Pathogenesis of mouse hepatitis virus-induced demyelination, *J. Neurovirol.* **2**: 361–376.

Koup, R., 1994, Virus escape from CTL recognition, *J. Exp. Med.* **180**: 779–782.

Kyuwa, S. and Stohlman, S. A., 1990, Pathogenesis of a neurotropic murine coronavirus, strain JHM in the central nervous system of mice, *Sem. Virol.* **1**: 273–280.

Lane, T. E. and Buchmeier, M. J., 1997, Murine coronavirus infection: a paradigm for virus-induced demyelinating disease, *Trends Microbiol.* **5**: 9–14.

Meyerhans, A., Cheynier, R., Albert, J., Seth, M., Kwok, S., Sninsky, J., Morfeldt-Manson, L., A., B. , and Wain-Hobson, S., 1989, Temporal fluctuations in HIV quasispecies in vivo are not reflected by sequential HIV isolations, *Cell* **58**: 901–910.

Parker, S. E., Gallagher, T. M., and Buchmeier, M. J.,1989, Sequence analysis reveals extensive polymorphism and evidence of deletions within the E2 glycoprotein gene of several strains of murine hepatitis virus, *Virology* **173**: 664–673.

Perlman, S., Schelper, R., Bolger, E., and Ries, D., 1987, Late onset, symptomatic, demyelinating encephalomyelitis in mice infected with MHV-JHM in the presence of maternal antibody, *Microb. Pathog.* **2**: 185–194.

Pewe, L., Wu, G., Barnett, E. M., Castro, R., and Perlman, S., 1996, Cytotoxic T cell-resistant variants are selected in a virus-induced demyelinating disease, *Immunity* **5**: 253–262.

Pewe, L., Xue, S., and Perlman, S., 1997, Cytotoxic T cell-resistant variants arise at early times after infection in C57Bl/6 but not in SCID mice infected with a neurotropic coronavirus, *J. Virol.*

Pircher, H., Moskophidis, D., Rohrer, U., Burki, K., Hengartner, H., and Zinkernagel, R., 1990, Viral escape by selection of cytotoxic T cell-resistant virus variants in vivo, *Nature* **346**: 629–633.

Weidt, G., Deppert, W., Utermohlen, O., Heukeshoven, J., and Lehmann-Grube, F., 1995, Emergence of virus escape mutants after immunization with epitope vaccine, *J. Virol.* **69**: 7147–7151.

Xue, S., Jaszewski, A., and Perlman, S., 1995, Identification of a CD4+ T cell epitope within the M protein of a neurotropic coronavirus, *Virology* **208**: 173–179.

Zinkernagel, R. M., 1996, Immunology taught by viruses, *Science* **271**: 173–178.

67

USING A DEFECTIVE-INTERFERING RNA SYSTEM TO EXPRESS THE HE PROTEIN OF MOUSE HEPATITIS VIRUS FOR STUDYING VIRAL PATHOGENESIS

Xuming Zhang,[1,5] David Hinton,[1,2] Sungmin Park,[3] Ching-Len Liao,[3] Michael M. C. Lai,[1,3,4] and Stephen Stohlman[1,3]

[1]Department of Neurology
[2]Department of Pathology
[3]Department of Molecular Microbiology and Immunology
[4]Howard Hughes Medical Institute
University of Southern California School of Medicine
Los Angeles, California 90033
[5]Department of Microbiology and Immunology
University of Arkansas for Medical Sciences
Little Rock, Arkansas 72205

1. ABSTRACT

We have developed a defective-interfering (DI) RNA of mouse hepatitis virus (MHV) as a vector for expressing a variety of cellular and viral genes including the chloramphenicol acetyltransferase (CAT), hemagglutinin'esterase (HE), and gamma interferon. Here, we used the HE-expressing DI RNA for examining the role of HE protein in viral pathogenesis. The pseudorecombinant virus containing an expressed HE protein was generated by infecting cells with MHV-A59, which does not express HE, and transfecting the *in vitro*-transcribed DI RNA containing the HE gene. These pseudorecombinant viruses (DE-HE A59) were then inoculated intracerebrally into mice. Viruses recovered from cells infected with A59 and transfected with DI RNA expressing the CAT gene (DE-CAT A59) were used as a control. At various time points after inoculation, mice were observed for clinical symptoms. Tissues (brains and livers) were obtained for determining the replication of DI RNA by RT-PCR, virus replication by plaque assay, antigen expression by immunohistochemistry, and pathological changes. Results showed that all mice infected with DE-CAT A59 succumbed to infection by 9 days postinfection (d p.i). These

Coronaviruses and Arteriviruses, edited by Enjuanes *et al.*
Plenum Press, New York, 1998

data are identical to the pathogenesis of the parental A59 virus, demonstrating that inclusion of the DI RNA did not by itself alter pathogenesis. In contrast, only 40 % of mice infected with DE-HE A59 succumbed to infection. The subgenomic mRNAs transcribed from the DI vector were detected at 1 and 2 d p.i. but not at subsequent time points, indicating that the genes in the DI vector were expressed only at an early stage of viral infection. No significant difference in virus replication in the brains was detected between these two groups of mice, suggesting that virus replication in brains was not affected by the expression of the HE. Histopathological examination showed only a small increase in the extent of inflammatory cell infiltration and reduced viral antigen in the mice infected with DE-HE A59. There was no difference in virus replication in the livers at 2 and 4 d p.i., but a 3 $\log_{10}$ reduction was detected in the livers of mice infected with DE-HE A59 at 6 d p.i. Histological examination showed a significant reduction in viral antigen, inflammation and necrosis in mice infected with DE-HE A59. These results indicate that the expression of HE from the DI vector altered the viral pathogenesis. This study thus demonstrates the usefulness of this system in studying the role of viral or cellular genes expressed locally at the sites of viral infection in viral pathogenesis.

2. INTRODUCTION

Mouse hepatitis virus, the prototype murine coronavirus, is a member of the *Coronaviridae*. It contains a single-strand, positive-sense RNA genome of 31 kb (Lee et al., 1991). Due to the unusual large genome size, it has not been able to make an infectious cDNA clone of any coronaviruses to date. Thus, studies of coronavirus pathogenesis at the molecular level have been largely hampered. The development of a defective-interfering RNA as an expression system provides an alternative genetic approach for defining the roles of individual viral gene products in viral pathogenesis.

Viral defective-interfering (DI) particles are usually generated under certain evolutionary pressure from infected hosts or tissue cultures. They require helper virus for their replication; but they in turn interfere with the replication of a co-infected helper virus. Many DI particles of MHV have been isolated, two, which have been derived from MHV strain JHM, are well-characterized (Makino et al., 1988). DIssE contains an RNA genome of 2.2 kb and DIssF 3.5 kb. They contain three and five fragments derived from discontinuous regions of the parental viral genome, respectively. These DI RNAs also contain essential cis-acting signals for replication, thus behaving like a minigenome. Upon transfection, they replicate efficiently in the presence of a helper virus. However, sinve these DI RNAs do not contain intergenic (IG) sequences, which serve as transcription initiation sites or promotors, they are not able to transcribe subgenomic mRNAs. Insertion of an IG sequence into the DI RNA allows transcription of a subgenomic mRNA from the IG site (Makino et al., 1991). Because of the small size as well as the presence of all *cis*-acting replication signals, DI RNAs have become an alternative yet powerful genetic approach for studying replication, transcription, recombination and pathogenesis of coronaviruses (Lai, 1992; Lin et al., 1993; Liao and Lai, 1992; Liao and Lai, 1994; Zhang et al., 1994).

MHV contains four or five structural proteins. The spike (S) protein, which forms the characteristic peplomers, is involves in virus attachment to the receptors of permissive cells, elicitation of neutralizing antibodies, and cell-fusion. The membrane (M) protein and the small envelope (E) protein are essential for virion assembly, morphogenesis, and formation of the viral envelope. The nucleocapsid (N) protein is associated with viral genomic RNA to form the nucleocapsid. The hemagglutinin/esterase (HE) protein is an optional glycoprotein, present only in certain MHV strains (Yokomori et al., 1989). It contains both the receptor-binding (binding to sialic acid-containing receptor) and receptor-destroying (acetylesterase) activities (Yokomori and Lai, 1989), similar to the HEF protein of influenza C virus (Herrler et al., 1985a,b). In bovine coronavirus, which is from the same antigenic subgroup as MHV, HE appears to be important for virus infection, as monoclonal antibodies against the HE have been shown to neutralize bovine coronavirus infectivity (Deregt and Babiuk, 1987; Deregt et al., 1989). This is in contrast to MHV-DVIM, in which the S protein was prevented from interacting with its cell receptor but which had a functional HE protein, which was unable to initiate a productive infection (Gagneten et al., 1995). However, the presence of HE in the neuropathogenic strain JHM and the preservation of the function of HE known in other viruses raise an interesting question: Does the HE play a role in pathogenesis, particularly in neuropathogenesis, of the neuropathogenic MHV strain JHM in mice? Passive immunization of mice with HE monoclonal antibodies protected mice from lethal infection and altered the pathogenicity (Yokomori et al., 1992). JHM variant, which expresses abundant HE, exhibited more neurovirulent in C57BL/6 mice than one which expresses little HE (Yokomori et al., 1995). These findings suggest that the HE protein contributes to viral neuropathogenicity by influencing either the rate of virus spread and/or cell tropism (Yokomori et al., 1995). However, the contribution of other viral gene products cannot be rigorously ruled out. Thus, the exact role of the HE protein in MHV neuropathogenicity remains to be illucidated.

Recently, we developed the DI RNA expression system, in which the bacterial chloramphenicol acetyltransferase (CAT) gene (Liao and Lai, 1994), the MHV HE gene (Liao et al., 1995) and the mammalian cellular gene interferon gamma (Zhang et al., 1997) were placed behind an IG sequence as a promoter and were expressed in cell culture. In the present study, we used the HE gene as an example to to test the idea of whether a viral structural gene can be efficiently expressed locally in the central nervous system (CNS) using this expression system. Our results showed that the HE gene was expressed in the CNS, and that the expression of HE altered viral pathogenesis. This study thus demonstrates the feasibility of using the DI expression system to express a foreign gene in the loci of virus replication *in vivo* and provides a model for studying viral pathogenesis.

3. MATERIALS AND METHODS

3.1. Mice

C57BL/6 mice were purchased from Jackson Laboratories (Bar Harbor, MA) at 6–7 weeks of age. Mice were housed locally in micro isolator cages and were used within 7 days of receipt. None was found to have preexisting anti-MHV antibodies prior to use. Mice were infected intracranially (i.c.) with 1×10^5 plaque forming units (PFU) of virus contained in 30 μl of phosphate buffered saline (PBS).

3.2. Viruses and Cells

MHV strain A59 was used throughout the study. It was plaque-purified twice and further passaged twice in cell cultures to enrich its infectivity before further use. The murine astrocytoma cell line DBT (Hirano et al., 1974) was used for virus propagation, plaque assay and preparation of DI stocks.

3.3. Construction of DI Vector

The HE gene was cloned into the DI cDNA of p25CAT to replace the CAT gene, resulting in pDE-HE, as described previously (Liao et al., 1995).

3.4. Preparation of DI Stock Viruses

For making DI stock viruses, plasmid DNAs of pDE-HE and p25CAT were linearized with restriction enzyme *Xba*I and subjected to *in vitro* transcription as described (Zhang et al., 1994). The *in vitro* transcribed RNA was then transfected into MHV A59-infected DBT cells with the DOTAP method (Boehringer-Mannheim). At 12 to 14 h posttransfection, culture medium was harvested and cell debri were clarified by low speed centrifugation. Clarified supernatants were used as DI stocks, designated DE-HE A59 or DE-CAT A59.

3.5. Determination of Virus Titers in Tissues

Virus titers in tissue were determined by homogenization of 1/2 the brain or 1/2 lobe of liver in 4.0 ml of Dulbecco's PBS, pH7.4 using Tenbrock tissue homogenizers. The remainder was processed for histopathology (see below). Following centrifugation at 1,500 g for 20 min at 4 °C supernatants were assayed immediately, or frozen at -70 °C. Virus titers were determined by plaque assay using monolayers of DBT cells as previously described (Yokomori et al., 1992). Data presented are the average titer per gram of tissue for groups of three or more mice.

3.6. RNA Isolation and RT-PCR Analysis

For detection of the expression of HE and CAT genes in mice following infection with DI viruses, RNAs were isolated from tissues with Trizol Reagent according to the manufacturer's instruction (Life Technologies), and were subjected to reverse transcription (RT) and polymerase chain reaction (PCR) using primer pairs 5'-L9 (specific to the leader RNA)/3'-HE74 for HE gene (Zhang and Lai, 1994) and 5'-L9/3'CAT542 (Zhang et al., 1994) for the CAT gene.

3.7. Histopathology

For routine histopathological analysis mice were sacrificed by CO_2 asphyxiation. Brains were removed and dissected in the midcoronal plane. Brains, spinal cords and livers were fixed for 3 h in Clark's soloution (75% ethanol and 25% glacial acetic acid) and embedded in paraffin. Sections were stained with hematoxylin and eosin for routine examination. The distribution of viral antigen was examined using immunoperoxidase staining (Vectastain-ABC kit, Vector Laboratories, Burlingame, CA) and anti-JHMV mAb J.3.3 specific for the carboxy terminus of the N protein (Fleming et al., 1983). All slides were read in a blinded manner.

4. RESULTS AND DISCUSSION

4.1. Expression of the HE Gene in the CNS Via an MHV DI Vector

To determine the expression of the HE gene in the CNS, RNA was extracted from the brains of DE-HE A59 infected mice at various time points and analyzed by RT-PCR.

The RT-PCR specifically detects HE-containing subgenomic mRNA which is transcribed from the IG site and contains a leader sequence at the 5'-end. Thus, subgenomic mRNA can be distinguished from the original input DI RNA. The subgenomic HE-containing mRNA was undetectable at 12 h p.i. However, expression of the HE mRNA could be detected at 24 and 48 h p.i., but not at subsequent time points. Similarly, in the control experiment, in which mice were infected with DE-CAT A59, the CAT mRNA could also be detected only at these time points. These data suggest that the expression of DI RNA following infection is limited to the first 2 days of the infection.

4.2. Expression of the HE Gene Alters Viral Pathogenesis

To determine if the expression of the HE gene from a DI vector could influence the pathogenesis of A59 virus infection, mice were infected i.c. with 1×10^5 PFU of DE-HE A59, DE-CAT A59, and parental A59. All mice infected with DE-CAT A59 succumbed to infection by day 9 p.i, similar to mice infected with parental A59 alone. These data demonstrate that inclusion of the DI RNA in the virus pool did not by itself alter the outcome of A59 infection. In contrast to these groups of mice, mice infected with the identical dose of DE-HE A59 showed identical signs of encephalitis by day 6 p.i.; however, 60% of these mice survived infection. Clinical symptoms resolved in all surviving mice by 12 to 14 d p.i. and no mice succumbed during the 30 days of observation. These data demonstrate that expression of the HE gene from the DI RNA alters the pathogenesis of A59 infection.

4.3. Viral Replication and Pathological Changes in the CNS of Infected Mice

To determine the basis for the reduction in mortality, virus titers in the brains of mice infected with DE-HE A59 were compared to mice infected with DE-CAT A59. As shown in Fig. 1A, a slight, but not statistically significant reduction in virus titers within the CNS was found at day 6 p.i. in the mice infected with DE-HE A59 compared to mice infected with DE-CAT A59 or parental A59 (data not shown). No difference was detected at either day 1 or day 2 p.i. when HE gene expression was detected in the CNS. Brains and spinal cords were also examined for histopathological changes at days 1, 2, 4, and 6 p.i. No differences were detected in either the extent or distribution of either mononuclear cell infilrates or viral antigen at days 1, 2 and 6 p.i (data not shown). At day 4 p.i. there was a small increase in the extent of inflammatory cell infiltration and reduced viral antigen in the mice infected with DE-HE A59 (data not shown). These data suggest that expression of the HE gene within the CNS results in a transient increase in CNS inflammation and reduction of viral antigen within the CNS.

To examine whether the altered pathogenesis of DE-HE A59 was due to the transient expression of the HE gene, virus was recovered from the brains of DE-HE A59 infected mice at 6 d p.i. This time point was chosen based on the absence of detectable HE gene expression in the CNS. All of the mice infected with this recovered virus (A59-R) with 1×10^5 PFU succumbed to infection by 9 d p.i.. The amount of virus present in the CNS of these mice was similar to that in mice infected with DE-HE A59, DE-CAT A59 or parental A59. These data suggest that the difference in mortality following infection with DE-HE A59 is not due to alterations in either virus replication within the CNS, cellular tropism or extent of inflammatory changes. However, the pathogenesis of the recovered A59 suggests that the reduced mortality is due to the transient expression of the HE gene from the DI RNA.

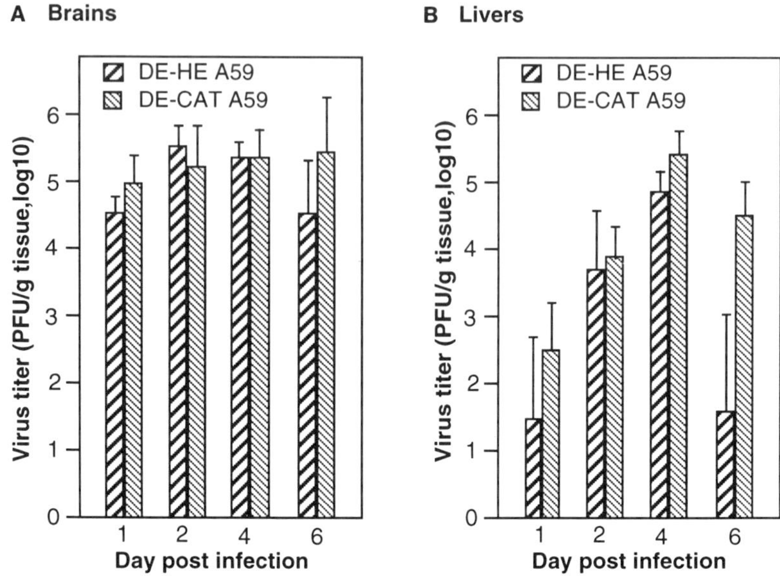

Figure 1. Viral replication in mice after intracrainial infection. C57BL/6 mice were inoculated intracraniallly with 1×10^5 PFU of DE-HE A59 and DE-CAT A59. There were at least 3 mice in each group. At various time points after infection, mice were sacrificed, and brains and livers were collected for virus isolation. Virus titer was determined by plaque-assay and is indicated as PFU/gram tissue. (A) Virus titers in the brains. (B) Virus titers in the livers.

4.4. Viral Replication and Pathological Changes in the Livers of Infected Mice

The A59 strain of MHV is both neurotropic and hepatotropic (Lavi et al., 1986). To explore the possibility that the decreased mortality of mice infected with DE-HE A59 was due to altered hepatotropism, virus replication and pathological changes in the livers of mice infected with DE-HE A59 and DE-CAT A59 were compared. Fig. 1B shows that at day 1 p.i. there was approximately 1 log less infectious virus recovered from the livers of mice infected with DE-HE A59 compared to mice infected with DE-CAT A59. This initial difference was transient, however, since there was no difference in virus titer at day 2 or 4 p.i. In contrast, by day 6 p.i., when the DE-CAT A59 infected mice were beginning to succumb to infection, there was a substantial 3 $\log_{10}$ difference in infectious virus as compared to mice infected with DE-CAT A59, parental A59, or A59-R (data not shown). Histological examination of the livers of the mice infected with DE-CAT A59 showed increased viral antigen and inflammatory changes beginning at day 2 p.i. By day 6 p.i., the livers of these mice showed evidence of extensive necrosis with prominent viral antigen. No differences were noted in the extent of inflammatory changes, necrosis or viral antigen in the livers of the mice infected with DE-HE A59 at day 1 or 2 p.i. However, by day 4 p.i., less viral antigen and less necrosis was noted in the mice infected with DE-HE A59. By day 6 p.i., there was a significant reduction in the amount of viral antigen, consistent with the decrease in recovery of infectious virus and substantially less necrosis compared to the almost confluent necrosis found in the mice infected with either parental A59 or DE-CAT A59. Similar to the data presented above, mice infected i.c. with virus recovered from mice initially infected with DE-HE A59 showed extensive liver necrosis by day 6 p.i. Taken together, these data indicate that the reduced mortality found in mice infected with DE-HE A59 is due to reduced hepatitis and is a consequence of the transient expression of the HE gene.

ACKNOWLEDGMENTS

We would like to thank Wenqiang Wei for preparing histological slides and Steve Ho for plaque assay, and Dr. Tom Kienzle for reading the manuscript. This work was supported by the Public Health Research Grant NS 18146 from the National Institute of Health. M.M.C.L. is an Investigator of the Howard Hughes Medical Institute.

REFERENCES

Deregt, D., and Babiuk, L. A., 1987, Monoclonal antibodies to bovine coronavirus: Characteristics and topographical mapping of neutralizing epitopes on the E2 and E3 glycoproteins, *Virology* **161**: 410–420.

Deregt, D., Gifford, G.A., Khalid Ijaz, M., Watts, T.C., Gilchrist, J.E., Haines, D.M., and Babiuk, L.A., 1989, Monoclonal antibodies to bovine coronavirus glycoproteins E2 and E3: demonstation of in vivo neutralizing activity, *J. Gen. Virol.* **70**:993–998.

Fleming, J.O., Stohlman, S.A., Harmon, R.C., Lai, M.M.C., Frelinger, J.A., and Weiner, L.P., 1983, Antigenic relationship of murine coronaviruses: analysis using monoclonal antibodies to JHM (MHV-4) virus, *Virology* **131**: 296–307.

Gagnetan, S., Gout, O., Dubois-Dalcq, M., Rottier, P., Rossen, J., and Holmes, K.V., 1995, Interaction of mouse hepatitis virus (MHV) spike glycoprotein with receptor glycoprotein MHVR is required for infection with an MHV strain that expresses the hemagglutinin-esterase glycoprotein, *J. Virol.* **69**: 889–895.

Herrler, G., Rott, R., Klenk, H.D., Müller, H.P., Shukda, A.K., and Schauer, R., 1985, The receptor-destroying enzyme of influenza C is virus neuraminidate-O-acetyl-esterase, EMBO J. **4**:1503–1506.

Herrler, G., Rott, R., and Klenk, H.D., 1985, Neuraminic acid is involved in the binding of influenza C virus to erythrocytes, *Virology* **141**: 144–147.

Hirano, N., Fujiwara, K., Hino, S., and Matsumoto, M., 1974, Replication and plaque formation of mouse hepatitis virus (MHV-2) in mouse cell line DBT culture, *Arch. Gesamte Virusforsch.* **44**, 298–302.

Lai, M.M.C., 1992, RNA recombination in animal and plant viruses, *Microbiol. Rev.* **56**: 61–79.

Lavi, E., Gilden, D.H., Highkin, M.K., Weiss, S.R., 1986, The organ tropism of mouse hepatitis virus A59 in mice is dependent on dose and route of inoculation, *Lab. Ani. Sci.* **36**:130–135.

Lee, H.-J., Shieh, C.-K., Gorbalenya, A. E., Koonin, E. V., La Monica, N., Tuler, J., Bagdzyahdzhyan, A., and Lai, M. M. C., 1991, The complete sequence (22 kilobases) of murine coronavirus gene 1 encoding the putative proteases and RNA polymerase, *Virology* **180**: 567–582.

Liao, C.-L., and Lai, M. M. C., 1994, Requirement of the 5'-end genomic sequence as an upstream cis-acting element for coronavirus subgenomic mRNA transcription, *J. Virol.* **68**: 4727–4737.

Liao, C.-L., Zhang, X. M., and Lai, M. M. C., 1995, Coronavirus defective-interfering RNA as an expression vector: the generation of a pseudorecombinant mouse hepatitis virus expressing hemagglutinin-esterase, *Virology* **208**: 319–327.

Lin, Y.-J., and Lai, M. M. C, 1993, Deletion mapping of a mouse hepatitis virus defective interfering RNA reveals the requirement of an internal and discontiguous sequence for replication, *J. Virol.* **67**: 6110–6118.

Makino, S., Shieh, C.-K., Soe, L. H., Baker, S. C., and Lai, M. M. C., 1988, Primary structure and translation of a defective-interfering RNA of murine coronavirus, *Virology* **166**: 550–560.

Makino, S., Joo, M., and Makino, J. K., 1991, A system for study of coronavirus mRNA synthesis: A regulated, expressed subgenomic defective-interfering RNA results from intergenic site insertion, *J. Virol.* **65**: 6031–6041.

Yokomori, K., La Monica, N., Makino, S., Shieh, C.-K., and Lai, M.M.C., 1989, Biosynthesis, structure, and biological activities of envelope protein gp65 of murine coronavirus, *Virology* **173**: 683–691.

Yokomori, K., Baker, S.C., Stohlman, S.A., and Lai, M.M.C., 1992, Hemagglutinin-esterase (HE)-specific monoclonal antibodies alter the neuropathogenicity of mouse hepatitis virus, *J. Virol.* **66**:2865–2874.

Yokomori, K., Asanaka, M., Stohlman, S.A., Makino, S., Shubin, R.A., Gilmore, W., Weiner, L.P., Wang, F.I., and Lai, M.M.C., 1995, Neuropathogenicity of mouse hepatitis virus JHM isolates differing in hemagglutinin-esterase protein expression, *J. Neurovirol.* **1**:330–339.
Zhang, X. M., Liao, C.-L., and Lai, M. M. C., 1994, Coronavirus leader RNA regulates and initiates subgenomic mRNA transcription both in *trans* and in cis, *J. Virol.* **68**: 4738–4746.
Zhang, X.M., Hinton, D.R., Cue, D., Stohlman, S., and Lai, M.M.C., 1997: Expression of gamma interferon by a coronavirus defective-interfering RNA vector and its effect on viral replication, spread and pathogenicity, *Virology.* In press.

68

THE MOUSE HEPATITIS VIRUS A59 SPIKE PROTEIN IS NOT CLEAVED IN PRIMARY HEPATOCYTE AND GLIAL CELL CULTURES

Susan T. Hingley,[1] Isabelle Leparc-Goffart,[2] and Susan R. Weiss[2]

[1]Department of Microbiology and Immunology
Philadelphia College of Osteopathic Medicine
Philadelphia, Pennsylvania 19131
[2]Department of Microbiology
University of Pennsylvania
Philadelphia, Pennsylvania 19104

1. ABSTRACT

Mouse hepatitis virus strain A59 (MHV-A59) produces mild meningoencephalitis and severe hepatitis during acute infection. To determine whether an *in vitro* system could be established which would mimic *in vivo* replication of the virus, we examined the ability of MHV-A59 to replicate in primary cultures of hepatocytes derived from C57BL/6 mice. Infection of hepatocytes with MHV-A59 resulted in low levels of replication, with virus remaining cell associated. Maximum viral yield was observed at 24 hours postinfection, while occasional syncytia were observed at 48 hours postinfection. Primary glial cell culture represents a potential *in vitro* system representing the second main target of MHV-A59, namely the brain. It is known that MHV-A59 produces a productive, but nonlytic infection in these cultures. Since cell-to-cell fusion is associated with the cleavage of S, the observation of little or no syncytia following MHV-A59 infection of both hepatocytes and glial cells prompted us to examine the cleavage of the spike protein (S) by Western blot analysis. The cleavage of S is inefficient in MHV-A59 infected hepatocytes and in glial cells. Furthermore, no cleavage of this protein was detected in liver homogenates from C57BL/6 mice infected with MHV-A59. These data suggest that cleavage of the MHV-A59 S protein, and by inference cell-to-cell fusion, does not seem to be essential for entry and spread of the virus *in vivo* and *in vitro*.

Coronaviruses and Arteriviruses, edited by Enjuanes *et al.*
Plenum Press, New York, 1998

2. INTRODUCTION

Mouse hepatitis virus strain A59 (MHV-A59) is a murine coronavirus which is both hepatotropic and neurotropic. In the central nervous system (CNS), MHV-A59 produces an acute meningoencephalitis, as well as chronic demyelinating disease in animals which survive acute infection (Lavi *et al.*, 1988; Lavi *et al.*, 1990).

MHV-A59 produces a lytic infection with extensive syncytia formation in continuously cultured cell lines. However it produces a productive, nonlytic infection in primary mixed glial cell cultures (Lavi, *et al.*, 1987). Since the two main sites of MHV-A59 pathology in C57BL/6 mice are the CNS and the liver, we were interested in determining what type of infection the virus would produce in primary hepatocyte cultures. MHV-A59 infection of primary cell cultures may be more reflective of the process of infection that occurs *in vivo* in infected animals than infection of continuously cultured cell lines.

MHV infectivity is determined, at least in part, by the spike or S protein. The S protein is an envelope glycoprotein responsible for viral attachment and virus-induced cell-to-cell fusion. Variants of both MHV-A59 and the JHM strain of MHV, with mutations in the S protein and an altered fusion phenotype, were attenuated *in vivo* and *in vitro*, suggesting that this viral protein is an important mediator of pathogenicity (Gallagher *et al.*, 1991; Hingley *et al.*, 1994). However, the fusion phenotype does not necessarily correlate with virulence. For instance, fusion-competent revertants of the MHV-A59 mutants were still attenuated *in vivo* (Hingley *et al.*, 1994). Furthermore, for the JHM fusion mutant as well, it appears that pathogenicity is not linked to the fusion phenotype (M. Buchmeier, personal communication). Therefore, the ability to induce cell-to-cell fusion may not be important in determining the virulence of MHV.

It has been reported that cleavage of the MHV S protein into its S1 and S2 subunits results in more efficient induction of syncytia than that induced by uncleaved S (Bos *et al.*, 1995; Frana *et al.*, 1985; Gombold *et al.*, 1993; Sturman *et al.*, 1985). While it is well established that the S proteins of many coronaviruses are cleaved *in vitro* by continuously cultured cell lines (Spaan *et al.*, 1988), it has not been shown whether cleavage of MHV S proteins occurs *in vivo*. The lack of syncytia observed in primary glial cell cultures implies that S may not be efficiently cleaved by this cell type.

Results presented here indicate that there is little or no cleavage of the S protein of MHV-A59 following viral replication in primary cultures of murine glial cells or hepatocytes. Furthermore, the S protein present in homogenates of livers taken from MHV-A59-infected mice also was not cleaved, suggesting that cleavage of S may not occur *in vivo*. This observation implies that efficient cell-to-cell fusion may not be required for viral virulence *in vivo*.

3. METHODS AND MATERIALS

3.1. Virus and Cells

MHV-A59 was obtained from L. Sturman (Albany, NY), and propagated in mouse L2 cells. Virus was plaque purified by three passages through L2 cells prior to use.

L2 cells, a mouse fibroblast cell line, were grown in Dulbecco's modified Eagles's medium (DMEM) containing 10% fetal bovine serum (FBS).

Primary cells included mixed glial cell cultures and hepatocyte cultures. Glial cell cultures were established as described previously (Gombold and Weiss, 1992; Lavi *et al.*,

1987). Primary hepatocyte cultures were established using the technique of Seglen (1997). Hepatocytes were isolated from 4–5 week old C57BL/6 mice by *in situ* perfusion of the liver with collagenase. After perfusion, the liver was removed, gently dissociated by pipetting and then filtered through nylon mesh (100μm). The cell suspension was centrifuged, the pellet resuspended and hepatocytes purified by low-speed centrifugation of the cell suspension through a Percoll medium having a density of 1.06 g/ml. This results in the separation of single and viable hepatocytes (parenchymal cells) from dead, aggregated hepatocytes, debris and non parenchymal cells (including Kupffer and endothelial cells; Kreamer *et al.*, 1986). The viable hepatocytes were washed, counted and plated on collagen coated dishes and grown in Hepatocyte-SFM medium (GibcoBRL). The purity of hepatocytes was assessed by the distinctive large size and morphology of the cells, and appeared to be nearly 100% hepatocytes by this criterion.

3.2. Infection of Primary Hepatocytes

Confluent monolayers of hepatocytes were infected with virus at a multiplicity of infection (MOI) of 5 pfu/cell. After 1 hour of adsorption at 37°C, the cells were washed three times and incubated in Hepatocyte-SFM medium. At various times postinfection, supernates were removed and the cells lysed by freeze/thawing twice in one ml of PBS. Viral titers in both supernates and samples from lysed cells were determined by plaque assay on L2 cells as described previously (Hingley *et al*, 1994).

3.3. SDS Page and Western Blotting

Cell lysates used in SDS PAGE were obtained by treating infected monolayers with lysis buffer (50 mM Tris, pH 7.4; 150 mM NaCl; 1% Nonidet P-40) for 5 minutes with agitation. Lysates were diluted in 2X Laemmli sample buffer, electrophoresed into 6 or 8% SDS polyacrylamide gels following standard electrophoresis procedures, and transferred to nitrocellulose for Western blotting. In some instances, culture supernates from virus infected L2 or glial cell cultures, or liver homogenates from infected animals, were diluted in 2X Laemmli sample buffer and used in SDS PAGE as described above.

Following transfer to nitrocellulose, proteins were probed using a polyclonal goat anti-S antibody obtained from Dr. K. Holmes (Denver, Colorado). Bound antibody was detected by enhanced chemiluminescence (Amersham, Arlington Heights, IL) following the manufacturers written protocol.

4. RESULTS

4.1. Replication of MHV-A59 in Hepatocytes

A representative experiment of viral titers present in the supernates and cell fractions of primary cultures of hepatocytes infected with MHV-A59 is shown in Figure 1. MHV-A59 is capable of replicating in hepatocytes, but remains primarily cell associated. This differs from the pattern of replication seen in L2 cells or glial cells, where virus is released into the supernate and achieves titers several logs higher than what was observed in hepatocyte cultures. Only minimal syncytia formation was observed in hepatocyte cultures at 48 hours postinfection, after the peak of viral replication.

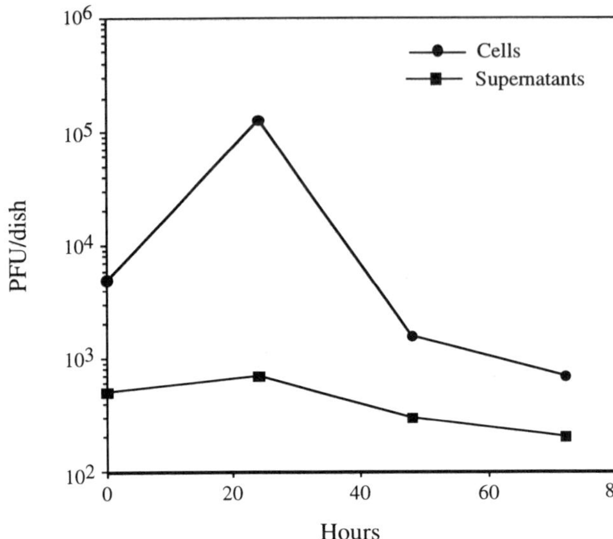

Figure 1. Replication of MHV-A59 in primary cultures of hepatocytes. Confluent monolayers of hepatocytes derived from C57BL/6 mice were infected with MHV-A59 at an MOI of 5 PFU/cell. Supernates and cells were obtained at 0, 24, 48 and 72 hours pi. Viral titers were determined by plaque assay on L2 cells.

4.2. Cleavage of MHV S Proteins

Previous studies have examined the replication of MHV-A59 in primary glial cell cultures (Gombold and Weiss, 1992; Lavi *et al.*, 1987). It was known that MHV-A59 produces a productive infection of glial cells with progeny virus released into the supernate and no observable syncytia formation. Similarly, very few syncytia were present in hepatocyte cultures, and not until 48 hours postinfection. Since cleavage of the S protein correlates with syncytia formation in many strains of coronavirus, we examined the extent of cleavage of S by primary cultures of glial cells and hepatocytes, as compared to that seen

Table 1. Cleavage of S and syncytia formation following infection of different cell types with MHV-A59

	Cleavage of S[4]	Syncytia formation
L2 cells	+	+
Hepatocytes[1]	–	–
Liver tissue[2]	–	unknown
Glial cells[3]	–	–

[1]Primary hepatocytes, derived from C57BL/6 mice were infected with virus at an MOI of 2.5 PFU/cell. Cells were collected and lysed at 24 hours post infection; proteins from cell lysates were visualized by Western blot analysis for the cleavage of S protein.

[2]C57BL/6 mice were inoculated intracranially with 5000 PFU of MHV-A59. At 4 days pi, animals were sacrificed, livers were removed and homogenized, and homogenates were used directly in Western blots for analysis of the cleavage of S.

[3]Infections (MOI of 2.5 PFU/cell) were carried out on primary glial cells derived from C57BL/6 mice. Supernates were collected at 24 hours pi and visualized by Western blot analysis for the cleavage of S.

[4]Cleavage of S was detected by Western blot analysis with a polyclonal monospecific antisera A04 (Dr. K. Holmes) at a dilution of 1/2000.

in L2 cells. In contrast to results from L2 cells, the MHV-A59 S protein showed little or no cleavage following viral replication in both types of primary cell cultures (Table 1).

Since it appeared that hepatocytes were unable to cleave MHV-A59 S protein, we looked at virions from liver homogenates of MHV-A59 infected mice. Consistent with the results from primary hepatocyte cultures, the MHV-A59 S protein was not cleaved in liver homogenates (Table 1).

5. DISCUSSION

MHV-A59 infects both the CNS and liver of susceptible mice. We examined the replication of this virus in primary cell cultures from both the brain and the liver in order to evaluate *in vitro* systems which may be relevant to the study of MHV-A59 pathogenesis. While it is known that MHV-A59 produces a productive, nonlytic infection in primary glial cell cultures, less is known about the ability of MHV-A59 to infect different cells in the liver. Here we examine the ability of MHV-A59 to replicate in primary cultures of hepatocytes, a cell type which may be involved in pathogenesis.

Results published previously demonstrate that MHV-A59 replication in glial cell cultures derived from C57BL/6 mice is comparable to that observed with continuously cultured cell lines (Lavi *et al.*, 1987). In contrast, this virus exhibits only limited replication in hepatocytes (derived from C57BL/6 mice), with titers increasing approximately 20-fold in 24 hours, then decreasing. In addition, unlike glial cells, where virus is released into the culture supernate, MHV-A59 remained cell-associated during infection of primary hepatocyte cultures.

The relatively poor ability of MHV-A59 to replicate in hepatocytes raises the possibility that hepatocytes are not a major site of replication for MHV-A59. It is possible that replication of MHV-A59 in other cell types, such as Kupffer cells or liver endothelial cells for example, contributes to the hepatitis observed during infection of C57BL/6 mice with this virus. In support of this hypothesis, other studies have compared the ability of hepatotropic (MHV-3) and nonhepatotropic (MHV-4) strains of MHV to replicate in different cell types isolated from livers of Balb/c mice (Joseph *et al.*, 1995; Martin *et al.*, 1994). These studies concluded that replication in Kupffer cells and/or hepatic endothelial cells, but not hepatocytes, correlates with hepatotropism. If hepatovirulence is associated with hepatotropism, it is quite possible that the ability of MHV to cause hepatitis is not determined by its ability to infect hepatocytes.

Many studies on the pathogenesis of MHV focus on the viral spike or S protein. The S protein is a surface glycoprotein responsible for both viral attachment and cell fusion. Cleavage of this protein correlates with an enhanced ability to induce cell-to-cell fusion *in vitro* (Bos *et al.*, 1995; Frana *et al.*, 1985; Gombold *et al.*, 1993; Sturman *et al.*, 1985), however the significance of cell fusion for virulence has not been established. For instance, the two neurotropic strains of MHV, MHV-A59 and JHM, have cleaved S proteins and can induce syncytia formation in continuously cultured cell lines. Fusion mutants of these strains of MHV were found to be attenuated *in vivo* (Gallagher *et al.*, 1991; Hingley *et al.*, 1994), however, attenuation is not necessarily associated with the altered fusion phenotype (Hingley *et al.*, 1994).

In contrast to the extensive viral-mediated cell-to-cell fusion which occurs in continuously cultured cell lines, in primary hepatocyte or glial cell cultures, minimal or no syncytia were observed during infection by MHV-A59. Since cleavage of S is associated with more efficient cell-to-cell fusion, this lack of syncytia formation suggests that the S protein of MHV-A59 may not be cleaved by these cell types. Indeed, the MHV-A59 S protein was not cleaved when the virus replicated in primary hepatocyte or glial cell cultures, although it is cleaved by L2 cells. Western blot analysis, therefore, is consistent with the observation that syncytia formation by MHV-A59 correlates with cleavage of the S protein. Furthermore, the lack of cleavage of MHV-A59 S in liver homogenates of infected C57BL/6 mice implies that *in vivo*, cell-to-cell fusion may not occur.

The association between cleavage of S, induction of cell-to-cell fusion and pathogenicity may not be as strong with MHV as with other viruses. Many viral fusion glycoproteins require proteolytic cleavage in order to be mediate both viral-cell fusion and cell-to-cell fusion (Guo *et al.*, 1990; Kawaoka and Webster, 1988; McCure *et al.*, 1988). For MHV, on the other hand, cleavage of S enhances fusogenicity, but is not absolutely required for either infectivity (*i.e.* viral-cell fusion) or syncytia formation (*i.e.* cell-to-cell fuison) (Bos *et al.*, 1995; Frana *et al.*, 1985; Gombold *et al.*, 1993; Sturman *et al.*, 1985). With other viruses, the ability to produce cell-to-cell fusion is important in determining viral pathogenicity and spread in the infected host (Bosch *et al.*, 1981; Glickman *et al.*, 1988; Klenk & Rott, 1988; Ohuchi *et al.*, 1989). With MHV, the observation that the fusion glycoprotein may not be cleaved *in vivo* and the subsequent implication that cell-to-cell fusion may not occur, suggests that cell-to-cell fusion may not be an important factor in viral virulence.

ACKNOWLEDGMENTS

This work is supported by public health service grant numbers NS-30606 (SRW) and NS-21954 (SRW).

REFERENCES

Bos, E.C.W., Heijnen, L., Luytjes, W., and Spaan, W.J.M., 1995, Mutational analysis of the murine coronavirus spike protein: effect on cell-to-cell fusion, *Virology* **214**:453–463.

Bosch, F. X., Garten, W., Klenk, H.-D., and Rott, R., 1981, Proteolytic cleavage of influenza virus hemagglutinins: primary structure of the connecting peptide between HA1 and HA2 determines the proteolytic cleavability and pathogenicity of avian influenza viruses, *Virology* **113**:725–735.

Frana, M.F., Behnke, J.N., Sturman,S., and Holmes, K.V., 1985, Proteolytic cleavage of the E2 glycoprotein of murine coronavirus: host-dependent differences in proteolytic cleavage and cell fusion, *J. Virol.* **56**:912–920.

Gallagher, T.M., Escarmis,C., and Buchmeier,M. J., 1991, Alteration of pH dependence of coronavirus-induced cell fusion: Effect of mutations in the spike glycoprotein, *J. Virol.* **65**:1916–1928.

Glickman, R.L., Syddall, R.J., Iorio, R.M., Sheehan, J.P., and Bratt, M.A., 1988, Quantitative basic residue requirements in the cleavage-activation site of the fusion glycoprotein as a determinant of virulence for Newcastle disease virus, *J. Virol.* **62**:354–356.

Gombold, J.L., and Weiss, S.R., 1992, Mouse hepatitis virus A59 increases steady state levels of MHC mRNAs in primary glial cell cultures and in the murine central nervous system, *Microbial Pathogenesis* **13**:493–505.

Gombold, J.L., Hingley, S.T., and Weiss, S.R., 1993, Fusion-defective mutants of mouse hepatitis virus A59 contain a mutation in the spike protein cleavage signal, *J. Virol.* **67**:4504–4512.

Guo, H.-G., diMarzo Veronese, F., Tschachler, E., Pal, R., Kalyanaraman, V.S., Gallo, R.C., and Reitz, Jr., M.S., 1990, Characterization of an HIV-1 point mutant blocked in envelope glycoprotein cleavage, *Virology* **174**:217–224.

Hingley, S.T., Gombold, J.L., Lavi, E., and Weiss, S.R., 1994, MHV-A59 fusion mutants are attenuated and display altered hepatotropism, *Virology* **200**:1–10

Joseph, J., Kim, R., Siebert, K., Lublin, F.D., Offenbach, C., and Knobler, R.L., 1995, Organ specific cell heterogenicity influences differential replication and cytopathogenicity of MHV-3 and MHV-4, *Adv. Exp. Med. Biol.* **380**:43–50.

Kawaoka, Y. and Webster, R.G., 1988, Sequence requirements for cleavage activation of influenza virus hemagglutinin expressed in mammalian cells, *Proc. Natl. Acad. Sci. USA* **85**:324–328.

Klenk, H.-D. and Rott, R., 1988, The molecular biology of influenza virus pathogenicity, *Adv. Vir. Res.* **34**:247–281.

Kreamer, B.L., Staecker, J.L., Sawada, N., Sattler, G.L., Hsia, M.T.S., and Pitot, H.C., 1986, Use of low-speed, iso-density percoll centrifugation method to increase the viability of isolated rat hepatocyte preparations, *In Vitro Cell. Devel. Biol.* **22**:201–211.

Lavi, E., Suzumura, A., Hirayama, M., Highkin, M.K., Dambach, D.M., Silberberg, D.H., and Weiss, S.R., 1987, Coronavirus mouse hepatitis virus (MHV)-A59 causes a persistent, productive infection in primary glial cell cultures, *Microbial Pathogenesis* **3**:79–86.

Lavi, E., Fishman, P.S., Highkin, M.K., and Weiss, S.R., 1988, Limbic encephalitis after inhalation of a murine coronavirus, *Lab. Investig.* **58**:31–36.

Lavi, E., Murray, E.M., Makino, S., Stohlman, S.A., Lai, M.M.C., and Weiss, S.R., 1990, Determinants of coronavirus MHV pathogenesis are localized to 3' portions of the genome as determined by ribonucleic acid-ribonucleic acid recombination, *Lab. Investig.* **62**:570–578.

Martin, J.P., Chen, W., Koehren, K., Pereira, C.A., 1994, The virulence of mouse hepatitis virus 3 as evidenced by permissivity of cultured hepatic cells toward escape mutants, *Res. Virol.* **145**:297–302.

McCune, J.M., Rabin, L.B., Feinberg, M.B., Leiberman, M., Kosek, J.C., Reyes, G.R., and Weissman, I.L., 1988, Endoproteolytic cleavage of gp160 is required for the activation of human immunodeficiency virus, *Cell* **53**: 55–67.

Ohuchi, M., Orlich, M., Ohuchi, R., Simpson, B.E.J., Garten, W., Klenk, H.-D., and Rott, R., 1989, Mutations in the cleavage site of the hemagglutinin alter the pathogenicity of influenza virus A/Chick/Penn/83 (H5N2), *Virology* **168**:274–280.

Seglen, P.O., 1976, Preparation of isolated rat liver cells, *Meth. Cell Biol.* **13**:29–83.

Spaan, W.J.M., Cavanagh, D., and Horzinek, M.C., 1988, Coronaviruses. Structure and genome expression, *J. Gen. Virol.* **69**:2939–2952.

Sturman, L.S., Ricardi, C.S., and Holmes, K.V., 1985, Proteolytic cleavage of the E2 glycoprotein of murine coronavirus: activation of cell fusing activity of virions by trypsin and separation of two different 90K cleavage fragments, *J. Virol.* **56**:904–911.

Taguchi, F., Kawamura, S., and Fujiwara, K., 1983, Replication of mouse hepatitis viruses with high and low virulence in cultured hepatocytes, *Infect. Immun.* **39**:955–959.

69

THE PATHOGENESIS OF MHV NUCLEOCAPSID GENE CHIMERIC VIRUSES

E. Lavi,[1] J. A. Haluskey,[1] and P. S. Masters[2]

[1]Division of Neuropathology
Department of Pathology and Laboratory Medicine
University of Pennsylvania, School of Medicine
Philadelphia, Pennsylvania 19104
[2]Wadsworth Center for Laboratories and Research
New York State Department of Health
Albany, New York 12201

1. ABSTRACT

A set of viruses in which various segments of the nucleocapsid (N) gene of MHV have been substituted with the corresponding segments of bovine coronavirus (BCV) by targeted recombination were analyzed for their biologic properties. Histology for organ pathology and plaque assay for viral titer analysis following intracerebral (IC) inoculation were studied. One chimeric virus (Alb85), in which only a small segment of the N gene was replaced, exhibited a phenotype similar to wild type MHV-A59. However, three of the chimeric viruses (Alb106, Alb112 and Alb100) produced acute encephalitis and demyelination but without hepatitis following IC inoculation. Intravenous (IV) and intrahepatic (IH) inoculations were able to restore the ability of these viruses to produce hepatitis. The common denominator of the three chimeric viruses with a different phenotype is a region between aa 306 and aa 386 in which 17 amino acids (aa) differences exist between the two strains. Thus this region may contain determinants which enable the virus to exit the brain and reach the blood stream.

2. INTRODUCTION

Coronavirus mouse hepatitis virus (MHV) strain A59 produces acute focal encephalitis and hepatitis and chronic demyelinating disease (Lavi et al., 1988; Lavi et al., 1984a; Lavi et al., 1986; Lavi et al., 1984b; Lavi and Weiss, 1989). The genome of MHV-A59 is

Coronaviruses and Arteriviruses, edited by Enjuanes *et al.*
Plenum Press, New York, 1998

nearly of 32kb in size, the largest among RNA viruses (Pachuk et al., 1989). The lack of an available infectious clone of the entire virus prevents us from studying the pathogenesis of MHV-A59 by simply exchanging portions of the virus between strains with different biologic properties. However, in recent years, targeted recombination between 3' segments of coronaviruses became available (Koetzner et al., 1992). To study potential determinants of pathogenesis within the N gene of MHV and to test whether chimeric viruses produced by targeted recombination are suitable for pathogenesis studies we used a set of viruses in which segments of the N gene of MHV have been substituted with the corresponding segments of bovine coronavirus (BCV).

The preparation of MHV-BCV chimeric viruses Alb 106, Alb 112, Alb 85, and Alb 100 has been previously described (Peng et al., 1995). Alb 106 has a replacement of aa 227–397 of the MHV-A59 N gene with aa 226–398 of BCV N gene. Alb 112 has a replacement of aa 123–187 and 307–397 with BCV N aa 120–184 and 306–398 respectively. Alb100 has a replacement of aa 304–397 with BCV 303–398, and Alb85 has a replacement of aa 384–397 with aa 386–398 of BCV.

3. METHODS AND MATERIALS

3.1. Viruses

The stock of MHV-A59 is a three times plaque purified virus with a titer of 2×10^7pfu/ml (Lavi et al., 1986; Lavi et al., 1984b). The chimeric viruses Alb 106, Alb 112, Alb 85, Alb 100 were previously described (Peng et al., 1995). Viruses were propagated and titered in L2 transformed murine fibroblasts in DMEM media supplemented with heat-inactivated 10% fetal calf serum, glutamine and antibiotics.

3.2. Mice and Inoculations

Four-week-old C57BL/6 virus free male mice were used (Jackson Laboratories, Bar Harbor, Maine). Injection of 20µl of virus or cell lysate as control, was done into the left hemisphere of mice lightly anesthetized with methoxyflurane. One LD50 dose for each virus was used (30,000 pfu for Albany viruses and 5000 pfu for wt MHV-A59).

3.3. Virulence Assay

The lethal dose of viruses that kills 50% of the animals (LD50) was used to determine the virulence of viruses. It was examined by intracerebral inoculation of groups of 4 week old mice with the viruses as previously described (Lavi et al., 1984b). Four week old C57Bl/6 mice were inoculated IC with 10 fold serial dilutions of recombinant and parental viruses. Ten mice were used per dilution and five 10 fold dilutions per virus were used. Mice were monitored daily for signs of disease, and mortality. The LD50 was calculated according to the standard Reed-Muench formula as we have previously described (Lavi et al., 1984b).

3.4. Viral Replication in Organs

To assess viral replication in organs viral titers in tissues were determined by plaque assay. Organs were aseptically removed at autopsy and while half of the organs were fixed in formalin for light microscopy another portion (half of the brain and equivalent portion of the liver) were frozen, homogenized, serially diluted in PBS and plaque-assayed in 6-well-plate agarose-based plaque assay in L2 transformed murine fibroblasts (Lavi et al., 1986; Lavi et al., 1984b).

3.5. Histopathology and Organ Tropism of the Viruses

Four week old C57Bl/6 mice were inoculated IC with 1 LD50 dose of the virus. Groups of 3–5 mice per time point (days 1,3,5,7,11,15,30,60 post inoculation) were sacrificed by anesthetic (methoxyflurane) overdose and perfused intracardially with PBS followed by 10% buffered formalin. Brain, spinal cord and liver were removed and fixed in 10% buffered formalin. Tissues are embedded in paraffin, sectioned and stained with H&E for light microscopy evaluation as previously described (Lavi et al., 1988; Lavi et al., 1986; Lavi et al., 1984b; Lavi et al., 1990).

3.6. Demyelination

Four week old C57Bl/6 mice were inoculated IC with 1 LD50 dose of the virus. At 30 days post infection, mice were perfused with PBS, followed by 10% buffered formalin. Brains and spinal cords were removed, tissue was embedded in paraffin, and sectioned for staining with luxol fast blue to detect plaques indicative of demyelination. Alternatively, mice were perfused and spinal cords were fixed in 4% glutaraldehyde, embedded in Epon and stained with toluidine blue (Gombold et al., 1995; Lavi et al., 1984b).

4. RESULTS

4.1. Growth Characteristics of Chimeric Viruses in Culture

To rule out the possibility that chimeric viruses have different biologic properties due to defects in replication we first tested the ability of viruses to replicate in culture. Growth kinetics of chimeric viruses was tested in susceptible L2 murine fibroblast cell cultures by a 48 hour growth curve with time points titers at 4, 8, 12, 16, 24, 48 hours post infection. All chimeric viruses had growth characteristics in tissue culture similar to MHV-A59.

4.2. Virulence and Organ Tropism of Chimeric Viruses in Mice

The LD50 dose of chimeric viruses was determined. The LD50 of Alb106, Alb112, Alb85 and Alb100 was approximately 3×10^4 pfu. One LD50 dose of Alb106, Alb112, Alb100, Alb85, MHV-A59 and mock-infection with L2 cell lysate were injected intracerebrally into 4-week old C57Bl/6 mice. Tissues were harvested and examined for liver and brain histology at 1,3,5,7,9,11,30 days post inoculation. Mock-infected mice had no pathologic changes in organs. All the chimeric viruses produced encephalitis similar to wt MHV-A59. Alb106, Alb112, Alb100 did not produce hepatitis as assessed by light microscopy of tissue sections stained with H&E. Alb85 produced hepatitis similar to wt MHV-A59.

4.3. Replication of Chimeric Viruses in Mouse Organs

Replication of viruses in organs was assessed by plaque assays following inoculation with one LD50 dose of Alb106, Alb112, Alb100, Alb85, MHV-A59 and mock-infection with L2 cell lysate injected intracerebrally into 4-week old C57Bl/6 mice. Tissues were harvested and examined for liver and brain titers at 1,3,5,7,9,11,30 days post inoculation. Mock-infected mice had no viral titers in organs. All the chimeric viruses produced viral titers in the brain similar to wt MHV-A59. Alb106, Alb112, Alb100 did not produce viral titers in the liver as assessed by plaque assays except for one mouse in which titers in the liver were below 2 logs of virus on day 5 post inoculation when brain titers reached 6–7 logs of virus. Alb85 produced viral titers in the liver as well as brain similar to wt MHV-A59. Thus the titer assays were consistent with the histopathological results.

4.4. Demyelination

Demyelination was assessed following IC inoculation with either Alb 112 or Alb106. Groups of 5 mice per virus were inoculated with 1LD50 dose of each one of the viruses. All three viruses produced demyelination in mice. Mice inoculated with Alb106 had demyelination in 2/6 mice. Mice inoculated with Alb112 had demyelination in 4/6 mice and mice inoculated with Alb100 had demyelination in 4/5 mice.

4.5. Comparative Pathology and Replication of Chimeric Viruses in Mouse Organs by IC, IV, and IH Routes of Inoculation

We then asked whether the reason for the inability to produce hepatitis was because of inability of the viruses to infect the liver or the inability of the viruses to reach the liver following IC inoculation. The main route of dissemination of MHV through the infected organism is presumed to be via hematogenous dissemination. Thus Alb106 and Alb112 and MHV-A59 were used for comparative analysis of different routes of inoculation. Each one of the viruses was injected IV or IH in separate experiments. Both Alb106 and Alb112 produced acute hepatitis and had high viral titers recovered from the livers under these conditions similar to MHV-A59. There was no evidence of encephalitis upon IV or IH inoculations with all three viruses, consistent with the inability of the virus to cross the blood brain barrier if not injected directly into the brain or into the nasal cavity where it is capable of using intraneuronal transport through the nerve endings of the olfactory nerves.

5. DISCUSSION

The study described here provides several new insights into the pathogenesis of MHV. It shows for the first time that targeted recombination can be used as a tool in studying MHV pathogenesis. The exchange of sequences between the two strains of viruses did not by itself change the phenotype as shown by the fact that the phenotype of Alb85 was similar to that of MHV-A59. Only viruses in which a specific region, presumably important for a certain function was exchanged, showed selective change of a specific biologic property. Three different chimeric viruses (Alb106, Alb112, and Alb100) which were produced under different recombination events, but contain this putative region, consistently showed the same phenotype in repeated experiments and in multiple animals.

The experiments shown here revealed the existence of viruses which were unable to produce hepatitis upon IC inoculation without effect on the ability of these viruses to produce encephalitis or demyelination. Thus the neurotropic properties of MHV are unrelated to the ability of the virus to cause hepatitis. This provides further support for the idea that the property of hepatitis in MHV-A59 should not interfere with the reliability of the study of the neurotropic properties of the virus.

The most important finding in the experiments described here show that a region between aa 306 and aa 386 of the MHV-A59 N gene contains sequences which control the ability of the virus to reach the blood stream following acute infection of the brain. This property may be related to the interaction between the virus and brain endothelial cells or other components of the blood brain barrier. The region in the N gene which contains this property contains differences in 17 aa between the two strains MHV-A59 and BCV. A sequence in that region (APTAGAFFF) has been linked to recognition of the virus by cytotoxic T cells derived from MHV-JHM infected mice (Bergmann and Stohlman, 1996) but other functions of this region are still not clear. Further investigation is necessary to determine the minimum aa sequence necessary for the change in phenotype and to determine the exact function of this region.

ACKNOWLEDGMENTS

This work was supported by a grant from the National Multiple Sclerosis Society RG-26151/2 (EL), as well as by a public health service (PHS) grant AI31622 (PSM) from the National Institute of Health (NIH).

REFERENCES

Bergmann, C. C., and Stohlman, S. A., 1996, Specificity of the H-2 L(d)-restricted cytotoxic T-lymphocyte response to the mouse hepatitis virus nucleocapsid protein, *J. Virol.* **70**: 3252–3257.

Gombold, J. L., Sutherland, R. M., Lavi, E., Paterson, Y., and Weiss, S. R., 1995, Mouse hepatitis virus A59-induced demyelination can occur in the absence of $CD8^+$ T cells, *Microb. Pathogen.* **18**: 211–221.

Koetzner, C. A., Parker, M. M., Ricard, C. S., Sturman, L. S., and Masters, P. S., 1992, Repair and mutagenesis of the genome of a deletion mutant of the coronavirus mouse hepatitis virus by targeted RNA recombination, *J. Virol.* **66**: 1841–1848.

Lavi, E., Fishman, S. P., Highkin, M. K., and Weiss, S. R., 1988, Limbic encephalitis following inhalation of murine coronavirus MHV-A59, *Lab. Invest.* **58**: 31–36.

Lavi, E., Gilden, D. H., Highkin, M. K., and Weiss, S. R., 1984a, Detection of MHV-A59 RNA by in situ hybridization, in *"Molecular biology and pathogenesis of coronaviruses"* (P. M. Rottier, B. M. A. van der Zeijst, W. J. M. Spaan, and M. C. Horzinek, Eds.), Vol. 173, pp. 247–258. Plenum Press, New York.

Lavi, E., Gilden, D. H., Highkin, M. K., and Weiss, S. R., 1986, The organ tropism of mouse hepatitis virus A59 is dependent on dose and route of inoculation, *Lab. Anim. Sci.* **36**: 130–135.

Lavi, E., Gilden, D. H., Wroblewska, Z., Rorke, L. B., and Weiss, S. R., 1984b, Experimental demyelination produced by the A59 strain of mouse hepatitis virus, *Neurology* **34**: 597–603.

Lavi, E., Murray, E. M., Makino, S., Stohlman, S. A., Lai, M. M., and Weiss, S. R., 1990, Determinants of coronavirus MHV pathogenesis are localized to 3' portions of the genome as determined by ribonucleic acid-ribonucleic acid recombination, *Lab. Invest.* **62**: 570–578.

Lavi, E., and Weiss, S. R., 1989, Coronaviruses, in *"Clinical and molecular aspects of neurotropic viral infections"* (D. H. Gilden, and H. L. Lipton, Eds.), pp. 101–139. Kluwer, Academic Publishers, Boston.

Pachuk, C. J., Bredenbeek, P. J., Zoltick, P. W., Spaan, W. J. M., and Weiss, S. R., 1989, Molecular cloning of the gene encoding the putative polymerase of mouse hepatitis coronavirus, strain A59, *Virology* **171**: 141–148.

Peng, D., Koetzner, C. A., McMahon, T., Zhu, Y., and Masters, P. S., 1995, Construction of murine coronavirus mutants containing interspecies chimeric nucleocapsid proteins, *J. Virol.* **69**: 5475–5484.

70

TARGETED RECOMBINATION BETWEEN MHV-2 AND MHV-A59 TO STUDY NEUROTROPIC DETERMINANTS OF MHV

E. Lavi,[1] L. Kuo,[2] J. A. Haluskey,[1] and P. S. Masters[2]

[1]Division of Neuropathology
Department of Pathology and Laboratory Medicine
University of Pennsylvania, School of Medicine
Philadelphia, Pennsylvania 19104
[2]Wadsworth Center for Laboratories and Research
New York State Department of Health
Albany, New York 12201

1. ABSTRACT

MHV-A59 produces acute encephalitis, acute hepatitis and chronic demyelination in infected mice. MHV-2 produces only hepatitis and mild meningitis but without encephalitis or demyelination. We have previously studied a set of recombinant viruses between these two strains. The common denominator of viruses that produced encephalitis was a membrane (M) gene derived from MHV-A59. Thus to study the potential contribution of the M gene to acute encephalitis, chimeric viruses were produced in which the M gene of MHV-A59 was substituted with the M gene of MHV-2 by targeted recombination. A control virus was produced in which the M gene of A59 was recombined back into an A59 background. Viruses were then analyzed for their biologic properties and compared with the phenotypes of MHV-A59 and MHV-2 by histopathology and plaque assays for viral titers in organs following intracerebral (IC) inoculation. All three chimeric viruses had a phenotype similar to MHV-A59. Thus, the replacement of the M gene of MHV-A59 with that of MHV-2 is insufficient to produce a phenotype that lacks encephalitis similar to MHV-2.

2. INTRODUCTION

MHV-A59 is a hepatotropic-neurotropic strain of MHV (Lavi et al., 1988; Lavi et al., 1984a; Lavi et al., 1986; Lavi et al., 1984b; Lavi and Weiss, 1989). It produces acute

Coronaviruses and Arteriviruses, edited by Enjuanes *et al.*
Plenum Press, New York, 1998

encephalitis, acute hepatitis and chronic demyelination. The pathologic hallmarks of MHV-A59 during chronic disease are similar to the human disease multiple sclerosis which makes it a useful model system to study the relationship between virus infection and induction of an immune mediated demyelinating disease. MHV-2, a strain closely related to MHV-A59 with presumably over 80% homology in the genes that have been sequenced so far, produces only hepatitis and mild meningitis but without encephalitis or demyelination (Lavi et al., 1995). To understand the properties of neurotropism, and the relationship between acute encephalitis and chronic demyelination we have previously studied a set of recombinant viruses between these two strains which were produced in Dr. Michael Lai's laboratory (Keck et al., 1988). The common denominator of all viruses that produced encephalitis was a membrane (M) gene derived from MHV-A59 although recently we found that several amino acid (aa) similarities also exist in all of these recombinant viruses within the E gene. Since the M region of the genome has been shown to contain a common denominator for all the recombinant viruses that produce acute encephalitis we sought to produce targeted recombination in which portions or the entire M gene of A59 will be replaced with the equivalent region of MHV-2.

3. METHODS AND MATERIALS

3.1. Construction of M Chimeric Viruses by Targeted Recombination

The construction of recombinants was done according to methods previously described (Peng et al., 1995). We used a mutant virus of A59 N gene deletion which renders it temperature sensitive and incapable of growing at nonpermissive temperatures. The ts mutant virus (Alb4) was used for infection of cultures that were co-transfected with an A59 DI construct DNA containing the 3' part of the virus within a vector. In this construct MHV-A59 the M gene was replaced with the M gene of MHV-2. The relevant parts of the recombined viruses were sequenced to confirm the extent of the recombinations.

3.2. Viruses

The stock of MHV-A59 is a three times plaque purified virus with a titer of 2×10^7 pfu/ml (Lavi et al., 1986; Lavi et al., 1984b). The chimeric viruses Alb132, Alb134 and Alb139 were prepared as described below. Viruses were propagated and titered in L2 transformed murine fibroblasts in DMEM media supplemented with heat-inactivated 10% fetal calf serum, glutamine and antibiotics.

3.3. Mice and Inoculations

Four-week-old C57BL/6 virus free male mice were used (Jackson Laboratories, Bar Harbor, Maine). Injection of 20µl of virus or cell lysate as control, was done into the left hemisphere of mice lightly anesthetized with methoxyflurane. One LD50 dose for each virus was used.

3.4. Virulence Assay

The lethal dose of viruses that kills 50% of the animals (LD50) was used to determine the virulence of viruses. It was examined by intracerebral inoculation of groups of 4 week old mice with the viruses as previously described (Lavi et al., 1984b). Four week old C57Bl/6 mice inoculated IC with 10 fold serial dilutions of recombinant and parental viruses. Ten mice were used per dilution and five 10 fold dilutions per virus were used. Mice were monitored daily for signs of disease, and mortality. The LD50 was calculated according to the standard Reed-Muench formula as we have previously described (Lavi et al., 1984b).

3.5. Viral Replication in Organs

To assess viral replication in organs viral titers in tissues were determined by plaque assay. Organs were aseptically removed at autopsy and while half of the organs were fixed in formalin for light microscopy another portion (half of the brain and equivalent portion of the liver) were frozen, homogenized in a tissue disemembranator, serially diluted in PBS and plaque-assayed in 6-well-plate agarose-based plaque assay in L2 transformed murine fibroblasts (Lavi et al., 1986; Lavi et al., 1984b).

3.6. Histopathology and Organ Tropism of the Viruses

Four week old C57Bl/6 mice were inoculated IC with 1 LD50 dose of the virus. Groups of 3–5 mice per time point (days 1,3,5,7,11,15,30,60 post inoculation) were sacrificed by anesthetic (methoxyflurane) overdose and perfused intracardially with PBS followed by 10% buffered formalin. Brain, spinal cord and liver were removed and fixed in 10% buffered formalin. Tissues are embedded in paraffin, sectioned and stained with H&E for light microscopy evaluation as previously described (Lavi et al., 1988; Lavi et al., 1986; Lavi et al., 1984b; Lavi et al., 1990).

3.7. Demyelination

Four week old C57Bl/6 mice were inoculated IC with 1 LD50 dose of the virus. At 30 days post infection, mice were perfused with PBS, followed by 10% buffered formalin. Brains and spinal cords were removed, tissue was embedded in paraffin, and sectioned for staining with luxol fast blue to detect plaques indicative of demyelination. Alternatively, mice were perfused and spinal cords were fixed in 4% glutaraldehyde, embedded in Epon and stained with toluidine blue (Gombold et al., 1995; Lavi et al., 1984b).

4. RESULTS

4.1. Targeted Recombination between MHV-A59 and MHV-2

The infection of Alb4 with co-transfection of the DI RNA produces recombination events in which the 3' parts of the ts mutant Alb4 are replaced with portions of the DI RNA thus correcting the ts mutation of Alb4 and allowing growth of virions in the temperatures that are non-permissive for growth of Alb4. Thus the only viruses that grow in this culture in nonpermissive temperatures are either revertants of tsAlb4 or recombinant viruses. The recombinant viruses can be identified based on restriction analysis of RT-PCR fragments. The entire gene of each recombinant was then analysed by direct RNA sequencing. We obtained two such recombinants in which the M gene ORF of A59 has been replaced with all or part of the M gene ORF of MHV-2. Alb132 is an A59 based virus containing the entire M ORF of MHV-2 replacing the M ORF of A59. Alb134 contains the carboxyl half of the M ORF of MHV-2 substituting for the carboxyl half of M ORF of A59. The crossover occurred between nucleotides 342 and 374, which are identical in both strains. Alb139 is a wild type A59 in which A59 M ORF was put back into an A59 background, thus it is an isogenic counterpart of Alb 132 for control purposes.

4.2. Growth Characteristics of Chimeric Viruses in Culture

To rule out the possibility that chimeric viruses have different biologic properties due to defects in replication we first tested the ability of viruses to replicate in culture. Growth kinetics of chimeric viruses was tested in susceptible L2 murine fibroblast cell cultures by a 48 hour growth curve with time points titers at 4, 8, 12, 16, 24, 48 hours post infection. All three chimeric viruses had growth characteristics in tissue culture similar to MHV-A59.

4.3. Virulence and Organ Tropism of Chimeric Viruses in Mice

The LD50 dose of chimeric viruses was determined. The LD50 of Alb132, Alb134 and Alb139 was found to be approximately 3×10^4pfu. One LD50 dose of Alb132, Alb134, Alb139, MHV-A59 and mock-infection with L2 cell lysate were injected intracerebrally into 4-week old C57Bl/6 mice. Tissues were harvested and examined for liver and brain histology at 1,3,5,7,9,11,30 days post inoculation. Mock-infected mice had no pathologic changes in organs. All the chimeric viruses produced encephalitis and hepatitis similar to MHV-A59 as assessed by light microscopy of tissue sections stained with H&E.

4.4. Replication of Chimeric Viruses in Mouse Organs

Replication of viruses in organs was assessed by plaque assays following inoculation with one LD50 dose of Alb132, Alb134 and Alb139, MHV-A59 and mock-infection with L2 cell lysate injected intracerebrally into 4-week old C57Bl/6 mice. Tissues were harvested and examined for liver and brain titers at 1,3,5,7,9,11,30 days post inoculation. Mock-infected mice had no viral titers in organs. All the chimeric viruses produced viral titers in the brain and liver similar to MHV-A59. Thus the titer assays were consistent with the histopathological results.

4.5. Demyelination

Demyelination was assessed following IC inoculation with either Alb132, Alb134 or Alb139. Groups of 5 mice per virus were inoculated with 1LD50 dose of each one of the viruses. All chimeric viruses produced demyelination in mice similar to MHV-A59. Alb134 produced demyelination in 2/2 surviving mice in 7/22 quadrants of spinal cord sections. Alb139 produced demyelination in 3/3 of surviving mice in 20/44 quadrants of spinal cord sections examined.

5. DISCUSSION

The purpose of this study was to examine the role of the M gene in producing acute encephalitis. MHV-A59 produces acute encephalitis of specific regions of the brain which are associated with the olfactory-limbic systems or physiologically connected with these systems (Lavi et al., 1988). MHV-2 lacks the ability to produce encephalitis. The pathologic hallmarks of acute encephalitis are mononuclear inflammatory cells, predominantly lymphocytes and macrophages, microglial proliferation, microglial nodules and neuronophagia. Immunohistochemistry during acute infection usually shows viral antigens in neurons and glial cells in these locations. Previous analysis of a set of recombinant viruses that were prepared between MHV-A59 and MHV-2 showed that all of the recombinants produced acute encephalitis similar to MHV-A59 even in those viruses in which there was predominantly MHV-2 sequences and only minimal MHV-A59 derived sequences. Initial analysis showed a common M gene in all these viruses but recently by direct sequencing of all the S, E, and N genes of the recombinant viruses we found minimal similarities in aa also in the E gene. Our study here shows that replacement of the M gene alone is insufficient to change the phenotype. Future studies will replace portions of the E gene or perhaps a combination of M and E genes from MHV-2 are necessary to change the phenotype. The M and E genes have been recently found to be closely linked in various functions related to viral replication and assembly.

ACKNOWLEDGMENTS

This work was supported by a grant from the National Multiple Sclerosis Society RG-26151/2 (EL), as well as by a public health service (PHS) grant AI31622 (PSM) from the National Institute of Health (NIH).

REFERENCES

Gombold, J. L., Sutherland, R. M., Lavi, E., Paterson, Y., and Weiss, S. R., 1995, Mouse hepatitis virus A59-induced demyelination can occur in the absence of CD8$^+$ T cells, *Microb. Pathogen.* **18**: 211–221.

Keck, J. G., Soe, L. H., Makino, S., Stohlman, S. A., and Lai, M. M. C., 1988, RNA recombination of murine coronavirus: recombination between fusion-positive mouse hepatitis virus A59 and fusion-negative mouse hepatitis virus 2, *J. Virol.* **62**: 1989–1998.

Lavi, E., Fishman, S. P., Highkin, M. K., and Weiss, S. R.,1988, Limbic encephalitis following inhalation of murine coronavirus MHV-A59, *Lab. Invest.* **58**: 31–36.

Lavi, E., Gilden, D. H., Highkin, M. K., and Weiss, S. R., 1984a, Detection of MHV-A59 RNA by in situ hybridization, in *"Molecular biology and pathogenesis of coronaviruses"* (P. M. Rottier, B. M. A. van der Zeijst, W. J. M. Spaan, and M. C. Horzinek, Eds.), Vol. 173, pp. 247–258. Plenum Press, New York.

Lavi, E., Gilden, D. H., Highkin, M. K., and Weiss, S. R., 1986, The organ tropism of mouse hepatitis virus A59 is dependent on dose and route of inoculation, *Lab. Anim. Sci.* **36**: 130–135.

Lavi, E., Gilden, D. H., Wroblewska, Z., Rorke, L. B., and Weiss, S. R., 1984b, Experimental demyelination produced by the A59 strain of mouse hepatitis virus, *Neurology* **34**: 597–603.

Lavi, E., Murray, E. M., Makino, S., Stohlman, S. A., Lai, M. M., and Weiss, S. R., 1990, Determinants of coronavirus MHV pathogenesis are localized to 3' portions of the genome as determined by ribonucleic acid-ribonucleic acid recombination, *Lab. Invest.* **62:** 570–578.

Lavi, E., Wang, Q., Hingley, S. T., and Weiss, S. R., 1995, Genomic regions associated with neurotropism identified in MHV by RNA-RNA recombination, *Adv, Exp. Med. Biol.* **380**: 51–56.

Lavi, E., and Weiss, S. R., 1989, Coronaviruses, in *"Clinical and molecular aspects of neurotropic viral infections"* (D. H. Gilden, and H. L. Lipton, Eds.), pp. 101–139. Kluwer, Academic Publishers, Boston.

Peng, D., Koetzner, C. A., McMahon, T., Zhu, Y., and Masters, P. S., 1995, Construction of murine coronavirus mutants containing interspecies chimeric nucleocapsid proteins, *J. Virol.* **69**: 5475–5484.

71

MOUSE HEPATITIS VIRUS RECEPTOR LEVELS INFLUENCE VIRUS-INDUCED CYTOPATHOLOGY

P. V. Rao and T. M. Gallagher

Department of Microbiology and Immunology
Loyola University Medical Center
Maywood, Illinois 60153

1. ABSTRACT

We developed human (HeLa) cell lines in which mouse hepatitis virus receptor (MHVR) levels could be regulated by addition of tetracycline. We used these cell lines to determine whether MHVR levels impact the degree of cytopathology induced by infection with the lytic MHV A59 strain. Two cultures were studied; HeLa-MHVR^{lo} (less than 3,000 molecules per cell) and HeLa-MHVR^{hi} (300,000 molecules per cell). Both supported synthesis of infective A59 virus. However, the MHVR^{lo} cells showed no virus-induced cytopathology while the MHVR^{hi} cells uniformly died within 14 hours after infection. This cell death was not related to virus-induced syncytium formation as it occurred even in subconfluent cells overlaid with fusion-blocking antiviral antibodies. MHV A59 spike proteins produced by vaccinia vectors also killed the MHVR^{hi} cells within 12 hours postinfection—MHVR^{lo} cells infected in parallel were intact as judged by trypan blue exclusion. Our current hypothesis is that the accumulation of intracellular complexes composed of spike and MHVR proteins leads to acute single cell lysis.

2. INTRODUCTION

Infection by MHV requires receptor (MHVR) proteins on the cell surface. The receptors, members of the biliary glycoprotein (Bgp) family (Williams *et al.*, 1991), are type I integral membrane glycoproteins whose extracellular structure is comprised of two or four immunoglobulin-like domains (Dveksler *et al.*, 1991). The receptors are functionally important at two temporal stages in the infection cycle. First, the membrane-distal immunoglobulin domain interacts with spike (S) glycoprotein projections on incoming virions, and virion:

Coronaviruses and Arteriviruses, edited by Enjuanes *et al.*
Plenum Press, New York, 1998

cell membrane fusion ensues (Dveksler *et al.*, 1993a). Second, the same domain interacts with S proteins displayed on the surface of infected cells. During infection by most MHV strains, this latter interaction leads to intercellular fusion and syncytium formation, a hallmark of infection both *in vivo* (Lavi *et al.*, 1986) and *in vitro* (Frana *et al.*, 1985).

Both the virion:cell and the infected cell:cell interactions that ultimately lead to membrane fusion require a critical density of MHVR (Dveksler et al., 1993b). The precise density is not yet known, nor is it known whether the virion:cell fusion requires MHVR levels distinct from those that enhance cell:cell fusion. These questions led us to develop a set of HeLa cell transfectants that vary only in their level of MHVR on the surface. Using these lines we found that the receptor densities required for infection by virions were in fact lower than the densities that promote syncytia. This result is discussed in the context of similar studies focussed on interactions between other enveloped viruses and their receptors. With these cell lines we made the additional observation that virus-induced cytopathology is strongly correlated with receptor levels. This interesting finding is described herein.

3. METHODS AND RESULTS

3.1. Regulated Expression of MHVR cDNA in HeLa-MHVR Cells

In HeLa-MHVR cells (Gallagher, 1996), expression of the MHVR cDNA requires binding of a tetracycline-controlled transactivator (tTA) to tetracycline operator DNA (Figure 1). Addition of tetracycline to the growth medium releases the tTA from operator DNA and consequently reduces MHVR expression (Gössen and Bujard, 1992). Parallel cultures of HeLa-MHVR cells (clone #5) were incubated with doses of the tetracycline-derivative doxycycline (Gossen et al., 1995) ranging from 10^{-3} to 10^{0} μg/ml. Using antireceptor antibody CC1 (kindly provided by Dr. K. Holmes) in FACS analyses, we found that one-week incubation periods were necessary to establish new steady-state surface MHVR levels. By the same FACS methods we also found that the collection of doxycycline-exposed cultures encompassed a 140-fold range of steady-state MHVR densities. Cells with the lowest receptor levels (10^{0} μg/ml doxycycline) were designated HeLa-MHVR^{lo}, intermediate levels (10^{-1} μg/ml doxycycline) were HeLa-MHVR^{int}, and highest levels (no doxycycline) were HeLa-MHVR^{hi}.

The average number of MHVR molecules per HeLa cell was estimated by a western immunoblot procedure in which the MHVR proteins were identified using Mab CC1 (data not shown). The signal intensities in these blots were compared to those generated by known quantities of a recombinant soluble MHVR (Gallagher, 1997). In this way we roughly estimated that actively growing HeLa-MHVR^{hi} cultures contained ~300,000 receptors per cell. Immunoblot signals were not detected among proteins from the corresponding

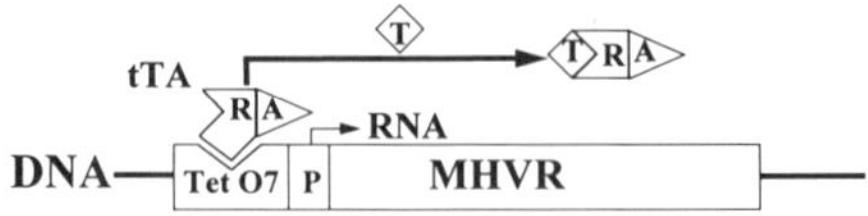

Figure 1. Schematic depiction of the tetracycline-controlled expression system. The tetracycline-controlled transactivator (tTA) is a recombinant protein comprising the carboxy-terminal domain of the herpes simplex virus VP16 (A) and the tetracycline repressor (R). Addition of tetracycline (T) to culture medium prevents binding of tTA to operator DNA (TetO7), thereby preventing transcription from the minimal cytomegalovirus promoter (P).

HeLa-MHVR^{lo} cells; our detection limit in these experiments was ~3,000 receptors per cell. Given that the FACS data indicated that MHVR^{lo} cells had only 0.7% as many surface receptors as MHVR^{hi}, we speculate that the MHVR^{lo} cells contained ~2000 receptors per cell.

3.2. HeLa-MHVR^{lo} Cells Are Resistant to Virus-Induced Syncytium Formation

To directly assess the importance of MHVR levels in virion infection and subsequent syncytium formation, we inoculated the MHVR^{hi} and MHVR^{lo} cultures with MHV strain A59. To identify individual infected cells, indirect immunoflourescence assays were performed using anti-matrix Mab 5A5.2 (Collins *et al.*, 1982). We found that 25% of the MHVR^{lo} cells contained the viral matrix protein. By microscopic examination, there was absolutely no evidence of syncytia in these cultures; this was verified by quantitative fusion assays (Gallagher, 1996; data not shown). In contrast, essentially all MHVR^{hi} cells were matrix-positive and part of multinucleated syncytia (Figure 2).

3.3. High Yields of Progeny Virions from Infected HeLa-MHVR^{lo} Cells

We and others have found that abundant MHVR synthesis inhibits the secretion of MHV-A59 from infected cells (Gallagher, 1995, Chen *et al.*, 1997). A complete mechanis-

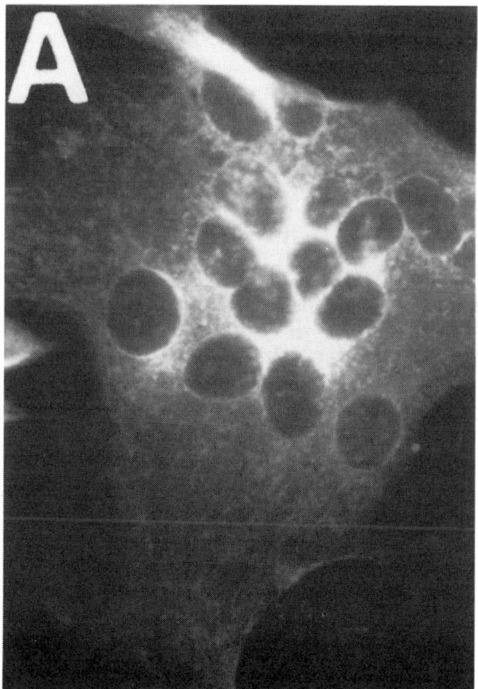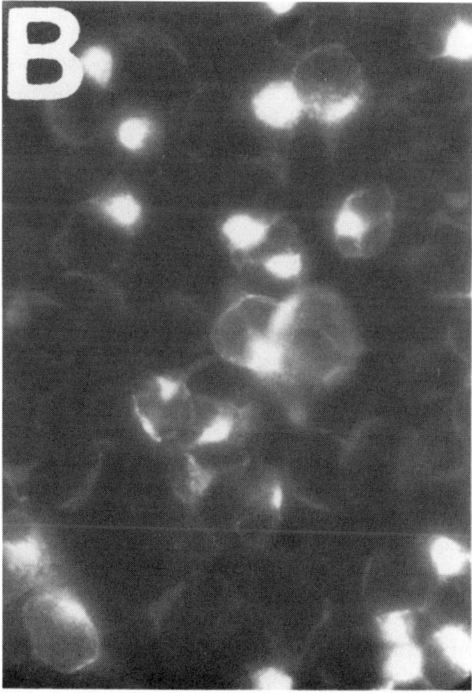

Figure 2. Identification of MHV A59 infection in HeLa-MHVR cells by indirect immunofluorescence. At 8 h post-MHV A59 infection (moi = 100 pfu/cell), cells were washed with PBS, acetone-fixed and incubated with mouse 5A5.2 ascites (1:250 in PBS-2% BSA). After the secondary incubation with FITC-conjugated goat antibody directed against mouse Ig, cells were rinsed, mounted and photographed using a Leitz fluorescence microscope. Panel (A); HeLa-MHVR^{hi}. Panel (B); HeLa-MHVR^{lo}.

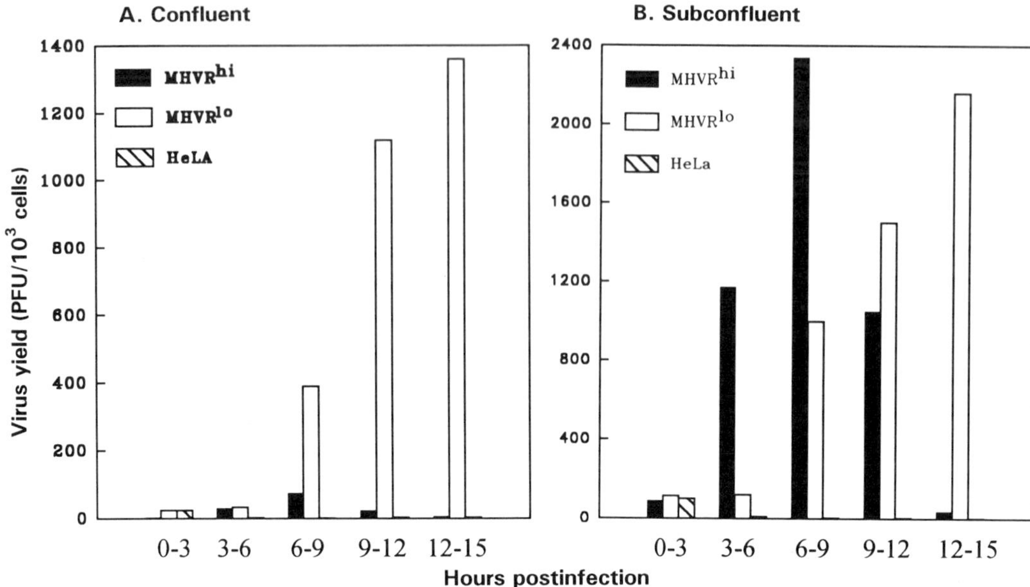

Figure 3. MHV A59 yields from MHVR^{lo} cells exceed those from MHVR^{hi} cells. Following inoculation with MHV A59 (moi = 10 pfu/cell), cultures were incubated at 37°C using 1 ml growth medium per 10^6 cells. At 3 h intervals, growth media were removed and replaced with fresh media. Viral infectivities in the spent media were determined by plaque assays using HeLa-MHVR^{hi} cells as indicators.

tic understanding of this inhibition is not yet available. To further investigate the role of MHVR levels during virus secretion, we monitored the time course of progeny virus development (Figure 3). As expected, our findings indicated that yields from MHVR^{hi} cultures were markedly lower than those from MHVR^{lo} cultures (Figure 3A). These differences were partly due to rapid induction of syncytia in the MHVR^{hi} cultures—this occurred at around 7 hours postinfection and destroyed the capacity to support virus secretion. However, syncytia were not solely responsible for the low yields. Even when MHVR^{hi} cultures were seeded at densities low enough to prevent any intercellular contact, yields at 9 to 15 hours postinfection remained relatively low (Figure 3B). This last finding suggests that the MHVR^{hi} cells—but not the MHVR^{lo} cells—were susceptible to a virus-induced "single cell death" that was not dependent on intercellular fusion.

3.4. A "Single Cell Death" Occurs following Infection of HeLa-MHVR^{hi} Cells

To measure single cell death following virus infection, we seeded cultures at subconfluent densities, infected them with MHV A59, then blocked any possibility of intercellular fusion by adding fusion-inhibiting polyclonal antiserum MK15 (a gift of Dr. Susan Baker). At 14 hours postinfection, viable and non-viable cells were discriminated by trypan blue exclusion.

In MHVR^{hi} cells, we found the proportion of infected cells matched the proportion of non-viable cells (Figure 4). This concordance remained even as the input multiplicity of infection was varied. In the MHVR^{int} and MHVR^{lo} cultures, the proportion of infected

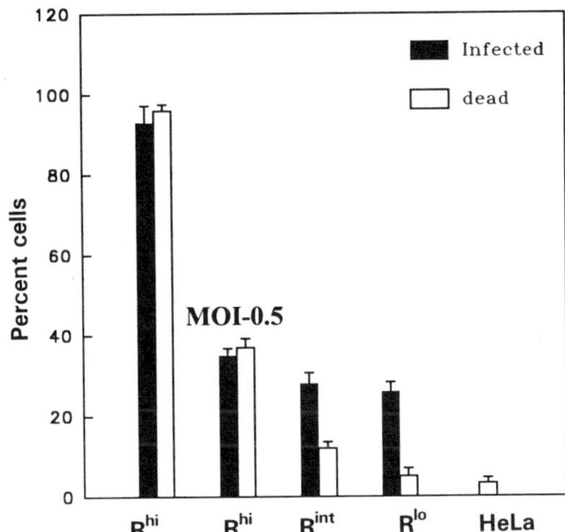

Figure 4. Rapid virus-induced cell death in MHVR^{hi} (but not MHVR^{lo}) cells. Infection of subconfluent cells with MHV A59 (moi = 100 pfu/cell) was followed 3 h later by addition of hyperimmune anti-A59 rabbit serum #MK15, to 1% final concentration. At 14 hours postinfection, cells were suspended using trypsin, and trypan blue was used to monitor cell lysis (white bars). Infected cells (black bars) were scored by indirect immunofluorescence, as described in the legend to Figure 2.

cells significantly exceeded the killed proportion, indicating survival of individual infected cells.

3.5. Spike Proteins Synthesized from Vaccinia Vectors Induce "Single Cell Death" in HeLa-MHVR^{hi} Cells

Mutations in the S gene modulate the pathogenesis of MHV. Additionally, S expression induces extreme cytopathology in tissue culture (Daya *et al.*, 1989). To find out whether S-mediated cytopathology is solely due to its known syncytium-inducing properties, we produced MHV A59 S proteins from vaccinia vectors in subconfluent MHVR^{lo} and MHVR^{hi} cells and monitored the resulting cell death as described above. At 12 hours postinfection, cell death in MHVR^{hi} cells was 25%; in contrast less than 4% death occurred in the MHVR^{lo} cells. We currently speculate that "single cell death" in MHVR^{hi} cells is due to accumulation of MHVR:S complexes within the lumenal cavities of intracellular organelles. Studies addressing this hypothesis are in progress.

4. DISCUSSION

Our results indicate that sensitivity to MHV A59-induced cytopathic effect increases with cellular MHVR levels. This was readily evident by microscopic examination of infected HeLa-MHVR cells. Inoculation of confluent MHVR^{hi} cultures induced complete (100%) syncytia by 9 hours postinfection while the corresponding MHVR^{lo} cells were entirely devoid of syncytia. Given that about 25% of the MHVR^{lo} cells were infected by A59, our results indicate that the MHVR threshold necessary for virion:cell fusion (virus entry) is lower than that required for cell:cell fusion (syncytia development). It is possible that virions fuse with MHVR^{lo} cells due to their closely-spaced spikes—infected cells by contrast may not display spikes at sufficient densities for fusion to occur. This hypothesis implies that a critical density of MHVR must be reached on the cell surface to congregate spikes into proximity close enough to form a "fusion pore". This general view of a fusion pore involving multiple, closely-spaced fusion proteins has support from careful biochemical studies of orthomyxo and retrovirus glycoproteins (White, 1994, Frey *et al.*, 1995).

In this study we observed a second type of cytopathic effect that increased with MHVR levels and was independent of syncytia development. Infected MHVR^{hi} cells died within 14 hours postinfection (as judged by trypan blue exclusion) even when maintained as subconfluent monolayers. MHVR^{lo} cells infected in parallel survived. We speculate that the "single cell death" occurring in MHVR^{hi} cells arises from the formation of MHVR:S complexes within the cell. We have indeed isolated intracellular MHVR:S complexes by co-immunoprecipitation (data not shown) and we suggest that these proteins have properties similar to CD4:gp160 complexes (Cao *et al.*, 1996)—namely that they destabilize intracellular membranous organelles. In this regard, the MHV matrix protein, which is normally localized to the Golgi appratus (Klumperman *et al.*, 1994), is disseminated throughout the cytoplasm in individual A59-infected MHVR^{hi} cells (data not shown).

Infected Hela-MHVR^{lo} cultures immediately supported persistent infection. No acute phase of infection was observed. Infected HeLa-MHVR^{hi} cultures also supported persistent infection, but only after an acute phase that destroyed all but the few cells producing low levels of MHVR, i.e., cells repopulating the culture after infection were MHVR^{lo}. Thus the maintanence of persistent infection depends on relatively low MHVR levels, as indicated previously by our colleagues (Chen and Baric, 1996; Sawicki *et al.*, 1995).

ACKNOWLEDGMENTS

We thank H.-M. Jäck, H. Gössen and J. Bujard for providing HeLa tTA cells and pUHD-10 expression vectors, B. Hsiang and E. Sethi for assistance in developing and maintaining the HeLa-MHVR cell lines. Special thanks go to K.V. Holmes for providing antireceptor Mab CC1. This research was supported by NIH grant NS31636 and by a grant from the Schweppe Foundation of Chicago.

REFERENCES

Cao, J., Park, I-W., Cooper, A., and Sodroski, J., 1996, Molecular determinants of acute single-cell lysis by human immunodeficiency virus type 1, *J. Virol.* **70**:1340–1354.

Chen, W., Madden, V.J., Bagnell, C.R. Jr., and Baric, R.S., 1997, Host derived intracellular immunization against mouse hepatitis virus infection, *Virology.* **228**:318–332.

Chen, W., and Baric, R.S., 1996, Molecular anatomy of mouse hepatitis virus persistence: Coevolution of increased host cell resistance and virus virulence, *J. Virol.* **70**:3947–3960.

Collins, A.R., Knobler, R.L., Powell, H., and Buchmeier, M.J., 1982, Monoclonal antibodies to murine hepatitis virus-4 (strain JHM) define the viral glycoprotein responsible for attachment and cell-cell fusion, *Virology.***119**:358–371.

Daya, M., Wong, F., Cervin, M., Evans, G., Vennema, H., Spaan, W., and Anderson, R., 1989, Mutation of host cell determinants which discriminate between lytic and persistent mouse hepatitis virus infection results in a fusion-resistant phenotype, *J. Gen. Virol.* **70**:3335–3346.

Dveksler, G.S.,. Pensiero, M.N., Dieffenbach, C.W., Cardellichio, C.B., Basile, A.A., Elia, P.E., and Holmes, K.V.,1993a, Mouse hepatitis virus strain A59 and blocking antireceptor monoclonal antibody bind to the N-terminal domain of cellular receptor, *Proc. Natl. Acad. Sci. USA.* **90**:1716–1720.

Dveksler, G.S., Dieffenbach, C.W., Cardellichio, C.B., McCuaig. K., Pensiero, M.N., Jiang, G-S., Beauchemin, N., and Holmes, K.V., 1993b, Several members of the mouse carcinoembryonic antigen-related glycoprotein family are functional receptors for the coronavirus mouse hepatitis virus-A59, *J. Virol.* **67**:1–8.

Dveksler, G.S., Pensiero, M.N., Cardellichio, C.B., Williams, R.K., Jiang, G-S., Holmes, K.V., and Dieffenbach, C.W., 1991, Cloning of the mouse hepatitis virus (MHV) receptor:Expression in human and hamster cell lines confers susceptibility to MHV, *J. Virol.* **65**:6881–6891.

Frana, M.F., Behnke, J.N., Sturman, L.S., and Holmes, K.V., 1985, Proteolytic cleavage of the E2 glycoprotein of murine coronavirus:host-dependent differences in proteolytic cleavage and cell fusion, *J. Virol.* **56**:912–920.

Frey, S., Marsh, M., Gunther, S., Pelchen-Matthews, A., Stephens, P., Ortlepp, S., and Stegmann, T., 1995, Temperature dependence of cell-cell fusion induced by the envelop glycoprotein of human immunodeficiency virus type 1, *J. Virol.* **69**:1462–1472.

Gallagher, T.M., 1997, A role for naturally occurring variation of the murine coronavirus spike protein in stabilizing association with the cellular receptor, *J. Virol.* **71**:3129–3137.

Gallagher, T.M., 1996, Murine coronavirus membrane fusion is blocked by modification of thiols buried within the spike protein, *J. Virol.* **70**:4683–4690.

Gallagher, T.M., 1995, Overexpression of the MHV receptor. Effect on progeny virus secretion, *Adv. Exp. Med. Biol.* **380**:331–336.

Gossen, M., Freundlieb, S., Bender, G., Muller, G., Hillen, W., and Bujard, H., 1995, Transcriptional activation by tetracyclines in mammalian cells, *Science.* **268**:1766–1769.

Gossen, M. and Bujard, H., 1992, Tight control of gene expression in mammalian cells by tetracycline-responsive promoters, *Proc. Natl. Acad. Sci. USA.* **89**:5547–5551.

Klumperman, J., Locker, J.K., Meijer, A., Horzinek, M.C., Geuze, H.J., and Rottier, P.J.M., 1994, Coronavirus M proteins accumulate in the golgi complex beyond the site of virion budding, *J. Virol.* **68**:6523–6534.

Lavi, E., Gilden, D.H., Highkin, M.K., and Weiss, S.R., 1986, The organ tropism of mouse hepatitis virus A59 in mice is dependent on dose and route of inoculation, *Lab. Anim. Sci.* **36**:130–135.

Sawicki, G.S., Lu, J-H., and Holmes, K.V., 1995, Persistent infection of cultured cells with mouse hepatitis virus (MHV) results from the epigenetic expression of the MHV receptor, *J. Virol.* **69**: 5535–5543.

White, J. M., 1994, Fusion of influenza virus in endosomes: Role of the hemagglutinin, in E. Wimmer (ed), *Cellular receptors for animal viruses.* Cold Spring Harbor laboratory Press, Cold Spring Harbor, N.Y, pp. 281–301.

Williams, R.K., Jiang, G-S., and Holmes, K.V., 1991, Receptor for mouse hepatitis virus is a membrer of the carcinoembryonic antigen family of glycoproteins, *Proc. Natl. Acad. Sci. USA.* **88**:5533–5536.

72

IS THE SIALIC ACID BINDING ACTIVITY OF THE S PROTEIN INVOLVED IN THE ENTEROPATHOGENICITY OF TRANSMISSIBLE GASTROENTERITIS VIRUS?

C. Krempl,[1] H. Laude,[2] and G. Herrler[1]

[1]Institut für Virologie
Philipps-Universität Marburg
35037 Marburg, Germany
[2]Unité de Virologie et Immunologie Moléculaires
Institut National de la Recherche Agronomique
Jouy-en-Josas, France

1. ABSTRACT

Transmissible gastroenteritis virus (TGEV) is able to recognize sialic acid on sialoglycoconjugates. Analysis of mutants indicated that single point mutations in the S protein (around amino acids 145–155) of TGEV may result both in the loss of the sialic acid binding acitivity and in a drastic reduction of the enteropathogenicity. From this observation we conclude that the sialic acid binding acitivity is involved in the enteropathogenicity of TGEV. On the basis of our recent results we propose that binding of sialylated macromolecules to the virions surface may increase virus stabiltiy. This in turn would explain how TGEV as an enveloped virus can survive the gastrointestinal passage and cause intestinal infections.

2. INTRODUCTION

Coronaviruses usually infect their host via the respiratory and/or enteric eptithelium. Though some viruses of this family may spread to other tissues or organs, the majority of diseases caused by coronaviruses are due to localized infections of the respiratory or enteric tract. In order to use the gastrointestinal tract as a portal of entry, viruses have to survive under the specific environmental conditions encountered in the stomach and in the

Coronaviruses and Arteriviruses, edited by Enjuanes *et al.*
Plenum Press, New York, 1998

Table 1. Enteropathogenic viruses that invade their
host via the gastrointestinal tract

	Genome	Lipid envelope
Coronaviruses	ssRNA	yes
Astroviruses	ssRNA	no
Caliciviruses	ssRNA	no
Picornaviruses	ssRNA	no
Rotaviruses	dsRNA	no
Parvoviruses	ssDNA	no
Adenoviruses	dsDNA	no

enteric tract. Factors that may be detrimental for viruses are (i) low pH, (ii) proteases, and (iii) bile salts. The detergent-like bile salts are expected to be an effective defence mechanism against enveloped viruses. Indeed, as shown in Table 1, most viruses that invade their host via the gastrointestinal route lack a lipid envelope. A notable exception are coronaviruses. How these enveloped RNA viruses survive the gastrointestinal passage has not been explained.

Porcine transmissible gastroenteritis virus (TGEV) is a typical enteropathogenic coronavirus. TGEV infections are especially severe in piglets resulting in diarrhea and death within a few days (Penseart et al., 1993). Interestingly there is a related virus, porcine respiratory coronavirus (PRCV), that has lost enteropathogenicity (Laude et al., 1993). Infections by PRCV are restricted to the respiratory tract. The nucleotide sequences of the genomes of TGEV and PRCV are very similar (Rasschaert et al., 1990). The most prominent difference is a large deletion within the S gene of PRCV. Because of this deletion a stretch of some 220 amino acids is missing in the S protein of PRCV compared to the respective protein of TGEV (illustrated in Fig. 1). Comparative studies on the enteropathogenic TGEV and the respiratory variant PRCV should help to elucidate the properties of TGEV that make this virus enteropathogenic. A functional difference between TGEV and PRCV has been reported recently: A sialic acid binding site is present on TGEV but absent from PRCV (Schultze et al., 1996). This binding actvitiy enables TGEV to agglutinate erythrocytes. In general the hemagglutination reaction is a convenient assay for the sialic acid binding activity. However, in contrast to other enveloped viruses with sialic acid binding activity, TGEV does not contain a "receptor-destroying enzyme". Influenza viruses, paramyxoviruses or coronaviruses from the BCV serogroup contain either a neuraminidase or an acetylesterase (reviewed by Herrler et al., 1995). These enzymes inactivate sialylated glycoconjugates that may bind to the virion surface and prevent the virus from agglutinating erythrocytes. TGEV does not contain such a "receptor-destroying enzyme" and, therefore, is sensitive to the action of inhibitors. For optimal hemagglutinating activity TGEV has to be pretreated with neuraminidase to inactivate virus-bound hemagglutination inhibitors (Schultze et al., 1996). The sialic acid binding activity of TGEV is most effective towards N-glycolylneuraminic acid, a type of sialic acid that is quite common on porcine cells. The comparison of TGEV with PRCV suggested that the sialic acid binding site of the TGEV S protein is located within that stretch of amino acids that are missing in PRCV because of the deletion in the S gene. This has been confirmed by studies with monoclonal antibodies. Only antibodies to a single epitope prevented TGEV from agglutinating erythrocytes (Schultze et al., 1996). This epitope is not found on PRCV because it is located in that portion of the TGEV S protein that is missing in the surface protein of PRCV.

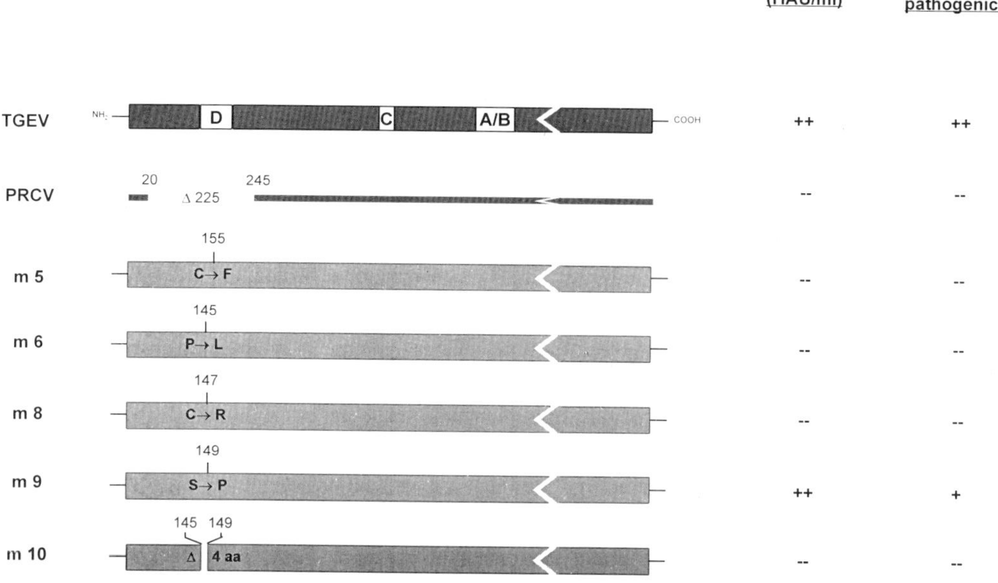

Figure 1. Schematic representation of the S proteins of TGEV, PRCV, and 5 site D mutants. In the S protein, the location of the antigenic sites is indicated. On the right it is indicated whether the viruses are able (++) or unable (--) to agglutinate chicken erythrocytes. It is also indicated whether the enteropathogenicity (++) is slightly (+) or strongly (--) reduced.

Here we summarize our recent data that indicate (i) that the hemagglutination-inhibitors that are often associated with TGEV are—at least in part—of cellular origin; (ii) that the sialic acid binding activity is involved in the enteropathogenicity of TGEV. We discuss the possibility that the sialic acid binding activity may increase the viral stability und thus may help to survive the gastrointestinal passage.

3. METHODS AND MATERIALS

The data described below have been obtained by methods that have been described elsewhere in detail (Schultze et al., 1996; Krempl et al., 1997).

4. RESULTS

We have previously reported that infection of cultured cells by TGEV results in a transient release of hemagglutinating virions. While hemagglutinating activity was detectable in the cell supernatant at 24 h.p.i., it decreased at later times p.i. and all the activity had disappeared by 72 h.p.i. The loss of the hemagglutinating activity was not due to a reduction of the virus titer, because high HA-titers were determined when the culture medium was treated with sialidase (neuraminidase). This result indicated that at the late stage of the infectious cycle sialylated glycoconjugates are blocking the sialic acid binding site and thus prevent the hemagglutination reaction. The nature of the inhibitors has not not been elucidated yet. The result described above does not allow a conclusion as to the origin of the inhibitory sialoglycoconjugates. However, a recent observation suggests that they are—at least in part—derived from the host cell. When LLC-PK1 (pig kidney) cells or ST (swine testicular) cells were first treated with neuraminidase and then, after removal of the enzyme, infected by TGEV, the virions released from these cells showed hemagglutinating activity that was preserved also at the late stage of infection (Krempl et al., 1997). In this case the neuraminidase only had contact with cellular glycoconjugates and nevertheless was able to inactivate the hemagglutination-inhibitors. Therefore, these inhibitors—or a substantial part of them—have to be derived from the host cells. They appear to be regenerated during the infectious cycle only to a limited extent, because the newly synthesized sialoglycoconjugates are not sufficient to prevent TGEV from agglutinating erythrocytes.

The comparison between TGEV and PRCV suggested that the hemagglutinating activity is involved in the enteropathogenicity of TGEV. The structural difference between both viruses involves a large and a few minor deletions affecting the S gene and the open reading frame of a nonstructural protein. Because of these differences it was not possible to strictly correlate the sialic acid binding activity of TGEV with the enteropathogenicity of this virus. Interestingly, mutants of TGEV have been described that were resistant to a monoclonal site D antibody due to point mutations or a deletion of twelve nucleotides in the S gene (Delmas et al., 1986). The resulting amino acid changes (amino acids 145–155) were located in that portion of the S protein that is missing in the PRCV S protein (illustrated in Fig. 1). As a consequence of these mutations, the enteropathogenicity of the mutants was strongly reduced (Bernard and Laude, 1995). We analyzed these mutants for hemagglutination activity and found that in addition to the enteropathogenicity they had lost the ability to agglutinate erythrocytes (Krempl et al., 1997; Fig. 1). Thus, a single point mutation in the S protein can affect both the sialic acid binding activity and the enteropathogenicity of TGEV.

5. DISCUSSION

Bernard and Laude (1995) reported that point mutations between amino acids 145 - 155 (within antigenic site D) of the S protein of TGEV resulted in a markedly reduced enteropathogenicity.This finding demonstrated that a critical determinant of the pathogenicity of TGEV is located on the S protein, within or close to site D. We have shown that these site D mutants were unable to agglutinate erythrocytes, i.e. they had lost the sialic acid binding activity. Therefore, we can assign a function to this region of the S protein. As a single point mutation can affect both the pathogenicity and the hemagglutinating activity, we conclude that the sialic acid binding activity is a determinant of the enteropathogenicity of TGEV.

An answer to the question how the sialic acid binding activity affects the enteropathogenicity of TGEV may come from our finding that the binding site for sialic acid is sensitive to competitive inhibitors. Virions that are released from infected cells interact with sialic acid containing substances and lose their hemagglutinating activity. Treatment with sialidase inactivates the inhibitors and restores the ability to agglutinate erythrocytes. The inhibitors appear to be of cellular origin as they are inactivated by enzyme treatment of cells prior to infection. Such inhibtors may be encountered by TGEV also during natural infection. Mucins that are covering the respiratory and the intestinal epithelium of pigs are rich in N-glycolylneuraminic acid and may interact with TGEV. Bound mucins would block the sialic acid binding site and prevent the virus from using this binding activity for attachment to cells. However, the cellular receptor for TGEV is aminopeptidase N (Delmas et al., 1992) and there is no evidence that the sialic acid binding activity is required for attachment to cells. On the other hand, binding of mucins or other sialylated substances to the virion surface may increase virus stability. A coat of sialylated macromolecules may render the virus less susceptible to the action of proteases or bile salts. An increased resistance against such substances should help to survive the passage through the gastrointestinal tract. In this way the sialic acid binding activity of TGEV might contribute to the enteropathogenicity and explain how TGEV as an enveloped virus can survive the gastrointestinal passage. Future work is directed to determine the importance of the sialic acid binding activity for the susceptibility of TGEV towards detergents and proteases.

REFERENCES

Bernard, S, and Laude, H., 1995, Site-specific alteration of transmissible gastroenteritis virus spike protein results in markedly reduced pathogenicity, *J. Gen. Virol.* **76**:2235–2241.

Delmas, B., Gelfi, J., and Laude, H., 1986, Antigenic structure of transmissible gastroenteritis virus. II. Domains in the peplomer glycoprotein, *J. Gen. Virol.* **67**:1405–1418.

Delmas, B., Gelfi, J., L'Haridon, R., Vogel, L.K., Sjöström, H., Noren, O., and Laude, H., 1992, Aminopeptidase N is a major receptor for the enteropathogenic coronavirus TGEV, *Nature* **357**:417–419

Herrler, G., Hausmann, J., and Klenk, H.-D., 1995, Sialic acid as receptor determinant of ortho- and paramyxoviruses, in *Biology of the Sialic Acids* (A. Rosenberg , ed.), Plenum Press, New York pp. 315–336.

Krempl, C., Laude, H., and Herrler, G., 1997, Point mutations in the S protein connect the sialic acid binding activity with the enteropathogenicity of transmissible gastroenteritis coronavirus, *J. Virol.* **71**:3285–3287.

Laude, H., van Reeth, K., and Pensaert, M., 1993, Porcine respiratory coronavirus: molecular features and virus - host interactions, *Vet. Res.* **24**:125–150.

Pensaert, M., Callebaut, P., and Cox, E., 1993, Enteric coronaviruses of animals, in *Viral infections of the gastrointestinal tract*, 2nd ed. (A.Z. Kapikian, ed.),Marcel Dekker, New York, pp. 627–696.

Rasschaert, D., Duarte, M., and Laude, H., 1990, Porcine respiratory coronaviurs differs from transmissible gastroenteritis virus by a few genomic deletions, *J. Gen. Virol.* **71**:2599–2607.

Schultze, B., Krempl, C., Shaw, L., Enjuanes, L., Schauer, R., and Herrler, G., 1996, Transmissible gastroenteritis coronavirus but not the related porcine respiratory coronavirus has a sialic acid (N-glycolylneuraminic acid) binding activity, *J. Virol.* **70**:5634–5637.

73

ISOLATION OF HEMAGGLUTINATION-DEFECTIVE MUTANTS FOR THE ANALYSIS OF THE SIALIC ACID BINDING ACTIVITY OF TRANSMISSIBLE GASTROENTERITIS VIRUS

C. Krempl,[1] M. L. Ballesteros,[2] L. Enjuanes,[2] and G. Herrler[1]

[1]Institut für Virologie
Philipps-Universität Marburg
35037 Marburg, Germany
[2]Centro Nacional de Biotecnologia
Department of Molecular and Cell Biology, CSIC
Campus Universidad Autonoma de Madrid
Canto Blanco, 28049 Madrid, Spain

1. ABSTRACT

The surface protein S of transmissible gastroenteritis virus (TGEV) has a sialic acid binding activity that enables the virus to agglutinate erythrocytes. A protocol is described that has been successfully applied to the isolation of hemgglutination-defective mutants. The potential of these mutants for the characterization of the sialic acid-binding site and the function of the binding activity is discussed.

2. INTRODUCTION

The ability to recognize sialic acid is shared by more viruses than any other of the viral binding acitivities elucidated so far. In addition to several nonenveloped viruses (rotaviruses, reoviruses, encephalomyocarditis virus, polyomavirus) a number of enveloped viruses attach to sialic acid residues present on glycoproteins or glycolipids (reviewed by Herrler et al., 1995). Most information is available about the sialic acid binding activity of enveloped viruses. Influenza viruses, several members of the paramyxovirus family, as well as bovine coronavirus and antigenically related human and porcine coronaviruses use

Coronaviruses and Arteriviruses, edited by Enjuanes *et al.*
Plenum Press, New York, 1998

                                              C. Krempl *et al.*

sialic acids on the cell surface as receptor determinants for attachment to cells. In the case of erythrocytes virus binding results in a hemagglutination reaction that serves as a convenient assay for the viral sialic acid binding activity. In the case of cultured cells, binding to sialylated surface glycoconjugates is the initial stage of the virus infection. As shown in Fig. 1, viruses may differentiate between different types of sialic acid. Influenza A and B as well as the paramyxoviruses have a preference for N-acetylneuraminic acid. Influenza C virus and bovine coronavirus recognize N-acetyl-9-O-acetylneuraminic acid. In addition, these viruses may have a preference for a certain linkage type between sialic acid and the neighboring sugar (e.g. α2,3 or α2,6-linked to galactose). Efficient virus binding also requires a multivalent virus-cell interaction. Therefore, only a limited number of surface sialoglycoconjugates appear to be suitable as virus receptors. In the case of MDCK cells, a single surface glycoprotein has been shown to be the major protein recognized by influenza C virus and bovine coronavirus (Zimmer et al., 1995; Schultze et al., 1996). A hall-

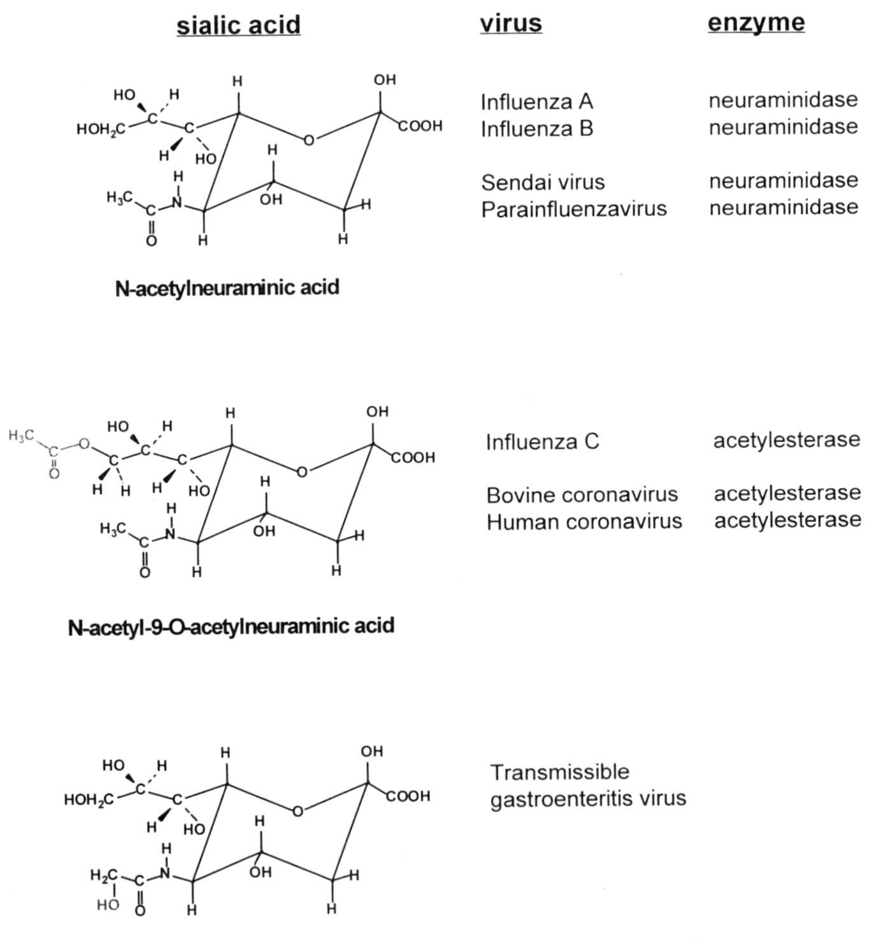

Figure 1. Different types of sialic acid that function as receptor-determinants for enveloped viruses. The viruses that recognize each of these sialic acids are indicated. Several of these viruses contain a receptor-destroying enzyme. On the right it is indicated whether the enzyme is a neuraminidase or an acetylesterase.

mark of enveloped viruses that use sialic acid to initiate infection is the presence of a receptor-destroying enzyme on the virion surface (Fig. 1). This enzyme is a neuraminidase in the case of influenza A and B viruses as well as of paramyxoviruses (Sendai virus and parainfluenza viruses). With viruses that recognize 9-O-acetylated sialic acids—influenza C virus and bovine coronavirus—the receptor-destroying enzyme is an acetylesterase releasing the acetyl group from position C-9 of sialic acid. It is assumed that these viruses require a receptor-destroying enzyme to prevent interactions with sialoglycoconjugates that are unfavorable for infection. Available evidence indicates that lack of such an enzyme results in the fromation of virus aggregates on the surface of infected cells and thus affects the spread of virus to other cells (Liu et al., 1995; Höfling et al., 1996).

Coronaviruses are exceptional, because within one genus they contain both viruses with and viruses without a receptor-destroying enzyme. While viruses of one genetic lineage (bovine coronavirus, human coronavirus OC43, and porcine hemagglutinating encephalomyelitis virus) contain an acetylesterase (Vlasak et al., 1988; Schultze et al., 1990), two other genetic lineages lack such an enzyme. The latter two genetic groups are represented by porcine transmissible gastroenteritis virus (TGEV) and avian infectious bronchitis virus (IBV). Though both TGEV and IBV do not contain a receptor-destroying enzyme, they nevertheless are able to recognize sialic acid (Schultze et al. 1992, 1996). The sialic acid binding activity of these two viruses can also be assayed in a hemaglutination assay. However, in the course of infection there is usually only a transient hemagglutinating activity detectable in the cell supernatant (Schultze et al., 1996). The disappearance of the hemagglutinating activity late in infection is due to competitive inhibitors that interact with the sialic acid binding site and prevent the virus from agglutinating erythrocytes. The hemagglutinating activity of viruses with a receptor-destroying enzyme is usually observed throughout the infectious cycle, because these viruses are able to inactivate potential competitive inhibitors by the viral acetylesterase or neuraminidase, respectively. In order to reproducibly detect hemagglutinating activity with TGEV, competitive inhibitors have to be inactiviated by the addition of an exogenous receptor-destroying enzyme. Neuraminidase has been shown to be effective in this respect (Schultze et al., 1992,1996). Following neuraminidase treatment of the cell supernatant of infected cells, the hemagglutinating activity of TGEV is no longer a transient event but is as stable as that of influenza viruses or bovine coronavirus.

Results obtained with mutants of TGEV indicate that point mutations in the surface protein S of TGEV may result in the loss of both the hemagglutinating activity and the enteropathogenicity (Krempl et al., 1997). These findings correlate the sialic acid binding activity with the enteropathogenicity of TGEV.

3. EXPERIMENTAL OUTLINE

In order to strengthen the experimental data that correlate the sialic acid binding activity of TGEV with the enteropathogenicity of this virus, it is desirable to isolate hemagglutination-defective mutants of TGEV. For this purpose we have applied a selection procedure that is schematically shown in Fig. 2. In a first step a preparation of TGEV was treated with neuraminidase (150 mU of the *Vibrio cholerae*-enzyme in a final volume of 3 ml PBS, 30 min, 37°C) in order to inactivate potential inhibitors that may block the sialic acid binding site. In this way virions are obtained that show optimal hemagglutinating activity (Schultze et al., 1996). Neuraminidase was removed by ultracentrifugation (150.000xg, 1 h, 4°C). The virus sediment was resuspended in 3 ml of PBS and filtrated to

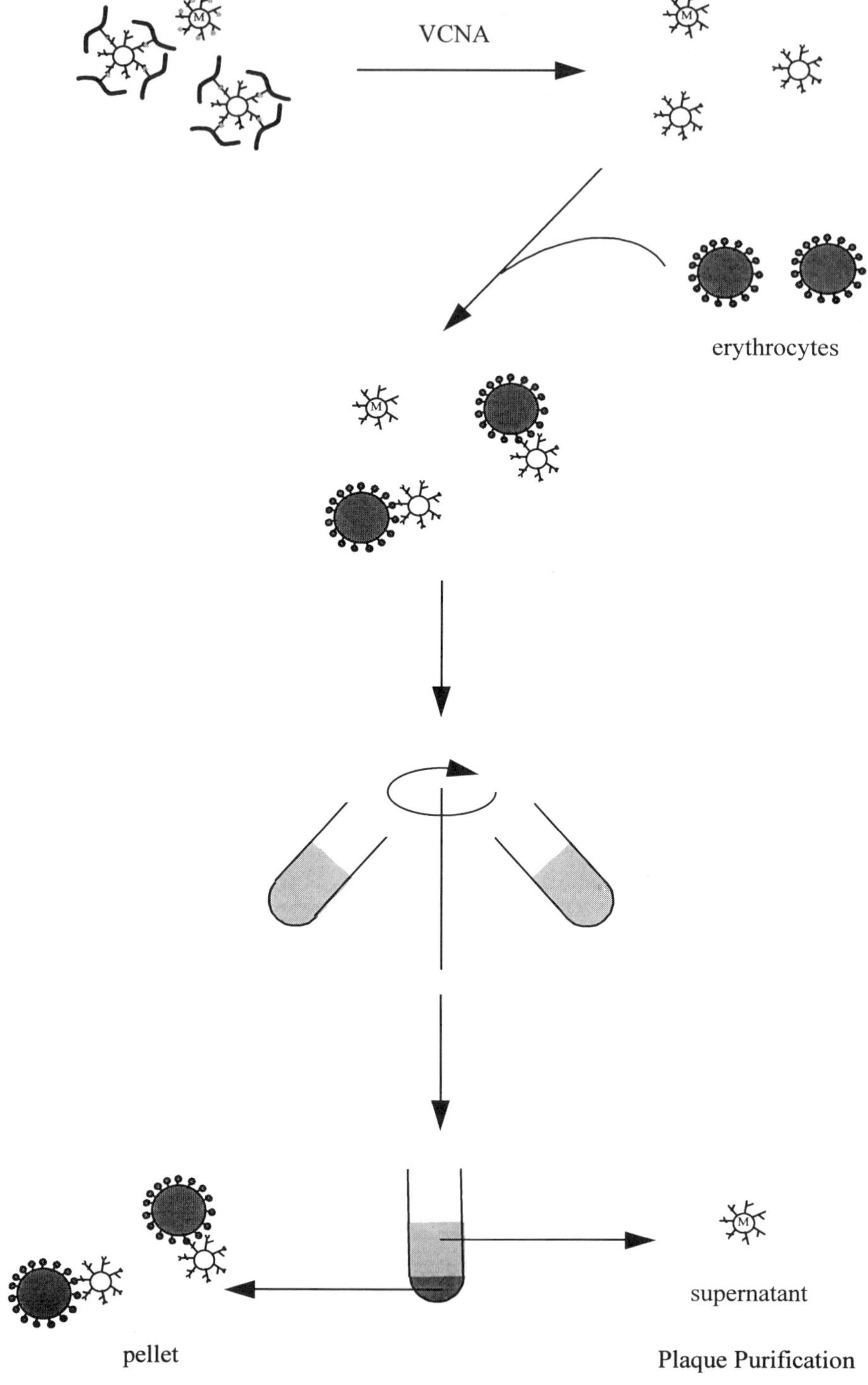

Figure 2. Protocol for the selection of hemagglutination-defective mutants of TGEV. Virus was treated with neuraminidase from *Vibrio cholerae* to inactivate competive inhibitors (sialoglycoconjugates) that may block the sialic acid-binding site of the S protein. The virus was then incubated with glutaraldehyde-fixed erythrocytes. Virions with intact sialic acid-binding site will bind to the erythrocytes and can be sedimented by low speed centrifugation while hemagglutination-defective mutants (designated M) will remain in the supernatant.

Table 1. Infectivity and hemagglutinating activity of
TGEV after different rounds of the selection procedure

	Virus grown in ST cells	
Step of the selection procedure	Infectivity PFU/ml	HA-activity HAU/ml
Before selection	3.2×10^8	1024
After first round	1.5×10^7	8
After second round	4.2×10^8	< 2
After third round	3.0×10^8	< 2

remove microbial contaminants. To the sterile virus suspension, glutaraldehyde-fixed bovine erythrocytes were added to a final concentration of 2.5%. The latter cells are more sensitive to the hemagglutinating activity of TGEV than the erythrocytes from other species tested (Schultze et al., 1996). This suspension was incubated for 1 h on ice to allow the virions to attach to the erythrocytes. Following centrifugation at 2.500xg for 10 min, the supernatant was subjected to two further rounds of incubation with bovine erythrocytes. The supernatant derived from the third batch of erythrocytes was expected to be enriched in natural variants of TGEV that were defective in the sialic acid-binding activity. It was used to infect swine testicular (ST) cells. At about 18 h.p.i., when a cytopathic effect was detectable, virions were harvested. The virus was subjected to the selection procedure (neuraminidase treatment, three incubation rounds with bovine eryhtrocytes, growth in ST cells) two more times. Finally, individual viruses were isolated by two rounds of plaque purification. Stock virus was grown in ST cells and analyzed for hemagglutination activity as described previously (Schultze et al., 1996).

4. RESULTS

The selection protocol described above was applied to the PUR46-MAD (Sanchez et al, 1990) strain of TGEV. After the first round of the selection procedure (neuraminidase treatment, 3 incubations with bovine erythrocytes), the final supernatant was grown in ST cells. As shown in Table 1, virus recovered from the infected cells had a reduced hemagglutinating activity compared to the parental virus. This virus was again subjected to the selection protocol. This time, virus was recovered that had the same infectivity as the original virus with no hemagglutinating activity detectable. After having run through the selection procedure for a third time, virus was plaque-purified. Eight individual plaques were picked to grow stock virus. All of them were devoid of hemagglutinating activity.

5. DISCUSSION

Hemagglutination-defective mutants are interesting in several respects. It is expected that most of these viruses differ from the parental virus by mutations affecting the sialic acid binding site. Comparison of the sequences of the respective S genes should reveal amino acids that are involved in the receptor-binding site. The available evidence indicates that the binding site or an essential part of it is located between amino acids 20 and 245 (Schultze et al., 1996). This stretch of amino acids is missing in the S gene of a related virus, porcine respiratory coronavirus (PRCV), that is unable to agglutinate erythrocytes. The portion of the TGEV S protein that is deleted in the corresponding PRCV protein contains an antigenic site and monoclonal antibodies directed to this epitope have been found to inhibit the hemagglutinating activity of TGEV very efficiently. Several antibody-resistant variants have been reported to lack hemagglutinating activity (Krempl et al., 1997). These variants differed from the parental virus by point mutations or a deletion of four amino acids. All changes were located between amino acids 145 and 155. These amino acids appear to be located at or close to the sialic acid binding site of TGEV. Sequence analysis of the hemagglutination-defective viruses described above should help to further identify amino acids of the S protein that are involved in the sialic acid binding site of TGEV.

Hemagglutination-defective viruses may also be useful in elucidating the functional role of the sialic acid binding site. One concept is that binding of sialylated macromolecules to the virion surface increases the resistance of TGEV to detrimental agents such as detergents and proteases (Schultze et al., 1996; Krempl et al., 1997). In this way the sialic acid binding activity may help TGEV to survive the gastrointestinal passage. In future work we will analyze whether the hemagglutination-defective virions are more sensitive to the action of detergents and proteases than is the parental virus.

REFERENCES

Herrler, G., Hausmann, J., and Klenk, H.-D., 1995, Sialic acid as receptor determinant of ortho- and paramyxoviruses, in *Biology of the Sialic Acids* (A. Rosenberg , ed.), Plenum Press, New York pp. 315–336.

Höfling, K., Brossmer, R., Klenk, H.-D. and Herrler, G., 1996; Transfer of an esterase-resistant receptor analog to the surface of influenza C virions results in reduced infectivity due to aggregate formation, *Virology* **218:** 127–133.

Krempl, C, Laude, H., and Herrler, G., 1997, Point mutations in the S protein connect the sialic acid binding activity with the enteropathogenicity of transmissible gastroenteritis coronavirus, *J. Virol.* **71**: 3285–3287.

Liu, C., Eichelberger, M.C., Compans, R.W. and Air, G.M., 1995, Influenza type A virus neuraminidase does not play a role in viral entry, replication, assembly, or budding, *J. Virol.* **69**:1099–1106.

Sanchez, C. M., Jimenez, G., Laviada, M. D., Correa, I., Sune. C., Bullido, M. J., Gebauer, F., Smerdou, C., Callebaut, P., Escribano, J. M., and Enjuanes, L., 1990, Antigenic homology among coronaviruses related to transmissible gastroenteritis virus, *Virology* **174**: 410–417.

Schultze, B., Groß, H.J., Brossmer, R., Klenk, H.-D. and Herrler, G. 1990, Hemagglutinating encephalomyelitis virus attaches to N-acetyl-9-O-acetylneuraminic acid-containing receptors: comparison with bovine coronavirus and influenza C virus, *Virus Res.* **16**: 185–194.

Schultze, B., Cavanagh, D. and Herrler, G. 1992, Neuraminidase treatment of avian infectious bronchitis virus reveals a hemagglutinating activity that is dependent on sialic acid containing receptors on erythrocytes, *Virology* **189**: 792–794.

Schultze, B., Krempl, C., Shaw, L., Enjuanes, L. and Herrler, G., 1996, Transmissible gastroenteritis coronavirus but not the related porcine respiratory coronavirus has a sialic acid (N-glycolylneuraminic acid) binding activity, *J. Virol.* **70**: 5634–5637.

Schultze, B., Zimmer, G. and Herrler, G., 1996, Virus entry into a polarized epithelial cell line (MDCK): similarities and dissimilarities between influenza C virus and bovine coronavirus, *J. Gen. Virol.* **77**: 2507–2514.

Vlasak, R., Luytjes, W., Spaan, W. and Palese, P., 1988, Human and bovine coronaviruses recognize sialic acid-containing receptors similar to those of influenza C viruses, *Proc. Natl. Acad. Sci. U.S.A.* **85**: 4526–4529.

Zimmer, G., Klenk, H.-D. and Herrler, G., 1995; Identification in a MDCK cell line of a 40 kDa cell surface sialoglycoprotein with the characteristics of a major influenza C virus receptor, *J .Biol. Chem.* **270**: 17815–17822.

74

ROLE OF MOUSE HEPATITIS VIRUS-A59 RECEPTOR Bgp1a EXPRESSION IN VIRUS-INDUCED PATHOGENESIS

Catherine Godfraind,[1] Kathryn V. Holmes,[3] and Jean-Paul Coutelier[2]

[1]Laboratory of Pathology
[2]Unit of Experimental Medicine
Catholic University of Louvain
1200 Bruxelles, Belgium
[3]Department of Microbiology
University of Colorado Health Sciences Center
80262 Denver, Colorado

1. ABSTRACT

Expression of Bgp1a, a glycoprotein that serves as receptor for mouse hepatitis virus-A59 has been analyzed in various mouse tissues and correlated with the pathogenicity that this virus induces in the corresponding organs. Expression of Bgp1a was observed in many cells of epithelial origin, including hepatocytes and endothelial cells. It was also shown on macrophages and B lymphocytes. Bgp1a localization may easily explain infection and lysis of some cell types like hepatocytes. In contrast, other cell types that express the viral receptor are not infected after *in vivo* inoculation with mouse hepatitis virus-A59, which may be due to inaccessibility of the receptor to the virus during mouse infection, or to resistance to this virus in some cell types. This may account for the ability of the blood-brain barrier to prevent mouse hepatitis virus-A59 spreading into the central nervous system. In other organs, the virus may induce pathogenesis indirectly, resulting in the destruction of cells that do not express Bgp1a, like thymic lymphocytes, or else impair cell functions such as cytokine and immunoglobulin production by macrophages and B lymphocytes, respectively.

2. INTRODUCTION

A wide range of pathologies may develop in mice infected with mouse hepatitis virus (MHV), depending on the host genetic background and virus strain. Frequent consequences of MHV-A59 infection of sensitive BALB/c, C57BL/6 or CBA/Ht mice include hepatitis, diarrhea, transient or chronic demyelinating lesions of the central nervous system and various alterations of the immune system, including polyclonal activation of B lymphocytes, impairment of macrophage functions and thymus involution.

Bgp1a, a biliary glycoprotein with properties of an adhesion molecule, member of the carcinoembryonic family, has been recently identified as the major receptor for MHV-A59 (Dveksler *et al.*, 1993a; 1993b; 1991). Using antibodies raised against this protein, its expression has been localized in various mouse tissues. This paper summarizes the data reported so far on Bgp1a localization and on the relationship between expression of this viral receptor and pathogenicity of MHV-A59.

3. MATERIALS AND METHODS

Bgp1a expression was analyzed at the mRNA level by RT-PCR (Coutelier *et al.*, 1994), or at the protein level by immunohistochemistry (Godfraind *et al.*, 1995b) or by flow cytometry analysis (Coutelier *et al.*, 1994). For the latter techniques, both MAb-CC1 monoclonal antibody recognizing the N-terminal domain of murine Bgp1a (Dveksler *et al.*, 1993b) and Ab-655 polyclonal rabbit anti-mouse Bgp1 antibody (Dveksler *et al.*, 1991) were used. Functional ability of Bgp1a to bind MHV-A59 was tested on cryostat tissue sections with a new virus binding assay (Godfraind *et al.*, 1995b). After intraperitoneal inoculation into BALB/c or CBA/Ht mice of approximately 10^4 - 10^5 tissue culture infectious doses ($TCID_{50}$) of the MHV-A59 grown in NCTC 1469 cells, infected cells were detected by immunohistochemistry using a polyclonal goat antibody directed against the spike glycoprotein of MHV-A59, isolated from detergent-disrupted virions (anti-S) (Sturman *et al.*, 1980) or by *in situ* hybridization (Godfraind *et al.*, 1995a). Apoptosis was analyzed by electron microscopy and *in situ* end labelling of fragmented DNA (Godfraind *et al.*, 1995a) and cytokine expression was measured by RT-PCR.

4. RESULTS

4.1. Bgp1a Expression May Lead to Cell Infection and Lysis

Analysis of Bgp1a localization by immunohistochemistry indicated that this molecule is expressed on epithelial cells, including enterocytes, ciliated and non-ciliated respiratory epithelial cells, hepatocytes, and a wide range of epithelial cells in the bile ducts, kidney tubules, adrenal gland, and thyroid follicles (Godfraind *et al.*, 1995b). It was also found on endothelial cells in all tissues tested. This distribution is similar to that reported for rat (Odin *et al.*, 1988) and human Bgp glycoproteins (Prall *et al.*, 1996). However, it differs from the localization reported for Bgp1a in embryonic mouse (Huang *et al.*, 1990), indicating age-dependent regulation of its expression.

In the liver, Bgp1a is preferentially expressed around the bile canaliculi and on the membranes facing the space of Disse. Its ability to bind MHV-A59 virions was demonstrated by an *in situ* virus binding assay performed on tissue sections (Godfraind *et al.*, 1995b). Therefore, it is not surprising that intraperitoneal infection of BALB/c mice with MHV-A59 leads to infection, followed by lysis of large foci of hepatocytes (Godfraind *et al.*, 1995b). Although other mechanisms may also be involved, virus-triggered destruction of target cells expressing the viral receptor seems to be a major cause of hepatitis developing in infected animals.

4.2. Resistance of Some Bgp1a-Expressing Cells to MHV-A59 Infection and Prevention of Passage of the Virus through the Blood-Brain Barrier

In contrast to hepatocytes, some cell types that express Bgp1a, like enterocytes or thyroid follicle cells are not infected after intraperitoneal inoculation of MHV-A59. This is probably because the localization of the viral receptor does not allow contact with virions in the general blood circulation. Interestingly, endothelial cells of the central nervous system (CNS) express Bgp1a at their apical pole and thus could come in contact with blood-borne virions, but resist MHV-A59 infection (Godfraind, manuscript in preparation). The virus binding assay indicated that the receptor on CNS endothelial cells could effectively bind MHV-A59 virions. However, after brief intravenous treatment with a detergent, virus replication was detected in some brain endothelial cells, suggesting that the resistance to infection developed by this cell subpopulation affects viral entry (Godfraind, manuscript in preparation). Therefore, the efficient prevention of MHV-A59 spreading from the blood circulation into the CNS probably relies on specific mechanisms of resistance to viral infection evolved by endothelial cells that form the blood-brain barrier.

Once MHV-A59 has penetrated into the CNS, it can infect glial cells, although Bgp1a expression could not be detected on these cells by immunohistochemistry. We therefore postulate that either Bgp1a is expressed on glial cells at a density too low to allow detection by this method, or that virus binding to these cells involves another receptor, such as an isoform of Bgp1a (Yokomori and Lai, 1992), Bgp2 (Nédellec *et al.*, 1994) or bCEA (Chen *et al.*, 1995).

4.3. Indirect MHV-A59-Induced Death of Cells that Do Not Express Bgp1a and Thymus Involution

MHV-A59 infection of BALB/c mice is followed by a severe, but transient involution of the thymus, characterized by a preferential decrease of cortical immature double-positive CD4+CD8+ lymphocytes (Godfraind *et al.*, 1995a). Electron microscopy and *in situ* end labelling of fragmented DNA on thymus sections indicated that apoptosis of lymphocytes was involved in this thymus atrophy. Although no Bgp1a expression could be detected on T lymphocytes, this receptor was present on thymus epithelial cells and some of which were infected after MHV-A59 inoculation (Godfraind *et al.*, 1995a). Since thymic epithelial cells play a critical role in T lymphocyte maturation, we postulate that infection of these epithelial cells impairs a cell function required for T cell development, or causes inappropriate production of some soluble factor that induces apoptosis in immature T lymphocytes. Therefore, T cell death and the resulting thymus involution are an indirect consequence of infection of Bgp1a-bearing epithelial cells.

4.4. MHV-A59-Induced Impairment of Immune Cell Functions: Cytokine Production by Macrophages and B Lymphocyte Polyclonal Activation

In the mouse spleen, Bgp1a expression was found by flow cytometry on B lymphocytes and macrophages, but not on T cells (Coutelier *et al.*, 1994). MHV-A59 infection triggers strong polyclonal activation of B lymphocytes, leading to enhancement of concomitant non antiviral antibody responses, including autoantibody production, and to an isotypic preponderance of the IgG2a subclass (Lardans *et al.*, 1996; Coutelier *et al.*, 1991; Coutelier *et al.*, 1988). Since direct activation of B cells through virus/receptor interaction has been previously demonstrated for other viruses (Scalzo and Anders, 1985), it may also be postulated for MHV-A59. On the other hand, the IgG2a preponderance, which is a T helper-dependent phenomenon (Lardans *et al.*, 1996), may result from a cascade of events. Indeed, MHV-A59 induces production of cytokines such as interleukin-12 by macrophages (Coutelier *et al.*, 1995) and, as a probable effect of this cytokine secretion, the preferential differentiation of T helper lymphocytes towards the Th1 subpopulation (Coutelier *et al.*, 1991). Thus, MHV-A59-induced IgG2a-restricted B lymphocyte polyclonal activation requires functional alteration of two cell types that express Bgp1a and can interact with the virus, namely B lymphocytes and macrophages.

5. CONCLUSIONS

Expression of the MHV-A59 receptor Bgp1a is observed on a wide range of epithelial cells, including hepatocytes, on endothelial cells, on macrophages and on B lymphocytes. After MHV-A59 inoculation, cellular expression of this viral receptor may result in different outcomes (Figure 1): (a) infection and lysis of cells expressing Bgp1a leading to pathology in the corresponding organ, such as hepatitis; (b) indirect destruction of cells that do not express Bgp1a and that are not infected by MHV-A59, as a consequence of infection of another cell type. Such an indirect mechanism triggers apoptosis of thymic lymphocytes; (c) impairment of the function of cells bearing the MHV-A59 receptor. The combined effect of the virus on two cell subpopulations, i.e. macrophages and B lymphocytes may lead to pathological mechanisms such as IgG2a-restricted enhanced autoanti-

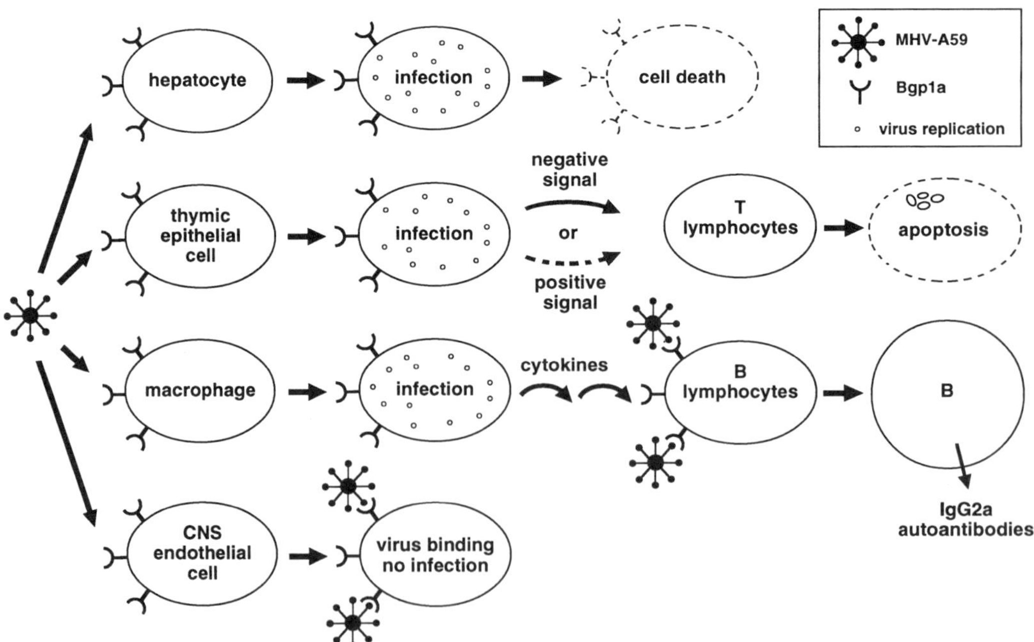

Figure 1. Relationship between MHV-A59 Bgp1a receptor expression and viral pathogenesis.

body production; (d) lack of infection of cells bearing the viral receptor, which may correspond to mechanisms of resistance to infection. At the level of CNS endothelial cells, this resistance prevents spreading of the virus through the blood-brain barrier. Therefore, MHV-A59 provides a useful model to analyze the relationship between cellular expression of a viral receptor and viral pathogenesis.

ACKNOWLEDGMENTS

This work was supported by the Fonds National de la Recherche Scientifique (FNRS), Fonds de la Recherche Scientifique Médicale (FRSM), Loterie Nationale and the State-Prime Minister's Office–S.S.T.C. (interuniversity attraction poles, grant n°44) and the French Community (concerted actions, grant n° 88/93–122), Belgium, by the association "Recherche & Partage", Paris, France, and by NIH grant #25231. J.-P. C. is a senior research associate with the FNRS.

REFERENCES

Chen, D.S., Asanaka, M., Yokomori, K., Wang, F.-i., Hwang, S.B., Li, H.-P., and Lai, M.M.C., 1995, A pregnancy-specific glycoprotein is expressed in the brain and serves as a receptor for mouse hepatitis virus, *Proc. Natl. Acad. Sci. USA* **92:** 12095–12099.

Coutelier, J.-P., van der Logt, J.T.M., Heessen, F.W.A., Vink, A., and Van Snick, J., 1988, Virally induced modulation of murine IgG antibody subclasses, *J. Exp. Med.* **168:** 2373–2378.

Coutelier, J.-P., van der Logt, J.T.M., and Heessen, F.W.A., 1991, IgG subclass distribution of primary and secondary immune responses concomitant with viral infection, *J. Immunol.* **147:** 1383–1386.

Coutelier, J.-P., Godfraind, C., Dveksler, G.S., Wysocka, M., Cardellichio, C.B., Noël, H., and Holmes, K.V., 1994, B lymphocyte and macrophage expression of carcinoembryonic antigen-related adhesion molecules that serve as receptors for murine coronavirus, *Eur. J. Immunol.* **24:** 1383–1390.

Coutelier, J.-P., Van Broeck, J., and Wolf, S.F, 1995, Interleukin-12 gene expression after viral infection in the mous, *J. Virol.* **69:** 1955–1958.

Dveksler, G.S., Pensiero, M.N., Cardellichio, C.B., Williams, R.K., Jiang, G.S., Holmes, K.V., and Dieffenbach, C.W., 1991, Cloning of the mouse hepatitis virus (MHV) receptor: expression in human and hamster cell lines confers susceptibility to MH, *J. Virol.* **65:** 6881–6891.

Dveksler, G.S., Dieffenbach, C.W., Cardellichio, C.B., McCuaig, K., Pensiero, M.N., Jiang, G.S., Beauchemin, N., and Holmes, K.V., 1993a, Several members of the mouse CEA-related glycoprotein family are functional receptors for murine coronavirus MHV-A59, *J. Virol.* **67:** 1–8.

Dveksler, G.S., Pensiero, M.N., Dieffenbach, C.W., Cardellichio, C.B., Basile, A.A., Elia, P.E., and Holmes, K.V., 1993b, Mouse hepatitis virus strain-A59 and blocking antireceptor monoclonal antibody bind to the N-terminal domain of cellular receptor, *Proc. Natl. Acad. Sci. USA* **90:** 1716–1720.

Godfraind, C., Holmes, K.V., and Coutelier, J.-P., 1995a, Thymus involution induced by mouse hepatitis virus A59 in BALB/c mice, *J. Virol.* **69:** 6541–6547.

Godfraind, C., Langreth, S.G., Cardellichio, C.B., Knobler, R., Coutelier, J.-P., Dubois-Dalcq, M., and Holmes, K.V., 1995b, Tissue and cellular distribution of an adhesion molecule in the carcinoembryonic antigen family that serves as a receptor for mouse hepatitis virus, *Lab. Invest.* **73:** 615–627.

Huang, J.Q., Turbide, C., Daniels, E., Jothy, S., and Beauchemin, N., 1990, Spatiotemporal expression of murine carcinoembryonic antigen (CEA) gene family members during mouse embryogenesis, *Development* **110:** 573–588.

Lardans, V., Godfraind, C., van der Logt, J.T.M., Heessen, F.W.A., Gonzalez, M.-D., and Coutelier, J.-P, 1996, Polyclonal B lymphocyte activation induced by mouse hepatitis virus A59 infection, *J. Gen. Virol.* **77:** 1005–1009.

Nédellec, P., Dveksler, G.S., Daniels, E., Turbide, C., Chow, B., Basile, A.A., Holmes, K.V., and Beauchemin, N., 1994, Bgp2, a new member of the carcinoembryonic antigen-related gene family, encodes an alternative receptor for mouse hepatitis viruses, *J. Virol.* **68:** 4525–4537.

Odin, P., Asplund, M., Busch, C., and Öbrink, B., 1988, Immunohistochemical localization of CellCAM 105 in rat tissues: appearance in epithelia, platelets, and granulocytes, *J. Histochem. Cytochem.* **36:** 729–739.

Prall, F., Nollau, P., Neumaier, M., Haubeck, H.-D., Drzeniek, Z., Helmchen, U., Löning, T., and Wagener, C., 1996, CD66a (BGP), an adhesion molecule of the carcinoembryonic antigen family, is expressed in epithelium, endothelium, and myeloid cells in a wide range of normal human tissues, *J. Histochem. Cytochem.* **44:** 35–41.

Scalzo, A. A., and Anders, E.M, 1985, Influenza viruses as lymphocyte mitogens. II. Role of I-E molecules in B cell mitogenesis by influenza A viruses of the H2 and H6 subtypes, *J. Immunol.* **135:** 3524–3529.

Sturman, L.S., Holmes, K.V., and Behnke, J.N., 1980, Isolation of coronavirus envelope glycoproteins and interaction with the viral nucleocapsid, *J. Virol.* **33:** 449–462.

Yokomori, K., and Lai, M.M.C., 1992, Mouse hepatitis virus utilizes two carcinoembryonic antigens as alternative receptors, *J. Virol.* **66:** 6194–6199.

75

PERSISTENT INFECTION OF NEURAL CELL LINES BY HUMAN CORONAVIRUSES

Nathalie Arbour and Pierre J. Talbot

Laboratory of Neuroimmunovirology
Institut Armand-Frappier
Université du Québec
Laval, Québec, Canada H7V 1B7

1. ABSTRACT

Human coronaviruses (HCV) have been associated mainly with infections of the respiratory tract. Accumulating evidence from *in vitro* and *in vivo* observations is consistent with the neurotropism of these viruses in humans. To verify the possibility of a persistent infection within the central nervous system (CNS), various human cell lines of neural origin were tested for their ability to maintain chronic infection by both known strains of HCV, OC43 and 229E. Production of infectious progeny virions was monitored by an immunoperoxydase assay on a susceptible cell line and viral RNA was observed after RT-PCR. Astrocytic cell lines U-373 MG and U-87 MG did not sustain a persistent HCV-229E infection, even though they were susceptible to an acute infection by this virus. On the other hand, these two cell lines could maintain a persistent infection by HCV-OC43 for as many as 25 cell passages (about 130 days of culture). Relatively stable titers of infectious viral particles, as well as apparently constant amounts of viral RNA were detected throughout the persistent infection of U-87 MG cells. However, persistent infection of U-373 MG cells was accompanied by the detection of infectious viral particles from passage 0 to passage 13 and then from passage 20 to the end of the experiment. This gap in the production of infectious virions was correlated by a drop in the apparent amount of viral RNA detected at passages 15 and 20. These results confirm the ability of HCV-OC43 to persistently infect cells of an astrocytic lineage and, together with our previous observations of HCV infection of primary cultures of human astrocytes and the detection of HCV RNA in human brains, are consistent with the possibility that this human coronavirus could persist in the human CNS by targeting astrocytes.

2. INTRODUCTION

Human coronaviruses (HCV) cause up to one-third of common colds (Myint 1994). Neonatal nosocomial respiratory coronavirus infections appear to be frequent and may be

Coronaviruses and Arteriviruses, edited by Enjuanes *et al.*
Plenum Press, New York, 1998

associated with apnea (Sizun *et al.* 1995). Other pathologies have sporadically been associated with HCV such as pneumonia, perimyocarditis, meningitis, radiculitis (Riski and Hovi, 1980), as well as diarrhea (Resta *et al.* 1985). Increasing evidence from animal models and studies in humans suggest a neurotropism for these viruses.

The murine counterpart of HCV, murine hepatitis virus (MHV), has been studied as an animal model of a virus-induced disease of the central nervous system (CNS) (ter Meulen *et al.* 1989). After intranasal inoculation, MHV enters the CNS of mice via the olfactory nerve and then spreads to the brain (Barnett and Perlman 1993). Moreover, a neurotropic MHV was shown to enter the CNS of primates after peripheral inoculation (Cabirac *et al.* 1994).

Salmi *et al.* (1982) detected intrathecal antibody synthesis to HCV in humans, particularly in multiple sclerosis patients, suggesting a CNS infection. Moreover, HCV RNAs have repeatedly been detected in human brains (Stewart *et al.* 1992, Murray *et al.* 1992, Arbour and Talbot 1997). Finally, we have recently shown that HCV could infect primary cultures of human neural cells (Bonavia *et al.* 1997). We speculate that HCV could persistently infect human neural cells since MHV RNA could be detected in the brains of infected mice for a long period of time after the initial infection (Adami *et al.* 1995, Rowe *et al.* 1997) and that HCV were detected in human brains and are able to infect human neural cells both in primary and immortalized cultures. Since primary cultures do not allow us to perform a long time-scale study that is involved for viral persistence, we have used cell lines representative of different neural cell types to verify the possibility of persistent HCV infections. Viral RNA and infectious viral particles were monitored during a prolonged infection cycle.

3. MATERIALS AND METHODS

3.1. Virus and Cell Lines

Both HCV strains (229E and OC43) were originally obtained from the American Type Culture Collection (ATCC; Rockville, MD), plaque-purified twice and grown on either L132 cells (229E) or HRT-18 cells (OC43), as described previously (Jouvenne *et al.* 1992, Mounir and Talbot 1992). The third passage of HCV-229E with a titer of 5.5×10^5 $TCID_{50}$/ml and the fourth passage of HCV-OC43 with a titer of 5.15×10^5 $TCID_{50}$/ml, from laboratory stocks kept at -90°C, were used for all experiments. Human neuronal and glial cell lines: H4, SK-N-SH, MO3.13, U-373 MG, U-87 MG, were obtained and cultured as already described (Talbot *et al.* 1994). Cell monolayers at 60–80% confluence were infected in 25 cm^2 culture flasks with 0.5 ml of viral suspension and incubated 2 hours at 33 °C, with periodical agitation. Cell monolayers were then washed with PBS and grown in the regular cell culture medium at 37°C. Cells were passaged by trypsination every 4–8 days. Samples of supernatants and cells were kept for viral titration at each passage and cells kept at each fifth passage for RNA extraction.

3.2. Immunoperoxidase Assay for Quantitation of Infectious Virus Titers

The immunoperoxydase assay for quantitation of infectious virus titers was performed as described (Bonavia *et al.* 1997). Antiviral mouse monoclonal antibody 1.10-C.1 (anti-HCV-OC43) was produced in our laboratory by standard hybridoma technology.

3.3. Preparation of RNA and Reverse-Transcription Polymerase Chain Reaction

To extract total cellular RNA from infected or controls cells, cell monolayers were washed twice with PBS and then kept at -90 °C until RNA extraction. Cells were lysed with GIT buffer (4 M guanidine isothiocyanate, 2.5 mM Na acetate, 12 mM 2-mercaptoethanol). Lysates were passed through a 26G needle at least 6 times and then layered onto a cesium chloride cushion (5.7 M cesium chloride, 2.5 mM Na acetate) for a 12 to 20 hours centrifugation at 150 000 x g. The supernatant was removed and the RNA pellet resuspended in sterile dd H_2O. The pair of primers used for HCV-OC43 amplification as well as the one for control gene (glyceraldehyde-3-phosphate dehydrogenase) are described in Table 1. A mixture of 40 pmol (4 µl) of the inverse-complementary primer and about 5 µg of total cellular RNA was incubated at 65 °C for 5 min to denature RNA, followed by a slow cool down to 37 °C for annealing. Reverse transcription with Moloney murine leukemia virus reverse transcriptase (50 U; Boehringer Mannheim) was performed at 42 °C for 90 min in the presence of 60 U of RNA Guard (Pharmacia), 0.4 mM of each dNTP (Boehringer Mannheim), 1X Reverse Transcriptase buffer [50 mM Tris-HCl pH 8.3, 40 mM KCl, 5 mM $MgCl_2$ 0.5% Tween® 20 (v/v)] and 10 mM dithiothreitol (Boehringer Mannheim). For PCR, one fifth of the synthesized cDNA was incubated in the presence of 20 pmol of the O1 and O3 primers, or 50 pmol of the GAPDH-H and GAPDH-I primers, 2.5 mM $MgCl_2$ (1.5 mM in the case of GAPDH) (BIO/CAN), 1X PCR buffer (10 mM Tris-HCl pH 8.3, 1.5 mM $MgCl_2$, 50 mM KCl) (Boehringer Mannheim) and 0.4 mM of each dNTP (Boehringer Mannheim) at 94°C for 5 min and at 60°C (50°C for GAPDH) for another 5 min. After addition of the *Taq* DNA polymerase (2.5 U; Boehringer Mannheim), 30 cycles of 2 min at 72 °C, 1 min at 94 °C and 2 min at 60°C (50°C for GAPDH) were performed with a final elongation step of 10 min at 72 °C. The DNA amplicons were separated by electrophoresis in a 1.5% (w/v) agarose gel containing 1 µg/ml ethidium bromide.

4. RESULTS and DISCUSSION

4.1. Quantitation of Infectious Progeny Virions

Although the human astrocytic glial cell lines U-373 MG and U-87 MG were previously shown to be acutely infectable by HCV-229E (Talbot *et al.* 1994), no infectious viral particles could be detected after the 5th cell passage during a persistent infection (data not shown). Therefore, it appears that astrocytic cell lines did not sustain a persistent infection by HCV-229E. These results could be easily compared with the ones obtained with

Table 1. Primers for RT-PCR

RNA amplified	Primers-corresponding region	
HCV-OC43 nucleoprotein[1]	5′ - CCCAAGCAAACTGCTACCTCTCAG - 3′	O1-sense: 215- 238
.	5′ - GTAGACTCCGTCAATATCCCTGCC - 3′	O3- antisense: 497-520
GAPDH[2]	5′ - GTGAAGGTCGGAGTCAACG - 3′	GAPDH-H-sense: 10-68
	5′ - CACCTGGTGCTCAGTGTAGC - 3′	GAPDH-I-antisense: 824-843

[1](Stewart *et al.* 1992); (Kamahora *et al.* 1989)
[2](Ercolani *et al.* 1988)

primary cultures of human astrocytes where the HCV-229E infection could only be detected by RT-PCR and not by immunofluorescence or by production of infectious progeny virions (Bonavia *et al.* 1997). Therefore, it is possible that HCV-229E could infect astrocytic cells but that the infection is rapidly cleared. These results suggest that the presence of HCV-229E in human brains may not involve astrocytes. However *in situ* hybridization assays from HCV-229E RNA positive human brain samples will be necessary before such a conclusion can definitively be reached.

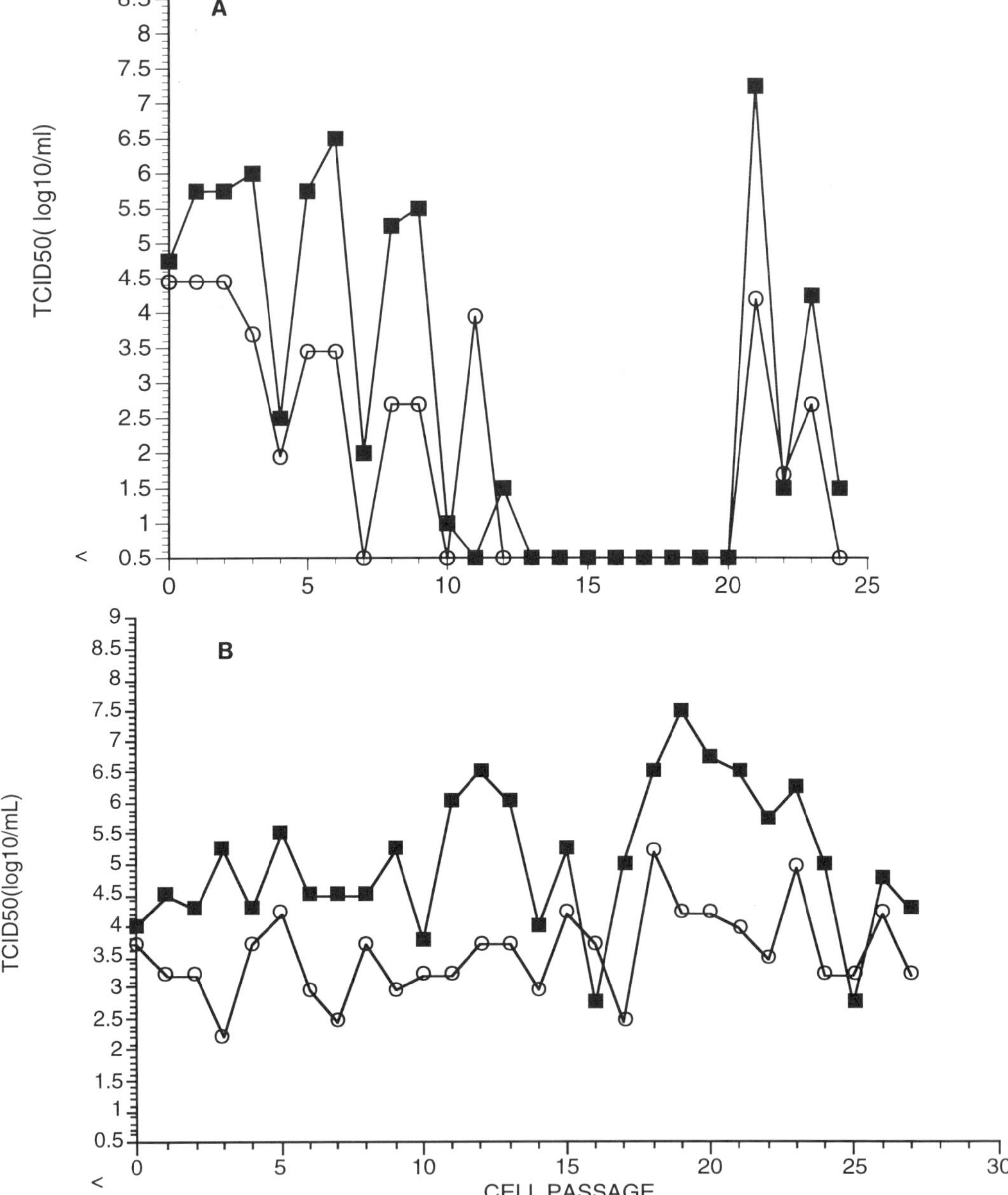

Figure 1. Infectious virus titers from HCV-OC43 persistent infections of astrocytic cell lines. Panel A: U-373 MG; Panel B: U-87 MG. ■: extracellular, ○: intracellular.

Persistent infection of the same astrocytic cell lines by HCV-OC43 did result in the production of infectious virions as late as the 25th cell passage (about 130 days in culture). Figure 1 shows the infectious viral titers obtained from persistently infected U-373 MG and U-87 MG cell lines (panels A and B, respectively). In both cases, viral titers obtained for the extracellular component were higher than for the intracellular components. These astrocytic cell lines did produce infectious virions in amounts comparable to that obtained after infection of the reference cell line HRT-18 (data not shown). Approximately 10% of U-87 MG cells were infected in the culture, as estimated by immunofluorescence (data not shown). This percentage of infected cells appears to be sufficient to maintain the presence of virus in the culture. Collins and Sorensen (1986) previously showed that U-87 MG cells could be persistently infected with HCV-OC43 and release infectious virions, although infection was monitored on a much shorter time scale of 28 days.

No virus was detected between passages 13 and 20 of the persistently infected U-373 MG cell culture, despite the presence of viruses at the beginning and at the end of the culture, particularly in the extracellular compartment (Fig. 1, panel A). At the end of the culture, about 5 % of U-373 MG cells were infected, as shown by immunofluorescence (data not shown). Persistent infections by HCV-229E and HCV-OC43 of other neural cell lines representative of human neurons (H4) and oligodendrocytes (MO3.13) were also observed, including production of infectious virions after more than 100 days of culture (data not shown).

4.2. Detection of HCV-OC43 RNA

Figure 2 shows RT-PCR results obtained from HCV-OC43 persistent infection of U-373 MG and U-87 MG cells. The RT-PCR assay was performed on RNA extracted from acutely infected cells (lanes A) and non-infected cells (lanes N). GAPDH RNA could be amplified in each case (data not shown), indicating that RNA was suitable for amplification even for non-infected cells, where no viral RNA could be detected. Viral RNA could be detected at every passage we looked at : P2, P5, P10, P15, P20 and P25 (both cell lines) and P28 (U-87 MG only). Apparently comparable amounts of viral RNA were detected in

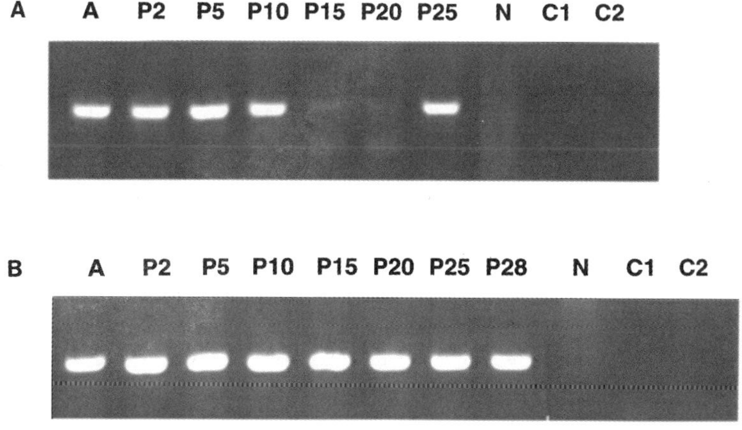

Figure 2. Detection of HCV-OC43 RNAs in persistently infected cells by RT-PCR visualized on a 1.5% (w/v) agarose gel containing ethidium bromide. Panel A: U-373 MG, Panel B: U-87 MG. A: acute infection, Px, cell passage number, N: non-infected cells, C1: RT negative control and C2: PCR negative control.

acute and persistently infected cells. For U-87 MG cells, comparable amounts of viral RNA could be amplified at each passage, which is consistent with a more or less constant amount of viral RNA in the culture. Other authors have been able to detect viral RNA during a persistent HCV-OC43 infection of U-87 MG cells (Collins and Sorensen 1986). During the persistent infection of U-373 MG cells, we observed that apparently less viral RNA was detectable at P15 and P20, even though the same amount of total cellular RNA was used for RT-PCR in each case. When these results are compared to those of the immunoperoxydase experiments (Fig. 1, panel A), we observe that passages where no infectious virus was detected (13–20) correspond to low viral RNA levels.

We have recently shown that HCV-OC43 could infect primary cultures of human fetal astrocytes, adult microglia and adult astrocytes, as monitored by the immunofluorescence detection of viral proteins (Bonavia *et al.* 1997). It is therefore not surprising that astrocytic cell lines also supported a HCV-OC43 infection. Interestingly, in mice, the murine strain MHV-JHM infects astrocytes after an intranasal inoculation (Sun and Perlman 1995). Moreover during the chronic demyelination induced in infected mice, astrocytes are the predominant cells expressing inflammatory cytokines such as IL-1β and IL-6, as well as nitric oxide (Sun *et al.* 1995) probably involved in tissue damage, such as demyelination. In addition, *in situ* hybridization combined with immunohistochemistry performed on monkey brains after intracerebral infection has shown that astrocytes are the target cells in white matter during acute infection by MHV-JHM (Murray *et al.* 1997). Our results, combined with the work of other research teams, are consistent with a potential role of astrocytes in CNS infection by human coronaviruses.

ACKNOWLEDGMENTS

We thank Francine Lambert for excellent technical assistance. N. Arbour is grateful to the Institut Armand-Frappier for studentship support and P. J. Talbot to the *Fonds de la recherche en santé du Québec* for a senior scholarship. This work was supported by an operating grant from the Medical Research Council (MRC) of Canada to P.J.T.

REFERENCES

Adami, C., Pooley, J., Glomb, J., Stecker, E., Fazal, F., Fleming, J.O., and Baker, S.C., 1995, Evolution of mouse hepatitis virus (MHV) during chronic infection: quasispecies nature of the persisting MHV RNA, *Virology*, **209**:337–346.

Arbour, N., and Talbot, P.J., 1997, unpublished observations.

Barnett, E.M., and Perlman, S., 1993, The olfactory nerve and not the trigeminal nerve is the major site of CNS entry for mouse hepatitis virus, strain JHM, *Virology* **194**:185–191.

Bonavia, A., Arbour, N., Yong, V.W., and Talbot, P.J., 1997, Infection of primary cultures of human neural cells by human coronaviruses 229E and OC43, *J. Virol.* **71**:800–806.

Cabirac, G.F., Soike, K.F., Zhang, J.Y., Hoel, K., Butunoi, C., Cai, G.Y., Johnson, S., and Murray, R.S., 1994, Entry of coronavirus into primate CNS following peripheral infection, *Microb. Path.* **16**:349–357.

Collins, A.R., and Sorensen, O., 1986, Regulation of viral persistence in human glioblastoma and rhabdomyosarcoma cells infected with coronavirus OC43, *Microb. Path.* **1**: 573–582.

Ercolani, L., Florence, B., Denaro, M., and Alexander, M., 1988, Isolation and complete sequence of a functional human glyceraldehyde-3-phosphate dehydrogenase gene, *J. Biol. Chem.* **263**:15335–15341.

Jouvenne, P., Mounir, S., Stewart, J.N., Richardson, C.D., and Talbot, P.J., 1992, Sequence analysis of human coronavirus 229E mRNAs 4 and 5: evidence for polymorphism and homology with myelin basic protein, *Virus Res.* **22**:125–141.

Kamahora, T., Soe, L.H., and Lai, M.M.C., 1989, Sequence analysis of nucleocapsid gene and leader RNA of human coronavirus OC43, *Virus Res.* **12**:1–9.

Mounir, S., and Talbot, P.J., 1992, Sequence analysis of the membrane protein gene of human coronavirus OC43 and evidence for *O*-glycosylation, *J. Gen. Virol.* **73**:2731–2736.

Murray, R.S., Brown, B., Brian, D., and Cabirac, G.F., 1992, Detection of coronavirus RNA and antigen in multiple sclerosis brain, *Ann. Neurol.* **31**:525–533.

Murray, R.S., Cai, G.Y., Soike, K.F., and Cabirac, G.F., 1997, Further observations on coronavirus infection of primate CNS, *J. Neurovirol.* **3**:71–75.

Myint, S.H., 1994, Human coronaviruses-a brief review, *Rev. Med. Virol.* **4**:35–46.

Resta, S., Luby, J.P., Rosenfeld, C.R., and Siegel, J.D., 1985, Isolation and propagation of a human enteric coronavirus, *Science* **229**:978–981.

Riski, H., and Hovi, T., 1980, Coronavirus infections of man associated with diseases other than the common cold, *J. Med. Virol.* **6**:259–265.

Rowe, C.L., Baker, S.C., Nathan, M.J., and Fleming, J.O., 1997, Evolution of mouse hepatitis virus: detection and characterization of spike deletion variants during persistent infection, *J. Virol.* **71**:2959–2969.

Salmi, A., Ziola, B., Hovi, T., and Reunanen, M., 1982, Antibodies to coronaviruses OC43 and 229E in multiple sclerosis patients, *Neurology* **32**:292–295.

Schreiber, S.S., Kamahora ,T., and Lai, M.M.C., 1989, Sequence analysis of the nucleocapsid protein gene of human coronavirus 229E, *Virology* **169**:142–151.

Sizun, J., Soupre, D., Legrand, M.C., Giroux, J.D., Rubio, S., Cauvin, J.M., Chastel, C., Alix, D., and de Parscau, L., 1995, Neonatal nosocomial respiratory infection with coronavirus: a prospective study in a neonatal intensive care unit, *Acta Paediatr.* **84**:617–620.

Stewart, J.N., Mounir, S., and Talbot, P.J., 1992, Human coronavirus gene expression in the brains of multiple sclerosis patients, *Virology* **191**:502–505.

Sun, N., and Perlman, S., 1995, Spread of a neurotropic coronavirus to spinal cord white matter via neurons and astrocytes, *J. Virol.* **69**:633–641.

Sun, N., Grzybicki, D., Castro, R.F., Murphy, S., and Perlman, S., 1995, Activation of astrocytes in the spinal cord of mice chronically infected with a neurotropic coronavirus, *Virology* **213**:482–493.

Talbot, P.J., Ékandé, S., Cashman, N.R., Mounir, S., and Stewart, J.N., 1994, Neurotropism of human coronavirus 229E, *Adv. Exp. Med. Biol.* **342**:339–346.

ter Meulen, V., Massa, P.T., and Dörries, R., 1989, Coronaviruses, in: *Handbook of clinical neurology: viral disease, revised series,* Volume 12 (P.J. Vinken, G.W. Bruyn, and H.L. Klawans, eds.), Elsevier, New-York, pp. 439–451.

76

NEUROPATHOGENICITY AND SUSCEPTIBILITY TO IMMUNE RESPONSE ARE INTERDEPENDENT PROPERTIES OF LACTATE DEHYDROGENASE-ELEVATING VIRUS (LDV) AND CORRELATE WITH THE NUMBER OF N-LINKED POLYLACTOSAMINOGLYCAN CHAINS ON THE ECTODOMAIN OF THE PRIMARY ENVELOPE GLYCOPROTEIN

Zongyu Chen,[1] Kehan Li,[1] Raymond R. R. Rowland,[2] and
Peter G. W. Plagemann[1]

[1]Department of Microbiology
University of Minnesota Medical School
Minneapolis, Minnesota 55455
[2]Department of Biology and Microbiology
South Dakota State University
Brookings, South Dakota 57007

1. ABSTRACT

We have developed differential RT-PCR methods to distinguish different isolates of
LDV and have purified several quasispecies by repeated end point dilution in mice. They
fall into two groups, each possessing two or more members. Group A viruses are non-
neuropathogenic, highly resistant to in vitro neutralization by antibodies and efficient in
establishment of a life-long, persistently viremic infection in mice despite a detectable immune response. Group B viruses, on the other hand, are neuropathogenic, much more sensitive to antibody neutralization and have an impaired ability to establish a high viremia
persistent infection in immune competent mice. These properties seem to be interdependent and correlate with the number of N-glycosylation sites on the short (about 30 amino
acid long) ectodomain of the primary envelope glycoprotein, VP-3P, which probably is
part of the attachment site for the LDV receptor on permissive cells and harbors an epi-

Coronaviruses and Arteriviruses, edited by Enjuanes *et al.*
Plenum Press, New York, 1998

tope(s) reacting with neutralizing antibodies. Group A viruses possess three closely spaced N-linked polylactosaminoglycan chains, whereas group B viruses lack the two N-terminal ones. We postulate that lack of these polylactosaminoglycan chains endows group B viruses with the ability to interact with a receptor on anterior horn neurons resulting in neuropathogenesis. At the same time, it increases an interaction with neutralizing antibodies thus impeding the infection of macrophages newly generated during the persistent phase of infection which is essential for the continued rounds of replication of the virus.

2. INTRODUCTION

Lactate dehydrogenase-elevating virus (LDV) is a member of the newly established family Arteriviridae (Canavagh et al., 1997). It is very closely related to another arterivirus, porcine reproductive and respiratory syndrome virus (PRRSV), in both genome size and sequence homology (Fig. 1; Meulenberg, et al., 1993; Murtaugh et al., 1995). In fact, for several viral proteins, the homology between LDV and PRRSV is comparable with that between the European and North American PRRSV isolates (Plagemann, 1996). However, LDV and PRRSV cannot reciprocally infect each other's hosts (Plagemann, 1996). LDV infection in its identified only host, the mouse, inevitably results in a life-long, asymptomatic, viremic infection in spite of a significant host immune response (Plagemann 1996; Plagemann et al., 1995). Under specific conditions, infection of mice with some LDV isolates can cause age-dependent poliomyelitis (ADPM), a fatal paralytic disease similar to amyotrophic lateral sclerosis of humans, through interaction with a normally nonpathogenic endogenous murine retrovirus (Plagemann 1996; Anderson et al., 1995b). LDV infection therefore represents a unique experimental model for studying the mechanisms of virus persistence and pathogenesis and may yield important information relevant to PRRSV biology and patho-

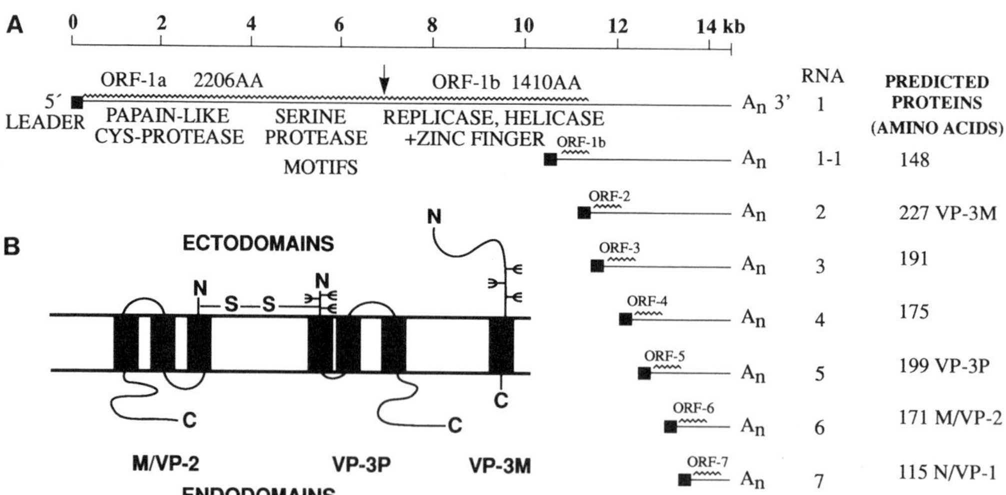

Figure 1. LDV genome organization and subgenomic mRNA expression (A). Proposed topology of the three viral structural proteins in the LDV-P virion (B). N and C denote the amino and carboxy termini of the proteins, respectively. S-S represents the disulfide bond(s) between VP-3P and M/VP-2 which is essential for viral infectivity. N-linked polylactosaminoglycan chains are indicated by ψ. VP-3M is present in a very small quantity in the virion.

genesis which are currently being extensively researched in order to curtail the huge economic loss brought about by the recent outbreaks of PRRSV infection.

We have recently discovered the coexistence, in neuropathogenic LDV isolates, of different quasispecies with different neuropathogenicity and susceptibility to antibody neutralization. Characterization of these quasispecies indicate that they fall into two groups. Group A viruses are non-neuropathogenic, highly resistant to in vitro neutralization by antibodies and efficient in establishment of a life-long, persistently viremic infection in mice despite a detectable immune response. Group B viruses, on the other hand, are neuropathogenic, much more sensitive to antibody neutralization and have an impaired ability to establish a high viremia persistent infection in immune competent mice. These biological features of LDV seem to be interdependent and correlate with the number of N-linked polylactosaminoglycan chains on the short ectodomain of the primary envelope glycoprotein, VP-3P. We postulate that lack of the two N-terminal chains endows group B viruses with the ability to interact with a receptor on anterior horn neurons resulting in neuropathogenesis. At the same time, it increases an interaction with neutralizing antibodies thus impeding the infection of macrophages newly generated during the persistent phase of infection which is essential for the continued rounds of replication of the virus.

3. MATERIALS AND METHODS

3.1. LDV

All neuropathogenic isolates of LDV came from C58/M mice carrying transplantable Ib leukemia cells (Murphy et al., 1983; 1987). During the repeated passages through C58/M mice, the tumor cells apparently became contaminated with LDV of unknown origin as a passenger virus. LDV concentrations were estimated by an end point dilution assay in FVB mice (Plagemann et al., 1963; Chen and Plagemann, 1997). LDV titers are expressed as 50% infectious dose (ID_{50}). Stocks of LDV consisted of plasma harvested from groups of 2 or more mice 1 day after inoculation with about 10^6 ID_{50} of virus. The plasma contained 10^9 to 10^{10} ID_{50}/ml. For measuring viremia, plasma was obtained from mice by the orbital bleeding method using heparinized blood collection tubes (Fischer Scientific, Pittsburgh, PA).

For ADPM experiments, C58/M mice about 2 months of age or older were injected intraperitoneally (i.p.) with about 10^6 ID_{50} of LDV. The mice were also injected i.p. with 200 mg cyclophosphamide per kg of body weight 1 day before and in recent experiments also at weekly intervals p.i. to continually suppress the formation of anti-LDV immune responses (Anderson et al., 1995a). The mice were subsequently monitored for paralytic symptoms for about 6 weeks p.i..

3.2. Sequence Analysis of the Primary Envelope Glycoprotein (VP-3P)

LDV genomic RNA extraction, RT-PCR amplification of ORF 5 (encoding VP-3P), cloning and sequencing have been described elsewhere (Chen and Plagemann, 1997; Chen et al., 1997a and b).

3.3. Epitope Mapping by Synthetic Peptide ELISA

The synthetic peptides and the procedure have been described elsewhere (Li et al., 1997).

```
LDV

  10899                A1509/
  P    TACAAACGTGGGCCATCCACCTATACCACAAGTAATTTTTGCCCGCCTTGTAAAAGATA.CTGCAGTACCCGTTGGCTGT
  C    --T-GG---------------------------------CA--------T-G--T--G--C-t----T--.-----------C


                                                                          ORF2->
  P    AAAGGCAGTGGGTACATGTTTCCCAAGTAATGGGTAATGTTGCAACTCTTATAAACAAGGCTATTGAAGATGCATTTAT
  C    -----A--C-------------A----G-----C--CA-------A---G-T--T--A--C--A-------G---G-

       11506                                    /A1584
  P    ......CAGTGTATAATGTAACCATCAAGTTTAGCAATGAAACCATAGCCATCAGCTTTGGACCATCAAACACATCCCT
  C    ......----T--------------------TTA---C----AG-GCTG--TGAT--C--C--G-----TT-T-----
                                                /A1583


PCR product size: 650bp
```

Figure 2. Positions and sequences of the common 5' sense oligonucleotide A1509/ and the differential 3' antisense oligonucleotides /A1583 and /A1584 used for differentiation of LDV-C and LDV-P, respectively. The sequence motifs for leader primed synthesis of subgenomic mRNAs 2 and 3 (Chen et al., 1993) are in bold face, so is the ORF 2 translational initiation codon. The expected PCR products are 650 bp.

3.4. Differential RT-PCR

The procedure has been described elsewhere (Chen and Plagemann, 1997; Chen et al., 1997a and b).

4. RESULTS

We (Chen et al., 1997a) and others (Murphy et al., 1983) have found that neuropathogenic LDV isolates lost neuropathogenicity during persistent infection regardless of ADPM susceptibility of the host mice. Sequence analyses have shown that the predominant LDV quasispecies that grew out of the neuropathogenic isolates during persistent infections are different from those in the original isolates (Chen et al., 1997a and b). Taking advantage of the sequence differences, we have developed differential RT-PCR assays using common 5' sense oligonucleotides and differential 3' antisense oligonucleotides (Chen et al., 1997a and b; e.g. Fig. 2 for nonneuropathogenic LDV-P and neuropathogenic LDV-C) to distinguish and study the various LDV quasispecies at different stages of infection. Using the differential RT-PCR assays, we have observed in the mouse plasma a dramatic switch in the dominance of LDV quasispecies after infection

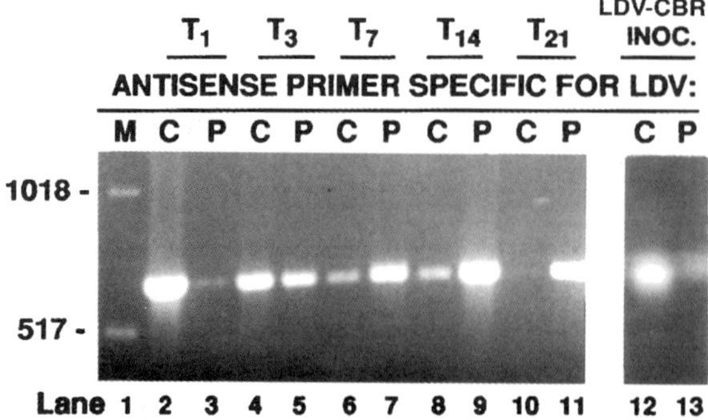

Figure 3. Detection of LDV-C and LDV-P by differential RT-PCR. Plasma samples were collected from FVB mice at different times p.i. with the neuropathogenic isolate LDV-CBR. LDV genomic RNA was extracted from these samples as well as the original LDV-CBR inoculum and subjected to differential RT-PCR analysis. The predicted PCR products are 650 bp.

Table 1. Comparison of the biological and molecular properties of various LDV quasispecies

	LDV	Neuro-pathogenicity *	In vitro Ab neutralization** ($\log_{10}$ ID$_{50}$)	Viremia during persistent infection (ID$_{50}$/ml)	Polylactosamino-glycan chains on VP-3P ectodomain
Group B	C	80-100%	2-3	Low 10^2-$10^{3.5}$	1
Group B	v	80-100%	2-3	Low 10^2-$10^{3.5}$	1
Group A	P	0	0.5-1.5	High 10^5-10^6	3
Group A	vx	0	0.5-1.5	High 10^5-10^6	3
Group A	a	0	0.5-1.5	High 10^5-10^6	3
Group A	363	0	0.5-1.5	High 10^5-10^6	3
	C-T163	30%	0.5-1.5	10^3-10^4	3

* Measured in 2 month old and older C58/M mice immunosuppressed by injections with cyclo-phosphamide at -1, 7, 14, 21 days p.i.
** Virus incubated with neutralizing Ab in vitro and residual infectivity titrated in mice

with neuropathogenic LDV isolates, regardless of the MHC of the mice (Chen et al., 1997a and b; Fig. 3). For example, when FVB mice were injected with the neuropatho-genic isolate LDV-CBR (which consists of LDV-C and a lower concentration of LDV-P; Fig. 3 lanes 12 and 13), LDV-C was rapidly replaced as the predominant quasispecies in the plasma. In fact, LDV-C became hardly detectable by 21 days p.i.. The switch coin-cided with the loss of neuropathogenicity of the LDV reisolated from the plasma (Chen et al., 1997a and b). We have biologically cloned the neuropathogenic LDV quasispecies from their original isolates by repeated end point dilutions (EPD) and confirmed their neuropathogenicity (Chen and Plagemann, 1997; Chen et al., 1997a and b). We have also found that the neuropathogenic LDV-C and LDV-v are incapable of maintaining a high viremia persistent infection and are much more sensitive to in vitro neutralization by im-mune mouse plasma (IMP) and neutralizing mAbs (Chen et al., 1997a and b; Li et al., 1997; Table 1). These properties are in sharp contrast to the nonneuropathogenic quasispe-cies all of which are capable of a high viremia persistent infection and are very resistant to in vitro neutralization by IMP and neutralizing mAbs (Chen et al., 1997b; Li et al., 1997; Table 1). It therefore seems that LDVs can be classified into two groups: group A viruses are non-neuropathogenic, highly resistant to neutralization by antibodies and capable of high viremia persistent infection, whereas group B viruses are neuropathogenic, much more sensitive to antibody neutralization and have an impaired ability to establish a high viremia persistent infection in immune competent mice. There also appears to be a close relationship between neuropathogenicity and susceptibility to host immune response, sug-gesting that these two properties might be determined by the same molecular feature.

We have examined the primary envelope glycoprotein (VP-3P) as a candidate for the molecular determinant of neuropathogenicity and susceptibility to immune response be-cause it has been suggested to be involved in binding to the receptor on permissive cells (Faaberg et al., 1995a; Faaberg and Plagemann, 1995) and contains probably the only neu-tralization epitope(s) for LDV (Coutelier et al., 1986; Coutelier and van Snick, 1988; Ca-

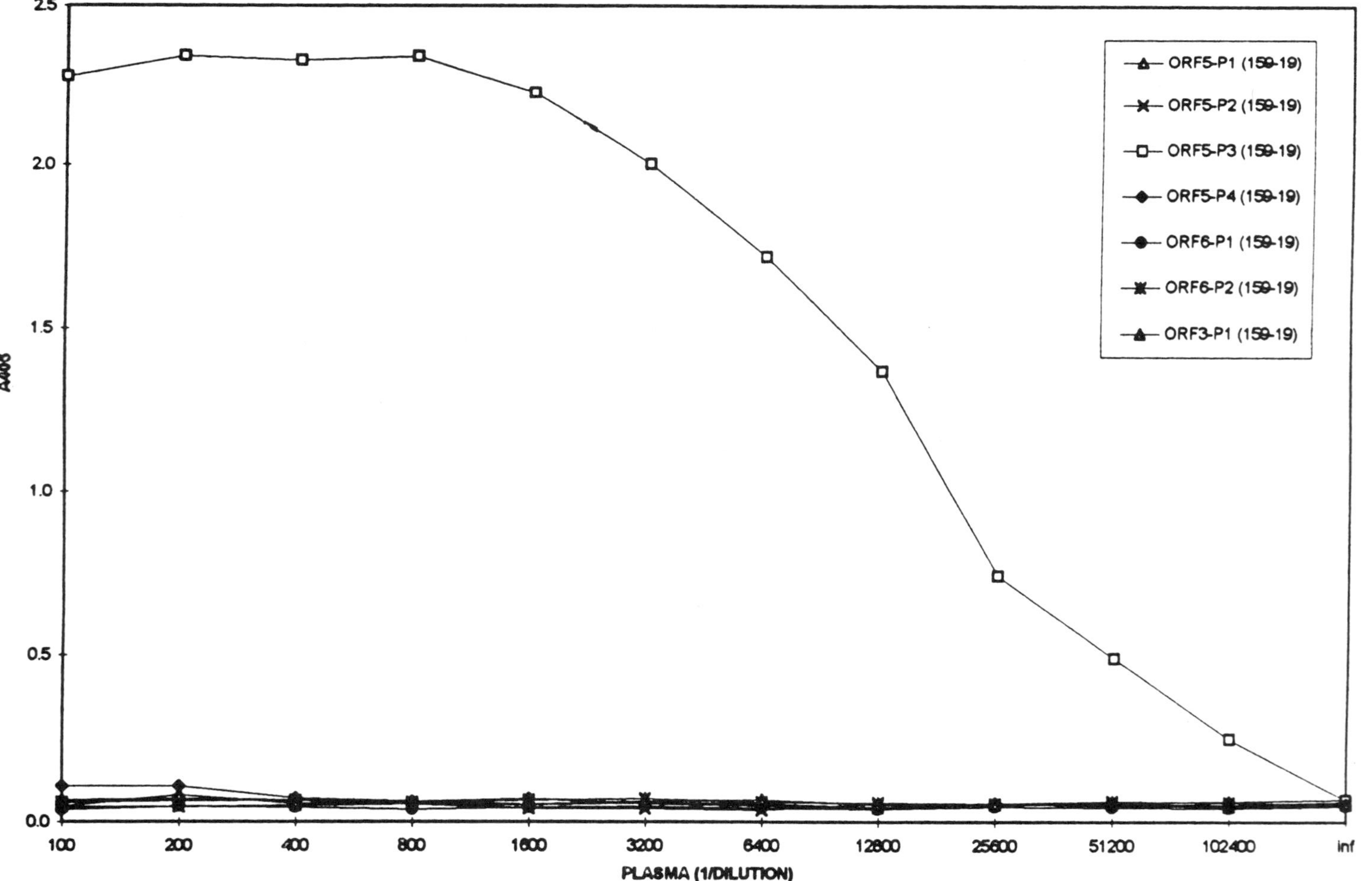

Figure 4. Mapping of the LDV neutralization epitope by ELISA using synthetic peptides. The results are for the neutralizing mAb 159–19. Similar results have been obtained for other neutralizing mAbs. The ELISA procedure has been described elsewhere (Li et al., 1997). For locations and sequences of the peptides, refer to text and the above reference.

fruny et al., 1986; Harty and Plagemann, 1988). We have developed an ELISA assay using synthetic peptides for epitope mapping and have demonstrated that all neutralizing mAbs (e.g. mAb 159-19) as well as IMP bound specifically to a synthetic peptide representing the ectodomain of VP-3P (ORF5-P3; see Fig. 4). None of them bound to peptides derived from other LDV proteins (ORF3-P1, ORF6-P1 and ORF6-P2), or other regions of VP-3P (ORF5-P1 and ORF5-P2), or the same ectodomain peptide with 4 amino acid substitutions (ORF5-P4; also see Li et al., 1997). These results strongly support the ectodomain of VP-3P as a candidate for the molecular determinant of susceptibility to antibody neutralization and possibly neuropathogenicity.

Comparison of the predicted amino acid sequences of VP-3P ectodomain of a variety of LDV quasispecies has revealed a consistent correlation between the number of N-glycosylation sites and the biological properties of the quasispecies. Whereas group A viruses possess three closely spaced N-glycosylation sites, group B viruses lack the two N-terminal ones (Fig. 5; Table 1). Because these sites have been shown to be indeed glycosylated at least for LDV-P (Faaberg et al., 1995b), we postulate that the three closely spaced N-linked polylactosaminoglycan chains on the ectodomain of group A viruses may help mask the only neutralization epitope thus rendering these viruses resistant to antibody neutralization whereas the lack of the two N-terminal polylactosaminoglycan chains in group B viruses unmask the neutralization epitope hence susceptibility to the host immune response. In addition, the lack of the two chains in group B viruses could endow these viruses with the ability to interact with a receptor on anterior horn neurons resulting in neuropathogenesis.

In order to test our hypothesis, we are in the process of isolating and studying mutants of group B viruses that may have changed their phenotypes to those of group A viruses and consequently become selected during persistent infection. Preliminary results from studying one such mutant (LDV-C-T163) have demonstrated that it had become largely resistant to in vitro antibody neutralization and capable of maintaining a higher persistent viremia than its parental quasispecies and at the same time possessed reduced neuropathogenicity (Table 1). However, it has not been determined whether the remaining neuropathogenicity of LDV-C-T163 is the property of the mutant itself or is due to the parental LDV-C which could be potentially coexisting and is indistinguishable from the mu-

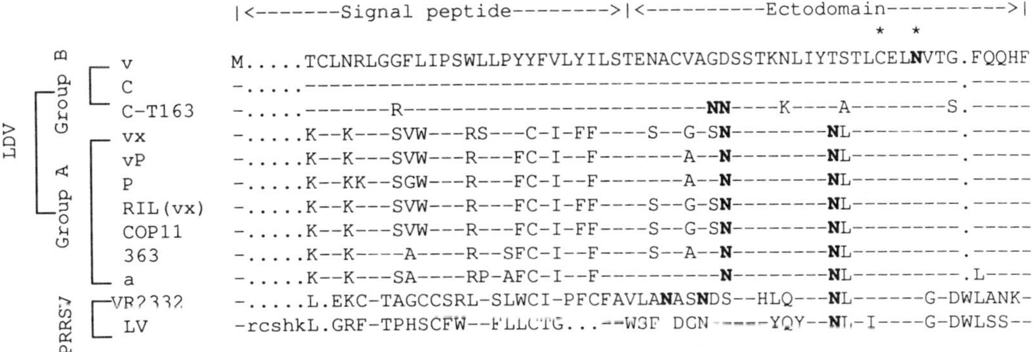

Figure 5. Comparison of the predicted N-terminal amino acid sequences of the primary envelope glycoproteins of various LDV quasispecies and PRRSV isolates. The signal peptide and ectodomain regions are indicated. N-glycosylation sites are shown in bold face. The conserved cysteine residue and the C-terminal N-glycosylation site are marked by asterisks.

tant by the differential RT-PCR used for characterization. Analysis of the predicted amino acid sequence of its VP-3P ectodomain shows that it has gained two new contiguous N-glycosylation sites, one located at exactly the same position as the N-terminal one found in group A viruses (Fig. 5). These findings are consistent with our hypothesis which we hope to confirm with the results from more mutant analyses.

5. DISCUSSION

We have discovered the coexistence, in neuropathogenic LDV isolates, of quasispecies that differ drastically in neuropathogenicity and the ability to establish a high viremia persistent infection. Although the coexistence of multiple quasispecies in an RNA virus is not surprising (Domingo and Holland, 1994; Holland et al., 1992), the dramatic differences in the biology of the LDV quasispecies are quite unusual. The selection and maintenance of the neuropathogenic quasispecies may have resulted from passaging over many years the LDV-contaminated Ib leukemia cells in immunosuppressed C58 mice (Murphy et al., 1983, 1987). The specific condition in these mice may have created an environment suitable for the selection of the neuropathogenic quasispecies (Chen et al., 1997b).

Analyses of the biological properties of the different LDV quasispecies have revealed a close relationship between neuropathogenicity and susceptibility to host immune response (specifically antibody neutralization). Whereas the nonneuropathogenic group A viruses are resistant to host immune response, the neuropathogenic group B viruses are much more susceptible. The finding that the LDV-C-T163 mutant derived from a parental group B virus had increased resistance to host immune response and reduced neuropathogenicity further supports the interdependence of these two biological properties. Such interdependence suggests that the two properties may be determined by the same molecular feature.

We have found a consistent correlation between the number of N-glycosylation sites on the ectodomain of the major envelope glycoprotein (VP-3P) and the biological properties of LDV. The identification of the ectodomain of VP-3P as a potential determinant for neuropathogenicity and susceptibility to immune response is not unexpected since it has been postulated to be involved in binding to LDV receptors on permissive cells (Faaberg et al., 1995a; Faaberg and Plagemann, 1995). Furthermore, the ectodomain of VP-3P has been shown to contain probably the only neutralization epitope(s) for LDV (Coutelier et al., 1986; Coutelier and van Snick, 1988; Cafruny et al., 1986; Harty and Plagemann, 1988; Li et al., 1997). We were initially somewhat concerned about the feasibility of using synthetic peptides and ELISA to identify the neutralization epitope(s) since antibodies usually recognize complex conformational structures in native proteins rather than "linear" structures like peptides. Our results, however, have proved the feasibility of our approach, most likely because the ectodomain of VP-3P consists of only about 30 amino acids and therefore probably lacks any complex conformational structures. We postulate that the neutralization epitope(s) in the ectodomain of VP-3P may be well masked by the three closely-spaced polylactosaminoglycan chains present in group A viruses leading to their resistance to Ab neutralization whereas the lack of two N-terminal chains in group B viruses may render the neutralization epitope(s) less protected, making these viruses more susceptible to suppression by host immune response. In addition, the lack of the two chains in group B viruses may endow these viruses with the ability to interact with a receptor on anterior horn neurons resulting in neuropathogenesis. Our preliminary results from characterization of the mutant LDV-C-T163 provides further support to our postulation. However, the fact that LDV-C-T163 could not maintain a persistent viremia as high as those of group A viruses suggests that the location of the N-glycosylation sites may be also important. Furthermore, the question of whether the low level of neuropathogenicity of LDV-C-T163 is attributed to the mutant itself or the potentially coexisting parental group B virus still needs to be resolved perhaps by end point dilution analysis. The confirmation of our postulation still requires the characterization of more mutants.

The importance of the ectodomain of VP-3P in determining LDV persistence and pathogenicity may have interesting implications for PRRSV research. Because LDV and PRRSV are highly related viruses, their corresponding viral proteins may have similar biological functions. In fact, there is significant sequence homology between the two viruses in the ectodomain of VP-3P (Fig. 5). The positions of a cysteine residue proposed to form disulfide bond with the M/VP-2 protein, which is essential for LDV infectivity (Faaberg et al., 1995a), and the C-terminal N-glycosylation site are conserved in all LDVs and PRRSVs examined. The middle N-glycosylation site, which is conserved in all non-neuropathogenic LDVs, is also present at the exact location in the PRRSVs examined. There is a significant variation in the number and location of the N-terminal N-glycosylation site in the PRRSVs. Whether the differential N-glycosylation of the VP-3P ectodomain of PRRSV are related to persistence and pathogenicity may be interesting to investigate.

ACKNOWLEDGMENT

We thank Dr. Robert Wohlhueter for providing synthetic peptides, Dr. Grant Anderson and Steve Wietgrefe for helpful discussions. Z.C. was supported by USPH training grant CA 09138.

REFERENCES

Anderson, G.W., Even, C., Rowland, R.R.R., Palmer, G.A., Harty, J.T., and Plagemann, P.G.W., 1995a, C58 and AKR mice of all ages develop motor neuron disease after lactate dehydrogenase-elevating virus infection but only if antiviral immune responses are blocked by chemical or genetic means or as a result of old age, *J. Neurovirol.* **1**: 244–252.

Anderson, G.W., Palmer, G.A., Rowland, R.R.R., Even, C., and Plagemann, P.G.W., 1995b, Infection of central nervous system cells by ecotropic murine leukemia virus in C58 and AKR mice and in utero infected CE/J mice predisposes mice to paralytic infection by lactate dehydrogenase-elevating virus, *J. Virol.* **69**: 308–319.

Cafruny, W.A., Chan, S.P.K., Harty, J.T., Yousefi, S., Kowalchyk, K., McDonald, D., Foreman, B., Budweg, G., and Plagemann, P.G.W., 1986, Antibody response of mice to lactate dehydrogenase-elevating virus during infection and immunization with inactivated virus, *Virus Res.* **5**: 357–375.

Cavanagh, D., Brian, D.A., Brinton, M., Enjuanes, L., Horzinek, M.C., Lai, M.M.C., Laude, H., Plagemann, P.G.W., Siddell, S., Spaan, W.J.M., and Talbot, P.J., 1997, Nidovirales: a new order comprising Coronaviridae and Art-Chen, Z., Kuo, L., Rowland, R.R.R., Even, C., Faaberg, K.S. and Plagemann, P.G.W., 1993, Sequence of 3'-end of genome and of 5'-end of ORF 1a of lactate dehydrogenase-elevating and common junction motifs between 5'-leader and bodies of subgenomic mRNAs, *J. Gen. Virol.* **74**: 643–660.

Chen, Z., and Plagemann, P.G.W., 1997, Detection of lactate dehydrogenase-elevating virus in transplantable mouse tumors by biological assay and RT-PCR assays and its removal from the tumor cells, *J. Virol. Meth* In press.

Chen, Z., Rowland, R.R.R., Anderson, G.W., and Plagemann, P.G.W., 1997a, Coexistence in lactate dehydrogenase-elevating virus pools of variants that differ in neuropathogenicity and ability to establish a persistent infection, *J. Virol.* **71**: 2913–2920.

Chen, Z., Li, K., Rowland, R.R.R., Anderson, G.W., Schuler, T.W., and Plagemann, P.G.W., 1997b, Neuropathogenicity and low viremia during persistent infection are related properties of lactate dehydrogenase-elevating virus and are associated with lacking two of three N-glycosylation sites on the ectodomain of primary envelope glycoprotein, *submitted to Virology.*

Coutelier, J.P. and van Snick, J., 1988, Neutralization and sensitization of lactate dehydrogenase-elevating virus with monoclonal antibodies, *J. Gen. Virol.* **69**: 2097–2100.

Coutelier, J.P., van Roost, F., Lambotte, P., and van Snick, J., 1986, The murine antibody response to lactate dehydrogenase-elevating virus, *J. Gen. Virol.* **67**: 1099–1108.

Domingo, E. and Holland, J.J., 1994, Mutations and rapid evolution of RNA viruses, in *The Evolutionary Biology of Viruses,* pp 161–184. Edited by S.S. Morse. New York: Raven Press.

Faaberg, K.S., and Plagemann, P.G.W., 1995, The envelope proteins of lactate dehydrogenase-elevating virus and their membrane topography, *Virology* **212**: 512–525.

Faaberg, K.S., Even, C., Palmer, G.A., and Plagemann, P.G.W., 1995a, Disulfide bonds between two envelope proteins of lactate dehydrogenase-elevating virus are essential for viral infectivity, *J. Virol.* **69**: 613–617.

Faaberg, K.S., Palmer, G.A., Even, C., Anderson, G.W., and Plagemann, P.G.W., 1995b, Differential glycosylation of the ectodomain of the primary envelope glycoprotein of two strains of lactate dehydrogenase-elevating virus that differ in Neuropathogenicity, *Virus Res.* **39**: 331–340.

Harty, J.T., and Plagemann, P.G.W., 1988, Formalin inactivation of the lactate dehydrogenase-elevating virus reveals a major neutralizing epitope not recognized during natural infection, *J. Virol.* **62**: 3210–3216.

Holland, J.J., de la Torre, J.C. and Steinhauer, D.A., 1992, RNA virus populations as quasispecies, *Curr. Top. Microbiol. Immunol.* **176**: 1–18.

Li, K, Chen, Z and Plagemann, P.G.W., 1997, Differences in sensitivity to in vitro neutralization between different lactate dehydrogenase-elevating virus quasispecies correlates with differences in number of polylactosaminoglycan chains associated with neutralizing epitope on the ectodomain of primary envelope glycoprotein, *submitted to Virology.*

Meulenberg, J.J.M., Hulst, M.M., de Meijer, E.J., Moonen, P.J.L.M., den Besten, A., de Kluyver, E.P., Wensvoort, G., and Moormann, R.J.M., 1993, Lelystad virus, the causative agent of porcine epidemic abortion and respiratory syndrome (PEARS), is related to LDV and EAV, *Virology* **192**: 62–72.

Murphy, W.H., Nawrocki, J.F., and Pease, L.R., 1983, Age-dependent paralytic viral infection in C58 mice: Possible implication in human neurologic disease, *Prog. Brain Res.* **59**: 291–303.

Murphy, W.H., Mazur, J.J. and Fulton, S.A., 1987, Animal model for motor neuron disease, in *Clinical Neuroimmunology*, pp 135–155. Edited by W.M.H. Behan, P.O. Behan & J.A. Aarli. Oxford, England: Blackwell Scientific Publications.

Murtaugh, M.P., Elam, I.R., and Kakach, L.T., 1995, Comparison of the structural protein coding sequences of the VR2332 and Lelystad virus strains of the PRRS virus, *Arch. Virol.* **140**: 1451–1460.

Plagemann, P.G.W., 1996, Lactate dehydrogenase-elevating virus and related viruses, in *Virology*, 3rd Edition, pp 1105–1120. Edited by B.N. Fields, D.M. Knipe, and P.M. Howley. New York: Raven Press.

Plagemann, P.G.W., Gregory, K.F., Swim, H.E. and Chan, K.K.W., 1963, Plasma lactic dehydrogenase elevating agent of mice: Distribution in tissues and effect on lactic dehydrogenase isozyme patterns, *Can. J. Microbiol.* **9**: 75–86.

Plagemann, P.G.W., Rowland, R.R.R., Even, C., and Faaberg, K.S., 1995, Lactate dehydrogenase-elevating virus-an ideal persistent virus? *Semin. Immunopathol.* **17**: 167–186.

77

ARTERIVIRUS PRRSV

Experimental Studies on the Pathogenesis of Respiratory Disease

Jörg Beyer,[1] Dieter Fichtner,[1] Horst Schirrmeier,[1] Harald Granzow,[1] Ulf Polster,[1] Emilie Weiland,[2] Angela Berndt,[3] and Helmut Wege[1]

[1]Federal Research Centre for Virus Diseases of Animals
Friedrich-Loeffler-Institutes
D-17498 Isle of Riems
[2]and D-72076 Tübingen
[3]Federal Institute for Health Protection of Consumers and Veterinary Medicine
D-07743 Jena, Germany

1. ABSTRACT

Pigs were infected with the porcine respiratory and reproductive syndrome virus (PRRSV) by the oronasal route. We studied the development of histological lesions, sites of virus infection and of inflammatory infiltrates by quantitative evaluation of reactive cells. The animals developed a multifocal interstitial pneumonia. Clinical signs of pneumonia were observed from day 7 to 21. In the first stage, an acute alveolitis was found, which was characterised by a hyperplasia of type II pneumocytes within the septa and an accumulation of macrophages in the alveolar spaces. Within 2–4 days p.i., virus infected cells were prominent in lymphatic organs, but their number declined rapidly during the following days. In the following period, the number of virus antigen positive cells increased in the lung. An interesting discrepancy existed between the relatively small number of virus specific cells and the degree of intensive pneumonia. As a first step to analyse mechanisms leading to the induction of pneumonia, we studied transcriptional expression of cytokines and other immunomodulatory molecules by semiquantitative RT-PCR.

2. INTRODUCTION

The porcine reproductive and respiratory syndrome (PRRS) is a viral disease of the swine of increasing economical importance (reviewed by Meredith 1995). Retrospective

Coronaviruses and Arteriviruses, edited by Enjuanes *et al.*
Plenum Press, New York, 1998

studies indicate, that this syndrome can be followed back to the early 1980s in the USA and spread rapidly until 1990. The syndrome comprises a spectrum of symptoms ranging from early farrowings, late abortions and a highly prevalent respiratory disease. In particular the respiratory syndrome is often accompanied by a number of other pathogenic bacterial or viral agents. In the field PRRS is of increasing concern. PRRS was described under different nomenclature and the incertainity concerning its true etiology complicated the diagnosis. In Europe the first PRRS cases were noticed 1990 and the disease spread rapidly from Germany to the Netherlands, Belgium and England. An important progress was the first isolation and definition of this virus ("Lelystad virus", Wensvoort et al. 1991, Terpstra et al., 1991). The virus is highly variable in terms of virulence. In addition, between the strains from USA and Europe remarkable serological and genetic differences were found (Kapur et al., 1996; Suárez et al., 1996; Wieczorek-Krohmer, 1996; Halbur et al., 1995 Rossow et al., 1995). Nowadays, live vaccines are available but are debated because of their potential to induce persistent infections and reproductive failures. It prooved to be difficult to reproduce the disease experimentally. The knowledge on the immune response and the pathogensis of PRRS is limited. Only contradictory data are available on the suspected synergism with other potential pathogens (Molitor et al., 1997; Van Reeth, 1997). Furthermore, PRRSV infections offer an opportunity to study inflammatory mechanisms and immunological defense in the lung as a model for human respiratory diseases. In order to investigate the pathogenesis of pneumonia in conventionally raised pigs, we initiated experimental studies with PRRSV (Fichtner et al., 1993). Here we describe the pathological changes, virus replication and transcriptional expression of proinflammatory cytokines during a kinetic study.

3. MATERIALS AND METHODS

Conventionally raised piglets (6–7 weeks old) were infected by the oronasal route with an European strain (Cobbelsdorf) of PRRSV. At day 2, 3, 4, 7, 14, 21, 28 and 35 p. i. each time one uninfected control animal and two infected piglets were euthanized and slaughtered for dissection. Organ samples were shock frozen and stored at -70°C before processing for immunofluorescence on cryostate sections or RT-PCR. Immunostaining was performed with mAbs specific for PRRSV ORF 7 protein (Wieczorek-Krohmer, 1996) and alveolar macrophages. Organ samples for histology and immunohistology were fixed in 4% neutral formalin and embedded in paraffin. For RT-PCR, RNA extracted from organ samples were reversely transcribed to cDNA (oligo dT primed). The optimal conditions for PCR were established for each primer set. The primers were delineated from published sequence informations and data banks (Murtaugh et al., 1994). The correct identity of the PCR-product was monitored by Southern-blot hybridisation. Isolation of infectious virus was performed by inoculation of MARC 145 cell monolayers with homogenised tissue samples.

4. RESULTS

4.1. Pathomorpholcgical Findings

Histological changes typical for a mild, multifocal interstitial lobular pneumonia were noticed first at the third day post infection. Macroscopic lesions were first observed on day 4 p.i. and developed further from day 7 to 21 p.i., but in general they remained mild to moderate (Figure 1). By contrast, the affected lobuli displayed pronounced histo-

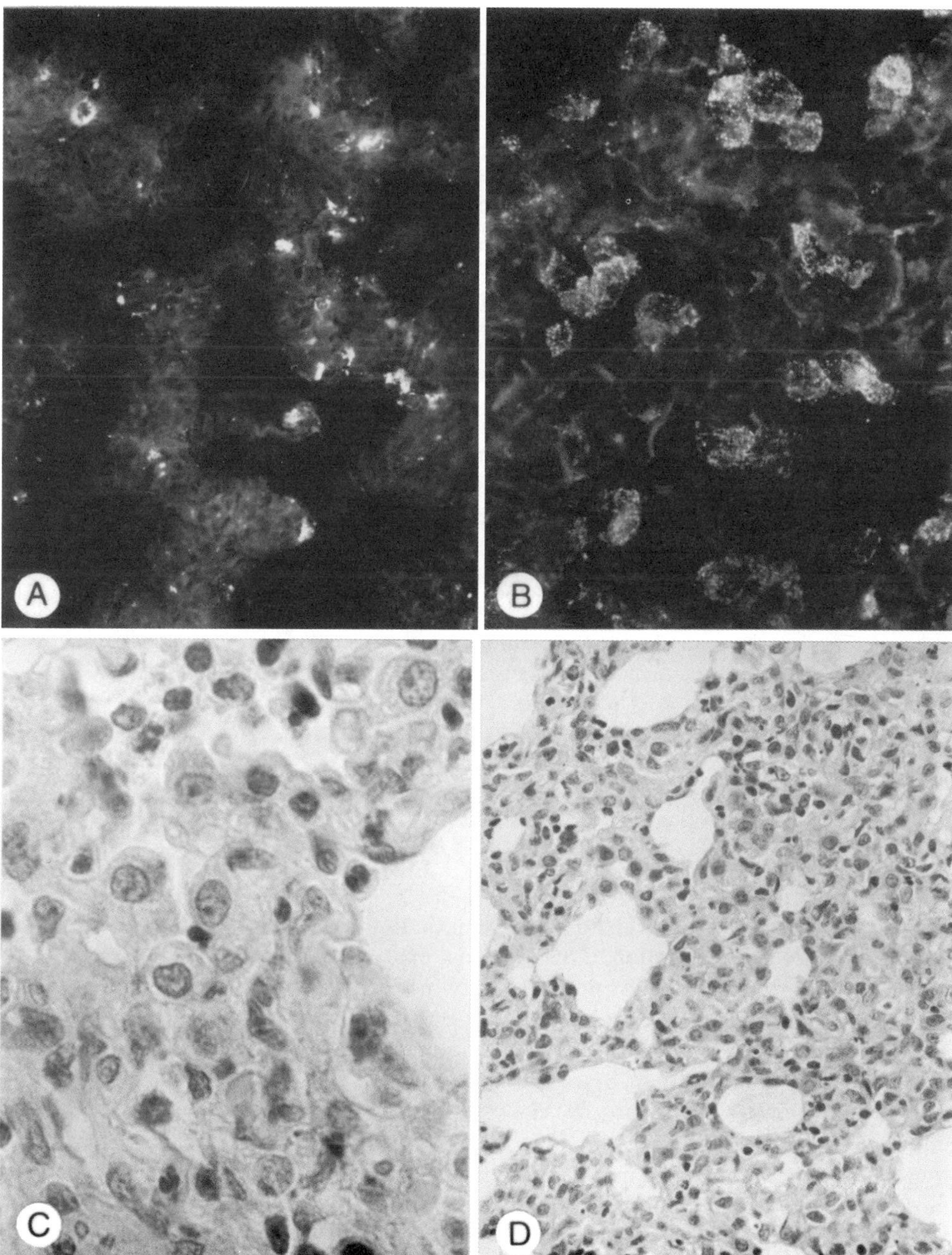

Figure 1. Typical features of PRRSV induced pneumonia. A) PRRSV - antigen in alveolar macrophages, 7 days p. i. Indirect immunofluorescence on cryostate section, anti-PRRSV mAb P3/27. 235:1. B) Accumulation of alveolar macrophages in an early stage of pneumonia, 21 days p. i. Indirect immunofluorescence on cryostate section, anti-macrophage mAb 2G6. 470:1. C) Interstitial pneumonia.Type II pneumocytes displaying hyperplasia and hypertrophy, alveolar macrophages are scattered in between. 14 days p. i., HE staining, 960:1. D) Pronounced interstitial pneumonia, 21 days p.i., thickening of septal walls. HE staining, 300:1.

logical lesions. Early stages are histologically characterised by an acute alveolitis. Within these lesions, the number of alveolar macrophages increased and a hyperplasia of type II pneumocytes was visible. In some areas, the alveolar macrophages appeared destroyed. Advanced stages of pneumonia displayed a thickening of the alveolar septa and alveolar spaces were often filled with cell detritus. As a consequence of the strong hyperplasia and hypertrophy of type II pneumocytes, the alveolar spaces were increasingly restricted. Desquamated, decaying type II pneumocytes and alveolar macrophages accumulated in the remaining alveoli. Solid pneumonic lesions were first seen on day 7 p. i. and consisted of proliferating type II pneumocytes including a quite variable proportion of macrophages scattered in between. In localised areas, necrotic type II pneumocytes and aggregations of decaying alveolar macrophages were encountered.

4.2. Kinetics of Virus Spread and Cell Tropism

The time course of virus spread and the infiltration process was analysed employing virus specific monoclonal antibodies, immunofluorescence and RT-PCR technology. In lymphatic organs, a high number of virus infected cells was found within 2–4 days p.i., which declined rapidly within the following days. At the same time, only a few virus positive cells were noted in lung tissue. From day 7–21 p.i., the number of virus antigen positive cells increased in the lung tissue (Figure 1 A). It is noteworthy, that the degree of intensive pneumonia did not correlate with the relatively small number of virus specific cells. The virus antigen was located predominantly in alveolar macrophages. In addition, cells which were not yet unequivocally identified displayed virus antigen.

4.3. Transcriptional Expression of Virus RNA and Mediators of Inflammation

As a first step towards an analysis of local immunity and inflammatory mechanisms we monitored the organ specific transcriptional expression of viral RNA in parallel to proinflammatory cytokines IL-1a, IL1–1b, IL-6, IL-8, AMCFII (alveolar macrophage chemotactic factor), TNFa and TNFb. In correlation to the results of virus detection by immunofluorescence and reisolation, transcriptional expression of viral RNA was already detectable within 2 days p. i. in tonsils and lymph nodes. Starting at 4 days p. i. also lung tissue, spleen and thymus were highly positive. Strong involvement was typical for tonsils and at variable degrees for thymus and spleen until the experiment was terminated 35 days p. i. However, in lung tissue of both healthy control animals and infected pigs, a strong transcriptional expression of the cytokines IL-1a, IL1–1b, IL-8, TNFa and TNFb was found, with not much variation throughout the time kinetics. Since conventional raised pigs were employed, this result may reflect the reactivity of the mucosal defence system to bacteria. Attempts of semiquantitation by taking a series of samples at different PCR-cycle numbers and comparing to standardised probes did not reveal significant differences between infected and uninfected controls.

Remarkable correlations between PRRSV induced pneumonia and the transcriptional expression of the cytokines IL-6 and AMCFII were noticed. Table 1 illustrates the kinetics of transcriptional expression throughout the experiment concerning virus, IL-6, IL-8 and AMCFII. From day 4 p. i. onwards IL-6 expression in lung tissue was significantly enhanced compared to uninfected controls. The transcriptional expression of IL-6 was also increased in organ samples from lymphnodes, spleen, thymus and tonsils, indicating a systemic upregulation. Furthermore, the transcriptional expression of AMCFII

Table 1. Transcriptional expression of PRRSV and cytokines: kinetics in lung tissue

Days p.i.	Status	Organ	PRRSV	AMCF II	IL-6	IL-8
2	control	lung	–	±	±	++++
2	infected	lung	±	±	±	++
2	infected	lung	–	+	±	+++
3	control	lung	–	±	+	+++
3	infected	lung	++	±	±	++++
3	infected	lung	–	+	+	+++
4	control	lung	–	±	+	++++
4	infected	lung	++++	±	+	+++
4	infected	lung	++	±	+	+++
7	control	lung	–	–	+	+++
7	infected	lung	++++	+++	+++	+++
7	infected	lung	+	±	+	+++
14	control	lung	–	+	+	++++
14	infected	lung	+++	++	+++	++++
14	infected	lung	+++	++	+	+++
21	control	lung	–	±	±	+++
21	infected	lung	++++	++	+	+++
21	infected	lung	++++	++	++	+++

was selectively upregulated in PRRSV-infected lung tissue, beginning from day 7 throughout the following time points studied. Rather independent of the stage of inflammation this chemokine was strongly expressed in tonsils and at a variable degreee in thymus and spleen.

5. DISCUSSION

Field observations and experimental data indicate, that the pathogenicity and virulence of PRRSV strains differ widely and are strongly influenced by host factors (Kapur et al., 1996; Suárez et al., 1996; Halbur et al., 1995; Rossow et al., 1995). The European strains in general appear to induce a milder disease than strains from the USA. Unrecognised subacute and persistent infections with PRRSV may lead to long lasting health problems in pig herds. Therefore we started to analyse the pathogenesis of pneumonia employing deliberately conventionally raised pigs and an European strain of PRRSV to obtain informations reflecting the situation in the field (Fichtner et al., 1993).

It is not yet clear, whether the virus infection spreads to the septal walls of the lung strictly via the respiratory route or if a hematogeneous transmission is involved. Alveolar macrophages are obviously the major target, but by morphological criteria type II pneumocytes are probably involved. We observed widespread inflammation in advanced stages of pneumonia, although the number of infected cells remained rather small. A number of results indicate, that the interactions between PRRSV and macrophage populations play a pivotal role determining the outcome of infection (Molitor et al., 1997). The balance between tissue injury and protection strongly depends on the pattern of proinflammatory cytokines produced by alveolar macrophages (Murtaugh et al., 1994 and 1996). Furthermore, chemokines which attract mononuclear phagocytes and neutrophils may have a great impact on the disease process.

We employed the RT-PCR to obtain preliminary information on the expression of mRNA specific for cytokines and immunomodulatory substances. The protein levels of many cytokine genes are regulated by transcription. Therefore, RT-PCR technology can complement bioassays or immunoassays both in terms of sensitivity and specificity, especially when no other options are available. However, RT-PCR results should be interpreted with caution. First, for some cytokines transcriptional expression does not correlate with biological activity. Secondly, only limited conclusions can be drawn because of the lack of localisation and insufficient information on the morphological context. This ambiguity is illustrated by the observation, that by RT-PCR potentially harmful cytokines appeared to be transcriptionally expressed in healthy control pigs. Further studies are in progress to clarify these points.

PRRSV replicated in early stages in lymph organs, later lung tissue was involved. It is an open question, whether the infection results directly in immunosuppression because of a functional and quantitative impairment of alveolar macrophages. In other studies, at least a temporal disturbance of lung defence mechanisms against secondary infections were observed. However, the diminished number of lymphocytes and macrophages recovered within a few weeks p. i. (Molitor et al., 1997; Shimizu et al., 1996). An infection with PRRSV may possibly booster the immune response induced by vaccinations possibly through activation of macrophages. Furthermore, it was observed that certain antibodies to PRRSV can even enhance the uptake of virus by alveolar macrophages (Yoon et al., 1997).

IL-6 was transcriptionally upregulated in infected lung tissue as well as in other organs. This cytokine is produced by several cell types. It mediates an early protection by acute phase proteins and enhances the antibody response. A most interesting observation was the selective induction of the alveolar macrophage chemotactic factor AMCF II, a chemokine important for attraction of neutrophils. The peptide is related to the chemokine family GRO and CINC (Goodman et al., 1991 and 1992). It is structurally different from the chemokine IL-8, which belongs to the integrine family but has similar biological properties. In our studies, IL-8 was upregulated independently of AMCFII. Further studies are of interest to analyse the relations of PRRSV infection to signal mediated induction of immunomodulatory substances.

The clinical course of PRRSV induced interstitial pneumonia was relatively mild. The pathology differs clearly from the pleuropneumonia induced by some gram-negative bacteria, which may be triggered by the immunological effects of lipopolysaccharides. One striking difference concerns the kinetics of disease. Only a few hours following infection of swine by Pasteurella multocida pronounced infiltrations were observed (Berndt and Müller, 1995). Furthermore, following endotracheal infection with Actinobacillus pleuropneumoniae lavage cells expressed already within 2 hours p. i. proinflammatory cytokines IL-1 and IL-8 followed later by IL-6 (Baarsch et al., 1995). It is conceivable, that mutual interactions between viral and bacterial agents occur, which finally precipitate or exacerbate a clinical disease. PRRSV infections may thus complicate the enzootic pneumonia of pigs, which represents a multifactorial disease associated with a number of bacterial infections (Molitor et al., 1997; Van Reeth, 1997). This health problem is of increasing importance for animal welfare. We hope, that further studies of PRRS will help to understand more precisely the relations between inflammation, virus infection and the immune response.

ACKNOWLEDGMENTS

We thank Gabriele Czerwinski, Brigitte Nolting, Regine Schröder and Hanna Wege for excellent technical assistance.

REFERENCES

Baarsch, M.J., Scamurra, R.W., Burger, K., Foss, D.L., Maheswaran, S.K., and Murtaugh, M.P., 1995, Inflammatory cytokine expression in swine experimentally infected with Actinobacillus pleuropneumoniae, *Infect. Immunol.* **6**: 3587–3594.

Berndt, A., and Müller, G., 1995 Occurence of T lymphocytes in perivascular regions of the lung after intratracheal infection of swine with Pasteurella multocida, *Vet. Immunol. Immunopath.* **49**: 143–159.

Fichtner, D., Beyer, J., Leopoldt, D., Schirrmeier, H., Bergmann, S., and Fischer, U., 1993, Experimentelle Reproduktion der respiratorischen Verlaufsform der Infektion mit dem Erreger des Seuchenhaften Spätabortes der Schweine, *Berl. Münch. Tierärztl. Wschr.* **106**: 145–149.

Goodmann, R.B., Forstrom, J.W., Osborn, S.G., Chi, E.Y., Martin, T.R., 1991, Identification of two neutrophil chemotactic peptides produced by porcine alveolar macrophages, *J. Biol. Chem.* **266**: 8455–63.

Goodmann, R.B., Foster, D.C., Mathewes, S.L., Osborn, S.G., Kuijper, J.L., Forstrom, J.W. and Martin, T.R., 1992, Molecular cloning of porcine alveolar macrophage-derived neutrophil chemotactic factors I and II: Identification of porcine IL-8 and another intercrine-α protein, *Biochemistry* **31**: 10483–10490.

Halbur, P.G., Paul, P.S., Frey, M.L., Landgraf, J., Eernisse, K., Meng, X.-J., Lum, M.A., Andrews, J.J. and Rathje, J.A., 1995, Comparison of the pathogenicity of two US porcine reproductive and respiratory syndrome virus isolates with that of the Lelystad virus, *Vet. Pathol.* **32**: 648–660.

Kapur, V., M.R. Elam, T.M. Pawlovich and Murtaugh, M.P., 1996, Genetic variation in porcine reproductive and respiratory syndrome virus isolates in the midwestern United States, *J. Gen. Virol.* **77**: 1271–1276.

Meredith, M.J., 1995, Porcine and resproductive and respiratory syndrome (PRRS), *Boehringer Ingelheim*, ISBN 0–9520409–7–2.

Molitor, T.W., Bautista, E.M. and Choi, C.S., 1997, Immunology to PRRSV: Double-edged sword, *Vet. Microbiol.* **55**: 265–276.

Murtaugh, M.P., 1994, Porcine cytokines, *Vet. Immunol. Immunopathol.* **43**: 37–44.

Murtaugh, M.P., Baarsch, M.J., Zhou, Y., Scamurra, R.W., Lin, G., 1996, Inflammatory cytokines in animal health and disease, *Vet. Immunol. Immunopathol.* **54**: 45–55.

Rossow, K.D., Collins, J.E., Goyal, S.M., Nelson, E.A., Christopher-Hennings, J. and Benfield, D.A., 1995, Pathogenesis of porcine reproductive and respiratory syndrome virus infection in gnotobiotic pigs, *Vet. Pathol.* **32**: 361–373.

Shimizu, M., Yamada, S., Kawashima, K., Ohashi, S., Shimizu, S. and Ogawa, T., 1996, Changes of lymphocyte subpopulations in pigs infected with porcine reproductive and respiratory syndrome (PRRS) virus, *Vet. Immunol. Immunopath.* **50**: 19–27.

Suárez, P., Zardoya, R., Martin, M.J., Prieto, C., Dopazo, J., Solana, A. and Castro, J.M., 1996, Phylogenetic relationships of European strains of porcine reproductive and respiratory syndrome virus (PRRSV) inferred from DNA sequences of putative ORF-5 and ORF-7 genes, *Virus Res.* **42**: 159–165.

Terpstra, C., Wensvoort, G., Pol, JMA., 1991, Experimental reproduction of porcine epidemic abortion and respiratory syndrome (mystery swine disease) by infection of Lelystad virus: Koch's postulates fullfilled, *Vet. Quarterly* **13**: 121–130.

Van Reeth, K., 1997, Pathogenesis and clinical aspects of a respiratory porcine reproductive and respiratory syndrome virus infection, *Vet. Microbiol.* **55**: 223–230.

Wensvoort, G., Terpstra, C., Pol, J.M.A., ter Laak, E.A., Bloemraad, M., de Kluyver, E.P., Kragten, C., van Buiten, L., den Besten, A., Wagenaar, F., Broekhuijsen, J.M., Moonen, P.L.J.M., Zetstra, T., de Boer, E.A., Tibben, H. J., de Jong, M.F., van't Veld, P., Groenland, G.J.R., van Gennep, J.A., Voets, M.Th., Verheijden, J.H.M., Braanskamp, J., 1991, Mystery swine disease in the Netherlands: the isolation of Lelystad virus, *Vet. Quarterly* **13**: 121–130.

Wieczorek-Krohmer, M., Weiland, F., Conzelmann, K., Kohl, D., Visser, N., van Woensel, P., Thiel, H.-J. and Weiland, E., 1996, Porcine reproductive and respiratory syndrome virus (PRRSV): Monoclonal antibodies detect common epitopes on two viral proteins of European and U.S. isolates, *Vet. Microbiol.* **51**: 257–266.

Yoon, K.-J., Wu, L.-L., Zimmermann, J.J. and Platt, K.B., 1997, Field isolates of porcine reproductive and respiratory syndrome virus (PRRSV) vary in their susceptibility to antibody dependent enhancement (ADE) of infection, *Vet. Microbiol.* **55**: 277–287.

SPREAD OF SWINE HEMAGGLUTINATING ENCEPHALOMYELITIS VIRUS FROM PERIPHERAL NERVES TO THE CNS

N. Hirano,[1] K. Tohyama,[2] and H. Taira[3]

[1]Department of Veterinary Microbiology
[3]Department of Bioscience and Technology
Iwate University
Morioka, Japan
[2]Department of Cell Biology and Neuroanatomy
Iwate Medical University
School of Medicine
Morioka, Japan

1. ABSTRACT

Swine hemagglutinating encephalomyelitis virus (HEV) strain 67N was inoculated into the sciatic nerve or the right leg crural muscle of rats. In both cases, the virus was isolated first from the caudal half of the spinal cord on day 2 after inoculation, and from the rostral half of the spinal cord and the brain on day 3. The virus titers in the brain reached a maximum when the infected rats developed CNS symptoms on day 5.

Using confocal laser scanning microscope, fluorescent positive cells were first found in the lumbar dorsal root ganglion (DRG) and spinal cord ipsilateral of the inoculated leg on day 3. Antigen positive neurons were found bilaterally in the lumbar DRG and spinal cord on day 4. On day 5 specific fluorescence was observed in the neurons of the cerebral cortex, hippocampus, brainstem and Purkinje cells in the cerebellum.

2. INTRODUCTION

The HEV strain 67N causes encephalomyelitis or vomiting and wasting disease in piglets (Andries et al., 1980, Mengeling ct al., 1972; Roe et al., 1958). In experimental oronasal infection of piglets, the virus spreads to the CNS predominantly via the nerve pathways (Andries et al., 1980; Andries et al., 1981). In our experimental studies of HEV

Coronaviruses and Arteriviruses, edited by Enjuanes *et al.*
Plenum Press, New York, 1998

strain 67N, the virus caused encephalomyelitis in mice when inoculated by intracerebral (i.c.), intranasal (i.n.), intraperitoneal (i.p.) or subcutaneous (s.c.) routes, and was propagated mainly in the nerve cells. However, 20-day-old or older mice were resistant to the virus inoculated by i.n., i.p. or s.c. routes (Hirano et al., 1995). In experimental infections of rats, 4-week-old rats died of encephalitis after intravenous (i.v), i.p. and s.c. as well as i.c. and i.n. routes. However, the rats inoculated by s.c. route died a few days earlier than those i.p. and i.v. inoculated and the virus was isolated only from the spinal cord and the brain, but not from the spleen and liver. By immunohistochemistry, viral antigen positive neurons were found in the spinal cord and brain on day 4. These findings suggest that the virus might spread to the CNS by neural pathways rather than through the blood stream (Hirano et al., 1993).

To follow the virus spread from the peripheral nerve to the CNS, the virus was inoculated s.c. into the foot pad of the right leg. Rats were operated on to cut the sciatic nerve after virus inoculation. The fatal infection of HEV was aborted by cutting the sciatic nerve within 6 hours but not after 12 or more hours. When the virus was directly inoculated into the sciatic nerve, rats were protected from encephalitis by cutting the proximal segment to the inoculated site of the sciatic nerve within 1 hour following inoculation. These studies strongly suggest the crucial role of neural spread of the virus in the induction of fatal encephalitis with HEV in rats (Hirano et al., 1995).

In the present study, we attempted to analyze the virus spread to the CNS from the peripheral nerve after virus inoculation into the sciatic nerve or skeletal muscle using immunohistochemical techniques.

3. MATERIALS AND METHODS

Virus. Plaque-purified HEV 67N strain was propagated and assayed for infectivity in SK-K cells as described previously (Johnson et al., 1965). The infectivity titers were expressed in plaque-forming units (PFU).

Animals and inoculation: Four- to 8-week-old Wistar male rats were obtained from a commercial breeder colony, which was serologically negative for murine coronavirus infections. The virus (1×10^4 PFU/0.02 ml) was inoculated directly into the sciatic nerve under surgical operation, or into the crural muscle of right leg of rats anesthetisized with inhaled halothane.

Infectivity assay of the brain and spinal cord: Ten percent tissue homogenates were prepared from the brain and spinal cord and assayed for infectivity in SK-K cells (Johnson et al., 1965). The spinal cord was divided into 2 pieces, anterior and posterior half for each assay.

Immunohistochemistry: Rats were perfused with a fixative containing 4% paraformaldehyde in 0.1M phosphate buffer (pH 7.4) and the brain and spinal cord were cut out and processed for histopathological and immunohistochemical studies. For histopathological studies, serial sections of the CNS were stained by hematoxylin and eosin (HE). For immunostaining, coronal sections were obtained on a freezing microtome, treated with anti-HEV 67N mouse antibody (1:1000) at 4^0C overnight, and labeled with horse radish peroxidase (HRP)- or FITC-conjugated anti-mouse IgG goat serum at room temperature for 2 hours. HRP-labeled specimens were observed after visualization with the DAB reaction. FITC-labeled specimens were examined under a confocal laser scanning microscope.

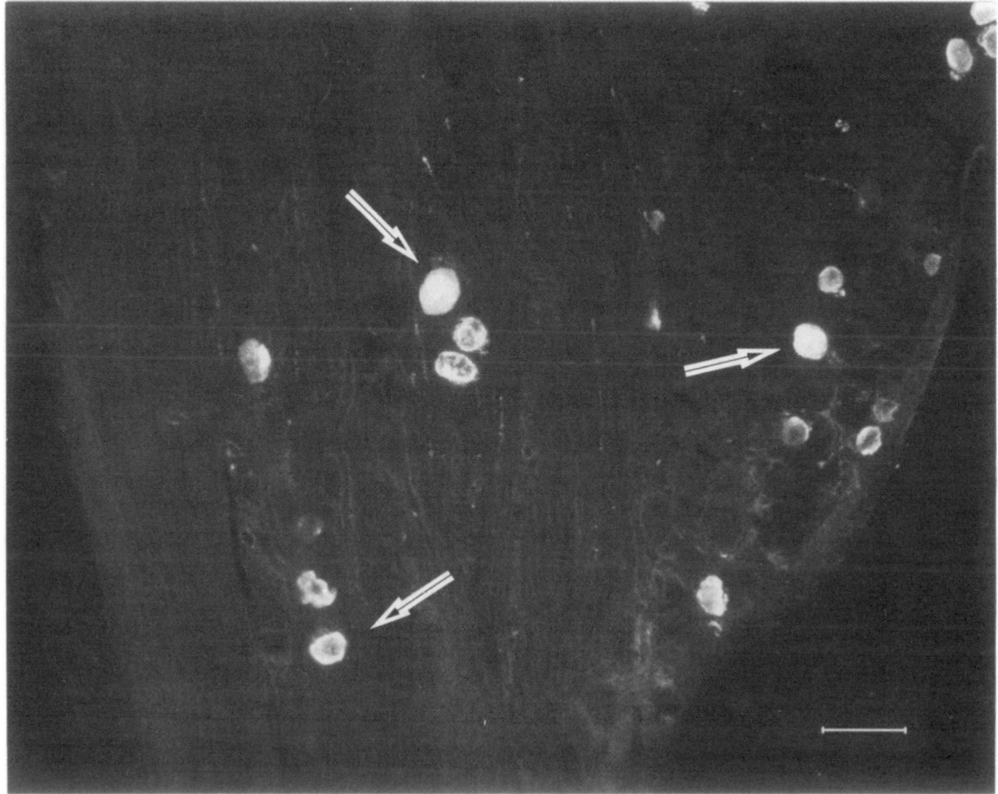

Figure 1. Ipsilateral L4 DRG on day 3 after viral inoculation. Many antigen positive neurons (arrows) are shown. FITC-labeling. Scale bar: 100 µm.

4. RESULTS

Infectivity in the spinal cord and brain: After inoculation with the virus (1×10^4 PFU) into the sciatic nerve or muscle of right leg, the virus was first isolated from the posterior half of the spinal cord at an infectivity titer of 10^5 to 10^6 PFU/0.2g. On day 3 the virus was detectable from the anterior half of the spinal cord and the brain. On day 5 the virus titer in the brain reached a maximum of 10^6 to 10^7 PFU/0.2g when the animals developed CNS signs consisting of ataxia, flapping ears and hypersensitivity. No virus was detected from the spleen, liver, and blood.

By HE staining a few neurons were pyknotic and the typical perivascular infiltration was not found in the spinal cord nor in the brain.

Immunofluorescence: On day 3 after inoculation, specific fluorescence was first found in the lumbar DRG neurons (Fig. 1) and lumbar spinal cord (Fig. 2) ipsilateral to the inoculated right leg. On day 4 antigen positive neurons were bilaterally distributed in the spinal cord, cerebral cortex, hippocampus, cerebellum and brainstem. In the spinal cord many neurons in the ventral and dorsal horn were antigen positive as were neurons in the contralateral DRG. At this time, antigen positive neurons were found in the thoracic and cervical spinal cord but the number of these cells in this area was smaller than in the lumbar spinal cord. In the cerebral cortex, pyramidal cells were infected in both supragranular

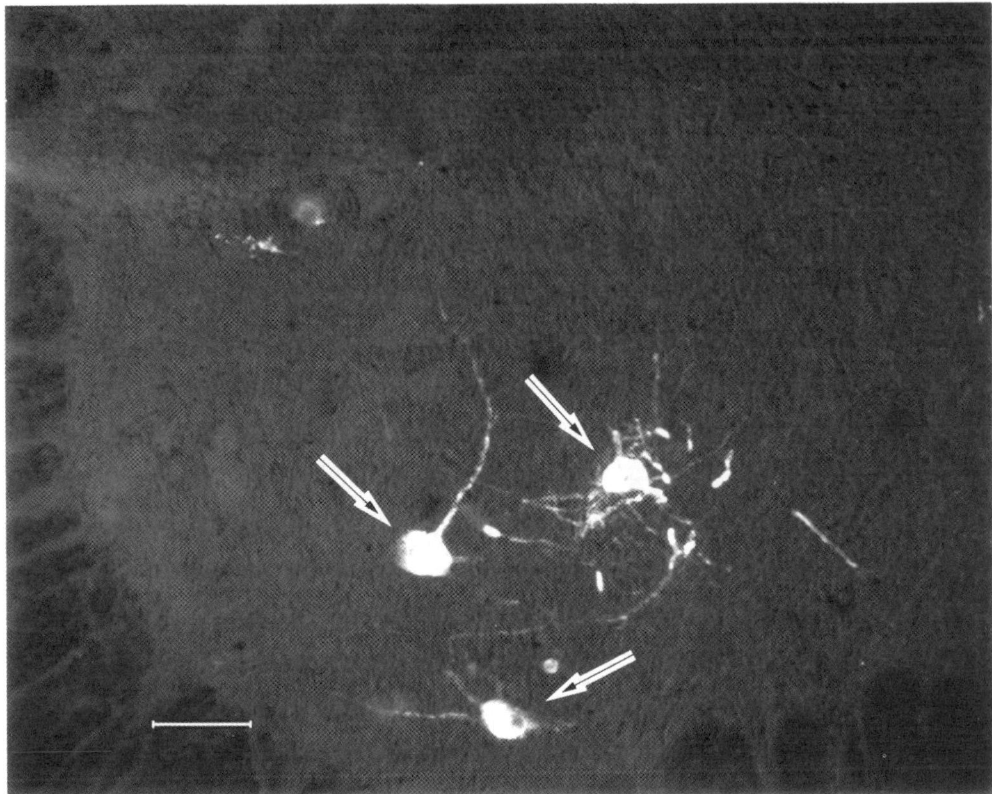

Figure 2. Ventral horn of ipsilateral lumbar spinal cord (L4) on day 3. A couple of HEV-positive motorneurons (arrows) are shown. FITC-labeling. Scale bar: 100 μm.

and infragranular layers, whereas neurons in the layer IV were generally not infected (Fig. 3). Clusters of antigen positive neurons were distributed in a patch-like manner. In the cerebellum, only a few Purkinje cells were antigen positive, but not neurons in the granular and molecular layers (Fig. 4). No fluorescence was detected in the choroid plexus and ependymal cells in the ventricle, and in extraneural tissues.

5. DISCUSSION

In the present study the experimental infection of rats with HEV was characterized by a selective vulnerability of neurons, apparently synchronous development of the virus growth within these neurons with a paucity of histopathological reactions, and a spread of virus from peripheral nerves or extraneural tissues to the CNS along neural pathways. These findings were different from those of HEV infection in rats by i.c. inoculation (data not shown). In rats infected by the i.c. route, virus antigen was widely found not only in neurons in the brain but also in the ependymal cells and in the choroid plexus in the ventricle. In the present study, antigen positive neurons were distributed in a patch-like manner in the cerebral cortex, but the ependymal cells and choroid plexus were negative. These findings suggest that virus infection in the CNS of i.c. inoculated rats might be established by cell to cell spread or via ventricles, and that HEV infection in this study might be established by virus spread through neural connections.

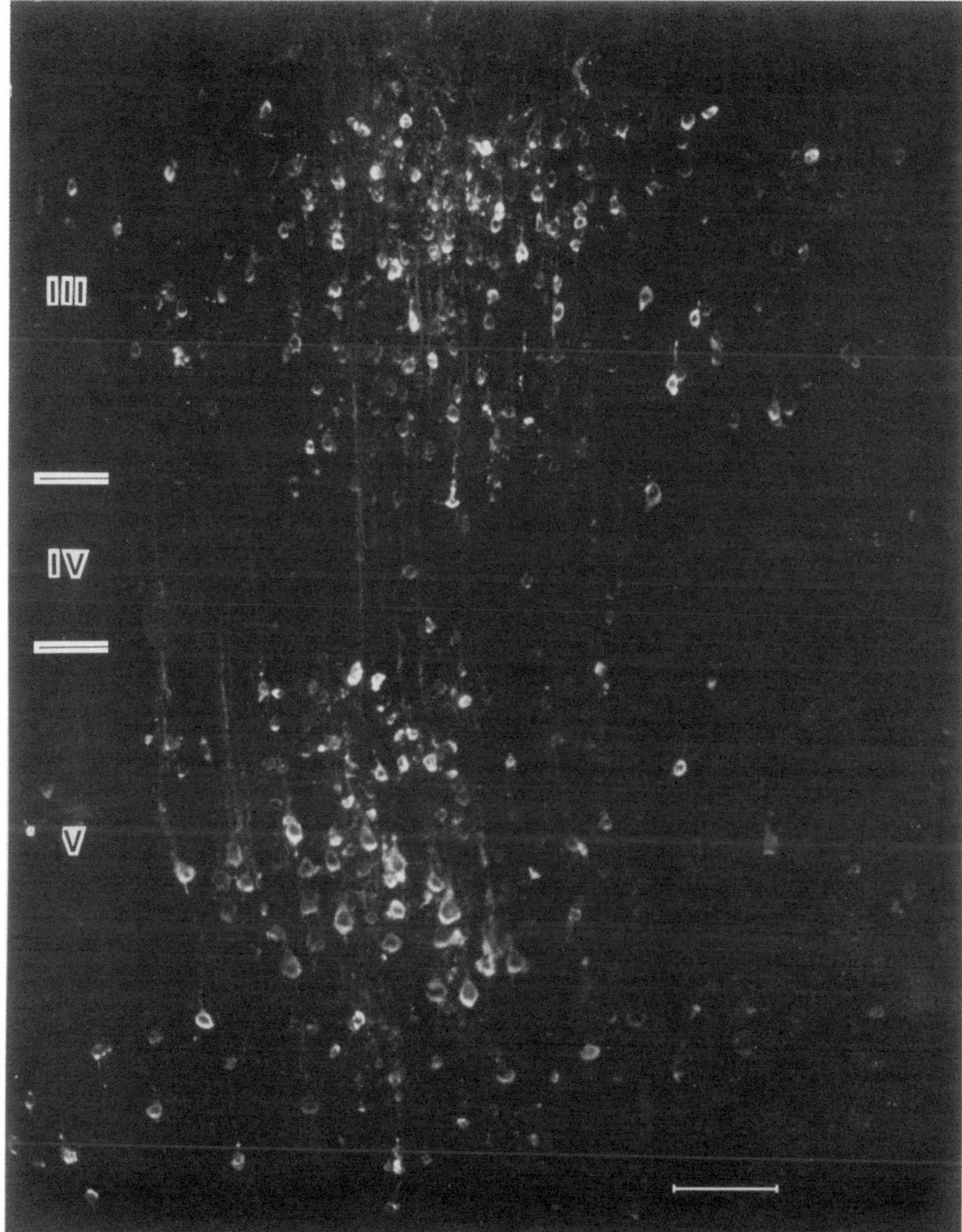

Figure 3. Cerebral cortex on day 4. Clusters of pyramidal neurons in the layer III and V are HEV-positive whereas neurons in layer IV were negative. FITC-labeling. Scale bar: 100 μm.

Our previous experiments showed that the infection of HEV was aborted by cutting the ipsilateral sciatic nerve within 6 hours after s.c. inoculation of the virus into the foot pad (Hirano et al., 1993). Furthermore, when the virus was directly inoculated into the sciatic nerve, only cutting of the proximal, but not the distal segment of the nerve within one hour post-infection prevented the rats from developing encephalitis (Hirano et al., 1995).

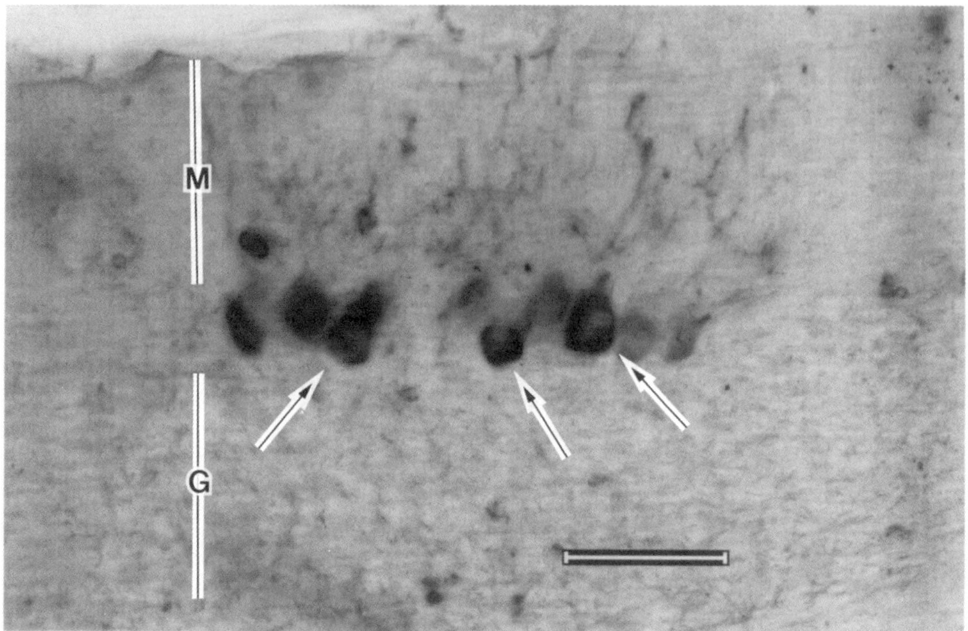

Figure 4. Cerebellar cortex on day 4. A cluster of HEV-positive Purkinje cells (arrows) are shown whereas no antigen positive neurons in the molecular (M) and granular (G) layers. DBA reaction. Scale bar: 50 μm.

In the present study, after virus inoculation into the sciatic nerve or the muscle, the virus was detectable from the lumbar spinal cord on day 2. By immunohistochemistry, specific antigen was found in the lumbar DRG neurons and in the ventral and dorsal horn neurons of the lumbar spinal cord ipsilateral to the inoculated leg on day 3. On day 4, specific fluorescence was also found in the contralateral DRG. On day 5, antigen positive cells were found bilaterally in the thoracic and cervical spinal cord, cerebral cortex and cerebellum. In the present study only neurons were found to be susceptible to HEV infection and, furthermore, only selective neuron populations. This was best demonstrated in the cerebellum where the Purkinje cells were infected, while there was no evidence of infection of any cells in the adjacent molecular or granular layers.

These findings are similar to those of experimental infection of mice with fixed rabies virus by the s.c. route (Roe et al., 1958).

This study clearly demonstrated that the virus spread via nerve pathways to the CNS from the sciatic nerve or extraneural tissue (gatrocnemius muscle) when inoculated directly with HEV, as demonstrated with rabies virus infections in mice (Johnson et al., 1965). From a viewpoint of neuroanatomy, HEV is useful as a transneural tracer for analysing neural connections in the CNS.

ACKNOWLEDGMENTS

The authors thank Dr., K. Sano, Department of Anatomy, Iwate Medical University, for his critical advises. This study was partly supported by Grants-in-Aids (No. 07660088 and 08556052) from the Ministry of Education and Cultures of Japan.

REFERENCES

Andries, K., and Pensaert, M. B., 1980, Immunofluorescence studies on the pathogenesis of hemagglutinating encephalomyelitis virus infection in pigs after oronasal inoculation, *Am. J. Vet. Res.* **41**: 1372–1385.

Andries, K., and Pensaert, M. B., 1981, Vomiting and wasting disease, a coronavirus infection of pigs, *Adv. Exp. Med. Biol.* **142**: 399–408.

Hirano, N., Haga, S., and Fujiwara, K., 1993, The route of transmission of hemagglutinating encephalomyelitis virus (HEV) 67N strain in 4-week-old rats, *Adv. Exp. Med. Biol.* **342**: 333–335.

Hirano, N., Nomura, T., Tawara, K., Ono, K., and Iwasaki, Y., 1995, Neuronal spread of swine hemagglutinating encephalomyelitis virus (HEV) 67N strain in 4-week-old rats, *Adv. Exp. Med. Biol.* **380**: 117–119.

Hirano, N., Ono, K., Takasawa, H., and Haga, S. J., 1990, Replication and plaque formation of swine hemagglutinating encephalomyelitis virus (67N) in swine cell line, SK-K culture, *Virol. Meth.* **27**: 91–100.

Johnson, R. T. J., Studies of cellular vulnerability and pathogenesis using fluorescent antibody staining, 1965, *Neuropathol. Exp. Neurol.* **24**: 662–674.

Mengeling, W. L., Boothe, A. D., and Richite, A. E., 1972, Characteristics of a coronavirus (Strain 67N) of pigs, *Am. J. Vet. Res.* **33**: 297–308.

Roe, C. K., and Alexander, T. J. L., 1958, A disease of nursing pigs previously unreported in Ontario, *Cand. Comp. Med.* **22**: 305–307.

Yagami, K., Hirai, K., and Hirano, N. J., Pathogenesis of haemagglutinating encephalomyelitis virus (HVE) in mice experimentally infected by different routes, 1986, *Comp. Pathol.* **96**: 645–657.

79

EXPRESSION OF THE *fgl2* AND ITS PROTEIN PRODUCT (PROTHROMBINASE) IN TISSUES DURING MURINE HEPATITIS VIRUS STRAIN-3 (MHV-3) INFECTION

J. W. Ding,[1] Q. Ning,[1] M. F. Liu,[1] A. Lai,[1] K. Peltekian,[2] L. Fung,[1] C. Holloway,[3] H. Yeger,[4] M. J. Phillips,[4] and G. A. Levy[1]

[1]Department of Multi Organ Transplantation Program and Medicine
 Toronto Hospital
[3]Department of Surgery
 Women's College Hospital
[4]Department of Pathology
 The Hospital for Sick Children
 University of Toronto
 Toronto, Ontario
[2]Department of Medicine
 Dalhousie University
 Halifax, Nova Scotia, Canada

1. ABSTRACT

Murine Hepatitis Virus Strain 3 (MHV-3) produces fulminant hepatitis with 80–90% mortality in Balb/cJ mice. Previous studies in our laboratory have shown that peritoneal macrophages from MHV-3 infected mice produce a procoagulant (PCA) which has the ability to cleave prothrombin to thrombin (prothrombinase) encoded by the gene *fgl2* located on chromosome 5. PCA accounts for sinusoidal thrombosis and hepatic necrosis and the necrosis and mortality can be prevented by treatment of animals with a monoclonal antibody to PCA. These present studies were designed to examine the expression of this gene (mRNA by Northern analysis and *in situ* hybridization) and the gene product PCA (immunochemistry) in tissues recovered from MHV-3 infected Balb/cJ mice in an attempt to explain the liver specific nature of MHV-3 disease. *Fgl2* gene expression was detected as early as 8 hours after MHV-3 infection which persisted to 48 hours in the liver, spleen and lungs whereas no gene expression was seen in the brain or kidneys despite the fact

Coronaviruses and Arteriviruses, edited by Enjuanes *et al.*
Plenum Press, New York, 1998

that equivalent viral titers were detected in all tissues at all times. In the liver, *fgl2* gene expression was confined to endothelial and Kupffer cells with no expression in hepatocytes. Immunochemistry localized the PCA protein to Kupffer cells and endothelial cells and necrotic foci within the liver. No PCA protein was detected by immunochemistry in any other tissues at any time during the course of MHV-3 infection.

These results explain the liver specific nature (fulminant hepatitis) of MHV-3 infection and provides further evidence for the role of PCA in the pathogenesis of fulminant hepatitis. MHV-3 induces selective transcription of the gene *fgl2* and only hepatic reticuloendothelial cells produce functional protein (PCA) which is known to account for fulminant hepatic failure produced by MHV-3.

2. INTRODUCTION

Studies from this laboratory have demonstrated a pivotal role for a procoagulant with prothrombin cleaving activity (prothrombinase) in murine hepatitis virus strain 3 induced hepatic necrosis (Levy, 1981). Macrophages isolated from susceptible (Balb/c) but not resistant (A/J) animals express large amounts of this prothrombinase *in vitro* and *in vivo*. Of great interest, inhibition of the prothrombinase with high titered neutralizing monoclonal antibody prevents hepatic necrosis and prolongs survival in MHV-3 susceptible mice (Li, 1992). Recently, this laboratory reported the isolation and functional expression of complementary DNA clones for this prothrombinase, termed *fgl2* (Parr, 1995). The molecular clone corresponded to a mouse fibrinogen like protein (musfiblp). The murine gene for the prothrombinase *fgl2* was localized to proximal regions of mouse chromosome 5. We and others have shown that MHV-3 replicates in many tissues including lung, bowel, brain, kidney and spleen in addition to the liver and yet disease as defined by fibrin deposition and cellular necrosis is confined to the liver (Yuwaraj, 1996; Houtman, 1986). The present study demonstrates that following MHV-3 infection of susceptible Balb/c mice, *fgl2* mRNA transcripts are strongly induced in macrophage-rich organs, such as the liver, spleen and lungs, however, immunoreactive *fgl2* (*fgl2* prothrombinase) was only expressed in the liver. Furthermore, both *fgl2* mRNA transcripts and *fgl2* prothrombinase were only expressed by Kupffer cells and endothelial cells and not hepatocytes of the liver after MHV-3 infection. The expression of the *fgl2* prothrombinase was promptly followed by fibrin deposition and massive hepatic necrosis, hallmarks of fulminant liver disease.

3. MATERIALS AND METHODS

3.1. Mice and Virus

Female Balb/c mice (Jackson Lab, ME and Charles River, Quebec, Canada), 8 weeks, were infected with MHV-3 by intraperitoneal injection. MHV-3 was plaque purified on monolayers of DBT cells. It was grown to a titre of 1.5×10^7 plaque forming units (PFU) per ml in 17 CL1 cells. Viral titres were determined on monolayers of L2 cells in a standard plaque assay as described elsewhere (Parr, 1995). Susceptible Balb/c mice were infected by intraperitoneal administration of MHV-3 and tissues were removed at various time points following MHV-3 infection for a variety of analyses including Northern blot analysis, *in situ* hybridization, immunohistochemistry, viral titration as well as rountine histology.

3.2. Hepatic Non-Parenchymal Cell Isolation

The method for isolating non-parenchymal cells was described elsewhere (Janousek, 1993; Seglen, 1973). Briefly, normal liver and the liver 24 hours after MHV-3 infection were perfused through portal vein with 0.02% collagenase B and 0.001% DNAse I, and the liver capsule and cells were gently removed. Cell suspension was then centrifuged in 30% metrizamide following incubation with 0.02% pronase and 0.01% DNAse. Non-parenchymal cell band was collected and washed free of metrizamide with HBSS, and non-parenchymal cells and hepatocytes were subjected to total RNA extraction and Northern blot analysis.

3.3. Northern Analysis

Methods for macrophage and tissue total RNA extraction were described previously (Parr, 1995; Ribaudo, 1991) Both macrophage (10 μg) and tissue total RNA (80 μg) was loaded and electrophoresed in 1% agarose gel with 1.7% formaldehyde and then transferred to Hybond-N$^+$ membrane (Amersham International plc, UK) in 20 × SSC. RNA was immobilized by UV crosslinking and hybridized in 50% formamide, 5 × SSPE, 5 × Denhardt's solution, 0.5% SDS and 100 µg/ml denatured salmon sperm DNA at 42 °C with ~ 1.3 kb fragment of *fgl2* gene cDNA labelled with [α-^{32}P] dCTP for 18 hours. The membrane was washed with 1 × SSPE and 0.1% SDS at room temperature for 15 min and then autoradiographed by exposing the membrane to Kodak film (X-OMAT^{TM}; Eastman Kodak Company, Rochester, NY).

3.4. In Situ Hybridization

The method used in the experiment has been described previously (Wang, 1996) with slight modification. Digoxigenin-UTP labelled-RNA probes (antisense and sense) were synthesized by T7 and T3 RNA polymerases following subcloning a 650 bp *fgl2* cDNA fragment into EcoRI site in pBluescript plasmid. The hybridization mixture consisted of 50% deionized formamide, 5% dextran sulfate, 250 μg/ml salmon sperm DNA and 2 μg/ml Digoxigenin-labelled ribo-probe in 2×SSC, and hybridization was allowed to persist overnight at 42 °C following prehybridization of tissue sections in 50% deionized formamide and 2×SSC for 1 hour at room temperature. Digoxigenin-labelled RNA probes were detected by polyclonal anti-Digoxigenin F$_{ab}$ fragment, conjugated to alkaline phosphatase in a dilution of 1:500 in Tris-HCl buffer (pH 7.5) (Boehringer). Purple colour develops after incubation with alkaline phosphatase substrates, 5-bromo-4-chloro-3-indolyl-phosphate (BCIP) and nitro blue tetrazolium (NBT), at room temperature.

3.5. Immunohistochemical Staining

a) Immunofluoresence for detection of the fgl2 prothrombinase. Immunohistochemical stain of snap-frozen tissue specimens including liver, spleen, lungs, kidneys and brain using rabbit polyclonal antibody against PCA was performed. The tissue cryosections were blocked with 10% normal horse serum prior to incubation with rabbit polyclonal antibodies against *fgl2* prothrombinase, followed by incubation with FITC conjungated goat IgG fraction to rabbit IgG F$_c$. Tissue slices were then air dried, mounted with 90% glycerol and cover slips, and photographed with ultraviolet Leitz Laborlax S microscope (Ernst Leitz Welzlar GmbH, Germany). b) Immunoperoxidase staining of fibrin. Liver tis-

sue from Balb/c mice obtained at 0, 24, 36 and 48 hours after MHV-3 infection were examined for presence of fibrin. The primary antibody used was a rabbit anti-human fibrinogen, a purified immunoglobulin fraction of rabbit antiserum (DAKO, CARPINTERIA,
CA). The antibody is known to react with native human fibrinogen as well as with the fibrinogen of other species and with fibrin in human tissues and in the mouse. The technique utilized for detection of fibrin was the standard avidin-biotin complex method. The
biotinylated secondary antibody was against rabbit IgG linked to peroxidase reacted with
DAB chromagen followed by counterstaining with hematoxylin. Serial sections were
stained with hematoxylin and eosin.

3.6. Statistical Analysis

Quantitative data were expressed as mean±SD. Statistical analysis was carried out
using one way analysis of variance (ANOVA) and a p value less than 0.05 was considered
statistically significant.

4. RESULTS

4.1. Histology and Viral Titers in the Organs after MHV-3 Infection

Tissues including liver, spleen, lungs, kidneys and brain were examined from Balb/c
mice after MHV-3 infection as described above. Histopathological changes were only found
in the liver but not in the other organs (data not shown). High titers of virus were found in
all tissues examined. The viral titers in the liver detected 24 hours following MHV-3 infection were significantly higher than those in the lungs, kidneys and brain (p<0.001); but there
was no difference in viral titers between liver and spleen (p>0.05) (Figure 1).

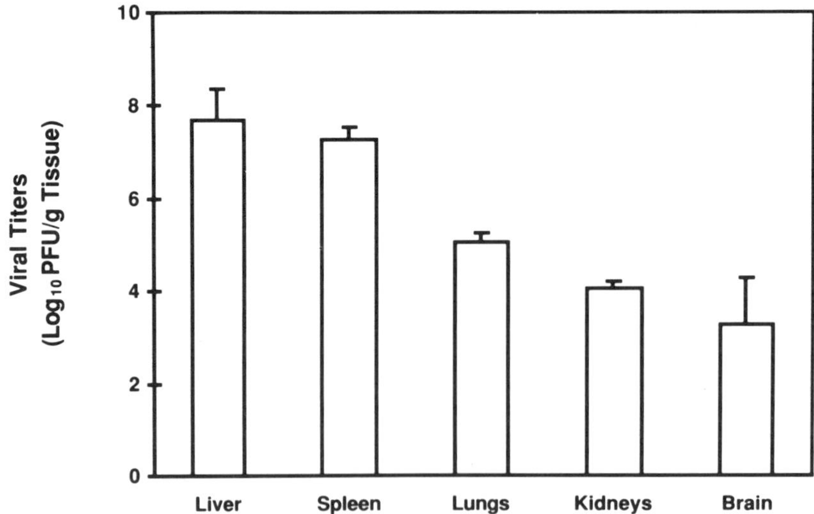

Figure 1. Viral titer determination in organs 24 hours following MHV-3 infection. Viral titer in the liver was significantly higher than those in the lung, kidney and brain (p<0.001), but did not significantly differ from the spleen
(p>0.05).

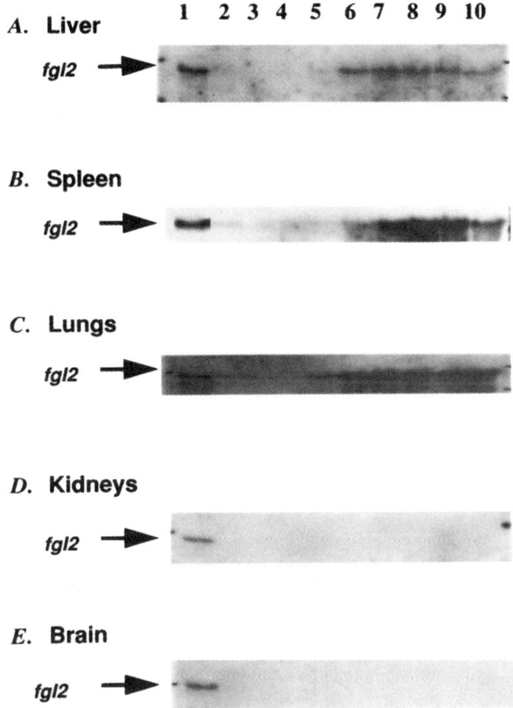

Figure 2. Tissue expression of *fgl2* mRNA transcripts post MHV-3 infection. Eighty (80) μg of total RNA was loaded in each well and hybridized with an *fgl2* cDNA. Lane 1: RNA (15μg) from MHV-3 infected peritoneal macrophages (positive control); Lane 2: RNA from organs immediately after MHV-3 infection (negative control); Lanes 3–10: RNA extracted 2, 4, 6, 8, 16, 24, 36 and 48 hours after MHV-3 infection. (A) MHV-3 induced *fgl2* mRNA transcription in the liver started at 8 hours and persisted to 48 hours whereas expression of *fgl2* mRNA transcripts was first detected at 6 hours in spleen (B) and lungs (C) and persisted for 48 hours. In contrast, *fgl2* mRNA transcripts were not detected in kidneys (D) and brain (E) after MHV-3 infection. A *GAPDH* cDNA was used to ensure equal loading of RNA in all lanes (data not shown).

4.2. *Fgl2* mRNA Expression following MHV-3 Infection in Tissues

Northern blot analysis of total cellular RNA extracted from uninfected and infected organs at various time points after intraperitoneal administration of MHV-3 is shown in Figure 2. The mRNA for *fgl2* was not constitutively expressed in the tissues examined in this study. In the liver, *fgl2* mRNA transcripts were detected at 8 hours and persisted to 48 hours after MHV-3 injection , the last time point studied (Figure 2A). mRNA transcripts for *fgl2* were first detected at 6 hours in the spleen (Figure 2B) and lungs (Figure 2C) and persisted to 48 hours after MHV-3 infection, whereas *fgl2* mRNA transcripts could not be detected in the kidney or brain at all time points studied (Figures 2D and 2E).

4.3. *Fgl2* mRNA Expression in Cellular Subpopulations of the Liver

Northern blot analysis of total RNA isolated from parenchymal (hepatocytes) and non-parenchymal (Kupffer and endothelial) demonstrated that *fgl2* gene transcripts were only expressed in non-parenchymal cells (Figure 3).

4.4. In Situ Hybridization

Fgl2 mRNA transcripts were evident in endothelial cells and Kupffer cells in infected Balb/c mouse liver 24 hours following MHV-3 (Figure 4). Furthermore, *fgl2* mRNA expression was seen in splenic reticuloendothelial cells and macrophages and was also detected in bronchiolar, alveolar duct epithelial cells and alveolar macrophages (data not shown). Low level expression of *fgl2* was seen in the kidney and the brain following

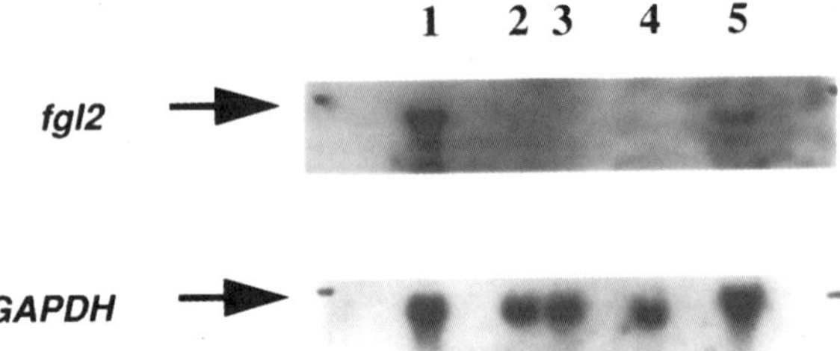

Figure 3. Northern blot analysis of total cellular RNA from both uninfected and infected hepatocytes and non-parenchymal cells 24 hrs after MHV-3 infection. Lane 1: RNA from MHV-3 infected peritoneal macrophages (positive control); Lanes 2 and 3: RNA from uninfected hepatocytes and non-parenchymal cells; Lanes 4 and 5: RNA from hepatocytes and non-parenchymal cells 24 hours after MHV-3 infection. Twenty (20) μg of total RNA was added to each lane and hybridized with a *fgl2* cDNA. A *GAPDH* cDNA was used to ensure equal loading of RNA in all lanes.

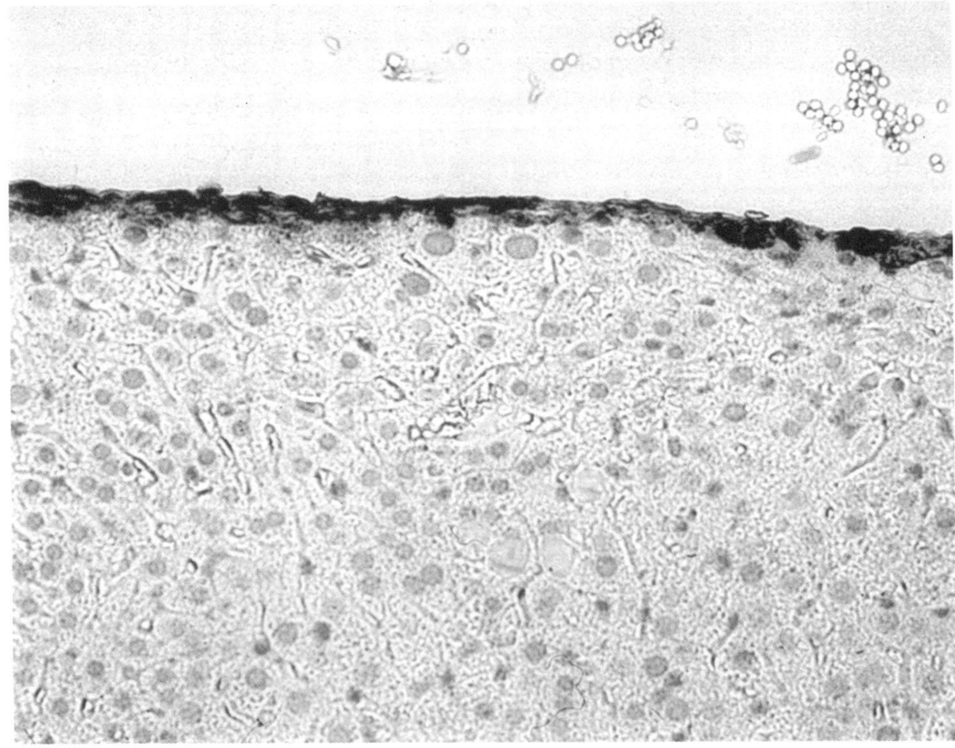

Figure 4. *In situ* hybridization detecting *fgl2* gene expression in the liver. *Fgl2* gene expression was detected by an antisense *fgl2* probe, to detect sense mRNA. *Fgl2* transcripts were found in endothelial cells of large veins and sinusoids and Kupffer cells 24 hours after MHV-3 infection. Alkaline phosphatase stain, ×400.

MHV-3 infection (data not shown). In the brain, *fgl2* transcripts were found in subcortical neurons and in the kidney, the *fgl2* transcripts were confined to tubular epithelium of the inner medulla. Tissues from uninfected Balb/c mice hybridized with antisense or sense probes did not reveal any constitutive *fgl2* mRNA expression (data not shown).

5. DISCUSSION

The recent molecular characterization of the *fgl2* prothrombinase now allows for definitive studies to determine the cellular mechanisms involved in the pathogenesis of murine hepatitis virus strain 3 (MHV-3) induced fulminant hepatic failure. mRNA transcripts for murine *fgl2* gene were seen in reticuloendothelial cells of the liver 8 hours after MHV-3 infection persisting to 48 hours. *Fgl2* prothrombinase was detected in the liver 24 hours following MHV-3 infection and was expressed in endothelial cells lining large hepatic veins, the hepatic sinusoids and in the areas of focal necrosis (Figure 5). Concomitant fibrin deposition in the hepatic sinusoids and focal hepatocellular necrosis with *fgl2* prothrombinase expression 24 hours following MHV-3 infection were observed, and these changes progressed to more extensive fibrin deposition and confluent hepatic necrosis 36–48 hours following MHV-3 infection (Figure 6). Although by Northern blot analysis and *in situ* hybridization *fgl2* mRNA transcripts were detected in other organs, no protein product (*fgl2* prothrombinase) or disease was seen outside of the liver despite the presence of high viral titers. These results indicated that the pattern of the disease is organ-specific. Studies from our and other laboratories have suggested that host factors may be more critical than viral replication in the pathogenesis of MHV disease (Pope, 1995; Yuwaraj, 1996; Houtman, 1996). In this and previous studies MHV-3 replication significantly differed in each organ examined, only in the liver is pathology seen following MHV-3 infection. The absence of disease in the spleen which had similar high viral titres to those found in the liver supports the concept that viral replication alone will not explain MHV-3 disease.

Fibrin deposition in the liver is a unique feature of MHV-3 induced fulminant liver failure. It is well documented that MHV-3 disease is localized to the liver and other organs are relatively spared (Piazza, 1969). The fact that no *fgl2* prothrombinase and no pathological changes were seen outside of the liver had further suggested the importance of this molecule in the MHV-3 induced liver disease. The importance of the immune coagulation system resulting in fibrin deposition in human fulminant liver failure is controversial (Levy, 1984). Mori et. al. (1981; 1986) have reported that fibrin is a classical feature of fulminant hepatitis but this has not been substantiated by others (Portmann, 1975; Zimmerman, 1982). The reason for not finding fibrin in tissue sections may rest with the methodologies used. In routine hematoxylin and eosin (H & E) stained sections, fibrin is not easily recognized and its presence is further obscured by hepatic necrosis. Even the use of special stains for fibrin are not always conclusive in our experience. However, using recently available specific antibodies such as those used in this study, fibrin is clearly evident in the hepatic lesions of MHV-3 infected livers from susceptible mice.

In conclusion, *fgl2* mRNA transcripts were seen in reticuloendothelial rich organs including liver, spleen and the lungs following MHV-3 infection. However, *fgl2* prothrombinase was only evident in the liver. The expression of *fgl2* prothrombinase in the hepatic sinusoids and large portal veins initiates microthrombosis and fibrin deposition which leads to perturbations in microvascular hemodynamics and subsequent liver cell necrosis. These studies firmly establish the pivotal role of *fgl2* and its protein product in the pathogenesis of fulminant liver failure caused by MHV-3.

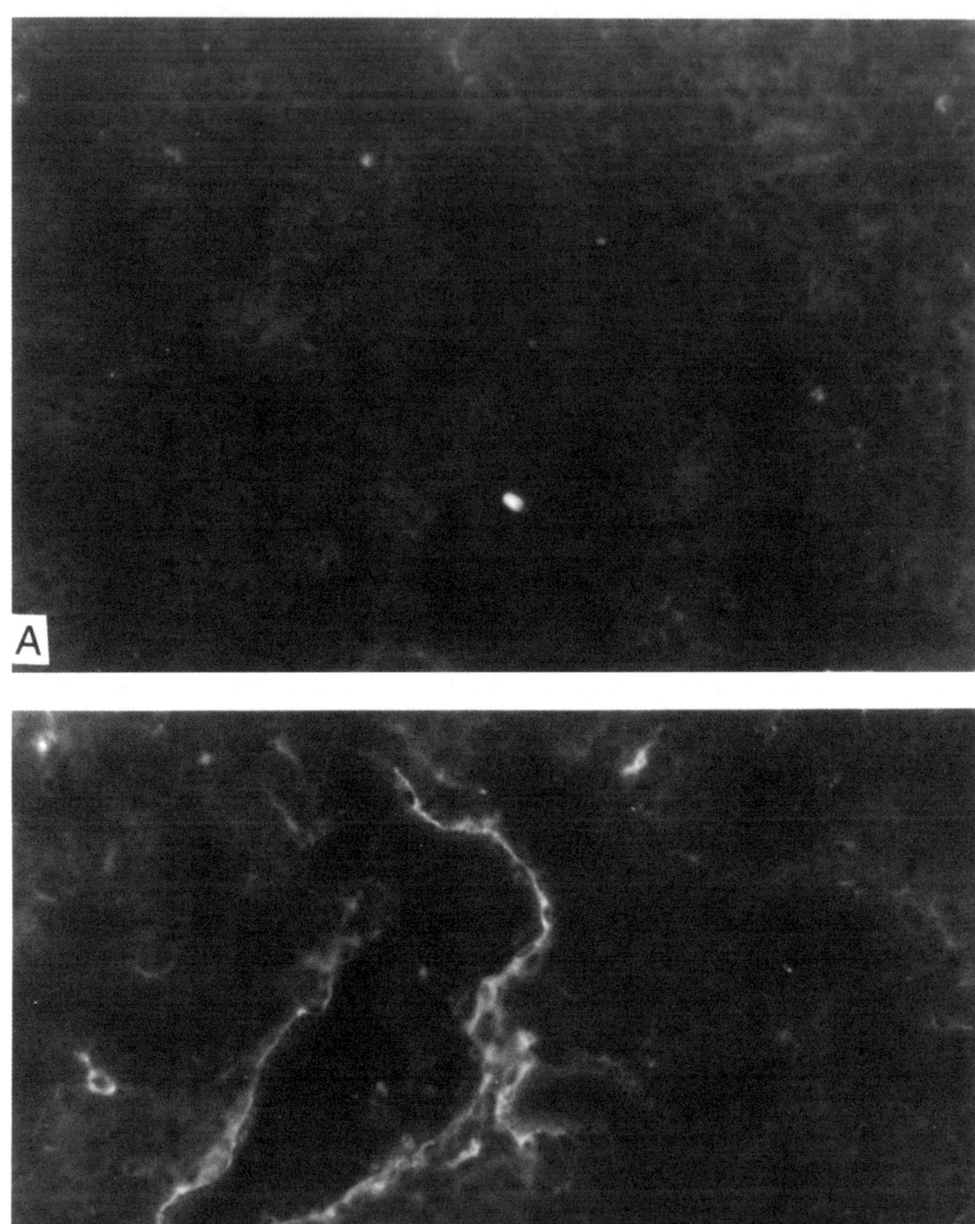

Figure 5. Immunohistochemical staining of the liver post MHV-3 infection using a rabbit polyclonal antibody against the *fgl2* prothrombinase. (A) Absence of staining in a normal control (×100); *Fgl2* prothrombinase was detected in (B) the large veins and sinusoids (×100) and at 48 hours (C) in sinusoids and necrotic foci (×400), following MHV-3 infection by immunofluorescence.

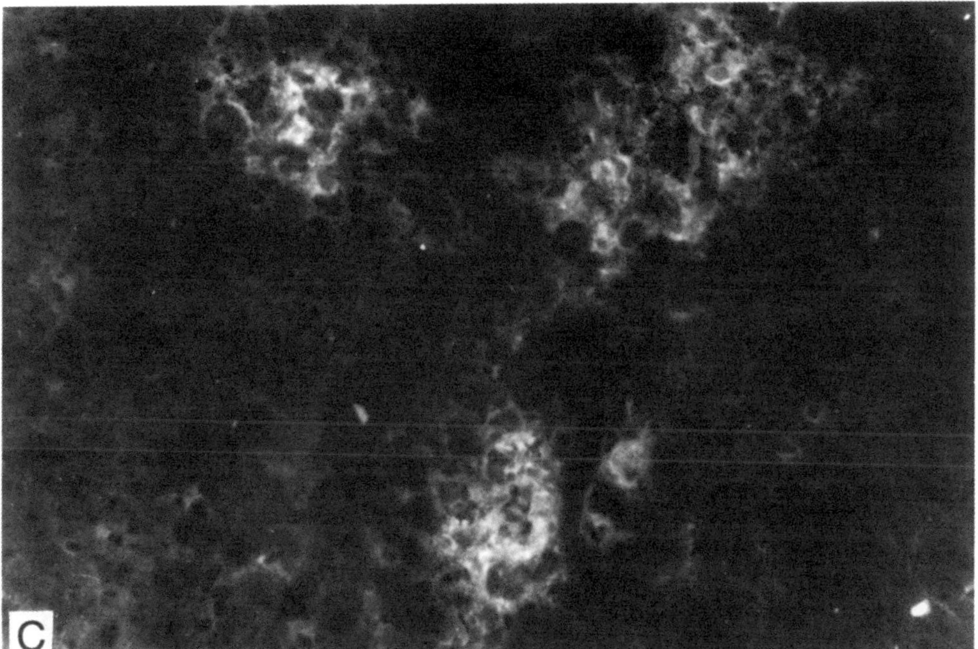

Figure 5C.

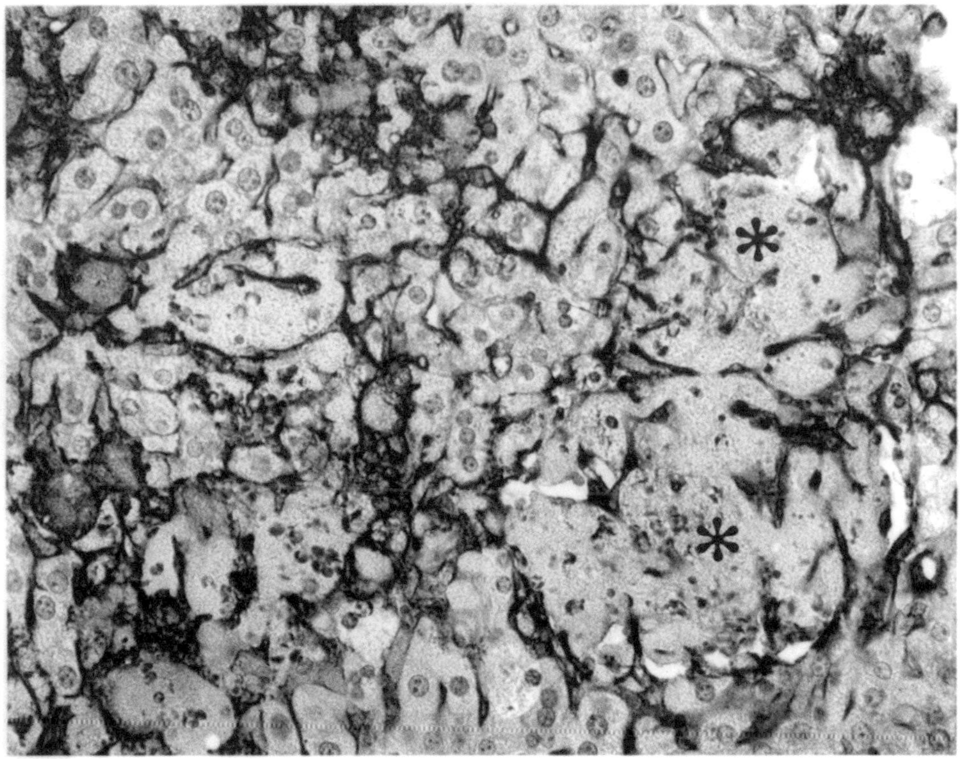

Figure 6. Immunohistochemical staining of the liver for fibrin. Widespread fibrin deposition is seen (dark areas) in sinusoids frequently outlining areas of hepatocellular necrosis (✳). Larger focal areas of fibrin deposition are also seen with loss of liver cells. Immunoperoxidase stain, ×400.

ACKNOWLEDGMENT

We are grateful to Charmaine Beal for her excellent secretarial assistance. This work was supported by a group grant PG 11810 from the Medical Research Council of Canada.

REFERENCES

Houtman, J. J., and Fleming, J. O., 1996, Pathogenesis of mouse hepatitis virus-induced demyelination, *J. Neurovirol.* **2**:361–270.

Janousek, J., Strmen, E., and Gervais, F., 1993, Purification of murine Kupffer cells by centrifugal elutriation, *J. Immunol. Methods* **164**:109–117.

Levy, G. A., Leibowitz, J. L., and Edgington., T. S., 1981, The induction of monocyte procoagulant activity by murine hepatitis virus MHV-3 parallels disease susceptibility in mice, *J. Exp. Med.* **154**:1150–1163.

Levy, G. A., Helin, H., and Edgington, T. S., 1984, The pathobiology of viral hepatitis and immunologic activation of the coagulation protease network, *Sem. Liver Dis.* **4**:59–68.

Li, C., Fung, L. S., Crow, A., Myers-Mason, N., Phillips, M. J., Leibowitz, J. L., Cole, E., Ottaway, C. A., and Levy, G. A., 1992, Monoclonal antiprothrombinase (3D4.3) prevents mortality from murine hepatitis virus (MHV-3) infection, *J. Exp. Med.* **176**:689–697.

Mori, W., Shiga, J., and Irie, H., 1986, Schwartzman reaction as a pathogenetic mechanism in fulminant hepatitis, *Sem. Liver Dis.* **6**:267–269.

Mori, W., and Naoto, A., 1981, Acute hepatic cell necrosis experimentally produced by viral agents in rabbits, *Am. J. Pathol.* **103**:31–35.

Parr, R. L., Fung, L., Reneker, J., Myers-Mason, N., Leibowitz, J. L., and Levy, G.A., 1995, Association of mouse fibrinogen-like protein with murine hepatitis virus-induced prothrombinase activity, *J. Virol.* **69**:5033–5038.

Piazza, M. (ed.) 1969, *Experimental viral hepatitis*, Charles C. Thomas, Springfield, Illinois.

Pope, M., Rotstein, O., Cole, E., Sinclair, S., Parr, R., Cruz,B., Fingerote, R., Chung, S., Gorczynski, R., Fung, L., Leibowitz, J., Rao,Y. S., and Levy, G. A., 1995, Pattern of disease after murine hepatitis virus strain 3 infection correlates with macrophage activation and not viral replication, *J. Virol.* **69**:5252–5260.

Portmann, B., Talbot, I. C., Day, D. W., Davidson, A. R., Murray-Lyon, I. M., and Williams, R, 1975, Histopathological changes in the liver following a paracetamol overdose: correlation with clinical and biochemical parameters, *J. Pathol.* **117**:169–181.

Qureshi, S. T., Clermont, S., Leibowitz, J. L., Levy, G. A. and Malo, D., 1995, Mouse hepatitis virus-3 induced prothrombinase (*Fgl2*) maps to proximal chromosome 5, *Genomics* **29**:307- 309.

Ribaudo, R., Gilman, M., Kingston, R. E., Chomczynski, P., and Sacchi, N., 1991, Single-step RNA isolation from cultured cells or tissues, in: *Current Protocols in Immunology (Supplement 3)* (J. E. Coligan, A. M. Kruisbeek, D.H. Margulies, E.M. Shevach and W. Strober, eds.), John Wiley & Sons, Inc., pp. 10.11.7–10.11.14.

Seglen, P. O., 1973, Preparation of rat liver cells, *Exptl. Cell Res.* **82**:391–398.

Wang, D., Yeger, H., and Cutz, E., 1996, Expression of gastrin releasing peptide receptor gene in developing lung, *Am. J. Repir. Cell Mol. Biol.* **14**:409–416.

Yuwaraj, S., Cattral, M., Pope, M. and Levy, G. A. 1996, Murine hepatitis virus: Molecular biology and pathogenesis, *Viral Hepatitis* **2**:125–142.

Zimmerman, H. J., and Ishak, K. G, 1982, Valproate-induced hepatic injury: analysis of 23 fatal case, *Hepatology* **2**:591–597.

80

THE PATTERN OF INDUCTION OF APOPTOSIS DURING INFECTION WITH MHV-3 CORRELATES WITH STRAIN VARIATION IN RESISTANCE AND SUSCEPTIBILITY TO LETHAL HEPATITIS

Michail Belyavskyi,[1] Gary A. Levy,[2] and Julian L. Leibowitz[1]

[1]Department of Pathology and Laboratory Medicine
Texas A&M University College of Medicine
208 Reynolds Building
College Station, Texas 77843-1114
[2]Multi Organ Transplant Program
The Toronto Hospital–University of Toronto
621 University Ave.
NU-10-151, Toronto, Ontario

1. ABSTRACT

In the present study we have investigated the possibility that strain specific differences in the induction of apoptosis in macrophages could play a role in the resistance of strain A/J mice to MHV-3 induced hepatitis. MHV-3 infected macrophages from Balb/c and A/J mice were analyzed at various time points after infection. Apoptosis in A/J macrophages could be detected at 8 h post infection and increased significantly by 12 h, when almost 50–70% of the infected cells were undergoing apoptosis. In Balb/c macrophages, apoptotic changes were less pronounced and were observed in only 5–10% of the cells. MHV-3 induced apoptosis was inversely correlated with the ability of this virus to induce expression of fgl-2 prothrombinase protein and syncytia formation. Infected macrophages from A/J mice did not express fgl-2 protein and did not form syncytia. In contrast, infection of Balb/c derived macrophages resulted in fgl-2 expression and extensive syncytia formation. These data fit a model in which apoptosis of virally infected cells is a protective response which eliminates cells whose survival might be harmful for the whole organism.

Coronaviruses and Arteriviruses, edited by Enjuanes *et al.*
Plenum Press, New York, 1998

2. INTRODUCTION

Host protective non-specific responses are important factors in the pathogenesis of viral infections. Macrophages can limit viral infections by phagocytosing and killing virus particles, as well as by elaborating interferons and pro-inflammatory cytokines which can limit virus replication. The ability of viruses to replicate efficiently in macrophages is an important determinant in the pathogenesis of many diseases, including the hepatitis induced in mice during infection with mouse hepatitis virus (MHV) (Piazza, 1969). In fully susceptible strains of mice (Balb/c), MHV-3 infection produced a fulminant hepatitis, characterized by abnormalities of the hepatic microcirculation (Levy, et al., 1983). Strain A mice (A/J) are resistant to the development of disease even though MHV-3 replicates in their livers (Dindzans et al., 1985). The induction of a unique macrophage prothrombinase (PCA) in response to MHV-3 infection correlated with the severity of infection. This unregulated elaboration of PCA during infection is a major contributor to the pathogenesis of the lethal hepatitis in susceptible strains of mice (Dindzans et al., 1986). Apoptosis can be viewed as a protective mechanism to eliminate infected cells, whose survival might be harmful for the organism. Apoptosis is characterized by distinctive morphological and biochemical changes including nucleolytic degradation of chromosomal DNA, compaction and fragmentation of chromatin, cellular shrinkage, cytoplasmic blebbing, and fragmentation.

In the present work we have examined a possible role of apoptosis in the differences in response to MHV-3 infection. Macrophages derived from strain A/J mice undergo rapid apoptosis after infection with MHV-3, whereas macrophages from Balb/c mice do not. We propose that the rapid induction of apoptosis in macrophages from resistant mice is a key protective non-specific host response to MHV-3 infection. It prevents what would otherwise be an injurious prothrombinase response and eliminates infected macrophages which might be harmful for the whole organism.

3. MATERIALS AND METHODS

3.1. Virus and Cells

Stocks of MHV-3 were grown as described by Levy et al (1981). Thioglycollate elicited peritoneal macrophages were prepared from mice and infected as described (Parr et al., 1995). At various times post infection (p.i.) cells were harvested with a rubber policeman, resuspended in PBS containing 2% paraformaldehyde, incubated at 4°C for 10 minutes, washed and fixed in ice cold 70% ethanol. The samples were stored at -20°C before analysis by flow cytometry.

3.2. BrdUrd Labeling

Macrophages were infected with MHV-3 and labeled in media containing 10μM BrdUrd (Belyavskyi, 1994). At the indicated time points macrophages were harvested with a rubber policeman, washed twice in PBS, fixed in 70% ethanol, and kept at -20°C until analyzed.

3.3. Flow Cytometry

Fluorescence measurements were performed on a Becton Dickinson FACS flow cytometer as described previously (Belyavskyi, 1994). Before analysis cells were passed through a 23-gauge needle to break up clusters of cells. Fixed macrophages were washed twice in PBS (pH 7.0) and resuspended in 0.5 ml of propidium iodide (PI) solution for 20 minutes at 37°C. The data were gated to eliminate particles not the correct size for intact cells, and a two-parameter histogram of light scatter versus red fluorescence was plotted. At least 5×10^4 cells were counted in each assay. Relative anti-BrdUrd fluorescence intensity (RFI) was calculated as described (Schutte et al., 1987).

3.4. TUNEL Assay

TUNEL assays were performed on cells grown on chamber slides using a kit purchased from Promega Corp. TUNEL-positive cells were visualized by fluorescence microscopy.

3.5. Western Blotting

Cytoplasmic extracts were prepared from A/J and Balb/c macrophages with 0.5% NP-40. Proteins (20 µg) were resolved by SDS-PAGE, transferred to Immobilon membranes, and probed with antibodies to the fgl-2 or the MHV N proteins and detected by ECL (Amersham).

4. RESULTS

4.1. MHV-3 Induced Apoptosis

The majority of Balb/c macrophages infected with MHV-3 developed pronounced cytopathic effects, namely cell fusion with syncytial giant cell formation, after which the cells die. In A/J macrophages there was no detectable syncytia formation up to 8–12 hpi.

To examine the relationship between macrophages which were dying by apoptosis after MHV-3 infection and those that were incorporated into or form syncytia, we used the in situ TUNEL assay. A/J and Balb/c macrophages were mock or MHV-3 infected and TUNEL staining was performed 12hpi and visualized by fluorescence microscopy.

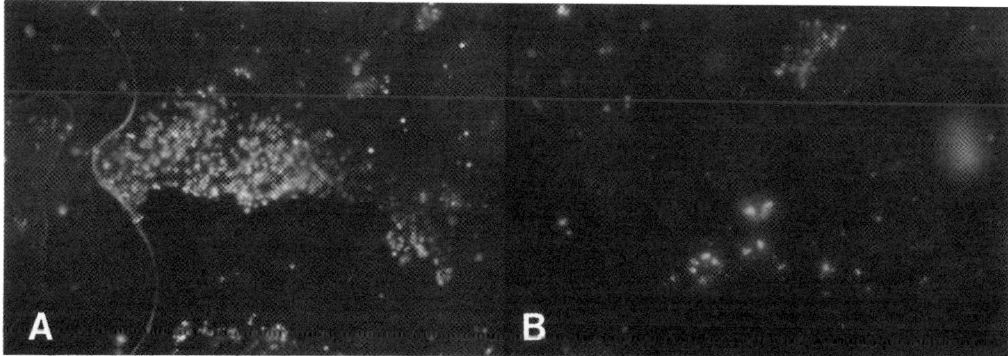

Figure 1. Detection of apoptosis by TUNEL assay. Macrophages from A/J and Balb/c mice were infected with MHV-3 and processed for TUNEL assay at 8 hpi. Representative fluorescence photomicrographs are shown for A/J (A) and Balb/c (B) macrophages.

Mock-infected A/J or Balb/c derived cells had the typical morphology of intact macrophages and contained few TUNEL-positive apoptotic cells (not shown). The majority of MHV-infected Balb/c macrophages underwent cell fusion forming large syncytia. TUNEL positive cells were observed only in a small fraction of the macrophages (about 10%), and these cells were not incorporated into syncytia. Quite different results were observed when A/J macrophages were infected with MHV-3. There was no visible syncytia formation during the first 12 h after infection. Cell counts indicated that the number of TUNEL-positive cells made up 60–70 % of the population. Based on these data we conclude that TUNEL-positive macrophages were not incorporated into or form syncytia during MHV-3 infection.

To further address the possibility that only single cells which were not incorporated into syncytia underwent apoptosis we used flow cytometric analysis. Sample preparation and gating paremeters for the flow cytometer exclude syncytial giant cells from the analysis. Thus macrophages were infected with MHV-3 and at various times after infection analyzed for DNA content by flow cytometry. At 4 hpi the majority of infected A/J and Balb/c macrophages show a typical G0+G1 DNA profile, similar to that of uninfected cells (not shown). Between 6–8 hpi a subpopulation of cells with less than G0+G1 DNA content appeared (about 20%), which subsequently increased (up to 50%) between 8–12 hpi (data not shown). The timing of this change in DNA profile was closely correlated with an increase in the proportion of shrunken cells detected by 90° light scatter (data not shown). DNA fragmentation with a resultant decrease in cellular DNA content and decreasing cell size are the hallmarks of apoptosis (Wyllie et al., 1980).

4.2. Fgl-2 Expression

We have previously shown that the appearence of fgl-2 mRNA after MHV-3 infection is delayed by about 2 h in A/J macrophages (5–6 hpi) relative to its appearence in Balb/c macrophages (3–4 hpi). To determine if fgl-2 protein was synthesized in MHV-3 infected A/J macrophages, cell lysates were prepared at various times after infection and analyzed by western blotting (Fig. 2A). Fgl-2 protein was not detected in A/J macrophages at any time examined, while in Balb/c macrophages fgl-2 protein was detected at 8 hpi and increased by 12 hpi. To verify that differences in fgl-2 protein expression were not related to variation in infectivity between A/J and Balb/c macrophages the same samples were probed with anti-MHV nucleocapsid antibodies (Fig. 2B). The nucleocapsid protein was detected in both A/J and Balb/c macrophages.

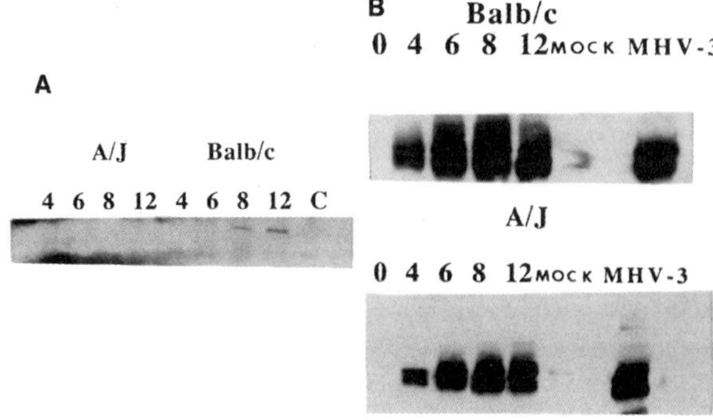

Figure 2. Western blot analysis of fgl-2 and nucleocapsid proteins. A/J and Balb/c macrophages were infected with MHV-3 and at the indicated times, cytoplasmic extracts were prepared and analyzed by immunoblotting with antibodies recognizing fgl-2 (A) and MHV nucleocapsid (B).

4.3. BrdUrd Incorporation into Macrophages after MHV-3 Infection

A comparison of DNA histograms from uninfected and MHV-3 infected A/J and Balb/c macrophages revealed that starting at 2–4 hpi a small fraction (5–15%) of cells contained typical S phase DNA content (data not shown). In uninfected macrophages the number of cells with S phase DNA content was 1–2%. To confirm this observation we directly measured the initiation of DNA replication by BrdUrd labeling infected and uninfected macrophages. At various times after infection the amount of incorporated BrdUrd was measured by flow cytometry (Fig. 3). Over the first 4 h of MHV-3 infection of A/J macrophages there was no detectable BrdUrd labeling above the background level. Between 4–6 hpi the amount of BrdUrd labeling started to increase, and then dropped back to the background level after 8 hpi. The pattern of BrdUrd labeling of infected Balb/c macrophages was similar. Labeling began between 2–4 hpi and then decreased, reaching background levels at 8 hpi. During the course of these experiments the infected cells were constantly incubated in media containing BrdUrd. If the proportion of the cells in which DNA replication was stimulated by MHV-3 remained fairly constant at various times after infection, and/or those infected cells proceeded normally to the next stage of the cell cycle, then the amount of incorporated BrdUrd would be proportional to the post infection incubation time. The data indicate that after a transient increase, the amount of BrdUrd incorporated into infected macrophages decreases with increasing hpi. One possible explanation of these results is that cells incorporating BrdUrd were being eliminated from the culture. A possible mechanism for this process is the induction of apoptosis in the cells with "unscheduled" DNA replication.

5. DISCUSSION

In the present study the basic mechanisms underlying the difference in response in fully susceptible and resistant strains of mice to MHV-3 infection were examined. To our

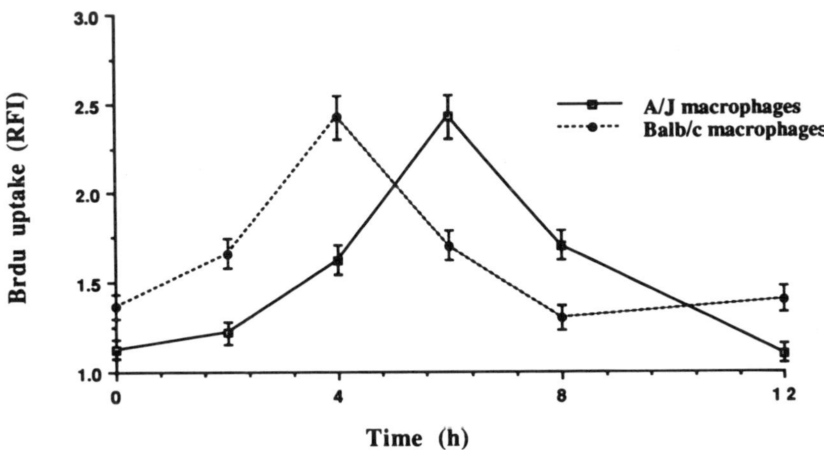

Figure 3. BrdUrd uptake in A/J and Balb/c macrophages infected with MHV-3. Macrophages were infected with MHV-3, labeled with BrdUrd, and analyzed by flow cytometry. Each data point is the mean value of three independent experiments. The error bars indicate the standard errors of the mean.

knowledge this study is the first demonstration of MHV-3 induced apoptosis, and raises the possibility that this is a key protective non-specific host response to the infection.

Our data, using two different methods to detect apoptosis, reveals that only single cells not incorporated into syncytia underwent apoptosis. This was true in both Balb/c and A/J macrophages, although apoptosis was much greater for the A/J derived cells. The mechanism by which MHV-3 induces apoptosis in macrophages is unknown. Our flow cytometric analysis of BrdUrd incorporation into A/J and Balb/c macrophages after MHV-3 infection demonstrated a temporary increase in BrdUrd labeling between 4–6 hpi. The amount of BrdUrd incorporated into cellular DNA subsequently dropped down to the background level, though the cells were continuously incubated in media with BrdUrd. Terminally differentiated cells, including macrophages, do not normally incorporate BrdUrd unless they are stimulated to initiate DNA synthesis. Cells which incorporated BrdUrd after MHV-3 infection should have been present in the infected cultures, unless they were specifically and constantly eliminated. This suggests that like several DNA viruses, infection of macrophages with MHV-3 overcomes the multiple host cell checkpoints that detect and prevent uncontrolled initiation of DNA synthesis. This observation provides an indication that the same pathways to apoptosis which are triggered by some DNA viruses (Cuff and Ruby, 1996) might also be involved in the induction of apoptosis by MHV-3. The difference in the outcome of MHV-3 infection of macrophages amongst the two strains could be due to the somewhat more rapid replication of MHV-3 in Balb/c macrophages than in A/J macrophages (Macnaughton and Patterson, 1980). This could commit infected Balb/c macrophages to high levels of S protein expression, with resulting cell fusion, prior to the cells acting on the apoptotic signal. In this scenario, in Balb/c mice MHV-3 simply outruns the apoptotic response, in A/J mice the response is effective. Other possible explanations of the strain-related differences in the ability of MHV-3 to induce apoptosis include variations in bcl-2 levels or other proteins important in decreasing programmed cell death.

Macrophages from A/J mice do not elaborate fgl-2 protein synthesis in response to MHV-3 infection, although there is a slow (compared to that seen in Balb/c macrophages) induction of prothrombinase mRNA. This suggests the possibility that in A/J macrophages, the rapid development of apoptosis or the signals leading to apoptosis during MHV-3 infection could supervene and inhibit fgl-2 protein synthesis before it is well established, thus contributing to the resistance of this strain of mice to MHV-3 induced hepatitis.

ACKNOWLEDGMENTS

This work was supported by NIH grant AI31069 and Canadian MRC program project grant PPG11810. We thank Drs. Helmut Sauer and Van Wilson for critical comments on this manuscript.

REFERENCES

Belyavskyi, M., Miller, J., and Wilson, V., 1994, The bovine papillomavirus E1 protein alters the host cell cycle and growth properties, *Virology* **204**:132–143.
Cuff, S., and Ruby, J., 1996, Evasion of apoptosis by DNA viruses, *Immun.Cell Biol.* **74**: 527–537.
Dindzans, V., MacPhee, P., Fung, L., Leibowitz, J., and Levy, G., 1985, The immune response to mouse hepatitis virus: expression of monocyte procoagulant activity and plasminogen activator during infection *in vivo*, *J. Immunol.* **135**:4189–4197.

Dindzans, V., Skamene, E., and Levy, G., 1986, Susceptibility/resistance to mouse hepatitis virus strain 3 and macrophage procoagulant activity are genetically linked and controlled by two non-H2-linked genes, *J. Immunol.* **137**:2355–2360.

Levy, G. A., J. L. Leibowitz, and T. S. Edgington., 1981, The induction of monocyte procoagulant activity by murine hepatitis virus MHV-3 parallels disease susceptibility in mice, *J. Exp. Med.* **154**:1150–1163.

Levy, G., MacPhee, P., Fung, L., Fisher, M., and Rappaport, A., 1983, The effect of mouse hepatitis virus infection on microcirculation of the liver, *Hepatology* **3**: 964–973.

Macnaughton, M., and Patterson, S., 1980, Mouse hepatitis virus strain 3 infection of C57, A/Sn and A/J strain mice and their macrophages, *Arch. Virol.* **66**:71–75.

Parr, R. L., Fung, L. S., Reneker, S. J., Myers-Mason, N., Leibowitz, J. L., and Levy, G. A., 1995, Association of mouse fibrinogen like protein (musfiblp) with murine hepatitis virus induced prothrombinase activity, *J. Virol.* **69**:5033–5038.

Piazza, M., 1969, "*Experimental viral hepatitis*". CA Thomas, Springfield, IL.

Schutte, B., Reynders, M., van Assche, C. L. M., Hupperets, P. S. G. J., Bosman, F. T., and Blijham, G. H., 1987, An improved method for the immunocytochemical detection of bromodeoxyuridine labeled nuclei using flow cytometry, *Cytometry* **8**:372–376.

Wyllie, A. H., Kerr, J. F. R., and Currie, A. R., 1980, Cell death: The significance of apoptosis, *Int. Rev. Cytology* **68**:251–306.

81

THE C12 MUTANT OF MHV-A59 IS VERY WEAKLY DEMYELINATING AND HAS FIVE AMINO ACID SUBSTITUTIONS RESTRICTED TO THE SPIKE AND REPLICASE GENES

Isabelle Leparc-Goffart,[1] Susan T. Hingley,[2] Xinhe Jiang,[1] Ming Ming Chua,[1] Ehud Lavi,[3] and Susan R. Weiss[1]

[2]Department of Microbiology
[3]Department of Pathology
University of Pennsylvania
Philadelphia, Pennsylvania 19104
[1]Department of Microbiology and Immunology
Philadelphia College of Osteopathic Medicine
Philadelphia, Pennsylvania 19131

1. ABSTRACT

C12, an attenuated, fusion defective, very weakly hepatotropic mutant of MHV-A59 has been further characterized. Analysis of C12 *in vivo* in C57BL/6 mice has shown that despite the fact that this virus replicates in the brain to titers at least as high as wild type and causes acute encephalitis similar to wild type, this virus causes minimal demyelination. Thus acute encephalitis is not sufficient for induction of demyelination by wild type MHV-A59. We have previously shown that C12 has two amino acid substitutions relative to wild type virus in the spike gene, Q159L (in the receptor binding domain of S1) and H716D (in the signal sequence for cleavage of S). We have now sequenced the rest of the 31 kb genome of C12 and compared it to wild type virus. Only three additional amino acids substitutions were found, all within the replicase gene, one in the predicted papain like proteinase (PLP)-2 domain and one in the predicted helicase domain. Thus, determinants of virulence, hepatotropism, and demyelination may map to the replicase gene as well as to the spike gene.

Coronaviruses and Arteriviruses, edited by Enjuanes *et al.*
Plenum Press, New York, 1998

2. INTRODUCTION

The murine coronavirus, mouse hepatitis virus strain A59 (MHV-A59), produces both hepatitis and neurological disease in susceptible mice. Neurological disease includes both acute meningoencephalitis and chronic demyelinating disease. *In vitro*, MHV-A59 causes a lytic infection in mouse fibroblast cells, but a persistent productive, non-lytic infection of cultured glial cells.

We have previously characterized fusion-defective mutants of MHV-A59 which were derived from persistently infected glial cell cultures (Hingley *et al.*, 1994; Gombold *et al.*, 1993). One of these, C12, has multiple phenotypes different from wild type MHV-A59. We have previously demonstrated that C12 induced fusion only at very late times after infection compared to wild type virus and that this phenotype maps to the H716D mutation in the signal sequence for cleavage of the spike glycoprotein (S) (Gombold *et al.*, 1993). C12 is also attenuated *in vivo* and has greatly reduced capacity to replicate in the liver and cause hepatitis (Hingley *et al.*,1994). We demonstrate here a further difference in pathogenesis between wild type virus and C12. Despite the fact that C12 replicates in the brain and causes encephalitis to a degree greater than or equal to that of wild type virus, C12 is a very weak inducer of demyelination.

It is presumed that the spike glycoprotein is a major determinant of viral tropism. Thus we compared the sequence of wild-type MHV-A59 and C12 to identify candidate amino acid substitutions that might effect pathogenesis. Aside from the cleavage site mutation (H716D), the S gene of C12 contained one, and only one, additional point mutation in S1, Q159L (Hingley *et al.*,1994). This mutation is located in the receptor binding domain of S1. This remains a likely candidate to effect pathogenic properties.

Earlier studies also demonstrated that viruses with the identical sequence of the S gene had different virulence and hepatotropism phenotypes. For example the mutant B11 appeared to be blocked in its spread from the central nervous system (CNS) to the liver, while the mutant C12 was inhibited in its ability to replicate in hepatocytes (Hingley *et al.*, 1994). The B12 mutant, while weakly hepatotropic like C12, retained its virulence. Thus, we concluded that these pathogenic properties must also involve genes outside of the spike. To determine which additional genes may influence pathogenesis, we sequenced the entire C12 genome. In addition to the two mutations found in the S gene, there were surprisingly few other substitutions, only three, all within the replicase gene. This demonstrates that mutations in the replicase gene, as well as in the S gene, are likely to affect pathogenesis.

3. METHODS AND MATERIALS

3.1. Virus

MHV-A59 was obtained originally from Dr. Lawrence Sturman (Albany, NY). C12 was isolated from a persistently infected glial cell culture at 16 weeks after infection and characterized as previously described (Hingley *et al.*, 1994; Gombold *et al.*, 1993).

3.2. Demyelination

C57BL/6 mice were inoculated intracranially with 10-fold serial dilutions of wild type or mutant MHV-A59, 5 mice/dilution. At 30 days post infection, infected mice were perfused with PBS, followed by 10% buffered formalin. Brains and spinal cords were removed, tissue was embedded in parafin, and sectioned for staining with luxol fast blue to detect plaques indicative of demyelination. For each of 5 mice at each dose of virus (or less if at a very high dose relative to the LD_{50}), demyelination was quantitated by examining 16–20 sections from three levels of spinal cord (Gombold *et al.*, 1995).

3.3. Genome Sequencing

Sequencing of the S gene was previously described (Hingley *et al.*, 1994). For sequencing of the rest of the genome, RT/PCR was carried out on cytoplasmic RNA extracted from L2 cell monolayers infected with either wild type MHV-A59 or C12. This approach did not include a cloning step and thus was designed to obtain a consensus sequence. MHV-A59-specific primers were used both to generate DNA templates for sequencing, and in sequencing reactions. Each fragment (about 600 base pairs) was sequenced in both directions. Automated sequencing was performed using the Taq dye terminator procedure according to the manufacturer's protocol (Taq DyeDeoxy Terminator Cycle Sequencing kit, Applied biosystems).

4. RESULTS

4.1. C12 Has a Greatly Reduced Ability to Induce Demyelination

We have demonstrated previously that C12 has the ability to demyelinate (Hingley *et al.*1994). However, in a more recent study using a low dose of virus, C12 appeared to demyelinate less well than wild type virus (Gombold *et al.*, 1995). To assess the difference in demyelination more quantitatively, we measured the amount of demyelination (as described in Materials and Methods) induced by wild type and C12 viruses as a function of dose of virus inoculated. As shown in Table 1 with a dose of 5000 PFU (approx $2LD_{50}s$) of wild type virus, all the animals infected had demyelination (5/5) and demyelination was present in approximately 63% of the spinal cord quadrants examined. At the same dose C12 cause no detectable

Table 1. Pathogenic properties of wild type (WT) MHV-A59 and the C12 mutant

Virus	Virulence (LD_{50})[1]	Brain replication/ encephalitis[2]	Demyelination[3]
WT	++++	++++	++++
C12	+	++++	+

[1] Virus was inoculated intracranially at 10 fold dilutions. LD_{50} values were calculated as previously described (Hingley *et al.*, 1994). C12 had an LD_{50} approx 2.7 log10 higher than wild type virus.

[2] Virus (5000 PFU or approx $2LD_{50}s$ of wild type virus) was inoculated intracranially; replication was assayed by plaquing of brain homogenates on L2 cell monolayers. Encephalitis was assayed by staining of tissue sections with hematoxylin and eosin and also staining for viral antigen all as previously described (Hingley *et al.*, 1994).

[3] Various dilution of virus were inoculated intracranially; demyelination was measured by luxol fast blue staining of brain and spinal cord sections as described in the text. At 5000 PFU, wild type caused extensive demyelination in all animals and demyelination was undetectable in C12 infected animals.

demyelination. Even at 100 fold higher doses of inoculation, demyelination was observed in 3/5 of C12 infected animals and in only 12.8 % of the quadrants. Thus, despite the fact that C12 replicates to high titer in the brain and causes encephalitis equivalent to wild type virus (Hingley *et al.*, 1994), it is a very poor inducer of demyelination.

4.2. Sequence of the C12 Genome and Comparison to the Wild Type MHV-A59 Genome

We have sequenced the entire 31 kb genome of C12. For sequencing our approach was seen to use RT-PCR to amplify regions of the C12 genome as described in the Materials and Methods section. We compared our C12 sequence with the published sequence (much of it from our lab) for wild type A59. When we found a difference we did a second independent cDNA synthesis and PCR amplification of the C12 genome to verify the C12 sequence. For each mutation we verified that the change was not also in the genome of our wild type virus from which the mutants were selected. Changes found in both mutant and wild type were eliminated from further study. The C12 genome had surprisingly few amino acid substitutions (five) compared to the wild type genome after 16 weeks of culture in glial cells. Table 2 shows the comparison of the two genomes.

With the exception of the two previously discussed mutations in the S gene, H716D and Q159L (see above), there are no amino acid substitutions in the deduced sequences of any of the structural proteins of C12 as compared to our wild type MHV-A59. This includes the S, M, N and Sm proteins as well as the ORFs for the I protein, encoded within, but out of frame with, the N gene (Fischer *et al.*, 1997) and the HE gene, which is present but not expressed in MHV-A59 (Luytjes *et al.*, 1988).

The deduced sequences of the nonstructural proteins encoded in ORFs 2a, 4 and 5a are also identical for C12 and wild type virus. In the 21 kb sequence of ORF1a and ORF1b of the replicase gene of the C12 genome, we have found three amino acid substitutions compared to wild type virus. There are two mutations in ORF1a (P1699S and M2196K) and one in ORF1b, (R1130S). This small number of mutations in the replicase gene of C12 are highly likely to play a role in pathogenesis.

We have also compared the non-coding regions of C12 and wild type MHV-A59. We have found no differences between C12 and wild type in the 5' or 3' non-coding regions.

Table 2. Sequence differences between the genomes of wild type MHV-A59 and the C12 mutant

Gene	Functional domain[1]	Amino acid/Ntd substitution[2]
Pol (ORF1a)	PLP-2[3]	P1699S
Pol (ORF1a)	Unknown	M2196K
Pol (ORF1b)	Helicase	R1330S
Spike (S1)	Receptor binding	Q159L
Spike (S1)	Cleavage site	H716D
Intergenic genes 5b/M	Promoter (?) Transcription (?)	CAAACto UAAAC

[1]Mutations were mapped to domains in the replicase gene using published sequences (Lee *et al.*1991; Bonilla *et al.*1994) as were the positions of the receptor binding domain of S (Kubo *et al.*1994) and intergenic sequences (Budzilowicz *et al.*1985).

[2]Amino acid substitutions are listed at the position within each open reading frame (ORF); the nucleotide change is shown for the intergenic region.

[3]Papain-like proteinase-2.

Among the intergenic sequences, there is only one nucleotide difference between C12 and wild type. In the intergenic region preceding the M gene of C12 the sequence is AAU-CUAAAC instead of AAUCCAAAC present in wild type MHV-A59.

5. DISCUSSION

5.1. Demyelination Phenotype

The mechanism of MHV induced demyelination is to this date still not understood. There is mounting evidence that both immune mediated components as well as direct infection of virus are factors in the establishment of demyelination (Houtman and Fleming, 1996). It has been shown that while infectious virus cannot be recovered from the CNS after the acute infection, viral RNA persists in the white matter of the CNS probably for the lifetime of the mouse (Lavi *et al.,* 1984; Adami *et al.,* 1995).

Since C12 was isolated from persistently infected glial cells, it might be expected that this mutant would have an enhanced ability to persist in the CNS and cause demyelination. Therefore, it was a somewhat surprising observation that C12 had lost the ability to induce demyelination efficiently despite the fact that they replicate to high titer in the brain and induce encephalitis at least to the level of wild type virus. Thus, in the case of MHV-A59, encephalitis does not necessarily lead to demyelination. Similar results were obtained for B11 another attenuated variant isolated from a parallel infection from C12 (data not shown).

Several other groups have reported mutants of the JHM strain of MHV, in which encephalitis and demyelination are also dissociated. Two examples are the attenuated JHM2.2-V-1 mutant (Fleming *et al.,* 1986; Fleming *et al.,* 1987) and the MHV-4 mutant 5A13.5 (Dalziel *et al.,* 1986; Fazakerley *et al.,* 1992). These mutants differ from C12 and B11 in that they cause little encephalitis compared to wild type but do demyelinate efficiently. The mutants we have selected from persistently infected glial cells (C12 and B11 for example) are the first examples of MHV isolates, to our knowledge, in which encephalitis occurs without subsequent demyelinating disease. We are currently investigating whether there is a difference in the ability of C12 RNA to persist in the brain.

5.2. Mapping of Pathogenic Properties

We have previously shown that C12 as well as the other fusion defective, weakly hepatotropic MHV-A59 mutants isolated from persistently infected glial cells have two amino acid substitutions relative to wild type in the gene encoding the spike glycoprotein S. The cleavage site amino acid substitution at position 716 correlates with fusogenicity (Gombold *et al.,* 1993). The other substitution Q159L appears to correlate with the loss of hepatotropism in B11, B12 and C12 (Hingley *et al.,* 1994). Experiments are underway to determine whether the demyelination phenotype exhibited by C12 and B11 also correlates with the Q159L mutation. The Q159L mutation is a good candidate to play a role in tropism and pathogenesis as it is the predicted receptor binding domain (Kubo *et al.,* 1994).We are currently carrying out experiments in collaboration with Dr. K. Holmes to determine if S protein bearing this mutation interacts differently with various forms of the MHV receptor as compared with wild type virus.

The observation that C12, B11 and B12 have different pathogenic properties but the same sequence of S suggests that mutations outside of S must influence the pathogenesis of MHV-A59. In the comparison of the C12 and wild type genomes, we found surprisingly few mutations considering the C12 was purified from the supernatant of a glial cell culture that had been persistently infected for 16 weeks. Other than the two amino acid substitutions found in the S gene of C12 and described above, there were no other differences between the wild type and C12 in any of the other structural proteins, M, N, Sm, HE or the I protein (Fischer *et al.,* 1997). There were several differences between C12 and the published sequences for A59 but all these differences were found in our wild type MHV-A59. For example sequence of HE in the genome of our wild type virus, as well as in the C12 genome, differs from the published sequence for MHV-A59 (Luytjes *et al.,* 1980) by a one nucleotide deletion resulting in a frame shift near the end of the ORF, resulting in the addition of 8 amino acids to the end of the ORF. Thus, our wild MHV-A59, has an additional mutation in the HE ORF; this is consistent with the notion that this protein is not essential in the MHV-A59 life cycle *in vitro* or *in vivo.*

While ORFs 2a, 4 and 5a, encoding non structural proteins were identical for C12 and wild type A59, there were three amino acid substitutions in the replicase gene. There are two mutations in ORF1a (P1699S and M2196K) and one in ORF1b (R1130S). The P1699S substitution is in (PLP-2), the second of two predicted papain-like proteinase domains and R1130S, is in the predicted helicase domain (Bredenbeek *et al.,* 1990, Lee *et al.,* 1991). It is difficult to predict whether the mutation in PLP-2 is likely to be significant. The B11 mutant, another attenuated, poorly demyelinating glial cell mutant also has an amino acid substitution, albeit a different one from C12, in PLP-2. Attempts in our lab so far to detect an activity for PLP-2 have not been successful. Thus if this PLP-2 is not an active proteinase, it may be a region of the replicase gene in which mutations do not have a phenotype. However if this is a necessary activity for replication these mutations may be significant. The C12 mutation in ORF 1b is in the helicase domain. The helicase activity has not yet been demonstrated from MHV but it is likely that such a polypeptide might interact with host cell proteins and thus influence tropism. The third mutations in ORF 1a, lies within a region that does not yet have a predicted function. This small number of mutations in the replicase gene of C12 are likely to play a role in pathogenesis.

We have found no differences between C12 and wild type in the 5' or 3' non-coding regions. With the exception of the intergenic region preceding the M gene there were also no differences between C12 and wild type MHV-A59 intergenic sequences. In C12 the sequence is AAUCUAAAC instead of AAUCCAAAC present in wild type MHV-A59. The AAUCUAAAC sequence observed preceding the M gene of C12 is the sequence found in several intergenic regions in the wild type genome (Budzilowicz *et al.,* 1985). Thus it is clear that this sequence can function as an intergenic region. It is however possible that this nucleotide change could effect the transcription of the C12 mRNA encoding the M gene.

We are currently attempting to use targeted recombination to introduce each of the two mutations (H716D and Q159L) in the S gene individually into the MHV-A59 genome (Fischer *et al.,* 1997) to determine whether they influence the pathogenic properties of C12, in particular the ability to induce demyelination. Our long term goal is to be able to use recombination to determine the phenotypes induced by the three amino acid substitutions in the replicase gene.

ACKNOWLEDGMENTS

This work was supported by public health service grant numbers NS-30606 (SRW) and NS-21954 (SRW) as well as grants RG-2585A4/1 (SRW) and RG-26151/2 (EL) from the National Multiple Sclerosis Society.

REFERENCES

Adami, C., Pooley, J., Glomb, J., Stecker, E., Fazal, F., Fleming, J. O., and Baker, S. C., 1995, Evolution of mouse hepatitis virus (MHV) during chronic infection: quasispecies nature of the persisting MHV RNA, *Virology* **209**: 337–346.

Bonilla, P. J., Gorbalenya, A. E., and Weiss, S. R., 1994, Mouse hepatitis virus strain A59 RNA polymerase gene ORF 1a: heterogeneity among MHV strains, *Virology* **198**: 736–740.

Bredenbeek, P.J., Pachuck, C.J., Noten, A.F.H., Charité, J., Luytjes, W , Weiss, S.R., and Spaan, W.J.M., 1990, The primary structure and expression of the second open reading frame of the polymerase gene of the coronavirus MHV-A59; a highly conserved polymerase is expressed by an efficient ribosomal frameshifting mechanism, *Nucleic Acids Res.* **18**:1825–1832.

Budzilowicz, C. J., Wilczynski, S. P., and Weiss, S. R., 1985, Three intergenic regions of coronavirus mouse hepatitis virus strain A59 contain a common nucleotide sequence that is homologous to the 3' end of the viral mRNA leader sequence, *J. Virol.* **53**: 834–840.

Dalziel, R. G., Lampert, P. W., Talbot, P. J., and Buchmeier, M. J., 1986. Site-specific alteration of murine hepatitis virus type 4 peplomer glycoprotein E2 results in reduced neurovirulence, *J. Virol.* **59**:463–471.

Fazakerley, J. K., Parker, S. E., Bloom, F., and Buchmeier, M. J., 1992, The V5A13.1 envelope glycoprotein deletion mutant of mouse hepatitis virus type-4 is neuroattenuated by its reduced rate of spread in the central nervous system, *Virology* **187**: 178–188.

Fischer, F., Peng, D., Hingley, S. T., Weiss, S. R., and Masters, P. S., 1997, The internal open reading frame within the nucleocapsid gene of mouse hepatitis virus encodes a structural protein that it not essential for viral replication, *J. Virol.* **71**: 996–1003.

Fleming, J. O., Trousdale, M. D., Bradbury, J., Stohlman, S. A., and Weiner, L. P., 1987, Experimental demyelination induced by coronavirus JHM (MHV-4): molecular identification of a viral determinant of paralytic disease, *Microbial Pathogenesis* **3**: 9–20.

Fleming, J. O., Trousdale, M. D., El-Zaatari, F. A. K., Stohlman, S. A., and Weiner, L. P., 1986, Pathogenicity of antigenic variants of murine coronavirus JHM selected with monoclonal antibodies, *J. Virol.* **58**: 869–875.

Gombold, J. L., Hingley, S. T., and Weiss, S. R., 1993, Fusion-defective mutants of mouse hepatitis virus A59 contain a mutation in the spike protein cleavage signal, *J. Virol.* **67**: 4504–4512.

Gombold, J. L., Sutherland, R. M., Lavi, E., Paterson, Y., and Weiss, S. R., 1995, Mouse hepatitis virus A59-induced demyelination can occur in the absence of CD8+ T cells. *Microbial Pathogenesis* **18**: 211–221.

Hingley, S. T., Gombold, J. L., Lavi, E., and Weiss, S. R., 1994, MHV-A59 fusion mutants are attenuated and display altered hepatotropism, *Virology* **200**: 1–10.

Houtman, J. J., and Fleming, J. O.,1996, Pathogenesis of mouse hepatitis virus-induced demyelination, *J. Neurovirol.* **2**: 361–376.

Kubo, H., Yamada, Y. K., and Taguchi, F., 1994, Localization of neutralizing epitopes and the receptor binding site within the amino-terminal 330 amino acids of the murine coronavirus spike protein, *J. Virol.* **68**: 5404–5410.

Lavi, E., Gilden, D. H., Highkin, M. K., and Weiss, S. R.,1984, Persistence of MHV-A59 RNA in a slow virus demyelinating infection in mice as detected by *in situ* hybridization, *J. Virol.* **51**: 563–566.

Lee, H. J., Shieh, C. K., Gorbalenya, A. E., Koonin, E. V., LaMonica, N., Tuler, J., Bagdzhadzhyan, A., and Lai, M. M. C., 1991, The complete sequence of the murine coronavirus gene 1 encoding the putative protease and RNA polymerase, *Virology* **180**: 567–582.

Luytjes, W., Bredenbeek, P. J., Noten, A. F., Horzinek, M. C., and Spaan, W. J. M., 1988, Sequence of mouse hepatitis virus A59 mRNA 2: indications for RNA recombination between coronaviruses and influenza C virus, *Virology* **166**: 415–422.

82

HUMAN MACROPHAGES ARE SUSCEPTIBLE TO CORONAVIRUS OC43

Arlene R. Collins

Department of Microbiology
School of Medicine and Biomedical Sciences
State University of New York at Buffalo
Buffalo, New York 14214

1. ABSTRACT

Adherent adult and cord blood macrophages were infected with human coronavirus OC43 at a multiplicity of 1–1.5 and washed twice to remove unbound virus. Virus progeny was detected in the supernatant on day 1 and peaked at 2–3 days at an average titer of $5\pm3.9 \times 10^6$ pfu/ml from seven samples. Viral RNA was detected by nested set RT-PCR in infected macrophages incubated for 48 hr. Intracellular viral nucleocapsid was detected in 15% of the cells and surface staining for viral spike antigen was observed using monoclonal antibodies. Amplification of infectious virus and detection viral RNA and antigen synthesis in macrophages *in vitro* indicates susceptibility to OC43 virus.

2. INTRODUCTION

Human coronaviruses are the second most frequent viruses after the rhinoviruses, isolated from individuals with common colds. Two major serotypes, 229E and OC43 are found. Very little is known about the cytopathology of OC43 virus in the nasopharyngeal epithelium because of its fastidious growth requirements (McIntosh,1996). In human tracheal organ cultures, OC43 shows a slow patchy destruction of ciliated epithelial cells that is reflected in a loss of beating cilia. It is likely that this destructive effect has a role in the pathogenesis of disease caused by human coronaviruses. Beyond this however, there is little information concerning disease mechanisms (Tyrrell and Bynoe, 1965). OC43 causes a persistent infection in U87-MG glioblastoma cells (Collins, 1987). The purpose of this work is to determine the susceptibility of primary human monocytes/macrophages to infection with OC43.

Coronaviruses and Arteriviruses, edited by Enjuanes *et al.*
Plenum Press, New York, 1998

3. MATERIALS AND METHODS

3.1. Cells and Viruses

Primary monocyte/macrophage cells were isolated from six cord blood samples and one adult peripheral blood. Samples were collected in vacutainer tubes with EDTA (Becton Dickinson). The leucocyte layer was separated by gradient centrifugation at 800xg through Nycoprep solution (Accurate Chemical) and cultured in RPMI 1640 medium (GIBCO/BRL) with 10%fetal bovine serum (fbs) (Atlanta Biologicals), 2mM glutamine and 0.05mg/ml Gentamycin. Nonadherent cells were removed during exchange with fresh medium. BAES-2B cells, derived from adenovirus transformed human bronchial epithelium, were grown and maintained in LHC-9 serum-free defined medium (Becker et al.,1993). After five to seven days in culture, cells were infected with a multiplicity of 1–1.5 pfu per cell in 25 cm^2 flask containing 1–1.5 × 10^6 cells. After adsorption at 37°C, for 1 hr, cells were washed twice with medium alone, and the first sample of medium was taken for virus assay. Samples were taken daily and stored at -70°C. Virus yield was determined by plaque assay on MRC-5 human diploid lung cell monolayers (ViroMed, Minneapolis MN) and expressed as plaque forming units per milliliter (pfu/ml) (Collins, 1995).

3.2. Detection of Viral RNA

RNA was isolated from 1 × 10^6 infected and uninfected cells using PUREscript RNA isolation system (Gentra). For RT-PCR, the nested PCR assay of Myint et al. (1994) was used except that for first strand cDNA synthesis in which downstream primer 5'TGTGATTCTTCCAATTGGCC3' (Kamahora et al, 1989) was used.

3.3. Immunocytochemistry

Macrophages were placed in ice cold, phosphate buffered saline 0.1M, pH 7.2 (PBS) for 15 min followed by scraping and resuspension in fresh medium. Cells were seeded in 60 mm Petri dishes containing glass coverslips. After attachment, the cells were infected with virus at a multiplicity of 1–1.5. After five days, infected cells were washed twice in PBS, fixed in cold acetone for 10 min, air dried and stained with monoclonal antibody (mAb) 4B6.2 (1:500) to viral nucleocapsid antigen, followed by horse anti-mouse antibody conjugated to FITC (1:50,Vector, Burlingame CA).

For cell surface antigens, infected cells after washing in PBS, were incubated with rat mAb I16 to OC43 spike antigen (1:.25, J Vautherot, INRA, Juoy-en-Josas, France) or mAb B1–1G6 to β2 microglobulin, light chain of HLA class I (Immunotech, 1:50), followed by biotin-conjugated goat anti-mouse IgG(1:40, Vector), then by avidin FITC (1:50, Vector). Cells were post-fixed in methanol/acetone 1:1 for 10 min, visualized at 400x magnification in a Leitz Dialux20 UV fluorescence microscope and photographed using Kodak P3200 black and white film. The number of fluorescent cells was evaluated in 200–500 cells per culture.

4. RESULTS

Monocytes/macrophages recovered from healthy donor blood were infected with OC43 virus. Virus replication was monitored by plaque assay of the culture supernatant. In Table 1, the virus titers from monocyte/ macrophage cultures obtained from seven donors,

Table 1. Susceptibility of human cord blood macrophage cultures
to coronavirus OC43

	Virus titer, PFU/ml of supernatant medium on day			
Sample	0^c	1	2	3
214	1.2×10^2	–	1.8×10^7	–
217	2.8×10^2	3.9×10^4	2.2×10^5	5.2×10^5
218	2.2×10^2	4.5×10^4	1.2×10^6	5.7×10^5
226^a	–	1.0×10^6	1.5×10^6	3.0×10^6
231	–	7.5×10^5	2.5×10^6	3.8×10^5
239	1.6×10^2	–	3.2×10^6	2.0×10^6
241^b	–	–	6.3×10^6	1.5×10^5

[a] Cells were infected at a multiplicity of 1.5; all others were infected at a multiplicity
of 1.

[b] Adult peripheral blood monocytes; all the other cells are from cord blood.

[c] Supernatant medium was sampled after washing the monolayer twice to remove unbound virus.

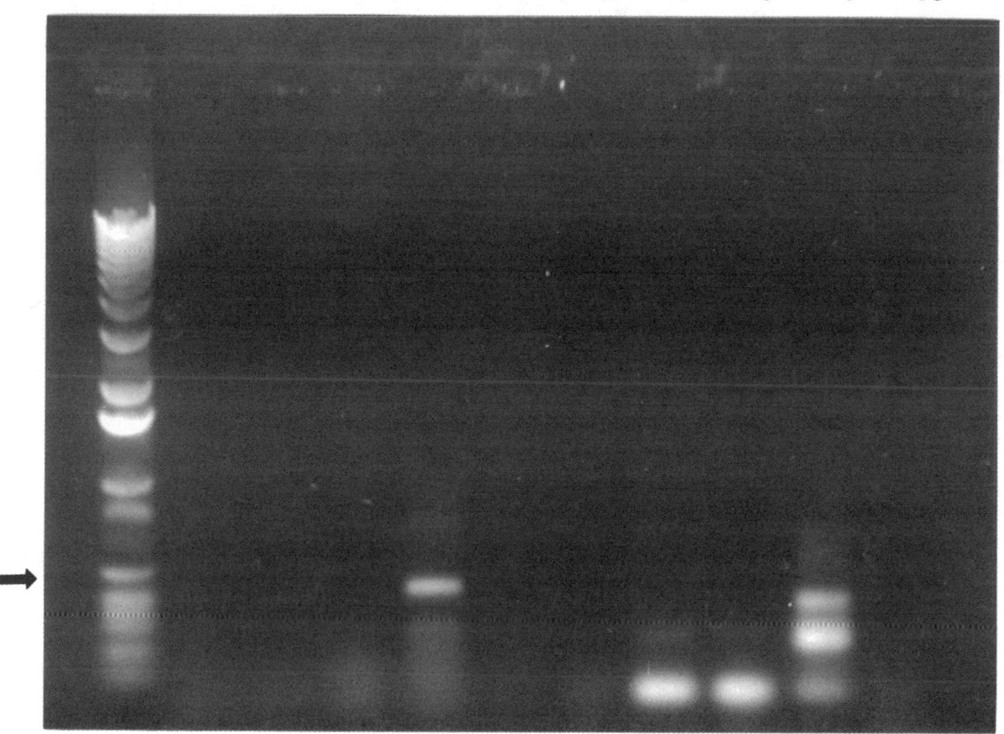

Figure 1. Agarose gel electrophoresis of products from RT-PCR. On left, molecular size markers, arrow shows
500 bp. Lanes 1 and 5, first round products. Lanes 6 and 10, second round products. Lanes 1 and 6 uninfected
macrophages. Lanes 2 and 7 infected macrophages. Lanes 3 and 8 infected BAES-2B cells Lanes 4 and 9 infected
MRC-5 cells. Lanes 5 and 10, uninfected MRC-5 cells.

six cord blood and one adults are shown. After 48hr, an average of $5\pm3.9 \times 10^6$ infectious virus particles per ml of supernatant medium was consistently recovered.

RT-PCR was applied to monocyte/macrophage cells from the eighth donor cord blood after infection with OC43 virus and incubation for 48 hr. Similarly infected MRC-5 and BAES-2B cells and uninfected macrophages were used as controls. Figure 1 shows that RNA from infected MRC-5 cells was positive for coronavirus in the first PCR cycle and gave a product of the expected size, 450 base pairs (bp). Macrophages and BAES-2B cells were positive after the second cycle. The product was a 100 base pair fragment, as expected. The second amplification from MRC-5 cells gave two additional products, one of 450bp, identical to that seen in the first round and another of 300bp which was observed previously (Myint et al. 1994).

After immunocytochemical staining, microscopic observation and counting showed that 15% of macrophages that stained positive with anti-nucleocapsid mAb on day five. Nucleocapsid staining was a punctate or granular cytoplasmic fluorescence, Figure 2A. Macrophages stained with secondary antibody alone did not give this appearance, Figure 2B. Viral spike antigen on the surface of infected monocytes was also detected (Figure 2C). Unstained cells lacking spike antigen were also visible. Figure 2D shows that cells stained fairly uniformly positive for surface β2m. Presence of surface staining for OC43 spike corroborated the intracytoplasmic staining for nucleocapsid antigen.

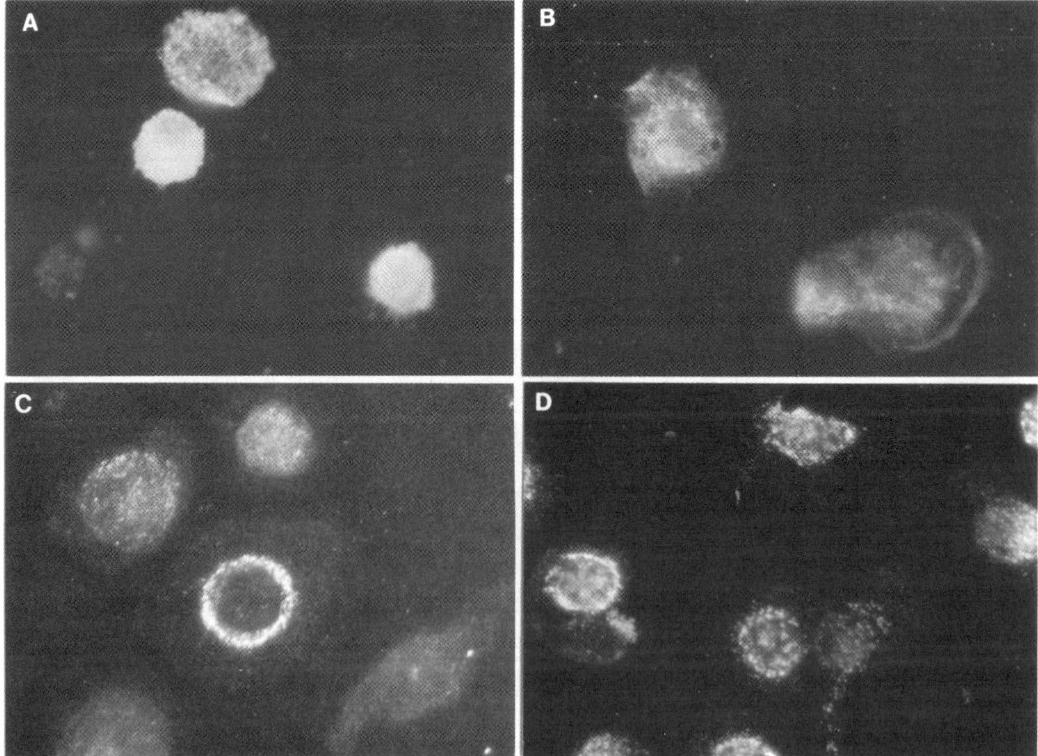

Figure 2. Immunofluorescent staining of OC43 infected monocyte/macrophages. A. day five after infection, stained with anti-nucleocapsid mAb. Cytoplasmic granules are stained. B. Same as A, except stained with secondary antibody, horse anti-mouse IgG (H and L). C. day three after infection, stained as unfixed cells with anti-spike mab. D. Same as C, except stained with anti-β2m mAb. Magnification 400x.

5. DISCUSSION

Plaque assay, RT-PCR and immunocytology strongly indicate that OC43 virus replicated in human monocyte/macrophage cells *in vitro*. The estimated number of cells with intracytoplasmic nucleocapsid antigen was 15%, suggesting that a subpopulation of macrophages was infected. Susceptibility of a subset of macrophages to respiratory syncytial virus has also been demonstrated (Midulla et al., 1989).

OC43 causes the common cold with cough and hyperemia, and reinfection is common. OC43 also can infect neurons (Collins, 1995; Bonavia et al., 1996) and viral RNA was found in the brain of multiple sclerosis patients (Talbot, 1997). These are reasons to look further into the pathogenesis of OC43 virus.

REFERENCES

Becker, S., Koren, H.S., and Henke, D.C., 1993, Interleukin-8 expression in normal nasal epithlelium and its modudaltion by infection with respiratory syncytial virus and cytokines tumor necrosis factor, interleukin-1 and interleukin-6, *Amer. J. Resp. Cell. Mol. Biol.* **8**: 20–27.

Bonavia, A., Arbour, N., Yong, V.W., Talbot, P.J., 1997, Infection of primary cultures ofhuman neural cells by human coronavirus 229E and OC43, *J. Virol.* **71**: 800–6.

Collins, A.R., and Sorenson, O., 1986, Regulation of viral persistence in human glioblastoma and rhabdosarcoma cells infected with coronavirus OC43, *Microb. Path.* **1**: 573–582.

Collins, A.R., 1995, Interferon gamma potentiates human coronavirus infection of neuronal cells by modulation of HLA class I, *Immunol. Invest.* **24**: 977–986.

Kamahora, T., Soe, L.H., and Lai, M.M.C., 1989, Sequence analysis of nucleocapsid gene and leader RNA of human coronavirus OC43, *Virus Res.* **12**: 1–9.

Mc Intosh, K., 1996, Coronaviruses, in: *Fundamental Virology*, (B.N.Fields, D.M. Knipe, R.M. Chanok, M. S. Hirsch, J.L. Melnick, T.P. Monath, B. Roizman, eds.), Raven, Philadelphia, pp.1093–1103.

Midulla, F., Huang, Y.T., and Gilbert, I.A., 1989, Respiratory syncytial virus infection of human cord and adult blood monocytes and alveolar macrophages, *Amer Rev. Res. Dis.* **140**: 771–777.

Myint, S., Johnson,S., Sanderson,G., and Simpson, H., 1994, Evaluation of nested polymerase chain methods for the detection of human coronavirus 229E and OC43, *Molec. Cell. Probes* **8**:357–361.

Talbot, P.J. Cote G., Arbour, N., 1997, Human coronavirus OC43 and 229E persistence in neural cell cultures and human brains, *Adv. Exp. Med. Biol.* In press.

Tyrrell, D.A.J., and Bynoe, M.L., 1965, Cultivation of a novel type of common cold virus in organ culture, *Br. Med. J.* **1**: 1467–1470.

83

TRANSMISSIBLE GASTROENTERITIS CORONAVIRUS CARRIER SOW

Roger D. Woods and Ronald D. Wesley

Virology Swine Research Unit
National Animal Disease Center
USDA, Agricultural Research Service
2300 Dayton Avenue, P.O. Box 70
Ames, Iowa 50010

1. ABSTRACT

A sow infected with virulent transmissible gastroenteritis virus (TGEV) shed virulent virus in her feces for 18 months. The virus was isolated from rectal swabs beginning 2 days postexposure (PE) and continued at irregular intervals. Virus shedding was detected on 24 separate occasions. The titer of the virus shed ranged from $<1 \times 10^2$ pfu/ml to 7.2×10^3 pfu/ml, while the duration of the shedding ranged from 1 to 5 consecutive days. Inoculation of 3-day-old piglets with TGEV isolated from the sow proved the virus was virulent throughout the study. Virulent TGEV was isolated from the spleen, mesenteric lymph nodes, and the liver of the sow 544 days PE. This study demonstrates an apparently healthy sow can be a reservoir and shed virulent TGEV for an extended period of time.

2. INTRODUCTION

Transmissible gastroenteritis (TGE) is an economically important disease of swine (Mousing *et al.*, 1988). While considerable information is available on the virus and the disease process, significant information is lacking on natural virus reservoirs, maintenance of the virus through non-seasonal periods, and a vaccine that protects greater than 90% of piglets nursing vaccinated sows and challenged with virulent TGE virus (TGEV) (Saif and Wesley, 1992). Dogs, foxes, flies, and starlings have been shown to be mechanical carriers of the TGEV for periods up to 14 days, but they are not believed to be natural reservoirs or involved in maintenance of the virus through non-seasonal periods (Gough and Jorgenson, 1983; Pilchard, 1965; Haelterman, 1962). Herds with enzootic TGE (sows and feeder

Coronaviruses and Arteriviruses, edited by Enjuanes *et al.*
Plenum Press, New York, 1998

pigs) are the most likely source or reservoir for virulent TGEV (Pritchard, 1985; Morin, 1978 *et al.*). The addition of susceptible animals to the herd (either through farrowing or new breeding stock) would assist in maintaining the virus throughout the year (Maes and Haelterman, 1979). Chronically infected swine also offer the potential of being reservoirs for the virus. Using pigs less than 7 days of age for virus isolation, Kemeny (1978) and Cook *et al.* (1991) detected virulent TGEV in pharyngeal swabs, muscle, and lymph node specimens collected from sows at local abattoirs. In experimentally infected pigs, virus can be recovered from the lungs and small intestines up to 14 days PE (Pensaert *et al.*, 1970; Kemeny, 1978). Under experimental conditions, virus has been isolated from intestinal tissues up to 8 weeks PE (Lee, 1954) and from lung tissue up to 104 days PE (Underdahl *et al.*, 1975). In contrast, Wiseman et al (1988) added sentinel pigs to a herd 3, 4, and 5 months after a TGEV outbreak and none of the pigs developed clinical signs of an infection or seroconverted to TGEV. Harris *et al.* (1987) reported the spread of virulent virus could be stopped in an infected herd by a "space" between swine groups. The space was designed to allow a thorough cleaning of the contaminated facilities and all associated areas and then leaving the pens free of swine for about 1 month. After the porcine respiratory coronavirus (PRCV) became widespread in the European swine population, epizootic TGE decreased (Pensaert and Cox, 1989). However, in the U.S. swine population, the PRCV has not spread as widely and therefore a similar reduction in outbreaks of TGE has not been observed.

The purpose of this report is to present information on a long term carrier sow from which TGEV was isolated for a period of one and a half years.

3. MATERIALS AND METHODS

3.1. Cells and Virus

The stock strain of TGEV, virulent Miller #3, was used in this study. Its history and growth characteristics have been previously reported (Frederick and Bohl, 1976). The McClurkin swine testicular (ST) cell line was used for virus isolations, to determine viral plaque titers (VT) and 50% virus plaque reduction titers (VN) (Woods *et al.*, 1988).

3.2. Animals and Experimental Protocol

One-year-old sows (30 & 32), sero-negative for TGEV antibodies by VN, were contact challenged (nursing piglets exposed to virulent Miller #3 TGEV). Both sows were given two additional doses [1×10^6 plaque forming units (pfu)/dose] of virulent TGEV: the first oral/nasally (O/N) 3 weeks PE and the second intravenously (IV) 6 weeks PE. Serum VN titers were determined monthly. Rectal swabs were taken daily from each sow for 3 months. Rectal swabs were taken 5 to 7 times a week from sow 30 for an additional 1 year and 3 months. After collection, the swabs were placed in a snap cap vial containing 1 ml of cold cell culture medium and 5% v/v swine sera containing antibodies against rotaviruses and enteroviruses, placed in an ice bucket, and returned to the lab for processing.

Multiple tissue samples for virus isolation were collected from sow 30 544 days PE. Samples included tonsils, lungs, heart, liver, spleen, lymph nodes draining mammary glands, mammary tissues, muscle, mesenteric lymph nodes, ovaries, kidneys, large and small intestines, and cecum. Each tissue sample was weighed and a 10% suspension (w/v) was made in tissue culture medium. The material was placed in a glass beaker surrounded with ice, minced with scissors, and homogenized. The material was centrifuged at 4000 x g for 5 minutes to remove cell debris and the supernatant fluids were dispensed in 1.5 ml aliquots and frozen at -75° C.

3.3. Virus Isolation

The swabs were wrung, fluids transferred to Eppendorf tubes, and centrifuged at 4000 x g for 5 minutes at 4 C to pellet debris. One hundred μl of fluid from each swab and processed tissue specimen was inoculated onto confluent cultures of ST cells in 60 × 15 mm dishes, incubated at 37° C for 1 hr, inoculum removed, and the cell sheet flooded with 3 ml of culture medium. The cultures were returned to the CO_2 incubator and monitored twice a day for a cytopathic effect (CPE). If CPE was observed, the culture was frozen at -75° C and held for virus identification. Isolation of TGEV was confirmed by a monoclonal antibody based indirect fluorescent antibody (IFA) staining procedure and the VN assay (Woods, 1988). In the absence of CPE by 48 hrs, a second pass (100 μl of culture fluids) was made in ST cells. If a third pass was necessary, it was made on a 4-chambered slide confluent with ST cells. The slide culture was incubated for 2 days, fluids were removed, and the cells fixed with acetone/methanol. Two chambers of the slide were stained by the direct FA procedure with a fluorescein labeled conjugate prepared with hyper immune serum from a sow exposed to the virulent Miller #3 TGEV and two chambers were stained by the IFA procedure with MAb 4F6 (Woods *et al.*, 1988) and stained with goat anti-mouse fluorescein labeled $F(ab)_2$.

3.4. Virulence of Virus Isolated from Swabs and Tissue Samples

All animal experiments were conducted in an AAALAC-certified laboratory. Thirty-two caesarian-derived colostrum-deprived (CDCD) piglets from sows sero-negative for TGEV antibodies by VN were obtained from the Animal Resources Unit of the National Animal Disease Center, Ames, Iowa. The piglets were housed in individual plexiglass isolators in a room maintained at 34 to 36° C under slightly negative air pressure (11 air changes/hour). The pigs were fed a sow's milk replacer three times/day (piglets initially received 180 ml/day which was increased to 560 ml/day by the time they were 2 weeks old). At 3 days of age, 24 of the pigs were used for virulence testing of the tissue samples. Each piglet was given 25 ml of homogenized tissue fluids (2.5 grams equivalents) via stomach tube. The piglets were observed four times a day for clinical signs of disease. Piglets exhibiting clinical signs typical of a virulent TGEV infection (vomiting, diarrhea, fever, inappetence, dehydration, and rough hair coat) were euthanized and a portion of the ileum was removed for virus isolation. At 21 days PE, surviving piglets were bled and their TGEV VN titer was determined.

Eight piglets were used to test the virulence of TGEV isolated from the rectal swabs. Virus from the 3rd, 6th, 9th, 12th, 15th, 18th, 21st, and 24th isolates were tested. Each piglet was given 1×10^6 pfu of virus from the isolate's second cell culture pass. The piglets were observed four times a day for clinical signs of disease. Piglets with clinical signs typical of a virulent TGEV infection were euthanized and a portion of the ileum was removed for virus isolation.

4. RESULTS

Clinical signs of a TGEV infection (off feed on PE days 2, 3, and 4 and an elevated temperature on PE days 1, 3, and 4) were observed in sow 32. By 5 days PE sow 32 was

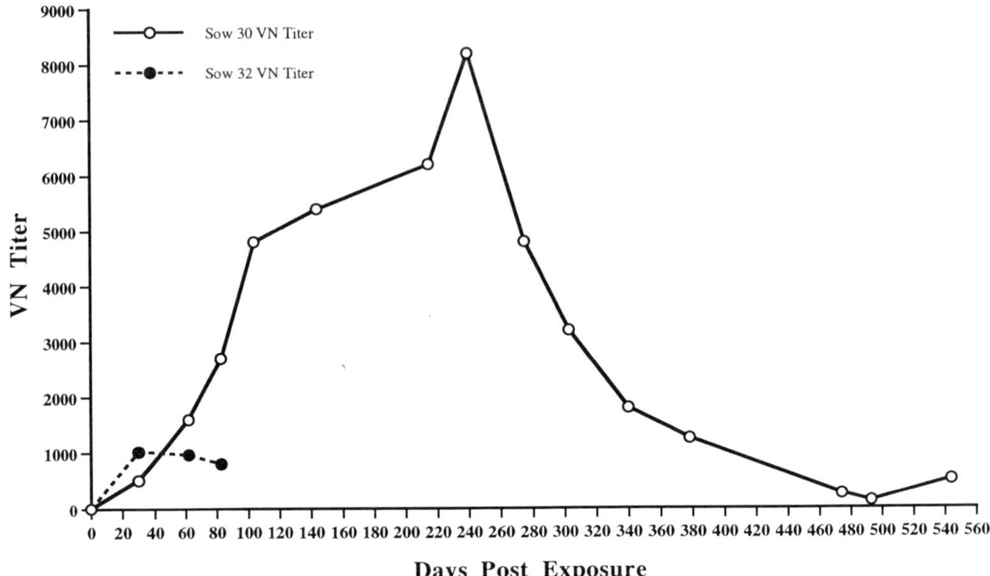

Figure 1. Fifty percent VN plaque reduction titers of sow 30 and sow 32 against Miller #3 TGEV. Sows oral/nasally exposed 21 days PE (1 × 10⁶ pfu) and intravenously exposed 42 days PE (1 × 10⁶ pfu). Sow 32 euthanized 81 days PE. Sow 30 euthanized 544 days PE.

free of clinical signs and remained healthy until the end of the experiment. Clinical signs of a TGEV infection (inappetence on PE days 2, 3, 4 and 5, diarrhea on PE days 2 and 3, an elevated temperature on PE days 4 and 5, and a rough hair coat for approximately three months) were observed in sow 30. By 6 days PE, sow 30 was free of clinical signs (except hair coat) and remained healthy for the duration of the experiment.

Serum VN titers for TGEV were elevated in both sows following multiple exposures (Fig. 1). Sow 32 had a VN titer of 1:800 56 days PE. However, at 80 days PE, her VN titer had declined to 1:640 and she was euthanized to obtain a large quality of hyper immune sera. Sow 30's VN titer reached a peak of 1:8200 by 8 months PE. During the next 8 months, her VN titer declined to a low of 1:128 at 493 days PE. At the time she was euthanized (544 days PE), her VN antibody titer was 1:512.

TGEV was isolated from 3 of 90 rectal swab samples taken from sow 32. The virus titer was 3 × 10³ pfu/ml on day 2 and <1.0 × 10² pfu/ml on days 4 and 5. TGEV was isolated from rectal swabs taken from sow 30 beginning 2 days PE and continued at irregular intervals for a period of 18 months (Fig. 2). The titer of virus shed and duration of the shedding was sporadic. Beginning 80 days PE, virus was detected in cell culture on 24 separate occasions. The number of days virus was isolated varied from a minimum of 1 day (9 of the 24) to a maximum of 5 days (Fig. 2). On four occasions TGEV was isolated for 2 days, and on one occasion 1 day separated two isolations.

Piglets given TGE virus isolated from rectal swabs collected from sow 30, 80 to 544 days PE, developed clinical signs typical of a virulent TGEV infection within 72 hrs. To prevent pain and distress, the piglets were euthanized within 24 hrs after developing severe clinical signs (inappetence for 24 hrs and dehydration). TGEV was isolated from the ileum of each exposed piglet. Piglets given homogenized tissues samples of liver, spleen, and mesenteric lymph nodes from sow 30 developed clinical signs typical of a virulent

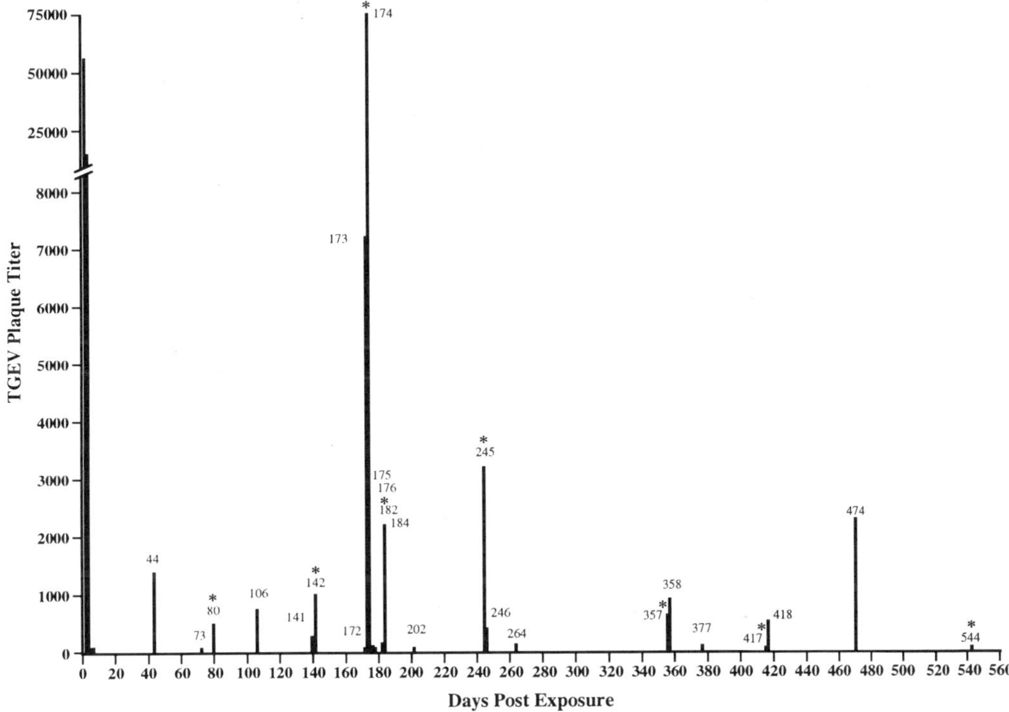

Figure 2. Isolation of TGEV from a long term carrier sow from rectal swabs. Virus isolations were attempted 5 to 7 times per week. Days PE as listed on peaks are the dates virus was isolated. Height of bar indicates number of plaque forming units per ml isolated (range= <100 to 7200). * = virus virulent for 3-day-old CDCD piglets (all isolates tested were virulent).

TGEV infection within 48 hrs PE that increased in severity until the piglets were euthanized. TGEV was isolated from the ileum of each euthanized piglet. Piglets given homogenized tissue samples from the other 21 locations remained clinically normal. Serum samples from piglets surviving 21 days were negative for TGEV VN antibodies.

5. DISCUSSION

During production of hyper immune antisera in sows against virulent TGEV, differences were observed in the clinical and antibody responses of the sows. Sow 30 developed clinical signs of diarrhea, transient fever, inappetence, and a rough hair coat. In addition, she had a rising VN titer and virus was sporadically isolated from rectal swabs beyond the usual 10 to 14 days PE. In contrast, sow 32 developed mild clinical signs of a TGEV infection and a moderate VN titer. The reason(s) for the difference in the sows responses are unknown. Some possibilities include susceptibility of the sows, amount of virus each animal received, the duration of exposure to the virus, and the health of the sow at the time of exposure. Because the sows were originally exposed through contact with their TGEV challenged piglets, it is not possible to define any of the initial exposure parameters other than the natural O/IN exposure route. Because sow 30 had a rising VN antibody response and sporadically shed TGE virus for more than 14 days after her initial exposure, she was monitored to determine the duration of viral shedding.

Following exposure to virulent virus, both sows developed TGEV VN antibodies. By 80 days PE, sow 32 had a waning TGEV VN titer and was exsanguinated. In contrast, the VN titer of sow 30 continued to rise for the next 8 months. Based on the sporadic isolation of TGE virus from this sow, it is probable that she received additional antigenic stimuli on several occasions. Her highest VN titer of 1:8200 was detected 240 days PE. By 303 days PE, her VN titer had declined to 1:3200 and continued to decline until a low of 1:128 was recorded at PE day 493. One factor contributing to this decline may have been the loss of the antigenic stimuli. During the months of June, July, and August (264 to 357 days PE), virus was not isolated from this sow. Thus, lacking the antigen stimuli, the sow's immunological system reduced the production of antibodies for TGEV. After TGEV was recover from the sow 357 days PE, her VN titer slowly increased. At the end of the experiment, the VN titer was 1:512.

All TGEV isolates from rectal swabs tested for virulence in young piglets produced clinical signs of a virulent TGEV infection. TGEV isolated from sow 30 544 days PE was as virulent for young piglets as the original Miller #3 virus. Because this sow was maintained in isolation and never bred, it was not possible to determine if piglets nursing this animals or other swine in contact with her would have seroconverted with antibodies to the TGEV or developed a TGEV infection.

In this study, it is uncertain if the additional O/N and IV doses of TGEV had any effect on establishing the carrier and shedder patterns observed in sow 30. The doses were given to increase the serum VN titers of the sows. Sow 32 remained free of clinical signs and virus was not isolated from any rectal or pharyngal swab collected after the initial exposure. However, sow 30 was shedding TGEV and had mild clinical signs (weight loss, rough hair coat) of an infection at the time of the second O/N dose and the third IV dose. Additionally, the sows had TGEV VN antibodies that would have formed an antigen/antibody complex with new TGEV and this complex should have been destroyed by the sows immune system. *In vitro* studies have demonstrated alveolar macrophages support the growth of attenuated TGEV (Laude *et al.*, 1984). Thus, it is possible that the lymphatic system of sow 30, but not sow 32, became infected during the initial O/N exposure and the virus remained in the system for the duration of the experiment. Support for this concept occurs in the post-mortem evaluation of sow 30 as virulent TGE virus was isolated from the spleen, liver and mesenteric lymph nodes but not from the kidneys. Woods (1978) demonstrated a persistent TGEV infection could be established in a leukocyte cell line and infectious TGEV was chronically shed from the infected cells into the culture medium. Underdahl isolated TGEV from an experimentally infected animal 3 months PE (Underdahl *et al.*, 1975). This study extends that time interval to 18 months.

These results do not agree with those of Wiseman *et al.* (1988) and Harris *et al.* (1987). They reported that infected animals did not shed TGEV beyond 30 days after clinical signs disappeared. Although seroconversion was not detected in their sentinel animals, the potential of intermittent virus shedding was not eliminated. Their results failed to report the contact of the sentinels to the feces (slatted floors or solid floors) and the season of the year. In our study two variations were noted, sow 30 was maintained on a solid floor and TGEV was not isolate during June, July and August. Peak virus shedding was detected during late winter (172 to 184 PE) which agrees with most published literature concerning seasonality of the disease (Saif, 1992). While this report involves a single sow, the results demonstrated an apparently healthy sow can maintain virulent TGEV for an extended period of time and intermittently shed infectious virus to the environment. These findings indicate the lymphatic system of swine seropositive for TGEV could be reservoirs for the virus.

REFERENCES

Cook D., Hill H., and Taylor J., 1991, Oral transmission of transmissible gastroenteritis virus by muscle and lymph node from slaughtered pigs, *Aust. Vet. J.* **68**:68–70.

Frederick G.T., and Bohl E.H., 1976, Local and systemic cell-mediated immunity against transmissible gastroenteritis, an intestinal viral infection of swine, *J. Immunol.* **116**:1000–1004.

Gough P.M., and Jorgenson R.D., 1983, Identification of porcine transmissible gastroenteritis virus in house flies (Musca domestica Linneaus), *Am. J. Vet. Res.* **44**:2078–2082.

Haelterman E.O., 1962, Epidemiological studies of transmissible gastroenteritis of swine, in Proceeding,. *U.S. Livest. Sanit. Assoc.* **1962**:305–315.

Harris D.L., Bevier G.W., and Wiseman B.S., 1987, Eradication of transmissible gastroenteritis virus without depopulation, *Proc. 18th Am. Assoc. Swine Pract.* **18**:557–561.

Kemeny L.J., 1978, Isolation of transmissible gastroenteritis virus from pharyngeal swabs obtained from sows at slaughter, *Am. J. Vet. Res.* **39**:703–705.

Kemeny L.J., Wiltsey V.L., and Riley J.L., 1975, Upper respiratory infection of lactating sows with transmissible gastroenteritis virus following contact exposure to infected piglets, *Cornell Vet.* **65**:352–362.

Laude H., Charley B., and Bonnardiere C., 1984, Interactions of porcine enteric coronavirus TGEV with macrophages and lymphocytes, *Adv. Exp. Med. Biol.* **173**:385–386.

Lee K., Moro M., and Baker J., 1954, Transmissible gastroenteritis in pigs, *Am. J. Vet. Res.* **15**:364–372.

Maes R.K., and Haelterman E.O., 1979, A sero-epizootiological study of five viruses in a swine-evaluation station, *Am. J. Vet. Res.* **40**:1642–1645.

Morin M., Solorzano R.F., Morehouse L.G., and Olson L.D., 1978, The postulated role of feeder swine in the perpetuation of the transmissible gastroenteritis virus, *Can. J. Comp. Med.* **42**:379–384.

Mousing J., Vagsholm I., Carpenter T.E., Mousing J., Gardner I.A., and Hird D.W., 1988, Financial impact of transmissible gastroenteritis in pigs, *J. Am. Vet. Med. Assoc.* **192**:756–759.

Pensaert M.B., and Cox E., 1989, Porcine respiratory coronavirus related to transmissible gastroenteritis virus, *Agri-Pract.* **10**:17–21.

Pensaert M.B., Haelterman E.O., and Burnstein T., 1970, Transmissible gastroenteritis of swine; virus-intestinal cell interactions. I. Immunofluorescence, histopathology and virus production in the small intestine through the course of infection, *Arch. Gesamte Virusforsch* **31**:321–334.

Pilchard E.I., 1965, Experimental transmission of transmissible gastroenteritis by starlings, *Am. J. Vet. Res.* **26**:1177–1179.

Pritchard G.C., 1985, Transmissible gastroenteritis in endemically infected breeding herds of pigs in East Anglia, 1981–1985, *Vet. Rec.* **120**:226–230.

Saif L.J., and Wesley R.D., 1992, Transmissible gastroenteritis, in: *Diseases of Swine,* 7th Edition, (A.D., Leman, B.E. Straw, W.L. Mengeling, S. D'Allaire, and D.J. Taylor, eds.), Iowa State University Press, Ames, Iowa, pp. 362–386.

Underdahl N.R., Mebus C.A., and Torres-Medina A., 1975, Recovery of transmissible gastroenteritis virus from chronically infected experimental pigs, *Am. J. Vet. Res.* **36**:1473–1476.

Wiseman B.S., Harris D.L., and Curran B.J., 1988, Elimination of transmissible gastroenteritis virus from a herd affected with the enzootic form of the disease, *Proc 19th Am Assoc Swine Pract* **19**:145–149.

Woods R.D., 1978, Small plaque variant transmissible gastroenteritis virus, *J. Am. Vet. Med. Assoc.* **173**:643–647.

Woods R.D., Wesley R.D., and Kapke P.A., 1988, Neutralization of porcine transmissible gastroenteritis virus by complement-dependent monoclonal antibodies, *Am. J. Vet. Res.* **49**:300–304.

84

EQUINE VIRAL ARTERITIS

Current Status in Finland

Anita Huovilainen and Christine Ek-Kommonen

Virology Unit
National Veterinary and Food Research Institute
P.O. Box 368
00231 Helsinki, Finland

1. SUMMARY

A serological study for antibodies against equine arteritis virus (EAV) in Finland was performed during 1996. All equine sera delivered to the Virology Unit at the National Veterinary and Food Research Institute were tested with a micro-neutralization test, using the Arvac strain as antigen. The study also included imported horses to evaluate EAV circulation in the countries of origin.

Nucleocapsid gene sequences of 2 Finnish equine semen isolates were amplified with RT-PCR and sequenced. The genetic relationships of those isolates with strains isolated elsewhere in the world were analyzed. The Finnish isolates shared 98.2% nucleotide identity, and the closest relatives to the Finnish strains were isolated from the semen of 2 Norwegian horses in 1988 and 1989.

2. INTRODUCTION

All known EAV strains belong to a single serotype with a worldwide distribution (Mumford et al. 1985). The clinical signs of infection vary from mild respiratory disease, fever, edema of legs, genitalia and abdomen, general weakness and ataxia to abortion in pregnant mares. It has been shown that subclinical infections are up to 6 times more common than clinical infections (Timoney & McCollum 1988). Virus spreads mainly by aerosol via the respiratory tract (McCollum et al. 1971), but about one third of infected stallions shed virus in their semen for at least 1–2 years and consequently infect mares during breeding (Timoney et al. 1986). Two EAV vaccines are commercially available: a live, attenuated vaccine and a killed vaccine (Arvac and Artervac, respectively, Fort

Coronaviruses and Arteriviruses, edited by Enjuanes *et al.*
Plenum Press, New York, 1998

Dodge Laboratories). In Finland no EAV vaccines are currently in use, thus all seropositive animals have been vaccinated abroad or have gone through natural infection.

Strain variation has been demonstrated by oligonucleotide fingerprinting (Murphy et al.1992) and by serological studies (Fuginaga et al. 1994). The reference strain (Bucyrus) has been completely sequenced (Den Boon et al. 1991). Genetic relationships of European and North American EAV field isolates have been studied based on G_S and G_L gene sequences (Hedges et al. 1996, Balasuriya et al. 1995a). Recent evidence indicates that the major epitope involved in virus neutralization is located in the G_L (Balasuriya et al. 1995b, Chirnside et al. 1995). Partial sequence data covering the 3' end of the genome encoding the structural proteins M and N have also been published (Chirnside et. al. 1994).

3. MATERIALS AND METHODS

3.1. Serological Studies

A total of 729 equine sera were tested with the micro-neutralization test for the presence of EAV antibodies. Briefly, the sera reacted first with the EAV Arvac strain (kindly provided by Dr Berndt Klingeborn, Uppsala, Sweden) for 1 h at +37°C, RK cells were added and the cytopathic effect was followed by microscope. The sera were tested in duplicate, using serial dilutions from 1:4 to 1:512, a titer $\geq 1:4$ was considered a positive result.

In the group diagnosis (Table 1) samples were taken from horses with symptoms diagnostic of the presence of EVA. Imported horses originated mainly from Estonia (246 sera) and Russia (54 sera), but also from other Eastern European countries such as Latvia, Belorussia, Slovakia, and Ukraine (9, 6, 5, and 2 sera, respectively), in 16 cases the origin was not given. The reasons for testing in the group other included the following: travel, stud service, competition, research, or unknown.

3.2. Genetic Analysis

Two EAV strains were isolated in Finland during the 1990s, the 1st in 1992 (318/92) and the second in 1995 (752/95). Both strains are semen isolates from stallions which arrived in Finland from the United States via Sweden. RNA was purified from EAV-positive cell cultures with SDS and proteinase K treatment. After phenol extraction and ethanol precipitation the RNA was dried, resuspended in RNAse-free water, and RNAase inhibitor was added. RT-PCR was modified from the 1st part of diagnostic-nested RT-PCR developed by Belak et al. (1994), which yields a product of 395 bases from the 3' end of the EAV genome. To obtain

Table 1. Reasons for serological testing, number of positive sera and the percentage of those positive in each group

Reason	Number of sera	Number of positive sera	Percentage of those positive
Diagnosis	229	31	13.5
Import	338	37	10.9
Export	59	11	18.6
Other	103	21	20.4
Total	729	100	13.7

more target DNA for sequencing, 50 pmol of each primer per 50 µl reaction was used and the concentration of dNTPs was increased to 300 µM each. Two independent PCRs of 35 cycles were performed for both strains, in which one of the primers was biotinylated in each reaction. The capture of biotin-labelled amplicons with avidin and DNA sequencing were performed at the DNA Synthesis and Sequencing Laboratory, Institute of Biotechnology, University of Helsinki. The genetic relationships of the Finnish strains with strains isolated elsewhere (Chirnside et al. 1994) were studied based on the nucleocapsid gene sequence. Nucleotide identities were calculated with the GAP program and the dendrogram was created with the Pileup program of the Genetics Computer Group Package (Devereux et al. 1984).

4. RESULTS

The serological data were divided into 4 groups. The total number of samples and the number and percentage of positive sera in each group are given in Table 1. The import group can be divided further: horses arriving from Estonia were seropositive in 7.3% of cases, from Russia in 14.8%, and from other countries (Latvia, Belorussia, Slovakia, Ukraine, or unknown) in 28.9% of cases.

Sequence analysis revealed that the closest relatives to the Finnish strains were strains isolated in Norway in 1988 and 1989 (Fig. 1). The nucleotide sequence identity between strains 318/92 and 752/95 was 98.2%. All nucleotide differences between these two strains were silent and their amino acid sequences were identical with the Bucyrus reference strain. The Arvac vaccine strain shared 97.3 % nucleotide identity with both Finnish strains.

5. DISCUSSION

The prevalence of EAV antibodies among domestic horses implies moderate EAV circulation in Finland. The domestic horses tested for reasons other and export were seropositive more often than those tested for diagnostic purposes. This probably indicates that horses belonging to these two groups are transported more widely and potentially more exposed to EAV infections. It is also possible that some have been vaccinated abroad. Horses imported from Estonia and Russia do not appear to be a health risk to Finnish horses as far as EAV infection is concerned. From an epidemiological standpoint, viruses arriving with foreign horses may posses different antigenic characteristics and thus serve as potential agents for outbreaks among Finnish horses.

The micro-neutralization test is widely used and considered to be reliable when EAV antibodies are measured, however, for example corticosteroids interfere with the EAV neutralization test (Chrinside 1992). In the present study, the medication for sick horses was unknown and in some cases might have influenced the results.

Molecular epidemiology of EAV strains has been studied based on G_L, G_S, M and N gene sequences, and no recombination between strains has been demonstrated so far. The N gene is highly conserved and therefore the scale in the dendrogram is smaller than it would be e.g. in the dendrogram based on G_L gene sequences. Sequence analysis revealed that both Finnish (originally American) stallions were most probably not infected in the United States, but rather during their stay in Sweden prior to entering Finland, or later in Finland. Sequencing of the G_L gene will show whether these Norwegian and Finnish isolates belong to the European E_2 group (Balasuriya et. al. 1995) or form a group of their own.

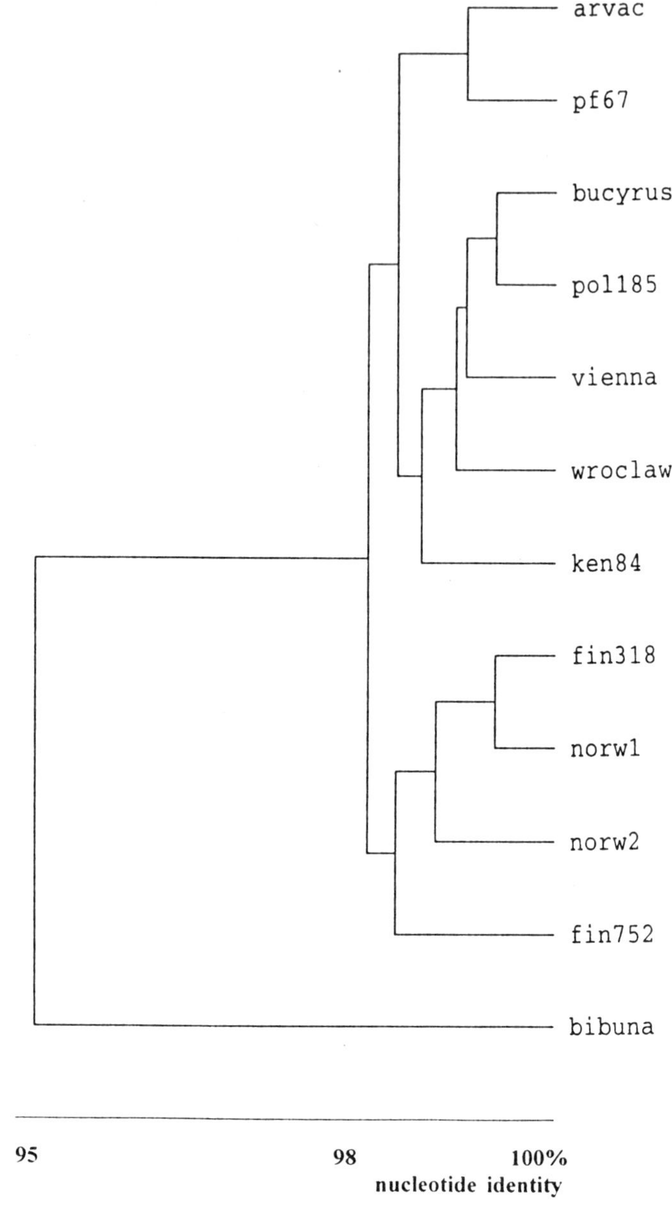

Figure 1. Dendrogram based on nucleotide identity of the EAV nucleocapsid gene.

REFERENCES

Balasuriya U. B. R., Timoney P. J., McCollum W. H. and MacLachlan N. J., 1995a, Phylogenetic analysis of open reading frame 5 of field isolates of equine arteritis virus and identification of conserved and nonconserved regions in the GL envelope glycoprotein, *Virology*, **214**: 690–697.

Balasuriya U. B. R., Maclachlan N.J., de Vries A. A. F., Rossitto P. V. and Rottier P. J. M., 1995b, Identification of a neutralization site in the major envelope glycoprotein (G_L) of equine arteritis virus, *Virology* **207**: 518–527.

Belak S., Ballagi-Pordany A., Timoney P. J., McCollum W.H., Little T. V., Hyllseth B. and Klingeborn B., 1994 , Evaluation of a nested PCR assay for the detection of equine arteritis virus, in: *Proc. 7th Internat. Conf. Equine Infect. Dis.* , Tokyo. (H. Nagajima and W. Plowright, eds). R&W Publications, Ltd., Newmarket. pp 33–38.

Chirnside E. D., 1992, Equine arteritis virus: an overview, *Br. Vet. J.* **148**: 181–197.

Chirnside E. D., Wearing C. M., Binns M. M. and Mumford J. A., 1994, Comparison of M and N gene sequences distinguishes variation amongst equine arteritis virus isolates, *J. Gen. Virol.* **75**: 1491–1497.

Chirnside E. D., de Vries A. A. F., Mumford J. A. and Rottier P. J. M., 1995, Equine arteritis virus neutralizing antibody in the horse is induced by a determinant on the large envelope glycoprotein GL, *J. Gen. Virol.* **76**: 1989–1998.

Den Boon J. A., Snijder E. J., Chirnside E. D., de Vries A. A. F., Horzinek M. C. and Spaan W. J. M., 1991, Equine arteritis virus is not a togavirus but belongs to the Coronaviruslike superfamily, *J. Virol.* **65**: 2910–2920.

Devereux J., Haeberli P. and Smithies O., 1984, A comprehensive set of sequence analysis programs for the VAX, *Nucleic. Acids Res.* **12**: 387–395.

Fugunaga Y., Matsumura T., Sugiura T., Wada R., Imagava H., Kanemaru T. and Kamada M., 1994, Use of the serum neutralization test for equine viral arteritis with different virus strains, *Vet. Rec.* **134**: 574–576.

Hedges J. F., Balasuriya U. B. R., Timoney P. J. McCollum W. H. and MacLachlan N. J., 1996, Genetic variation in open reading frame 2 of field isolates and laboratory strains of equine arteritis virus, *Virus Res.* **42**: 41–52.

McCollum W. H., Prickett M. E. and Bryans J. T., 1971, Temporal distribution of equine arteritis virus in respiratory mucosa, tissues and body fluids of horses infected by inhalation, *Res. Vet. Sci.* **2**: 459–464.

Mumford J. A., 1985, Preparing for equine arteritis, *Equine Vet. J.* **17**: 6.

Murphy T. W., McCollum W. H., Timoney P. J., Klingeborn B. W., Hyllseth B., Golnik W. and Erasmus B., 1992, Genomic variability among globally distributed isolates of equine arteritis virus, *Vet. Microbiol.* **32**: 101–115.

Timoney P. J. and McCollum W. H., 1988, Equine viral arteritis: epidemiology and control, *J. Equine Vet. Sci.* **8**: 54–59.

Timoney P. J., McCollum W. H., Roberts A. W. & Murphy T.W., 1986, Demonstration of the carrier state in naturally acquired equine arteritis virus infection in the stallion, *Res. Vet. Sci.* **41**: 279–280.

Strategies to Control Coronavirus-Induced Diseases

85

ISOLATION AND RECOMBINANT EXPRESSION OF AN MHV-JHM NEUTRALISING MONOCLONAL ANTIBODY

Andreas F. Kolb, Monika Lechermaier, Angelien Heister, Atiye Toksoy, and Stuart G. Siddell

Institute of Virology and Immunology
University of Würzburg
Versbacherstr. 7
D-97078 Würzburg, Germany

1. ABSTRACT

The monoclonal antibody A1 (mab A1) efficiently neutralises the infection of susceptible cells by the murine hepatitis virus MHV-JHM in vitro and in vivo (Wege et al., 1984). The variable regions of mab A1 were amplified from mRNA of the respective hybridoma cell line by RT-PCR and integrated into different eukaryotic expression vectors. The biological function of the recombinant antibody constructs was verified by virus neutralisation assays. Whereas a complete recombinant antibody (mab A1rec.) expressed in transfected murine myeloma cells inhibited the MHV-JHM infection as well as the parental antibody, a single-chain Fv derived from mab A1 did not show any neutralising activity.

2. INTRODUCTION

Coronaviruses are enveloped, positive-stranded RNA viruses causing diseases of economical importance in animals and humans. Generally, coronavirus induced diseases show a pronounced pathogenicity in new-born animals. The marked susceptibility of new-born animals is presumably a consequence of their still immature immune system. Lactotropic transfer of maternal, antiviral antibodies via the placenta and milk efficiently protects the new-born animals against fatal consequences of acute coronaviral infections. Administration of neutralising antibodies via an intraperitoneal injection and also by oral application has emerged as potential tool of interference with coronavirus infections in offspring of non-immune mothers (Wege et al., 1984; Enjuanes and van der Zeijst, 1995).

Coronaviruses and Arteriviruses, edited by Enjuanes *et al.*
Plenum Press, New York, 1998

Generation of transgenic animals expressing neutralising antibodies directed against coronaviral pathogens in the lactating mammary gland may therefore provide new-born animals with an efficient protection against fatal infections during a critical period.

The murine hepatitis virus (MHV) strain JHM leads to fatal neurological disorders, especially in new-born animals. MHV-JHM infections can be efficiently inhibited in vitro and in vivo by neutralising antibodies and CD4+ T-cells (Wege et al., 1984; Flory et al., 1993). The majority of neutralising mabs are directed against the viral surface glycoprotein (S). Mab A1 (Wege et al., 1984) is one of the most potent antibodies with regard to virus neutralisation and inhibition of virus induced cell-to-cell fusion. It interacts with the S1 subunit of the MHV-JHM S-protein. We have isolated the variable regions of mab A1 and transferred them into different eukaryotic expression vectors. The biological function of the recombinantly expressed antibodies was assessed by MHV-JHM neutralisation assays.

3. MATERIALS AND METHODS

3.1. Cells and Viruses

DBT cells were grown as described (Kolb et al., 1996) NS0 mouse myeloma cells (NS0, ECACC 85110503) were cultivated at 37°C in RPMI 1640 medium (Sigma, Deisenhofen, Germany) supplemented with 10% FBS, non-essential amino acids, glutamine and antibiotics. The virus strain used for these studies was described previously (Routledge et al., 1991). Transfections were done as described (Kolb and Siddell, 1996). For virus neutralisation assays 500μl of the respective MHV-JHM dilution was mixed with 500μl of tissue culture supernatant from antibody expressing cells or control cells and incubated for 1h at 37°C. Subsequently, the virus-antibody mixture was added to confluent DBT cells and incubated for one further hour at 37°C. The virus containing supernatant was then removed. The cells were washed twice with PBS and overlaid with cell culture medium containing 1% agarose. Virus plaques were counted after an incubation period of 16h at 37°C.

3.2. DNA Cloning

The isolation and cloning of the RT-PCR products corresponding to the variable regions of the mab A1 heavy and light chain cDNAs was performed as described (Orlandi et al., 1989). The single-chain Fv derived from mab A1 consists of 1. a 216bp fragment of the murine Metallothionein promoter, 2. 400bp of the human growth hormone (hGH) gene encoding the hGH signal peptide, 3. the mab A1 heavy and light chain variable regions, 4. an 38nm spacer region (3 consecutive stretches of the amino acid sequence GGGGS; Huston et al., 1988) which is inserted between the heavy and light chain variable domains, 5. a peptide tag recognised by the monoclonal antibody 30B (Routledge et al., 1991) (amino acid sequence: IPRSRQIDLQIG) (Fig. 3a) and 6. 275bp of the 5th exon of the hGH gene containing the polyadenylation signal. The protein encoding DNA fragments were assembled in one open reading frame which encodes a protein with a predicted molecular mass of 26500 Da.

3.3. Protein Analysis

Western blot analyses were performed as described (Kolb et al., 1996). Development of the blots was done by using 4-Chloronaphtol or the Amersham ECL system.

4. RESULTS

4.1. Isolation of the Variable Regions of Mab A1

The monoclonal antibody A1 efficiently neutralises the infection of MHV-JHM into susceptible cells. To express a recombinant antibody with the same biological properties we isolated the variable regions of the mab A1 by reverse transcription and PCR by using degenerate primers corresponding to conserved regions of the variable domains (Orlandi et al., 1989). The sequence of the mab A1 heavy and light chain variable regions are available under the gene bank accession numbers AF 001575 and AF 001576, respectively. The PCR products of 322bp and 347bp derived from the light and heavy chain cDNAs were subsequently cloned into the expression vectors Lys17 and Lys30 (Orlandi et al., 1989), respectively. These vectors carry IgG constant domains of human origin. Additionally, the vectors contain the hygromycin-phosphotransferase and the xanthin-guanosine-phosphoribosyl-transferase genes as selectable markers.

4.2. Recombinant Expression of the Mab A1

The expression plasmid Lys17-A1L was then transfected into NS0 myeloma cells. After selection in medium containing 100µg/ml hygromycin B, 10 of the 52 resulting cell clones were analysed for the expression of the IgG kappa chain. 8 of the 10 cell clones secreted high amounts of light chain protein into the tissue culture supernatant. One of these clones was then re-transfected with the plasmid Lys30-A1H. After selection in medium containing 1µg/ml mycophenolic acid one cell clone (designated NS0 Lys17-A1L/Lys30-A1H) could be recovered which expressed detectable amounts of heavy chain (Fig. 1). The amount of light chain protein present in the supernatant clearly exceeds the amount of heavy chain (Fig. 1). However, after purification of the supernatant over a protein G-sepharose column equimolar amounts of heavy and light chain protein can be detected in a Western blot analysis (Fig. 1). Comparison to a human IgG protein standard reveals that NS0 Lys17-A1L/Lys30-A1H cells (1×10^7 cells in 14 days) can accumulate to about 3mg/ml of recombinantly expressed light chain in their supernatant (Fig. 1). The amount of heavy chain protein appears to be tenfold lower (Fig. 1).

The biological activity of the recombinantly expressed antibody (mab A1rec.) was analysed by virus neutralisation assays. The tissue culture supernatant of NS0 cells transfected with the parental plasmids Lys17 and Lys30 encoding an irrelevant antibody was used as a negative control. As expected, only the supernatant of NS0 Lys17-A1L/Lys30-A1H cells was capable of neutralising MHV-JHM infection of DBT cells (Fig. 2a). The efficiency of virus neutralisation was compared between the supernatants of A1 hybridoma cells and NS0 Lys17-A1L/Lys30-A1H cells (Fig. 2b). The plaque reduction efficiency was in the same order of magnitude for both supernatants indicating that the correct variable region were cloned and expressed.

4.3. Expression of an scFv Derived from Mab A1

We next investigated whether a single-chain Fv (scFv) fragment carrying the mab A1 variable regions would also possess the highly neutralising activity of its parental counterpart. Single-chain Fvs offer several advantages over complete antibody molecules, in that they are made up of only one peptide chain, so that only one transcriptional unit would have to be transferred in a transgenic approach. Inherent with the nature of the scFv

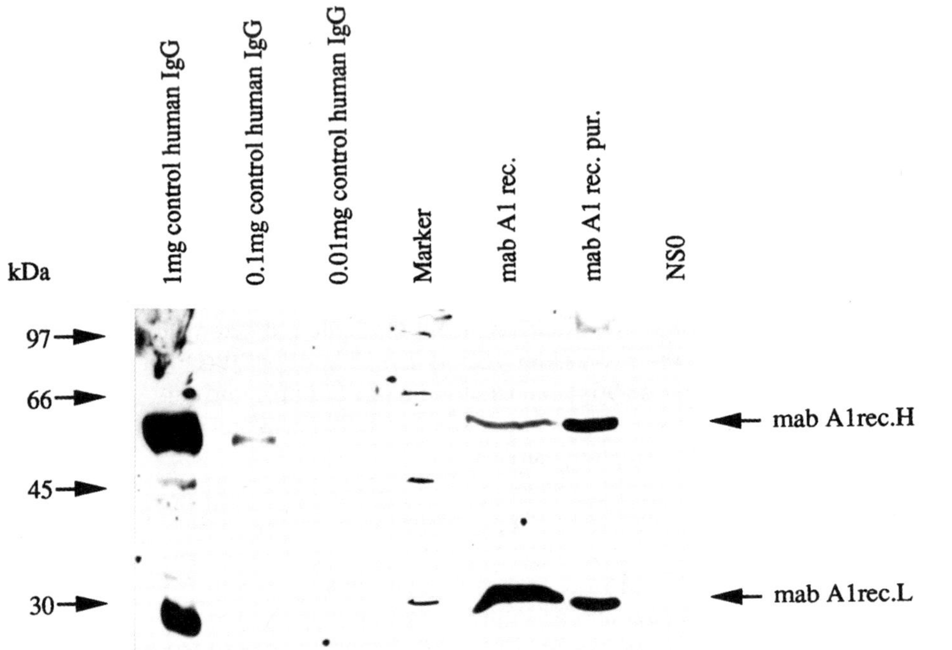

Figure 1. Western blot analysis of the recombinantly expressed mab A1 (mab A1rec.). Supernatants of NS0 cells and NS0 cells stably transfected with the expression vectors Lys17-A1L and Lys30-A1H and 3 different dilutions of a human IgG protein standard (Behring-Werke, Marburg, Germany) were separated on a 15% polyacrylamide gel and blotted to nitro-cellulose. The molecular masses of the molecular weight marker proteins and the positions of the IgG heavy (mab A1rec.H) and light chain proteins (mab A1rec.L) are indicated.

molecule, heavy and light chain sequences are present in equimolar amounts. Finally, scFvs may also be produced in bacterial expression systems, so that large amounts of an antiviral compound can be synthesised.

We generated an scFv expression cassette (Fig. 3a) in which the heavy and light chain variable regions of mab A1 were separated by a linker consisting of 3 consecutive stretches of the amino acid sequence GGGGS (Huston et al., 1988). In order to allow for the secretion of the scFv, the human growth hormone (hGH) signal peptide (sp) was linked to the N-terminus of the predicted peptide chain. A peptide tag recognised by the monoclonal antibody 30B (Routledge et al., 1991) was added at the C-terminus of the scFv open reading frame to permit the immunological detection of the protein. The expression cassette was linked the strong constitutive murine Metallothionein (MT) promoter and the hGH polyadenylation signal (Fig. 3a). The expression vector (designated MThGH-A1scFv-tag) was transfected into DBT cells together with the plasmid pSV2-neo carrying a selectable neomycin-phosphotransferase marker gene. After selection in medium containing 400μg/ml G418, the resulting clones were analysed for the expression of the A1-scFv by Western blotting. The highest amounts of the A1-scFv could be detected in cytoplasmic extracts and the supernatant of the cell clone DBT MThGh-A1scFv-tag-6 (Fig. 3b).

Again the biological activity of the recombinantly expressed antibody derivative was assessed by a virus neutralisation assay. In a first experiment DBT MThGh-A1scFv-

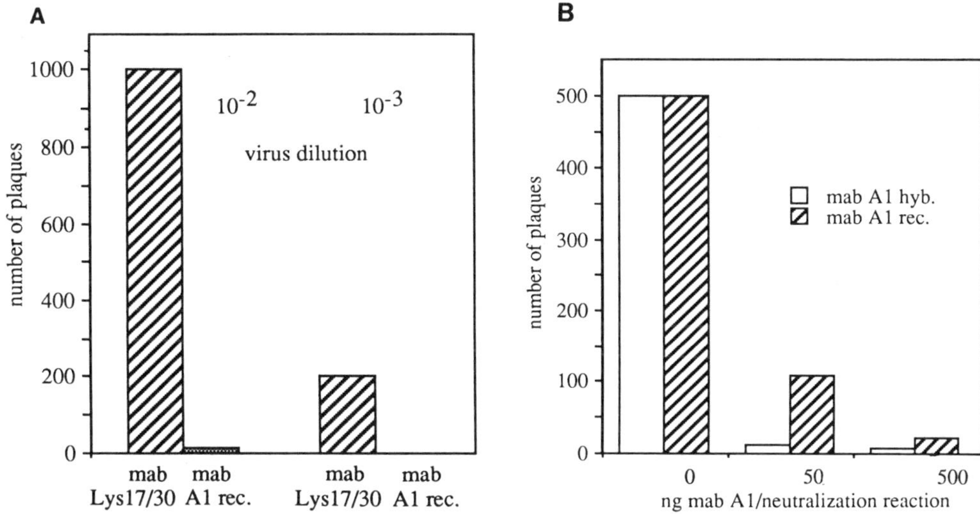

Figure 2. Virus neutralisation assay. A) Tissue culture supernatants derived from the cell clone NS0 Lys17-A1L/Lys30-A1H or the negative control cell clone NS0 Lys17/Lys30 (stably transfected with the plasmids Lys17 and Lys30 which encode an irrelevant mab) were analysed in a virus neutralisation assay. The number of virus plaques obtained in a representative experiment with virus dilutions of 10^{-2} and 10^{-3} are presented. B) The supernatants of A1 hybridoma and NS0 Lys17-A1L/Lys30-A1H cells were diluted in MEM cell culture medium. The amounts of antibody present in the supernatant was estimated by calibration against murine and human antisera in a Western blot assay. The number of virus plaques obtained are presented in relation to the total amount of mab present in the supernatants applied to the neutralisation reaction.

tag-6 cells were infected with different dilutions of MHV-JHM. As a negative control un-transfected DBT cells and DBT cells transfected with the control plasmid pBK-CMV (Stratagene) were infected in parallel. However, no differences in the number and size of virus-induced plaques could be detected (Fig. 3c). This suggests that the intracellular ex-pression of the A1-scFv does not lead to an inhibition of MHV-JHM infection. One reason for this result might be that the A1-scFv is not present in the same cellular compartments as the MHV-JHM surface glycoprotein (S) and therefore fails to inhibit the S-protein func-tion in virus entry and virus maturation. We therefore analysed the ability of the secreted A1-scFv to neutralise MHV-JHM infection. The supernatant of DBT MThGh-A1scFv-tag-6 cells, however, was also unable to inhibit MHV-JHM entry into DBT cells (Fig. 3d).

5. DISCUSSION

We have isolated the variable regions of a monoclonal antibody which neutralises MHV-JHM infections into susceptible cells. A recombinant version of the mab A1 was successfully expressed in NS0 myeloma cells. An scFv carrying the variable regions of mab A1, however, failed to inhibit MHV-JHM infection, indicating that an scFv derived from a monoclonal antibody does not necessarily display the same biological function. A number of reasons could account for the inability of the A1-scFv to neutralise MHV-JHM and future experiments will preferentially address the question as to whether the A1-scFv is able to physically interact with the MHV-JHM surface glycoprotein.

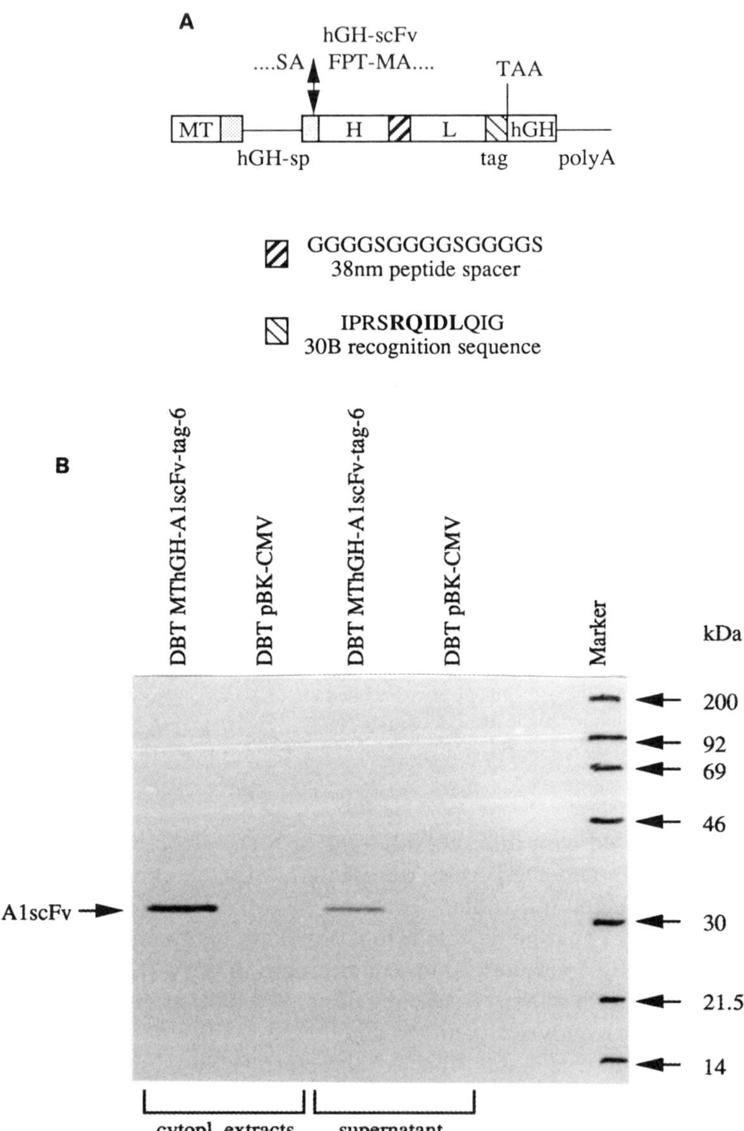

Figure 3A and B. A) Schematic representation of the single-chain Fv derived from mab A1. The murine Metallothionein promoter (MT), heavy (H) and light chain (L) variable regions and the 5th exon of the human growth hormone (hGH) are marked as open boxes. The hGH signal peptide, the 38nm spacer segment and the 30B amino acid tag sequence are represented as dotted, heavily striped and striped boxes, respectively. The translational stop codon (TAA) and the hGH polyadenylation signal (polyA) are indicated. The junction between the hGH signal peptide and the scFv open reading frame is marked with a dash (-). The site at which the hGH signal peptide is predicted to be cleaved is marked with a double arrow. B) Western blot analysis of A1 scFv expression. Cytoplasmic extracts and tissue culture supernatants of DBT cells transfected with pBK-CMV and of the cell clone DBT MThGh-A1scFv-tag-6 carrying the A1 scFv expression plasmid were separated on a 15% polyacrylamide gel and blotted to nitro-cellulose. scFv protein was detected by using the antibody 30B.

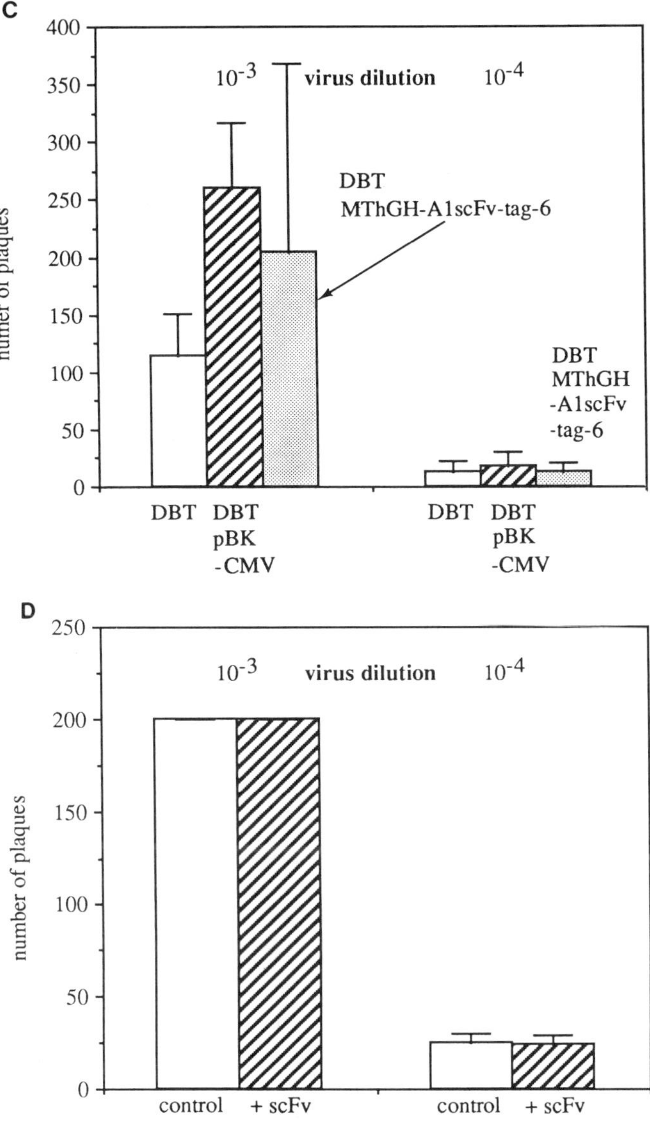

Figure 3C and D. C) Virus neutralisation assay. DBT cells, DBT cells transfected with the plasmid pBK-CMV and DBT MThGh-A1scFv-tag-6 cells were analysed for their susceptibility to MHV-JHM infection. The average number of virus plaques scored after 3 independent infections with 2 different virus dilutions are presented. D) Virus neutralisation assay. The supernatants of DBT pBK-CMV cells and DBT MThGh-A1scFv-tag-6 cells were analysed for ability to neutralise MHV-JHM infection into DBT cells. The average number of virus plaques scored after 3 independent infections with 2 different virus dilutions are presented.

REFERENCES

Enjuanes, L. and van der Zeijst, B.A.M., 1995, Molecular basis of transmissible gastroenteritis virus epidemiology. in: *The Coronaviridae* (S. G. Siddell, ed.), Plenum Press, New York, pp. 337–376.

Flory, E., Pfleiderer, M., Stuhler, A., and Wege, H., 1993, Induction of protective immunity against coronavirus-induced encephalomyelitis: evidence for an important role of CD8+ T cells in vivo, *Eur. J. Immunol.* **23:** 1757–61.

Huston, J.S., Levinson, D., Mudgett-Hunter, M., Tai, M.-S., Novotny, J., Margolies, M.N., Ridge, R.J., Bruccoleri, R.E., Haer, E., Crea, R., and Oppermann, H., 1988, Protein engineerig of antibody binding sites: Recovery of specific activity in an anti-digoxin single-chain Fv analogues produced in Escherichia coli, *Proc. Natl. Acad. Sci. USA* **85:** 5879–5883.

Kolb, A.F. and Siddell, S.G., 1996, Genomic targeting with an MBP-Cre fusion protein, *Gene* **183:** 53–60.

Kolb, A.F., Maile, J., Heister, A., and Siddell, S.G., 1996, Characterization of functional domains in the human coronavirus HCV 229E receptor, *J. Gen. Virol,* 2515–2521.

Orlandi, R., Güssow, D.H., Jones, P.T., and Winter, G., 1989, Cloning immunoglobulin domains for expression by the polymerase chain reaction, *Proc. Natl. Acad. Sci. USA* **86:** 3833–3837.

Routledge, E., Stauber, R., Pfleiderer, M., and Siddell, S.G., 1991, Analysis of murine coronavirus surface glycoprotein functions by using monoclonal antibodies. *J. Virol.* **65:** 254–262.

Wege, H., Dörries, R., and Wege, H., 1984, Hybridoma antibodies to the murine coronavirus JHM: characterization of epitopes on the peplomer protein E2, *J. Gen. Virol.* **65:** 1913–1941.

INTERFERENCE OF CORONAVIRUS INFECTION BY EXPRESSION OF IgG OR IgA VIRUS NEUTRALIZING ANTIBODIES

Isabel Sola, Joaquín Castilla, and Luis Enjuanes

Centro Nacional de Biotecnología
Consejo Superior de Investigaciones Científicas (CSIC)
Department of Molecular and Cell Biology
Campus Universidad Autónoma, Cantoblanco
28049 Madrid, Spain

1. ABSTRACT

Mouse immunoglobulin gene fragments encoding the variable modules of the heavy (VH) and light (VL) chains of a transmissible gastroenteritis coronavirus (TGEV) neutralizing monoclonal antibody (MAb) have been cloned and sequenced. The selected MAb recognizes a highly conserved viral epitope and does not lead to the selection of neutralization escape mutants. Chimeric immunoglobulin genes with the variable modules from the murine MAb and constant modules of human gamma 1 and kappa chains were constructed using RT-PCR. These chimeric immunoglobulins were stably or transiently expressed in murine myelomas and COS cells, respectively. The secreted recombinant antibodies had radioimmunoassay (RIA) titers higher than 10^3 and reduced the infectious virus more than 10^4-fold. Recombinant dimeric IgA showed a 50-fold enhanced neutralization of TGEV relative to a recombinant monomeric IgG1 which contained the identical antigen binding site. Epithelial cell lines stably-transformed with these constructs and expressing either recombinant IgG or IgA TGEV neutralizing antibodies reduced virus production by >10^5-fold after infection with homologous virus, although a residual level of virus production (<10^2 PFU/ml) remained in less than 0.1 % of the cells.

2. INTRODUCTION

The mucosal immune system and its predominant effector, secretory immunoglobulin A (IgA), provide the initial immunologic barriers against most pathogens that invade

Coronaviruses and Arteriviruses, edited by Enjuanes *et al.*
Plenum Press, New York, 1998

the body at a mucosal surface (Mazanec et al., 1995; McGhee et al., 1989; Mestecky and McGhee, 1987). This is especially true for viruses, since resistance to infection has been strongly correlated with the presence of specific IgA antibody in mucosal secretions (Armstrong et al., 1990). Traditionally, the neutralization of viruses by immunoglobulins is thought to result from the binding of antibody to virion attachment proteins, thereby preventing adherence to epithelial cells surfaces. In addition, mucosal antibody interacts intracellularly with viruses preventing their replication, possibly by interfering with virus assembly (Mazanec et al., 1992).

New strategies, including the introduction of antibody genes into cells have been recently explored in model tissue culture systems. This approach may potentially be applied to *in vivo* protection of mucosal surfaces by gene therapy with antibody-encoding genes since monoclonal antibodies (MAbs) are now available against a vast range of viruses. Virus neutralizing MAbs may protect mucosal tissues against viral infections, however it is not known whether antibody secreting cells will provide protection to the neighboring tissues.

Transmissible gastroenteritis coronavirus (TGEV) infects both enteric and respiratory tissues and causes a mortality close to one hundred percent when newborn animals are infected (Enjuanes and Van der Zeijst, 1995). Full protection against TGEV can be provided by lactogenic immunity from immune sows (Saif and Wesley, 1992; Stone et al., 1977). Investigations by our laboratory into the mechanisms of TGEV neutralization (Suñé et al., 1990) and of antigenic and genetic variability (Sanchez et al., 1992; Sanchez et al., 1990) have led to the identification of a mouse MAb which neutralized all TGEV isolates tested, and also TGEV-related coronaviruses. No neutralization escape mutants (*mar* mutants) appeared when this MAb was employed (Gebauer et al., 1991).

We have studied the protection of epithelial cell monolayers against TGEV infection using expression plasmids encoding virus neutralizing MAbs. We found that cell lines stably-transformed with these MAb-producing vectors were substantially protected against TGEV challenge, however a small fraction of these cells continued to produce low levels of viral progeny.

3. MATERIALS AND METHODS

3.1. Cell and Viruses

Swine testis (ST) (McClurkin and Norman, 1966) and SV40 transformed monkey kidney COS-1 cells (ATCC CRL-1650), non secreting murine myeloma Sp2/0 (ATCC, CRL-1581), and MAb 6A.C3 and MAb 1G.A7-secreting murine hybridomas (Correa et al., 1988; Jiménez et al., 1986) were grown as described (Castilla et al., 1997). TGEV PUR46-MAD strain (Gebauer et al., 1991) was grown, purified, and titrated in ST cells as described (Jiménez et al., 1986). Vesicular stomatitis virus (VSV) was grown and titrated as described (Bullido et al., 1989).

3.2. Immunofluorescence Microscopy

Sp2/0 and ST cells expressing recombinant mouse-human (rMH) antibodies, recombinant mouse-swine (rMS) antibodies, and the MAb 6A.C3 secreting hybridoma were grown in microslide culture chambers (Miles Scientific). Immunofluorescence to detect rMAb expression was performed as described (Castilla et al., 1997). ST cells expressing rMH and rMS antibodies, or untransformed ST cells, were infected with TGEV. To detect recombinant antibodies and viral proteins in the same cells, double immunofluorescence was performed as previously described (Castilla et al., 1997).

3.3. Radioimmunoassay (RIA), Virus Neutralization, and Western Blot Analysis

The procedures for RIA, virus neutralization and Western blot have been described (Correa et al., 1988).

3.4. RNA Extraction

Total cytoplasmic RNA from hybridoma 6A.C3 was prepared as described (Castilla et al., 1997) PolyA$^+$ mRNA was isolated using the PolyATtract mRNA Isolation System (Promega).

3.5. Sequencing, Characterization of 6A.C3 Variable (V)-Modules RNA, and Synthesis of MAb 6A.C3 cDNA Encoding the Variable Module

MAb 6A.C3 polyA$^+$ mRNA was sequenced as previously described (Castilla et al., 1997).

3.6. Construction of Immunoglobulin Expression Plasmids

The expression plasmids pINLC6A and pINHC6A containing the recombinant light and heavy chains respectively were constructed as described (Castilla et al., 1997).

3.7. Interference of TGEV Infection in ST Cell Lines Expressing Recombinant TGEV Neutralizing Antibodies

Cloned and uncloned ST cell lines expressing recombinant TGEV neutralizing antibodies with IgG$_1$ or IgA isotype were infected with TGEV as described (Castilla et al., 1997). Supernatants from infected and non-infected cells were analyzed at 1 to 6 days post infection (p.i.). Infections with vesicular stomatitis virus (VSV) were performed in parallel as a control.

4. RESULTS

4.1. Sequence of Genes Encoding Variable Modules of a TGEV Neutralizing MAb

In order to protect cell monolayers from virus infection using MAbs it is convenient to use an antibody with a high titer in virus neutralization, that recognizes an epitope present in all virus isolates, and which does not lead to the selection of neutralization escape mutants. These conditions were fulfilled by MAb 6A.C3 (Gebauer et al., 1991; Jiménez et al., 1986).

The sequence of MAb 6A.C3 V-modules was determined. The typical L and H chain organization was identified (data not shown). The sequences of MAb 6A.C3 L and H chain variable modules showed high homology with kappa (99.4%) and gamma 1 (92.7%) sequences of subgroup V and subgroup IIIC immunoglobulin genes, respectively. The J regions of the L and H chains belong to the J_2 type.

4.2. Generation of rMAbs 6A.C3

Human κ and γ constant modules were flanked by the SV-40 early promoter and polyadenylation signals and were subcloned into plasmids pINLC6A and pINHC6A, respectively, which carry the mouse immunoglobulin enhancer at the 5' end of the expression cassettes. The engineering of a rIgA immunoglobulin with the same variable modules than the $rIgG_1$ has been described (Sola et al., 1997).

4.3. Functional Analysis of Recombinant MAbs with Gamma 1 and Alpha Isotypes

Western-blot analysis performed under non-reducing conditions demonstrated that MAb 6A.C3 with the IgG_1 isotype was monomeric (molecular mass 150 kDa), while rIgA presented a dimeric structure of about 300 kDa (results not shown). The secreted chimeric immunoglobulins expressed both in COS and SP2/0 cells bound TGEV by RIA with titers up to 10^4, and neutralized virus infectivity around 10^4-fold (neutralization indices, NIs, around 4) (Table 1).

The final aim of this work is to express the rMAbs in the swine enteric tract to examine their protective effect on mucosal surfaces against viral challenge. In these experiments IgA isotype antibodies are known to be more stable than those with an IgG isotype

Table 1. Functional characterization of recombinant antibodies

	Titer determined by	
Antibody[a]	RIA[b]	Neutralization index[c]
MAb-6A.C3	10^4	>4
rIgG-Sp2/0 cells	10^2–10^3	2–3
rIgA-Sp2/0 cells	10^2–10^3	4
rIgG-COS cells	10^2–10^3	4
rIgA-COS cells	10^3–10^4	3
rγ	5	<0.3

[a]Chimeric immunoglobulins rIgG1 and rIgA expressed in COS monkey kidney cells and in Sp2/0 myeloma cells were analyzed by RIA and virus neutralization assays. MAb-6A.C3, antibody secreted by the original hybridoma. rIgG-Sp2/0 and rIgA-Sp2/0, recombinant mouse-human and mouse-swine antibodies, respectively, secreted by transformed Sp2/0 myeloma cells. rIgG-COS and rIgA-COS, recombinant mouse-human and mouse-swine antibodies, respectively, secreted by transformed COS cells. rγ, recombinant mouse-human gamma chain.

[b]RIA titer, highest dilution giving 3-fold the background. Similar results were obtained in more than five independent evaluations of the antibodies secreted by the selected cell lines.

[c]Neutralization index, $\log_{10}$ of the ratio of the PFU after incubating the virus in the presence of medium or the indicated MAb. Similar results were obtained in the evaluation of more than five independent cultures of transiently transformed COS cells.

(Lamm et al., 1995). To compare the neutralizing activity of $rIgG_1$ and rIgA, supernatants containing recombinant antibodies of the same RIA titer were used in neutralization assays. Recombinant IgA neutralized TGEV 50-fold more effectively than $rIgG_1$ as expected for a dimeric immunoglobulin.

4.4. Generation of ST Cells Expressing $rIgG_1$ and rIgA and Evaluation of Their Resistance to TGEV Infection

Porcine epithelial (ST) cells susceptible to infection with TGEV were transfected with DNA constructs encoding the chimeric H and L chains to produce either the $rIgG_1$ or the rIgA TGEV specific antibodies. RIA titers of the recombinant MAbs in supernatants from the cell lines ranged from 10^2 to 10^3. Immunofluorescence microscopy studies using immunoglobulin-specific antibodies revealed that, before the stably transformed cells were cloned, 10 to 15 % of the cells of each line expressed the recombinant MAbs (results not shown). Cell lines expressing $rIgG_1$ or rIgA TGEV-specific antibodies were infected with TGEV. Transformed cell lines were apparently resistant to TGEV infection (Figure 1A) since no cytopathic effect was observed 48 h post-infection while untransformed ST cell monolayers were completely lysed. TGEV-infected ST cells produced 10^7 PFU/ml while IgG1 and IgA-transformed ST cell supernatants had titers which dropped to 10^3 PFU/ml (Figure 1B). This inhibition in viral synthesis was specific, since VSV grew to the same titer in transformed and untransformed cells (Figure 1B). While there was a significant (10^4-fold) reduction in virus synthesis in the antibody-producing cells, they were not fully resistant to TGEV infection since a residual level of virus synthesis (10^3 PFU/ml) persisted in the absence of any discernible cytopathic effect.

4.5. Interference of TGEV Infection in Cloned ST Cell Lines Expressing Recombinant Antibodies

To determine whether the residual infection of transformed ST cell lines by TGEV was due to the presence of a large proportion of cells that did not produce the antibody, two cell lines producing the highest levels of either $rIgG_1$ or rIgA antibodies were cloned by limiting dilution and infected. All cloned cells expressed the expected rMAbs, either $rIgG_1$ or rIgA as determined by immunofluorescence microscopy (Figure 2a and 2c) and >99.9% were free of viral antigens (Figure 2b and 2d). No cytopathic effect was observed at 72 h p.i. in the transformed cells, in contrast to the complete lysis observed in the untransformed ST cell monolayer, similarly to what was observed (Figure 1) in the uncloned cells (results not shown).

Next, the kinetics of antibody production, viral synthesis, and cytopathic effects were determined (Figure 3). Virus production in the supernatant of transformed cultures was reduced from approximately $10^{7.5}$ PFU/ml to $<10^2$ PFU/ml. Cytopathic effects were not detected in rIgG-producing cells (Figure 3C) and only minor effects were observed at 72 h p.i. in rIgA-producing cells. A high reduction in the virus titer was detected ($>10^5$-fold) but a residual infectivity ($<10^2$ PFU/ml) persisted in the supernatants. Low levels of infectious virus particles were produced by the transformed ST cells in the presence of virus neutralizing antibody (Figure 3A and 3B). To test whether a neutralization resistant virus had been selected, the virus produced at 1, 2, 3, and 6 d p.i. was studied. Neutralization assays by different MAbs demonstrated that a neutralization escape mutant was not present.

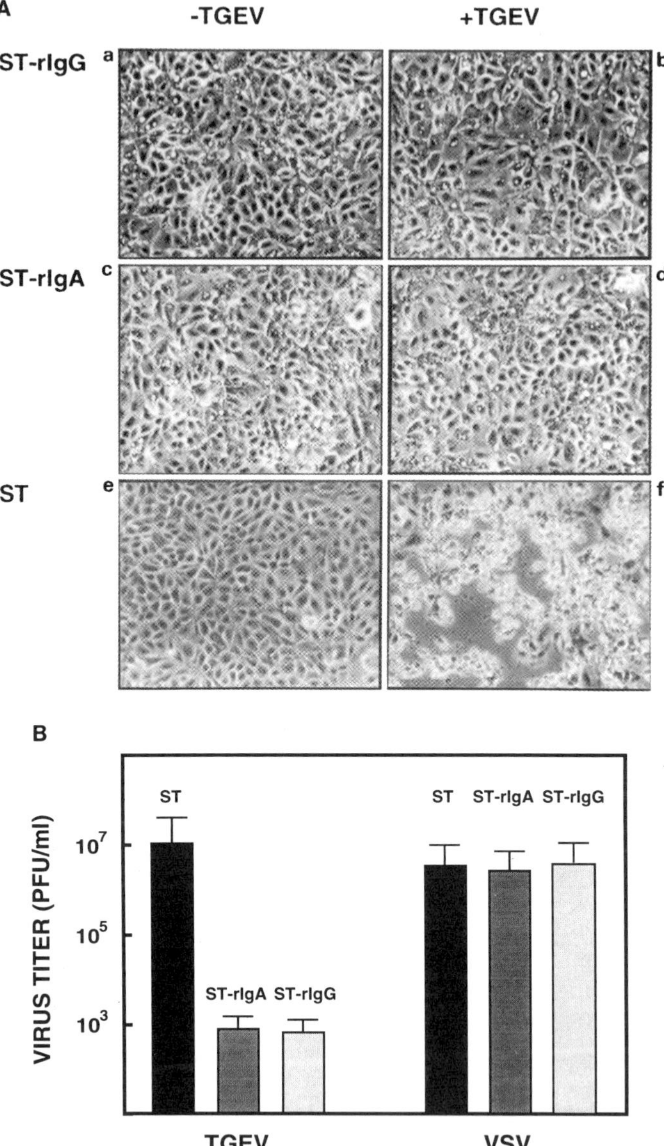

Figure 1. Resistance of rMAb producing ST cells to virus infection. (A) Phase contrast microscopy of rIgG1 (a, b) or rIgA (c, d) producing ST cells, or of untransformed cells (e, f). Cells were infected with TGEV (MOI 0.5) (b, d, f) or uninfected (a, c, e). (B) Specificity of the virus growth inhibition in antibody producing cell lines. Untransformed ST cells or ST cells producing TGEV neutralizing rIgA or IgG1 antibodies derived from MAb 6A.C3 were infected with TGEV or VSV. Supernatants were harvested 48 h p.i. and titrated on ST cells. Thin vertical bars, standard error. ST, untransformed cells. ST-rIgG and ST-rIgA, cells producing rIgG or rIgA neutralizing TGEV, respectively. The mean ± the standard deviation of three experiments is shown.

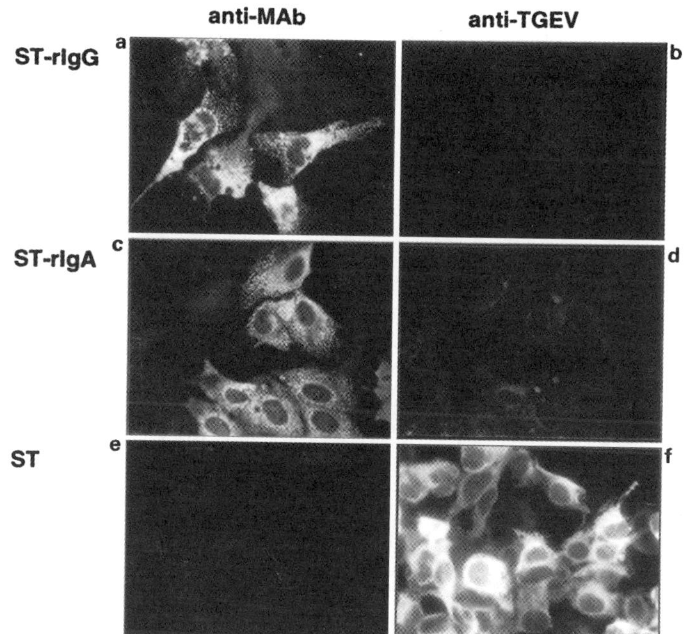

Figure 2. Immunofluorescence of TGEV-infected ST cells expressing neutralizing MAbs. Control ST cells or cells expressing MAbs were infected with TGEV. At 48 h p.i. cells were fixed with methanol and stained by two color immunofluorescence to detect antibodies and viral proteins. Recombinant antibodies were detected in transformed ST cells (a and c) but not in untransformed cells (e). Conversely, viral antigens were detected in untransformed cells (f) but not (<0.1% of cells) in rIgG (b) or rIgA (d) transformed cells. ST, untransformed cells. ST-rIgG and ST-rIgA, cells producing rIgG or rIgA neutralizing TGEV, respectively. Anti-MAb and anti-TGEV, cells stained by immunofluorescence using antibodies specific for immunoglobulins or TGEV.

5. DISCUSSION

The genetic engineering and the anti-viral efficacy of two TGEV neutralizing MAbs with the same binding site and IgG1 and IgA isotypes are described. The dimeric rMAb with an IgA isotype had a 50-fold higher efficiency in virus neutralization assays than the monomeric IgG1 isotype. Transformation of ST cells with plasmids encoding TGEV-neutralizing rMAbs with IgG_1 or IgA isotype reduced virus infectivity $>10^5$-fold and prevented the appearance of cytopathic effects *in vitro*. A low level of virus production was detected in a few antibody producing cells.

A high level of functional chimeric antibody was produced in immunoglobulin-gene transiently-transformed COS cells and in stably-transformed myeloma cell lines (Sp2/0) as were the epithelial (ST) cells. This indicates that in principle, these DNA constructs could be used to transform other epithelial cells such as those found at porcine mucosal cell surfaces making them prime candidates for use as a somatic cell gene therapy-based vaccination strategy.

A highly significant ($>10^5$-fold) reduction in the virus recovery in rIgG- or rIgA-producing cell lines has been demonstrated. This reduction in virus production in conjunction with other natural antiviral mechanisms including non-specific immune effector molecules such as interferon may be sufficient to eliminate the residual virus or the virus producing cells to provide protection against virus infection *in vivo*.

ST cell lines expressing a TGEV-neutralizing rMAb, in which only 10 to 15% of the cells were transformed, were largely protected against TGEV infection. Nevertheless, protection was not complete. Supernatants of transformed ST cell clones had virus titers $<10^2$ PFU/ml, very low in comparison with the 10^7 PFU/ml observed in untransformed cells under the same conditions. The decrease in virus titer probably was not due to a non-specific reduction in the capacity of the transformed cells to produce virus, since the transformed

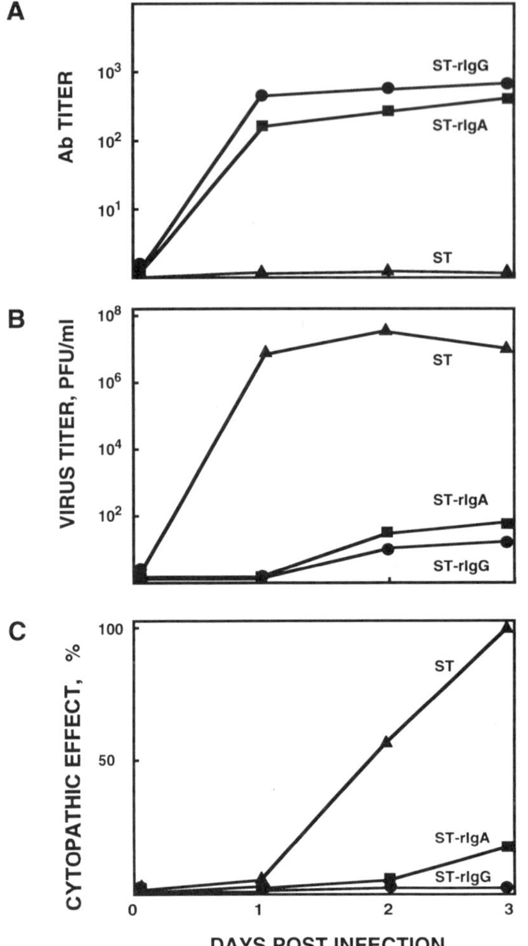

Figure 3. Kinetics of antibody production, virus titer, and cytopathic effects in recombinant antibody producing cells. Untransformed ST cells or cell lines producing recombinant IgG1 or IgA were infected with TGEV. (A) Antibody titers for TGEV were determined by RIA. (B) Virus titers were determined using ST cells. (C) The cytopathic effect was calculated by counting the cells that remained bound to culture plates and was expressed as the percentage of the cells that were detached. ST, untransformed cells. ST-rIgG and ST-rIgA, cells producing rIgG or rIgA neutralizing TGEV, respectively. One representative experiment selected from three assays with similar results is shown.

ST cells produced VSV as efficiently as did the untransformed cells. In addition, the TGEV titer was similar in the supernatant of cell lines in which only 10 to 15% of the cells were transformed, than in the cloned transformed cell lines in which 100% of the cells were antibody producers. The most likely explanation for this residual virus persistence is that the antibody-virus neutralization complex is reversible and, at the antibody concentrations present in the supernatants of the rMAb transformed cells, a small virus fraction (10^2–10^3 PFU) remains as free infectious virus, while most of the virus (10^4 to 10^5 PFU) is in the form of non-infections virus-antibody complexes (Suñé et al., 1990).

Extracellular neutralization is likely to be responsible for the results of this study, because the antibody is continuously released into the medium, even during virus infection, and in the immunoglobulin gene-transformed uncloned cells about 85 to 90% of the cells did not produce the antibody yet they were still protected from TGEV infection. Intracellular neutralization of virus in cultured cells has been demonstrated in the influenza and Sendai virus systems during transcytosis of dimeric IgA (dIgA) (Mazanec et al., 1995). This mechanism may also contribute part of the observed protection in antibody-producing ST cells infected with TGEV since during virus and antibody synthesis these proteins could co-localize leading to an intracellular interference of protein transport as described in other systems (Mazanec et al., 1992).

A neutralization escape mutant was not selected for because the virus produced at 1 and 6 days p.i. was neutralized with the same efficiency by MAb 6A.C3, produced by the transformed ST cells, and by MAb 1G.A7, which binds to a different antigenic subsite of the TGEV spike protein (Correa et al., 1988; Gebauer et al., 1991). This is in agreement with our previous results indicating that the epitope recognized by MAb 6A.C3 is an inter-species-conserved epitope which seems to be essential for virus replication, and for which neutralization escape mutants have never been observed (Sanchez et al., 1990).

The transient expression of virus neutralizing antibodies in mucosal surfaces could be used to provide immediate protection of these tissues against viral infection. This type of somatic cell gene therapy may particularly be efficacious to protect newborn animals. This new vaccination strategy may also be advantageous both in the prevention of virus infections, and therapeutically, following the possible ingestion of a variety of viruses, where fast immune intervention is required in a defined tissue, since viral vectors could express antibody genes within two to three hours following delivery.

ACKNOWLEDGMENTS

This work has been supported by grants from the Consejo Superior de Investigaciones Científicas, the Comisión Interministerial de Ciencia y Tecnología (CICYT), La Consejería de Educación y Cultura de la Comunidad de Madrid, and Laboratorios Sobrino (Cyanamid Group) from Spain, and the European Communities (Projects Biotechnology and FAIR). JC and IS received fellowships from the Department of Education, University and Research of the Gobierno Vasco and the Consejo Superior de Investigaciones Científicas (Spain), respectively.

REFERENCES

Armstrong, S. J., Outlaw, M. C., and Dimmock, N. J., 1990, Morphological studies of neutralization of influenza virus by IgM, *J. Gen. Virol.* **71:** 2313–2319.

Bullido, M. J., Correa, I., Jiménez, G., Suñé, C., Gebauer, F., and Enjuanes, L., 1989, Induction of transmissible gastroenteritis coronavirus-neutralizing antibodies *in vitro* by virus-specific T helper cell hybridomas, *J. Gen. Virol.* **70:** 659 - 672.

Castilla, J., Sola, I., and Enjuanes, L., 1997, Interference of coronavirus infection by expression of immunoglobulin G (IgG) or IgA virus-neutralizing antibodies, *J. Virol.* **71:** 5251–5258.

Correa, I., Jiménez, G., Suñé, C., Bullido, M. J., and Enjuanes, L., 1988, Antigenic structure of the E2 glycoprotein from transmissible gastroenteritis coronavirus, *Virus. Res.* **10:** 77–94.

Enjuanes, L., and Van der Zeijst, B. A. M., 1995, Molecular basis of transmissible gastroenteritis coronavirus epidemiology, in: *The Coronaviridae* (S. G. Siddell, eds.), Plenum Press, New York, pp. 337–376.

Gebauer, F., Posthumus, W. A. P., Correa, I., Suñé, C., Sánchez, C. M., Smerdou, C., Lenstra, J. A., Meloen, R., and Enjuanes, L., 1991, Residues involved in the formation of the antigenic sites of the S protein of transmissible gastroenteritis coronavirus, *Virology* **183:** 225–238.

Jiménez, G., Correa, I., Melgosa, M. P., Bullido, M. J., and Enjuanes, L., 1986, Critical epitopes in transmissible gastroenteritis virus neutralization, *J. Virol.* **60:** 131–139.

Marasco, W. A., Haseltine, W. A., and Chen, S., 1993, Design, intracellular expression, and activity of a human anti-human immunodeficiency virus type 1 gp120 single-chain antibody, *Proc. Natl. Acad. Sci. USA* **90:** 7889–7893.

Mazanec, M. B., Coudret, C. L., and Fletcher, D. R., 1995, Intracellular neutralization of influenza virus by immunoglobulin A anti-hemagglutinin monoclonal antibodies, *J. Virol.* **69:** 1339–1343.

Mazanec, M. B., Kaetzel, C. S., Lamm, M. E., Fletcher, D., and Nedrud, J. G., 1992, Intracellular neutralization of virus by immunoglobulin A antibodies, *Proc. Natl. Acad. Sci. USA* **89**: 6901–6905.

McClurkin, A. W., and Norman, J. O., 1966, Studies on transmissible gastroenteritis of swine. II. Selected characteristics of a cytopathogenic virus common to five isolates from transmissible gastroenteritis, *Can. J. Comp. Vet. Sci.* **30**: 190–198.

McGhee, J. R., Mestecky, J., Elson, C. O., and Kiyono, H., 1989, Regulation of IgA synthesis and immune response by T cells and interleukins, *J. Clin. Immunol.* **9**: 175–199.

Mestecky, J., and McGhee, J. R., 1987, Immunoglobulin A (IgA): molecular and cellular interactions involved in IgA biosynthesis and immune response, *Adv. Immunol.* **40**: 153–245

Saif, L. J., and Wesley, R. D., 1992, Transmissible gastroenteritis, in: *Diseases of Swine*, (A. D. Leman, B. E. Straw, W. L. Mengeling, S. D'Allaire, and D. J. Taylor, eds.), Wolfe Publishing Ltd, Ames. Iowa, pp. 362–386.

Sanchez, C. M., Gebauer, F., Suñé, C., Méndez, A., Dopazo, J., and Enjuanes, L., 1992, Genetic evolution and tropism of transmissible gastroenteritis coronaviruses, *Virology* **190**: 92–105.

Sanchez, C. M., Jiménez, G., Laviada, M. D., Correa, I., Suñé, C., Bullido, M. J., Gebauer, F., Smerdou, C., Callebaut, P., Escribano, J. M., and Enjuanes, L., 1990, Antigenic homology among coronaviruses related to transmissible gastroenteritis virus, *Virology* **174**: 410–417.

Sola, I., Castilla, J., Pintado, B., Sánchez-Morgado, J. M., Whitelaw, B., Clark, J., and Enjuanes, L., 1998, Transgenic mice secreting coronavirus neutralizing antibodies into the milk. *J. Virol.* **72**:3762–3772.

Stone, S. S., Kemeny, L. J., Woods, R. D., and Jensen, M. T., 1977, Efficacy of isolated colostral IgA, IgG, and IgM(A) to protect neonatal pigs against the coronavirus transmissible gastroenteritis, *Am. J. Vet. Res.* **38**: 1285–1288.

Suñé, C., Jiménez, G., Correa, I., Bullido, M. J., Gebauer, F., Smerdou, C., and Enjuanes, L., 1990, Mechanisms of transmissible gastroenteritis coronavirus neutralization, *Virology* **177**: 559–569.

87

LACTOGENIC IMMUNITY IN TRANSGENIC MICE PRODUCING RECOMBINANT ANTIBODIES NEUTRALIZING CORONAVIRUS

J. Castilla,[1] I. Sola,[1] B. Pintado,[2] J. M. Sánchez-Morgado,[1] and L. Enjuanes[1]

[1]Department of Molecular and Cell Biology
Centro Nacional de Biotecnología, CSIC
Campus Universidad Autónoma, Canto Blanco
28049 Madrid, Spain
[2]Department of Animal Reproduction. INIA
Carretera La Coruña km 5.9
28040 Madrid, Spain

1. ABSTRACT

Protection against coronavirus infections can be provided by the oral administration of virus neutralizing antibodies. To provide lactogenic immunity, eighteen lines of transgenic mice secreting a recombinant IgG_1 monoclonal antibody ($rIgG_1$) and ten lines of transgenic mice secreting recombinant IgA monoclonal antibodies (rIgA) neutralizing transmissible gastroenteritis coronavirus (TGEV) into the milk were generated. Genes encoding the light and heavy chains of monoclonal antibody (MAb) 6A.C3 were expressed under the control of regulatory sequences derived from the mouse genomic DNA encoding the whey acidic protein (WAP) and beta-lactoglobulin (BLG), which are highly abundant milk proteins. The MAb 6A.C3 binds to a highly conserved epitope present in coronaviruses of several species. This MAb does not allow the selection of neutralization escaping virus mutants. The antibody was expressed in the milk of transgenic mice with titers of one million as determined by RIA, and neutralized TGEV infectivity by one million fold corresponding to immunoglobulin concentrations of 5 to 6 mg per ml. Matrix attachment regions (MAR) sequences were not essential for $rIgG_1$ transgene expression, but co-microinjection of MAR and antibody genes led to a twenty to ten thousand-fold increase in the antibody titer in 50% of the $rIgG_1$ transgenic animals generated. Co-microinjection of the genomic BLG gene with rIgA light and heavy chain genes led to the generation of transgenic mice carrying the three transgenes. The highest antibody titers were produced by transgenic mice that had integrated the antibody and BLG genes, al-

Coronaviruses and Arteriviruses, edited by Enjuanes *et al.*
Plenum Press, New York, 1998

though the number of transgenic animals generated does not allow a definitive conclusion on the enhancing effect of BLG co-integration. Antibody expression levels were transgene copy number independent and integration site dependent. The generation of transgenic animals producing virus neutralizing antibodies in the milk could be a general approach to provide protection against neonatal infections of the enteric tract.

2. INTRODUCTION

The mucosal immune system and one of its predominant effectors, immunoglobulin, provide the initial immunologic barriers against most pathogens that invade the body at mucosal surfaces (Mazanec *et al.*, 1996). This is especially true for viruses, since resistance to infection has been frequently correlated with the presence of specific antibody in mucosal secretions. Of particular interest is the lactogenic immunity by which the antibodies in the milk confer immunity to newborns. The oral administration of virus neutralizing MAbs may protect mucosal tissues against viral infections that invade enteric areas (Enjuanes and Van der Zeijst, 1995; Saif and Wesley, 1992).

Transgenic technology now makes possible to express novel proteins in the milk under the control of regulatory sequences controlling the expression of abundant milk proteins such as the whey acidic protein (WAP) and the β-lactoglobulin (Castilla *et al.*, 1997; Sola *et al.*, 1997). This approach could be extended to the expression of virus neutralizing antibodies in order to provide protection against enteric infections.

The immune response to TGEV has been characterized and full protection against TGEV can be provided by lactogenic immunity. It has also been shown that the passive oral administration of serum elicited by recombinant adenoviruses expressing the spike protein completely protects piglets against virulent virus challenge (Torres *et al.*, 1995). Bottle feeding of newborn animals may target the antibody to epithelial surfaces providing protection against enteric virus infections, but this approach is not practical. Alternatively, transgenic animals secreting virus neutralizing antibodies in their milk during lactation should protect piglets against coronavirus infection. To explore this approach, the expression of a recombinant TGEV neutralizing MAb with a human IgG_1 and swine IgA isotypes under the control of regulatory sequences controlling the expression of abundant milk proteins was investigated.

TGEV infects both enteric and respiratory tissues and causes a mortality close to one hundred percent when newborn animals are infected (Enjuanes and Van der Zeijst, 1995; Saif and Wesley, 1992). The major antigenic sites of TGEV involved in the induction of virus neutralizing antibodies are located in the globular portion of the spike (S) protein (Correa *et al.*, 1990; Delmas and Laude, 1990; Gebauer *et al.*, 1991). Investigations by our laboratory into the mechanisms of TGEV neutralization and of antigenic and genetic variability (Sanchez *et al.*, 1992; Sanchez *et al.*, 1990) have led to the identification of mouse MAb 6A.C3 which neutralized all the TGEV isolates tested, and also TGEV-related coronaviruses which infect at least three animal species: porcine, canine, and feline. This MAb probably binds to an epitope essential for virus replication since no neutralization escape mutants were generated (Gebauer *et al.*, 1991) and has been used in the construction of the transgenic mice.

Transgenic mice were constructed carrying two expression cassettes containing the cDNA sequences encoding the light and heavy chains of a chimeric immunoglobulin. $rIgG_1$ cDNAs were co-microinjected with MAR sequences that increase transgene expression by flanking the transcription units and isolating transgenes from the negative influences of surrounding regions or by altering chromatin structure (Bonifer *et al.*, 1990). rIgA cDNAs were co-microinjected with gene expression regulatory sequences derived from the BLG gene. Transgenic mice that secrete high titer virus neutralizing $rIgG_1$ and rIgA into their milk have been obtained.

3. MATERIALS AND METHODS

3.1. Cells, Viruses, and Antibodies

Swine testis (ST) cells, COS-1 cells, non secreting murine myeloma Sp2/0, MAb 6A.C3, and S2.1 IgA secreting porcine hybridoma were grown as previously described (Castilla, Sola, and Enjuanes, 1997). TGEV PUR46-MAD strain was grown, purified, and titrated in ST cells as described previously (Jiménez *et al.*, 1986). Antibody titers were determined by radioimmunoassay (RIA) using purified TGEV as the antigen following a reported procedure (Jiménez *et al.*, 1986). Heavy chain titers were determined by a double antibody sandwich (DAS) similar to the RIA (Castilla *et al.*, 1997).

3.2. Western Blot Analysis

The procedures for Western blot have been described (Correa *et al.*, 1988). The antiserum used to develop the RIA assays to detect $rIgG_1$ RIA (Castilla *et al.*, 1997) was rabbit anti-human IgG (Cappel). To detect rIgA rabbit anti-swine IgA (Bethyl Laboratories, Inc. Texas. USA) was used.

3.3. Construction of Transgenes

cDNAs encoding the constant modules of human *gamma* and *kappa* chains (Castilla, Sola, and Enjuanes, 1997) were expressed using regulatory sequences from the WAP gene. The two cDNA fragments were blunt ended and inserted separately at the unique EcoRV site of the plasmid pWAP7K (Hennighausen and Sippel, 1982) carrying the WAP gene (Fig. 1A).

cDNAs encoding the constant modules of porcine *alpha* and *kappa* chains (Sola *et al.*, 1997) were expressed using a BLG construct pSS1tgXS. To generate the expression cassette pBJ41 (Fig. 1B), an EcoRV cloning site was inserted between exons 1 and 5 of BLG gene. Introns 5 and 6 were removed. The cDNA fragments encoding the chimeric light and heavy chains of rIgA were cloned separately at the unique EcoRV site of the plasmid pBJ41.

3.4. Functional Analysis of Transgenes

In order to study whether the resulting plasmids carrying the immunoglobulin genes under the control of WAP and BLG regulatory sequences were functional and hormone inducible, mouse mammary gland epithelial cells (HC11) were used. Plasmids encoding immunoglobulin light and heavy chains under the control of WAP or BLG gene regulatory elements were linearized and were co-transfected into 2×10^7 HC11 cells by electroporation. Antibody expression levels from transformed HC11 cells were analyzed in the supernatant by RIA after hormone induction.

3.5. Generation of Transgenic Mice

Transgenic mice were generated essentially as described (Castilla *et al.*, 1997; Sola *et al.*, 1997).

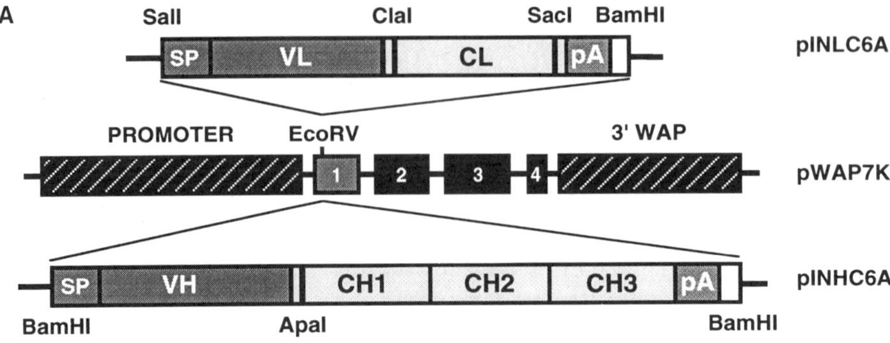

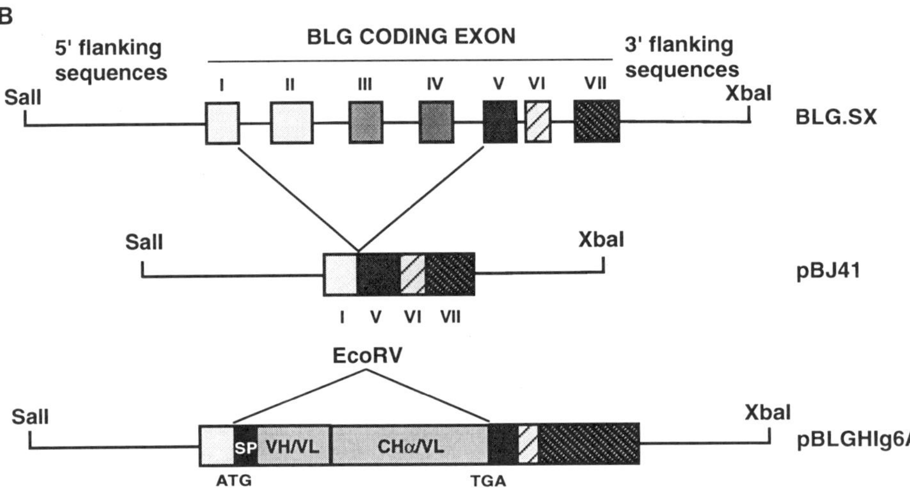

Figure 1. Cloning of immunoglobulin light and heavy chain cDNAs into expression cassettes. (A) The two cDNA fragments encoding the chimeric light and heavy chains of rIgG$_1$, were extracted from the plasmids pINLC6A and pINHC6A respectively and inserted separately at the unique EcoRV site of the plasmid pWAP7K carrying the WAP gene. SP, signal peptide. VL and VH, variable modules. CL, CH1, CH2 and CH3, constant modules. pA, SV40 polyadenylation signal. Key restriction endonuclease sites used in the cloning are indicated. (B) Structure of the BLG gene (BLG.SX) indicating the introns deleted to generate the pBJ41 plasmid carrying the intronless BLG-minigene. The ATG present in the BLG was removed in pBJ41 plasmid and an EcoRV cloning site was created by linkers between exons I and V. Exon VII does not have an open reading frame. CAAT, TATAA and AATAAA regulatory signals are indicated. Bold lines indicate 5'- and 3'-flanking sequences and intronic sequences of BLG; boxes, BLG exon sequences. VH or VL and CHα or CL, cDNA encoding immunoglobulin chains. All constructs comprise identical 5'- and 3'-regulatory sequences. Relevant restriction enzyme cutting sites are shown. SP, signal peptide.

3.6. Screening of Founders

DNA from tail biopsies was prepared as described (Castilla *et al.*, 1997; Sola *et al.*, 1997). The presence of WAP or BLG-immunoglobulin transgenes and MAR sequences were identified by a PCR assay using primers specific for the WAP or BLG promoters, the immunoglobulin variable modules and MAR sequences (Castilla *et al.*, 1997; Sola *et al.*, 1997).

To determine transgene copy number, PCR-positive animals were examined by Southern blot analysis using a probe that detected the expression cassette for the L and H immunoglobulin genes and also the single copy endogenous WAP gene. Genomic DNA was digested with BglII and three fragments corresponding to endogenous WAP, WAP-LC, and WAP-HC genes were detected by hybridization. The radiolabeled DNA fragment used to probe the blot was specific to the common WAP promoter allowing a comparison of the transgene copy number in each animal on the same Southern blot.

In the case of BLG constructs, EcoRI-cleaved DNA was analyzed by agarose gel electrophoresis and Southern blotting. The DNA fragment used to probe the blot was specific for the common BLG promoter and therefore detected the light and heavy immunoglobulin chain genes and the BLG gene simultaneously, allowing the comparison of the copy numbers of the three on the same Southern-blot. Transgene copy number was determined by comparison with known amounts of reference plasmids.

3.7. Characterization and Functional Activity of the Recombinant Antibody in Mouse Milk

Milk from four to six month old transgenic mice was collected as described (Castilla *et al.*, 1997).

4. RESULTS

4.1. Engineering TGEV Neutralizing Recombinant Antibodies

The construction of the chimeric recombinant antibodies required the fusion of MAb 6A.C3 light (L) or heavy (H) variable modules to human or porcine constant modules as previously reported (Castilla *et al.*, 1997; Sola *et al.*, 1997). Immunoglobulin variable modules of the MAb 6A.C3 were cloned by RT-PCR and joined to the constant module of human or porcine antibodies. To verify the functionality of $rIgG_1$ and rIgA, COS-1 and murine Sp2/0 myeloma cells were transiently and stably transformed, respectively, with constructs encoding the chimeric L and H chains. Antibody expression levels ranging between 20 to 40 µg/ml were detected in the supernatant. The secreted chimeric immunoglobulins bound TGEV by RIA with titers up to 10^3, and neutralized virus infectivity around 10^4-fold (i.e., neutralization indices, NIs, around 4), indicating that the recombinant immunoglobulins had the expected biological activity.

In order to generate transgenic mice expressing mouse-human $rIgG_1$ 6A.C3 or the rIgA 6A.C3 in the milk, rMAbs were expressed under the control of WAP or BLG gene regulatory sequences, respectively by inserting immunoglobulin gene cDNAs encoding L and H chains of MAb 6A.C3 into exon 1 of the WAP gene (Fig. 1A) or between exons 1 and 5 of a BLG derived expression vector (Fig. 1B).

To study whether antibody expression under the control WAP or BLG regulatory sequences was hormone inducible in epithelial HC11 cells, these cells were stably transformed with the WAP or BLG-immunoglobulin constructs. Antibody expression was lactogenic hormone dependent (results not shown) indicating that these mouse mammary gland epithelial cells synthesized, assembled and secreted functional recombinant antibodies.

4.2. Obtention of $rIgG_1$ Transgenic Mice

An equimolar mix of the WAP gene-based expression cassettes encoding L and H chains were co-microinjected into the pronucleus of fertilized ova. These expression cassettes were co-microinjected with an equimolar amount of MAR sequences as indicated. To identify founder animals carrying transgenes, genomic DNA isolated from tail biopsies of animals born from microinjected fertilized eggs was screened by PCR for the presence of mouse WAP-immunoglobulin transgenes using primer pairs that hybridized with the WAP gene (5' primer) and with the immunoglobulin gene (3' primer). From embryos microinjected using the immunoglobulin genes without MAR sequences (MAR⁻), 13 transgenic mice were obtained. From embryos microinjected with the immunoglobulin gene with MAR sequences (MAR⁺) 33 transgenic mice were selected. The majority of transgenic mice (35 out of 46) had co-integrated the transgenes encoding both the light and heavy chains of rIgG$_1$. All 27 animals that were generated from embryos microinjected with DNAs encoding the light and heavy chain of the recombinant antibody with MAR sequences had integrated the three DNAs (data not shown). Transgenic founder animals carrying MAR sequences, light, and heavy genes were bred for further analysis.

A high number of animals (8 MAR⁻ and 27 MAR⁺) that had integrated both the heavy and light immunoglobulin gene chains transmitted all of these genes to their progeny for at least four generations (data not shown), strongly suggesting that the light and the heavy chains co-integrated in the nineteen selected mice and co-segregated as a single Mendelian character.

4.3. Construction of rIgA Transgenic Mice

Analysis of DNA prepared from tail biopsies showed that 23 of the 93 generation zero (G0) mice had integrated at least one of the transgenes. Transgene integration in the genome of a modified animal does not guarantee its expression, since it may be integrated in a silent chromosome region. It has been previously reported that it is possible to enhance the efficiency of transgene expression by co-integrating the expression cassette with a genomic clone of BLG. To attempt an increase of antibody expression, transgenic (BLG⁺) mice were produced by co-injection of the BLG gene with the expression cassettes for the light (BLG-SLC) and heavy (BLG-SHC) chains. In all (16 out of 16) of the BLG⁺ transgenic mice in which one or both immunoglobulin genes were integrated, the BLG gene was also integrated (data not shown). The integration of expression cassettes and of genomic BLG was determined by PCR. After screening more than 250 progeny mice, derived from the 16 founder mice, the co-segregation of immunoglobulin and BLG genes was observed in a high proportion of transgenic animals (>68%), indicating that in general both immunoglobulin and BLG genes had been integrated in the same chromosomal locus. The majority of the BLG⁻ and BLG⁺ transgenic mice (17 out of 23) had co-integrated the transgenes encoding the light and heavy rIgA chains. Twelve of the BLG⁺ transgenic mice carrying both light and heavy chains had integrated the BLG gene.

At least 10 out of 17 transgenic founders carrying both SLC+SHC transmitted both transgenes to their progeny suggesting that they have been co-integrated in a single site in each line.

4.4. Expression of rIgG$_1$ in Milk

Milk was collected from the generation one (G1) of each transgenic mouse line which transmitted both transgenes. rIgG$_1$ was detected by RIA in the milk using TGEV as an antigen. Transgenic (MAR⁻) and (MAR⁺) produced high titer TGEV specific antibodies in the milk (Fig. 2A). Five out of the five MAR⁻ (two producing the whole antibody and three secreting the heavy chain) secreted TGEV specific antibodies with titers of around 1×10^2 as

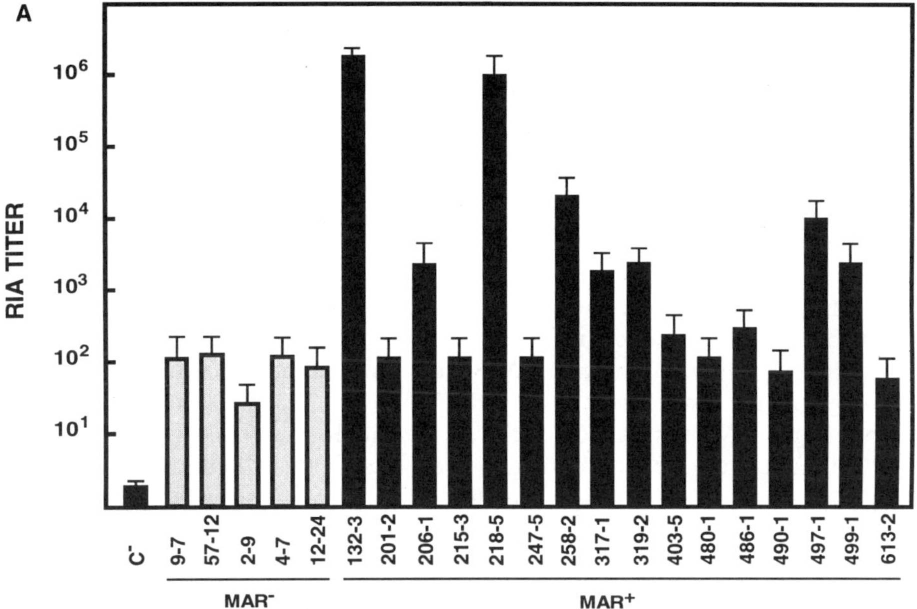

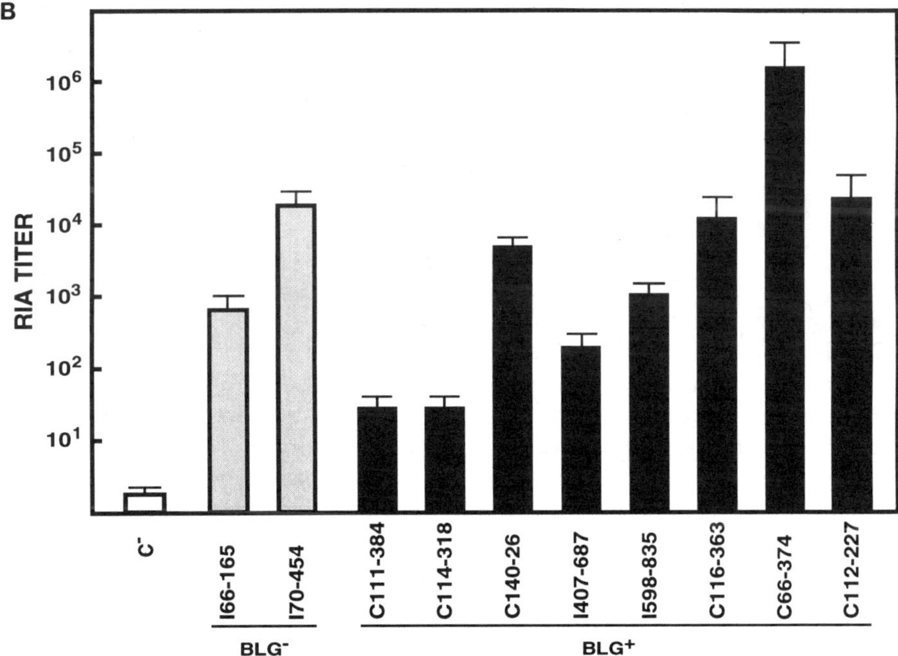

Figure 2. Analysis of recombinant immunoglobulin expression in the milk of transgenic mice. (A) IgG1 expression. Milk from four transgenic mice that were generated by microinjection of immunoglobulin gene expression cassettes without the MAR sequences (MAR⁻) and sixteen mice in which immunoglobulin genes were co-microinjected with MAR sequences (MAR⁺) were evaluated for the presence of TGEV binding antibodies by RIA. Four to ten mice of each line were analyzed and the mean ± the standard deviation of the maximum titers obtained during lactation are shown. C⁻, milk from non-transgenic mice. (B) IgA expression. The RIA titers for rIgA in the milk of BLG⁻ and BLG⁺ transgenic mice are shown as the mean of the antibody titers in three to six milk samples of the same transgenic mice collected at different days through lactation. C⁻, milk from non-transgenic mice.

determined by RIA (Fig. 2A), while 16 out of the 17 transgenic MAR⁺ mice expressed rIgG$_1$ in the milk with titers ranging from 1×10^2 to 1×10^6. Within each line of transgenic mice no significant variation in rIgG$_1$ expression levels was observed. The two MAR⁻ mice and the sixteen MAR⁺ transgenic mice that transmitted both light and heavy chain genes to the progeny for at least two generations secreted virus neutralizing antibodies into the milk (results not shown).

The co-integration of MAR sequences was not an essential requirement for the expression of immunoglobulin genes (results not shown). The titers of the secreted TGEV specific antibody or immunoglobulin chain produced in five of the MAR⁻ mice were around 10^2. The antibody titers in the milk of these MAR⁻ mice were determined by direct RIA in the two lines of transgenic mice secreting both the light and heavy antibody chains, and using a double antibody sandwich (DAS) technique in the three lines of transgenic mice secreting only the heavy chain of the same MAb (Fig. 2A). Interestingly, in the mice that had co-integrated MAR sequences antibody titers in the milk ranged between 10^2 and 10^6. In 50% of these mice (8 out of 16) antibody expression levels were increased between twenty and ten thousand-fold in relationship to the antibody levels produced by the MAR⁻ mice.

The milk from selected MAR⁺ transgenic females 132-3 and 218-5 had TGEV specific antibodies with RIA titers close to 10^6. The milk of these mice showed a reduction in TGEV infectivity around one million-fold (results not shown). These results indicated that the rIgG$_1$ synthesized in the mammary gland and secreted into the milk of transgenic mice was active in TGEV neutralization. Recombinant IgG$_1$ production ranged from 0.005 mg/ml to approximately 5 mg/ml in the different mice.

4.5. Expression of rIgA in Milk

Milk was collected from G0 females or female progeny of mice which transmitted the transgenes for both the heavy and light genes. Recombinant IgA was detected by RIA in the milk of the two BLG⁻ transgenic lines, with titers ranging from 8×10^2 to 3×10^4 (Fig. 2B). Eight out of 12 BLG⁺ transgenic founders expressed rIgA in the milk with titers in RIA ranging from 5×10^1 to 5×10^6.

Neutralization assays with milk samples showed that virus infectivity was reduced around one million-fold using milk with the highest titers (mouse C66–374). These results indicated that the rIgA synthesized in the mammary gland and secreted into the milk of transgenic mice was biologically active in TGEV neutralization.

No significant differences in transgene expression frequency were observed between BLG⁻ and BLG⁺ transgenic mice (Fig. 2A). Nevertheless, higher Ab titers ($>10^6$) were obtained in mice that had cointegrated immunoglobulin and BLG genes (Fig. 2A).

The rIgA concentrations in transgenic mouse milk were measured by RIA using a purified rIgA standard and ranged from 0.005 mg/ml to 6 mg/ml.

4.6. Analysis of Transgene Integration

To determine whether antibody expression levels were transgene copy number dependent, transgenic mice carrying rIgG$_1$ and rIgA genes were analyzed by Southern blot using a probe specific for a promoter sequence present in the expression cassette for the light and heavy immunoglobulin chain genes and in the endogenous WAP or BLG gene. DNA extracted from the tails of transgenic mice was digested with convenient restriction endonucleases to identify specific fragments (Fig. 3A and 3C) which corresponded to the

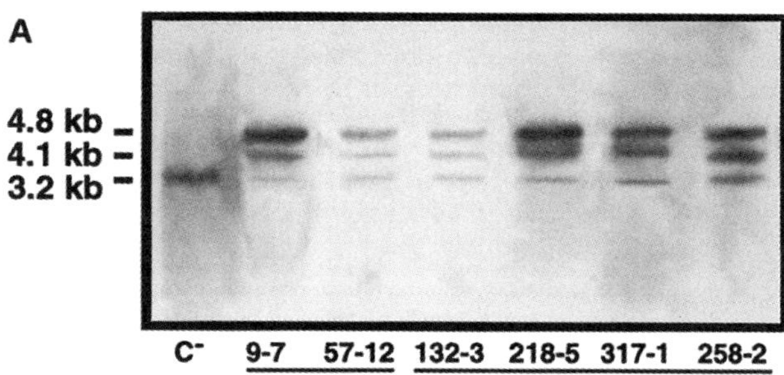

B

Transgenic mice		Copy number[a]		Expression[b]
		Light Chain	Heavy Chain	
MAR⁻	9-7	10	40	$1\text{-}5\times10^2$
	57-12	1	3	$1\text{-}5\times10^2$
MAR⁺	132-3	1	3	$1\text{-}8\times10^6$
	218-5	3	12	$1\text{-}5\times10^5$
	317-1	2	3	$1\text{-}5\times10^3$
	258-2	2	2	$1\text{-}5\times10^4$

a. Transgene copy number.
b. Titer, highest dilution giving a binding three-fold higher than the background.

Figure 3. Relationship between transgene copy number and antibody expression levels. (A) Southern blot analysis of WAP-immunoglobulin transgene integration. Genomic DNA from transgenic mice were hybridized with a WAP-specific probe. The bands shown correspond to WAP-heavy transgene (4.8 kb), WAP-light transgene (4.1 kb), and endogenous WAP gene (3.2 kb) sequences. (B) Antibody expression levels in the milk of transgenic mice with variable numbers of transgene copy numbers. MAR⁺, with MAR sequences integrated. MAR⁻, without MAR sequences integrated.

genes encoding the light and heavy immunoglobulin chains and the endogenous genes (WAP and BLG). The intensity of the endogenous gene band was taken as the reference to estimate the integrated immunoglobulin gene copy number. A large variation in the transgene copy number (from 1 to 40) was detected in the transgenic lines analyzed (Figure 3B and 3D) and no correlation was observed between antibody expression levels and transgene copy number.

5. DISCUSSION

Transmisible gastroenteritis virus produces high mortality in suckling piglets. In order to effectively protect offspring, it is necessary to produce high levels of neutralizing antibodies in the milk of lactating sows. To this end, we have constructed transgenic mice expressing the genes encoding the chimeric TGEV neutralizing MAb 6A.C3 with human constant module (IgG_1 isotype) and porcine constant module (IgA isotype) under the control of abundant milk protein (WAP and BLG) gene regulatory elements, respectively.

Transgenic mice were constructed that secreted the $rIgG_1$ and rIgA MAb into their milk with titers up to 5×10^6 as determined both by RIA and neutralization. Chimeric IgA antibodies have been shown to be more efficient in virus neutralization than the recombinant antibodies with identical specificity and the IgG isotype, probably because of the tetravalency of dimeric rIgA.

The expression levels of functional TGEV-specific $rIgG_1$ and rIgA in the milk of several transgenic mice (up to 6 mg/ml) represents one of the highest expression levels ob-

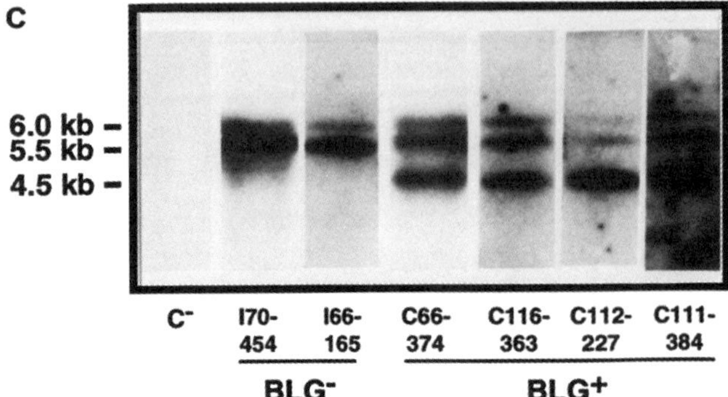

D

Transgenic mice		Copy number[a]		Expression[b]
		Light Chain	Heavy Chain	
BLG⁻	I70-454	8	2	3×10^4
	I66-165	6	2	$4\text{-}8 \times 10^2$
BLG⁺	C66-374	4	2	$1\text{-}5 \times 10^6$
	C116-363	5	3	$1\text{-}8 \times 10^3$
	C112-227	4	2	$1\text{-}5 \times 10^4$
	C111-384	6	2	50

a. Transgene copy number.
b. Titer, highest dilution giving a binding three-fold higher than the background.

Figure 3. *(Continued)* (C) Southern-blot analysis of BLG⁻ and BLG⁺ transgenic lines. C⁻, non-transgenic mouse DNA. Tail DNA samples were cleaved with EcoRI and analyzed by electrophoresis and Southern-blot. Filters were probed with 5' flanking sequences common to the three transgenes. The 6.0 kb BLG-SHC, 5.5 kb BLG-SLC and 4.5 kb BLG⁻ transgene specific EcoRI fragments are shown. (D) Comparison of rIgA titers in BLG⁻ and BLG⁺ transgenic lines carrying different transgene copy numbers. The expression levels are reported as the RIA titers of the highest dilution giving a binding three-fold the background. The RIA titers of rIgA in the milk of BLG⁻ and BLG⁺ transgenic mice were calculated as the mean of antibody titers of three to six milk samples from the same mouse.

served in a complex recombinant protein in any mammalian expression system, including transgenic mice. This result was particularly interesting since it indicated that the epithelial cells of the mammary gland successfully produced both the light and heavy immunoglobulin chains and provided the adequate environment for the assembly of a functional IgG$_1$ and IgA molecule.

Recombinant immunoglobulin expression levels in the milk of sows, similar to those produced by the transgenic mice described in this manuscript, may suffice to protect piglets against TGEV infection. The recombinant immunoglobulin that has been used contains the variable modules of the MAb 6AC3, that neutralizes very efficiently all known TGEV strains and does not lead to the selection of neutralization escape mutants. This fact, the continuous intake of virus neutralizing recombinant antibodies from the milk of trangenic sows during lactation and the natural defenses should provide *in vivo* protection against TGEV infection.

In the experimental model system used, the transgenic lines of mice that were not co-microinjected with MAR sequences also expressed rIgG$_1$. Consequently, we could not assert the effect of MAR sequence inclusion on the frequency of antibody-expressing mice. Among the mice that co-integrated MAR sequences with rIgG$_1$ genes, no direct relationship between transgene copy number and the rIgG$_1$ expression levels in the milk was found. This suggests that transgene integration site determines transcriptional levels.

In contrast with the findings of other investigators (Clark *et al.*, 1993), the co-microinjection of BLG sequences with antibody genes has not led to a significant increase of the average antibody expression levels, since in the absence or in the presence of BLG gene, aproximately the same antibody titers were obtained in the milk of transgenic mice. Nevertheless, it is interesting to note that maximum antibody expression levels were obtained when BLG and antibody sequences were co-microinjected, although, since a small number of transgenic is presented, the significance of BLG co-integration to achieve high antibody expression levels can not be definitively concluded. The requirement for introns to achieve an efficient transgene expression, probably due to the presence of cis-acting elements, is well documented (Whitelaw *et al.*, 1991). However, in our system, cDNAs encoding rIgA light and heavy chains were inserted into BLG intronless constructs and an efficient expression of rIgA was observed in the milk of the transgenic animals, suggesting that sequences within these cDNAs can also favor expression. One possibility is that some sequences present in the variable module of MAb 6A.C3 light and heavy chains may enhance expression, since this MAb was selected from two thousand independently generated MAbs because its specificity and high expression levels. In this context, transgene rescue by co-microinjecting the BLG gene with the transgenes (Clark *et al.*, 1993; Clark *et al.*, 1992) may not enhance the efficiency of expression as dramatically as in those cases with a very inefficient expression of the intronless transgenes.

Normal development in the mice secreting high titer coronavirus neutralizing antibodies in the milk was observed, indicating that the production of pathogen neutralizing antibodies in the milk could be a useful approach to prevent enteric infections of the newborn.

The modular approach to obtain recombinant antibodies (i.e., the fusion of variable to constant immunoglobulin domains) described in this manuscript could easily be applied to other antibodies with different therapeutic purposes. Transgenic swine expressing TGEV-neutralizing recombinant immunoglobulins are currently being made using the same expression cassettes described in this paper. This new system will allow us to directly test the lactogenic immunity provided by the transgenic sows to neonates following challenge with TGEV. Since MAbs specific for many enteric viruses are available, this strategy could be a general procedure to generate animals resistant to numerous viral infections of the enteric tract.

REFERENCES

Bonifer, C., Vidal, M., Grosveld, F., and Sippel, A. E., 1990, Tissue specific and protein position independent expression of the complete gene domain for chicken lysozyme in transgenic mice, *EMBO J.* **9**: 2843–2848.

Castilla, J., Pintado, B., Sola, I., Sánchez-Morgado, J. M., Hennighausen, L., and Enjuanes, L., 1998, Engineering lactogenic immunity in transgenic mice secreting virus neutralizing antibodies in the milk, *Nat. Biotech.* **16**: 349–353.

Castilla, J., Sola, I., and Enjuanes, L., 1997, Interference of coronavirus infection by expression of immunoglobulin G (IgG) or IgA virus-neutralizing antibodies, *J. Virol.* **71**: 5251–5258.

Clark, A. J., Archibald, A. L., McClenaghan, M., Simons, J. P., Wallace, R., and Whitelaw, C. B. A., 1993, Enhancing the efficiency of transgene expression, *Transg. Res. Soc. Lond.* **339**: 225–232.

Clark, A. J., Cowper, A., Wallace, R., Wright, G., and Simons, J. P., 1992, Rescuing transgene expression by co-integration, *Biotechnology* **10**: 1450–1454.

Correa, I., Gebauer, F., Bullido, M. J., Suné, C., Baay, M. F. D., Zwaagstra, K. A., Posthumus, W. P. A., Lenstra, J. A., and Enjuanes, L., 1990, Localization of antigenic sites of the E2 glycoprotein of transmissible gastroenteritis coronavirus, *J. Gen. Virol.* **71**: 271–279.

Correa, I., Jiménez, G., Suñé, C., Bullido, M. J., and Enjuanes, L., 1988, Antigenic structure of the E2 glycoprotein from transmissible gastroenteritis coronavirus, *Virus. Res.* **10**: 77–94.

Delmas, B., and Laude, H., 1990, Assembly of coronavirus spike protein into trimers and its role in epitope expression, *J. Virol.* **64**: 5367–5375.

Enjuanes, L., and Van der Zeijst, B. A. M., 1995, Molecular basis of transmissible gastroenteritis coronavirus epidemiology. *In* "The Coronaviridae" (S. G. Siddell, Ed.), pp. 337–376. Plenum Press, New York.

Gebauer, F., Posthumus, W. A. P., Correa, I., Suñé, C., Sánchez, C. M., Smerdou, C., Lenstra, J. A., Meloen, R., and Enjuanes, L., 1991, Residues involved in the formation of the antigenic sites of the S protein of transmissible gastroenteritis coronavirus, *Virology* **183**: 225–238.

Hennighausen, L. G., and Sippel, A. E., 1982, Mouse whey acidic protein is a novel member of the family of four-disulfide core proteins, *Nuc. Acid. Res.* **10**: 2677–2684.

Jiménez, G., Correa, I., Melgosa, M. P., Bullido, M. J., and Enjuanes, L., 1986, Critical epitopes in transmissible gastroenteritis virus neutralization, *J. Virol.* **60**: 131–139.

Mazanec, M. B., Huang, Y. T., Pimplikar, S. W., and Lamm, M. E., 1996, Mechanisms of inactivation of respiratory viruses by IgA, including intraepithelial neutralization, *Sem. Virol.* **7**: 285–292.

Saif, L. J., and Wesley, R. D., 1992, Transmissible gastroenteritis. Seventh ed. In *"Diseases of Swine"* (A. D. Leman, B. E. Straw, W. L. Mengeling, S. D'Allaire, and D. J. Taylor, Eds.), pp. 362–386. Wolfe Publishing Ltd, Ames. Iowa.

Sanchez, C. M., Gebauer, F., Suñé, C., Méndez, A., Dopazo, J., and Enjuanes, L., 1992, Genetic evolution and tropism of transmissible gastroenteritis coronaviruses, *Virology* **190**: 92–105.

Sanchez, C. M., Jiménez, G., Laviada, M. D., Correa, I., Suñé, C., Bullido, M. J., Gebauer, F., Smerdou, C., Callebaut, P., Escribano, J. M., and Enjuanes, L., 1990, Antigenic homology among coronaviruses related to transmissible gastroenteritis virus, *Virology* **174**: 410–417.

Sola, I., Castilla, J., Pintado, B., Sánchez-Morgado, J. M., Whitelaw, B., Clark, J., and Enjuanes, L., 1998, Transgenic mice secreting coronavirus neutralizing antibodies into the milk, *J. Virol.* **72**: 3762–3772.

Torres, J. M., Sánchez, C. M., Suñé, C., Smerdou, C., Prevec, L., Graham, F., and Enjuanes, L., 1995, Induction of antibodies protecting against transmissible gastroenteritis coronavirus (TGEV) by recombinant adenovirus expressing TGEV spike protein, *Virology* **213**: 503–516.

Whitelaw, C. B. A., Archibald, A. L., Harris, S., McClenaghan, M., Simons, L. P., and Clark, A. J., 1991, Targeting expression to the mammary gland: intronic sequences can enhance the efficiency of gene expression in transgenic mice, *Transgenic Res.* **1**: 3–13.

88

UTILISING A DEFECTIVE IBV RNA FOR HETEROLOGOUS GENE EXPRESSION WITH POTENTIAL PROPHYLACTIC APPLICATION

S. A. Evans, K. Stirrups, K. Dalton, K. Shaw, D. Cavanagh, and P. Britton

Division of Molecular Biology
Institute for Animal Health, Compton Laboratory
Compton, Newbury, Berkshire, RG20 7NN, United Kingdom

1. ABSTRACT

Based on the natural ability of coronaviruses to undergo homologous RNA recombination, we are working to produce infectious bronchitis virus (IBV) recombinants using RNA generated from recombinant fowlpox viruses (FPV). The aim is to replace the spike (S) gene of an existing IBV vaccine strain with the S gene of a heterologous strain. CD-61 is an IBV defective RNA (D-RNA) derived from a naturally occurring IBV D-RNA (CD-91). CD-61 D-RNA is being investigated as an RNA vector for the expression of heterologous genes. T7-derived RNA transcripts of CD-61 can be replicated and passaged in the presence of helper virus, following electroporation into IBV-infected cells. CD-61 cDNA was modified by the addition of the hepatitis delta virus ribozyme plus T7 terminator downstream of the 3' UTR. This allowed the synthesis of discreet RNA transcripts. The complete cassette was cloned into an FPV transfer vector (pEFL10) for generating recombinant fowlpox viruses. FPV/CD-61 recombinants will be assessed for D-RNA production in IBV-infected cells. The luciferase reporter gene sequence has been inserted into the modified CD-61, under the control of the IBV transcription associated sequence (TAS) from gene 5. Luciferase has been successfully expressed from CD-61 in helper virus-infected cells.

2. INTRODUCTION

The coronavirus, infectious bronchitis virus (IBV), is an enveloped virus with a 27.6kb single-stranded, positive sense RNA genome. The virus is both a welfare and economic problem in the chicken industry. A major contributory factor is that many antigenic variants of IBV exist, resulting from sequence differences in the spike (S) protein. The S protein forms

Coronaviruses and Arteriviruses, edited by Enjuanes *et al.*
Plenum Press, New York, 1998

the characteristic peplomer on the virion surface and is responsible for binding virus particles to cell surface receptors (Cavanagh, 1983). It is also involved in the generation of neutralising antibodies and cell-mediated immunity. In order to study pathogenicity, protein function and antigenicity of IBV we propose to replace the S gene (target gene) of an existing IBV vaccine strain with the S gene of a heterologous strain. Our approach to the production of recombinant IBV genomes is based on their tendency to undergo homologous RNA recombination by template switching, as demonstrated experimentally for IBV (Kottier *et al.*, 1995).

Coronaviruses have the largest genomes of any known RNA virus and the production of full length cDNA clones has not yet been achieved. For this reason, defective RNAs (minigenomes) of IBV have been employed as an RNA vector system for increasing the probability of recombination. Important features of these minigenomes include sequences for the replication and for the packaging of the RNA into virus particles. The helper IBV provides the functional polymerase and structural proteins. Interestingly, RNA recombination can occur between the coronavirus genome and synthetic transcripts from cDNA copies of D-RNAs. A naturally occurring 9.1kb IBV D-RNA (CD-91) was identified and characterised following repeated high-multiplicity passage of the Beaudette strain in chicken kidney (CK) cells. CD-91 consists of three discontinuous regions of the genome, with a long ORF encoding a potential product of 2,155 amino acids (Pénzes *et al.*, 1994). The replicating D-RNA CD-61 was generated from CD-91 by deletion of 3kb from the region of the polymerase gene (domain II). We have demonstrated that RNA produced from CD-61 cDNA, under the control of the T7 promoter, can be rescued (replicated and packaged into virions) by the Beaudette strain of IBV.

In this paper, we describe the construction of a modified cDNA clone of CD-61, containing a ribozyme/termination signal at the 3' end. Moreover, utilising a unique restriction site within a potentially inert region of the CD-61, a luciferase reporter gene was inserted into the modified cassette. We show the replication and packaging of modified CD-61 (CD-61+) RNA, following electroporation of *in vitro*-transcribed CD-61+ RNA into helper virus-infected cells. Consequently, CD-61+ constructs were used to prepare recombinant transfer plasmids. The presence of fowlpox virus (FPV)-specific flanking sequences in these shuttle vectors enabled the isolation of recombinant fowlpox viruses (rFPV). Co-infection of selected rFPV/CD-61+ with a rFPV expressing T7 RNA polymerase (Britton *et al.*, 1996) and IBV into cells, resulted in expression of luciferase and rescue of CD-61 RNA. Our observations indicate that rFPV/CD-61+ can be used to generate RNA for recombination with IBV; an important step towards isolating recombinant coronaviruses.

3. MATERIALS AND METHODS

3.1. Cells and Viruses

The Beaudette strain of IBV was propagated in primary chicken kidney (CK) cells as described by Pénzes *et al.* (1994). Fowlpox viruses; FP9 and rFPV/T7 (Britton *et al.* 1996) were grown in chicken embryo fibroblast (CEF) cells using 199 medium supplemented with 2% new born calf serum (NBCS).

3.2. Construction of Plasmids Containing Modified CD-61

Briefly, the hepatitis delta ribozyme (HδR) plus T7 RNA polymerase termination signal (Tφ) was isolated as a discreet fragment from pVec2,0 (donated by A. Ball) and was ligated to CD-61 cDNA. The resulting construct, T7/CD-61/HδR/Tφ (CD-61+), has a unique *Pma* CI restriction site within the N gene sequences of CD-61, which was utilised for the cloning of the reporter gene luciferase. The luciferase (LUC) gene, under the control of an IBV transcription associated sequence (TAS) from Beaudette gene 5, was inserted into CD-61+ to give CD-61+LUC. The CD-61+ and CD-61+LUC cassettes were excised from their respective plasmids for insertion into the unique *Sca* I site of the FPV transfer vector pEFL10. Recombinant plasmids EFCD and EFCD/LUC were sequenced to confirm the orientation of each insert.

3.3. Rescue of Modified CD-61 by IBV Beaudette

Plasmids containing CD-61+ and CD-61+LUC were transcribed *in vitro* by T7 RNA polymerase (Promega). The RNA transcripts were electroporated into IBV-infected CK cells as described by Pénzes *et al.* (1996). Infected supernatants were inoculated onto fresh CK cells ~ 24 hours post-IBV infection; this was termed passage 1 (P1). Infected cell sheets were harvested at each passage. Cell pellets were divided equally for a luciferase assay (Promega) and total cellular RNA extraction (Pénzes *et al.* 1994). RNA isolated from infected cell lysates were analysed by Northern blotting and hybridisation to ^{32}P-labelled N gene or 3' UTR PCR DNA.

Recombinant transfer vectors pEFCD and pEFCD/LUC were also analysed for rescue of CD-61. In this case, intact plasmid DNA was electroporated into CK cells dually infected with rFPV/T7 and IBV. Virus harvest was passaged on fresh CK cells and the infected cell sheets were treated as before.

3.4. Isolation of CD-61+ Recombinant Fowlpox Viruses

The shuttle vector pEFL10 contains sequences from the termini of FPV ORF 1, flanking the unique cloning site, to allow homologous recombination with the FPV genome. The β-Galactosidase (Lac Z) gene is also present which enables definitive selection of potential rFPVs. Transfer vectors pEFCD and pEFCD/LUC were transfected into FPV/FP9-infected CEF cells via lipofectin. Harvested viruses were plaque titrated on CEF cells in the presence of X-Gal. Blue-stained plaques were purified three times by titration. Isolated rFPV were analysed for LUC expression and rescue of CD-61 in cells infected with both rFPV/T7 and IBV.

4. RESULTS

The approach we have taken towards the production of IBV recombinant viruses involves incorporating the CD-61 vector, with target gene, into the genome of FPV such that IBV-infected cells co-infected with rFPV/CD-61 and rFPV/T7 will express CD-61 RNA. Before preparation of recombinant FPV could commence, CD-61 was manipulated at the cDNA level. The modification of CD-61 cDNA was achieved by fusing the hepatitis delta virus ribozyme (HδR) and T7 RNA polymerase termination sequence (Tφ) to the end of the IBV sequences (see Fig. 1). This allows termination of RNA synthesis at a specific sequence and for the cleavage of the RNA portion comprising HδR/Tφ by the ribozyme, to give the CD-61 3' end.

In vitro T7-transcripts from pCD-61/HδR/Tφ were electroporated into IBV-infected cells and subsequently passaged. Northern blot analysis of RNA extracted from infected

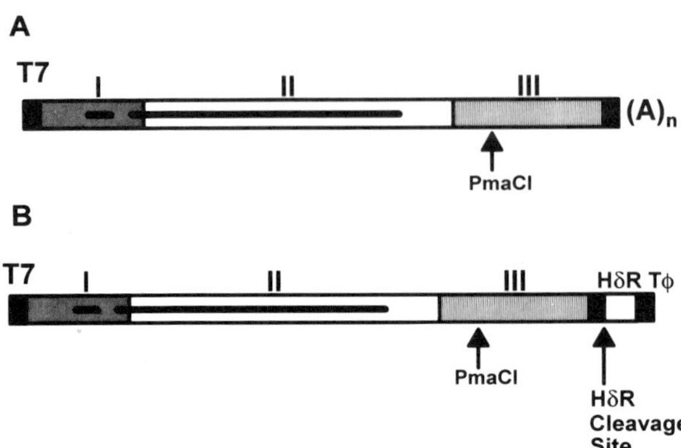

Figure 1. Modification of CD-61 cDNA. Salient features of CD-61 are shown in part (A). Three domains (I, II and III) depict discontinuous regions of the IBV genome followed by the poly (A) tail. The presence of a T7 promoter and unique *Pma* CI site are indicated. Sequences corresponding to the hepatitis delta ribozyme (HδR) and T7 terminator (Tφ) fused to the 3' end of CD-61 are shown in part (B). The ribozyme self cleavage site is indicated by an arrow. The unbroken line represents the truncated ORF 1a and CD-61-specific ORFs.

cell lysates showed rescue of CD-61 D-RNA from the modified minigenome (see Fig. 2). The new CD-61 RNA was detected by passage 4 (P4) of the virus. We concluded that RNA from CD-61+ could be replicated and packaged in the presence of helper virus, correlating with CD-61. This demonstrates that the additional HδR/Tφ sequences had not adversely affected CD-61, presumably because they had been successfully removed by self-cleavage by the ribozyme. A functional minireplicon cassette containing a ribozyme /termination signal is important for the generation of recombinant fowlpox viruses, since it will allow the production of discreet T7 RNA transcripts; preventing transcription of FPV genes downstream of CD-61 once inserted into the FPV genome.

For the purposes of expression of heterologous genes from the minigenome, a reporter gene was required to test the system. The luciferase (LUC) gene, directed by an IBV TAS, was cloned into the CD-61 cassette at the unique *Pma* CI site. Rescue of CD-61+LUC was confirmed by a luciferase assay, although not by Northern blotting.

Once the modified CD-61 cassettes were introduced into pEFL10, it was necessary to confirm that the minireplicon was still functional. This was achieved by transient expression of CD-61+ RNA with subsequent rescue attempts by IBV. Plasmid EFCD and EFCD/LUC DNAs were electroporated into CK cells co-infected with rFPV/T7 and IBV. Harvested virus was passaged on 24 hrs p.i. and cell lysates were assessed for luciferase activity and the presence of CD-61 D-RNA.

As can be seen from Figure 3, CD-61 was rescued from passage 4 (P4) following electroporation of pEFCD. This result demonstrated that CD-61 RNA generated *in situ*, by T7 polymerase, was replicated by helper IBV. Moreover, CD-61+LUC DNA within infected cells resulted in luciferase expression, indicating that transcription of the minireplicon by T7 polymerase and subsequent translation of the inserted reporter gene had occurred.

The shuttle vectors containing functional CD-61 minireplicons were used to prepare recombinant fowlpox viruses. Potential rFPV were analysed for expression of luciferase after co-infection of cells with rFPV/T7 and IBV. Two rFPV/CD-61+LUC exhibited luciferase activity, with higher luciferase activity than for EFCD+LUC DNA electroporated cells. This result suggested that the T7 polymerase from rFPV/T7 was able to transcribe the CD-61 minireplicon from rFPV/CD-61, which in the presence of IBV was replicated and transcribed the luciferase gene. We also needed to determine whether the RNA transcripts produced from rFPV/CD-61 could be rescued by the helper IBV. Preliminary experiments involving co-infection of rFPV/CD-61+, rFPV/T7 and IBV in CK cells and subsequent passages indicated the presence of CD-61 RNA in cell extracts by P4. This suggests that CD-61 RNA could be rescued by IBV from rFPV/CD-61+.

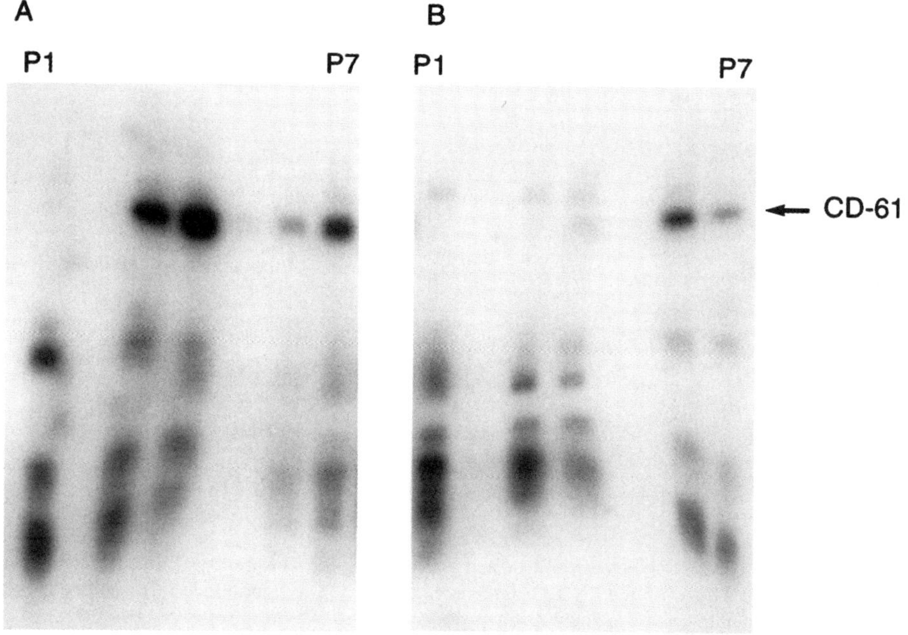

Figure 2. Northern blot analysis showing rescue of CD-61+ by IBV. T7-transcripts from CD-61 constructs were electroporated into IBV-infected cells. Harvested virus was passaged onto new cells up to seven times, RNA extracted from infected cell lysates was analysed by hybridisation to ^{32}P-labelled IBV cDNA (tracks P1 to P7). The position of CD-61 RNA is indicated by an arrow. Unmodified CD-61 electroporated into cells shows rescue of CD-61 RNA from passage 3 (P3; panel A). CD-61, modified with the ribozyme/ T7 terminator, shows rescue of CD-61 RNA at P4, after electroporation into cells (panel B).

5. DISCUSSION

Working towards improved methods of control for avian IBV requires indepth understanding of the molecular biology of the pathogen. Such knowledge will allow manipulation of the virus genome for the development of novel prophylactics. It is hoped that using the IBV D-RNA CD-61, developed as a minigenome, for the generation of recombinant IBV viruses, will promote studies on the basis of pathogenicity and protective immunity. The large surface spike (S) protein is a prime candidate as a determinant of pathogenicity and thus an important target gene for our studies. By using CD-61 as an RNA vector, we will attempt to introduce the S gene of the nephropathogenic IBV strain into the genome of a non-nephropathogenic strain. Moreover, this will enable us to assess the role of the spike protein in the determination of tissue tropism of IBV strains.

To increase the opportunity for recombination in IBV, we have inserted the CD-61 vector, with reporter gene, into the genome of fowlpox virus. Such that IBV-infected cells co-infected with rFPV/CD-61+ and rFPV/T7 were shown to express CD-61 RNA *in situ*. This should result in more cells producing CD-61 and in larger quantities than for the usual electroporation approach. The advantage of using a replicating minigenome is that the amount of D-RNA increases during replication and during passaging, which increases the chance of recombination between the D-RNA and the genomic RNA. Success of this

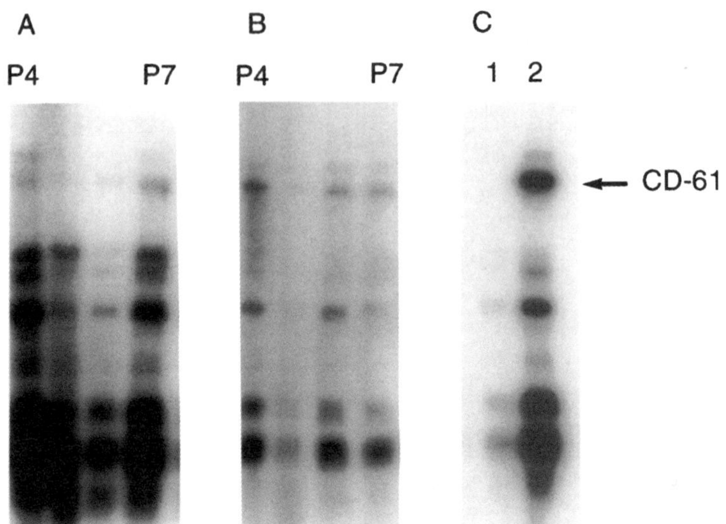

Figure 3. Northern blot analysis showing rescue of CD-61 from EFCD cDNA. Intact plasmid DNA constructs were electroporated into cells co-infected with IBV and a recombinant FPV expressing T7 RNA polymerase. Harvested virus was passaged up to P7 and RNA extracted at each passage was hybridised to ^{32}P-labelled IBV cDNA. CD-61 RNA is highlighted by an arrow. Panel A shows rescue of CD-61 (P4 to P7) after pCD-61+ DNA electroporation, whereas, panel B indicates rescue of CD-61 RNA following pEFCD electroporation. Track 1 (panel C) shows RNA from IBV-infected cells only. A positive control, with CD-61 T7-transcripts electroporated into IBV-infected cells, is shown by track 2 (panel C).

approach would also permit the development of recombinant IBV's for vaccine purposes. By introducing the S gene of a heterologous IBV strain into a vaccine strain, the scope of the original vaccine would be extended.

ACKNOWLEDGMENTS

We thank Andrew Ball for the gift of plasmid Vec2,0 (with ribozyme/termination sequences). This work was funded by a ROPA grant.

REFERENCES

Britton, P., Green, P., Kottier, S., Mawditt, K. L., Pénzes, Z., Cavanagh, D. and Skinner, M. A., 1996, Expression of bacteriophage T7 RNA polymerase in avian and mammalian cells by a recombinant fowlpox virus, *J. Gen. Virol.* **77**:963–967.

Cavanagh, D., 1983, Coronavirus IBV: structural characterisation of the spike protein, *J. Gen. Virol.* **64**:2577–2583.

Kottier, S. A., Cavanagh, D. and Britton, P., 1995, Experimental evidence of recombination in coronavirus infectious bronchitis virus, *Virology* **213**:569–580.

Pénzes, Z., Tibbles, K., Shaw, K., Britton, P., Brown, T.D.K. and Cavanagh, D., 1994, Characterisation of a replicating and packaged defective RNA of avian coronavirus infectious bronchitis virus, *Virology* **203**:286–293.

Pénzes, Z., Wroe, C., Brown, T.D.K., Britton, P. and Cavanagh, D., 1996, Replication and packaging of coronavirus infectious bronchitis virus defective RNAs lacking a long open reading frame, *J. Virol.* **70**(12):8660–8668.

89

INTRAMUSCULAR INJECTION OF PLASMID DNA EXPRESSING mRNA7 CODING THE NUCLEOCAPSID PROTEIN OF JHMV PARTIALLY PROTECTED MICE AGAINST ACUTE INFECTION WITH JHMV

M. Hayashi,[1] K. Ishida,[2] A. Maeda,[2] Y. Kon,[2] T. Mizutani,[2] T. Watanabe,[2] S. Arai,[1] and F. Okada[3]

[1]Department of Veterinary Radiology
Rakuno Gakuen University
Ebetsu 069, Japan
[2]Laboratory of Experimental Animal Science
Graduate School of Veterinary Medicine
[3]Laboratory of Pathology
Cancer Institute, School of Medicine
Hokkaido University, Sapporo 060, Japan

1. ABSTRACT

We constructed a plasmid expressing mRNA7 coding the nucleocapsid (N) protein of JHM strain of mouse hepatitis virus (JHMV) under the control of Rous sarcoma virus LTR, referred to as pRSV-mRNA7. When C57BL/6 mice injected intramuscularly (i.m.) with control plasmid DNA which contained no viral sequence were infected with JHMV, necrotic figures of neural cells and diffuse immersion of lymphatic cells in the pedunculus cerebri were observed. In the hypothalamus, vascular cuffing consisting of lymphatic cells was observed. In contrast, no histological change was observed throughout these areas of the brains in the JHMV-infected mice after injection with pRSV-mRNA7. These results showed that the injection with plasmid DNA expressing mRNA7 of JHMV partially protected mice against acute infection with JHMV in the brain.

The plasmid DNA was i.m. injected into mice and the cytolytic activity of spleen cells from the mice was assessed by ^{51}Cr-release assay. The spleen cells from the mice administrated with pRSV-mRNA7 showed a significant level of cytolytic activities against the transfected cells expressing the viral N protein even though the spleen cells were not

Coronaviruses and Arteriviruses, edited by Enjuanes *et al.*
Plenum Press, New York, 1998

cocultivated with stimulator cells. When the spleen cells from administrated mice with pRSV-mRNA7 were cocultivated with stimulator cells, higher cytolytic activity was observed against the transfected cells, compared with the activity without stimulation.

2. INTRODUCTION

It has been reported that an intramuscular injection of plasmid DNA in mice results in the uptake of DNA by the muscle cells and expression of the protein encoded by the plasmid DNA (Ascardi et al., 1991; Wolff et al., 1990). Plasmids are shown to be maintained episomally and do not replicate. This technique can be a method of inducing antibodies against viral proteins and introducing viral proteins into the antigen-processing pathway that results in the generation of virus-specific CTLs (Martins et al., 1995; Xiang et al., 1994; Ulmer et al., 1993).

Infection of susceptible strains of mice with the neurotropic JHMV produces a panencephalitis accompanied by primary demyelination with no or little evidence of hepatitis (Stohlman and Weiner, 1981). This infection is studied as a model of virus-induced central nervous system (CNS) demyelination. The immune response during the actute phase of the disease appears to play a pivotal role in determining the eventual outcome of the infection. The major components of the immune system, including antiviral antibody, natural killer cells, monocytes and CD4$^+$ and CD8$^+$ T cells, are present within CNS during the acute infection (Williamson et al., 1991; Kyuwa and Stohlman, 1990; Williamson and Stohlman, 1990).

In the present study, we investigated whether intramuscular injection of plasmid DNA can protect mice against the infection with JHMV. A nucleocapsid (N) protein is translated from mRNA7 of JHMV and the most abundant viral protein in the infected cells (Stohlman and Lai, 1979). Furthermore, neutralizing monoclonal antibodies to the N protein protect mice against a lethal challenge to mouse hepatitis virus (Nakanaga et al., 1986). Therefore, we constructed a plasmid expressing mRNA7, which codes the N protein of JHMV.

3. MATERIALS AND METHODS

3.1. Cells and Virus

Murine astrocytoma-derived DBT cells (Hirano et al., 1974) were cultured in Eagle's minimum essential medium (MEM) with 5% calf serum. C57BL/6 mouse fibrosarcoma-derived QR32 cells (Okada et al., 1990) were maintained in Eagle's MEM supplemented with 8% fetal calf serum, 1mM sodium pyruvate, non-essential amino acids and 2 mM L-glutamine (MEM$^+$). The JHM strain of MHV (Makino et al., 1983) was propagated in DBT cells.

3.2. Mice

Pathogen-free C57BL/6//JCL (B6) mice were purchased from CLEA, Japan at 4 weeks of age. Serum samples obtained from representative mice were tested for JHMV specific antibody by enzyme-linked immunosorbent assay (ELISA) and found to be negative. Research was conducted according to the principles defined in the "Guide for the Care and Use of Laboratory Animals" prepared by Rakuno Gakuen University.

3.3. Construction of Plasmid and Transfection of Cells

A full length of cDNA fragment of mRNA7, which codes the N protein of JHMV, was ligated into the expression vector pRSV-CAT after removal of chloramphenicol acetyltransferase gene. The plasmid was referred to as pRSV-mRNA7. QR32 cells were cotransfected with 10 µg of pRSV-mRNA7 and pSV2-neo DNA by standard calcium precipitation procedure and selected in MEM$^+$ containing G418 (1 mg/ml). The transfected cell clone expressing mRNA7 was referred to as QR-mRNA7. An expression of the mRNA7 and the N protein in QR-mRNA7 cells was confirmed by Northern and Western blot analyses, respectively. Northern and Western blot analyses were carried out as described previously (Maeda, et al., 1995).

3.4. Histopathological Examination

B6 mice were injected in the quadriceps muscle with 100 µg of a closed circular form of pRSV-mRNA7 DNA. The mice were injected with the plasmid DNA once a week at one-week intervals, receiving a total of four injections. The mice were inoculated intraperitoneally with 10^5 PFU of JHMV at the next day after the fourth injection. At 10 days postinfection, the mice were anaesthesized with ether and sacrificed by cervical dislocation. Following the fixation of tissue samples in Carnoy's solution, the samples were dehydrated and embedded in paraffin. Five µM-thick sections were then stained with hematoxylin and eosine for histopathological examinations.

3.5. Mixed Lymphocyte Cell Cultures and ^{51}Cr-Release Assay

B6 mice were administrated with the plasmid DNA described above. The next day after the fourth injection, spleens were aseptically removed from the mice and homogenized in loose fitting glass homogenizer. The suspensions of spleen cells were then passed through four layers of gauze. The QR-mRNA7 cells to be used as stimulators were treated with mitomycin C (100 µg/ml) for 1 h at 37C in a 5% CO_2 atmosphere. Spleen cell suspensions (2×10^7 cells) were cultured with the mitomycin C-treated stimulator cells (5×10^5) as described previously (Okada et al., 1990). After 5 days of incubation, the stimulated spleen cells were collected and their cytolytic activity was assessed.

Labeling of the cells with sodium [^{51}Cr] chromate and ^{51}Cr-release assay were carried out as described previously (Okada et al., 1990).

4. RESULTS

We constructed a plasmid expressing mRNA7, which codes the N protein of JHMV under the control of RSV LTR, referred to as pRSV-mRNA7. When mice i.m. injected with control pRSV-CAT plasmid DNA were infected with JHMV, necrotic figures of neural cells and diffuse immersion of lymphatic cells in the pedunculus cerebri were observed (Fig. 1a). In the hypothalamus, vascular cuffing consisting of lymphatic cells was observed (Fig. 1b). In the control uninjected mice, similar histological changes were observed after infection with JHMV (data not shown). In contrast, no histological change was observed throughout these areas of the brains in the JHMV-infected mice after injection with pRSV-mRNA7 (Fig. 1c and d). These results showed that injection with the plasmid DNA expressing mRNA7 of JHMV partially protected mice against acute infection

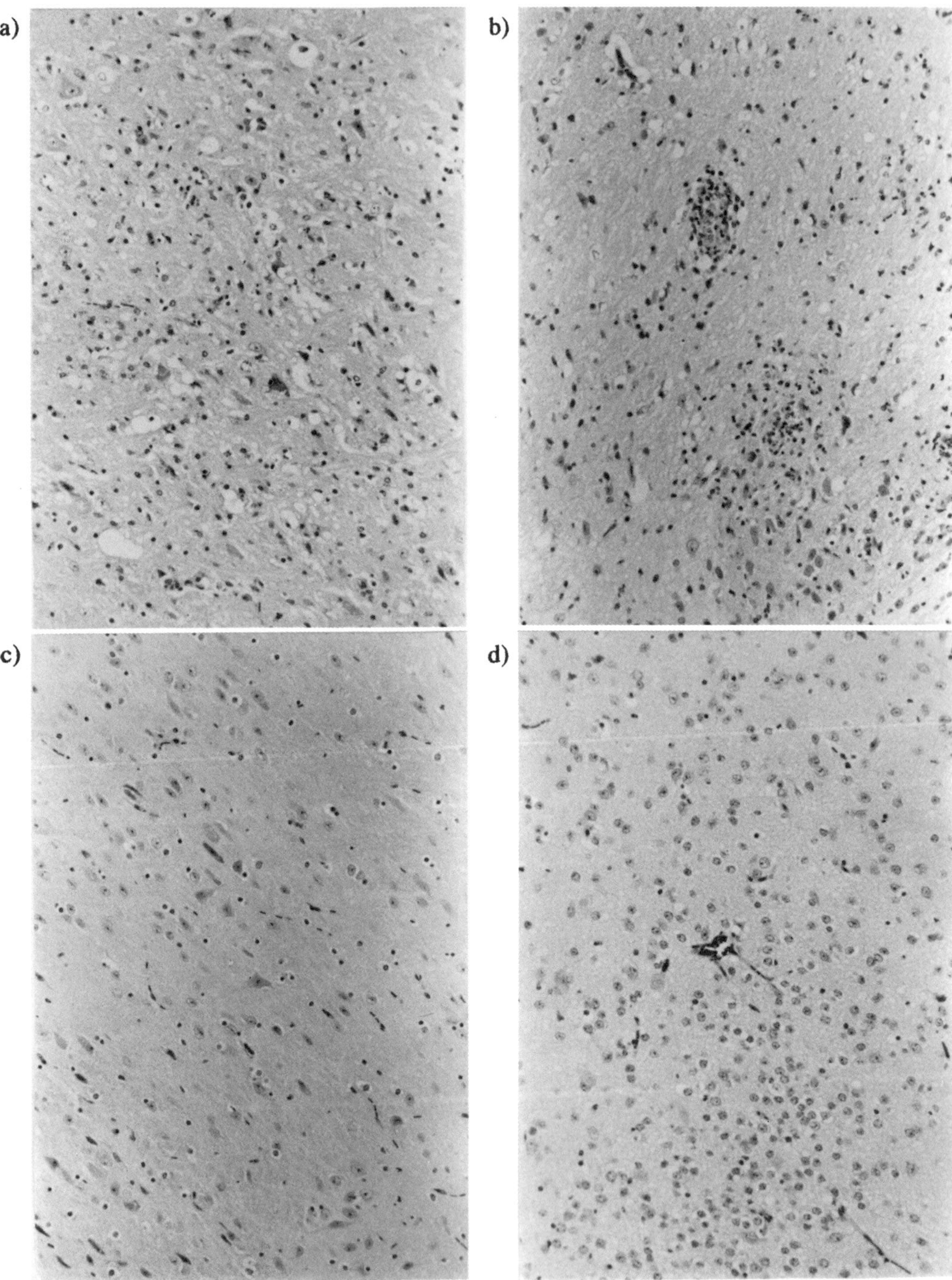

Figure 1. Histopathological figures of brains after the infection of JHMV at 10 days postinfection. Cerebral peduncle (a and c) and hypothalamus (b and d) of the mice after intramuscular injection with pRSV-CAT (a and b) and pRSV-mRNA7 (c and d). (magnification X 165).

with JHMV in the brain. There was no obvious histological change in other tissues, including the liver, of JHMV-infected mice (data not shown).

In the serum samples from the mice administrated with pRSV-mRNA7 DNA, the antibody against the N protein was not observed by ELISA (dat not shown). Therefore, the cytolytic activities of spleen cells of the mice i.m. injected with pRSV-mRNA7 were assessed by ^{51}Cr-release assay. The B6 mouse fibrosarcoma-derived QR 32 cells were transfected with pRSV-mRNA7, referred to as QR-mRNA7 cells, and the QR-mRNA7 cells were used as stimulator and target cells. When pRSV-mRNA7 was i.m. injected into B6 mice, cytolytic activity was bserved using QR-mRNA7 cells as a target, even though the spleen cells were not cocultivated with stimulator cells (Fig. 2a). No cytolysis was observed when the untransfected QR 32 cells were used as a target. The spleen cells from the mice receiving pRSV-CAT plasmid DNA showed no cytolytic activity against QR 32 and QR-mRNA7 cells (data not shown). Therefore, the intramuscular injection of pRSV-mRNA7 DNA induced specific cytolytic activity against the cells expressing the viral N protein. When the spleen cells from the mice injected with pRSV-mRNA7 were cocultivated with QR-mRNA7 cells, higher cytolytic activity was observed using QR-mRNA7 cells as a target, compared with the activity without stimulation (Fig. 2b). However, cytolysis of the untransfected QR 32 cells used as a target was also observed. The levels of cytolysis of QR-mRNA7 cells were significantly higher than those of the untransfected QR 32 cells.

5. DISCUSSION

In the present study, we showed that the intramuscular injection with the plasmid DNA expressing mRNA7, which codes the N protein of JHMV protected mice against acute infection with JHMV in the brain. Although the antibody against the N protein was not observed by ELISA in the serum samples from the mice administrated with pRSV-mRNA7 DNA (dat not shown), the spleen cells from the mice i.m. injected with pRSV-mRNA7 showed a significant level of cytolytic activities against QR-mRNA7 cells.

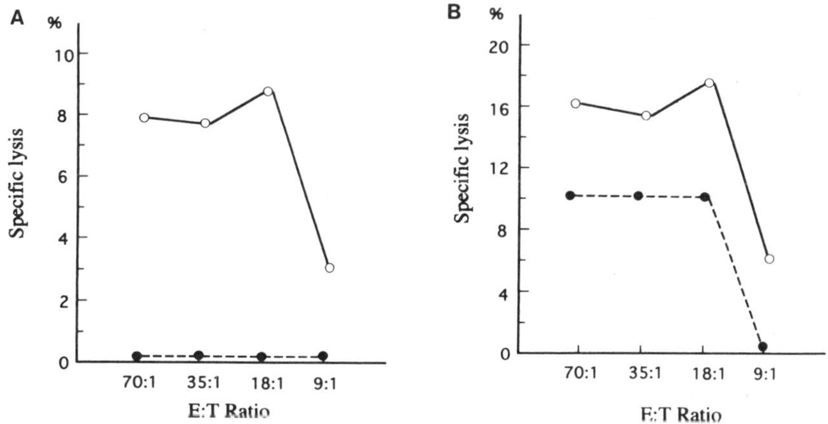

Figure 2. Cytolytic activities of spleen cells from the mice administrated with pRSV-mRNA7. pRSV-mRNA7 was i.m. injected into B6 mice. Specific lysis of control untransfected QR32 cells (●) and QR-mRNA7 cells (○) used as targets by spleen cells which cultivated without stimulator cells (a) and with QR-mRNA7 cells (b) was assayed. Each point represents the average value of 3 separate experiments.

Therefore, the protection of the mice against acute infection with JHMV by the intramuscular injection with pRSV-mRNA7 DNA might result from the induction of cytolytic activity against the cells expressing the viral N protein. Immunization of BALB/c mice with JHMV induces CTLs that are specific for the N protein (Stohlman et al., 1992). However, in the case of B6 mice the dominant epitopes for CTLs are reported to be located within the viral spike protein of JHMV (Bergmann et al., 1996; Castro and Perlman, 1995). Therefore, the immune response and the nature of the induced cells that showed the cytolytic activity in the present study may be different from those in natural infection of B6 mice with JHMV. Further characterization of the induced cells showing cytolytic activity should be necessary.

The reason why the cytolysis was observed by the spleen cells cocultivated with stimulator cells against the untransfected QR32 cells as a target remains unknown. It has been reported that the QR32 cells are resistant to cytolysis caused by natural killer cells but sensitive to that caused by CTLs and lymphokine activated killer cells (LAK) (Okada et al., 1990). Therefore, killer cells nonspecific for antigen such as LAK may be generated.

A nucleic acid vaccination with naked plasmid DNA encoding specific virus antigens offers a promising means to induce immunity against several important pathogens (Xiang et al., 1994; Cox et al., 1993; Ulmer et al., 1993). Furthermore, it has been reported that nanogram quantities of naked DNA effectively generate antibodies and cytotoxic T cell responses using gene gun (Pertmer et al., 1995). Thus, administration of DNA that codes the viral proteins may be a potential method of vaccination.

REFERENCES

Ascardi, G., Dickinson, G., Love, D.R., Jani, A., Walsh, F.S., Gurusinghe, A., Wolff, J.A., and Davis, K.E., 1991, Human dystrophin expression in mdx mice after intramuscular injection of DNA constructs, *Nature* **352**: 815–818.

Bergmann, C.C., Yao, Q., Lin, M., and Stohlman, S.A. 1996, The JHM strain of mouse hepatitis virus induces a spike protein-specific D^b-restricted cytotoxic T cell response, *J. Gen. Virol.* **77**: 315–325.

Castro, R.F., and Perlman, S., 1995, $CD8^+$ T-cell epitopes within the surface glycoprotein of a neurotropic coronavirus and correlation with pathogenicity, *J. Virol.* **69**: 8127–8131.

Cox, G.J., Samb, T., and Babiuk, L.A., 1993, Bovine herpesvirus 1: immune responses in mice and cattle injected with plasmid DNA, *J. Virol.* **67**: 5664–5667.

Hirano, N., Fujiwara, K., Hino, S., and Matumoto, M., 1974, Replication and plaque formation of mouse hepatitis virus (MHV-2) in mouse cell line DBT culture, *Arch. Gesamte Virusforsch.* **44**: 298–302.

Kyuwa, S., and Stohlman, S.A., 1990, Pathogenesis of a neurotropic murine coronavirus, strain JHM in the central nervous system of mice, *Semin. Virol.* **1**: 273–280.

Maeda, A., Hayashi, M., Ishida, K., Mizutani, T., Watanabe, T., and Namioka, S., 1995, Characterization of DBT cell clones derived from cells persistently infected with the JHM strain of mouse hepatitis virus, *J. Vet. Med. Sci.* **57**: 813–817.

Makino, S., Taguchi, F., Hayami, M., and Fujiwara, K., 1983, Characterization of small plaque mutants of mouse hepatitis virus, JHM strain, *Microbiol. Immunol.* **27**: 445–454.

Martins, L.P., Lau, L.L., Asano, M.S., and Ahmed, R., 1995, DNA vaccination against persistent viral infection, *J. Virol.* **69**: 2574–2582.

Nakanaga, K., Yamanouchi, K., and Fujiwara, K., 1986, Protective effect of monoclonal antibodies on lethal mouse hepatitis virus infection in mice, *J. Virol.* **59**: 168–171.

Okada, F., Hosokawa, M., Hasegawa, J., Ishikawa, M., Chiba, I., Nakamura, Y., and Kobayashi, H., 1990, Regression mechanisms of mouse fibrosarcoma cells after in vitro exposure to quercetin: diminution of tumorgenicity with a corresponding decrease in the production of prostaglandin E2, *Cancer Immunol. Immunother.* **31**: 358–364.

Pertmer, T.M., Eisenbraun, M.D., McCabe, D., Prayage, S.K., Fuller, D.H., and Haynes, J.R., 1995, Gene gun-based nucleic acid immunization: Elicitation of humoral and cytotoxic T lymphocyte responses following epidermal delivery of nanogram quantities of DNA, *Vaccine* **13**: 1427–1430.

Stohlman, S.A., and Lai, M.M.C., 1979, Phosphoproteins of murine hepatitis virus, *J. Virol.* **32**: 672–675.

Stohlman, S.A., Kyuwa, S., Cohen, M., Bergman, C., Polo, J.P., Yeh, J., Anthony, R., and Keck, J.G., 1992, Mouse hepatitis virus nucleocapsid protein-specific cytotoxic T lymphocytes are L^d restricted and specific for the carboxy terminus, *Virology* **189**: 217–224.

Stohlman, S. A., and Weiner, L.P., 1981, Chronic nervous system demyelination in mice after JHM virus infection, *Neurology* **31**: 38–44.

Ulmer, J.B., Donnelly, J.J., Parker, S.E., Rhodes, G.H., Felgner, V.J., Dwarki, V.J., Gromkowski, S.H., Deck, R.R., DeWitt, C.M., Friedman, A., Hawe, L.A., Leadner, K.R., Martinez, D., Perry, H.C., Shiver, J.W., Montogomery, D.L., and Liu, M.A., 1993, Heterologous protection against influenza by injection of DNA encoding a viral protein, *Science* **259**: 1745–1749.

Williamson, J.S.P., and Stohlman, S.A., 1990, Effective clearance of mouse hepatitis virus from the central nervous system requires both CD4[+] and CD8[+] T cells, *J. Virol.* **64**: 4589–4592.

Williamson, J.S.P., Sykes, K.C., and Stohlman, S.A., 1991, Characterization of brain-infiltrating mononuclear cells during infection with mouse hepatitis virus strain JHM, *J. Neuroimmunol.* **32**: 199–207.

Wolff, J.A., Malone, E.W., Williams, P., Chamg, W., Ascardi, G., Jani, A., and Felgner, P.L., 1990, Direct gene transfer into mouse muscle in vivo, *Science* **247**: 1465–1468.

Xiang, Z.Q., Sptalnik, S., Tran, M., Wunner, W.H., Chang, J., and Ertl, H.C.J., 1994, Vaccination with plasmid vector carrying the rabies virus glycoprotein gene induces protective immunity against rabies virus, *Virology* **199**: 132–140.

90

INHIBITORY EFFECTS OF MODIFIED OLIGONUCLEOTIDES COMPLEMENTARY TO THE LEADER RNA ON THE MULTIPLICATION OF MOUSE HEPATITIS VIRUS

M. Hayashi,[1] A. Maeda,[2] M. Kihara,[1] S. Arai,[1] K. Hanaki,[3] and T. Nozaki[3]

[1]Department of Veterinary Radiology
Rakuno Gakuen University, Ebetsu 069
[2]Laboratory of Experimental Animal Science
Graduate School of Veterinary Medicine
Hokkaido University, Sapporo 060
[3]The Chemo-Serotherapeutic Research Institute
Kumamoto 869-12, Japan

1. ABSTRACT

Phosphorothioate oligonucleotides (PS-oligo) and PS-oligos with cholesterol conjugates (ChPS-oligo) complementary to the leader RNA of strain JHM of mouse hepatitis virus (JHMV) were more effective inhibitors of viral multiplication than natural oligodeoxynucleotides (PO-oligo) in JHMV-infected DBT cells. PS- and ChPS-oligos were 1,000 times more potent than unmodified PO-oligo. No significant difference was observed in the inhibitory efficiency between PS-oligo and ChPS-oligo. Sequence-dependent inhibition of viral multiplication was shown at low concentrations (0.001–0.1 M) of antisense PS-oligo and ChPS-oligo. Phosphorothioate oligodeoxycytidine, PS-(dC)$_{20}$, and PS-(dC)$_{20}$ with cholesterol conjugates, and PS- and ChPS-oligo which have no significant homology to the JHMV sequences, showed inhibitory effects on JHMV multiplication at concentrations higher than 0.5 M. These results showed that PS-oligo and ChPS-oligo were more potent than PO-oligo in the inhibition of JHMV multiplication, and that PS-oligo and ChPS-oligo may inhibit JHMV multiplication by two different mechanisms, that is by sequence-dependent and sequence-independent manners.

Coronaviruses and Arteriviruses, edited by Enjuanes *et al.*
Plenum Press, New York, 1998

2. INTRODUCTION

Antisense oligodeoxynucleotides (ODNs) within cells targeted towards the RNA transcript of a specific gene can inhibit the expression or promote the degradation of the transcript, resulting in suppression of the function coded for by the gene (Cohen, 1991; Colman, 1990). Antisense ODNs have also been widely used as tools for inhibiting viral replication. However, the susceptibility of the phosphodiester linkage to degradation by nucleases in natural ODNs (PO-oligos) appears to reduce their persistence in vivo and their potency as antiviral agents (Wickstrom, 1986). Therefore, several analogues have been developed in order to obtain chemically stable ODNs which are nuclease-resistant, but retain their Watson-Crick base pairing specificity. Among these, phosphorothioate ODNs (PS-oligos) are now widely used. However, PS-oligos show low efficiency in uptakes into cells mainly due to their negative charges (Crooke, 1992; Stein, 1992). Furthermore, when PS-oligos are added to culture medium, they tend to accumulate in lysosomal compartments (Tonkinson et al., 1994; Iversen et al., 1992). Such intracellular localization might affect an interaction between antisense ODNs and virus mRNA. It has been reported that cholesterol conjugation improves the uptake of ODNs into cells and antiviral activities (Kreig et al., 1993; Letsinger et al., 1989). Although a large number of studies concerning the antiviral activities of a variety of modified ODNs have been reported, most of them were targeted to viruses that replicate in the nucleus, such as human immunodeficiency virus and hepatitis B virus (Korba and Gerin, 1995; Agrawal et al., 1988). The replication of many RNA viruses, including mouse hepatitis virus (MHV), occurs only in the cytoplasm of infected cells. Therefore, intracellular localization might be an important factor in determining the efficacy of the antiviral activities of antisense ODNs for RNA viruses. Previously we reported that a PO-oligo complementary to a leader RNA containing the conserved sequence, UC(U/C)AAAC, of MHV (Shieh et al., 1987) inhibits viral multiplication (Mizutani et al., 1992). Therefore, we selected the leader sequence containing the conserved sequence as a target region for antisense ODNs and investigated the inhibitory effects of PS- and ChPS-oligos on MHV multiplication.

3. MATERIALS AND METHODS

3.1. Cells and Virus

Mouse asterocytoma-derived DBT cells (Hirano et al., 1974) were cultured in Eagle's minimum essential medium (MEM) supplemented with 5% calf serum (CS). The JHM strain of MHV (Makino et al., 1983) was used throughout this study.

3.2. Oligodeoxynucleotides

PS-oligo and ChPS-oligo were obtained from Sawadii Tech. Co. AL-oligo contained a sequence complementary to the conserved sequence of the leader RNA of JHMV (Fig. 1). ML-oligo contained a sequence with 70% homology to AL-oligo. Random-oligo contained no significant homology with any reported MHV sequences. Oligodeoxycytidine, $(dC)_{20}$ was also synthesized.

3.3. Inhibition of Virus Multiplication by ODNs

DBT cells were infected with JHMV at a multiplicity of infection (m.o.i.) of 0.1 in the absence or presence of ODNs at concentrations from 0.001 to 10 M for 1 h at 37C in

60   70   80

5'-AUCUAAUCUAAUCUAAACUUUA-3'

Oligonucleotides	Sequences 5' - 3'
AL-oligo	AAAGTTTAGATTAGATTAGA
ML-oligo	AAAGTTAATGTAATGTTAGA
Random-oligo	ACCCGTAGCTAATGAGATAT
$(dC)_{20}$	CCCCCCCCCCCCCCCCCCCC

Figure 1. Sequences of the leader RNA containing the conserved sequence UC(U/C)AAA, PS- and ChPS-oligos.

CS-free MEM. After incubation, ODNs and virus were removed, and the cells were washed twice with CS-free MEM. After the addition of MEM with 5% CS, plaque assays were performed to titrate infectious progeny at varions time points post infection (p.i.) according to the method of Hirano et al. (1974).

4. RESULTS

The treatment of DBT cells with PS- and ChPS-oligos had no apparent detrimental effect on cell viability of (data not shown). The yields of infectious virus particles at 12 h p.i. from the cells treated with AL-oligo and ML-oligo at 0.001 M were reduced significantly compared to the yields from control cells which were not treated with ODNs (Fig. 2). At concentrations higher than 0.1 M, the viral multiplication was inhibited more than 95%.

Random-oligo and $(dC)_{20}$ showed inhibitory effects on viral multiplication at concentrations higher than 0.1 M for ChPS-oligos and at 0.5 M for PS-oligos. At concentra-

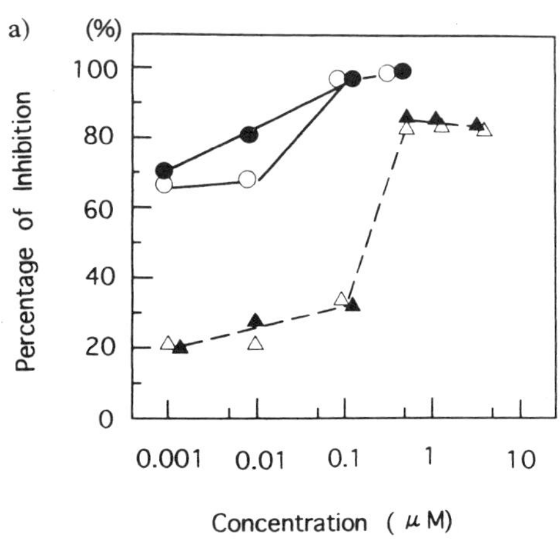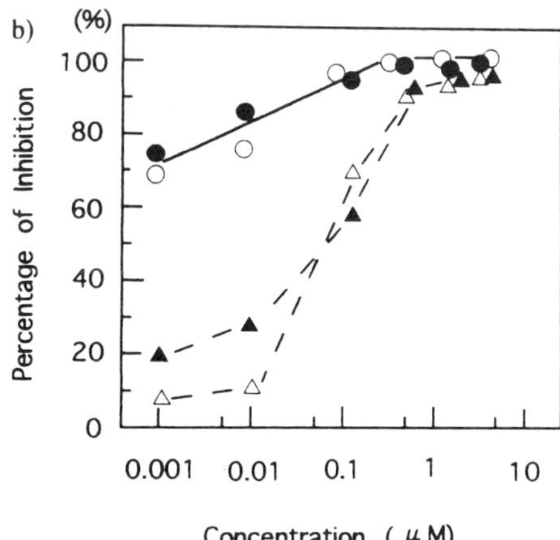

Figure 2. The effects of PS-oligos (a) and ChPS-oligos (b) on JHMV multiplication. DBT cells were infected with JHMV (0.1 m.o.i.) in the presence of AL-oligo (O), ML-oligo (●), Random-oligo (Δ), and $(dC)_{20}$ (▲) for 1 h at 37C. Plaque assays were performed to titrate infectious progeny in the supernatants at 12 h p.i. Each point represents an average of 3–4 separate experiments.

tions higher than 1.0 M, viral multiplication was inhibited more than 80% as, compared to the yields from control cells (Fig. 2).

5. DISCUSSION

Previously we showed that at 12 h p.i. the yields of infectious virus particles from cells treated with natural PO-oligo complementary to the leader RNA (PO-AL-oligo) at concentrations from 5 to 25 M is reduced to 25 to 65% as compared to infected untreated control cells, and that no inhibitory effect on the viral multiplication was observed at 1 M after treatment with PO-AL-oligo (Mizutani et al., 1992). The present results showed that virus multiplication was significantly inhibited at 0.001 M by treatment with PS- and ChPS-AL-oligos, and that the viral multiplication was inhibited more than 95% at concentrations higher than 0.1 M. Thus, PS- and ChPS-oligos were 1,000 times more potent inhibitors of JHMV multiplication than unmodified PO-oligos. PS-oligos are more resistant to nuclease digestion in cells and in the whole body than PO-oligos (Agrawal et al., 1991). Therefore, PS-oligos might inhibit viral multiplication more effectively than PO-oligos. It has been reported that cholesterol conjugation improves the uptake of ODNs into cells and antiviral activities (Kreig et al., 1993; Letsinger et al., 1989). Indeed, we obtained the results that ChPS-oligo was incorporated into the cells 4–5 times more eficiently than PS-oligo (data not shown). Although no significant difference was observed in the inhibitory effects between PS-oligo and ChPS-oligo in the present study, Northern blot analysis showed that ChPS-oligo more effectively inhibited the synthesis of viral RNA than PS-oligo did at early times postinfection (data not shown). We also obtained preliminary results that ChPS-oligos inhibited MHV multiplication in the infected mice, but not PS-oligos (data not shown). Thus, PS- and ChPS-oligos showed potent antiviral activities against JHMV whose replication cycle occurs only in the cytoplasm.

At concentrations higher than 1.0 M, viral multiplication was inhibited more than 80% by modified random-oligo and $(dC)_{20}$. The percentages of inhibition by random-oligo and $(dC)_{20}$ were significantly lower than those by AL- and ML-oligo at low concentrations. Thus, sequence-dependent inhibition was observed at low concentrations of PS- and ChPS-oligos, and sequence-independent inhibition was observed at high concentrations. The sequence-independent inhibition of HIV multiplication is observed at a high concentration of PS-oligos (Agrawal et al., 1989). The present study showed that such a sequence-independent inhibitory effect on viral multiplication by PS- and ChPS-oligos was not specific to HIV. A study of mechanism of the sequence-independent inhibition is now in progress.

REFERENCES

Agrawal, S., Goodchild, J., Civeira, M.P., Thornton, A., Sarin, P.S., and Zamecnik, P.C., 1988, Oligodeoxynucleoside phosphoramidates and phosphorothioates as inhibitors of human immunodeficiency virus, *Proc. Natl. Acad. Sci. U.S.A.* **85**: 7079–7083.

Agrawal, S., Ikeuchi, T., Sun, D., Sarin, P.S., Konopka, A., Maizel, J., and Zamecnik, P.C., 1989, Inhibition of human immunodeficiency virus in early infected and chronically infected cells by antisense oligodeoxynucleotides and their phosphorothioate analogues, *Proc. Natl. Acad. Sci. U.S.A.* **86**: 7790–7794.

Agrawal, S., Temsamani, J., and Tang, J.Y., 1991, Pharmacokinetics, biodistribution, and stability of oligodeoxynucleotide phophorothioates in mice, *Proc. Natl. Acad. Sci. U.S.A.* **88**: 7595–7599.

Cohen, J.S., 1991, Antisense oligodeoxynucleotides as antiviral agents. *Antiviral Res.* **16**: 121–133.

Colman, A., 1990, Antisense strategies in cell and developmental biology, *J. Cell Sci.* **97**: 399–409.

Crooke, S.T., 1992, Therapeutic applications of oligonucleotides, *Biotechnology* **10**: 882–886.

Hirano, N., Fujiwara, K., Hino, S., and Matumoto, M., 1974, Replication and plaque formation of mouse hepatitis virus (MHV-2) in mouse cell line DBT culture, *Arch. Gesamte Virusforsch.* **44**: 298–302.

Iversen, P.L., Zhu, S., Meyer, A., and Zon, G., 1992, Cellular uptake and subcellular distribution of phosphorothioate oligonucleotides into cultured cells, *Antisense Res. Dev.* **2**: 211–222.

Korba, B.E., and Gerin, J.L., 1995, Antisense oligonucleotides are effective inhibitors of hepatitis B virus replication in vitro, *Antiviral Res.* **28**: 225–242.

Kreig, A.M., Tonkinson, J., Matson, S., Zhao, Q., Saxon, M., Zhang, L.M., Bhanja, U., Yakubov, L., and Stein, C.A., 1993, Modification of antisense phophodiester oligodeoxynucleotides by a 5' cholesteryl moiety increases cellular association and improves efficacy, *Proc. Natl. Acad. Sci. U.S.A.* **90**: 1048–1052.

Letsinger, R.L., Zhang, G.R., Sun, D.K., Ikeuchi, T., and Sarin, P.S., 1989, Cholesteryl-conjugated oligonucleotides: synthesis, properties, and activity as inhibitors of replication of human immunodeficiency virus in cell culture, *Proc. Natl. Acad. Sci. U.S.A.* **86**: 6553–6556.

Makino, S., Taguchi, F., Hayami, M., and Fujiwara, K., 1983, Characterization of small plaque mutants of mouse hepatitis virus, JHM strain, *Microbiol. Immunol.* **27**: 445–454.

Mizutani, T., Hayashi, M., Maeda, A., Yamashita, T., Isogai, H., and Namioka, S., 1992, Inhibition of mouse hepatitis virus multiplication by an oligonucleotide complementary to the leader RNA, *J. Vet. Med. Sci.* **54**: 465–472.

Shieh, C.K., Soe, L.H., Makino, S., Chang, M.F., Stohlman, S.A., and Lai, M.M.C., 1987, The 5'-end sequence of the murine coronavirus genome: implications for multiple fusion sites in leader-primed transcription, *Virology* **156**: 321–330.

Stein, C.A., 1992, Anti-sense oligodeoxynucleotides—promises and pitfalls, *Leukemia* **6**: 967–974.

Tonkinson, J.L., and Stein, C.A., 1993, Antisense nucleic acids—prospects for antiviral intervention, *Antiviral Chem. Chemother.* **4**: 193–200.

Wickstrom, E., 1986, Oligodeoxynucleotide stability in subcellular extracts and culture media, *J. Biochem. Biophys. Meth.* **13**: 97–102.

91

ADAPTATION AND SERIAL PASSAGE OF BOVINE CORONAVIRUS IN AN ESTABLISHED DIPLOID SWINE TESTICULAR CELL LINE AND SUBSEQUENT DEVELOPMENT OF A MODIFIED LIVE VACCINE

Mark W. Welter

Oragen L.C.
Des Moines, Iowa 50322

ABSTRACT

A virulent bovine coronavirus isolate (newborn calf diarrheal) was adapted and serially passaged in an established diploid swine testicular cell line (ST cells). The same cells have been used to produce modified live porcine rotavirus and coronavirus vaccines that are federally licensed and sold worldwide. Growth of the bovine coronavirus resulted in cytopathic effect characterized by cellular stranding and subsequent cell lysis. Virus yields were relatively high in the ST cells and active replication was confirmed by immune electron microscopy and immunofluorescence. Adaptation of bovine coronavirus to a diploid swine cell line has not been previously reported.

Different cell culture passage levels of bovine coronavirus were evaluated by oral inoculation of clean-catch, colostrum-deprived calves. A passage level of bovine coronavirus was identified that multiplied in the calf without the clinical signs of disease associated with virulent passages. The modified live bovine coronavirus vaccine remained safe and efficacious even after 5-backpassages in calves. Further efficacy studies have shown that the modified live bovine coronavirus vaccine significantly protected calves from highly virulent challenges with either winter dysentery or newborn calf diarrheal coronavirus isolates.

INTRODUCTION

Coronaviruses are a major cause of gastroenteritis in newborn calves and adult cattle worldwide. It has been estimated that economic losses due to neonatal diarrhea approach

Coronaviruses and Arteriviruses, edited by Enjuanes *et al.*
Plenum Press, New York, 1998

$500 million annually for calves in North America alone (Saif and Theil, 1990). On a worldwide basis, this translates into significant economic losses in animal production. Although commercial vaccines are available their efficacy is limited in both experimental and field conditions. This is directly attributed to the strain or biotype of coronavirus used in the vaccines.

For years it had been assumed that all coronavirus infections in cattle were the same, but it has now been shown that two biotypes of bovine coronavirus exist (Millane, 1995). Winter dysentery (WD) is one type of disease due to bovine coronavirus infection. This disease is caused by pneumoenteric bovine coronaviruses that replicate in the epithelium of the upper respiratory tract and the enterocytes of the intestinal tract. The acute disease in adult cattle has been characterized by dark bloody watery diarrhea. These clinical outbreaks frequently result in depression and dehydration which can lead to a decreased milk production. Winter dysentery spreads rapidly to cattle of all ages with infections in young calves resulting in clinical diarrhea and frequently rhinitis (Saif and Theil, 1990). Winter dysentery infections result in high morbidity but low mortality.

Another type of disease due to bovine coronavirus infection has been described as newborn calf diarrheal (NCD). These coronaviruses infect the differentiating epithelial cells of the small intestine and colon (Saif and Theil, 1990). Virus infections in young calves result in villous atrophy leading to watery diarrhea with high morbidity and moderate mortality rates. Researchers have shown that both of these coronavirus biotypes can be differentiated by hemagglutination assays performed at 4° C and 37° C (Millane, 1995).

Recent work has shown that animals that have recovered from an acute infection of WD are susceptible to a subsequent challenge with NCD. Animals that have recovered from an acute infection with NCD are not susceptible and are completely immune to subsequent challenges with WD. The current vaccines on the market have been reported to have limited field efficacy and it has been suggested that the strain of virus used in the products will not sufficiently protect against all bovine coronavirus infections.

A bovine coronavirus vaccine has been developed and is briefly described in this presentation. The vaccine contains a NCD strain of bovine coronavirus and studies have confirmed that this vaccine will prevent bovine coronavirus diseases due to either WD or NCD infections. The modified live vaccine described in this paper remained safe and efficacious even after 5-backpassages in clean-catch, colostrum-deprived calves. This new and unique vaccine will provide the worldwide cattle industry with a safe and most effective means of controlling bovine coronavirus infections.

METHODS AND RESULTS

Growth of BCV (NCD Strain) in STs

A field isolate of bovine coronavirus (NCD strain) was adapted and serially passaged in an established diploid swine testicular cell line (ST cells). Virus infectivity is best described as a typical cytopathic effect (CPE) resulting in a moderate fusogenic characteristic leading to cellular stranding and complete death within 2 days after inoculation. This cytopathology was compared to the reference Mebus strain and found to be characteristically different with the Mebus strain of BCV being very weakly fusogenic. Active replication and identity of the BCV (NCD strain) were confirmed by immunofluorescence, immune electron microscopy and differential hemeagglutination assays. Virus yields are relatively high with titers ranging from 10^8 to 10^9 $TCID_{50}$/ml. The ST cell line that was used is a federally licensed cell line that has been used to produce the modified live porcine rotavirus and coronavirus vaccines that are sold worldwide. This ST cell line is different than other ST cell lines that are commercially available. The diploid chromosome number of this ST cell line has been shown to remain stabile even after 40 passages from master stock. Unlike the commercially available ST cells which are at a much higher passage level and are aneuploidy.

Selection of a BCV Vaccinal Candidate and Preparation of Master Seed

Development of a modified live bovine coronavirus (NCD strain) vaccine was achieved by serial passage of the virus in ST cells. All vaccine safety evaluations were conducted in clean catch, colostrum-deprived calves (cc-calves). These calves were housed in germfree bubbles and fed a medicated milk replacer that was previously determined to be devoid of BCV antibodies. This animal model was used to conduct the differential virulence studies since cc-calves are the most susceptible and thereby the most sensitive model available. Different passage levels of BCV were used to orally inoculate cc-calves. The BCV cell culture passages were classified into three categories based on their infectivity for the cc-calves. The first category was defined as those BCV cell culture passage that were still virulent when inoculated into cc-calves as determined by clinical observations. The second category was defined as those BCV cell culture passages that were not virulent for cc-calves yet still infectious as measured by virus shedding and seroconversion. The third category was defined as those BCV cell culture passages that did not replicate in cc-calves and were therefore considered to be over-attenuated. After these studies were completed a BCV (NCD) cell culture passage level was identified and labeled as a vaccine candidate. The virus was then plaque purified three times and a large pool of virus was generated and designated as master seed virus (MSV). The master seed has passed all purity, identity and safety tests as outlined by 9CFR guidelines including extensive adventitious virus tests.

Backpassage of BCV (NCD) in Clean Catch Calves

The reversion to virulence study was conducted using the BCV MSV as the original inoculum. Two cc-calves were used for initial inoculation and one cc-calf for each subsequent backpassage of the BCV vaccinal strain. All calves were orally inoculated between 5 and 7 days of age. Animals were sacrificed and samples of duodenum, jejunum and ileum were collected for gross histopathology and virus isolation. It was determined that the best time to sacrifice animals for the purpose of virus recovery was between 72 and 96 hours post inoculation. To prepare backpass inoculum, a 50% (weight per volume) small intestinal tissue extract (sie) was prepared from the remainder of the small intestine. A portion of the extract was then orally fed to the next calf. This procedure was continued until 5-consecutive backpassages in cc-calves had been completed. All extracts were titrated and the amount of virus used in each backpassage inoculum was determined. Virus isolation/titration was performed by inoculating confluent human rectal tumor cells (HRTs) and staining by indirect immunoflourescence at 72 hours post-inoculation. All cc-calf tissue samples appeared normal and no signs of villous atrophy were observed in any of the five backpassages. BCV (NCD) was isolated and identified from the small intestinal samples of all of the backpassage inoculated animals, although the amount of virus recovered dropped significantly after the 2nd backpassage.

Table 1. Comparison of BCV (NCD) calf backpassage
number 5 to virulent NCD challenge

Virus inoculation	Clinical observations	Mortality
Backpass - 5	0/5 (0 %)	0/5 (0 %)
NCD Challenge	5/5 (100 %)	3/5 (60 %)

To confirm the safety of the BCV (NCD) vaccine the 5th backpassage sie was used
to orally inoculate 5 cc-calves. These animals were compared to 5 cc-calves that were
orally inoculated with a virulent intestinal extract (standard challenge). Animals were between 7 and 10 days of age and were housed in germfree bubbles, fed medicated milk replacer and observed for 14-days post-inoculation. Both viruses were titrated prior to oral
inoculation. The virulent virus was diluted so as to have the same titer for inoculation as
backpass #5. All five backpassage #5 inoculated cc-calves remained clinically normal.
Two of the five animals developed a creamy diarrhea that persisted for 3 days, but it was
at 10-days after inoculation and this was not considered a clinical sign due to inoculation.
Only a non-pilliated beta-hemolytic E. coli was isolated from the fecal samples collected
from these calves. All of the backpass #5 animals significantly gained weight at a greater
rate than the virulent inoculated animals. All five virulent inoculated animals developed
severe watery diarrhea within 72 hours post-inoculation which persisted for up to 10 days
Three of the five virulent-inoculated calves were moribund and had to be sacrificed prior
to the end of the test. Autopsy results were typical of a severe bovine coronavirus infection with the villi in the jejunum and ileum appearing to be completely denuded. BCV was
isolated from the small intestine of each of the sacrificed animals.

BCV (NCD) Vaccine Protects against a Winter Dysentery Challenge

Six cc-calves between the ages of 7 to 10 days were used in this study. Three animals were orally inoculated with the BCV (NCD) vaccine and the other three animals remained as nonvaccinated controls. All animals were orally challenged with a WD virulent
intestinal extract at 24 days of age and were observed for an additional 14-days. Fecal
samples were collected at time of challenge, 3 days post-challenge (3-DPC), 5-DPC, 7-
DPC, 10-DPC and 14-DPC. The vaccinated animals had normal feces and remained clinically normal and gained weight at a greater rate than the nonvaccinated controls during
the post-challenge period. At 72 hours post-challenge all nonvaccinated controls had developed a severe watery diarrhea that persisted for 5 to 7 days. No virus was detected in
the vaccinated group post-challenge compared to the control group in which virus was detected in fecal samples from 3-DPC to 10-DPC.

Table 2. Effectiveness of BCV (NCD) vaccine against a virulent winter dysentery
challenge

Group	Morbidity incidence (%)	Fecal shedding post-challenge (%)				
		3-DPC	5-DPC	7-DPC	10-DPC	14-DPC
Vaccinates	0	0	0	0	0	0
Controls	100	100	100	100	33	0

DISCUSSION

In conclusion, a vaccine for bovine coronavirus infections has been developed using a newborn calf diarrheal isolate (NCD). The virus was adapted and serially passaged in an established diploid swine testicular cell line. A vaccine candidate was determined in a series of studies where cc-calves were orally inoculated with different cell culture passage levels of BCV (NCD). These various BCV cell culture passaged virus isolates should prove to be useful for determining changes on a molecular level. Master seed has been prepared and passed all testing as outlined by 9CFR. The vaccine strain master seed remained safe even after 6 backpassages in calves as determined in a side by side study with virulent NCD inoculated calves.

This BCV (NCD) vaccine is unique in that it will prevent both WD and NCD infections. The BCV (NCD) vaccine was able to prevent clinical signs and inhibit virulent virus shedding associated with a severe WD challenge.

This BCV (NCD) vaccine will provide the worldwide cattle industry with means to control prevalent bovine coronavirus infections and thereby satisfy a demand for a more efficacious product in the field.

REFERENCES

Babiuk, L.A., Sabara, M., and Hudson, G.R., 1985, Rotavirus and coronavirus infections in animals, *Prog. Vet. Microbiol. Immunol.* **1**: 80–120.

Millane, G., Michaud, L., and Dea, S., 1995, Biological and molecular differentiation between coronaviruses associated with neonatal calf diarrhoea and winter dysentery in adult cattle, *Adv. Exp. Med. Biol.* **380**: 29–33.

Saif, L. J., and Theil, K. T. (eds), 1990, *Viral diarreheas of man and animals*. CRC Press, Boca Raton, Florida.

92

EUROPEAN SEROTYPE PRRSV VACCINE PROTECTS AGAINST EUROPEAN SEROTYPE CHALLENGE WHEREAS AN AMERICAN SEROTYPE VACCINE DOES NOT

P. A. M. van Woensel, K. Liefkens, and S. Demaret

Intervet International B.V.
Wim de Körverstraat 35
5831 AN Boxmeer, The Netherlands

1. SUMMARY

Pigs were either vaccinated with an American serotype Porcine Reproductive and Respiratory Syndrome Virus (PRRSV) vaccine or with a European serotype vaccine. A control group of was left unvaccinated. At four weeks after vaccination the PRRSV-specific antibody titres were determined and one third of each group was challenged with a Spanish, one third with a German and one third with a Dutch PRRSV wild type strain. The serological responses, measured at 4 weeks after vaccination, confirmed that both vaccines were of a different serotype. It was demonstrated that vaccination with an American serotype vaccine slightly reduced the amount of viraemia after challenge with European PRRSV wild type strains. Only after challenge with the Spanish PRRSV strain a moderate, and statistically significant, reduction in viraemia was observed. This is in contrast to vaccination with a European vaccine strain, where viraemia was completely suppressed after challenge with the German PRRSV isolate and almost completely suppressed after challenge with the Spanish and Dutch isolates.

2. INTRODUCTION

The Porcine Reproductive and Respiratory Syndrome virus, a member of the Arteriviruses, has become endemic in pigs since its discovery in 1991. Although structurally similar, PRRSV isolates can be classified into two distinct serotypes; the North-American serotype (further refered to as the American serotype) and the European serotype (Wens-

Coronaviruses and Arteriviruses, edited by Enjuanes *et al.*
Plenum Press, New York, 1998

voort *et al.,* 1992). The serological differences (Wensvoort *et al.,* 1992; Nelson *et al.,* 1993; Drew *et al.,* 1995) are complemented by extensive sequence data (Meng *et al.,* 1994; Mardassi *et al.,* 1995; Murtaugh *et al.,* 1995; Kapur *et al.,* 1996; Suarez *et al.,* 1996) all supporting the distinction between the American and the European serotypes. Now the first vaccines against PRRSV have been brought onto the market it has become quite relevant to investigate the extent of cross-protection between both serotypes.

3. MATERIALS AND METHODS

Two groups, each containing 30 five-weeks old fattening pigs were used for vaccination. One group was vaccinated with Porcilis PRRS, a European based vaccine strain, the other group with a commercial available vaccine based on an American strain. Six sentinel pigs were placed with each group to monitor the spreading of the vaccines. Vaccinations were all done according the manufactures instructions. In addition one group of twelve non-vaccinated pigs was used as a control group. At 4 weeks after vaccination all groups were challenged. One third of each group was challenged with a Spanish (S1), one third with a German (DL3) and one third with a Dutch (I2) PRRSV wild type strain.

The spreading of both vaccine virus strains was determined by the serological examination of the sentinels for the presence of PRRSV specific antibodies at 4 weeks post vaccination. The harvested serum was placed in 2-fold serial dilutions on plates of PRRSV-infected MA104 cells. To prepare the plates MA104 cells were infected with either a European or an American PRRSV isolate and at 24 to 30 hours after infection, fixed with 96% ethanol of -70°C for period 30 min at room temperature. Before the plates were incubated, 2 wash steps with PBS were performed to remove the ethanol. After the sera had been incubated on the plates for 60 min at 37°C, 3 wash steps with PBS were performed and a FITC labelled anti-swine conjugate was added to the plates. After 1 hour of incubation and 3 new washes with PBS, 50 µl of 70% (v/v) glycerol in PBS pH 8.4, was added and the end points determined by fluorescence microscopy. Based on the number of seroconverted sentinels, the reproduction ratio was calculated according the method of Becker.

The efficacy of both vaccines was determined by the amount of viraemia after challenge.

Blood taken from the jugular vein was centrifuged for 10 min. at 1000xg and the serum harvested. The serum was incubated in 10-fold serial dilutions on primary porcine macrophages. At 7 days post-inoculation, cells were scored for the presence of CPE and the 50% tissue culture infectious dose was determined. From these data the average titre per group on each sampling day were determined and an estimation was made of the total amount of virus excreted in the blood. The total amount of viraemia was estimated by calculating the area of the average virus titres per day. All date were analyzed with an ANOVA using Fishers least significance test.

4. RESULTS

At challenge, 4 weeks post-vaccination, an average PRRSV-specific antibody titre of 11.7 ($\log_2$) was found in the animals vaccinated with the European serotype and an average titre of 8.6 ($\log_2$) in the pigs vaccinated with the American serotype. However, when an American serotype virus was used in the test system, an average PRRSV specific anti-

Table 1. Average PRRSV specific antibody titres ($\log_2$) per group determined on MA104 cells infected with either a European or American serotype virus

Vaccine	Average PRRSV specific antibody titres at different weeks after vaccination in the test system		
	European serotype virus		American serotype virus
	0	4[*]	4[*]
European serotype vaccine	0	11.7	8.3
American serotype vaccine	0	8.6	12.4
Controls	0	0	0

*At challenge. Bold, Statistically significantly better than the competitor vaccine (α = 0.05).

body titre of 8.3 ($\log_2$) was found in the animals vaccinated with the European serotype and 12.4 ($\log_2$) in the pigs vaccinated with the American serotype. PRRSV-specific antibody titres of the vaccinated groups, determined in the homologous system were significantly (α = 0.05) higher than titres found in the heterologous system (Table 1). At challenge the controls were all negative for PRRSV-specific antibodies.

With regard to the number of seroconverted sentinel pigs, significantly more sentinels were seroconverted in the American serotype vaccinated group (5 out of 6) than in the European serotype vaccinated group (1 out of 6). In this experiment is was also demonstrated that the use of either serotype in the test system will lead to a number of false negatives (Table 2).

At the time of challenge no vaccine virus could be detected. The average challenge virus titres in the blood after challenge and the total amount of challenge virus shed in the blood stream are given in Table 3. The total amount of challenge virus shed in the blood, visualised in Figure 1, was strongly reduced by vaccination with the European serotype vaccine and were significantly less than that found in both the controls and in the American serotype-vaccinated pigs. This is in sharp contrast to the results with the American serotype-vaccinated pigs. Only after challenge with the Spanish strain a moderate and statistically significant reduction in viraemia was found. After challenge with the two other European challenge viruses no statistically significant differences were observed between the American serotype-vaccinated animals and the non-vaccinated animals.

Table 2. Number of seroconverted sentinel pigs per group determined by seroconversion at 4 weeks post-vaccination using MA104 cells infected with a either a European or an American serotype virus and the resulting R_o for each vaccine

Vaccine	Number of seroconverted sentinels from a total of 6 in the test system		Reproduction ratio vaccine strains* (R_o [95% interval])
	European serotype virus	American serotype virus	
European serotype vaccine	1	1	0.032 [0;0.08]
American serotype vaccine	2	5	$\geq$0.14 [0.09;0.19]

*Reproduction ratios were calculated by the method of Becker.

Table 3. Statistical analysis of the average PRRSV challenge virus titres ($\log_{10}$) found in serum per group and the total virus excretion during the experiment ($\log_{10}$ x day)

Vaccine	Challenge virus	Titre ($\log_{10}$) at days post-challenge				Total ($\log_{10}$ x days)
		3	7	10	14	
European serotype	I2	0.2	**0.2**	**0**	0.5	**2.3**
American serotype	I2	1	1.8	1.5	0.5	15.9
Controls	I2	0.6	2.2	2.4	1.1	20.3
European serotype	S1	**0.1**	**0.2**	**0.2**	**0.6**	**3.1**
American serotype	S1	1.2	1.2	1.4	0.3	13.7
Controls	S1	2.1	1.3	2.6	1.9	26.2
European serotype	DL3	0	**0**	**0**	0	**0.0**
American serotype	DL3	1.2	2	1.2	0.1	15.9
Controls	DL3	1.5	2.2	2.1	0.5	21.0

Bold: significantly better than the competitor vaccine ($\alpha = 0.05$); underlined: significantly better than the controls ($\alpha = 0.05$).

5. DISCUSSION

Due to the fact that clinical signs of PRRSV are quite variable and hard to reproduce in fattening pigs we chose to assess the efficacy by determining the reduction in viraemia. It is obvious that a reduction in viraemia will proportionally decrease the clinical signs caused by PRRSV. Additionally, the reduction of viraemia is quite indicative for the curtailing of viral spread.

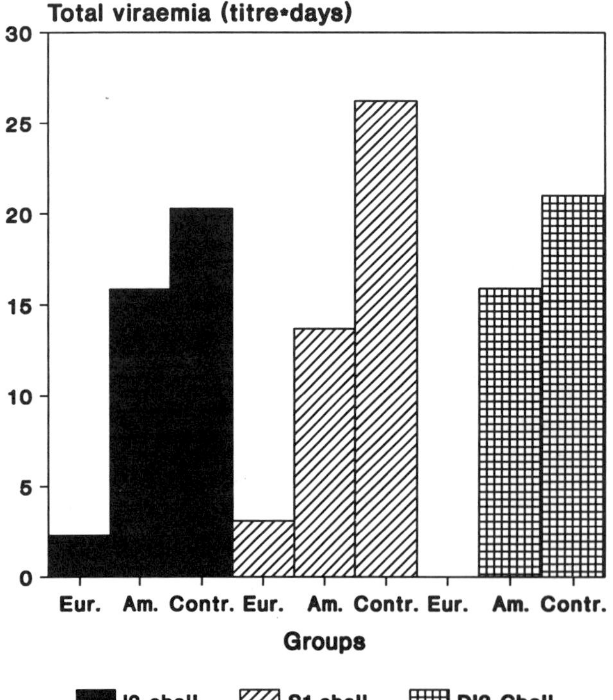

Figure 1. Total amount of viraemia (titre x day) during the experiment after challenge with 3 different European PRRSV strains found in the controls and pigs vaccinated with either an American or a European serotype PRRSV strain.

From serology data and sequence data it was already evident that large differences were present between the European and the American PRRSV strains, suggesting that both types belong to distinct genotypes. The reported serological differences between both PRRSV types were confirmed in this experiment. Sera of pigs vaccinated with the American serotype, incubated on cells infected with a European serotype, had significantly lower end points than the sera from European serotype-vaccinated pigs and vice versa. The necessity to use a test system with a reporter virus of the same serotype was demonstrated by the sentinels included in the test. The relatively low titres of 3 sentinels placed with the American serotype-vaccinated pigs were not detected in the immunofluorescence test using a European reporter virus. Only when a reporter virus of the same serotype was used, these animals were scored as being seropositive. The fact that the American serotype vaccine spreaded significantly more than the European serotype vaccine indicates that the American strain used in this vaccine was less attenuated than the European strain used in this experiment.

With regard to the efficacy it was clearly demonstrated that the reported serological differences and the genomic differences between the American and the European isolates have serious practical consequences with regard to vaccination. Vaccination with an American serotype vaccine will certainly not prevent a European strain from replicating in a vaccinated host. Although not statistically significant in 2 out of 3 challenges, vaccination with an American strain seems to suppress the challenge virus replication to some extent. However, when compared to the European serotype-vaccinated pigs a far more effective suppression and even a total suppression of challenge virus replication in the host could be obtained. Inevitably, after vaccination with an American serotype virus, European wild type viruses are still able to spread substantially more than after using a European serotype vaccine. The demonstrated lack of cross protection shows that both PRRSV types not only differed in their previously reported serological response but that their immunologically distinct. The American and European PRRSV strains therefore, represent two separate PRRSV types with their own unique immunological characteristics. With regard to the practical consequences of this observation, due to the absence of clinical signs in the none vaccinated controls after challenge, it could not be determined if vaccination with an American serotype vaccine might result in a reduction of clinical signs. However, based on the reduction in viraemia, it is clear that a European-serotype vaccine probably will be proportionally more efficacious in the prevention of clinical signs caused by European wild type strains. These results therefore, show that the control of PRRSV in Europe can only be achieved by using European serotype-based vaccines.

REFERENCES

Becker N.g., 1989, Analysis of infectious disease data. *Chapman and Hall Ltd,* London, New York.

Drew, T.W., Meulenberg, J.M., Sands, J., and Paton, D.J., 1995, Production, characterization and reactivity of monoclonal antibodies to porcine reproductive and respiratory syndrome virus, *J. Gen. Virol.* **76**: 1361–1369.

Kapur, V., Elam, M.R., Pawlovich, T.M,. and Murtaugh, M.P., 1996, Genetic variation, in porcine reproductive and respiratory syndrome virus isolates, in The Midwestern United States, *J. Gen. Virol.* **77**: 1271–1276.

Meng, X.-J., Paul, P.S., Halbur, P.G., and Lum, M.A., 1994, Phylogenetic analyses of the putative M (ORF 6) and N (ORF 7) genes of porcine reproductive and respiratory syndrome virus (PRRSV); implications for the existence of two genotypes of PRRSV in the U.S.A. and Europe, *Arch. Virol.* **140**: 745–755.

Mardassi, H., Mounir, S., and Dea, S., 1995, Molecular analysis of the orfs 3 to 7 of porcine reproductive and respiratory syndrome virus, Quebec Reference Strain. *Arch. Virol.* **140**: 1405–1418.

Murtaugh, M.P., Elam, M.R., and Kakach, L.T., 1995, Comparison of structural protein coding Sequences of the Vr-2332 and Lelystad virus strains of the PRRS virus. *Arch. Virol.* **140**: 1451–1460.

Nelson, E.A., Cristopher-Hennings, J., Drew, T., Wensvoord, G., Collins, J.E., and Benfield, D.A., 1993, Differentiation of U.S. and European isolates of porcine reproductive and respiratory syndrome virus by monoclonal antibodies, *J. Clin. Microbiol.* **31**: 3184–3189.

Suarez, P., Zardoya, R., Jesus-Martin, M., Prieto, C., Dopazo, J., Solana, A., and Castro, J.M., 1996, Phylogenic relationships of European strains of porcine reproductive and respiratory syndrome virus (PRRSV) inferred from DNA sequences of putative orf-5 and orf-7 genes. *Virus Res.* **42**: 159–165.

Wensvoort, G., M., DeKluyver, E.P., Luijtze, E.A., DenBesten, A., Harris, L., Collins, J.E., Christianson, W.T., and Chladek, D., 1992, Antigenic comparison of Lelystad virus and swine infertility and respiratory syndrome (SIRS) virus, *J. Vet. Diag. Investi.,* **4**: 134–138.

Variability and Evolution

93

POPULATION DYNAMICS IN THE EVOLUTION OF RNA VIRUSES

Esteban Domingo, Cristina Escarmís, Noemí Sevilla, and Eric Baranowski

Centro de Biología Molecular "Severo Ochoa" (CSIC-UAM)
Universidad Autónoma de Madrid
Cantoblanco, 28049 Madrid, Spain

1. ABSTRACT

RNA virus quasispecies are subjected to processes of positive Darwinian selection, to a very active and continuous negative selection and to random genetic drift. The course of RNA virus evolution is often unpredictable, and recent results suggest that even highly conserved motifs, once regarded as essential for infectivity, may be rendered dispensable by singular evolutionary events. An immediate consequence of the quasispecies genetic organization of RNA viruses is a surprising ability to gain fitness once a minimal replication ability is established in a biological enviroment. The unique features of RNA genetics should not be underestimated since they are at the basis of the emergence of new viral diseases and of the current difficulties to control many diseases associated variable viruses.

2. INTRODUCTION

Work in several laboratories during the last decade has firmly established that RNA viruses replicate as complex collections of closely related genomes that have been termed viral quasispecies (recent reviews in Domingo and Holland, 1997; Domingo *et al.* 1995; Holland *et al.* 1992). The molecular basis of viral genome complexity at the population level is the limited copying fidelity exhibited by viral replicases including the RNA-dependent RNA polymerases and RNA-dependent DNA polymerases or retrotranscriptases. The absence or very low activity of proofreading-repair activity in these enzymes has been supported by biochemical (Steinhauer *et al.* 1982) and structural (Sousa, 1996) studies. The latter have evidenced the absence in the viral replicases analyzed to date of any domain that could be equated with a 3' to 5' exonuclease, the activity responsible for proofreading in several cellular DNA polymerases. Mutation rates during viral replication have

Coronaviruses and Arteriviruses, edited by Enjuanes *et al.*
Plenum Press, New York, 1998

been estimated using a variety of genetic and biochemical methods, and most figures are in the range of 10^{-3} to 10^{-5} substitutions per nucleotide (s/nt) and round of copying (several examples were compiled by Domingo and Holland, 1994).

With mutation rates of about 10^{-4} s/nt, a viral genome of 10,000 nucleotides will produce an average of one mutation per template copied. That is, if an infected cell accumulates in the order of 10^5 viral RNA molecules ready to be encapsidated and released from the cell (a reasonable estimate for picornaviruses for example), most of them will be single mutants. Of course in viral infections not only one cell but hundreds of cells may be infected, and often at different compartments in the organism. If one assumed a cascade of infections of cells by the prior progeny, after as little as 10 rounds of infection one would end with viral population averaging 10 substitutions per genome relative to the genome that initiated the infection. Soon viral genomes would diverge so much that they would not maintain their structural and biological identity. Obviously there must be some force counteracting accumulation of mutations due to error-prone replication. This force is called negative selection. A balance between mutational imput, negative selection, and the effects of population bottlenecks constitute important parameters that define the population dynamics of RNA virus populations. This is the topic of this review.

3. SELECTION DETERMINES A DIFFERENCE BETWEEN MUTATION RATE AND MUTATION FREQUENCY

The invented example of Fig. 1 illustrates the difference between mutation rate and mutation frequency. In the example the template adenosine (A) in genomic position 2001 is miscopied into C in the complementary strand at a higher rate (10^{-4} s/nt) than into G (10^{-6} s/nt). This means that C at this position will be represented one hundred times more frequently than G in the population of complementary strands. However, assume that the template activity of molecules with G is vastly superior than the template activity of molecules with C, for example because G is more acceptable than C in the interaction with a host factor needed for replication. In this case many more progeny genomes will be produced with C-2001 than with

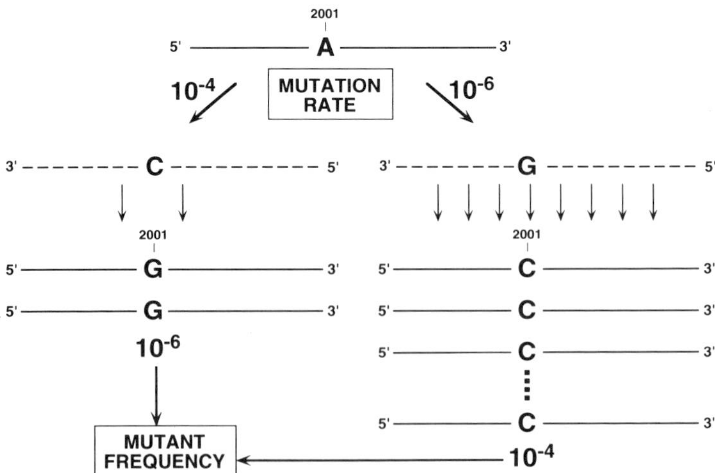

Figure 1. The difference between mutation rate and mutation frequency. Positive strands are depicted as continuous lines and negative strands (replication intermediates) as discontinuous lines (see text).

G-2001. The mutant frequency of A→G may be hundred-fold lower than the frequency of A→C, and these values would be unrelated to the mutation rates acting to produce their respective ancestor complementary strands (Figure 1). Obviously, any steps in the infectious cycle of a virus may influence the proportion of the different types of mutants that populate viral quasispecies. The mutation rate reflects a biochemical event at the level of misincorporation during template copying while the mutation frequency is the proportion of mutations found in the population under analysis once selective forces have acted.

For a virus well adapted to its biological environment (for example to a type of cell in culture) deleterious mutations will be more frequent than advantageous mutations. This may explain why determinations of mutation frequencies by biochemical procedures (such as the error oligonucleotide method of Steinhauer and Holland, 1986) tend to yield slightly higher values than procedures based on genetic analyses (compare Ward and Flanegan, 1992; Sedivy *et al.* 1987; White and McGeoch, 1987; Steinhauer and Holland, 1986; review in Domingo and Holland, 1994). In the error oligonucleotide method, the proportion of molecules lacking selected T1 RNase-cleavable G residues was determined in clonal populations of vesicular stomatitis viruses passaged to the minimal extent feasible to permit the chemical detection of mutant molecules. The limited number of virus passages probably did not allow selective forces to modulate the proportion of different mutants, a feature of most methods based on genetic analysis.

Mutation frequencies for RNA viruses are generally in the range of 10^{-5} to 10^{-3} s/nt (Domingo and Holland, 1994), the larger values being generally observed during prolonged persistent infections in heterogeneous environments. For example, Nájera *et al.* (1995) measured mutation frequencies of up to 10^{-2} in the *pol* gene of HIV-1 genomes present within infected individuals. We know now that an array of different cell types expressing a variety of coreceptors may be the target of HIV-1 variants in the course of the infection.

4. SELECTION AND DRIFT

In contrast to negative selection which tends to eliminate or keep at low proportion unfit variants, positive or Darwinian selection promotes the rapid dominance of certain variants in evolving quasispecies. Examples are the selection of mutant viruses resistant to monoclonal or polyclonal antibodies, to soluble receptors, or to antiviral agents. The distinction between positive and negative selection can at times become rather fuzzy; negative selection acting on a subset of genomes can be viewed as positive selection acting on the complement subset of biologically competent variants. This argument could be regarded as anecdotal were it not for the fact that part of the controversy accompanying the views of neutralists and selectionists in molecular evolution stems from an imprecise distinction between positive and negative selection (for a review of these concepts from the point of view of two virologists see Domingo and Holland, 1994; for the ultimate defense of neutralism to interpret RNA virus evolution, see Kimura, 1989).

Genetic drift is a modification of the genetic composition of a population due not to positive selection but to the random sampling and multiplication of a variant genome. Virologists are very familiar with genetic drift since it may occur when one or a few infectious particles from a viral quasispecies initiate an infection. Plating a virus on a cell monolayer and the subsequent isolation of virus from a plaque is the most severe form of a genetic bottleneck since the entire population is the progeny of a single genome. Tolerated mutations present in this ancestor genome will be maintained in further viral amplifications not as a result of their selective value, but as the passive outcome of their being present in the plaque that was isolated at random.

Wain-Hobson (1994) has emphasized that activation of a lymphocyte harboring a HIV-1 provirus will result in the production of viral progeny irrespective of the fitness of the virus being made. There is very little solid evidence to support that either positive selection or genetic drift play the main role in virus evolution (Domingo and Holland, 1994). Very often we do not understand the subtle ways in which single or multiple mutations affect viral fitness in particular environments.

Remarkable Genetic Changes in Foot-and-Mouth Disease Virus

A recent example using foot-and-mouth disease virus (FMDV) serves to illustrate the unexpected course that evolution of RNA viruses may take. FMDV is an economically important pathogen of the *Picornaviridae* family. Our laboratory has used FMDV as a model system to analyze the evolution of RNA viruses and also to define the molecular basis of antigenic variation as one of the main problems for the efficacy of synthetic vaccines. FMDV has a mobile, protruding loop on the surface of its capsid of icosahedral symmetry (Acharya *et al.* 1989). This loop is involved in at least two recognized biological activities. It is a receptor-binding site, and it uses a typical Arg-Gly-Asp domain to recognize integrin $a_v b_3$ or a closely related molecule (Berinstein *et al.* 1995). It is also one of the sites of binding of neutralizing antibodies, and the residues from the integrin-recognition triplet (mainly the aspartate) play a critical role in antibody binding (Verdaguer *et al.* 1995, 1996, and submitted for publication). The Arg-Gly-Asp motif is highly invariant in FMDV and site-directed mutations affecting this triplet (but not neighbouring residues) generally produced noninfectious virus (Mason *et al.* 1994; Leippert *et al.* 1997; an exception was a Asp®Glu substituion in FMDV of serotype O which rendered infectious virus [Leippert *et al.* 1997]). In our studies with FMDV of serotype C monoclonal antibody-escape mutants showed amino acid substitutions at positions around the Arg-Gly-Asp motif but never within it (Mateu *et al.* 1989, 1990), consistently with the findings that alterations of this motif were lethal. Interestingly and surprisingly, a clonal FMDV population that had been passaged one hundred times in BHK-21 cells showed a radical change of behavior. The repertoire of escape mutants of the virus at passage one hundred was substantially expanded, and some mutants showed amino acid substitutions at the Arg-Gly-Asp, and grew normally. Tolerance to replacements at the critical receptor-binding site may be associated to substitutions at a few capsid sites around the receptor binding loop, but the exact mechanism that renders the Arg-Gly-Asp dispensable remains unknown (Martínez *et al.* 1997). Had such additional changes occured without knowledge of the altered (and increased) potential of the virus to evade neutralization they would be given the rank of neutral and with little if any selective value. Experiments are now in progress to try to elucidate whether such FMDV mutants recognize alternative receptors. It is worth stressing that all the FMDV changes took place upon serial passage in BHK-21 cells that did not coevolve with the virus (since they were periodically thawed from the same cell stock to mimmize cell variation). It is not the same situation of FMDV coevolution with host cells during persistent infections (de la Torre *et al.* 1988) or the alterations in cellular receptor choice of the coronavirus murine hepatitis virus in its coevolution with murine astrocytoma cells (Chen and Baric, 1996).

5. THE SEARCH FOR FITNESS OPTIMA

Eigen (1992) emphasized the concept of sequence space as applied to simple replicons and viruses. Sequence space indicates all possible nucleotide sequences that a viral genome could theoretically use. This number is extremely large ($4^{30,000}$ for an RNA virus of 30,000 nucleotides, a number impossible to comprehend). Most of the theoretical positions in sequence space cannot be occupied by extant viruses because they make "no biological sense". A virus is a coordinated ensemble of interacting nucleic acid and protein domains kept in close check by cellular structures and activities. Thus, movements in sequence space are necessarily very limited. No matter how great are the mutational and selective pressures triggering diversification of HIV-1, the occupation of sequence space is still so limited that we can classify HIV-1 isolates as belonging to groups M or O and subtypes A, B, C, D, E, F, G, H, I, J, etc. Evolution is a walk in sequence space impulsed by genetic change, guided by selection, and in search of fitness optima.

As studied and visualized in the classical work of Wright (1931), a fitness landscape is made of mountains, peaks, valleys and pits. Mutations and selection guide viruses to climb to fitness peaks. The quasispecies nature of RNA viruses predicts that the numbers of genomes subjected to competitive selection will very much influence fitness gain or fitness loss. This has been documented experimentally by quantitating average fitness losses when RNA viruses have been subjected to repeated bottleneck events, achieved by serial plaque-to-plaque transfers (Chao, 1990; Duarte *et al.* 1992; Escarmís *et al.* 1996). In contrast, massive serial infections where mutant genomes are allowed to compete, result in rapid increases of viral fitness (Novella et al. 1995a). Repeated bottlenecks do not permit an optimization of the viral quasispecies whereas massive infections lead to selection of the most competent (fit or viable) viruses (reviews in Domingo *et al.* 1996; Domingo and Holland, 1997).

A fitness landscape varies with the environment in which viral replication takes place. Studies with vesicular stomatitis virus (VSV) revealed that after its persistence in insect cells, viral fitness reached values which were million-fold greater in insect cells than in mammalian cells (Novella *et al.* 1995b). Highly debilitated viruses very frequently find their way to fitness peaks. This is one of the most direct demonstrations of adaptability of RNA viruses. In turn, the dynamics of fitness change in RNA viruses cannot be explained except by considering the quasispecies nature of RNA viruses since adaptability lies in the selective amplification of subsets of genomes. In the course of an HIV-1 infection of humans, the virus evolves to use expanded sets of coreceptors. At early stages of infection when disease symptoms are still limited, HIV-1 strains are often of a non-syncytium inducing (NSI) phenotype and use a coreceptor termed CCR5. In patients which develop AIDS, viruses expand their use of coreceptors to include CCR3, CCR2b and CXCR-4. The latter, also termed fusin, is used by virus depicting a syncytium-inducing (SI) phenotype. The use of different coreceptors correlates with changes in a highly variable V3 loop located on the surface glycoprotein gp120. That is, in the course of a natural infection, HIV-1 increases its fitness to enter additional cells by a variety of mechanisms, but this does not ensure that the same SI viruses are equally fit than their NSI ancestors at the early stages of infection. In fact many infections are initiated with viruses predominantly of a NSI phenotype which later evolve to a SI phenotype. Viral persistence is a manifestation of virus adaptability.

6. CONCLUDING REMARKS

Genetic variation is a general feature of all living creatures since it is the very foundation of adaptability and survival. This generality cannot be used as an argument to deny novelty to quasispecies as a descriptor of RNA viruses. This would be a damaging mistake since many urgent problems in viral diseae emergence and control stem from some very distinct features of RNA virus variation as compared to cellular variation (see Domingo, 1996). In essence the critical difference is that considering the mutation frequencies, the genome sizes and the population numbers (total numbers, not calculated effective numbers) of RNA viruses, their potential for exploration of sequence space is vastly superior than for cellular organisms (Domingo *et al.* 1995; Domingo, 1996). If we deny differences between mutability of RNA genomes and cells we will underestimate the evolutionary potential for new RNA viral pathogens to arise and to evolve. As J. Lederberg rightly pointed out "The survival of the human species is not a preordained evolutionary program. Abundant sources of genetic variation exist for viruses to learn new tricks, not necessarily confined to what happens routinely or even frequently". In fact one of the most striking and also disconcerting features of RNA virus evolution is that often each individual infection (in a cell culture or in a live host organism) is a unique evolutionary episode. Each infection is somewhat a new start in a fleeting ceaseless successions of evolutionary events where extinctions (failures) are far more abundant than creations (successes). However, one creation was sufficient to trigger the AIDS plague.

REFERENCES

Acharya, R., Fry, E., Stuart, D., Fox, G., Rowlands, D. and Brown, F., 1989, The three-dimensional structure of foot-and-mouth disease virus at 2.9Å resolution, *Nature* **337**: 709–716.

Berinstein, A., Roivainen, M., Hovi, T., Mason, P.W. and Baxt, B., 1995, Antibodies to the vitronectin receptor (integrin avb3) inhibit binding and infection of foot-and-mouth disease virus to cultured cells, *J. Virol.* **69**: 2664–2666.

Chao, L., 1990, Fitness of RNA virus decreased by Muller's ratchet, *Nature* **348**: 454–455.

Chen, W. and Baric, R.S., 1996, Molecular anatomy of mouse hepatitis virus persistence: Coevolution of increased host cell resistance and virus virulence, *J. Virol.* **70**: 3947–3960.

de la Torre, J.C., Martínez-Salas, E., Díez, J., Villaverde, D., Gebauer, F., Rocha, E., Dávila, M. and Domingo, E., 1988, Coevolution of cells and viruses in a persistent infection of foot-and-mouth disease virus in cell culture, *J. Virol.* **62**: 2050–2058.

Domingo, E., 1996, Biological significance of viral quasispecies, *Viral Hepatitis Reviews*, **2**: 247–261.

Domingo, E. and Holland, J.J., 1994, Mutation rates and rapid evolution of RNA viruses, *Evolutionary Biology of viruses*, S.S. Morse, ed., Raven Press, New York, pp. 161–184.

Domingo, E. and Holland, J.J., 1997, RNA virus mutations and fitness for survival, *Annu. Rev. Microbiol.* **51**: 151–178.

Domingo, E., Holland, J.J., Biebricher, C. and Eigen, M., 1995, Quasispecies: The concept and the word, in: *Molecular Evolution of the Viruses*, A. Gibbs, C. Calisher and F. García-Arenal, eds. Cambridge University Press, pp. 171–180.

Domingo, E., Escarmís, C., Sevilla, N., Moya, A., Elena, S.F., Quer, J., Novella, I.S. and Holland, J.J., 1996, Basic concepts in RNA virus evolution, *FASEB J.* **10**: 859–864.

Duarte, E., Clarke, D., Moya, A., Domingo, E. and Holland, J.J., 1992, Rapid fitness losses in mammalian RNA virus clones due to Muller's ratchet, *Proc. Natl. Acad. Sci. USA* **89**: 6015–6019.

Eigen, M., 1992, Steps towards life. Oxford University Press.

Escarmís, C., Dávila, M., Charpentier, N., Bracho, A., Moya, A. and Domingo, E., 1996, Genetic lesions associated with Muller's ratchet in an RNA virus, *J. Mol. Biol.* **264**: 255–267.

Holland, J.J., de la Torre, J.C. and Steinhauer, D.A., 1992, RNA virus populations as quasispecies, *Curr. Top. Microbiol. Immunol.* **176**: 1–20.

Kimura, M., 1989, The neutral theory of molecular evolution and the world of the neutralists, *Genome* **31**: 24–31.

Leippert, M., Beck, E., Weiland, F. and Pfaff, E., 1997, Point mutations within the bG-bH loop of foot-and-mouth disease viurs O_1K affect virus attachment to target cells, *J. Virol.* **71**: 1046–1051.

Martínez, M.A., Verdaguer, N., Mateu, M.G. and Domingo, E., 1997, Evolution subverting essentiality: Dispensability of the cell attachment Arg-Gly-Asp motif in multiply passaged foot-and-mouth disease virus, *Proc. Natl. Acad. Sci. USA.* In press.

Mason, M.W., Rieder, E. and Baxt, B., 1994, RGD sequence of foot-and-mouth disease virus is essential for infecting cells via the natural receptor but can be bypassed by an antibody dependent enhancement pathway, *Proc. Natl. Acad. Sci. USA* **91**: 1932–1936.

Mateu, M.G., Martínez, M.A., Capucci, L., Andreu, D., Giralt, E., Sobrino, F., Brocchi, E. and Domingo, E., 1990, A single amino acid substitution affects multiple overlapping epitopes in the major antigenic site of foot-and-mouth disease virus of serotype C, *J. Gen. Virol.* **71**: 629–637.

Mateu, M.G., Martínez, M.A., Rocha, E., Andreu, D., Parejo, J., Giralt, E., Sobrino, F. and Domingo, E., 1989, Implications of a quasispecies genome structure: affect of frequent, naturally occurring amino acid substitutions on the antigenicity of foot-and-mouth disease virus, *Proc. Natl. Acad. Sci. USA* **86**: 5883–5887.

Nájera, I., Holguín, A., Quiñones-Mateu, M.E., Muñoz-Fernández, M.A., Nájera, R., López-Galíndez, C. and Domingo, E., 1995, The *pol* gene quasispecies of human immunodeficiency virus. Mutations associated with drug resistance in virus from patients undergoing no drug therapy, *J. Virol.* **69**: 23–31.

Novella, I.S., Duarte, E.A., Elena, S.F., Moya, A., Domingo, E. and Holland, J.J., 1995a, Exponential increases of RNA virus fitness during large population transmissions, *Proc. Natl. Acad. Sci. USA* **92**: 5841–5844.

Novella, I.S., Clarke, D.K., Quer, J., Duarte, E.A., Lee, C.H., Weaver, S.C., Elena, S.F., Moya, A., Domingo, E. and Holland, J.J., 1995b, Extreme fitness differences in mammalian and insect hosts after continuous replication of vesicular stomatitis virus in sandfly cells, *J. Virol.* **69**: 6805–6809.

Sedivy, J.M., Capone, J.P., Raj Bhandary, U.L. and Sharp, P.A., 1987, An inducible mammalian amber suppressor: propagation of a poliovirus mutant. *Cell* 50: 379–389.

Sousa, R., 1996, Structural and mechanistic relationships between nucleic acid polymerases, *TIBS* **21**: 186–190.

Steinhauer, D.A. and Holland, J.J., 1986, Direct method for quantification of extreme polymerase error frequencies at selected single base sites in viral RNA, *J. Virol.* **57**: 219–228.

Steinhauer, D., Domingo, E. and Holland, J.J., 1992, Lack of evidence for proofreading mechanisms associated with an RNA virus polymerase, *Gene* **122**: 281–288.

Verdaguer, N., Mateu, M.G., Andreu, D., Giralt, E., Domingo, E. and Fita, I., 1995, Structure of the major antigenic loop of foot-and-mouth disease virus complexed with a neutralizing antibody: direct involvement of the Arg-Gly-Asp motif in the interaction, *EMBO J.* **14**: 1690–1696.

Verdaguer, N., Mateu, M.G., Bravo, J., Domingo, E. and Fita, I., 1996, Induced pocket to accomodate the cell attachment Arg-Gly-Asp motif in a neutralizing antibody against foot-and-mounth disease virus, *J. Mol. Biol.* **256**: 364–376.

Wain-Hobson, S., 1994, Is antigenic variation of HIV important for AIDS and what might be expected in the future? in: *The Evolutionary Biology of Viruses*, S.S. Morse, ed., pp. 185–209. Raven Press, Ltd., New York.

Ward, C.D. and Flanegan, J.B., 1992, Determination of the poliovirus RNA polymerase error frequency at eight sites in the viral genomes, *J. Virol.* 66: 3784–3793.

White, B.T. and McGeoch, D.J., 1987, Isolation and characterization of conditional lethal amber nonsense mutants of vesicular stomatitis virus. *J. Gen. Virol.* **68**: 3033–3044.

Wright, S., 1931, Evolution in Mendelian populations, *Genetics* **16**: 97–159.

94

DOES IBV CHANGE SLOWLY DESPITE THE CAPACITY OF THE SPIKE PROTEIN TO VARY GREATLY?

D. Cavanagh,[1]* K. Mawditt,[1] A. Adzhar,[1] R. E. Gough,[2] J.-P. Picault,[3] C. J. Naylor,[1] D. Haydon,[4] K. Shaw,[1] and P. Britton[1]

[1]Institute for Animal Health
Compton Laboratory
RG20 7NN, United Kingdom
[2]Central Veterinary Laboratory
Weybridge, KT15 3NB, United Kingdom
[3]Laboratoire Central de Reserches Avicole et Porcine
22440 Ploufragan, France
[4]Department of Zoology
University of Oxford, OX1 3PS

1. SUMMARY

We have sequenced that part of the spike protein (S) gene which encodes the amino-terminal and most variable quarter (hypervariable region, HVR) of the S1 subunit of 28 isolates of the 793/B (also known as CR88 and 4/91) serotype of infectious bronchitis virus (IBV) and the whole of S1 for nine of them. The isolates were from France and Britain between the years 1985 (first isolation) and 1996. The maximum nucleotide and amino acid differences between the first isolate and the others were 4.1% and 7.6%, respectively, for the whole of S1 and 7.1% and 14.6%, respectively, in the HVR. Analysis within clearly recognisable subgroups suggested that even in the HVR the nucleotide mutation rate was only 0.3 to 0.6% per year. However, there was no evidence that mutations had become fixed in a progressive manner; this serotype did not appear to be evolving. Strains isolated several years apart could be more similar than those isolated in a given year. It is likely that the amino acid changes are largely at positions where amino acid differences are tolerated rather than as a consequence of immune pressure. Reasons for this conclusion are discussed.

Coronaviruses and Arteriviruses, edited by Enjuanes *et al.*
Plenum Press, New York, 1998

2. INTRODUCTION

IBV is known to exist as many serotypes, defined by haemagglutination-inhibition (HI) or virus neutralization (VN) tests, and doubtless many more await discovery. The epitopes which define serotype are formed largely by amino acids within the first and third quarters of the S1 subunit (Kant et al., 1992). It is in these regions that most amino acid differences occur between serotypes and within serotypes (Cavanagh, 1995). Serotypes frequently, but not invariably, differ by 20 to 25% of S1 amino acids. The amino-terminal quarter of S1 is the most variable, this hypervariable region (HVR) differing between serotypes by 35% or more of amino acids. Such differences might suggest that IBV mutates rapidly.

In the winter of 1990/91 in Britain chickens suffered respiratory disease, and unusual pathology, caused by a serotype of IBV not previously detected in Britain (Gough et al., 1992; Parsons et al., 1992). The serotype was named 793/B (also 4/91) after one of the earliest British isolates. It differs from all known IBV serotypes at 21 to 25% of S1 amino acids (Adzhar et al., 1997). It subsequently transpired that a large number of strains isolated in France in 1988 (prefixed by CR88) were of the same serotype (Picault et al., 1995). Retrospective serological analysis revealed that one French isolate of 1985 was also of this serotype. These epidemiological studies presented us with an opportunity to examine the manner and extent to which changes were occurring in this serotype over a decade.

3. MATERIALS AND METHODS

The French isolates (number examined in parenthesis) were from years 1985 (1), 1988 (3) and 1994 (6) and the British isolates were from years 1991 (7), 1993 (6), 1995 (2) and 1996 (4).

The S1 part of the S gene of the 1991 and 1993 isolates was amplified in a reverse transcriptase polymerase chain reaction (RT-PCR) using 'universal' IBV oligonucleotides as primers (Adzhar et al., 1997). The DNA products were cloned prior to sequencing. For all other isolates sequencing was performed directly on the PCR products. The HVR was amplified using two primers specific for the 793/B genotype.

The phylogenetic relationships between the strains were investigated using the DNAML (maximum likelihood) package of Felsentein (1993); 11 'hill climbs' were perfomed. Estimates of the mean corrected (Jukes and Cantor, 1969) numbers of synonymous changes per synonymous site (d_s) and non-synonymous changes per non-synonymous site site (d_{ns}) per sequence pair were obtained using the method of Nei and Gojobori (1986).

4. RESULTS

4.1. Nucleotide Sequence Comparisons

It is not suggested that the British strains had developed directly from the French strains in this study or, indeed, from any other French strains. It is highly likely that this serotype was prevalent in Continental Europe, but not in Britain, by 1990 (Cook et al., 1996). Notwithstanding, we have used CR85131, the earliest isolate of the 793/B serotype, as a reference point.

The sequence of that part of the S gene encoding the whole of the S1 subunit was determined for nine of the isolates. Comparison of each nucleotide sequence with that of

Table 1. Percentage differences between the nucleotide and amino acid sequence of the whole of S1 of the earliest 793/B isolate (CR85131) with those of eight other strains of the same serotype

Isolate	Year	Country	Difference (%) from whole S1 sequence of CR85131	
			Nucleotides	Amino acids
CR88061	1988	France	3.0	5.6
2/91	1991	Britain	3.3	6.5
3/91	1991	Britain	3.1	5.9
5/91	1991	Britain	3.2	5.8
7/91	1991	Britain	3.3	6.5
7/93	1993	Britain	3.9	7.6
CR94047	1994	France	4.1	6.7
1233-95	1995	Britain	3.7	6.5

the earliest isolate (CR85131) of the 793/B serotype showed that the strains differed at 3.0 to 4.1% of nucleotides (Table 1).

In order to get a more thorough picture of the sequence differences exhibited by this genotype we sequenced the HVR (approximately 450 nucleotides) of an additional 19 isolates. Relationships among the 28 isolates in this region are illustrated by a maximum likelihood phylogeny unrooted tree in Figure 1. A number of subgroups can be discerned.

The best estimate of the rate at which these viruses are mutating can be deduced from the subgroup at the top of Figure 1. This subgroup comprises 13 strains isolated in Britain between 1991 and 1996. The greatest difference between two of the isolates, isolated in 1991 and 1996, respectively, was only 1.5% of nucleotides. This is very low when one considers that these figures refer to the HVR. It is likely, therefore, that all these isolates have a common origin. If that is the case then there has been a change of only 1.5 % of nucleotides in five years, or 0.3%/year. The rate for the whole of the S1 part of the S gene would be approximately half this figure.

The other British subgroup comprises five isolates. The greatest difference in the HVR, between a 1991 and a 1995 isolate, was 2.6%. If the 1995 isolate had developed from the 1991 isolate then this indicates a rate of change of 0.6%/year.

The five 1994 French isolates which are on the same branch as the earliest (1985) isolate (CR85131) have differences of 2.6 to 3.9% from this isolate. If these strains developed directly from CR85131 then this difference over nine years suggests a mutation rate of 0.3 to 0.4% in the HVR. This is similar to the rate estimated for the two British subgroups and supports the phylogenetic analysis that these French 1994 isolates have developed from CR85131.

The three 1988 French isolates differed from CR85131 by 4.7 to 5.8%. Given the mutation rates estimated above this would suggest that these three isolates were introduced into France independently of CR85131.

4.2. Amino Acid Sequence Analysis

The amino acid sequence of S1 of CR85131 differed from that of the other eight strains for which the whole of S1 was sequenced by 5.6 to 7.6% (Table 1). The ratio (% amino acid difference:% nucleotide differences) was approximately 1.8.

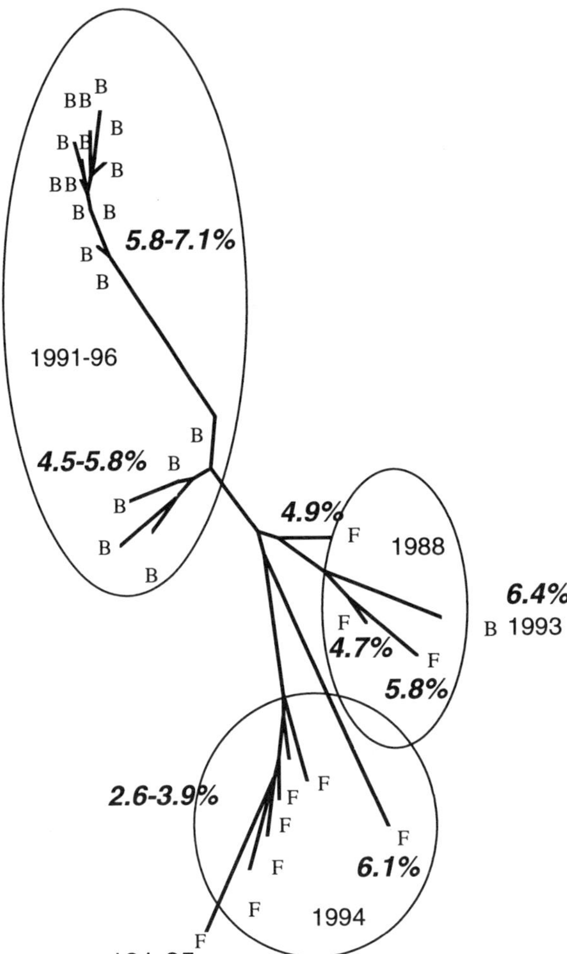

Figure 1. Relationships among 28 isolates of the 793/B serotype, using data corresponding to the hypervariable region, illustrated by a maximum likelihood phylogeny unrooted tree. The French and British isolates are indicated by 'F' and 'B', respectively. The ellipses encircle strains isolated in the years shown. The differences (%) between the nucleotide sequence of the HVR of the oldest isolate, CR85131, and the others is shown.

In the HVR the range of amino acid differences from CR85131 was 7.6 to 14.6%. The ratio (% amino acid difference:% nucleotide difference) was in the range 1.8 to 2.3 for the majority of strains, indicating a generally higher proportion of non-synonymous mutations in the HVR than in the S1 gene as a whole.. The exceptions were four of the isolates in the same branch as CR85131 in Figure 1, where the ratio was 2.6 to 2.8, indicative of a higher proportion of non-synonymous mutations than in the other isolates.

We have analysed the ratio of synonymous to non-synonymous mutations in the HVR region of 12 British 793/B strains isolated in 1991–93 and compared it to the same ratio obtained from an analysis of a 360 nucleotide partial sequence from the nucleocapsid protein (N) gene of 21 strains of IBV of many serotypes (Zwaagstra et al., 1992). The N protein is much less variable than the S1 protein. The expected value of d_s/d_{ns} (see Materials and Methods) for the N gene was 6.48 (standard error 1.36) and for the HVR of S1 gene was 1.35 (standard error 0.46). These ratios are significantly different from each other ($p<0.0002$), demonstrating that non-synonymous substitutions were relatively more common with respect to synonymous substitutions in the S1 gene than the N gene. Indeed the value of 1.35 demonstrates that a high proportion of mutations in the HVR of S1 led to changes in amino acid composition.

5. DISCUSSION

A d_s/d_{ns} ratio of 1.0 is often cited as evidence for the presence of positive Darwinian selection for amino acid changes (Hughes and Hughes, 1995; Seibert et al., 1995). A ratio of slightly greater than 1.0 could result either from a selection pressure arising from immune surveillance, or as a consequence of reduced structural constraints on the protein structure thereby permitting a wider diversity of tolerated amino acid substitutions. While both of these factors probably interact to determine the observed d_s/d_{ns} ratio, we favour the latter—low structural constraints—as being the major factor.

In Britain alone there are approximately 700 000 000 meat-type chickens in a year. Most of these live for only 6 to 8 weeks and are quickly replaced with day-old chicks. Throughout the country the chicks receive a similar vaccination regime, typically one, sometimes two, vaccinations with live IBV vaccine of the Massachusetts serotype. Chicks vaccinated with Massachusetts vaccine develop clinical signs when challenged with IBV of the 793/B type (Parsons et al., 1992). If a 793/B-type strain can successfully infect a flock of meat-type chickens, and our epidemiological studies show that this is common, there is little immune pressure on the S1 protein to change as the replacement flock will have virtually the same immune status. Therefore it is likely that many of the amino acid differences, which arise from random mutations, are simply tolerated because of low structural constraints in this part of the protein. It is the equivalent part of the S protein of porcine transmissible gastroenteritis virus (TGEV), canine coronavirus, feline coronavirus and human coronavirus 229E which shows the greatest variation among those species. Indeed, in the respiratory variant of TGEV known as porcine respiratory coronavirus, the corresponding part of the S protein is missing (Cavanagh, 1995).

Egg-type chickens number approximately 60 000 000 in Britain. They live for a year or more, sufficient time for more than one infection by IBV. Immunity developed to infection by a 793/B strain during the first few weeks of life might be expected to have some protective effect against subsequent infections by strains of the same serotype. Changes in the S protein during the first infection might give some selective advantage when the virus infects a chicken that has previously been infected with the 793/B serotype. Further changes in the S protein may occur during this second infection and be selected.

Our analysis of the S1 gene sequences of the 28 strains in this study suggest that the mutation rate is not especially high. While many of the mutations are non-synonymous there is no obvious fixation of amino acid changes. Rather, different amino acids can be tolerated at some locations, especially within the first and third quarters of S1. The maximum amino acid difference between two 793/B isolates over the whole of S1 was 7.6%, approximately one-third of the difference between many IBV serotypes. It may take a considerable time for an IBV to accumulate a change of 20% of its S1 amino acids. A live 793/B vaccine was introduced recently. If this becomes widely used two scenarios may be envisaged. Firstly, the virus may accumulate amino acid changes in S1, and possibly other proteins, more quickly, in response to immune pressure. Secondly, the vaccine may reduce the amount of 793/B in the environment to such an extent that a 'vacuum' is created, paving the way for another serotype to become dominant. The dominant serotype/genotype (D274) in Continental Europe during the early/mid 1980s decreased greatly in the later part of the decade, during which time D274-specific vaccines were introduced. Concomitant with the decline of D274 was the rise of 793/B.

ACKNOWLEDGMENTS

We acknowledge the financial support of the British Chicken Association and the Ministry of Agriculture, Fisheries and Food, Great Britain.

REFERENCES

A. Adzhar, R. E. Gough, D. Haydon, K. Shaw, P. Britton & D. Cavanagh., 1997, Molecular analysis of the 793/B serotype of infectious bronchitis virus in Great Britain. *Avian Pathol.* 27. In press.

Cavanagh, D., 1995, The coronavirus surface glycoprotein, in: S.G. Siddell (Ed) *The Coronaviridae*, 73–114, New York, Plenum Press.

Cook, J.K.A., Orbell, S.J., Woods, M.A. & Huggins, M.B., 1996, A survey of the presence of a new infectious bronchitis virus designated 4/91 (793B*). Vet. Rec.* **138**: 178–180.

Felsentein, J., 1993, PHYLIP (phylogeny inference package), version 3.5c. Department of Genetics, University of Washington, Seattle.

Gough, R.E., Randall, C.J., Dagless, M., Alexander, D.J., Cox, W.J. & Pearson, D., 1992, A 'new' strain of infectious bronchitis virus infecting domestic fowl in Great Britain. *Vet. Rec.* **130**:493.

Hughes, M.K. & Hughes, A.L., 1995, Natural selection on plasmodium surface proteins. *Mol. Biochem. Parasitol.* **71**: 99–113.

Jukes, T.H. & Cantor, C.R., 1969, Evolution of protein molecules, in: Munro, H.N. (Ed) *Mammalian Protein Metabolism*, 21–132.

Kant, A., Koch, G., Van Roozelaar, D.J., Kusters, J.G., Poelwijk, F.A.J. & Van Der Zeijst, B.A.M., 1992, Location of antigenic sites defined by neutralizing monoclonal antibodies on the S1 avian infectious bronchitis virus glycopolypeptide. *J. Gen. Virol.* **73**: 591–596.

Nei, M. & Gojobori, T., 1986, Simple methods for estimating the numbers of synonymous and non-synonymous nucleotide substitutions. *Mol. Biol. Evol.* **3**: 418–426.

Parsons, D., Ellis, M.M., Cavanagh,D. & Cook, J.K.A., 1992, Characterisation of an infectious bronchitis virus isolated from vaccinated broiler breeder flocks. *The Vet. Rec.*: **131**: 408–411.

Picault, J.P., Drouin, P., Lamande, J., Allee, C., Toux, J.Y., Le Coq, H., Guittet, M. & Bennejean, G., 1995, L'epizootie recente de bronchite infectieuse aviaire en France: importance, evolution et etiologie. 1[eres] Journee de la Recherche Avicole, Centre de Congres, d'Angers.

Seibert, S.A., Howell, C.Y., Hughes, K.A., & Hughes, K.L., 1995, Natural selection on the gag, pol and env genes of human immunodeficiency virus 1 (HIV-1). *Mol. Biol. Evol.*, **12**: 803–813.

95

SELECTION IN PERSISTENTLY INFECTED MURINE CELLS OF AN MHV-A59 VARIANT WITH EXTENDED HOST RANGE

Jeanne H. Schickli,[1] David E. Wentworth,[1] Bruce D. Zelus,[1] Kathryn V. Holmes,[1] and Stanley G. Sawicki[2]

[1]Department of Microbiology
University of Colorado Health Sciences Center
Denver, Colorado 80262
[2]Department of Microbiology
Medical College of Ohio
Toledo, Ohio 43699

1. ABSTRACT

Murine coronavirus MHV-A59 normally infects only murine cells *in vitro* and causes transmissible infection only in mice. In the 17 Cl 1 line of murine cells, the receptor for MHV-A59 is MHVR, a biliary glycoprotein in the carcinoembryonic antigen (CEA) family of glycoproteins. We found that virus released from the 600th passage of 17 Cl 1 cells persistently infected with MHV-A59 (MHV/pi600) replicated in hamster (BHK-21) cells. The virus was passaged and plaque-purified in BHK-21 cells, yielding the MHV/BHK strain. Because murine cells persistently infected with MHV-A59 express a markedly reduced level of MHVR (Sawicki, *et al.*, 1995), we tested whether virus with altered receptor interactions was selected in the persistently infected culture. Infection of 17 Cl 1 cells by MHV-A59 can be blocked by treating the cells with anti-MHVR MAb-CC1, while infection by MHV/BHK was only partially blocked by MAb-CC1. MHV/BHK virus was also more resistant than wild-type MHV-A59 to neutralization by purified, recombinant, soluble MHVR glycoprotein (sMHVR). Cells in the persistently infected culture may also express reduced levels of and have altered interactions with some of the Bgp-related glycoproteins that can serve as alternative receptors for MHV-A59. Unlike the parental MHV-A59 which only infects murine cells, MHV/BHK virus was able to infect cell lines derived from mice, hamsters, rats, cats, cows, monkeys and humans. However, MHV/BHK was not able to infect all mammalian species, because a pig (ST) cell line and a dog cell line (MDCK I) were not susceptible to infection. MHV/pi600 and MHV/BHK replicated in murine cells more slowly than

Coronaviruses and Arteriviruses, edited by Enjuanes *et al.*
Plenum Press, New York, 1998

MHV-A59 and formed smaller plaques. Thus, in the persistently infected murine cells which expressed a markedly reduced level of MHVR, virus variants were selected that have altered interactions with MHVR and an extended host range. *In vivo*, in mice infected with coronavirus, virus variants with altered receptor recognition and extended host range might be selected in tissues that have low levels of receptors. Depending upon the tissue in which such a virus variant was selected, it might be shed from the infected animal or eaten by a predator, thus presenting a possible means for initiating the transition of a variant virus into a new host as a model for an emerging virus disease.

2. INTRODUCTION

MHV is a murine coronavirus that naturally infects only mice, causing either inapparent infection or hepatitis, diarrhea, splenolysis or neurological disorders. The virus is transmitted via the respiratory and enteric routes, and spreads rapidly through mouse colonies (Compton, *et al.*, 1993). Though infection *in vivo* is usually limited to mice, a few exceptions have been reported. Following intracerebral inoculation, MHV/JHM can infect brain of rats and owl monkeys (Cabirac, *et al.*, 1994; Murray, *et al.*, 1992; Wege, *et al.*, 1982). MHV infection *in vitro* is also usually limited to mice. The principal receptor used by MHV-A59 to infect murine 17 Cl 1 cells, MHVR (also called Bgp1[a]), is a biliary glycoprotein (Bgp) in the carcinoembryonic antigen (CEA) family (Dveksler, *et al.*, 1991). MHVR is expressed at high levels in mouse colon, small intestine, liver, trachea and smaller airways of the lung, kidney, and in lower levels on B lymphocytes and macrophages (Godfraind, *et al.*, 1995; Coutelier, *et al.*, 1994). Splicing of mRNA generates MHVR isoforms with either 2 or 4 Ig-like domains and either long or short cytoplasmic tails. Several isoforms can be co-expressed in murine cells. SJL/J mice are highly resistant to MHV-A59 infection. DNA sequence encoding MHVR is absent in SJL/J mice, but the genome encodes the Bgp1[b] (mmCGM2) glycoprotein in place of MHVR (Dveksler, *et al.*, 1993a). In tissue culture, Bgp1[b] and some additional murine Bgp-related glycoproteins can serve as MHV-A59 receptors if the recombinant proteins are expressed at high levels in cells of an MHV-A59 resistant species such as BHK or COS-7 cells. These receptors are Bgp2, which is a biliary glycoprotein with 2 Ig-like domains, and bCEA, a pregnancy specific glycoprotein (PSG) that is a soluble protein containing 3 Ig-like domains and lacking a membrane anchor (Chen, *et al.*, 1997; Chen, *et al.*, 1995; Nedellec, *et al.*, 1994).

Species specificity of MHV-A59 is determined, in part, by the virus-receptor interaction. MHV-A59 does not infect the BHK-21 line of hamster cells (called BHK below), but BHK cells transfected with cDNA encoding MHVR are susceptible to infection by MHV-A59. Anti-MHVR MAb-CC1 binds to the N-terminal domain of MHVR that is present in both the 4-domain and 2-domain isoforms of MHVR, and blocks MHV-A59 infection of murine cells and MHVR-transfected BHK cells (Dveksler, *et al.*, 1993b). The level of expression of MHVR is markedly reduced in murine 17 Cl 1 cells persistently infected with MHV-A59 and the cells are resistant to superinfection with wild type MHV-A59 (Sawicki, *et al.*, 1995). As summarized below, we found that virus from the persistently infected culture utilized MHVR in a different manner than MHV-A59 and may also utilize an additional receptor on persistently infected 17 Cl 1 cells (Schickli, *et al.* submitted). We also found that the virus from the persistently infected culture was able to infect cells of several non-murine species. We suggest that persistent infection of murine 17 Cl 1 cells with MHV-A59 causes a marked reduction in the expression of MHVR, which in turn selects for virus with altered interactions with MHVR and an extended host range.

3. RESULTS AND DISCUSSION

Cells from the 17 Cl 1 line of spontaneously transformed BALB/c mouse fibroblasts were inoculated with MHV-A59 at a high multiplicity of infection (100 PFU/cell) and then the cells were serially passaged (Sawicki, *et al.*, 1995). At passage 600, virus (MHV/pi600) from the supernatant medium was plaque-purified on 17 Cl 1 cells. Unlike the species-specific MHV-A59, MHV/pi600 could infect hamster (BHK) cells. This virus was serially passaged 12 times in BHK cells, then plaque-purified twice on BHK cells to yield MHV/BHK virus. Both MHV-A59 and MHV/BHK caused fusion of 17 Cl 1 cells, though fusion induced by MHV/BHK was less extensive (Figure 1). The plaques produced on mouse 17 Cl 1 cells by MHV/BHK were considerably smaller than those of MHV-A59 (0.5–1mm and 4–5mm, respectively).

To determine if MHV/BHK infects 17 Cl 1 cells by interacting with MHVR, we used anti-MHVR MAb-CC1 to block virus binding to MHVR. The cells were incubated for one hour at 37°C with supernatant medium from hybridomas producing either MAb-CC1 or a control MAb and then inoculated with either MHV-A59 or MHV/BHK at 3 PFU/cell. After three hours at 37°C, the inocula were removed and the cells were refed with medium containing either MAb-CC1 or control MAb. Virus released into the supernatant medium at 4, 9.5, and 21 hours after virus inoculation was titered in 17 Cl 1 cells using a $TCID_{50}$ assay.

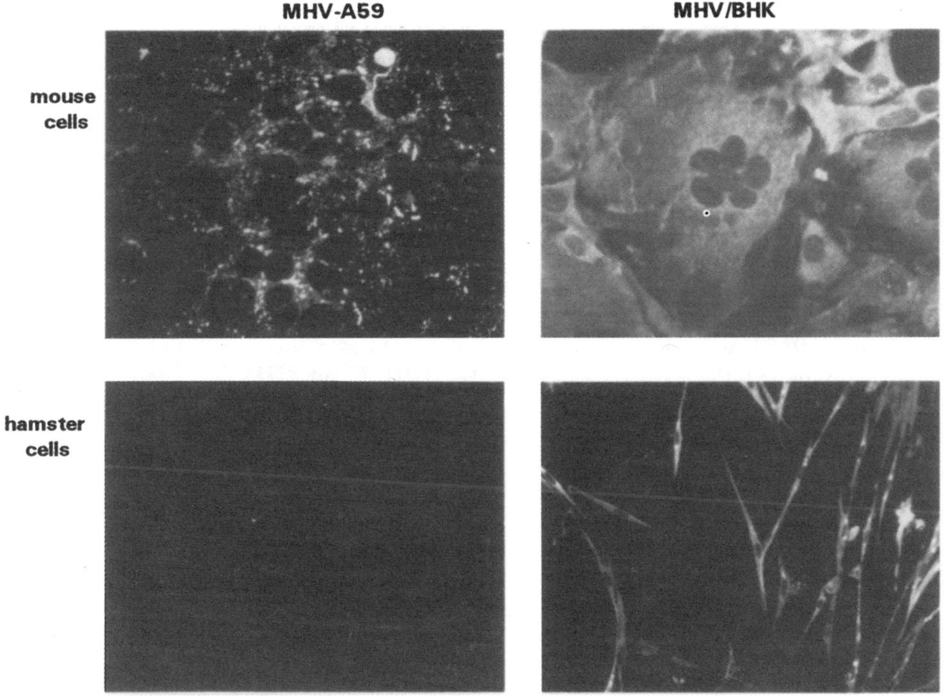

Figure 1. Susceptibility of murine (17 Cl 1) cells and hamster (BHK) cells to infection with MHV-A59 or MHV/BHK. Cells were inoculated with either MHV-A59 or MHV/BHK derived from persistently infected murine cells at 3 to 5 PFU/cell and fixed in acetone at 10 hr. after inoculation for mouse cells or 24 hours for hamster cells. Viral nucleocapsid protein was detected in the cytoplasm of infected cells by immunolabeling. MHV/BHK infected both mouse and hamster cell lines, while the parental MHV-A59 virus infected only mouse cells.

Table 1. Yields of MHV-A59 and MHV/BHK from murine 17 Cl 1 cell
treated with anti-MHVR MAb-CC1. Virus yields of MHV-A59 or
MHV/BHK at the indicated time points, in the presence of either
anti-MHVR MAb-CC1 or a control MAb, were determined using
a $TCID_{50}$ assay in 17 Cl 1 cells

Time of harvest**	MHV-A59		MHV/BHK	
	Control MAb	MAb-CC1	Control MAb	MAb-CC1
4.0	1.5*	0	2.5	0
9.5	4.8	0	3.5	1.5
21.0	6.5	0	6.3	5.3

*$\log_{10} TCID_{50}$
**Hours

While MAb-CC1 completely blocked MHV-A59 infection, it only partially inhibited infection of 17 Cl 1 cells by MHV/BHK (Table 1). This indicates that MHV/BHK can use MHVR as a receptor, but in a different manner from MHV-A59 and/or that MHV/BHK may also use an alternative receptor to initiate infection of 17 Cl 1 cells.

We also incubated MHV-A59 or MHV/BHK with varying concentrations of purified recombinant, soluble MHVR containing all four Ig-like domains (sMHVR) and then assayed for virus infectivity. The sMHVR neutralized MHV-A59 more than 100 fold more effectively on a micromolar basis than it neutralized MHV/BHK virus. Because MHV/BHK was neutralized by high concentrations of sMHVR , the MHV/BHK virus is apparently able to utilize MHVR, although much less effectively than MHV-A59.

Genomes of all MHV strains contain a gene encoding a hemagglutinin esterase (HE) glycoprotein that can bind to 9-O-acetylated neuraminic acid residues on membrane macromolecules (Vlasak, *et al.*, 1988). In wild type MHV-A59, although the HE gene is present in the viral genomic RNA, no HE protein is expressed due to 3 different mutations (Yokomori, *et al.*, 1991). We addressed the possibility that the mutant MHV/BHK virus had acquired the ability to express HE which would allow it to bind to 9-O-acetylated sialylic acid residues on host cells, possibly facilitating virus entry. 17 Cl 1 cells were inoculated with 10-fold serial dilutions of either MHV-DVIM which expresses HE (Gagneten, *et al.*, 1995; Sugiyama, *et al.*, 1986), wild type MHV-A59 which does not express HE, or MHV/BHK. After 1 hour at 37°C, the virus inocula were removed and gel overlay was applied. At 2 days (MHV-A59) or 3 days (MHV-DVIM and MHV/BHK) after inoculation

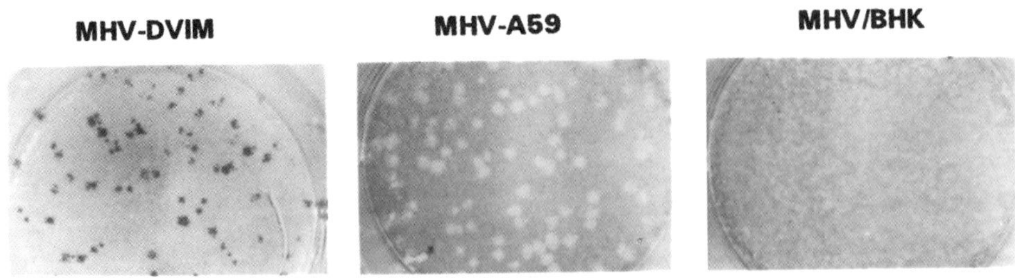

Figure 2. Acetyl esterase activity of hemagglutinin esterase (HE) glycoprotein in MHV plaques. Virus plaques on 17 Cl 1 cells were fixed 2 days after inoculation with MHV-A59 and 3 days for MHV-DVIM and MHV/BHK. The red product of the esterase activity was seen in plaques of MHV-DVIM, but not in plaques of MHV-A59 or MHV/BHK.

Table 2. Host range of MHV/BHK virus obtained from murine cells persistently infected with murine coronavirus MHV-A59

	Susceptibility to infection with	
Cell lines	MHV-A59	MHV/BHK
Mouse	+	+
Hamster	–	+
Rat	–	+
Cat	–	+
Cow	–	+
Human	–	+
Dog	–	+/–
Pig	–	+

the gel was removed, and the cells were fixed in cold formalin/acetone and tested for esterase activity as described by Wagaman , et al. (1989) (Figure 2). Plaques of the positive control, MHV-DVIM, showed deep red staining on the infected cells, while plaques of MHV-A59, which served as a negative control, showed no esterase activity. Plaques of MHV/BHK also did not show esterase activity. Thus, the HE gene in this virus is not expressed and HE glycoprotein is absent from the envelope of MHV/BHK virions so it cannot bind to an alternative carbohydrate moiety as a receptor.

MHV/BHK developed the ability to infect hamster cells during persistent infection in mouse 17 Cl 1 cells without prior exposure to hamster cells. We investigated whether cell lines of species other than mouse and hamster were also susceptible to MHV/BHK infection. Infection by MHV/BHK was assayed by immunofluorescence. The cells were fixed in -20°C acetone at 10, 17, or 24 hours after inoculation, and nucleocapsid antigen was detected with anti-N MAb and FITC-labeled goat anti-mouse IgG as previously described (Dveksler, et al., 1991). The results are summarized in Table 2. We found that rat (RIE), cat (Fcwf), cow (MDBK) and human (L-132 and HeLa) cell lines were susceptible to infection by MHV/BHK. In these cell lines, 5% to 100% of the cells were infected by 24 hours after virus inoculation. One dog cell line (A72) was only slightly susceptible to MHV/BHK infection, as <1% of the cells contained viral antigen at 24 hours after inoculation. The dog kidney (MDCK I) and the swine testis (ST) cell lines were completely resistant to MHV/BHK infection.

Infection of the non-murine cell lines by MHV/BHK did not induce extensive cell fusion. In hamster cells, <1% of the cells were binucleated by 24 hours after inoculation. No fusion was seen in any of the non-murine cell lines tested (Figure 1). However, MHV/BHK was able to induce fusion in murine 17 Cl 1 cells, though to a lesser extent and more slowly than MHV-A59. OBLV60, an acid-dependent variant of MHV-4 (JHM-MHV) was selected during persistent infection in OBL21a cells that are resistant to fusion (Gallagher, et al., 1991). OBLV60 enters OBL21a and Sac- cells via the endocytic pathway. In contrast, MHV-4 virus grown in Sac- cells that are not resistant to fusion enters Sac- cells via both endocytic and non-endocytic pathways (Nash and Buchmeier, in press). Possibly, MHV/BHK infection of murine cells also can be initiated both by the endocytic pathway and by direct fusion at the plasma membranes, while infection of the non-murine cells may only occur via the endocytic pathway.

In addition to the differences in cell fusion activity between MHV-A59 and MHV/BHK, differences in the rates of viral RNA synthesis were observed. The rate of viral RNA synthesis was determined by measuring Actinomycin D-resistant RNA synthesis as described by Sawicki and Sawicki (1986) in 17 Cl 1 cells or BHK cells infected with either MHV-A59, MHV/pi600 or MHV/BHK (Schickli, *et al.*, submitted). In 17 Cl 1 cells infected with MHV-A59, viral RNA synthesis was maximal at 7 hours after inoculation, while in cells infected with either MHV/pi600 or MHV/BHK, the rate of viral RNA synthesis peaked at 10 hours and was 3-fold and 2-fold lower, respectively, than MHV-A59 (data not shown). In BHK cells, the synthesis of MHV/pi600 or MHV/BHK viral RNA was considerably slower than for the same viruses in murine 17 Cl 1 cells. The rate of MHV/BHK RNA synthesis at 24 hours after inoculation of BHK cells was roughly 3-fold less than that of 17 Cl 1 cells infected with MHV-A59 at 7 hours (data not shown). Perhaps MHV-A59 has adapted to maximize viral RNA synthesis in murine cells.

The MHV/BHK virus formed small plaques on both 17 Cl 1 and BHK cells, had altered interactions with MHVR, was less fusogenic in murine cells and had an extended host range compared to MHV-A59. MHV/BHK virus was selected in persistently infected murine cells that express low levels of MHVR (Sawicki, *et al.*, 1995). Alternative MHV receptors such as Bgp2 may also be expressed on these cells, so mutant virus that can utilize the alternative receptor(s) better than MHV-A59 may have a selective advantage in the persistently infected murine cell culture. Murine tissues differ markedly in their level of expression of MHVR (Godfraind, *et al.*, 1995). High levels of MHVR are found in respiratory and enteric epithelium, portals of entry and shedding of MHV, while other tissues such as spleen and brain express much less MHVR. *In vivo*, in MHV-infected murine tissues that express low levels of MHVR, mutant virus that has an extended host range may be selected as in the persistently infected murine cell culture. Mutant virus would probably not be shed effectively from these tissues, although it might play a role in the disease process within the infected mouse. If a mouse infected with such a host range virus variant were ingested by a predator or scavenger, the mutant virus might gain the opportunity to initiate infection in a new host species, possibly leading to the emergence of a new virus disease.

REFERENCES

Cabirac, G. F., Soike, K. F., Zhang, J. Y., Hoel, K., Butunoi, C., Cai ,G. Y., Johnson, S., and. Murray, R. S., 1994.,Entry of coronavirus into primate CNS following peripheral infection, *Microb. Pathog.* **16**:349–357.

Chen, D. S., Asanaka, M., Chen, F.S., Shively, J.E., and Lai, M.M.C., 1997, Human carcinoembryonic antigen and biliary glycoprotein can serve as mouse hepatitis virus receptors, *J. Virol.* **71**:1688–1691.

Chen, D. S., Asanaka, M., K. Yokomori, K., Wang, F. I., Hwang. S.B., Li, H.P., and Lai, M.M.C., 1995, A pregnancy-specific glycoprotein is expressed in the brain and serves as a receptor for mouse hepatitis virus, *Proc. Natl. Acad. Sci. USA* **92**:12095–12099.

Compton, S.R., Barthold, S.W., and Smith, A.L., 1993, The cellular and molecular pathogenesis of coronaviruses, *Lab. Anim. Sci.* **43**:15–28.

Coutelier, J-P, Godfraind, C., Dveksler, G.S., Wysocka, M.,. Cardellichio, C.B, Noel, H., Holmes, K.V., 1994, B lymphocyte and macrophage expression of carcinoembryonic antigen-related adhesion molecules that serve as receptors for murine coronavirus, *Eur. J. Immunol.* **24**:1383–1390.

Dveksler, G. S., Pensiero, M. N., Cardellichio, C. B., Williams, R. K., Jiang, G. S., Holmes, K. V., and Dieffenbach, C. W., 1991, Cloning of the mouse hepatitis virus (MHV) receptor: expression in human and hamster cell lines confers susceptibility to MHV, *J. Virol.* **65**:6881–6891.

Dveksler, G.S., Dieffenbach, C.W., Cardellichio, C.B., McCuaig, K., Pensiero, M.N., Jiang, G., Beauchemin, N., and Holmes, K.V., 1993a, Several members of the mouse carcinoembryonic antigen-related glycoprotein family are functional receptors for the coronavirus mouse hepatitis virus-A59, *J. Virol.* **67**:1–8.

Dveksler. G.S., Pensiero, M.N., Dieffenbach, C.W., Cardellichio, C.B., Basile, A.A., Elia, P.E., and Holmes, K.V., 1993b, Mouse hepatitis virus strain A59 and blocking antireceptor monoclonal antibody bind to the N-terminal domain of cellular receptor, *Proc. Natl. Acad. Sci. USA* , **90**:1716–1720.

Gagneten, S., Gout, O., Dubois-Dalcq, M., Rottier, P., Rossen, J., and Holmes, K. V., 1995, Interaction of mouse hepatitis virus (MHV) spike glycoprotein with receptor glycoprotein MHVR is required for infection with an MHV strain that expresses the hemagglutinin-esterase glycoprotein, *J. Virol.* **69**:889–895.

Gallagher, T.M., Escarmis, C., and Buchmeier, M.J., 1991, Alteration of the pH dependence of coronavirus-induced cell fusion: Effect of mutation in the spike glycoprotein, *J. Virol.* **65**:1916–1928.

Godfraind, C., Langreth, S. G., Cardellichio, C. B., Knobler, R.,. Coutelier, J. P., Dubois-Dalcq, M., and Holmes, K. V., 1995, Tissue and cellular distribution of an adhesion molecule in the carcinoembryonic antigen family that serves as a receptor for mouse hepatitis virus, *Lab. Invest.* **73**:615–627.

Murray, R. S., Cai, G. Y., Hoel, K., Zhang, J. Y., Soike, K. F., and Cabirac, G. F., 1992, Coronavirus infects and causes demyelination in primate central nervous system, *Virology.* **188**:274–284.

Nash, T.C., Buchmeier, M.J., 1997, Entry of mouse hepatitis virus into cells by endosomal and nonendosomal pathways, *Virology.* in press.

Nedellec, P., Dveksler, G. S., Daniels, E., Turbide, C., Chow, B., Basile, A. A., Holmes, K. V., and Beauchemin, N., 1994, Bgp2, a new member of the carcinoembryonic antigen-related gene family, encodes an alternative receptor for mouse hepatitis viruses, *J. Virol.* **68**:4525–4537.

Sawicki, S. G., Lu ,J. H., andHolmes, K. V., 1995, Persistent infection of cultured cells with mouse hepatitis virus (MHV) results from the epigenetic expression of the MHV receptor, *J. Virol.* **69**:5535–5543.

Sawicki, S. G., and Sawicki, D. L., 1986, Coronavirus minus-strand RNA synthesis and effect of cycloheximide on coronavirus RNA synthesis, *J. Virol.* **57**:328–334.

Schickli, J.H., Zelus, B.D., Wentworth, D.E., Sawicki, S.G., and Holmes, K.V., 1997, Murine coronavirus MHV-A59 from persistently infected murine cells exhibits an extended host range, submitted.

Sugiyama, K., Ishikawa, R. and Fukuhara, N., 1986, Structural polypeptides of the murine coronavirus DVIM, *Arch. Virol.* **89**:245–254.

Vlasak, R., Luytjes, W., Leider, J., Spaan, W., and Palese, P., 1988, The E3 protein of bovine coronavirus is a receptor-destroying enzyme with acetylesterase activity, *J. Virol.* **62**:4686–4690.

Wagaman, P. C., Spence, H. A., and O'Callaghan, R. J., 1989, Detection of influenza C virus by using an in situ esterase assay, *J. Clin. Microbiol.* **27**:832–836.

Wege ˉ ., Siddell, S., and Meulen, V. ter., 1982, The biology and pathogenesis of coronaviruses, *Curr. Top. Microbiol. Immunol.* **99**:165–200.

Yokomori, K., Banner, L. R., and Lai, M. M., 1991, Heterogeneity of gene expression of the hemagglutinin-esterase (HE) protein of murine coronaviruses, *Virology* **183**:647–657.

96

RECEPTOR HOMOLOGUE SCANNING FUNCTIONS IN THE MAINTENANCE OF MHV-A59 PERSISTENCE IN VITRO

Wan Chen, Boyd Yount, Lisa Hensley, and Ralph S. Baric

Departments of Epidemiology, Microbiology and Immunology
University of North Carolina at Chapel Hill
Chapel Hill, North Carolina 27599-7400

1. ABSTRACT

Cell lines and viruses were isolated from mouse hepatitis virus (MHV-A59) persistently-infected DBT cells at different times postinfection. Cloned cell lines had cured virus infection, displayed low levels of MHVR receptor expression and were progressively more resistance to MHV infection. MHV persistence was likely maintained by epigenetic expression of the MHVR receptor in subsets of these resistant cells and by the emergence of persistent viruses characterized by high affinity MHVR receptor usage. Persistent viruses also displayed higher affinity for alternative biliary glycoprotein receptors suggesting that receptor homologue scanning functioned in the maintenance of persistence. Importantly, persistent viruses isolated after 210 days postinfection efficiently replicated in human HepG2 cells, a hepatocarcinoma cell line, suggesting that persistence promotes the interspecies transfer of MHV.

2. INTRODUCTION

Mouse hepatitis viruses produce acute infections that are confined to the mouse and are often accompanied by persistence that may last for the life of the host. In animals, enteric and respiratory infections are typically cleared from survivors. In contrast, infections in the brain and spinal cord are oftentimes not cleared providing an insightful model for studying virus persistence in the CNS. Although persistent MHV infections can be rapidly established in vivo and in vitro, the molecular mechanisms by which cytoplasmic RNA viruses evolve and persist are largely unknown. Recent studies in our laboratory and others have suggested that the crucial element in establishing MHV persistence is the emergence

Coronaviruses and Arteriviruses, edited by Enjuanes *et al.*
Plenum Press, New York, 1998

of resistant host cells that express little if any MHVR receptor, a biliary glycoprotein (Bgp) (Chen and Baric,1996; Sawicki *et al.*, 1996). The emergence of these resistant cells subsequently selected for the coevolution of more virulent virus variants that displayed increased affinity for both the and Bgp1[b] biliary glycoprotein receptors for docking and entry (Chen and Baric, 1996). Persistence following reovirus, parvovirus and poliovirus infection may also occur at the level of entry by similar coevolutionary mechanisms (de la Torre *et al.*, 1988; Dermody *et al.*, 1993; Martin-Hernandez et al., 1994).

In this manuscript we demonstrate that progressively more resistant host cell populations evolved during the first 210 days of persistence. Epigenetic expression of the MHVR receptor functioned in the maintenance of MHV-A59 persistence. In addition, persistent viruses replicated more efficiently in these resistant cells and also displayed increased affinity for MHVR, Bgp1[b] and Bgp2 murine biliary glycoprotein receptors. MHV persistence is likely established and maintained at the level of Bgp receptor expression and also by the emergence of persistent viruses that contain mutations which promote receptor homologue scanning. Surprisingly, in vitro persistence also promoted the interspecies transfer of MHV since persistent viruses isolated at 210 days postinfection replicated efficiently in human cell lines.

3. MATERIAL AND METHODS

3.1. Virus and Cells

Persistent DBT cell clones were isolated from day 30 (PDBT5) and 210 days postinfection (P51–1,4) using techniques previously described in our laboratory entry (Chen and Baric, 1996). Hamster cell lines expressing the MHVR (BHK-mmCGM1) or Bgp1[b] (BHK-mmCGM2; CHO-mmCGM2) receptors were kindly provided by Dr. K.Holmes (Holmes and Dveksler, 1994). These cell lines were maintained in alpha minimum essential medium containing 10% fetal calf serum, 10% tryptose phosphate broth, 1% penicillin and streptomycin and 800 ug/ml geneticin (G418-sulfate). Human HepG2 cells were maintained in MEM containing 10% FCS, 5% tryptose phosphate broth and 0.05 µg/ml gentamicin and 0.25 µg/ml kanamycin at 37°C.

Persistent viruses were isolated from passages 8 (V8), 10 (V10), 19 (V19) 30 (V30) and 51 (V51) and plaque purified twice in DBT cells. V30 has been previously described as V16 (Chen and Baric, 1996). Virus growth curves and plaque assays were performed as previously described by our laboratory.

3.2. Blockade of the MHVR Receptor

Different cell lines were seeded at densities of 5×10^4 in 8-chambered LabTek slides overnight and then incubated with 100 µl of a 1:4 dilution of anti-MHVR monoclonal antibody CC1 for 1 hr at 37°C. The CC1 monoclonal antibody was collected from hybridoma culture supernatants, and recognizes the N-terminal domain of MHVR (Holmes and Dveksler, 1994). After incubation, the antibodies were removed and the cells challenged with MHV-A59 or different persistent viruses at an MOI of 5 for 1 hr at room temperature. The virus inoculate was then removed, the cultures washed twice with PBS to remove residual virus, and 400 µl of complete medium containing a 1:16 dilution of CC1 added to the cultures. Virus samples were harvested at different times postinfection. As a control, cells were pretreated with an equivalent amount of a monoclonal antibody (EDDA or R501) with the same IgG isotype as CC1, but directed against an irrelevant antigen.

4. RESULTS

4.1. Coevolution of Increased Host Cell Resistance and Persistent Viruses

Acute MHV-A59 infection in DBT cells is characterized by extensive syncytium formation and cell death within the 24 hrs of virus infection. Surviving cells (~4–5%) replicate to confluence over the next few days, and virus persists in carrier cultures that displayed progressively decreased levels of virus-induced syncytium formation and cell death (Chen and Baric, 1996). To determine if progressively more resistant cell lines evolved, we cloned several cell lines (P51–1,4) from 210 day old persistently infected cultures (Figure 1). Consistent with earlier findings, all P51 clones had cured virus infection, as no infectious virus, viral RNA or viral protein was detected (data not shown). Importantly by RT-PCR, no MHVR (Bgp1) mRNA transcripts were present in the P51 cell clones consistent with previous findings that persistent clones expressed little MHVR.

To study the phenotypes of persistent viruses which had evolved during persistence, we cloned several persistent viruses from different passages corresponding to 30, 38, 75, 119 and 210 days postinfection (Figure 1). The parental MHV-A59 strain and all the persistent viruses replicated efficiently in our laboratory stock of DBT-9 cell clones. In all cases, greater than 90% of the cells were positive for viral antigen by FA. In the PDBT5 cell clone isolated from 30 day old persistent cultures, the parental MHV-A59 virus replicated to tiers of ~10^5 PFU/ml (Table 1). In contrast, the V10, V30 and V51 persistent viruses replicated more efficiently in these cells with peak titers ranging from ~0.05 to 6 × 10^6 PFU/ml(Table 1). In PDBT5 cultures, only 1–4% of the cells infected with MHV-A59 were viral antigen positive while about 15–20% of the cells infected with V10 and V30 were positive for viral antigen by FA. In comparison, V51 could replicate in more than 40% of these cells. In the P51–1 clone that was isolated from 210 day old cultures, MHV-A59 replication was almost completely ablated with titers at 24 hrs postinfection approaching 5 × 10^3 PFU/ml (Table 1). By FA, significantly less than 0.1% of the cells were infected with the wildtype MHV-A59 strain. While the V10 and V30 persistent viruses replicated to titers ranging from 0.2 to 3 × 10^5 PFU/ml and infected about 4% and 10% of the cells as measured by FA, respectively. The V51 variant viruses replicated most effi-

Isolation of cell lines

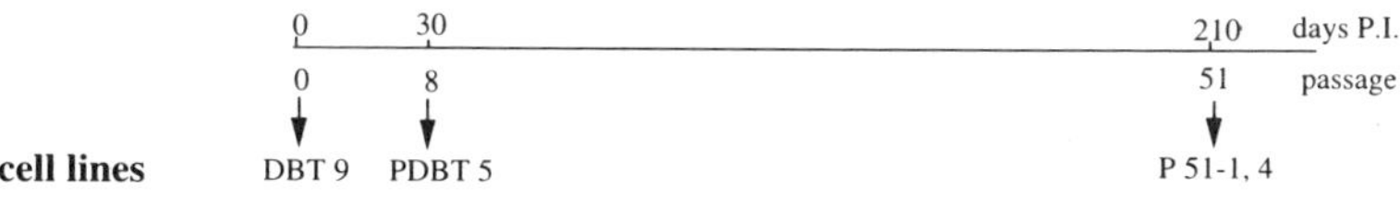

Isolation of variant viruses

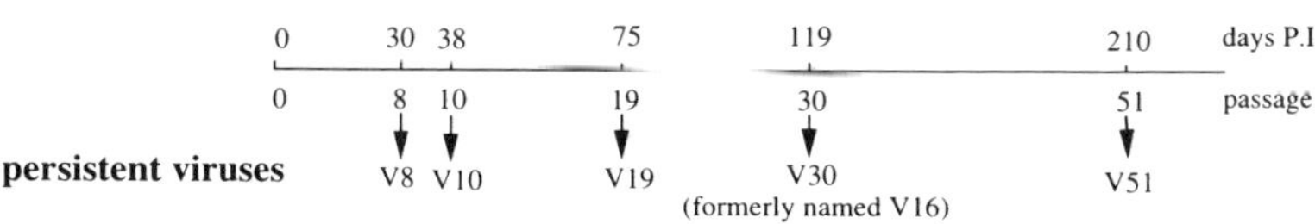

Figure 1. Passage history of the isolation of cell lines and variant viruses.

Table 1. Replications of persistent viruses in different cell lines

Virus strain	Passage number	Days post-infection	Virus titers in		
			DBT-9[1]	PDBT5[1]	P51-4[1]
A59	0	0	1.5×10^7	3.0×10^4	3.5×10^3
V8	8	30	1.0×10^7	2.8×10^4	1.0×10^4
V10	10	38	1.0×10^7	4.5×10^4	2.5×10^4
V19	19	75	1.5×10^7	6.5×10^4	1.0×10^5
V30	30	119	2.5×10^8	1.0×10^6	3.0×10^5
V51	51	210	3.0×10^8	6.5×10^6	1.2×10^7

[1]Titers at 24 hrs postinfection.

ciently with peak titers approaching 1×10^7 PFU/ml. By FA, between 20–30% of the P51–1 cells were infected by V51. These data indicated that progressively more resistant host cells had evolved during persistent MHV infection.

4.2. Epigenetic Expression of MHVR Functions in MHV Persistence

If epigenetic expression of MHVR is functioning in the maintenance of MHV persistence in DBT cells, CC1 monoclonal antibodies that bind to the N-terminal domain of MHVR should block or significantly reduce virus infection in these highly resistant clones (Sawicki *et al.*, 1996). Cultures of the P51–1 cells were treated with a 1:4 dilution of the CC1 monoclonal antibody and infected with MHV-A59, V30 or V51 persistent viruses at an MOI of 5 for 1 hr. CC1-treatment completely ablated the replication of wildtype MHV-A59 (Table 2), and no cells were positive for viral antigen by FA. Peak yields of V30 and V51 were also reduced following CC1 blockade (Table 2). Following CC1-treatment, about 0.2–0.3% of the P51–1 cells were infected with V30 while 6–10% of these cells infected with V51 were positive for viral antigens by FA. These data are consistent with the hypothesis that epigenetic expression of MHVR functions in the maintenance of MHV persistence in highly resistant cell clones isolated at 210 days postinfection (Table 2). Although epigenetic expression of MHVR clearly functions in maintaining MHV persistence in DBT cells in vitro (Sawicki *et al.*, 1996), persistent virus replication was not completely ablated indicating that additional mechanisms must function to maintain MHV persistence.

4.3. Persistent Viruses Display Increased Affinity for MHVR

To compare affinity of persistent viruses for MHVR, blockade of MHVR was performed. Cultures of DBT cells were treated with a 1:5 dilution of CC1 monoclonal anti-

Table 2. Epigenetic expression of MHVR functions in the maintenance of MHV persistence

Virus strain	Passage number	Days post-infection	Virus titers[1] in	
			P51-1	CC1-treated P51-1
A59	0	0	7.0×10^3	$<2.0 \times 10^2$
V30	30	119	1.5×10^5	2.0×10^4
V51	51	210	6.0×10^6	1.0×10^6

[1]Virus titers at 24 hrs postinfection.

Table 3. Resistance of persistent viruses to CC1 blockade

Virus strain	Passage number	Virus titers[1] in			
		Untreated		CC1-treated	
		DBT-9	BHK-mmCGM1	DBT-9	BHK-mmCGM1
A59	0	2.4×10^7	5.8×10^7	1.0×10^4	1.5×10^3
V10	10	2.9×10^7	2.9×10^7	7.5×10^6	3.5×10^3
V30	30	4.2×10^7	1.2×10^8	9.0×10^6	9.5×10^4
V51	51	1.0×10^8	6.0×10^7	1.3×10^8	7.8×10^6

[1]Virus titers taken at 30 hrs postinfeciton.

body for 1 hr, and then infected with MHV-A59, V8, V10, V30 or V51 at an MOI of 10 (Table 3). Variants had evolved within the first 38 days of postinfection, since V10 was extremely resistant to CC1 blockade. While V30 appeared more resistant to blockade than V10, the V51 variant (210 days) was the most resistant and replicated to titers equivalent to untreated controls (Table 3). To determine the mechanism of persistent virus resistance to CC1 blockade, cultures of BHK-mmCGM1 cells which are BHK cells expressing MHVR were pretreated with CC1, and infected with the persistent viruses (Table 3). All persistent viruses were capable to recognize MHVR as a receptor for entry, since they replicated to high titers in BHK-mmCGM1 cells. MHV-A59 and V10 were highly sensitive to CC1 blockade in BHK-mmCGM1 cells.. Increasing resistance to CC1 blockade was noted with the V30 and V51 viruses and V51 exhibited equivalent resistance as seen in DBT cells. These data suggested that resistance to CC1 blockade likely occurred by two distinct mechanisms. Initial resistance in DBT cell lines likely occurred by virus recognition of alternative receptors (Bgp1[b], Bgp2) for entry. Variant virus resistance to CC1 blockade in BHK-mmCGM1 cells developed later suggesting that V30 and V51 displayed a higher affinity for MHVR, or recognized spatially distinct domains that were less sensitive to CC1 blockade.

4.4. Receptor Homologue Scanning Functions in the Maintenance of MHV Persistence

Since DBT cells were isolated from outbred CD1 mice and express MHVR, Bgp1[b] and Bgp2 and other CEA receptors (Chen *et al.*, 1997; Sawicki *et al.*, 1995), we next determined if persistent viruses recognized Bgp1[b] for docking and entry. Two different hamster cell lines, BHK-mmCGM2 and CHO-mmCGM2 cells, were chosen since they likely expressed different levels of the Bgp1[b] receptor. Cultures of cells were infected with MHV-A59, V8, V10, V30 and V51 at an MOI of 5 for 1 hr. Samples were harvested at the indicated times and assayed by plaque assay. Consistent with previous reports, the murine *Bgp1[b]* glycoprotein gene served as a functional, but poor receptor for MHV-A59 docking and entry (Table 4) (Chen and Baric, 1996; Gallagher, 1997). About 2% of the BHK-mmCGM2 cells infected with MHV-A59 were positive for viral antigens by FA. The V8 and V10 persistent viruses recognized the Bgp1[b] glycoprotein as a receptor for docking and entry more efficiently than MHV-A59 as evidenced by higher titers of infectious virus and higher percentages of infected cells (Table 4). Greater than 40% of the BHK-mmCGM2 cells infected with persistent viruses were positive for viral antigen by FA. In contrast, BHK cells lacking the receptor did not support MHV-A59 or persistent viruses replication. These data were consistent with the notion that early resistance to CC1 blockade was mediated by

Table 4. Persistent virus usage of Bgp1[b]

Virus strain	Passage number	Virus titers[1] in	
		CHO-mmCGM2	BHK-mmCGM2
A59	0	3.0×10^5	2.0×10^3
V8	8	6.5×10^5	4.5×10^5
V10	10	8.0×10^6	1.9×10^6
V19	19	1.0×10^7	1.1×10^6
V30	30	8.0×10^6	7.4×10^6
V51	51	6.0×10^7	6.0×10^6

[1]titers at 24 hrs in CHO-mmCGM2 and 30 hrs in BHK-mmCGM2 cells.

alternative biliary glycoprotein receptor usage. Importantly, the V30 and V51 variant viruses were even more efficient at recognizing the Bgp1[b] receptor for docking and entry into BHK-mmCGM2 cells and infected greater percentages of cells as compared with MHV-A59, the V8 and the V10 persistent viruses. These data were consistent with the hypothesis that progressively increased affinity for the murine Bgp1[b] receptor had also evolved during the course of MHV-A59 persistence. Similar findings were also noted in CHO expressing Bgp2 cell lines (data not shown). Because persistent viruses had evolved increased affinity for both the Bgp1[b] and Bgp2 receptors for docking and entry, receptor homologue scanning also functions in the maintenance of MHV persistence in DBT cell lines.

4.5. Persistence Promotes Interspecies Transfer of MHV-A59

Since human and murine CEA genes are highly homologous and wildtype MHV-A59 may recognize human CEAs as weak receptors for entry (Chen *et al.*, 1997), persistent viruses evolving increased affinity for different murine Bgp receptors may also evolve increased affinity for human Bgp or CEA receptors. Cultures of BHK and HepG2 cells were infected with V8, V10, V19, V16, V51, and MHV-H2 at an MOI of 10. Virus titers clearly demonstrate that the persistent viruses could not replicate in BHK cell lines (Table 5). Human HepG2 cell lines were extremely resistant to infections with MHV-A59, and the persistent viruses V8, V10, V19, and V30. The V51 persistent viruses, however, replicated efficiently in HepG2 cells, demonstrating that MHV persistence in DBT cells selected for variant viruses that bridged the species barrier.

5. DISCUSSION

MHV-A59 persistence in vitro provides a rare opportunity to identify sites of virus-receptor interaction that regulate the molecular mechanisms governing persistence, host susceptibility/resistance to virus infection at the cellular level, virulence, receptor homologue scanning and virus interspecies transfer. We have demonstrated that coevolution of increasing host cell resistance and persistent viruses that replicate more efficiently in these resistant cells functions in establishing and maintaining persistent infections. In persistently infected 17CL-1 and DBT cells, epigenetic expression of the MHVR receptor likely functions in the maintenance of MHV persistence by creating productive cellular targets that are susceptible to persistent virus infection. These findings are also consistent with earlier reports that indicated that a small subpopulation of susceptible cells was essential for the maintenance of

Table 5. Persistence promotes interspecies transfer of MHV-A59 into human cell lines

			Virus titers[*] in	
Virus strain	Passage number	Days post-infection	BHK cells[1]	HepG2 cells[1]
A59	0	0	$<1.0 \times 10^2$	3.8×10^3
V8	8	30	$<1.0 \times 10^2$	5.0×10^3
V10	10	38	5.0×10^2	1.0×10^3
V19	19	75	ND	3.9×10^3
V30	30	119	5.0×10^3	5.5×10^3
V51	51	210	1.5×10^4	3.2×10^6

[1]Virus titers at 30 hrs postinfection.
[*]Titers at time 0 averaged between $1\text{-}5 \times 10^4$ PFU/ml in these cell lines.
ND=not done.
NA= not applicable.

MHV persistence in vitro (Macintyre *et al.*, 1989). Low level and/or epigenetic expression of the MHVR receptor has most likely selected for the coevolution of viral mutants that recognized different Bgp receptors for entry and displayed increased affinity for MHVR. Such persistent viruses would most likely bind quickly to cells expressing low levels of receptor, replicate and be retained in the culture. The evolving "viral receptor mutants" would subsequently select for host cell lines which were more resistant to persistent virus infection causing additional rounds of coevolution and selection between the host and virus. In support of this hypothesis, elegant studies measuring differences in binding on and off characteristics using soluble receptor have indicated that naturally occurring MHV strains display wide ranges in MHVR binding in biochemical assays (Gallagher, 1997). Although the genetic targets are less well characterized, similar phenomena have also been observed following foot and mouth disease and reovirus persistent infections in vitro (de la torre *et al.*, 1988; Dermody *et al.*, 1993; Martin-Hernandez *et al.*, 1994).

Persistence is a complex phenomena likely requiring alterations in a variety of viral and cellular genes. In addition to the epigenetic expression of MHVR, we have shown that virus receptor homologue scanning functions in the maintenance of MHV persistence. Persistent viruses may evolve the more efficient recognition of Bgp1[b] and Bgp2 by using a common binding domain in different Bgp receptors, tolerating increased heterogeneity within a common binding domain, or binding to different portions of different biliary glycoprotein receptors. Receptor homologue scanning may also occur following HIV infection in humans since HIV variants rapidly emerge which recognize different chemokine co-receptors for entry into macrophages and T-lymphocytes (Wimmer, 1994). Although the molecular mechanisms mediating receptor homologue scanning will require additional study, it seems likely that this phenomena has also expanded MHV host range by virus recognition of closely homologous CEA glycoproteins in different species. Preliminary findings in our laboratory indicate that V51 recognizes human biliary glycoproteins as receptors for entry into human cells. It will be of interest to determine if outbred mice persistently-infected with MHV-A59 or MHV-JHM evolve the capacity to replicate in alternative host species.

REFERENCES

Chen, W. and Baric, R.S., 1996, Molecular anatomy of MHV persistence: Coevolution of increased host resistance and virus virulence, *J. Virol.* **70**: 3947–3960.

Chen, D.S., Asanaka, M., Yokomori, K., Wang, F-I, Hwang, S.B., Li, H-P., and Lai, M.M.C., 1995, Pregnancy-specific glycoprotein is expressed in the brain and serves as a receptor for mouse hepatitis virus, *Proc. Natl. Acad. Sci. USA* **92**: 12095–12099.

Chen, DS, Asanaka, M., Chen, FS, Shively, JE and Lai, MMC., 1997, Human carcinoembryonic antigen and biliary glycoprotein can serve as mouse hepatitis virus receptors, *J.Virol.* **71**:1688–1691.

de la Torre, J.C., Martinez-salas, E., Diez, J., Villaverde, A., Gebauer, F., Rocha, E., Davila, M., and Domingo, E., 1988, Coevolution of cells and viruses in a persistent infection of foot-and mouth disease virus in cell culture, *J. Virol.* **62**: 2050–2058.

Dermody, T.S., Nibert, M.L., Wetzel, D., Tong, X., and Fields, B.N., 1993, Cells and viruses with mutations affecting viral entry are selected during persistent infections of L cells with mammalian reoviruses, *J. Virol.* **67**: 2055–2063.

Gallagher, T.M., 1997, A role for naturally occurring variation of the murine coronavirus spike glycoprotein in stabilizing association with the cellular receptor, *J. Virol.* **71**: 3129–3137.

Holmes, K.V. and Dveksler. G.S., 1994, Specificity of Coronavirus receptor interactions. p 403–443. In: E. Wimmer (ed.). *Cellular Receptors for Animal Viruses.* Cold Spring Harbor Laboratory Press, Cold Spring Harbor, NY.

Macintyre, G., Wong, F., and Anderson, R., 1989, A model for persistent murine coronavirus infection involving maintenance via cytopathically infected cell centers, *J. Gen. Virol.* **70**: 763–768.

Martin-Hernandez, A.M., Carrillo, E.C., Sevilla, N., and Domingo, E., 1994, Rapid cell variation can determine the establishment of a persistent viral infection, *Proc. Natl. Acad. Sci. USA* **91**: 3705–3709.

Sawicki, S.G., Lu, J.-H., and Holmes, K.V., 1995, Persistent infection of cultured cells with mouse hepatitis virus (MHV) results from the epigenetic expression of the MHV receptor, *J. Virol.* **69**: 5535–5543.

Wimmer, E., 1994, Cellular Receptors for Viruses. p. 1–14 in: E. Wimmer (ed.). *Cellular Receptors for Animal Viruses.* Cold Spring Harbor Laboratory Press, Cold Spring Harbor, NY.

97

VIRAL EVOLUTION AND CTL EPITOPE STABILITY DURING JHMV INFECTION IN THE CENTRAL NERVOUS SYSTEM

C. C. Bergmann,[1,2] E. Dimacali,[1] S. Stohl,[1] N. Marten,[1] M. M. C. Lai,[2,3] and S. A. Stohlman[1,2]

[1]Department of Neurology
[2]Microbiology and Immunology
[3]Howard Hughes Medical Institutes
University of Southern California School of Medicine
Los Angeles, California 90033

1. ABSTRACT

The JHM strain of mouse hepatitis virus (JHMV) establishes a persistent infection in the murine central nervous system (CNS) associated with chronic ongoing demyelination in the absence of detectable virus. To distinguish between immune and replication associated mechanisms of persistence, brains from acutely and persistently infected mice were analyzed for viral RNA mutations in the encapsidation sequence (ECS) and regions encoding either the transmembrane domains of the matrix (M) protein or a protective CTL epitope in the nucleocapsid (N) protein. Detection of the ECS to 120 days post infection (p.i.) indicated low levels of replication. The ECS remained stable whereas the fragment encoding the CTL epitope revealed extensive diversity with mutation frequencies in the order of 2.0 per 1000 nts. The M gene also remained stable despite random mutations during the acute phase. Mutations in the N gene were random and not selected for during persistence, with the exception of a single prominent Pro363 to Ser substitution in a region not associated with any known regulatory function or immune response. Mutations within the CTL epitope affecting CTL recognition were found early in responder BALB/c mice (H-2^d), but also in non-responder C57BL/6 (H-2^b) mice, suggesting that CTL escape variants play no significant role in establishing persistence.

Coronaviruses and Arteriviruses, edited by Enjuanes *et al.*
Plenum Press, New York, 1998

2. INTRODUCTION

Infection of the murine CNS with JHMV results in an acute demyelinating panencephalomyelitis with virus replication in oligodendroglia, astrocytes, ependyma, microglia and, to a lesser extent, neurons (Kyuwa and Stohlman, 1990). Although the cellular immune response provides protection and reduces viral titers, the inability to achieve sterile immunity results in viral RNA persistence associated with chronic ongoing primary demyelination in the absence of infectious virus. Attempts to understand the relationships between JHMV replication, the CNS, the immune response and persistence are severely hampered by the inability to recover infectious virus after 21 days (d) p.i. and the large genome size. Studies have therefore focused on the Spike (S) protein, which is not only the ligand for the cell surface receptor (Lai and Cavanagh, 1997), but also contains both neutralizing antibody (Kyuwa and Stohlman, 1990) and CTL epitope domains (Pewe et al., 1996). Furthermore, attenuated viral phenotypes are associated with a hypervariable region known to contain point mutations and deletions (Fleming et al., 1986).

To dissect the contribution of immune or replication mediated parameters in driving JHMV persistence, this study examined RNA variability within regions comprising a dominant, CTL epitope in BALB/c mice and in regions critical for viral assembly. The crucial role of CTL in protection and JHMV clearance from the CNS is documented by CD8 depletion, adoptive transfer, and immunization studies (Stohlman et al., 1995; Williamson and Stohlman, 1990) and recently by the emergence of CTL escape mutants in suckling mice nursed by immunized dams (Pewe et al., 1996). As the CTL response in BALB/c mice is primarily targeted against a single epitope comprising aa 318–326 of the N protein (Bergmann et al., 1993), mutants leading to escape from CTL recognition provide a potentially potent mechanism contributing to viral persistence. To assess a role of mutations leading to reduced viral assembly, the ECS (Fosmire et al., 1992) and sequences encoding the membrane-spanning region of the M protein (Rottier et al., 1986) were chosen.

3. MATERIALS AND METHODS

3.1. Mice, Viruses, and CTL Assays

BALB/c (H-2^d) and C57BL/6 (H-2^b) mice were obtained from the Jackson Laboratory, Bar Harbor MA and housed in isolator cages. Mice were infected at 6–7 wks of age intracranially with plaque purified JHMV-DS (Stohlman et al., 1982) using 300 PFU for C57BL/6 and 600 PFU for BALB/c mice. This dose resulted in 50 % fatalities 7 to 9 d p.i.. Antigenicity of peptides was tested using pN-specific T cell clones at an effector to target ratio (E:T) of 5:1 and preincubating labeled J774.1 target cells with peptide as described (Bergmann and Stohlman,1996). Peptide pN318–326 (APTAGAFFF) and the derivative (APTAGAF*L*F) were purchased from Chiron Mimotopes (Clayton, Australia).

3.2. RNA Extraction and RT-PCR Analysis

Brains were removed and one brain half per mouse disrupted in a Tenbrock homogenizer containing 2 ml GIT solution. The suspension was overlaid on a 2.5 ml CsCl cushion, centrifuged at 30 K r.p.m. overnight in a Beckman SW50.1 rotor and the RNA pellet resuspended in DEPC treated water. For reverse transcription (RT) reactions 2 µg total RNA and 1µg random primer (Promega, Madison, WI) were mixed, heat denatured for 5 min at 72° C and then chilled on ice. Synthesis of cDNA was carried out for 60 min at 42° C in 25 µl buffer containing 50 mM Tris-Cl (pH 8.3), 10 mM $MgCl_2$, 50 mM KCl, 0.5 mM spermidine, 10 mM DTT, 0.4mM dNTPs, 14 U RNAsin and 10 U AMV-Reverse Transcriptase (Promega, Madison, WI). The cDNA was amplified using primers EC5' (ATT TTG GTT TGC TAT GCG TAG AG) and EC3' (ATC GAG GTG AGA AAG AGG AC) spanning ECS nucleotides (nts) 20286–20479 (Fosmire et al., 1992), JN-984 (5-CCC GCT AGC CAT GGT AAA GCT TGG AAC TAG TGA TCC) and JN1250 (ACA GGT ACC TAC ATC TGC ACC ACC ATC TTG) spanning N gene nts 984–1250 (Skinner and Siddell, 1983) and JM67 (TTA TGA GTA GTA CCA CTC AGG CC) and JM418 (GCT CCA CCA GCT ACC AGT CC) spanning M gene nts 67–418 (Pfleiderer et al., 1986). Amplification was carried out in 35–40 cycles using 20–50 % of the reverse transcription mixture in buffer containing 10 mM Tris-Cl (pH 8.3), 2 mM $MgCl_2$, 50 mM KCl, 0.1mM dNTPs, 0.25 µg of each primer and 1.25 U Amplitaq DNA Polymerase (Perkin Elmer). Oligonucleotides were purchased from Genosys Biotechnologies (The Woodlands, TX).

3.3. cDNA Cloning and Sequencing

PCR products were purified using Gene clean (BIO 101, Inc., Vista, CA) and using the ZERO Background cloning kit (Invitrogen, San Diego, CA). Viral inserts were directly sequenced by the Sanger dideoxy chain termination method using Sequenase enzyme 2.0 (Sequenase DNA Kit version 2.0, U.S. Biochemicals) and flanking T7 or SP6 primers. Some samples were sequenced using the dsDNA cycle sequencing system (Gibco BRL) following PCR amplification of the insert with SP6/T7 primers. Bulk PCR products were directly analyzed by automated sequencing using dye deoxy terminators and virus-specific primers (USC Norris Cancer Center Microchemistry Core Laboratory).

4. RESULTS

4.1. The ECS and RNA Encoding the M Protein Are More Conserved Than RNA Encoding the N Protein during Persistence

To explore the possibility that defective viral assembly or CTL escape mutants contribute to persistence, brains from infected BALB/c and C57BL/6 mice were analyzed for the presence of the ECS on genomic length RNA in addition to both the M and N protein coding sequences. PCR products were amplified from individual mice at 6, 60, 90, and 120 d p.i. and 5–18 clones sequenced per mouse. All 3 sequences were detectable in persistently infected, but not uninfected mice. ECS clones from 6 d infected mice were all wt, and only 2 out of 22 clones from 60 d persistently infected mice had a single mutation, resulting in a low mutation frequency of 0.6 per 10^3 nts (Table 1). The respective mutations, C20326 to T and C20337 to T, were located outside of the hairpin structure essential for encapsidation (Fosmire et. al., 1992) and therefore unlikely to result in a phenotype. Furthermore, no mutations were detected in clones isolated from a 90 d persistently infected mouse.

M protein coding sequences were analyzed for sequence variations in the N-terminal domain to detect potential changes in membrane topology leading to impaired budding. Of 17 clones spanning nts 67–418 (aa 1–117), 8 carried variations at d 6. All clones contained

Table 1. Mutation frequencies of JHMV ECS, M and N gene sequences per 1000 nts

Viral sequence	BALB/c				C57BL/6			
	Time p.i.	Clones seq.	Mutated clones	*Mut freq.	Time p.i.	Clones seq.	Mutated clones	*Mut freq.
EC	6-A	10	0	0	6A	8	0	0
nts 20286-20479								
	60-A	13	1	0.5	6B	3	0	0
	60-B	10	1	0.7				
	90-B	10	0	0				
M	6-A	17	8	2.3				
nts 67-418								
	60-A	6	0	0				
	60-B	5	5	0.6				
N	6-A	7	3	2.0	6-A	16	4	1.7
nts984-1250								
	6-B	7	3	2.0	6-B	3	1	1.5
	60-A	15	9	1.9	120-A	20	7	1.6
	90-A	18	10	2.1				
	90-B	10	5	2.3				

Clones were sequenced from individual mice (A,B) at the indicated time point p.i.

*Calculations exclude sequences comprising the primers and mutations at the same position in different clones.

a silent A190 to G transition, which was also present in the input virus. In contrast to the ECS, this M protein domain exhibited a mutation frequency of 2.3 per 1000 nts (Table 1). A high number of mutations (92%) resulted in aa changes with an apparent bias from hydrophobic to polar or charged residues (72%). Although all mutations were localized to different positions within a mRNA population, their polar nature may lead to instability in the membrane thereby decreasing viral progeny. However, analysis of 60 d persistently infected mice, revealed that none of the original mutations were selected. Whereas only wt sequences were recovered from one mouse, a Phe86 to Leu mutation was found in all clones of the second mouse. Furthermore, no prominent mutations were evident in bulk PCR products from additional 2 and 4 mice 90 and 135 d p.i., respectively. The relative stability of the ECS and M coding region throughout persistent infection suggested that these domains are not contributing to viral persistence.

To account for the potential emergence of CTL escape mutants, sequences comprising the coding region for the 9 aa epitope (APTAGAFFF) were amplified from brains of responder BALB/c mice 6, 60 and 90 d p.i.. To verify a role for CTL pressure, the identical region was analyzed from CTL nonresponder C57BL/6 mice. As CTL escape in the form of antagonism only requires the presence of small populations of mutant peptide (Klenerman et al., 1996) 10–17 clones were sequenced per timepoint from individual mice (Table 1). Cloned PCR products revealed extensive diversity with mutation frequencies in the order of 2 per 1000 nts. The distribution of mutations was *random* throughout the sequence with 63% leading to aa changes in the form of substitutions, stop codons or frame shifts (Fig. 1). One prominent C toT transition at nt 1170 resulting in a Pro363 to Ser mutation accumulated in all persistently infected mice suggesting haplotype independent evolution (Fig. 1). The transition from wt (d 6), to predominantly mutant (d 60 and d90) was clearly supported by automated sequencing of bulk PCR products, suggesting a selective advantage of this mutant during persistence.

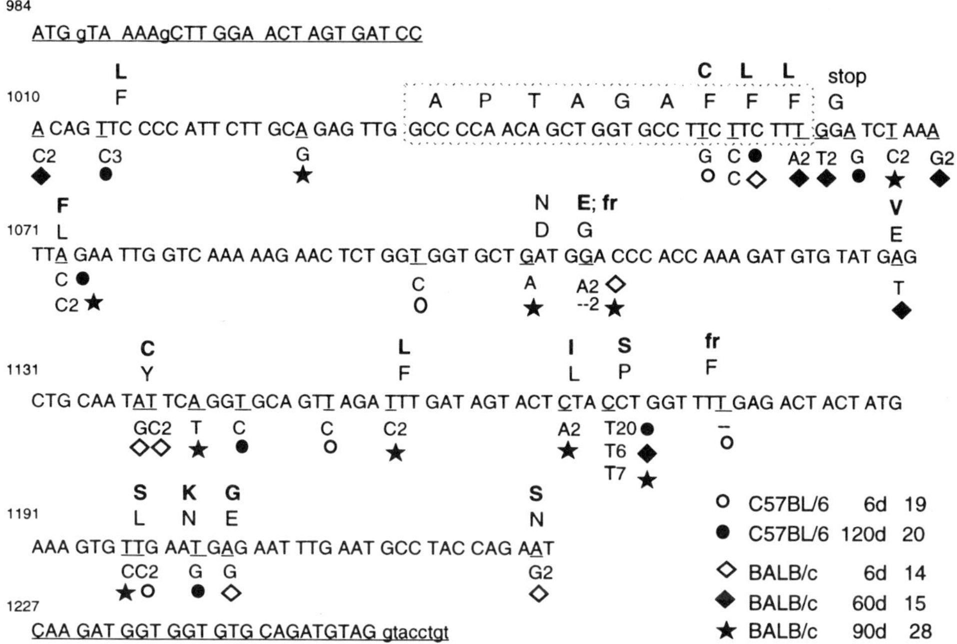

Figure 1. Distribution of mutations in the N region spanning nts 984–1250 (aa 310–382). Symbols represent groups of mice at 6, 60, 90 and 120 d p.i.; numbers indicate total clones sequenced per group. The CTL epitope is boxed. Mutated nts are underlined, and individual mutations listed together with the symbol representing the mouse group in which they occurred and the number of clones containing this mutation. Mutations leading to aa changes are indicated by the respective substitution in bold; fr indicates frameshift mutations. 5' and 3' priming sequences are underlined.

4.2. Persistence in BALB/C Mice Is Not Associated with CTL Escape Mutants

Mutations within and adjacent to the CTL epitope were identified early in individual mice, but there was *no* evidence for the *prevalent* emergence or *accumulation* of mutations within or proximal to the epitope (Fig. 1). Despite their limited number, epitope mutations were specifically in positions affecting class I binding (positions 2 and 9) and TcR recognition (positions 7 and 8). Analysis of the antigenicity of mutant peptides revealed that the conservative F326 to L variant in position 9 enhanced L^d binding and CTL recognition (Bergmann and Stohlman, 1996). By contrast, the F325 to L mutation in the subdominant TcR contact residue is only recognized by a subset of CTL clones derived from the CNS of acutely infected mice (Fig. 2). Although this may be interpreted as a decrease in the overall CTL response, the infrequent occurrence of this mutation suggested no significant contribution in CTL-mediated clearance. This was further supported by the inability to inhibit CTL function via an antagonistic mechanism (Klenerman et al., 1996). The fact that PCR products from C57BL/6 mice, which do not mount a CTL response to the N epitope, also revealed mutations in the TcR contact residues (Fig. 1), further indicated that the epitope mutations in BALB/c mice were not CTL driven.

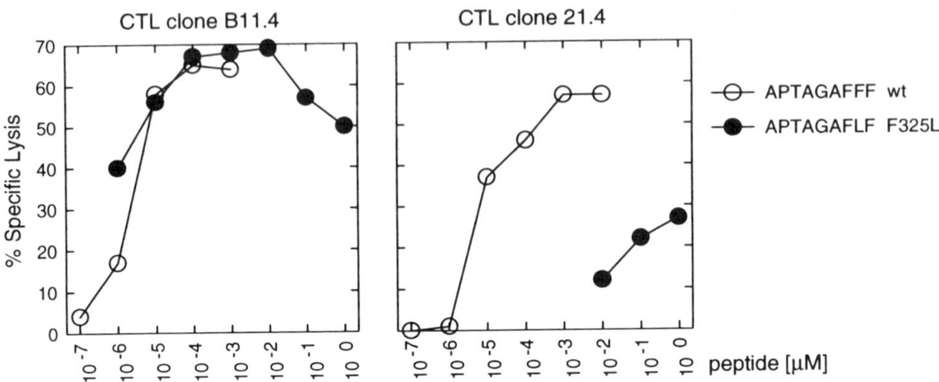

Figure 2. Differential recognition of a mutated CTL epitope by N-specific CTL clones.

5. DISCUSSION

MHV genomes evolve into quasispecies both *in vivo* and *in vitro,* with an estimated mutation frequency of 0.5–5 per 1000 nts (Adami et al, 1994; Wege, pers. communication). ECS stability compared to the N and M gene sequences during acute infection suggests that the viral genome may be more stable than subgenomic mRNA, reflecting the relative amounts of these RNA subsets (Lai and Cavanagh, 1997). The high mutation frequencies during acute infection with no evidence of increasing mutations with time support this notion. By contrast, cDNA clones generated from the CNS of 2.2.v-1 persistently infected C57BL/6 mice demonstrated high and increasing mutation rates in both the N and S genes with time (Adami et al., 1994). Screening of a large panel of clones further revealed an increased frequency of deletions in the hypervariable region of the S gene (Rowe et al., 1997). Different still, analysis of RNA isolated from diseased mice in the protected suckling mouse model exposed mutations solely in the sequence encoding the dominant S protein CTL epitope (Pewe et al., 1996). Taken together these studies indicate that viral RNAs may exhibit divergence in the prevalence and areas of mutations depending on individual MHV variants and differing mouse models.

The diverse and random nature of mutations found in the biphasic disease may reflect a multitude of attenuated genomes giving rise to an overall low replication phenotype; alternatively, it may reflect the sum of individually dominant mutations localized to different foci of persistent infection. The size of the MHV genome and complexity of mutations predict that factors affecting replication may substantially contribute to persistence *in vivo.* This may be supported by the finding of an accumulative dominant mutation in the N protein; despite its unknown significance, the Pro to Ser substitution may impair protein folding as well as alter N protein regulatory functions via differential phosphorylation (Stohlman et al., 1983). Attenuated replication as a mechanism of persistence is also favored by the finding that persistence in BALB/c mice is *not* associated with CTL-escape mutants, although CTL within the CNS comprise the major mechanism of JHMV clearance. This contrasts with the finding of numerous CTL escape epitopes in diseased C57BL/6 suckling mice from immunized dams (Pewe et al., 1996). A likely explanation may be that CTL escape mutants to JHMV and other viruses primarily evolve in the presence of persistent infections associated with ongoing virus release.

ACKNOWLEDGMENTS

This work was supported by U.S. Public Health Service Research Grants NS18146 and T32NS07149.

REFERENCES

Adami, C., Pooley, J., Glomb, J., Stecker, E., Fazal, F., Fleming, J.O., and Baker, S.C., 1995, Evolution of mouse hepatitis virus (MHV) during chronic infection: Quasispecies nature of the persisting MHV RNA, *Virol.* **209**:337.

Bergmann, C. and Stohlman, S., 1996, Specificity of the Ld restricted CTL response to the mouse hepatitis virus nucleocapsid protein, *J. Virol.* **70**:3252.

Bergmann, C., McMillan, M., and Stohlman, S., 1993, Characterization of the Ld restricted CTL epitope in the mouse hepatitis virus nucleocapsid protein, *J. Virol.* **67**: 7041.

Fleming, J. O., Trousdale, M.D., El-Zaatari, F.A.K., Stohlman S.A., and Weiner L.P., 1986, Pathogenicity of antigenic variants of murine coronavirus JHM selected with monoclonal antibodies, *Virology* **58**: 869.

Fosmire, J. A., Hwang K. and Makino S., 1992, Identification and characterization of a coronavirus packaging signal, *J. Virol.* **66**: 3522.

Klenerman, P., Phillips, R., and McMichael, A.,1996, Cytotoxic T-cell antagonism in HIV-1, *Seminars Virol.* **7**:31.

Kyuwa, S., and Stohlman, S.A., 1990, Pathogenesis of a neurotropic murine coronavirus, strain JHM in the central nervous system of mice, *Seminars Virol.* **1**: 273.

Lai, M., and D. Cavanagh., 1997, Molecular biology of coronaviruses, *Adv. Virus Res.* In press.

Pewe, L., Wu ,G.F., Barnett, E.M., Castro, R.F., and Perlman S., 1996, Cytotoxic T cell-resistant variants are selected in a virus-induced demyelinating disease. *Immunity* **5**:253.

Pfleiderer, M., Skinner, M.A., and Siddell, S.K., 1986, Coronavirus MHV-JHM: nucleotide sequence of the mRNA that encodes the membrane protein, *Nucl. Acids Res.* **14**:6338.

Rottier, P. J., Welling, G. E., Welling-Webster, S., Niesters, H. G., Lenstra, J. A., and van der Zeijst, B.A.M., 1986, Predicted membrane topology of the coronavirus protein E1, *Biochem.* **25**: 1335.

Rowe, C.L., Baker, S.C., N. M.J., and Fleming, J.O., 1997, Evolution of mouse hepatitis virus: Detection and characterization of spike deletion variants during persistent infection, *J. Virol.* **71**:2959.

Skinner, M.A., and Sidell, S.G., 1983, Coronavirus JHM: nucleotide sequence of the mRNA that encodes nucleocapsid protein, *Nucl. Acids Res.* **11**:5045.

Stohlman, S.A., Bergmann, C., van der Veen, R., and Hinton, D., 1995, Mouse hepatitis virus-specific cytotoxic T lymphocytes protect from lethal infection without eliminating virus from the central nervous sytem, *J. Virol.* **69**:684.

Stohlman, S. A., Brayton P.R., Fleming, J.O., Weiner, L.P. and Lai, M.M.C., 1982, Murine coronaviruses: isolation and characterization of two plaque morphology variants of the JHM neurotropic strain, *J. Gen. Virol.* **63**:265.

Stohlman, S.A., Fleming, J.O., Patton,C.D., and Lai, M.M.C., 1983, Synthesis and subcellular localization of the murine coronavirus nucleocapsid protein, *Virology* **130**:527.

Williamson, J.S.P., and Stohlman S.A.S., 1990, Effective clearance of mouse hepatitis virus from the CNS requires both CD4$^+$ and CD8$^+$ T cells, *J. Virol.* **64**:4590.

98

QUASISPECIES DEVELOPMENT BY HIGH FREQUENCY RNA RECOMBINATION DURING MHV PERSISTENCE

C. L. Rowe,[1] S. C. Baker,[1] M. J. Nathan,[2] J.-Y. Sgro,[3] A. C. Palmenberg,[3,4] and J. O. Fleming[2]

[1]Department of Microbiology and Immunology and Molecular Biology
   Program
Loyola University of Chicago
Stritch School of Medicine
Maywood, Illinois 60153
[2]Departments of Neurology and Medical Microbiology
University of Wisconsin and William S. Middleton Veterans Hospital
Madison, Wisconsin
[3]Institute for Molecular Virology
[4]Department of Animal Health and Biomedical Science
University of Wisconsin
Madison, Wisconsin

1. ABSTRACT

Recent studies suggest that infectious viruses and particularly persisting viral RNAs often exist as diverse populations or "quasispecies". We have developed an approach to characterize populations of the murine coronavirus mouse hepatitis virus (MHV) generated during persistent infection which has allowed us to begin to address the role of the viral quasispecies in MHV pathogenesis. We analyzed the population of persisting viral RNAs using reverse-transcription polymerase chain reaction amplification (RT-PCR) of the S1 "hypervariable" region of the spike gene followed by differential colony hybridization to identify spike deletion variants (SDVs) from acute and persistently infected mice. Sequence analysis revealed that mice with the most severe chronic paralysis harbored the most complex quasispecies. Mapping of the SDVs to the predicted RNA secondary structure of the spike RNA revealed that an isolated stem loop structure is frequently deleted. Overall, these results are consistent with high frequency recombination at sites of RNA

Coronaviruses and Arteriviruses, edited by Enjuanes *et al.*
Plenum Press, New York, 1998

secondary structure contributing to expansion of the viral quasispecies and persisting viral pathogenesis.

2. INTRODUCTION

Viruses with RNA genomes rapidly adapt to changing environments. This is because they exist as a "quasispecies" of replicating RNAs (Novella et al., 1995). Two major mechanisms contribute to the viral quasispecies: mutation and RNA recombination. Mutations are rapidly incorporated into the population because RNA viruses have no known proof-reading ability, and therefore are unable to correct random mutations introduced into their viral genome. In addition, RNA viruses have been shown to undergo RNA recombination which may radically alter or delete portions of the RNA virus genome. Copy-choice RNA recombination, as described for poliovirus, occurs when partially synthesized RNAs dissociate from one template and then re-join the same or another template at either the same site (homologous recombination) or a different site (aberrant-homologous or non-homologous recombination) (Lai 1992). Both mutation and RNA recombination contribute to quasispecies expansion during persistent infection of MHV and may significantly contribute to viral pathogenesis.

A number of investigators have identified infectious MHV variants during the sub-acute stage of disease which harbor deletions in the S1 "hypervariable region" (nt 1200–1800) (reviewed in Rowe et al, 1997a). However, the significance of these variants in chronic disease is complex and requires further investigation. During the chronic stage of disease there is a significant reduction in infectious virus, however persisting viral RNA can be detected by RT-PCR amplification (Adami et al., 1995). We wanted to know if genetic changes in the viral RNA might contribute to persistence and chronic disease. To begin to understand the significance of genetic changes in the spike gene, we monitored the genetic changes in the S1 hypervariable region and determined if these genetic changes were associated with more chronic and severe disease.

3. MATERIALS AND METHODS

As reported in Rowe et al. 1997a and 1997b, C57BL/6 mice were inoculated intracerebrally with 10^3 p.f.u. MHV-JHM 2.2-V-1 and periodically monitored for clinical signs of demyelination. Mice were sacrificed at 4, 42 and 100 days post-inoculation (PI), and total RNA was isolated from the brains and spinal cords of individual mice. An 848 bp S1 region of the spike gene was amplified by RT-PCR from total RNA with primers S1–1 and S1–2. To identify individual isolates containing SDVs, the S1 products were cloned into pGEM-T (Promega, Madison, WI), transformed into competent *E. coli* and analyzed by differential colony hybridization using an 837 bp wild-type (WT) probe and small 40–80 bp probes to the core (C) of the hypervariable region and to flanking regions A, B and D. Clones which hybridized with the wild-type probe but did not hybridize with probes A-D were sequenced by the sanger dideoxy chain termination method.

Using MFOLD version 2.2, the RNA secondary structure of the 4,139 nt MHV-JHM 2.2-V-1 spike RNA [(accession # D10235) according to Taguchi et al. (Taguchi et al., 1992) with mutations incorporated according to Wang et al. (Wang et al., 1992)] was folded (Rowe et al., 1997b).

4. RESULTS AND DISCUSSION

We were interested in determining the extent of the diversity in the population of persisting MHV RNAs. We also wanted to determine if changes in the viral RNA were associated with chronic disease and what mechanisms might be responsible for putative changes in the persisting viral RNA. To address these questions, we amplified the spike "hypervariable" region from acute and persistently infected mice (Figure 1). We identified SDVs using differential colony hybridization with S1 subdomain probes (A-D) and a full length probe (WT) (Rowe et al., 1997b). As reported (Rowe et al., 1997a, b), we identified 25 different patterns of SDVs. Interestingly, we found distinct quasispecies populations in the CNS of individual mice. For example mouse 42–8, which recovered from paralysis, had no detectable SDVs in the brain, only wild-type sequences and only a very low level of a single type of SDV was detected in the spinal cord (Figure 1B). In contrast mouse 100–1, which displayed chronic severe paralysis, harbored a higher frequency of SDVs both in the brain and the spinal cord and a more diverse quasispecies. In fact, our screening identified four distinct types of SDVs in the spinal cord of mouse 100–1 alone. We consistently found this pattern, where mice with the highest frequency and most diverse quasispecies had the most severe disease (Rowe et al, 1997a). These observations are con-

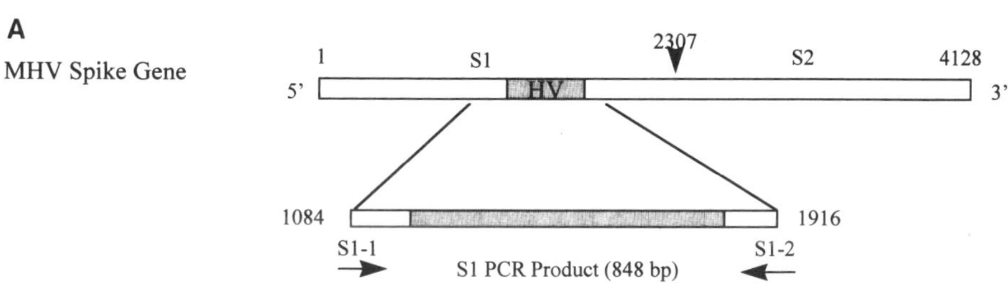

Figure 1. A) Schematic diagram of the position of the S1 primers used to amplify SDVs from the brain and spinal cord of acute and persistently infected mice. B) Examples of SDV patterns identified in mice with different clinical outcomes. Mice were monitored and scored on a scale from 0 (normal) to 4 (severe paralysis) every four days after inoculation. *Clinical score at day 24.

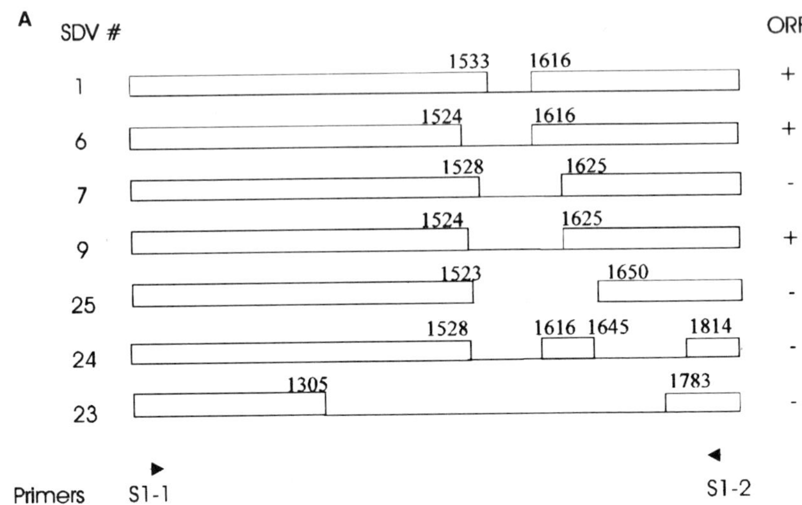

Figure 2A. Schematic diagram of the SDV patterns identified by differential colony hybridization with probes A, B, C, D and WT.

sistent with those seen in chronic infections with hepatitis C virus (Martell et al., 1992) and measles virus (Baczko et al., 1993). Thus, by studying the dynamics of MHV persistence in mice we may be better able to understand how persistent viral infections may contribute to chronic diseases such as MS.

We also wanted to determine how MHV quasispecies arise. RNA secondary structure has been shown to be important in directing recombination in a number of RNA viruses such as tomato bushy stunt virus (TBSV) and turnip crinkle virus (TCV) (White et al, 1994; Cascone et al., 1993). In fact, Simon and colleagues has shown that single base mutations that disrupt a stem loop eliminate recombination events in TCV (Cascone et al., 1993). To analyze the role of RNA secondary structure in the generation of coronavirus SDVs, we mapped a subset of the SDV patterns onto the MHV 4 kb spike RNA secondary predicted by the computer program MFOLD (Figure 2).

We noted that the SDVs frequently map to the base of an isolated stem loop structure, suggesting that this region may be prone to RNA recombination. In addition, the SDV deletion endpoints were frequently proximal to one another, suggesting that these SDVs may have been generated by an intramolecular recombination event in which the polymerase encounters the stem loop and instead of reading through it, simply jumps across to a nearby site, thereby deleting the stem. By modeling the MHV spike RNA and mapping of spike deletion variants we identified a putative RNA secondary structure which may play an important role in the high frequency generation of SDVs during persistence.

Figure 3 highlights factors which may affect MHV quasispecies populations. Potential selective pressures include: 1) altered binding of variants to host specific receptors, 2) replication advantages of variants, 3) altered recognition of variants by cytotoxic T lymphocytes, 4) reduced capacity for fusion of variants, and 5) reduced cytopathic effects of variants. Interestingly, a spike protein epitope (Db-restricted; a.a 510–518) recognized by CD8+ cytotoxic T lymphocytes (CTLs) is deleted in all of the SDVs that we identified. CTL escape may be one way that the virus persists and causes disease. However, we also identified wild-type epitopes which persisted in some mice, suggesting that deletion of the CTL epitope is not the only way that MHV persists.

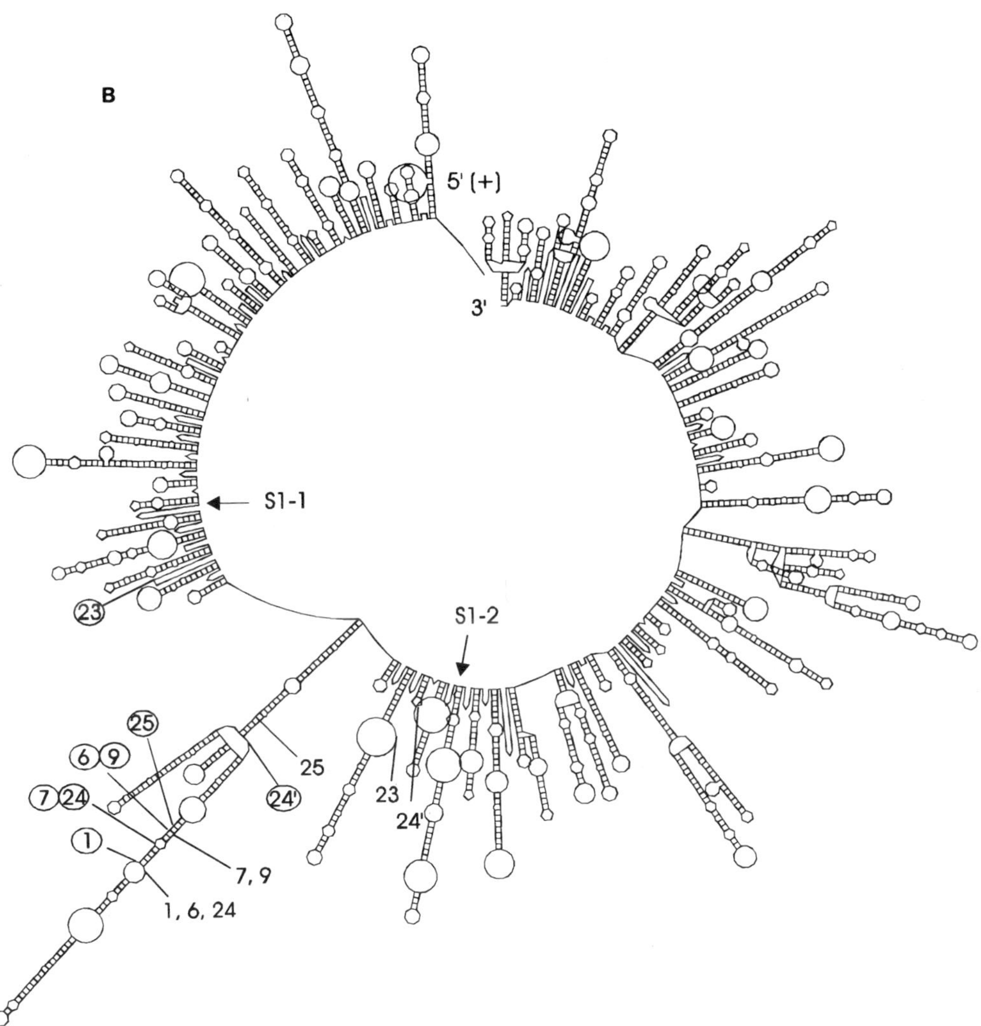

Figure 2B. Mapping of SDV patterns from (A) to the predicted RNA secondary structure of the MHV spike RNA. The left (circled) and right crossover sites of the SDVs are indicated.

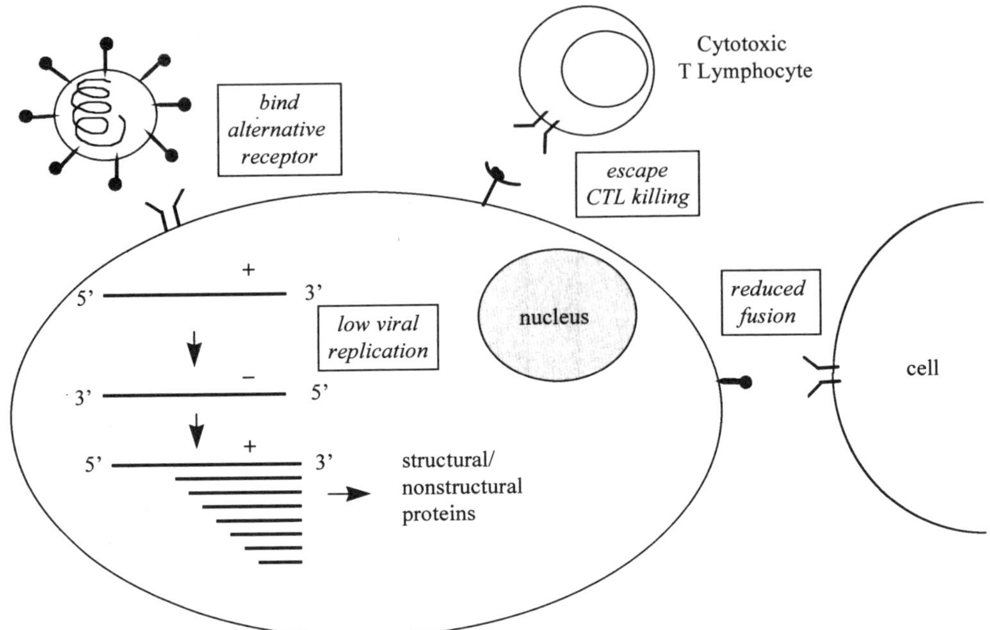

Figure 3. Factors which may contribute to selection of persisting virus or viral RNAs.

REFERENCES

Adami, C., J. Pooley, J. Glomb, E. Stecker, F. Fazal, J. O. Fleming, and S. C. Baker, 1995, Evolution of mouse hepatitis virus (MHV) during chronic infection: quasispecies nature of the persisting MHV RNA, *Virology* 209:337–346.

Baczko, K., J. Lampe, U. G. Liebert, U. Brinckmann, V. ter Meulen, I. Pardowitz, H. Budka, S. L. Cosby, S. Isserte, and B. K. Rima, 1993, Clonal expansion of hypermutated measles virus in a SSPE brain, *Virology* 197:188–195.

Bergmann, C. C., Q. Yao, M. Lin, and S. A. Stohlman, 1996, The JHM strain of mouse hepatitis virus induces a spike protein-specific Db-restricted cytotoxic T cell response, *J. Gen. Virol.* 77:315–325.

Cascone, P. J., T. F. Haydar, and A. E. Simon, 1993, Sequences and structures required for recombination between virus-associated RNAs, *Science* 260:801–805.

Castro, R. F., and S. Perlman, 1995, CD8+ T-cell epitopes within the surface glycoprotein of a neurotropic coronavirus and correlation with pathogenicitiy, *J. Virol.* 69:8127–8131.

Lai, M. M., 1992, RNA recombination in animal and plant viruses, *Microbiol. Rev.* 556:61–79.

Martell, M., J. I. Esteban, J. Quer, J. Genesca, A. Weiner, R. Esteban, J. Guardia, and J. Gomez., 1992, Hepatitis C virus (HCV) circulates as a population of different but closely related genomes: quasispecies nature of HCV genome distribution, *J. Virol.* 66:2547–2556.

Novella, I. S., E. Domingo, and J. J. Holland, 1995, Rapid viral quasispecies evolution - implications for vaccine and drug strategies, [Review]. *Molecular Medicine Today* 1:248–253.

Parker, S. E., T. M. Gallagher, and M. J. Buchmeier, 1989, Sequence analysis reveals extensive polymorphism and evidence of deletions within the E2 glycoprotein gene of several strains of murine hepatitis virus, *Virology* 173:664–673.

Rowe, C. L., S. C. Baker, M. J. Nathan, and J. O. Fleming, 1997a, Evolution of Mouse Hepatitis Virus: Detection and Characterization of Spike Deletion Variants During Persistent Infection, *J. Virol.* 71:2959–2969.

Rowe, C. L., J. O. Fleming, J.-Y. Sgro, A. C. Palmenberg and S. C. Baker, 1997b, Generation of coronavirus spike deletion variants (SDVs) by high frequency recombination at regions of predicted RNA secondary structure, *J. Virol.*, in press.

Taguchi, F., T. Ikeda, and H. Shida, 1992, Molecular cloning and expression of a spike protein of neurovirulent murine coronavirus JHMV variant cl-2 [published erratum appears in *J Gen Virol* 1992 Oct; 73(Pt10):2767], *J. Gen. Virol.* **73**:1065–1072.

Wang, F. I., J. O. Fleming, and M. M. Lai, 1992, Sequence analysis of the spike protein gene of murine coronavirus variants: study of genetic sites affecting neuropathogenicity, *Virology* **186**:742–749.

White, K. A., and T. J. Morris, 1995, RNA determinants of junction site selection in RNA virus recombinants and defective interfering RNAs. *RNA* **1**:1029–1040.

99

CORONAVIRUS INFECTION AND DEMYELINATION

Sequence Conservation of the S-Gene during Persistent Infection of Lewis–Rats

Helmut Wege,[1] Albert Stühler,[2] Hans Lassmann,[3] and Hanna Wege[1]

[1]Institute of Diagnostic Virology
Federal Research Centre for Virus Diseases of Animals
Friedrich-Loeffler-Institutes
D-17498 Isle of Riems
[2]Ludwig Institute for Cancer Research
London W2 1PG, England
[3]Institute of Neurology
A-1090 Vienna, Austria

1. ABSTRACT

Coronaviruses display a large phenotypic variability, which may be an important factor for diversification and selection. Previous studies have demonstrated that the S-protein is an essential determinant of virulence and pathogenicity. Therefore we studied the S-gene as an indicator molecule for selection processes employing two different MHV-JHM variants. First, Lewis–rats were infected with MHV-JHM-Pi, a variant that causes demyelinating disease after several weeks p. i. It was not possible to isolate infectious MHV-JHM-Pi from such rats, although viral proteins were expressed. The S-gene was rescued directly from brain tissue employing RT-PCR technology. The amplicons were sequenced in bulk or at the level of single clones. We detected no evidence for an increase of S-gene mutants during the length of time. Only few mutations were found at the clonal level. The changes were distributed throughout the analysed S-gene fragments without a predilection in their location. The frequency of mutation remained low within a range of 0.03 to 0.5 mutations per thousand nucleotides. As a second approach, we sequenced the S-genes of viruses isolated from brain tissue infected with MHV-JHM-ts43. Infection of adult Lewis rats with that mutant resulted several weeks to months p.i. in demyelinating encephalomyelitis. The S-gene of this virus contains an insertion of 423 bp in the S1 region, which is identical to a polymorphic region described for MHV-4. In contrast to JHM-Pi, infectious MHV-JHM-ts43 was readily to iso-

Coronaviruses and Arteriviruses, edited by Enjuanes *et al.*
Plenum Press, New York, 1998

late from brain tissue. The S-gene sequences of virus isolated 45–106 days p.i. from diseased rats were identical with that of the input virus. These results show, that during a persistent infection of Lewis–rats the S-gene was highly conserved.

2. INTRODUCTION

The single stranded, positive sense RNA genome of Coronaviridae comprises a size of 31 kb. Transcription and gene expression are mediated by 5–8 subgenomic RNAs, which are produced by a discontinuous RNA synthesis process. Because of the infidelity of the RNA polymerase newly generated viral RNA can display mutations, deletions and changes are introduced by recombination (Rowe et al., 1997a; Kottier et al., 1995; LaMonica et al., 1991; Banner et al., 1990; Parker et al., 1989). The resulting virus population in an infected host may therefore comprise a heterogenous but defined distribution of molecular variants, termed as "quasispecies". Experimental results and field observations led to the hypothesis, that a quasispecies distribution of RNA virus populations may provide the basis for diversification (Smith et al., 1997; Holland, 1992).

Infections of mice and rats with murine coronaviruses provide fascinating experimental systems to study demyelination and persistence in the central nervous system (Wege, 1995; Kyuwa & Stohlman, 1990). The mouse hepatitis virus MHV-JHM in particular causes in Lewis–rats a subacute demyelinating encephalomyelitis (SDE) characterized by inflammation and selective loss of myelin which resemble the lesions characteristic for multiple sclerosis of humans (Barac-Latas et al., 1997; Wege et al., 1984 and 1983). Coronaviruses are composed of up to five major structural proteins. These are the nucleocapsid protein (N), the membrane protein (M), the small membrane protein (sM), the hemagglutinin esterase protein (HE), and the large surface glycoprotein (S or spike). Results from a number of studies have shown, that both the S- and HE-protein are important for the neuropathogenicity of the virus and for the ability to infect specific cell types (LaMonica et al., 1991; Parker et al., 1989). The S-protein is very important for receptor binding, cell fusion and induction of the immune response. In tissue culture, virus variants had been selected in the presence of neutralizing antibodies which displayed a significant loss of virulence. The S-gene of such escape variants are characterised by mutations and deletions (Gallagher et al., 1990; Wege et al., 1988; Dalziel et al., 1986; Fleming et al., 1986). It is conceivable, that coronavirus induced chronic infection and demyelination is promoted by evasion of humoral and cellular immune responses by antigenic variation. Based on these observations a number of recent studies focused on the S-protein gene as an indicator molecule for selection processes in vivo (Rowe et al., 1997a,b; Stühler et al., 1997; Pewe et al., 1996; Adami et al., 1995). The association between chronic disease and the role of genetic heterogeneity in virus persistency is an interesting but controversial discussed issue. We investigated the S-gene variability during persistent infection in Lewis–rats employing two different MHV-JHM variants and alternative experimental approaches. As an unexpected result we found that under the constraints of a persistent infection the S-gene sequence was highly conserved and no evidence for a quasispecies distribution was found.

3. MATERIALS AND METHODS

MHV-JHM-Pi was derived from a persistently infected cell line (Baybutt et al., 1984). MHV-JHM-ts43 is a temperature sensitive mutant selected after mutagenesis with fluoruracil

(Wege et al., 1984 and 1983). Both variants are closely related to MHV-JHM wildtype (Schmidt et al., 1987). Specific pathogen free Lewis–rats were intracerebrally inoculated, either with MHV-JHM-Pi at an age of 4–5 days or with MHV-JHM-ts43 at an age of 3–5 weeks. For histology, animals were perfused with 4% paraformaldehyde in 0.1 M phosphate buffer. Parts of brain and spinal cord were embedded in paraffin. To obtain specimens for RT-PCR, animals were dissected and left half of the brain frozen in liquid nitrogen before storage at -80 °C. The right half of the brain and spinal cord was further processed for histology and immunocytochemistry as described previously (Barac-Latas et al., 1997; Wege et al., 1984 and 1983). Processing of tissue for RT-PCR, selection of primer pairs for defined regions of the S-gene, cloning and sequencing was performed according to Stühler et al., 1997).

4. RESULTS

4.1. Neurovirulence of MHV-JHM-Pi and MHV-JHM-ts43 for Lewis–Rats

The tissue culture adapted MHV-JHM wild-type strain usually causes within 6–12 days p.i. an acute encephalomyelitis, independent of the age of the rat at time of infection (Wege et al., 1983 and 1984). By contrast, the variant MHV-JHM-Pi replicates in the brain of adult Lewis–rats without causing disease (Baybutt et al., 1984). It was also not virulent for adult mice. If 4–5 days old rats were infected, about 25% of the animals developed subacute demyelinating encephalomyelitis (SDE) within 20 to 50 days p.i., the majority of rats remained healthy (Stühler et al., 1997). The brain tissue of rats with SDE contained high amounts of virus antigen and developed demyelinating lesions in the white matter of the central nervous system (Barac-Latas et al., 1997). It was not possible to isolate infectious virus from brain homogenates of rats with SDE, although it was easy to isolate virus within 6–8 days p.i. during the acute phase.

The temperature sensitive mutant MHV-JHM-ts43 was used for infection of adult Lewis–rats (Wege et al. 1983 and 1984). These animals developed SDE as late as several weeks to months p.i. and revealed pronounced lesions of primary demyelination. It was most interesting, that a number of rats developed a relapse of SDE after recovery from disease. In contrast to the results with MHV-JHM-Pi, infectious virus was easily isolatable from tissue of brain and spinal cord.

4.2. Sequencing of the S-Gene of the Variant MHV-JHM-Pi and MHV-JHM-ts43

In order to detect changes and to examine variability in the S-genes isolated from infected animals we first determined the consensus sequence of MHV-JHM-Pi from three individual plaques of the input virus. Exactely as described for the MHV-JHM wild-type, the S-gene of MHV-JHM-Pi comprises 3705 nucleotides coding for a protein of 1235 amino acids (Schmidt et al., 1987). In comparison to MHV-JHM wild-type, the S-gene of MHV-JHM-Pi contains 14 nucleotide substitutions (point mutations) which all do result in amino acid changes. These changes were distributed throughout the S-protein and not limited to domains. In comparison to the JHM strain MHV-4 (Parker et al., 1989) both the S-gene of MHV-JHM-Pi and the MHV-JHM wild-type have a deletion of 423 nucleotides starting at position 1359 and extending through to nucleotide 1781.

The complete S-gene of the MHV-JHM-ts43 used for infection of Lewis rats was sequenced for comparison with the other MHV-strains. The S-gene consensus sequence of MHV-JHM-ts43 appeared to be almost identical with MHV-4. This sequence is 4128 nucleotides long. In comparison to MHV-JHM and MHV-JHM-Pi this virus contains 423 additional nucleotides starting at position 1359 and extending through position 1781 of the S-gene (Parker et al., 1989). Position 763, 764 and 765 were found to be GCT instead of CTG in agreement with the sequences published by Schmidt et al. (1987). The position 3990 was changed from C to T. This mutation results only in a silent mutation (amino acid Ala→Ala).

4.3. S-Gene Sequence Variation in Lewis–Rats Persistently Infected with MHV-JHM-Pi

Rats were infected with MHV-JHM-Pi and brain tissue of animals suffering from subacute demyelinating encephalomyelitis collected for the analysis of the S-gene without any cell culture passage. The RNA was extracted, transcribed to cDNA and the S-gene was amplified by PCR in three overlapping fragments covering the entire gene. In total, brain tissue from six rats was examined for this analysis (Stühler et al., 1997). The amplified S-gene fragments derived fom brain tissue were exactely of the same size than corresponding S-gene amplicons from the input virus propagated in tissue culture. The consensus sequences of the S-genes derived from brain tissue were identical with the sequence of the input virus.

Because individual mutations at the molecular level might not influence the consensus sequence, sequencing studies of cloned PCR products were performed to reveal a possible quasispecies nature of the S-gene. According to results from previous studies, those gene fragments were selected for clonal analysis which should encompass regions involved in virus neutralization and cell fusion. The brain tissue from six rats with SDE was employed for this analysis. From each animal 20 individual molecular clones per each fragment were sequenced. The RT-PCR amplification products were cloned and then sequenced. We detected some mutations at the clonal level, but the total number of nucleotide substitutions was very low (51 mutations in 360 clones). The distribution of the detected mutations gave no indication for "hot spots" or "clustering" in their location. In total, 181.2 kb were sequenced (30.2 kb per animal and 10 kb per single fragment). The number of nucleotide substitutions was by statistical analysis smaller than the theoretical error rate of Taq polymerases (0.5 mutations per thousand nucleotides). As described in detail by Stühler et al. (1997), even attempts to drive selection by application of a neutralizing antibody in vivo did not influence the rate of mutations. Their frequency remained always within a range between 0.03 to 0.5 mutations per thousand nucleotides.

4.4. S-Gene Sequence Variability of MHV-JHM-ts43 Isolates from Diseased Lewis–Rats

As an alternative approach we investigated the S-gene variability of MHV-JHM-ts43 reisolated from persistently infected rats. In contrast to MHV-JHM-Pi, this variant is virulent for adult mice and rats, and infectious virus was readily isolatable from brain tissue. Furthermore, its S-gene contains the same insert of 423 nucleotides than MHV-4, which appears to be polymorphic and hypervariable (Gallagher et al., 1990). We concentrated on five MHV-JHM-ts43 isolates from brain tissue of Lewis–rats, which developed SDE after an incubation time between 45, 48, 54, 95, and 106 days p. i. respectively (Wege et al., 1984 and 1983). The virus stocks propagated for anaylysis of the S-gene were kept within 2–4 cell passages at most. These virus stocks displayed both the temperature sensitive phenotype and neurovirulence as the input virus. The complete S-gene was sequenced in eight overlapping RT-PCR amplicons. In summary, we could not detect any diversity of the consensus sequence of the S-gene of these reisolates in comparison to the MHV-JHM-ts43 input virus. Again, these results are implicating, that the S-gene of coronavirus MHV-JHM is strongly conserved under the conditions of a persistent infection in vivo. In total, 20.6 kb of information were sequenced.

5. DISCUSSION

It is well documented by a number of studies that at least the S-gene and HE-gene of coronaviruses are very polymorphic and this variability may play an important role for changes in virulence and during pathogenesis (LaMonica et al., 1991; Gallagher et al., 1990; Wege et al., 1988; Dalziel et al., 1986; Fleming et al., 1986). According to the sequence data on variability the S-gene appears to be an interesting indicator molecule to reveal a quasispecies distribution within the "virus swarm" and it may be a major target for selection based on an enhanced mutant frequency. At least in the mouse system, good evidence had been obtained for an association of genetic variability and persistency in vivo (Rowe et al., 1997a,b; Pewe et al., 1996; Adami et al., 1995). However, some controversy remains in the interpretation of the biological meaning of the observed variability.

The observed genetic stability of the MHV-JHM-Pi S-gene in vivo may reflect a bottleneck situation valid as an exception for this persistent infection. First, MHV-JHM-Pi was only infectious for newborn rats. The immune system of this animals was not fully developed at the timepoint of infection and although virus antigen was detectable in the brain tissue of diseased rats, it was not possible to isolate infectious virus. It is conceivable, that the lack of proper receptors on the host cells of adult rats allows only the survival of the already well adapted S-protein of MHV-JHM-Pi fitting to the rat variant of the MHV-receptor. Second, coronavirus persistence requires a continuous synthesis of viral proteins and the virus may spread slowly within the infected organs. Under these conditions the number of replication cycles may be relatively small and a high mutation frequency of the multifunctional S-protein may disrupt this balanced virus-host relationship. Third, MHV-JHM-Pi is derived from the MHV-JHM wild-type strain which contains a S-gene differing from the JHM strain MHV-4 by a deletion of 423 nucleotides (Rowe et al., 1997b; Parker et al., 1989; Schmidt et al., 1987). It had been shown recently, that S-protein deletion mutants accumulate during persistent infection of mice and that this "hypervariable region" is involved in forming a secondary stem-loop structure, which may be a "hot spot" for RNA recombination (Rowe et al., 1997a,b). We were interested to analyse the S-gene variability of MHV-JHM-ts43 because it contains this 423 nucleotide insert as described for MHV-4 (Parker et al., 1989). Furthermore, in contrast to MHV-JHM-Pi this variant can infect adult rats with a mature immune system and infectious virus could be isolated from diseased rats. However, at least at the level of the consensus sequence, no S-gene changes between the input virus and the isolated virus following a long period of persistency was found.

It has been shown by a number of studies that neutralizing antibodies can drive selection of escape variants, which display an attenuated virulence (Wege et al., 1988; Dalziel et al., 1986; Fleming et al., 1986). In MHV-4 the "hypervariable region" mentioned above behaved highly polymorphic in tissue culture experiments under the selection pressure by neutralizing antibodies (Gallagher et al., 1990). Following antibody selection, the S-gene of escape variants displayed deletions in this region. In our case, even by mimicking an immunological selection pressure with a neutralizing antibody in vivo no increase in the frequency of mutations was found (Stühler et al., 1997). On the other hand, depending on the host genetic context the cellular immune response may be a much stronger factor driving selection during chronic disease. In a mouse model for MHV-JHM induced demyelination, mutations were defined especially in the region coding for an immunodominant epitope recognized by cytotoxic T-cells (Pewe et al., 1996).

Taken together, our results appear to be controversial in view of the genetic heterogeneity described in the studies employing mice. On the other hand, only about 50 % of the mice harbored S-gene mutants with deletions, but all animals were persistently infected (Rowe et al., 1997b). Therefore, the impact of genetic heterogeneity of an individual virus population for the outcome of disease and the emergence of variants may differ with the particular virus-host combination. Our results illustrate that a genetically quite heterogenuous virus may remain in an evolutianary silent state during persistence in vivo. RNA replication may be reduced while persisting in infected animals in contrast to acute infections or during replication in tissue culture. These findings are compatible with a possibly high rate and frequency S-gene mutations during natural spread by acute infections in large host populations. The background of neutral mutations may be an unpredictible factor allowing evolution and emergence of new variants involving recombination, deletions and selection by host factors (Baric et al., 1997; Kottier et al., 1995; LaMonica et al., 1991; Gallagher et al., 1990). The study of virus-host relations during persistency and the implications for health and disease remains a fascinating topic.

ACKNOWLEDGMENTS

The work was supported by the Deutsche Forschungsgemeinschaft and Hertie Stiftung.

REFERENCES

Adami, C., Pooly, J., Glomb, J., Stecker, E., Fazal, F., Fleming, J.O. and Baker, S.C., 1995, Evolution of mouse hepatitis virus (MHV) during chronic infection: Quasispecies nature of the persisting MHV RNA, *Virologi* **209**: 337–346.

Banner, L.R., Keck, G.K. and Lai, M.M.C., 1990, A clustering of RNA recombination sites adjacent to a hypervariable region of the peplomer gene of murine coronavirus, *Virol.* **175**: 548–555.

Barac-Latas, V., Suchanek, G., Breitschopf, H., Stühler, A., Wege, H. and Lassmann, H., 1997, Patterns of oligodendrocyte pathology in coronavirus induced subacute demyelinating encephalomyelitis in the Lewis rat, *Glia* **19**: 1–12.

Baric, R., Yount, B., Hensley, L., Peel, S.A. and Chen, W., 1997, Episodic evolution mediates interspecies transfer of a murine coronavirus, *J. Virol.* **71**: 1946–1955.

Baybutt, H.N., Wege, H., Carter, M.J. and ter Meulen, V., 1984, Adaptation of coronavirus JHM to persistent infection of murine Sac(-) cells, *J. Gen. Virol.* **65**: 915–924.

Dalziel, R.G., Lampert, P.W., Talbot, P.J. and Buchmeier, M.J., 1986, Site-specific alterations of murine hepatitis virus type 4 peplomer glycoprotein E2 results in reduced neurovirulence, *J. Virol.* **59**: 463–471.

Fleming, J.O., Trousdale, M.D., El-Zaatari, F.A.K., Stohlman, S.A. and Weiner, L.P., 1986, Pathogenicity of antigenic variants of murine coronavirus JHM selected with monoclonal antibodies, *J. Virol.* **58**: 869–875.

Gallagher, T.M., Parker, S.E. and Buchmeier, M.J., 1990, Neutralization-resistant variants of a neurotropic coronavirus are generated by deletions within the aminoterminal half of the spike glycoprotein, *J. Virol.* **64**: 731–741.

Holland, J.J. (editor, 1992). Genetic diversity of RNA viruses. *Curr. Top. Microbiol. Immunol.* Springer-Verlag, Berlin-Heidelberg-New York **176.**

Kyuwa, S. and Stohlman, S.A., 1990, Pathogenesis of a neurotropic murine coronavirus, strain JHM in the central nervous system of mice, *Semin. Virol.* **1**: 273–280.

Kottier, S., Cavanagh, D. and Britton, P., 1995, Experimental evidence of recombination in coronavirus infectious bronchitis virus, *Virologi* **213**: 569–580.

Lamonica, N., Banner, L.R., Morris, V.L. and Lai, M.M.C., 1991, Localization of extensive deletions in the structural genes of two neurotropic variants of murine coronavirus JHM, *Virol.* **182**: 883–888.

Parker, S.E., Gallagher, T.M. and Buchmeier, M.J., 1989, Sequence analysis reveals extensive polymorphism and evidence of deletions within the E2 glycoprotein gene of several strains of murine hepatitis virus, *Virologi* **173**: 664–673.

Pewe, L., Wu, F.G., Barnett, E.M., Castro, R. and Perlman, S., 1996, Cytotoxic T cell-resistant variants are selected in a virus-induced demyelinating disease, *Immunity* **5**: 253–262.

Rowe, C.L., Fleming, J.O., Nathan, M.J., Sgro, J.-Y., Palmenberg, A. and Baker, S.C., 1997a, Generation of coronavirus spike deletion variants by high-frequency recombination at regions of predicted RNA secondary structure, *J. Virol.* **71**: 6183–6190.

Rowe, C.L., Baker, S.C., Nathan, M.J. and Fleming, J.O., 1997b, Evolution of mouse hepatitis virus: Detection and characterization of spike deletion variants during persistent infection, *J. Virol.* **71**: 2959 -2969.

Schmidt, I., Skinner, M.A. and Siddell, S.G., 1987, Nucleotide sequence of the gene encoding the surface projection glycoprotein of the coronavirus MHV-JHM, *J. Gen. Virol.* **68**: 47–56.

Smith, D.B., McAllister, J., Casino, C. and Simmonds, P., 1997, Virus "quasispecies": making a mountain out of a molehill? *J. Gen. Virol.* **78**: 1511–1519.

Stühler, A., Flory, E., Wege, H., Lassmann, H. and Wege, H., 1997, No evidence for quasispecies populations during persistence of the coronavirus mouse hepatitis virus JHM: sequence conservation within the surface glycoprotein gene S in Lewis rats, *J. Gen. Virol.* **78**: 747–756.

Wege, H., 1995, Immunopathological aspects of coronavirus infections, *Springer Semin. Immunopathol.* **17**: 133–148.

Wege, H., Watanabe, R. and ter Meulen, V., 1984, Relapsing subacute demyelinating encephalomyelitis in rats in the course of coronavirus JHM infection, *J. Neuroimmunol.* **6**: 325–336.

Wege, H., Koga, M., Watanabe, R., Nagashima, K. and ter Meulen, V., 1983, Neurovirulence of murine coronavirus JHM temperature sensitive mutants in rats, *Infect. Immun.* **39**: 1316–1324.

Wege, H., Winter, J. and Meyermann, R., 1988, The peplomer protein E2 of coronavirus JHM as a determinant of neurovirulence: definition of critical epitopes by variant analysis, *J. Gen. Virol.* **69**: 87–98.

100

PROKARYOTIC EXPRESSION OF PORCINE EPIDEMIC DIARRHOEA VIRUS ORF3

A. Schmitz, K. Tobler, M. Suter, and M. Ackermann

Institute of Virology
University of Zurich
Winterthurerstrasse 266a
CH 8057 Switzerland

1. ABSTRACT

Wild type (wt) and cell culture adapted (ca) strains of the coronavirus PEDV differ in their ability to cause diarrhea in neonate piglets: the wt strains are virulent; the ca strains are attenuated. Comparison of the available nucleotide sequences obtained from the different viral isolates revealed almost complete sequence identity with the exception of variations and truncations in open reading frame 3 (ORF3) observed exclusively in ca-PEDV isolates. In order to study the biological function(s) of the putative ORF3 product, the molecule was expressed as a heterodimeric fusion protein in *E.coli*. ORF3 was fused in frame to the alkaline phosphatase gene. Simultaneously, the construct was designed to form specific heterodimers by inclusion of the well known leucine zipper motiv of Jun and Fos. The heterodimerization partner contained the *E.coli* heat-labile enterotoxin subunit B (LTB) to allow specific binding to the eukaryotic cell receptor GM1. Our results indicate that heterodimeric fusion protein containing a truncated form of ORF3 was produced in high amounts, carried the expected ORF3 epitope, showed phosphatase activity, and was able to bind to the GM1 receptor. In contrast, a fusion protein containing the entire sequence of the ORF3 product was produced in minute amounts, indicating that it may have biological activity in prokaryotes, which led to the reduction of the amounts of proteins expressed.

2. INTRODUCTION

Comparison of the available nucleotide sequences obtained from wild type (wt) and either the corresponding or a different cell culture adapted (ca) strain of the coronavirus

Coronaviruses and Arteriviruses, edited by Enjuanes *et al.*
Plenum Press, New York, 1998

PEDV revealed almost complete sequence identity with the exception of variations and truncations in the open reading frame 3 (ORF3) observed exclusively in ca-PEDV isolates (Duarte et al., 1994; Tobler and Ackermann, 1995). A Spanish wild type isolate of PEDV acquired mutations in ORF3 during adaptation to growth in cell culture (Carvajal, Utiger, and Ackermann, unpublished observations).

Attempts to express ORF3 in eukaryotic cells, either by transient expression or in recombinant viruses, e.g. Baculovirus or Vacciniavirus, were unsuccessful. In contrast, it was possible to translate ORF3 mRNA *in vitro* or to produce ORF3 in the cytosol of *E.coli* (Utiger et al., 1995).

To study the functions of the putative ORF3 product, we planned to express ORF3 as a heterodimeric fusion protein in *E.coli*. The N-terminal part of the fusion protein was designed to contain the pelB leader to guide the polypeptide to the periplasmic space of the bacteria in order to allow correct folding of the aminoacid (aa) chain. Fused to this component the entire ORF3 or a fragment thereof was arranged to follow. A third component of the fusion protein consisted of *fos* sequences, and alkaline phosphatase and was designed to represent the C-terminus of one part of the heterodimeric protein. The second part was composed to contain again the pelB leader, followed by the *E.coli* heat-labile enterotoxin subunit B (LTB) and the *jun* polypeptide. A well known leucine zipper motiv of Jun and Fos is known to cause specific dimerization of the two components (Crameri and Suter, 1993). The LTB module was included in order to allow binding of the fusion protein to the eukaryotic cell receptor GM1 (Holmgren, 1973).

Here we show that the PEDV ORF3 product may have biological activity in prokaryotes, which led to the reduction of the amounts of proteins expressed. Heterodimeric fusion proteins containing truncated forms of ORF3 were produced in high amounts, they carried the expected ORF3 epitope, showed phosphatase activity, and were able to bind to the GM1 receptor.

3. MATERIALS AND METHODS

3.1. Plasmid Constructions

To construct pAS1, the sequence encoding the *E.coli* heat-labile enterotoxin subunit B (LTB) was amplified by PCR from pLTB0 (Aitken et al., 1994), and integrated as a *Sal*I/*Hind*III fragment into pJuFo (Crameri and Suter, 1993) containing the gene encoding the *E.coli* alkaline phosphatase (phoA). For the construction of pAS6, pAS1 was partially digested with *Nco*I and cut with*Sac*I before the complete ORF3 was inserted. The plasmid pAS12 was created by digestion of pAS6 with *Sac*I and religation, deleting the 3' last 574 nucleotides of ORF3. Gene expression in all constructs was under the control of *lacZ* promoters. The nucleotide sequence of the plasmid inserts was verified by sequencing.

3.2. Protein Expression and Extraction

For the induction of protein expression, bacteria were cultured at 37°C for 3h in LB containing 0.2% glucose, 100µg/ml ampicillin, and 1mM IPTG.

After the induction period, the cells were centrifuged and resuspended in 1/50 volume of 20% sucrose 50mM Tris pH 8.0. 1/1000 volume of 10mg/ml lysozyme in 50mM Tris pH 8.0 was added and the cells were kept on ice for 15min. 1/1000 volume of 0.5M EDTA pH 8.0 was added and incubation on ice continued for 10min. 1/500 volume of 10%

Brij58 in 50mM Tris pH8.0 and 1/1000 volume of 10mM PMSF were added and the cells were incubated on ice for an additional 30min. The cells were centrifuged and the supernatant containing the periplasmic extract was removed.

3.3. Nitrocellulose Test

Serial dilutions of the recombinant proteins in PhoA buffer (100mM Tris pH 9.5, 100mM NaCl, 50mM MgCl2) were sptted onto nitrocellulose sheets. PhoA activity was monitored by addition of substrate solution (0.45mg/ml of 4-nitroblue tetrazolium chloride and 0.175mg/ml of 5-bromo-4-chloro-3-indolyl-phosphate, Boehringer, in PhoA buffer). Crude bacterial alkaline phosphatase (Sigma) bound in defined concentrations to the same membrane served as a standard for evaluation of the enzymatic activity.

3.4. Immunodetection of Proteins on Western Blots

Periplasmic isolates were solubilized in loading buffer, boiled and separated on standard 10% polyacrylamide gels containing SDS before being transferred to nitrocellulose by electroblotting. For the immunodetection of proteins, the Western blots were first blocked in washing buffer (50mM Tris pH 7.4, 140mM NaCl, 5mM EDTA, 0.05% Nonidet P40, 0.25% gelatine) containing 10% skimmed milk (s.m.), then incubated with either a rabbit anti-peptide serum (1:100) specific for an N-terminal epitope of the putative ORF3product (Utiger et al., 1995) or the corresponding preimmune serum. Then, the blots were washed, incubated with a horseradish peroxidase-conjugated goat anti-rabbit IgG (Nordic Immunological Laboratories), and washed again before being developed by addition of the substrate solution containing 0.01% 4-chloro-1-naphtol and 0.015% H_2O_2.

Subsequently, the same blots were probed for PhoA using a mAb (Anawa) specific against this protein and the glucose oxidase mouse IgG kit (ABC-GO) and the GO-subrate kit III (Vector Laboratories). Blots were incubated and developed by addition of the GO substrate containing INT tetrazoium salt in Tris pH 9.5 essentially according to the protocols of the manufacturer.

3.5. GM1-Enzyme-Linked-Assay (GM1-ELA)

ELISA plates were coated over night at 4°C with 100μl of 1μg/ml GM1 (Sigma) in PBS pH 7.4. The plates were blocked for 1h at 37°C with 100μl of 3% BSA in PBS. 100μl of undiluted periplasmic extracts were applied to the wells and incubated for 1h. The plates were washed three times before the substrate (1.5mg/ml of 4-nitrophenyl phosphate, Sigma, in 1M diethanolamine buffer pH 9.8 containing 0.5M MgCl2) was added and PhoA activity measured at 405nm after 15, 75 and 800 min.

4. RESULTS

4.1. Construction of Vectors for the Expression of ORF3 Fusion Proteins

We attempted to express ORF3 in prokaryotes in order to analyse the protein in vitro and to deliver the product into eukaryotic cells by the LTB fusion approach.

Complete and truncated ORF3 constructs were integrated into a cassette (pAS1) consisting of the E.coli LTB and alkaline phosphatase, linked together by the leucine zipper formation of jun and fos sequences, which were also part of the constructs. A plasmid termed pAS6 contained the entire ORF3 coding sequence within the cassette. For the construction of pAS12, a SacI fragment was removed from pAS6. Thus, the new plasmid encoded, apart from the LTB-jun fusion protein, a truncated second polypeptide consisting of the 33 aa nearest to the N-terminus of the putative ORF3 product fused to fos and the alkaline phosphatase. The nt sequence of the constructs was confirmed by sequencing (data not shown) before attempts were made to express the corresponding proteins in bacteria following induction with IPTG. Periplasmic protein extracts were harvested and serial dilutions thereof were spotted on nitrocellulose sheets. The relative amounts of recombinant proteins were monitored by measuring phosphatase activity on the nitrocellulose sheets. Interestingly, considerable phosphatase activity was observed with the periplasmic extracts of bacteria transformed with pAS1 or pAS12 but only minute activity was seen with pAS6 (data not shown).

4.2. Analysis of Prokaryotically Expressed ORF3 Fusion Proteins

In order to verify that the product obtained from pAS12 contained both alkaline phosphatase and the PEDV ORF3 tag, Western blot analyses were performed. The results are shown in Fig. 1. Alkaline phosphatase, visualized by monoclonal antibody and an anti-mouse glucose-oxidase conjugate (red color), was present in extracts of both pAS12 and the positive control. In contrast, the PEDV ORF3 tag was labelled exclusively in pAS12 extracts, using a rabbit anti-peptide serum (but not preimmune serum) specific for the N-terminus of the putative ORF3 product and an anti-rabbit peroxidase conjugate (purple color). Neither phosphatase nor the ORF3 tag was detected in extracts obtained from control bacteria.

The sheet containing lanes 1 through 3 was then re-incubated using preimmune rabbit serum (PI), whereas a rabbit anti-peptide serum (I) specific for the N-terminus of the putative ORF3 product was used for the sheet containing lanes 5 through 7. The reactions were visualized using anti-rabbit peroxidase conjugate and chloronaphtol.

The circles in the left panel refer to bands stained unspecifically by the preimmune rabbit serum. These bands allow inference on the abundancy of protein loaded in each lane. The arrow points to the only band in the figure which was stained both by the monoclonal antibody against phosphatase and the rabbit anti-ORF3 serum. The numbers on the right of the figure indicate the relative mobility of the proteins of the molecular weight marker (MW, lane 4).

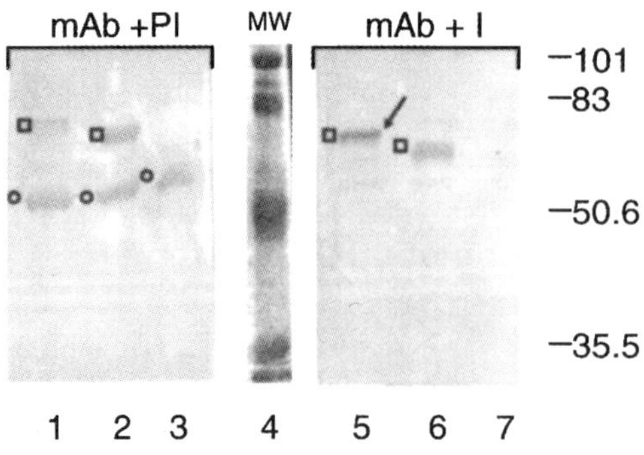

Figure 1. A double stained Western immunoblot is shown. Periplasmic extracts of *E.coli* transformed with pAS12 (lanes 1 and 5), pAS1 (lanes 2 and 6) or not transformed (lanes 3 and 7) were separated on 10% SDS polyacrylamide gels and transferred to nitrocellulose. The nitrocellulose sheets were first stained for the detection of alkaline phosphatase by a monoclonal antibody (mAb) and the mouse ABC-GO system as described in Materials and Methods. Since colors cannot be discriminated in this Figure, the red-stained bands are indicated by squares.

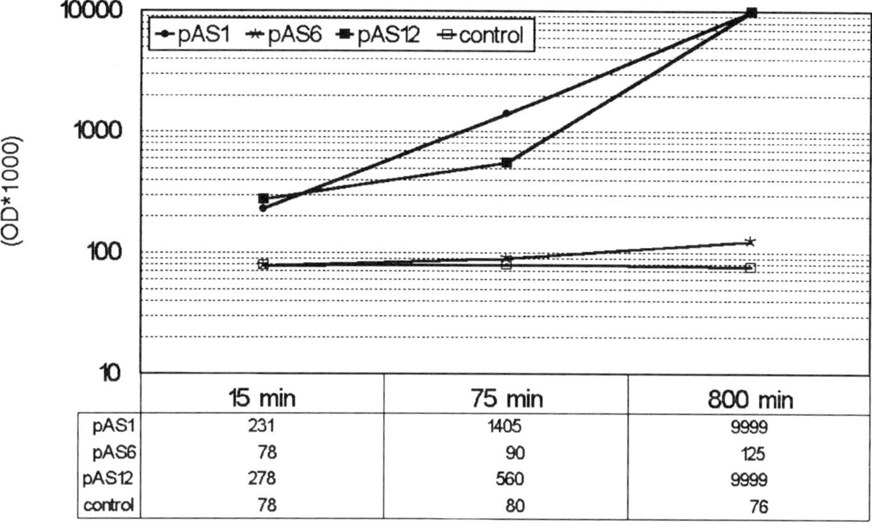

	15 min	75 min	800 min
pAS1	231	1405	9999
pAS6	78	90	125
pAS12	278	560	9999
control	78	80	76

Figure 2. Enzyme-linked binding assay (ELA). Microtiter plates were coated with GM1 before being incubated with periplasmic extracts of *E.coli* transformed with either pAS1 (dots), pAS6 (asteriks), pAS12 (filled squares) or a control plasmid (open squares). After washing, binding of LTB-containing fusion proteins was visualized and quantified by the addition of phosphatase substrate. Absorption (OD405 nm) was measured in an ELISA reader at 15, 75, and 800 minutes after the addition of substrate. The values for each construct at each time point are indicated in the table below the figure.

4.3. Functional Analysis of the LTB-Binding Domain

In order to test the functional capacity of the LTB-domain to bind to the eukaryotic cell receptor GM1, extracts of bacteria transformed with pAS1, pAS6, pAS12, or control plasmids were incubated on GM1 coated microtiter plates and bound phosphatase activity was quantified at different time points after the addition of substrate (Fig. 2). Extracts of both, pAS1 and pAS12 showed almost linear increase of phasphatase activity over time, whereas no phosphatase activity was observed with the negative control. As expected, slight but reproducibly measurable phosphatase activity was detected in association with GM1 when the product was expressed from plasmid pAS6.

5. DISCUSSION

Several conventional eukaryotic expression systems, e.g. transient expression in transfected Vero cells, as well as the recombinant baculovirus- and vaccinia virus systems, were inadequate to provide sufficient quantities of the Porcine Epidemic Diarrhoea coronavirus putative ORF3 product to study its properties in detail (Tobler and Ackermann, unpublished observations). On the other hand, there was clear evidence from *in vitro* translation (Sirejols, Tobler, and Ackermann, unpublished) and cytosolic expression in *E.coli* (Utiger et al., 1995) that the ORF3 gene encoded a product which could be translated.

Therefore, our plan was to generate the ORF3 product in a prokaryotic system in order to deliver it to eukaryotic cells. To this aim, a heterodimeric fusion protein was constructed (Crameri and Suter, 1994). One part of the heterodimer consisted of the pelB-leader followed by either full length or truncated ORF3 sequences, and fused to fos and the alkaline phosphatase. The second part consisted of the pelB-leader, followed by the *E.coli* heat-labile enterotoxin subunit B (LTB) and jun at the carboxy terminus. The pelB-leader was used to guide the growing amino acid chains into the periplasmic space of *E.coli* in order to allow correct folding of the protein. The jun-fos leucine zipper was included to allow heterodimerization of the two fusion proteins. Phosphatase activity permitted a read-out for the correct reading frame and translation of the ORF3 fusion protein, and LTB was added as a means to bind the heterodimeric protein to the eukaryotic cell receptor GM1 for internalization (Aitken et al., 1994).

Our results clearly indicated that all of the components, pelB-leader, phosphatase, LTB, and the jun-fos leucine zipper were intact when expressed from pAS12, which encoded a truncated ORF3 product. Since pAS12 was a deletion mutant produced from pAS6, which encoded the entire ORF3, we conclude that the differences seen between the pAS12 and pAS6 expression products must be due to the biological activities of the ORF3 product. It appears as if ORF3 was able to reduce the net amounts of protein produced in *E.coli*. On the other hand, the work presented here confirmed the specificity of the rabbit anti-peptide serum for the N-terminus of the ORF3 protein (Utiger et al., 1995) as well as the possibility to use the N-terminal 33 amino acids of the ORF3 product as a tag in fusion proteins. The leucine zipper of jun-fos allowed heterodimerization of as large macromolecules as LTB and alkaline phosphatase without disturbing each others functions. These constructs will now be used to fine map the biological activities of the ORF3 product in more detail.

ACKNOWLEDGMENTS

These studies were supported by the Swiss National Science Foundation, grant #31-43503.95.

REFERENCES

Aitken R., Brock J., Sinclair M.C., 1994, Receptor-binding subunits from cholera toxin and heat-labile enterotoxin as immunological carriers, in: *Bacterial protein toxins*, Zbl. Bakt. Suppl. **24**, (Freer et al., eds), Gustav Fischer, Stuttgart, pp 467–473.

Crameri R., Suter M., 1993, Display of biologically active proteins on the surface of filamentous phages: a cDNA cloning system for selection of functional gene products linked to the genetic information responsible for their production, *Gene* **137**: 69–75.

Duarte M., Tobler K., Bridgen A., Rasschaert D., Ackermann M. and Laude H., 1994, Sequence analysis of the porcine epidemic diarrhea virus genome between the nucleocapsid and spike protein genes reveals a polymorphic ORF, *Virology* **198**: 466–476.

Holmgren J., Lonnroth I., Svenerholm, L., 1973, Tissue receptor for cholera exotoxin: postulated struture from studies with GM1 ganglioside and related glycolipids, *Infect. Immun.* **8**: 208–14.

Tobler K. and Ackermann M., 1995, PEDV leader and junction sites. In: Corona- and Related Viruses (P.J. Talbot and G.A. Levy, eds.), *Adv. Exp. Med. Biol.* **380**: 541–542.

Utiger A., Tobler K., Bridgen A., Suter M., Singh M., Ackermann M., 1995, Identification of proteins specified by porcine epidemic diarrhoea virus. In: Corona- and Related Viruses (P.J. Talbot and G.A. Levy, eds.), *Adv. Exp. Med. Biol.* **380**: 287–290.

101

FURTHER ANALYSIS OF THE GENOME OF PORCINE EPIDEMIC DIARRHOEA VIRUS

A. Bridgen,[1,2] R. Kocherhans,[1] K. Tobler,[1] A. Carvajal,[1] and M. Ackermann[1]

[1]Institute of Virology
University of Zurich
Winterthurerstrasse 266a
CH 8057 Switzerland
[2]Institute of Virology
Church Street
Glasgow G11 5JR, Scotland

1. ABSTRACT

We report here the continued determination and analysis of the nucleotide sequence of both wild type (wt) and cell culture adapted (ca) porcine epidemic diarrhoea coronavirus (PEDV). These studies were undertaken with two objectives in mind: the identification of common and divergent features in the genomic sequences of wt and ca PEDV which can explain the differences in virulence of these isolates and the further exploration of the relationship of PEDV to other coronaviruses.

2. INTRODUCTION

PEDV is, as its name implies, responsible for causing diarrhoea in pigs, particularly in neonates. We have previously completed the sequencing of the M, sM, ORF 3 and N genes of the CV777 strain of PEDV. The S, M, sM, ORF 3 and N genes of a British strain of PEDV have also been sequenced (Duarte *et al.*, 1994; Duarte and Laude, 1994 and Bridgen *et al.*, 1993). These results indicated that PEDV occupies a position intermediate between the two well characterised members of the group 1 coronaviruses, transmissible gastroenteritis virus (TGEV) and human coronavirus (HCV) 229E.

Our original cloning approach involved designing primers based on conserved regions of the coronavirus M and N genes to amplify the equivalent as then unknown PEDV sequence (Bridgen *et al.*, 1994). In this study, we were interested in knowing whether this

Coronaviruses and Arteriviruses, edited by Enjuanes *et al.*
Plenum Press, New York, 1998

technique could be used to clone the ORF 1 of PEDV. Such a method is useful for viruses which do not grow to high titre, avoids lengthy screening of clones and could potentially be applied to the cloning of any group 1 coronavirus. However, the large size of the ORF (more than 20,000 nucleotides) and the paucity of sequence data from other coronaviruses make this an ambitious objective. Only two group 1 coronaviruses have been sequenced completely, HCV229E and TGEV (Herold *et al.*, 1993 and Eleouet *et al.*, 1995). In addition, two strains of MHV, JHM and A59 and also IBV have also been completely sequenced. A number of conserved functional domains have been identified in the predicted ORF 1 products, but these domains are mainly located in the ORF 1b region and leave large regions of the 1a product with no known function and only a low level of sequence conservation between different coronavirus genomes.

Wild type (wt) and cell culture adapted (ca) PEDV exhibit remarkably different phenotypes with regard to virulence in piglets (Bernasconi et al., 1995). It was therefore of interest to compare the sequences of the two virus types at the nucleotide and predicted amino acid levels.

3. MATERIALS AND METHODS

3.1. Growth of PEDV and Preparation of Viral DNA

Growth of ca PEDV was performed essentially as has been described previously (Bridgen *et al.*, 1993), except that virus-infected cells were harvested at approximately 18 h pi. Cells were freeze-thawed three times and cell debris removed by low speed centrifugation. Virus was pelleted by centrifugation for 2 h at 22,000 rpm and 4°C in a SW28 rotor. Virus pellets were resuspended in 0.5 ml Trizol^{TM}, and RNA was prepared as recommended by the manufacturer. Pre-existing cDNA clones (Tobler and Ackermann, 1995) derived from RNA of gut tissue of pigs infected with either Belgian (Bernasconi et al., 1995) or Spanish (Carvajal et al., 1995) strains of PEDV were used for the determination wt PEDV sequences.

3.2. cDNA Synthesis and PCR Amplification of Viral Sequences

RNA prepared from two 175 cm^2 flasks of virus-infected cells was denatured for 10 min at 65°C and first strand cDNA was prepared in a 50 μl total reaction volume for 70 min at 42°C using Superscript^{TM} (Gibco-BRL), with conditions as recommended by the manufacturer. Each 50 μl reaction contained up to six anti-genome sense primers (Schmidheini AG) based on PEDV sequences or conserved regions of the HCV229E and TGEV genomes (see Table 1). Samples were treated with RNaseH for 30 min at 37°C prior to PCR amplification in a Perkin Elmer DNA Thermal Cycler 480. Amplifications were performed using pfu polymerase (Stratagene) in a 30 μl reaction volume containing 0.75 μM each primer. Reaction conditions varied with the primer pair but were typically one cycle of 95°C for 5 min, 50°C for 5 min, 74°C for 5 min followed by thirty five cycles of 95°C for 50 sec, 50°C for 50 sec, 74°C for 2 min.

3.3. Cloning of PCR Fragments

PCR fragments were phenol/chloroform extracted and ethanol precipitated prior to phosphorylation for 30 min at 37°C using T4 polynucleotide kinase in the presence of

Table 1. Primers used for RT-PCR amplification of the ORF 1a (1st column) and ORF 1b regions (2nd column) of the PEDV genome

Primer[a]	Sequence[b]	Primer[c]	Sequence[d]
o199	ggaacatttgttattgacatg	o203	tatgatgattctttctgatga
o200	ccatatggttgtgccatttt	o204	tgctgtgaacaaaattcatg
o205	gacaagttttacaagcgcatt	o207	tctcgtaatggcaacgttaa
o216	aggtttkccatcrgcrccaca	o208	tttgtcttgtgggattatgaa
o217	taaaaccggtccaatcwccaac	o209	tcttcttccattgtgctgcga
o238	cattwgwsagcatytttacgcagtt	o210	tgtcccamccgccataaaactt
o246	gagccacttgctactatcta	o211	catactattcttatatgatgcct
o258	gmgttwatccarcarttrttrtc	o226	caccaaatatatcactcttaac
o261	gatkttattgtwaatgctgcwaa	o239	gttgtmgtkgatgaggtstctatg
o263	ggtaagttytgyaaraarca	o240	ttrttttggaaytgtaatgtrga
o266	ctgcctgaataatacttgtact	o251	aaccaattcattcctgtgcac
o267	agctgtccwattgtwgtwgg	o284	tgacctacagcgagtatcaaa
o269	tgagttacaattcaactctaca		
o272	gcagctatgtayaargargc	o214[e]	gcaaccagaagtaaattaaagac
o286	ccaaacgtctcaacataccaaa	o117[e]	gttagctcttttctagacc
o287	aaataaacacctgctgggatacc	o241[e]	ggatagttagctcttttctag

[a]Designation and [b]nucleotide sequence of primers used for the amplification of the ORF 1a region.
[c,d]Primers designated for the amplification of the ORF 1b region.
[e]Primer o214 is in the S gene and primers o117 and o241 in the leader sequence.
Abbreviations are standard, ie w=a/t; s=c/g; r=a/g; y=c/t; m=a/c ; k=g/t.

500 µM ATP. Blunt-ended vector was prepared by digestion of pBS KS+ with *Eco* RV, dephosphorylation and phenol extraction to remove the alkaline phosphatase. Both PCR fragment and vector were electrophoresed through a 1% agarose gel, excised and purified by gene clean.

4. RESULTS

4.1. RT-PCR Cloning of the PEDV ORF 1

In order to clone and determine the sequences for the PEDV ORF 1, the predicted amino acid sequences of the HCV229E and TGEV polymerase open reading frames were aligned and homologous regions identified. The HCV229E and TGEV ORFs were sufficiently closely related to allow complete alignment of the predicted expression products. In contrast, the MHV and IBV sequences were much more divergent, and could only be aligned with the group 1 sequences in some of the conserved regions. Conservation between the HCV229E and TGEV sequences was much higher in the pol 1b than the pol 1a ORF, in which considerable regions of the predicted polypeptides showed under 30% homology.

A mixture of degenerate and non-degenerate primers were designed from amino acids conserved between the predicted products of the HCV229E and TGEV and, where possible, the MHV and IBV ORF 1. These primers were used both to prime reverse transcription and for the PCR amplification. Initially, 3 primer pairs yielded products. Two of them mapped to the pol 1b region and one to pol 1a. Sequencing of the amplification products allowed the design of PEDV-specific primers and confirmed the expected homology between the PEDV and the HCV229E and TGEV sequences. Continued use of this strategy using the primers described in Table 1 eventually led to RT-PCR amplification and cloning of almost the entire ORF 1 of PEDV (Figure 1).

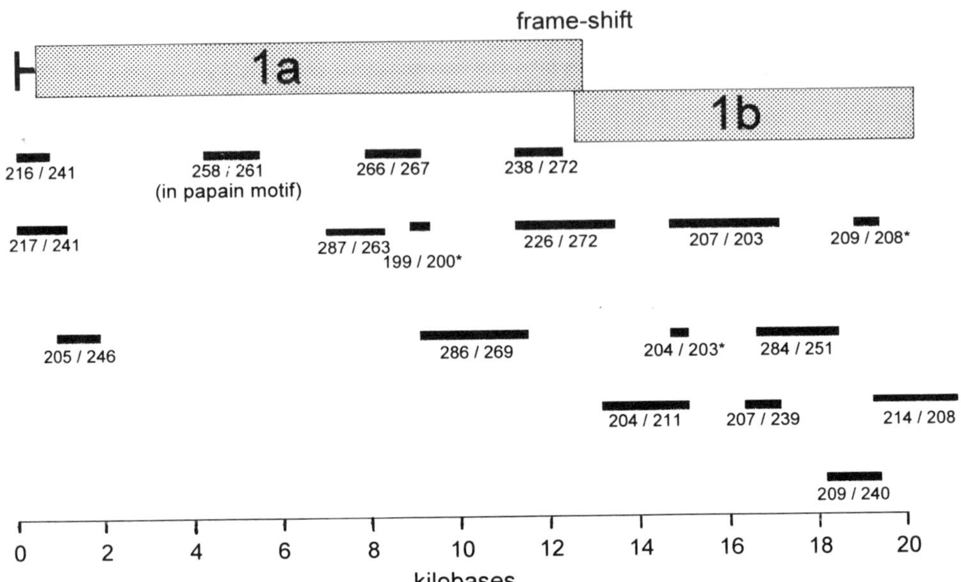

Figure 1. RT-PCR amplification of the PEDV ORF 1 using primers designed from conserved coronavirus sequences. The figure is drawn to scale, with the exception of the frame shift region and 5' leader sequence. PCR products are numbered according to the primers used for the amplification. Stars indicate the first products obtained by RT-PCR.

Analysis of the PEDV sequences continued to confirm the close relationship of PEDV with both TGEV and HCV229E. Similar results were obtained when the PEDV sequence was compared with both HCV229E and TGEV: a homology of approximately 70% and 25–30% at the amino acid level was determined for the pol 1b and pol 1a products, respectively. However the PEDV polymerase product showed a consistent slightly greater homology to the corresponding HCV229E than the TGEV product.

4.2. Sequence Comparisons of Wild Type and Cell Culture Adapted PEDV

Comparison of the two wt and CV777 ca PEDV genomes remarkably revealed no differences in either the coding or the non-coding sequences between the 3' end of the ORF 3 gene and the 3' end of the genome. Differences, however, between wt and ca strains were found within ORF 3 and within the S gene.

The sequence variations and truncations which were observed in ORF 3 of both a ca Belgian and a ca British strain (Duarte *et al.*, 1994) were not detected in wt isolates of either Belgian or Spanish origin. In contrast, the maximum possible length of the ORF 3, which had never been detected in a ca isolate, was the only form of ORF 3 which could be detected in the wt isolates. This observation suggested that an intact ORF 3 was maintained in wt viruses, whereas it seemed to be incompatible with replication of PEDV in cell culture.

The sequence of the S gene was determined only from one strain, CV777, the wt parent virus of our Belgium ca PEDV. The sequence comparison was made to the se-

Table 2. Five nucleotide changes within the S gene cause five differences between the predicted amino acid sequences of wild type and cell culture adapted PEDV

Nucleotide	From[a]	To[b]	Amino acid	From[a]	To[b]	Remarks
347	T	C	116	I (aliphatic)	T (hydrophilic)	creates new potential glycosylation site in ca PEDV
1168	C	T	390	P (imino acid)	S (hydrophilic)	
1264	A	T	422	N (hydrophilic)	Y (aromatic)	destroys potential glycosylation site in ca PEDV
3224	C	T	1075	S (hydrophilic)	L (hydrophobic)	S at this position is typical of group 1 coronaviruses
4123	C	G	1375	Q (neutral)	E (acidic)	

[a]Sequence in wt PEDV
[b]Sequence in ca PEDV

quence of the British ca PEDV isolate, which had been published previously (Duarte and Laude, 1995). Most interestingly only 5 nucleotide changes were observed within the coding sequences for S. All of those changes, however, were meaningful and led to changes in the predicted amino acid sequence. The results are summarised in Table 2. The overall number of 29 potential glycosylation sites remained unchanged. One of the changes in the amino acid sequence destroyed such a site, another change created a new one at a different location. A potential cleavage site for S1-S2 cleavage, which has been described for other coronaviruses, was not detected in either the wt or the ca sequences.

5. DISCUSSION

The technique of RT-PCR cloning the PEDV ORF 1 using primers based on conserved coronavirus sequences has proved successful, despite the lack of conserved sequences in the ORF 1a, and we would therefore recommend this technique as a means of cloning other group 1 ORF 1 sequences.

The sequence results obtained for the ORF 1 of PEDV revealed the presence of conserved domains typical of all the coronaviruses sequenced to date, such as the ribosomal frame shift site and polymerase and helicase domains. The 5' end of the genome was successfully amplified using a primer based on the leader sequence of PEDV (Tobler and Ackermann, 1995), thus confirming the relationship between these two sequences. In addition, the sequence data confirmed the homology of PEDV to HCV229E and TGEV.

The comparison of the ca and wt PEDV sequences has identified changes in both ORF 3 and the S gene between these isolates. However, in the absence of a reverse genetics system to study coronavirus genomes, completion of the entire sequence of both the wt and ca isolates will be necessary before we know which changes are responsible for the different biological properties of the isolates.

ACKNOWLEDGMENTS

These studies were supported by the Swiss National Science Foundation, grant #31–43503.95.

REFERENCES

Bernasconi C., Guscetti F., Utiger A., Van Reeth K., Ackermann M., and Pospischil A., 1995, Experimental infection of gnotobiotic piglets with a cell culture adapted porcine epidemic virus: clinical, histopathological and immunohistochemical findings, in: *Immunobiology of viral infections* (M. Schwyzer, M. Ackermann, G. Bertoni, R. Kocherhans, K. McCullough, M. Engels, R. Wittek, and R. Zanoni, eds.), Fondation Marcel Merieux, Lyon, France, pp 542–546.

Bridgen A., Duarte M., Tobler K., Laude H. and Ackermann M., 1993, Sequence determination of the nucleocapsid protein gene of the porcine epidemic diarrhoea virus confirms that this virus is a coronavirus related to human coronavirus 229E and porcine transmissible gastroenteritis virus, *J. Gen. Virol.* **74**: 1795–1804.

Bridgen A., Tobler K. and Ackermann, M., 1994, Identification of coronaviral conserved sequences and application to viral genome amplification. In: Coronaviruses: Molecular Biology and Virus-Host Interactions (H. Laude and J-F. Vautherot, eds.), *Adv. Exp. Med. Biol.* **342**: 81–82.

Carvajal A., Lanza I., de Arriba M.L., Rubio P., del Pozo M., Ackermann M., and Carmenes P., 1995, Preliminary characterisation of a Spanish isolate of porcine epidemic diarrhoea coronavirus. IVth National Virology Congress, Madrid, 21–23 September 1995.

Duarte M. and Laude H., 1994, Sequence of the spike protein of the porcine epidemic diarrhoea virus, *J. Gen. Virol.* **75**: 1195–1200.

Duarte M., Tobler K., Bridgen A., Rasschaert D., Ackermann M. and Laude H., 1994, Sequence analysis of the porcine epidemic diarrhea virus genome between the nucleocapsid and spike protein genes reveals a polymorphic ORF, *Virology* **198**: 466–476.

Eleouet J.F., Rasschaert D., Lambert P., Levy L., Vende P. and Laude H., 1995, Complete sequence (20 kilobases) of the polyprotein-encoding gene 1 of transmissible gastroenteritis virus, *Virology* **206**: 817–822.

Herold J., Raabe T., Schelle-Prinz B. and Siddell S.G., 1993, Nucleotide sequence of the human coronavirus 229E RNA polymerase locus, *Virology* **195**: 680–691.

Tobler K. and Ackermann M., 1995, PEDV leader and junction sites. In: Corona- and Related Viruses (P.J. Talbot and G.A. Levy, eds.), *Adv. Exp. Med. Biol.* **380**: 541–542.

102

GENETIC VARIATION IN THE PRRS VIRUS

Michael P. Murtaugh, Kay S. Faaberg, Judy Laber, Margaret Elam, and
Vivek Kapur

Department of Veterinary PathoBiology
University of Minnesota
1971 Commonwealth Avenue
St. Paul, Minnesota 55108

1. ABSTRACT

Porcine reproductive and respiratory syndrome (PRRS) is characterized by late-term
abortions and stillbirths in sows and respiratory difficulties in nursery pigs. The disease
appeared in Europe and North America at approximately the same time between 1985 and
1990. The PRRS virus was isolated shortly thereafter and demonstrated unexpectedly profound differences between European (Lelystad) and North American (VR2332) isolates as
measured by serological crossreactivity and nucleotide sequence similarity. In order to determine the amount of genetic variation in the PRRS virus and to understand the molecular mechanisms of viral evolution, nucleotide sequences of PRRS virus strains were
determined. Comparisons among ten U.S. strains showed that variation in primary nucleotide sequence between isolates ranged from 2.5% to 7.9% for ORFs 2–7. In contrast, Lelystad virus was, on average, 35% different from US clones. These results provided direct
molecular evidence that US and European PRRSV isolates represented genetically distinct
groups of the same viral family. A further analysis of more than 150 isolates in the United
States and Canada demonstrated that the PRRS virus in North America represents a single
large and diverse genetic group that is distinct from European forms of the virus.

2. INTRODUCTION

A new viral disease of pigs was detected in North America in 1987 (Hill, 1990), and
in Europe in 1990 (Paton et al., 1991). The disease, now known as porcine reproductive
and respiratory syndrome (PRRS), is characterized by late-term abortions and stillbirths in
sows, and respiratory difficulties in nursery pigs (Pol et al., 1991; Wensvoort et al., 1991;
Collins et al., 1992). The virus is recovered primarily from alveolar macrophages of in-

Coronaviruses and Arteriviruses, edited by Enjuanes *et al.*
Plenum Press, New York, 1998

fected swine and is a small, enveloped positive-stranded RNA virus (Benfield et al., 1992; Wensvoort et al., 1992a). It is classified in the Arteriviruses based on morphology, genome organization, transcriptional regulation, and macrophage specificity (Plagemann and Moennig, 1992). The Arterivirus genus includes the PRRS virus, equine arteritis virus (EAV, den Boon et al., 1991), lactate dehydrogenase-elevating virus (LDV, Plagemann and Moennig, 1992) and simian hemorrhagic fever virus (SHFV, Godeny et al., 1993).

The PRRS virus and other Arteriviruses share the genome organization and expression strategy of the Coronaviruses (Spaan et al., 1988, Plagemann and Moennig, 1992, Meulenberg et al., 1993, Murtaugh et al., 1995). Seven proteins are expressed from a nested set of RNA transcripts with overlapping 3' ends. The first two open reading frames (ORFs 1a and 1b) encode the viral RNA polymerase. ORFs 2, 3, and 4 encode proteins of unknown function, ORFs 5 and 6 appear to encode structural proteins associated with viral membranes, and ORF 7 is believed to encode a nucleocapsid protein.

The American and European isolates, VR-2332 and Lelystad virus (LV), respectively, show a high level of molecular and antigenic variation. In a comparison of 24 field sera and seven viral isolates from Europe and North America, Wensvoort et al. (1992b) found that the European and American isolates were immunologically distinct and had no common antigens. The complete 15 kb nucleotide sequence of a European isolate (Lelystad strain) of the PRRS virus was determined by Meulenberg et al. (1993), and the 3'-hydroxy terminal region was recently characterized from a U.S. isolate (VR-2332) by Murtaugh et al. (1995). These studies showed that while European and U.S. isolates of the PRRS virus may have evolved from a common ancestor, there was substantial variation at the nucleotide and amino acid levels. For instance, there was only 55% identity in the amino acid sequence of ORF 5 between the U.S. and European isolates (Murtaugh et al., 1995).

In order to determine the amount of genetic variation among isolates recovered from infected swine in the U.S., and to understand the molecular mechanisms that govern the evolution of the PRRS virus, nucleotide sequences from ORFs 2 through 7 of ten U.S. isolates were characterized. In addition, a 1102 base nucleotide sequence encoding ORF 5 and ORF 6 was determined from more than 150 additional U.S. and Canadian isolates in order to evaluate the population genetic structure of the PRRS virus in North America. The results revealed substantial nucleotide diversity among the individual isolates, and suggested that the virus is evolving by processes other than the simple accumulation of random neutral mutations. Moreover, the data indicate that intragenic recombination has played an important role in the genetic diversification of PRRS virus isolates, and that evolutionary pressure has acted to maintain primary amino acid structure of individual ORFs. Furthermore, the PRRS virus in North America represents a single large and diverse genetic group that is distinct from European forms of the virus.

3. MATERIALS AND METHODS

3.1. Viruses and Cells

The VR-2332 strain of the PRRS virus was a fourth cell culture passage of ATCC VR-2332 grown on CL2621 cells (Collins et al. 1992). The other virus strains were field isolates cloned by the University of Minnesota Veterinary Diagnostic Laboratory by limiting dilution (Bautista et al., 1993) or isolated on MA-104 cells or porcine alveolar macrophages (see Acknowledgment). Cells were cultured in MEM supplemented with 4–10% fetal calf serum in a 5% humidified CO_2 atmosphere at 37°C.

3.2. Isolation of RNA, RT-PCR, and DNA Sequencing

Total RNA from infected cell supernatants was isolated by acid guanidine phenol extraction as described (Chomczynski and Sacchi, 1987) using equal volumes of cell supernatant containing virus and the guanidine lysis solution. Reverse transcription was performed using random primers as directed by the manufacturer (Perkin-Elmer Cetus, Norwalk, CT). For PCR reactions, primers were synthesized based on VR-2332 sequence and designed to amplify regions encompassing each ORF. Briefly, 20 µl of cDNA was amplified in a 100 µl reaction volume as directed by the manufacturer. Nucleic acids were denatured at 94°C, followed by 30 cycles at 55°C for 30 seconds, 72°C for 45 seconds, and 93°C for 45 seconds. Amplified fragments were polished at 72°C for 10 minutes, and purified using SpinBind® columns (FMC, Rockland, ME). Automated DNA sequencing reactions were performed according to the directions provided with the Taq DyeDeoxy™ Terminator Cycle Sequencing Kit (Applied Biosystems Inc., Foster City, CA) using a PE 480 Thermocycler (Perkin Elmer, Foster City, CA). Sequencing primers were selected from sequences within the PCR fragments.

3.3. Analysis of Nucleotide Sequences

Sequence data were assembled and analyzed with the computer programs SEQMAN and MEGALIGN (DNASTAR Inc., Madison, WI). Phylogenetic analysis of the compiled sequences was performed with distance-matrix, maximum-likelihood, and parsimony methods, with PHYLIP 3.4 (J. Felsenstein, Department of Genetics, University of Washington, Seattle, WA), the programs NJTREE and NJBOOT (T.S. Whittam, Institute of Molecular Evolutionary Genetics, Pennsylvania State University), and PAUP 3.1.1 (Swofford, 1991). The robustness of the phylogenetic analysis and significance of the branch order was determined by bootstrap analysis (Felsenstein, 1985), with 1000–2000 replications. Statistical analysis for clustering of polymorphic sites was performed by Sawyer's method (Sawyer, 1989), and individual ORFs were analyzed for pairwise comparisons of the number of synonymous substitutions per synonymous site (d_S) and coding substitutions per nonsynonymous site (d_N) by the method of Nei and Gojobori (1986).

4. RESULTS

4.1. Properties of PRRS Virus Gene Sequences

Complete nucleotide sequences (range 3256–3260 bp) of ORFs 2 through 7 were determined for 10 PRRSV isolates (Kapur et al., 1996). Alignment of the full length nucleotide sequences resulted in a consensus sequence of 3266 bp with a total of 611 polymorphic nucleotide sites. The total nucleotide diversity ([1]) of individual ORFs ranged from 0.038 ± 0.003 for ORF 6 to 0.097 ± 0.007 for ORF 5, with a mean nucleotide diversity for all six ORFs of 0.060 ± 0.008. This indicates that there are substantial differences in the levels of total nucleotide diversity among the individual ORFs, and suggests that the virus is evolving by mechanisms other than the simple accumulation of random neutral mutations. For instance, it is possible that host selective pressure may play a role in enhancing the nucleotide substitution rate for specific ORFs. An alternate hypothesis is that intragenic recombination has occurred in specific regions of the viral genome, resulting in an elevated or depressed nucleotide substitution rate in these regions.

4.2. Patterns of Nucleotide Substitution

One method of determining whether there is evolutionary pressure on specific genes to mutate in response to host selection or increase the fitness of the organism is to obtain the ratio of the number of nucleotide substitutions that result in amino acid replacement (nonsynonymous substitutions) to those that do not (synonymous substitutions). Under the neutral theory of evolution, the levels of synonymous substitution are expected to be high compared with those that result in amino acid replacements. On the other hand, if there were a selective advantage to maintain or create amino acid polymorphisms, the level of synonymous substitutions should be lower than nonsynonymous substitutions. Since the number of nucleotide substitutions between all possible pairs of sequences is large, and in many instances there are two or more nucleotide differences within a single codon, it is difficult to distinguish between synonymous and nonsynonymous substitutions. Therefore, the numbers of synonymous changes per synonymous site (d_S) and coding changes per nonsynonymous site (d_N) were calculated by the unweighted pathway method of Nei and Gojobori (1986). The results indicate a three- to five-fold excess of synonymous over nonsynonymous substitutions in the six ORFs. For instance, in ORF 5, the d_S values were 0.213 ± 0.022 while the d_N values were 0.063 ± 0.006, a $d_S{:}d_N$ ratio of 3.38:1. Similarly, for ORF 6, the $d_S{:}d_N$ ratio was 5.6:1. This indicates that there is evolutionary pressure to conserve the primary structure of these viral proteins.

4.3. Phylogenetic Relationships among PRRSV Isolates

Pairwise comparisons of the 10 aligned nucleotide sequences revealed mean nucleotide divergence ranging from a low of 2.5% for the isolates NE1 and VR2332 to a high of 7.9% (Kapur et al., 1996). An unrooted bootstrap phylogenetic tree for the complete region sequenced indicates that the ten US isolates grouped together. Phylogenetic trees constructed with the neighbor-joining and maximum-likelihood methods gave similar results. When the nucleotide sequence from the corresponding region of the single European PRRS virus isolate (Lelystad strain) previously characterized (Muelenberg et al., 1993) was included in the analysis, it always clustered as an outgroup with a branch length of, on average, more than an order of magnitude longer than the distance between any two of the U.S. isolates.

4.4. Intragenic Recombination in PRRS Virus Genes

Prompted by the finding of heterogeneity in nucleotide substitution rates among the PRRSV gene sequences, we examined the pattern of distribution of polymorphic sites in individual ORF alleles for evidence of intragenic recombination or gene conversion. An analysis of the 611 polymorphic nucleotide sites in the aligned sequences by the method of Sawyer (1989), revealed strong evidence for intragenic recombination in the PRRSV genome. In addition, a plot of the distribution of observed gap lengths and the probability of occurrence of the same gap length in 10,000 randomly permuted sequences reveals the presence of several fragments between 9 and 68 nucleotides in length that may represent intragenic recombination events.

In order to identify specific genes of the PRRS virus genome in which intragenic recombinational events may have occurred, the six individual ORFs were analyzed by Sawyer's method. The data provide statistical evidence for intragenic recombination in ORFs 2, 3, 4, 5, and 7, but not ORF 6. Further, an examination of the pattern of variation among the 10 PRRSV sequences by the construction of a haplotype plot revealed several possible intragenic recombinational events (Kapur et al., 1996). This indicates that some regions of the PRRS virus genome may have different evolutionary histories. Simply put, the PRRSV genome may represent a mosaic of RNA fragments from several distinct genetic backgrounds.

4.5. Taken Together, the Data Indicate that a Majority of the Genes in the PRRS Virus Genome Are Likely to Have Had One or More Episodes of Intragenic Recombination

An important implication of these results is that the construction of phylogenetic trees of ORFs 2, 3, 4, 5 and 7, all of which have evidence of recombination, may not be of value in determining the evolutionary history of this virus. It is likely that regions such as ORF 6 will be more informative for the purposes of reconstructing the true evolutionary history of the species.

4.6. Genetic Variation in North America

ORFs 5 and 6 comprise a total of approximately 1100 bases, or about 10% of the viral genome. They encode the two major proteins on the surface of the virus, existing as a disulfide-linked heterodimer, and thus are important targets for the swine immune system. In addition, they are interesting genetically since the ORF 5 protein product (the major envelope glycoprotein) is the most variable protein in the virus, and ORF 6 (encoding the integral membrane protein) is the most conserved protein and shows no statistical evidence of genetic recombination. Since it does not show evidence of recombination, the ORF 6 sequence is particularly useful for analysis of phylogenetic relationships among isolates.

Initial assessments of relatedness using PAUP 3.1.1 revealed that about 75 sequences were closely related to VR2332 and these were removed from the dataset since they might not be independent isolates and could be represented by VR2332. Other sequences which were obviously related by virtue of genetic similarity and isolated from the same location at the same time were also removed. The remaining 82 isolates which were thought to be independent were analyzed by bootstrap comparison of the ORF 5 and the ORF 6 regions, separately. The resulting dendrograms showed relatedness trees in which less than 10% of the branches appeared in 80% or more of the replicates. The fact that other branching patterns were obtained at a high frequency indicates that the isolates are interrelated. The detailed branching pattern is not identical for each ORF, also indicating that the PRRS virus in North America comprises a genetically related population of isolates in which virtually any isolate could suitably represent a typical PRRS virus. As observed previously, the phylogenetic analyses clearly showed that the PRRS virus in North America were related to each other and distinctly different from European isolates. Pairwise comparisons of North American isolates showed in most cases 10% or less sequence variation in ORF 5 and 5% or less in ORF 6. By contrast, the difference between Lelystad virus and VR2332 is about 45% in ORF 5 and 20% in ORF 6.

To gain further insight into the stability of the viral genome and the mechanisms by which the PRRS virus acquires genetic variation, the ORFs 5 and 6 genes were sequenced after an isolate was passaged approximately 70 times in cell culture and after it was sequentially passed six times through pigs. Nucleotide substitutions were observed at 5 locations, two in ORF 5 and three in ORF 6, and all of the nucleotide changes resulted in amino acid changes. The high rate of nonsynonymous changes was not expected, particularly in the ORF 6 gene (Kapur et al. 1996). Interestingly, an analysis of two related isolates which were obtained from the same farm at different times also revealed a total of 6 nucleotide substitutions, all nonsynonymous, in ORF 5 and ORF 6, in which four of the changes were at the same residues as observed previously. The biological significance of these substitutions is not known, since the in vitro growth characteristics of the isolates are the same, and there is no correlation between amino acid sequence and pathogenicity in animals.

5. DISCUSSION

How old is the PRRS virus? Epidemiologic studies suggest that the PRRS virus has only been recently introduced into pigs (Hill, 1990, Paton et al., 1991). However, the age of the PRRS virus, that is, the time since it diverged from its most recent ancestor, remains unknown. Studies of molecular evolution have shown that the rate of nucleotide substitutions in genes and genomes may be used to provide estimates of the time of divergence of a species (Nei, 1987). For the influenza virus, the rates of synonymous substitution sites per year of the hemagglutinin, neuraminidase, and nonstructural protein genes were estimated to be 0.014, 0.011, and 0.009, respectively (Hayashida et al., 1985). Based on the assumption that the mutation rate of ORF 6 represents the overall mutation rate for the virus in the absence of recombination, and that the overall synonymous nucleotide substitution rate is the same for influenza and PRRS viruses, we estimate that the total amount of nucleotide variation in the sample of ten isolates that were examined can be accounted for by 7 to 15 yr of virus evolution (Kapur et al., 1996). These results are consistent with the epidemiologic observations that suggest that the virus emerged as a swine pathogen approximately a decade ago. Estimates of the time of divergence of U.S. from European strains are more difficult to obtain since the levels of variation between the isolates are extremely high. For instance, the ORF 6 gene for the Lelystad virus and the US strains has, on average, only 45% sequence identity. Moreover, the degree and nature of genetic diversity among European PRRS virus isolates is also unknown.

In sum, our results indicate that PRRS virus isolates recovered from swine in the midwestern U.S. are evolving by both the accumulation of random neutral mutations, as well as by intragenic recombination. Further, the data indicate that there is substantial nucleotide variation between U.S. and European PRRS virus isolates. The striking differences in nucleotide sequence which were observed between the European Lelystad virus and the North American VR2332 ORFs 2 through 7 were unexpected when the first analyses were performed. The two isolates appeared at approximately the same time as the causative agent of a new disease in pigs. They shared similar, if not the same, biological characteristics, infection or immunization of pigs with one has a protective effect against challenge with the other, and there is some serological crossreactivity. Since the two isolates were presumed to be representative members of a new species of virus, now referred to as the PRRS virus, the genetic analysis raised the possibility that the virus was extremely heterogeneous with these two examples perhaps representing extremes in its genetic variation, or that there were genetically recognizable populations of the virus on different continents. With the additional nucleotide sequence information from a large sample of North American PRRS isolates, it now appears that the PRRS virus is represented by two subpopulations in which the North American isolates are all interrelated.

ACKNOWLEDGMENT

The authors thank the following individuals and institutions for providing PRRS virus isolates or nucleic acid sequences which were invaluable for the study: Dr. Beverly Schmitt of the National Veterinary Service Laboratory, Ames, Iowa, USA, Drs. Serge Dea and Carl Gagnon of the Institut Armand Frappier, Montreal, Quebec, Canada, Dr. John Harding of Animal Management Services, Humboldt, Saskatchewan, Canada, Dr. Michael Roof, NOBL Laboratories, Ames, Iowa, USA, and Drs. Sagar Goyal and Elida Bautista of the University of Minnesota Veterinary Diagnostic Laboratory, St. Paul, Minnesota, USA.

REFERENCES

Benfield, D. A., Nelson, E., Collins, J. E., Harris, L., Goyal, S. M., Robison, D., Christianson, W. T., Morrison, R. B., Gorcyca, D. E., and Chladek, D. W., 1992, Characterization of swine infertility and respiratory syndrome (SIRS) virus (isolate ATCC VR-2332), *J. Vet. Diagn. Invest.* **4**: 127–133.

Bautista, E. M. Goyal, S. G., Yoon, I. J., Joo, H.S.,. and Collins, J. E., 1993, Comparison of porcine aveolar macrophages and CL 2621 for the detection of porcine reproductive and respiratory syndrome (PRRS) virus and anti-PRRS antibody, *J. Vet. Diagn. Invest.* **5**:163–165.

Chomczynski, P. and Sacchi, N., 1987, Single-step method of RNA isolation by acid guanidinium thiocyanate-phenol-chloroform extraction, *Anal. Biochem.* **162**:156.

Collins, J. E., Benfield, D. A., Christianson, W. T., Harris, L., Hennings, J. C., Shaw, D. P., Goyal, S. M., McCullough, S., Morrison, R. B., Joo, H. S., Gorcyca, D. E., and Chladek, D. W., 1992, Isolation of swine infertility and respiratory syndrome virus (isolate ATCC VR-2332) in North America and experimental reproduction of the disease in gnotobiotic pigs, *J. Vet. Diagn. Invest.* **4**:117–126.

den Boon, J. A., Snijder, E. J., Chirnside, E. D., de Vries, A. A. F., Horzinek, M. C., and Spaan, W. J. M., 1991, Equine arteritis virus is not a togavirus but belongs to the coronavirus superfamily, *J. Virol.* **65**:2910–2920.

de Vries, A. A. F., Chirnside, E. D., Horzinek, M. C., and Rottier, P. J. M., 1992, Structural proteins of equine arteritis virus, *J. Virol.* **66**:6294–6303.

Felsenstein, J., 1985, Confidence limits on phylogenies: an approach using the bootstrap, *Evolution* **39**:783–791.

Godeny, E. K., Speicher, D. W., and Brinton, M. A., 1990, Map location of lactate dehydrogenase-elevating virus (LDV) capsid protein (Vp1) gene, *Virology* **177**:768–771.

Godeny, E. K., Zeng, L., Smith, S. L., and Brinton, M. A., 1993, in: *Proceedings of the 9th International Congress of Virology*, p. 22, August 8–13, Glasgow, Scotland.

Hayashida, H., Toh, H., Kikuno, R., and Miyata, T., 1985, Evolution of influenza virus genes, *Mol. Biol. Evol.* **2**:289–303.

Hill, H., 1990, Overview and history of mystery swine disease (swine infertility and respiratory syndrome), in: *Proceedings of the Mystery Swine Disease Committee meeting*, October 6, Denver CO, pp. 29–30. Livestock Conservation Institute, Madison, WI.

Kapur, V., Elam, M.R., Pawlovich, T. M., and Murtaugh, M.P., 1996, Genetic variation in porcine reproductive and respiratory syndrome virus isolates in the midwestern United States, *J. Gen. Virol.* **77**:1271–1276.

Kuo, L., Harty, J. T., Erickson, L., Palmer, G. A., and Plagemann, P. G. W., 1991, A nested set of eight mRNAs is formed in macrophages infected with lactate dehydrogenase-elevating virus, *J. Virol.* **65**:5118–5123.

Meulenberg, J. J. M., Hulst, M. M., de Meijer, E. J., Moonen, P. L. J. M., den Besten, A., de Kluyver, E. P., Wensvoort, G., and Moormann, R. J. M, 1993, Lelystad virus, the causative agent of porcine epidemic abortion and respiratory syndrome (PEARS), is related to LDV and EAV, *Virology* **192**:62–72.

Murtaugh, M. P., Elam, M. R., and Kakach, L. T, 1995, Comparison of the structural protein coding sequences of the VR-2332 and Lelystad strains of the PRRS virus, *Arch. Virol.* **140**:1451–1460.

Nei, M., and Gojobori, T., 1986, Simple methods for estimating the numbers of synonymous and nonsynonymous substitutions, *Mol. Biol. Evol.* **3**:418–426.

Paton, D. J., Brown, I. H., Edwards., S. and Wensvoort, G., 1991, Blue ear disease of pigs, *Vet. Record* **128**:617.

Plagemann, P. G. W. and Moennig, V., 1992, Lactate dehydrogenase-elevating virus, equine arteritis virus and simian hemorrhagic fever virus: a new group of positive-strand RNA viruses, *Adv. Virus Res.* **41**: 99–192.

Pol, J. M. A., van Dijk, J. E., Wensvoort, G., and Terpstra, C., 1991, Pathological, ultrastructural, and immunohistochemical changes caused by Lelystad virus in experimentally induced infections of mystery swine disease (synonym: porcine epidemic abortion and respiratory syndrome (PEARS), *Vet. Quarterly* **13**:137–143.

Spaan, W. J. M., Cavanagh, D. and Horzinek, M. C., 1988, Coronaviruses: structure and genome expression, *J. Gen. Virol.* **69**:2939–2952.

Sawyer, S., 1989, Statistical tests for detecting gene conversion, *Mol. Biol. Evol.* **6**:526–538.

Swofford, D.L., 1991, PAUP 3.1.1: *Phylogenetic analysis using parsimony*. Illinois Natural History Survey. Champaign, Ill. University of Illinois.

Wensvoort, G., Terpstra, C., Pol, J. M. A., ter Laak, E. A., Bloemraad, M., de Kluyver, E. P., Kragten, C., Van Buiten, L., den Besten, A., Wagenaar, F., Broekhuijsen, J. M., Moonen, P. L. J. M., Zetstra,T., de Boer, E. A., Tibben, H. J., de Jong, M. F., van't Veld, P., Groenland, G. J. R., van Gennep, J. A., Voets, M. T., Verheijden, J. H. M., and Braamskamp, J., 1991, Mystery swine disease in the Netherlands: the isolation of Lelystad virus, *Vet. Quarterly* **13**:121–130.

Wensvoort, G., de Kluyver, E. P., Pol, J. M. A., Wagenaar, F., Moormann, R. J. M., Hulst, M. M. Bloemraad, R., den Besten, A., Zetstra, T. and Terpstra, C. 1992a, Lelystad virus, the cause of porcine epidemic abortion and respiratory syndrome: a review of mystery swine disease research at Lelystad, *Vet. Microbiol.* **33**:185–193.

Wensvoort, G., de Kluyver, E. P., Luijtze, E. A., den Besten, A., Harris, L., Collins, J. E., Christianson, W. T. and Chladek, D., 1992b, Antigenic comparison of Lelystad virus and swine infertility and respiratory syndrome (SIRS) virus, *J. Vet. Diagn. Invest.* **4**:134–138.

103

SEQUENCE ANALYSIS OF THE NUCLEOCAPSID PROTEIN GENE OF THE PORCINE REPRODUCTIVE AND RESPIRATORY SYNDROME VIRUS TAIWAN MD-001 STRAIN

L. L. Chueh,[1] K. H. Lee,[1] F. I. Wang,[1] V. F. Pang,[1] and C. N. Weng[2]

[1]Department of Veterinary Medicine
National Taiwan University
142 Chou San Road, Taipei, Taiwan
[2]Pig Research Institute, Taiwan
Chunan, Miaoli, Taiwan

1. ABSTRACT

The 3'-portion of the genome from a Taiwan isolate of porcine reproductive and respiratory syndrome (PRRS) virus, strain MD-001, was cloned and sequenced. The resultant 549 nucleotides contained an open reading frame with a coding capacity of 123 amino acids (predicted Mr 13 600). The predicted protein corresponds to the nucleocapsid protein, the gene product of ORF7. Comparative sequence analysis of several known PRRSV strains indicated that this protein showed the highest degree of amino acid similarity to the US VR2332 and the Canadian IAF-Exp91 strains (92.7%) and the least to the Dutch Lelystad strain (56.5%). The phylogenic trees constructed on the basis of the known PRRSV nucleotide sequences indicated that MD-001 strain belongs to the North American strain cluster and that it is distinct from the European virus.

2. INTRODUCTION

Porcine reproductive and respiratory syndrome has emerged in the 1990s as an important new viral disease of swine. The disease, characterized by reproductive failure in sows and respiratory symptoms in pigs of all ages, has spread rapidly and caused alarm in the swine industry worldwide (Goyal, 1993; Done *et al.*, 1996). The causative agent, a small enveloped RNA virus, was first identified in 1991 in the Netherlands (Wensvoort *et*

Coronaviruses and Arteriviruses, edited by Enjuanes *et al.*
Plenum Press, New York, 1998

al.). Based on the size of the virion, viral genome organization, and replication strategy, the PRRS virus (PRRSV) has been provisionally classified in the Arterivirus group (Meulenberg *et al.*, 1993; Plagemann and Moennig, 1992; den Boon *et al.*, 1991; Spaan *et al.*, 1988).

The nucleocapsid (N) protein is encoded by ORF 7 which is located at the 3' end of the PRRSV genome. This protein is highly immunogenic (Yoon *et al.*, 1995). PRRS was first identified from an outbreak in Taiwan in 1991. The causative virus was isolated (Chang *et al.*, 1993a) and the disease subsequently reproduced in pathogen/free piglets (Chang *et al.*, 1993b). The present study reports the cloning and nucleotide sequence analysis of the nucleocapsid gene of this Taiwan isolate and its genetic relationship to the known PRRSV strains.

3. MATERIALS AND METHODS

3.1. Virus and Cells

Virus strain, MD-001 (Chang *et al.*, 1993), was grown in a continuous cell line (MARC-145). Virus stocks were prepared following three rounds of purification by limited dilution. After complete degeneration of the monolayers, supernatant fluid was clarified and the viral particles were concentrated by ultracentrifugation through a sucrose gradient.

3.2. RNA Preparation, cDNA Synthesis, and Cloning

The viral genomic RNA was isolated according to the method described (Chomczynki and Sacchi, 1987). To clone the 3' end of the viral genome, oligo (dT) primed cDNA was produced using the Gibco BRL cDNA synthesis kit. *Eco*RI /*Not*I adaptor ligation and phosphorylation were performed. Purified cDNA was cloned into the pUC19 plasmid vector, expanded in bacteria, purified and used directly for Southern hybridization.

3.3. Southern Hybridization

DIG-labeled DNA probes (Boehringer Mannheim) were prepared using RT-PCR from the viral genomic RNA as a template. Primers (F1: 5'GGGAATGGCCAGCCAGT-CAATCAACTGT3' and R1: 5'TGTAGAAGTCACGCGAATCAGGCGCACT3') targeted to nucleocapsid gene were modified from Suarez *et al.* (1994). Colony hybridization to screen the cDNA library was performed as described (Sambrook *et al.*, 1989). Positive colonies were selected and subjected to sequence analysis.

3.4. DNA Sequencing

cDNA clones were sequenced on both strands by the dideoxynucleotide chain termination method (Sanger *et al.*, 1977) using T7 DNA polymerase and an Automated Laser Fluorescent DNA sequence analyser (Pharmacia LKB). Sequence analysis was performed on an Apple Macintosh computer using the Seqman and Megalign sequence analysis programs (DNASTAR).

4. RESULTS AND DISCUSSION

The nucleotide sequence of the 3'-terminal 549 nt of the MD-001 strain of PRRSV was determined. This genomic region encompassed a large ORF encoding a polypeptide of 123 amino acids with a predicted Mr of 13 600, consistent with the estimated Mr of the nucleocapsid (N) protein (Meulenberg *et al.*, 1995; Mardassi *et al.*, 1994). Comparative sequence analysis with other known PRRSV strains indicated that the highest degree of amino acid similarity with the US VR2332 and the Canadian IAF-Exp91 strains (92.7%). The Japanese EDRD-1 together with the US VR2385 strains showed 91.9% amino acid similarity to the N protein of the MD-001 strain whereas all the European strains exhibited only 56.5 % similarity. Such relatively high divergence resulted from a number of nucleotide substitutions, insertions or deletions, making the MD-001 N protein five amino acids shorter than that of the European strains (Fig.1). Two amino acid streches, TAPM and QGAS, situated at the N-terminal and C-terminal regions respectively of the N protein of four analysed European isolates, were from the MD-001 N protein. These two missing amino acid domains were also observed in the American VR2332, Canadian IAF-exp91 and Japanese EDRD-1 strains (Fig.1).

The Clustal method was used to construct a phylogenetic tree of nine PRRSV isolates. The results indicated that these isolates can be divided into two proups. The first genotype was represent by European Lelystad, Boxmeer, No 1 and Olot/91 strains. Asian strains, MD-001 together with EDRD-1 strains, were in the same group as the North American strains and represented the second genotype (Fig. 2). The genetic relatedness of our PRRSV to the North American strain may be due to the importation of the carrier breeder pigs from those geographical areas.

ACKNOWLEDGMENTS

This work was supported by National Science Council of Republic of China project grant NSC85-2321-B-002-066-A13.

```
MD-001      MPNNNGKQQN KKK****GDG QPVNQLCQML GKIIAQQSQS RVKGPGRKNK KKNPEKPHFP LATEDDVRHH
IAF-exp91   ------R--K ---------- ---------- -------N-- -G----K--- ---------- ----------
VR2332      --------TE E--------- ---------- -------N-- -G----K--- ---------- ----------
VR2385      ----T----K R--------- ---------- -----H-N-- -G----K--- ---------- ----------
EDRD-1      --------K R-T-----N- ---------- ---------- -G----N--- ---------- ----Y-----
No1         -AGK-QS-KK --STAPMGN- --------L- -AM-KS-*** -QQPR-GQA- --K------- --A---I---
Olot/91     -AGK-QS-KK --SAAPMGN- --------L- -AM-KS-*** -QQPR-GQA- --K------- --A---I---
Boxmeer10   -AGK-QS-KK --STAPMGN- --------L- -AM-KS-*** -QQPR-GQA- --K------- --A---I---
Lelystad    -AGK-QS-KK --STAPMGN- --------L- -AM-KS-*** -QQPR-GQA- --K------- --A---I---


MD-001      FTPSERQLCL SSIQTAFNQG AGTCILSDSG RISYTVEFSL PTHHTVRLIR VTAPPSA   123
IAF-exp91   ---------- ---------- ----T----- ----A----- ---------- ---S---   123
VR2332      ---------- ---------- ----T----- ---------- ---------- ---S---   123
VR2385      ---------- ---------- ----T----- ---------- ---------- ---S---   123
EDRD-1      ---------- ---------- ----T----- ---------- ---------- ---S---   123
No1         L-QT--S--- Q--------- ---AS--S-- KV-FQ---M- -VA------- --STSASQGA S 128
Olot/91     L-QT--S--- Q--------- ---AS--S-- KV-FQ---M- -VA------- --STSASQGA S 128
Boxmeer10   L-QT--S--- Q--------- ---AS- S   KV-FQ---M- -VA------- --STSASQGA S 128
Lelystad    L-QT--S--- Q--------- ---AS--S-- KV-FQ---M- -VA------- --STSASQGA S 128
```

Figure 1. Alignment of amino acid sequences of the N gene of PRRSV. MD-001 sequence, strain IAF-exp91 (Mardassi *et al.*, 1995), VR2332 (Murtaugh *et al.*, 1995), VR2385 (Meng *et al.*, 1994), EDRD-1 (Saito *et al.*,1996), No1 (Drew *et al.*, 1996), Olot/91 (Plana *et al.*, 1996), Boxmeer 10 (Conzelmann *et al.*, 1993), and Lelystad (Meulenberg *et al.*, 1993). Deletions are indicated by (★) and identical residues are indicated by (−).

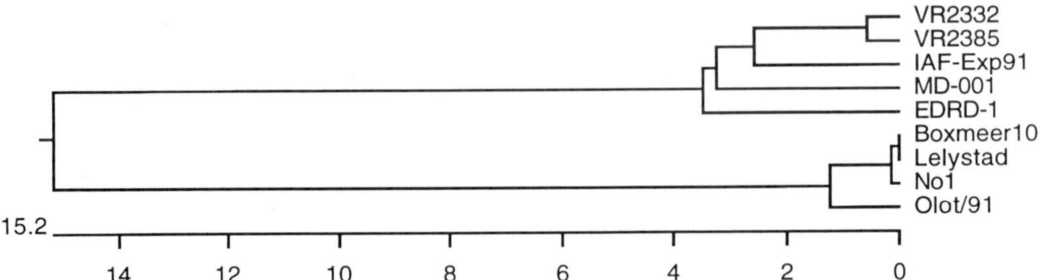

Figure 2. Phylogenetic trees based on the analysis of nucleotide sequences of the N gene of PRRSV. The tree was derived using the Clustal method and nucleotide sequence data obtained from EMBL+GeneBank for the American VR2332 (U00153) and VR2385 (U03040), Canadian IAF-Exp 91 (L40898), Taiwanese MD-001, Japanese EDRD-1 (D45852), Dutch Boxmeer 10 (L04493) and Lelystad (M96262), British No1 (L77924) and Spanish Olot/91(X92942) strains.

REFERENCES

Chang C.C., Chung W.B., Lin W.M., Weng C.N., Yang P.C., Chiu Y.T., Chang W.F. and Chu R.M., 1993a, Porcine reproductive and respiratory syndrome (PRRS) in Taiwan I. Viral isolation, *J. Chin. Soc. Vet. Sci.* **19**:268–276.

Chang C.C., Chung W.B., Lin W.M., Yang P.C., Weng C.N., Chiu Y.T., Chang W.F. and Chu R.M., 1993b, Porcine reproductive and respiratory syndrome (PRRS) in Taiwan II. Experimental reproduction of PRRS in specific-pathogen free (SPF) pigs, *J. Chin. Soc. Vet. Sci.* **19**:277–284.

Chomczynski P. and Sacchi N., 1987, Single-step method of RNA isolation by acid guanidium thiocyanate-phenol-chloroform extraction, *Anal. Biochem.* **162**:156–159.

Conzelmann K.K., Visser N., Woensel P.V. and Thiel H.J., 1993, Molecular characterization of porcine reproductive and respiratory syndrome virus, a member of the Arterivirus group, *Virology* **193**:329–339.

Den Boon J.A., Snijder E.J., Chirnside E.D., De Vries A.A.F., Horzinek M.C. and Spaan W.J.M., 1991, Equine arteritis virus is not a togavirus but belongs to the coronavirus-like superfamily, *J. Virol.* **65**:2910–2920.

Done S.H., Paton D.J. and White M.E.C., 1996, Porcine reproductive and respiratory syndrome (PRRS): A review, with emphasis on pathological, virological and diagnostic aspects, *Br. Vet. J.* **152**:153–174.

Drew T.W., Lowings J.P. and Yapp F., 1996, Variation in the open reading frame ORF3 and ORF7 of porcine reproductive and respiratory syndrome virus isolates in Great Britain. *Vet. Microbiol.* (in press).

Goyal S.M., 1993, Review article: Porcine reproductive and respiratory syndrome, *J. Vet. Diagn. Invest.* **5**: 656–664.

Mardassi H., Mounir S. and Dea S., 1994, Identification of major differences in the nucleocapsid protein genes of a Quebec strain and European strains of porcine reproductive and respiratory syndrome virus, *J. Gen. Virol.* **75**:681–685.

Mardassi H., Mounir S. and Dea S., 1995, Molecular analysis of the ORF 3 to 7 of porcine reproductive and respiratory syndrome virus, Quebec reference strain, *Arch. Virol.* **140**:1405–1418.

Mardassi H., Massie B. and Dea S., 1996, Intracellular synthesis, processing, and transport of proteins encoded by ORFs 5 to 7 of porcine reproductive and respiratory syndrome virus, *Virology* **221**:98–112.

Meng X.J., Paul P.S.and Halbur P.G., 1994, Molecular cloning and nucleotide sequencing of the 3'-terminal genomic RNA of the porcine reproductive and respiratory syndrome virus, *J. Gen. Virol.* **75**:1795–1801.

Meulenberg J.J.M., Hulst M.M., De Meijer E.J., Moonen P.L.J.M., Den Besten A., De Kluyver E.P., Wensvoort G. and Moormann R.J.M., 1993, Lelystad virus, the causative agent of porcine epidemic abortion and respiratory syndrome (PEARS), is related to LDV and EAV, *Virology* **192**:62–72.

Meulenberg J.J.M., Petersen-Den Besten A., De Kluyver E.P., Moonen R.J.M., Schaaper W.M.M. and Wensvoort G., 1995, Characterization of proteins encoded by ORF 2 to 7 of Lelystad virus, *Virology* **206**:155–163.

Murtaugh M.P., Elam M.R. and Kakach L.T., 1995, Comparison of the structural protein coding sequences of the VR-2332 and Lelystad virus strains of the PRRS virus, *Arch. Virol.* **140**:1451–1460.

Plagemann P.G.W. and Moennig V., 1992, Lactate dehydrogenase-elevating virus, equine arteritis virus and simian hemorrhagic fever virus, a new group of positive strand RNA virus, *Adv. Virus Res.* **41**:99–192.

Saito A., Kanno T., Murakami Y., Muramatsu M. and Yamaguchi S., 1996, Characteristics of major structural protein coding gene and leader-body sequence in subgenomic mRNA of porcine reproductive and respiratory syndrome virus isolated in Japan, *J. Vet. Med. Sci.* **58**:377–380.

Sambrook J., Fritsch E.F. and Maniatis T., 1989, *Molecular cloning: a laboratory manual.* 2nd ed., Cold Spring Harbor Laboratory Press, New York.

Sanger F., Nicklen S. and Coulson A.R., 1977, DNA sequencing with chain-terminating inhibitors, *Pro. Natl. Acad. Sci. USA* **74**:5463–5467.

Spaan W., Cavanagh D. and Horzinek M.C., 1988, Coronaviruses: structure and genome expression, *J. Gen. Virol.* **69**:2939–2952.

Suarez P., Zardoya R., Prieto C., Solana A., Tabares E., Bautista J.M. and Castro J.M., 1994, Direct detection of the porcine reproductive and respiratory syndrome (PRRS) virus by reverse polymerase chain reaction (RT-PCR), *Arch. Virol.* **135**: 89–99.

Wensvoort G., Terpstra C., Pol J.M., Ter Laak E.A., Bloemraad M., De Kluyver E.P., Kragten C., Van Butten L., Den Besten A., Wagenaar F., Broekhuijsen J.M., Moonen P.L.J.M., Zetstra T., De Boer E.A., Tibben H.J., De Jong M.F., Van't Veld P., Groenland G.J.R., Van Gennep J.A., Voets M.T., Verheijden J.H.M. and Braamskamp J., 1991, Mystery swine disease in the Netherlands: The isolation of Lelystad virus, *Vet. Quart.* **13**:121–130.

Yoon K.J., Zimmerman J.J., Swenson S.L., McGinley M.J., Eernisse K.A., Brevik A., Rhinehart L.L. Frey M.L., Hill H.T. and Platt K.B., 1995, Characterization of the humoral immune response to porcine reproductive and respiratory syndrome (PRRS) virus infection, *J. Vet. Diagn. Invest.* **7**:305–312.

104

COMPARISON OF THE DI- AND TRINUCLEOTIDE FREQUENCIES FROM THE GENOMES OF NINE DIFFERENT CORONAVIRUSES

Kurt Tobler* and Mathias Ackermann

Institute of Virology
Vet.-med. Faculty
University of Zurich
Winterthurerstrasse 266a, 8057 Zurich, Switzerland

1. ABSTRACT

As an alternative to protein alignments for the comparison of sequences, the reiterations of mono- di- and trinucleotide frequencies were used for the comparison of coronavirus sequences. The relative abundance of the di- and trinucleotide frequencies within the 3' part from nine coronavirus genomes were determined.

The patterns of dinucleotide frequencies and the trinucleotide frequencies showed some common features for all coronaviruses but also differences between the groups formerly defined on the base of antigenic relatedness.

The normalised dinucleotide frequencies were further used to calculate the distances between coronavirus sequences. Based on the dinucleotide frequency distances, coronaviruses can be divided into two groups which roughly reflect the taxonomic groups. In this kind of evaluation, however, IBV occupies a position different to the one that it would take based on most protein sequence comparisons. Based on similarities within coding sequences and antigenic properties IBV occupies a place outside of both groups. Based on the dinucleotide frequencies IBV gained a position in between of the TGEV-related and the MHV-clustered coronaviruses.

---

* Present address: BCBMB, Northwestern University, 2153, North Campus Drive, Evanston Illinois, 60208-3500.

Coronaviruses and Arteriviruses, edited by Enjuanes *et al.*
Plenum Press, New York, 1998

2. INTRODUCTION

Evolutionary comparisons of genomes are generally based on amino acid sequence comparisons of specific proteins. Since some proteins are more conserved than others the calculated relatedness of the sequences and the implied evolutionary relationship of the organisms are strongly dependent on the sequences used for the alignment.

An alternative for the evolutionary comparison of genomes is to analyse the frequencies of di- tri- or higher order oligonucleotides. Early biochemical experiments established that the set of dinucleotide frequencies is a remarkable stable property of the DNA of an organism (Josse et al., 1961).

The frequencies of the dinucleotides were shown to be similar in related and different in non-related organisms (Burge et al., 1992; Karlin et al., 1994). These kinds of genomic compositional inhomogeneities are widely recognised. The comparison of the frequencies mirrors the correlation of the sequences. A phylogenic relationship can be constructed on the base of these values. Since the function of a genome is not confined to encoding genes, the noncoding and untranslated regions turn out to be important as well (e.g. elements used for transcription and genome replication). These parts of the genome are considered by making dinucleotide comparisons.

3. MATERIALS AND METHODS

The published genomic nucleotide sequences of representative coronaviruses (Porcine epidemic diarrhoea virus (PEDV), Human coronavirus 229E (229E), Transmissible gastroenteritis virus (TGEV), Porcine respiratory coronavirus (PRCV), Canine coronavirus (CCV), Bovine coronavirus (BCV), Human coronavirus OC43 (OC43), Murine hepatitis virus A59 (A59), and Infectios bronchitis virus strain Beaudette (IBV)) beginning from the start codon of the S protein to the 3' end (but excluding the poly(A) tail) were included in this study. Then the frequencies of the mononucleotides, the dinucleotides, and the trinucleotides were determined. The dinucleotide and trinucleotide frequencies were then normalized by dividing them by the frequencies of the mononucleotides contained in the dinucleotides or trinucleotide, respectively. Values of ≤ 0.78 are considered as extreme under-representation, whereas values of ≥ 1.23 represent extreme over-representation of a given dinucleotide frequency (Karlin et al., 1994).

The sixteen normalized dinucleotide frequencies and the sixty-four trinucleotide frequencies were used in the further calculations. The equation below describes the distances of two nucleotide sequences g and f:

$$\delta(f, g) = \frac{1}{N} \sum \mathrm{abs}\left[\rho_{XY}(f) - \rho_{XY}(g)\right] \tag{1}$$

In this equation, the $\rho_{XY}(f)$ is the normalized frequency of the dinucleotide XY of the sequence f, and $\rho_{XY}(g)$ the normalized dinucleotide frequency from the dinucleotide XY of the sequence g, respectively. N is 16 for the number of dinucleotides. Similarly, a formula for the trinucleotide could be used. There, the sum of the positive differences of the corresponding trinucleotides is divided by 64. The resulting values describe the distances between two sequences based on the dinucleotide frequencies or trinucleotide frequencies. Lower values represent closer related sequences whereas higher values represent more distinct sequences.

Table 1. Selected ρ-values of dinucleotide and trinucleotide frequencies

	PEDV	229E	TGEV	PRCV	CCV	BCV	OC43	A59	IBV
CpG	0.60732	0.52635	0.49934	0.55922	0.50655	0.42774	0.46673	0.55559	0.51246
CpA	1.30596	1.31342	1.32916	1.27434	1.36635	1.16178	1.17393	1.15487	1.18162
CpC	0.84724	0.87599	0.87260	0.85933	0.89349	1.21401	1.16510	1.26549	1.08822
GpGpC	1.38944	1.37411	1.05993	1.19245	1.26368	1.14126	1.17165	1.42088	1.00900
CpCpC	0.66591	0.71401	0.67870	0.60824	0.82380	1.55631	1.40551	1.50988	1.08555
ApCpG	0.69475	0.72917	0.63374	0.77147	0.68903	0.38848	0.37223	0.41751	0.59202
ApCpA	1.42879	1.64026	1.66573	1.64538	1.66452	1.16011	1.14785	0.93200	1.18569
ApCpC	1.12254	1.23814	1.26219	1.11744	1.24829	1.45204	1.31410	1.43669	1.30236

The theoretical values for a random sequence would be 1.0 for every ρ-value. Values of [2]0.78 are considered as extreme underrepresentation, whereas values of [3]1.23 represent extreme over-representation of a given dinucleotide frequency.

4. RESULTS AND DISCUSSION

The dinucleotide frequencies and some specifically chosen trinucleotide frequencies are shown in Table 1. For all sequences included in the analysis, the normalized dinucleotide frequencies revealed a constantly high value for the dinucleotide UpG and a low abundance for the dinucleotide CpG. The normalised frequencies for the dinucleotide CpC are specifically high (around 1.2) for BCV, HCV OC43, and MHV A59 but low for the viruses PEDV, HCV 229E, TGEV, PRCV, and CCV (around 0.9). In contrast, the frequencies for the dinucleotide CpA are specifically high for the viruses PEDV, HCV 229E, TGEV, PRCV, CCV but lower for BCV, HCV OC43 and MHV A59.

Similar observations could be made by analysing the trinucleotide frequencies. The trinucleotides GpGpG, UpCpG, CpGpG, and CpGpU are underrepresented in all coronavirus genomes included in this analysis. In contrast, an overrepresentation of the trinucleotides GpUpG, GpCpA, CpApG, and CpApA, CpCpA is common for all coronavirus genomes. The values for the trinucleotides GpCpC, ApCpC and CpCpC are significantly higher in the genomes of BCV, HCV OC43, and MHV A59 than in the genomes of PEDV, HCV 229E, TGEV, PRCV, and CCV. In contrast, the trinucleotide frequencies of ApCpG and ApCpA are more abundant in the genomes of PEDV, HCV 229E, TGEV, PRCV, and CCV compared to the genomes of BCV, HCV OC43, and MHV A59.

In a consecutive analysis, the distances between the relative abundance of di- and trinucleotide frequencies of nine different coronaviruses were calculated. The distances based on the dinucleotide frequencies are presented in Table 2. Based on the distances of the dinucleotide frequencies coronaviruses can be divided into two groups which interestingly reflect the taxonomic groups. The group related to TGEV includes HCV229E, CCV, PRCV and PEDV while the other group related to MHV includes BCV and HCV OC43. In this kind of evaluation IBV occupies a position different to the one that it would take based on most protein sequence comparisons. Interestingly, based on similarities and antigenic properties IBV occupies a place outside both; but based on the dinucleotide frequencies the distances between IBV and the members of the MHV-related group are less than the distances to the members of the TGEV-clustered group and vice versa. According to the dinucleotide analyses, the ancestors of IBV might have branched later from the ancestors of the MHV-related group than the ancestors of the TGEV-clustered group.

It was shown by S. Karlin that the sequence lengths used for the analyses have only a minor effect on the results of the dinucleotide frequencies (Karlin and Burge, 1995).

Table 2. The δ-values calculated between coronavirus genomes

	PEDV	229E	TGEV	PRCV	CCV	BCV	OC43	A59	IBV
PEDV	0	0.058	0.049	0.044	0.061	0.090	0.080	0.102	0.078
229E		0	0.048	0.049	0.051	0.116	0.105	0.122	0.087
TGEV			0	0.020	0.026	0.079	0.072	0.100	0.061
PRCV				0	0.037	0.083	0.078	0.096	0.065
CCV					0	0.089	0.084	0.103	0.075
BCV						0	0.023	0.061	0.069
OC43							0	0.051	0.059
A59								0	0.079
IBV									0

The sequence from the start codon of the S gene to the poly(A) tail were used for the calculation of the ρ-values. Value interpretation: closely related ($\delta < 0.05$); moderately related (δ range from 0.05 to 0.075); weakly related (δ range from 0.075 to 0.1); distantly related ($\delta > 0.1$).

The same was observed for the coronavirus sequences. The normalized dinucleotide frequencies for the N gene of the viruses, for the entire genome if available, or the part of the genome from the S gene to the poly(A) tail are similar (data not shown). The longer the sequence used for calculating the frequencies, the more accurate are the values of the frequencies. The stretches of sequence used for the dinucleotide analyses of the coronaviruses are small compared to the amount of sequence taken for the comparisons of large genomes such as herpesviruses, bacteria, or eukaryotes.

ACKNOWLEDGMENTS

These studies were supported by the Swiss National Science Foundation, grant #31–43503.95.

REFERENCES

Burge C., Campbell A.M. and Karlin S., 1992, Over- and under-representation of short oligonucleotides in DNA sequences, *Proc. Natl. Acad. Sci. USA* **89**: 1358–1362.

Josse J., Kaiser A.D. and Kornberg A., 1961, *J. Biol. Chem.* **236**: 864–875.

Karlin S. and Burge C., 1995, Dinucleotide relative abundance extremes: a genomic signature, *Trends in Genetics* **11**: 283–290.

Karlin S., Ladunga I. and Blaisdell B.E., 1994, Heterogeneity of genomes: measures and values, *Proc. Natl. Acad. Sci. USA* **91**: 12837–12841.

105

SEQUENCE DETERMINATION AND GENETIC ANALYSIS OF THE LEADER REGION OF VARIOUS EQUINE ARTERITIS VIRUS ISOLATES

A. Kheyar, G. St-Laurent, M. Diouri, J. Dufresne, and D. Archambault

Université du Québec à Montréal
Département des Sciences Biologiques
C. P. 8888, Succursale Centre-Ville
Montréal, Québec, Canada, H3C 3P8

1. ABSTRACT

The entire leader sequence of ten equine arteritis virus (EAV) isolates including the Bucyrus reference strain was determined and analyzed at the primary nucleotide and secondary structure levels. The leader sequence of eight EAV isolates was determined to be 206 nucleotides (nt) in length, whereas those of the 86AB-A1 and 86NY-A1 isolates were found to be 205 and 207 nt in length, respectively. The sequence identity of the leader sequences between the different isolates and the Bucyrus reference strain ranged from 94.2 to 98.5%. An AUG start codon found at position 14 in all EAV isolates could initiate an open reading frame (ORF) that could produce a polypeptide of 37 amino acids, except for the 86NY-A1 isolate where the intraleader polypeptide would contain 54 amino acids. Five patterns of computer-predicted RNA secondary structures were identified in the ten EAV leader regions analyzed. All EAV isolates showed three conserved stem-loops (designated A, B and C). An additional conserved stem-loop (D) was observed in six EAV isolates, including the Bucyrus reference strain. Based on the presence or absence of stem-loop D, all EAV isolates analyzed in this study could be tentatively classified into two genogroups (I and II). The significance of the intraleader ORF and the predicted secondary structures has yet to be determined.

2. INTRODUCTION

Equine arteritis virus (EAV) is the etiologic agent of equine viral arteritis. The clinical outcome following EAV exposure varies widely from subclinical infection to EAV sys-

Coronaviruses and Arteriviruses, edited by Enjuanes *et al.*
Plenum Press, New York, 1998

temic disease of variable intensity, indicating that virulence varies among EAV isolates (McCollum and Swertzek, 1978). Abortion is common when pregnant mares become infected (McCollum and Swertzek, 1978).

EAV genome is a positive, polyadenylated, single-stranded RNA of 12.7 kb in length (Den Boon et al., 1991). During EAV replication, a 3' end-coterminal nested set of seven virus-specific RNAs (ORFs 1 to 7) is produced. Each ORF is preceded by the sequence motif 5'-UCAAC- 3', designated the leader-body junction site, which is involved in the formation of six subgenomic EAV RNAs with a common leader sequence derived from the extreme 5' end of the viral genome (De Vries et al., 1990). ORFs 1a/1b encode for the viral replicase, and ORF 3- and 4-encoded products are believed to be glycosylated non-structural proteins (Den Boon et., 1991; De Vries et al., 1992). Four viron structural proteins are produced from ORFs 2, 5, 6 and 7. ORFs 6 and 7 encode an unglycosylated membrane (M) protein of 16 kDa and a 14 kda nucleocapsid (N) protein, respectively. ORFs 2 and 5 encode the glycosylated 25 kDa small (G_S) membrane protein and the heterogeneously glycosylated 30- to 42-kDa large (G_L) membrane protein, respectively (De Vries et al., 1992).

Very little information is available for the EAV leader region. We recently determined the sequence of the terminal 5' leader region of four isolates of EAV, including the Bucyrus reference strain (Kheyar et al., 1996). We, and other investigators (Van Dinten et al., 1997), showed that the leader region is 206 nucleotides (nt) in length (not including a putative 5' cap structure-associated nt). We also identified an intraleader ORF of 111 nt in length (Kheyar et al., 1996).

Leader sequences of several other families of RNA positive-sense viruses such as picornaviruses (Duke et al., 1992; Haller et al., 1996), flaviviruses (Fukushi et al., 1994) or coronaviruses (Chen and Baric, 1995; Hofmann et al., 1993) have been shown to be involved in the regulation of critical viral functions (such as replication, transcription and/or translation) by the involvement of either intraleader ORFs or sequences that, by intramolecular base pairing, form unique secondary and tertiary structures. These considerations prompted us to extend our analysis of the EAV leader region. The primary goal of this sudy was to determine the complete leader sequence of the ATCC Bucyrus strain of EAV and of nine EAV field isolates originating from both different geographic regions (United States, n = 3; Canada, n = 5; and Austria, n = 1) and years of isolation to allow a comparative genetic analysis. Second, we wished to compare the predicted RNA secondary structures in the leader region of these isolates and consider how these structures might have a role in EAV biogenesis.

3. MATERIALS AND METHODS

The EAV isolates used in this study are listed in Table 1. Virus propagation and EAV genomic RNA extraction of each isolate were performed as described (Kheyar et al., 1996). The Rapid Amplification of cDNA ends (RACE) method based on the single strand ligation to single-stranded cDNA (SLIC) was used to obtain sequences of the 5' end of each EAV isolate genome (Kheyar et al., 1996). EAV genomic RNA was reverse transcribed using the procedure described in the Amplifinder Race Kit (Clontech Laboratories, Palo Alto, CA). The reverse transcription reaction was primed with the antisense oligonucleotide primer PEV-L1, which is complementary to nt 306 to 323 of the EAV genome (Den Boon et al., 1991). After RNA hydrolysis, the single-stranded cDNA was purified and ligated to the 3' end blocked AmpliFINDER anchor sequence with T4 RNA ligase.

Table 1. Characteristics of the EAV isolates used in this study

Isolate	Origin (year of isolation)	Source	Passage history*
Bucyrus**	Ohio, USA, (1953)	Fetus lung	Horse P14/LLC-MK2, P1
87AR-A1	Arizona, USA, (1987)	Semen	RK, P4/V, P7
86AB-A1	Alberta, Canada, (1986)	Fetus	RK, P4/V, P6
86NY-A1	New-York, USA, (1986)	Semen	ED, P4/V, P2
84KY-A1	Kentucky, USA(1984)	Nasal swab	RK, P5
Vienna	Austria (1968)	Nasal swab	ED, P1/RK, P2
19933	Guelph, Canada, (1992)	Semen	RK, P5
15492	Guelph, Canada, (1991)	Semen	RK, P5
11958	Guelph, Canada, (1990)	Semen	RK, P5
T1329	Guelph, Canada, (1988)	Neonatal lung***	RK, P5

*Cells: ED; equine dermis, HK; primary horse kidney, LLC-MK2; Rhesus monkey kidney, RK; rabbit kidney-13, V; Vero, P; refers to passage number.
**ATCC number: VR-796.
***EAV was isolated from 5 days old standardbred foal.

The single stranded ligation product was then amplified in the polymerase chain reaction (PCR) procedure by using the antisense primer PEV-L02, which is complementary to an internal sequence (nt 165 to 145) of the EAV leader sequence (Den Boon et al., 1991), and the sense AmpliFINDER anchor primer (AFAP). A second set of primers (sense, nt 1 to 18; antisense, complementary to nt 165 to 145) was selected according to den Boon et al. (1991) and was used to obtain, by PCR, overlapping cDNA fragments containing the remainder of the leader sequence at the 3' end. PCR amplifications were performed with the Taq polymerase (Promega, Madison, WI). The gel-purified PCR cDNA products for each isolate were cloned into Sma I-cleaved pBluescript II KS+ (Stratagene, La Jolla, CA) or PCR™II (Invitrogen, San Diego, CA) plasmid vectors. Two or more clones of each PCR products (both strands) were sequenced using the Sanger dideoxynucleotide chain termination method (Sanger et al., 1977).

Comparison and multiple alignments of nucleic acid sequences were carried out with the University of Wisconsin Genetics Computer Groups software package (GCG, version 8.1). Putative secondary structures of the leader sequence of all EAV isolates exhibiting minimum-free energies were predicted with the FOLD and SQUIGGLES graphics output programs of the GCG software package.

4. RESULTS

The complete nucleotide sequences of the leader region of all EAV isolates including that of the ATCC Bucyrus strain are depicted in Figure 1. The RACE method successfully identified the extreme 5' ends of the genomes of all of the EAV isolates studied and revealed the presence of 17 additional nt for nine EAV isolates, not previously identified, located upstream of the published sequence (Den Boon et al., 1991). In the case of the 86AB-A1 isolate, the first of these 17 nt was missing. The 17 base sequence in the T1329 isolate was identical to that of the Bucyrus reference strain. A G→A substitution at position 1 was observed for the 87AR-A1, Vienna and 15492 isolates. Additional base substitutions of A→G at positions 6 (86AB-A1 isolate) or 7 (86AB-A1, 84KY-A1, 86NY-A1, 11958 and 19933 isolates), and C→G or T at position 2 (86AB-A1 and 19933 isolates, respectively)

```
BUCYRUS   GCTCGAAGTG TGTATGGTGC CATATACGGC TCACCACCAT ATACACTGCA    50
87AR-A1   A......... .......... .......... .....G.... ..G.......
86AB-A1   -G...GG... .......... .......... .......... ..G.......
86NY-A1   ......G... .......... .......... .......... ..........
84KY-A1   ......G... .......... .......... .......... ..........
VIENNA    A......... .......... .......... ........G. ..........
19933     .T....G... .......... .......... ..C..G.... ..........
15492     A......... .......... .......... .....G.... ..........
11958     ......G... .......... .......... .......... ..........
T1329     .......... .......... .......... .....G.... ..........

BUCYRUS   AGAATTACTA TTCTTGTGGG CCCCTCTCGG TAAATCCTAG AGGGCTTTCC   100
87AR-A1   .......... .......... .......... .......... ..........
86AB-A1   .......... .......... .......... .......... ..........
86NY-A1   .......... .......... .......... .......... .........T.
84KY-A1   .......... .......... .......T.. .......... ..........
VIENNA    .......... .......... .......T.. ...T...... ..........
19933     .......... .......... .......... .......... ..........
15492     .......... .......... .......... .......... ..........
11958     .......... .......... .......... .G........ ..........
T1329     .......... .......... .......T.. ...C...... ..........

BUCYRUS   TCTCGTTATT GCGAGATTCG TCGTTAGATA ACGGCAAGTT -CCCTTTCTT   150
87AR-A1   .......... ........T. .......... .......... -.........
86AB-A1   .......... .......... .......... .......... -.........
86NY-A1   .......C. .......... .....GT... .......... C.........
84KY-A1   .......... .......... .......... .......... -.........
VIENNA    .......... .......... .......T.. .....-.T.. C..TA.....
19933     .......... .......... .....A... .......... -.........
15492     .......... .......... .......... .........C -..T......
11958     .......... .......... .......... .......... -........A
T1329     .......... .......... .......... ......A.. -.........

BUCYRUS   ACTATCCTAT TTTCATCTTG TGGCTTGACG GGTCACTGCC ATCGTCGTCG   200
87AR-A1   .......... .......... .......... .......... ..........
86AB-A1   .......... .......... .......... .......... ..........
86NY-A1   ...T...... .......... .......... .......... ..........
84KY-A1   ...T...... .......... .......... .A........ ..........
VIENNA    .TCT...... .......... .......... A ......... ..........
19933     ...T...... .......... .......... .......... ..........
15492     ...T...... .......... .......... .......... ..........
11958     .......... .......... .......... .......... ..........
T1329     ...T...... .......... .......... .......... ..........

BUCYRUS   ATCTCTA
87AR-A1   .......
86AB-A1   .......
86NY-A1   .......
84KY-A1   .......
VIENNA    .......
19933     .......
15492     .......
11958     .......
T1329     .......
```

Figure 1. Alignment of the nucleotide sequences of the leader region of nine arteritis virus isolates with the Bucyrus reference strain. AUG start codons are double-underlined and in-frame stop codons are also underlined. The GenBank accession numbers of the different isolates are the following: AF001259 (ATCC Bucyrus); U46946 (87AR-A1); U65723 (86AB-A1); U46945 (86NY-A1); U65724 (84KY-A1); U46947 (Vienna); U65728 (19933); U65727 (15492); U65726 (11958); U65725 (T1329).

were also observed by comparaison with the Bucyrus reference strain. The identification of these additional nt brings the length of the leader sequence to 206 nt without the junction site motif and the putative 5' cap structure-associated nt for most EAV isolates, except the 86AB-A1 isolate which is 205 nt. Another exception is the 86NY-A1 isolate, which has a leader region of 207 nt in length due to a C insertion at position 141.

When the leader sequences were aligned and compared to that of the Bucyrus reference strain, a total of 29 base positions varied among isolates (14% of the 206 nt positions) (Figure 1). Beside the substitutions mentioned above that occurred at the extreme 5' end of the leader region, base substitutions were common at particular positions. For instance, substitutions of A→G at positions 36 and 43, C→T at position 78, and A→T at position 154 were observed with regularity. Such base mutations are common in RNA viruses (Domingo et al., 1985).

The levels of identity for the leader sequence of the nine EAV field isolates to the Bucyrus reference strain ranged from 94.2 (Vienna isolate) to 98.5% (11958 isolate) (data not shown). When the leader sequences of the EAV field isolates were compared to each other, both the 86AB-A1 and 11958, and the 84KY-A1 and T1329 isolates were the most closely related at 97.6% identity. Sequences of the North American isolates (five Canadian and four American, including the Bucyrus strain) were 95.2 to 98.5% identical when compared to each other, while all were less closely related to the Vienna European isolate where the identity of the sequence ranged from 92.2 (86NY-A1 and 19933 isolates) to 94.6% (T1329 and 15492 isolates).

A striking feature in all of the EAV leader sequences was the presence of a AUG start codon located at positions 14 to 16. This start codon has not been previously identified because it is located upstream of the published sequence (den Boon et al., 1991). The presence of an amber (UAG) or an ochre (UAA) stop codon for all EAV isolates, except the 86NY-A1 isolate, at positions 125 to 127 in frame with the first start codon predicts an ORF of 111 nt that could encode a small polypeptide of 37 amino acids. In contrast, the 86NY-A1 isolate contains an opal (UGA) stop codon at positions 176 to 178, so that its intraleader ORF is 162 nt in length and the predicted polypeptide would include 54 amino acids. When the sequences of the predicted 37 amino acid peptides for all EAV isolates were aligned and compared to that of the Bucyrus reference strain, only a few amino acid substitutions were observed and, in most cases, were common among the EAV field isolates (for instance, Arg→His at position 8, Met→Ile at position 10, and Ser→Leu at position 22) (data not shown). The significance of these variations is not known.

The putative secondary structures of the leader sequence of all EAV isolates are depicted in Figure 2. Based on the number and locations of predicted stem-loop structures, five different secondary structure patterns were obtained. Three of the predicted stem-loop structures, which would form in the first half of the leader region at nt 7 to nt 100, designated A, B and C, were found in all of the EAV isolates. Moreover, a fourth stem-loop, designated D, was predicted for the six isolates comprising patterns 1 and 2. This suggests that the EAV isolates might segregate into two major groups (I and II) based on the presence or absence of stem-loop D.

Stem-loop A (41 nt in length) forms by base-pairing nucleotides from nt 7 to nt 47, stem-loop B (20 nt in length) uses nt 48 to nt 67, stem-loop C (33 nt in length) uses nt 68 to nt 100 (except for the 86NY-A1 isolate where a shorter stem-loop C is formed from nt 72 to nt 94), and stem-loop D (16 nt in length) uses nt 101 to nt 116. For most isolates, these sets of three or four stem-loop structures were branched out from an internal loop designated I. Although some nucleic acid substitutions were observed within the se-

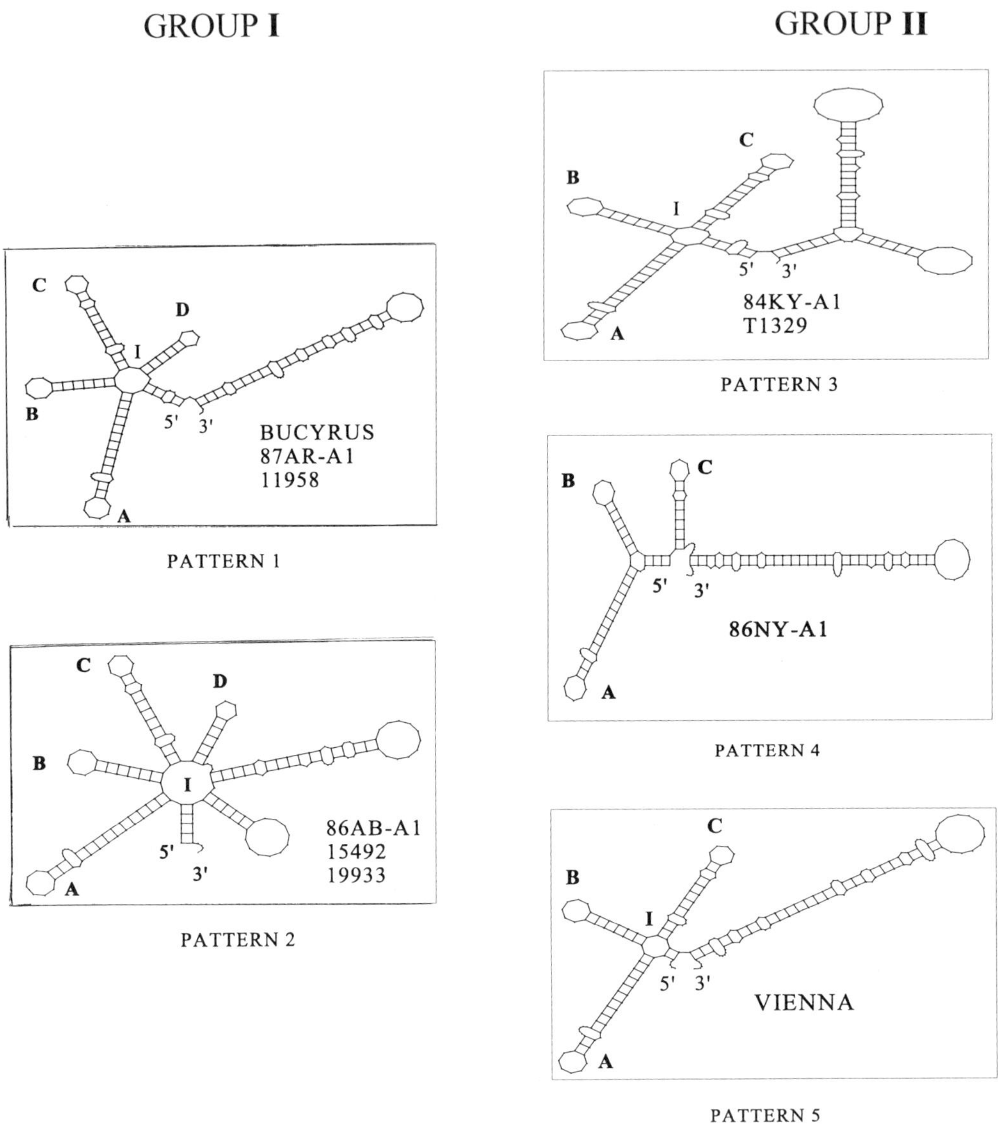

Figure 2. Predicted secondary structure of the leader sequence of ten EAV isolates using the FOLD program of the GCG software package. The conserved stem-loops (A to D) and the internal loop (I) are designated with upper cases in parenthesis.

quence of all four of these stem-loops, the general form and the length of these stem-loops were unaltered regardless of the mutations (data not shown) indicating a high degree of structural conservation for the stem-loop structures of both groups I and II. Such conservation of the general form of these stem-loop structures indicates that most mutational changes are not tolerated and suggests an important functional role of these structural domains in EAV replication or biogenesis. Furthermore, the presence of stem-loop D in six EAV isolates is of interest because it might alter or modulate EAV replication and/or virulence.

5. DISCUSSION

In this study, the complete leader region sequences of ten EAV isolates, including the ATCC Bucyrus strain were determined.The results demonstrated a high degree of nucleic acid conservation in the leader sequences among EAV isolates. Moreover, an intraleader ORF associated to a AUG start codon at residues 14 to 16 was found in all EAV isolates. A second AUG codon located at residues 41 to 43 was also observed in two EAV isolates (87AR-A1 and 86AB-A1) (Figure 1). This second AUG codon observed was in frame with the first start codon, so that a shorter ORF (nt 41 to nt 125) could theoritically produce a 28 amino acid polypeptide. However, the primary sequence in the immediate vicinity of an AUG initiation codon is important for initiation of translation (Kozak, 1991), and it is unlikely that the AUG codon at positions 41 to 43 of the EAV leader sequence would act as a start codon because, with both positions -3 and +4 occupied by a C residue, it is unlikely that translation could be initiated. In contrast, the AUG codon at positions 14 to 16 is likely to act as a functional start codon because, with the -3 and +4 positions occupied by U and G residue, respectively, it is capable, albeit at a suboptimal level, of initiating translation. Experiments to test for the presence of functional intraleader ORFs within the EAV leader region with the use of antisera to synthetic oligopeptides are in progress in our laboratory. This work may be of some interest because, as shown for coronaviruses, an intraleader ORF was reported to modulate virus translation efficiency either by attenuating (in bovine coronavirus, Hofmann et al., 1993) or by enhancing (in murine coronavirus, Chen and Baric, 1995) the translation rate of downstream ORFs.

Analysis of putative secondary structures predicted the presence of three stem-loop structures (A, B and C) in all EAV isolates. Moreover, we could theoritically segregate the EAV isolates into two major groups based on the presence or absence of the fourth stem-loop D. The nucleotide sequences and predicted secondary structures downstream from either stem-loop C (from nt 100 for group I) or stem-loop D (from nt 117 for group II) were found much more variable than the 5' end (Figures 1 and 2). This variability is seen in the sequence following the stop codon at nt 125–127 through nt 154, and is reflected by a variable pattern of additional possible stem-loop structures that would vary from one isolate to another in terms of position and sequence. Variability in this portion of the leader sequence, in marked contrast to the conserved sequence and limited variability in the secondary structures predicted for the 5' end, suggests a lesser role for this region of the genome in regulation of EAV replication. Nevertheless, such variation might also be associated with differences in EAV virulence, as seen elsewhere where variations in the secondary structure of the leader sequence of echovirus and Sabin poliovirus reflected variation in neurovirulence and the level of virus attenuation in tissue cultures, respectively (Macadam et al., 1992; Romero and Rotbart, 1995).

The role of secondary structures in leader sequences in the initiation of translation of eukaryotic mRNAs has been well documented (Kozak, 1991). Moreover, in picornaviruses (e.g. Theiler's virus), mutational analyses demonstrated that neurovirulence of certain strains was associated with a specific secondary structure in the leader sequence (Haller et al., 1996). Therefore, replication and translation strategies, as well as other features in EAV biogenesis such as virulence, might well be influenced by variations of the primary sequence, or RNA secondary structure within the leader region. Experiments with appropriate cDNA mutant clones are under way to investigate how the variation in EAV leader sequence secondary structure impacts on subgenomic mRNA translation efficiency.

ACKNOWLEDGMENTS

This work was supported by an operating grant from the National Sciences and Engineering Research Council of Canada to D. Archambault. A. Kheyar is supported by a graduate student fellowship from Université de Montréal. D. Archambault is the holder of a research scholarship from the Fonds de la Recherche en Santé du Québec (FRSQ). We thank Peter J. Timoney and William H. McCollum (Gluck Equine Research Center, Lexington, Kentucky), and Susy Carman (Veterinary Laboratory Services, Guelph, Ontario) for providing the EAV laboratory strains and field isolates. We are also grateful to John Dahlberg for reviewing the manuscript.

REFERENCES

Chen W. and Baric R.S., 1995, Function of a 5'-end genomic RNA mutation that evolves during persistent mouse hepatitis virus infection in vitro, *J. Virol.* **69:** 7529–7540.

Den Boon J.A., Snijder E.J., Chirnside E.D., De Vries A.A.F., Horzinek M.C. and Spaan W.J.M., 1991, Equine arteritis virus is not a togavirus but belongs to the coronaviruslike superfamily, *J. Virol.* **65:** 2910–2930.

De Vries A.A.F., Chirnside E.D., Bredenbeek P.J., Gravestein L.A., Horzinek M.C. and Spaan W.J.M., 1990, All subgenomic mRNAs of equine arteritis virus contain a common leader sequence, *Nucleic Ac. Res.* **18:** 3241–3247.

De Vries A.A.F., Chirnside E.D., Horzinek M.C. and Rottier P.J.M., 1992, Structural proteins of equine arteritis virus, *J. Virol.* **66:** 6294–6303.

Domingo E., Martinez-Salas E., Sobrino F., de la Torre J.C., Portela A., Ortin J., Lopez-Galindez C., Perez-Brena P., Villanueva N. and Najera R., 1985, The quasispecies (extremely heterogeneous) nature of viral RNA genome populations: biological relevance, a review, *Gene* **40:** 1–8.

Duke G.M., Hoffman, M.A. and Palmenberg A.C., 1992, Sequence and structural elements that contribute to efficient encephalomyocarditis virus RNA translation, *J. Virol.* **66:** 1602–1609.

Fukushi S., Katayama K., Kurihara C., Ishiyama N., Hoshini F.B., Ando T. and Oya A., 1994, Complete 5' noncoding region is necessary for the efficient internal initiation of hepatitis C virus RNA, *Biochem. Biophys. Res. Com.* **199:** 425–432.

Haller A.A., Stewart S.R. and Semler B.L., 1996, Attenuation stem-loop lesions in the 5' noncoding region of poliovirus RNA: neuronal cell-specific translation defects, *J. Virol.* **70:** 1467–1474.

Hofmann M.A., Senanayake S. and Brian D.A., 1993, A translation attenuating intraleader open reading frame is selected on coronavirus mRNAs during persistent infection, *Proc. Natl. Acad. Sci. USA* **90:** 11733–11737.

Kheyar A., St-Laurent G. and Archambault D., 1996, Sequence determination of the extreme 5' end of the equine arteritis virus leader region, *Virus Genes* **12:** 291–295.

Kozak M., 1991, An analysis of vertebrate mRNA sequences: intimations of translational control, *J. Cell Biol.* **115:** 887–903.

Macadam A.J., Ferguson G., Burlison J., Stone D., Skuce R., Almond J.W. and Minor P.D., 1992, Correlation of RNA secondary structure and attenuation of Sabin vaccine strains of poliovirus in tissue culture, *Virology* **189:** 415–422.

McCollum W.H. and Swertzek T.W., 1978, Studies of an epizootic of equine arteritis virus in racehorses, *J. Equine Med. Surg.* **2:** 459–464.

Romero J.R. and Rotbart H.A., 1995, Sequence analysis of the downstream 5' nontranslated region of seven echovirus with different neurovirulence phenotypes, *J. Virol.* **69:** 1370–1375.

Sanger F., Nicklen S. and Coulson A.R., 1977, DNA sequencing with chain-terminating inhibitors, *Proc. Natl. Acad. Sci. USA* **74:** 5463–5467.

Van Dinten, L.C., Den Boon J.A., Wassenaar A.L.M., Spaan W.J.M. and Snijder E.J., 1997, An infectious arterivirus cDNA clone: identification of a replicase point mutation that abolishes discontinuous mRNA transcription, *Proc. Natl. Acad. Sci. USA* **94:** 991–996.

106

GENETIC VARIATION AND PHYLOGENETIC ANALYSIS OF OPEN READING FRAMES 3 AND 4 OF VARIOUS EQUINE ARTERITIS VIRUS ISOLATES

D. Archambault, G. Laganière, and G. St-Laurent

Université du Québec à Montréal
Département des Sciences Biologiques
C.P. 8888, Succursale Centre-Ville
Montréal, Québec, Canada, H3C 3P8

1. ABSTRACT

The genetic variation in equine arteritis virus (EAV) nonstructural (NS) protein-encoding open reading frames (ORF) 3 and 4 genes was investigated. Nucleotide and deduced amino acid sequences from seven different EAV isolates (one European, one American and five Canadian isolates) and the Arvac vaccine strain were compared with those of the Bucyrus reference strain. ORF 3 nucleotide and amino acid sequence identities amongst these isolates (including the Arvac vaccine strain) and the Bucyrus reference strain ranged from 85.6 to 98.8%, and 85.3 to 98.2%, respectively, whereas ORF 4 nucleotide and amino acid sequence identities ranged from 90.4 to 98.3%, and 90.8 to 97.4%, respectively. Phylogenetic tree analysis based on the ORF 3 nucleotide sequences showed that the European Vienna isolate could be classified into a genetically divergent group from all other isolates and the Arvac vaccine strain. In contrast, a phylogenetic relationship among all EAV isolates and the Arvac vaccine strain based on the ORF 4 nucleotide sequences was observed.

2. INTRODUCTION

Equine arteritis virus (EAV) is the causative agent of equine viral arteritis, a debilitating respiratory disease which may cause abortion in pregnant mares (Timoney and McCollum, 1993). However, the clinical outcome following EAV exposure varies widely

Coronaviruses and Arteriviruses, edited by Enjuanes *et al.*
Plenum Press, New York, 1998

from subclinical infection to systemic EAV disease, indicating that virulence varies among EAV isolates (see Timoney and McCollum, 1993). The EAV genome is a polyadenylated positive single-stranded RNA of 12.7 kb in length, and eight open reading frames (ORFs) have been identified in EAV (Den Boon et al., 1991). ORF 1a/1b encoded for the viral polymerase, and ORFs 2, 5, 6, and 7 coded for EAV structural proteins (De Vries et al., 1992; Den Boon et al., 1991). These include a 14 kDa nucleocapsid protein (N, ORF 7), a 16 kDa unglycosylated membrane protein (M, ORF 6), a heterogeneously glycosylated 30 to 42 kDa large membrane protein (G_L, ORF 5), and a glycosylated 25-kDa small membrane protein (G_S, ORF 2). The ORF 3- and 4-encoded products are believed to be glycosylated nonstructural (NS) proteins (Den Boon et al., 1991; De Vries et al., 1992).

There is only one serotype identified in EAV (see Timoney and McCollum, 1993). However, RNA oligonucleotide fingerprint analysis has indicated genetic variation among various EAV isolates (Murphy et al., 1992). Comparison of M, N, G_L and G_S nucleotide sequences, whose products induce an antibody response (De Vries et al., 1992; Chirnside et al., 1995), has also demonstrated variation among EAV isolates (Chirnside et al., 1994; Balasuriya et al., 1995; Hedges et al., 1996; Lepage et al., 1996; St-Laurent et al., 1997). However, no sequence data are yet available for the EAV NS protein-encoding genes. Because the NS proteins are less likely to be exposed to immunological pressure, the ORF 3 and 4 genes might be relevant indicators of EAV random evolution. In this study, we report the characterization of the genetic variation and evolutionary relationships of various EAV isolates, including the Bucyrus reference (Den Boon et al., 1991) and the Arvac vaccine (Timoney and McCollum, 1993) strains, by comparison of the nucleotide sequences of ORFs 3 and 4. The deduced amino acid sequences of ORF 3 and 4-encoded proteins were also compared.

3. MATERIALS AND METHODS

The EAV isolates originated from Canada (n = 5), the United States (n = 1) and from Europe (n = 1) (Table 1). The Arvac vaccine EAV strain which is the attenuated form of the virulent Bucyrus strain was also included in this study. Virus propagation in cell cultures and EAV genomic RNA extraction of each isolate/strain were performed essentially as described (Lepage et al., 1996; St-Laurent et al., 1997).

Table 1. Characteristics of the equine arteritis virus (EAV) isolates used in this study

Isolate	Origin (year of isolation)	Source	Passage history*
T1329	Ontario, Canada (1988)	Neonatal lung**	RK, P5
11958	Ontario, Canada (1990)	Semen	RK, P5
15492	Ontario, Canada (1991)	Semen	RK, P5
19933	Ontario, Canada (1992)	Semen	RK, P5
Vienna	Vienna, Austria (1968)	Nasal swab	ED, P1/RK, P2
84KY-A1	Kentucky, U.S.A. (1984)	Nasal swab	RK, P5
86AB-A1	Alberta, Canada (1986)	Fetus	RK, P3/V, P1
Arvac***	Fort Dodge Laboratories	–	HK, p131/RK, p111/ED, p24/V, P2

*Cells: RK; rabbit kidney-13, ED; equine dermis, HK; primary horse kidney, V; Vero, P; refers to passage number.
**EAV was isolated from 5 day old standard bred foal.
***Vaccine strain of EAV.

The sense and antisense primer pairs [PEV-20 and PEV-21; (9807–9824, 10503–10486); PEV-30 and PEV-31 (10289–10305, 10783–10766); PEV-40 and PEV-41 (10683–10700, 11165–11148)] were used for reverse transcription-PCR (RT-PCR) amplification in order to obtain overlapping cDNA fragments which represent the entire nucleotide sequence of ORFs 3 and 4 (with an expected length of 492 and 459 bp, respectively). These primers were selected according to the nucleotide sequence of the EAV Bucyrus reference strain genome (Den Boon et al., 1991). RT-PCR assays were carried out as previously described (St-Laurent et al., 1994). The resulting amplified cDNA fragments of each EAV isolate were cloned, and sequenced using the Sanger dideoxynucleotide chain termination method (Sanger et al, 1977). Two or more independent cDNA clones were sequenced from RT-PCR products for each EAV isolate/strain.

Comparison and multiple alignments of nucleotide and deduced amino acid sequences were carried out with the University of Wisconsin Genetics Computer Groups software package (GCG, version 8.0). Phylogenetic analyses based on the analysis of ORF 3 and 4 nucleotide sequences were performed using the DNADIST and FITCH programs of the Phylogenic Inference Package (PHYLIP, version 3.5c) (Felsenstein, 1993). The ORFs 3 and 4 of lactate dehydrogenase-elevating virus (LDV) (Godeny et al., 1993) were used to generate outgroup-rooted trees. Bootstrap analysis was carried out on 1000 replicate datasets to assess the confidence limits of the branch pattern.

4. RESULTS

4.1. Nucleotide and Amino Acid Sequence Analysis of ORF 3

A total of 311 nucleotide mutations were observed within the ORF 3 sequence of all the EAV isolates the analyzed. Of the nucleotide substitutions observed, 45.3% lead to amino acid changes. The expected 492 bp ORF 3 gene was observed for all EAV isolates, except for the 11958 isolate and the Arvac vaccine strain whose ORF 3 length was determined to be 483 and 507 bp, respectively. This was due to nucleotide substitutions at position 481 (11958 isolate) which generates a stop codon, and at position 490 (Arvac vaccine strain) which abolishes the stop codon observed for the Bucyrus strain and all other EAV isolates, rendering the amino acid sequence in frame with a downstream stop codon located at nucleotide positions 505 to 507 (data not shown).

The levels of identity for ORF 3 nucleotide sequences between each EAV isolate/strain and the Bucyrus reference strain ranged from 85.6 (Vienna isolate) to 98.8% (Arvac vaccine strain) (Table 2). Deduced amino acid identities for the ORF 3-encoded product between each EAV isolate/strain and the Bucyrus strain ranged from 85.3 (Vienna isolate) to 98.2% (Arvac vaccine strain) (data not shown). When the EAV isolates were compared to each other, the Vienna and 19933 isolates showed the lowest levels of nucleotide sequence and amino acid identities at 82.7 and 76.1%, respectively. When the Canadian EAV isolates were compared to each other, the T1329 and 19933 isolates showed the lowest level of nucleotide sequence identity (87.8%), whereas the lowest level of amino acid identity was shown between the 19933 and 15492 isolates at 81.0%. The 11958 and 86AB-A1 Canadian isolates were found more closely related to each other with nucleic and amino acid identities of 96.7 and 94.4%, respectively. When the Arvac vaccine strain was compared to all EAV isolates, the lowest levels of nucleotide sequence and amino acid identities were shown with the Vienna isolate at 85.0 and 83.4%, respectively, whereas the highest were observed with the American 84KY-A1 isolate at 93.1 and 91.4%.

Table 2. Nucleotide sequence identities (%) of ORF 3 and 4 genes of equine arteritis virus (EAV) isolates*

Isolate	Bucyrus	T1329	11958	15492	19933	Vienna	84KY-A1	86AB-A1	Arvac**
Bucyrus		93.3	90.1	93.3	89.2	85.6	94.3	92.1	98.8
T1329	98.3		88.4	93.7	87.8	86.2	95.3	90.6	92.1
11958	93.2	94.1		90.1	92.1	83.4	90.3	96.7	89.0
14592	93.2	94.5	92.2		88.6	86.0	94.1	91.3	92.1
19933	93.9	94.3	96.5	93.5		82.7	88.0	93.9	88.0
Vienna	90.4	91.7	89.5	90.2	89.8		86.0	84.3	85.0
84KY-A1	94.8	96.1	94.5	95.4	95.9	91.5		91.3	93.1
86AB-A1	94.3	94.8	97.2	93.2	98.0	90.0	96.1		90.8
Arvac	98.0	97.2	92.6	92.2	93.2	90.0	94.1	93.7	

Upper section: ORF 3

Lower section: ORF 4

*The ORF 3 and 4 nucleotide sequence data reported for these isolates have been deposited in the Genbank Database under accession numbers: AF001079 to AF001094.

**Vaccine strain of EAV.

4.2. Nucleotide and Deduced Amino Acid Sequence Analysis of ORF 4

No insertions or deletions were found in the ORF 4 sequence for any of the EAV isolates including the Arvac vaccine strain. A total of 201 nucleotide changes were observed. Only 34.3% of all substitutions observed lead to amino acid changes. Nucleotide sequence identities with the Bucyrus reference strain ranged from 90.4 (Vienna isolate) to 98.3% (T1329 isolate) (Table 2). Deduced amino acid identities for the ORF 4-encoded product between each EAV isolate/strain and the Bucyrus strain ranged from 90.8 (Vienna isolate) to 97.4% (T1329 isolate) (data not shown). When the EAV isolates and Arvac vaccine strain were compared to each other, the Vienna and 11958 isolates were found to be the most divergent at the nucleotide level at 89.5% of identity, whereas the 19933 and 86AB-A1 Canadian isolates were found to be the most closely related with an identity of 98.0%. The Vienna isolate and Arvac vaccine strain showed the lowest level of amino acid identity at 91.4%, whereas the highest identity level (97.4%) was observed between the 11958 and 86AB-A1 Canadian isolates, and the 15492 and 84KY-A1 isolates. When the Canadian EAV isolates were compared to each other, nucleotide sequence identities ranged from 92.2 (11958 and 15492 isolates) to 98.0% (19933 and 86AB-A1 isolates), whereas amino acid identities between these isolates ranged from 94.7 (15492 and 11958 isolates, and 86AB-A1 and 15492 isolates) to 97.4% (11958 and 86AB-A1 isolates). The Arvac vaccine strain ORF 4 sequence showed nucleotide and amino acid sequence identities ranging from 90.0 (Vienna isolate) to 97.2% (T1329 isolate), and from 91.4 (Vienna isolate) to 95.4% (T1329, 19933 and 86AB-A1 isolates), respectively.

4.3. Phylogenetic Analysis of EAV Isolates Based on Their ORFs 3 and 4 Nucleotide Sequences

The phylogenetic analysis based on nucleotide sequence of ORF 3 (Figure 1) showed that the European Vienna isolate evolved independently and formed a distinct phylogenetic group (group E) from all other isolates and the Arvac vaccine strain. The North American isolates and the Arvac vaccine strain were clustered into a large clade (group A) which could be further subdivided into two distinct monophylogenic lineages. However, the confidence value obtained by bootstrap analysis of the lineage that includes the Bucyrus reference and Arvac vaccine strains and three other isolates (84KY-A1, T1329 and 15492) was low (37.3%). This indicates that the order of descent of this lineage was not fully resolved by this analysis, and suggests that the North American isolates and the Arvac vaccine strain are likely to represent a large phylogenetically-related group on the basis of the ORF 3 nucleotide sequences.

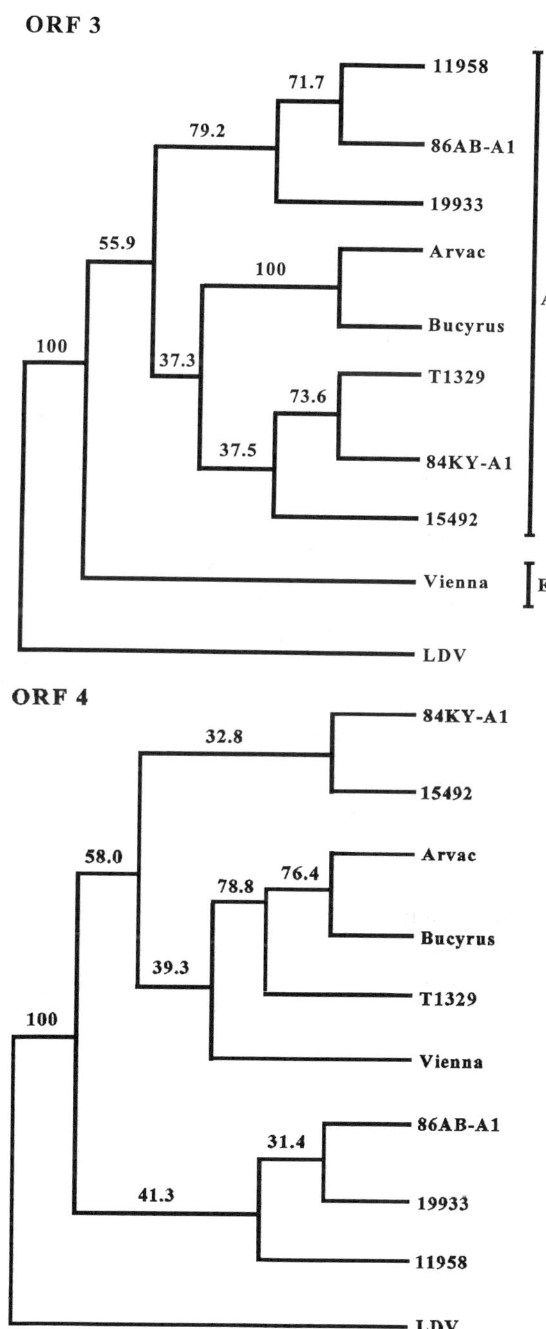

Figure 1. Phylogenetic relationship of nine equine arteritis virus (EAV) isolates including the Bucyrus reference and Arvac vaccine strains based on their ORF 3 or ORF 4 nucleotide sequences. The ORFs 3 and 4 of lactate dehydrogenase-elevating virus (LDV) were used as outgroups. The numbers indicate the branching node confidence values (%) obtained by bootstrap analyses.

In contrast, the phylogenetic tree based on the nucleotide sequence of ORF 4 (Figure 1) showed that the European Vienna isolate could not be clustered in a distinct group from the North American isolates. Although three Canadian isolates (11958, 19933 and 86AB-A1) might be classified as a distinct monophylogenic group from all other isolates, the order of descent of this lineage was not fully resolved by this analysis (confidence value of 41.3%). Indeed, the high levels of nucleotide and amino acid sequence identities described above indicate a close relationship between the ORF 4 sequences in all analyzed EAV isolates.

5. DISCUSSION

The results presented in this study demonstrate a high degree of nucleotide and deduced amino acid sequence conservation in the EAV ORF 4 among the analyzed EAV isolates. In contrast, the EAV ORF 3 gene was less conserved and was found to be genetically divergent such that the European Vienna isolate and the North American isolates could be classified into distinct phylogenetic groups on the basis of nucleotide sequences.

The results also showed that the EAV ORF 3 appears to be more variable genetically, based on percentages of identity of the nucleotide and amino acid sequence, than the ORF 2- ,5- , 6- and 7-encoded EAV structural proteins (Chirnside et al., 1994; Balasuriya et al., 1995; Hedges et al., 1996; Lepage et al., 1996; St-Laurent et al., 1997), the elicitors of antibody responses in EAV-infected horses (De Vries et al., 1992; Chirnside et al., 1995). This is particularly surprising for the ORF 5-encoded G_L protein which expresses the EAV neutralizing epitopes (Deregt et al., 1994; Balasurya et al., 1995; Chirnside et al., 1995; Glaser et al., 1995), and which, therefore, is likely to be the primary target of selective pressure by the host immune system.

The significance of the genetic variation observed in EAV ORF 3 has yet to be determined. The ORF 3-encoded product is believed to be an NS protein (Den Boon et al., 1991). However, the structural nature of the ORF 3-encoded product can not be ruled out. Genetic variation has also been reported with NS proteins of other RNA viruses, the conserved regions of which are believed to be important in the regulation of virus replication (Rao et al., 1995; Oberste et al., 1996). Analysis of the deduced amino acid sequence alignment of the ORF 3-gene product of all EAV isolates with the reference Bucyrus strain revealed the presence of two variable regions that encompassed residue positions 3 to 41 in the N-terminal region, and positions 98 to 121 in the central portion of the predicted 163 amino acid ORF 3-encoded product (data not shown). The remaining amino acid sequence of the protein was relatively well conserved among EAV isolates. Thus, it is likely that these conserved regions within the ORF-3 encoded protein might also have an important role in EAV pathogenesis. More in-depth studies are needed to resolve the molecular basis of EAV ORF 3 and 4 genetic variation and to determine the biological function of ORFs 3 and 4 in EAV biogenesis.

ACKNOWLEDGMENTS

This work was supported by an operating grant from the National Sciences and Engineering Research Council of Canada to D. Archambault. D. Archambault is the holder of a research scholarship from the Fonds de la Recherche en Santé du Québec (FRSQ). We also thank Peter Timoney and William McCollum (Gluck Equine Research Center, College of Agriculture, University of Kentucky, Lexington, KY, USA) for providing the Canadian 86AB-A1, American and European EAV isolates, and the Arvac vaccine strain, and Susy Carman (Veterinary Laboratory Services, Guelph, Ontario, Canada) for providing the Canadian EAV field isolates.

REFERENCES

Balasuriya U.B.R., MacLachlan N.J., De Vries A.A.F., Rossito P.V. and Rottier P.J.M., 1995, Identification of a neutralization site in the major envelope glycoprotein (G_L) of equine arteritis virus, *Virology* **207**: 518–527.

Balasuriya U.B.R., Timoney P.J., McCollum W.H. and MacLachlan N.J., 1995, Phylogenetic analysis of open reading frame 5 of field isolates of equine arteritis virus and identification of conserved and nonconserved regions in the G_L envelope glycoprotein, *Virology* **214**: 690–697.

Chirnside E.D., Wearing C.M., Binns M.M., Mumford J.A., 1994, Comparison of M and N gene sequences distinguishes variation amongst equine arteritis virus isolates, *J. Gen. Virol.* **75**: 1491–1497.

Chirnside E.D., De Vries A.A.F., Mumford J.A. and Rottier P.J.M., 1995, Equine arteritis virus-neutralizing antibody in the horse is induced by a determinant on the large glycoprotein G_L, *J. Gen. Virol.* **76**: 1989–1998.

Den Boon J.A., Snijder E.J., Chirnside E.D., De Vries A.A.F., Horzinek M.C. and Spann W.J.M., 1991, Equine arteritis virus is not a togavirus but belongs to the coronaviruslike superfamily, *J. Virol.* **65**: 2910–2920.

Deregt D., De Vries A.A.F., Raamsman M.J.B. and Elmgren L.D., 1994, Monoclonal antibodies to equine arteritis virus protein identify the G_L protein as a target for virus neutralization. *J. Gen. Virol,* **75**: 2439–2444.

De Vries A.A.F., Chirnside E.D., Horzinek M.C. and Rottier P.J.M., 1992, Structural proteins of equine arteritis virus, *J. Virol.* **66**: 6294–6303.

Felsenstein J., 1993, PHYLIP (Phylogenic Inference Package) 3.5c Manual, University of Washington, Department of Genetics SK-50, Seattle, WA 98195.

Glaser A.L., De Vries A.A.F., Dubovi E.J., 1995, Comparison of equine arteritis virus isolates using neutralizing monoclonal antibodies and identification of sequence changes in G_L associated with neutralization resistance, *J. Gen. Virol.* **76**: 2223–2233.

Godeny E.K., Chen L., Kumar S.N., Methven S.L., Koonin E.V. and Binton M.A., 1993, Complete genomic sequence and phylogenetic analysis of the lactate dehydrogenase-elevating virus (LDV), *Virology* **194**: 585–596.

Hedges, J.F., Balasuriya U.B.R., Timoney P.J., McCollum W.H. and MacLachlan N.J., 1996, Genetic variation in open reading frame 2 of field isolate and laboratory strains of equine arteritis virus, *Vir. Res.* **42**: 41–52.

Lepage N., St-Laurent G., Carman S. and Archambault D., 1996, Comparison of nucleic and amino acid sequences and phylogenic analysis of the Gs protein of various equine arteritis virus isolates, *Virus Genes* **13**: 87–91.

Murphy T.W., McCollum W.H., Timoney P.J., Klingeborn B.W., Hyllseth B., Golnik W. and Erasmus B., 1992, Genomic variability among globally distributed isolates of equine arteritis virus, *Vet. Microbiol.* **32**: 101–115.

Oberste M.S., Parker M.D. and Smith J.F., 1996, Complete sequence of Venezuelan equine encephalitis virus subtype IE reveals conserved and hypervariable domains within the C terminus of nsP3, *Virology* **219**: 314–320.

Rao C.D., Das M., Ilango P., Lalwani R., Bhargavi S.R. and Gowda K., 1995, Comparative nucleotide and amino acid sequence analysis of the sequence-specific RNA-binding rotavirus nonstructural protein NSP 3, *Virology* **207**: 327–333.

Sanger F., Nicklen S. and Coulson A.R., 1977, DNA sequencing with chain-terminating inhibitors, *Proc. Natl. Acad. Sci.* USA **74**: 5463–5467.

St-Laurent G., Morin G. and Archambault D., 1994, Detection of equine arteritis virus following amplification of structural and nonstructural viral genes by reverse transcription-PCR, *J. Clin. Microbiol.* **32**: 658–665.

St-Laurent G., Lepage N., Carman S. and Archambault D., 1997, Genetic and amino acid analysis of the G_L protein of Canadian, American and European equine arteritis virus isolates, *Can. J. Vet. Res.* **61**: 72–76.

Timoney P.J. and McCollum W.H., 1993, Equine viral arteritis, *Vet. Clinics North Amer.: Equine practice* **9**: 295–309.

Participants of the VIIth International Symposium on Coronaviruses and Arteriviruses, held May 10–15, 1997, in Segovia, Spain.

INDEX